28th International Laser Radar Conference 2017 (ILRC 28)

EPJ Web of Conferences Volume 176 (2018)

Bucharest, Romania
25-30 June 2017

Editors:

D. Nicolae
A. Vassilis
A. Behrendt
F. Gibert
M. P. McCormick
I. Veselovskii
A. Makoto
D. Balis
A. Comeron
E. Landulfo
C. Senff
U. Wandinger

ISBN: 978-1-5108-6146-6

Printed from e-media with permission by:

Curran Associates, Inc.
57 Morehouse Lane
Red Hook, NY 12571

Some format issues inherent in the e-media version may also appear in this print version.

Printed by Curran Associates, Inc. (2018)

For additional information, please contact EDP Sciences – Web of Conferences
at the address below.

EDP Sciences – Web of Conferences
17, Avenue du Hoggar
Parc d'Activité de Courtabœuf
BP 112
F-91944 Les Ulis Cedex A
France

Phone: +33 (0) 1 69 18 75 75
Fax: +33 (0) 1 69 28 84 91

contact-edps@webofconferences.org

Additional copies of this publication are available from:

Curran Associates, Inc.
57 Morehouse Lane
Red Hook, NY 12571 USA
Phone: 845-758-0400
Fax: 845-758-2633
Email: curran@proceedings.com
Web: www.proceedings.com

TABLE OF CONTENTS

ICLAS President
Alexandros Papayannis

International Coordination-group for Laser Atmospheric Studies ICLAS

ILRC 28 BUCHAREST

Conference Chair
Doina Nicolae
nnicol@inoe.ro

International Laser Radar Conference

25 - 30 June 2017
Bucharest, Romania

organized under the patronage of National Authority for Scientific Research and Innovation ANCSI

For a half of century, ILRC has been the largest forum for exchanging ideas, fostering collaborations, and reporting cutting edge research on Lidar techniques, technologies and applications. Organized under the auspices of the International Coordination-group for Laser Atmospheric Studies (ICLAS), this biennially conference has promoted generations of scientists and stimulated the advancement of the domain beyond expectations. Today, lidars are operating on ground, airborne and satellite platforms; continental and even global networks are setting up; weather services and airports are in the process of replacing classic sounding with remote sensing; lidars are taking over the outer space.

1st January 2017

- Submission of the 4-pages paper
- Submission of the travel grant request
- Submission of the request for invitation letter
- Payment of the the early registration fee for exhibitors

1st April 2017

- Payment of the early registration fee for authors
- Hotel registration (through the conference service)

1st June 2017

- Payment of the late registration fee

This edition will be co-hosted and supported by:

- National Institute of R&D for Optoelectronics
- "Politehnica" University of Bucharest
- National Institute of Aerospace Research "Elie Carafoli"
- University of Bucharest
- "Babes-Bolyai" University of Cluj-Napoca
- "Al. I. Cuza" University of Iasi

All sessions will be organized at the Conference Center of the "Politehnica" University of Bucharest.

The conference will follow the typical set up of the ILRCs, with plenary-only oral sessions (limiting the number of oral presentations to one per author), and correlated poster sessions.

Tentative topics are:

1. Recent advances in lidar technology
2. Spaceborne-Airborne lidar missions
3. Middle and upper atmospheric lidar
4. Lidar applications in weather/meteorology, air quality and climate-global change
5. Lidars applications in atmospheric structure and composition: aerosols, clouds, trace gases
6. Lidar applications in atmospheric dynamics and surface exchanges: boundary layer, winds and turbulence
7. Lidar applications in ecosystems research
8. Synergy of lidars and other sensors
9. Lidar networks
10. Large-scale campaigns
11. Commercial lidar systems, ceilometers and critical components.

Side events including a one-day tutorial for young scientists, exhibition for manufacturers and integrators of instruments, thematic round-tables, etc. will also be accommodated in the program.

A welcome reception (during the first day of the conference), and a banquet party coupled with a trip to historical places in the vicinity of Bucharest will be organized for all participants. Optional tours and events will be organized for the accompanying persons during the conference.

Conference website: http://ilrc28.inoe.ro
General contact: nnicol@inoe.ro

Venue

"Politehnica University of Bucharest"
the Conference Centre

ilrc 28
BUCHAREST

Conference fees

Early registration students	250 EUR	1st April 2017
Late registration students	350 EUR	1st June 2017
Early registration regular participant	400 EUR	1st April 2017
Late registration regular participant	500 EUR	1st June 2017
Early registration exhibitor	1500 EUR	1st January 2017
Late registration exhibitor	2500 EUR	1st June 2017

http://ilrc28.inoe.ro

Organizers

National Institute of R&D for Optoelectronics

"Politehnica" University of Bucharest

"Babes-Bolyai" University of Cluj-Napoca

National Institute of Aerospace Research "ELIE CARAFOLI"

"Al. I. Cuza" University of Iasi

Steering Committee

Doina Nicolae* (Romania)
Alexandros Papayannis* (Greece)
Andreas Fix* (Germany)
Eduardo Landulfo* (Brazil)
Upendra Singh* (USA)
Adolfo Comeron (Spain)
Albert Ansmann (Germany)
Anatolii Chaikovsky (Belarus)
Gelsomina Pappalardo (Italy)
Nobuo Sugimoto (Japan)
Vassilis Amiridis (Greece)

** ICLAS member*

edp sciences EPJ Web of Conferences

Statement of Peer review

In submitting conference proceedings to EPJ Web of Conferences, I certify to the Publisher that I adhere to the **Policy on Publishing Integrity** of the journal in order to safeguard good scientific practice in publishing.

1. All articles have been subjected to peer review administered by the proceedings editors.
2. Reviews have been conducted by expert referees, who have been requested to provide unbiased and constructive comments aimed, whenever possible, at improving the work.
3. Proceedings editors have taken all reasonable steps to ensure the quality of the materials they publish and their decision to accept or reject a paper for publication has been based only on the merits of the work and the relevance to the journal.

Title, date and place of the conference

INTERNATIONAL LASER RADAR CONFERENCE 28, 25-30 June 2017, Bucharest, Romania

Proceedings editor(s):

Doina Nicolae, Abo Makoto, Amiridis Vassilis, Dimitrist Balis, Andreas Behrendt, Adolfo Comeron, Fabien Gibert, Eduardo Landulfo, M. Patrick McCormick, Christoph Senff, Igor Veselovskii, Ulla Wandinger

Date and editor's signature

15/02/2018

PERFORMANCES OF A HGCDTE APD BASED DIRECT DETECTION LIDAR AT 2 μM. APPLICATION TO DIAL MEASUREMENTS.

Fabien Gibert[1]*, Arnaud Dumas[1], Johan Rothman[2], Dimitri Edouart[1], Claire Cénac[1], Jessica Pellegrino[1]

*[1]LMD, CNRS/ Ecole Polytechnique, Université de Paris-Saclay, France, *gibert@lmd.polytechnique.fr*
[2]LETI-CEA, Campus Minatech, France

ABSTRACT

A lidar receiver with a direct detection chain adapted to a HgCdTe APD based detector with electric cooling is associated to a 2.05 μm Ho :YLF pulsed dual wavelength single mode transmitter to provide the first atmospheric lidar measurements using this technology. Experiments confirm the outstanding sensitivity of the detector and hightligth its huge potential for DIAL measurements of trace gas (CO2 and H2O) in this spectral domain. Performances of coherent vs direct detection at 2.05 μm is assessed.

1 INTRODUCTION

CO2 is an unprecedented challenge for optic based remote sensing techniques, as its variation is rather small in the atmosphere. Typical natural variations due to its diurnal cycle (uptake by the vegetation, boundary layer dynamics) are around 5 % close to the surface but decrease to 0.1 % for turbulent linked fluctuations [1]. CO2 DIAL measurements requires indeed a high laser single shot signal to noise ratio (SNR). Although CO2 DIAL experiments emerged in the last decade [2,3], the technology faces two main challenges to get it: (i) a powerful dual wavelength single mode emitter and (ii) an efficient photodetector in the near infrared. Two spectral domains at 1.6 μm and 2 μm are relevant although higher optical depth at 2 μm enables more precision, flexibility and optimization of CO2 DIAL/IPDA measurements, ultimately from space.

In the framework of greenhouse gases studies, the LMD has developed a 2 μm DIAL emitter. Details of the laser may be found in [4]. The laser benefits from both fiber laser technology for compactness and reliability and free space operation to get high-energy pulses. LMD emitter delivers tens of millijoules and tens of nanosecond Fourier transform and diffracted limited beam pulses and can be operated at low or high repetition rates (100 Hz-10 kHz) depending the application.

To overcome the detector issue at 2 μm, LMD emitter was first associated with a coherent detection to make ground-based CO2 DIAL profiling in the atmospheric boundary layer [5]. However, recent advances in the field of HgCdTe APD may open the way to efficient direct detection [6,7]. The remarkable properties of HgCdTe APD (large gain, high efficiency, low dark current, close to unity excess noise factor, wide spectral range) make this technology especially suitable for very low intensity lidar signal detection. A dedicated APD for lidar application has been manufactured at CEA-LETI. This detector uses thermo-electric cooling (TEC) for space application. Some performances of the detector have been assessed in a precedent paper [8]. In the present paper we implement the detector in the LMD DIAL system to achieve first atmospheric lidar measurements at 2 μm using this technology. Preliminary CO2 DIAL measurements are presented and analyzed.

2 EXPERIMENTAL SET UP

2.1 HgCdTe APD detector prototype characteristics

The HgCdTe detector prototype developed by CEA-LETI is presented in Figure 1. The main characteristics are displayed in Table 1. The APD is located on several Peltier stages for electric cooling and positioned close to the transimpedance amplifier (TIA) to reduce parasitic capacitance. A dedicated substract on the APD surface acts as a high optical pass filter (> 800 nm). A quartz window serves as an interface between the detector vacuum box and lidar direct detection chain optical components.

The relevant detector characteristics for lidar detection are: a 180 μm active surface, a NEP of

75 fW/Hz$^{1/2}$ for a gain of 23 when polarized at 12 V and a total bandwidth (APD+TIA) of 20 MHz.

Table 1 HgCdTe APD main characteristics. NEP: Noise equivalent power, TEC: Thermoelectric cooler, TIA: transimpedance amplifier

APD	Diameter	180 µm
	Gain	23
	Operating voltage	12 V
	Operational temperature	180 K
	Equivalent capacitance	2 pF
	Excess noise factor	1.2
APD +TIA	Bandwidth APD APD +TIA	 > 60 MHz 20 MHz
	TIA gain	3.10^5 V/A
	NEP	75 fW/Hz$^{1/2}$

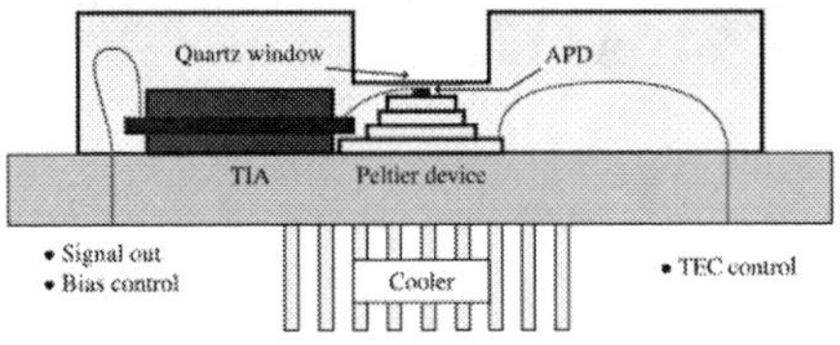

Figure 1 Sideview of the HgCdTe APD detector developed at CEA-LETI.

2.2 Direct detection DIAL experimental set-up at LMD

The APD was integrated in a lidar direct detection chain including a Cassegrain telescope, a diaphragm to control the field of view of the detection chain and different lens to image the telescope primary mirror on the detector surface. A 2-nm optical bandpass filter was also included to reduce the radiometric noise although the detector surface substrate already removed a big part of it for wavelength lower than 800 nm. The total transmission of the optical chain including telescope, lens and without the filter was estimated to be only 0.56 mainly due to the absence of dedicated coating an the telescope mirrors (telescope transmission is 0.8) and the optics. Given the size of the detector, the alignment of the optical components and the imaging on the detector surface is difficult to achieve although this is critical to obtain an operation in good conditions. This direct detection chain was associated with the LMD 2 µm dual wavelength single mode pulsed Ho:YLF emitter (Fig. 2). Some details may be found in Table 2. The ON wavelength is usually locked on the center of the R30 CO2 absorption line at 2050.967 nm although off-center locking is also possible using an additional DFB laser. The whole DIAL system was installed on the second floor of the LMD lab and the line of sight was practically horizontal.

Table 2 HgCdTe APD main characteristics. NEP: Noise equivalent power, TEC: Thermoelectric cooler, TIA: transimpedance amplifier

Emitter	Thulium fiber pump	
	Ho:YLF oscillator	
	Pulse Energy / Rate	10 mJ/ 1 kHz
	On/ Off wavelengths	2050.97/ 2050.26 nm
	Laser divergence after beam expander	0.3 mrad
Receiver	Telescope diameter	0.2 m
	Field of view	0.5 mrad
	Filter bandwidth	2 nm
	Detector	HgCdTe APD

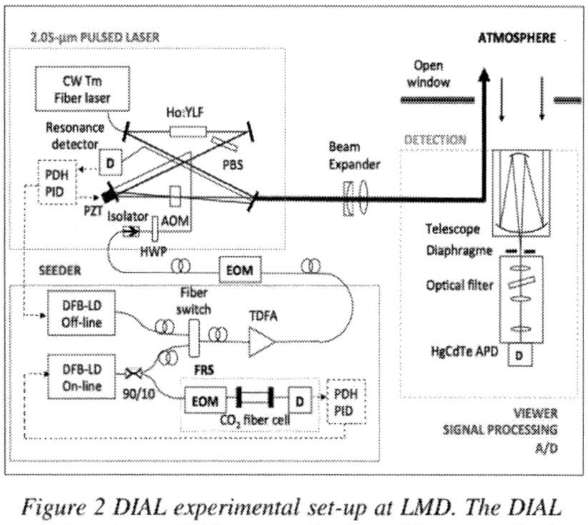

Figure 2 DIAL experimental set-up at LMD. The DIAL consists in a 2.05 µm pulsed oscillator, a dual-wavelength seeding module locked to a frequency reference system (FRS) and a direct detection. EOM: electro-optical modulator; AOM: acousto optical modulator; PID: proportional integral derivative; PDH: Pound Drever Hall; AOFS: acousto optic frequency shifter; TDFA: Thulium doped fiber amplifier; PBS: polarizer beam splitter; HWP: half wave plate; QWP: quarter wave plate; PZT: piezoelectric transducer..

3 PERFORMANCES

3.1 Lidar signal to noise ratio

Assuming the speckle noise is negligible (usually true in DIAL configuration) the lidar signal to noise ratio (SNR) may be written [6]:

$$SNR = \frac{P}{\sqrt{2h\ FB\left(P+P_{Bg}\right)+NEP^2 B}} \quad (1)$$

where P is the useful signal, P_{Bg} is the background signal which results mainly from the scattered solar radiation but also from some thermal radiation by components around the detector. F is the noise factor and B is the detection chain bandwidth.

The different sources of noise (shot noise, background, detector) were calculated for a single shot and compared with the lidar signal (background corrected) obtained in horizontal measurements (Fig. 3). For this particular case, there was no optical filter. Other measurements showed that the background noise is actually negligible when such a filter is used. Thus, Figure 3 shows that the single shot lidar SNR is detector NEP limited for a range longer than 0.6 km. A SNR of 0 dB is estimated at 3 km. It is worth pointing out that the SNR may vary by more than a factor of two during the day just due to the change in the atmospheric conditions and the particles content.

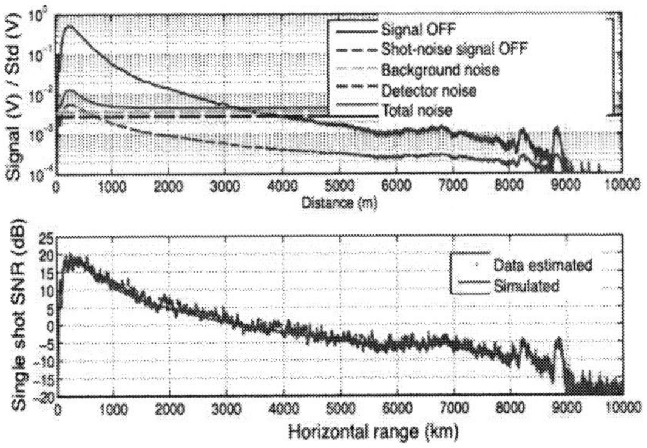

Figure 3 Different source of noise standard deviation (Std) estimated from the lidar signal and detector data. The lidar signal is averaged over 9000 shots for visualisation. The resulting SNR is calculated and compared with data estimated single shot SNR.

A longer data set was used to verify successfully that the SNR increases with the square root of the number of shots at least up to 4000 averaged shots.

3.2 Preliminary CO2 DIAL measurements

The DIAL configuration of the lidar was preliminary tested with the ON wavelength positioned in the center of the R30 CO2 absorption line at 2050.967 nm. Theoretically, this results in an optical depth of almost 0.65 for 1 km (and for 400 ppm of CO2 in standard conditions of temperature and pressure). Figure 4 displays the results.

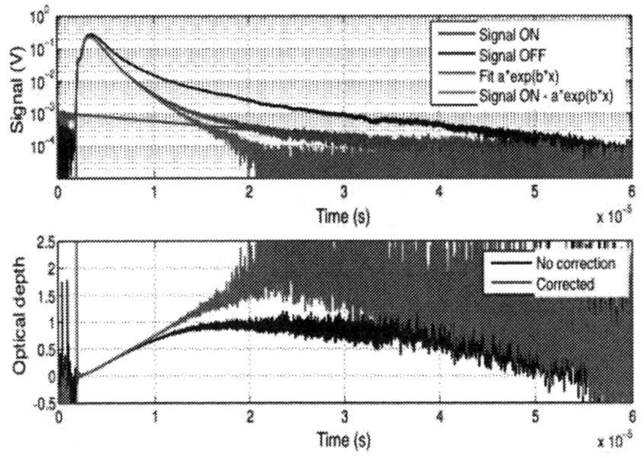

Figure 3 Lidar signals for the OFF and ON (CO2 line center) wavelengths and resulting optical depth due to CO2 absorption. An exponential fit of the ON-line data is used to correct from the slow decay of the signal with time. For range estimation, 1 µs corresponds to 150 m.

The main result is that, after 900 m, the calculated optical depth deviates significantly from an expected linear trend. An exponential decay of the signal is observed with a characteristic time of 16 µs. This parasitic signal is unlikely linked to the APD itself as its theoretical response time does not match [9] but seems rather to be linked to the optical set-up and an illumination on the edge of the detector. This may be corrected as shown in Figure 4 given that the ON-line signal is negligible after 3 km (97% of signal extinction with an absorption of 0.65 km^{-1}).

The DIAL system was also tested successfully with an ON-wavelength positioning on the wing of the CO2 absorption line (3 GHz off-set locking). For each case, the relative error on CO2 absorption coefficient was estimated and compared to previous results obtained in coherent detection [5]. The time and space resolution were fixed to 4 s and 100 m, respectively (Fig. 5).

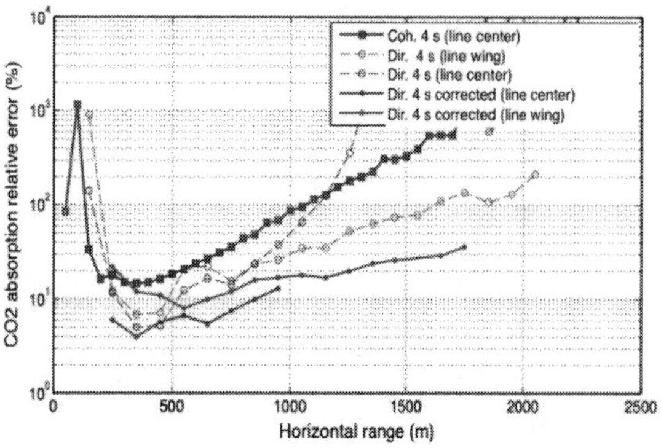

Figure 5 Relative error on CO2 absorption coefficient for a 100 m range gate and 4 s of time averaging (4000 shot-pair). The corrected data used the exponential correction described in Figure 4.

A reduction of the absorption enables to reach a longer range and a slower increase of the error with the distance as expected. Meanwhile the ON lidra signal is less sensitive to the parasitic signal. The direct detection shows a lower error at least by a factor of 5 than coherent detection (CO2 line center positioning).

4 CONCLUSIONS

A new direct detection chain at 2 µm using a HgCdTe APD prototype has been implemented and associated with a dual wavelength pulsed Ho:YLF oscillator. The performances of the lidar in terms of signal to noise ratio have been assessed. For the displayed data set, a SNR of 1 is obtained at 3 km mainly limited by the NEP of the detector. Preliminary DIAL measurements have been tested. The error budget on CO2 absorption coefficient estimate shows already a clear advantage of direct detection compared to coherent detection.

ACKNOWLEDGEMENTS

The authors thank the French national space agency (CNES) for supporting this work.

References

[1] Gibert F., Koch G., Davis K.J., beyond J.Y., Hilton T., Andrews, A., Flamant P.H., Singh U.N. 2011: Can CO2 turbulent flux be measured by lidar? A preliminary study., *J. Atmos. Ocean. Tech.,* **28**, 365-377

[2] Koch G.J., Barnes B.W., Petros M., Beyon J.Y., Amzajerdian F., Yu J., Davis R.E., Ismail S., Vay S., Kavaya M.J., Singh U.N. 2004: Coherent differential absorption lidar measurements of CO2, *Appl. Opt.* **43**, 5092-5099.

[3] Sakaizawa D., Nagazawa C., Nagai T., Abo M., Shibata Y., Nakasato M., Sakai T. 2009: Development of a 1.6 µm differential absorption lidar with a quasi-phase matching optical parametric oscillator and photon counting detector for the vertical CO2 profile, *Appl. Opt.*, 48, 748-757

[4] Gibert F., Edouart D., Cénac C., Le Mounier F., 2014: 2-µm high power multiple-frequency single mode Q-switched Ho:YLF laser for DIAL application, *Appl. Phys. B* **116** (4), 967-976.

[5] Gibert F., Edouart D., Cénac C., Le Mounier F., Dumas A. 2015: 2-µm Ho emitter-based coherent DIAL for CO2 profiling in the atmosphere, *Opt. Lett.* **40** (16), 3093-3096.

[6] Beck J., Welsch T., Mitra P., Reiff K., Sun X., Abshire J. 2014: A highly sensitive multi-element HgCdTe e-APD detector for IPDA lidar application, *J. Electronic Materials.* **43** (8), 2970-2977.

[7] Foubert K., Lasfargues G., Mathieu L., Benahmed S., Vejetta G., Rothman J., Benoit à la Guillaume Q., Calvo V., Picot-Clemente J., Le Mounier F., Gibert F. 2013: Development of HgCdTe single-element APD based detectors for low flux short wave infrared applications, *Proc. SPIE 8621, Optical Components and Materials X*, 86210F

[8] Dumas A., Rothman J., Gibert F., Lasfargues G., Zanatta J.-P., Edouart D. 2015: Performances of HgCdTe APD based detector with electric cooling for 2-µm DIAL/IPDA applications, *Proc. International Lidar Radar Conference (ILRC27)*

[9] Rothman J., Foubert K., Lasfargues G., Lergeron C. 2014: Response time measurements in short-wave infrared HgCdTe e-APDs, *J. Electronic Materials.* **43** (8), 2947-2954.

PERFORMANCE OF CHARM-F – THE AIRBORNE DEMONSTRATOR FOR MERLIN

Axel Amediek[1]*, Gerhard Ehret[1], Andreas Fix[1], Martin Wirth[1], Christian Büdenbender[1], Mathieu Quatrevalet[1], Christoph Kiemle[1]

[1]Deutsches Zentrum für Luft- und Raumfahrt (DLR) Oberpfaffenhofen, Germany
**axel.amediek@dlr.de*

ABSTRACT

In 2015, the new airborne lidar CHARM-F for the measurement of the greenhouse gases carbon dioxide and methane was set into operation. It is an integrated path differential absorption (IPDA) lidar designed for the use onboard the German research aircraft HALO. Due to its high technological similarity to the MERLIN space lidar, it is also a demonstrator for the space system. Here, we report on CHARM-F's technology, the experiences with the IPDA lidar method and the achieved performance.

1 INTRODUCTION

Better understanding of the carbon fluxes in the Earth's atmosphere is a current research issue. This topic is of particular importance, because carbon dioxide (CO_2) and methane (CH_4) are the two most relevant anthropogenic greenhouse gases contributing to global radiative forcing. The concentrations of both gases have significantly increased since the beginning of the industrial era (CO_2 from 280 ppm to 400 ppm and CH_4 from about 700 ppb to 1847 ppb). There is still a lack of knowledge related to sources and sinks, which limits the accuracy of climate change projections.

Active remote sensing has the potential to close the gaps. The integrated path differential absorption (IPDA) lidar technique can reach the stringent requirements to be met for significant contribution here. The aim is to generate high resolved data of CO_2 and CH_4 gradients to be used as inputs for inverse modeling.

CHARM-F (CO_2 and CH_4 atmospheric remote monitoring - Flugzeug) is an IPDA lidar for the simultaneous measurement of CO_2 and CH_4, designed for operation onboard the German research aircraft HALO (high altitude and long range research aircraft). Its purpose is to quantify concentration gradients and surface fluxes of CO_2 and CH_4, both, over point sources and large-scale sources.

Due to the high technological similarity, it is an airborne demonstrator for the French-German satellite mission MERLIN (methane remote sensing lidar mission), which recently has entered its project phase C. Furthermore, it is planned to use CHARM-F for the validation of future MERLIN data.

In 2015, CHARM-F was flown for the first time during a test campaign. For spring 2017, the first scientific campaign CoMet [1] is planned.

2 THE IPDA LIDAR PRINCIPLE

The key factor of the IPDA method is the fact that the strong laser light backscatter from a defined surface is measured instead of the atmospheric backscatter as in case of classical DIAL systems (see Figure 1). This surface is called "hard target" and can be the ground, the sea surface or a sharply edged cloud surface. The reasons are the comparatively weak atmospheric backscatter in the 1.6 µm wavelength region, which is on the other hand favorable for the measurement of CO_2 and CH_4, and the required high measurement precision to resolve the existent concentration variations of these gases. The derived quantity is the column-averaged dry-air mixing ratio of the trace gas *Xgas* [2]:

$$Xgas = \frac{DAOD}{\int_{p\_aircr}^{p\_target} WF(p,T)\cdot dp} \qquad (1)$$

The differential absorption optical depth *DAOD* is calculated from the measured signals S^{target} (hard

 https://doi.org/10.1051/epjconf/201817601002

target backscatter) and E^{ref} (the energy reference inside the instrument) for online and offline:

$$DAOD = \frac{1}{2} \cdot \ln \frac{S_{off}^{target} / E_{off}^{ref}}{S_{on}^{target} / E_{on}^{ref}} \qquad (2)$$

The weighting function *WF* is integrated in (1) from the hard target to the aircraft (here expressed in pressure coordinates). It is calculated as follows:

$$WF(p,T) = \frac{\sigma_{on}(p,T) - \sigma_{off}(p,T)}{g \cdot (m_{dry-air} + m_{H_2O} \cdot q_{H_2O})} \qquad (3)$$

using the absorption cross sections σ for online and offline, the normal gravity g and the molecular masses $m_{dry\text{-}air}$ of an average dry-air molecule and m_{H2O} of a water molecule, with q_{H2O} as the dry-air volume mixing ratio of water vapor.

For calculating *WF*, external data are required, such as the temperature *T*, the pressure *p* and q_{H2O} along the column (e.g. from numerical weather prediction like ECMWF) as well as spectroscopic data for σ (e.g. from databases like HITRAN).

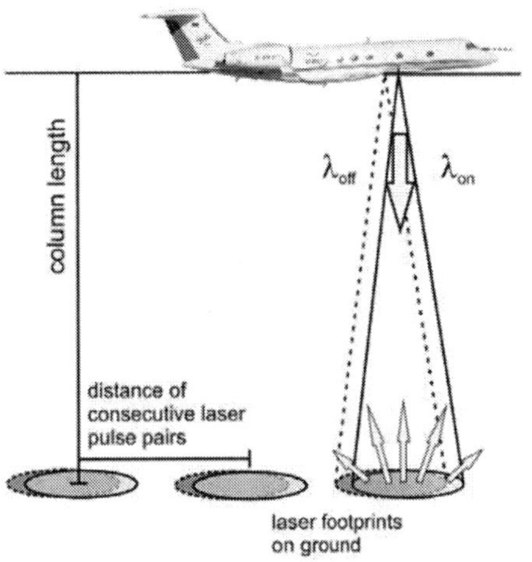

Figure 1. CHARM-F's measurement principle: Laser double pulses (online and offline) are emitted and backscattered by the ground (or cloud tops or the sea surface)

The instrument CHARM-F operates with short laser double pulses (20 ns each with 500 µs separation at 50 Hz), which leads to a precise ranging capability and high SNRs for each single shot measurement. Due to the double pulses, the illuminated spots on ground are almost identical for the online and offline measurements, which minimizes the impact of variable ground reflectivities. Main instrument specifications are listed in Table (1). The instrument, installed in the aircraft, is shown in Figure 2.

Table 1. CHARM-F system specifications

Parameter	Value
Lidar type	Integrated path differential absorption (IPDA) lidar using 2 wavelengths (online and offline) per trace gas
Receiving method	direct detection
Measured trace gases	atmospheric CO_2 and CH_4 (column averaged dry-air mixing ratios)
Laser type	Nd:YAG pumped OPOs
Laser specifications	10 mJ pulse energies, 50 Hz double pulses (500 µs pulse separation), 20 ns pulse width
Wavelengths	1572 nm (CO_2) 1646 nm (CH_4)
Detector types	InGaAs PINs and APDs
Reference gas cell	36-m multipass, CO_2 and CH_4 filled

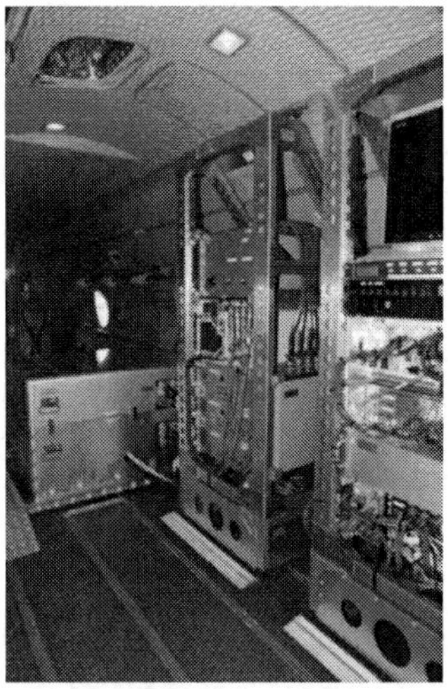

Figure 2. CHARM-F system installed in HALO

3 RESULTS

In the following, important results of the 2015 test flight campaign are presented.

3.1 System performance

One key objective of the instrument is to achieve a high measurement precision (low statistical noise) to resolve small gradients in the trace gas concentrations. The results of the first flight campaign were analyzed to that effect. Figure 3 (CO_2) and Figure 4 (CH_4) show the decreasing of the mixing ratio standard deviations along with averaging.

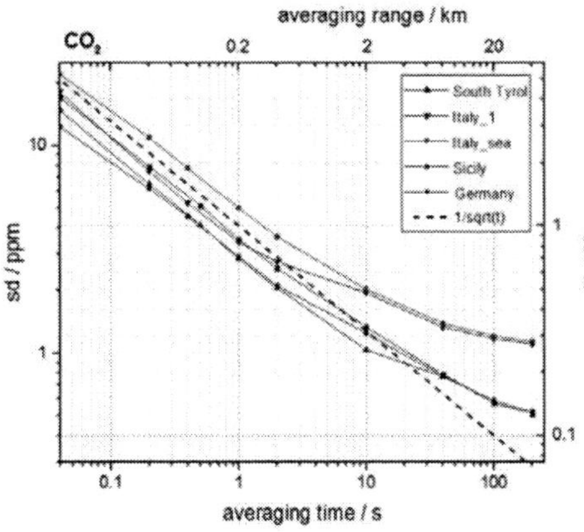

Figure 3. Noise statistics for CO_2 measurements from different flight legs along with data averaging.

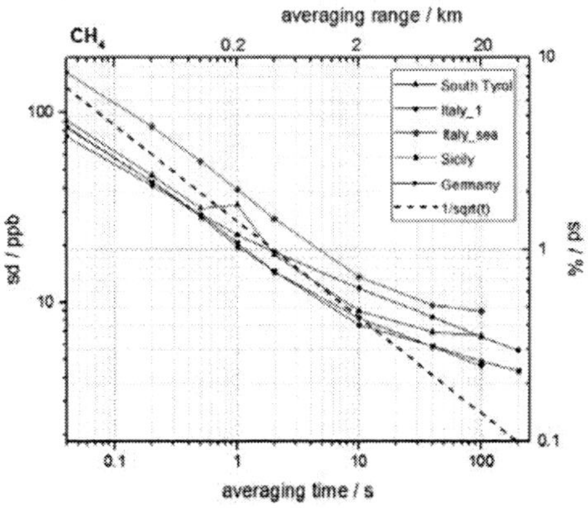

Figure 4. Noise statistics for CH_4 measurements from different flight legs along with data averaging.

Each figure contains plots from different regions including the sea surface. It is shown that during the test flights a precision of below 1% was reached after few seconds of averaging and after 100 s (i.e. 20 km horizontal averaging) a mixing ratio precision of below 0.3% (CO_2) and 0.5% (CH_4).

The use of laser double pulses reduces the "reflectivity noise" effectively, which occurs if the ground reflectivity for online and offline laser footprints varies. This is a special challenge resulting from the IPDA lidar approach. However, due to the short distance (in time and space) between online and offline measurements, the impact of these unavoidable reflectivity variations is minimized for the CHARM-F system. This is clearly shown by the flight measurements.

The absolute values retrieved from the measurements show offsets, if in-situ profile measurements are taken as a basis for comparisons, which were obtained by an onboard cavity ring down spectrometer [3] during the campaign. These offsets can be partly explained by beam misalignments. The reasons and possible improvements are currently under investigation. This issue has to be resolved before the next flights. In case of CH_4 another reason could play a decisive role: The accuracy of the spectroscopic parameters in combination with the applied absorption line shape model for the calculation of the absorption cross sections, as used in Eq. (3). It is possible that the calculated cross section for online (within a "trough" formed by the overlay of six methane absorption lines) is imprecise. Studies using different data and models are in progress.

3.2 Broken clouds

A major advantage of active remote sensing of trace gases from aircraft or satellite is the ability to retrieve meaningful data even under conditions of broken clouds, since lidars operate with small and defined footprints compared to passive sensors. The CHARM-F aircraft measurements demonstrate that the separation of clouds and ground can be done easily shot-by-shot using the ranging information. The Figure 5 and Figure 6 show the alternating *DAOD*s for clouds and ground, as expected. This allows also the retrieval of column parts, e.g. mixing ratio values for the boundary layer separated from the whole column.

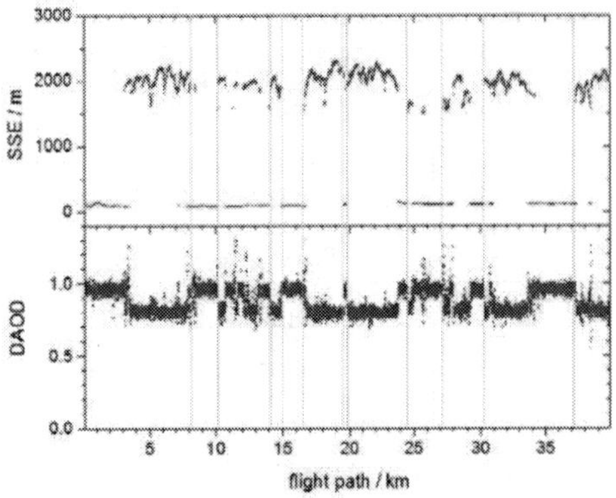

Figure 5. DAOD single-shot measurements of CO_2 for a broken cloud scenery. Cloud tops and the ground can be distinguished by the SSE levels (surface scattering elevation)

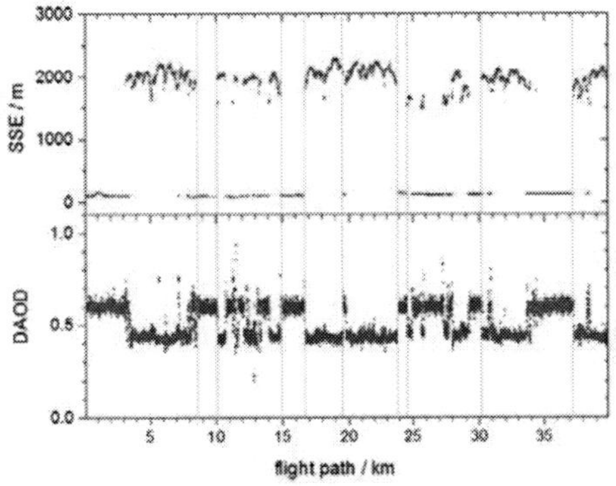

Figure 6. DAOD single-shot measurements of CH_4 for a broken cloud scenery (the same track section as shown in Figure 5).

3.3 Point source emissions

During the test flight campaign, known strong point sources of CO_2 and CH_4 were overflown. In one case a CH_4 plume from a hard coal mine ventilation shaft could be identified and for CO_2, the plume of a huge coal fired power plant was detected.

By means of the Gaussian plume approximation and the budget approach, the emission rates of the point sources were estimated. They agree well with reported numbers. For more information, see the paper presented by Ehret et al. [4].

4 CONCLUSIONS

The first flight measurements using CHARM-F on the German research aircraft HALO show promising results. The IPDA-lidar approach using two wavelengths (online and offline) emitted as laser double pulses with short separation allows high measurement precisions of less than 0.5% for 100 s averages (i.e. 20 km horizontal averaging).

However, shortcomings with respect to the absolute accuracy of the retrieved mixing ratios were found when the results were compared to independent data. Probably laser beam misalignments partly cause the deviations. Additionally, the spectroscopic modelling of the trace gas absorption could cause offsets.

The capability to handle broken cloud conditions was demonstrated.

The crossings of CH_4 and CO_2 plumes, emitted by strong point sources were clearly detectable in the DAODs. Estimations of the emission rates using these data resulted in reasonable values compared to reported data.

For spring 2017 the first scientific aircraft measurement campaign "CoMet" with CHARM-F on HALO is planned.

References

[1] Fix, A., Amediek, A., Bovensmann, H., Ehret, G., Gerbig, C., Gerilowski, K., Pfleilsticker, K., Roiger, A., Zöger, M., 2017: CoMet: An airborne mission to simultaneously measure CO_2 and CH_4 using lidar, passive remote sensing, and in-situ techniques, *Proceedings of the 28th International Laser Radar Conference* (ILRC 28)

[2] Ehret, G., Kiemle, C., Wirth, M., Amediek, A., Fix, A., Houweling, S., 2008: Space-borne remote sensing of CO_2, CH_4, and N_2O by integrated path differential absorption lidar: a sensitivity analysis, *Applied Physics B*, **90**(3-4), 593-608.

[3] Filges, A., Gerbig, C., Chen, H., Franke, H., Klaus, C. and Jordan, A., 2015: The IAGOS-core greenhouse gas package: a measurement system for continuous airborne observations of CO_2, CH_4, H_2O and CO, *Tellus B*, **67**(0)

[4] Ehret, G., Amediek, A., Wirth, M., Fix, A., Quatrevalet, M., 2017: Quantification of greenhouse gas emission of strong point sources by airborne IPDA-lidar measure-ments, *Proceedings of the 28th International Laser Radar Conference* (ILRC 28)

INELASTIC HYPERSPECTRAL LIDAR FOR AQUATIC ECOSYSTEMS MONITORING AND LANDSCAPE PLANT SCANNING TEST

Guangyu Zhao[1,*], Elin Malmqvist[2], Klas Rydhmer[2], Alfred Strand[2], Giuseppe Bianco[3], Lars-Anders Hansson[3] Sune Svanberg[1,4], Mikkel Brydegaard[2,3,5]

[1] *Center of Optical and Electromagnetic Research, South China Normal University, Guangzhou, China*
[2] *Combustion Physics Division, Department of Physics, Lund University, Sweden*
[3] *Centre for Animal Movement, Department of Biology, Lund University, Sweden*
[4] *Atomic Physics Division, Department of Physics, Lund University, Sweden*
[5] *Norsk Elektro Optikk A/S, Norway*
* guangyu.zhao@coer-scnu.org

ABSTRACT

We have developed an aquatic inelastic hyperspectral lidar with unrestricted focal-depth and enough sensitivity and spatio-temporal resolution to detect and resolve position and behavior of individual sub-millimeter aquatic organisms. We demonstrate ranging with monitoring of elastic echoes, water Raman signals and fluorescence from chlorophyll-bearing phytoplankton and dye tagged organisms. The system is based on a blue CW diode laser and a Scheimpflug optical arrangement.

1 BACKGROUND

Underwater suspended particle monitoring has an important role in aquatic ecosystem research. For instance, a specialized zooplankton – Daphnia is a crucial and dominant species of the Arctic zone, and can serve as an indicator of conditions in polar area environments, which are particularly influenced by global change. There is considerable diversity in the layered vertical structures of salinity, illumination, and temperature as well as biomass distribution which exhibits rapid changes, e.g., the diel vertical migration of zooplankton shows changes even on a scale of one hour [1]. When it comes to monitoring these phenomena, remote sensing and real-time methods provide advantages over conventional methods, such as water sampling, which is laborious and intrusive.

Individual submillimeter organism detection is difficult to accomplish by sonar which is the most common and mature technique used in aquatic monitoring [2]. Laser optical particle counters (LOPC) and in-line digital holography are capable of detecting underwater sub-millimeter organisms with high precision [3, 4]. However,

the monitoring area of these methods is limited, and such equipment is commonly installed on underwater vehicles to accomplish large area measurements. Light detection and ranging (lidar) techniques are readily used for atmospheric fauna and particle detection, as well as for aquatic monitoring with airborne systems. Recently, our group has developed atmospheric fauna lidar systems [5], which are based on high- power continuous-wave (CW) laser diodes rather than pulsed lasers which are used in time-of-flight (TOF) lidar systems. TOF lidar has been widely used in atmospheric and topography monitoring with kilometer distance detection. However, rapidly varying particle monitoring is limited by the pulse width and repetition rate of the pulsed lasers used in TOF lidars [6]. In addition, lidar techniques are little developed for detailed feature assessment, e.g. of vertical structure, sub-millimeter organism size distribution, and fluorescence features. We here introduce a novel lidar technique, which provides hyperspectral monitoring and spatio-temporal resolution with a reasonable sensitivity [7].

2 SYSTEM DESCRIPTION

The system is based on a blue CW diode laser and a special optical arrangement. Range resolution can be achieved by using a Scheimpflug arrangement and employing the Hinge rule rather than the familiar time-of-flight approach. Based on the Scheimpflug lidar technique, hyperspectral detection is combined in the system to achieve range and spectral resolution simultaneously. The system provides

enough sensitivity for underwater Raman as well as algae detection. The system shows advantages in cost, size and weight. A 1 W laser diode at 445 nm (3W is commercially available) with a 1×20 μm chip element size was used in the current system. 445 nm is efficient to induce fluorescence of chlorophyll of phytoplankton (algae), and has potential to also detect constituents of zooplankton. A standard video objective (ø12 mm, f_{exp}= 20 mm) with adjustable focus and iris is used to collimate the laser light to focus the laser beam on the beam termination. The fast axis of the laser-diode chip is aligned in parallel with the baseline of the instrument as is indicated in Fig. 1. The smaller beam width (fast axis) is chosen to achieve a smaller beam width at the far range to compensate for the degraded range resolution. The whole transmission part is mounted with two rotational stages to adjust the orientation of the laser beam (shown in Fig. 1). To be able to achieve online background subtraction, the laser is modulated on and off, as triggered by a strobe signal from the detector,
Backscattered light is collected by an achromatic duplet lens (L1) and passes a 100 μm slit. The slit is tilted by 45°. Then the light is collimated by a lens (L2) after the slit. A prism-grating-prism (PGP) optical assembly will disperse the light and image the light distribution on the sagittal plane of the imaging sensor. A lens (L3) is mounted after the PGP to focus the image on the detector. Correspondingly, the range information will be distributed in the meridian plane of the imaging sensor. Therefore, a 2D array (CCD) is used, which is a 12-bit industrial GigE camera with an active size of 7x5 mm. An appropriate long-pass filter is used to suppress the short wavelengths down to 1% to balance the relative intensities of the elastic and inelastic light. In the current system, the long-pass filter is inserted in the PGP component. The schematic of the system is shown in Fig. 1.

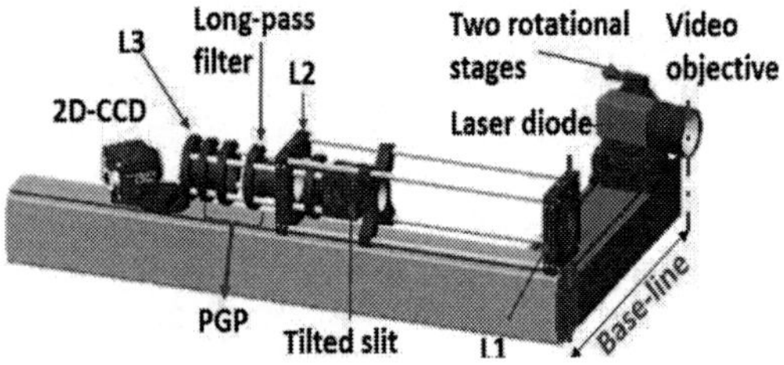

Fig. 1. Schematic of the inelastic hyperspectral lidar with its components.

The PGP is able to disperse the light in the spectral range 430-700 nm onto the CCD in the current system. The current system has 61 spectral bands, achieved even with a relatively wide slit, which allows detailed analysis of the spectra, e.g. Raman backscatter from liquid water. The wide spectral range gives the opportunity also for chlorophyll fluorescence monitoring. An experiment to demonstrate detection of phytoplankton, and labeled and unmarked zooplankton was performed. The advantage in low cost and small size could allow the system to be widely implemented for different layers of the ocean, as well as for shallow waters and polar environments. The system provides a novel non-intrusive tool for detecting aquatic plankton and suspended particulate matters.

3 EXPERIMENT

In order to evaluate the performance of the novel instrument, a tank which is 5 m long and supplied with transparent walls was used to provide a horizontal aquatic shooting range. The experimental arrangement is presented in Fig. 2.

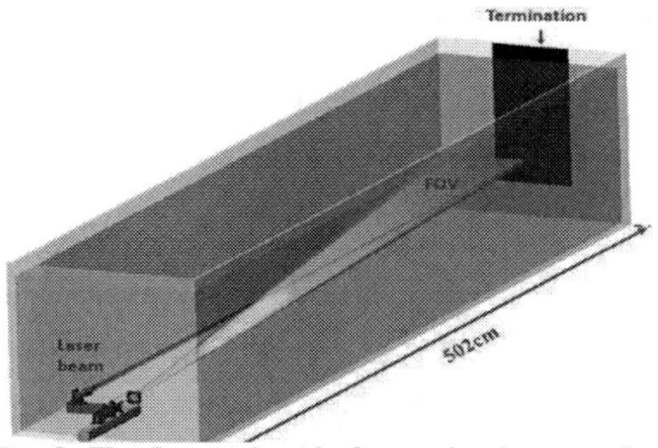

Fig. 2. The 5 m tank with the overlap between the laser beam and the FOV (field of view) indicated.

The tank was filled with clean freshwater. A black antistatic foam was used to terminate the laser beam with a resulting low intensity of the termination echo. The beam is focused on the termination from the 12 mm video objective aperture to a diffraction limited image of the laser chip. The volumetric backscatter signal of clean water and the H_2O Raman signal are useful to help adjusting the best overlap between beam and FOV. The termination echo is used to calibrate the range [8]. A second-order polynomial fit on 3 laser lines (445, 532, and 660 nm) has been used to calibrate the spectral distribution, and this calibration was double-checked by using the water O-H stretch Raman signal at 524 nm, corresponding to a Raman shift of about 3400 cm^{-1} from the 445 nm laser line. A white halogen tungsten lamp, along with an integrated sphere, was used to calibrate the spectral response.

4 RESULTS

A 3D visualization of a seven-second recording is given in Fig. 3 which demonstrates the capability of the instrument.

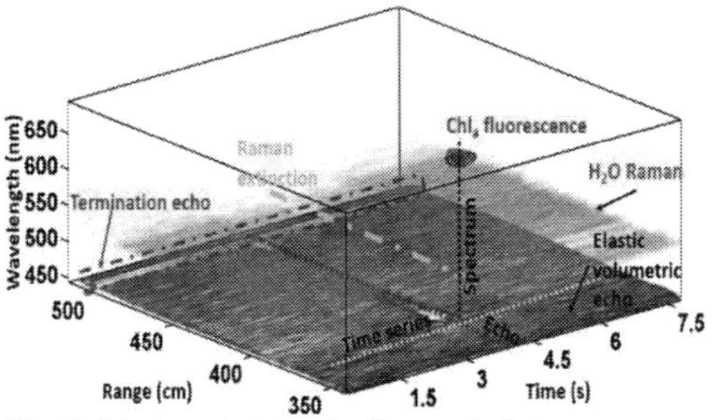

Fig. 3. 3D visualization of a 7 seconds data stream.

In Fig. 3, a drop of water containing phytoplankton was released close to the laser beam and the recording, which is represented by a data cube with an isosurface chosen with a strategic isolevel. The elastic, Raman and fluorescence signals are represented simultaneously. The volumetric elastic echo is shown as a blue layer, and the termination echo is observed and indicated in the figure. A Raman echo and fluorescence echo are shown by a green and a red layer, separately. Correspondingly, the termination echo could not be seen in the Raman and fluorescence bands. The red fluorescence of chlorophyll could be observed (red layer) when injecting the drop of water containing phytoplankton. In addition, corresponding elastic signals and the associated optical extinction (white hole in blue layer) are presented. The H_2O Raman signal also shows a decrease due to the optical extinction of the phytoplankton from the injecting position to the termination. The spectral signals were recorded simultaneously, and a continuous monitoring could be achieved.

Fig. 5 shows the elastic, Raman and chlorophyll fluorescence curves of the static clean water and for water containing phytoplankton. The increase of the elastic echo and fluorescence signal due to

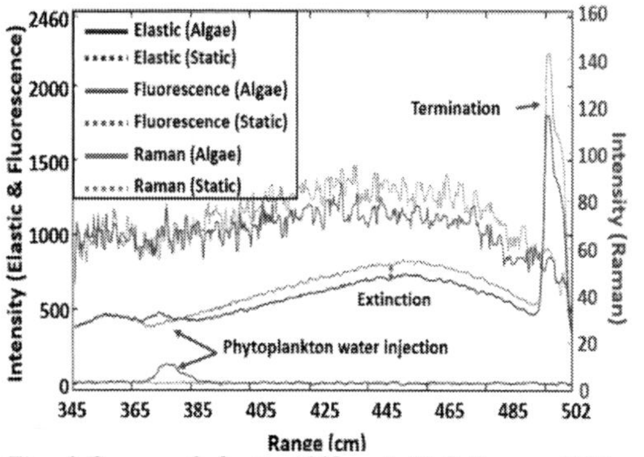

Fig. 4 Curves of elastic (445 nm), H_2O Raman (524 nm) and chlorophyll fluorescence (680 nm). Water containing phytoplankton was injected 370 cm from the entrance of the beam into the tank.

the algal release is noted. The optical extinction induced by the algae, which is noted above, is also represented. Consequently, the H_2O Raman echo and the following decrease of the Raman signal are shown, from which it is possible to estimate the range resolved extinction coefficient.

In Fig. 5, unstained Daphnia and labeled Daphnia were injected into the lidar transect at distinct ranges and times. Fig. 5. C) shows the maximum, median and minimum echoes. These indicate rare zooplankton, static water volumetric echo and the noise level, respectively [5, 8, 9]. The time series of the three panels in Fig. 5. C) present the two unstained Daphnia and labeled Daphnia with yellow fluorescence.

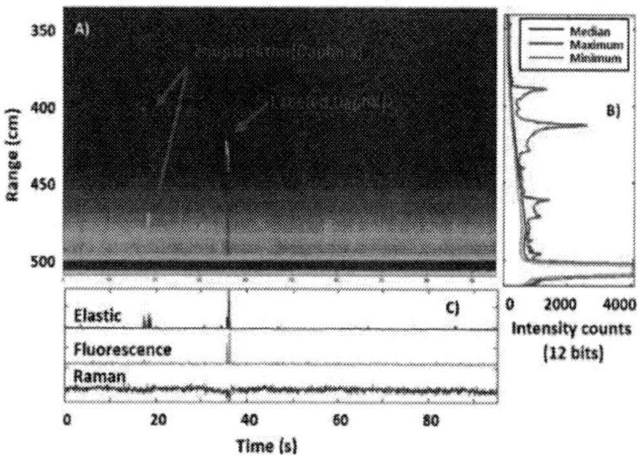

Fig. 5. A). Time-range map of the elastic echo of three Daphnia observations at different ranges and times. B) Minimum echo, median echo and maximum echo in a 100 s time window. C) Time-intensity curves of three spectral bands (445, 524 and 580 nm).

5 DISCUSSION, SUMMARY AND PERSPECTIVES

An inelastic hyperspectral lidar has been developed for an aquatic environment to detect and resolve the position and behavior of individual sub-millimeter aquatic organism. We demonstrate the performance of the instrument with monitoring of elastic echoes, water Raman signals and fluorescence from phytoplankton and dye labeled zooplankton with good range resolution. Hyperspectral detection is combined with continuous-wave lidar technique in the system to achieve range and spectral detection simultaneously. The system shows advantages in low cost, small size and light weight. The instrument provides a novel tool for non-intrusive monitoring in plankton research.

In addition, we are looking forward to using the device for canopy scanning in landscape. A coarse test of canopy monitoring has been performed on a small vegetation branch, and an example of the result is shown in Fig. 6. The elastic and chlorophyll fluorescence signals are displayed in different colors. 3-D shape and spectral information are achieved.

Fig. 6 3-D visualization of a scanned branch of Thuja.

ACKNOWLEDVHGEMENTS

The equipment could be acquired by an award from Edmund Optics. The work is supported by Lund University and the Overseas Study Program of Guangzhou Elite Project, and the Guangdong Province Innovation Research Team Program (No. 201001D0104799318).

REFERENCES

[1]. Lampert, W. Ultimate causes of diel vertical migration of zooplankton: new evidence for the predator-avoidance hypothesis. Diel Vertical Migration of Zooplankton, 39, 79-88, 1993.

[2]. Lemon, D., Johnston, P., Buermans, J., Loos, E., Borstad, G., & Brown, L. (2012, October). Multiple-frequency moored sonar for continuous observations of zooplankton and fish. In Oceans, (pp. 1-6). IEEE, 2012.

[3]. Herman A W, Beanlands B, Phillips E F. The next generation of optical plankton counter: the laser-OPC. Journal of Plankton Research, 26(10): 1135-114, 2004.

[4]. Jericho, S. K., Garcia-Sucerquia, J., Xu, W. B., Jericho, M. H. & Kreuzer, H. J. Submersible digital in-line holographic microscope. Rev. Sci. Instr. 77, doi:10.1063/1.2193827, 2006.

[5]. Brydegaard, M. Towards quantitative optical cross sections in entomological laser radar – Potential of temporal and spherical parameterizations for identifying atmospheric fauna. PLoS One 10, e0135231, 2015.

[6]. Wright, C. W., Hoge, F. E., Swift, R. N., Yungel, J. K. & Schirtzinger, C. R. Next-generation NASA airborne oceanographic lidar system. Applied Optics 40, 336-342, 2001.

[7]. Zhao, G., Ljungholm, M., Malmqvist, E., Bianco, G., Hansson, L. A., Svanberg, S., & Brydegaard, M. Inelastic hyperspectral lidar for profiling aquatic ecosystems. Laser & Photonics Reviews, 10(5), 807-813, 2016.

[8]. Mei, L. & Brydegaard, M. Continuous-wave differential absorption lidar. Laser & Photonics Reviews 9, 629-636, doi:10.1002/lpor.201400419, 2015.

[9]. Brydegaard, M., Gebru, A., Kirkeby, C., Åkesson, S. & Smith, H. Daily evolution of the insect biomass spectrum in an agricultural landscape accessed with lidar, in ILRC 27. New York, USA, 2015.

Measurements of Particle Backscatter, Extinction, and Lidar Ratio at 1064 nm with the Rotational Raman Method in Polly-XT

Ronny Engelmann[1]*, Moritz Haarig[1], Holger Baars[1], Albert Ansmann[1], Michael Kottas[2], Eleni Marinou[2]

[1]*Leibniz Institute of Tropospheric Research (TROPOS), Leipzig, Germany, *Email: ronny@tropos.de,*
[2]*National Observatory of Athens (NOA), Greece*

ABSTRACT

We replaced a 1064-nm interference filter of a Polly-XT lidar system by a 1058-nm filter to observe pure rotational Raman backscattering from atmospheric Nitrogen and Oxygen. Polly-XT is compact Raman lidar with a Nd:YAG laser (20 Hz, 200 mJ at 1064 nm) and a 30-cm telescope mirror which applies photomultipliers in photon-counting mode.

We present the first measured signals at 1058 nm and the derived extinction profile from measurements aboard RV Polarstern and in Leipzig. In combination with another Polly-XT system we could also derive particle backscatter and lidar ratio profiles at 1064 nm.

1 INTRODUCTION

The rotational Raman lidar (RRL) technique has been used for a long time, especially for temperature [1] and also for aerosol extinction [2] measurements in the atmosphere. Recently it was presented by Veselovskii et al. [3] that nowadays the availability of specially designed interference filters can enable multiwavelength lidars to use the RRL technique as a replacement for the more traditional vibrational Raman lidar technique at 532 nm wavelength [4]. This method is intriguing because on the one hand the RRL results in a factor of 20 more signal strength and on the other hand the assumption of the Angström exponent between elastic and inelastic vibration-rotational scattering is not needed which can reduce errors in backscatter and extinction retrieval methods.

A disadvantage, however, is that the RRL filters must have a much larger bandwidth than vibrational Raman filters in order to transmit the lidar return signals independently of temperature. As a consequence, the noise from sky background light reduces the advantage of the higher rotational backscattering cross section during daytime.

Up to now, measurements of the extinction coefficient and of the lidar ratio at 1064 nm are not very common. Most so called 3+2 multiwavelength aerosol lidars measure the backscatter coefficient at three wavelengths (355, 532, and 1064 nm) and the extinction coefficient at two wavelengths (355 and 532 nm) only. Haarig et al. [5] showed for the first time the use of RRL at 1064 nm in order to independently measure the extinction coefficient at 1064 nm. However, this study was performed with a high-power lidar system [6] and presented in a first step a cirrus case.

In the following work, we employed the same filters that Haarig et al. [5] used in one of our Polly-XT systems in order to test if the RRL method would also be applicable for our smaller and automated systems.

2 DESIGN OF INTERFERENCE FILTERS

It can be calculated that the backscatter coefficient of the sum of both rotational-Raman wings of atmospheric gases at 1064 nm is of the order of 0.0023 $Mm^{-1}sr^{-1}$ for normal pressure and a temperature of 300 K. The interference filters that we use are centered on the Anti-Stokes lines to be able to angle-turn them to shorter wavelengths in case the manufacturing margins were too high. In this way, the required suppression of the elastically scattered light should still be achievable. Observing only the Anti-Stokes lines means that the rotational Raman backscatter coefficient which has to be measured is on the order of 0.001 $Mm^{-1}sr^{-1}$. In contrast, the backscatter coefficient of the Cabannes line is 0.091 $Mm^{-1}sr^{-1}$ for normal pressure and a temperature of 300 K [1]. Typical backscatter coefficients at 1064 nm are on the order of 1-5 $Mm^{-1}sr^{-1}$ for aerosols and of 500 $Mm^{-1}sr^{-1}$ for water clouds. Hence, a suppression of the fundamental laser wavelength of 10^{-6} would only result in equal signals for the rotational Raman backscattering and for the elastic backscattering in the case of clouds. About four more orders of magnitude of suppression would be desirable to apply the Raman method [1, 2, 3, 4].

Figure 1 shows the transmission curve of our 1058-nm filters and the location of the rotational-Raman-scattering lines of Nitrogen and Oxygen from a fundamental laser wavelength of 1064.15 nm. The filters were manufactured by Alluxa and have a peak transmission >95%. The full width at half maximum is 8 nm, and was chosen so that the temperature dependence of the rotational Raman scattering is minimized. By theoretical design each filter should have an attenuation of the fundamental laser wavelength of 10^{-6}. The measured attenuation of the finished filters however is only on the order of 10^{-4} as given by Alluxa.

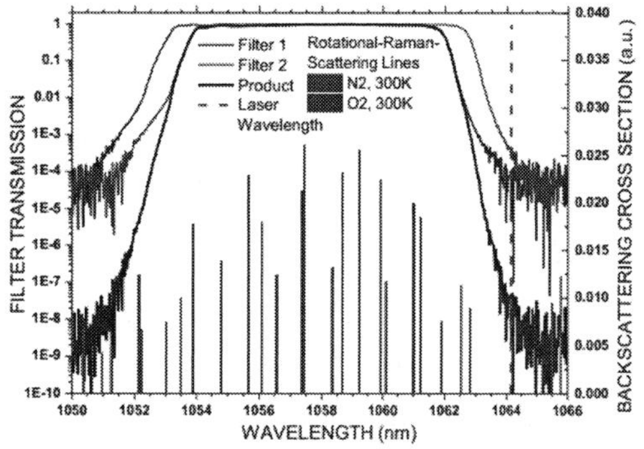

Figure 1: Transmission of both 1058-nm filters and their product. The locations of the laser wavelength and of the Anti-Stokes rotational Raman lines at 300 K are shown in addition.

As proposed by Veselovskii et al. [3] we used two filters to achieve a better suppression of the fundamental laser wavelength. In our case the theoretical suppression of 10^{-12} could not be achieved and is now only between 10^{-7} and 10^{-8}. It can be seen, that both filters are indeed slightly shifted in their transmission window and Filter 2 should be a bit tilted for future measurements. For the following study however, we operated both filters at 90° incident angles.

3 THE LIDAR POLLY-XT

The automated multiwavelength Raman polarization lidar Polly-XT has been developed at TROPOS [7, 8]. Nine of these lidar systems exist meanwhile. For this study we employed the lidars of the mobile ocean facility (OCEANET) of TROPOS and the Polly-XT of the National Observatory of Athens (NOA). Both systems are installed in standard 20-feet containers for more stable operations in maritime environments. In November 2016 Polly-XT-OCEANET was aboard the research vessel (RV) Polarstern crossing the Atlantic Ocean. In January 2017 both lidars were simultaneously operated at TROPOS, Leipzig.

In terms of the optical setup the lidar is completely build on a carbon fiber board which gives a light-weight but sturdy framework for the components. Figure 2 shows the schematically setup of the optical parts. The flash-lamp pumped and frequency doubled and tripled Nd:YAG laser emits radiation of about 200, 100, and 50 mJ at 1064, 532, and 355 nm wavelengths at 20 Hz. The beam is expanded 8 times before it is transmitted into the atmosphere with a divergence of 0.2 mrad.

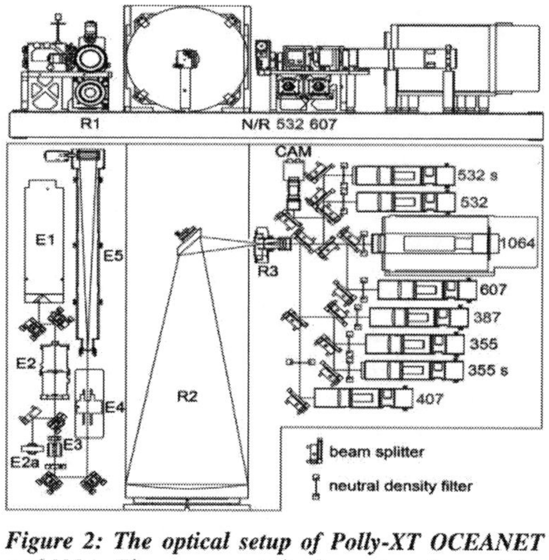

Figure 2: The optical setup of Polly-XT OCEANET and NOA. The upper part displays a top view and the lower part the front view of the system.

The receiving telescope (R2) has a diameter of 300 mm, a field of view of 1 mrad, and a full overlap at around 800 m above the lidar. An automated depolarization calibration unit (R3) is placed in front of the field stop. By use of the Δ90° method absolute calibration is possible on a regular basis. Various dichroic beam splitters are used to separate all paths for the detection channels shown in Fig. 2. The detection is performed in photon-counting mode in all channels. For 1064 nm we operate a R3236 photomultiplier tube (PMT, Hamamatsu) cooled down to −35 °C. This PMT has a quantum efficiency of only 0.03 % at 1064 nm wavelength, but a very small dead-time of around 2.5 ns. We employ a data acquisition that has been developed by Holger Linné of the Max-Planck Institute, Hamburg, Germany.

4 FIRST RESULTS

During our regular transects aboard RV Polarstern (cruise no. PS102) we occasionally exchanged the interference filter at 1064 nm of Polly-XT OCEANET by the two RRL filters. During the night from 25 to 26 November 2016 a marine boundary layer of 600 m height and on top a polluted dust layer up to 3.5 km height prevailed over the Atlantic Ocean about 500 km west of the coast of Guinea at 10 °N 20 °W. Figure 3 presents the range-corrected signal and volume linear depolarization ratio at 532 nm from that night. The white lines indicate the almost 6-hour cloud-free time period (0015 – 0600 UTC) which we used to average the weak RRL signal.

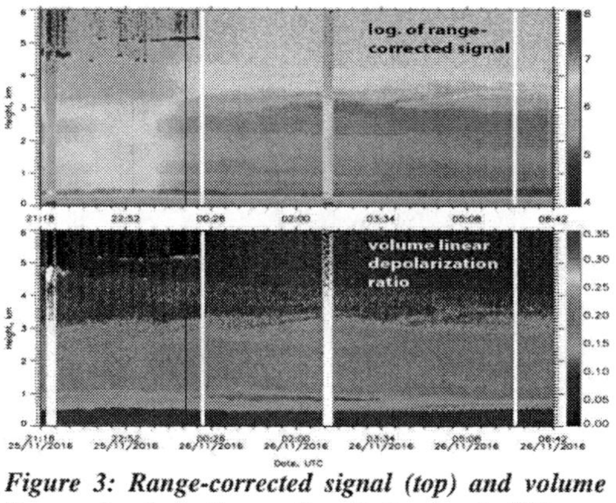

***Figure 3:** Range-corrected signal (top) and volume linear depolarization ratio (bottom) at 532 nm measured at 10 °N, 20 °W during the night from 25 to 26 Nov. 2016. At 0230 UTC an automated depolarization-calibration routine took place.*

Figure 4 shows the optical products calculated from that period. The profiles of the lidar ratio (S) and the particle linear depolarization ratio (δ_P) show an occurrence of a marine layer (S and δ_P rather low) at the surface and a dust layer with enhanced depolarization above. In the dust layer $S > 60$ sr at 355 nm and $\delta_P < 21$ % which indicates that the dust is not "pure" but most likely mixed with smoke. Further analysis is not performed here.

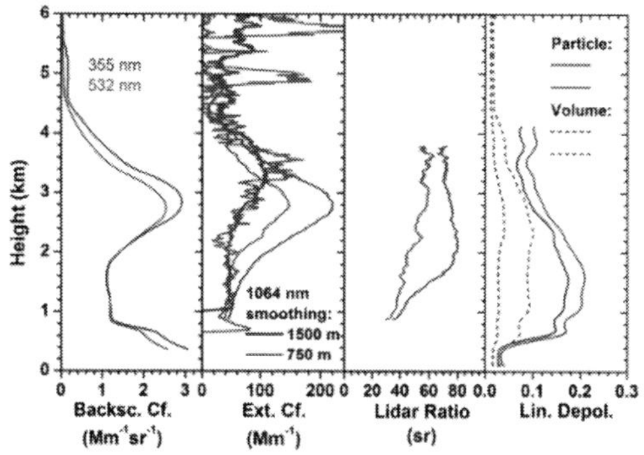

***Figure 4:** Optical products calculated from the 6-hour measurement shown in Fig. 3. The height resolution is 750 m for all profiles. The 1064-nm extinction is additionally shown at 1500 m height resolution.*

The focus of this work lies on the 1064-nm particle extinction. Figure 4 shows the extinction calculated from the weak RRL signal with the Raman method for two different smoothing lengths (750 and 1500 m, red and purple lines). Between 1 and 2 km height an extinction coefficient of 50 Mm^{-1} is obtained which coincides with the values at 355 and 532 nm in the layer with the highest depolarization, i.e. in the dust-dominated layer the Angström exponent approaches 0. The backscatter coefficient at 1064 nm could not be derived because this channel was replaced by the RRL channel.

It can be noted that the maximum of the 1064-nm extinction is found above the extinction maxima at 355 and 532 nm. Moreover, this maximum is found where the backscatter gradient is strongest for 355 and 532 nm at around 3.5 km height. This fact suggests that the suppression of the elastically backscattered light is not sufficient and a significant crosstalk is still present. Consequently, the additional contribution of 1064-nm light affects the derivative of the logarithm of the RRL signal and thus affects the calculated extinction profile.

In a second case, we present a measurement with two Polly-XT systems performed on 13 January 2017 in Leipzig. One of the lidars (OCEANET) was equipped with the RRL filters the other one (NOA) measured with its 1064 nm filter. With this combination we were able to measure the backscatter coefficient with the Raman method and thus the lidar ratio at 1064 nm, too. Figure 5 shows the derived products at 1064 nm for the first time measured with Polly-XT.

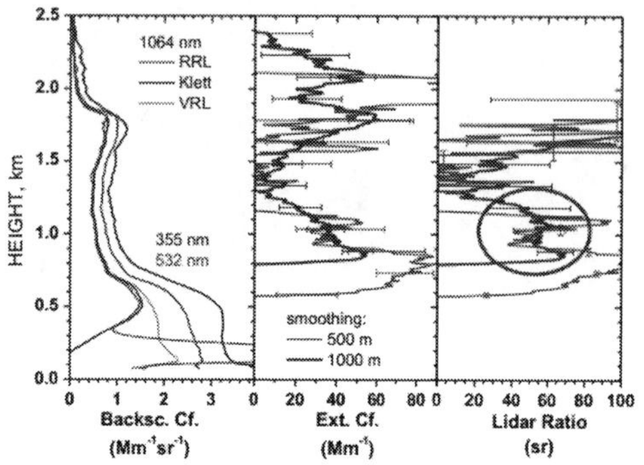

***Figure 5:** Lidar profiles measured with Polly-XT on 13 Jan. 2017 from 2035 to 2130 UTC in Leipzig. Left: backscatter coefficient at 355, 532, and 1064 nm. Center: extinction coefficient at 1064 nm with two different smoothing lengths. Right: corresponding lidar ratios. The green circle indicates a "trustworthy" height region. Standard signal-noise errors are shown in addition.*

The backscatter coefficients at 355, 532, and the orange 1064 nm profile were calculated with the vibrational-rotational Raman lidar (VRL) method. For 1064 nm we used the 607 nm signal. The Klett solution (brown) and the RRL solution (red) are also shown. The RRL profile at 1064 nm is erroneous below 600 m height because the two different lidars don't show the same overlap function. As seen in the previous case, the RRL filters show crosstalk with the elastically backscattering so that the extinction and lidar ratios are not yet trustworthy were gradients in the backscatter profiles are present. The green circle indicates a height region at which no gradient is present and a lidar ratio of 50-60 sr is obtained.

5 OUTLOOK

We found that the interference filter based RRL method at 1064 nm could be even applicable in a relatively small system like Polly-XT. However, the signals are very weak due to the low quantum efficiency of the 1064-nm PMT and an averaging time period of several hours is needed. Further improvements would include a more sensitive detector as well as additional blocking of the cross talk from elastically backscattered light at 1064 nm.

ACKNOWLEDGEMENT

We gratefully thank Igor Veselovskii for his work on the calculation and the initial order of the RRL filters from Alluxa. The research leading to these results has received partly funding from the European Union Seventh Framework Programme (FP7/2007-2013) under grant agreement n° 262254. The Polarstern cruise was supported by the Alfred-Wegener Institute (AWI_PS102_00). Polly-XT NOA received funding from the European Union's Seventh Framework Programme for research, technological development and demonstration (in the framework of BEYOND funded under: FP7-REGPOT-2012-2013-1) under grant agreement n° 316210.

References

[1] Wandinger, U., 2005: Raman Lidar, in: Lidar – Range-Resolved Optical Remote Sensing of the Atmosphere, edited by: C. Weitkamp, *Springer – Series in Optical Sciences*, New York, 241–271.

[2] Behrendt, A., T. Nakamura, M. Onishi, R. Baumgart, and T. Tsuda, 2002: Combined Raman lidar for the measurement of atmospheric temperature, water vapor, particle extinction coefficient, and particle backscatter coefficient. *Appl. Opt.*, **41** (36), 7657–7666.

[3] Veselovskii, I., D. N. Whiteman, M. Korenskiy, A. Suvorina, D. Pérez-Ramírez, 2015: Use of rotational Raman measurements in multiwavelength aerosol lidar for evaluation of particle backscattering and extinction, *Atmos. Meas. Tech.*, **8**, 4111–4122.

[4] Ansmann, A. and D. Müller, 2005: Lidar and atmospheric aerosol particles, in: LIDAR – Range-resolved optical remote sensing of the atmosphere, edited by: C. Weitkamp, *Springer Series in Optical Sciences*, New York, 105–141.

[5] Haarig, M., R. Engelmann, A. Ansmann, I. Veselovskii, D. N. Whiteman, D. Althausen, 2016: 1064 nm rotational Raman lidar for particle extinction and lidar-ratio profiling: cirrus case study, *Atmos. Meas. Tech.*, **9**, 4269–4278.

[6] Althausen, D., D. Müller, A. Ansmann, U. Wandinger, H. Hube, E. Clauder, and S. Zörner, 2000: Scanning 6-Wavelength 11-Channel Aerosol Lidar, J. *Atmos. Oceanic Technol.*, **17**, 1469–1482.

[7] Engelmann, R., T. Kanitz, H. Baars, B. Heese, D. Althausen, A. Skupin, U. Wandinger, M. Komppula, I. S. Stachlewska, V. Amiridis, E. Marinou, I. Mattis, H. Linné, A. Ansmann, 2016: The automated multiwavelength Raman polarization and water-vapor lidar PollyXT: the neXT generation, *Atmos. Meas. Tech.*, **9**, 1767–1784.

[8] http://polly.tropos.de (1 February 2017).

LIDAR USING THE BACKSCATTER AMPLIFICATION EFFECT

Igor A. Razenkov*, Victor A. Banakh

*V.E. Zuev Institute of Atmospheric Optics, Tomsk, 634055, RUSSIA, *Email: lidaroff@iao.ru*

ABSTRACT

Experimental data proving the possibility of lidar measurement of the refractive turbulence strength based on the effect of backscatter amplification (BSA) are reported. It is shown that the values of the amplification factor correlate with the variance of random jitter of optical image of an incoherent light source depending on the value of the structure constant of the air refractive index turbulent fluctuations averaged over the probing path. This paper presents the results of measurements of the BSA factor in comparison with the simultaneous measurements of the BSA peak, which is very narrow and only occurs on the laser beam axis. It is constructed the range-time images of the derivative of the amplification factor gives a comprehensive picture of the location of turbulent zones and their temporal dynamics.

1. INTRODUCTION

It is known that, on double passage paths in random media, the mean intensity of the return optical wave in the strictly backward direction to the wave source can exceed the intensity of the wave reflected by a target in the same direction in a free space [1–4]. An increase of the mean intensity of the reflected wave in the strictly backward direction in random media as compared with that in a free space is referred to as the effect of backscatter amplification (BSA) [5-7]. This effect is caused by the correlation of the forward and backward waves traversing the same random inhomogeneities of the medium.

The BSA lidar is an aerosol micro pulse lidar (MPL) with one transmitting channel and two receiving (axial and off-axis) channels [8,9]. The axial receiving channel coincides with the transmitting channel. It records the echo signal in the direction of the axis of the probing beam. The receiving aperture of the off-axis channel is separated from the receiving aperture of the axial channel by the distance exceeding the correlation scale of intensity fluctuations in the probing beam. The ratio of the powers of lidar echo signals recorded in the axial and off-axis receiving channels determines the amplification *N*-factor [9].

This paper presents the results of measurements of the BSA factor in comparison with the simultaneous measurements of the variance of random jitter of optical image of an incoherent light source, whose value is directly proportional to the strength of refractive turbulence at the probing path. The results of scanning the BSA-peak and space-time mapping of the *N*-factor derivative are given.

2. LIDAR SETUP

The design of the lidar is shown schematically in Fig. 1. A laser beam 5 mm in diameter from fiber laser 1 (532 nm) is directed by mirror 2 to 50/50 beam splitter 3. After beam splitter 3, the beam is launched into the atmosphere through mirrors 4 and 5, afocal telescope 6-7, and round hole 9 (45 mm in diam.) in screen 8.

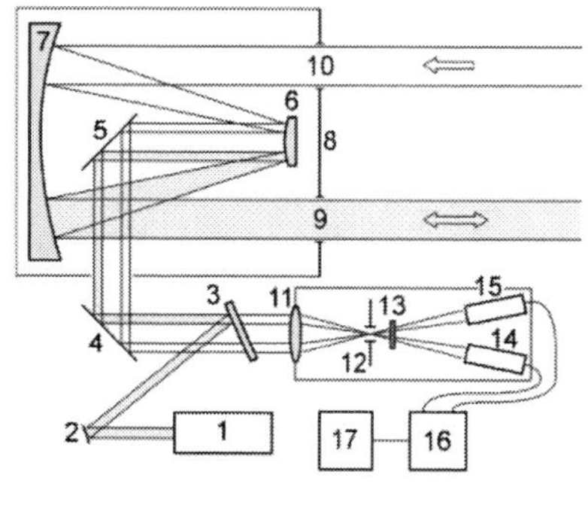

Fig. 1. Block diagram of the micro pulse BSA lidar.

The radiation backscattered in the atmosphere comes to the axial receiving channel through the same hole 9 in screen 8, the probing beam goes out. The backscattered radiation comes into the off-axis receiving channel through another hole 10 in screen 8, which is placed symmetric about the telescope axis.

After telescope 6-7, mirrors 5 and 4, and beam splitter 3, the scattered radiation of the axial and off-axis channels comes into the receiving box with a focusing lens 11 at an entrance. Aperture diaphragm 12 is set at a focus of lens 11. The diameter of this diaphragm determines the lidar field-of-view (350 μrad). The diaphragm is followed by interference filter 13 and PMT photo detectors of the axial 14 and off-axis 15 receiving channels. The repetition frequency of laser pulses with duration of 5 ns and energy of 10 μJ was 50 kHz. Accumulation time was 5 min, and the spatial resolution after averaging bins (1 bin =2 m) was 50 m.

For the additional experiment we used the third receiving channel served for recording of the profile of the backscatter amplification factor across the axial receiving channel [10]. For this purpose we set up beam-splitting plate (80/20) into the outgoing laser beam and small 10 mm objective was placed in the reflected incoming beam and installed on a computer-controlled movable platform. Platform allows one to scan the power distribution of the scattered radiation from hard target across the axial receiving channel. The use of the third receiving channel was possible only in the nighttime, when background noise was low.

3. EXPERIMENT

The experiment was conducted in Tomsk city at a 2-km path between two buildings 5 and 6 (Fig. 2). The measurements were carried out in winter time of 2016 at negative open-air temperatures. The BSA lidar operated continuously in the automatic mode. The recorded profiles of the amplification factor started at a distance of 800 m and ended at 1900 m. Simultaneously with measurements of the amplification factor, we measured the variance of arrival angles of the light wave from a white disk 80 mm in diameter on black 3 being at a distance of 2 km. For registration of the disk image a CCD camera with Maksutov objective was used. This additional setup is called as "image jitter sensor" (IJS). The focal length F of the objective was 1 m, and the diameter of the entrance aperture was D=100 mm. Objective 4 of the IJS was close to the axis of the transceiving telescope of lidar 1 at a 30 cm apart.

For the time of accumulation of the lidar signal (300 sec), the laser made 15 000 000 shots, and the receiver detected from hundred to several thousands of photons in every bin (13.3 ns) depending on the distance. The CCD camera recorded images with a frequency of 15 Hz. For the time of signal accumulation 4500 frames were recorded. During the measurement, the coordinates of energy centroid of the image of the white diffuse disk were determined for every frame in real time, and then the variances σ_y^2 and σ_z^2 of turbulence-induced random displacements of the energy centroid of the image along axes y and z were calculated.

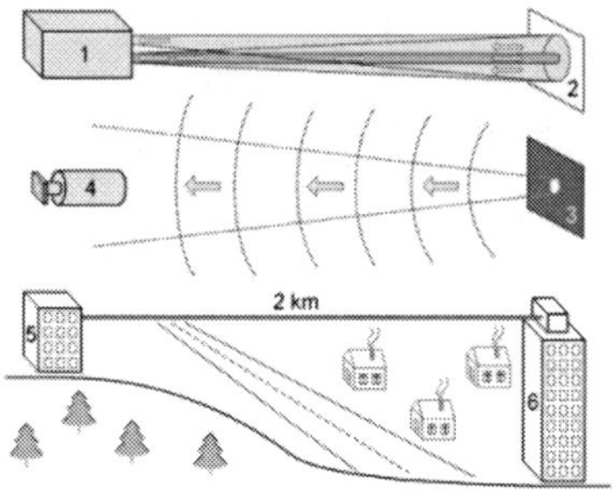

Fig. 2. Schematic of the experiment: lidar 1 and topographic target 2 (top), image jitter sensor: diffuse luminous white disk 3 on black and teleobjective with CCD camera 4 (center), probing path between buildings 5 and 6, and rough site plan (bottom).

4. BSA PEAK RESULTS

Based on the results [11], for the variance of random image jitter of an incoherent light source (diffuse disk) $\sigma_c^2 = \sigma_y^2 + \sigma_z^2$ we can obtain

$$\sigma_c^2 = c\, C_n^2\, D^{-\frac{1}{3}}\, L, \qquad (1)$$

where c is numerical constant, C_n^2 is the structure constant of turbulent fluctuations of the refractive index of air, L is the path length. In common case the variance of jitter σ_c^2 depends on the integral value of the air refractive index structure constant C_n^2 along the propagation path, and can be used for monitoring the refractive turbulence strength on the probing path.

Figure 3a shows the temporal profile of the amplification factor N at a distance of 1900 m. During this experiment was run the scanner operated continuously. Figure 3b depicts the value of the scanner signals for three scans within aria shown by double arrow on the Figure 3a, when the amplification factor was maximal. In all the scans, we can clearly see the profile of the amplification factor in the cross section of the axial receiving channel. The position of the transceiver aperture ("beam") is shown by the horizontal bar in Fig. 3(b). Half-width of the BSA peak is equal ~15..20 mm.

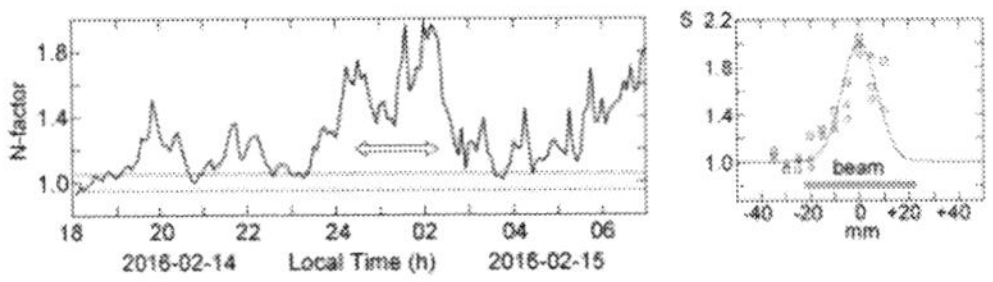

(a) (b)

Fig. 3. Temporal profile of the backscatter amplification factor N at a distance of 1900 m (a) and the normalized scanner signal S at the night (b) of February 14-15 of 2016 for three scans (○, □ and Δ) corresponding to the time interval on the left marked with an double arrow. The Gaussian approximation of experimental values is shown by the solid curve.

5. BSA-LIDAR vs. IMAGE JITTER SENSOR

In Fig. 4 the temporal profile of the amplification factor N for the same period as in Fig. 3a is compared with the root-mean-square (rms) deviation of the angular random displacements of the image energy centroid $\sigma=\sigma_c/F$. From Fig. 4 it is seen that the variations of the amplification N-factor correlate with the variations of σ. During our experiments the correlation coefficient was from 0.6 to 0.8. The temporal profile of the rms deviation σ is close to the variation of the amplification factor N in time. Thus, the value of the amplification factor depends on the refractive turbulence strength on the probing path.

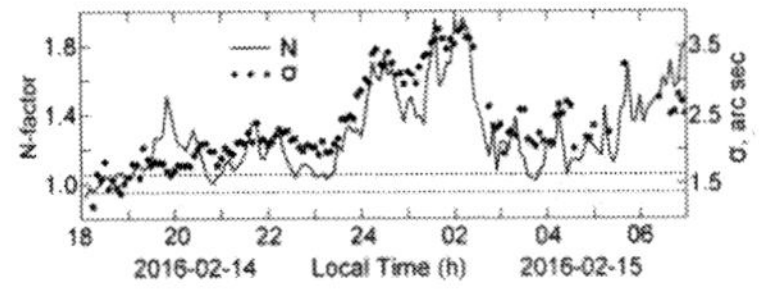

Fig. 4. Temporal variations of the amplification factor N at a distance of 1900 m (solid curve) and rms deviation of random image jitter σ (circles). Local time. Data of February 14-15 of 2016.

6. SPACE-TIME BSA-MAPPING

The function of the N-factor is a monotonic function with the positive first derivative [9,10]. If we differentiate this function, then the result is the rate of N increase along the sensing path. It allows represent the whole data array as a single range-time image of dN/dz (map). The derivative was determined from five values (dz=250 m). The procedure of the range-time mapping was applied to the data obtained on March 7 to 8 of 2016.

The data are depicted in Figure 5, and it consists of three parts. The figures show (a) integral values of the BSA factor N at the end of the 2-km sensing path, (b) standard deviation of the centroid of IJS disk image in arc seconds, and (c) range-time distribution of the derivative dN/dz. The depicted data reflect the dynamics of the process and allow to localize the areas with the weak and increased turbulence intensity.

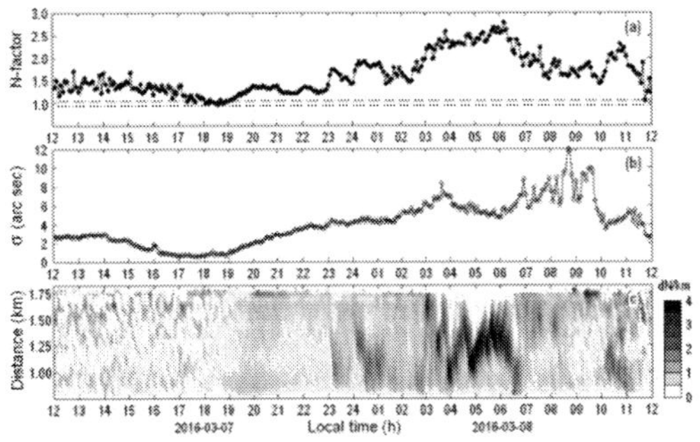

Fig. 5. Diurnal time series (a) of the integral backscatter amplification N-factor at a distance of 2 km, (b) variance of jitter of the disk image centroid at the 2-km path, and (c) range-time distribution of the derivative of the BSA factor. The correlation coefficient is equal to 0.65. Record from 12am of March 7 of 2016 to 12am of March 8 of 2016.

The comparison of the daytime and nighttime behavior of the N-factor shows that the space-time distribution is more homogeneous in the daytime, when the atmospheric temperature stratification is unstable. At 7 p.m. the neutral temperature stratification followed by the monotonic increase of the N-factor was observed. We observed the significant increase of the BSA factor in nighttime, which can be explained by appearance of significant temperature gradients.

The range-time distribution of dN/dz in Fig. 5c also allows us to explain the difference between the data of the BSA-lidar and the IJ-sensor, which can be seen in Figs. 5a and 5b. The point is that the image jitter sensor recorded the maximal values of the variance near 8-9 a.m. (Fig. 5b), while the data of the BSA-lidar by this time decreased and showed the intensification at a level lower than two (Fig. 5a). If we refer to Fig. 5c, then we can see that by 7 a.m. the zone of intense turbulence became weaker and shifted toward the path end. However, for the IJ-sensor, just this section of the path is a major relative contributor to the sensor readings. Consequently, the discrepancy between the data of the BSA-lidar and the IJ-sensor is the caused by the inhomogeneous distribution of the turbulence intensity along the sensing path.

7. CONCLUSIONS

Thus, the paper has presented brief experimental results demonstrating the possibility of the BSA lidar measurement of the refractive turbulence strength averaged over the probing path. The incorporation of the third receiving channel into the BSA lidar has allowed us to obtain experimentally the profiles of the backscatter amplification factor over the receiving aperture.

The measured values of the amplification factor have been compared with the measured variance of random jitter of the optical image of the incoherent light source, which is directly proportional to the integral value of the structure constant of turbulent fluctuations of the refractive index of air on the probing path. The variations of the amplification factor in time are synchronous with the variations of the rms deviation of random image jitter. The high correlation between the amplification factor and the synchronously measured jitter rms deviation proves the possibility of lidar measurement of the refractive turbulence strength based on the effect of backscatter amplification.

Winter field lidar experiment in Tomsk city was carried out at the 2-km path to record the profile of the backscatter amplification coefficient. The differentiation of individual profiles of the BSA factor and drawing of their range-time image have allowed us to visualize the processes occurring in the atmosphere. It became possible to localize the zones with the increased turbulence intensity and to follow their temporal dynamics. The inhomogeneous emission of heat at night from heating of private houses leads to the smooth shift of turbulent zones, which can be followed remotely with the backscatter amplification lidar.

Under conditions of fine cloudless weather, the behavior of the BSA factor in daytime is quite predictable and correlates exactly with the amount of heat incoming to the surface. In the nighttime, the behavior of the BSA factor becomes hard to predict, because of a few different factors influencing turbulence in the air surface layer.

ACKNOWLEDGMENTS

The authors are grateful to A.S. Gurvich for initiation of the study, attention, and support, as well as to A.I. Nadeev for the help in design of instrumentation and consultations. Russian Scientific Foundation for Maintenance and Development, Agreement #14.607.21.0151.

REFERENCES

1. A.G. Vinogradov, Yu.A. Kravtsov, and V.I. Tatarskii, Radiophysics and Quantum Electronics **16**, 818 (1973).
2. A.S. Gurvich, and S.S. Kashkarov, Radiophysics and Quantum Electronics **20**, 547 (1977).
3. V. A. Banakh, and V.L. Mironov, *Lidar in a turbulent atmosphere* (Artech House. Boston & London. 1987. 185 p.)
4. A.S. Gurvich, Izvestiya, Atmospheric and Oceanic Physics **48**, 585 (2012).
5. V.A. Banakh, "Enhancement of the laser return mean power at the strong optical scintillation regime in a turbulent atmosphere", Atmospheric and Oceanic Optics **26**, 90–95 (2013).
6. I.N. Smalikho, "Calculation of the backscatter amplification coefficient of laser radiation propagating in a turbulent atmosphere using numerical simulation", Atmospheric and Oceanic Optics **26**, 135–139 (2013).
7. Yu.A. Kravtsov, A.I. Saichev, Physics-Uspekhi, **25**, 494 (1982).
8. I.A. Razenkov, V.A. Banakh, F.I. Nadeev. Russian Federation Patent No.153460, 2015.
9. V.A. Banakh, I.A. Razenkov, and I.N. Smalikho, Applied Optics, **54** (24), 7301 (2015).
10. V.A. Banakh and I.A. Razenkov. Optics Letters, **41** (19), 4429 (2016).
11. S. Manning, B.A. Clare, K.J. Grant, and K.A. Mudge, Optical Engineering **54** (11), 114104-1 (2015).

NEW CAPABILITY FOR OZONE DIAL PROFILING MEASUREMENTS IN THE TROPOSPHERE AND LOWER STRATOSPHERE FROM AIRCRAFT

Johnathan Hair[1], Chris Hostetler[1], Anthony Cook[1], David Harper[1], Anthony Notari[1], Marta Fenn[2], Mike Newchurch[3], Lihua Wang[3], Shi Kuang[3], Travis Knepp[2], Sharon Burton[1], Richard Ferrare[1], Carolyn Butler[2], Jim Collins[2], Amin Nehrir[1]

[1]*NASA Langley Research Center Hampton, VA USA: Johnathan.W.Hair@nasa.gov*
[2]*Science Systems and Applications, Inc., Hampton, VA USA,* [3]*University of Alabama in Huntsville, Huntsville, AL USA*

ABSTRACT

Recently, we successfully demonstrated a new compact and robust ozone DIAL lidar for smaller aircraft such as the NASA B200 and the ER-2 high-altitude aircraft. This is the first NASA airborne lidar to incorporate advanced solid-state lasers to produce the required power at the required ultraviolet wavelengths, and is compact and robust enough to operate nearly autonomously on the high-altitude ER-2 aircraft. This technology development resulted in the first new NASA airborne ozone DIAL instrument in more than 15 years. The combined ozone, aerosol, and clouds measurements provide valuable information on the chemistry, radiation, and dynamics of the atmosphere. In particular, from the ER-2 it offers a unique capability to study the upper troposphere and lower stratosphere.

1 INTRODUCTION

NASA Langley's current airborne lidar Ozone and Aerosol Lidar instrument provide profiles of ozone via the Differential Absorption Lidar (DIAL) technique and aerosol and cloud optical properties via the High Spectral Resolution Lidar (HSRL) technique. This lidar has been a much relied-upon facility-class instrument for deployment on the larger aircraft platforms (i.e., DC-8) over the last three decades, having been deployed on 33 major field experiments focused on investigation of regional- and global-scale processes within tropospheric and stratospheric, satellite validation, and model assessments [1]. Ozone lidar measurements continue to be a requirement on many NASA-sponsored airborne process studies and validation campaigns such as the recent and successful SEAC4RS and KORUS campaigns and will be important for future satellite (e.g. TEMPO and SAGE-ISS) validation campaigns.

This abstract provides a technical overview of major subsystems of this new instrument, an overview of potential new investigations, and highlight a few examples of ozone and aerosol profiles acquired during flights in January 2016 and April 2016 from the B200 and ER-2 aircraft platforms, respectively. In addition, the ozone profiles measured by the lidar are compared with profiles from ozonesondes and ground-based TOLNet ozone lidar systems.

2 METHODOLOGY

This new system is designed to combine aerosol/cloud and the trace gas ozone measurements into a single integrated instrument. The driving requirement was to develop a lidar system that is much smaller, lighter, and uses less power compared to previous generations. In addition, to fly on the ER-2 aircraft the system had to operate autonomously or with very limited remote control.

2.1 Laser Subsystem

A critical development was the laser systems. The laser system consists of a Nd:YAG pump laser built by Fibertek Inc. and a nonlinear optics module (NLO) that was built by IIT Industries with modifications by NASA LaRC and Welch Mechanical Designs, LLC. The NLO consists of two Optical Parametric Oscillators (OPOs) and two crystals for Sum Frequency Generation (SFG). The pump laser is a seeded single frequency 1064nm laser that has a single oscillator and a single preamplifier. The light from the pump laser is then split to two amplifier paths. This results in two output beams with 100mJ per pulse at 200Hz (20W) in each output. One of the amplifier outputs is frequency doubled (SHG) and then frequency tripled (THG). The other amplifier output is used to pump the two OPOs that generate 1608nm and 1945nm which are then mixed with 355nm in two SFG crystals to produce the 290nm and 300nm ozone DIAL wavelengths.

Nd:YAG Laser (bottom in *Figure 1*)
(Fibertek Inc.)

- 40W laser, 200Hz pulse rate, seeded single frequency
- 20W leg to pump dual Optical Parametric Oscillators (OPO)
- 20W leg to generate 355nm, 532nm, 1064nm
- Aerosol and cloud measurements are conducted at 355, 532, 1064nm

Nonlinear Optics Module (top in *Figure 1*)
(ITT Industries, Welch Mechanical Designs, NASA LaRC)

- 1W @ 290nm and 0.5W @ 300 nm, 200Hz
- Dual OPOs: 1608nm and 1945nm
- Sum frequency generation (SFG): mix OPO and 355nm to generate Ozone DIAL wavelengths

Summary of Transmitted Wavelengths

- Transmit single frequency 3.6W @ 1064nm, 2.5W @532nm, and 3.5W @ 355nm wavelengths required for HSRL measurements.
- Transmit 1W @ 290nm and 0.5W @ 300nm

Figure 1. Picture of the Nd:YAG laser (bottom) and the nonlinear optics module (top) that contain dual OPOs and SFG mixers.

2.2 Receiver Subsystem

The receiver consists of multiple channels for the aerosol/cloud and ozone measurements. The system incorporates the HSRL technique at both 532nm and 355nm using an iodine vapor filter and Michelson Interferometer, respectively. The 1064nm channels are detected using the standard backscatter approach with calibration from the 532nm channel. The 532nm and 1064nm retrievals are implemented similar to Hair et al. (2010) [2]. The system also measures the linear depolarization ratio for all three wavelengths [3]. The system has flown on multiple campaigns with the aerosol/cloud measurement channels since 2012. Here we report additional channels that are included in the system to enable ozone profiling through the Differential Absorption Lidar (DIAL) technique focused on measurements in the lower stratosphere and troposphere. These ozone wavelengths are transmitted simultaneously and are split in the receiver with a dichroic filter where each channel has 1nm wide interference filter before the photomultiplier detectors.

2.3 Overall System Design

As noted above, a main objective was to reduce the overall size, weight, and power requirements of this new instrument compared to the current airborne ozone DIAL system that Langley has flown over the past 3 decades. This was accomplished mainly by the laser system approach and design along with the supporting custom electronics and chillers. In addition, considerable effort was put forth in the design of the receiver and instrument structure. The telescope incorporates a 40cm diameter all-metal telescope and a dual-sided aft-optics breadboard. A comparison of the two system sizes is provided in Figure 2 and Table 2. The new system is configured to view downward (e.g. from ER-2) in contrast to the current DIAL system, which simultaneously views in the nadir and zenith directions (e.g. from DC-8).

Table 2. Comparison of size, weight, and power of the new system compared to the current system.

	UV DIAL	HSRL/ DIAL	Change
Volume(m^3)	3.7	0.34	11x
Weight(kg)	907	135	7x
Power(kW)	9.75	1.7	6x

 https://doi.org/10.1051/epjconf/201817601006

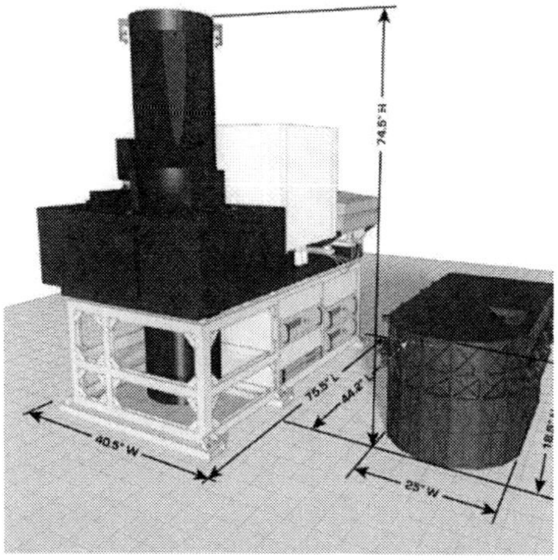

Figure 2. Comparison of the size of the current (left) and the new (right) lidar transceivers.

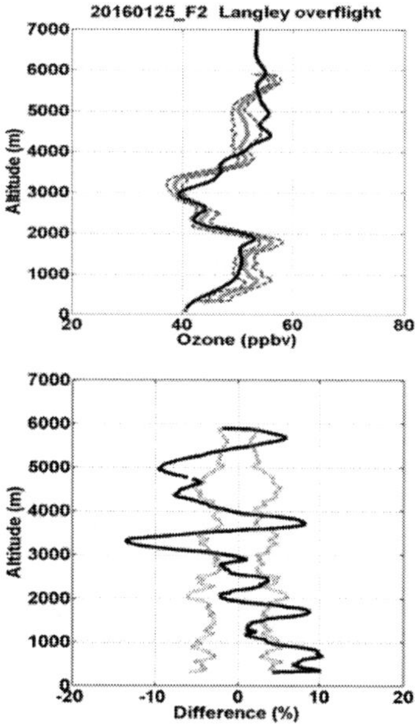

Figure 3. Comparison of lidar and ozonesonde profiles measured 25 January 2016. The top plot shows the sonde (black) and the lidar (green) ozone mixing ratio. The red lines show the ozone standard deviation over the measured interval for the lidar. The bottom plot shows the difference (black) in the ozone profile and the standard deviation over the averaging interval of the lidar.

3 RESULTS

Two series of test flights were conducted on the NASA B200 aircraft from Hampton, VA in January 2016 (5 flights) and on the ER-2 aircraft from Palmdale CA in April 2016 (7 flights). The flights conducted from the B200 focused on evaluating the measurement performance of the ozone measurement and included overpasses of three TOLNet ground-based lidar systems: LMOL in Hampton VA, TropOz in Greenbelt MD, and RO_3QET in Huntsville AL. Ozonesondes were launched at three locations (Hampton and Wallops Island VA and Huntsville AL) during overpasses of the B200 aircraft. Airborne lidar measurements from 25 January 2016 are compared to measurements from an ozonesonde launched in Hampton, VA (Figure 3) showing a typical ozone profile. The lidar profile (green) compared favorably with the sonde profile (black) with a maximum and minimum difference of -13.5% and 10% and the mean standard deviation over the profile of 5.9%. There was no mean bias over the profile for this particular sonde comparison. The average of all six ozonesonde comparisons resulted in a mean profile bias of -1.2% and a mean profile standard deviation of 5.7%.

The seven ER-2 flights focused mainly on operational performance of the instrument, including flying at altitudes up to ~20km to assess the quality of the measurements within the lower stratosphere and near the tropopause. A flight conducted on 13 April 2016 flew from Southern California (lat=34.6N) south to 15N over the Pacific Ocean. The ozone curtain measured during the transit to this southern point is shown in Figure 4. The profiles extended down to the cloud tops near 2km altitude after the ER-2 descended to 18km. The unique capability of the lidar clearly observes the large-scale features in the ozone horizontal and vertical distribution. The lidar ozone profiles revealed a large tropopause fold that extended nearly over the entire flight track. The small aerosol layers, shown in Figure 4 (bottom) correlated with the increased ozone in the mid-troposphere. A sudden decrease in ozone is observed while crossing the Inter-Tropical Convergence Zone (ITCZ). The data also highlights the transport of lower ozone due to the

https://doi.org/10.1051/epjconf/201817601006

large-scale dynamics related to convection in the tropics and the Brewer-Dobson circulation.

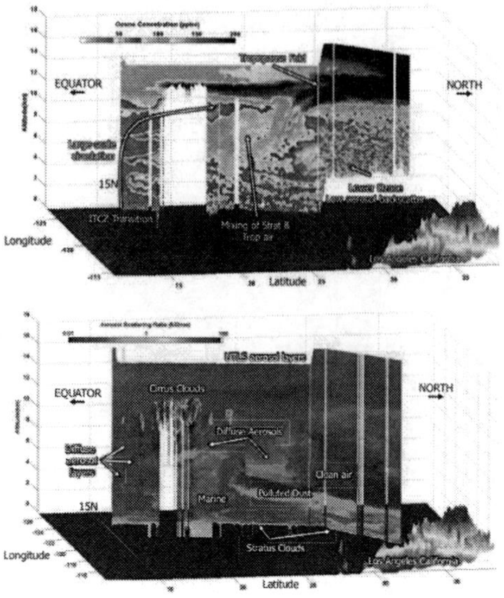

Figure 4. Ozone distribution (top) and the 532nm aerosol to molecular backscatter ratio (bottom) measured on flight from 13 April 2016.

4 CONCLUSIONS

The first NASA airborne ozone lidar developed to fly on both small aircraft (B-200) and high-altitude aircraft (ER-2) has successfully demonstrated. The instrument performed well on all the 12 flights conducted to date and demonstrated reliable and hands-off performance of dual OPO-based UV lasers at 290nm and 300nm. These flights provided the first opportunity to assess the measurement performance. The lidar ozone profiles compared well with profiles from coincident ozonesonde launches to within 1.2% with a variation of 5.7% on average. The airborne lidar profiles also compared well with profiles from the ground-based TOLNet DIAL lidar at University of Alabama in Huntsville. Comparisons to the other TOLNet sites have yet to be analyzed. The flights from the ER-2 demonstrated the lidar performance in a near autonomous operation and provided unique profiles of the lower stratosphere and upper troposphere ozone and aerosol distributions. In addition, current plans are to implement a recently developed Fibertek Inc. OPO laser system that operates at 304 and 316nm. This laser system is currently being integrated with the same pump laser described above and will enable profiling from 20km to the surface at all latitudes.

ACKNOWLEDGEMENTS

This technology development supported through the NASA Earth Science Technology Office (ESTO) and the NASA Science Mission Directorate (SMD), Research and Development Program.

References

[1] Browell, E. V., S. Ismail, and W. B. Grant. "Differential absorption lidar (DIAL) measurements from air and space." *Applied Physics B: Lasers and Optics* 67.4 (1998): 399-410.

[2] Hair, Johnathan W., et al. "Airborne high spectral resolution lidar for profiling aerosol optical properties." *Applied optics* 47.36 (2008): 6734-6752.

[3] Burton, S. P., et al. "Observations of the spectral dependence of linear particle depolarization ratio of aerosols using NASA Langley airborne High Spectral Resolution Lidar." *Atmospheric Chemistry and Physics* 15.23 (2015): 13453-13473.

CO_2-TEA PULSE CLIPPING USING PULSED HIGH VOLTAGE PRE-IONIZATION FOR HIGH SPATIAL RESOLUTION I.R. LIDAR SYSTEMS

Taieb Gasmi
Division of Science & Engineering, Saint Louis University-Madrid Campus
Avenida del Valle, 34, Madrid 28003, Spain. E-mail: tgasmi@slu.edu

ABSTRACT

An extra-cavity CO_2-TEA laser pulse clipper for high spatial resolution atmospheric monitoring is presented. The clipper uses pulsed high voltage-to facilitate the breakdown of the gas within the clipper cell. Complete extinction of the nitrogen tail, that degrades the range resolution of LIDARS, is obtained at pressures from 375 up to 1500 Torr for nitrogen and argon gases whereas an attenuation coefficient of almost 10^2 is achieved for helium. Excellent energy stability and pulse width repeatability were achieved using high voltage pre-ionized gas technique.

1. INTRODUCTION

The TEA transversely excited atmospheric CO_2 laser pulse temporal profile typically has a tall, sharp spike followed by a long, drawn out nitrogen tail. Although the amplitude of the nitrogen tail is not nearly as high as the initial spike, the tail is drawn out over a time interval of 2–5 μs and holds up to two-thirds of the total energy of the pulse. The nitrogen tail occurs due to the long lifetime of the nitrogen molecule vibrational energy level used to excite the CO_2 molecules and initiate lasing. A sufficient number of nitrogen molecules remain excited after the initial spike to continue the lasing process at a lower power level for the duration of the tail. If there is some low level of near-field scatter into the receiver, this effect can be quite significant, as the aerosol backscatter return is of the order of 10^{-4} of the transmitted power within this wavelength interval. In this case, the unclipped near-field scattered signal can interfere with the weak signal backscattered from the atmospheric aerosols. This type of pulse limits the range resolution of LIDARS to some hundreds of meters unless pulse-deconvolution techniques [1] are adopted. Short laser pulses can also be obtained by various techniques such as mode locking, free induction decay, pulse slicing with electro-optic switched [2-5]. However, output pulses from these require further amplification for any useful application due to their very low energy content. Moreover, laser self-induced gas breakdown clippers are unpredictable due to the large jitter characterizing their operations. They also tend to generate clipped pulses of unstable energies and pulse widths. The use in this work of a HV pulse to pre-ionize the sealed clipper gas prior to the arrival of the laser pulse achieved excellent energy stability and pulse width repeatability of the clipped pulse. The clipper hence achieves high range-resolved remote sensing of atmospheric constituents.

2. RESULTS AND DISCUSSION

The laser beam is usually focused inside the sealed clipper to high field intensities to create self-induced plasma which acts as a pulse clipper. Both higher breakdown levels and long-term stability operations of the clipper dictated the choice of helium, argon, and nitrogen as filling gases [6-8]. The sketch of the CO_2-TEA laser pulse clipper is shown in figure 1. The HeNe laser is used for alignment purposes.

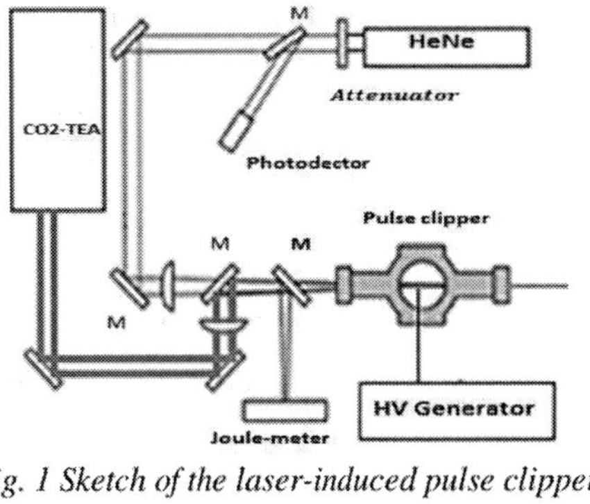

Fig. 1 Sketch of the laser-induced pulse clipper.

The breakdown of the filling gas is hence provided by two separate channels. The first channel is purely electrical and provides a high voltage pulse whose amplitude is just below the gas breakdown threshold but is sufficient to facilitate the plasma initiation by providing starting electrons for the avalanche ionization process for gas breakdown. The second channel is optical provided by the focused CO_2-TEA laser pulse, whose field intensity is far superior to reach or even exceed the threshold of the gas breakdown in the clipper. This electrical-optical sequenced mechanism also achieves fast triggering of the plasma with minimal jitter. Laser-induced clipping in helium and nitrogen is shown in Fig. (2, 3) at the 9P(20) and for argon Fig. 4 at the 10P(20) laser lines. Attenuation of the clipped portion of the laser pulse within 500 ns of the breakdown is approximately 10^2. Average clipping efficiency for helium at pressures of 600–900 Torr is approximately 75%. The latter is far superior than that reported by other authors [2] for a range of pressures extending from 200 to 300 Torr. Complete extinction of the nitrogen tail is obtained at pressures extending from 375 up to 1500 Torr for the nitrogen and argon gases. Attenuation of the clipped portion of the laser pulse is approximately 10^5, which is three orders of magnitude higher than that of helium. No pulse recovery occurs after clipping takes place; nevertheless, the nitrogen and argon plasma clipping drains out almost 40% and 50% of the laser energy, respectively. The evolution the pulse width versus cell pressure for three gases is illustrated in Fig. 5. The pulse width decreases gradually as the pressure of helium and nitrogen tends to 300 Torr and then decreases rapidly between 300–600 Torr to reach a nearly asymptotic value at pressures higher than 700 Torr. As for argon, the pulse decreases very sharply from a nitrogen-tailed pulse to a clipped pulse at pressures as low as 150 Torr. Figure 6 shows the LIDAR return signals where the solid-line curve represents the clipped pulse characterized both by the normal exponential decay of the single gain-switched spike and total absence of the hump, which was initially associated with the nitrogen tail of the TEA–CO_2 pulse. The dotted curve represents the return of the raw pulse.

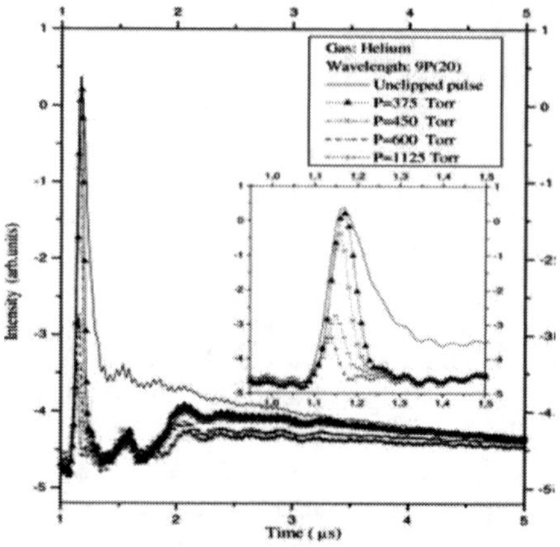

Fig. 2: Temporal profiles of clipped laser pulse versus helium cell pressure.

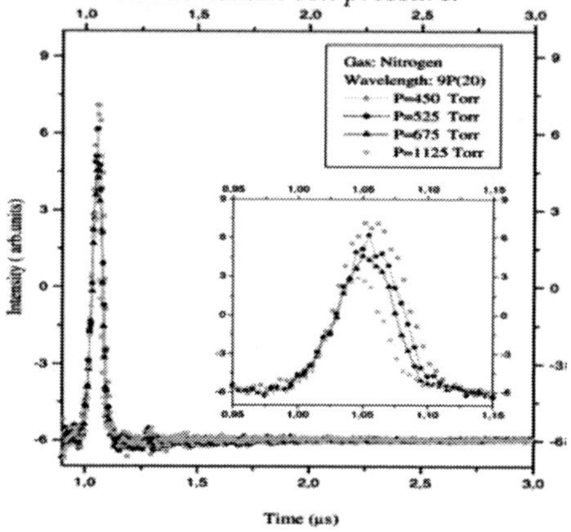

Fig. 3: Temporal evolution of the clipped laser pulses in presence of nitrogen at the various pressures.

Figure 7 shows the threshold breakdown intensities versus the 75–1500 Torr pressure range of helium, argon, and nitrogen. Gas breakdown intensity for the 10(P20) line have been plotted on a log(I_{th}) versus log(P) scale and fitted to a straight line with 1%–8% standard deviation of the fits. For helium and argon, the threshold intensities extend from 3×10^{11} Wcm^{-2} –5×10^{12} Wcm^{-2} and from 2×10^{11}Wcm^{-2}–2×10^{12}Wcm^{-2}, respectively. The plot shows no evidence for the plateau that illustrates the multiphoton ionization (MPI) processes in gases [3] but does confirms the predominance of the collision ionization process due to the IR emission band of TEA–CO_2 lasers where the laser-induced breakdown is expected to be driven by an avalanche ionization mechanism [4]. For nitrogen, the threshold breakdown intensities extend from 3×10^{13}Wcm^{-2}–2×10^{14} Wcm^{-2} which is rather higher than that of helium and argon due to the additional vibrational energy losses inherent to the nitrogen molecule.

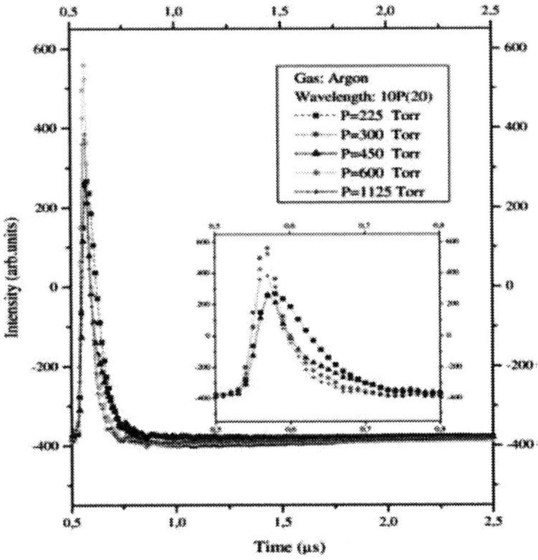

Fig. 4: Temporal profiles of the clipped laser pulse in the presence of argon as a function of pressure.

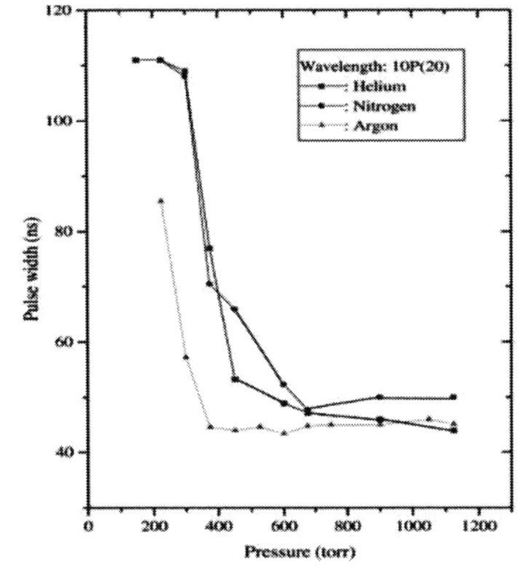

Fig.5: Evolution in time of the clipped pulse width as a function of pressure for helium, nitrogen and argon

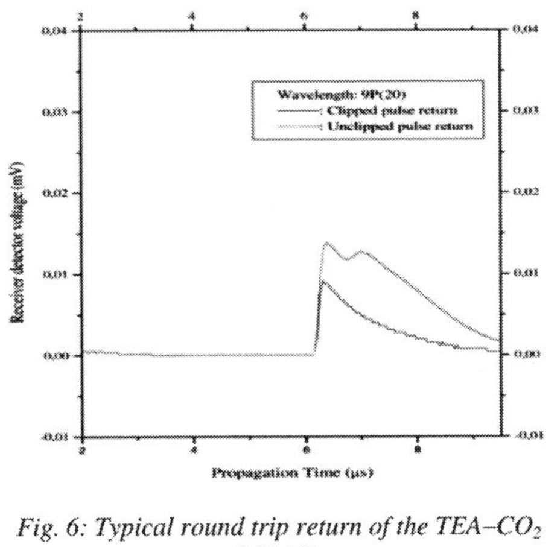

Fig. 6: Typical round trip return of the TEA–CO_2 LIDAR.

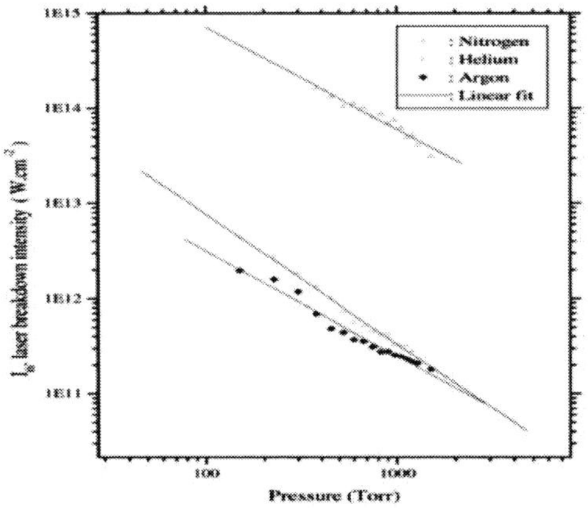

Fig.7: Threshold breakdown intensities for nitrogen, helium and argon versus cell pressure.

Laser pulse energy and stability were studied and it was found that the mean pulse energy was 490 mJ; the standard deviation in energy was 9 mJ as shown in Fig. 8.

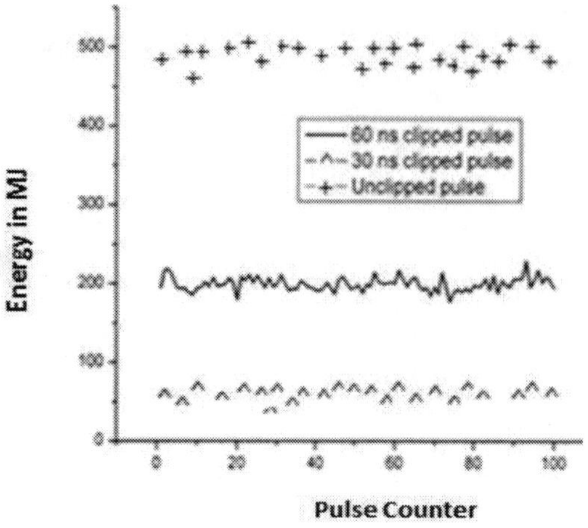

Fig.8: *Pulse stability over 100 consecutive pulses for 30 and 60 ns clipped pulses and the unclipped pulse.*

3. CONCLUSION

We present a high stability and energy-efficient laser-induced plasma clipper to clip the nitrogen tail of CO_2-TE based LIDAR. Optimum pressures for helium, argon, and nitrogen, that provide the best stability of the transmitted energy and complete extinction of the nitrogen tail, are located within the 450–600 Torr interval. Excellent range resolutions can now be achieved with TEA–CO_2 laser-based LIDAR systems.

References

[1] Y.J. Park, H.J. Kong; 1997: "iterative deconvolution algorithm for improving range resolution of a CO_2 laser based lidar system", Optical Review, Springer

[2] N. Hurst, S. S. Harilal, 2009: Pulse shaping of CO_2-TEA laser using a simple plasma shutter, Rev. Sci. Ins. 80.

[3] Y. Qu; X. Hu; D. Ren, B. Zhou, W. Song, 2001: Plasma shutter for pulse shaping of CO_2-TEA laser. Proc.of SPIE 4914, High-Power Lasers and Applications II, 273.

[4] I. Tanaka, 1990: High-efficiency, all-solid-state exciters for high-repetition-rated, high-power CO_2–TEA lasers, Rev. Sci. Instrum. 61: 2092–2096.

[5] T. Donnelly et. al, 2012: CO_2 laser pulse shortening by laser ablation of a metal target. Review of Scientific Instruments, 83.

[6] V. Parthasarathy, S. Nad, K. Rao, S.K. Sarkar, 1998: Intensity and pressure effects in CO_2 laser induced carbon-13 enrichment with temporally modified pulses, J. Photochem. Photobiol. A: Chem. 115, 1.

[7] C.L.M. Ireland, C.G. Morgan, 1973: Laser-induced breakdown of gases, J. Phys. D 6, 720.

[8] R. Tambay, R.K. Thareja, 1991: Laser induced breakdown of Ar, N_2 and O_2 gases using 1.064, 0.532, 0.355 and 0.266μm radiation, J. Appl. Phys. 70, 2890.

LIDAR SPRECTROSCOPY INSTRUMENT (LISSI): AN INFRASTRUCTURE FACILITY FOR CHEMICAL AEROSOL PROFILING AT THE UNIVERSITY OF HERTFORDSHIRE

Matthias Tesche[1,*], Boyan Tatarov[1], Youngmin Noh[2] and Detlef Müller[1,2]

[1]*University of Hertfordshire, Hatfield, United Kingdom, *m.tesche@herts.ac.uk*
[2]*Gwangju Institute of Science and Technology, Gwangju, South Korea*

ABSTRACT

The lidar development at the University of Hertfordshire explores the feasibility of using Raman backscattering for chemical aerosol profiling. This paper provides an overview of the new facility. A high-power Nd:YAG/OPO setup is used to excite Raman backscattering at a wide range of wavelengths. The receiver combines a spectrometer with a 32-channel detector or an ICCD camera to resolve Raman signals of various chemical compounds.

The new facility will open new avenues for chemical profiling of aerosol pollution from measurements of Raman scattering by selected chemical compounds, provide data that allow to close the gap between optical and microphysical aerosol profiling with lidar and enables connecting lidar measurements to parameters used in atmospheric modelling.

1 INTRODUCTION

Conventional Raman or high spectral resolution lidar is used for deriving profiles of aerosol optical properties and, if proper input data are provided, retrieving aerosol microphysical properties through inversion techniques. The lidar development at the University of Hertfordshire (UH) is centred on exploring the feasibility of using Raman backscatter of various target species to obtain mass concentration profiles of selected chemical compounds. This marks the logical next step in aerosol profiling with lidar and will allow for defining aerosol types according to their chemical composition rather than optical properties. Ultimately, this will unify the treatment of aerosol types in lidar remote sensing with that of in-situ measurements and atmospheric modelling. The feasibility of the underlying methodology has been proven for measurements of Raman backscattering at silicone dioxide molecules within Asian dust layers [1, 2, 3].

2 OBJECTIVES

The lidar spectroscopy instrument (LiSsI) will enable profiling of trace gases, chemical components in particles, and bio-aerosols in atmospheric aerosol pollution through the combination of different nonlinear spectroscopy techniques (photoluminescence, fluorescence, Raman and coherent anti-Stokes Raman spectroscopy) in a single measurement platform. The work at UH includes (1) the development of an end-to-end simulator that allows us to model the processes of inelastic photoluminescence and Raman scattering by aerosol particles and gases in the context of lidar remote sensing, (2) the identification of the luminescence and Raman scattering characteristics for a set of key aerosol types and gases (of natural and anthropogenic origin) by means of laboratory experiments, (3) test measurements and case studies as a proof of concept of the technique under laboratory conditions, and (4) the characterisation and development of a mobile prototype inelastic lidar spectrometer receiver for field deployment at established lidar sites.

3 LABORATORY SETUP

Figure 1 presents a sketch of the setup of the LiSsI facility. The main components are a high power Nd:YAG laser (Continuum Powerlite Furie LD), an Optical Parametric Oscillator (Horizon OPO), a motorised beam combiner, a multiwavelength depolarisation Raman lidar receiver, a Horiba 1250M spectrometer that can be used with an intensified CCD camera (ICCD, Princeton Instruments PI-MAX4 1024i-HBf) or a 32-channel Licel PMT (32PMT), an Olympus BX51TRF-6 Raman/flourescence microscope, and several custom-made gas chambers.

An overview of the properties of the transmitter is provided in Table 1. The combination of a high-power Nd:YAG laser and an Optical Parametric Os-

 https://doi.org/10.1051/epjconf/201817601008

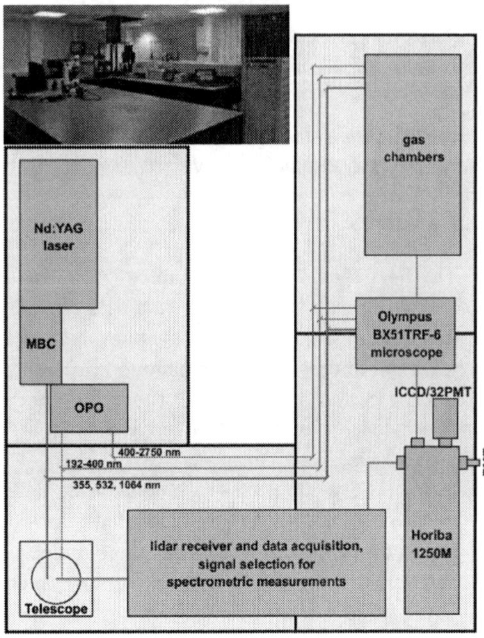

Figure 1: Sketch of the laboratory setup including a high-power Nd:YAG laser, a motorised beam combiner (MBC), an Optical Parametric Oscillator (OPO), an advanced lidar receiver, a spectrometer, an intensified CCD camera (ICCD), a 32-channel Licel PMT (32PMT), a Raman/flourescence microscope, and several gas chambers.

cillator allows for exciting Raman backscatter for a wide range of excitation wavelengths. The novel lidar receiver setup has been designed and simulated using ZEMAX and will combine the spectral resolution of spectrometers with a 32-channel detector to resolve Raman signals of a variety of chemical compounds. An overview of the configuration and properties of the receiver and data acquisition is provided in Table 2. Aerosol chambers will enable measurements of a range of chemical species to test the methodology for known compounds and to assess the respective detection limit with respect to applied laser power and resolvable particle concentrations. A Raman microscope will be used to measure the Raman scattering cross-sections of the selected target species—information that is needed to transform the detected signals into mass concentrations.

Table 1: Properties of the transmitter.

Continuum Powerlite Furie LD	
Laser type	Nd:YAG, injection seeded
Wavelength and pulse energy	7500 mJ @1064 nm 5000 mJ @532 nm nm 2500 mJ @355 nm
Beam divergence	0.5 mrad, 0.1 mrad after 5-fold beam expansion
Repetition rate	10 Hz
Linewidth	<0.003 cm^{-1}
Pulse duration	<15 ns
Horizon Optical Parametric Oscillator	
Pumping wavelength	355 nm
Wavelength and pulse energy	from 192 to 2750 nm scanning step 0.01 nm 120 mJ @400 nm 60 mJ @600 nm 25 mJ @300 nm
Beam divergence	<2 mrad (both axes)
Repetition rate	10 Hz
Linewidth	2-6 cm^{-1}, injection seeded
Pulse duration	<15 ns

The aim of the facility is to carry out precise inelastic spectroscopy experiments that target measurements of photoluminescence, fluorescence, and Raman spectra of aerosol and gas samples. These measurements will include the identification of spectra, absolute values of fluorescence, and Raman cross-sections that are currently poorly known or unknown.

4 APPLICATIONS IN LIDAR AND SPECTROSCOPY

The LiSsI facility has been designed to allow for comprehensive laboratory experiments as well as for atmospheric observations. The laser beam can be released into the atmosphere through a hatch in the roof of the laboratory. The backscattered light is collected with a 14-inch Schmidt-Cassegrain telescope and guided to the different components of the receiver (Table 2). Depending on the setup of the experiment, LiSsI can be used as multi-wavelength elastic backscatter lidar for measurements of aerosols and temperature from the troposphere to the mesosphere, as multi-channel spectroscopic Raman lidar (using Stokes and anti-Stokes, rotational, and rotational-vibrational Raman scattering), multi-channel spectroscopic photoluminescence/fluorescence lidar, high spectral resolution li-

Table 2: Properties of the receiver and data acquisition system.

Schmidt-Cassegrain telescope	
Focal length Field of view	3910 mm (14 inch) 0.5-4.0 mrad (variable)
HORIBA 1250M Research Spectrometer	
Focal length Aperture Spectral range Grating size Dispersion @500 nm Accuracy Repeatability	1.25 m F/9 0-1500 nm mechanical range (1200 g/mm grating) 110 mm × 110 mm 0.65 nm/mm ±0.15 nm ±0.005 nm
Gratings and resolutions @313.183 nm	2400 g/mm blaze 250 nm, max resolution 0.003 nm 1800 g/mm blaze 400 nm, max resolution 0.004 nm 1200 g/mm blaze 330 nm, max resolution 0.006 nm 600 g/mm blaze 500 nm, max resolution 0.012 nm
Detection	
Mie and Rayleigh scattering	355 nm, PMT HV-R9880U-20, bandwidth 1.3 nm 532 nm, PMT HV-R9880U-20, bandwidth 1.3 nm 1064 nm, APD InGaAs50, bandwidth 4 nm
Spectroscopic 1 Licel SP32-20	Hamamatsu H7260-20, 0.8 mm×7 mm×32 anodes spectral response 300-920 nm
Spectroscopic 2 Princeton Instruments PI-MAX4 ICCD camera	1024×1024 imaging pixels; 12.8 μm×12.8 μm pixels Gen III filmless intensifier Sensitive range 290-710 nm QE>20% in range 355-700 nm; QE>40% in range 410-640 nm
Data acquisition system	
Mie and Rayleigh scattering	Licel transient recorders, 16 bit, 20 MHz A/D converters and photon-counters maximum count rate 250 MHz, variable range resolution
Multi-anode PMT	Single-photon counting system, maximum count rate 100 MHz, 50 ns resolution
ICCD	Digitization 16 bit, 32 MHz, minimum gate width 2 ns

dar (HSRL), polarization lidar, or infrared absorption and differential absorption lidar (DIAL).

In addition, LiSsI opens possibilities for a wide range of applications in spectroscopy: general spectroscopy (attenuation, transmission, and reflectance spectroscopy), high-spectral resolution spectroscopy, Stokes and anti-Stokes Raman spectroscopy of gas and solid material including applications involving a microscope, fluorescence spectroscopy of gas and solid material including microscope applications, laser-induced breakdown spectroscopy, coherent-anti-Stokes Raman spectroscopy, time-resolved spectroscopy, infrared gas analysis and materials processing.

5 FIRST RESULTS

The LiSsI laser has been installed in September 2017. Figure 2 shows one of the first measurements performed between 1026 and 1330 UTC on 22 September 2017 using only a 532-nm elastic channel and a laser power of about 30 mJ. The analog counting signal is shown in the figure. Despite the low laser power, a rather small telescope and

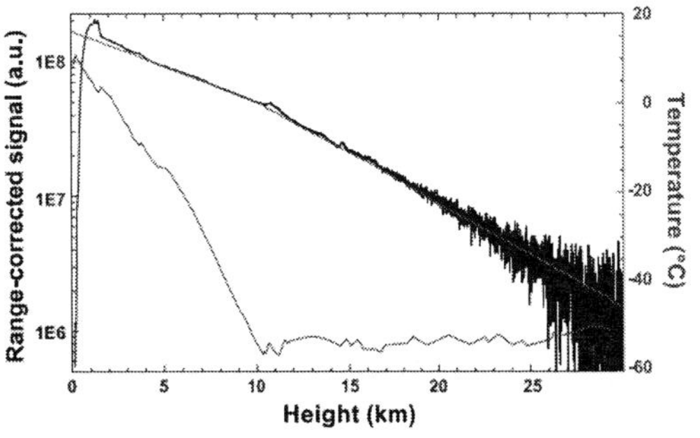

Figure 2: First measurement at 532 nm on 22 September 2017 (black) together with a molecular (red) and a temperature (blue) profile from a sounding launched at Nottingham at 0000 UTC on 22 September 2017.

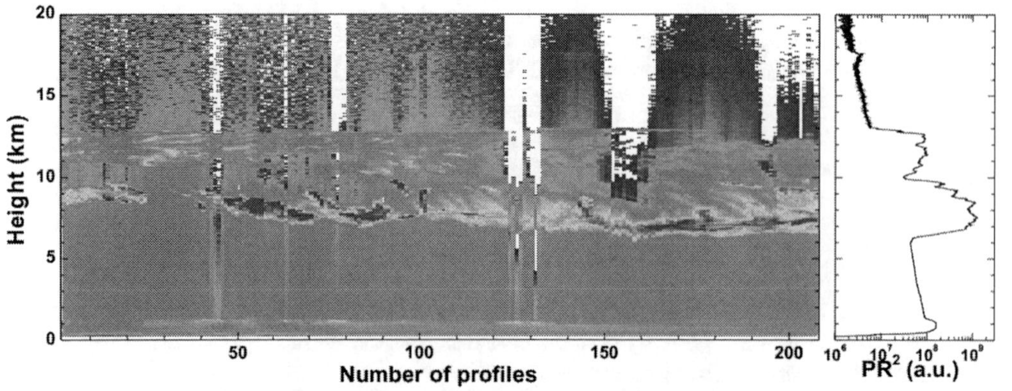

Figure 3: Time-height display (left) and average profile (right) of the range-corrected signed at 532 nm detected during a measurement between 1644 and 2010 UTC on 28 September 2017 with a temporal resolution of one minute.

observations being performed during daytime, reasonable signals can be obtained up to a height of 20 km. Aerosol can be seen in the lowermost 4 km as well as in layers at 7, 11, 14, 15, and 16 km height. The temperature profile of a radiosonde launched at Nottingham (about 150 km north of Hatfield) shows the tropospause at about 10 km height, putting the upper aerosol layers—likely originating from pyro-cumulonimbus events in Canada several weeks earlier—well within the stratosphere.

Figure 3 presents data from the first nighttime measurement performed with LiSsI. The configuration has been the same as during the measurements presented in Figure 2. Cirrus clouds are clearly visible between 6.5 and 13.0 km. The stratospheric aerosol layer from Figure 2 is still present between 17 and 18 km height.

6 NEXT MILESTONES

The coming milestones in developing the LiSsI facility are aligned with increasing the complexity of the optical setup. After the initial test measurements (Figure 2), we intend to perform atmospheric measurements of elastically and inelastically backscattered light at the different laser wavelength using full power. After that, the spectrometer will be integrated into the lidar receiver to allow for multispectral measurements in the atmosphere and using the gas and aerosol chambers in the laboratory.

ACKNOWLEDGEMENTS

The LiSsI facility has been funded by the University of Hertfordshire through capital investment. Boyan Tatarov is supported by a Marie Skłodowska-Curie Action Fellowship of the European Commission (CAPABLE, H2020-MSCA-IF-2015).

References

[1] Tatarov, B. and Sugimoto, N., 2005: Estimation of quartz concentration in the tropospheric mineral aerosols using combined Raman and high spectral resolution lidars, *Opt. Lett.* **30**, 3407-3409.

[2] Müller *et al.*, 2010: Mineral quartz concentration measurements of mixed mineral dust/urban haze pollution plumes over Korea with multiwavelength aerosol Raman-quartz lidar, *Geophys. Res. Lett.* **37**, 2010GL044633.

[3] Tatarov *et al.*, 2011: Lidar measurements of Raman scattering at ultraviolet wavelength from mineral dust over East Asia, *Opt. Expr.* **19**, 1569-1581.

DEVELOPMENT OF AN ADVANCED TWO-MICRON TRIPLE-PULSE IPDA LIDAR FOR CARBON DIOXIDE AND WATER VAPOR MEASUREMENTS

Mulugeta Petros[1]*, Tamer F. Refaat[1], Upendra N. Singh[1], Jirong Yu[1], Charles Antill[1], Ruben Remus[1], Bryant D. Taylor[2], Teh-Hwa Wong[2], Karl Reithmaier[2], Jane Lee[2], Syed Ismail[3] & Kenneth J Davis[4]

*[1]NASA Langley Research Center, Hampton, VA 2381, USA, * mulugeta.petros-1@nasa.gov*
[2]Science Systems and Applications, Inc., Hampton, VA 23666, USA,
[3]Analytical Services and Materials, Inc., Hampton, VA 23666, USA,
[4]Pennsylvania State University, University Park, PA 16802, USA

ABSTRACT

An advanced airborne triple-pulse 2-µm integrated path differential absorption (IPDA) lidar is under development at NASA Langley Research Center that targets both carbon dioxide (CO_2) and water vapor (H_2O) measurements simultaneously and independently. This lidar is an upgrade to the successfully demonstrated CO_2 2-µm double-pulse IPDA. Upgrades include high-energy, high-repetition rate 2-µm triple-pulse laser transmitter, innovative wavelength control and advanced HgCdTe (MCT) electron-initiated avalanche photodiode detection system. Ground testing and airborne validation plans are presented.

1 INTRODUCTION

Understanding the process of CO_2 exchange in the context of global climate change has become the topic of vital importance to the scientific community. Knowledge of the spatial and temporal distribution of CO_2 sources and sinks, and transport in a global scale is crucial to predict and possibly manage the carbon cycle process. In response to these challenges, laser-based active remote sensing instruments are being developed at NASA Langley Research Center (LaRC) [1-2]. Active remote sensors have a number of advantages including night, seasonal, high latitude, and tropical ecosystems measurements. LaRC researchers have envisioned a 2-µm triple-pulse integrated path differential absorption (IPDA) lidar that can measure atmospheric column concentrations of CO_2 and H_2O simultaneously and independently. This lidar operates on both ground and airborne platforms. The 2-µm wavelength chosen for this instrument provides attractive spectroscopic features that include strong CO_2 absorption, low temperature dependence and minor interference from other atmospheric gases. In addition, the 2-µm favors low tropospheric weighting functions that focus on CO_2 near-surface dynamics. Pulsed laser operation enables a direct measurement of absorption column length. This paper presents the technical methodology and technology progress for this 2-µm IPDA lidar.

Utilizing meteorological data, this unique 2-µm triple-pulse IPDA lidar is used to measure atmospheric dry-air column mixing ratios of CO_2 (XCO_2) and H_2O (XH_2O) with projected precision of 0.2% and 0.5%, respectively [1]. Three wavelengths are carefully selected to achieve independent and simultaneous measurements of CO_2 and H_2O as shown in figure 1. The first and second, λ_1 and λ_2, are strongly absorbed by H_2O and CO_2, respectively. The third, λ_3, is tuned where both have low absorption. The ratio of transmitted energy, E, and received signals power, P, reflected off a hard target for on- and off-line wavelength provides the measured differential optical depth, dOD, according to

$$dOD = \ln\left[\frac{P_{off}\cdot t_{off}}{E_{off}} \Big/ \frac{P_{on}\cdot t_{on}}{E_{on}}\right] \qquad (1)$$

which is the primary IPDA product, where t is the hard target return pulse width. Meteorology and spectroscopic data base (HITRAN) define the range-based weighting function from

$$WF(r) = \Delta\sigma(r)\cdot N_{dry}(r) \qquad (2)$$

where $\Delta\sigma$ is the differential absorption cross section, N_{dry} is the dry-air number density and r is the range. The dry-air weighted-column mixing ratio, X, is given by

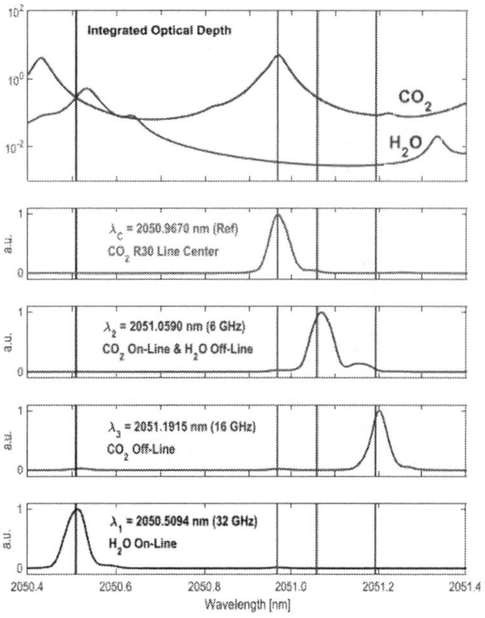

Figure 1 Integrated optical depth spectra of CO_2 and H_2O as compared to the selected 2-μm triple-pulse IPDA operating wavelengths [1] (top). Wavelengths are tuned and locked with respect to CO_2 R30 line center (bottom). Observed broadening is due to spectrometer resolution.

$$X = dOD \Big/ \int_{R_A}^{R_G} WF(r) \cdot dr \qquad (3)$$

where R_A and R_G are the instrument altitude and target elevation, respectively.

2 TECHNOLOGY DEVELOPMENT

In this IPDA the transmitter emits three pulses separated by 200 μs at 50 Hz repetition rate. The IPDA transmitted pulses time separation maximizes the measured column volume-overlap. These pulses are seeded at selected wavelengths that are generated by an offset with respect to the CO_2 R30 line center. The transmitter is co-axially aligned to the receiver through bore-sighting. The IPDA receiver system is similar to the successfully demonstrated 2-μm CO_2 double-pulse IPDA lidar [3].

2.1 Laser Transmitter Module Design

The IPDA 2-μm laser transmitter is based on Ho:Tm:YLF crystal with an emitted radiation wavelengths that cover strong CO_2 absorption features. The long upper laser level lifetime of Tm and Ho provides the capability of generating three Q-switched pulses using a single pump pulse. The crystal is end pumped using 792 nm to excite Tm. The excited Tm pumps Ho which generates the 2-μm emission. The co-doped approach has the advantage of providing close to 2 quantum efficiency in the Tm, and overall system simplicity [4]. Table 1 lists the main parameters of the IPDA laser transmitter. Figure 2 shows the temporal variation of the triple-pulse generation event. Stable pulse energy output is observed as shown in figure 3. Heat dissipation is essential for high intensity pump where high repetition rate operation is required. This required additional mitigation strategies for in laser head design.

A wavelength control unit provides three locked seeding wavelengths for the triple-pulse laser. The wavelength control consists of a single cw semiconductor laser diode [5]. The laser diode wavelength reference is locked to the CO_2 R30 line center through a gas cell. The three seeding wavelengths, shown in Figure 1, are generated by frequency offset with respect to the locked center line.

Custom designed electronics modules are used to synchronize and control timing between various transmitter components and IPDA receiver. The transmitter is housed in a sealed 67×16.5×26 cm^3 enclosure purged with dry-air. The laser bench is temperature controlled to avoid thermal induced misalignment. All optical mounts are designed to be adjustable, lockable and flight hardened.

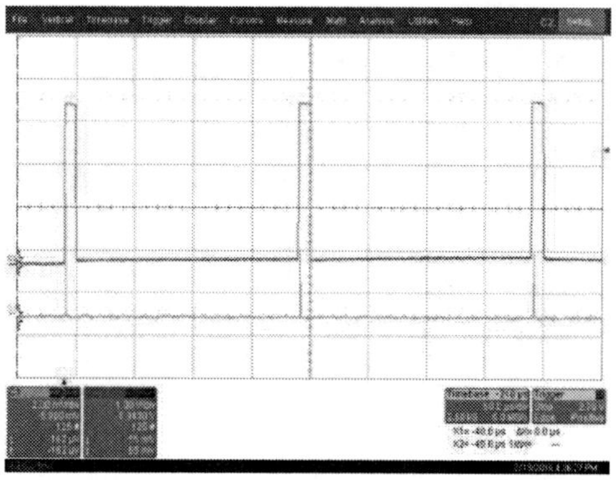

Figure 2 Single-shot temporal profile of the triple-pulse generation (yellow) as compared to Q-switch timing (red).

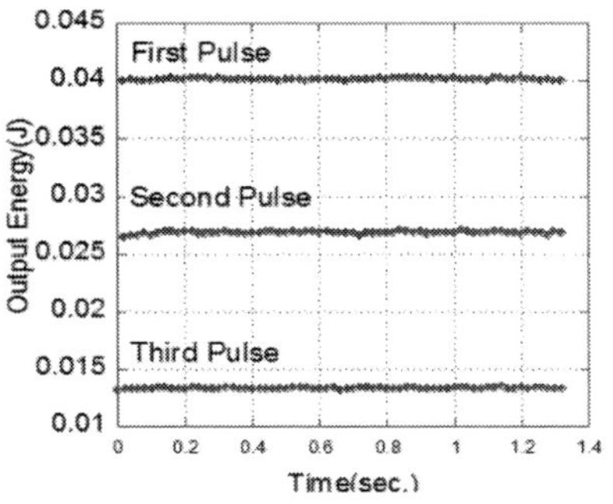

Figure 3 Shot by shot variation of the triple-pulse generated energies.

Table 1 IPDA transmitter parameters

Operating Wavelength	λ_1	2050.5094 nm
	λ_2	2051.0590 nm
	λ_3	2051.1915 nm
Output Energy	λ_1	40 mJ
	λ_2	26 mJ
	λ_3	14 mJ
Pulse Width	λ_1	22 ns
	λ_2	32 ns
	λ_3	60 ns
Line Width	Transform Limit	
Beam Quality	< 2	
Frequency Accuracy	< 1 MHz	

2.2 Receiver Design

Figure 4 shows the main receiver components which includes a telescope, an aft-optics box and detection systems. A 0.4 m (F2.3) diameter aluminum, Newtonian telescope is custom designed for this lidar. The primary mirror is designed with hyperbolic curvature to minimize aberration. The secondary mirror is a two-surface flat, dichroic fused silica. The top surface turns the return radiation 90° to the aft-optics and the bottom surface transmits the expanded laser beam down coaxially. The telescope field-of-view is 300 µrad and focus spot size diameter is 300-µm. The telescope output radiation is coupled to a field-imaging aft-optics. The aft-optics collimates and splits the return radiation into two 90/10 detection channels.

The 90% return is focused onto InGaAs-based pin detector, similar to the double-pulse IPDA [3]. The other 10% is directed, using an optical fiber, to an advanced HgCdTe (MCT) electron-initiated avalanche photodiodes (e-APD), developed at DRS Technologies. This detector has a 4×4 pixel format with integrated trans-impedance amplifier (TIA) for each pixel. MCT e-APD detection system integration was conducted at NASA GSFC and includes cryogenic cooling, TIA buffering and summing electronics, and 2×2 pixel custom focusing optics. This detection channel exhibits high sensitivity with a typical 6 MHz bandwidth noise-equivalent-power (NEP) of 1.4 $fW/Hz^{1/2}$ per pixel. Operation of this detection system, in terms of pixel activation, bias voltage and gain selection is fully controlled with interfacing software [6].

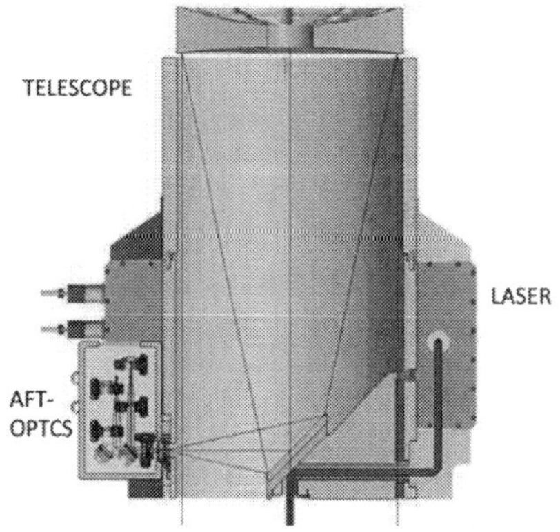

Figure 4 Schematic of the IPDA receiver with two-channel aft-optics for high-gain (90%) InGaAs pin and low-gain (10%) MCT e-APD detection systems.

2.3 Data Acquisition and Processing

IPDA data acquisition is based to two advanced 12-Bits, 1 GS/s, two-channel digitizers. One digitizer is dedicated for transmitted laser energy monitors. The other digitizer is dedicated for the two return signals channels. Both digitizers are enabled by the pump pules and triggered using the laser Q-switch driver. Several aspects of the digitizers were characterized and resulted in desirable performance in terms of dynamic range, overload recovery, linearity and noise. Total average noise was less than 1 mV. For a 2 V peak-to-peak input range, this translates to a maximum signal-to-noise ratio of 2000. Figure 5 shows a snapshot of the data acquisition Graphical User Interface (GUI) during ground testing. The GUI shows the high and low gain channels return signals and one active energy monitor channel. Real-time data analysis will be tested during final phases of integration.

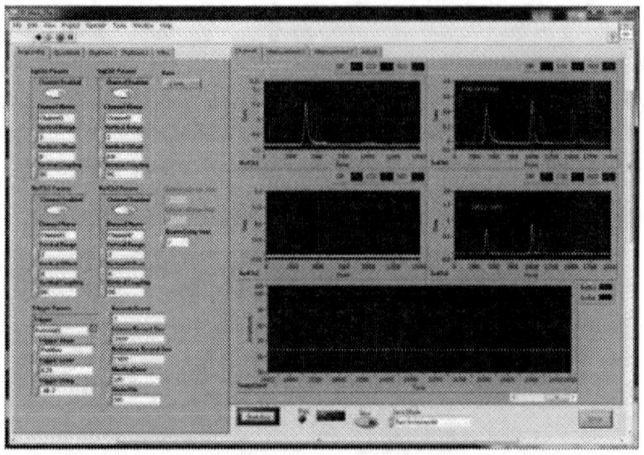

Figure 5 Data acquisition software GUI showing the triple-pulse energy monitor and return signals using InGaAs and MCT e-APD detection systems.

3 VALIDATION PLANS

The 2-μm triple-pulse IPDA lidar validation is important to assess measurement capability of XCO_2 and XH_2O. IPDA ground validation starts after integration in a mobile lidar laboratory. This facility enables IPDA horizontal measurement using calibrated targets and vertical measurements using clouds. Ground validation objectives are to check the IPDA lidar operational readiness and data processing algorithms before aircraft deployment.

The design of the 2-μm IPDA lidar was set to meet the payload size, weight and power consumption requirements of the NASA-B200 aircraft. Initial engineering flights will focus on instrument operational reliability and initial performance testing. Later flights will focus on comparing CO_2 and H_2O measurements against correlative *in-situ* sensors and models. This includes evaluation of systematic and random errors. This will rely on onboard sensors as well as coordinated measurements with other independent sensors, such as NOAA air-sampling. Airborne validation will be planned to test the performance under different conditions including surface reflectivity changes; day and night background; clear, cloudy and broken cloud conditions; variable surface elevation; urban pollution; and plume detection.

4 CONCLUSIONS

A Novel IPDA lidar system that is capable of measuring both atmospheric CO_2 and H_2O independently and simultaneously from airborne platform is under development. All the different sub systems have been designed, built and characterized. IPDA integration and ground testing is planned to be completed soon. Airborne measurements are planned in conjunction with operation of other CO_2 sensors such as NOAA tall tower network and airborne flask air-sampling for the validation of the IPDA. These evaluations will be conducted in various environment and ground reflectivity conditions such as vegetation, snow, sand and ocean during day and night. This IPDA lidar development opens a path toward space-based active remote sensing.

ACKNOWLEDGEMENTS

This work was funded and supported by NASA Earth Science Technology Office.

References

[1] T. Refaat, U. Singh, J. Yu, M. Petros, S. Ismail, M. Kavaya, K. Davis, 2015: Evaluation of an airborne triple-pulsed 2 μm IPDA lidar for simultaneous and independent atmospheric water vapor and carbon dioxide measurements, *Appl. Opt.* **54**, 1387-1398.

[2] U. Singh, B. Walsh, J. Yu, M. Petros, M. Kavaya, T. Refaat, N. Barnes, 2015: Twenty years of Tm:Ho:YLF and LuLiF laser development for global wind and carbon dioxide active remote sensing, *Opt. Mater. Exp.* **5**, 827-837.

[3] T. Refaat, U. Singh, M. Petros, R. Remus, J. Yu, 2015: Self-calibration and laser energy monitor validations for a double-pulsed 2-μm CO_2 integrated path differential absorption lidar application, *Appl. Opt.* **54**, 7240-7251.

[4] B. Walsh, N. Barnes, M. Petros, J. Yu, U. Singh, 2004; Spectroscopy and modeling of solid state lanthanide lasers: Application to trivalent Tm and Ho in YLiF and LuLiF, *J. Appl. Phys.* **95**, 3255–3271.

[5] M. Bagheri, G. Spiers, C. Frez, S. Forouhar, F. Aflatouni, 2015; Linewidth measurement of distributed-feedback semiconductor lasers operating near 2.05 μm, *IEEE Phot. Tech. Let.* **27**, 1934-1937.

[6] X. Sun, J. Abshire, J. Beck, 2014: HgCdTe e-APD detector arrays with single photon sensitivity for space lidar applications, *Proc. of SPIE* **9114**, 91140K.

REMOTE SENSING OF METHANE EMISSIONS BY COMBINING OPTICAL SIMILITUDE ABSORPTION SPECTROSCOPY (OSAS) AND LIDAR

Sandrine Galtier[1], Christophe Anselmo[1], Jean-Yves Welschinger[2], Jean-Pierre Cariou[3], Jean-François Sivignon[1], Alain Miffre[1] and Patrick Rairoux[1*]

[1] *Institut Lumière Matière, UMR5306 Université Lyon 1-CNRS, Université de Lyon, 69622 Villeurbanne, France, patrick.rairoux@univ-lyon1.fr*

[2] *Institut Camille Jordan, UMR 5208 Université Lyon 1-CNRS, Université de Lyon, 69622 Villeurbanne cedex, France*

[3] *Leosphere France, 14-16 rue Jean Rostand, 91400 Orsay, France*

ABSTRACT

Monitoring the emission of gases is difficult to achieve in industrial sites and in environments presenting poor infrastructures. Hence, robust methodologies should be developed and coupled to Lidar technology to allow remote sensing of gas emission. OSAS is a new methodology to evaluate gas concentration emission from spectrally integrated differential absorption measurements. Proof of concept of OSAS-Lidar for CH_4 emission monitoring is here presented.

1 INTRODUCTION

Lidar remote sensing has found a large interest in dedicated field studies on atmospheric trace gases and high concentrated atmospheric compounds for air quality control, industrial emission surveys and for long-term evaluation of climate change [1-3]. For this purpose, methodologies development and technical solutions are well documented in the literature [4, 5]. Among these methodologies, differential absorption Lidar (DIAL) and Raman scattering Lidar (RSL) are appropriate for atmospheric trace gases and water vapor active remote sensing. In the past fifteen years, improvements in laser technology and new methodology developments have been achieved, making DIAL and RSL a standard in the environment and climate survey [6]. However, due to the weakness of Raman scattering cross-sections, the RSL-methodology is often limited to the remote sensing detection of high concentrated gases such as water vapor. Moreover, to achieve accurate DIAL measurements, a tunable laser, having a narrow spectral line width, is required, with a high spectral resolution detector. Hence, when applying the DIAL-methodology, limitations may occur when the spectral width of the emission laser does not match the target gas absorption line [7]. Conversely, when the absorption line width is in the same range as the laser spectral width (or even narrower than), effective absorption cross section should be taken into account. It follows that a measurement uncertainty is added as for the detection of greenhouse gases such as water vapor (H_2O), methane (CH_4) or carbon dioxide (CO_2) [8] in the mid-infrared spectral region. In this case, other methodologies, based on a broadband emission laser and a high spectral resolution detector, have been successfully applied, as we formerly published [9].

In this contribution, a new methodology is proposed to remotely address atmospheric traces gases, with no particular consideration on the spectral width of the emission laser. It is based on coupling Lidar remote sensing with OSAS [10]. Our study underscores that OSAS-Lidar can overcome the possible mismatch between the spectral width of the laser emission and the TG-absorption line, which may occur for atmospheric greenhouse gases remote sensing. Pioneer work in this field has been published by H. Edner et al. [11] On Hg-containing atmosphere. More recently, Thomas et al. [12] proposed a more appropriate approach to solve the problem of non-spectrally resolved absorption lidar measurements, which was the basis of OSAS-Lidar. This contribution is organized as follows: in section 2, the OSAS-Lidar methodology is presented by highlighting the mathematical formalism. OSAS experimental verification is shown, followed by a proof of concept. This contribution ends with a discussion and proposes outlooks.

https://doi.org/10.1051/epjconf/201817601010

2 OSAS LIDAR METHODOLOGY

2.1 OSAS Principle

In the OSAS-methodology [10], the power spectral density (PSD) of a broadband light source (i.e. broader than the spectral width of one or several molecular absorption lines) is set to be similar with the TG-gas absorption spectrum (the active channel), while a second PSD is set in a less absorbing spectral region (the reference channel). The OSAS-principle is shown in Fig. 1 when considering methane in its $2\nu_3$ absorption band. The active channel $P_{0;1}(\lambda)$, which overlaps the Q-branch absorption lines, undergoes light absorption along the optical path, while the reference channel $P_{0;2}(\lambda)$ experiences a lower light extinction. Hence, two different signals, S_1 and S_2, are detected after light absorption.

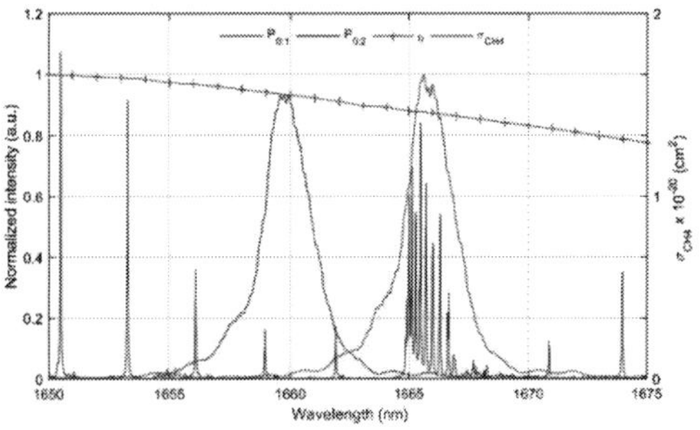

Figure 1: OSAS-principle applied to the $2\nu_3$ Methane gas absorption band. The PSD of a broadband light source is spectrally shaped matching (red curve) or not (blue curve) the absorption cross-section spectrum $\sigma_{CH4}(\lambda)$ (green lines) In purple, the opto-electronic efficiency spectrum $\eta(\lambda)$ of the OSAS detection part.

2.2 Target gas concentration retrieval

The inversion algorithm used to retrieve the target gas concentration from spectrally-integrated OSAS-lidar signals P_1(z) and P_2(z) at altitude z is a particular application of what has been published in [10]. The starting point is the single-scattering elastic lidar equation:

$$P_i(z) = \frac{K}{z^2}\int_{\Delta\lambda} O(z,\lambda)P_{0,i}(\lambda)\beta(z,\lambda) \quad (1)$$
$$T_{TG}^2(z,\lambda)T_{\text{atm}}^2(z,\lambda)\eta(\lambda)\,\mathrm{d}\lambda$$

where $O(z, \lambda)$ is the overlap function and $\beta(z, \lambda)$ the total backscattering coefficient. In the transmission term, the extinction from the target gas (TG) and from other atmospheric compounds are differently expressed, the latter being defined as follows at range z and wavelength λ:

$$T_{\text{atm}}(z,\lambda) = \exp\left[-\int_0^z \alpha(z',\lambda)\,\mathrm{d}z'\right] \quad (2)$$

The extinction coefficient α(z',λ) is the sum of four contributions, one per atmospheric main component (aerosols, molecules, except the target gas) considered in each physical process (light scattering, light absorption) [5]. TG-absorption is considered in the lidar equation by its transmission:

$$T_{TG}(z,\lambda) = \exp\left[-\int_0^z \sigma_{TG}(z',\lambda)N_{TG}(z')\,\mathrm{d}z'\right] \quad (3)$$

Where N_{TG}(z') is the number density of the absorbing molecule of interest whose cross-section is σ_{TG}. Practically, the OSAS-lidar signals must – as in conventional lidar, be corrected from the sky background signal and electronic offsets. As in DIAL formalism, we made the assumptions that: (i) the overlap function $O(z)$ is the same for both wavelengths, (ii) the wavelength dependence of the extinction and backscatter coefficients $\alpha(z, \lambda)$ and $\beta(z, \lambda)$ of the atmosphere are negligible over the wavelength interval Δλ. In this way, the signals are modeled with the f_i-functions, which are function of the path-integrated concentration or cumulative concentration parameter x at distance z:

$$f_i(z,x) = \int_{\Delta\lambda} P_{0,i}(\lambda)\exp\left[-2\sigma_{TG}(\lambda)x(z)\right]\eta(\lambda)\,\mathrm{d}\lambda \quad (4)$$

Following the formalism developed by Anselmo et al. [10], defining a g-function depending on x allows retrieving the cumulative concentration $CC(z)$ as follows:

$$g(z,x) = S_2 f_1(z,x) - S_1 f_2(z,x) \quad (5)$$

where S_i are the lidar measurements. Because the g-function depends on the range z, the cumulated concentration $CC(z)$ appears to be the first zero of the g-function. This zero value can be retrieved iteratively at each range z by applying, as for example, the Newton-Raphson root-finding algorithm. Finally, the TG density number N at range z is retrieved by derivating $CC(z)$ with respect to z:

$$N_{TG}(z+\Delta z/2) = \frac{dCC(z)}{dz} \quad (6)$$

where Δz is the spatial resolution. Prior to applying the OSAS formalism, a verification that the ratio $f_1(x)/f_2(x)$ is monotonic on the concentration interval $[0, x_{max}]$ should be made. Values of x_{max} can reach up to 10^{27} mol·m^{-3} [10].

3 OSAS-LIDAR SETUP

This section presents the experimental arrangement of the OSAS-lidar methodology, built up to verify the proposed methodology for methane concentration retrievals. The general setup is presented in figure 2. For this purpose, the emitter part is a OPO pumped fs Ti:Sapphire laser emitting in the NIR with a pulse repetition frequency of 1 kHz. The spectral switch between the active channel $P_{0;1}(\lambda)$ at 1665.5 nm and the reference channel $P_{0;2}(\lambda)$ at 1660 nm is achieved with an acousto-optic programmable dispersive filter. Laser pulses with energy of 30 µJ and with a 2 nm spectral width are achievable. As the laser amplitude modulation is performed at the emission, the lidar receiver configuration is quite simple. The light is collected with a Newtonian telescope having a focal length of 450 mm and a primary mirror diameter of 115 mm. The NIR detector consists in an InGaAs based APD mounted at the telescope's focus point. The field of view 0.44 mrad is governed by the 200 µm detector diameter. The signal acquisition system is a 12 bits digitizing oscilloscope.

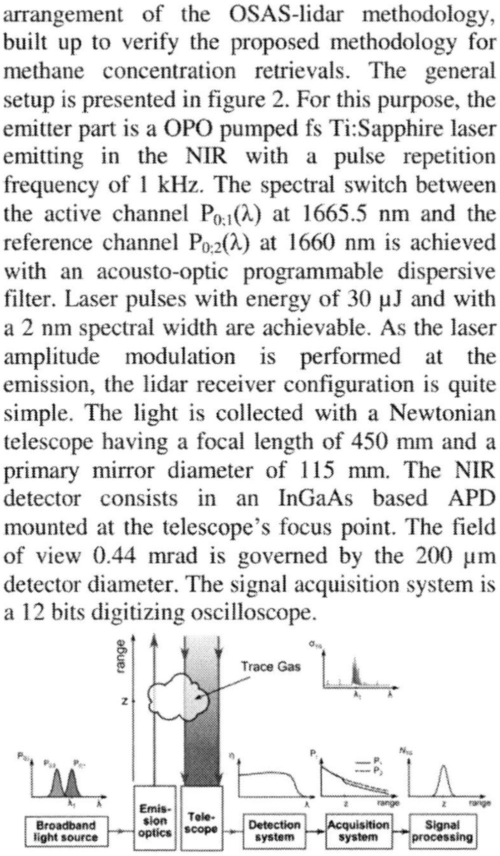

Figure 2: Principle of OSAS-lidar. Two different laser pulses are alternatively generated. Using a lidar system, these two emitted laser pulses through the atmosphere undergo different absorptions due to the TG presence, giving rise to the output OSAS-lidar signals $S_1(z)$ and $S_2(z)$, after collection by a lidar receiver and detection on a photoreceptor.

For the first experimental achievement, a gas leak has been made artificially by using open gas cell located at a horizontal distance of 90 meters from the telescope. The open gas cell is 6 meters long with diameter of 60 centimeters. Hence, laser beam is able to pass through the gas cell without major interaction on his mount. To overcome the problem of low overlap between the receiver field-of-view and the laser transmitter in the near field, a biaxial geometry has been adopted.

4 RESULTS

The first lidar backscattering profiles with 30 µJ operating at 1665.5 nm for P_1 and at 1660 nm for P_2 are presented in Fig. 3. The strong and rapid signal increase observed at a z = 110 meters distance is due to the reflection of a Teflon based hard target. The low signal increase at 90 meters range over a distance is due to the presence of particles injected in the open gas cell. Due to the turbulence highlighted by the fast particles concentration change (not visible in the figure), monitoring a CH_4 with open path gas cell is found to be difficult within the signal to noise ratio related to fast acquisition. Then, TG concentration retrieval based on hard target reflection approach was therefore envisaged. The signal-to-noise ratio on the hard target reflection reach than 10^3 (compared to SNR = 2 at z = 50 m).

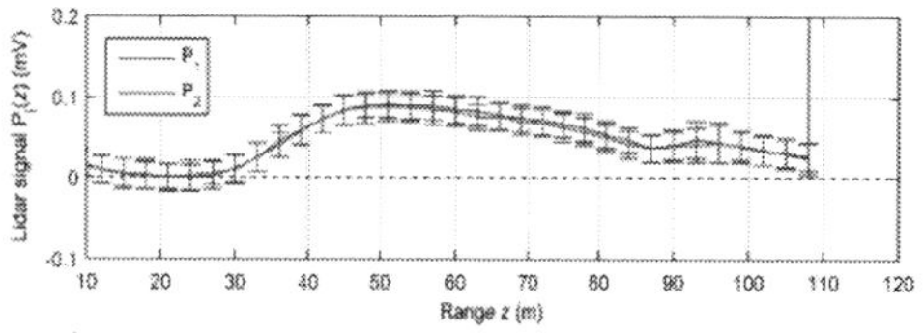

Figure 3: OSAS-Lidar signal for the active channel (red) and the reference channel (blue). The background noise signal is numerically subtracted. Each signal corresponds to a 1 second average time.

Remote Methane detection with OSAS-lidar is there achieved by filling out the open cell with air-methane. The first results on methane path integrated concentration (PIC) retrievals are presented in Fig. 4. When opening the gas cell after CH_4 filling, the measurements show a methane PIC above 2.5×10^4 ppm.m. The dispersion of the methane plume contained into the open gas cell is very quick as it is expected. The results seem to be in agreement with the

https://doi.org/10.1051/epjconf/201817601010

maximal value expected if we consider the 1.8 % of methane mixing ratio released into a 6 meters long gas cell (10.8×10^4 ppm.m). Since the gas cell is not perfectly sealed and its pressure remains at 1 atm, a part of methane has necessarily leaked around during the filling.

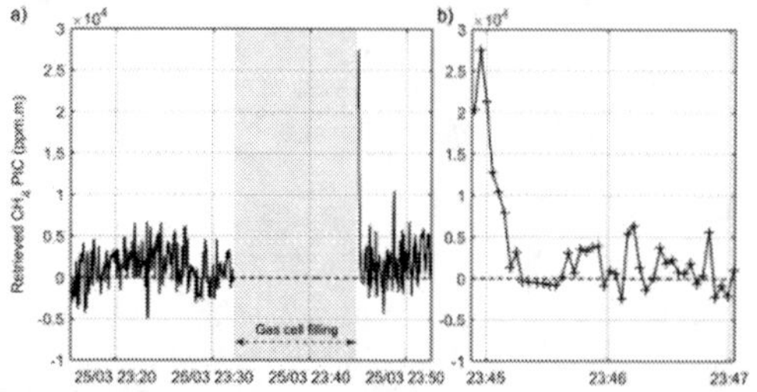

Figure 4: a) Retrieval of methane path integrated concentrations. The gas cell was closed during methane filling. b) Methane concentration dispersion. Once the gas cell opened, the methane concentration decreases quickly to the detection limit in less than 30 seconds.

An in-situ methane sensor placed within the open cell would have been very valuable to compare the two CH_4 measurements. However, the detection limit (1σ) is estimated at 2.2×10^3 ppm.m.. As expected, this detection limit value is above the one obtained in laboratory ($\sigma = 0.65 \times 10^3$ ppm.m).

5 CONCLUSION

These preliminaries results appear promising though the estimated limit of detection is rather low. The possible outlooks of this work are numerous since these experimental results have been performed with a non-specifically dedicated laser light source. This work also settles the basements towards the use of the OSAS-methodology to improve DIAL retrievals using broadband laser sources.

References

Numbered in the order they appear in the text, with brackets. Example of references:

[1] G. Ehret, C. Kiemle, M. Wirth, M. A. Amediek, A. Fix,S. Houweling, 2008: Space-borne remote sensing of CO_2, CH_4, and N_2O by integrated path differential absorption lidar: a sensitivity analysis , *Appl. Phys. B*, **90**, 593-608.

[2] C. Stephan, M. Alpers, B. Millet, G. Ehret, P. Flamant, and C. Deniel, 2011: MERLIN: a space-based methane monitor, *Proc. SPIE* **8159**, 815908.

[3] L. Fiorani, S. Santoro, S. Parracino, M. Nuvoli, C. Minopoli, and A. Aiuppa, 2015: Volcanic CO_2 detection with a DFM/OPA-based lidar, *Opt. Lett.*, **40,** 1034–1036.

[4] R. M. Measures, 1992: Laser Remote Sensing, Fundamentals and applications, *Krieger Ed.*

[5] C. Weitkamp, 2005: Lidar, Range-Resolved Optical Remote Sensing of the Atmosphere, *Springer Ed*., Berlin, New-York.

[6] P. Di Girolamo, M. Cacciani, D. Summa1, A. Scoccione, B. De Rosa1, A. Behrendt, and V. Wulfmeyer, 2017: Characterisation of boundary layer turbulent processes by the Raman lidar BASIL in the frame of HD(CP)2 Observational Prototype Experiment, *ACP*, **17**, 745–767.

[7] V.D. Burlakov, 2010: A differential-absorption lidar for ozone sensing in the upper atmosphere-lower stratosphere, *Instruments and Experimental Techniques*, **53**, 886–889.

[8] S. Ismail et al., 2010: Laser measurements of water vapor, aerosol, and cloud distributions in Saharan air layers and tropical disturbances, *J. Atmos. Sci.*, **67**, 1026–1047.

[9] P. Rairoux et al. , 2000: Remote sensing of the atmosphere using ultrashort laser pulses, *Appl. Phys. B*, **71**, 573–580.

[10] C. Anselmo, J.Y. Welschinger, J.P. Cariou, A. Miffre, P. Rairoux, 2016: Gas concentration measurement by optical similitude absorption spectroscopy: methodology and experimental demonstration, *Opt. Expr.*, **24**, 12588-12599.

[11] H. Edner, S. Svanberg, L. Unéus, W. Wendt, 1984: Gas correlation lidar, *Opt. Lett.*, **9**, 493.

[12] B. Thomas et al., 2013: Remote sensing of atmospheric gases with optical correlation spectroscopy and lidar: first experimental result on water vapor profile measurements, *Applied Physics B*, **113**, 265-275.

FIRST APPLICATION OF THE OPTIMAL ESTIMATION METHOD TO RETRIEVE TEMPERATURE FROM PURE ROTATIONAL RAMAN SCATTER LIDAR MEASUREMENTS

S. Mahagammulla Gamage[1,*], A. Haefele[2,1] and R.J. Sica[1,2]

[1]*Department of Physics and Astronomy, The University of Western Ontario, Canada*
[2]*Federal Office of Meteorology and Climatology, MeteoSwiss, Payerne, Switzerland*

ABSTRACT

We present the application of the Optimal Estimation Method (OEM) to retrieve atmospheric temperatures from pure rotational Raman (PRR) backscatter lidar measurements. A forward model (FM) is developed to retrieve temperature and tested using synthetic measurements. The OEM offers many advantages for this analysis, including eliminating the need to determine temperature calibration coefficients.

1 INTRODUCTION

The Raman lidar technique uses the pure rotational Raman (PRR) spectra of atmospheric molecules to measure the temperature. The PRR spectrum contains two branches called Stokes and anti-Stokes. Both branches have approximately the same amplitude and they are positioned symmetrically in wavenumber on either side of the excitation line. The traditional Raman lidar temperature retrieval method, introduced by Cooney in 1972, used the ratio of two signals from the lidar's PRR channels [1]. The traditional retrieval algorithm consists of smoothing raw lidar measurements, determining the lidar calibration function coefficients using ancillary measurements such as those from radiosondes, and then calculating temperatures and estimating the absolute and relative uncertainties [2].

The most important capability of the Optimal Estimation Method (OEM) is that it can directly retrieve the lidar constant along with the temperatures from the raw lidar measurements. The OEM also allows us to estimate both random and systematic uncertainties on a profile-by-profile basis.

We use lidar measurements from the Raman Lidar for Meteorological Observations (RALMO) located in Payerne, Switzerland (46°48′N, 6°56′E). RALMO is a fully automated lidar developed by the Federal Insitute of Technology in Lausane, EPFL, and operated by MeteoSwiss. It has two PRR channels that measure the Raman-shifted backscatter arising from rotational energy state transitions in atmospheric nitrogen and oxygen molecules due to excitation at the laser wavelength of 354.7 nm. The High Quantum number channel (JH) in the RALMO measures the backscattered signals from the energy exchange that occur in the high quantum states in both Stokes and anti-Stokes branches while the Low Quantum number channel (JL) measures the signals from the energy exchange occurring in the low quantum states. Detailed description of the RALMO PRR system is given in Dinoev et al.[3].

Implementation of the OEM will be presented in section 2.3. In section 3 retrievals from the synthetic measurements are presented. Conclusions and future work are given in section 4.

2 METHODOLOGY

2.1 Basics of the OEM

The OEM is a regularized matrix inverse method based on Bayes' theorem. It requires a knowledge of a forward model capable of predicting data if the model parameters are already known. The variables that need to be estimated are called state variables, and they are assembled into a state vector $\boldsymbol{x}$. Similarly, measurements are assembled into a measurement vector $\boldsymbol{y}$. Then the relationship between $\boldsymbol{x}$ and $\boldsymbol{y}$ is described by the forward model $\boldsymbol{F}$, which con-

tains all the physical and instrumental factors describing the measurements.
The general form of the forward model $\boldsymbol{F}$ is,

$$\boldsymbol{y} = \boldsymbol{F}(\boldsymbol{x},\boldsymbol{b}) + \varepsilon \tag{1}$$

where $\boldsymbol{b}$ is a parameter vector including all model variables and ε is an uncertainty vector including contributions from random uncertainties in the measurements, in the forward model, and in the model parameters. The state vector and the model parameters describe the atmosphere and the instrument measuring it. The state vector contains all the parameters that are to be retrieved. The model parameters contain all other parameters that are needed to fully model the measurement but are not retrieved. In the presence of measurement uncertainties ($\varepsilon \neq 0$), a statistical estimate for the state vector ($\widehat{\boldsymbol{x}}$, retrieved parameters) or the optimal solution of $\boldsymbol{x}$, can be obtained using an estimate of the *a priori* state vector ($\boldsymbol{x_a}$) before the measurements were made.
The cost function associated with the OEM includes the measurement $\boldsymbol{y}$, and its covariance $\boldsymbol{S}_y$, the forward model, the retrieved state, model parameters, the *a priori* and its covariance $\boldsymbol{S}_a$:

$$\begin{aligned} cost = &[\boldsymbol{y} - \boldsymbol{F}(\boldsymbol{x},\boldsymbol{b})]^T \boldsymbol{S}_y^{-1} [\boldsymbol{y} - \boldsymbol{F}(\boldsymbol{x},\boldsymbol{b})] \\ &+ [\boldsymbol{x} - \boldsymbol{x_a}]^T \boldsymbol{S}_a^{-1} [\boldsymbol{x} - \boldsymbol{x_a}]. \end{aligned} \tag{2}$$

If the forward model is non-linear the Marquardt-Levenberg method can be used iteratively to minimize the cost. Further details of the OEM are given in Rodgers [4].

2.2 Forward model

We use a forward model based on the Raman lidar equation, with the number density replaced by pressure and temperature via the ideal gas law [5]. The forward model for true counts from both PRR channels is:

$$\begin{aligned} N_{RR,t}(z) = \frac{C^*_{RR}}{z^2} \frac{P(z)}{kT(z)} \Theta(z) \sigma_{RR}(T,z) \times \\ \Gamma^2_{atm}(z) + B^*_{RR}. \end{aligned} \tag{3}$$

C_{RR} is the lidar calibration constant that includes transmitted photons, detector efficiency, height resolution and the area of the telescope. $\Theta(z)$ is the overlap function, $P(z)$ is pressure, $T(z)$ is temperature. $\Gamma^2_{atm}(z)$ is the atmospheric transmission, B_{RR} is the background noise and z is the altitude. The backscatter coefficient $\sigma_{RR}(T,z)$ is given by:

$$\sigma_{RR}(T,z) = \sum_{i=O_2,N_2} \sum_{Ji} \Gamma_{RR}(J_i^*) \eta_i (\frac{d\sigma}{d\Omega})_{\Pi}^{RR,i}(J_i^*). \tag{4}$$

$\Gamma_{RR}(J_i)$ is the transmission of the receiver at the wavelength of the RR line J_i, $\eta(i)$ is the relative volume abundance of N_2 and O_2, $(\frac{d\sigma}{d\Omega})_{\Pi}^{RR,i}(J_i)$ is the differential back-scatter cross section (depends on temperature), and the corresponding quantum state is J.
For each channel RR can be replaced by JL and JH and C^*_{RR}, B^*_{RR} and J^* have different values. For a channel with a height-dependent background, B_{RR} would then be a function of altitude, and the additional coefficients would be included in the forward model.
The two lidar calibration constants C_{JL} and C_{JH} can be related by:

$$R = \frac{C_{JH}}{C_{JL}}. \tag{5}$$

The ratio, R, is constant over long time periods where C_{JL} and C_{JH} are stable over hours only. It is also independent of temperature and allows us to retrieve only one lidar calibration constant. R can be derived using:

$$R = \left(\frac{N_{t,JH} - B_{JH}}{N_{t,JL} - B_{JL}} \right) \Big/ \left(\frac{\sigma_{JH}(T,z)}{\sigma_{JL}(T,z)} \right) \tag{6}$$

Based on the assumption of non-paralyzable detectors, the true count profile can be transformed into the observed height profile using the dead-time correction in Eq.(7):

$$N_{true} = \frac{N_{obs}}{1 - N_{obs}\gamma}. \tag{7}$$

Currently we retrieve temperature, background noise (that includes photomultiplier shot noise, and sky background) and the lidar calibration constant C_{JL}. As the measurements and retrieval grid spacings are relatively small compared to the height of the atmosphere, we use linear interpolation between the two grids.

2.3 Implementation of the OEM

For this study we use the Qpack software package in Matlab implemented by Eriksson et al.[6]. Qpack is a part of the ARTS model used by the radiometer and microwave community. We use the Marquardt–Levenberg method in Qpack to do non-linear retrievals. This also allows us to retrieve parameters using an *a priori* considerably far from the true solution [7].
For the preliminary testing of the forward model we use the US standard atmospheric model as *a priori* knowledge. As discussed in Eriksson et al. [6], when there is some correlation between the various quantities the respective covariance matrix should have non zero off-diagonal elements. The correlation is provided by using correlation functions with a correlation length appropriate to the physical situation [7]. We use a tent function with a covariance length of 1 km for the temperature covariance.
We use measurements from both JL and JH channels in the measurement vector and the covariance matrix $\boldsymbol{S_y}$ is a diagonal matrix. With the use of the Qpack OEM non-linear solver the optimal solution is obtained minimizing the cost.

3 RESULTS

We present the initial test results of the OEM using the synthetic measurements.The US standard atmospheric model temperature and pressure profiles were used in the forward model to generate the synthetic measurements shown in Fig. 1. Poisson noise and a constant background were added to the synthetic measurements to make those similar to the 4 hours of integrated nighttime RALMO measurements obtained on 28th June 2011.
The averaging kernel in Fig. 2, shows the sensitivity of the retrieved temperatures to the true temperatures. We can retrieve real temperatures with 1 km vertical resolution up to about 15 km. However, the sensitivity decreases with the increasing height. The sum of the averaging kernel shown in red should equal to unity where the retrievals are accurate.

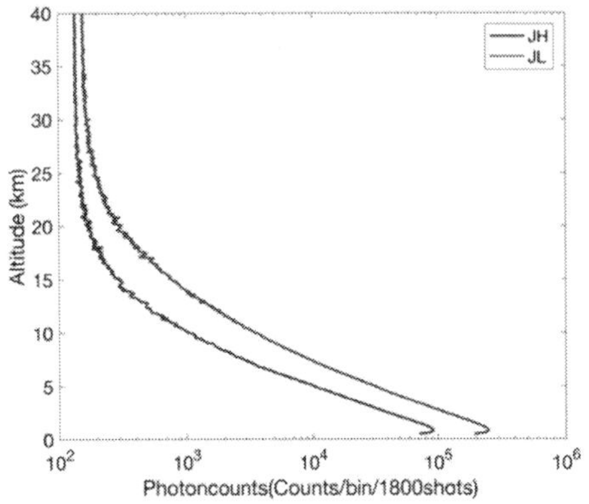

Figure 1: Synthetic measurements of JH and JL channels, generated using the forward model in Eq.(3).

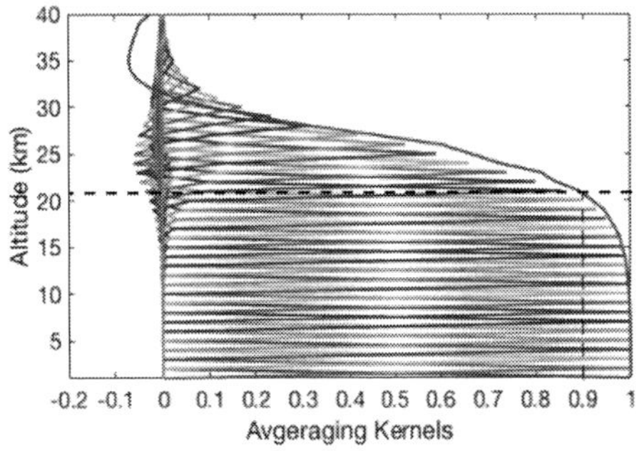

Figure 2: The averaging kernel for temperature. The red curve shows the sum of the kernels.

Fig. 3, shows the percent difference between the synthetic measurement and the retrievals. The difference between the measurements and model is within the Poisson photoncount uncertainty. The lidar calibration constant was retrieved with a percent difference less than 0.5% and the percent difference for the retrieved background is less than 3%.

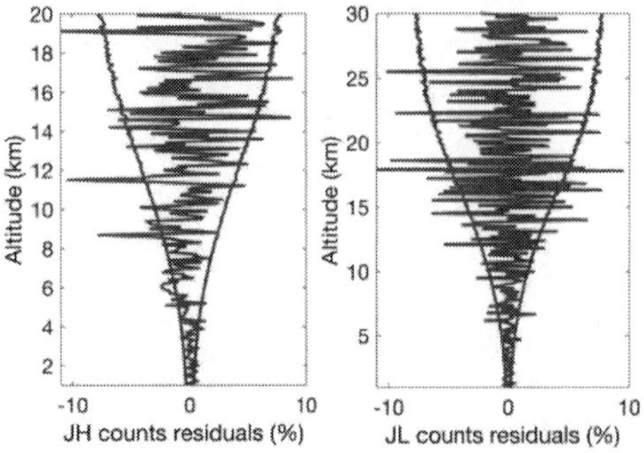

Figure 3: Residuals between the forward model and the measurements (blue curves). The red curves show the variance of the measurements.

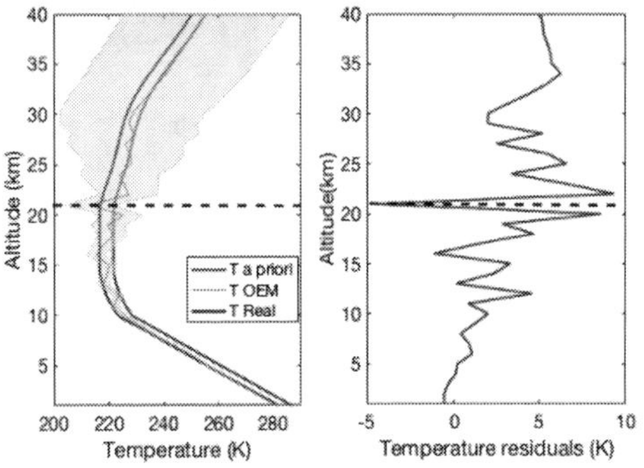

Figure 4: Figure on left shows the retrieved temperatures(red curve). The blue and green curves shows the true and the a priori temperatures respectively. Figure on the right shows the temperature residuals between retrieved and true states. Dashed line shows the cutoff height of the retrievals

Fig. 4 (left), shows a comparison of the retrieved temperatures to the *a priori* and true temperatures. In Fig. 4 (right), the difference between the retrieved and the true temperature profiles are shown. The uncertainty and temperature residuals increase with height. The retrieved and the true temperatures up to 15 km agree within 3 K. As was seen in Fig. 3, the sensitivity of the retrievals decrease after 15 km and reach zero eventually. The retrieved temperatures start to fall back to the *a priori* state from 21 km height. Therefore, the cutoff height of the retrieval can be set around 21 km where the averaging kernels is 0.9.

These results were produced by the OEM solver after completing 5 iterations. The initial and the final costs of the retrievals are 5.2566 and 0.6037 respectively.

4 CONCLUSIONS AND OUTLOOK

The OEM has been shown to be able to successfully retrieve temperatures from the synthetic, but realistic, pure rotational Raman-scattering-lidar measurements. The temperature retrievals agree within 3 K up to 15 km with the true profile. The retrieval uncertainty and vertical resolution increase as shown by the averaging kernels. Currently background noise and the lidar calibration constant for the JL channel can also be retrieved using the OEM with less than 3% uncertainty. We are currently in the process of extending the retrieval to include dead time, transmission and other parameters in addition to temperature, with the goal of then applying the retrieval to RALMO measurements and validating them.

ACKNOWLEDGEMENTS: We thank the Natural Sciences and Engineering Research Council (NSERC), Canadian Space Agency (CSA) and MeteoSwiss for supporting this project.

References

[1] Cooney, J., 1972. Measurement of atmospheric temperature profiles by Raman backscatter. Journal of applied meteorology, 11(1), pp.108-112.

[2] Zuev, Vladimir V., et al. Tropospheric temperature retrievals using nonlinear calibration functions in the pure rotational Raman lidar technique. Atmos. Meas. Tech. Discuss., doi 10.

[3] Dinoev, T. S., et al. Monitoring of Eyjafjallajökull ash layer evolution over payerne Switzerland with a Raman lidar. Proceedings of the TECO (2010).

[4] Rodgers, C.D., 2000. Inverse methods for atmospheric sounding: theory and practice (Vol. 2). World scientific.

[5] Behrendt A. Temperature measurements with lidar. InLidar 2005 (pp. 273-305). Springer New York.

[6] Eriksson, P., Jiménez, C. and Buehler, S.A., 2005. Qpack, a general tool for instrument simulation and retrieval work. Journal of Quantitative Spectroscopy and Radiative Transfer, 91(1), pp.47-64.

[7] Sica, R.J. and Haefele, A., 2015. Retrieval of temperature from a multiple-channel Rayleigh-scatter lidar using an optimal estimation method. Applied optics, 54(8), pp.1872-1889.

EPJ Web of Conferences **176**, 01012 (2018) https://doi.org/10.1051/epjconf/201817601012
ILRC 28

SHORT-WAVE INFRARED ATMOSPHERIC SCHEIMPFLUG LIDAR

Mikkel Brydegaard*[1,2,3], Jim Larsson[2], Sandra Török[2], Elin Malmqvist[2], Guangyu Zhao[4], Samuel Jansson[2], Mariam Andersson[2], Sune Svanberg[4,5], Susanne Åkesson[3], Fredrik Laurell[6], Joakim Bood[2]

[1] *Norsk Elektro Optikk A/S, Norway, * mikkel.brydegaard@fysik.lth.se*

[2] *Combustion Physics Division, Department of Physics, Lund University, Sweden*

[3] *Centre for Animal Movement Research, Department of Biology, Lund University, Sweden*

[4] *Center for Optical and Electromagnetic Research, South China Normal University, Guangzhou, China*

[5] *Atomic Physics Division, Department of Physics, Lund University, Sweden*

[6] *Department of Applied Physics, Royal Institute of Technology, Stockholm, Sweden*

Atmospheric dual-band Scheimpflug lidar is demonstrated at 980 and 1550 nm. Signals are compared during three weather conditions, and the spatio-temporal resolution of the atmospheric structure is considered. The potential for aerosol classification is evaluated, and future directions are discussed.

1 SCHEIMPFLUG LIDAR

In recent years we have developed the Scheimpflug lidar concept [1]. We have demonstrated the implementation for modulation spectroscopy on aerofauna [2], multiband aerosol monitoring [3, 4], polarimetric differential absorption lidar [5] and fluorescence and Raman lidar [6]. The method is based on high power CW laser diodes rather than pulsed solid-state lasers used in time-of-flight (ToF) lidars. Ranging is done by imaging the illuminated volume onto a tilted linear array. Utilizing the Scheimpflug condition and the Hinge rule, focus is maintained throughout the probe volume. Some advantages of Scheimpflug lidar are that resolution and range coverage can be scaled by scaling the optics. Also, whereas signal decays by r^{-2} and range resolution is constant in ToF lidar, this is the opposite for Scheimpflug lidar.

2 SHORT-WAVE INFRARED LIDAR

Until now, all implementations of Scheimpflug lidar have been based on Si CCD or CMOS arrays in the UV-NIR spectral region (405-808 nm). In this report we demonstrate that the method is feasible in SWIR (900-1700 nm) by using an InGaAs array. Atmospheric ToF lidar in the SWIR region is generally challenging due to the low backscatter and the need for complex laser systems with Raman cells [7], OPO conversion [8] or fiber lasers with low peak power. In this report we show how atmospheric echoes can be retrieved in the SWIR range using the Scheimpflug lidar method. Spectral bands can be chosen arbitrarily within the detection band, by replacing diode lasers for a cost comparable to that of a spectral filter. Lidar in the SWIR region has several benefits. The region has higher power density limits in relation to eye safety. The complex refractive index of liquid and frozen water has a detailed structure which can be indexed differentially through selecting appropriate laser wavelengths. Eumelanin is the most common pigment of the aerofauna, but because of its dull absorption spectrum ($\lambda^{-3.48}$ from 280 to 1064 nm) it is challenging to quantify without positioning a λ_{off} band in the SWIR region. This method could potentially quantify melanization as function of wingbeat with kHz sampling [9]. Several atmospheric bulk and trace gases, such as H_2O, CO_2, CO and CH_4 have absorption bands in the SWIR region. Scheimpflug SWIR lidar can potentially map these species at high spatio-temporal resolution which can be realized with narrow-band seeded fiber amplifiers [10].

3 SETUP AND TEST RANGE

The instrument previously described [3, 11] was modified and equipped with a 2048 pixel (12x250μm) InGaAs array. Two multimode (100μm wide) diode lasers at 980 nm and 1550 nm were superimposed with a dichroic beam splitter. The total peak power was measured to 4.2W and 2.5W, respectively (half of available output). The lasers are operated at 33% duty cycle synchronized with the line rate of the detector. Some ~50% of the laser light was collected and expanded by a ø2", F/3, SWIR coated lens. The beam was transmitted from a laboratory window over the university campus and terminated on a roof at 980 m distance. The near limit was 35 m and at 70 m various particles were released. Backscattered light was collected by a ø8" Newtonian reflector with a InGaAs array tilted in the Scheimpflug condition. Normally, a dual bandpass filter is installed in front of the detector, but because of a mismatch the system was operated without filters. This limited the system to night-time operation or reduced exposure time (thus also reduced SNR) during daytime. In Fig. 1 data from three different atmospheric conditions is seen. As the scattering increases, so does the extinction. During clear conditions, flat atmospheric echoes are seen all way to the termination; this is characteristic for Scheimpflug lidars. The anomalies at r < 50m are caused by stray light and partial overlap of the beam and field-of-view.

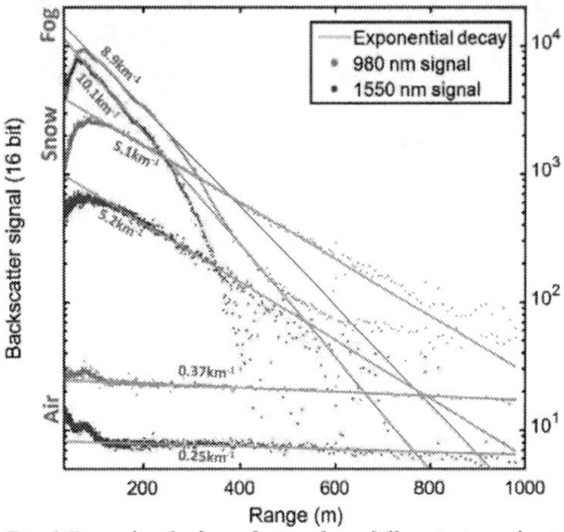

Fig. 1 Example of echoes during three different atmospheric conditions. The curves are normalized to the exposure time of 1ms for comparison. For the highly scattering conditions, light is extinguished halfway to the termination.

Table 1. Atmospheric conditions for the various scenarios presented.

Date/ time	Condition	Δt	T	V	RH	μ_{ext} 980nm	μ_{ext} 1550nm	Experiments
2016 31 Dec 10..11	Fog Daylight	500 µs	+2 °C	SE 6m/s	95 %	8.9 km^{-1}	10.1 km^{-1}	Wind Homogeneity
2017 5 Jan 16..18	Clear Darkness	2 ms	-9 °C	NW 3m/s	48 %	0.37 km^{-1}	0.25 km^{-1}	Particle release
2017 7 Jan 9..10	Snow Daylight	500 µs	-3 °C	E 2m/s	95 %	5.1 km^{-1}	5.2 km^{-1}	Sparse particles

4 PARTICLE RELEASE

Several types of aerosols and smoke were released at a test range of 70 m. The plumes were identified when signal exceeded the static air scatter plus the noise [12]. The most distinct cases are displayed in Fig. 2 with further details in Table 2. The distribution displays large covariance. The general conception is that concentration (which is unknown in our experiment) alters the absolute scatter cross

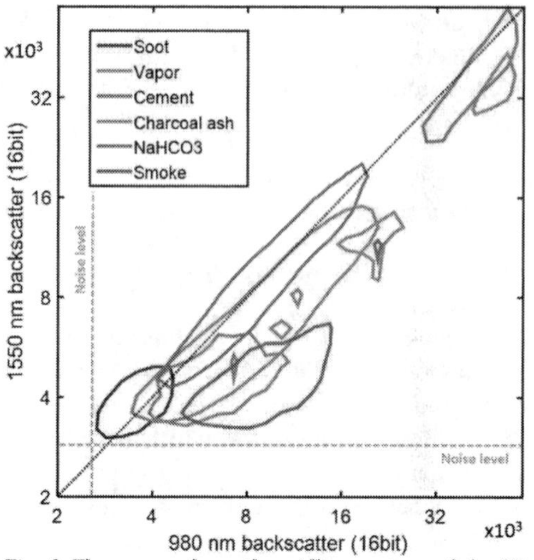

Fig. 2 The curves shows the midline contour of the 2D distribution from hundreds of thousand time-range pixels from each plume.

section, whereas the type of aerosol alters the spectral slope. Large non-absorptive salts such as $NaHCO_3$ align on the diagonal. New smoke with small particle sizes display strong scattering of the shorter band. Vapor, from the human breath, shows lower reflectance at 1550 nm where water absorbs. As can be expected the distributions heavily overlap and few aerosols occupy the upper triangle of the parameter space in Fig. 2. This could in principle be improved by placing strategic bands on the water absorption spectrum so that the shorter band coincides with higher water absorption. From the termination echo at 980 m range plume opacity was derived. However, backscatter from the plume saturates in most cases when significant extinction is detected.

Table 2. The various released particles plumes.

Species	Release method	Plume peak opacity	Spectral slope $\sigma_{1550}/\sigma_{980}$
Soot	Inverse vacuum	39%	110%
Vapor	Exhalation	10%	57%
Cement	Inverse vacuum	56%	84%
Ash	Inverse vacuum	87%	70%
NaHCO3	Inverse vacuum	27%	100%
Smoke	Tire burning	25%	48%

5 RESOLVING ATMOSPHERIC STRUCTURES IN SPACE AND TIME

One particular feature of Scheimpflug lidar is that it can reach spatial resolution down to a few cm (or less for other systems [13]) and temporal resolution of several kHz [12]. The idea of a homogeneous- or a granular atmosphere much depends on the spatio-temporal scale in question.

 https://doi.org/10.1051/epjconf/201817601012

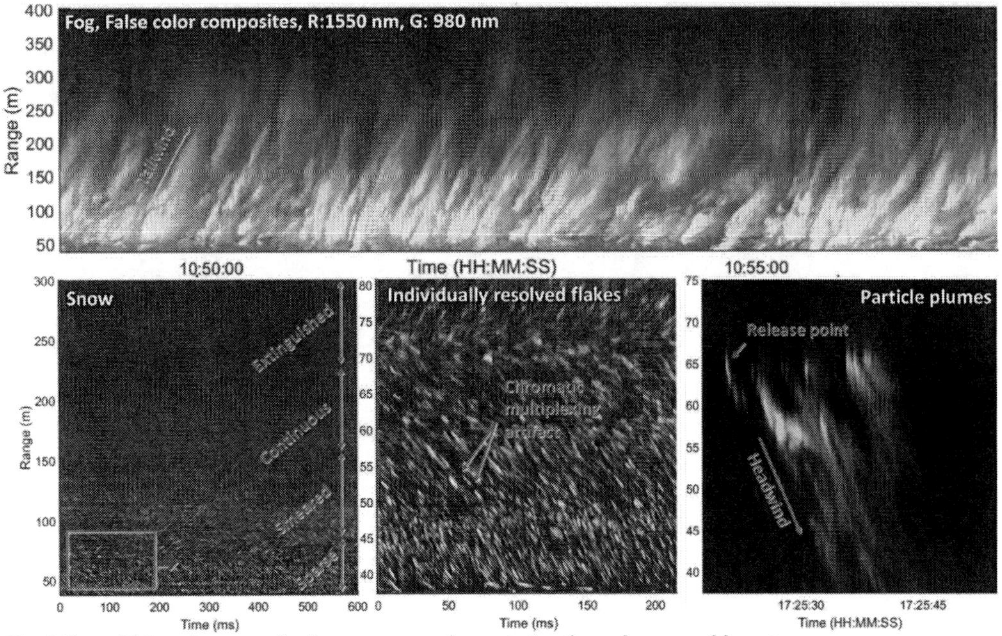

Fig. 3. Several false color composite time-range maps demonstrating the performance of the system.

Fig. 3 presents several false color composite time range maps for the days listed in Table 1. In the top of Fig. 3, a 5 minute data stream of the first 400 m is observed where turbulent fluffy features blow through the probe volume at 6m/s tail wind. At a range of 400 m both wavelengths are essentially entirely extinguished in fog; compare to Fig.1 with the echo profiles of the same fog data. In contrast to snowy and clear conditions, fog displays distinct extinction for the two spectral bands. In the lower left case of snowy weather we observe a second data stream with a speckled pattern in the closest range. At farther ranges the speckles smear out, eventually becoming a continuously distributed back scatterer which is ultimately attenuated. The extinction during snowfall is roughly half of that during fog. When zooming into the closest range, a peculiar pattern of individual snowflakes is seen. Over longer time periods consistent structures, where the flakes blow in opposite directions at the same time but at different ranges can be observed. In Fig. 3 a simple background subtraction is employed. Because of the fast changing scenery during snow, both the background and the two bands are not measured instantaneously but in a time multiplexing sequence. It can be observed that flakes are red when entering the probe volume and green when they leave. This can be improved with more advanced interpolation techniques [3].

The last subfigure in Fig. 3, shows a plume from a teaspoon of $NaHCO_3$ powder, which was released at 65 m range using an inverted vacuum cleaner. The time and range pixels for similar plumes were cropped out automatically and their scatter (in addition to the static scattering from clear air) was analyzed, see Fig. 2. The upper part of Fig. 4 shows how the statistical distribution and skewness varies with range (thus resolution) for different situations. Clean air produces Gaussians regardless of distance. Snow and fog display significant skewness when the spatial variance is resolved; However, the curvature is very different. The curves are presented for the 1550 nm scatter cross section, but similar findings can been obtained for 980 nm. The temporal dynamics of the three situations are displayed in the lower row of Fig. 4. The clean air shows a static power spectrum on all ranges. Even when resolved, the fluffy fog only changes the scattering in the probe volume this fast and decays quickly with frequency. The sparse snowflakes, on the other hand, enter and leave the probe volume rapidly, and when spatially resolved, they show a broad frequency distribution. The overall common dampening towards the Nyquist frequencies is governed by the system bandwidth, which is a sinc function.

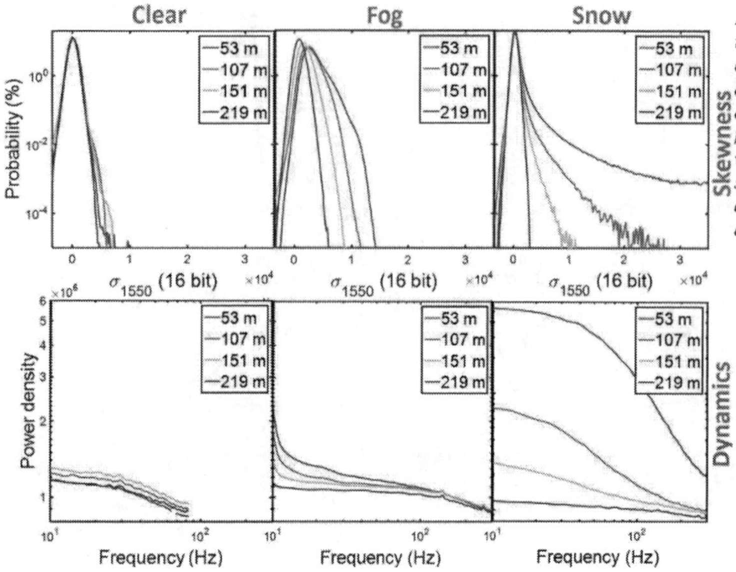

Fig. 4. When the spatial resolution increases in the near field, clear air, fog and snow show both distinct skewness of the statistical distribution and distinct frequency content. This aspects would affect the results of a system with lower resolution or averaging based on Gaussian assumption.

6 CONCLUSION AND OUTLOOK

We have shown that the atmospheric Scheimpflug lidar concept can be extended into the short-wave infrared range. Flat echoes from clean air (μ_{att}=0.32km^{-1}) were retrieved with SNR=1 at 32 ms exposure single shot, but the laser power could potentially be increased by a factor five. The system was tested in wintertime, and we are unsure if noise levels and line rates are sufficient for lidar entomology. However, some opportunities of high spatio-temporal resolution lidar were demonstrated. We hope to improve the system with dual-band pass filters, increased power, reduced stray light and to improve aerosol specificity by rearranging the bands. Further, we hope to pursue DIAL with seeded fiber lasers in the near future.

ACKNOWLEDGEMENTS

We appreciate the support from the Physiographical Society in Lund, from Norsk Elektro Optikk AS, from Lund University, from a Linnaeus grant to the Lund Laser Centre, and from the Swedish Research Council.

REFERENCES

[1] M. Brydegaard, A. Gebru, and S. Svanberg, "Super resolution laser radar with blinking atmospheric particles – application to interacting flying insects.," *Prog. Electromagn. Res.*, vol. 147, pp. 141-151, 2014.

[2] M. Brydegaard, "Towards quantitative optical cross sections in entomological laser radar – Potential of temporal and spherical parameterizations for identifying atmospheric fauna," *PLoS One*, vol. 10, p. e0135231, 2015.

[3] G. Zhao, E. Malmqvist, S. Török, P.-E. Bengtsson, S. Svanberg, J. Bood, *et al.*, "Dual-band continuous-wave lidar system employed for particle classification," *To appear*, 2017.

[4] L. Mei and M. Brydegaard, "Atmospheric aerosol monitoring by an elastic Scheimpflug lidar system," *Optics Express*, vol. 23, pp. A1613-A1628, 2015.

[5] L. Mei and M. Brydegaard, "Continuous-wave differential absorption lidar," *Laser & Photonics Reviews*, vol. 9, pp. 629-636, 2015.

[6] G. Zhao, M. Ljungholm, E. Malmqvist, G. Bianco, L.-A. Hansson, S. Svanberg, *et al.*, "Inelastic Hyperspectral Lidar for Profiling Aquatic Ecosystems," *Laser & Photonics Reviews*, vol. 10, pp. 807–813, 2016.

[7] W. P. Hooper and G. M. Frick, "Lidar detected spike returns," *J. Appl. Remote Sensing*, vol. 4, p. 4(1) 043549, 2010.

[8] K. Numata, H. Riris, S. Li, S. Wu, S. R. Kawa, M. Krainak, *et al.*, "Development of optical parametric amplifier for lidar measurements of trace gases on Earth and Mars," in *SPIE Optical Engineering+ Applications*, 2011, pp. 81590E-81590E-8.

[9] A. Gebru, M. Brydegaard, E. Rohwer, and P. Neethling, "Probing insect backscatter cross-section and melanization using kHz optical remote detection system," *SPIE, Journal of Applied Remote Sensing*, vol. 16611P, 2016.

[10] P. Zeil, V. Pasiskevicius, and F. Laurell, "Efficient spectral control and tuning of a high-power narrow-linewidth Yb-doped fiber laser using a transversely chirped volume Bragg grating," *Optics Express*, vol. 21, pp. 4027-4035, 2013/02/25 2013.

[11] L. Mei and M. Brydegaard, "Development of a Scheimpflug lidar system for atmospheric aerosol monitoring," presented at the International Laser Radar conference 27, NYC, 2015.

[12] E. Malmqvist, S. Jansson, S. Torok, and M. Brydegaard, "Effective parameterization of laser radar observations of atmospheric fauna," *Selected Topics in Quantum Electronics, IEEE Journal of*, vol. PP, pp. 1-1, 2015.

[13] E. Malmqvist, M. Brydegaard, M. Aldén, and J. Bood, "CW-Laser Radar for Combustion Diagnostics," in *Imaging and Applied Optics 2016*, Heidelberg, 2016, p. LW2G.2.

A SCANNING SCHEIMPFLUG LIDAR SYSTEM DEVELOPED FOR URBAN POLLUTION MONITORING

Yang Yang[1], Peng Guan[1], & Liang Mei[1]*

[1] *School of Optoelectronic Engineering and Instrumentation Science, Dalian University of Technology, Dalian 116024, China, *meiliang@dlut.edu.cn*

ABSTRACT

A scanning Scheimpflug lidar system based on the Scheimpflug principle has been developed by employing a high power multimode 808 nm laser diode and a highly integrated CMOS sensor in Dalian University of Technology, Dalian, Northern China. Atmospheric scanning measurements in urban area were performed for the studies of particle emission sources.

1 INTRODUCTION

Air pollution, especially particulate pollution, has been a serious environmental problem as the rapid development of industrialization and urbanization during past decades in China. By transmitting a nanosecond-scale pulsed laser (e.g., Nd:YAG) beam into atmosphere and capturing the backscattered light from particles/molecules via e.g., photomultiplier tubes (PMT), conventional pulsed lidar techniques have been widely employed for atmospheric particle and gas monitoring since 1960s [1-5]. As a remote sensing technique, lidar is capable of real-time monitoring of atmospheric particle distribution and variation for a large area, e.g., urban area pollution monitoring, while local point monitoring instruments are only able to measure localized pollutants. Extensive work has been pursued by utilizing scanning atmospheric lidar systems for remote monitoring of atmospheric pollutants and/or emission sources. For instance, Scott M. Spuler *et. al.*, presented a field-deployable scanning lidar system with a blind range of about 500 m based on an eye-safe laser (1.54 μm) [6]. T. Y. He, *et. al.*, presented a scanning lidar system employing a pulsed Nd:YAG laser, which was capable of tracking two-dimensional spatial and temporal particle distribution with an angular resolution of 0.1° [7]. However, the near-range particle distribution could not be resolved due to the long blind range, i.e., 0.8 km of the lidar system.

The lidar systems discussed above mainly employ high-cost and sophisticated laser sources, e.g., Nd: YAG lasers and require a data acquisition unit with high sampling rate and large dynamic range to achieve high spatial resolution and long detection range. Furthermore, a common issue of the pulsed lidar systems mentioned above is of long blind range because of incomplete geometric overlap between the transmitter and receiver. Recently, a new lidar technique, referred to as the Scheimpflug lidar has been demonstrated for atmospheric remote sensing [8-13]. The Scheimpflug lidar technique is based on the Scheimpflug principle, which implies that the backscattered light of the entire aerosol volume illuminated by the laser beam can be detected by a tilted image sensor if the image sensor plane, the lens plane and the laser beam plane intersect with each other. The pixels of the CCD or CMOS sensors correspond to the distances of the illuminated volume. The Scheimpflug lidar technique can achieve a short blind range less than 100 m or even up to 30 m by employing a long senor (e.g., 28 mm length).

In this work, we developed a scanning Scheimpflug lidar system which was capable of scanning in both azimuth and elevation angle with 1° angular resolution and short blind range. Field campaign was performed by monitoring particle distribution in Dalian, Northern China during a haze weather condition. The performance and feasibility of employing the scanning Scheimpflug lidar system for tracking emission sources is investigated.

2 METHODOLOGY

The schematic of the scanning Scheimpflug lidar system is shown in Figure 1. A C-mount 808-nm laser diode with 4 W output power is employed as the light source. The laser diode chip is a wide-stripe with an emission area of 230 μm (slow axis)

×1 µm (fast axis). As has been discussed in [9], the fast axis should be oriented along the plane of the optical layout in order to optimize the effective spatial resolution. To suppress the background noise, an 808-nm interference filter with 3 nm full width at half maximum (FWHM) and a high-pass color filter (RG780) with 780 nm cut-off wavelength are employed. Besides, multimode laser diodes typically have large uncertainties (e.g., ±3 nm) in terms of center emission wavelength. To optimize the transmission through the interference filter and stabilize the system performance, the 808-nm laser diode is housed by a home-made mount with large-capacity thermoelectric cooler to control the case temperature of the laser diode.

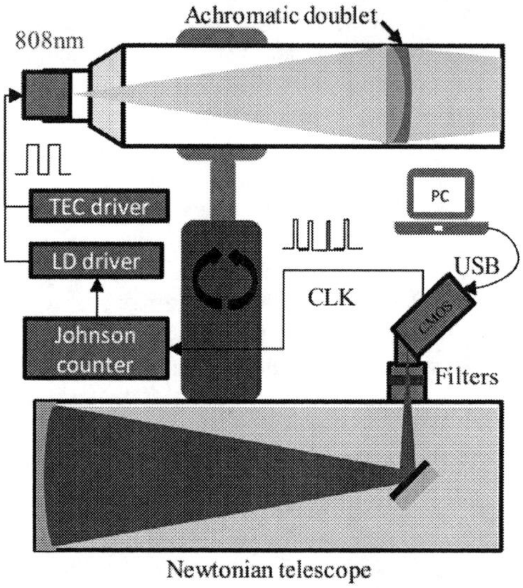

Figure 1 Scanning Scheimpflug lidar system.

A Newtonian telescope (Skywatcher, focal length 800 mm, Ø200 mm) with approximately 806 mm separation to the laser beam collimator is employed to collect the backscattering light of the laser beam. A 2D camera (CMOSIS CMV2000, 2048×1088, 5.5 µm), is employed to image the profile of the laser beam. A trigger signal is generated by the camera and fed to a Johnson counter that is employed as a two-frequency divider. Thus, the laser diode is turned on and off, while synchronizing with the exposure of the camera. The camera can then record the images of the laser beam and the background alternatively. The background as well as the dark current of the camera can thus be subtracted from the recorded raw images. The whole setup is assembled on an equatorial mount (Skywatcher, EQ6) which is capable of scanning the atmosphere in vertical and horizontal directions. A LabVIEW-based program was developed by our group to alternatively scan the lidar system and perform the measurements, according to user-set configurations.

3 MEASUREMENTS AND RESULTS

Field campaign was performed to scan the atmosphere of the south area of Ganjingzi District in Dalian on November 4th 2016. The scanning Scheimpflug lidar system was placed on a roof of a tall building (7 floors) in order to obtain a wide field of view. The elevation angle of the system was set to 3° to avoid some tall buildings within the field of view. The scanning range covers 90° with angular resolution of 1°, and the system was programed to work periodically and continuously. Before the measurement, the Scheimpflug lidar system is calibrated first in order to obtain the precise relationship between the position of pixel and the measuring distance, as discussed before. Figure 2 shows the scanning region and the location of the Scheimpflug lidar system and the calibration target.

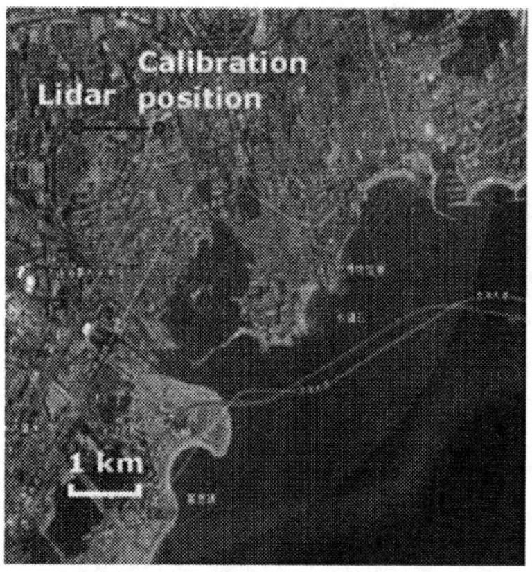

Figure 2. Top view of the scanning region (the blue sector) and the Scheimpflug lidar site.

Scanning lidar signals were recorded over a few hours. Linear interpolation and median analysis were employed to process the recorded lidar curves, in order to eliminate the influence of the swift variation of atmosphere and extremely strong sunlight background, as discussed in previous work. Since atmospheric aerosol distribution is changing all the time, high scanning rate is necessary to capture the spatial and temporal particle distributions. However, high scanning rate often implies a short measurement time (less signal-averaging times) and thus lower signal-to-noise ratio (SNR). To improve the SNR and thus achieve a relatively higher scanning rate, a low-pass finite response (FIR) filter with a Hanning window function was employed to suppress the solar radiation (shot) noise, readout noise and dark current noise. As shown in Figure 3, the SNR ratio was improved by a factor of 6 after employing the low-pass filter. The scanning speed of the Scheimpflug lidar system is 12 min/90° with an angular resolution of 1°, allowing fast monitoring of the urban particle distributions.

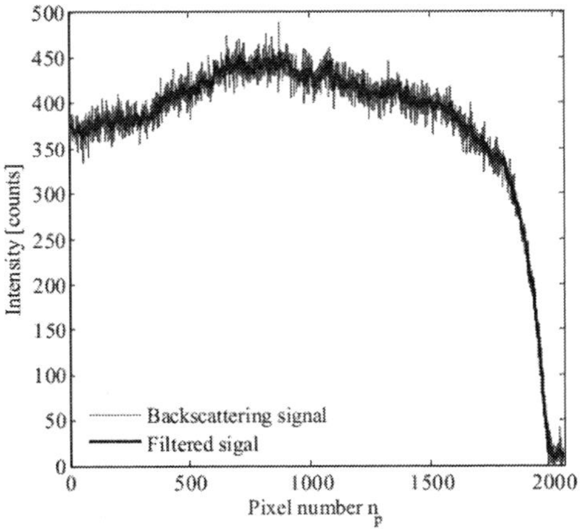

Figure 3. Backscattering signal (50 times averaging, approximately 2 seconds) recorded by the image sensor at a certain angle and the signal after applying a low-pass filter.

Figure 4 shows 2D maps of atmospheric backscattering intensities measured by the scanning Scheimpflug lidar system. As can be seen, two emission sources with substantially stronger atmospheric backscattering signal were identified, which were figured out to be two heating plants. From the backscattering signal evolution in 7 hours, it can be concluded that the maximum emission of the two plants occurred at around 16:00. Heating plants are actually one of the main sources of atmospheric pollution and causing haze weather in northern China during winter.

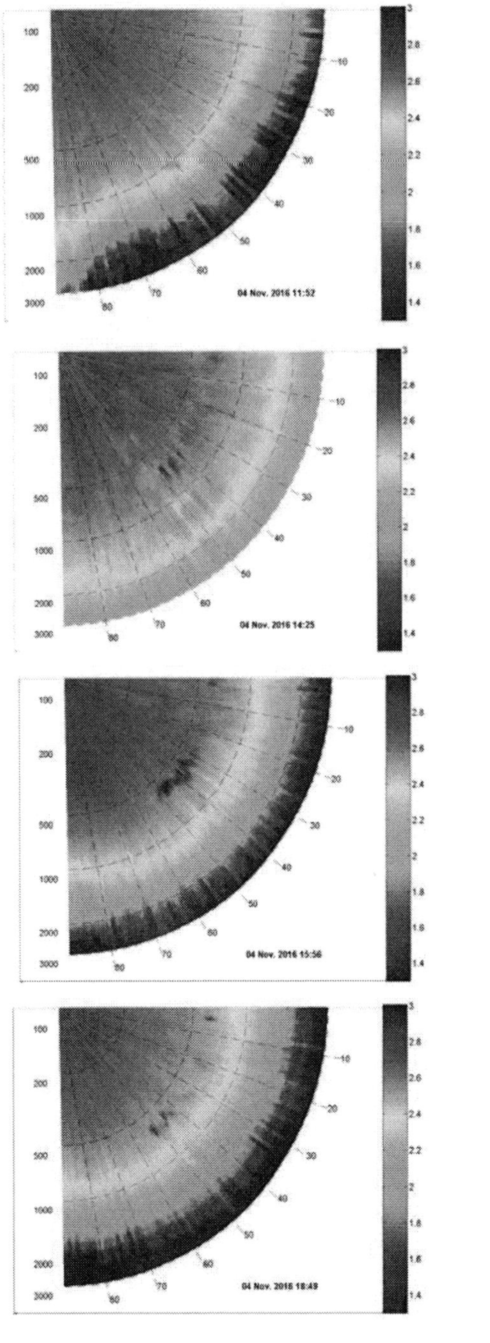

Figure 4. 2D maps of the backscattering intensities measured by the scanning lidar system. The intensities are plotted in log-scale, and y-axis is distance in meter.

4 CONCLUSIONS

In urban areas, aerosol sources could be difficult to localize by single point ground-based aerosol monitoring stations due to the spatial and temporal dispersions of particles. Since the aerosol has been increasingly severe in big cities, aerosol sources identification is highly necessary for pollution control and managements. The scanning Scheimpflug lidar system developed in this work is capable of tracking the locations of the emission sources in urban area, and it is of low cost by employing continuous-wave laser diodes and image sensors.

A field campaign was performed to monitor urban particle emissions in Ganjingzi District, Dalian on November 4th 2016. The scanning speed of the scanning Scheimpflug lidar system is 12 min/90° with an angular resolution of 1°. During the campaign, two pollutant sources are identified clearly from the two-dimension figures. The blind range of the scanning Scheimpflug lidar system is very short, i.e., 80 m and thus near-range particles can be detected as well. During the campaign, we found out that the most time-consuming procedure for the measurement is actually the movement of the equatorial EQ6. To achieve faster scanning and higher time and spatial resolution, the scanning mechanics should be furthered improved. Field campaign with improved performance is planned in the near future.

ACKNOWLEDGEMENTS

This work was supported by the National key research and development program of China (2016YFC0200600); Fundamental Research Funds for the Central Universities (DUT15RC(3)107); Natural Science Foundation of Liaoning Province, China (201602163).

References

[1] A. Papayannis, D. Balis, V. Amiridis, G. Chourdakis, G. Tsaknakis, C. Zerefos, A. D. A. Castanho, S. Nickovic, S. Kazadzis, and J. Grabowski, 2005: Measurements of Saharan dust aerosols over the Eastern Mediterranean using elastic backscatter-Raman lidar, spectrophotometric and satellite observations in the frame of the EARLINET project, Atmos. Chem. Phys. **5**, 2065-2079.

[2] Z. G. Guan, P. Lundin, L. Mei, G. Somesfalean, and S. Svanberg, 2010: Vertical lidar sounding of atomic mercury and nitric oxide in a major Chinese city, Appl. Phys. B **101**, 465-470.

[3] J. Lenoble, L. Remer, and D. Tanre, *Aerosol Remote Sensing*. (Springer, Verlag Berlin Heidelberg, 2013).

[4] L. Mei, G. Y. Zhao, and S. Svanberg, 2014: Differential absorption lidar system employed for background atomic mercury vertical profiling in south China, Opt. Laser Eng. **55**, 128-135.

[5] M. R. Perrone, F. De Tomasi, and G. P. Gobbi, 2014: Vertically resolved aerosol properties by multi-wavelength lidar measurements, Atmos. Chem. Phys. **14**, 1185-1204.

[6] S. M. Spuler and S. D. Mayor, 2004: Scanning Eye-Safe Elastic Backscatter Lidar at 1.54 μm, Journal of Atmospheric & Oceanic Technology **22**, 696.

[7] T. Y. He, S. Stanic, F. Gao, K. Bergant, D. Veberic, X. Q. Song, and A. Dolzan, 2012: Tracking of urban aerosols using combined LIDAR-based remote sensing and ground-based measurements, Atmos. Meas. Tech. **5**, 891-900.

[8] M. Brydegaard, A. Gebru, and S. Svanberg, 2014: Super resolution laser radar with blinking atmospheric particles - Application to interacting flying insects, PIER **147**, 141-151.

[9] L. Mei and M. Brydegaard, 2015: Atmospheric aerosol monitoring by an elastic Scheimpflug lidar system, Opt. Exp. **23**, 247841.

[10] L. Mei and M. Brydegaard, 2015: Continuous-wave differential absorption lidar, Laser & Photon. Rev. **9**, 629-636.

[11] L. Mei and P. Guan, 2017: Development of an atmospheric polarization Scheimpflug lidar system based on a time-division multiplexing scheme, Opt. Lett. **42**, 3562-3565.

[12] L. Mei, P. Guan, and Z. Kong, 2017: Remote sensing of atmospheric NO_2 by employing the continuous-wave differential absorption lidar technique, Opt. Exp. **25**, A953-A962.

[13] L. Mei, P. Guan, Y. Yang, and Z. Kong, 2017: Atmospheric extinction coefficient retrieval and validation for the single-band Mie-scattering Scheimpflug lidar technique, Opt. Exp. **25**, A628-A638.

ATMOSPHERIC AEROSOL MEASUREMENTS BY EMPLOYING A POLARIZATION SCHEIMPFLUG LIDAR SYSTEM

Liang Mei[1]*, Peng Guan[1], & Yang Yang[1]

[1] *School of Optoelectronic Engineering and Instrumentation Science, Dalian University of Technology, Dalian 116024, China, *meiliang@dlut.edu.cn*

ABSTRACT

A polarization Scheimpflug lidar system based on the Scheimpflug principle has been developed by employing a compact 808-nm multimode high-power laser diode and two highly integrated CMOS sensors in Dalian University of Technology (DLUT), Dalian, China. The parallel and orthogonal polarized backscattering signal are recorded by two 45 degree tilted image sensors, respectively. Atmospheric particle measurements were carried out by employing the polarization Scheimpflug lidar system.

1 INTRODUCTION

Lidar (light detection and ranging) techniques have been widely employed for atmospheric remote sensing of particles/aerosols and gases since decades. Conventional lidar techniques measure atmospheric backscattering echoes based on the time-of-flight principle, by employing high-power nanoscale lasers - and photomultiplier tubes (PMT) [1-5]. Since 1980s, polarization lidar, capable of detecting the parallel and orthogonal polarized backscattering light respectively, has been employed as an effective tool for the studies of atmospheric cloud phase, ice crystals, and particle classification [6-9], etc. Nevertheless, high power pulsed Nd:YAG lasers and PMTs are to some extent complicated and of high cost. Besides, both analog and single photon counting (SPC) detection schemes are often utilized to increase the dynamic range as the lidar signal decreases over the square of the distance. All these high technical demands increase the system cost, complexity and maintenance requirement, which limit the applications of conventional pulsed lidar techniques mainly in scientific fields.

Recently, a new atmospheric lidar technique, referred to as Scheimpflug lidar, is developed for atmospheric gas and aerosol monitoring [10-13]. The Scheimpflug lidar technique is based on the Scheimpflug principle, which describes the relationship between the object, image and lens planes when they are non-paralleled in an imaging system. Briefly, if a collimated laser beam is transmitted into atmosphere, the entire laser beam over km-range can be clearly imaged by a titled image sensor while using a large aperture lens/telescope if the image sensor plane, the laser beam plane and the lens plane intersect with each other. The atmospheric backscattering signal can then be measured, and the sensor pixel corresponds to the distance. The range-resolved lidar signal is in this case angular-resolved, while it is time-resolved in conventional pulsed lidar techniques [10]. The fundamental difference of the Scheimpflug lidar technique enables the usage of low-cost, compact high-power continuous-wave laser diodes and highly-integrated CMOS/CCD image sensors, which greatly reduces the system complexity and cost.

In this work, we demonstrate a polarization Scheimpflug lidar system for the polarization property studies of atmospheric aerosols. A ground-based polarization Scheimpflug lidar system, based on low-cost, compact laser diodes and CMOS sensors, was developed in Dalian University of Technology, China. Atmospheric remote sensing measurements have been performed by using the polarization Scheimpflug lidar system to validate the system feasibility.

2 METHODOLOGY

The system schematic of the polarization Scheimpflug lidar is shown in Figure 1. A 4 W 808 nm multimode (3 nm bandwidth) laser diode is employed as light source. The laser beam is collimated by a F6 achromatic telescope (Focal length 600 mm, Ø100 mm) for transmission. A half-wave plate is placed before the achromatic lens for rotating the polarization of the laser beam, which can be used to calibrate the polarization-dependent system constant. The atmospheric backscattering light is collected by a F4 Newtonian telescope (focal length 800 mm, Ø200 mm). The received backscattering light first passes through an absorption filter (RG780, Edmund Optics) and an interference filter (808 nm, 3 nm FWHM, Edmund

Optics) before separated by a polarization beam splitter cube. Two titled CMOS sensors (CMOSIS, 2088×1088 pixels, 5.5 μm pixel width) are utilized to simultaneously detect the orthogonal and parallel polarized backscattering light. A trigger output signal from one of the CMOS sensor is fed to a Johnson counter, which then generate a square signal to modulate the laser diode for dynamic background subtraction. The equipment mentioned above are first assembled and then carried by the two arms of an equatorial mount (Skywatcher, EQ6) in accordance with the Scheimpflug principle.

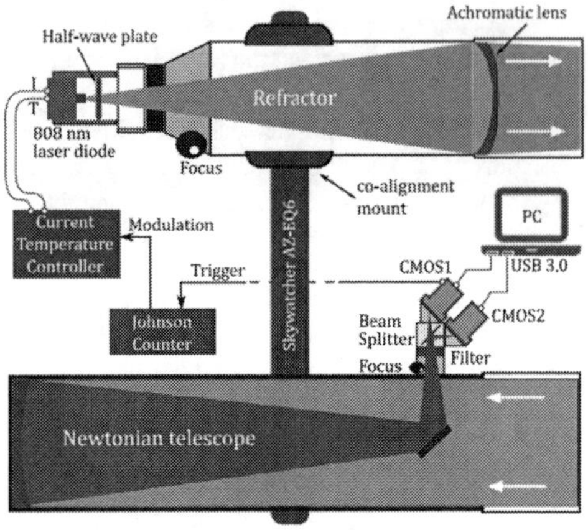

Figure 1. The schematic of the Scheimpflug Lidar system. The co-alignment mount is capable of tuning the refractor in two dimensions for the field-of-view overlap between the Newtonian telescope and the refractor. The half-wave plate can be rotated manually.

The laser beam image recorded by the CMOS sensor after subtraction of a background image is vertically (perpendicular to the laser beam transmission direction) binned in a LabVIEW-based program developed by our research group. The pixel binning process leads to a single pixel-intensity curve. After the system calibration with a hard target at a known distance [10], the pixel-distance relationship can be calculated and the distance-intensity curve, i.e., the range-resolved lidar signal is then retrieved. According to the Scheimpflug principle and geometrical optics, the lidar signal detected by the Scheimpflug lidar technique is given by [10]

$$P(\lambda,z) = K\beta(\lambda,z)\exp[-2\int_0^z \alpha(\lambda,z')dz']. \quad (1)$$

Here, λ is the laser center wavelength, z is the measurement distance , $\beta(\lambda,z)$ and $\alpha(\lambda,z)$ are the backscatter and extinction coefficients at the distance z . From the above equation, we note that the lidar signal does not decrease over the square of the distance which is the case in conventional pulsed lidar techniques. This greatly facilitates the detection scheme of the Scheimpflug lidar technique, and improves the signal-to-noise ratio (SNR) for the far distance. Following Eq. (1), the depolarization ratio $\delta(z)$ can be given by [14]

$$\delta(z)=\frac{k_p}{k_s}\frac{P_s(\lambda,z)}{P_p(\lambda,z)}=\mathrm{k}\frac{P_s(\lambda,z)}{P_p(\lambda,z)}. \quad (2)$$

Here $P_p(\lambda,z)$ and $P_s(\lambda,z)$ are the parallel and orthogonal polarized signal. k_p and k_s are the system constant in the parallel and orthogonal channels of the Scheimpflug lidar system, and $k = k_p/k_s$ is a polarization-dependent system constant of the Scheimpflug lidar system. The system constant k can be calibrate by, e.g., non-polarized light source method, air molecular method and half wave-plate method [15]. In this article, we employ the half wave-plate method to determine the system constant k during the measurement campaign.

3 MEASUREMENTS AND RESULTS

Depolarization measurements of atmospheric aerosol were performed on December 14th, 2016 from 12:30 – 21:00. The local temperature was about -6° to -3°. The laser beam was pointed to the south. Atmospheric backscattering signal was measured with 20 ms exposure time, and each lidar curve was generated from the mode-average of 5000-time recordings (about 4 minutes). The time-range maps of the parallel and orthogonal polarized light are shown in Figure 2. The system constant factor was calibrated by the half-wave plate method. Since atmospheric backscattering signal is angular resolved for the Scheimpflug lidar technique, the system constant factor was found to be variable for different pixels/distances in this case. In the present work, an averaged value of the polarization dependent system constant was estimated, i.e., 1.3. The Depolarization ratio map can then be retrieved. The high depolarization ratio, observed in Figure 3, should be attributed to the highway maintenance at around 200-300 m which generated significant amount of dust, and the wind from north (Level 3-4).

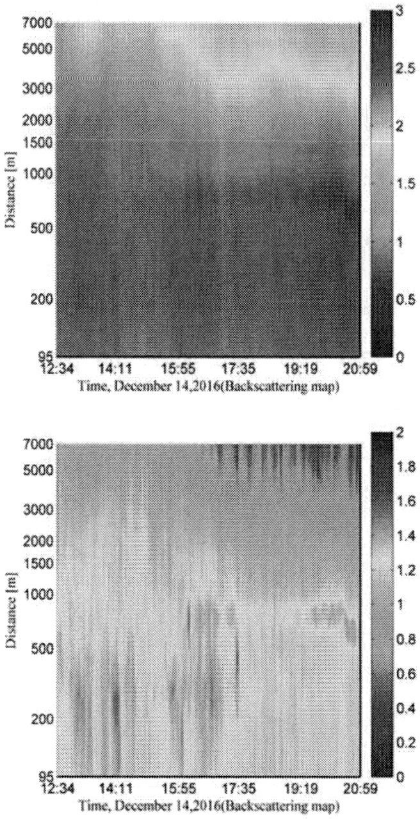

Figure 2. Parallel (up) and orthogonal (Bottom) polarized atmospheric backscattering time – range map measured by the Scheimpflug lidar system.

4 CONCLUSIONS

A polarization Scheimpflug lidar system as well as the preliminary results were presented in this work. As has been demonstrated, atmospheric depolarization ratio can be retrieved over several-kilometer range by the Scheimpflug lidar technique even during day time. Detailed analysis about the system constant calibration accuracy, atmospheric vertical measurements will be presented in the near future to further validate the capability of the polarization Scheimpflug lidar technique.

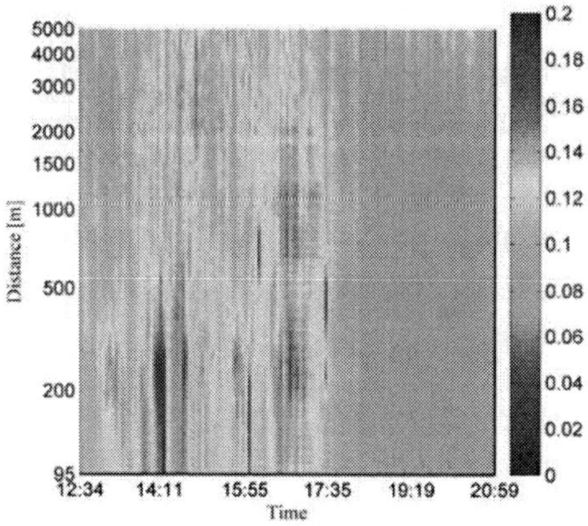

Figure 3. Depolarization ratio time – range map measured by the Scheimpflug lidar system.

ACKNOWLEDGEMENTS

National key research and development program of China (2016YFC0200600); Fundamental Research Funds for the Central Universities (DUT15RC(3)107); Natural Science Foundation of Liaoning Province, China (201602163) .

References

[1] M. Sicard, F. Molero, J. L. Guerrero-Rascado, R. Pedros, F. J. Exposito, C. Cordoba-Jabonero, J. M. Bolarin, A. Comeron, F. Rocadenbosch, M. Pujadas, L. Alados-Arboledas, J. A. Martinez-Lozano, J. P. Diaz, M. Gil, A. Requena, F. Navas-Guzman, and J. M. Moreno, 2009: Aerosol lidar intercomparison in the framework of SPALINET-The Spanish lidar network:methodology and results, IEEE Trans. Geoscience and Remote Sensing **47**, 3547-3559.

[2] Z. G. Guan, P. Lundin, L. Mei, G. Somesfalean, and S. Svanberg, 2010: Vertical lidar sounding of atomic mercury and nitric oxide in a major Chinese city, Appl. Phys. B **101**, 465-470.

[3] J. Lenoble, L. Remer, and D. Tanre, *Aerosol Remote Sensing*. (Springer, Verlag Berlin Heidelberg, 2013).

[4] L. Mei, G. Y. Zhao, and S. Svanberg, 2014: Differential absorption lidar system employed for background atomic mercury vertical profiling in south China, Opt. Laser Eng. **55**, 128-135.

[5] G. Pappalardo, A. Amodeo, A. Apituley, A. Comeron, V. Freudenthaler, H. Linne, A. Ansmann, J. Bosenberg, G. D'Amico, I. Mattis, L. Mona, U. Wandinger, V. Amiridis, L. Alados-Arboledas, D.

Nicolae, and M. Wiegner, 2014: EARLINET: towards an advanced sustainable European aerosol lidar network, Atmos. Meas. Tech. **7**, 2389-2409.

[6] S. Pal and A. Carswell, 1978: Polarization properties of lidar scattering from clouds at 347 nm and 694 nm, Appl. Opt. **17**, 2321-2328.

[7] V. Noel, D. M. Winker, M. McGill, and P. Lawson, 2004: Classification of particle shapes from lidar depolarization ratio in convective ice clouds compared to in situ observations during CRYSTAL - FACE, Journal of Geophysical Research: Atmospheres **109**,

[8] J. A. Shaw, N. L. Seldomridge, D. L. Dunkle, P. W. Nugent, L. H. Spangler, J. J. Bromenshenk, C. B. Henderson, J. H. Churnside, and J. J. Wilson, 2005: Polarization lidar measurements of honey bees in flight for locating land mines, Opt. Exp. **13**, 5853-5863.

[9] S. S. Matthew Hayman, and Bruce Morley, 2014: Polarization lidar observations of backscatter phase matrices from oriented ice crystals and rain, Opt. Exp. **22**, 16976-16990.

[10] L. Mei and M. Brydegaard, 2015: Atmospheric aerosol monitoring by an elastic Scheimpflug lidar system, Opt. Exp. **23**, 247841.

[11] L. Mei and M. Brydegaard, 2015: Continuous-wave differential absorption lidar, Laser & Photon. Rev. **9**, 629-636.

[12] P. Guan and L. Mei, "Range-resolved atmospheric aerosol monitoring by a Scheimpflug lidar system based on a continuous-wave laser diode," Asia Communications and Photonics Conferene (ACP), Wuhan, China, AS1F.5, (2016).

[13] L. Mei and M. Brydegaard, "Development of a Scheimpflug lidar system for atmospheric aerosol monitoring," EPJ Web of Conferences (ILRC 27), 119, 27005, (2016).

[14] K. Sassen, "Polarization in lidar," in *Lidar*, edited by C.Weitkamp (Springer, New York, 2005), pp. 19-42.

[15] Z. Wang, R. Chi, B. Liu, and J. Zhou, 2008: Depolarization properties of cirrus clouds from polarization lidar measurements over Hefei in spring, Chinese Optics Letters **6**, 235-237.

CW LASER RADAR FOR COMBUSTION DIAGNOSTICS

Elin Malmqvist[1], Mikkel Brydegaard[1,2], Marcus Aldén[1], Joakim Bood[1]*

[1] *Div. of Combustion Physics, Dep. of Physics, Lund University, Sweden*
[2] *Norsk Elektro Optikk A/S, Norway*
* *joakim.bood@forbrf.lth.se*

ABSTRACT

A CW-laser radar system developed for combustion diagnostics is described. The system is based on triangulation to attain range information. A portable system has been constructed and here we show some result from measurements in various flames, for example Rayleigh scattering thermometry and monitoring of particle distributions with high temporal and spatial resolution. The concept can equally well be based on pulsed lasers, allowing suppression of background emission through gated detection.

1 INTRODUCTION

Lidar is an interesting concept for practical combustion diagnostics, since it allows single-ended measurements, i.e. only one optical access is needed. This feature opens up for range-resolved *in-situ* measurements inside devices with intractable geometries, such as full-scale furnaces, boilers, and power plants. Our research group has in recent years developed a lidar system based on a picosecond laser as transmitter and an ultrafast detector (streak camera or MCP-PMT) to achieve adequate range resolution (~1 cm) [1]. The ps-lidar system has been demonstrated for Rayleigh thermometry (including two-dimensional) in flames [1] and full-scale room fire experiment [2], quantitative species detection based on differential absorption lidar (DIAL) [3,4], and measurements of soot volume fraction in flames [5].

Laser-induced fluorescence (LIF) is a very sensitive and highly species-selective technique, which therefore has been used extensively for combustion studies. Unfortunately LIF is difficult to utilize in ps-lidar. Since ps-lidar attains range resolution from the time-of-flight principle, the finite fluorescence lifetime (typically on the nanosecond scale in combustion environments), prevents adequate range resolution. This problem can, in principle, be circumvented if the recorded signal is deconvoluted with the fluorescence decay curve, but since the fluorescence lifetime generally is unknown and also varies along the measurement path, this is not a viable solution.

Based on the pioneering work by Brydegaard and co-workers, who has developed a novel continues-wave (cw) lidar concept based on the Scheimpflug principle [6,7], we intend to pursue this technique for combustion diagnostics. In cw-lidar, range resolution is achieved through triangulation instead of utilizing a time-of-flight approach as in conventional lidar. By positioning the detector, collection optics, and the laser beam in a configuration which fulfils the Scheimpflug condition, infinite focal depth can be obtained and the laser beam can thus be sharply imaged onto the detector. Since range resolution in this lidar configuration is achieved by imaging the laser-induced emission, instead of utilizing time-resolved detection, LIF is fully applicable. This new approach also opens up for the use of small inexpensive diode lasers, which may reduce the size, weight, and cost of the system, thereby also facilitating its mobility.

2 METHODOLOGY

The Scheimpflug-lidar setup currently under development in our laboratory is illustrated in Fig. 1. The detector is a linear array CMOS camera (Glaz-S, Synertronic Designs) with 2048 pixels (200×14 μm pixel size), providing a maximum scan rate of 4 kHz. The kHz-scan rate enables online background subtraction by modulating the laser output on and off in synchronization with the camera exposure. The high scan rate also opens up the possibility to study fast combustion processes. The signal is collected with a 2-inch spherical lens (f = 200 mm), and a bandpass filter is used in front of the detector to suppress background radiation. A tele-xenar objective (1:2.8/100, Schneider Kreuznach) expands and transmit the laser beam towards the region of interest. Diode lasers (O-like lasers) emitting at 445 nm (3-W output power) and 405 nm (0.5-W

 https://doi.org/10.1051/epjconf/201817601015

output power) are currently available for measurements.

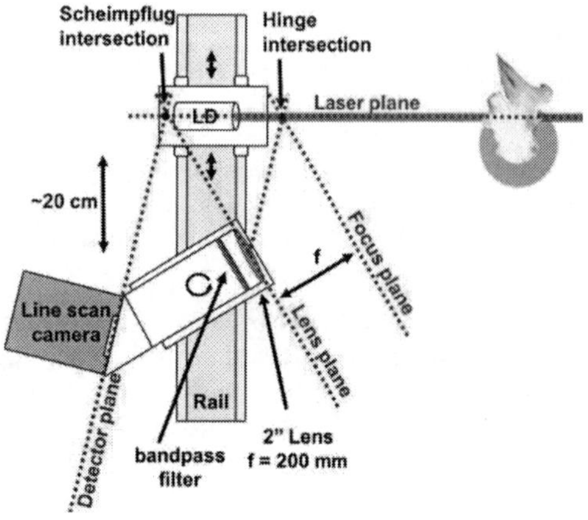

Fig. 1 Schemtaic setup of cw-lidar in Scheimpflug configuration for combustion diagnostics. The laser is a diode laser and the camera is linear array CMOS camera allowing a maximum scan rate of 4 kHz.

Figure 2 shows a photograph of the lidar system in operation. The bright blue spot, visible in the upper-middle part of the figure, is scattering from the beam termination trap.

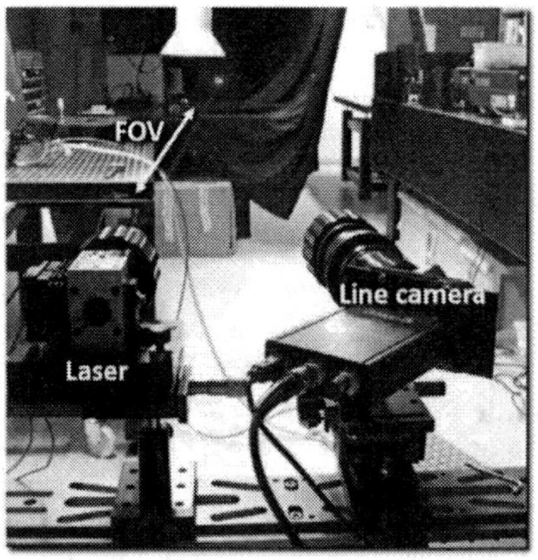

Fig. 2 Photograph of the cw-lidar system.

3 RESULTS

3.1 Elastic scattering from sooty flames and smoke

Elastic scattering from two highly sooty candle flames, placed in the lidar transect at 230 and 352 cm, was studied with the cw-lidar system. The data was collected with a 2 kHz scan rate and 400 μs integration time. Figure 3 shows the attained time-range map, after background subtraction. The mean intensity over the time window for each range is shown at the left of each map. As expected, strong echoes, due to scattering from soot particles in the flames, give rise to the distinct peaks at 230 and 352 cm. A closer view of the signal from the closest flame, i.e. the signal confined by the blue rectangle, is shown beneath the full time-range map. It is clearly seen that the instability of the flame can be resolved in time and space by the system.

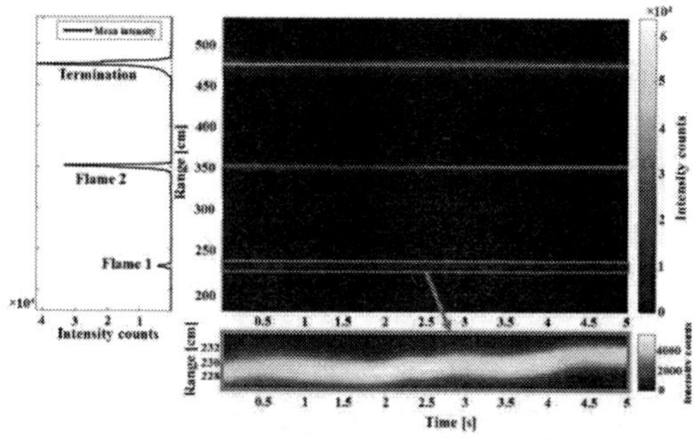

Fig. 3 Time-range map, covering a 5-second time window and a range interval of 187 – 530 cm, recorded with two sooty flames present in the lidar transect. The mean intensity over the full time window for each range is shown in the left panel, while the bottom panel is a zoom-in of the signal inside the blue rectangle.

In order to study particle distributions in smoke, the candle flame located at 350 cm was abruptly blown out, which resulted in the time-range map depicted in Fig. 4. Prior to extinguishing the flame, i.e. in the time interval 0 to 4 sec, the observed signal is elastic scattering from soot particles in the burning flame. After the flame has been blown out it is possible to monitor how the smoke is moving in time and space. This result suggests that cw-lidar in Scheimpflug configuration may deliver particle distributions

https://doi.org/10.1051/epjconf/201817601015

with high temporal and spatial resolution, crucial information in combustion and environmental research.

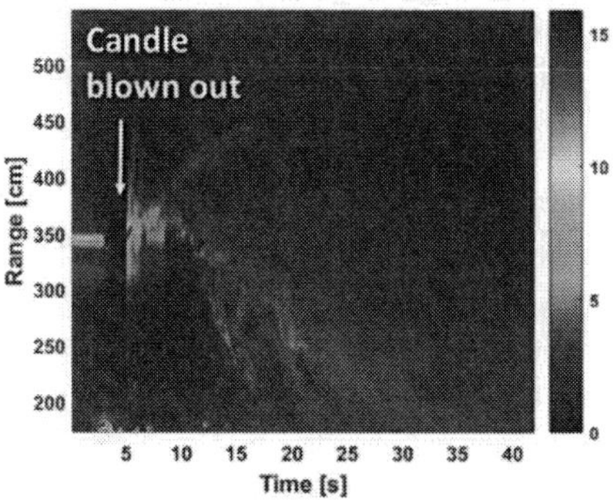

Fig. 4 Time-range map acquired in an experiment where a sooty flame was extinguished (at approximately the 4-sec mark). The spatio-temporal dynamics of the smoke is revealed by the recorded data.

3.2 Elastic scattering from non-sooty flames – thermometry

Two non-sooty flames were positioned in the lidar transect. A conical flame on a Bunsen burner was positioned at 181 cm, while a flat flame on a McKenna burner was placed at 194 cm. Since no soot is present in the flames, the scattering of the laser beam was pure molecular Rayleigh scattering. An example of a recorded lidar curve is shown in Fig. 5.

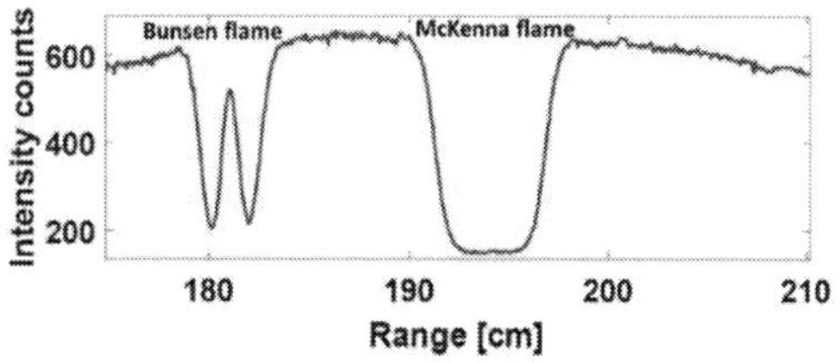

Fig. 5 Elastic lidar signal recorded with a non-sooty conical flame present at 181 cm and a non-sooty flat flame located at 194 cm.

The Rayleigh scattering intensity is proportional to the number density of molecules. From the ideal gas law it then follows that the Rayleigh scattering intensity is inversely proportional to the temperature, which explains the lower lidar signal intensities at the locations of the two flames. At the location of the McKenna burner, the lidar signal is fairly constant through the flame, which is expected since the flame is flat (i.e. temperature and species concentrations do not vary across the flame). At the location of the Bunsen burner two intensity dips are observed, which is expected since this flame is conical. The dips correspond to the reaction zone, i.e. the transition region between the unburnt reactants and the product gases. The lidar signal is significantly higher in the center of the flame, i.e. at 181 cm, actually only slightly lower than the signal level for the surrounding air. This is also expected since cold reactants (fuel and air) are present in the center of the flame.

Using the same McKenna burner, but positioned at a slightly shorter distance, a lidar curve was recorded through a slightly fuel-lean premixed methane/air flame, which corresponds to the blue curve in the left panel of Fig. 6. With identical system settings, a lidar curve was recorded with the flame turned off and room-temperature air flowing through the burner, which corresponds to the red curve in Fig. 6.

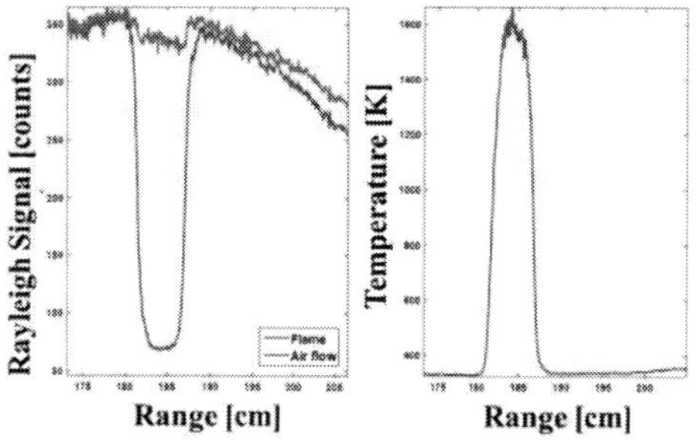

Fig. 6 Left: Lidar curves corresponding to molecular Rayleigh scattering in a slightly fuel-lean methane/air flame on a McKenna burner (blue curve) and with only air flowed through the burner (red curve). Right: Evaluated flame temperature profile.

The flame temperature can now be determined using Equation (1), where σ_{flame} and σ_{air} are Rayleigh cross sections of the flame and air, respectively, I_{flame} and I_{air} are the measured signal intensities in the flame and air, respectively, and T_{air} is the air temperature (295 K).

$$T_{flame} = \frac{\sigma_{flame}}{\sigma_{air}} \frac{I_{air}}{I_{flame}} T_{air} \quad (1)$$

Since the major flame species and their concentrations are known, the Rayleigh cross sections can be determined (relative values are sufficient). Using these cross sections and the data plotted in the left panel of Fig. 6, the temperature profile shown in the right panel of Fig. 6 was extracted. A flame temperature of ~1600 K is in fairly good agreement with previously reported results [1]. The reason why our measurement slightly underpredicts the flame temperature is, at least to some extent, due to uncertainty in the recording of the reference lidar curve (red curve in Fig. 6), which is related to difficulties in preventing dust particles from penetrating into the probe volume and stray light.

4 FUTURE WORK

Currently cw-lidar experiments utilizing LIF is under development in our laboratory. If successful, stand-off range-resolved detection of trace species with high temporal resolution can be added to the assets demonstrated here, making cw-lidar a very versatile tool for combustion diagnostics.

The Scheimpflug lidar concept is not restricted to cw lasers, but it can equally well be based on pulsed lasers. In fact, a pulsed laser source together with a gated intensified CCD camera is probably the best setup for most applied combustion experiments, as strong background radiation is anticipated. Therefore, such lidar configurations will soon be developed and investigated in our laboratory.

5 CONCLUSIONS

A cw-lidar system based on a Scheimpflug configuration has been successfully demonstrated in sooty and non-sooty flames. The results show that the method has the potential to monitor particle distributions with high temporal and spatial resolution and to remotely measure temperature profiles in non-sooty flames with high range- and time resolution. It is anticipated that the portable lidar system constructed will be a very useful tool not only for fundamental combustion studies, but in particular in numerous field campaigns at industrial sites.

ACKNOWLEDGEMENTS

The authors would like to thank the Swedish Energy Agency for financial support through the Center for Combustion Science and Technology (CECOST). Also acknowledged is the funding through the Advanced Grant TUCLA of the European Research Council (ERC).

References

[1] B. Kaldvee, A. Ehn, J. Bood, M. Aldén, Development of a picosecond lidar system for large-scale combustion diagnostics, Appl. Opt. **48**, B65-B72 (2009).

[2] B. Kaldvee, J. Wahlqvist, M. Jonsson, C. Brackmann, B. Andersson, P. van Hees, J. Bood, M. Aldén, Room Fire Characterization Using Highly Range-Resolved Picosecond Lidar Diagnostics and CFD Simulations, Comb. Sci. Tech. **185**, 749-765 (2013).

[3] B. Kaldvee, C. Brackmann, M. Aldén, J. Bood, Highly range-resolved ammonia detection using near-field picosecond differential absorption lidar, Opt. Expr. **20**, 20688-20697 (2012).

[4] T. Leffler, C. Brackmann, A. Ehn, B. Kaldvee, M, Aldén, M, Berg, J. Bood, Range-resolved detection of potassium chloride using picosecond differential absorption light detection and ranging, Appl. Opt. **54**, 1058-1064 (2015)

[5] B. Kaldvee, C. Brackmann, M. Aldén, J. Bood, LII-lidar: range-resolved backward picosecond laser-induced incandescence, Appl. Phys. B **115**, 111-121 (2014).

[6] M. Brydegaard, A. Gebru, S. Svanberg, Super resolution laser radar with blinking atmospheric particles – application to interacting flying insects, Prog. Electromagn. Res. **147**, 141-151 (2014).

[7] L. Mei and M. Brydegaard, Continuous-wave differential absorption lidar, Laser Photon. Rev. **9**, 629-636 (2015).

HIGH SPECTRAL RESOLUTION LIDAR USING SPHERICAL FABRY-PEROT TO MEASURE AEROSOL AND ATMOSPHERIC MOLECULAR DENSITY

Caraty Yann, Hauchecorne Alain, Keckhut Philippe, Mariscal Jean-François, Dalmeida Eric

Affiliation main author: LATMOS, France, yann.caraty@latmos.ipsl.fr
Affiliation second, third and fourth author: LATMOS, France

ABSTRACT

In theory, the HSRL method should expand the validity range of the atmospheric molecular density and temperature profiles of the Rayleigh LIDAR in the UTLS below 30 km, with an accuracy of 1 K, while suppressing the particle contribution. We tested a Spherical Fabry-Perot which achieves these performances while keeping a big flexibility in optical alignment. However, this device has some limitations (thermal drift and a possible partial depolarisation of the backscattered signal).

1 INTRODUCTION

Temperature is a main parameter for the radiative, dynamic and chemical processes of the atmosphere. The measure of the vertical profile of this variable is therefore a major challenge in the meteorological and climatic context.

Analysis of the temperature parameter requires a global and complete spatial coverage, a good vertical resolution and a temporal continuity of the measurements. Various techniques for analysing the vertical profile of atmospheric temperature exist. The LIDAR (Light Detection and Ranging) allows the measure of this profile in accordance with the previous requirements through the Rayleigh technique [1] [2].

This technique involves vertically emitting a short light pulse via a laser towards the atmosphere at a wavelength corresponding to the atmospheric constituents scattering. The method involves then collecting the photons of the pulsed laser beam which are backscattered by the atmospheric particles by means of a telescope pointing to the zenith in the emission direction.

In the upper stratosphere (above 30 km), the backscattered signal comes only from Rayleigh or molecular scattering by air molecules (there are no aerosols). The number of photons recorded is directly proportional to the atmospheric density. This one can be used to trace back to the atmospheric temperature vertical profile (via the ideal gas law and the assumption of an atmosphere in hydrostatic equilibirum with a constant air mass [3]).

In the upper troposphere and the lower stratosphere (the UTLS below 30 km), aerosols contaminate the backscattered signal which now comes from two scatterings: molecular scattering (by air molecules) and Mie or particle scattering (by aerosols). The Rayleigh LIDAR method hence becomes inapplicable.

The technique developed involves using spectral filtering to eliminate the particle scattering contribution within the LIDAR signal below 30 km. This is made possible by the Doppler broadening of the Rayleigh line shape, the linewidth of the particle scattering being equivalent to the linewidth of the laser [4]. Thus, the High Spectral Resolution Lidar (HSRL) method aims to expand the validity range of the atmospheric molecular density and temperature profiles of the Rayleigh LIDAR in the UTLS [5]. The subject of this article is therefore to introduce an atmospheric molecular density and temperature measurement technique using an innovative HSRL method developed around a Spherical Fabry-Perot (SFP) [6].

In a first part, we describe the HSRL technique with this SFP and discuss its implementation in the optical box of the Rayleigh LIDAR in operation at the Observatory of Haute-Provence (OHP). Then, we present and analyse the initial results. Finally, we conclude with the performances and limitations of this method and discuss its potential for improvements.

2 METHODOLOGY

2.1 The principle of the HSRL technique

The HSRL method is based on using the line shapes of the light beam backscattered by the atmospheric constituents. In the UTLS, a backscattered line shape (by molecular and particle scatterings) is represented by the superposition of two curves focusing on the same wavelength [7]: a Voigt profile similar at a Gaussian curve (molecular contribution) whose broadening depends on the temperature and pressure values superimposed with a thin central peak (aerosols contribution) whose broadening is conditioned by the characteristics of the laser [4] (Fig. 1).

The principle of the HSRL method is to use a filtering device (as the SFP) whose transmission function makes it possible to discriminate between the Rayleigh and particle contributions within the backscattered signal in the UTLS. The objective is to measure the atmospheric temperature profile from the analysis of the molecular portion of the backscattered line shape. The transmission function selects a spectral line finer than the Rayleigh line shape but broader than the particle line shape (Fig. 1).

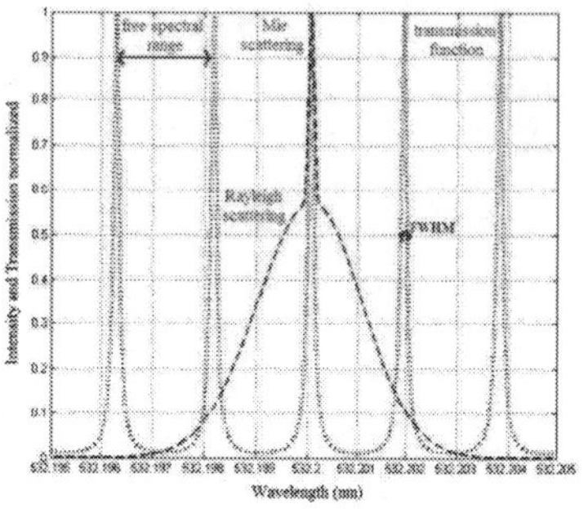

Figure 1:
Simulation of the backscattered line shape at 532.2 nm by molecular and particle scatterings at T ≈ 220 K (red). Simulation of the SFP transmission function (blue).

According to the specifications from the company *Spectra Physics* (Nd:YAG laser pulsed at $\lambda_0 = 532.2$ nm and single-mode through injection) and taking into account an average temperature of T ≈ 220 K within the UTLS, the SFP should make it possible to isolate a $\Delta\lambda_{Mie} < 0.168$ pm linewidth from a background width of $\Delta\lambda_{Rayleigh} \approx 2.12 \pm 0.10$ pm (Fig. 1).

The objective is to retain the aerosols contribution by eliminating almost all of the molecular component. If the characteristics of the filtering system are know, it is then possible to isolate the Rayleigh and particle signals through calculation. Finally, this estimate of the molecular scattering unpolluted makes it possible to infer the temperature profile through the Rayleigh LIDAR method as in the upper stratosphere [3] [5].

2.2 Description and general properties of the SFP

The principle of the SFP was developed by Pierre Connes in 1958 to specifically study hyperfine structures and line shapes as small as 10^{-3} cm^{-1}.

This is a unit-magnification afocal system consisting of two identical concave spherical blades which are each centred on top of one another. These blades are restricted by circular diaphragms of diameter D. They are both connected by two flat blades produced from a material having a zero expansion coefficient. So, the system is stiffened by molecular adhesion (Fig. 2.a).

In this study, the SFP is in a hermetic metal enclosure equipped with an adjustable screw so that the volume and pressure can be mechanically modified, and in turn, the refractive index of internal air n (Fig. 2.b). So, we can spectrally adjust this filtering device on the λ_0 laser wavelength by changing the pressure between the blades.

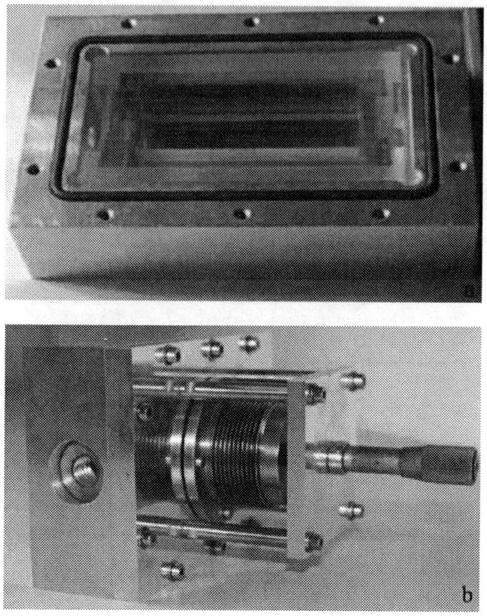

Figure 2.a and 2.b:
Photograph of the SFP in its metallic box (a). Photograph of the external appearance of the SFP (b).

However, the peculiarity of the SFP concerns the light paths in the device and at its output (Fig. 3).

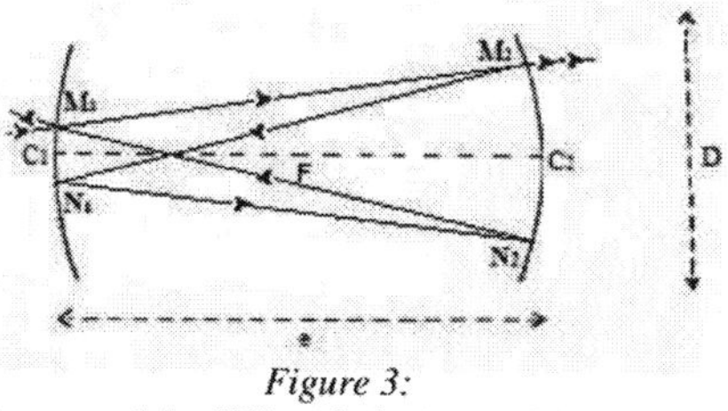

Figure 3:
Diagram of the SFP with the internal light paths (projection on a main section plane).

This specificity results in the angular independence of the δ optical path difference ($\delta = 4.n.e$ with the thickness e) and the transmission of the SFP unlike the case of the Plane Fabry-Perot (PFP).

This filtering device is characterised by its thickness e which can be equated to the curvature radius. However, the thickness of this filtering system is fixed and cannot be modified unlike for the PFP. The spherical interferometer also has characteristics similar to those of a double thickness PFP. It has the same transparency, contrast and finesse F. The R_0 theoretical resolving power, the δ_0 resolution limit and the free spectral range also are identical [6].

It is thus possible to define the SFP transmission function thanks to the PFP theory with a double reflection and the angular independence (Fig. 3). Here, we developed a SFP of thickness $e=75.10^{-3}$m equivalent to a single thickness PFP with a finesse of F ≈ 30 and a power reflection coefficient of R ≈ 0.900. The characteristics of this SFP are summarized in table 1.

Table 1: Summary of the main technical parameters of the SFP.

Main technical parameters of the SFP	Numerical values
Thickness	$e = 75.10^{-3}$ m
Optical path difference	$\delta \approx 0.149$ m
Finesse	$F \approx 14.97$
Free Spectral Range	$FSR \approx 1.88 \pm 8.10^{-3}$ pm
FWHM (resolution)	$\Delta\lambda_{FPS} \approx 0.126 \pm 8.10^{-3}$ pm

Finally, with $\Delta\lambda_{FPS}$=FWHM≈0.126 ± 8.10^{-3} pm, this spherical interferometer is able to select a spectral line finer than the Rayleigh line shape ($\Delta\lambda_{Rayleigh}$≈2.12±0.10pm) and as broad as the particle line shape ($\Delta\lambda_{Mie}$<0.168 pm). Its transmission function thus has the ability to differentiate the molecular and aerosols contributions within the backscattered signal (Fig. 1 and table 1). So, the SFP is a filter transmitting only the energy emitted by the laser source within its bandwidth.

2.3 Characterisation of the optical set-up

The filtering system is implemented in the optical analysis box of the Rayleigh LIDAR in operation at the OHP whose the optical set-up is investigated in Fig. 4 and [3].

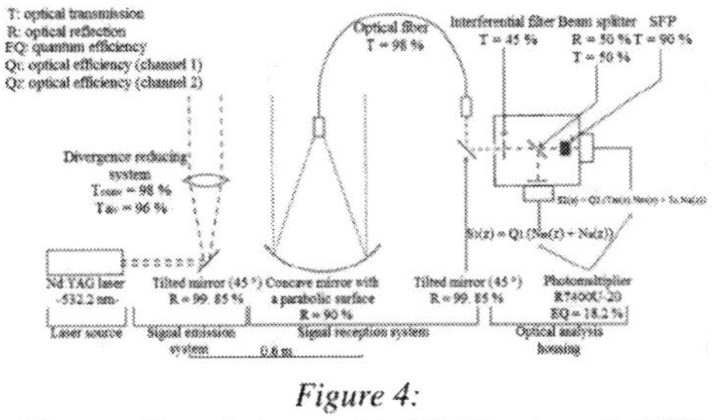

Figure 4:
Diagram of the optical set-up of the LIDAR station at the OHP.

The Nd:YAG laser vertically emits a short monochromatic light pulse towards the atmosphere at $\lambda_0 = 532.2$ nm. The laser beam first passes through the emission system which reduces its divergence through an afocal optical device. The backscattered photons are then collected by the reception telescope pointing to the zenith in the direction of the emission (Fig. 4).

An optical fiber placed at the main mirror's focal point ensures the transmission of the signal to the optical box (Fig. 4). Inside this, the light beam begins by passing through an interference filter that eliminates atmospheric noise (Fig. 4). Finally, the signal is dissociated into two channels (with the SFP on the second channel) by a beam splitter 50/50 and sensed by two photomultipliers (PMs) R7400U-20 in "photon counting" mode (Fig. 4).

3 INITIAL RESULTS

We present the results for the night of 22nd to 23rd July 2015 between 21:35 UT and 22:35 UT with the presence of low-lying cirrus (Fig. 5). The profiles are represented between 682 m (altitude of the station) and 40 km with a vertical resolution of $\Delta z = 150$ m.

The optical set-up analyses two output signals (Fig. 4): the $S_1(z)$ signal corresponding to the channel without SFP (channel 1) and the $S_2(z)$ signal corresponding to the channel with the SFP (channel 2). $S_1(z)$ is the total of the laser photons backscattered by air molecules ($N_m(z)$) and by almost all the aerosols ($N_a(z)$). $S_2(z)$ is proportional to the sum of the photons backscattered by air molecules ($T_m(z).N_m(z)$) and by the aerosols ($T_a.N_a(z)$) transmitted by the SFP.

These signals show that the aerosols distribution is generally diffuse within the UTLS with the exception of areas with relatively dense cloud cover between 7 km and 9 km and between 10 km and 11 km where we detect high particle concentrations which result in an increase in the number of the backscattered photons (intensity peaks of the $S_1(z)$ and $S_2(z)$ profiles).

The combination of $S_1(z)$ and $S_2(z)$ signals leads to the $N_m(z)$ molecular and $N_a(z)$ particle density vertical profiles (equations (1) and (2)):

$$N_m(z) = \frac{\frac{Q_1}{Q_2}.S_2(z) - T_a.S_1(z)}{T_m(z) - T_a} \qquad (1)$$

$$N_a(z) = S_1(z) - N_m(z) \qquad (2)$$

Analysis of $N_a(z)$ confirms that the aerosols contribution to the backscattered signal in the UTLS is to be taken into account only at the level of cloud covers.

The acquisition range of the $N_m(z)$ profile spans the interval [2.5 ; 20] km (Fig. 5). The number of the photons backscattered by air molecules decreases exponentially as altitude increases (Fig. 5). This decrease is coherent with a diminishing molecular concentration as altitude increases and a particle distribution which is generally diffuse within the UTLS. However, the main interest of this profile lies in the flattening of intensity peaks in ranges [7 ; 9] km and [10 ; 11] km (Fig. 5).

The partial ([7 ; 9] km) or total ([10 ; 11] km) suppression of these peaks means that we have succeeded in considerably limiting the aerosols contribution to the backscattered signal in the UTLS.

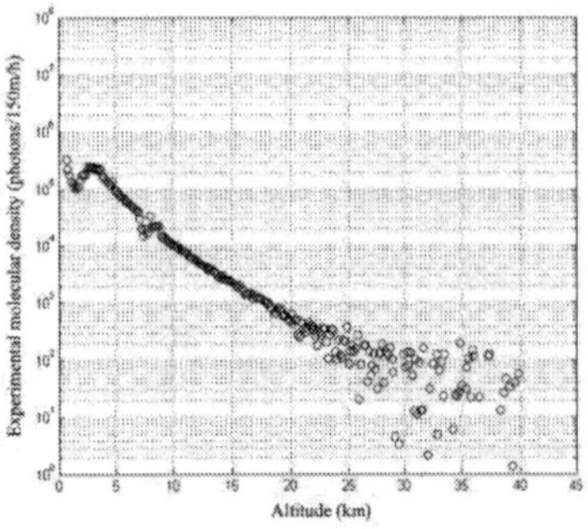

Figure 5:
Representation of the $N_m(z)$ experimental molecular density profile in the UTLS for the night of 22nd to 23rd July 2015 between 21:35 UT and 22:35 UT at the OHP.

Ultimately, the HSRL method with the SFP improves the capabilities of the Rayleigh LIDAR by extending its validity range to the UTLS between 2.5 km and 20 km altitude. However, the reliability of this technique decreases when the particle concentration is too high (here between 7 km and 9 km altitude). The aerosols contribution is then only minimised and there remain a particle residue of about 15 % which cannot be corrected (Fig. 5).

Two phenomena may restrict these performances. This could be a problem of partial depolarisation of the backscattered signal associated with the operating mode of the optical box. One solution would be to retain only one polarisation in the optical box through a polarising beamsplitter cube at its input. The spectral adjustment of the laser source with the SFP has also to be considered. Indeed, we note a $\Delta\lambda_{déc} \approx 0.10 \pm 8.10^{-3}$ pm spectral shift of the device relative to λ_0 at the end of the experimental phase. This, because of a thermal drift of the laser and/or the SFP. The solution is to control their internal temperature through an automatic monitoring system.

4 CONCLUSIONS

Finally, with a resolution of $\Delta\lambda_{FPS} \approx 0.126 \pm 8.10^{-3}$ pm, the SFP is able to select a $\Delta\lambda_{Mie} < 0.168$ pm linewidth from a background width of $\Delta\lambda_{Rayleigh} \approx 2.12 \pm 0.10$ pm. This filtering device therefore seems well adapted to separate the molecular and aerosols contributions within the backscattered laser signal in the UTLS. Thus, the HSRL method with this SFP improves the capabilities of the Rayleigh LIDAR at the OHP by extending its validity range to the UTLS between 2.5 km and 20 km altitude.

The SFP offers many advantages relative to the cumbersome parameterisation of the PFP (thickness > 8 cm and diameter > a few centimeters) with regard to achieving the resolution levels necessary within the framework of these studies of hyperfine structures from 10^{-3} cm^{-1}. Despite its performances, these capabilities decrease when the particle concentration is too high. The problems stem in particular from the possible partial depolarisation of the backscattered signal and the probably thermal drift of the laser source and/or the SFP.

Improvements in this filtering technique (integration of a polarising beamsplitter cube and an automated temperature control device) will soon form the basis for the design of a new instrument of higher precision.

ACKNOWLEDGEMENTS

This work has drawn heavily on research conducted by Pierre Connes in 1958. Virginia Ciardini contributed to this study during her masters degree in 2005.

REFERENCES

[1] Hauchecorne A., Chanin M. L., 1980: Density and temperature profiles obtained by Lidar between 35 and 70 km, *Geophys. Res. Lett.*, **7**, 565-568.

[2] Chanin M. L., Hauchecorne A., 1984: Lidar studies of temperature and density using scattering, *Map Handb.*, **13**, 87-89.

[3] Keckhut P., Hauchecorne A., Chanin M. L., 1993: A Critical Review of the Database Acquired for the Long-Term Surveillance of the Middle Atmosphere by the French Rayleigh Lidars, *J. Atm. Ocean. Tech.*, **10**, 850-867.

[4] Fiocco G., Benedetti-Michelangeli G., Maischberger K., Madonna E., 1971: Measurement of Temperature and Aerosol to Molecule Ratio in the Troposphere by Optical Radar, *Nature Physical Science*, **299**, 78-79.

[5] Hauchecorne A., Chanin M. L., Keckhut P., Nedeljkovic D., 1992: Lidar monitoring of the temperature in the middle and lower atmosphere, *Appl. Phys.*, **B55**, 29-34.

[6] Connes P., 1958: L'étalon de Fabry-Pérot Sphérique, *Le journal de physique et le radium*, **19**, 262-269.

[7] Fiocco G., De Wolf J. B., 1968: Frequency Spectrum of Laser Echoes from Atmospheric Constituents and Determination of the Aerosol Content of Air, *J. Atmos. Sci.*, **25**, 488-496.

CHARACTERIZATION OF HIGH TEMPORAL RESOLUTION PRR ACQUISITION BY FAST COMTEC CARD: DEADTIME, PRR DESATURATION, TEMPERATURE CALIBRATION AND RETRIEVAL.

Giovanni Martucci[1]*, Valentin Simeonov[2], Ludovic Renaud[1], and Alexander Haefele[1]

[1] *Federal Office of Meteorology and Climatology, MeteoSwiss, Payerne, Switzerland,*
**Giovanni.martucci@meteoswiss.ch*

[2] *Ecole Polytechnique Federale de Lausanne - EPFL, CH-1015 Lausanne, Switzerland*

ABSTRACT

RAman Lidar for Meteorological Observations (RALMO) is operated at MeteoSwiss and provides continuous measurements of water vapor and temperature since 2010. While the water vapor has been acquired by a Licel acquisition system since 2008, the temperature channels have been migrated to a Fastcom P7888 acquisition system, since August 2015. We present a characterization of this new acquisition system, namely its dead-time, desaturation, temporal stability of the Pure Rotational Raman signals and the retrieval of the PRR-temperature.

1 INTRODUCTION

The real time and continuous monitoring of the temperature in the troposphere and lower stratosphere has both a meteorological and climatological importance. Accurate measurements of the atmospheric temperature are needed for numerical weather prediction and on the long term to understand the changes and trends of temperature for a climatological applications. The temperature can also be used, in combination with the water vapor, to calculate the relative humidity, a variable of prime importance to understand the dynamics of cloud formation and the hygroscopic growth of aerosols in the boundary layer.

In collaboration with the École Polytechnique Fédérale de Lausanne (EPFL), the Federal Office of Meteorology and Climatology MeteoSwiss operates a Raman LIDAR to measure water vapor and temperature in the troposphere and lower stratosphere. The RAman Lidar for Meteorological Observations (RALMO) is operated at the Aerological station at Payerne (46°480 N, 6°560 E, 491ma.s.l.) by MeteoSwiss and provides continuous measurements of water vapor and temperature almost uninterruptedly since 2008 [1],[2]. While the water vapor has been acquired by the Licel acquisition system (licel.com/) since 2008, the Pure Rotational Raman (PRR) channels have been migrated to a new and faster acquisition system, the FAST ComTec P7888 (www.fastcomtec.com), since August 2015. This study presents a characterization of the acquisition system, namely its deadtime, the desaturation procedure of the high (JH) and low (JL) rotational quantum number PRR channels based on the non-paralyzable condition and the temporal stability of the PRR signals. The retrieval procedure of the PRR-based temperature profile using different parameterizations accounting for higher-degree terms and its calibration based on radiosonde data are also discussed

2 INSTRUMENT DESCRIPTION

When RALMO has been designed, the main objectives of the project helped defining the system parameters. The main purpose was to measure water vapor and temperature profiles in the middle and high troposphere during day and night continuously. In order to achieve this target with profiles of high quality, the system has been created based on to the following specifications: a narrow receiver's field of view, a narrowband grating polychromator that isolates the ro-vibrational Q-branches allowing to minimize the contamination by solar background and to maximize the signal-to-noise ratio, a near real-time data availability, long-term data consistency and system stability, autonomous operation, minimal maintenance and eye safety to ensure the possibility to measure without interfering with the surroundings. Such a system ensures high accuracy and precision of the measured parameters, the traceability of the measurement and high data availability since 2008. Table 1 summarizes the system parameters as it is described more in detail by Dinoev et al., in 2013 [1]. RALMO laser's source emits at 1064 nm its fundamental harmonic, but uses the third harmonic at 355 nm, the tripled harmonic is then optically directed to the LIDAR transceiver optics. A beam expander reduces the beam divergence at the instrument's output and the expanded beam is transmitted into the atmosphere. The light backscattered by the atmosphere is collected by the LIDAR's telescope (4 mirrors of 30 cm diameter) and transmitted by optical fibers separately to the H_2O and Aerosol/Temperature polychromators for the wavelength separation. The photo multipliers detect the optical signals and convert them into electrical signals, which are acquired and stored for the data analysis.

Table 1 RALMO system parameters

Transmitter	Receiver
Nd:YAG laser – third harmonic:	*Four fiber-coupled parabolic mirrors:*
Wavelength: 354.7 nm	Focal length: 1m
Spectral line width: 0.7 cm−1	Diameter: 0.3m
Rep. rate: 30 Hz	Axial displacement: 235mm
Pulse energy: 300 mJ	FOV: 3 × 0.20 and 1 × 0.22 mrad
Pulse duration: 8 ns	
Beam expander – Galilean type:	
Expansion ratio 15×	*Two-stage, grating Polychromator*
Transmitted beam:	
Divergence: calculated 0.06 mrad	*Photodetectors (Hamamatsu)*
measured 0.09 ± 0.02 mrad	R12421
Diameter: 140 mm	

3 FAST ComTec P7888

The Model P7888 Series is one of the fastest commercially available multiple-event time digitizer with up-to four inputs. P7888 has been installed to acquire the PRR RALMO channels in August 2015 to reduce significantly the acquisition system's deadtime and consequently to minimize the saturation effects in the photon-counting channel. Full specifications of the FAST ComTec P7888 acquisition card can be found at www.fastcomtec.com. Compared to the Licel acquisition system, the P7888 acquires the PRR channels with higher range resolution and with about twice shorter deadtime, τ. The new acquisition system uses solely the photon counting method with two transmissivity channels (high and low) for JH and JL. The high-transmission channel corresponds to the 90% of the signal intensity, while the low-transmission corresponds to the remaining 10%. The low-transmission channels ($JH^{10\%}$, $JL^{10\%}$) are theoretically never saturated and can act as reference unsaturated channel when they are linearly compared with the high-transmission channels ($JH^{90\%}$, $JL^{90\%}$). In order to calculate the count rate threshold at which saturation starts to occur and the corresponding deadtime, we have studied the departure from linearity of the high-transmission channel. For over a year of data acquired by the P7888 card we have characterized the values of τ, when the saturation was directly observable.

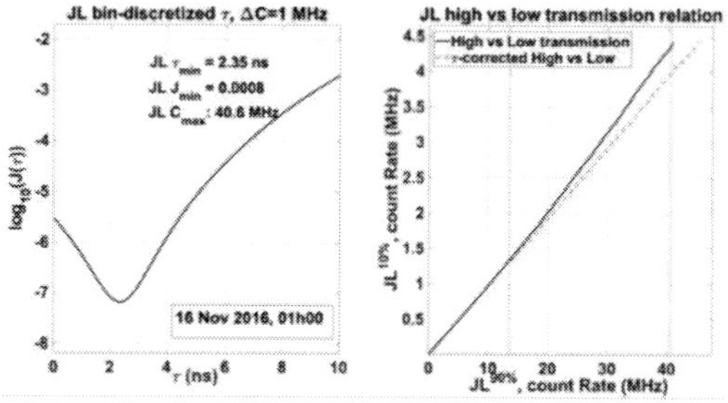

Figure 1 Deadtime calculation and high-transmission channel deadtime correction.

By applying the non-paralyzable assumption [3], the two signals $JH^{90\%}$ and $JL^{90\%}$show saturation values as soon as the signal exceeds the threshold counts of about 35 MHz (~10 MHz for Licel) with values of τ of 2 ns and 1.35 ns, respectively. The method adopted to calculate τ and to desaturate the PRR channels is based on the non-linear relation between an unsaturated and a saturated channel and has been described by previous authors, e.g. [3], [4]. Figure 1 shows an example of calculation of the deadtime τ for the $JL^{90\%}$ signal obtained by minimization of the least-squares matrix ***J*** (left panel) and the desaturation by use of the non-paralyzable assumption (right panel). The matrix ***J*** depends on the fit-error normalized differences between the deadtime-corrected count rates and the virtual count rates.

4 PRR-BASED TEMPERATURE

In addition to the Rayleigh light backscattered by the aerosols and molecules at the same frequency as the incident one, the O_2 and N_2 molecules returns a frequency-shifted Raman light to the LIDAR receiver. The returned signal is shifted in frequency due to the rotational and vibrational Raman effect. In this study only the pure rotational part of the spectrum around the Rayleigh frequency is analyzed (Fig. 2).

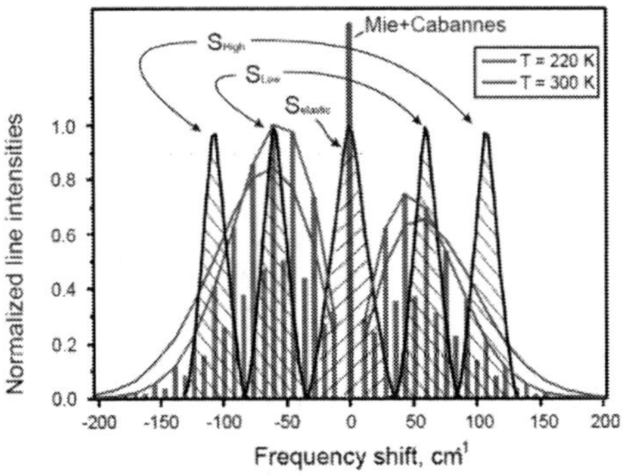

Figure 2 Temperature dependence of the pure-rotational Raman spectrum of N_2 [6]

The high-frequency and low-frequency shifted signals in the Stokes and anti-stokes Q-branches depend on the temperature of the probed atmospheric volume. The temperature T at distance z is derived from the ratio of the intensity of the PRR signals JL and JH at z, using the following approximate relationship [5], where the constants A and B are determined during the LIDAR calibration.

$$T(z) = \frac{A}{\log\left(\frac{JL(z)}{JH(z)}\right) + B} \quad (1)$$

4.1 Nonlinear parameterizations

The temperature retrieval accuracy and number of coefficients as well as the degree of the terms depending on the PRR ratio depends on the degree of the expanded Taylor's series of the log(JL/JH) ratio. The generally-used parameterization has a linear dependence on the logarithmic term JL/JH, which simplifies the retrieval and the calibration of the temperature, but that ignores the broadening of individual atmospheric N_2 and O_2 PRR lines, contributing to the overall uncertainty of the retrieval. Moreover, the theoretical expression of eq.(1) refers strictly to a single line of the PRR spectrum, and the fact of using it for multiple spectral lines introduces an additional uncertainty. The overall uncertainty related to the coupled system retrieval + instrument, where the instrument contributes with the measurement noise, is on the order of ±1 K [7], [8]. The pressure broadening of N_2 and O_2 PRR lines dominates over other types of broadening in the troposphere, and it should not be neglected when higher-degree terms are taken into account for the retrieval of the temperature. When expanding the Taylor's series at higher degrees other two parameterizations can be used as suggested by previous authors, i.e., [8],[9].

$$T = \frac{A}{\log\left(\frac{JL}{JH}\right)^2 + B\log\left(\frac{JL}{JH}\right) + C} \quad (2)$$

$$T = \frac{\log\left(\frac{JL}{JH}\right)}{A\log\left(\frac{JL}{JH}\right)^2 + B\log\left(\frac{JL}{JH}\right) + C} \quad (3)$$

Especially Eq. (3) has proved to provide better calibrations of the temperature profiles [8] and a reduced related error. We have tested over one year of RALMO PRR data, retrieving the temperature using equations 1-3: the results proved a good agreement with the findings of [9], showing better calibrations using equations 2 and 3, but an unsatisfactory reproducibility of the calibration coefficients over the days and weeks following a calibration.

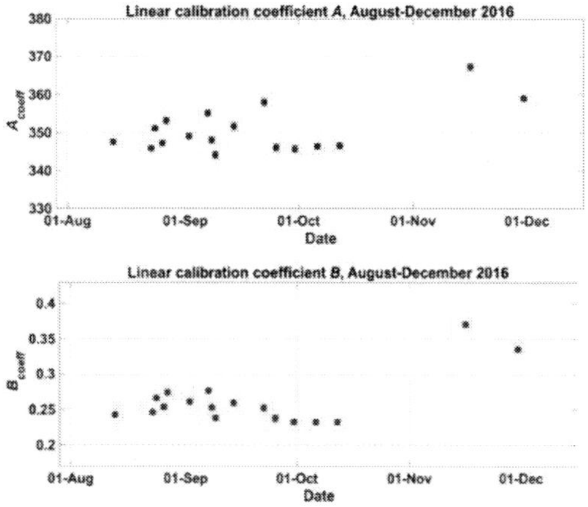

Figure 3 A timeseries of and B calibration coefficients

Finally eq. (1) leads to a slightly larger related error in the temperature retrieval (~1K) compared

https://doi.org/10.1051/epjconf/201817601017

to eqs. 2-3, but the retrieved calibration coefficients can be used robustly over several weeks, allowing for a more stable and homogeneous temperature time series. We have then decided to use only eq. (1) to retrieve *T*.

In Figure 3 we show the variability of the two calibration coefficients *A* and B based on eq. (1) over 3 months. The standard deviation of *A* and *B* with relation to their mean values is 1.75 % and 14.24 %, respectively. The *T* profiles retrieved from RALMO profiles using eq.(1) and compared with the radiosounding (RS) after calibration have provided, for the period September-December 2016 (19 profiles), a statistical bias (RALMO-RS) and standard deviation (stdv) that are shown in Figure 4.

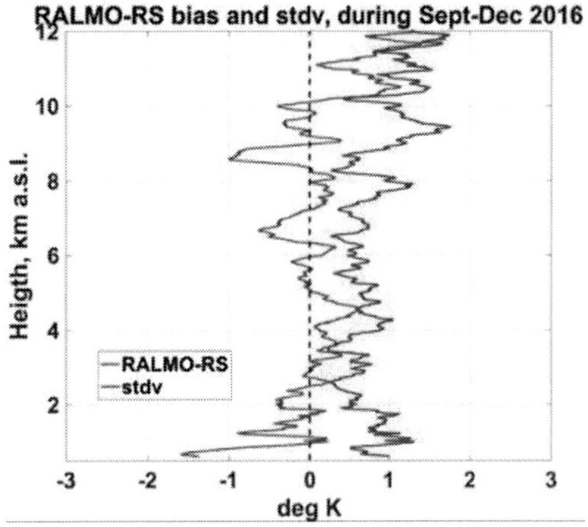

Figure 4 Bias and standard deviation (stdv) values of RALMO-RS during September-December 2016.

The mean bias does not exceed the [-1, +1] K interval, showing a cold bias of -1 K in the first few hundred meters. The RMSE is about 1 K up to 8 km and reaches 2 K between 8 and 12 km.

5 CONCLUSIONS

The high temporal resolution PRR acquisition card by FAST ComTec P7888, has been characterized for its deadtime for both the JH and JL PRR channels. The deadtime, compared to the Licel deadtime is shorter, which implies a reduced saturation of the PRR signals. Different parameterizations have been tested to calibrate the temperature profile retrieved from the PRR signals. The linear parameterization in eq. (1) provides the best compromise between accuracy and reproducibility of the calibration coefficients. Finally a 3-month statistic of stability of the calibration coefficients and the bias and standard deviation of the temperature values based on the radiosounding temperature profiles has been performed. The statistics show that the bias is restricted within the interval [-1, +1] K.

References

[1] Dinoev, T., Simeonov, V., Arshinov, Y., Bobrovnikov, S., Ristori, P., Calpini, B., Parlange, M., and van den Bergh, H.: Raman Lidar for Meteorological Observations, RALMO – Part 1: Instrument description, *Atmos. Meas. Tech.*, **6**, 1329–1346, doi:10.5194/amt-6-1329-2013, 2013

[2] Brocard, E., Philipona, R., Haefele, A., Romanens, G., Ruffieux, D., Simeonov, V., and Calpini, B.: Raman Lidar for Meteorological Observations, RALMO – Part 2: Validation of water vapor measurements, *Atmos. Meas. Tech.*,**6**, 1347–1358, doi:10.5194/amt-6-1347-2013, 2013.

[3] Whiteman, D. N.: Raman lidar system for the measurement of water vapor and aerosols in the Earth's atmosphere, *Appl. Opt.*, **31**, 3068–3082, 1992.

[4] Newsom R K, Turner D D, Mielke B, et al, 2009: Simultaneous analog and photon counting detection for Raman lidar. *Applied optics*, 2009, **48(20)**, 3903-3914.

[5] Arshinov, Y. F., Bobrovnikov, S. M., Zuev, V. E., and Mitev, V. M.: Atmospheric temperature measurements using a pure rotational Raman lidar, *Appl. Optics*, **22**, 2984–2990, 1983.

[6] Dinoev, T. S., V. B. Simeonov, B. Calpini, and M. Parlange, "Monitoring of Eyjafjallajökull ash layer evolution over payerne Switzerland with a Raman lidar," in Proceedings of the TECO, Helsinki, Finland, 2010.

[7] Behrendt, A., T. Nakamura, M. Onishi, et al.: *Appl. Opt.*, **41**, 7657, 2002.

[8] Zuev, V. V., Gerasimov, V. V., Pravdin, V. L., Pavlinskiy, A. V., and Nakhtigalova, D. P.: *Atmos. Meas. Tech.*, **10**, 315-332, doi:10.5194/amt-10-315-2017, 2017.

[9] Behrendt A., and J. Reichardt, *Appl. Opt.* **39**, 1372–1378, 2000.

ASSESSMENT OF CAPABILITIES OF LIDAR SYSTEMS IN DAY- AND NIGHT- TIME UNDER DIFFERENT ATMOSPHERIC AND INTERNAL-NOISE CONDITIONS

Ravil Agishev[1], Adolfo Comerón[2]

[1] *Kazan State Power Engineering University, 51, Krasnoselskaya St., Kazan 420066, Russia*
[2] *Universitat Politècnica de Catalunya, Jordi Girona 1-3, 08034 Barcelona, Spain*

ABSTRACT

As an application of the dimensionless parameterization concept proposed earlier for the characterization of lidar systems, the universal assessment of lidar capabilities in day and night conditions is considered. The dimensionless parameters encapsulate the atmospheric conditions, the lidar optical and optoelectronic characteristics, including the photodetector internal noise, and the sky background radiation. Approaches to ensure immunity of the lidar system to external background radiation are discussed.

1 INTRODUCTION

When conducting atmospheric measurements, the echo signals carrying information about the medium properties are received in presence of sky background radiance [1-4]. The signal-to-noise ratio (SNR) ρ of the generalized lidar system of direct photodetection with internal amplification in presence of internal noise and external background radiation can be written as follows [1-3]

$$\rho = \sqrt{N} \cdot P_s / \sqrt{P_q(P_s + P_b) + P_n^2} \qquad (1)$$

where P_s and P_b are respectively the signal and background powers reaching the photodetector, $P_q = 2 \cdot h \cdot c \cdot F \cdot \Delta f / \eta \cdot \lambda$, with h the Planck's constant, c the speed of light, F the photodetector excess noise factor, Δf the photoreceiver electrical bandwidth, η the photodetector quantum efficiency, and λ the wavelength, and $P_n = NEP \cdot \Delta f^{1/2}$; N is the number of accumulation cycles.

Our aim in the application and promotion of the methodology of dimensionless parametric modeling of lidar systems [5-8] is to show how the sensitivities of lidars using different types of photoreceivers with different noise parameters are impaired under the background of daytime sky and the variability and interdependent effects of atmospheric optical conditions, as well as to illustrate the practical possibilities for efficient and fairly visual comparison of different lidars.

2 METHODOLOGY

The dimensionless parametric methodology of analysis and comprehensive comparison of lidar systems [5-7], including generalization, modification and simplification of traditionally used algorithms to assess the SNR and measurement accuracy, facilitates a quantitative comparison of different lidars.

Following this approach, we introduce a reference echo-signal P_{s0} as the signal that would attain the lidar photodetector from the reference range R_0 in a standard reference molecular atmosphere. We also introduce a reference background power P_{b0} on the photodetector. The application of algorithms [8] normalizing the individual SNR components of (1) to the reference echo signal and comparing normalized system noises and atmosphere conditions refines the approach of the methodology presented in [5-7].

2.1 Partial SNRs and degrees of quantum noise exceeding

The normalized lidar signal P_s/P_{s0} can be presented as follows [5-7]:

$$\Pi \equiv P_s/P_{s0} = Q \cdot W^2 \cdot r^{-2}, \qquad (3)$$

where for aerosol backscattering lidar:

$$Q = \frac{\beta_\pi(\lambda,R)}{\beta_{0\pi}(\lambda,R_0)} \qquad W = \frac{T(\lambda,\alpha,R)}{T(\lambda,\alpha_0,R_0)} \qquad r = R/R_0, \qquad (4)$$

$\beta_\pi(\lambda,R)$ and $\beta_{0\pi}(\lambda,R_0)$ being the backscatter coefficient at the λ wavelength at respectively the range R in the actual conditions and at the reference range R_0 in the reference atmosphere. Likewise $T(\lambda,\alpha,R)$ and $T(\lambda,\alpha_0,R_0)$ are the atmospheric transmittances from the lidar respectively to range R in actual conditions and to range R_0 in the reference atmosphere. We will use the partial signal-to-noise ratios S_q, S_{b0}, S_n as specific indicators of potentialities and sensitivity decreasing of lidar sounding with respect to the reference atmosphere. The sensitivity reductions are caused by shot noise of signal and background clutter and by internal noise of the photodetector:

$$S_q = P_{s0}/P_q; \quad S_n = P_{s0}/P_n; \quad S_{b0} = P_{s0}/P_{b0} \quad (5)$$

The emphasis on the partial relations S_q, S_n, S_{b0} and the Π-parameter allows to consider different sources of internal and external noise inherent to a particular system, and to associate them with parameters of the sounded media [8]. In this case, the system noises and their influences can be considered not "in themselves" and independently, but in comparison with the reference echo-signal from the reference range, which is formed as a result of sounding the reference atmosphere by particular/concrete lidar system under the background radiation of the reference brightness.

Equation (1) can be rewritten using the normalized echo-signal of Eq. (3) and the partial SNRs S_q, S_{b0}, S_n of Eq. (5) [8]:

$$\rho = \frac{\sqrt{N}\cdot\Pi}{\sqrt{S_q^{-1}\cdot(\Pi+b_{rel}/S_{b0}) + S_n^{-2}}} \quad (6)$$

where the relative background brightness $b_{rel} = B_\lambda/B_{0\lambda}$ is the ratio of the actual sky brightness B_λ and the reference one $B_{0\lambda}$.

We'll use the generalized A-parameter [8] as a dimensionless characteristic of the lidar noises in presence of background radiation:

$$A = S_q/S_n^2 + b_{rel}/S_{b0} \quad (7)$$

Then the lidar SNR can be written as [8]:

$$\rho = \frac{\Pi\cdot\sqrt{N}\cdot\sqrt{S_q}}{\sqrt{\Pi+A}} = \sqrt{N\cdot\Pi\cdot S_q}\cdot\sqrt{\frac{\Pi}{\Pi+A}} \quad (8)$$

and it can be applied to any lidar instrument.

When sounding the night sky at the background brightness being considered minimal, i.e. when the ratio of the "current" to night sky background brightness $b_{rel}=B_\lambda/B_{0\lambda}=1$, it is useful [8] to select an important component A_0 within the A-parameter,

$$A_0 = S_q/S_n^2 + S_{b0}^{-1} \quad (9)$$

since $$A = A_0 + (b_{rel} - 1)/S_{b0} \quad (10)$$

It should be underlined, that taking into account S_{b0}^{-1} along with S_q/S_n^2 is especially important, when using photodetectors with low level of dark current.

2.2 L₀ as potentialities prediction parameter

Following Eq. (8), the dimensionless system parameter L_0 [8], defined as the SNR achieved by sounding the reference atmosphere at the reference range at reference night-sky background without accumulation (N=1) is:

$$L_0 \equiv \rho(\Pi = 1, b_{rel} = 1, N = 1) = \sqrt{S_q/(1 + A_o)} \quad (11)$$

Thus the L_0-parameter in advance gives a forecast for the particular/concrete lidar's SNR, when receiving the echo-signal from the reference range of the standard molecular atmosphere. By setting the L_0-parameter as combination of the partial SNRs of various natures, which characterize the system noises of a specific lidar, one can evaluate the potentialities and compare different lidar systems at the preliminary stage, even before the remote measurements.

2.3 Lidar excess noise factor introduction

The last multiplier of (8) is a factor of SNR reduction by excess noise Φ_{ex} [8], which caused by the detector's excess noise and the background radiation

$$\Phi_{ex} = \sqrt{\frac{\Pi}{\Pi+A}} = (1 + A/\Pi)^{-1/2}, \quad (12)$$

and indicates that the decrease of system capabilities can be caused by the interrelated influence of parameters of lidar system noises, background ligth and sounded medium.

2.4 Block diagram of system parameters' formation and interrelations

As follows from (8), (11) and (12), the system SNR for number of accumulation cycles N=1 is

$$\rho = \Pi\cdot L_0\cdot\sqrt{\frac{1+A_0}{\Pi+A}}, \quad \text{or} \quad \rho = \sqrt{\Pi\cdot S_q}\cdot\Phi_{ex} \quad (13)$$

when remembering that $\Pi = Q\cdot W^2\cdot r^{-2}$. Therefore, both forms of expression (13) are represented as the product of dimensionless factors having their own nature.

Fig.1 represents a block diagram of the model with relationships between normalized system parameters S_q, S_n, S_b, A, A_0, L_0, Φ_{ex}, Π and b_{rel}, which determine the system potentialities ρ.

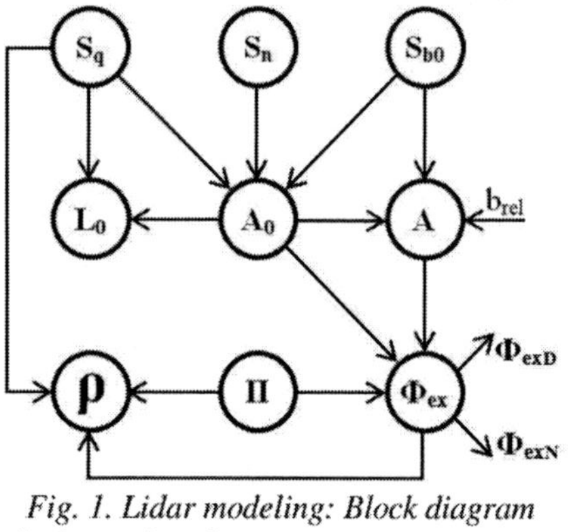

Fig. 1. Lidar modeling: Block diagram of relationships between system parameters.

 https://doi.org/10.1051/epjconf/201817601018

3 RESULTS

3.1 Degree of SNR reduction in day-time

In order to assess the lidar system capabilities under different atmospheric and internal-noise conditions, let's estimate the extent Φ of the SNR decrease when sounding in day- and night- time:

$$\Phi = \frac{\Phi_{exD}}{\Phi_{exN}} \equiv \frac{\Phi_{ex}(b_{rel})}{\Phi_{ex}(b_{rel}=1)} = \sqrt{\frac{\Pi+A_0}{\Pi+A}} \quad (14)$$

Fig. 2 shows the extent to which the potentialities of lidar in day-time compared to night-time can decrease by representing $\Phi = \Phi_{exD}/\Phi_{exN}$ as a function of the ratio $A/A_0 = 1 + (b_{rel} - 1)/S_{b0}$ for a wide range of optical and physical sensing conditions ($\Pi = 10^{-2}...2{\cdot}10^{2}$) and noise properties of photodetectors ($A_0 = 10^{-2}...10^{1}$) (see also Table 1).

Table 1. System parameters considered.

SYSTEM PARA-METERS	RANGE of CHANGES			CALCU-LATIONS	RANGE of CHANGES			
					absolute		moderate/ reasonable	
	min	max			min	max	min	max
P_{s0}, W	$3{\cdot}10^{-9}$	$3{\cdot}10^{-5}$		S_q	3	$3{\cdot}10^{6}$	10	10^{5}
P_q, W	10^{-11}	10^{-9}		S_n	1	$3{\cdot}10^{7}$	10	10^{6}
P_n, W	10^{-12}	$3{\cdot}10^{-9}$		S_{b0}	30	$3{\cdot}10^{7}$	1	10^{4}
P_{b0}, W	10^{-12}	10^{-10}		S_q/S_n^2	10^{-10}	300	10^{-3}	300
b_{rel}	1	300		A_0	$3{\cdot}10^{-8}$	300	10^{-4}	100
Π	10^{-6}	10^{3}		L_0	10^{-1}	$2{\cdot}10^{3}$	1	200

The 3-D-graphs of Fig. 2 illustrate the degree of SNR reduction for low-noise photodetectors with A_0<<1 (a), for detectors with A_0~1 (b) and for "noisy" detectors with A_0>>1 (c).

Fig. 2 is illustrates the fact that the impact of the detector's internal noise and of the background radiation are not absolute, but they depend on the optical and physical properties of the sounded atmosphere, combined into the Π-parameter.

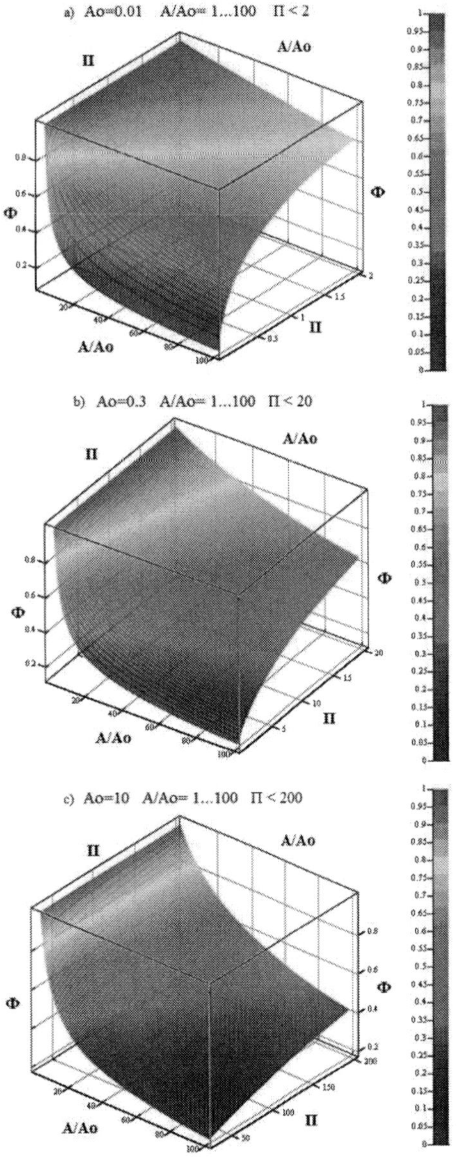

Fig. 2. The degree of SNR reduction as a function of the background noise intensity A/A_0, at wide variations of the normalized echo signal: $\Pi = 10^{-2}... 2{\cdot}10^{2}$, and the photodetector's excess noise: $A_0 = 10^{-2}$ (a), $A_0 = 0.3$ (b), $A_0 = 10$ (c).

3.3 Real lidar systems comparison

Results of the comparison of lidars A and B [9,10] are represented in Fig. 3. Their dimensionless parameters calculated are given in Table 2.

Table 2. Modeling parameters of Lidars A and B.

Modeling parameters	S_q	S_n	S_{b0}	A_0	L_0
Lidar A	$9.7{\cdot}10^{3}$	$1.5{\cdot}10^{3}$	$1.8{\cdot}10^{3}$	$5{\cdot}10^{-3}$	98.4
Lidar B	$4.0{\cdot}10^{2}$	$4.3{\cdot}10^{4}$	$3.5{\cdot}10^{4}$	$3{\cdot}10^{-5}$	20.1

The generalized parameters L_{0A} and L_{0B}, which predict the potentialities of lidars at sounding the reference medium, are determined at Π=1. The values b_{rel} =1, 30 and 300 correspond to results of A and B lidars simulation, when sounding the atmosphere at weak, average and bright sky background under wide variations of its optical and physical parameters (Π = var).

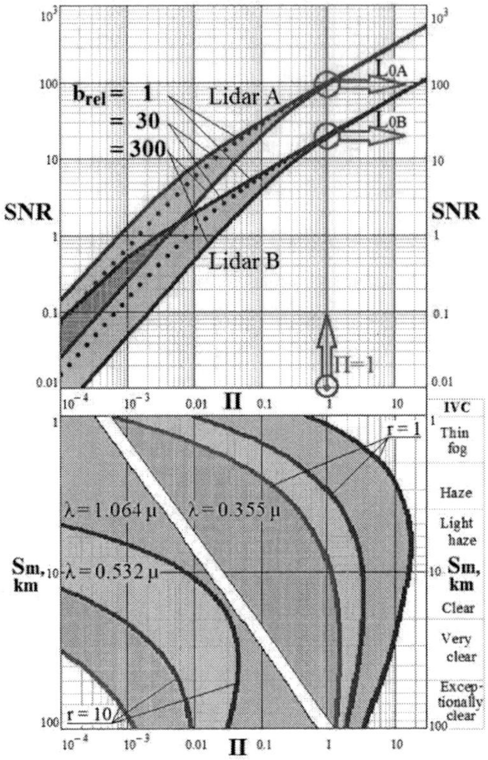

Fig. 3. Comparison of A and B lidars potentialities on the "map" SNR = f(Π). Lidars are considered sounding the same atmosphere for different values of the background noise brightness.
Here: Sm is the visibility, km; IVC is the International Visibility Code.

The generalized parameters L_{0A} and L_{0B}, which predict the potentialities of lidars at sounding the reference medium, are determined at Π=1. The values b_{rel} =1, 30 and 300 correspond to results of A and B lidars simulation, when sounding the atmosphere at weak, average and bright sky background under wide variations of its optical and physical parameters (Π = var).

As it can be seen from Fig. 3, the lidar A has the greater potentialities compared to the lidar B ($L_{0A} \approx 100$; $L_{0B} \approx 20$). It is seen that in day-time, when working under intense sky background (e.g. b_{rel} = 300), the A-lidar is able to receive approx. 10 times weaker echo-signals than B-lidar with a given accuracy (e.g. SNR=10).

References

[1] Rees W. Physical Principles of Remote Sensing / *Cambridge University Press*, 2012. – 470 p.
[2] Osche G. Optical Detection Theory for Laser Applications / *New York, Wiley*, 2011. - 442 p.
[3] Kovalev V., Eichinger W. Elastic Lidar: Theory, Practice, and Analysis Methods / *New York, Wiley-Interscience*, 2004. – 615 p.
[4] Saleh B., Teich M. Fundamentals of Photonics / *New York, Wiley*, 2011. - 965 p.
[5] Agishev R., Comeron A., Gross B., Gilerson A. et al. Application of the method of decomposition of lidar signal-to-noise ratio to the assessment of laser instruments for gaseous pollution detection / *Applied Physics B*, 2004, vol. 79, No. 2, 255–264.
[6] Agishev R., Comeron A., Rodriguez A. et al. Dimensionless parameterization of Lidar for laser remote sensing of the atmosphere and its application to systems with SiPM and PMT detectors / *Applied Optics*, 2014, vol. 53, No. 12, pp. 3164-3175.
[7] Agishev R. Lidar Monitoring of the Atmosphere / *Moscow, Phys-Math-Literature Press*, 2009. - 316 p.
[8] Agishev R. Potentialities of laser remote sensing systems of atmosphere monitoring at wide variability of optical physical characteristics / *Quantum Electronics*, 2017, vol. 47, No. 2, pp.140-152.
[9] Comerón A., Agishev R. Dimensionless parameters for lidar performance characterization / in: Remote Sensing of Clouds and the Atmosphere // *Proceedings of SPIE*, 2014, vol. 9242, pp. 1-10.
[10] Agishev R., Comeron A., Gilerson A. About the potential of LIDARs with different photodetectors under daytime sky radiation / *European Physical Journal: Laser Radar series*, 2016, vol. 119, pp. 2509-2513.

HYDROGEN GAS CONCENTRATION MEASUREMENT IN SMALL AREA USING RAMAN LIDAR MEASUREMENT TECHNNOLOGY

Sachiyo Sugimoto [1]*, Ippei Asahi [1], and Tatuso Shiina[2]

[1]*Shikoku Research Institution, INC, Japan, *s-sugimoto@ssken.co.jp*
[2]*Chiba university, Japan*

ABSTRACT

When change of hydrogen(H_2) gas concentration in a certain point is measured, non-contact measurement technology with high temporal and spatial resolution is necessary. In this study, H_2 concentration in the small area of <1cm^2 under the gas flow was measured by using a Raman lidar. Raman scattering light at the measurement point of 750mm ahead was detected by the Raman lidar. As a result, it was proved that the H_2 concentration of more than 100ppm could be successfully measured.

1 INTRODUCTION

Hydrogen is expected to become a clean energy in next-generation. Various proof examinations are carried out to create the hydrogen society. From such a background, the demand for hydrogen gas measurement is drastically increasing.

When change of hydrogen gas concentration in a certain point is measured, a gas sensor is fixed at the measurement point, or a syringe is utilized to sample the gas, and it is analyzed by a gas chromatography. However, the gas sensor and the syringe have influence on gas flow and its measurement values. Non-contact measurement technology with high temporal resolution and high spatial resolution is necessary to measure at such a small area.

It is effective to use a Raman lidar, as a non-contact measurement technique of hydrogen gas concentration distribution[1][2]. Hydrogen gas measurement by the Raman lidar doesn't have to install a sensor and a syringe into the gas flow, concentration distribution of hydrogen gas can be measured without interference to the gas flow at a measurement point. When the measurement area is very small of <1cm^2, however, the spatial resolution of the measurement delimits its accuracy. This is because the spatial resolution depends on the pulse width of the laser, and the hydrogen gas concentration measured within the pulse width is the mean value distributed over the range.

Therefore, in this study, for the purpose of non-contact measurement of the hydrogen gas concentration in the small area under the gas flow, a specialized Raman lidar was developed. A laser beam (wavelength 349nm, pulse energy 120μJ, PRF 1kHz) with the beam diameter of 1mmϕ at the measurement spot was irradiated to the hydrogen gas. Raman scattering light at the measurement point of 750mm ahead was focused by a Fresnel lens, which is arranged at the orthogonal direction again the transmitting laser beam. As a result, it was proved that Raman signal intensity showed good linear correlation with hydrogen density of more than 100ppm. The gas concentration in a small area could be quantitatively estimated without interference of the gas flow.

2 METHODOLOGY

2.1 Principle

The remarkable absorption wavelength of hydrogen molecules is less than 110nm[3] and does not have clear absorption lines from ultraviolet to the infrared wavelength. Therefore, although absorption spectroscopy is general technique for gas measurement, it is difficult to adapt it for hydrogen gas measurement. On the other hand, the hydrogen molecule shows strong Raman effect[4]. The Raman scattering is the inelastic scattering, which is caused by the interaction between the incident photon and target gas molecules. The Raman scattering shifts the scattering wavelength to the longer wavelength (Stokes light). When ω_P is an angular frequency of the irradiation light, and ω_S is an angular frequency of the Stokes light, and $\Delta\omega$ is the Raman shift of measurement gas molecules, from the law of the conservation of energy,

$$\omega_s = \omega_p - \Delta\omega \qquad (1)$$

is established. The energy difference of the photon is equal with internal energy level differences of molecules of $\Delta\omega$, and it is peculiarity for molecular species[5]. The Raman scattering light intensity is proportional to molecular concentration. Therefore, molecular species are clarified by Raman scattering wavelength, and gas concentration is measured quantitatively by Raman scattering intensity. Raman shift of hydrogen, nitrogen and oxygen, and Raman scattering wavelengths when laser light of 349nm is irradiated towards those gases are shown in table 1.

Table 1 Raman shifts and Raman scattering wavelength when using a laser wavelength 349nm

Molecule species	H_2	N_2	O_2
Raman shift [cm^{-1}]	4160	2331	1556
Raman scattering wavelength [nm]	408.3	377.9	369.0

The Raman scattering light intensity observed from the vertical direction against the laser optical axis is expressed by eq.(2)[6].

$$P_R = \frac{\lambda}{\lambda_R} P_L \chi \rho \sigma_R L N \eta_D \tag{2}$$

P_R : Intensity of Raman scattering
λ : Wavelength of laser irradiation
λ_R : Wavelength of Raman scattering
P_L : Output power of laser
χ : Molar function
ρ : Number density of molecules
σ_R : Raman scattering cross section
L : Pass length
N : Effective number of passes of the laser beam in field of view
η_D : Efficiency of the collecting and detecting optical system

From eq.(2), the Raman scattering light intensity is proportional to a molar function, namely the gas concentration of target gas. When concentration of hydrogen gas is measured, Raman scattering light intensity of the atmospheric nitrogen gas is also measured as reference. The Raman scattering light intensities of hydrogen gas (subscript H) and the nitrogen gas (subscript N) are expressed like eq.(3), eq.(4) using eq.(2) [1][2][7].

$$P_{RH} = \frac{\lambda}{\lambda_{RH}} P_L \chi_H \rho \sigma_{RH} L N \eta_{DH} \tag{3}$$

$$P_{RN} = \frac{\lambda}{\lambda_{RN}} P_L \chi_N \rho \sigma_{RN} L N \eta_{DN} \tag{4}$$

When the ratio of both Raman scattering lights is taken and the equation is expressed the molar function of the hydrogen gas, eq.(5) is provided.

$$\chi_H = \frac{P_{RH}}{P_{RN}} \frac{\lambda_{RH}}{\lambda_{RN}} \frac{\sigma_{RN}}{\sigma_{RH}} \frac{\eta_{DN}}{\eta_{DH}} \chi_N \tag{5}$$

Nitrogen gas concentration in the atmosphere is 78%, λ_H and λ_N are Raman scattering wavelength shown in Table 1, σ_H and σ_N are known[5], and η_H and η_N are values by the optical efficiency of the collecting and detecting system. Therefore, concentration conversion constant C is decided, and eq.(6) is provided.

$$\chi_H = C \frac{P_{RH}}{P_{RN}} \tag{6}$$

Therefore, the molar function that represents the concentration of hydrogen gas is calculated.

2.2 Experimental configuration

The experimental configuration in this study is shown in Figure 1.

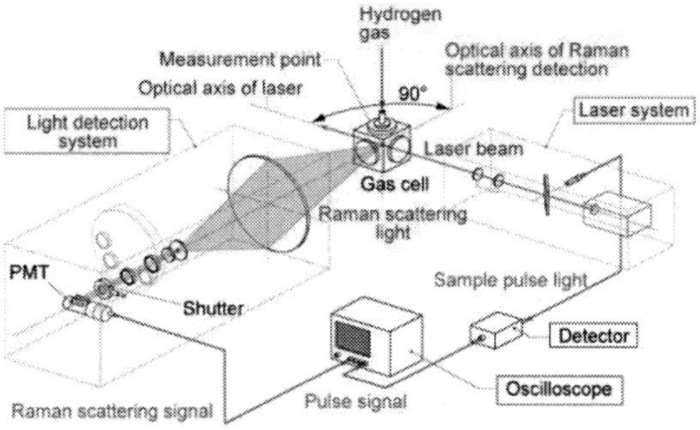

Figure 1 Experimental configuration of hydrogen gas measurement.

Nd:YLF laser (Explorer349 made in Spectra-physics, Wavelength 349nm, Pulse energy 120μJ, Pulse width 5ns, Repetition rate 1kHz) was used as a light source. The beam diameter is 1mmϕ at the measurement spot. Raman scattering light was focused by a Fresnel lens, which is arranged at the orthogonal direction again the transmitting laser beam, and the scattering intensity was measured by a photomultiplier tube (R12829 made in Hamamatsu Photonics). The aperture diameter and the focal distance of the Fresnel lens are 170mm and 230mm, respectively.

The light receiving optical system is shown in Figure 2.

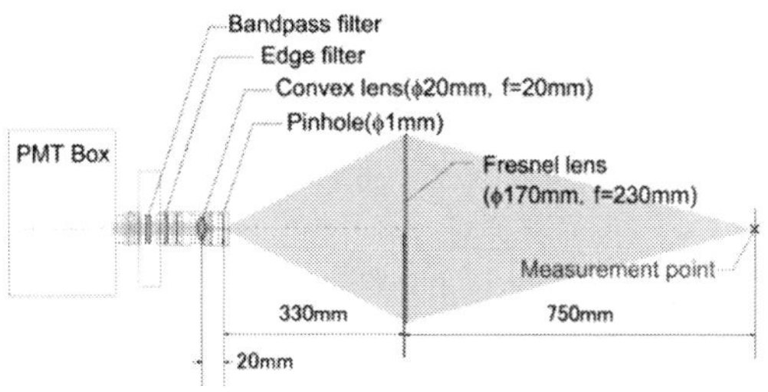

Figure 2 Optical components arrangement of light detection unit.

The light receiving optical system was Kepler type telescope using two convex lenses. The distance between the Fresnel lens and the pinhole was calculated by focus distance of the Fresnel lens of f_1=230mm and the distance to the measurement spot L_1=750mm. The pinhole of aperture diameter 1mmϕ was arranged at 330mm from the Fresnel lens.

The Raman scattering light was delivered to be parallel by a convex lens which was located behind a pinhole, and it was introduced into a photomultiplier tube through an edge filter (Transmissivity $<10^{-6}$ at 349nm, and 95% at 360nm or more) and a band pass filter (Center wavelength 410nm, FWHM 10nm).

The diameter of the receiver's field of view at the measurement spot was calculated as A_M=2.3mm when the pinhole aperture diameter A_P was 1mm, the distance to the measurement spot of L_1=750mm, the distance between Fresnel lens and the pinhole of L_2=330mm and their relationship with the image magnification shown in eq.(7).

$$\frac{A_P}{A_M} = \frac{L_2}{L_1} \tag{7}$$

The gas concentration measurement area of this experimental configuration which was observed from the Fresnel lens side was within the width of 2.3mm, height of 1mm, and depth of 1mm, because the beam diameter of the laser at the measurement spot is 1mmϕ and A_M=2.3mm.

3 RESULTS

In experimental configuration shown in Figure 2, low concentration hydrogen mixed gas (nitrogen gas balance) of known concentration were filled into a gas cell, and its Raman scattering light intensity was measured. Hydrogen gas of the concentration of 1,000ppm was used as standard, and the gas density in the cell was adjusted by changing the filling pressure into the gas cell. Measurement results are shown in Figure 3 and Figure 4.

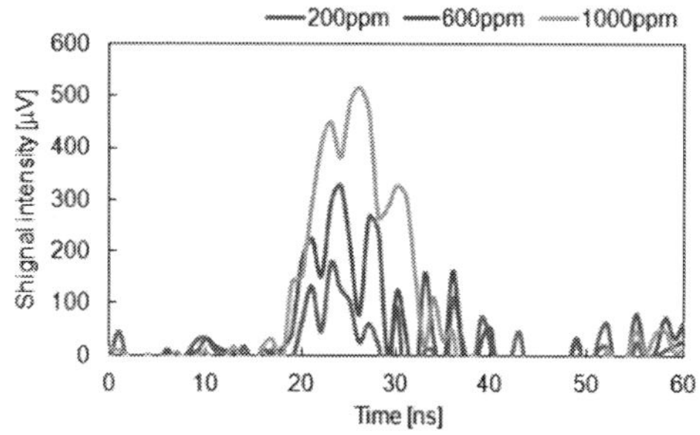

Figure 3 Time responses of Raman scattering intensity.

Figure 3 shows time responses of–Raman scattering signals from hydrogen gas which are measured by the photomultiplier tube. Pulse width of the light source was 5ns. FWHMs (Full Width Half Maximum) of Raman scattering signals were 10ns, it agreed with the time response of the pulsed light to pass through the measurement point.

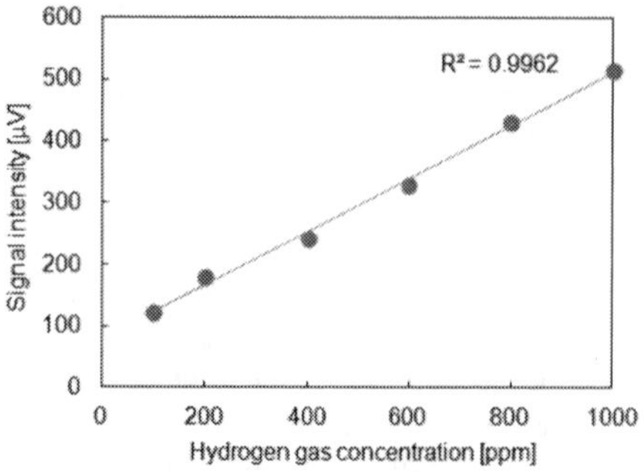

Figure 4 Dependence of Raman scattering intensity on the hydrogen concentration in the gas cell.

Figure 4 shows the peak value of the Raman scattering signal when the hydrogen gas concentration in the gas cell was changed. The Raman scattering signal intensity showed good linear correlation with hydrogen gas concentration of 100ppm or more. It was confirmed that the hydrogen gas concentration measurement in the certain small point was possible with the developed Raman lidar.

4 CONCLUSIONS

Remote measurement technique of gas concentration in the certain small point under its flow was considered, and the experimental system was developed. It was shown that low-concentrated hydrogen gas at the distance of 750mm was measured successfully with the liner correlation up to 100ppm.

From these results, it was shown that the gas concentration remote measurement ~~in~~ at the certain small leak point was possible without interfering the gas flow.

This result can be applied to monitoring of gas diffusion under a case that a gas tank was damaged and gas was leaked. In addition, other gases including oxygen and nitrogen etc. can be also measured by adjusting the observation Raman wavelength.

References

[1] H. Ninomiya, I. Asahi, S. Sugimoto and Y. Shimamoto, 2009: Development of remote sensing technology for hydrogen gas concentration measurement using Raman scattering effect, *IEEJ Trans. FM*, Vol.129, No.7, pp.1181-1185 (in Japanese).

[2] I. Asahi, H. Ninomiya and S. Sugimoto, 2010: Remote sensing of hydrogen concentration by low power laser, *IEEJ Trans. FM*, Vol.130, No.7, pp.1145-1150 (in Japanese).

[3] G. Herzberg, 1989: Molecular Spectra and Molecular Structure Vol. I : Spectra of Diatomic Molecules, 2nd Ed., *Krieger*, Malabar, Florida.

[4] J. C. White, 1987: Stimulated Raman Scattering", in "Tunable Lasers", L. F. Mollenauer and J. C. White, eds., *Springer-Verlag*, Berlin, pp.115-120.

[5] R. M. Measures, 1984: Laser Remote Sensing, John Wiley and Sons, New York, p.108.

[6] S. M. Adler-Golden, N. Goldstein, F. Bien, M. W. Matthew, M. E. Gresh, W. K. Gersh, W. K. Cheng and F. W. Adams, 1992: Laser Raman sensor for measurement of trace-hydrogen gas, *Appl. Opt.*, 31, pp.831-835.

[7] J. Kiefer, T. Seeger, S. Steuer, S. Schorsch, M. C. Weikl and A. Leipertz, 2008: Design and characterization of a Raman-scattering-based sensor system for temporally resolved gas analysis and its application in a gas turbine power plant, *Meas. Sci. Technol.*, 19, pp. 1-9.

COMPACT HIGH-POWER SHIPBORNE DOPPLER LIDAR BASED ON HIGH SPECTRAL RESOLUTION TECHNIQUES

Songhua Wu[1,2]*, Bingyi Liu[1], Guangyao Dai[1,3], Shenguang Qin[4], Jintao Liu[1], Kailin Zhang[1], Changzhong Feng[1], Xiaochun Zhai[1], Xiaoquan Song[1,2]

[1]*Ocean Remote Sensing Institute, Ocean University of China, Qingdao, China, *wush@ouc.edu.cn*
[2]*Laboratory for Regional Oceanography and Numerical Modeling, Qingdao National Laboratory for Marine Science and Technology, China*
[3]*Leibniz Institute for Tropospheric Research, Liepzig, Germany*
[4]*Seaglet Environmental Technology, Qingdao, China*

ABSTRACT

The Compact High-Power Shipborne Doppler Wind Lidar (CHiPSDWiL) based on high-spectral-resolution technique has been built up at the Ocean University of China for the measurement of the wind field and the properties of the aerosol and clouds in the troposphere. The design of the CHiPSDWiL including the transceiver, the injection seeding, the locking and the frequency measurement will be presented. Preliminary results measured by the CHiPSDWiL are provided.

1 INTRODUCTION

Atmospheric wind field has significant impact on atmospheric circulation, atmospheric carbon cycle, marine-atmosphere circulation and aerosol activities. Aerosols in the whole troposphere play a key role in climate change and air quality due to its direct, semi-direct and indirect effects on the radiation budget. It is the main driving force of various weather phenomena. To measure the characteristics of the optical properties, dynamics and the thermodynamics in the whole troposphere, the shipborne Doppler lidar based on High Spectral Resolution techniques is developed. Shipley [1] and Sroga [2] describe one of the first HSRLs. Eloranta [3] presents the principles of high spectral resolution lidar systems. Chanin et al. [4] presents the use of a dual Fabry-Perot interferometer (FPI) for Doppler frequency analysis. Korb et al. performed analyzes wind measurements with a single FPI [5] and a dual FPI [6-7]. Liu et al. [8] proposed the use of iodine vapor frequency discriminator for frequency analysis of backscattered light from both aerosol and air molecules for tropospheric wind measurements. Later on, a mobile ground-based incoherent Doppler wind lidar (MIDWiL) with an iodine filter was developed for the winds profiles from the signals of Mie and Rayleigh backscattering together in 2000. A van-based Doppler lidar using an injection-seeded diode-pumped Nd:YAG pulsed laser with a high repetition rate was established to measure the sea surface wind with high spatial and temporal resolution for the Sailing Game. [9] The purpose of this paper is to introduce a Compact High-Power Shipborne Doppler Wind Lidar (CHiPSDWiL) based on High-Spectral-Resolution technique established at the Ocean University of China for the offshore measurement of the wind field and the properties of the aerosol and clouds in the troposphere.

2 METHODOLOGY

The atmospheric wind field can be retrieved from the combined Mie and Rayleigh backscattering by use of a direct-detect technology. The use of the second harmonic of Nd:YAG wavelength at 532 nm, with adequate response to both Mie and Rayleigh scattering and a molecular filter as a frequency discriminator with 2 GHz bandwidth. The wind measuring sensitivity is proportional to the slope of the filter transmission at an operating point. Another unique feature of the CHiPSDWiL system is the capability to monitor the aerosol scattering ratio, with which the radial wind can be calculated according to the aerosol load. With the aerosol-scattering ratio independently measured, the lidar returns passing through the frequency discriminator can be processed to retrieve Doppler wind profiles under a varied mixture of Rayleigh and Mie scattering.

Figure 1 illustrates a schematic diagram of CHiPSDWiL, consisting of five major function blocks: telescope and scanner system, laser and seed injection system, frequency locking system, optical frequency discriminator system, and transceiver switch. All of the modules are integrated in a small container of 2m×2m×2.4m, excepting for the scanner, for the field experiment in a research vessel with limited spaces (Figure 2).

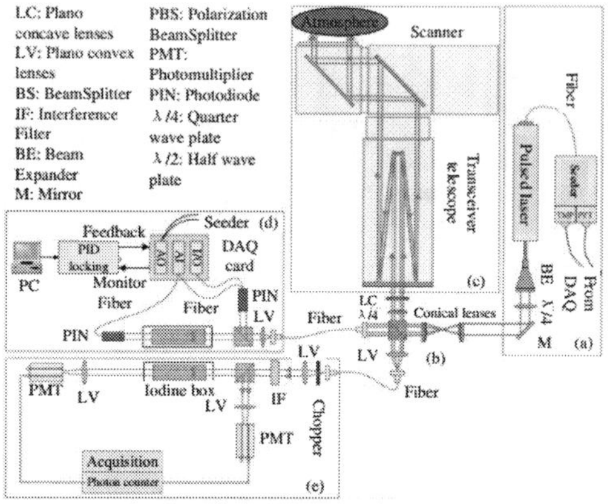

Figure 1 Schematic diagram of the CHiPSDWiL

Figure 2 CHiPSDWiL installed on the research vessel Dongfanghong 2

The lidar transmitter provides high-peak-power and frequency-stabilized laser pulses at 532 nm by a diode-pumped, frequency-doubled, pulsed Nd:YAG laser, injection seeded by a seed laser and stabilized to an iodine absorption line. The laser repetition rate is 100 Hz, and the single pulse energy is ~125 mJ with pulse width of 10 ns. The seed laser is a continuous None-Planer-Ring-Oscillator Nd:YAG laser and its 1064 nm output is injected into the pulsed Nd:YAG laser. The outgoing laser beam is firstly expanded to 250 μrad divergence by a 2X beam expander, and then directed into the transceiver telescope. A 300 mm diameter Cassegrain Dall Kirkham telescope with focal length of 3572 mm is the common part of the transmitting and the receiving optical path. Due to the structure blocking of the telescope, only the light beam in the annular space near the optical axis can be normally emitted, leading to much of the energy wasted. In order to solve this problem, for the first time to our knowledge, we use a pair of axicon lens together to produce an annular non diffractive Bessel beam to maximize the transmittance of laser beam and to minimize the unwanted back reflection and stray light. The annular beam transceiver simplifies the alignment between the transmitted beam and the receiver field of view and permits the lidar to operate with a 45-μrad FOV. Figure 3 shows the photos of transmitting annular beam at the surface of scanner and in the air.

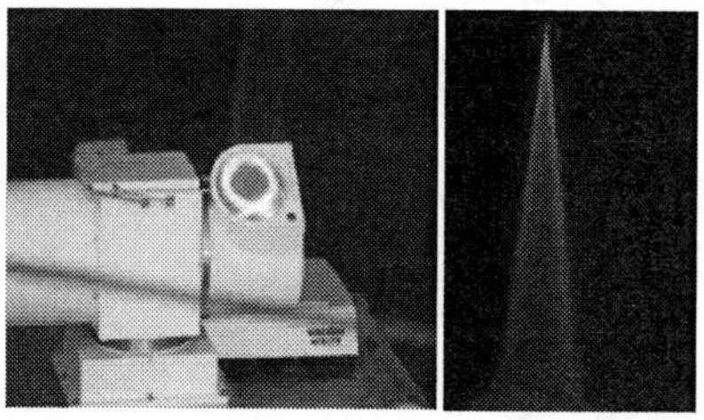

Figure 3 Lidar Scanner and the annular laser beam

A passive polarization transmit-receive switch consisting of a polarization beamsplitter and a 1/4-wave plate is employed. The telescope directs the received photons through the quarter-wave plate, which are converted into linear polarization with its axis of polarization perpendicular to that of the transmitted photons. So that the received photons pass through the polarizing cube. The received photons are then focused by lens onto the 200-μm-diameter multimode fiber that defines the receiver FOV. The fiber delivers the scattered light onto the high spectral resolution receiver system. After collimation by lens, scattered light passes through a skylight background filter comprised of a 0.1-nm interference filter followed by a non-polarizing beamsplitter. The signals are then split into two channels: one is detected directly by a photomultiplier tube (PMT) as a reference channel, and another goes through an iodine filter before being detected by another PMT in the measurement channel.

A small portion of leaked power from the transmitted laser beam through the polarizing beamsplitter is guided to a single mode fiber, which is directed to the second identical iodine cell and is locked to the 50% absorption level of the I2 line 1109 to stabilize the seed laser frequency. The frequency stability of the seed and amplified pulsed lasers are 200 kHz and 1 MHz, respectively. The spectral linewidth of the laser pulses is around 35 MHz.

The two-axis scanner provides coverage over an entire hemisphere (0-360° azimuth and 0-180° elevation). This scanner has the clear aperture of 320 mm and can accommodate co-axial transmission and reception with continuous scanning speed of 1-10°/s and directional accuracy of 0.1°. The scanning mirror has active stabilizing mechanism, which can compensate the pointing error caused by ship shake according to the information of orientation and attitude provided by the inertial measurement unit and GPS.

To measure the very large dynamic range of backscatter signal, an analog data acquisition and multi-channel scaler photon counting device is developed and used for the processes the data in real time with 10-m resolution for the altitude range from 0 m to 20 km to retrieve wind products. It also controls the laser trigger, the optical chopper, the scanner.

3 RESULTS

After the CHiPSDWiL system was built up, we conducted atmospheric measurements for verifying its capability. Aiming at validating the wind direction and speed measured by the CHiPSDWiL, the concurrent observation by the CHiPSDWiL and radiosonde was operated. One of the comparison results is shown in Figure 4.

The measurement of wind profile compared with radiosonde on January 16, 2015 from 19:47 local time (LT) to 20:36 LT at the OUC Laoshan campus (36.16°N 120.50°E). The Radiosonde data was from the Qingdao Observatory (36.06°N 120.33°E, Station identifier of ZSQD and Station number of 54857) and measurement time began at 20:00 LT. The wind profile by lidar was compatible with the corresponding by radiosonde when the observation height was below 3km. Considering the range difference (16.7 km) between the CHiPSDWiL and radiosonde, the slight deviations of the wind direction and speed are acceptable.

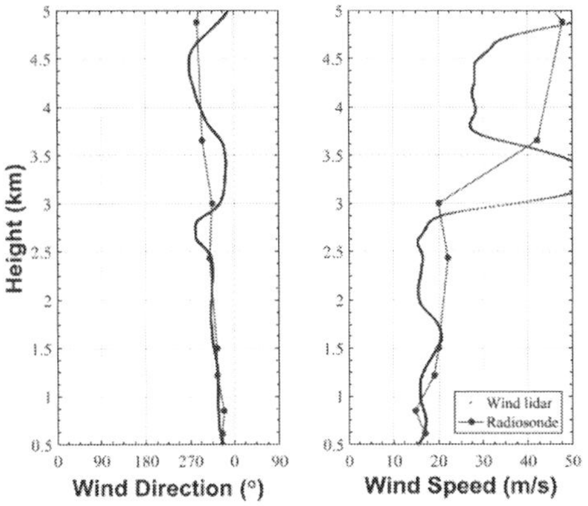

Figure 4 Comparison of profile on January 16, 2015. Lidar measurements were from 1947 local time (LT) to 2036 LT, and the radiosonde operated from 2000LT.

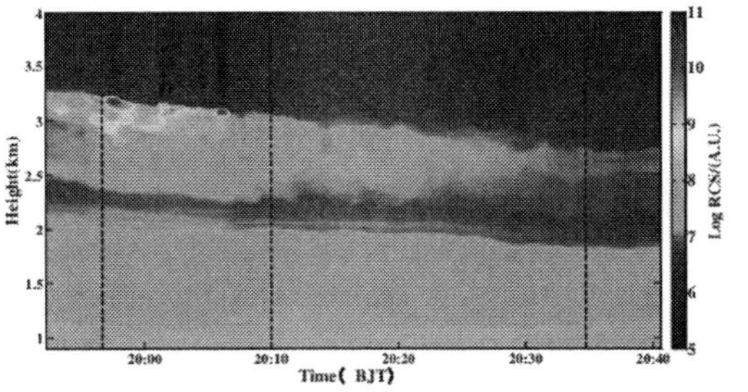

Figure 5 Range corrected atmospheric backscatter signal by CHiPSDWiL during 19:52~20:40 on 9

April 2015

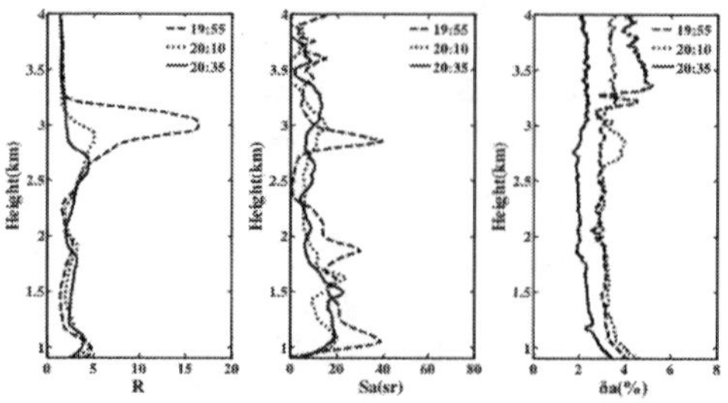

Figure 6 Profiles of aerosol backscattering ratio (R), lidar ratio (Sa) and depolarization ratio (δa) retrieved from five minutes' pulse accumulation signal at 19:55\20:10\20:35 on 9 April 2015

Figure 5 shows a range-squared backscatter measurement over the period 19:52 to 20:40 on April 9, 2015, presenting apparent aerosol layers and clouds. There are obvious clouds in the vicinity of 3 km and then gradually disappeared. In order to facilitate the analysis of the corresponding atmospheric conditions at different times, we take the characteristics of the experimental time at 19: 55, 20:10, 20:35 (Figure 5 black dotted line), respectively, with the accumulation time of 5min. The profiles of aerosol backscattering ratio (R), lidar ratio (Sa) and depolarization ratio (δa) are shown in Figure 6. The value of Sa of the aerosol layer below 2 km is in the range of 10-25, and δa is lower than 5%, which is in accordance with the characteristics of marine aerosols. On the other hand, at an altitude of 3 km, the value of R is 15 and δa is around 5% at the beginning of this observation, indicating a water cloud layer.

4 SUMMARY

We introduce a newly developed Doppler wind lidar/HSRL for measurement of wind profile and aerosol particle optical properties. The lidar transceiver transmits a unique Bessel laser beam into the atmosphere, offering a combination of advantages: (1) a high-peak-power seed injected DPSS laser and a normal aluminum coating DK telescope can be used to avoid laser damage and to improve the daytime observation capability. (2) Transceiver greatly releases the maintaining alignment between the laser beam and the receiver field of view and permits the lidar to operate in an unstable shipborne campaign.

ACKNOWLEDGEMENTS

This work was supported by the National Natural Science Foundation of China (NSFC) under grant 41375016.

References

[1] Shipley, S. T., Tracy, D. H., Eloranta, E.W., Trauger, J. T., Sroga, J. T., Roesler, F. L., and Weinman, J. A. 1983: High spectral resolution lidar to measure optical scattering properties of atmospheric aerosols. 1: Theory and instrumentation. Applied Optics, 22(23), 3716-3724.

[2] Sroga, J. T., Eloranta, E. W., Shipley, S. T., Roesler, F. L., and Tyron, P. J. 1983: High spectral resolution lidar to measure optical scattering properties of atmospheric aerosols. 2: Calibration and data analysis. Applied Optics, 22(23), 3725-373.

[3] Eloranta, E. W., 2005: Lidar: Range-Resolved Optical Remote Sensing of the Atmosphere, Claus Weitkamp ed., Springer Science+Business Media Inc, New York, USA, 143-164pp.

[4] Chanin, M. L., Gariner, A., Hauchecorne, A., Porteneuve, J., 1989: A Doppler lidar for measuring winds in the middle atmosphere, Geophys. Res. Lett. 16, 1273–1276.

[5] Korb, C. L., Gentry, B., Weng, C., 1992: The edge technique: theory and application to the lidar measurement of atmospheric winds, *Appl. Opt.* **31**, 4202–4213.

[6] Korb, C. L., Gentry, B. M., Li, S. X., Flesia, C., 1998: Theory of the double-edge technique for Doppler lidar wind measurement, *Appl. Opt.* **37**, 3097–3104.

[7] Flesia C., Korb, C. L., 1999: Theory of the double-edge molecular technique for Doppler lidar wind measurement, *Appl. Opt.* **38**, 432–440.

[8] Chen, W. B. Zhang, T. L., Hair, J. W., She, C. Y., 1997: An incoherent Doppler lidar for ground-based atmospheric wind profiling, *Appl. Phys. B* **64**, 561–566.

[9] Liu, Z. S., Liu, B. Y., Wu S. H., Li Z. G., Wang Z. J., 2008: A high spatial and temporal resolution mobile incoherent Doppler lidar for sea surface wind measurements, *Optics Letters* **33**(13), 1485-1487.

https://doi.org/10.1051/epjconf/201817601021

RAMAN SPECTROSCOPY MEASUREMENT OF CH_4 GAS AND CH_4 DISSOLVED IN WATER FOR LASER REMOTE SENSING IN WATER

Toshihiro Somekawa[1]* and Masayuki Fujita[1,2]

[1]*Institute for Laser Technology, Japan, *somekawat@ilt.or.jp*
[2]*Institute of Laser Engineering, Osaka Univ., Japan*

ABSTRACT

We examined the applicability of Raman spectroscopy as a laser remote sensing tool for monitoring CH_4 in water. The Raman technique has already been used successfully for measurements of CO_2 gas in water. In this paper, considering the spectral transmittance of water, third harmonics of Q-switched Nd:YAG laser at 355 nm (UV region) was used for detection of CH_4 Raman signals. The Raman signal at 2892 cm^{-1} from CH_4 dissolved in water was detected at a tail of water Raman signal.

1 INTRODUCTION

Lidar (light detection and ranging) is a very promising technique for monitoring the vast seafloor, and it is very useful for exploring mineral and natural gas deposits on a seafloor, monitoring the environmental change and working condition within submarine industrial facilities [1-4]. Although most lidar techniques can be applied to atmospheric sciences [5], their applications in water are not straightforward, because water is a strong light absorber showing relatively high transmission only in the shorter wavelength region from UV to Green spectral range [6]. For example, CO_2 gas is usually detected by infrared absorption methods [7, 8], but conventional infrared absorption spectroscopy is not suited for sensing gasses in water due to strong IR absorption [9]. Therefore, we proposed Raman lidar using UV-Green laser with relatively high transmission in water.

We reported on progress toward developing a technique for the underwater remote detection of gases in water using Raman lidar in our previous papers [1-4]. We used CO_2 gas to demonstrate our approach because it is easy to handle. Firstly, its effectiveness was demonstrated with remote identification of CO_2 dissolved in water in a glass bottle located 20 m away [1]. Next, applications of our approach to the quantitative analysis of the CO_2 concentration in water were presented for comparison with CO_2 dissolved in water and CO_2 bubbles [2, 3]. Also, the effect of seawater on CO_2 Raman spectra has been evaluated [4].

In this paper, we report measurements of CH_4 as well as CO_2 dissolved in water by laser Raman spectroscopy. We used the laser at 532 nm as the light source in the previous experiments. However, the CH_4 Raman signal (2892 cm^{-1}) excited by a 532 nm laser is distributed in the wavelength range at around 628.7 nm, where the spectral transmittance of water is not good [6]. Therefore, in this study, we utilized the 355 nm laser source which has lower attenuation in water compared with the 532 nm laser. The Raman signal at 2892 cm^{-1} from CH_4 dissolved in water was detected at a tail of water Raman signal using the 355 nm laser source. Thus, we can apply our methods to CH_4 as well as CO_2 gases. We expect that the submarine Raman lidar application will significantly improve the monitoring the vast seafloor.

2 EXPERIMENT

A schematic diagram of the Raman spectroscopy is given in Fig. 1. We use a standard Q-switched Nd:YAG laser (Continuum, Surelite) operating at its third harmonic wavelength of 355 nm. This laser generates a pulse-width of 10 ns and a pulse energy of 60 mJ at 10 Hz repetition rate. The laser output power is variable by the combination of half-wave plate (λ/2) and a polarizer.

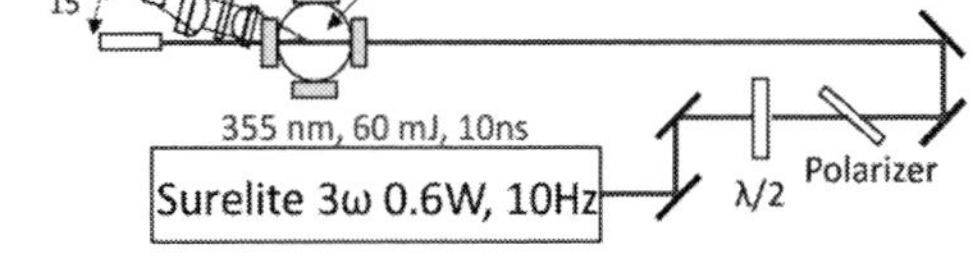

Figure 1 Schematic diagram of Raman spectroscopy experimental setup

The gas-water mixtures were realized by bubbling the CO_2 and CH_4 gases into the water in a high pressure chamber. This chamber is designed to withstand 10 MPa pressure, which is equivalent to a depth of 1000 m in the ocean. This chamber has four sapphire crystal windows and a rotation blade. The high-speed agitation by the rotation blade has efficiently achieved the equilibrium between the pressure in the water and the pressure in CO_2 and CH_4 gases. The pressure is monitored with both analog and digital pressure gauges. Also, a thermocouple located in water provided a measure of the water temperature.

The chamber is initially filled with about 180 mL deionized water, and CO_2 and CH_4 gases are introduced from a gas cylinder into the water. Each of the equilibrium steps is carried out for 5 minutes at about 22.5 degrees Celsius with rapid agitation (1000 rpm).

The Raman signals from samples are collected using an achromatic lens at an angle of 15° from the forward direction of the laser beam. This design provides a longer optical path length than a conventional detection geometry at 90°, offering an order of magnitude increase in Raman scattering intensity. After passing through the edge filter at 355 nm, the Raman signal is coupled into an optical fiber bundle by using an achromatic lens. The collected Raman signal was delivered to a spectrometer (Acton, SpectraPro-2300i) equipped with a thermoelectrically cooled charge-coupled device (CCD) camera (Princeton Instruments, PIXIS 400).

3 RESULTS AND DISCUSSION

Figure 2 shows Raman spectra of CO_2, CH_4 gases and CO_2, CH_4 dissolved in water. These Raman spectra were offset vertically for better visualization. The exposure time was 90 ms. The accumulation number was 3000 and it was repeated 5 times in each case. The entrance slit width of the spectrometer was set to 20 μm and the spectral resolution of this system was estimated to be 0.19 nm.

Figure 2(a) shows typical spectra of CO_2, CH_4 gases and CO_2, CH_4 dissolved in water at around 1 MPa. The broad spectra at 1635 and 3405 cm^{-1} are mainly the H_2O Raman signals detected in both samples of CO_2 and CH_4 dissolved in water.

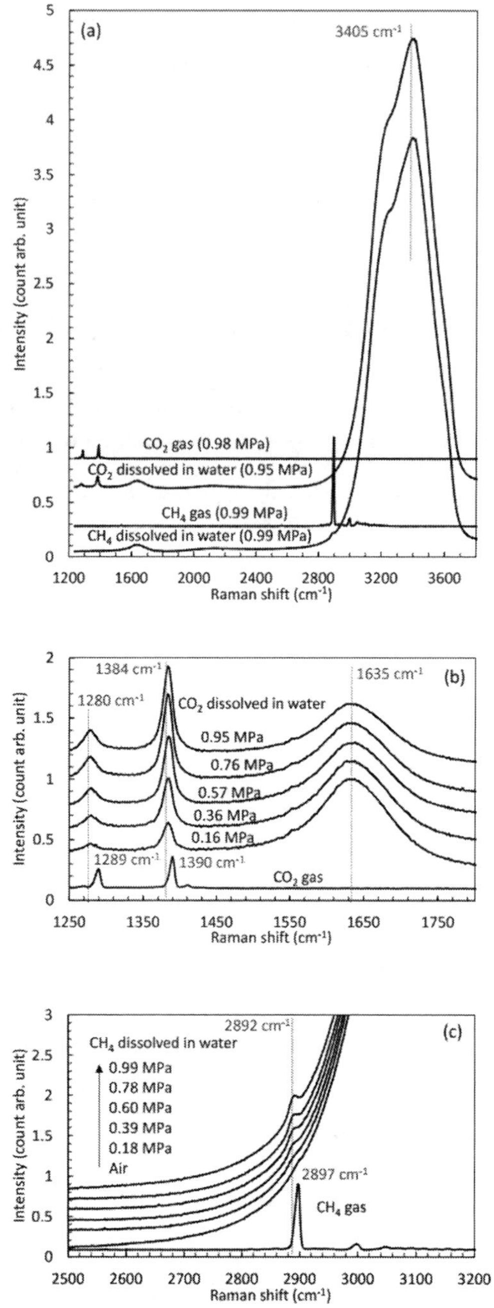

Figure 2 Raman spectra of CO_2, CH_4 gases and CO_2, CH_4 dissolved in water.

These water Raman signals are utilized as an internal standard for evaluating the dissolved gas concentration. In the CO_2 samples, the relatively sharp bands at 1280 and 1384 cm^{-1} are CO_2 Raman signals. On the other hand, the spectra of the CH_4 samples have the characteristic peak associated with CH_4 at 2892 cm^{-1}, which is located at a tail of water Raman signal. In the case of CO_2 and CH_4 gases at almost the same 1 MPa pressure, the CH_4 Raman signal is much stronger than the CO_2 Raman signal. This is because the Raman cross section of CH_4 is about five times larger than that of CO_2 [10]. However, in the case of CO_2 and CH_4 gases dissolved in water at almost the same ~1 MPa pressure, the CH_4 Raman signal is much weaker than the CO_2 Raman signal because of the difference in water solubility of these gases.

Figure 2(b) shows the spectra of CO_2 gas at different concentrations dissolved in water. These Raman spectra were normalized at 1635 cm^{-1} water Raman signal intensity peaks. The relatively sharp bands at 1280 and 1384 cm^{-1} are CO_2 Raman signals. When CO_2 gas is dissolved in water, the gas phase band position at 1390 cm^{-1} is shifted to 1384 cm^{-1} in water. This shift could be attributed to the variation of vibration modes in water.

Figure 2(c) shows the Raman spectra of CH_4 at different concentrations dissolved in water. These spectra have been normalized at ~3405 cm^{-1} peak value. The Raman signal of CH_4 dissolved in water is centered at 2892 cm^{-1}. When dissolved in water, the band position at 2897 cm^{-1} shifts to 2892 cm^{-1} due to the variation of CH_4 vibration modes in water.

Using 355 nm laser, we can detect the Raman signals of CH_4 dissolved in water, but its intensity is very weak. Also, the CH_4 Raman signal is located at a tail of water Raman signal, and it is difficult to distinguish between CH_4 and water Raman signals. However, Raman signals of CH_4 in gas phase is much stronger as shown in Fig. 2(a), and we have confirmed that our methods can be applied to the gas bubbles [3]. Therefore, we can expect that the Raman signals of CH_4 gases in water would be relatively easily detected at the coexistence areas of CH_4 gas dissolved in water and CH_4 bubbles.

4 CONCLUSIONS

We have demonstrated the applicability of Raman lidar technique for 3D laser mapping in water by measuring Raman signals of CO_2 and CH_4 gases in water. The Raman signals at 1280 and 1384 cm^{-1} from CO_2 in water are detected. The relatively strong Raman signal at 1384 cm^{-1} is a good candidate for doing Raman lidar measurements in water. Also, the CH_4 Raman signal at 2892 cm^{-1} was detected at a tail of water Raman signal, but it is very weak due to low water solubility of CH_4 gas. However, Raman signals of CH_4 in gas phase is much stronger, and we can expect that the Raman signals of CH_4 gases in water would be relatively easily detected at the coexistence areas of CH_4 gas dissolved in water and CH_4 bubbles.

ACKNOWLEDGEMENTS

This work was supported by JSPS KAKENHI Grant Number 15H05336.

References

[1] Somekawa, T., Tani, A., Fujita, M., 2011: Remote detection and identification of CO_2 dissolved in water using a Raman Lidar system, *Appl. Phys. Express,* **4**, 112401.

[2] Somekawa, T., Fujita, M., 2012: Development of Raman Lidar for Water Dissolved CO2 Detection, *Proc. of ILRC26*, S10-06.

[3] Somekawa, T., Takeuchi, T., Yamanaka, C., Fujita, M., 2014: Raman spectroscopy measurements of CO_2 dissolved in water and CO_2 bubbles for laser remote sensing in water, *Proc. of SPIE*, **9240**, 92400J.

[4] Somekawa, T., and Fujita, M., 2016: Raman Spectroscopic Measurements of CO2 Dissolved in Seawater For Laser Remote Sensing in Water, *EPJ Web of Conference*, **119** (7), 25017.

[5] Weitkamp C., 2005: Lidar Range-Resolved Optical Remote Sensing of the Atmosphere, *Springer*, New York.

[6] Smith R. C. and Baker, K. S., 1981: Optical properties of the clearest natural waters (200-800 nm), *Appl. Opt.*, **20** (2), 177-184.

[7] Somekawa T., Fujita, M., Izawa, Y., 2010: Direct Absorption Spectroscopy of CO2 Using a Coherent White Light Continuum, *Appl. Phys. Express*, **3**, 082401.

https://doi.org/10.1051/epjconf/201817601021

[8] Somekawa T., Manago, N., Kuze, H., Fujita, M., 2011: Differential optical absorption spectroscopy measurement of CO2 using a nanosecond white light continuum, *Opt. Lett.*, **36** (24), 4782-4784.

[9] Friedman D., 1969: Infrared Characteristics of Ocean Water (1.5-15μ), *Appl. Opt.*, **8** (10), 2073-2078.

[10] Fouche, D. G. and Chang, R. K., 1971: Relative Raman Cross Section for N_2, O_2, CO, CO_2, SO_2, and H_2S, *Appl. Phys. Lett.*, **18** (12), 579-580.

 https://doi.org/10.1051/epjconf/201817601022

POLARISATION EFFECTS IN LIDAR SIGNALS: DETERMINATION, CORRECTION, AND UNCERTAINTY CALCULATION

Volker Freudenthaler[1]*

*[1]Fakultät für Physik, Meteorologisches Institut, Ludwig–Maximilians–Universität, 80333 München, Theresienstrasse 37, Germany, *volker.freudenthaler@lmu.de*

ABSTRACT

An analytical framework, based on the Müller-Stokes formalism, is presented, which enables us to determine the impact of polarisation effects of the lidar optics on the lidar signals and on the measured atmospheric linear depolarisation. Furthermore, the equations enable us to investigate the accuracy of the gain ratio calibration of depolarisation channels with different calibration setups, to correct some of the polarisation effects, and to perform uncertainty analyses.

1 INTRODUCTION

The purpose of atmospheric depolarisation measurements with lidar is mainly to discern between more or less depolarising scatterers. In order to increase the accuracy, we have to analyse the impact of potential error sources, which is only possible for such sources that are included in the underlying analytical model of the lidar. While the consideration of errors due to random signal noise is largely standard, systematic errors in lidar signals very often have been neglected in the past or were included only for one or few parameters, supposedly because a complete mathematical model contains too many and often unknown parameters. This presentation gives an introduction to the mathematical lidar model recently published [1] to approach this problem regarding the polarization dependent systematic errors. For references or more detailed explanations please refer to this open access paper.

Polarization related systematic errors arise in lidar signals by three reasons: an uncertain state of polarization of the emitted laser beam, optical elements with polarization effects, and an inaccurate calibration of the gain ratio of the two lidar channels. The lidar model shown in Fig. 1 comprises all those parts, and at the same time it tries to reduce the number of parameters by combining the optics in three essential modules, which are the emitter optics ($\mathbf{M}_E$), the receiver optics ($\mathbf{M}_O$), and the polarization analysing optics ($\mathbf{M}_{T,R}$) that includes a transmitted ($_T$) and a reflected ($_R$) path of the polarizing beam splitter (PBS) with the subsequent optical elements.

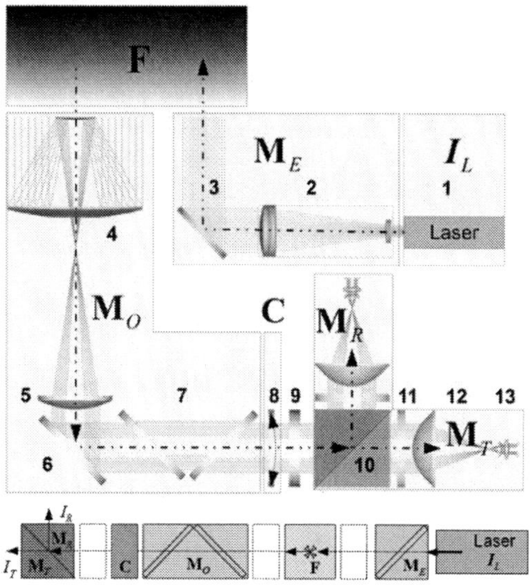

*Figure 1 Top: Exemplary depolarisation lidar setup with laser 1, beam expander 2, steering mirror 3, receiving telescope 4, collimator 5, folding mirror 6, dichroic beam-splitters 7, a rotating element for polarisation calibration 8, interference filter 9, and polarising beam-splitter cube 10 (PBS, polarising beam-splitter). The neutral density filters and cleaning polarisers 11, detector optics 12, and the detectors 13. The system can be subdivided in functional blocks which can be described with the Stokes-Müller formalism: $\mathbf{I}_L$ is the Stokes vector of the laser source, $\mathbf{M}_E$ is the Müller matrix of the the laser emitter optics, **F** of the atmospheric backscattering volume including depolarisation, $\mathbf{M}_O$ includes receiver optics as beam-splitters, **C** is the calibrator, and $\mathbf{M}_{T,R}$ is the polarising beam-splitter including the detector optics for the transmitted (T) and reflected (R) optical branches. Bottom: simplified schematic of the setup.*

This model cannot comprise all possible aerosol lidar systems but, we believe, a majority. Each module is described by a Müller matrix of a rotated retarding diattenuator, which is defined by the linear diattenuation parameter *D*, the retardance *Δ*, and the rotation *θ* of the module around the optical axis. The state of polarization of the emitted laser beam and the output beams in the transmitted and reflected channel are described by Stokes vectors $\boldsymbol{I}_L$ and $\boldsymbol{I}_{T,R}$, respectively, which are related by Eq. (1). The gain factors $\eta_{T,R}$ include the attenuation of optics and filters behind the PBS and the electronic amplification of each channel $_{T,R}$.

$$\boldsymbol{I}_{T,R} = \eta_{T,R}\mathbf{M}_{T,R}\mathbf{C}\mathbf{M}_O\mathbf{F}\mathbf{M}_E\boldsymbol{I}_L \qquad (1)$$

The equations for both signal intensities $I_{T,R}$, which is the first component of the Stokes vector, can be combined in one equation I_S:

$$I_S = \quad \eta_S \langle \mathbf{A}_S | \mathbf{C} | \boldsymbol{I}_{in} \rangle; \text{ with } S \in \{T,R\} \qquad (2)$$

The descriptions of the analyser part $<A_S|$ and of the input Stokes vector $|\boldsymbol{I}_{T,R}>$ are the same for all calibrators **C** (see below) and also for the standard signals without calibrator, wherefore we have to develop them only once. The atmospheric depolarisation is described by the Müller matrix for randomly oriented aerosol [2] as in Eq.(3)

$$\mathbf{F} = F_{11}\begin{pmatrix} 1 & 0 & 0 & 0 \\ 0 & a & 0 & 0 \\ 0 & 0 & -a & 0 \\ 0 & 0 & 0 & 1-2a \end{pmatrix} \qquad (3)$$

with the linear depolarisation ratio *δ* and the polarisation parameter *a* related by Eq. (4).

$$\delta = \frac{F_{11} - F_{22}}{F_{11} + F_{22}}; \; a = \frac{F_{11}}{F_{22}} \quad \Rightarrow a = \frac{1-\delta}{1+\delta} \qquad (4)$$

2 Standard measurements (0°)

The equations for the standard measurement intensities without calibrator have the form of Eq. (5) with the two new parameters G_S and H_S.

$$I_S = \eta_S T_S T_O T_{rot} F_{11} T_E I_L \left(G_S + aH_S\right) \qquad (5)$$

The GH-parameters allow us to formulate general expressions for *δ* in Eq. (6)

$$\delta = \frac{1-a}{1+a} = \frac{\delta^*\left(G_T + H_T\right) - \left(G_R + H_R\right)}{\left(G_R - H_R\right) - \delta^*\left(G_T - H_T\right)} \qquad (6)$$

with the calibrated signal ratio δ^* and the calibration factor *η* in Eq. (7),

$$\delta^* = \frac{1}{\eta}\frac{I_R}{I_T}(0°); \quad \eta \equiv \frac{\eta_R T_R}{\eta_T T_T} \qquad (7)$$

and for the atmospheric backscatter coefficient F_{11} in Eq. (8), which are valid for all lidar systems that can be described by the model in Fig. (1).

$$F_{11} \propto \eta H_R I_T - H_T I_R \qquad (8)$$

The GH-parameters are fully described by the Stokes vector of the emitted laser beam $\boldsymbol{I}_L$ and the optical properties of the modules $\mathbf{M}_E$, $\mathbf{M}_O$, and $\mathbf{M}_{T,R}$ in Fig. 1, i.e. the diattenuation parameters *D*, the retardances *Δ*, and the rotation *θ* of the modules around the optical axis.

3 Calibration measurements (±45°)

The calibration factor *η* (Eq. (7)) has to be determined by means of measurements with additional calibration elements (calibrator) that should distribute the incident light in equal intensities to both channels $_{T,R}$. For that purpose, four different calibrator types **C** can be inserted in our model at three locations as shown in Fig. 2.

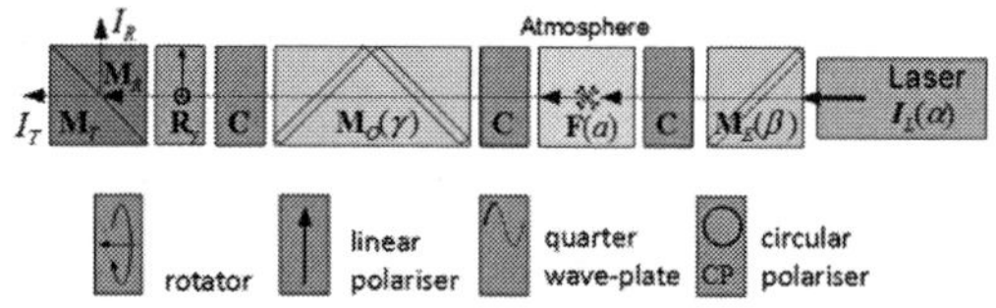

Figure 2 Different calibrator (red block) positions (top), and three options for the calibrator **C** *(bottom).*

We use as calibrator **C** the

- mechanical and λ/2-plate rotation ($\mathbf{M}_{rot}$(h)),
- a linear polariser ($\mathbf{M}_P$),
- a λ/4-plate ($\mathbf{M}_{QW}$), and
- a circular polariser ($\mathbf{M}_{CP}$).

These calibrators should be oriented at + or -45° with respect to the plane of incidence of the PBS in order to distribute the incident light equally, but as the absolute 45° orientation is difficult to achieve, an error ε is included in Eq. (9) for the gain ratio η^*. The calibration signals I_S in Eq. (9) include the same input Stokes vector and analyser part of Eq. (2) as the standard signals, but because of the different types of calibrators no general expression like Eq. (5) could be found for the calibration signals, and the expression for the calibration correction parameter K (Eq. (9)) has to be determined for each lidar setup individually.

$$\eta^*(\varepsilon) = \frac{I_R(\pm 45° + \varepsilon)}{I_T(\pm 45° + \varepsilon)} = K\eta \tag{9}$$

It can be shown, that the combination of the ±45° measurements in Eq. (9) at exactly 90° difference (Δ90-calibration), as in Eq. (10), reduces the uncertainty of the calibration.

$$\eta^*_{\Delta 90} \equiv \sqrt{\eta^*(+45° + \varepsilon)\eta^*(-45° + \varepsilon)} \tag{10}$$

4 Binary operators

In order to further reduce the number of equations, we introduce three binary operators, x, h, and y. With x (Eq. (11)) we combine the equations for the ±45° calibration measurements. With operator h (Eq. (12)) we combine the calibration equations for the mechanical and the λ/2-plate rotation ($\mathbf{M}_{rot}$(h)). The 90° rotation operator y (Eq. (13)) combines the equations for 90°-rotated PBSs (see Fig. (3)), because such a rotation is possible for some lidar systems.

$$\mathrm{x} = \pm 1: \ \Psi(\mathrm{x}, \varepsilon) = \mathrm{x}45° + \varepsilon \tag{11}$$

$$\mathrm{h} = \pm 1: \ \mathbf{M}_{rot}(\mathrm{x}, \varepsilon, \mathrm{h}) = \mathbf{R}(\mathrm{x}45° + \varepsilon)\mathrm{M}_{\mathrm{h}} \tag{12}$$

$$\mathrm{y} = \pm 1: \ \boldsymbol{I}_S = \eta_S \mathbf{M}_S \mathrm{R}(\mathrm{y}) \mathbf{C} \mathbf{M}_O \mathbf{F} \mathbf{M}_E \boldsymbol{I}_L \tag{13}$$

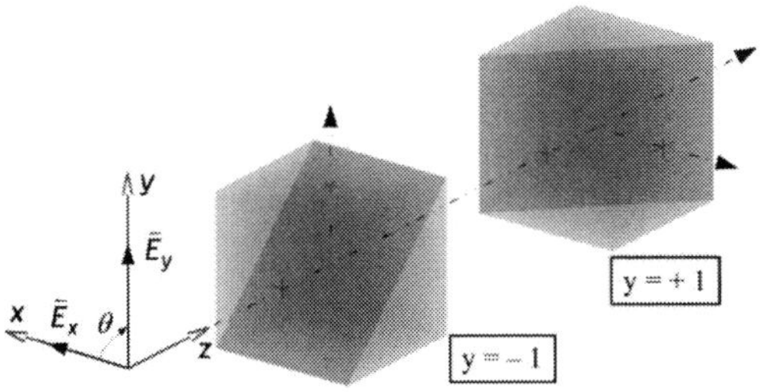

Figure 3 The orientation of the PBS with respect to the reference coordinate system for the two values of the binary operator y.

5 BENEFITS

The equations derived allow an analytical estimation of systematic error sources and their importance, and thus the optimization of the design of new and existing lidar systems with respect to their polarization properties. For example, it becomes evident that the Δ90-calibration largely removes errors due to a rotational misalignment ε of the calibrator setup and that the cleaning of the cross talk of the PBS by means of additional polarization filters, which is very cheap and quite easy, simplifies the calibration equations, especially for the Δ90-calibration. Furthermore, we can directly compare the advantages and disadvantages of different calibrators at different locations and choose the best combination for our special lidar setup. With calibrators at different locations it is possible to determine the diattenuation of the receiving optics. With a λ/4-plate calibrator it is possible to estimate the degree of circular polarization of the beam, and with a rotation-calibrator or a circular polarizing calibrator it is possible to determine the rotation of the plane of polarization of the laser beam.

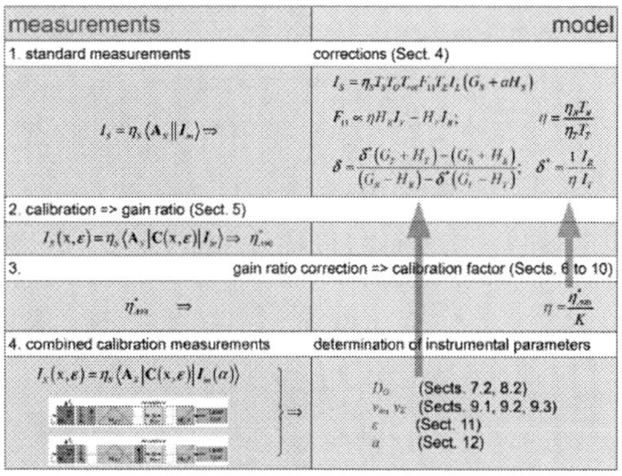

Figure 4 Four steps for calibrating and correcting the standard measurements for systematic errors by means of the model equations and additional calibration measurements. On the left side the measurements and their Müller matrix representations are listed and on the right side the scalar equations and their parameters, which can be retrieved from our model and from additional calibration measurements. References to sections are with respect to [1].

On the other hand, in case all optical parameters and rotational misalignments are known, it is possible to correct the measured signals with the provided equations. Fig. 4 summarizes the steps of this procedure.

In case the optical parameters and rotational misalignments are only known with uncertainties, a numerical complete search over the uncertainty space of all parameters (numerical error analysis) yields an overall systematic error estimation for the lidar signals and the retrieved linear depolarisation ratio and backscatter coefficient. Examples of such an error analysis for several EARLINET lidar systems with a Matlab code is shown by Bravo-Aranda et al. [3] (see their corresponding contribution at this conference), and an open source Python code including a real λ/2-plate calibrator with retardation and other additional details has been published recently [4].

ACKNOWLEDGEMENTS

The financial support for EARLINET in the ACTRIS Research Infrastructure Project by the European Union's Horizon 2020 research and innovation programme under grant agreement n. 654169 and previously under grant agreement n. 262254 in the 7th Framework Programme (FP7/2007-2013) is gratefully acknowledged.

References

[1] Freudenthaler, V., 2016: About the effects of polarising optics on lidar signals and the Δ90-calibration, *Atmos. Meas. Tech.* **9**, 4181-4255, doi: 10.5194/amt-9-4181-2016

[2] Mishchenko, M. I. & Hovenier, J. W., 1995: Depolarization of light backscattered by randomly oriented nonspherical particles *Opt. Lett., OSA,* **20,** 1356-1358, doi: 10.1364/OL.20.001356

[3] Bravo-Aranda, J., et al., 2016: Assessment of lidar depolarization uncertainty by means of \newline a polarimetric lidar simulator, *Atmos. Meas. Tech.* **9**, 4935-4953, doi: 10.5194/amt-9-4935-2016

[4] Freudenthaler, V., 2017: Open source Python code for polarization related error analysis of aerosol lidar signals: https://bitbucket.org/iannis_b/atmospheric_lidar_ghk

ADVANCES IN ATMOSPHERIC TEMPERATURE PROFILE MEASUREMENTS USING HIGH SPECTRAL RESOLUTION LIDAR

Ilya I. Razenkov, Edwin W. Eloranta

University of Wisconsin-Madison, 1225 W. Dayton St., Madison, WI, 53706, USA, razenkov@wisc.edu

ABSTRACT

This paper reports the atmospheric temperature profile measurements using a University of Wisconsin-Madison High Spectral Resolution Lidar (HSRL) and describes improvements in the instrument performance. HSRL discriminates between Mie and Rayleigh backscattering [1]. Thermal motion of molecules broadens the spectrum of the transmitted laser light due to Doppler effect. The HSRL exploits this property to allow the absolute calibration of the lidar and measurements of the aerosol volume backscatter coefficient. Two iodine absorption filters with different line widths are used to resolve temperature sensitive changes in Rayleigh backscattering for atmospheric temperature profile measurements.

1 INTRODUCTION

The atmospheric temperature is an important parameter characterizing the state of the atmosphere. Accurate temperature profiles are required for modeling atmospheric processes as well as for calibration of remote sensing instruments. Radiosonde measurements are usually available only twice a day and with very sparse ground coverage. Lidars are capable of measuring the atmospheric temperature with a relatively high spatial and temporal resolution. Several investigators have used High Spectral Resolution Lidar for temperature profiling [2, 3]. The reported techniques have low sensitivity to air temperature and the used instruments greatly exceed the eye-safe limit on the transmitted laser power.

The standard HSRL calibration requires the atmospheric temperature and pressure profile and typically uses radiosonde measurements. In order to enable temperature profile measurements, HSRL receiver has two iodine absorption filters with different absorption line widths [4]. This measurement capability makes the lidar intrinsically calibrated. The HSRL temperature measurement technique requires superior instrument performance. In order to achieve good measurement accuracy, the instrument has several modifications, which have improved instrument performance and data quality for standard HSRL measurements.

2 HIGH SPECTRAL RESOLUTION LIDAR SYSTEM

The simplified schematic of the HSRL lidar is shown in Figure 1. For a more detailed description of the instrument see "High Spectral Resolution Lidar at the University of Wisconsin-Madison" [5]. The transmitter uses a diode pumped Nd:YAG laser operating at 4kHz repetition rate. The average transmitted power is 200mW @ 532nm. The host laser cavity is injection seeded with the single mode distributed feedback diode (DFB) seed laser (1064nm). Active control loop is implemented to lock the seed laser frequency to the center of iodine absorption line #1109. The frequency locking setup uses a 100mm long iodine absorption cell and two detectors. The frequency locking accuracy is around ±15MHz with a frequency jitter of around ±15MHz amplitude.

A portion of the laser light is picked off from the beam by a wedged beamsplitter for frequency locking, laser spectrum quality control, and for the interferometer. The broadband beam splitter (R/T=65/35S) is used as a transceiver switch. Then, the beam is further expanded by 20x Dall-Kirkham telescope (400 mm aperture) to the eye-safe level and transmitted into the atmosphere. The telescope is shared between the transmitter and the receiver. The received light is split, filtered, and detected by the three channels: Combined, Molecular I2, and Molecular I2A. Signal in the combined channel contains photons scattered by both aerosols (Mie scattering) and by molecules (Rayleigh scattering). The first standard iodine filter (1.8 GHz Iodine line width) absorbs most of the aerosol signal photons (>99.98 %)

allowing around 25% of the molecular signal photons through this cell. The second iodine absorption filter contains an argon buffer gas, which broadens the iodine absorption line to 2.85 GHz. It allows around 50% of the first stage filtered molecular signal photons through this cell providing approximately equal count rate in both molecular channels.

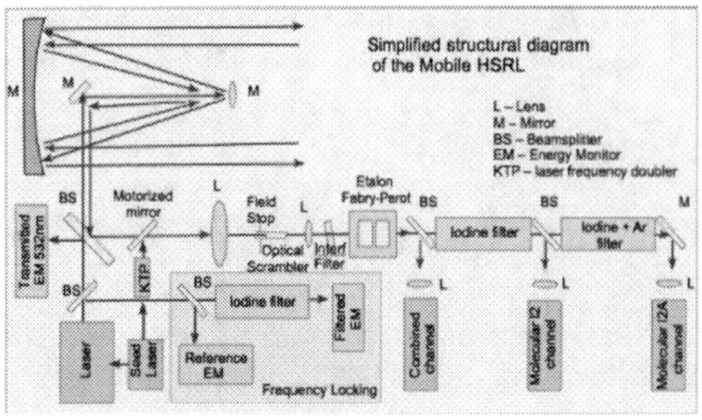

Figure 1. A simplified structural diagram of HSRL.

The argon-broadened filter minimum transmission in the center of the 1109 line is ~0.1% and cannot provide sufficient rejection of aerosol signal. For that reason a serial arrangement of the filters is used. Both absorption filters contain a fixed amount of iodine to provide stable filter properties. Heaters maintain all the iodine in the gaseous phase and ensure that no iodine condensation occurs on the windows of the cell. The cells temperature is around 33 degrees C. The pressure and the iodine amount in the argon gas buffered iodine cell were optimized to provide the instrument sensitivity of the signal to air temperature and transmission. The optimum pressure is found to be at around 110 Torr of the argon gas pressure providing the instrument temperature sensitivity coefficient $\frac{dR}{dT}$ around 0.145%/K.

In the previous instrument configuration, the difference in optical alignment of the two molecular channels introduced a range dependent differential geometry error in the temperature profile measurements. In order to eliminate this effect, a 25 mm long straight multimode optical fiber is installed behind the field stop with the core size equal to the field stop aperture (300 μm). The multimode fiber (optical scrambler) changes the intensity distribution of the received light and decreases the range dependency of the signals. It also decreases errors associated with non-uniform sensitivity of the detectors across the sensor area.

In order to measure the channels' transmission spectrum functions and the relative gain of the channels, we implemented seed laser frequency scans across the receiver bandwidth. The 95% duty cycle light pulses (~237μs pulse width) of the frequency doubled seed laser light (532nm) with a large signal levels significantly improved the accuracy of the measured channels' transmission spectrum functions. Previously, we used short (~50ns) pulses from the host laser at 4kHz pulse repetition rate. The seed laser scans also decreased the uncertainty in the laser output frequency and a laser spectral width. The newly measured channels' transmission functions do not require any adjustments for standard HSRL measurements. The temperature profile offsets due to scan-to-scan variations became smaller by approximately four times.

During instruments' operation the gain of the channels can change. In order to correct for these changes, we introduce high intensity green LED pulses (30 μs wide) into the receiver field-of-view 180 μs after the transmitted laser pulse. The ratio of the signals in molecular channels from the LED accounts for the relative gain change in the two channels: changes in iodine filter transmission and detector gain changes.

3 ATMOSPHERIC TEMPERATURE MEASUREMENT PRINCIPLES

The ratio R of the signals of the two molecular channels is a function of the atmospheric temperature and pressure. The received signals in the channels *Molecular I2* ($N_{m,1}$) and *Molecular I2A* ($N_{m,2}$), for $i=1$ *or* 2 are:

$$N_{m,i}(z) = N_0(z)f_{m,i}(z) + N_a(z)C_{am,i}(z),$$

where,

$$f_{m,1}(z) = \int \xi_1(\nu)F_1(\nu)\mathcal{R}(T(z),P(z),\nu)d\nu,$$

$$f_{m,2}(z) = \int \xi_2(\nu)F_1(\nu)F_2(\nu)\mathcal{R}(T(z),P(z),\nu)d\nu,$$

and $\int \mathcal{R}(T(z),P(z),\nu)d\nu = 1$, $N_0(z)$ – number of photons received from a distance z; $\xi_1(\nu), \xi_2(\nu)$ - the channels' gains; ν – frequency; $F_1(\nu), F_2(\nu)$ – transmission spectrums of a standard iodine and broadened iodine filters correspondingly; $\mathcal{R}(T,P,\nu)$ – spectrum of molecular return from

Tenti's s6 model; $T(z)$, $P(z)$ – atmospheric temperature and pressure; $N_a(z)C_{am,i}(z)$ - crosstalk terms allowing a leakage of the aerosol photons into the molecular channels, which are significant only for dense clouds and can be omitted for clear sky conditions. The $N_a(z)$ and $C_{am,i}(z)$ are derived from the standard HSRL inversion technique which uses the signal from the Combined channel. The channels' transmission spectrum functions $\xi_1(\nu)F_1(\nu)$, $\xi_2(\nu)F_1(\nu)F_2(\nu)$ are measured by performing a seed laser frequency scan across the receiver bandwidth. This provides the filter spectral band-pass and the relative sensitivity of the channels. The ratio of the signals in the two channels is the function of temperature and pressure:

$$R(z) = \frac{N_{m,1}(z) - N_a(z)C_{am,1}(z)}{N_{m,2}(z) - N_a(z)C_{am,2}(z)} = \frac{f_{m,1}(T(z),P(z))}{f_{m,2}(T(z),P(z))}$$

4 ATMOSPHERIC TEMPERATURE MEASUREMENT RESULTS

The atmospheric temperature profile measured on February 19th, 2017 from 0:00 to 5:00 UTC Madison, WI, and radiosonde observations from Green Bay, WI, are shown in Figure 2a. The vertical resolution is 300 meters. The difference of the measured profile and the radiosonde is shown in Figure 2b. For the profiles' correction the ratio of the signals in the molecular channels is measured with removed Argon buffered iodine filter. Then, the ratio is applied to the signals in one of the channels. A case for atmospheric temperature profile with inversion layer at the cloud top is shown in Figure 2c (February 15th, 2017 3:00-5:00UTC Madison, WI, and radiosonde Green Bay, WI, 150 meters vertical resolution). The site elevation is 330 meters. The HSRL measured profiles differ from the radiosonde observations in the lower 2 km. This difference is partially due to different physical locations of radiosonde and the lidar observations. Also, the earlier described optical scrambler is used to minimize the alignment errors. However, a residual range/time dependency error remains in the signals. We assume it is caused by insufficient optical scrambling, and the detectors' nonlinear response in presence of the large internally scattered laser pulse. The longer scrambler will be implemented in the future.

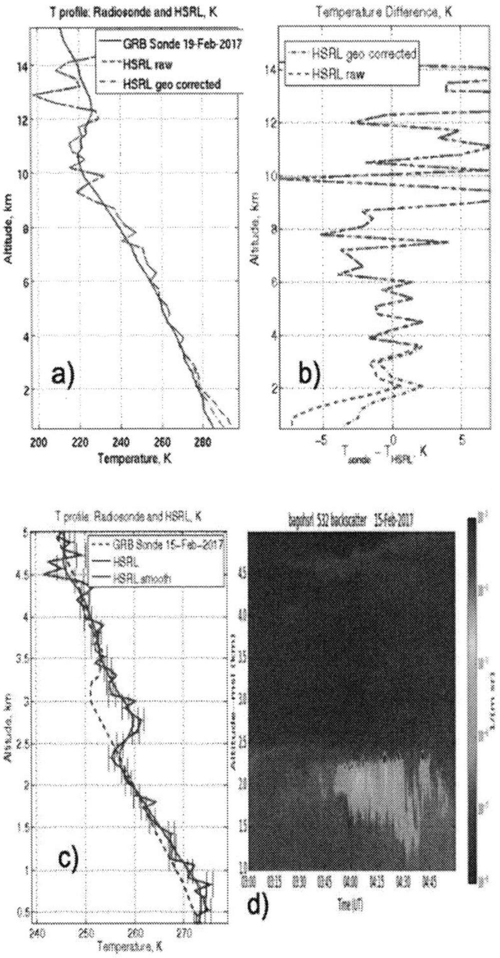

Figure 2. a) Temperature profiles: uncorrected-raw (red, dashed), and corrected (green, dashed); measured on February 19, 2017 from 0:00 to 5:00 UTC Madison, WI; and radiosonde profile (blue, solid) 0:00UTC Green Bay, WI; b) The difference of lidar and radiosonde profiles for uncorrected (red, dashed) and corrected profiles (green, dashed);
c) Temperature profile at 150 m range resolution (green, solid), smoothed profile at 75 m resolution (red, solid) on February 15, 2017 from 3:00 to 5:00 UTC Madison, WI; and radiosonde (blue, dashed) 0UTC Green Bay, WI; d) Range-time image for aerosol volume backscatter coefficient, 1/(m sr), corresponding to the temperature profile in Fig.c.

The HSRL is co-located with Atmospheric Emitted Radiance Interferometer instrument (AERI), which is capable of measuring the atmospheric temperature profile in the lower 4 km with ~200 m spatial resolution. This will be used for validation and for correction of the HSRL temperature profiles.

5 MEASUREMENT SENSITIVITY AND ERROR ANALYSIS

The temperature profile measured by the HSRL still exhibits an offset between around -5 and +5 K when compared with the radiosonde sounding. A correction on the relative gain of the two molecular channels must be applied for its correction. In order to explain the temperature offsets, we analyzed the sensitivity of the described temperature profile measurement method to the instrument calibrations. Some of the temperature measurement errors, instrument parameters, and calibration values are summarized in Table. 1.

Table 1 HSRL calibration and measurement errors.

Source	Estimate	Percent error in Signal Ratio	Temperature Error (.145%/K)
Laser Spectral purity	1:5000	0.001	<0.01 K (clear air)
Transmitted Spectral width	30MHz	0.10	.7 K
Frequency Lock Accuracy	±30 MHz	1.1	±7.5K
Etalon Drift Stability	±20MHz	0.08	±0.6K
Iodine filters' transmission (scan-to-scan variation)	20MHz	0.29	±2K
Detector gain stability and transmission functions	<0.45%	<0.45	3.1k
Statistical Noise at different ranges for 1-hr and 300-m integration	at 1 km at 5 km at 10 km at 15 km	.04 .1 .3 .7	0.3K 0.6K 2.0K 4.7K

The largest error is due to the frequency locking accuracy. The laser center frequency during the instrument operation can remain stable within 1 MHz for several hours. However, under instrument thermal disturbances, the center frequency experiences the offset up to 30 MHz from the peak absorption of the iodine line. Further improvements in the absolute frequency locking stability (~1 MHz) are required in order to decrease the offset to ~0.1 K. The calibration scans measurement error in the channels' transmission is another source of the temperature offsets. The transmissions are corrected by adjusting the relative gain of the channels. The correction remains constant if no changes in the transmission spectrum occur (i.e. etalon drift, iodine filters changes). The correction by LED pulses is used to adjust the channels' transmissions with around 0.01% accuracy. However, this technique requires more testing to confirm its efficiency.

6 CONCLUSIONS

The recent instrument modifications have improved the temperature measurement accuracy. However, it still requires a range correction in the lower 2 km and a precise laser frequency locking. In the near future we will improve the laser frequency locking accuracy by implementing an interferometer-based laser cavity controller. The future instrument improvements will make the temperature profile measurements robust and useful for meteorological applications.

ACKNOWLEDGEMENTS

This work is supported by the Space Science and Engineering Center U. of Wisconsin – Madison.

References

[1] Eloranta, E. W.: High Spectral Resolution Lidar, *in Lidar: Range-Resolved Optical Remote Sensing of the Atmosphere,* Klaus Weitkamp ed., Springer, NY, pp. 25-30.

[2] Hair J.W. et al, 2001: High spectral resolution lidar with iodine-vapor filters: measurement of atmospheric-state and aerosol profiles, *Applied Optics*, Vol. 40, Issue 30, pp. 5280-5294.

[3] Liu Z-S. et al 2009: Iodine-filter-based high spectral resolution lidar for atmospheric temperature measurements, *Optics Letters*, Vol. 34, Issue 18, pp. 2712-2714.

[4] Razenkov Ilya et al, 2015: Atmospheric temperature profile measurements performed by the University of Wisconsin-Madison High Spectral Resolution Lidar, Conference papers *ILRC27*.

[5] Razenkov Ilya et al, 2017: University of Wisconsin-Madison High Spectral Resolution Lidar, Conference papers *ILRC28*.

HIGH SPECTRAL RESOLUTION LIDAR AT THE UNIVERSITY OF WISCONSIN-MADISON

Ilya I. Razenkov, Edwin W. Eloranta

University of Wisconsin-Madison, 1225 W. Dayton St., Madison, WI, 53706, USA, razenkov@wisc.edu

ABSTRACT

This paper describes the modifications done on the University of Wisconsin-Madison High Spectral Resolution Lidar (HSRL) that improved the instrument's performance. The University of Wisconsin HSRL lidars designed by our group at the Space Science and Engineering Center were deployed in numerous field campaigns in various locations around the world. Over the years the instruments have undergone multiple modifications that improved the performance and added new measurement capabilities such as atmospheric temperature profile and extinction cross-section measurements.

1 INTRODUCTION

High Spectral Resolution Lidars developed in the University of Wisconsin-Madison have proven to be eye-safe and capable of continuous operation in remote areas requiring low maintenance. In the last 8 years our group has developed and built five HSRL instruments in four different configurations. The instruments were successfully deployed in 18 field campaigns around the globe including remote locations such as McMurdo Station in Antarctica, and Barrow, Alaska. This paper describes the latest version of our most advanced HSRL lidar.

2 HIGH SPECTRAL RESOLUTION LIDAR SYSTEM

2.1 Lidar transmitter

The HSRL technique uses spectral information from the lidar returns which requires a narrow high spectral resolution laser (<100MHz) and ~10 GHz laser frequency tuning capability [1]. The transmitter consists of the injection seeded diode pumped Nd:YAG laser from Photonics Industries Inc. The laser emits two wavelengths 532 nm and 1064 nm and operates at 4000 Hz pulse repetition rate. The single mode DFB seed laser diode (1064 nm) from QD-Lasers is used to injection-seed the host laser cavity. A custom controller was developed to provide a high frequency stability, smooth tuning, and good spectral purity of the laser. The host laser cavity end mirror is mounted on the piezo actuator and controlled by another custom-built controller. It adjusts the cavity length every four laser shots and tunes the cavity to the wavelength of the seed laser. The laser is tunable over 1 nm range at 1064 nm wavelength. The seed laser is injected into the host laser through a fast fiber-optic switch (~200 ns switching time) with two outputs. The switch is programmed to direct the light to the laser cavity ~10 ns before the host laser fires and switches off immediately after the laser pulse. The switch is implemented to avoid the amplification of the seed laser light. All the remaining time the fiber-optic switch directs the seed laser light to a second output, where the 1064 nm light is frequency doubled for use in the instrument calibration. The host laser cavity has a partially reflective mirror and an external KTP frequency doubling crystal mounted after the laser output coupler. KTP converts ~30% of the 1064nm light into the visible 532 nm light with the output power ratio ~700mW@532nm : 1400mW@1064nm.

The seed laser is frequency locked to the iodine absorption line #1109 using a frequency locking setup. The setup uses a 100 mm long iodine absorption cell and two energy monitors. The computer controlled DAQ applies voltage to the seed laser to control the frequency and does a slow frequency tuning of the laser. The laser cavity controller scans the host laser cavity end-mirror optimizing the cavity length to the resonant frequency, and measures the iodine line slope using the energy monitors' ratios (see Figure 1). The zero-slope corresponds to the peak absorption of the iodine line. The slope information is transferred to the computer providing an active feedback for frequency control. That allows the frequency locking accuracy of around ±15 MHz. The lidar transceiver parameters are summarized in Table 1.

Table 1 HSRL parameters

Transmitter	Laser Photonics Ind. Inc.	DS10-532/1064 SLM
	Wavelength	532/1064 nm
	Transmitted eye-safe power	0.2 Watt@532 0.6 Watt@1064
	Spectral width	<50MHz
	Spectral purity	1:5000
	Locking accuracy	30 MHz
	Pulse repetition rate	4 kHz
	Telescope Aperture	400 mm, Shared with receiver
	Transmitted beam divergence	~2 μrad
Receiver	Scanning telescope	400 mm diam. Dall-Kirkham
	Narrow field-of-view	100 μrad
	Wide field-of-view	1 mrad
	Spectral bandwidth	8 GHz
	Iodine filters (line #1109)	1.8 GHz (narrow iodine filter) 2.85 GHz (Argon broadened)
	Detectors	Geiger Mode APD SPCM-AQR-H-12

2.2 Optical design

In our HSRL instrument the transmitter and the receiver are located on opposite sides of the breadboard. The temperature difference of the breadboard skins can deform it and deflect the laser beam. Locating the optics on both sides of the breadboard compensates these beam deflection errors. The breadboard is positioned horizontally. The part of the transmitter's optics, along with frequency control optics and interferometer, are on the bottom side of the breadboard, and the receiver is on the top. The laser is mounted under the breadboard using a kinematic mount. Figure 1 shows the complete optical diagram of the High Spectral Resolution Lidar.

Some portion of the laser light is picked off from the beam by a wedge for locking setup, laser spectrum quality control and the interferometer. Then light passes through the glan polarizer and is expanded by the 10x beam-expanding telescope. The broadband non-polarizing beam splitter (R/T=65/35S) is used as a transceiver switch. Because the beamsplitter substrate is a 30 arcmin wedge, a compensating wedge is used to correct for beam's chromatic deflections due to glass dispersion at 532 and 1064 nm. The beamsplitter enables a detection of both parallel- and cross-components of the lidar returns and allows to alternate the polarization of the transmitted light pulses using a Pockels cell in the transmitter. It improves calibration of the cross-polarized channel and simplifies the optics alignment.

An achromatic half-wave plate is mounted on the telescope motor with a 1:2 gear reduction mechanism. It is used to compensate for the rotation of a polarization plane during telescope rotation. Then, the beam is further expanded by 20x Dall-Kirkham telescope (400 mm aperture) and transmitted into the atmosphere. The telescope is shared between the transmitter and the receiver. The telescope is mounted on a stepper motor, which enables elevation angle scanning capability. The polarization of the transmitted light pulses can be alternated, that allows a detection of oriented ice crystals.

The received light is focused by an apochromatic lens (EFL = 150 mm) into the field stop with an optical scrambler behind it. The optical scrambler is a 25mm long straight multimode optical fiber (300 μm core), which homogenizes the intensity distribution of the received light. Then, the second apochromatic lens collimates the light, and the 1064 and 532 nm lidar returns are separated by a dichroic beamsplitter. The background noise from the sun is rejected by interference filters (0.35 nm FWHM @ 532 nm and 1nm FWHM @ 1064 nm) and Fabry-Perot etalons (FWHM=7.5 GHz, FSR=375 GHz, Tpeak = 76%, Fe = 49 @ 532nm, FWHM=7GHz, FSR=207 GHz, Tpeak = 94%, Fe = 28 @ 1064nm). The 532nm channels use pressure tuned Fabry-Perot etalons, and the 1064 nm channel has a solid etalon (fused silica glass) mounted in the temperature controlled kinematic mount. The etalon thickness is trimmed to match the etalon peak transmission at 30 C to the wavelength, corresponding to a doubled wavelength of the #1109 iodine absorption line.

The 20 % of the received signal in HSRL channel is reflected by 20/80 beamsplitter to the combined channel, which detects photons scattered by aerosols and molecules. The remaining portion of light passes through the narrow iodine filter (FWHM 1.8 GHz) where the aerosol signal is rejected. Then 28% of that light is reflected into the first molecular channel and the remaining light passes through the iodine filter with argon buffer

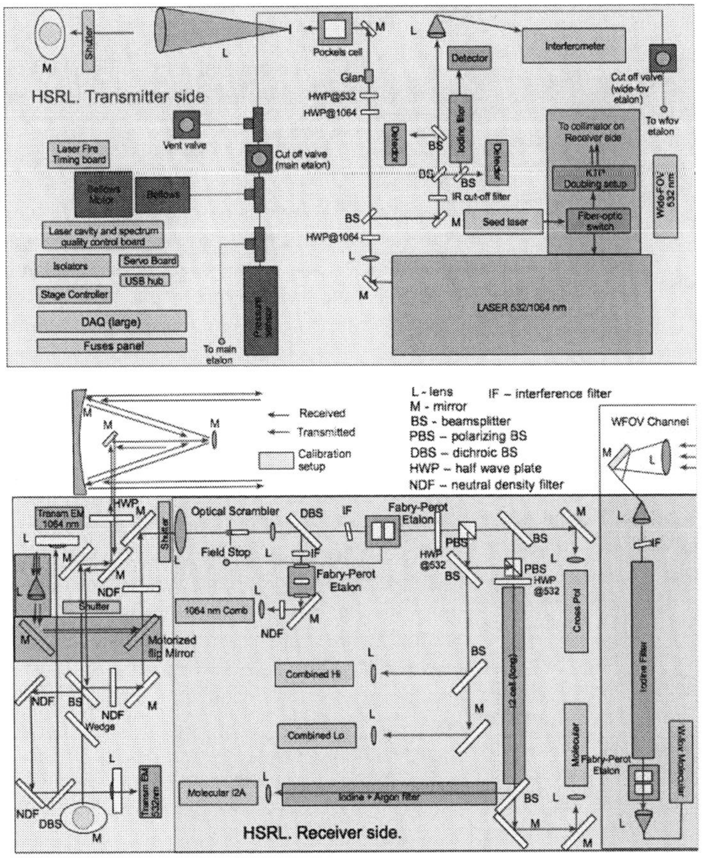

Figure 1 Schematic of the HSRL system: Top figure is the Transmitter side (bottom side of the optics breadboard); Bottom figure is the Receiver side (top side of the optics breadboard).

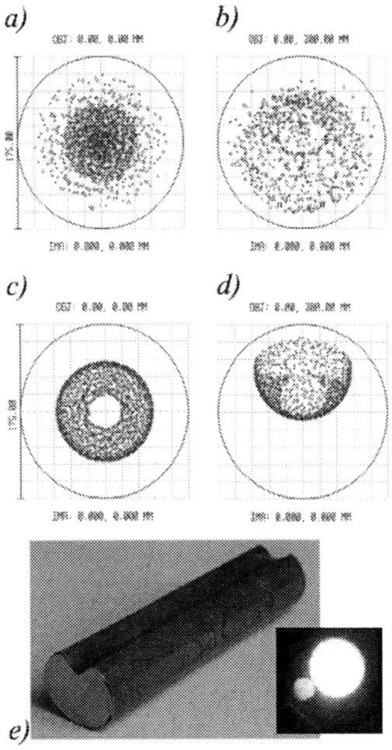

Figure 2 Zemax model for the image of point sources on the sensor at 2.5km. Figures a) and b) are a case with scrambler for on-axis and 200 mm off-axis point sources correspondingly. Figures c) and d) are a case without scrambler. A circle on the plots represents a 175 μm sensor scale. Figure 2e is the photo of the optical scrambler.

gas (FWHM 2.85GHz). The second Iodine absorption filter buffered with Argon enables atmospheric temperature profile measurements. HSRL temperature profiling is described in work [2].

When the polarization switching is enabled, parallel and cross-polarized channels are alternated depending on the polarization state of transmitted light. Due to two polarization-splitting cubes in the receiver, molecular channels always receive 80% of the signal in any mode.

The instrument is also equipped with a wide-field-of-view (WFOV) channel at 532 nm. A WFOV folded telescope is mounted on the secondary mirror support structure of the large telescope and is coaxial with the transceiver. An aspherical lens f=100 mm with f#=2 and a multimode fiber (105 μm core) are used to construct that telescope. The fiber delivered signal is filtered with an interference filter, Fabry-Perot etalon with the same parameters as the main receiver and the iodine filter. The light is further coupled with a multimode fiber to the APD detector. The WFOV channel enables direct measurement of the instrument overlap function for the narrow-field-of-view channels. The measured overlap function is important for atmospheric extinction cross-section measurements.

2.3 Overlap function improvement for 1064nm channel

The HSRL instrument operates at two wavelengths and is capable of measuring volume backscatter coefficient and color ratio for 532 nm and 1064 nm wavelengths. However, these measurements require accurate overlap functions for both wavelengths. In the previous instrument design each of the wavelengths had a separate field-stop, and, as a result, different overlap functions. Due to the difference in optics between the channels, that scheme caused large difference (up to 50%) in overlap functions between 532 and 1064 nm channels. In order to eliminate overlap function errors, we implemented apochromatic lenses and a shared field-stop in the receiver for all narrow-field-of-view channels.

2.4 Optical scrambling

The intensity distribution in the transmitted beam's cross-section is Gaussian. Due to a small field-of-view of the instrument, any shifts in the intensity distribution and possible non-uniform detector sensitivity across the sensor area appear in the signals as changes in the overlap function. Moreover, these changes can produce a difference in overlap functions between the channels sharing the same field stop. This is essential for atmospheric temperature profile measurements with HSRL [2]. In order to eliminate the differential overlap effect, we implemented a 25 mm long section of the multimode fiber (300 μm core diameter) mounted behind the field-stop. The straight glass rod homogenizes the received beam by scrambling rays while preserving the polarization information in the signal. That allows the atmospheric depolarization coefficient measurements.

The multimode fiber is glued to a stainless steel rod with v-groove using a silicon rubber glue to avoid mechanical stress of the glass rod. In the first experimental samples we used epoxy, which caused stress and the polarization contrast decrease from 1:1000 (0.1%) to 40:100 (40%). With the silicon rubber glue the scrambler samples had depolarization as low as 2:1000 (.2%). For optics alignment purposes, a single mode fiber is glued in parallel to the multimode fiber. The single mode fiber has a length matching to the multimode one, and the fibers' axes skewness is around 200 μrad. The scrambler is mounted on a translation stage so that when shifted laterally, in order to substitute the 3 μm core single mode fiber for the 300 μm, the fiber is used for lens collimation.

2.5 Calibration scans

The HSRL instrument uses an injection seeded Nd:YAG laser that produces two wavelengths. Originally, the seed laser operated in CW mode. After the laser firing, the seed laser light was amplified by the YAG crystal between the shots. This amplified light was saturating the IR channel detector. To avoid that, the seed laser was switched off after each laser shot and turned on around 80 μs before the shot with simultaneous gating of the IR channel detector. That limited the maximum range of this channel and also produced a laser frequency chirp during the laser firing. In order to overcome these problems, a fast fiber-optic switch was added to the instrument. One of its outputs is connected to the host laser, and the second output is connected to the KTP assembly partially converting 1064 nm light into 532 nm. The multimode fiber delivers both wavelengths to the fiber collimator. A mirror mounted on the motorized flip mount directs the light to the receiver. That calibration setup is marked with pink boxes on Figure 1. For the instrument calibration, seed laser frequency scans across the receiver bandwidth are performed in order to measure the channels' transmission spectrum functions. The 95% duty cycle light pulses (~237 μs pulse width) of the frequency doubled seed laser light (532nm) along with the large signal levels significantly improved the calibration accuracy. It decreased photon counting error by ~100 times and the uncertainty in the laser output frequency. Previously, we used short ~50 ns -wide pulses from the host laser at 4kHz pulse repetition rate. The newly measured channels' transmission functions do not require any adjustments for standard HSRL measurements.

ACKNOWLEDGEMENTS

This work is supported by the Space Science and Engineering Center U. of Wisconsin – Madison.

References

[1] Eloranta, E. W.: High Spectral Resolution Lidar, *in Lidar: Range-Resolved Optical Remote Sensing of the Atmosphere,* Klaus Weitkamp ed., Springer, NY, pp. 25-30.

[2] Razenkov Ilya et al, 2015: Atmospheric temperature profile measurements performed by the U. of Wisconsin – Madison High Spectral Resolution Lidar, *ILRC27.*

EPJ Web of Conferences **176**, 01025 (2018)
ILRC 28
https://doi.org/10.1051/epjconf/201817601025

HOW TO APPLY THE OPTIMAL ESTIMATION METHOD TO YOUR LIDAR MEASUREMENTS FOR IMPROVED RETRIEVALS OF TEMPERATURE AND COMPOSITION

R. J. Sica[1,2,*], A. Haefele[2,1], A. Jalali[1], S. Gamage[1] and G. Farhani[1]

[1]*Department of Physics and Astronomy, The University of Western Ontario, Canada,*
**sica@uwo.ca*
[2]*Federal Office of Meteorology and Climatology, MeteoSwiss, Payerne, Switzerland*

ABSTRACT

The optimal estimation method (OEM) has a long history of use in passive remote sensing, but has only recently been applied to active instruments like lidar. The OEM's advantage over traditional techniques includes obtaining a full systematic and random uncertainty budget plus the ability to work with the raw measurements without first applying instrument corrections. In our meeting presentation we will show you how to use the OEM for temperature and composition retrievals for Rayleigh-scatter, Raman-scatter and DIAL lidars.

1 INTRODUCTION

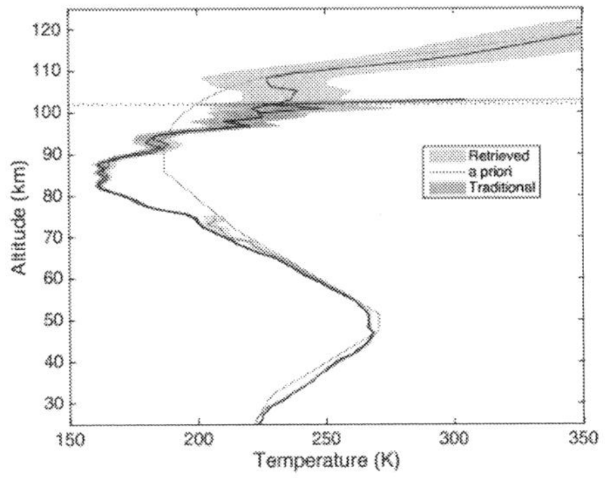

Figure 1: Two channel HSEQ temperature retrieval from Purple Crow Lidar measurements on 24 May 2012 (red curve) compared to the traditional method.

Rayleigh-scatter lidars are one of the best sources for temperature measurements in the middle atmosphere. Cooling in the middle atmosphere associated with warming in the lower atmosphere is an important measurement in assessing atmospheric change, as, although it is still complex, interpreting the middle atmospheric temperatures is simpler than interpreting changes in surface temperature. However, the magnitude of the changes is small, on the order of $\sim 1^\circ$ per decade. Thus, it is critical to have available an analysis technique that can perform a full uncertainty budget on a profile-by-profile basis. The OEM is a great choice for this application, and we applied it successfully to the Western Purple Crow Lidar [1]. Figure 1 shows a 2-channel temperature retrieval from Purple Crow Lidar measurements on 24 May 2012 (red curve). The figure also shows temperatures calculated using the traditional method of Hauchecorne and Chanin for the low-gain (green) and high-gain (blue) data channels [2]. The *a priori* temperature profile used for the retrieval is the U.S. Standard Atmosphere (cyan curve). The horizontal dotted line is the height above which the *a priori* temperature profile begins to make a significant contribution to the retrieval.

Figure 2 shows the uncertainty budget for the 2-channel temperature retrieval. The retrieval determines the statistical uncertainty (blue line), in addition to the systematic uncertainties: tie-on pressure (orange stars), ozone density (yellow +), ozone cross section (purple *), density

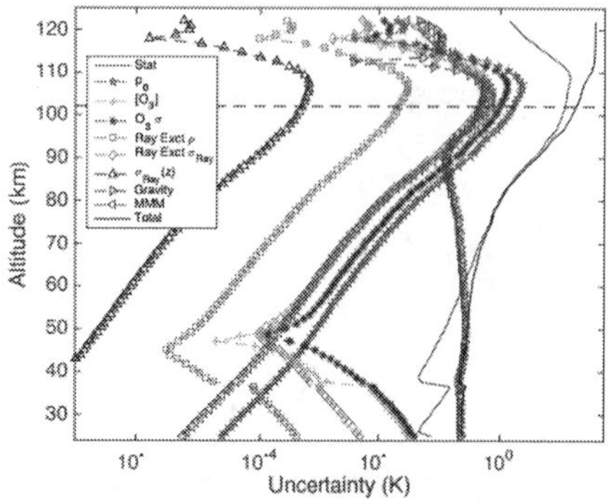

Figure 2: Uncertainty budget for the temperature retrieval in Figure 1. Details are given in the text.

profile for Rayleigh extinction (green squares), Rayleigh extinction cross section (blue diamonds), variation of Rayleigh-scatter cross section with height (red triangles), gravity model (blue triangles), variation of mean molecular mass with height (orange triangles) and total uncertainty (black line). The horizontal dashed line is the height above which the *a priori* temperature profile makes a significant contribution to the retrieval.

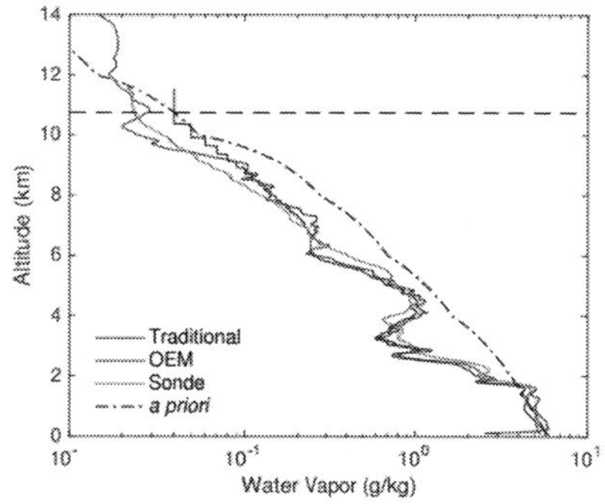

Figure 3: Four channel temperature retrieval of water vapour from the MeteoSwiss RALMO lidar on 5 September 2009 (red curve) compared to the traditional method.

After success with temperature retrieval, we applied the OEM to the retrieval of water vapor using measurements from the MeteoSwiss RALMO lidar [3]. The RALMO water vapor measurements, in terms of data quality and calibration, are among the best available in the world. In addition, ancillary instruments such as radiosondes and microwave radiometers are available in Payerne for validation of the RALMO measurements. Figure 3 shows the retrieved water vapor mixing ratio (red curve) using the OEM on 5 September 2009. The blue curve is the mixing ratio using the traditional analysis method. The green curve is the radiosonde measurement. The sonde is launched at the start of the 30 min RALMO average. The dot-dashed line is the *a priori* mixing ratio profile used by the OEM. The horizontal dashed line shows the height below which the retrieval is due primarily to the measurement and not the *a priori*.

In this paper we will discuss two technical aspects of applying OEM to lidar measurements, calculation of analytic forms for the Jacobians and practical considerations for working with non-linear counting systems.

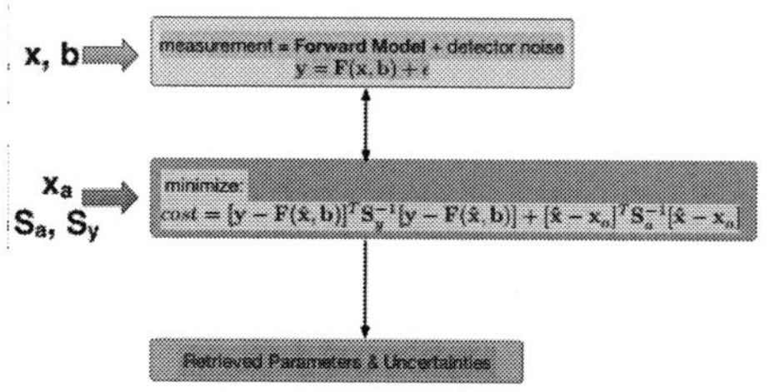

Figure 4: The OEM procedure. The retrieval is computed from an a priori state **a** *of the retrieval vector* **x**, *model parameters,* **b** *and covariances* **S** *using measurements,* **y**.

2 FORWARD MODELS

Figure 4 shows the basics of the Optimal Estimation Method [3]. Our retrievals are "first-principle retrievals;" that is, our forward model

(FM) includes all the instrumental and atmospheric parameters necessary to reproduce the raw measurements.

Our forward models are based on the lidar equation. For Rayleigh-scatter temperature retrievals we use the following form of the lidar equation for the true counts

$$N_t(z) = C\mathbb{T}^2 \frac{\psi(z)}{z^2} \frac{p(z)}{kT(z)} + B_t \tag{1}$$

where in this compact form all the instrumental parameters and constants are in the function ψ, except for the lidar constant C which is explicitly shown as it is often a retrieved quantity. The atmospheric transmission is $\mathbb{T}$. The true background, B_t, can be constant or the analytical form of the background appropriate to a given system. The pressure, $p(z)$ can be specified or computed from the temperature using the assumption of hydrostatic equilibrium [1].

Water vapour mixing ratio retrievals require a more complex set of equations, as many water vapour lidars will have 4 channels (2 digital and 2 analog) for nitrogen (N) and water vapour (H). For the linear case, applicable to analog systems or low-gain photomultipliers, the true counts are given by:

$$\begin{aligned} N_H &= \frac{O\mathbb{T}_L\mathbb{T}_H C_H n_{air}}{z^2} e^{q'} + B_H \\ N_N &= \frac{O\mathbb{T}_L\mathbb{T}_N C_N n_{N_2}}{z^2} + B_N \\ q' &= \ln q. \end{aligned} \tag{2}$$

Here the overlap function, O, varies with altitude, n is number density and the transmission reflects the inelastic scattering process. A log retrieval is used as water vapour mixing ratio, q, cannot be negative.

3 PRACTICAL JACOBIANS

The derivative of the FM, F, with respect to the retrieval vector, $\mathbf{x}$ is called the kernel, $\mathbf{K}$. The kernel is a $m \times k$ matrix whose elements are:

$$K_{ij} = \frac{\partial F_i(\mathbf{x})}{\partial x_j}. \tag{3}$$

Since $\mathbf{K}$ is a matrix of derivatives it can also be called the Jacobian (which is the term we will use, although in atmospheric science $\mathbf{K}$ is sometimes called the weighting function).

The size of the Jacobian, that is, the number of retrieval parameters k, in relation to the number of independent measurements m, relates to the regularization of the problem. When $m < k$ the problem is ill-posed (under-determined); when $m > k$ the problem is over-constrained (over-determined). Our lidar inversions have been restricted to the over-constrained situation.

To calculate the Jacobians, consider a retrieval of temperature from Rayleigh-scatter measurements for a system with 2 detection channels, which could be analog, digital or a combination of both and at different height resolutions. The data vector has m detector samples, y, defined at l_1 heights for the first channel and l_2 heights for the second channel, where $l_1 + l_2 = m$. We want to retrieve $x(k)$ quantities consisting of, for example, temperature at some number of retrieval heights $\leq k$, dead time(s) and background(s) depending on the number of channels.

The m columns of the Jacobian contain the sensitivity of a measurement to the retrieval vector. Parameters are retrieved on the retrieval grid, using the measurement vector specified on the data grid.

To visualize this consider a measurement vector comprised of a single channel of photocounts from which you want to retrieve temperature. The data grid is then a series of photocount measurements as a function of time (height) as shown in Figure 5.

For our specific case we retrieve a temperature at each height on the retrieval grid. Since a temperature has to be specified at all levels of the

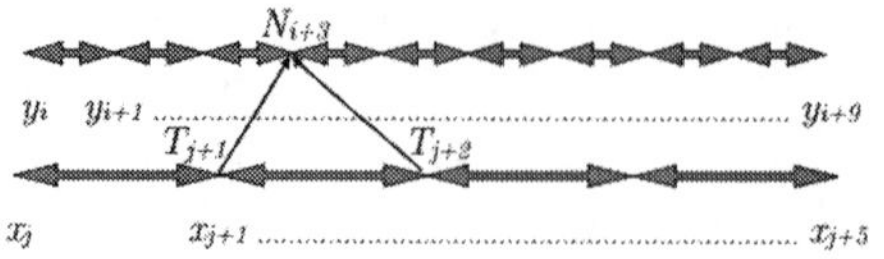

Figure 5: Interpolation of a retrieved parameter (here temperature, T) to photocounts, N, on the data grid.

data grid to evaluate the lidar equation, the FM performs an interpolation of temperature from the retrieval grid to the data grid. We choose to use linear interpolation. Linear interpolation is a reasonable choice for these retrievals as the both the data and retrieval grid spacing is much less than an atmospheric scale height. Using the Chain Rule we write the temperature Jacobian as:

$$\begin{aligned} K_{T_{ij}} &= \frac{\partial F_i(T)}{\partial T_i} \cdot \frac{\partial T_i}{\partial T_j} \\ &= \frac{\partial F_i(T)}{\partial T_i} \cdot \left(\frac{z_{j+1} - z_i}{z_{j+1} - z_j} \right) \end{aligned} \tag{4}$$

where the second term on the right hand side of Eq. (4) is due to the interpolation of the temperature on the retrieval grid to the data grid.

3.1 Analytical Derivatives

It is often not possible to calculate the analytical derivative of a model or retrieval parameter in the FM, in which case a numerical derivative can be calculated. Typically this derivative can be calculated using a simple finite difference scheme. However, if the analytical derivative can be determined, its exact form is a better choice than the numerical derivative. Some of the analytical derivatives are quite simple, such as for the lidar constant or a constant background. Others are not possible, such as the temperature dependence of the Rayleigh-scatter FM under the assumption of hydrostatic equilibrium, where T additionally appears in the integral used to determine pressure.

One Jacobian which can be determined analytically is the derivative of the FM with respect to density. Consider the simpler case where optical depth (or transmission) is not being retrieved, so transmission is specified on the data grid and an interpolation is not required. From Eq. (1) we see, using the Chain Rule, that the Jacobian of the FM with respect to density is:

$$\frac{\partial N_{t,\ell}}{\partial n_i} = \frac{\partial N_{t,\ell}}{\partial \mathbb{T}_\ell} \times \frac{\partial \mathbb{T}_\ell}{\partial n_i} \tag{5}$$

where ℓ and i range from 1 to m to form a $m \times m$ matrix. For discrete measurements the optical depth, τ, is given by

$$\tau_\ell = \sum_{i=1}^{\ell} \sigma n_i \Delta z \tag{6}$$

for equally spaced measurements and a constant cross section with altitude, σ, as in the case of water vapour and temperature retrievals. Note that τ can be a specific optical depth, e.g. aerosol optical depth, as required. The transmission is then

$$\begin{aligned} \mathbb{T}_\ell &= \exp\left(-\sum_{i=1}^{\ell} \sigma n_i \Delta z \right) \\ &= \prod_{i=1}^{\ell} \exp(-\sigma n_i \Delta z), \end{aligned} \tag{7}$$

where from the product above we see that the transmission Jacobian is:

$$\begin{aligned} \frac{\partial \mathbb{T}_\ell}{\partial n_i} &= -\sigma \mathbb{T}_\ell \Delta z \text{ for } i \leq \ell \\ &= 0 \text{ for } i > \ell. \end{aligned} \tag{8}$$

We can now calculate the Jacobian with respect to density using Eqs. (5) and (8):

$$\begin{aligned} \frac{\partial N_{t,\ell}}{\partial n_i} &= -2(N_{t,\ell} - B_{t,\ell}) \sigma \Delta z \text{ for } i \leq \ell \\ &= 0 \text{ for } i > \ell. \end{aligned} \tag{9}$$

This form of the Jacobian is fast computationally and avoids any numerical issues with the exponential quantities involved.

3.2 Effect of Detector Saturation

The FMs given in Eqs. (1) and (2) refer to the *true* count rate. Systems which measure in the daytime or have large dynamic range may have signals which are significantly nonlinear due to limitations in the detection system. For instance, for daytime water vapour retrievals the background count rate is extremely large and the observed background counts are not equal to the true (corrected) counts. Furthermore, in the specification of the parameters for the retrievals, the background term is estimated using the observed counts, not the true counts, so this difference must be accounted for in all quantities in the FM. Consider a non-paralyzable system where the observed counts are related to the true counts by

$$N_o = \frac{N_t}{1+\gamma N_t} \tag{10}$$

where γ is the counting system dead time. The derivative of the observed counts with respect to the true counts is then

$$\frac{\partial N_o}{\partial N_t} = (1-\gamma N_o)^2. \tag{11}$$

To apply our previous result for the Jacobian of the FM with respect to density for the nonlinear case, we must find the derivative of the observed photocounts with respect to n, using the observed photocounts N_o and the observed background, B_o. Using the Chain Rule and Eq. (9) we can show that

$$\frac{\partial N_{o,\ell}}{\partial n_i} = \frac{\partial N_{o,\ell}}{\partial N_{t,\ell}} \cdot \frac{\partial N_{t,\ell}}{\partial n_i} = (1-\gamma N_{o,\ell})^2 \cdot \left[-2\left(\frac{N_{o,\ell}}{1-\gamma N_{o,\ell}} - B_{o,\ell}\right)(\sigma\Delta z)\right]. \tag{12}$$

4 CONCLUSIONS

While each different type of scattering process requires a different FM, many of the retrieval and model parameters are the same. We demonstrated two tricks common to all retrievals, one an efficient analytical form for the transmission density Jacobian and the other inclusion of detector nonlinearity in the Jacobians. Currently, we are in the processes of developing general retrieval tools for use by the community in applying the OEM, which incorporate these results as well as some other finer points of the retrieval mechanics we have learned in developing these techniques.

ACKNOWLEDGEMENTS

We thank C. Dennison for advice on the calculation of the transmission Jacobian. This project has been supported by the Natural Sciences and Engineering Research Council (NSERC Discovery Grant & CREATE), Canadian Space Agency (CSA) and MeteoSwiss.

References

[1] Sica, R. J., and A. Haefele, 2015: Retrieval of temperature from a multiple-channel Rayleigh-scatter lidar using an optimal estimation method, *Appl. Opt.* **54** (8), 1872-1889.

[2] Hauchecorne, A., and M. Chanin, 1980: Density and temperature profiles obtained by lidar between 35 and 70 km, *Geophys. Res. Lett.*, **7**(8), 565-568.

[3] Sica, R. J., and A. Haefele, 2016: Retrieval of water vapor mixing ratio from a multiple channel Raman-scatter lidar using an optimal estimation method, *Appl. Opt.* **55** (4), 763-777.

[4] Rodgers, C. D., 2011: Inverse Methods for Atmospheric Sounding: Theory and Practice, *World Scientific*, Singapore.

WATER-VAPOUR MEASUREMENTS UP TO THE LOWER STRATOSPHERE — THE HIGH POWER RAMAN LIDAR AT THE SCHNEEFERNERHAUS

Lisa Klanner, Thomas Trickl and Hannes Vogelmann

Karlsruher Institut für Technologie, IMK-IFU, Garmisch-Partenkirchen, Germany
E Mail: thomas.trickl@kit.edu

ABSTRACT

A high-power Raman lidar system has been developed at the high-altitude research station Schneefernerhaus (Garmisch-Partenkirchen, Germany) at 2675 m, at the side of an existing differential-absorption lidar. It is based on a 180-W single-line XeCl laser and on two Newtonian telescopes (up to 1.5-m-diameter). In this way a vertical range up to more than 20 km and an accuracy level of the order of 10 % can be achieved for a measurement time of 1 h. Temperature measurements have been demonstrated to altitudes up to 54 km with just 1 % of the full 308-nm backscatter signal. Significantly higher altitudes are expected when using a chopper that cuts off the first 10 km or for 353 nm.

1 INTRODUCTION

Lidar measurements of water vapour in the climate-relevant upper troposphere (UT) and lower stratosphere (LS) are strongly motivated by the high spatial and temporal variability of this constituent [1, 2] that necessitates a reasonable time resolution. However, LS mixing ratios of the order of 5 ppm mean a considerable challenge for lidar sounding, given the additional burden of a backscatter signal strongly diminishing with altitude. During the past decade lidar measurements have been extended into the LS by groups of the Network for the Detection of Atmospheric Composition Change (NDACC, http://www.ndsc.ncep.noaa.gov/). Due to negligible light absorption in the troposphere the Raman lidar method has been preferred. A range up to about 20 km with uncertainties of the order of 20 % has been demonstrated for long averaging [2, 3]. This is an important step towards filling the existing gap in frequent accurate, vertically well-resolved measurements between 10 and 20 km.

In contrast to the DIAL method, Raman backscattering is background free during night-time and the concentrations can be directly related to the signal level. The quality of the H_2O data grows with the number of photons collected. As a consequence we have tried to improve the specifications by developing a Raman lidar system with roughly 50 times more backscatter signal than available in existing systems. More reliable results can be obtained in a substantially shorter data acquisition time. The new lidar system yields an ideal extension of the measurements with our differential-absorption lidar (DIAL) that provides accurate water-vapour profiles in most of the free troposphere [1,2,5-8]. Both systems are located in the same laboratory at the Schneefernerhaus high-altitude station (UFS) at an altitude of 2675 m.

2 SYSTEM DESCRIPTION

The details of the UFS Raman lidar were described in more detail in our previous ILRC contributions [9,10]. The entire set-up is shown in Fig. 1. Here, we just briefly summarize the most important properties.

2.1 Laser

The most powerful radiation sources in the ultraviolet are excimer lasers. Thus, a Lambda SX XeCl laser (Coherent, formerly Lambda Physik) with up to 1.2 J pulse energy at 308 nm and 350 Hz repetition rate was chosen. Since this laser system is normally used for industrial production we had to modify the optical layout to obtain linearly polarized single-line output with strongly reduced beam divergence. Under these conditions the peak average output power is 180 W.

The laser has been operated on its high-frequency component (λ = 307.955 nm in vacuum, $\Delta\lambda$ = 0.04 nm). All other contributions are very small. The spectral impurity mostly stays below 1 % at the highest repetition rates. The wavelength is controlled to within ±0.02 nm with a calibrated grating spectrometer (Ocean Optics, model HR4000).

After expanding the laser beam 35×35 mm^2 square it is focussed into a Raman shifter 3.6 m long with an f = 2 m lens. The Raman cell is filled with about 30 bar of hydrogen for generating a reference emission at 353.1 nm needed for ozone corrections and temperature measurements up to the mesosphere. The focal length will now be shortened to 1.75 m because of the frequently negligible conversion at high repetition rates.

From the focal point the beam expands towards an f = 10 m concave mirror used for collimation This means an overall beam expansion by a factor of 5.7 and a reduction of the beam divergence to less than 0.5 mrad, necessary for sufficiently tight focussing of the backscattered light in the two receivers.

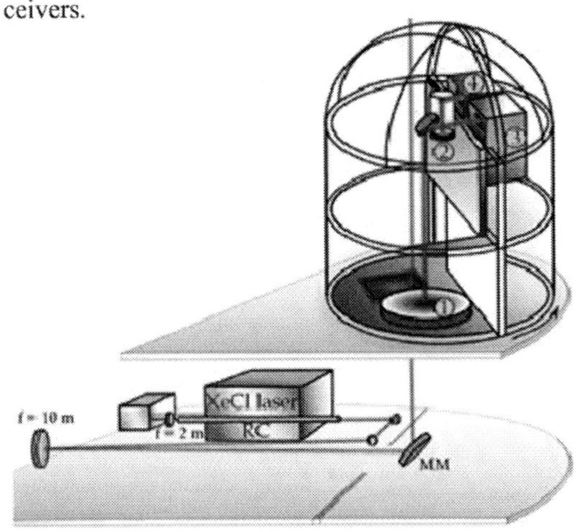

***Figure 1** (From Ref. 10) The radiation from both telescopes (1,2) is focussed into two six-channel polychromators shown in Ref. 9 (3, 4). A combination of polarization-sensitive optics and interference filters (Laseroptik G.m.b.H. and Materion Barr). The spectral widths are 0.25 nm (f.w.h.m.) for all channels except for H_2O where a 0.75-nm filter with 70 % transmittance is used. RC: Raman cell; MM: beam-steering mirror*

2.2 Receiver

Two separate Newtonian receivers for near-field and far-field detection are used (d = 0.38 m, f = 2 m, and d = 1.5 m, f = 5 m, respectively). Because of the long focal length of the large mirror the receivers are mounted in a tower on the terrace above the lidar laboratory (Fig. 1), covered by an astronomical dome suitable for the arctic conditions at the mountain site (Baader).The radiation is detected with Hamamatsu R7400 P-03 photomultiplier tubes (PMTs) with actively stabilized sockets yielding single-photon pulses without ringing (Romanski Sensors). The signal is processed by Licel 12-bit/20-MHz transient digitizers with a new ground-free input stage and a 5-GHz photon counting system (FAST ComTec).

3 RESULTS

3.1 Water Vapour

Results of the current testing period will be presented at the conference. During the past testing period in 2015 already a major step forward was achieved. The polychromator of the large telescope was shielded more carefully and, thus, the background light level could be reduced by a factor of 50 with respect to Ref. 10, to 150 counts per 15-m bin (Fig. 2). Although this is, still, far away from the targeted 0-2 counts per bin (see Secs. 3.2 and 3.3), but allowed us to judge the achievable performance in the stratosphere.

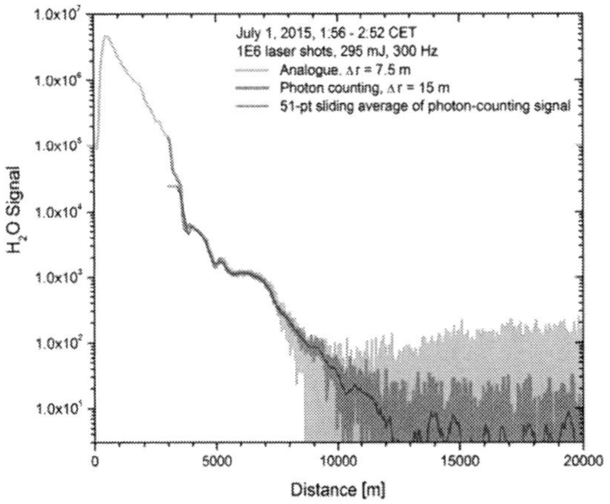

***Figure 2** H_2O backscatter signal from July 1, 2015 obtained with 1×10^6 laser shots (about 1 h) and just 90 W of laser power*

In Fig. 2, the backscatter signal for a water-vapour measurement shortly before dawn is shown. The peak analogue signal, measured with 25 mV full scale, was just about 3 mV, after a continuous drop in humidity until midnight. The noise amplitude of the photon-counting signal (adjusted to match the analogue signal) corresponds to about 1.6 nV. The noise of the analogue signal (7.5-m bins) is higher by one order of magnitude, which is excellent given the small signal voltages. A small electronic undershoot is seen around 11.5 km that is caused by the photon-counting discriminator. As one can see from the averaged signal the useful range in this case ends at 12.2 km above the lidar (14.9 km a.s.l.).

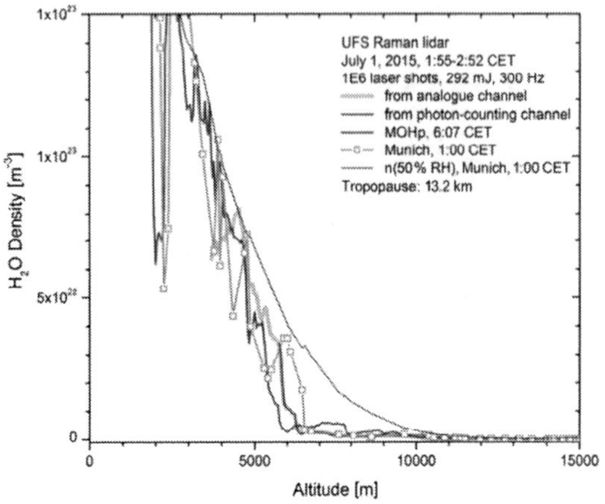

Figure 3 *H_2O density distributions obtained from the 1-h lidar measurement in Fig. 2*

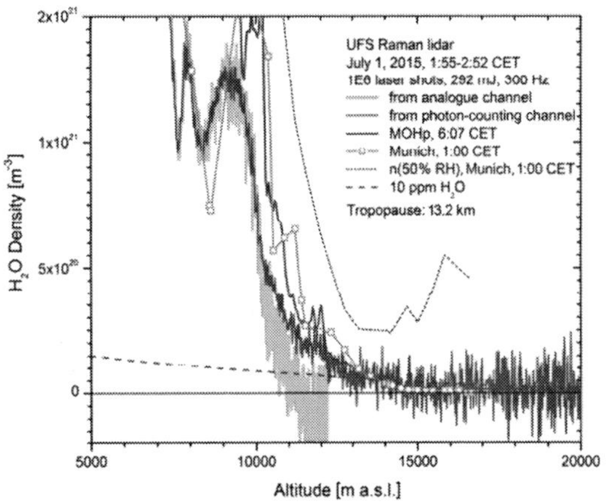

Figure 4 *Strongly expanded portion of Fig. 3, extended to 20 km: After smoothing the lidar data (not shown) mixing ratios of less than 5 ppm can be resolved to roughly 20 km.*

The water-vapour profile corresponding to Fig. 2 is shown for two different vertical scales in Figs. 3 and 4. For comparison, RS92 sonde measurements of the German Weather Service at Oberschleißheim ("Munich", 1:00 CET, distance 100 km) and (Hohenpeißenberg (MOHp), 6:07 CET, distance 38 km) are included that were used for approximate calibration of the lidar density profile. The agreement between lidar and sonde data is not good due to spatial inhomogeneity, which underlines the need for calibration with the DIAL. The DIAL has not been available recently due to a fatal laser damage.

It is interesting to note that there is reasonable agreement of the Hohenpeißenberg sonde and lidar above 12 km a.s.l. The two downward density steps coincide. The step above 14 km corresponds to a drop in (MOHp) mixing ratio from about 12 ppm to 2.5 ppm, with the well-known [6] cut-off at 1 % RH above 18.5 km. The low values around 15 km are tentatively explained by air-mass advection from the tropics (Caribbean Sea) where the UTLS can be very dry. Averaging the lidar values over ±25 data points (±375 m) yields positive values agreeing well with the sonde data up to about 21 km.

It is reasonable to assume a range extension to more than 20 km after a noise reduction by another two orders of magnitude as shown in Sec. 3.3.

3.2 Calibration

Because of the great advantage of side-by-side lidar sounding of the Raman lidar and the highly accurate [8] DIAL system a calibration can be achieved even under inhomogeneous condition. One example for a very demanding case with three dry stratospheric layers is given in Fig. 5. The agreement is acceptable despite slight differences due to a time shift between the two measurements.

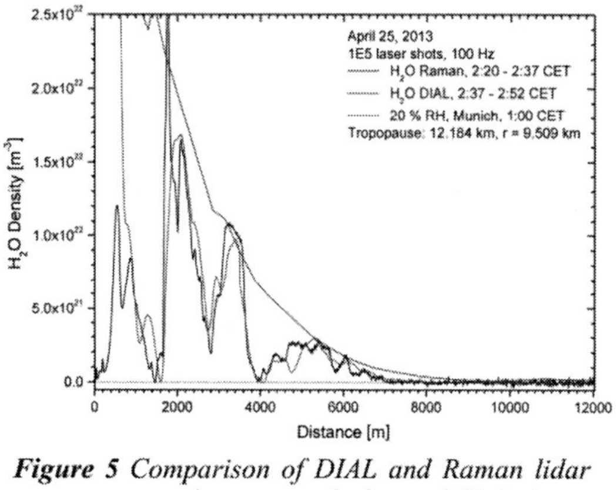

Figure 5 *Comparison of DIAL and Raman lidar measurements for a strongly layered H_2O distribution on April 25, 2013; at two altitudes zero humidity was found by both systems.*

3.3 Temperature

Temperature retrievals both from rotational Raman signals and Rayleigh profiles have been achieved by L. Klanner for her doctoral thesis (to be submitted in 2017) and by K. Höveler [11]. In the case of the measurement in Figs. 3 and 4 the 308-nm Rayleigh channel featured a background level of 0-4 counts per 15 m and hour. This resulted in an eight-decade dynamical range of the signal without smoothing, nine decades with smoothing. The air number density could be retrieved up to 65 km, the temperature up to 54 km (both profiles agreeing well with those from Hohenpeißenberg). During that measurement the backscatter signal was attenuated by more than two decades in order to avoid detector overload. After cutting off the first 8-10 km of signal with a chopper currently under development, this channel could be operated with full power, and temperature retrievals up to more than 80 km should be feasible. An even better performance is expected for the 353-nm after improving the stimulated Raman conversion of the 308 nm radiation to this wavelength (Sec. 2.1, [10]).

ACKNOWLEDGEMENTS

The authors thank H. P. Schmid for his support, W. Steinbrecht for making available data from the German Weather Service and the UFS team for their great help. The system development was funded by the Bavarian State Ministry of Environment and Health.

References

[1] Vogelmann H., R. Sussmann, T. Trickl, T. Borsdorff, 2011: Intercomparison of atmospheric water vapor soundings from the differential absorption lidar (DIAL) and the solar FTIR system on Mt. Zugspitze, *Atmos. Meas. Technol.*, **4**, 835-841

[2] Vogelmann, H., R. Sussmann, T. Trickl, A. Reichert, 2015: Spatiotemporal variability of water vapor investigated using lidar and FTIR vertical soundings above the Zugspitze, *Atmos. Chem. Phys.*, **15**, 3135-3148

[3] Whiteman D., K. Rush, S. Rabenhorst, W. Welch, W. Cadirola, G. McIntryre, F. Russo, M. Adam, D. Venable, R. Connell, I. Veselovskii, R. Forno, B. Mielke, B. Stein, T. Leblanc, S. McDermid, H. Vömel, 2010: Airborne and Ground-Based Measurements Using a High-Performance Raman Lidar, *J. Atmos. Oceanic Technol.*, **27**, 1781-1801

[4] Leblanc T., I. S. McDermid, T. D. Walsh, 2012: Ground-based water vapor Raman lidar measurements up to upper troposphere and lower stratosphere for long-term monitoring, *Atmos. Meas. Technol.*, **5**, 17-36

[5] Vogelmann H., T. Trickl, 2008: Wide-range sounding of free-tropospheric water vapor with a differential-absorption lidar (DIAL) at a high-altitude station, *Appl. Opt.*, **47**, 2116-2132

[6] Trickl, T., H. Vogelmann, H. Giehl, H. E. Scheel, M. Sprenger, A. Stohl, 2014: How stratospheric are deep stratospheric intrusions? Atmos. Chem. Phys., **14**, 9941-9961

[7] Trickl, T., H. Vogelmann, H. Flentje, L. Ries, 2015: Stratospheric ozone in boreal fire plumes – the 2013 smoke season over Central Europe, *Atmos. Chem. Phys.*, **15**, 9631-9649

[8] Trickl, T., H. Vogelmann, A. Fix, A. Schäfler, M. Wirth, B. Calpini, G. Levrat, G. Romanens, A. Apituley, K. M. Wilson, R. Begbie, J. Reichardt, H. Vömel, M. Sprenger, 2016: How Stratospheric Are Deep Stratospheric Intrusions into the Troposphere? LUAMI 2008, *Atmos. Chem. Phys.*, **16**, 8791-8815

[9] Klanner, L., T. Trickl, H. Vogelmann, 2012: On the Way to Combined DIAL and Raman-Lidar Sounding of Water Vapour on Mt. Zugspitze □ a Progress Report, pp. 853-826 in: *Reviewed and Revised Papers Presented at the 26th International Laser Radar Conference (ILRC 2012)*, Porto Heli (Greece), A. Papayannis, D. Balis, V. Amiridis, Eds., published in Greece on behalf of ICLAS

[10] Höveler, K., L. Klanner, T. Trickl, H. Vogelmann, 2016: The Zugspitze Raman Lidar: System Testing, 27th International Laser Radar Conference, New York City (U.S.A., 2015), B. Gross, F. Moshairy, M. Arend, Eds., Poster Session A2, „Lidar for Trace Gas Monitoring", A2.8, *EPJ Web of Conferences*, **199** (2016) 21004, 4 pp.

[11] Höveler, K., Entwicklung eines Nahbereichsempfängers zur Messung von Wasserdampf- und Temperaturprofilen für das Hochleistungs-Raman-Lidar am Schneefernerhaus, *Master Thesis*, Karlsruher Institut für Technologie, 129 pp., 2015 (in German)

NEW LASER DESIGN FOR NIR LIDAR APPLICATIONS

H. Vogelmann[1], T. Trickl[1], M . Perfahl[1], S. Biggel[2]

[1]*Karlsruhe Institute of Technology, IMK-IFU (Garmisch-Partenkirchen,Germany),, E-mail: hannes.vogelmann@kit.edu*
[2]*Optic Instiute Ulm, Germany*

ABSTRACT

Recently, we quantified the very high spatio-temporal short term variability of tropospheric water vapor in a three dimensional study [1]. From a technical point of view this also depicted the general requirement of short integration times for recording water-vapor profiles with lidar. For this purpose, the only suitable technique is the differential absorption lidar (DIAL) working in the near-infrared (NIR) spectral region. The laser emission of most water vapor DIAL systems is generated by Ti:sapphire or alexandrite lasers. The water vapor absorption band at 817 nm is predominated for the use of Ti:sapphire. We present a new concept of transversely pumping in a Ti:Sapphire amplification stage as well as a compact laser design for the generation of single mode NIR pulses with two different DIAL wavelengths inside a single resonator. This laser concept allows for high output power due to repetitions rates up to 100Hz or even more. It is, because of its compactness, also suitable for mobile applications.

1. INSTRUMENTS AND REQUIREMENTS

1.1. The Zugspitze DIAL

Our water vapor DIAL system is located just below Mt. Zugspitze (47.42°N, 10.98°E, 2962 m a.s.l.) which is by far the highest mountain on the northern rim of the Alps. The free troposphere above this site is representative of Central Europe. The mountain top is above the moist boundary layer for most of the year. Due to reduced absorption losses this site is ideal for sensitive spectroscopic measurements of water vapor throughout the free troposphere. Single absorption lines in the 817-nm band of H_2O are used for ground-based water vapor profiling in a vertical range from 2.95 km to roughly 12 km a.s.l.. Statistical measurement uncertainties are kept below about 5% to about 8 km by dynamically adapting the vertical resolution. The sensitivity limit is roughly 18 ppm at 10 km a.s.l. [2]. This DIAL system and the retrieval of water vapor profiles are described in more detail in [3]. Water-vapor profiles from the Zugspitze DIAL allow for retrieving IWV with a precision better than 0.1 mm [4]. For a suitable signal-to-noise ratio 10^4 laser shots of about 100 mJ have to be integrated each wavelength. At a repetition rate of 20 Hz of the current laser system, this takes almost 17 minutes overall. Our investigations [1] have shown, that an integration time of less than 5 minutes is desirable.

1.2. The mobile ATMONSYS lidar

Since 2012, a mobile scanning lidar for profiling water-vapor, aerosol and temperature in the planetary boundary layer (PBL) has been developed at IMK-IFU with in the ATMONSYS project. The new system will be used in field campaigns and can be transported to sites of special interest. Thus, the lidar is housed in a 20 feet shipping container and can easily be moved with a standard truck.

With respect to the high short-term variability in the PBL and the requirement of analyzing atmospheric short-term dynamics one major goal was to provide short integration times of the order of 10 s. Thus, a high repetition rate of 100 Hz and an infrared pulse energy of at least 50 mJ was aimed at 817 nm. Additional requirements for the ATMONSYS lidar system are robustness due to being transported on rough roads, compactness, and low power consumption due to the lack of high-power current at measurement sites somewhere in the hinterland.

1.3. Laser amplifier

Water-vapor measurements throughout the troposphere with the DIAL method require narrow-band laser emission with high spectral purity and high pulse energy. A high pulse energy is, in particular, essential during daytime since the lidar signal must be substantially larger than the solar background. Due to its advantageous wavelength range, covering the three important band systems of water-vapor around 725 nm, 817 nm and 935 nm, Ti:sapphire is ideal for humidity sounding and has, thus, been used as the laser medium in several successfully applied DIAL systems [5–9]. The design goal had been to provide widely tunable NIR radiation approaching the pulse-energy level of conventional fixed-frequency lasers of the order of 1 J. This could be achieved by directly pumping Ti:sapphire with flashlamps and this was our approach for the construction of Zugspitze DIAL [3]. Here, we could extract single-mode laser pulses with energies up to 250 mJ from an amplifier ring at a repetition rate of 20 Hz. This limitation was due to arcing inside the power supply above flashlamp load voltages of 26 kV. Increasing the average power by a higher repetition rate was limited by the performance of the power supply ($\approx$10 kW). For a safely routine operation we used a pulse energy of less than 150 mJ. After 11 years of lidar operation the power supply suffered a major damage without a chance of reparation or replacement. A new concept of laser amplification had to be found. A second motivation for the development of a new high power Ti:Sapphire laser amplifier was the construction of the mobile ATMONSYS lidar. A key requirement for quantitative DIAL sounding of water vapor is the spectral purity of the laser system [10]. The spectral purity of at least 99.9% was determined for the flashlamp pumped system and we expect the same for the transversely laser pumped setup.

Ordinary Ti:sapphire amplifiers work in a traveling wave configuration with multiple passes through a rod pumped longitudinally with a laser at 532 nm. Pumping Ti:sapphire longitudinally limits the length of the rod to the depth of penetration of the pump light which is typically about 10 mm at 532 nm. If pumped bidirectionally, a length of 20 mm is useful. Thus, it is not possible to increase the stored energy just by using a longer rod. In terms of single shots, this could partly be solved by enlarging the diameter of the rod and by enlarging the pump energy up to the limit of the damage threshold. In terms of average output power, this is also limited by thermal effects, because the cooling cylinder barrel grows only linear with the diameter of the rod, while the fluorescent area grows quadratically. With the introduction of diode pumped Nd:YAG lasers, pump sources with a moderate and appropriate pulse energy (100 mJ - 400 mJ), but with high repetition rates (100 Hz and more), have become available. The average output of such lasers is too much for conventional longitudinal pumping of 20-mm-long Ti:sapphire rods.

Experienced in transversely flashlamp pumping a Ti:Sapphire with an average pump power of more than 2 kW we started the promising development of a transversely laser pumped Ti:Sapphire amplifier stage. Transversely pumping allows for the use of a longer rod, because the penetration depth of the pump light does not limit its length. This yields the advantage of a much better ratio of cooled surface and fluorescent volume. Disadvantageous, on the other hand, are geometrical limitations in terms of a multi pass traveling wave arrangement with a geometrical setup as a dragonfly, because the ratio of aperture and length of the rod is much smaller than for longitudinally pumped rods. Thus, for high power applications, a long transversely pumped rod must be part of a traveling wave resonator with only one optical path. A principle setup is shown in Fig. 1.

The initially weak laser pulses from the oscillator are amplified in alternating sequence in the Ti:sapphire ring amplifier similar to the design by A. Kung [11]. The linearly polarized pulses enter the ring cavity where they are stored for amplification due to a rotation of the beam polarization by 90 degrees by a prism combination. After the first round trip the Pockels cell is switched on to counteract this rotation until the maximum amplification is reached after several round trips. After switching off the Pockels cell the amplified pulse is released. Therefore, the high voltage (6 kV) for the Pockels cell must be switched both on and off in a time significantly shorter than the circulation time of the laser pulses (e.g. 16 ns at a ring circumference of 4.8 m). We developed a pulse generator with both rise and fall times of 6 ns as well as a constant voltage level during the storage period.

1.4. Laser oscillators

The two DIAL wavelengths λ_{on} and λ_{off} of the Zugspitze DIAL are generated with two SLM optical parametric oscillators (OPO), which are based on a Littman cavity and a KTP (potassium titanyl phosphate) crystal as the OPO medium ([12]). This type of OPO provides a very wide tuning range from less than 700 nm to more than 1 μm with a rather constant output of about 0.5 mJ and is, therefore, highly suitable as the master oscillator for a tunable infrared laser system. The OPOs have been simultaneously pumped (10mJ each, 532 nm) by one common flashlamp-pumped injection-seeded and frequency-doubled Nd:YAG laser. The repetition rate of 20 Hz was adequate to the flashlamp pumped Ti:sapphire laser amplifier. By experiment, we showed, that a stable SLM operation is also possible at 100 Hz. After some improvements [3] we reduced the frequency noise of the OPOs from $\pm$140 MHz to $\pm$35 MHz and achieved a single-shot SLM emission of the OPOs almost at the Fourier-transform limit. A single shot bandwidth of 130 $\pm$ 15 MHz (4 ns) was measured. The lidar measurements have so far been carried out with the normal OPO pulse length of 2.0 ns resulting in an approximately doubled bandwidth. This is narrow enough to avoid errors by spectral contributions outside the absorption line center [9], considering the typical line widths of tropospheric water vapor between 2 GHz and 4 GHz.

In the ATMONSYS lidar, the two DIAL wavelength are generated with two tunable diode lasers (cw). The initial laser pulses are formed in a ring resonator following K. Ertel [8] with a longitudinal bidirectionally pumped Ti:sapphire rod with Brewster cut. This technique allows for a much more compact arrangement and provides much more mechanical stability due to the requirements of a mobile system. The laser pulses are generated by coupling the continuous emission from the diode lasers into the resonator in alternating sequence synchronously to the pump laser. A major challenge was the development of the electronics controlling the resonator in a way that both wavelength are resonant in alternating sequence while synchronized with the pump laser (100 Hz). The approach of Ertel et al. [8] using a backward and resonant signal processed by a personal computer was discarded because of being too slow for the operation with 100 Hz. Thus, we decided to use analog computers for controlling the resonator. Its length is modulated with a sine at a frequency of 200 Hz synchronized with the pump laser and with an amplitude covering at least two modes. Sample-and-hold circuits record the resonant mirror positions of both wavelengths while the pump laser is off in order to avoid a distortion by the pump light. By phase shifting the modulating sine, the two resonant mirror positions are tuned exactly to the time of the next pump pulse presuming that there is no significant drift of the two laser wavelengths during the period of about 5 ms. The stabilization of the mirror positions is achieved with a PID controller for each wavelength. The wavelengths the diode lasers are controlled with a WS 7 lambda meter (High Finesse).

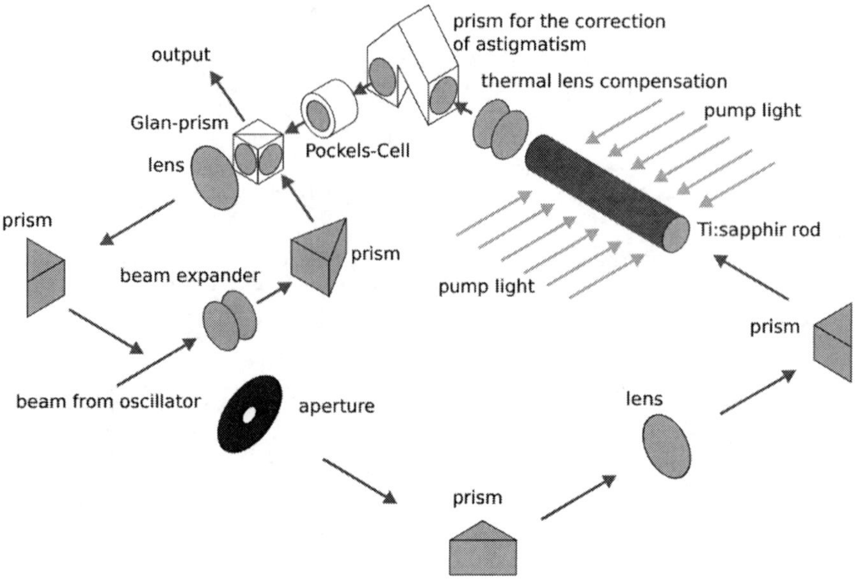

Figure 1: *Fundamental design of a traveling wave amplifier stage which is based on a ring resonator and suitable for a long transversely pumped Ti:sapphire rod. Because of the ring design, nanosecond pulses must actively be coupled in and out with a Pockels cell and a polarizing beam splitter. The two single lenses with the aperture in their focus work as spatial filter and as relays optic imaging the rod into itself. The thermal lensing of the Ti:sapphire rod is compensated by a concave-convex lens combination that is adjusted also for a conservation of the beam diameter. A major issue is the astigmatism of the thermal lens which causes problems with the beam quality without adequate compensation. A special combination of prisms in the upper corner solves this problem by rotating the beam by 90° around its axis. In this way the astigmatic distortion is self-corrected each round trip. This prism combination also rotates the polarisation about 90°, which is needed for the coupling operation with the Pockels cell.*

2. FIRST RESULTS OF TRANSVERSELY PUMPING TI:SAPPHIRE WITH A LASER

For experiments with transversely pumping we purchased several cylindrical Ti:sapphire rods (Ø 10 mm) with different lengths (50 mm, 100 mm, 150 mm) and different doping levels (0.7% and 0.1%).

In a very preliminary attempt we pumped a cylindrical 150 mm rod (Ø 10 mm) in a Fabry-Pérot resonator with 300 mJ (532 nm) from one side and extracted 7 mJ. Obviously, only a small fraction of the volume of the rod was pumped because the cylindrical barrel focussed the pump light to a narrow area at the opposite side. According to our former system pumped with four flashlamps, we tried to pump the rod in a Bethune arrangement [13] in 4 directions. This increased the pumped volume significantly, but, no laser emission was possible, even with a short rod of 50 mm. We concluded, that the polarisation of the pump light was not optimal because of the orientation of the c-axis of the Ti:sapphire crystal. Thus, the contribution to the laser gain was quite small for two transverse pumping directions, although the visible fluorescence seemed to be rather homogeneous across the rod (Fig. 2). In a next step we pumped the rod transversely from two opposite sides with an optimal orientation of the c-axis. Now we could extract 23 mJ from a Fabry-Pérot resonator and a cross-section of 25 mm^2. Nanosecond laser pulses from our OPOs could be amplified in a single pass with a gain of about 1.5 to 2.

3. OUTLOOK AND CONCLUSIONS

The next step will be the use of a rod with flat side walls. This should allow for a more homogeneous illumination of its volume. The rod is still in fabrication. For the operational use with high average power the rod needs to be cooled in a water floated bulb. By experiment we found that placing the rod in a water-filled glass tube slightly enhances the pump efficiency because of a better index matching. The results are rather promising. If pumping the entire cross-ection of 78 mm^2 with the same intensity one could expect a pulse energy of about 70 mJ from a 50 mm-rod and even more than 200 mJ from 150 mm-rod. A repetition rate of 100 Hz could perform an avarage power of about 7 W or even more than 20 W, respectively. For extracting 20 W, pumping with at least 100 W (532 nm) will be required, which is not available in our case. However, the method of transversely pumping has this great potential, and 100 W is still far from cooling

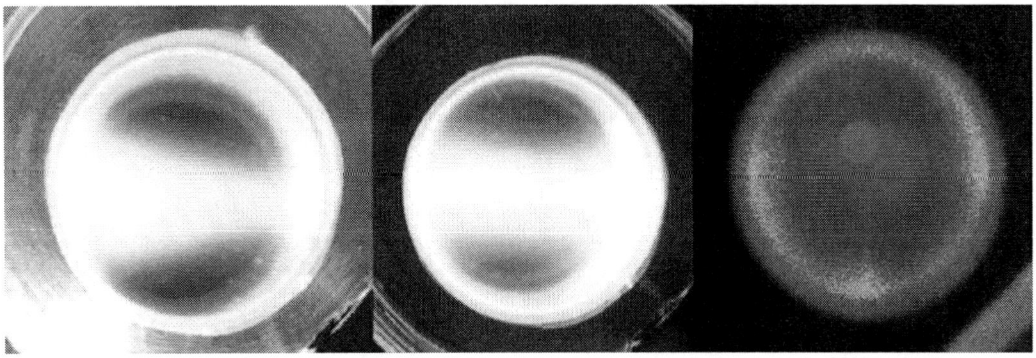

Figure 2: *Fluorescence of three different pumping configurations. Left: Single sided pumping from left. Middle: Two sided pumping from left and right. Right: Four sided pumping in a Bethune setup.*

limits if the rod is directly cooled by water. For narrowband lidar-soundings in the NIR this could contribute a significant progress, in particular for fast scanning applications or for observing short-term atmospheric processes.

ACKNOWLEDGMENTS

We thank Hans-Peter Schmid (KIT/IMK-IFU) for his continual interest in this work. We also acknowledge the funding by the German Federal Ministry of Economy within the program ZIM (Zentrale Innovation Mittelstand) and the assistance of our project partner Radiant Dyes Laser & Accessoires GmbH.

REFERENCES

1. Vogelmann, H., Sussmann, R., Trickl, T., and Reichert, A. Spatiotemporal variability of water vapor investigated using lidar and ftir vertical soundings above the zugspitze. *Atmos. Chem. Phys.*, 15:3135–3148, 2015.
2. Trickl, T., Vogelmann, H., Fix, A., Schäfler, A., Wirth, M., Calpini, B., Levrat, G., Romanens, G., Apituley, A., Wilson, K. M., Begbie, R., Reichardt, J., Vömel, H., and Sprenger, M. How stratospheric are deep stratospheric intrusions? LUAMI 2008. *Atmospheric Chemistry and Physics*, 16(14):8791–8815, 2016.
3. Vogelmann, H. and Trickl, T. Wide Range Sounding of Free Tropospheric Water Vapor with a Differential Absorption Lidar (DIAL) at a High Altitude Station. *Appl. Opt.*, 47(12):2116–2132, 2008.
4. Vogelmann, H., Sussmann, R., Trickl, T., and Borsdorff, T. Intercomparison of atmospheric water vapor soundings from the differential absorption lidar (DIAL) and the solar FTIR system on Mt. Zugspitze. *Atmos. Meas. Tech.*, 4(5):835–841, 2011.
5. Moore, A. S., Brown, K. E., Hall, W. M., Barnes, J. C., Edwards, W. C., Petway, L. B., Little, A. D., Luck, W. S., Jones, I. W., Antill, C. W., Browell, E. V., and Ismail, S. Development of the lidar atmospheric sensing experiment (lase) – an advanced airborne dial instrument. In A. Ansmann, R. Neuber, P. R. and Wandinger, U., editors, *Advances in Atmospheric Remote Sensing with Lidar*, pages 281–288. Springer-Verlag, 1996. Selected Papers of the 18th International Laser Radar Conference (ILRC), Berlin, 22-26 July 1996.
6. Browell, E. V., Ismail, S., and Grant, W. B. Differential absorption lidar (DIAL) measurements from air and space. *Appl. Phys. B*, B 67:399–410, 1998.
7. Nagasawa, C., Nagai, T., Abo, M., Shibata, Y., and Uchino, O. Developement of airborne dial for water vapor measurement. In Singh, U. N., Itabe, T., and Sugimoto, N., editors, *Lidar Remote Sensing for Industry and Environment Monitoring – Proceedings of SPIE - The International Society for Optical Engineering*, volume 4153, pages 599–606. 2001.
8. Ertel, K., Linné, H., and Bösenberg, J. Injection-seeded pulsed ti:sapphire laser with novel stabilization scheme and capability of dual-wavelength operation. *Appl. Opt.*, 44:5120–5126, August 2005.
9. Wulfmeyer, V. Ground-based differential absorption lidar for water-vapor temperatur-profiling: development and specifications of a high-performance laser transmitter. *Appl. Opt.*, 37(18):3804–3824, 1998.
10. Bösenberg, J. Ground-based differential absorption lidar for water-vapor and temperature profiling: methodology. *Appl. Opt.*, 37(18):3845–3860, 1998.
11. Kung, A. H. Regenerative amplification of a single-frequency optical parametric oscillator. *Opt. Lett.*, 18(23):2017 – 2019, April 1993.
12. Bosenberg, W. R. and Guyer, D. R. Broadly tunable, single-frequency optical parametric freqency-conversion system. *J. Opt. Soc. Am. B*, 10(9):1716–1722, 1993.
13. Bethune, D. S. Dye cell design for high-power low-divergence excimer-pumped dye lasers. *Appl. Opt.*, 20:1897–1899, June 1981.

SPACE- AND TIME-RESOLVED RAMAN AND BREAKDOWN SPECTROSCOPY: ADVANCED LIDAR TECHNIQUES

Gurlui Silviu[1]*, Cazacu Marius Mihai[1, 2], Timofte Adrian[1, 3],

Rusu Oana[1], Bulai Georgiana[1, 4], Dan Dimitriu[1]

*[1]Alexandru Ioan Cuza University of Iasi, Faculty of Physics, Atmosphere Optics, Spectroscopy and Lasers Laboratory, 700506 Iasi, Romania, *sgurlui@uaic.ro*

[2]Gheorghe Asachi Technical University of Iasi, Department of Physics, 700050 Iasi, Romania, marius.cazacu@tuiasi.ro

[3]National Meteorological Administration, Regional Forecast Center Bacau,3 Timpului Str., Bacau, Romania), timofte.adrian@gmail.com

[4] CERNESIM, Alexandru Ioan Cuza University of Iasi, 700506 Iasi, Romania

ABSTRACT

DARLIOES - the advanced LIDAR is based on space- and time-resolved RAMAN and breakdown spectroscopy, to investigate chemical and toxic compounds, their kinetics and physical properties at high temporal (2 ns) and spatial (1 cm) resolution. The high spatial and temporal resolution are needed to resolve a large variety of chemical troposphere compounds, emissions from aircraft, the self-organization space charges induced light phenomena, temperature and humidity profiles, ice nucleation, etc.

1 INTRODUCTION

Introduction of new technologies to investigate the physic-chemical phenomena related to the Earth's climate changes is highly required due to the increased number of pollutant sources (widely major relevant potential in the global warming effect) and their adverse greenhouse effects caused by compounds such as carbon dioxide, nitrous oxide, sodium chlorine, methane and high global warming potential gases like perfluorocarbons, sulfur hexafluoride, ozone substances etc. [1]. A large number of additional chemical processes involving aerosols have been postulated; however, a lack of information on the abundance, nature and reactivity of the various aerosol types (e.g., dust, soot, sulfate, nitrate, organic) has hindered an accurate assessment [2].

In the last decades, especially due to more frequent extreme weather events, worldwide Environmental Research concentrated its efforts toward development of climate change prediction and scenarios to improve the regional climate change predictions. It became necessary to use all available expertise and instrumentation in order to improve model representation of physical chemical and biological processes, to assess impacts of climate changes and to quantify and reduce the uncertainties in evaluation of the hydrological cycle and impacts on natural resources, socio-economic systems and human health. In order to accomplish this objective, a large amount of experimental data must be collected, processed, evaluated and analyzed

The study of the aerosols atmosphere optics of ultra-trace amounts such as organic or in-organic compounds plays a key impact factor in the earth physics also. To identify the physico-chemical species behaviours, their concentrations and human health risk, influence on climate changes, different sensors, both ground-based and satellites optical instruments are extensively used today. Moreover, the satellite spectral instruments, providing continuous, long-term data sets, environmental forecasting and global climate studies, have in particular a great potential to detect most of the atmospheric pollutants, water amount, brightness temperatures, etc. Thus, remote sensing from satellite evidences global coverage in latitude and longitude, but it may be hard to achieve any altitude selectively: time (magnitude of seconds)- space (magnitude of few km) resolution of measurements which are very limited in order to measure and understand fast chemical and optical phenomena induced from various physical atmospheric parameters.

2 DARLIOES SYSTEM

In order to better understand the fundamental of some critical physico-chemical transformation of the atmosphere compounds but also for applications point of view, a new optical instrument was developed in the framework of the STAR-ROSA. As in many scientific fields, in the atmosphere science, a major progress is driven through the introduction of new or improved techniques involving both analytical laboratory experiments and field campaigns. For example, the project "Fast laser imaging, detection and ranging of aerosol emissions in aircraft plumes"(**DARLIOES**) had the main objective to develop a new technique where coupled high resolution laser, passive and active remote optical emission spectroscopy instruments are used in order to study different aerosol emissions in aircraft plumes [DARLIOES was supported by the Romanian Space Agency (ROSA) within Space Technology and Advanced Research (STAR) Program (Project no. 98/ 29.11.2013), **http://spectroscopy.phys.uaic.ro/darlioes.html**]. Some advantage of the DARLIOES instrument are the following: time-space resolved temperature and humidity profiles (Raman scattering of oxygen 353 nm & 354 nm; real-time monitoring tool to measure concentration of sulphur dioxide injected in the troposphere from the volcanic eruption ash, water profiles and their dynamics, acid rain, chemical transformation/physical properties; DIAL H_2O and SO_2; UV fluorescence using the two stronger absorption emission lines of 218.9 nm and 220.8 nm, respectively; water vibrational Raman scattering at 407 nm; ice nucleation variability, physical and chemical impact, the influence upon the mixed-phase cloud; dust mineralogical composition dynamics influences upon the ice nucleation; Raman scattering of oxygen - 353 nm & 354nm; fast ICCD imagery (IR spectrum) of ice nucleation pattern clouds; studies of the ice nucleation and propagation in plants; UV fluorescence of PAH {Benzo(ghi)perylene Excitation (Ex-301 nm/ Emission Em-420 nm); Naphthalene (Ex-275 nm/ Em-325 nm); Anthracene (Ex-255 nm/Em-375 nm); Pyrene (Ex-270 nm/ Em-420 nm), etc)}.

Approaches of combining lidar and sun photometer data have been undertaken recently. A combination of spectrally resolved optical depth measured with sun photometer and backscatter coefficients measured with Raman lidar was used to explore the possibility of deriving important particle variables such as volume and surface-area concentration, effective radius, and single-scattering albedo [3]. Such synergetic retrieval results in additional enhancements in derived aerosol properties because the back-scattering observations by lidar improves the sensitivity in columnar property analysis of aerosol, while radiometric observations provide sufficient constraints on aerosol type and concentrations that are generally missing in lidar signals. Remote sensing methods, both active and passive, proved to be fruitful and convenient. A number of developed and launched space instruments [4], [5] provide global monitoring of aerosol properties [6], [7]. The retrieval of the physical, chemical and spectral aerosol properties from the optical measurements is a highly non-trivial and challenging task. It is often not possible to extract full information about the aerosol field from observations at fixed sites and for a limited set of wavelength channels. Besides that, the knowledge of the small-scale physical processes within the clouds and aerosols layers is also expected to improve. The used research infrastructure is shown schematically in Figure 1. The principle of operation is based on the interaction of short and high-energy laser pulses with aerosols and clouds and on the monitoring of radiation-induced effects through high speed spectral techniques (integration time in nanosecond range).

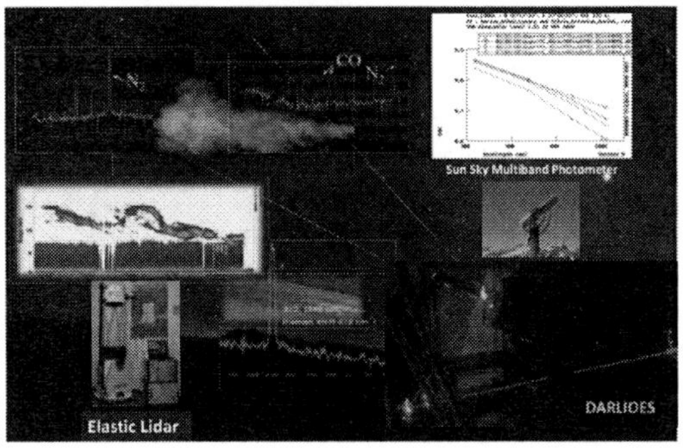

Figure 1: Approached concept of combining DARLIOES, lidar and sunphotometer

The basic lidar system described in previous papers [8, 9], was developed in the framework of ROmanian LIdar NETwork (ROLINET) and RADO (Romanian Atmospheric Research 3D Observatory) research projects.

 https://doi.org/10.1051/epjconf/201817601028

3 RESULTS

Preliminary measurements were performed on 27 January 2017. Meteorological data [http://www1.wetter3.de/] showed that on the 27th of January 2017, at the level of 500 hPa, field values were between 552 and 560 gpdam, highlighting the occurrence of a geopotential ridge but also thermal ridge (within 24 hours the temperature has evolved from approx. -26 to - 20 degrees Celsius). Wind direction passed in 24 hours from N to NE. At the ground level, the high pressure system was maintained with the same value at the centre- 1035 hPa, as in the previous day. In Figure 2 shows the RCS profile recorded with the elastic lidar at 532 nm. It can be observed that in the lower troposphere some intrusion occurs. Typically, the lidar data are correlated with complementary techniques such as the AERONET sunphotometers and with theoretical models (DREAM, HYSPLIT). Here, the preliminary results are approached by combining DARLIOES and lidar data.

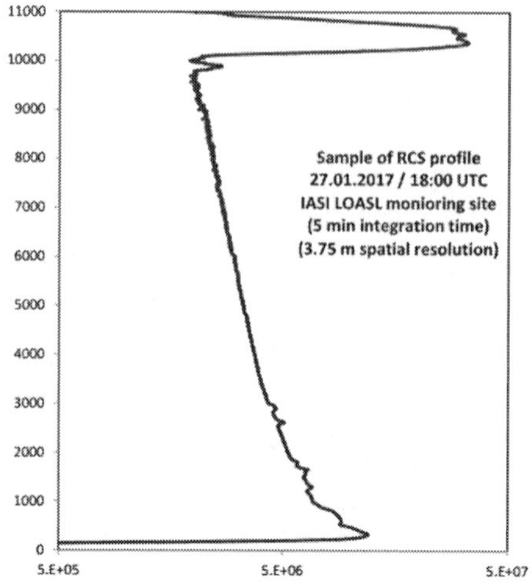

Figure 2: Sample of RCS profile

In order to analyse the atmosphere chemical compounds and their space-time dynamics optical emission spectra of the atmosphere were recorded (Figures 3 and 4).

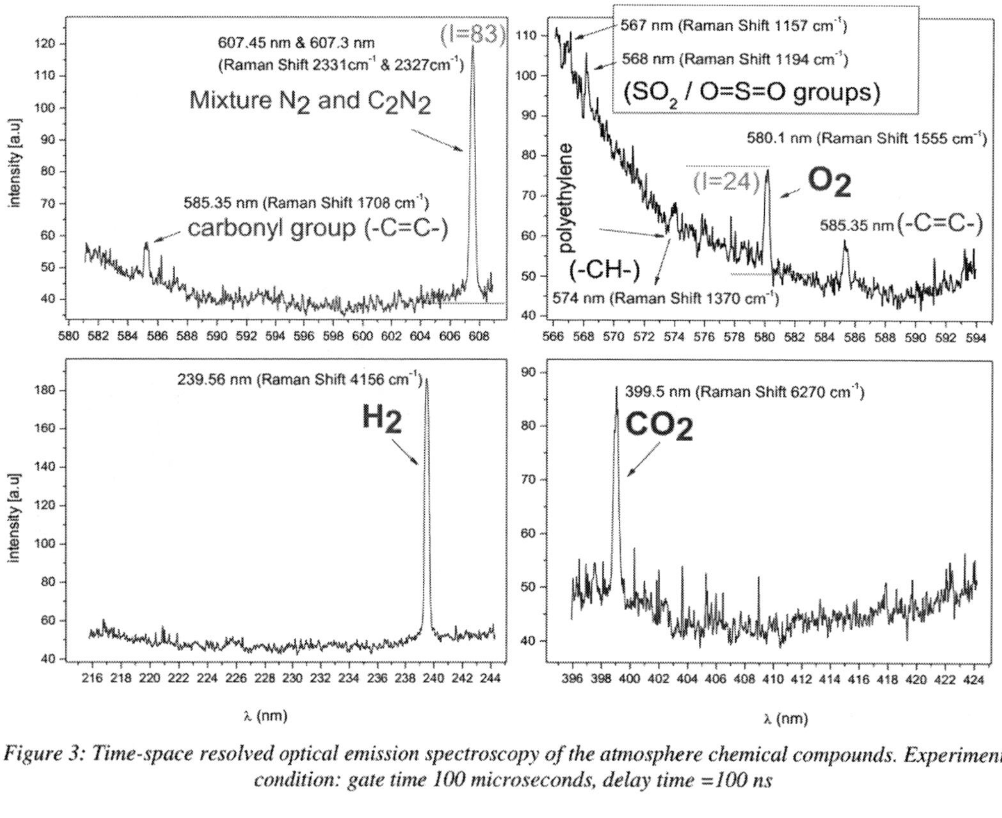

Figure 3: Time-space resolved optical emission spectroscopy of the atmosphere chemical compounds. Experimental condition: gate time 100 microseconds, delay time =100 ns

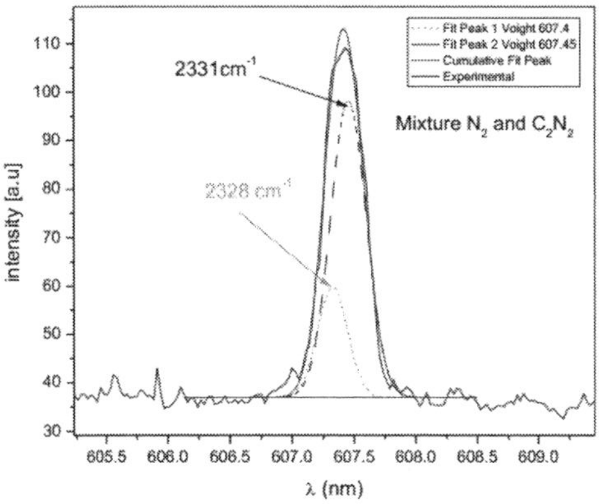

Figure 4: Time-space resolved optical emission spectroscopy of the nitrogen and cyanogens mixture.

Taking into account the wavelength of the excitation laser (532.1 nm) we calculated the Raman shifts for each spectral band. Therefore, we identified several known atmosphere compounds (nitrogen, oxygen, hydrogen, carbonate dioxide, sulphur dioxide, polypropylene, cyanogens, etc.). Moreover, the ratio of the two optical intensities of nitrogen and oxygen has been estimated. This is of about R = 3.5, in a good agreement with the ratio of the two volume gases (78% Nitrogen, 21% Oxygen; R= 3.7). In Figure 4 is showing the optical emission spectra and deconvolution of the recorded profile.

Two maxima have been evidenced at distance of 0.15 nm. This corresponds to nitrogen molecules and nitrogen derivatives (cyanogens). This is a very interesting result of the DARLIOES because it surpasses the disadvantage of the optical filter of the classical LIDAR where the difference of the two compounds may not be differenced yet. These are preliminary data and further measurements will be done in order to estimate the dynamics of several compounds with a time resolution of up to few nanoseconds. Moreover, RCS profiles with ~1 centimeter space resolution will be recorded in order to investigate the behavior of the trace gases compound, to estimate the altitude temperature, etc.

4 CONCLUSIONS

In order to better understand the fundamentals of some critical physico-chemical transformations of the atmospheric compounds and for practical purpose, our study relies on a new optical power instrument. It is able to capture a fast plume airborne image (2 ns gate time) and can be used to real-time investigate the behavior of several chemical compounds at a given point of the free atmosphere, with a spatial resolution of up to 1 cm. This new remote sensing optical instrument is based on the fast imaging and space-time resolved Raman Spectroscopy of both airborne and spaceborne plumes.

ACKNOWLEDGEMENTS

This work was financially supported by the Romanian Space Agency (ROSA) within Space Technology and Advanced Research (STAR) Program (Project no. 114/07.11.2016).

References

[1] Ciuraru, R. et al., 2011: Heterogeneous reactivity of chlorine atoms with sodium chloride and synthetic sea salt particles, *Phys. Chem. Chem. Phys.* **13** (43), 19460–70.

[2] IPCC, *Climate Change 2013: The Physical Science Basis*. 2013.

[3] Pahlow, M. et al., 2006: Retrieval of aerosol properties from combined multiwavelength lidar and sunphotometer measurements, *Appl. Opt.* **45** (28), 7429.

[4] Breon, F.M. et al., 2002: Scientific results from the POLarization and Directionality of the Earth's Reflectances (POLDER), *Adv. Sp. Res.* **30**, 2383.

[5] Winker, D.M. et al., 2007: Initial performance assessment of CALIOP, *Geophys. Res. Lett.* **34**, p. L19803,.

[6] King, M.D. et al., 1999: Remote sensing of Tropospheric aerosols from Space: Past, Present, and Future, *B. Am. Meteorol. Soc.* **80**, 2229.

[7] Kokhanovsky A. et al., 2007: Aerosol remote sensing over land: A comparison of satellite retrievals using different algorithms and instruments, *Atmos. Res.* **85**, 372–394,.

[8] Cazacu M.M. et al., 2012: Grimsvotn Volcano: atmospheric volcanic ash cloud investigations, modelling-forecast and experimental environmental approach upon the Romanian area, *J. Optoelectron. Adv. Mater.* **14**, (5–6), 517–522.

[9] Timofte A. et al., 2011: Romanian lidar investigation of the Eyjafjallajokull volcanic ash," *Environ. Eng. Manag. J.*, **10** (1), 91–97.

https://doi.org/10.1051/epjconf/201817601029

DEPOLARIZATION CHANNEL FOR BARCELONA LIDAR. IMPLEMENTATION AND PRELIMINARY MEASUREMENTS.

Alejandro Rodríguez-Gómez[1]*, Michaël Sicard[1,2], Constantino Muñoz-Porcar[1], Rubén Barragán[1,2], Adolfo Comerón[1], Francesc Rocadenbosch[1,2], Eric Vidal[3]

*[1]RSLab, BarcelonaTech University (UPC), Barcelona, Spain *alejandro@tsc.upc.edu*
[2]Ciències i Tecnologies de l'Espai - Centre de Recerca de l'Aeronàutica i de l'Espai / Institut d'Estudis Espacials de Catalunya (CTE-CRAE / IEEC), UPC, Barcelona, Spain
[3]UTC Building & Industrial Systems, Spain

ABSTRACT

A new depolarization channel has been implemented in the BarcelonaTech University (UPC) multi-wavelength lidar system. The optical and mechanical designs are presented. The special configuration of the total power channel is also detailed, with the relevant aspects in measurement inversion. Some preliminary measurements are presented for Saharan dust intrusion events.

1 INTRODUCTION

The capability of measuring depolarization of the backscattered light, together with multi-wavelength detection, improves the ability of a lidar system to identify the aerosols present in the atmosphere, as it has been pointed out by different authors [1]–[4].

The BarcelonaTech University (UPC) Remote Sensing Lab (RSLab) has implemented a new depolarization channel for its lidar system. The depolarization channel has been tested with measurements during Saharan dust intrusion events taking place on May and June 2016.

2 THE UPC LIDAR MAIN RECEIVER CONFIGURATION

The lidar system at UPC [5] presents a non-common receiver configuration that is sketched in Figure 1. A 14-inch telescope collects the backscattered light, which is delivered to the poly-chromator unit by means of a 3-m long multi-mode fiber bundle of 3 mm diameter. The poly-chromator delivers light to the 6 photo-receivers tuned at 1064, 532 and 355 nm (elastic backscattering) and 607, 407 and 387 nm (Raman channels).

According to fiber optics theory [6], a multi-mode fiber will not preserve the polarization of the guided light.

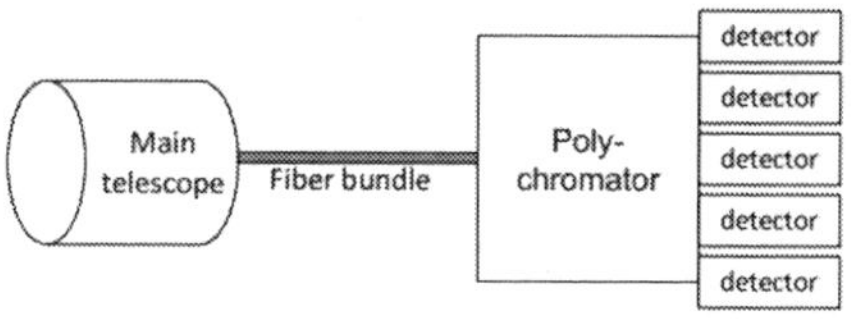

Figure 1 UPC lidar receiver configuration, including a depolarizing multi-mode fiber bundle.

This fact makes that the overall behavior of the receiver is non-sensitive to the polarization of the incoming light. So the main receiver becomes total-power collecting.

To check this feature, we measured the depolarization introduced by the fiber bundle, with a simple test set, presented in Figure 2. The laser beam passes through a linear polarizer before being fed to the fiber bundle.

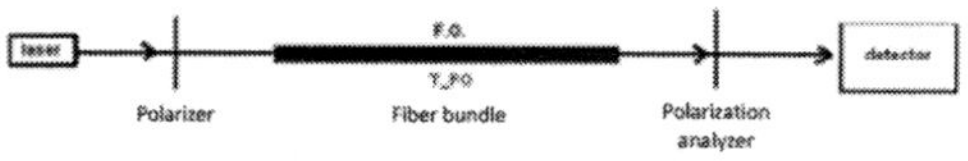

Figure 2 Test set for checking the polarization behavior of the fiber bundle

The polarization state of the output light was measured in two different situations: with the fiber bundle cable placed straight and rolled up. The results are plotted in Figure 3. The measured circularities are 0.96 for the cable straight and 0.93 for the cable rolled up.

This depolarization capability is also useful to make our system insensitive to the di-attenuation of the different beam-splitters used in the poly-chromator unit.

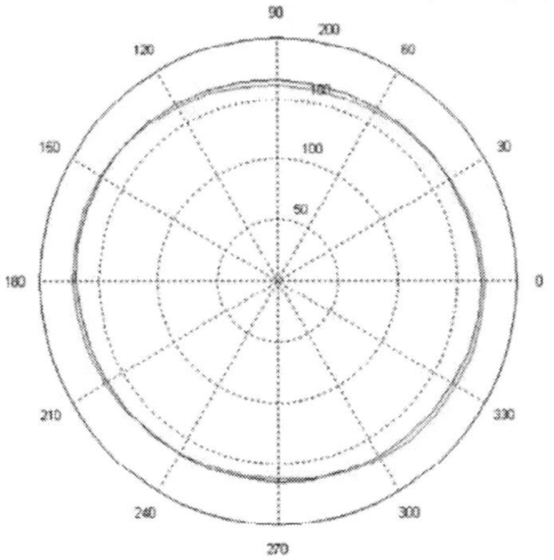

Figure 3 Polarization state measurement at the output of the fiber bundle cable for two situations: straight (blue line) and rolled up (red line) cable.[7]

3 NEW DEPOLARIZATION CHANNEL

The depolarization channel has been built by means of an auxiliary telescope attached to the lidar mechanical setup. The auxiliary telescope is, in fact, a 300-mm focal distance, 70-mm aperture diameter, TAIR-3S telephoto lens.

The optical setup, sketched in Figure 4, includes an eye-piece lens that produces an image of the telescope aperture onto the photo-receiver (a Licel R9880U PMT module) sensitive area. This arrangement allows distributing the collected light over the PMT sensitive surface, while a 1-mm field-stop iris limits the field of view to approx. 3.3 mrad.

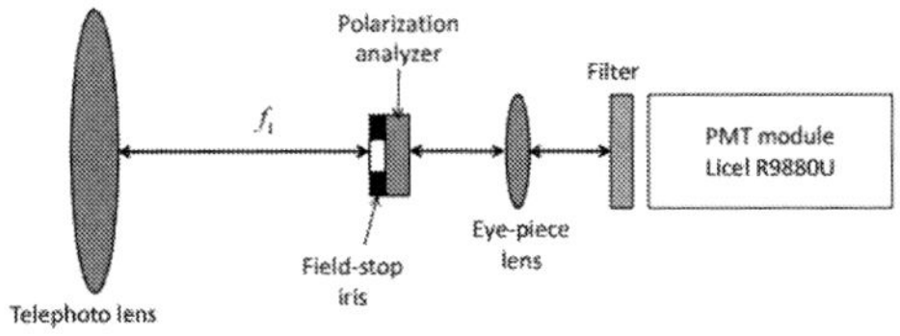

Figure 4 Optical setup of the depolarization channel

A polarization analyzer located at the focal plane of the telephoto lens selects the desired polarization of the collected light. When performing measurements, the position of the analyzer is oriented to collect the cross-polar component; when performing the needed calibration of the channel [4] the analyzer is oriented alternatively at +45 and -45 degrees from the nominal position.

Finally, an interference filter at 532 nm is located near the photo-receiver surface. According to this arrangement, the collected light rays are not parallel when passing through the filter. This fact implies that the filter effective spectral width is greater than the specified value (0.5 nm). This has implications especially in the molecular contribution to the depolarized backscattered light [8] and must be taken into account when processing the measurement results.

4 PRINCIPLE OF OPERATION [4]

The primary depolarization measurement is done by comparing the signals detected by the depolarization and total power channels

$$S_{Dep}(90^\circ, R) = V_{Dep}(R) \cdot P_\perp(R)$$
$$S_{Tot}(R) = V_{Tot}(R) \cdot P_{Tot}(R) \qquad (1)$$

where $P_\perp(R)$ and $P_{Tot}(R)$ are the cross-polar and total powers collected while $V_{Dep}(R)$ and $V_{Tot}(R)$ are the depolarization and total power channels responsivities, as a function of height over the lidar R. The primary observable is

$$\delta^*(90^\circ, R) = \frac{S_{Dep}(90^\circ, R)}{S_{TOT}(R)} \qquad (2)$$

In parallel to the measurement process, a channel calibration (with respect to the total power channel) must be performed to obtain the system channel ratio

$$V^*(R) = \frac{V_{Dep}(R)}{V_{Tot}(R)} \qquad (3)$$

The calibration is done by performing two short measurements with the polarization analyzer rotated +45 and -45 degrees from the nominal position, $S_{Dep}(90^\circ + 45^\circ, R)$ and $S_{Dep}(90^\circ - 45^\circ, R)$. The system calibration profile is then computed by comparing with the signal obtained simultaneously from the total power channel, $S_{Tot}(R)$:

https://doi.org/10.1051/epjconf/201817601029

$$V^*(R)=2\sqrt{\frac{S_{Dep1}(90^\circ-45^\circ,R)}{S_{Tot1}(R)}\cdot\frac{S_{Dep2}(90^\circ+45^\circ,R)}{S_{Tot2}(R)}} \quad (4)$$

Figure 5 shows the result of a calibration performed on December 2nd 2016. As it can be noted, a spatial average is performed

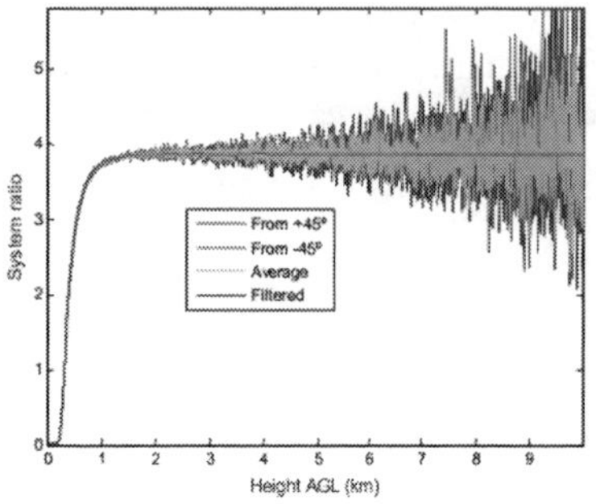

Figure 5 System channel-ratio calibration performed on December 2nd 2016.

The volume depolarization ratio is then computed

$$\delta^V(R)=\frac{\delta^*(90^\circ,R)}{V^*(R)-\delta^*(90^\circ,R)} \quad (5)$$

Finally, the particle depolarization ratio is computed by using concurrently the results of the molecular $\beta^m(R)$ and particle $\beta^p(R)$ backscattering coefficients of the atmosphere under measure, via an elastic or Raman inversion of the total power 532-nm channel:

$$\delta^p(R)=\frac{\left[1+\delta^m(R)\right]\delta^V(R)\rho(R)-\left[1+\delta^V(R)\right]\delta^m(R)}{\left[1+\delta^m(R)\right]\rho(R)-\left[1+\delta^V(R)\right]} \quad (6)$$

where:

$$\rho(R)=\frac{\beta^m(R)+\beta^p(R)}{\beta^m(R)} \quad (7)$$

and $\delta^m(R)$, the molecular volume depolarization ratio is usually considered a constant, which depends on the temperature and the spectral width of the photo-receiver [8].

5 RESULTS

Two Saharan dust intrusion events took place around Barcelona area in late May and early June 2016. Measurements were performed with the UPC lidar and simultaneously with a co-located Micro Pulse Lidar, associated to MPLnet [9]. This Micro Pulse Lidar provides the range-corrected total power and depolarization signals and calculates the volume depolarization in real time.

Figure 6 and Figure 7 show two consecutive measurements during May 26th. They show a Saharan dust layer around 3.7 km altitude, with particle depolarization ratio exceeding 20% (more than 15% is characteristic of this kind of aerosol). Both measurements show good agreement between the UPC lidar and the MPL.

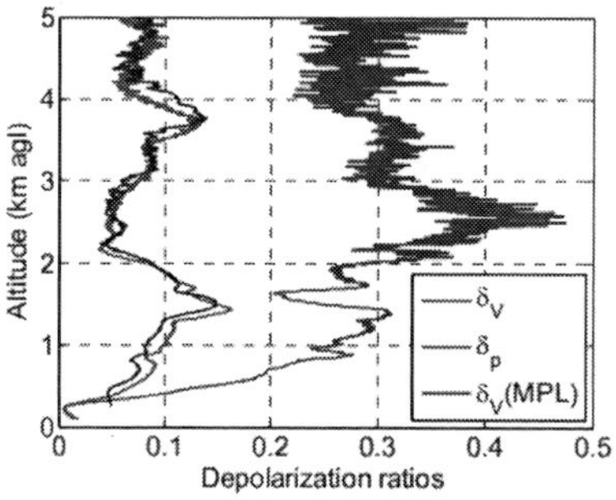

Figure 6 Measurement performed on May 26th 2016, 14:24 UTC

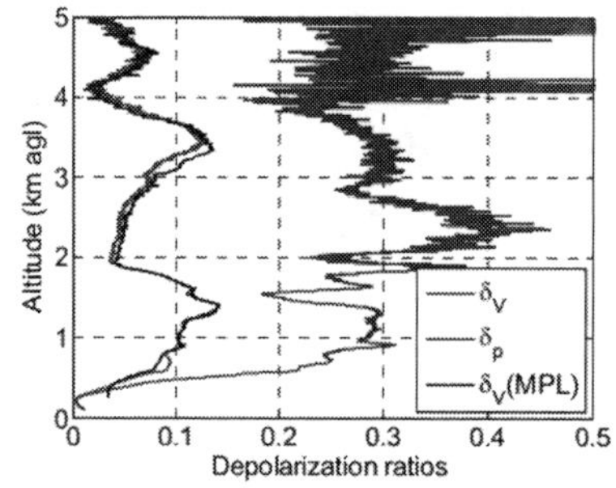

Figure 7 Measurement performed on May 26th 2016, 15:40 UTC.

Figure 8 and Figure 9 show measurements taken, respectively, in the beginning and in the maximum of an intrusion event in June 2016. Figure 8 shows a small value of volume depolarization up to 5 km, while Figure 9 shows a great mass of aerosol between 3 and 4 km height, also with particle depolarization over 20%. In this latest measurement, there is a noticeable difference with the MPL measurement, very likely due to some mechanical tolerance in the telescope pointing not considered because of the absence of a calibration on the day of the measurement.

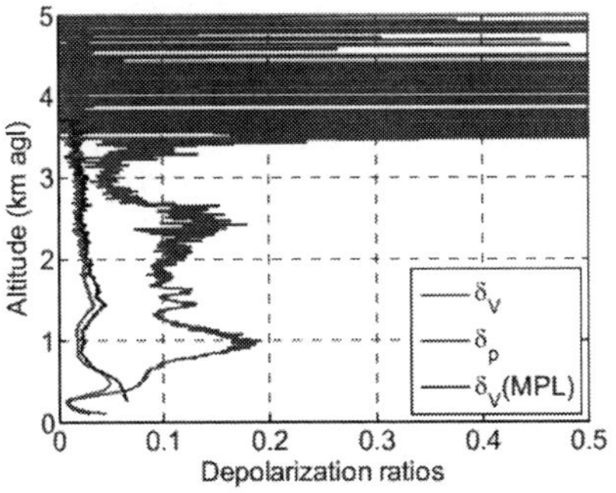

Figure 8 Measurement performed on June 7th 2016, 8:37 UTC.

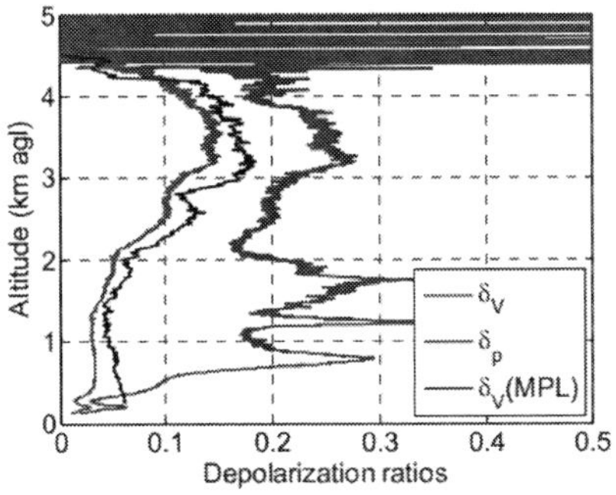

Figure 9 Measurement performed on June 10th 2016, 13:17 UTC

6 CONCLUSIONS

The new depolarization channel of the UPC lidar has been introduced with details of architecture and principle of operation. Measurements performed during two Saharan dust intrusion events are used to test the new channel. Particle depolarization ratios varying between 10 and 40 % are found in the dust layers. Comparisons of the volume depolarization with an MPL system show a good agreement between both systems and demonstrate the reliability of the new depolarization channel of the UPC multi-wavelength lidar.

ACKNOWLEDGEMENTS

This work has been financed by the European Union (H2020, grant 654169, ACTRIS-2), the European Fund for Regional Development, the Spanish Government (grant TEC2015-63832-P) and the Catalan Government (grant 2014 SGR 583).

The authors also acknowledge the priceless help of Joaquim Giner and Josep Pastor, both members of UPC technical staff.

References

[1] K. Sassen, "Polarization in Lidar," in *Lidar*, C. Weitkamp, Ed. New York: Springer-Verlag, 2005, pp. 19–42.

[2] K. Sassen, J. Zhu, P. Webley, K. Dean, and P. Cobb, "Volcanic ash plume identification using polarization lidar: Augustine eruption, Alaska," *Geophys. Res. Lett.*, vol. 34, no. 8, Apr. 2007.

[3] G. Wandinger, Ulla; Ansmann, Albert; Mattis, Ina; Müller, Detlef; Pappalardo, "Calipso and beyond: long-term ground-based support of space-borne aerosols and cloud lidar missions" in *24th International Laser Radar Conference*, 2008, pp. 715–718.

[4] V. Freudenthaler *et al.*, "Depolarization ratio profiling at several wavelengths in pure Saharan dust during SAMUM 2006," *Tellus, Ser. B Chem. Phys. Meteorol.*, vol. 61, no. 1, pp. 165–179, 2009.

[5] D. Kumar *et al.*, "Six-channel polychromator design and implementation for the UPC elastic/Raman lidar," *SPIE Int. Symp. - Remote Sens. Eur.*, vol. 8182, p. 81820W–1–81820W–10, 2011.

[6] G. P. Agrawal, *Fiber-Optic Communications Systems, Third Edition.*, vol. 6. 2002.

[7] E. Vidal, "Disseny d'un canal de despolarització a 532 nm per al lidar d'EARLINET de la UPC," BarcelonaTech, 2013.

[8] A. Behrendt and T. Nakamura, "Calculation of the calibration constant of polarization lidar and its dependency on atmospheric temperature," *Opt. Express*, vol. 10, no. 16, p. 805, Aug. 2002.

[9] NASA, "MPLNet," 2017. [Online]. Available: https://mplnet.gsfc.nasa.gov/. [Accessed: 11-Jan-2017].

REDUCTION OF MULTIPLE SCATTERING TO SECOND ORDER SCATTERING FOR DETERMINATION OF LIQUID-CLOUD MICROPHYSICAL PROPERTIES

Xiaoying Cao[1], Gilles Roy[2*] and Grégoire Tremblay[3]

[1]*Lidar Consultant, Ottawa, Canada;* [2*]*RDDC Valcartier 2459 de la Bravoure, Québec, Qc, Canada,* *gilles.roy@drdc-rddc.gc.ca*

[3]*AEREX Avionic, Levis, Canada;*

ABSTRACT

Under single scattering conditions, water droplets clouds do not depolarize the backscattered light. However, backscattered light from multiple scattering will be depolarized. The level of depolarization is a function of the droplets size and of the cloud extinction coefficient value and its profile; it has also a strong dependency on the lidar field-of-view (FOV) and azimuth angle [1]-[3]. It has been demonstrated in [4] that the multiple scattering cross polarization signal can be reduced to a second order scattering one using the azimuthal dependency of the cross polarization. In this paper we present how the multiple scattering cross polarization can be reduced to a second order scattering and how this signal is related to the droplet particle size. Monte Carlo simulations for clouds at different ranges and effective radius sizes are performed and compared with the results obtained with an analytical model.

1. AZIMUTHAL CROSS POLARIZATION PATTERN

Backscattered light from spherical particles does not change the polarization orientation state of a linearly polarized incident light. This particularity allows the discrimination of spherical particles from non-spherical particles. However, under multiple scattering conditions, even spherical particles do generate depolarized signals. In fact, the multiple scattering cross polarization shows an azimuthal pattern. This pattern is easily observed using a gated ICCD camera and a polarized beam splitter as shown in Fig. 1. A linearly polarized parallel laser beam is used. At the reception, after collimation, the cross polarization is imaged on the detector. An adjustable delay is used to image the backscattered light from the cloud from a specific cloud depth. The gate width of the ICCD is typically 10 to 20 ns. Please refer to [1] for relevant details.

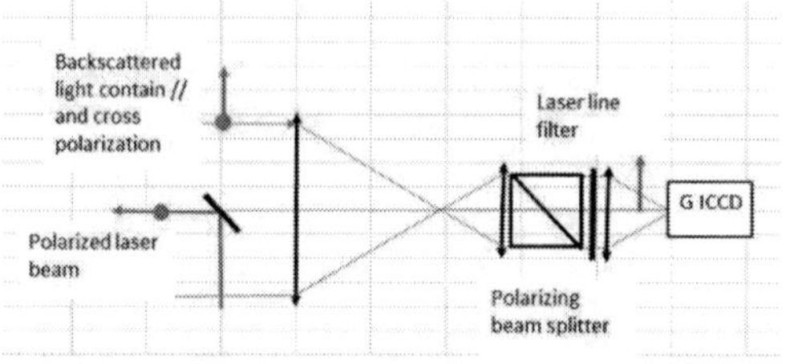

Figure 1. Lidar set up to capture the azimuthal cross polarization. The linearly polarized source is emitted on axis (lower left). The received signal is collimated through a polarizer and the cross polarization is imaged on the camera (right).

In order to better understand the mechanism of polarized multiple scattering, Monte Carlo (MC) simulation has been performed. The Undique MC simulator is a multithreaded software based on the Bohren and Huffman Mie scattering algorithm. The simulator reproduces the characteristics of a Flash Lidar system. It consists of an emitter/receiver system, a target and a propagation range including aerosols of various properties. The particularity of the Undique MC is its capability to image the scattered light on a detector array [5]. Fig. 2 shows the azimuthal pattern obtained for the simulation of a penetration of 50 m in a cloud of constant extinction 0.03 m^{-1} starting at 1000 m. The image at the top contains all the scattering orders, while the image at the bottom contains second order scattering only. The second order scattering cross polarization follows a $Cos(4\varphi)$ pattern; it means there is no second-order scattered energy for azimuths equal to 0°, 90°, 180°, 270°. In multiple scattering conditions (top part of Fig. 2) there clearly is a fair amount of energy at those angles. The subtraction of the higher order scattering energy found at those specific angles, when done at all azimuths, leads to a second order scattering pattern [4]. So, it appears that multiple scattering effects occurring in dense water cloud can be simplified to second

 https://doi.org/10.1051/epjconf/201817601030

order scattering. In the following, we will examine what are the mechanisms leading to depolarization of the lidar signal from second order scattering.

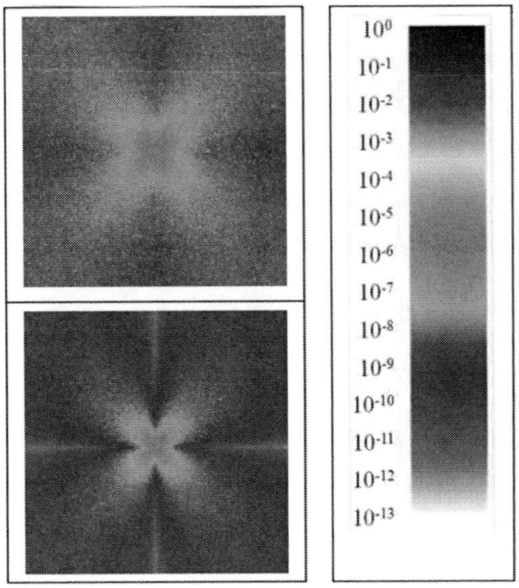

Figure 2. Monte Carlo simulation of cross polarization azimuthal pattern. The top image contains all orders scattering. The image at the bottom is for second order scattering only.

2. BACKSCATTERING AND DEPOLARIZATION

Lidars with polarization capabilities usually work at a specific FOV and detect the parallel and cross polarization components of the lidar return. The lidar signals will contain all scattering orders. In this paper, we want to reduce multiple scattering effects to second order scattering. To do so, it is necessary to have a look at the Mie scattering phase function and how the scattered light is depolarized.

The linear depolarization ratio (δ_{lin}) at a scattering angle β is defined as the ratio of the power integrated over the azimuthal angles φ in the plane of polarization orthogonal ($I_{\perp}$) and parallel ($I_{//}$) to the linearly polarized source:

$$\delta_{lin}(\beta) = \frac{<I_{\perp}>}{<I_{//}>} = \frac{\int_0^{2\pi} I_y(\beta,\varphi)d\varphi}{\int_0^{2\pi} I_x(\beta,\varphi)d\varphi}. \quad (1)$$

Using Mie theory, it can be shown that:

$$\delta_{lin}(\beta) = \frac{(P_2\cos^2\beta - 2P_3\cos\beta + P_1)}{(3P_2\cos^2\beta + 2P_3\cos\beta + 3P_1)}, \quad (2)$$

where in P_1, P_2 and P_3 are the Mie scattering phase matrix elements defined by Deirmendjian.

Fig. 3 shows the phase function and the depolarization ratio at a wavelength of 0.532 µm for linear incident waves on water-droplets of effective radius of 6 µm. The water droplets size distribution is a gamma distribution with $\alpha = 7$ and $\beta = 1.5$.

For depolarization, the major observations are:

1) The light is not depolarized in the forward scattering direction for angles smaller than 20°; 2) the depolarization is equal to 0 at 180°; 3) the depolarization increases very sharply around 180°.

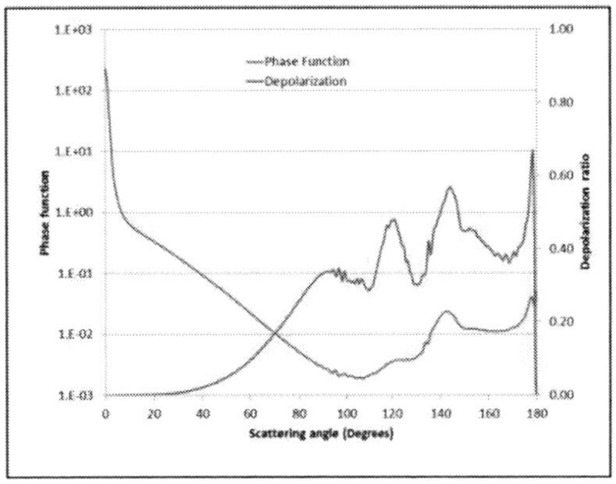

Figure 3. Phase functions and linear depolarization ratio for a water droplets gamma distribution with effective radius of 6 µm.

So, for a lidar, depolarization is mainly caused by a backscattering occurring after one or many forward scatterings as illustrated in Fig. 4.

A lidar with a given FOV (θ_i) will measure the backscattered light without any distinction regarding the forward scattering event position in space. The cross polarized second order scattering signal detected will be the sum of all those signals and it will be given by:

$$P_{D\perp}(z_c,\theta_i) = P(z_c)\int_0^{z_c}\int_0^{\beta_i} F(z,\beta)\sin\beta\, d\beta\, dz$$

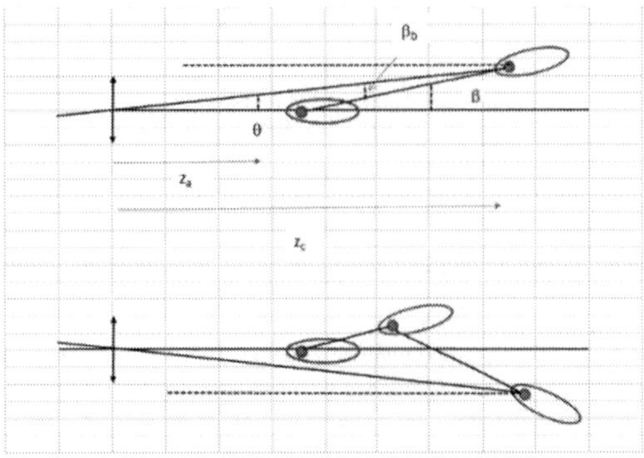

Figure 4. Double scattering processes and higher scattering processes leading to backscattering depolarization.

where $P(z_c) = P_0 \exp[-2\int_0^{z_c} \alpha(z)dz\ ]\dfrac{c\tau}{2}\dfrac{4\pi A}{z_c^2}$, and

$$F(z,\beta) = [\alpha_s(z)p(r,\beta)][\alpha_s(z_c)p_\perp(r,\beta_b)],$$

where, c, τ and A are respectively the speed of light, the pulse width, and the collecting optics area; α_s is the scattering extinction coefficient; $p(r,z\ ,\beta)$ is the value of the phase function for the forward scattering (β), and $p_\perp(r,z_c,\beta_b)$ the value of the perpendicular components of the phase function at backward scattering angles ($\beta_b = \pi - \beta + \theta$) for a particle of radius r; z_a is where the cloud starts; z_c is the distance where the scattered radiation is measured; the quantity $[\alpha_s(z)p(r,z\ ,\beta)]$ represents the forward scattering coefficient while $[\alpha_s(z_c)p_\perp(r,z_c,\beta_b)]$ represents the cross polarization backscattering coefficient. The FOV θ is easily related to the scattering angle β via: $\tan\beta = (z_c \tan\theta)/(z_c - z)$.

The forward scattering phase function is modelled as a summation of Gaussians [6]

$$p(r,\beta) = \frac{1}{2}\frac{1}{\pi\beta_d}\exp(-\beta^2/\beta_d^{\,2}) + \frac{A_g}{2}\frac{1}{\pi\beta_g}\exp(-\beta^2/\beta_g^{\,2})$$

$$\beta_d = 0.585(\lambda/2r_e);\ \beta_g = 0.481;\ A_g = 0.89;$$

r_e: effective radius, λ: wavelength.

There is no existing model for $p_\perp(r,\beta_b)$, the perpendicular components of the phase function. Our analysis has lead us rather to use the depolarization parameter ($D = p_\perp/(p_\perp + p_{//})$) rather than $p_\perp(r,\beta_b)$. Fig. 5 shows the definition of various parameters used to express the depolarization parameter as a function of scattering angles ranging from 170° to 180°. The complete derivation of the parameterization for r_e ranging from 2 to 12μm will be published in *Appl. Opt* in 2017. The mathematical representation is as follows:

$$\beta_{Max}(r_e) = 1.5845\ln(r_e) + 175.09^o$$

If, $\beta \geq \beta_{Max}(r_e)$

$$D(\beta,r_e) = A(r_e)[1 - \exp-((\pi-\beta)/(\beta_d(r_e)w(r_e))^4]$$

$$w(r_e) = 0.4625 + 0.10485r_e$$

If, $\beta < \beta_{Max}(r_e)$

$$D(\beta,r_e) = B_1(r_e)\cdot\exp(-B_3(0.944\pi - (\pi-\beta)) + B_2(r_e)$$

In short, $D(\beta,r_e)$ starts from a value of zero at 180° and increases quickly following a super-Gaussian. It needs to be noted here that the width of the super-Gaussian is directly related to the width of the forward scattering peak β_d. Past the maximum value, in the direction of decreasing angles, $D(\beta,r_e)$ decreases exponentially.

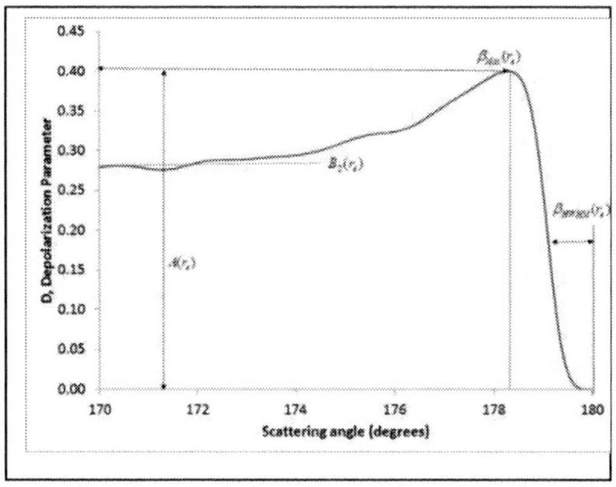

Figure 5. Parameterization of backscattering depolarization parameter. A water droplet cloud of effective radius of 6 μm is represented.

3. RESULTS

Fig. 6 shows the calculated and normalized detected energy in rings delimited by FOV θ_i

around the backscatter direction and for effective particle radius of 6 µm. It is obtained using the model previously described (see Fig. 5) for a wavelength of 532 nm, beam divergence: 0.3 mrad, pulse width: 5 ns, cloud base: 100 m, depth: 20 m, extinction coefficient equal to 0.25 m^{-1}.

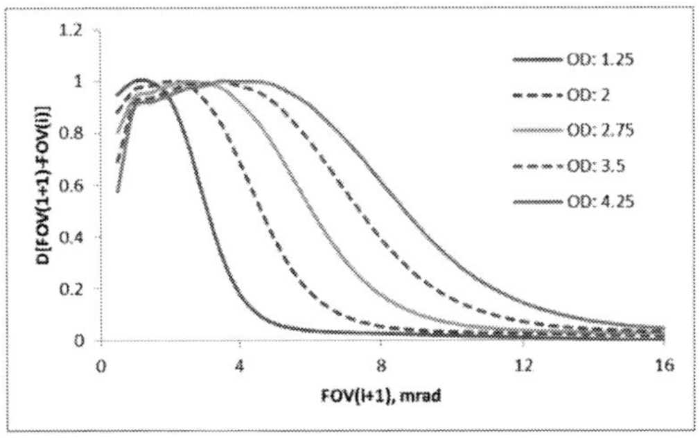

Figure 6. From 2nd scattering polarimetric model. Detected energy in rings delimited by FOV θ_i and θ_{i+1} for effective particle radius 6 µm.

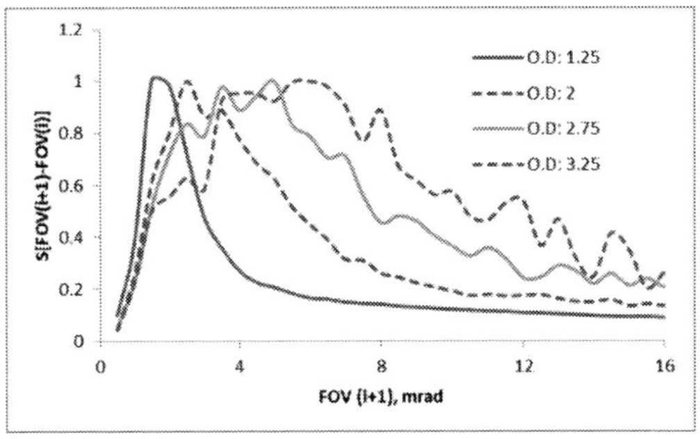

Figure 7. From MC simulation. Detected energy in rings delimited by FOV θ_i and θ_{i+1} for effective particle radius 6 µm

Using the same cloud parameters and lidar characteristics as for the model calculation displayed in Fig. 6, MC simulations were performed and images were analysed to recover second order scattering. Figure 7 shows the calculated and normalized detected energy in rings delimited by FOV θ_i and θ_{i+1} . Fig. 8 shows a clear dependency of the polarimetric signal over the droplet size

The MC results are noisier for the larger optical depths. The MC results and polarimetry double scattering model agree reasonably well in shape.

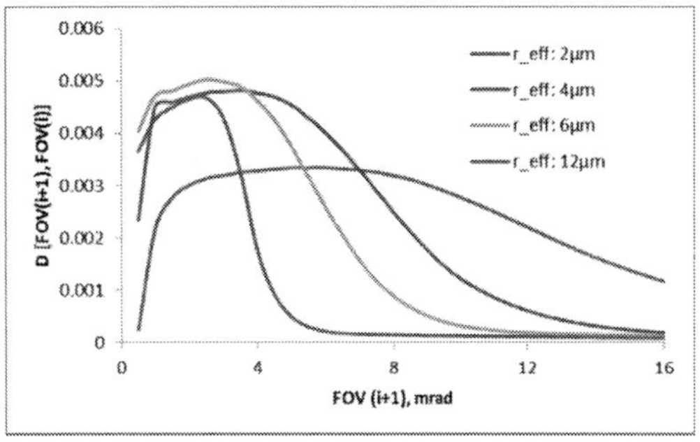

Figure 8. From 2nd scattering polarimetric model. Detected energy in ring delimited by FOV θ_i and θ_{i+1} for effective particle radius 2, 4, 6 & 12 µm. Same cloud extinction and position as Fig. 6 and Fig. 7. Penetration depth is of 11 m.

CONCLUSION

Monte Carlo simulations support experimental observations reported in [1]; i.e. multiple scattering cross polarization can be reduced to a second order scattering using the azimuthal dependency of the secondary polarization. A model of the polarimetric backscattering phase function has led to a mathematical expression for the polarimetric backscatter signal. This signal can be related to the droplet particle size. Future work will focus on the extraction of the droplet size.

REFERENCES

[1] S. R. Pal and A. I., Carswell, Polarization anisotropy in lidar multiple scattering from atmospheric clouds, *Appl. Opt.*, **24**, 3464-3471 (1985).

[2] V. Griaznov, I. Veselovskii, P. Girolamo, M. Korenskii and D. Summa, Spatial distribution of doubly scattered polarized laser radiation in the focal plane of a lidar receiver, *Appl. Opt.* **46**, 6821-6830, (2007)

[3] N. Roy, G. Roy, L. R. Bissonnette and J-R. Simard, Measurement of the azimuthal dependence of cross-polarized lidar returns and its relation to optical depth, *Appl. Opt.* **43**, 2777-2785, (2004)

[4] A. Doroshkevich, The effect of droplet cloudy microstructure on the polarization characteristics of double scattering lidar return, *SPIE* Vol. 10035, 1003550, (2016)

[5] G. Tremblay, G. Roy, High fidelity imaging algorithm for the Undique Monte Carlo simulator. *27rd International Laser Radar Conference,* New York, USA, (2015)

[6] Luc R. Bissonnette, Gilles Roy, Laurent Poutier, Stewart G. Cober, and George A. Isaac, Multiple-scattering lidar retrieval method: tests on Monte Carlo simulations and comparisons with in situ measurements, *Appl. Opt,* **41**, (2002)

EPJ Web of Conferences 176, 01031 (2018)
ILRC 28
https://doi.org/10.1051/epjconf/201817601031

VALIDATION OF DOUBLE-PULSE 1572 NM INTEGRATED PATH DIFFERENTIAL ABSORPTION LIDAR MEASUREMENT OF CARBON DIOXIDE

Juan Du[1,2], Jiqiao Liu[1*], Decang Bi[1], Xiuhua Ma[1], Xia Hou[1], Xiaolei Zhu[1], Weibiao Chen[1]

1Key Laboratory of Space Laser Communication and Detection Technology, Shanghai Institute of Optics and Fine Mechanics, CAS, No.390 Qinghe Road, Jiading, Shanghai 201800, China

2 University of Chinese Academy of Sciences, Beijing 100049, China

**Email: liujiqiao@siom.ac.cn*

ABSTRACT

A ground-based double-pulse 1572 nm integrated path differential absorption (IPDA) lidar was developed for carbon dioxide (CO_2) column concentrations measurement. The lidar measured the CO_2 concentrations continuously by receiving the scattered echo signal from a building about 1300 m away. The other two instruments of TDLAS and in-situ CO_2 analyzer measured the CO_2 concentrations on the same time. A CO_2 concentration measurement of 430 ppm with 1.637 ppm standard error was achieved.

1. INTRODUCTION

CO_2 is one of the main atmospheric greenhouse gases and plays an important role in global climate change. Global CO_2 concentrations with high precision measurements are significant for understanding the carbon cycle and improving the climate forecast model. A more accurate and effective technique to measure global atmospheric CO_2 concentrations has become strongly required. Differential absorption lidar (DIAL) technique is an effective method for trace gas concentrations measurement with high accuracy and sensitivity. In order to measure column CO_2 (XCO_2) concentrations from space, the integrated path differential absorption (IPDA) lidar is proposed [1-3.] . Usually 1.57 μm and 2 μm wavelength laser sources are acted as the preferred transmitter for IPDA lidar to measure XCO_2[4-6.]. Although the laser frequency stability is stricter for 1.57 μm than 2 μm laser source, the 1.57 μm lidar shows lower interference from water vapor and lower sensitivity from atmospheric temperature [7.]. In this paper, an IPDA lidar prototype system for CO_2 concentration measurement based on 1.57μm and some experiment results are presented. The optimized on-line and off-line wavelengths are 6361.2250 cm^{-1} and 6360.979 cm^{-1} respectively, which are shown in Figure 1. The lidar are compared with TDLAS and in-situ CO_2 analyzer (Los Gatos) in measuring the CO_2, and good agreements are shown.

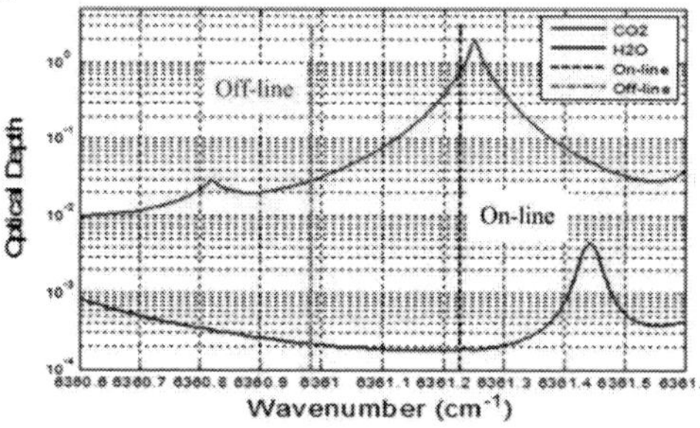

Figure 1 Operation wavelength of the IPDA lidar

2. SYSTEM CONFIGURATION

The block diagram of the IPDA lidar system is shown in Figure 2. This system consisted of the laser transmitter, the receiver, and the data acquisition and controller. The laser transmitter output 1572 nm and 1064 nm double-pulse lasers simultaneously. The double pulse's separation was 200 μs. The 1064nm laser was used to measure range to hard target. The 1572nm laser with on-line wavelength 6361.2250 cm^{-1} and off-line wavelength 6360.979 cm^{-1} was used to measure CO_2 concentrations. The 1572 nm laser was generated by the seeder injected optical parametric oscillator (OPO) method with single frequency 1064 nm laser acted as the pump laser pulse. The 1572 nm laser single pulse energy was about 1mJ with 50 Hz repetition rate and 20 ns pulse duration. In the laser transmitter, part of

1572 nm pulsed laser was received by an integrating sphere and detected by an InGaAs PIN photodiode. It was sampled and acted as the transmitting laser energy calibration reference. The receiver was composed of a 200 mm Schmidt-Cassegrain telescope, an 1572 nm detector channel and an 1064 nm detector channel. The 1572 nm detector channel signal was amplified by a 2 MHz bandwidth amplifier. Then the signal was sampled by an 100M/s AD card in the data acquisition and controller unit. The specifications of the system are summarized in the table 1.

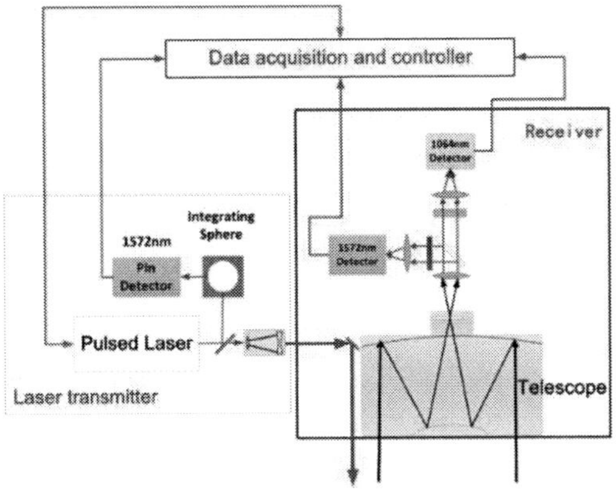

Figure 2 Block diagrams of the IPDA lidar system

Table1 Specifications of the lidar system

Parameters	value
Wavelength	1572 nm/1064 nm
Pulse energy	1 mJ@1572nm 5 mJ@1064nm
Pulse width	20 ns
Frequency stability	0.3MHz
Pulse repetition rate	50 Hz (double pulse)
Beam divergence	0.2 mrad
Telescope	Φ200 mm
FOV	0.5 mrad
Detector	Thorlabs PDA20CS-EC PIN Licel InGaAs APD module Licel Si APD module
AD	100M/s ,14bit, NI PCI5122

3. RESULTS

3.1 Retrieval method

The CO_2 differential absorption optical depth (DAOD) was calculated from the following expression:

$$DAOD = \int_0^R N_{CO_2}(r)\Delta\sigma(r)dr = \frac{1}{2}\ln\frac{P_{off}}{P_{on}}\frac{P_{on0}}{P_{off0}}, \quad (1)$$

Where $\Delta\sigma(r) = \sigma_{on} - \sigma_{off}$ was the differential absorption section, P_{on} and P_{off} were the echo signals from IPDA double pulses, P_{on0} and P_{off0} were the transmitting energy reference signals. Usually, multiple pulses were averaged to calculate DAOD with high accuracy.

The column CO_2 concentration (XCO_2) could be calculated from the DAOD and the integrated weighting function WF, and was expressed as:

$$XCO_2 = \frac{DAOD}{\int_{P_{TOA}}^{P_{surf}} WF(p,T)dp}, \quad WF(P,T) = \frac{\Delta\sigma_{CO_2}(P,T) * N_{air}}{1+\rho_{H_2O}}, \quad N_{air} = \frac{P\cdot NA}{RT}. \quad (2)$$

Where the weighting function $WF(p,T)$ was related to operational on-line and off-line wavelengths, and atmospheric parameters such as pressure and temperature and water vapor concentrations.

The XCO_2 relative random error (RRE) was expressed in the following as:

$$RRE = \frac{\Delta XCO_2}{XCO_2} = \frac{\Delta DAOD}{DAOD} = \frac{1}{2\cdot DAOD}\frac{1}{SNR}, \quad (3)$$

Where the SNR was the signal to noise ratio of total echo and energy monitor reference signals. So when the DAOD kept constant, RRE was only decided by the SNR. For lower RRE, higher SNR was required.

3.2 Experiments

Ground validation experiments were carried out for the IPDA lidar system compared with TDLAS and in-situ CO_2 analyzer instruments. Atmosphere pressure, temperature and water vapor concentration were recorded by the Davis weather station.

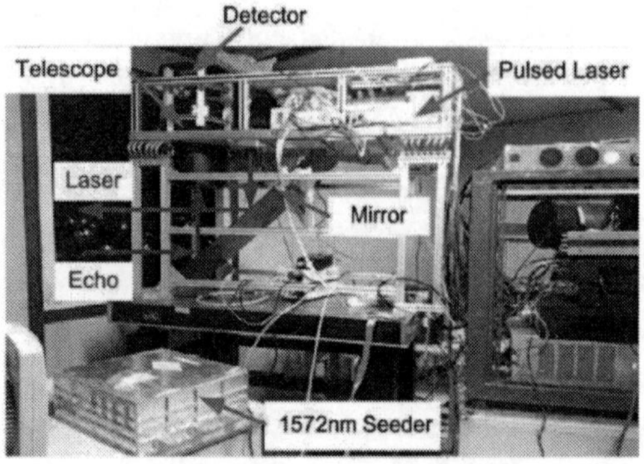

Figure 3 The IPDA lidar prototype

Figure 3 shows the IPDA lidar prototype. The transmitting laser beam was rotated 90 degrees by a mirror, and then passed the window in the lab to the atmosphere. The lab was on the top of five-story building. The lidar would receive the echo signal scattered by a building about 1300 m away. Then the XCO_2 in horizontal plane could be measured by the lidar. On the same time, the in-situ CO_2 analyzer could measure the CO_2 concentrations outside the lab with accuracy less than 0.3 ppm. The TDLAS instrument also was used to make comparisons. But it was different from the lidar and the path length was only about 120 m. Because the energy of its light was not enough to measure long-distance CO_2.

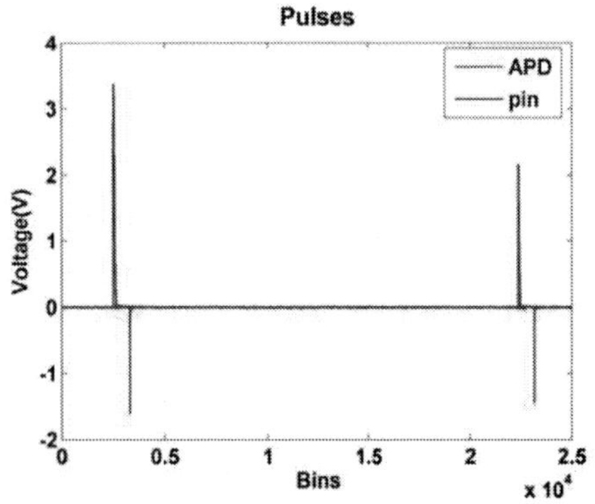

Figure 4 The lidar acquired double pulse echo and energy monitor signals

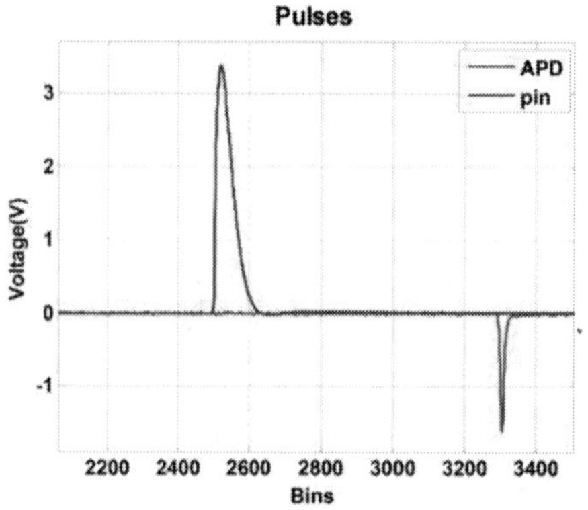

Figure 5 Single pulse echo and energy monitor signals waveform

The lidar acquired double pulse echo and energy monitor signals, shown in Figure 4. Energy monitor signals were received by a PIN detector with positive voltage. And the scattered echo signals were received by an APD detector with negative voltage output. The single pulse echo and energy monitor signals waveform are shown in Figure 5.

From the measured atmospheric parameters and calculated weighting function and background XCO_2 of about 400 ppm, we can calculate the theoretical value of DAOD which was 0.087. On-line and off-line normalized single echo signals are presented in Figure 6. When 4000 pulses averaged, the ratio of on-line and off-line normalized echo signal which is also the ratio in the log term of equation (1), shown in Figure 7. Then we got the DAOD, as shown in Figure 8. The mean DAOD was 0.0877, which agree with the theoretical DAOD.

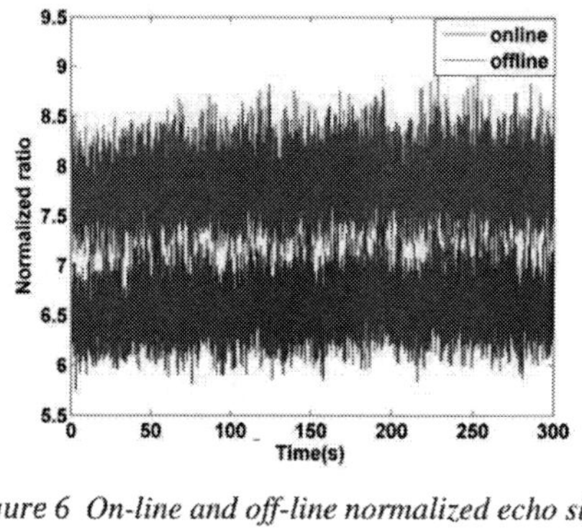

Figure 6 On-line and off-line normalized echo signal

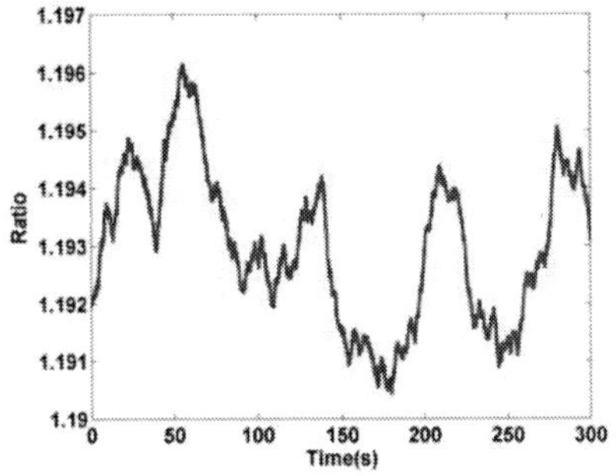

Figure 7 Ratio of on-line and off-line normalized echo signal

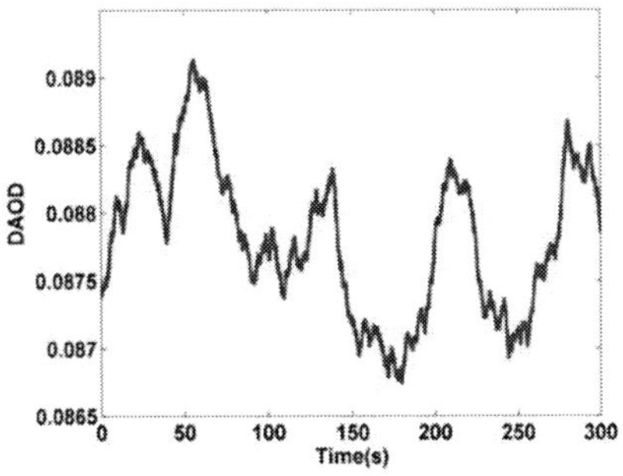

Figure 8 Calculated DAOD

The IPDA lidar retrieved XCO_2 compared with Los Gatos in-situ analyzer measured CO_2 concentration is shown in Figure 9. The mean XCO_2 from lidar is about 430 ppm with 1.637 ppm standard error. Both lidar and in-situ analyzer measured data agree well. On another day, the IPDA lidar retrieved XCO_2 is also compared with TDLAS, presented in Figure 10. The TDLAS path length was about 120 m and the cube corner retro-reflector target azimuth was different from the lidar scattered building wall. But both showed good correlation.

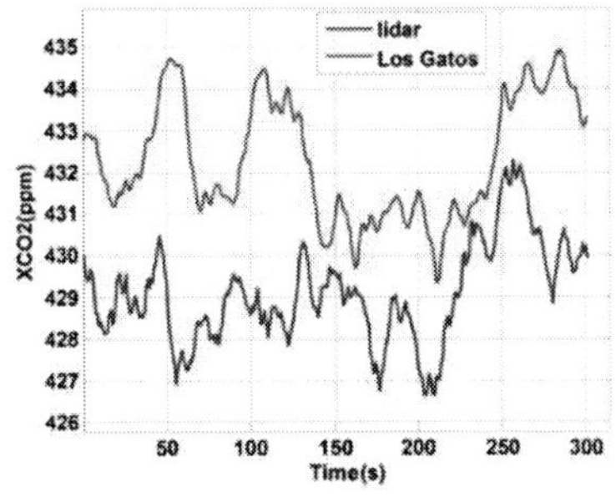

Figure 9 Comparison of XCO2 measured by lidar and Los Gatos in-situ analyzer

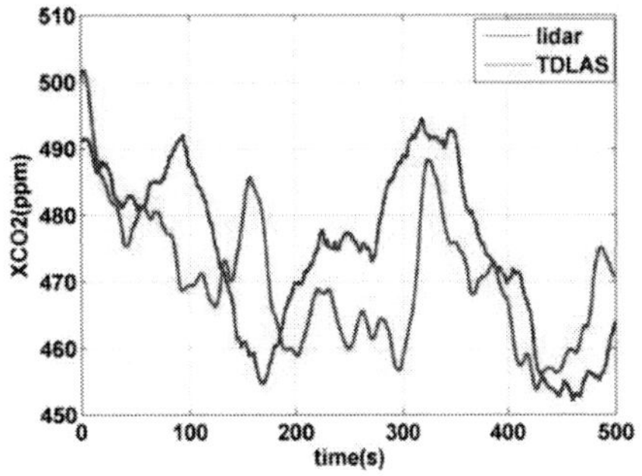

Figure 10 Comparison of XCO_2 measured by lidar and TDLAS

4. CONCLUSIONS

A ground-based double-pulse 1572 nm integrated path differential absorption (IPDA) lidar was developed for carbon dioxide (CO_2) column concentrations measurement. The lidar was compared with TDLAS and in-situ CO_2 analyzer to measure the column CO_2 concentrations. Good agreement was shown. When 4000 pulses averaged, the mean XCO_2 of 1300 m path length from lidar was about 430 ppm with 1.637 ppm standard error. This lidar will be upgraded soon for XCO_2 measurement in future airborne platform.

ACKNOWLEDGEMENT

This work is supported by Chinese atmosphere environment monitoring satellite project program.

REFERENCES

[1] ASCENDS (Active Sensing of CO2 Emissions over Nights, Days, and Seasons), http://decadal.gsfc.nasa.gov/ASCENDS.html.

[2] J. Caron, Y. Durand, J.-L. Bezy, and R. Meynart. 2009: Performance modeling for A-SCOPE, a space borne lidar measuring atmospheric CO2 . *Proc. SPIE*, 7479: 74790E.

[3] Ehret, G. und Kiemle, C. und Wirth, M. und Amediek, A. und Fix, A. und Houwrling, S. ,2008: Space-borne remote sensing of CO2, CH4, and N2O by integrated path differential absorption lidar: a sensitivity analysis. *Applied Physics B*, 90, p.593-608

[4]Amediek, A. Fix, M. Wirth, and G. Ehret, 2008: Development of an OPO system at 1.57 μm for integrated path DIAL measurement of atmospheric carbon dioxide, *Appl. Phys. B* ,92, 295–302

[5]G. Wagner and D. Plusquellic, 2016: Ground-based, integrated path differential absorption LIDAR measurement of CO2, CH4, and H2O near 1.6 μm, *Appl. Opt.* 55, 6292-6310

[6]Tamer F. Refaat, Upendra N. Singh, Jirong Yu, Mulugeta Petros, Ruben Remus, and Syed Ismail, 2016: Double-pulse 2-μm integrated path differential absorption lidar airborne validation for atmospheric carbon dioxide measurement, *Appl. Opt.* 55, 4232-4246

[7] Jérôme Caron and Yannig Durand, 2009: Operating wavelengths optimization for a spaceborne lidar measuring atmospheric CO2, *Appl. Opt.*48, 5413-5422 (2009)

Comparison between two Lidar methods to retrieve microphysical properties of liquid-water clouds

Cristofer Jimenez[1]*, Albert Ansmann[1], David Donovan[3], Ronny Engelmann[1], Jörg Schmidt[2] and Ulla Wandinger[1]

*[1]Leibniz Institute for Tropospheric Research (TROPOS), Leipzig, Germany, *jimenez@tropos.de*
[2]University of Leipzig, Institute for Meteorology (LIM), Leipzig, Germany
[3]Royal Netherlands Meteorological Institute (KNMI), De Bilt, The Netherlands

ABSTRACT

Since 2010, the Raman dual-FOV lidar system permits the retrieval of microphysical properties of liquid-water clouds during nighttime. A new robust lidar depolarization approach was recently introduced, which permits the retrieval of these properties as well, with high temporal resolution and during daytime. To implement this approach, the lidar system was upgraded, by adding a three channel depolarization receiver. The first preliminary retrieval results and a comparison between both methods is presented.

1 INTRODUCTION

The research on aerosol-cloud interaction [1] has intensified during the last years. Large uncertainties in the aerosol impact on liquid-water cloud evolution exist. To characterize this influence, accurate and simultaneous observations of the microphysical properties of the clouds and of aerosols below the clouds base are needed. Many scientist have attempted to estimate the microphysical properties of liquid clouds by analyzing the multiple scattering effects with lidar [2, 3, 4, 5, 6].

By measuring the nitrogen Raman signal returns with two FOVs, the retrieval of the single scattering extinction coefficient and the effective radius in thin layered liquid water clouds is possible [4, 5]. Since 2010, measurements of the Raman dual FOV lidar at TROPOS were used for this purpose. By combining the cloud results with vertical wind measurements from a Doppler lidar, a statistical analysis of the aerosol-cloud-dynamics relationship based on lidar measurements was possible [7,8]. However, for a deep statistical analysis, much more measurements are required, and since the dual-FOV is limited to nighttime measurements, an alternative method to extend the measurements to the daytime is needed.

Recently a robust optimal estimation procedure to retrieve this microphysical properties from lidar depolarization measurements was developed [6]. This is an innovative application, carrying on with previous work with the dual FOV techniques developed in the same group. The method makes use of the relationship between cloud multiple scattering effect and light depolarization by cloud droplets [9].

In order to implement this method, three telescopes for high accuracy depolarization measurements were implemented in the multi-wavelength Tropospheric Raman lidar for Temperature, Humidity, and Aerosol profiling (MARTHA). For first time, two different lidar-based methods to study the microphysics of liquid-water clouds are integrated in one lidar system, making possible a preliminary comparison of both methods in terms of the retrieved cloud microphysical and optical properties.

2 TWO LIDAR METHODS FOR CLOUD MICROPHYSICS OBSERVATIONS

A measurement case performed with the upgraded lidar MARTHA is presented in the Figure 1. A very stable stratocumulus layer developed in a polluted air mass. The time period analyzed was from 19:56 to 20:18 UTC. During this time period the cloud base remained stable (condition needed from both techniques).

2.1 Raman dual-FOV lidar method

This method is based on the measurement of nitrogen Raman backscatter signals (607 nm) with two FOVs, here with $\boldsymbol{FOV_1} = \mathbf{0.5}$ mrads (IN) and $\boldsymbol{FOV_2}$=2.0 mrads (OUT). Due to the weakness of the Raman signals, both profiles where averaged over 22 minutes.

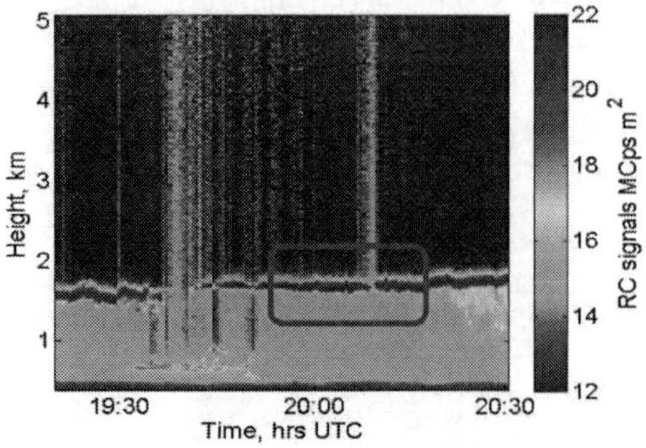

Figure 1: Height-time display of the range corrected elastic backscatter signal (532 nm) measured on 8 November 2016. The red rectangle shows the selected time period (19:56 to 20:18 UTC).

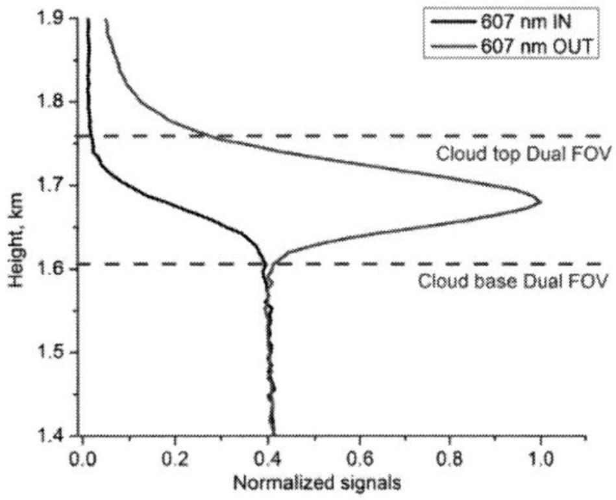

Figure 2: Normalized Raman signals from the two FOVs. Cloud base in 1.610 km.

The cloud base is directly obtained from the FOV_2 channel (red profile in Figure 2), since this channel only detects multiply scattered photons. In the Figure 2, the cloud base was set to 1.605 km and the cloud top to 1.76 km, the cloud was then divided into homogeneous layers.

An analytical relationship between the Raman lidar returns and the cloud extinction coefficient and effective radius is used to simulate both Raman signals [10], and to identify the best matching pair of cloud extinction coefficient and effective radius profiles [5].

The result is shown in the Figure 3. From the extinction coefficient and the effective radius profiles the liquid water content (LWC), is determined, and finally the cloud droplet number concentration (CDNC) is estimated [7].

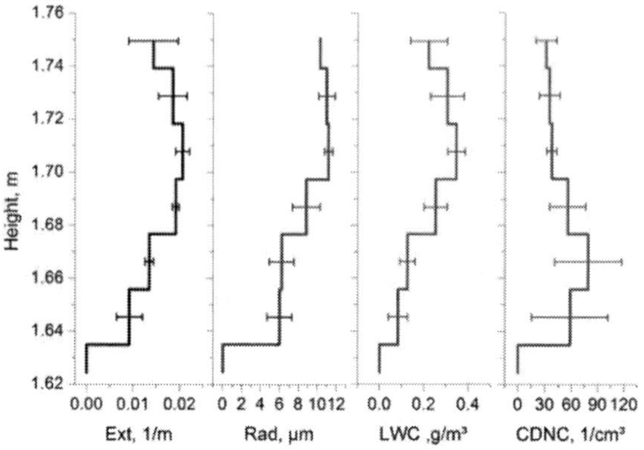

Figure 3: Results of the microphysical properties with the dual-FOV method.

2.2 Depolarization method

The same time period was selected to use the depolarization method [6], but the averaging is performed for 3-minute periods to obtain 7 profiles in total. The calibration for the polarization channels was based on a three signal procedure [11]. The found inter-channel calibration constant for this day is Cr=9.62 and a cross-talk factor $\delta_c = 0.04$.

The cloud base altitude for each profile considered in this inversion procedure is based on a threshold of β_{II}=0.05, where β_{II} is the attenuated parallel signal. As an average value the cloud base were set to 1.62 km. In this approach the cloud base region is considered as adiabatic, that means, a constant CDNC and a quasilinear LWC are assumed. At sub-adiabatic conditions $\alpha(z), R_{eff}(z), LWC(z)$ and $CDNC(z)$ are linked to each other [6].

A reference altitude of 100 meters above the cloud base was chosen. Look-up tables from the EarthCARE lidar Monte Carlo simulations were used for an optimal estimation minimization procedure. By means of the Look-up the attenuated calibrated parallel and cross polarized profiles can be computed as a function of cloud base altitude, FOV, α_{100} and $R_{eff,100}$.

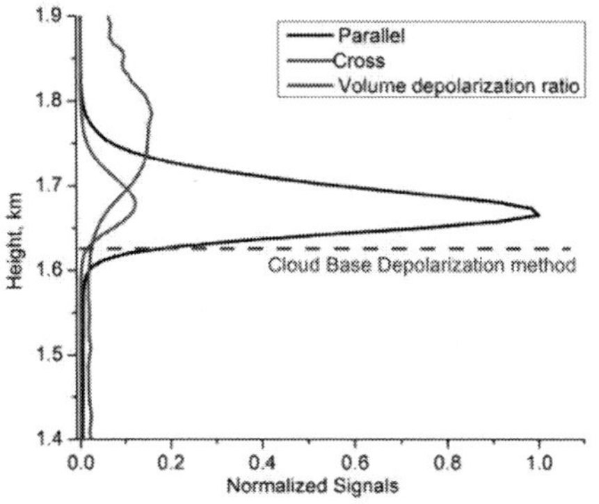

Figure 4: Averaged normalized polarization signals for the period from 19:56 to 20:18 UTC.

The inversion results of the depolarization method for 08 November 2016 from 19:56 to 20:18 UTC are shown in the Figure 5. On the bottom are the solutions for the extinction coefficient and the effective radius 100m above the cloud base, as a second product the LWC and the CDNC for this reference height were obtained [6].

3 COMPARISON BETWEEN BOTH METHODS

From Figures 2 and 4, a very good agreement between both criteria to define the cloud base height was observed. According to the dual-FOV method the cloud base was 1.605 km, the depolarization method detects the cloud base at 1.620 km height.

The dual FOV results (Figure 3) suggests that there is a quasi linear LWC and a constant CDNC as it has been assumed by the depolarization method, which makes this case a very good scenario for comparison.

Since the retrieval with the dual FOV method is given in one single profile for the 22-minute time period, for a first comparison the 7 retrievals obtained with the depolarization method are averaged. The values 100m above the cloud base (at 1.72 km) are presented in Table 1.

The depolarization method leads to higher values of the extinction coefficient and lower values for the effective radius. The large difference is then observed in the cloud droplet number concentration, which is about ten times higher when using the depolarization method.

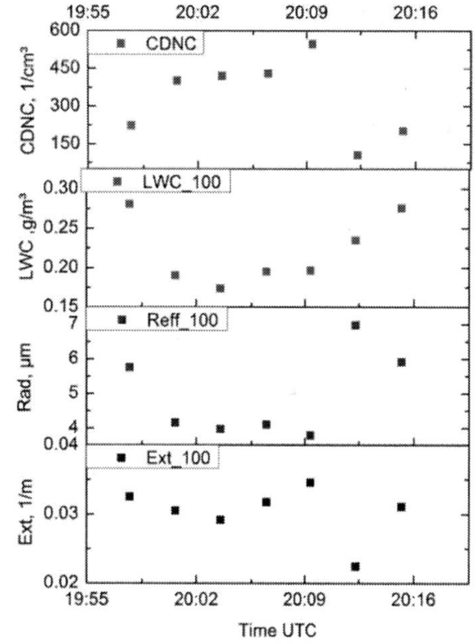

Figure 5: Inversion results from the depolarization measurements, by using the Donovan et al. method [6].

Table 1: Results 100 m above the cloud base.

	Dual FOV	*Depolarization*
α_{100}, m^{-1}	0.01845	0.03029
$R_{eff,100}, \mu m$	10.9768	4.9506
LWC_{100}, gm^{-3}	0.3037	0.2209
$CDNC_{100}, cm^{-3}$	34.81	332.68

4 CONCLUSIONS

A new depolarization-ratio method is implemented in the TROPOS MARTHA dual FOV Raman lidar to retrieve cloud microphysical properties as well as the cloud droplet extinction coefficient. First test measurements could be made. In the next phase, more cases need to be measured and analyzed in order to quantify the potential and accuracy of the new depolarization technique. These quality-assurance efforts are based on the comparison of the results obtained with the Raman dual-FOV lidar technique and the new depolarization technique.

ACKNOWLEDGEMENTS

The support of the lidar group of the Leibniz Institute for Tropospheric Research is acknowledged. The support of Dave Donovan of the KNMI is gratefully acknowledged. I acknowledge the financial support by DAAD.

References

[1] Twomey S., 1977: Influence of pollution on shortwave albedo of clouds, J. Atmos. Sci. 34, 1149–1152.

[2] Kim, D., Cheong, H., Kim, Y., Volkov, S., and Lee, J., 2010: Optical depth and multiple scattering depolarization in liquid clouds, Opt. Rev., 17, 507–512, doi:10.1007/s10043-010-0091-7.

[3] Roy G., Bissonnette L., Bastille C., and Valle´ e G., 1999: Retrieval of droplet-size density distribution from multiple-field-of-view cross-polarized lidar signals: theory and experimental validation. Aplied Optics Vol. 38, No. 24

[4] Malinka, A. V. and Zege, E. P.,2007: Possibilities of warm cloud microstructure profiling with multiple-field-of-view Raman lidar, Appl. Optics, 46, 8419–8427.

[5] Schmidt, J., Wandinger, U., and Malinka, A, 2013: Dual-field-of-view Raman lidar measurements for the retrieval of cloud microphysical properties, Appl. Optics, 52, 2235–2247.

[6] Donovan D. P, H. Klein Baltink, J. S. Henzing, S. R. de Roode, and A. P. Siebesma, 2015: A depolarisation lidar-based method for the determination of liquid-cloud microphysical properties.

[7] Schmidt J., Ansmann A., Bühl J., Baars Wandinger U., Müller D, and Malinka A. V, 2013: Dual-FOV Raman and Doppler lidar studies of aerosol-cloud interactions: Simultaneous profiling of aerosols, warm-cloud properties, and vertical wind, J. Geophys. Res. Atmos., 119, 5512–5527

[8] Schmidt J., Ansmann A., Bühl J, and Wandinger U., 2015: Strong aerosol–cloud interaction in altocumulus during updraft periods: lidar observations over central Europe.

[9] Hu, Y., Vaughan, M., Liu, Z., Lin, B., Yang, P., Flittner, D., Hunt, B., Kuehn, R., Huang, J., Wu, D., Rodier, S., Powell, K., Trepte, C., and Winker, D., 2007: The depolarization – attenuated backscatter relation: CALIPSO lidar measurements vs. theory, Opt. Express, 15, 5327–5332.

[10]Malinka A. V., Zege E. P., 2003: Analytical modeling of Raman lidar return, including multiple scattering. Applied optics vol 42, Nro 6.

[11]Reichardt J, Baumgart R. and McGee T. J., 2003: Three-signal method for accurate measurements of depolarization ratio with lidar. Applied optics, col. 42, No 24.

LIDAR TO DETERMINE THE FRACTIONS OF ICE, LIQUID AND WATER VAPOR IN POLAR TROPOSPHERIC CLOUDS

Gilberto J. Fochesatto[1]*, Oscar Galvez[2], Pablo Ristori[3], Douglas Keller[1] and Enzo Fochesatto[1]

[1]*Geophysical Institute, University of Alaska Fairbanks, AK 99775 USA*

[2]*Laboratorio de Fisica de Materiales, UNED. Madrid, España.*

[3]*Laboratorio Lidar. CITEDEF, Buenos Aires, Argentina.*

**Email: gjfochesatto@alaska.edu*

ABSTRACT

A new Lidar combining Raman spectroscopy and linear polarization analysis is presented. This new instrument identifies the fraction of ice, liquid, and water vapor in low level polar tropospheric clouds and provides the polarimetric S and P state of the backscattering 532 nm Lidar signal. An overview of the research applications is given followed by a theoretical estimation of the Lidar returns. The instrument concept and optical characteristics are discussed. First Lidar profiles and instrument evaluations will be provided during the conference.

1. INTRODUCTION

The thermodynamic state of water in the troposphere plays a central role describing aerosol-cloud microphysical and dynamical processes in polar atmospheres. Spectrally resolved information on the thermodynamic phases of water in terms of the amount of ice, liquid, and vapor can be achieved by Laser spectroscopy (i.e., spontaneous Raman scattering). Such an instrument would be able to provide key information on specific cloud and aerosol processes and also account for changes in the surrounding environment as cloud and aerosol layers evolve and interact. Measuring the Raman shift up to ~3800 cm^{-1} with spectral resolution better than 1 cm^{-1} allows selection of discrete Raman bands and vibrational lines to deduce the fractions of ice, liquid, and water vapor. Quantitative determination of concentration profiles requires knowing the Raman scattering cross section of the species under consideration. However, in this case this will be accomplished by referring all Raman returns to the well-known N_2 cross section [1]. Interpretation of the vibrational levels of water vapor does not represent a critical issue. However, interpretation of the Raman spectra during phase transitions that occur in a super-cooled state requires some further research work. This has been documented in the past by divergences between *ab initio* calculations, numerical approximations and laboratory experiments [2-8], and, of particular importance here, when cloud strata in the super-cooled state undergo homogenous or heterogeneous nucleation [9, 10]. In these cases, the intramolecular high frequency OH stretch mode shows significant changes around the isosbestic point in the spectral region 3000 to 3800 cm^{-1} as temperature decreases from ambient to -24 °C, as seen by the polarimetric sensitive Raman spectra [11]. And, as temperature lowers to -35 °C and even lower, the asymmetric hydrogen bond (<175 cm^{-1}) [12-15] is observed to decrease bond fluctuation, consistent with an increase in the formation of large structure polyhedral clusters in the region < 50 cm^{-1} [16]. (see Figure 1).

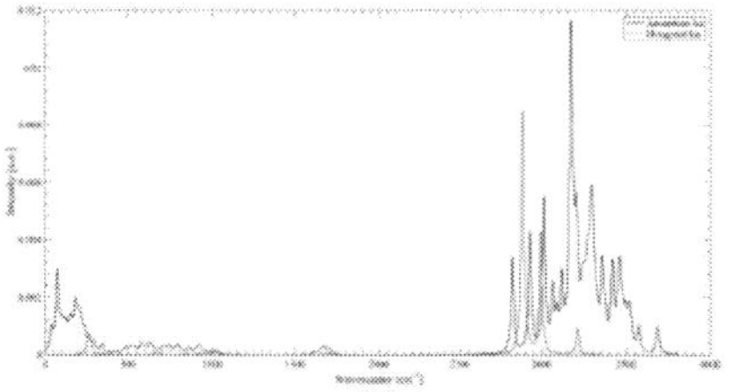

Figure 1. Theoretical Computation of Raman Spectrum of amorphous and hexagonal ice structures based on CASTEP. Lorentzian smearing of 20 cm^{-1} at 200 K and 10 cm^{-1} at 290 K respectively.

These complicated physicochemical processes that have been investigated by laser spectroscopy in laboratory settings are mature enough to be further studied in a Lidar configuration. This can provide more sophisticated aerosol-cloud freezing insight for model development.

Furthermore, the spectroscopic information provided by this new Raman Lidar will be combined with the determinations of dynamic and microphysical structure of aerosols and cloud layers by means of polarimetric backscatter determination [27, 28]. The simultaneous polarimetric-Raman capability will allow further identification of thermodynamic processes of water while in the presence of aerosols and clouds, identify non-spherical scatters, and detect ice formation, as well as estimate the crystal shape and habitat and their morphological changes, as they are also sensitive to the polarimetric observations from the laser line. Determination of atmospheric trace gas in the presence of aerosols is fundamental in studying physicochemical processes in multiphase and multicomponent atmospheric chemistry. These processes include aerosol seeding cloud formation, the role of aerosol surfaces assisting heterogeneous ice nucleation, and the chemical reactions on ice surfaces among others [17].

Since the inception of Raman Lidar methodology by Cooney [18], the technology has significantly evolved to a level that now allows simultaneous determination of several atmospheric chemical components using either Raman Lidars or combined Differential Absorption and Raman based Lidars to profile N_2, O_2 and H_2O, SO_2, O_3 and aerosols [19]. Most typical Raman Lidars [1, 20] involve the use of dedicated, custom made spectrometric units to spectrally resolve several molecular Raman features as a function of height. This currently available instrumentation operates in the visible and near UV spectra, normally stimulated by 532 and 355 nm pulsed lasers and in the UV-solar blind zone [19]. An example of the latter is the tropospheric ozone Lidar based on a 266 nm laser-pumped Raman multi-wavelength generation laser emission [21] and Raman Lidar using 266 nm as laser excitation [22].

On the other hand, Raman Lidars operating in the visible region are limited to nighttime minimal light conditions because the background solar radiation levels during the daytime surpass the Raman associated photon counting signal. In most of the cases, depending on the actual latitudinal location, the background radiation can top by several orders of magnitude the photon count level of the backscattering Raman signal. Some advances, including using Fabry-Perot etalons in the Lidar receiver, have partially overcome this issue [23]. Lidars operating in the UV solar blind region, in addition to being eye-safe, enhance the backscattering Raman cross section because of the short laser wavelength, typically <300 nm, but they experience strong signal absorption by ozone in the lower troposphere limiting their operation range [22]. Lidars operating in near-UV spectrum generally optimize the trade-off between maximizing the Raman cross section while minimizing background radiation. Still, they require the use of complicated optical arrangements, using multiple detectors and multichannel data acquisition receivers. A recent example [24] demonstrated the measurement of spectrally resolved water vapor and liquid Raman bands using a dispersive spectrometer coupled to a multichannel detector containing 32-fiber coupled integrated photomultipliers.

It is clear that in order to optimize the instrumental range -keeping the Raman cross section high enough while reducing background radiation- the laser stimulation has to be in near-UV spectrum. However, at shorter laser excitation wavelengths, Raman features narrow their spectral separation, making the receiver optics significantly more expensive. In this work, we therefore concentrate on demonstrating the feasibility of this Lidar spectroscopy, sacrificing signal-to-noise ratio and ranging capability by selecting the 532 nm wavelength as the laser excitation. Operating at this wavelength allows spectral requirements to be achieved within reasonable prices in the optics.

2. LIDAR SIMULATION

A simulation of the Raman Lidar returns, in MCPS (10^6 counts per second), was performed for the main Raman lines to ensure detection limits, maximum vertical resolution, and minimum averaging time. The simulation was initialized by radiosonde profile assuming specific atmospheric properties and optical characteristics in the Lidar emission and receiver. This simulation allows for the evaluation of the signals to be expected in the Raman channels. The Lidar equation was implemented, including the terms of spectral extinction α [m^{-1}], backscattering β [$sr^{-1}m^{-1}$] [25], and the Raman scattering cross sections of the main atmospheric features N_2, O_2 and H_2O (see Table 1). The following expression shows a

 https://doi.org/10.1051/epjconf/201817601033

compact form of the Raman-Lidar equation as a function of the detected wavelength (*X*).

$$S_X(z) = \frac{E.A}{z^2}.\eta_X.O(z).N_X(z).\sigma_X^{RAM}(\lambda^{RAY},\lambda_X^{RAM}).Tr(z,\lambda^{RAY}).Tr(z,\lambda_X^{RAM})$$

E is the laser energy per pulse, A is the telescope effective collecting area, z is the height, η_X is the optical throughput at the Raman wavelength (λ_X^{RAM}), $O(z)$ is the overlapping Lidar function, $N_X(z)$ is the concentration profile of the species of interest, $\sigma_X^{RAM}(\lambda^{RAY},\lambda_X^{RAM})$ is the Raman cross section of the species of interest depending on the laser stimulation λ^{RAY} and the Raman wavelength shift λ_X^{RAM}, $Tr(z,\lambda^{RAY})$ is the Rayleigh transmission of the atmosphere, including molecules and aerosols, at the laser wavelength, and $Tr(z,\lambda_X^{RAM})$ is the atmospheric transmission at the Raman wavelength in the time of flight returning to the telescope.

In this case, uniform aerosol vertical concentration profile in the atmospheric boundary layer (ABL) and wavelength scaling was considered, giving a specific Angstrom coefficient to simulate the spectral extinction due to aerosols [26]. The Lidar signal computation includes a random noise signal level based on a Poisson distribution to simulate real noise photon-detection conditions and the spectral gain curve associated with the detector. The MCPS photon count rate is given at the entrance of the spectroscopic device. The optical throughput of the device that produces the spectroscopy of the backscattered signal strongly changes depending on the spectral characteristics of the analyzer (e.g., dichroic mirrors and filters). The laser emission and receiver parameters, including the detector characteristics, are indicated in Table 1. Figure 2 illustrates the MCPS simulation of the molecular Raman channels. Additionally, theoretical ab initio calculations of Raman spectra based on hexagonal microphysical ice structure and amorphous ice consisting of 20 water molecules (for sub-freezing temperatures) and the theoretical Raman spectrum of amorphous ice consisting of 20 water molecules is shown in Fig. 1. This information, combined with that in Fig. 3, suggests that strong changes in a band around ~ 3100 cm^{-1} will indicate the formation and presence of ice, while liquid content should be determined after the isosbestic point where ice features are not prominent ~ 3370 cm^{-1}.

Table 1. 532 nm – Raman Lidar Simulation

Laser Emission Surelite SL II 10 Continuum Lasers	
Wavelength	1064.0 / 532 nm
Linewidth	1.0 cm^{-1}
Pulse repetition frequency	10 Hz
Energy per pulse	650 / 300 mJ
Lidar Receiver	
Smith-Cassegrain	F/5
Primary mirror diameter	50 cm
Focal	2.5 m
Custom Spectrometer	
Dichroic / Razor filters based	F/5 input
Optics and Fiber Optic coupler	Adaptor to F/2.5 coupled to fiber NA=0.2 – 1 mm core
Detection	
Hamamatsu PMT	H10720-20
Peak wavelength	630 nm
Cathode Radiant Sensitivity	78 mA/W
Cathode Quantum Efficiency	0.02
Detector Gain	10^6
Accumulation time	1 min.
Vertical resolution	1.5 m
Angle tuning Raman features	607 nm N_2 - 2331 cm^{-1} 643 ice - 3175.3 cm^{-1} 650 liquid - 3373.6 cm^{-1} 660 nm H_2O - 3651 cm^{-1}

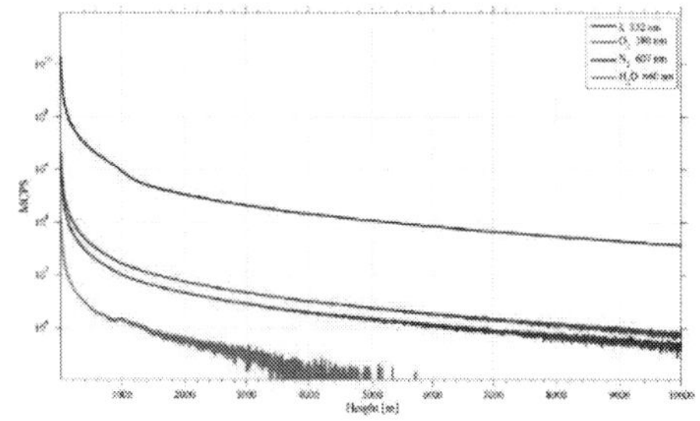

Figure 2. Raman Lidar Simulation. MCPS for 532 nm and Raman lines: O_2 at 580 nm; N_2 at 607 nm and H_2O at 660 nm. Thermodynamic profile from the NWS-PAFA June 21, 2006 at 00UTC (3 PM-AKT). ABL height was at 1 km, entrainment zone thickness of 200 m and ozone concentration set to 40 ppb both ABL and free troposphere. Aerosol optical thickness at 380 nm was 0.3, scaling wavelength Angstrom coefficient was 1.3 and a backscatter to extinction ratio of 0.012 sr^{-1}.

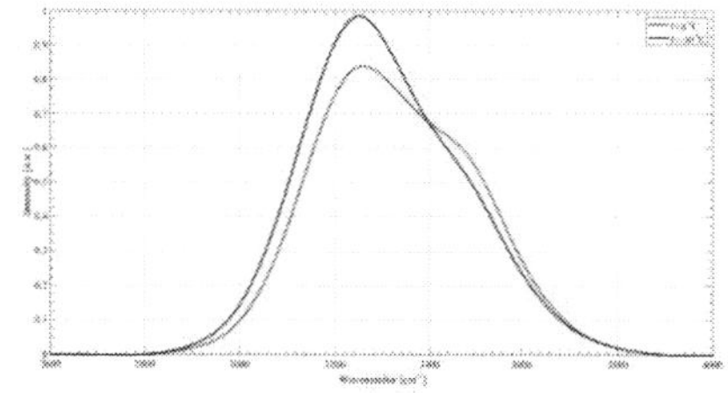

Figure 3. Raman Spectra of Water clusters, adapted from [11].

3. INSTRUMENT DESIGN

The Lidar emitter is based on a frequency doubled Nd:YAG pulsed laser operating at 532 nm. The Lidar receiver is based on a Smith-Cassegrain F/5 with a 0.5 m primary mirror diameter with

optimized optical throughput for visible spectroscopy. In the receiver, after splitting the linear polarization modes (S and P), the telescope is fiber coupled to a spectrometer designed to separate the vibrational Raman lines corresponding to N_2 and H_2O, and the Raman features corresponding to Raman bands of ice at 3175.3 cm^{-1} and liquid at 3373.6 cm^{-1}.
The instrument layout is shown in Figure 4.

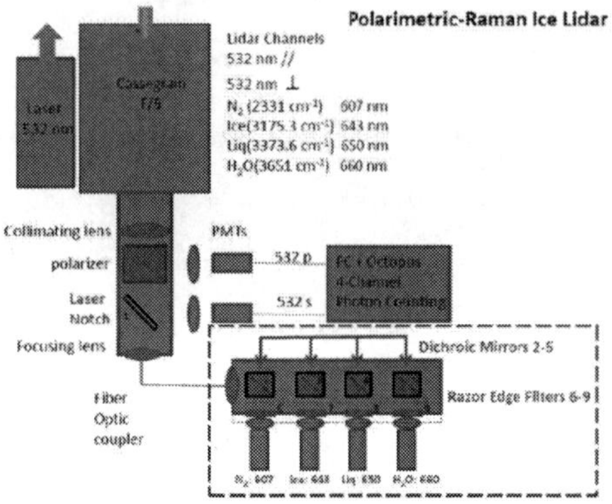

Figure 4. Polarimetric-Raman Ice Lidar Optical Layout.

4. CONCLUSIONS

A new Lidar system is being developed to investigate aerosol-cloud processes and composition in polar atmospheres. In this presentation a demonstration of feasibility will be presented illustrating several case examples in high latitude polar atmosphere.

ACKNOWLEDGEMENT

This project was funded by the US-NSF AGS-1443222. The authors recognize funding from the EU 2013-program Advanced Spectroscopy in Chemistry, Complutense University of Madrid and fruitful discussions with Prof. Rafael Escribano CSIC, Madrid, Spain.

REFERENCES

[1] Turner D. D. and D. N. Whiteman. 2002. Handbook of Vibrational Spectroscopy. Edited by J. M. Chalmers and P. R. Griffiths. John Wiley and Sons Ltd, Chichester, 2002.
[2] Sceats M., S. A. Rice and J.E. Butler. 1975. *J. Chem. Phys.*, 63, 12, 5390-5400.
[3] Irish D., T. Jary and C. Ratcliffe. 1982. *Applied Spectroscopy*, 36, 2, 137-140.
[4] Ratcliffe C. and D. Irish. 1982. *J. Phys. Chem.* 86, 4897-4905.
[5] Yeh Y., J.H. Bilgram and W. Kanzig. 1982. *J. Chem. Phys.*, 77(5), 2317-2321.
[6] Monosmith W. and G. Walrafen. 1984. *Chem. Phys.* 81, 2, 669-674.
[7] Lee Ch., et al. 1993. *Phys. Rev. B*, 47, 9, 4863-4872.
[8] Walrafen G. 2006. *J. Chem. Phys.*, 124, 184505; doi: 10.1063/1.2188942
[9] Angel C.A. *Ann. Rev. Phys. Chem.* 1983.34: 593-630
[10] Sokolov A., J. Hurst and D. Quitmann. 1995. *Physical Review B.*, 51, 18, 12865-12868.
[11] D'Arrigo G. et al. 1981. *J. Chem. Phys.* 75: 4264-4270.
[12] Krishnamurthy S., et al. 1983. *J. Chem. Phys.* 79, 12, 5863-5870.
[13] Kanno H., K. Tomikawa and O. Mishima. 1998. *Chem. Phys. Lett.,* 293, 412-416.
[14] Johari G. and O. Andersson. 2007. *Thermochimica Acta*, 461, 14-43.
[15] Suzuki H., et al., 2012. *J. Chem. Phys.*, 136, 234508; doi: 10.1063/1.4729476.
[16] Hare D.E. and S.M. Sorensen. 1990. *J. Chem. Phys.*, 93, 1, 25:33.
[17] Hoose C. and O. Möhler. 2012. *Atmos. Chem. Phys.*, 12, 9817–9854.
[18] Cooney J. 1969. *J. Appl. Meteor.* 9, 182-184.
[19] Weitkamp C. 2005. Springer. pp 455.
[20] Di Girolamo P. et al. 2004. *Geophys. Res. Lett.* 31, L01106, doi:10.1029/2003GL018342.
[21] Ancellet G. and F. Ravetta. 1998. *Appl. Opt.*, 37, 24, 5509-5521.
[22] Froidevaux, M., et al. 2013. A Raman lidar to measure water vapor in the atmospheric boundary layer. *Advances in Water Resources*, *51*, 345-356. DOI: 10.1016/j.advwatres.2012.04.008.
[23] Arshinov, Y. F., et al. 2003. Proceedings of the *ISTP* 2003, 142-144.
[24] Lui F. and F. Yi. 2013. *Appl. Opt.*, 52, 28, 6884-6895.
[25] Fochesatto, J., et al. 2004. *Advances in Space Research*, *34,* 10, 2227-2231.
[26] Ristori, P., et al. 2003. *Optics and Laser in Engineering, 40*, 91-104.
[27] Fochesatto, J., et al. 2005. *Proc. of SPIE, 5887,* doi: 10.1117/12.620970.
[28] Fochesatto, J., et al. 2008. *IJHSES. 18, 3, 713-726.*

SPACE-BASED ACTIVE OPTICAL REMOTE SENSING OF CARBON DIOXIDE COLUMN USING HIGH-ENERGY TWO-MICRON PULSED IPDA LIDAR

Upendra N. Singh[1]*, Tamer F. Refaat[1], Syed Ismail[2], Mulugeta Petros[1], Kenneth J. Davis[3], Stephan R. Kawa[4], and Robert T. Menzies[5]

*[1]NASA Langley Research Center, Hampton, VA 23681, USA, *Email: Upendra.n.singh@nasa.gov*
[2]Analytical Services and Materials, Inc., Hampton, VA 23666, USA,
[3]Pennsylvania State University, University Park, PA 16802, USA,
[4]NASA Goddard Space Flight Center, Greenbelt, MD 20771, USA,
[5]Jet Propulsion Laboratory, Pasadena, CA 91109, USA

ABSTRACT

Modeling of a space-based high-energy 2-µm triple-pulse Integrated Path Differential Absorption (IPDA) lidar was conducted to demonstrate carbon dioxide (CO_2) measurement capability and to evaluate random and systematic errors. A high pulse energy laser and an advanced MCT e-APD detector were incorporated in this model. Projected performance shows 0.5 ppm precision and 0.3 ppm bias in low-tropospheric column CO_2 mixing ratio measurements from space for 10 second signal averaging over Railroad Valley (RRV) reference surface.

1 INTRODUCTION

Sustained high-quality column CO_2 measurements from space are required to improve estimates of regional and global scale sources and sinks to attribute them to specific biogeochemical processes for improving models of carbon-climate interactions and to reduce uncertainties in projecting future change. Critical regions for CO_2 measurements are: high latitude ecosystems, tropical ecosystems, southern ocean, managed ecosystems, urban and industrial systems and coastal systems. The need for space-based IPDA measurements has been advocated by Active Sensing of CO2 Emission over Nights, Days, and Seasons (ASCENDS) [1] and Advanced Space Carbon and Climate Observation of Planet Earth (A-Scope) [2] studies. Space-based IPDA systems can provide sustained, high precision and low-bias column CO_2 in presence of thin clouds and aerosols. Technology developments are in progress to provide high pulse energy 2-µm IPDA that enables optimum, lower troposphere weighted column CO_2 measurements from space. This system provides simultaneous ranging; information on aerosol and cloud distributions, measurements over region of broken clouds, and reduces influences of surface complexities. Modeling the performance of a direct-detection high pulse energy 2-µm IPDA from space is presented in this paper [3].

2 MODEL BACKGROUND

The IPDA lidar transmitter is based on high-energy, tunable, triple-pulse 2-µm laser under development at NASA LaRC. The IPDA lidar receiver is based on an advanced HgCdTe (MCT) electron avalanche photodiode (e-APD) detection system demonstrated at NASA GSFC. IPDA model for simulating CO_2 Space-based measurements at 2-µm was conducted assuming 400 km altitude dawn-dusk orbit platform with US Standard atmospheric model and RRV as a reference surface. Water vapor (H_2O) is the dominant atmospheric molecule interfering CO_2 measurements. Figure 1 shows CO_2 (updated to 400 ppm) and H_2O double-path

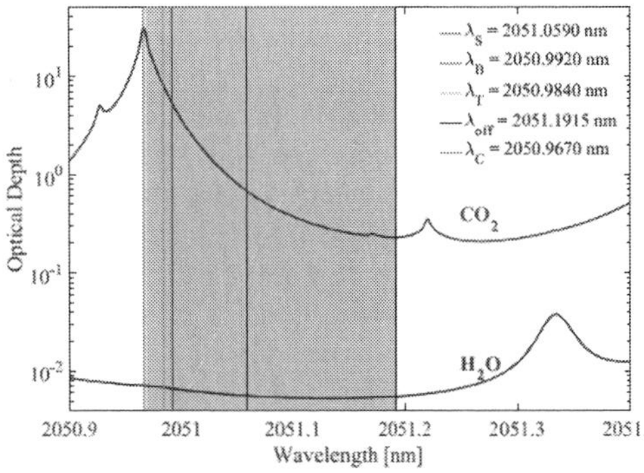

Figure 1 Integrated double-path differential optical depth spectra for CO_2 and H_2O obtained from 400 km altitude to mean sea level. Vertical lines mark selected on- and off-line wavelengths.

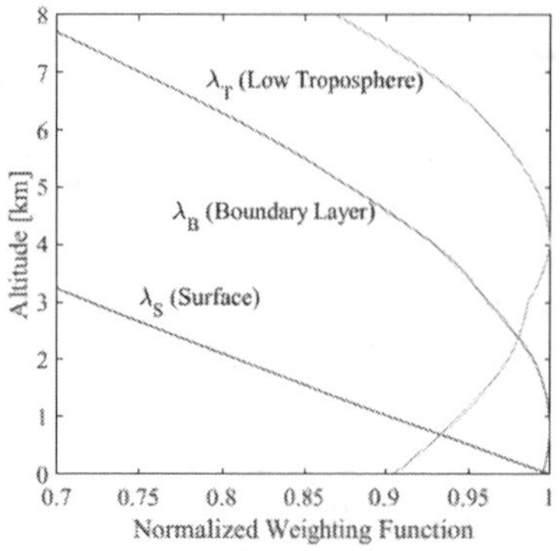

Figure 2 Pressure-based normalized weighting function for three selected on-line wavelengths, λ_S, λ_B and λ_T targeting surface, boundary layer and low tropospheric CO_2 column measurement.

optical depth spectrum. Vertical lines mark on-line positions for near surface (λ_S), boundary layer (λ_B), and tropospheric (λ_T) weighted column CO_2 measurements, as shown in figure 2, using common off-line wavelength (λ_{off}). Figure 2 indicates that an IPDA operating at the 2-μm wavelength region offers optimum CO_2 measurements in the lower troposphere and improves retrieval of the critically important surface flux estimates by a factor of two compared with a similar IPDA operating at 1.6-μm [2].

3 TECHNOLOGY DEVELOPMENTS

Progress in laser and detector technologies and airborne testing resulting in the development of a new aircraft based high-energy triple-pulsed 2-μm IPDA lidar. This IPDA promises a new avenue for remote sensing from space [4]. The 2-μm triple-pulse laser transmitter is the key component for this system. The performance of this transmitter is critical for achieving measurement sensitivity, accuracy, and range. The unique feature of this laser is the production of triple Q-switched pulses, separated by 150-200 μs, using a single pump pulse. The laser was developed at NASA LaRC and is based on Ho:Tm:YLF crystal technology [3]. Among other components, LaRC has been developing new technologies for the 2-μm lasers which include timing control, seeding, locking, and narrowing frequencies for both wind and CO_2 measurements [4]. Advanced wavelength control of the current triple-pulse laser uses a single seed laser and provides any offset-locked frequency within 32 GHz with respect to the CO_2 R30 line center (λ_C) shown in figure 1. The direct-detection IPDA receiver is based on the state-of-the-art, very low noise, 4×4 pixels MCT e-APD array [5]. This e-APD detection system has been developed and tested at GSFC for 1.6-μm pulsed IPDA lidar for different atmospheric trace gases such as CO_2 and CH_4. The projected 2-μm IPDA lidar parameters for the space-based system, and environmental parameters are listed in Table 1.

Table 1 2-μm triple-pulse IPDA lidar Parameters.

IPDA Lidar Transmitter	
On-line wavelength	Tunable#
Off-line wavelength	2051.1915 nm
Pulse energy	50 / 15 / 5 mJ
Pulse width	30 / 60 / 100 ns
Pulse separation	150 - 300 μs
Pulses repetition rate	50 Hz
Beam divergence	100 μrad
Laser line-width	Transform Limited
Frequency control accuracy	650 kHz
Wall-plug efficiency	2%
Beam expansion	×10
IPDA Lidar Receiver	
Optical efficiency	60%
Telescope diameter	1.5 m
Optical filter spectral width	1 nm
Field-of-view	150 μrad
Detection system NEP	1 $fW/Hz^{1/2}$
Detector responsivity	295.3 A/W
Detector gain	308
Detector excess noise factor	1.03
Detection bandwidth	10 MHz
Laser-to-telescope overlap	1.00
Settings and Environment	
Background solar irradiance	0.5 $mW/m^2 \cdot nm \cdot sr$
Atmospheric model	US Standard
Surface reflectivity*	0.51
Aerosols optical depth	0.0165
Platform altitude	400 km
Platform speed	28,000 km/hr
Integration time	10 s

* RRV surface reflectivity including the 24% lidar enhancement factor.

\# Tunable within 2050.97 nm to 2051.19 nm, as indicated by the shaded area of Figure 1.

4 IPDA RETURN POWER

The IPDA performance was estimated using the approach presented by Refaat et al. [3]. IPDA return signal power, total noise equivalent power (NEP) and signal-to-noise ratio (SNR) are shown in figure 3 as a function of on-line wavelength. The highest pair of transmitted energies (50 and 15 mJ) are assigned to on-line to account for the absorption loss in the measurement column. Off-line transmitted energies are set to either 15 or 5 mJ, which result in 185.55 or 37.11 nW return power, and 0.81 or 0.37 nW total NEP, respectively. Corresponding off-line SNR are 229 and 101. On-line wavelength selection close to R30 line center results in lower return power with total noise dominated by fixed background and electronic noise. Away from line center, on-line return power increases with shot-noise dominating total NEP and resulting in enhanced SNR. High energy and low detection noise combination allow short-time measurement over low reflectivity regions such as snow covered regions and ocean surface.

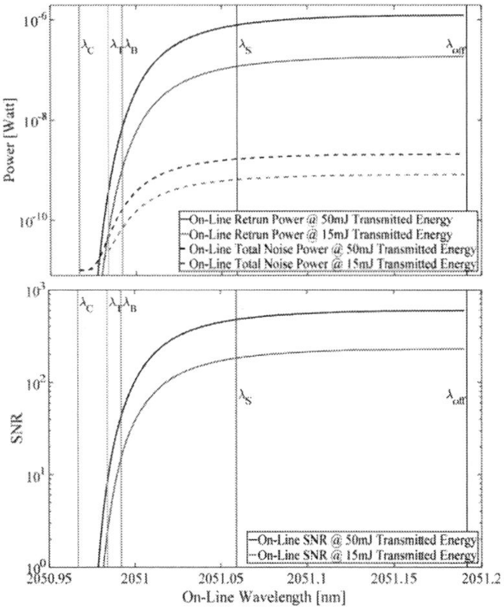

Figure 3 On-line surface return power detected by the IPDA form space and the corresponding integrated noise equivalent power using 50 and 15 mJ transmitted energies variation with on-line wavelength (top). Return power and noise are used to estimate SNR variations with on-line wavelength (bottom).

5 SENSITIVITY ANALYSIS AND ERROR

Analysis of both random error associated with IPDA, and systematic errors from atmospheric and instrument biases were estimated. Residual systematic errors in CO_2 measurement arise due to uncertainties in the knowledge of atmospheric and the IPDA instrument capabilities are shown in figure 4 with legend list in Table 2. The estimated CO_2 differential optical depth error from atmospheric effects include uncertainties in temperature (0.5°C), pressure (100 Pa), relative humidity (10%) and H_2O interference. A normally distributed random number generator was used to simulate the variability of these fields to evaluate the systematic error. CO_2 differential optical depth bias errors resulting from the IPDA transmitter uncertainties including on- and off-line laser position jitters (650 kHz) and laser spectral profiles were also estimated.

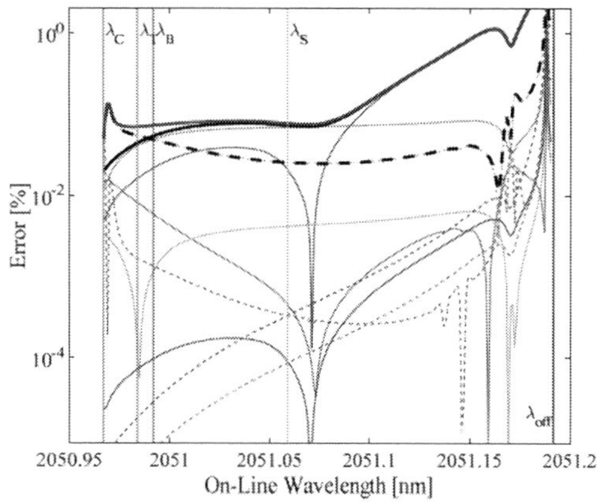

Figure 4 Atmospheric and laser transmitter errors and total systematic error estimation versus on-line wavelength for CO_2 measurement using the 2-μm IPDA lidar. Figure legend is in Table 2.

Table 2 Legends of figure 4.

———	H_2O Interference Error
———	Temperature Error
———	Pressure Error
———	Relative Humidity Error
———	Range Error
———	Total Atmospheric Errors
- - - - -	On-Line Jitter Error
- - - - -	Off-Line Jitter Error
- - - - -	On-Line Width Error
- - - - -	Off-Line Width Error
— — —	Total Transmitter Erros
———	Total Systematic Erros

Random errors for the combination of 50, 15, and 5 mJ on- and off-line energies and total (random + systematic) errors are shown in figure 5 with legend list in Table 3. Measurements with two weighting functions, at 50 Hz each, with the triple pulse system can be accomplished using two on-lines and a common off-line. The near optimum random error for each pair is <0.12% (<0.5 ppm), and the residual systematic error is <0.07% (0.3 ppm). Measurements can be optimized by tuning on-lines based upon ground target scenarios, environment and science objectives. With 10 MHz detection bandwidth, surface ranging with an uncertainty of <3 m can be achieved as demonstrated from earlier airborne flights [6].

6 CONCLUSIONS

High-energy triple-pulse 2-µm IPDA technology developments at LaRC are in progress to enable CO_2 column measurements from space. Currently

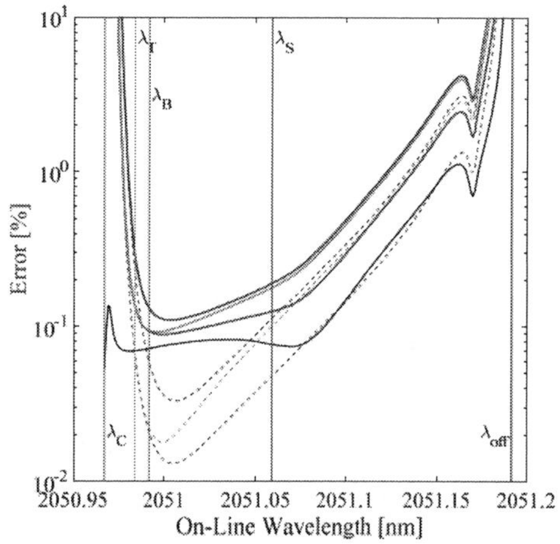

Figure 5 Random errors variation with on-line wavelength obtained using different on and off-line transmitted energy combinations. Total error for CO_2 measurement using the 2-µm IPDA lidar is obtained by adding random and total systematic errors. Figure legend is in Table 3.

Table 3 Legends of figure 5.

(dashed line)	Random Error, E_{on} = 50mJ & E_{off} = 15mJ
(dashed line)	Random Error, E_{on} = 15mJ & E_{off} = 5mJ
(dashed line)	Random Error, E_{on} = 50mJ & E_{off} = 5mJ
(solid line)	Total Systematic Error
(solid line)	Total Error, E_{on} = 50mJ & E_{off} = 15mJ
(solid line)	Total Error, E_{on} = 15mJ & E_{off} = 5mJ
(solid line)	Total Error, E_{on} = 50mJ & E_{off} = 5mJ

IPDA development is focused on measuring both H_2O and CO_2 simultaneously and independently. This allows measurement adaptability over a variety of atmospheric and target conditions. An airborne 2-µm IPDA lidar has demonstrated CO_2 measurements over land and ocean. Projected capability of high pulse energy 2-µm laser and an MCT e-APD detection system were incorporated for modeling the performance of a space-based IPDA. Random and systematic errors from the instrument and atmospheric effects including influences of H_2O were evaluated. The projected performance shows that high precision (0.5 ppm) and low bias (0.3 ppm) low tropospheric weighted dry-air column CO_2 mixing ratio measurements can be made from space with total error of <0.8 ppm for standard atmospheric condition and RRV reference with 10 second averaging.

ACKNOWLEDGEMENTS

This work was funded and supported by NASA Earth Science Technology Office.

References

[1] National Research Council, 2007: *Earth Science and Application from Space, National Imperatives for the Next Decade and Beyond.*

[2] European Space Agency Report for Assessment, 2008: A-SCOPE advanced space carbon and climate observation of planet Earth, SP-1313/1.

[3] T. Refaat, U. Singh, J. Yu, M. Petros, S. Ismail, M. Kavaya, K. Davis, 2015: Evaluation of an airborne triple-pulsed 2 µm IPDA lidar for simultaneous and independent atmospheric water vapor and carbon dioxide measurements, *Appl. Opt.* **54**, 1387-1398.

[4] U. Singh, B. Walsh, J. Yu, M. Petros, M. Kavaya, T. Refaat, N. Barnes, 2015: Twenty years of Tm:Ho:YLF and LuLiF laser development for global wind and carbon dioxide active remote sensing, *Opt. Mater. Exp.* **5**, 827-837.

[5] X. Sun, J. Abshire, J. Beck, 2014: HgCdTe e-APD detector arrays with single photon sensitivity for space lidar applications, *Proc. of SPIE* **9114**, 91140K.

[6] T. Refaat, U. Singh, J. Yu, M. Petros, R. Remus, S. Ismail, 2015: Double-pulse 2-µm integrated path differential absorption lidar airborne validation for atmospheric carbon dioxide measurement, *Appl. Opt.* **55**, 4232-4246.

SPACE-BORNE PROFILING OF ATMOSPHERIC THERMODYNAMIC VARIABLES WITH RAMAN LIDAR

Paolo Di Girolamo[1*], Andreas Behrendt[2], Volker Wulfmeyer[2]

[1]*Scuola di Ingegneria, Università degli Studi della Basilicata, Italy, *Email: digirolamo@unibas.it*

[2] *Institut für Physik und Meteorologie, Universität Hohenheim, Germany*

ABSTRACT

The performance of a space-borne water vapour and temperature Raman lidar has been simulated, with a specific attention to the Earth Explorer Missions in the frame of ESA's Living Planet Program. We report simulations under a variety of atmospheric scenarios, demonstrating the capability of a space Raman lidar to provide global-scale water vapour and temperature measurements in the troposphere with an accuracy fulfilling most observational requirements for numerical weather prediction (NWP) and climate research.

1 INTRODUCTION

An appropriate understanding and prediction of the Earth's temperature and water distributions is fundamental for a sustainable development of the Earth system. Unfortunately, our understanding of the water and energy cycles still shows critical gaps on all temporal and spatial scales. This is mainly due to a lack of measurement of water vapour and temperature profiles - hereafter called thermodynamic profiles - with high accuracy and high temporal-spatial resolution, especially in the lower troposphere.

Wulfmeyer *et al.* (2015) [1] demonstrated that the availability of global scale measurements of 3-dimensional thermodynamic profiles would have a revolutionary impact on our Earth system understanding in four key research areas: i) radiative transfer, with resulting implications on regional and global water and energy budgets, ii) land-atmosphere feedback including the surface energy balance closure in dependence of soil properties and land cover, iii) mesoscale circulations and convection initiation, and (iv) convective-scale data assimilation.

These observational gaps can only be closed by the development and operation of a new active remote sensing system in space based on the Raman lidar technique (Wulfmeyer *et al.*, 2015). In fact, combining vibrational and rotational Raman signals, simultaneous accurate and high-resolution measurements of water vapour and temperature profiles are possible. An instrumental concept is proposed here for a space-borne system based on the experience and know-how gained with the development and operation of several existing ground-based instruments ([2], [3]; [4], [5], [6], [7]; [8]) and airborne instruments [9]. Possible instrumental specifications for the different sub-systems are proposed to simulate the performances of a space-borne water vapour and temperature Raman lidar under a variety of environmental and atmospheric scenarios. For this purpose, different atmospheric reference models covering different climatic regions and seasons, as well as a variety of solar illumination conditions, were considered.

Observational requirements for NWP and climate research to be fulfilled by networks of satellite remote sensors, with a specific focus on the lower troposphere, have been identified by Wulfmeyer *et al.* (2015) [1]. The vertical resolution should be sufficiently high to resolve temperature and moisture gradients in the lower troposphere. Measurements are required at meso-beta to meso-gamma scale resolution. The water vapour/temperature bias should be 2–5% (0.2–0.5 K) and the noise errors smaller than 10% (1 K) in each single vertical range bin. In order to have a significant impact on all four above-mentioned primary application fields, space-borne lidar measurements with a vertical and horizontal resolution of 200 m and 50 km are required.

2. METHODOLOGY AND SYSTEM CONCEPT

An analytical lidar simulator, developed at Università degli Studi della Basilicata [5], has been applied in this paper to assess the expected performances of a space-borne water vapour and temperature Raman lidar system exploiting the vibrational and pure rotational Raman techniques in the ultraviolet (UV). The simulator allows to assess measurement quality in terms of both systematic and random measurement errors. The specifications of different lidar sub-systems have been assessed based on the application of the model.

Simulations consider a sun-synchronous low-Earth dusk/dawn orbit (inclination ~97 degrees), with an orbiting height and speed of 450 km and 7 km/s, respectively [10]. A dawn-dusk orbit characterized by an ascending node crossing time of 6 h was selected. A 3° off-nadir transmission of the laser beam is considered to avoid specular reflections from ice crystals.

The statistical uncertainty affecting water vapour mixing ratio and temperature measurements are determined through error propagation from the estimates of the noise errors affecting the Raman lidar signals, the latter being determined from the application of Poisson statistics to the signals. In this regard, it is to be specified that the application of Poisson statistics to the lidar signals is well suited in case of data acquired in photon-counting or in analogue mode, in the latter case after conversion of the analogue signals into "virtual" counts [3].

Atmospheric parameters considered in the simulation include vertical profiles of pressure, temperature, and humidity from three selected atmospheric reference models (tropical atmosphere, sub-Arctic winter, and U.S. Standard Atmosphere), considered for both summer and winter, and the median aerosol extinction data from the ESA Aerosol Reference Model of the Atmosphere (ESA-ARMA, 1999 [11]).

An analytical expression for the daylight sky background signals collected in the various channels has been formulated and tested in Di Girolamo *et al.* (2006 [5]). This expression includes three distinct contributions: namely, a cloud-free atmospheric contribution, a surface contribution, and a cloud contribution. The cloud-free atmospheric contribution accounts for the scattering of solar radiation by atmospheric constituents (molecular species and aerosols), the surface contribution is associated with Earth surface reflection of direct solar radiation, and the cloud contribution accounts for the cloud reflection of direct solar radiation. All three contributions depend on the solar zenith angle, with values largely increasing at small zenith angles. For this reason a sun-synchronous low Earth dusk-dawn orbit was selected for the simulations. Background contributions are proportional to the field of view of the receiving telescope (FOV) and this pushes in the direction to consider a very small FOV, compatibly with a reduced optical layout complexity and a large signal stability.

The Raman lidar is designed to collect four primary lidar signals: the water vapour vibrational Raman signal $P_{H_2O}(z)$, the high- and low-quantum-number rotational Raman signals, $P_{LoJ}(z)$ and $P_{HiJ}(z)$, and the elastic backscatter signal in the UV, $P_{354.7}(z)$. There is no need for the collection of the molecular nitrogen vibrational Raman signal $P_{N_2}(z)$ as in fact an alternative temperature-insensitive reference signal ($P_{ref}(z)$) can be obtained from the combination of the rotational Raman signals $P_{LoJ}(z)$ and $P_{HiJ}(z)$ [2]. This set of detected signals allows for independent measurements of atmospheric temperature, the water vapour mixing ratio (and consequently relative humidity) and the particle backscattering and extinction coefficients at 354.7 nm, the latter two such as delivered by EarthCARE mission. Besides the above mentioned parameters, very valuable additional products can be independently measured: the true atmospheric boundary layer depth over land and the oceans derived by the high-vertical resolution temperature profiles and the geometric and optical properties of clouds.

The lidar transmitter will consist of a frequency-tripled, diode-laser pumped Nd:YAG laser with efficiency of >10% and an average power in the UV (at 354.7 nm) of at least 250 W. The UV laser power required for the measurements can be achieved based on a new generation of pump chambers, with efficient pumping by diode lasers. The baseline to achieve the large UV laser powers required for these applications is represented by inclusion of several amplification stages. The implementation of diode laser pumping determines radiative or conductive cooling to be sufficient. Based on the electric power provided by the on-board solar arrays (2.5 kW) and an overall electric-to-355-nm efficiency in excess of 10 %, a laser power of >250 W at 355 nm can be generated.

The receiver must consist of a large-aperture telescope and an extremely stable and highly efficient receiving unit for the collection of the rotational and vibrational Raman signals. Simulations indicate the need for a telescope with a diameter of 8 m. The development of large-aperture telescopes may consider different technological solutions, as the use of segmented deployable or inflatable optics. New technological concepts for large aperture, lightweight telescopes, using thin deployable active mirrors have been demonstrated for lidar applications [12]. Several glass materials (e.g. Zerodur) have been tested and demonstrated to have the appropriate low weight and thermal stability characteristics for this type of space application. Simulations indicate the need for a receiving field-of-view (full width half maximum) of just 80 μrad. This requirement is a much more relaxed than the one considered for ADM (25 μrad), strongly reducing complexity and improving signal stability. The selection of the Raman scattered signals will be based on the use of interference filters; different options for the filters' specifications are considered in the simulations [4,13]. Simulation results illustrated in this paper are based on filters' specifications in [4]. The present simulations consider photon detection based on the use of photomultipliers and refer to the specifications of commercially available UV photomultipliers, with UV quantum efficiencies of approx. 30%. However, accumulation charge-coupled devices (ACCDs), as for example those developed for ADM, may also be used, thus allowing to achieve even higher quantum efficiencies (up to 85%) and to further improving measurement accuracy.

3 SIMULATION RESULTS

Simulations reported in the present paper refer to a vertical and horizontal resolution of 200 m and 50 km, respectively. Based on this horizontal resolution, more than 500 thermodynamic profiles/hour can be collected, i.e., ~12,000 profiles/day (as opposed to ~1200 radiosonde launches available from the international upper-air network). Figure 1 illustrates the vertical profile of the statistical uncertainty (precision) affecting water vapour mixing ratio measurements.

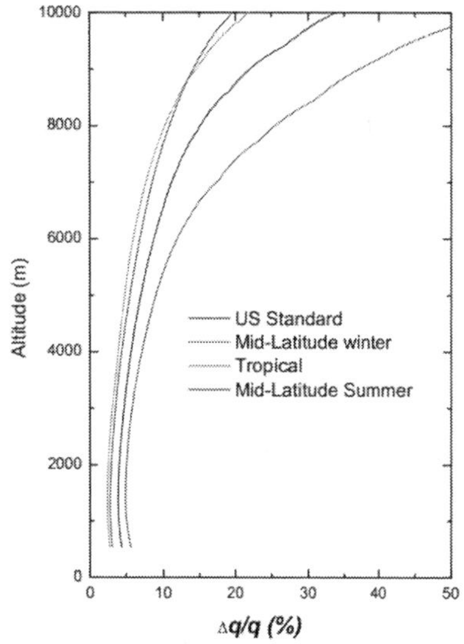

Figure 1: Vertical profile of the statistical uncertainty affecting water vapour mixing ratio measurements. The simulations consider four selected atmospheric reference models (tropical atmosphere, mid-latitude summer and winter, as well as U.S. Standard Atmosphere).

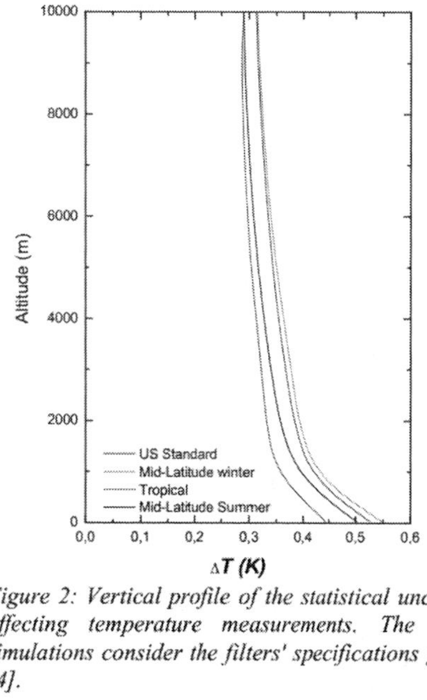

Figure 2: Vertical profile of the statistical uncertainty affecting temperature measurements. The present simulations consider the filters' specifications given in [4].

The simulations consider four different atmospheric reference models (tropical, mid-latitude summer and winter, as well as the U.S. Standard Atmosphere). Results in figure 1 reveal that the statistical uncertainty affecting water vapour mixing ratio has very small values throughout the boundary layer, with values not exceeding 5 % up to 3 km for three selected atmospheric reference models and < 10 % for all four. The statistical uncertainty is found to increase with altitude above 3 km, with values smaller than 20 % up to the upper troposphere. Results in figure 2 indicate that the statistical uncertainty affecting temperature measurements is only 0.5 K in the boundary layer, while it progressively decreases above the boundary layer to get down to 0.3 K at 10 km for all three selected atmospheric reference models. Correspondingly, the statistical uncertainty affecting relative humidity measurements ranges between 10 % in the boundary layer and 30 % in the upper troposphere. Simulations also reveal that the atmospheric boundary layer depth can be determined with a precision of 200 m, vertical profiles of particle backscatter & extinction with a precision of 5-20 %, while cloud geometrical properties and optical depth with a precision of 50-100 m and 5 %, respectively.

References

[1] Wulfmeyer, V., R. M. Hardesty, D. D. Turner, A. Behrendt, M. P. Cadeddu, P. Di Girolamo, P. Schlüssel, J. Van Baelen, and F. Zus, 2015: A review of the remote sensing of lower tropospheric thermodynamic profiles and its indispensable role for the understanding and the simulation of water and energy cycles, *Rev. Geophys.* **53**, 819–895.

[2] Behrendt, A., Nakamura, T., Onishi, M., Baumgart, R., and Tsuda, T., 2002: Combined Raman lidar for the measurement of atmospheric temperature, water vapor, particle extinction coefficient, and particle backscatter coefficient, *Appl. Opt.* **41**, 7657–7666.

[3] Behrendt, A.,V. Wulfmeyer, E. Hammann, S.K. Muppa, and S. Pal., 2015: Profiles of second to third order moments of turbulent temperature fluctuations in the convective boundary layer: First measurements with rotational Raman lidar, *Atmos. Chem. Phys.* **15**, 5485-5500.

[4] Di Girolamo, P., R. Marchese, D. N. Whiteman, and B. B. Demoz., 2004: Rotational Raman Lidar measurements of atmospheric temperature in the UV, *Geophys. Res. Lett.* **31**, L01106, ISSN: 0094-8276.

[5] Di Girolamo, P., A. Behrendt, and V.Wulfmeyer, 2006: Spaceborne profiling of atmospheric temperature and particle extinction with pure rotational Raman Lidar and of relative humidity in combination with differential absorption Lidar: performance simulations, *Appl. Opt.* **45**, 2474-2494.

[6] Di Girolamo, P., D. Summa, R. Ferretti, 2009: Multiparameter Raman Lidar Measurements for the Characterization of a Dry Stratospheric Intrusion Event, *J. Atmos. Ocean. Tech.* **26**, 1742-1762.

[7] Radlach, M., A. Behrendt, and V. Wulfmeyer, 2008: Scanning rotational Raman lidar at 355 nm for the measurement of tropospheric temperature fields. *Atmos. Chem. Phys.* **8**, 159–169.

[8] Hammann, E., A. Behrendt, F. Le Mounier, and V. Wulfmeyer, 2015: Temperature profiling of the atmospheric boundary layer with rotational raman lidar during the HD(CP)2 observational prototype experiment, *Atmos. Chem. Phys.* **15**, 2867-2881.

[9] Whiteman, D. N., R. Kurt, R. Scott, 2010: Airborne and Ground-Based Measurements Using a High-Performance Raman Lidar, *J. Atmos. Ocean. Technol.* **27**(11), 1781–1801.

[10] Wulfmeyer, V., H. Bauer, P. Di Girolamo, C. Serio, 2005: Comparison of active and passive water vapour remote sensing from space: An analysis based on the simulated performance of IASI and space borne differential absorption Lidar. Remote Sens. Environ. **95**, 211-230.

[11] ESA, ARMA Reference Model of the Atmosphere, Tech. Rep. APP-FP/99-11239/AC/ac (European Space Agency, 1999).

[12] Simonetti, F., A. Zuccaro Marchi, L. Gambicorti, V. Bratina, P. Mazzinghi, 2010: *Opt. Eng.* **49**, 073001.

[13] Hammann E., and A. Behrendt, 2015: Parametrization of optimum filter passbands for rotational Raman temperature measurements, *Opt. Express* **23**, 30767-30782.

CoMet: AN AIRBORNE MISSION TO SIMULTANEOUSLY MEASURE CO_2 AND CH_4 USING LIDAR, PASSIVE REMOTE SENSING, AND IN-SITU TECHNIQUES

Andreas Fix[1]*, Axel Amediek[1], Heinrich Bovensmann[2], Gerhard Ehret[1], Christoph Gerbig[3], Konstantin Gerilowski[2], Klaus Pfeilsticker[4], Anke Roiger[1], Martin Zöger[5]

[1] *DLR Institute of Atmospheric Physics, Oberpfaffenhofen, Germany, *andreas.fix@dlr.de*
[2] *Institute of Environmental Physics, University of Bremen, Germany*
[3] *Max Planck Institute for Biogeochemistry, Jena, Germany*
[4] *Institute of Environmental Physics, University of Heidelberg, Germany*
[5] *DLR Flight Experiments, Oberpfaffenhofen, Germany*

ABSTRACT

In order to improve our current knowledge on the budgets of the two most important anthropogenic greenhouse gases, CO_2 and CH_4, an airborne mission on board the German research aircraft HALO in coordination with two smaller Cessna aircraft is going to be conducted in April/May 2017. The goal of CoMet is to combine a suite of the best currently available active (lidar) and passive remote sensors as well as in-situ instruments to provide regional-scale data of greenhouse gases which are urgently required.

1 INTRODUCTION

Confronting Climate Change is one of the paramount societal challenges of our time. The main cause for global warming is the increase of anthropogenic greenhouse gases in the Earth's atmosphere. Together, carbon dioxide and methane, being the two most important greenhouse gases, globally contribute to about 81% of the anthropogenic radiative forcing [1].

However, there are still significant deficits in the knowledge about the budgets of these two major greenhouse gases (GHG) such that the ability to accurately predict our future climate remains substantially compromised. Different feedback mechanisms which are insufficiently understood have significant impact on the quality of climate projections. In order to accurately predict future climate of our planet and support observing emission targets in the framework of international agreements, the investigation of sources and sinks of the greenhouse gases and their feedback mechanisms is indispensable.

In the past years, inverse modelling has emerged as a key method for obtaining quantitative information on the sources and sinks of the GHG [2]. However, this technique requires the availability of sufficient amounts of precise and independent data on various spatial scales. Therefore, observing the atmospheric concentrations of the greenhouse gases is of significant importance for this purpose. In contrast to point measurements, airborne instruments are able to provide regional-scale data of greenhouse gases which are urgently required, though currently lacking.

Providing such data from remote sensing instruments (lidar and passive) supported by state-of-the-art in-situ sensors, and additionally comparing the results of the greenhouse gas columns retrieved from aircraft to the network of ground-based stations is the mission goal of the HALO CoMet campaign. The project also aims at preparing the validation activities for upcoming satellite missions such as the German-French Climate mission MERLIN [3], at developing new methodologies for GHG measurements, and promotes technological developments necessary for future Earth-observing satellites.

2 INSTRUMENTATION

The key instrumentation for the CoMet mission consists of a suite of the most sophisticated instruments currently available to measure atmospheric carbon dioxide and methane onboard aircraft. Remote sensing instruments, both active and passive, will be complemented by in-situ sensors to obtain maximum synergy. A list of the instruments is given in Table 1.

Table 1 CoMet airborne instrumentation

Instrument acronym	Description	Aircraft
CHARM-F	CO2 and CH4 Lidar (IPDA)	HALO
HALO_JIG	Cavity Ringdown Spectrometer	HALO
HALO_JAS	Flask sampler	HALO
miniDOAS	Differential Optical Absorption Spectroscopy	HALO
BAHAMAS	HALO basic data acquisition system	HALO
Dropsondes	Meteorological sondes	HALO
FOKAL	Frequency comb	HALO
MAMAP	NIR spectrometer	FUB-Cessna
METPOD	Cessna basic data acquisition system	DLR-Cessna
QCLS	Quantum Cascade Laser Spectrometer	DLR-Cessna
CRDS	Cavity Ringdown Spectrometer	DLR-Cessna
Sampler	Flask sampler	DLR-Cessna

2.1 Aircraft

Three aircraft are currently foreseen as part of the campaign to carry this scientific instrumentation. The German Research Aircraft HALO (Figure 1) serves as the flagship for CoMet. HALO is a Gulfstream G-550 aircraft specifically equipped with numerous provisions for in situ and remote sensing instruments. It enables measurements up to an altitude of 15 km within a range of 9000 km. HALO may carry an instrumental payload of up to 3 tons.

Figure 1 The German Research aircraft HALO (D-ADLR) operated by DLR

Co-ordinated HALO flights are planned with two smaller, low-flying aircraft (Figure 2). The first one is a Cessna T207A operated by Freie Universität Berlin (FUB) and providing a maximum scientific payload of up to 400 kg. Its typical speed during measurement flights is ~60 m/s at standard cruising altitudes of up to 3.5 km. During CoMet, this aircraft will be equipped with the passive spectrometer MAMAP [4] (see Section 2.3). The second one is a Cessna 208B Grand Caravan operated by DLR. This aircraft carries up to 500 kg of payload, can reach up to 7.6 km, and provides a maximum endurance of 1600 km. During CoMet, it will be equipped with in-situ instrumentation for greenhouse gases and an underwing meteorological sensor package that measures temperature, pressure and humidity as well as wind speed and direction (see Section 2.4).

Figure 2 Cessna C107 (D-EAFU, left) operated by FU Berlin and Cessna Grand Caravan (D-FDLR, right) operated by FU Berlin and DLR, respectively

Both Cessna aircraft play an important role in the EUFAR (EUropean Fleet for Airborne Research) initiative.

2.2 Lidar

Concerning active remote sensing, the Integrated Path Differential Absorption (IPDA) lidar technique using hard target reflection from the Earth's surface in the near IR has been identified in the last few years to measure the column averaged dry air mixing ratio of CO2 and CH4 with high accuracy and low bias [5]. The advantages of a lidar are that it does not require the sun as a light source, and can therefore provide both day and night, all-seasons and all latitude measurements. But more important, such an active instrument concept provides a direct measurement of the atmospheric path by means of runtime measurement and, thus, is comparatively weakly influenced by aerosol and thin clouds.

In the past years, an airborne IPDA system (CHARM-F) has been developed and successfully tested at DLR [6]. In contrast to other development in either the US or Japan, it uses a pulsed direct detection scheme at 1.6 μm. To our knowledge, CHARM-F currently is the only airborne IPDA capable to measure both greenhouse gases, CH_4 (at 1645 nm) and CO_2 (at 1572 nm) at the same time [6].

The lidar system is based on two optical parametric oscillators which are diode-pumped by means of injection seeded, Q-switched Nd:YAG lasers in a master-oscillator power-amplifier configuration. In order to fulfil the stringent requirements on frequency stability for the on-line and off-line wavelengths, a sophisticated locking scheme has been developed that is based on DFB lasers referenced to a multipass absorption cell and offset locking techniques. During CoMet, an airborne optical frequency comb (FOKAL) will be employed for the first time in order to monitor the frequency stability under in-flight conditions to highest accuracy. Previously, this was only possible in the laboratory.

2.3 Passive Remote Sensing

As a passive remote sensing instrument, the MAMAP airborne spectrometer system capable of direct and quantitative remote column-averaged measurements of atmospheric CH_4 and CO_2 complements the payload. This instrument measures reflected and scattered solar radiation in the short wave infrared (SWIR) and near-infrared (NIR) parts of the electro-magnetic spectrum and will be installed on the FUB Cessna. In recent campaigns in Europe [7] and the US, MAMAP data demonstrated that small gradients in total column CO_2 and CH_4 can be detected with sufficient precision to derive emissions from local sources (coal mine ventilation shafts, power plants, landfills, etc.).

The combination of CHARM-F and MAMAP offers the unique opportunity to further improve their individual data quality, for example, by minimizing large scale biases of MAMAP using co-located measurements with CHARM-F. On the other hand MAMAP data with its smaller ground scene measurements in comparison to CHARM-F can potentially help identifying emissions of smaller area sources below the detection limit of CHARM-F.

A second remote sensing instrument, mini-DOAS operated by Universität Heidelberg, will be installed on HALO to complement the passive remote sensing payload. Mini-DOAS is a 6-channel optical spectrometer for the detection of atmospheric stray-light in nadir and limb direction capable to detect various trace gases related to combustion in the UV/VIS/NIR spectral range and will contribute through characterization of the respective air masses.

2.4 In-Situ Instrumentation

In order to validate the remote sensing instruments and provide greenhouse gas profile information traceable to WMO scales, highly accurate in-situ instruments to measure CO_2 and CH_4 based on cavity-ringdown spectrometry (HALO_JIG) will complement the core payload on HALO. Also part of the HALO payload is a flask sampler (HALO_JAS) which will collect air samples for subsequent laboratory analysis, providing supplemental information on isotopic composition and other tracers correlated with the emission of GHGs. Along with the measurement of GHGs, the basic HALO measurement system (BAHAMAS) is going to provide important meteorological and aircraft attitude data indispensable for the retrieval of the remote sensors.

In parallel, onboard the DLR Cessna, an instrument package consisting of a quantum cascade laser spectrometer (QCLS), CRDS, and a flask sampler - similar to the one installed on HALO - will be deployed. The basic meteorological sensor package that measures temperature, pressure and humidity as well as wind speed and direction will provide ancillary data to enable budgeting of localized sources such as coal mine ventilation shafts, landfills, or urban agglomerations using the mass balance approach.

3 CAMPAIGN OBJECTIVES

For CoMet, an intensive measurement period of 4 weeks is planned at the beginning of the growing season in April-May 2017 in Central Europe. During this period, optimized HALO research flights will comprise extended latitudinal transects to capture the GHG gradients, flights over known regions of strong emissions as well as comparison overflights over the ground-based remote sensing sites of the Total Carbon Column Observing Network (TCCON). While HALO will provide a larger scale picture, the two Cessna aircraft will concentrate on two regions of prime interest, the Upper Silesian coal belt in Poland and the greater Berlin area. In these areas, additional ground-based remote sensing capabilities as well satellite products (i.e. GOSAT) will support the CoMet mission. These targets, characterized by significant anthropogenic CH_4 and CO_2 emissions, have been selected in order to demonstrate that the innovative airborne instrumentation can distinguish between anthropogenic and natural sources of GHGs.

For this purpose, modelling is an equally important component of the CoMet mission. The goal is to use the measurement results to determine regional scale fluxes of the GHGs more precisely than possible today. To do so, results from chemistry-climate models on various scales are required and need to be linked to regional transport models. It will also be possible to validate the transport models using such independent, non-surface, atmospheric CO_2 and CH_4 measurements. By selectively using different data streams in a regional inversion, we will be able to assess the benefit from high-resolution measurement of the dry-air columns.

Finally, the project also aims at preparing the validation activities for upcoming greenhouse gas satellites such as Sentinel 5-P, GOSAT-2, and MERLIN as well as at developing new, combined methodologies for greenhouse gas measurements from space using lidar and passive remote sensing.

OUTLOOK

Currently, a successor mission is already being planned that will extend the CoMet study from Central Europe to the highly relevant Arctic and Tropical wetlands and that also aims at validation of the German-French climate mission MERLIN that is scheduled for launch in ~2021.

At that time, MAMAP will have been upgraded into a 2-dimensional imaging spectrometer system providing improved detection sensitivity with certification for HALO in order to measure from the same platform as CHARM-F.

ACKNOWLEDGEMENTS

CoMet is partly supported by DFG within its Priority Program HALO-SPP 1294. We would like to thank the following individuals and institutions for scientific, technological, or logistical support: J. Marshall, MPI; P. Jöckel, and A. Butz, DLR; J. Swolkień and J. Nęcki, AGH University, Krakow; A. Kuze, JAXA; DLR Flight Experiments; Institute for Space Sciences, FU Berlin; and Menlo Systems GmbH.

References

[1] Stocker, T.F. et al.: Climate Change 2013: The Physical Science Basis. Contribution of Working Group I to the Fifth Assessment Report of the IPCC, Cambridge University Press.

[2] Gurney, K.R., et al., 2002: Towards robust regional estimates of CO2 sources and sinks using atmospheric transport models," Nature 415, 626-630.

[3] Pierangelo, C. et al., 2015: MERLIN (Methane Remote Sensing Lidar Mission): An Overview, in Proc. 27th International Laser Radar Conference, ILRC, New York, USA, (2015).

[4] Gerilowski, K. et al., 2011: MAMAP – a new spectrometer system for column-averaged methane and carbon dioxide observations from aircraft: instrument description and performance analysis, Atmos. Meas. Tech., 4, 215-243.

[5] Ehret, G. et al., 2008: Space-borne remote sensing of the greenhouse gases CO2, CH4, N2O by integrated path differential absorption lidar: a sensitivity analysis, Appl. Phys. B90, 593–608.

[6] Amediek, A. et al., 2017: CHARM-F – a new airborne integrated-path differential-absorption lidar for carbon dioxide and methane observations: measurement performance and methodology to quantify strong point source emission, Appl. Opt., submitted.

[7] Krings, T., et al., 2013: Quantification of methane emission rates from coal mine ventilation shafts using airborne remote sensing data, Atmos. Meas. Tech., 6, 151–166.

AIRBORNE TESTS OF AN OAWL DOPPLER LIDAR: RESULTS AND POTENTIAL FOR SPACE DEPLOYMENT

Michael Hardesty[1], Sara Tucker[2], Sunil Baidar[1], Mark Beubien[3]
[1]Cooperative Institute for Research in Environmental Sciences, USA, mike.hardesty@noaa.gov
[2]Ball Aerospace, USA
[3]Yankee Environmental Systems, USA

ABSTRACT

The 532 nm Green Optical Covariance Wind Lidar (GrOAWL) was flown on a NASA WB-57 research aircraft during the summer of 2016 to validate the instrument design and evaluate wind measurement capability and sensitivity. Comparisons with dropsondes and atmospheric models showed good agreement, demonstrating that a GrOAWL type instrument could provide high-value wind measurements from both airborne and space-based platforms

1 INTRODUCTION

Global wind profiles from space-based platforms are generally recognized as the most significant currently-missing observation needed to improve numerical weather predication [1]. Observing System Simulation Experiments (OSSEs) have shown that the assimilating wind data from a space-based Doppler lidar is likely to provide a significant increase in forecast skill [2], especially in the tropics and Southern Hemisphere where observations are currently limited.

The European Space Agency Aeolus Atmospheric Dynamics Mission, scheduled for launch in late 2017 or early 2018, will be the first lidar in space aimed at providing wind measurements. Operating in the ultraviolet, Aeolus will feature separate receiver channels to process lidar returns from both aerosol and molecular targets to measure radial wind profiles at a single fixed line of sight orthogonal to the polar orbital track.

As a follow-on to Aeolus, we have proposed the design-to-cost ATHENA-OAWL NASA Earth Venture Instrument (EVI) mission to measure two components of the wind from the International Space Station (ISS) using only aerosol/cloud atmospheric returns. Based on the Optical Autocovariance Wind Lidar (OAWL) concept, ATHENA would incorporate the 532-nm laser technology implemented on CALIPSO to reduce mission risk and cost.

Figure 1: AOVT geometry showing two looks at 45° nadir, 45° and 135° azimuth (relative to aircraft forward) on NASA WB-57

During summer 2016 we carried out the ATHENA-OAWL Venture-Technology (AOVT) aircraft flight program to demonstrate several key elements of the proposed ISS instrument, including instrument efficiency and wind-measuring precision, scaling to space, and two-look measurement capability. For AOVT the green OAWL (GrOAWL) instrument was mounted on a NASA WB-57, as shown in Figure 1, for a series of test flights out of Ellington Field in Houston, TX. Here we briefly describe the campaign and present some results and conclusions from the tests.

2 GrOAWL DESCRIPTION

The GrOAWL instrument flown during the AOVT campaign is a direct detection aerosol wind lidar built by Ball Aerospace that incorporates two

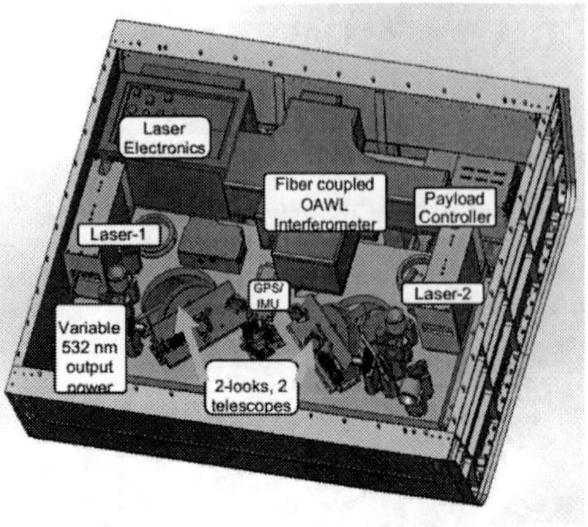

Figure 2: GrOAWL instrument shown in pressurized pallet, which mounts in the belly of the NASA WB-57 aircraft.

Nd:YAG lasers, both capable of operation at 532 and 355 nm, and a field-widened, quadrature-channel, Mach-Zender interferometer to measure the Doppler shifts [3]. For AOVT the system operated primarily at 532 nm with pulse energies of ~1.5 mJ transmitted at 200 Hz for each look. Two 29-cm diameter telescopes gathered the backscattered light at each look and directed it via optical fibers to the interferometer for processing. The fringes formed by the interferometer were sampled at 140 MHz (1.07 m); results were averaged both spatially and over multiple pulses to investigate wind estimate precision improvement through accumulation during the campaign. GrOAWL operated fully autonomously during each flight.

The GrOAWL system transmitted pulses alternately through each telescope, which were oriented at 45° nadir angle and 45° and 135° azimuth angle from aircraft forward as shown in Figure 1. The entire instrument, shown in Figure 2, was contained in a pressurized pallet that mounted in the WB-57 payload bay. Aircraft-specific designs were implemented to enable operation in a high altitude, high vibration environment; these included a vibration isolation layer underneath the instrument and kinematic mounting of the optical bench on that layer. The system was cooled by fluid loops running through external air-fluid heat exchangers. A GPS/IMU mounted on the optical bench provided information on pointing and platform motion.

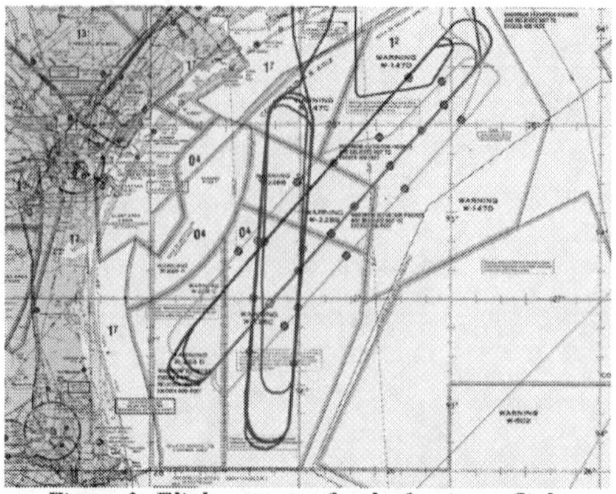

Figure 3: Flight patterns for the 3 science flights during the AOVT campaign. Dots indicate dropsonde launch positions.

3 MEASUREMENT CAMPAIGN

The AOVT measurement campaign consisted of 8 flights from Ellington Field, totaling approximately 38 flight hours during May-June, 2016. The flight tracks for the three science flights are shown in Figure 3. Flights were primarily over restricted areas, which provided flexibility to vary laser pulse energy, change altitude, avoid heavy clouds patterns, and minimize airspace coordination. The repeating racetrack flight patterns enabled multiple revisits of the same location, testing of the impact of multi-pulse averaging, and views of the same region of the atmosphere from opposite sides. An INMARSAT satellite link provided ground control and telemetry communication.

A key component of the campaign was the installation and utilization of the Yankee Environmental Systems dropsonde deployment system on the WB-57. A total of 21 dropsondes over three science flights were used for comparisons with the GrOAWL measurements. Radiosondes launched from Corpus Christi also provided wind data for comparison during the science flights.

4 SCIENCE RESULTS

After dealing with several platform-related issues such as fogging of the optical window and thermal control, we carried out three science flights during the latter part of the experiment. Figure 4 shows an example of the radial winds measured at both looks during the science flight on June 21 (blue track in

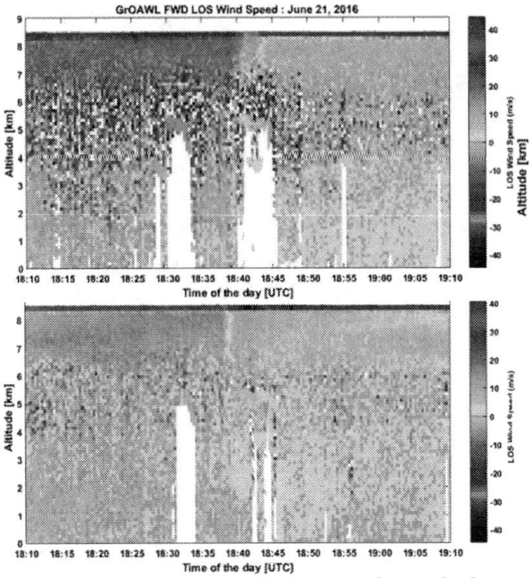

Figure 4: Radial velocity measurements for two looks on June 21. Color change at ~18:38 indicates a turn. Missing data areas are due to clouds.

Figure 3) over one hour of the racetrack pattern. In Figure 4 returns were averaged over 20 seconds in time and ~120 m vertically. Because data are collected and stored for every pulse, accumulation time can be varied to investigate wind estimate improvement. In Figure 4 the change in color is due to a turn at the end of the pattern, where the sign of the radial velocity changes. It is seen that the wind speeds for this day were quite low below 4 km altitude.

The profiles measured at each look were compared with the dropsonde observations. Figure 5 shows a comparison between the two looks and the dropsonde observations projected onto the line of sight for each look. Because the dropsondes fall more or less straight down from the aircraft, the lidar/sonde measurements don't overlap in space (see Figure 1). Consequently, to remove small-scale spatial variability in the wind field from the comparison we averaged the lidar wind results over 5 minutes for the comparison.

A scatterplot of the comparison between the dropsonde and the aft-pointing beam is shown in Figure 6 for the June 21 flight. Plots and comparison statistics have been generated for both looks for each flight as well as over the ensemble

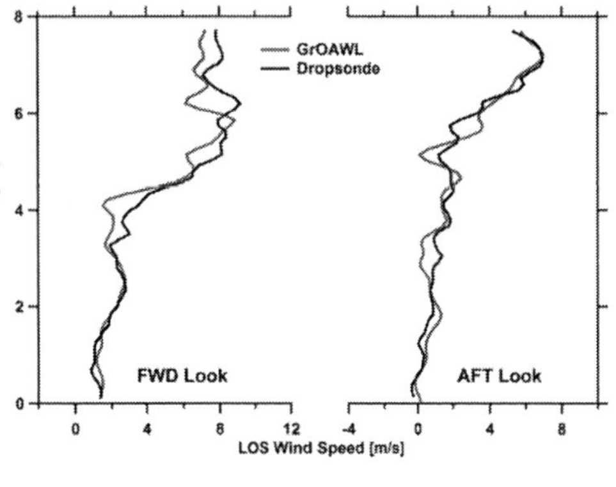

Figure 5: Example of a dropsonde comparison with two looks from GrOAWL on June 21.

of three science flights. The plot shows good agreement between the dropsonde and the GrOAWL measurements with very little bias. Bias is an important parameter for space-based observations; it must be low if the measurements are to have impact for numerical weather prediction.

Table 1 summarizes the dropsonde/GrOAWL comparisons from all three science flights (21 dropsondes in total) for three different altitude bins. Bias and standard deviation are low for all height regions and indicates excellent agreement between

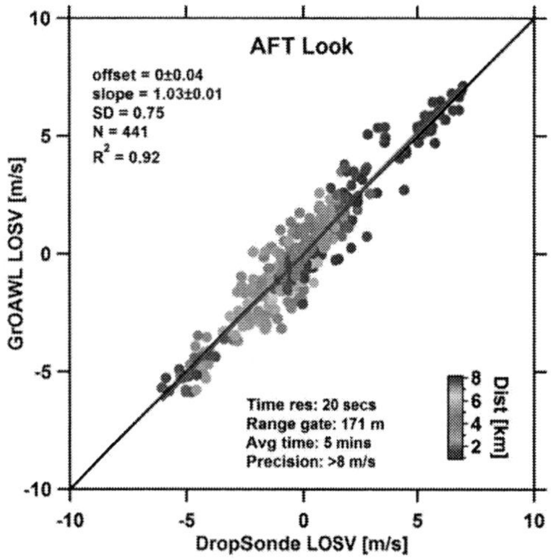

Figure 6: Scatter plot comparing GrOAWL aft look radial wind measurements with dropsonde observations. Color indicates dropsonde distance from the aircraft

Table 1: Summary of dropsonde/GrOAWL comparisons over the science flights for three altitude regions.

Altitude	V_{GrOAWL} - $V_{dropsonde}$ [m/s]	
	FWD Look	AFT Look
< 2 km	0.02 ± 0.76	0.02 ± 0.74
2-5 km	-0.10 ± 0.85	0.17 ± 0.84
> 5 km	-0.13 ± 0.87	0.13 ± 1.05

the GrOAWL measurements and the dropsonde observations over the course of the experiment.

Although numerical weather prediction models will likely assimilate line-of-sight velocities rather than wind vectors, it is useful to examine wind vectors computed by combining the two looks from the GrOAWL measurements. Figure 7 shows wind vectors from a portion of the June 21 flight computed from the two looks. Also shown are the dropsonde measurements represented by dots attached to the wind bars. A turn at the end of the racetrack pattern occurred between 18:40 and 18:50, as shown by the wind-field symmetry as the lidar re-samples the same region.

An important goal for the AOVT test program was validation of OAWL instrument sensitivity as represented in models used to predict performance of an OAWL-type instrument in space. To examine sensitivity, we attempted to characterize the backscatter environment for each flight from the GEOS-5 model analyses, Aeronet measurements, and lidar observations obtained at the same time of year during previous campaigns. Based on our estimates of backscatter during the flights while accounting for some degradation of the GrOAWL measurements from the harsh platform environment, we determined that our instrument model accurately estimates space-based performance for an OAWL instrument. Further work is currently being carried out in collaboration with NCAR to use HSRL measurements to more precisely characterize the scattering environment during GrOAWL surface-based operation.

5 DISCUSSION AND FUTURE WORK

Results from the AOVT campaign provide strong evidence that the OAWL concept is suitable for measurement of winds from space. Advantages of

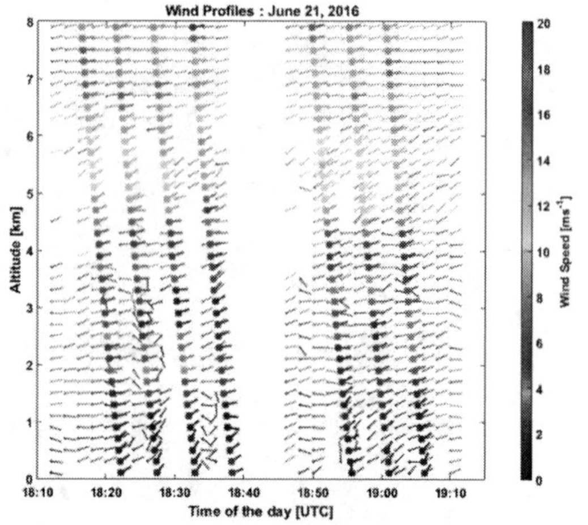

Figure 7: Wind vectors computed from June 21 GrOAWL and dropsonde observations.

the OAWL design include relaxed specifications on laser pulse-to-pulse frequency stability, high transmission, capability to extract very weak signals by averaging, potential for inclusion in a single transmitter/nested receiver design to measure winds from both aerosol and molecular scatterers, and potential for a combined HSRL/Doppler lidar. Projected next steps for GrOAWL include modification for deployment on the NASA DC-8, utilization for Aeolus Calibration and Validation, and application for airborne science flights studying dynamics and transport.

5 ACKNOWLEDGEMENTS

We thank the NASA Earth Science Venture Technology Program, the Ball hardware team, and the NASA WB-57 group for their valuable support.

References

[1] Baker, W. E., et al, 2013: Lidar-Measured Wind Profiles: The Missing Link in the Global Observing System, Bull. Amer. Meteor. Soc., **95**, 4, 543-564.

[2] Riishojgaard, L. P., et al, 2012: Observing System Simulation Experiments for a global observing wind sounder, Geophy. Res. Lett. **39**, Article L17805.

[3] Tucker, S. C., R. M. Hardesty, S. Baidar, and C. Weimer, 2016: The ATHENA-OAWL Venture Tech Instrument. Proc. 18th Coherent Laser Radar Conference, Boulder, CO, June 26-July 1.

EARLINET VALIDATION OF CATS L2 PRODUCT

Emmanouil Proestakis[1,2,*], Vassilis Amiridis[1], Michael Kottas[1], Eleni Marinou[1,3], Ioannis Binietoglou[1,4], Albert Ansmann[5], Ulla Wandinger[5], John Yorks[6], Edward Nowottnick[7], Abduvosit Makhmudov[8], Alexandros Papayannis[9], Aleksander Pietruczuk[10], Anna Gialitaki[1,11], Arnoud Apituley[12], Constantino Muñoz-Porcar[13], Daniele Bortoli[14], Davide Dionisi[15], Dietrich Althausen[5], Dimitra Mamali[16], Dimitris Balis[3], Doina Nicolae[4], Eleni Tetoni[1,11], Gian Luigi Liberti[15], Holger Baars[5], Iwona S. Stachlewska[17], Kalliopi-Artemis Voudouri[3], Lucia Mona[18], Maria Mylonaki[9], Maria Rita Perrone[19,20], Maria João Costa[14], Michael Sicard[13,21], Nikolaos Papagiannopoulos[18], Nikolaos Siomos[3], Pasquale Burlizzi[19], Ronny Engelmann[5], Sabur F. Abdullaev[8], Julian Hofer[5], Gelsomina Pappalardo[18]

*[1]IAASARS, National Observatory of Athens, Athens, 15236, Greece, *proestakis@noa.gr*
[2]Laboratory of Atmospheric Physics, Department of Physics, University of Patras, 26500, Greece
[3]Department of Physics, Aristotle University of Thessaloniki, Thessaloniki, 54124, Greece
[4]National Institute of R&D for Optoelectronics, Magurele, Romania
[5]Leibniz Institute for Tropospheric Research, Leipzig, 04318, Germany
[6]NASA Goddard Space Flight Center, Greenbelt, Maryland, USA
[7]GESTAR, Universities Space Research Association, Columbia, Maryland, US
[8]Physical Technical Institute of the Academy of Sciences of Tajikistan, Dushanbe, Tajikistan
[9]Laser Remote Sensing Unit (LRSU) National Technical University of Athens Physics Department, Zografou, Greece
[10]Institute of Geophysics PAS, Warsaw, Poland
[11]University of Athens, School of Physics, Division of Environmental Physics-Meteorology, 15784, Athens, Greece
[12]KNMI - Royal Netherlands Meteorological Institute, The Netherlands
[13]Remote Sensing Laboratory, Universitat Politècnica de Catalunya, Barcelona, Spain
[14]Departamento de Física, Instituto de Ciências da Terra, Escola de Ciências e Tecnologia, Universidade de Évora, Évora, Portugal
[15]Consiglio Nazionale delle Ricerche, Istituto Scienze dell'Atmosfera e del Clima (CNR-ISAC), Rome-Tor Vergata, Italy
[16]Department of Geoscience and Remote Sensing, TU Delft, Delft, The Netherlands
[17]Institute of Geophysics, Faculty of Physics, University of Warsaw (IGFUW), 02-093 Warsaw, Poland
[18]Istituto di Metodologie per l'Analisi Ambientale (CNR-IMAA), Tito scalo, Italy
[19]Dipartimento di Matematica e Fisica, Università del Salento, Lecce, Italy
[20]CNISM-Consorzio Nazionale Interuniversitario per le Scienze Fisiche della Materia, Lecce, Italy
[21]CiènInstitut d'Estudis Espacials de Catalunya (CTE-CRAE / IEEC), Universitat Politècnica de Catalunya, Barcelona, Spain

ABSTRACT

The Cloud-Aerosol Transport System (CATS) onboard the International Space Station (ISS), is a lidar system providing vertically resolved aerosol and cloud profiles since February 2015. In this study, the CATS aerosol product is validated against the aerosol profiles provided by the European Aerosol Research Lidar Network (EARLINET). This validation activity is based on collocated CATS-EARLINET measurements and the comparison of the particle backscatter coefficient at 1064nm.

1 INTRODUCTION

The Cloud-Aerosol Transport System (CATS), a lidar remote sensing payload onboard the International Space Station (ISS), provides near-real-time profiles of aerosols and cloud optical properties [1]. Operating since February 2015, CATS's main objective is to study the 4D distribution, evolution, and transport of aerosols and clouds. Furthermore, CATS envisages to bridge space-borne lidar observations, between the Cloud-Aerosol Lidar and Infrared Pathfinder Satellite Observations (CALIPSO) and future space-borne lidars (e.g ESA's ADM-Aeolus and EarthCARE, NASA's ACE). The retrieval algorithms of CATS are similar to CALIPSO. In contrast with CALIPSO, CATS main products are the backscatter coefficient and depolarization ratio at 1064 nm.

The aim of this study is to validate the CATS retrievals and establish the accuracy of the derived aerosol backscatter profiles. To this end, the lidar

https://doi.org/10.1051/epjconf/201817602005

systems of the European Aerosol Research Lidar Network (EARLINET) [2], provide the ideal reference dataset. EARLINET consists of a large number of sophisticated lidars, which provide long-term quality-assured multi-wavelength aerosol retrievals. The objective of EARLINET is to provide a climatologically-significant profiling aerosol dataset.

In this study, we present and discuss the methodology for the validation of the profiles of CATS L2 backscatter coefficient at 1064 nm (Section 2). The methodology is demonstrated using a case study of a single CATS-ISS overpass over the Polly^{XT} lidar of the National Observatory of Athens (NOA) on the 13th of January 2016 (Section 3.1). The approach followed is mostly based on the previous expertise and methodologies that have been developed and applied in EARLINET for the CALIPSO cal/val activity [3]. Finally, the first consolidated output from the comparison with a number of EARLINET stations is given in Section 3.2., while Section 4 provides our first conclusions.

2 DATA AND METHODOLOGY

At present, 13 EARLINET stations operating the 1064 nm channel contribute to this study. The permanent stations are located at: Athens (Greece), Barcelona (Spain), Belsk (Poland), Bucharest (Romania), Cabauw (Netherlands), Evora (Portugal), Lecce (Italy), Leipzig (Germany), Potenza (Italy), Rome (Italy) and Thessaloniki (Greece). Furthermore, the non-permanent EARLINET stations at Dushanbe (Tajikistan) and the lidar system of NOA (Athens) have been integrated in the study.

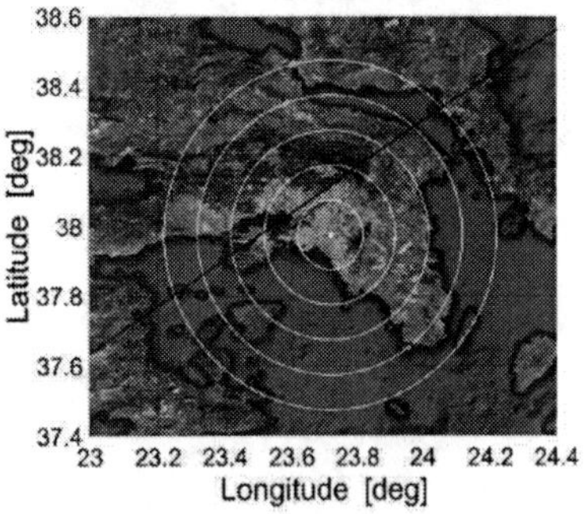

Figure 1. The nighttime ISS orbit over Athens on the 13th of January 2016 (black line). The concentric white circles denote regions of radius 10, 20, 30, 40 and 50 km from the position of NOA-Polly^{XT} lidar system.

For the purpose of demonstrating the methodology applied for the CATS validation, we use the NOA-Polly^{XT} system, which was deployed at the Thiseion site of Athens, between May 2015 and March 2016, in the framework of the ACTRIS JRA1 campaign. The nighttime study case of the CATS-ISS overpass on the 13th of January 2016 is shown in Fig.1.

2.1 CATS-ISS

CATS is a low cost payload onboard the ISS. Due to the ISS inclination orbit, CATS retrievals are confined in the geographical region between 51°S and 51°N. CATS operates at 415 km altitude. The orbit characteristics of ISS enable CATS to provide information on the vertical structure of the atmosphere at different local times each day.

The CATS lidar was designed to operate in three different modes and in four different Instantaneous Fields Of View (IFOV). Mode 1 and Mode 3 are not operational due to a failure of the electronics of laser1 and to a problem related to the optical path respectively. Mode M7.2 is limited to the 1064 nm channel. The vertically-resolved retrievals of aerosol and clouds are provided in different processing levels [4]. In this study, EARLINET is utilized for the validation and exploitation of CATS M7.2 L2 of the V1-05 version of data.

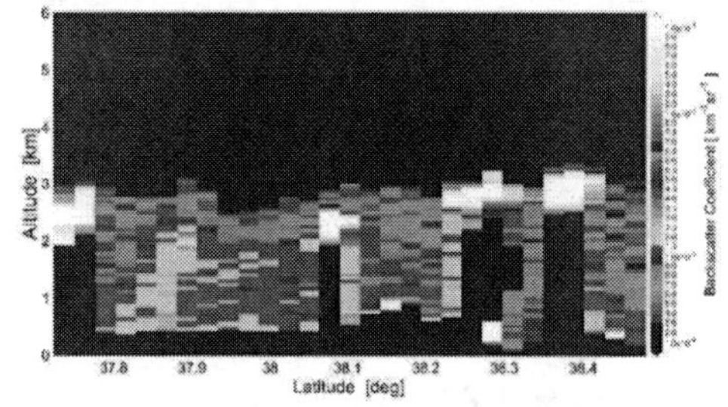

Figure 2. CATS Backscatter Coefficient 1064 nm of the overpass on the 2016-01-13, 01:16:38 UTC.

CATS L2 product provides information on aerosols and clouds along the ISS track, with 5 km horizontal and 60 m vertical resolution, for the altitudinal range between 28 and -2 km. L2 product includes the backscatter coefficient at

 https://doi.org/10.1051/epjconf/201817602005

1064 nm, the Feature Type, Aerosol Subtype and the confidence of the classification are provided. The total backscatter coefficient at 1064 nm of the CATS overpass of the study case on the 13th of January 2016 is shown in Fig.2.

For the comparison of the correlative backscatter coefficient dataset between CATS and ground-based lidars, cases that are not contaminated by clouds are selected. Regarding the accuracy of the classification of the cloud and aerosol layers, a Cloud-Aerosol Discrimination (CAD) score is assigned to the identified atmospheric layers.

2.2 A typical EARLINET system

Fig. 3 shows an example of EARLINET lidar quicklooks, providing the range-corrected signal time-height plot at 1064nm (13th of January 2016). In the example presented here, the sky above the Athens site was contaminated by clouds during the ISS overpass. Thus the time averaging window of the lidar signal is shifted for few minutes after the satellite overpass, in order to avoid cloud presence.

Figure 3. The range corrected signal time-series for the Polly^{XT} 1064nm channel. The red box depicts the temporal averaging of the lidar signals for the selected scene (01:30-02:28), while the white dashed line indicates the CATS overpass time.

3 RESULTS

3.1 EARLINET-CATS comparison: 2016-01-13, 01:16:38 UTC

CATS L2 backscatter profiles within a region of 50 km radius from the ground-based lidar are spatially averaged into a mean CATS profile. CAD score equal to -5 has been used for filtering out of the analysis the low confidence aerosol layers, which are possibly misclassified or contaminated by clouds. Accordingly, the mean backscatter profile is computed through the temporal averaging of the lidar signal. Fig.4 shows the comparison between the averaged Backscatter profiles of CATS and Polly^{XT} ground-based system. The CATS error-bars shown correspond to the standard deviation from spatially averaging the CATS scene. Regarding EARLINET, typical uncertainties are in the range of 5–10% for the backscatter coefficient retrieved with the Raman method.

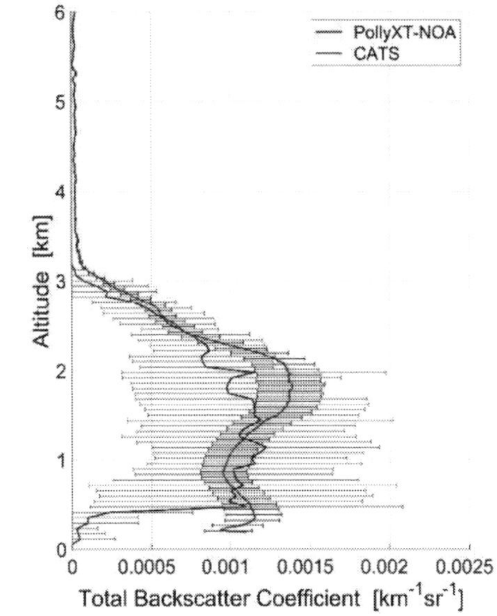

Figure 4. CATS (blue line) and EARLINET (red line) mean profiles of backscatter coefficient at 1064 nm.

From the analysis of the 13th of January 2016, it is observed that under homogeneous, relatively cloud free, nighttime conditions, the mean Backscatter Coefficient Profile at 1064 nm retrieved by CATS is in good agreement with the profile retrieved by the ground-based lidar. The comparison demonstrates a good nighttime performance of CATS. The observed disagreements to low altitudes (below 0.4 km) are most probably attributed to the effect of the orography in the spatial window used for averaging.

3.2 EARLINET-CATS comparison

The distribution of the absolute differences between CATS and the EARLINET stations of Leipzig (13 cases), Dushanbe (5 cases) and NOA

(4 cases), for nighttime ISS overpasses and for the altitudinal range between 0 and 10 km a.s.l., is shown in Fig.5. EARLINET measurements lower than the CATS minimum nighttime detectable backscatter limit (5e-05 $km^{-1}sr^{-1}$) are rejected from our analysis, for the same cases that CATS measurement provides values equal to zero as well. Based on the nighttime cases, the distribution of the absolute differences is characterized by mean value of -1.519e-04 $km^{-1}sr^{-1}$, median value of -9.94e-05 $km^{-1}sr^{-}1$ and standard deviation of 6.634e-04 $km^{-1}sr^{-1}$.

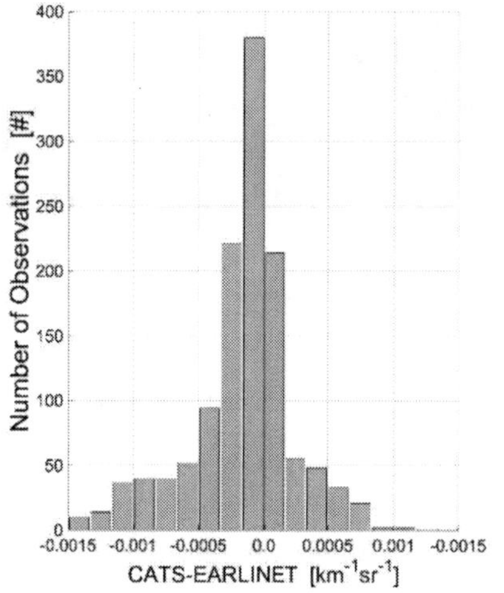

Figure 5. Distribution of the mean absolute differences between CATS and EARLINET backscatter measurements of Leipzig, Dushanbe and NOA stations and nighttime ISS overpasses.

The observed disagreements are most probably related to CATS minimum detectable backscatter, to the incomplete overlap of the ground-based lidars for lower altitudes, to the effect of the orography on CATS spatial averaging and to the horizontal distance between CATS footprint and the location of the EARLINET stations.

4 CONCLUSIONS

EARLINET is utilized for the validation study of CATS L2 backscatter profiles at 1064nm. Results demonstrate the good performance of CATS during nighttime conditions. The distribution of the absolute differences between CATS and EARLINET, for all the correlative measurements, is characterized by a mean value of -1.518e-04 $km^{-1}sr^{-1}$, a median value of -9.94e-05 $km^{-1}sr^{-}1$ and a standard deviation of 6.634e-04 $km^{-1}sr^{-1}$. Future work includes the integration of the entire EARLINET network in this study. Furthermore the study will be expanded to include the CATS-EARLINET daytime intercomparison.

ACKNOWLEDGEMENTS

The authors acknowledge ACTRIS-2 under grant agreement no. 654109 from the European Union's Horizon 2020 research and innovation programme. The research leading to these results has received funding from the European Union's Horizon 2020 Research and Innovation programme under grant agreement No 602014, project ECARS. The ISS NASA Research Office (NRO) funded the CATS instrument. CATS browse images and data products are freely distributed via the CATS web site at http://cats.gsfc.nasa.gov/data/.

References

[1] McGill, M. J., J. E. Yorks, V. S. Scott, A. W. Kupchock, and P. A. Selmer (2015), The Cloud Aerosol Transport System (CATS): A technology demonstration on the International Space Station, Proc. SPIE 9612, Lidar Remote Sensing for Environmental Monitoring XV, 96120A, doi:10.1117/12.2190841.

[2] Pappalardo, G., Amodeo, A., Apituley, A., Comeron A., Freudenthaler, V., Linné, H., Ansmann, A., Bösenberg, J., D'Amico, G., Mattis, I., Mona, L., Wandinger, U., Amiridis, V., Alados-Arboledas, L., Nicolae, D., and Wiegner, M.: EARLINET: towards an advanced sustainable European aerosol lidar network, Atmos. Meas. Tech., 7, 2389–2409, doi:10.5194/amt-7-2389-2014, 2014.

[3] Pappalardo, G., Wandinger, U., Mona, L., Hiebsch, A., Mattis, I., Amodeo, A., Ansmann, A., Seifert, P., Linné, H., Apituley, A., Alados Arboledas, L., Balis, D., Chaikovsky, A., D'Amico, G., De Tomasi, F., Freudenthaler, V., Giannakaki, E., Giunta, A., Grigorov, I., Iarlori, M., Madonna, F., Mamouri, R., Nasti, L., Papayannis, A., Pietruczuk, A., Pujadas, M., Rizi, V., Rocadenbosch, F., Russo, F., Schnell, F., Spinelli, N., Wang, X., and Wiegner, M.: EARLINET correlative measurements for CALIPSO: first intercomparison results, J. Geophys. Res., 115, D00H19, doi:10.1029/2009JD012147, 2010.

[4] Yorks, J. E., M. J. McGill, S. P. Palm, D. L. Hlavka, P. A. Selmer, E. P. Nowottnick, M. A. Vaughan, S. D. Rodier, and W. D. Hart (2016): An overview of the CATS level 1 processing algorithms and data products, Geophys. Res. Lett., 43, 4632–4639, doi:10.1002/2016GL068006.

ENHANCEMENTS TO THE CALIOP AEROSOL SUBTYPING AND LIDAR RATIO SELECTION ALGORITHMS FOR LEVEL II VERSION 4

A. Omar[1], J. Tackett[2], M-H. Kim[3], M. Vaughan[1], J. Kar[2], C. Trepte[1], D. Winker[1]

[1]*NASA Langley Research Center, Atmospheric Sciences, MailStop 475, Hampton, VA 23681 ali.h.omar@nasa.gov*

[2]*Science Systems Applications Inc (SSAI), Hampton, VA,*

[3]*NASA NPP (USRA)*

1. ABSTRACT

Several enhancements have been implemented for the version 4 aerosol subtyping and lidar ratio selection algorithms of Cloud Aerosol Lidar with Orthogonal Polarization (CALIOP). Version 4 eliminates the confusion between smoke and clean marine aerosols seen in version 3 by modifications to the elevated layer flag definitions used to identify smoke aerosols over the ocean. To differentiate between mixtures of dust and smoke, and dust and marine aerosols, a new aerosol type will be added in the version 4 data products. In the marine boundary layer, moderately depolarizing aerosols are no longer modeled as mixtures of dust and smoke (polluted dust) but rather as mixtures of dust and seasalt (dusty marine). Some lidar ratios have been updated in the version 4 algorithms. In particular, the dust lidar ratios have been adjusted to reflect the latest measurements and model studies.

2. INTRODUCTION

The version 4 aerosol subtyping algorithms are considerably improved over the earlier versions. They initially employ more accurately calibrated level 1 data and the MERRA-2 meteorological data. These discussed in detail in the data quality summaries (https://goo.gl/qVRmd8). The version 4 Level 2 aerosol subtyping algorithm is preceded by a much improved cloud aerosol discrimination algorithm and includes improvements in the tropospheric aerosol subtyping algorithm, and new schemes for stratospheric aerosol subtyping (c.f. [1]) and updated lidar ratios. Version 4 permits identification of all aerosol types over snow/tundra land surface types (Polar Regions) and therefore correctly accounts for transport of assorted aerosol types to polar regions, especially to the Arctic. Another significant improvement is that the version 4 algorithms account for the settling of dust plumes into the marine boundary layer which results in mixtures of dust and seasalt aerosols.

3. POLLUTED DUST AND DUSTY MARINE AEROSOL DEFINITION

Several comparisons of CALIOP's lidar ratios with High Spectral Resolution Lidar (HSRL) over

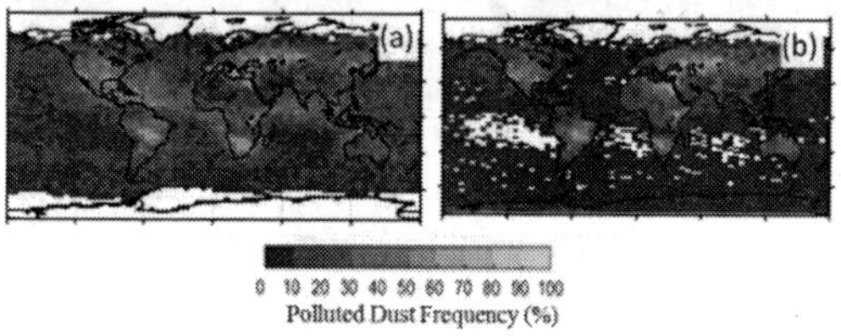

Figure 1. Polluted dust frequency for (a) version 3 and (b) version 4 for June-July-August 2008 nighttime layers.

the ocean show that the lidar ratio (S_a) estimates for dust and marine are in agreement. However, these same comparisons also show that the CALIOP lidar ratio assigned for polluted dust (55 sr) is typically much larger that the lidar ratio measured by the HSRL (~35 sr). [2]. Since S_a is a good proxy for aerosol type, this suggests that CALIOP's aerosol model is wrong for moderately depolarizing cases near the ocean surface and points to a need to define a more representative aerosol type. The version 4 aerosol subtyping algorithm now include a dusty marine (dust + marine) aerosol type, with a S_a value of 37 sr, for moderately depolarizing aerosols in and near the marine boundary layer (altitudes less than 2.5 km above the ocean surface). Figure 1 illustrates the improvement in the distributions of polluted dust over version 3. Though the distributions of polluted dust in version 4 are more reasonable, high frequencies in the northern Pacific and Indian Ocean persist. In version 3 smoke and polluted continental aerosol layers are occasionally misclassified as marine layers near continental outflows. The map in Figure 2(a) shows the altitude at which clean marine aerosols

 https://doi.org/10.1051/epjconf/201817602006

are most frequently found in version 3. There is a high frequency of elevated marine layers on the south west coast of Africa and the region stretching from the Kamchatka peninsula to the Sea of Japan. Figures 2 (b) – (c) show a case study of this misclassification arising from a number of bugs and incorrect elevated layer flag thresholds. Note the large number of elevated marine layers and attached polluted continental layers

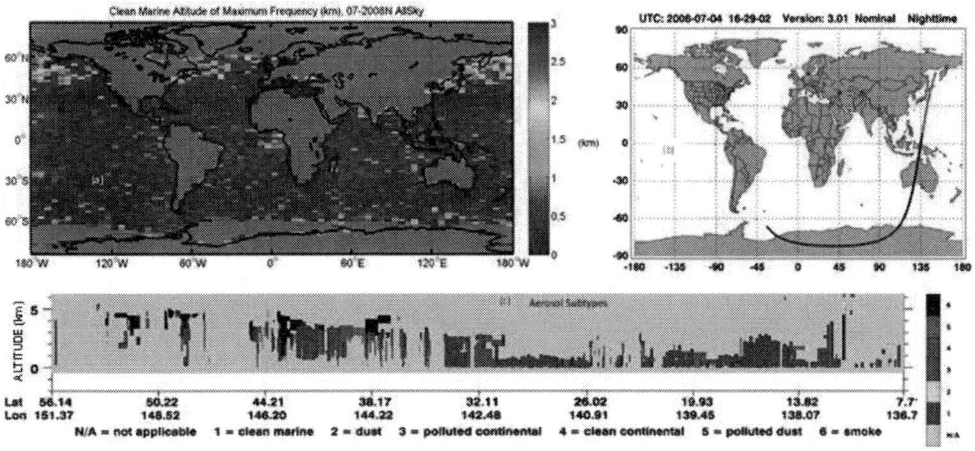

Figure 2. (a) The altitude of maximum frequency of marine layers in July 2007, (b) night time orbit and (c) the corresponding subtypes.

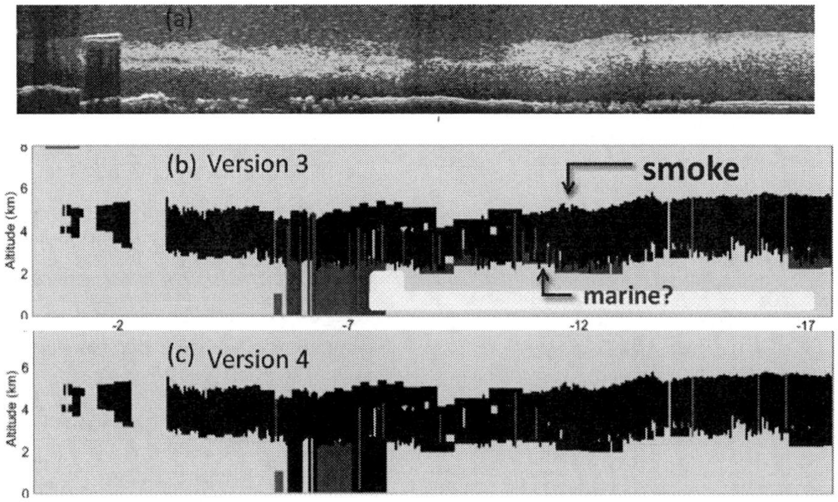

Figure 3(a) Browse image of an aerosol layer of the Gulf of Guinea, and (b) the version 3 and (c) version 4 classifications of the aerosol layer showing smoke in black, marine in blue and polluted dust in brown.

4. MARINE AND SMOKE TYPES

In version 3, smoke is a non-depolarizing and elevated aerosol. However, the definition also included 'elevated' included a threshold geometric thickness which resulted in spurious classifications and the large frequencies of smoke misclassified as marine shown in Figure 2 (a) above. Figure 3 shows a case study of a robust smoke layer (a) off the Gulf of Guinea and its version 3 classification (b). The version 4 algorithm, illustrated by Fig 3(c), identifies Smoke as non-depolarizing and elevated (i.e., layer top > 2.5 km) irrespective of the geometric thickness. In the version 4 subtyping products, fewer smoke layers are misclassified as marine and the 'striping' by marine layers in smoke plume is substantially reduced.

5. LIDAR RATIOS

[3] developed a technique that takes advantage of the relatively small variability of the lidar ratio for water clouds and uses opaque water clouds as a reference to determine the optical depth of overlying transparent aerosol or cirrus layers. [4] applied this technique to CALIOP level 1 measurements in dust region off the coast of W. Africa to retrieve, unambiguously, distributions of dust and smoke lidar ratios.

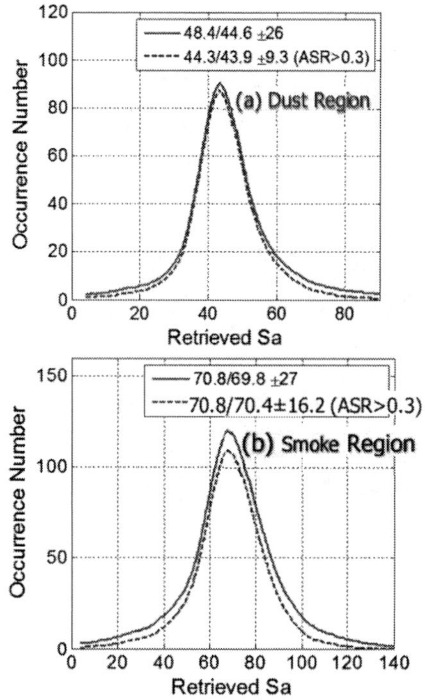

Figure 4. Multiyear retrieval of dust and smoke lidar ratios (Liu et al. 2015). Lidar ratios of dust and smoke are well constrained around 44 sr and 70 sr, respectively. The blue dotted lines are for aerosol scattering ratios (ASR) > 0.3.

These are shown in Figure 4 and are used to inform the adjustment of the dust S_a (44 sr) for version 4 products. While the smoke S_a remains unchanged from version 3, the uncertainty is reduced. The lidar ratios and uncertainties used for all version 4 aerosol subtypes are listed in Table 1. Version 4 assumes the 532-nm and 1064-nm lidar ratios are the same.

Table 1. 532 nm Lidar Ratios of Versions 3 and 4

Layer Type	532 nm Lidar Ratios (sr)	
	version 3	version 4
Dust	40 ± 20	44 ± 9 [A,B]
Smoke	70 ± 28	70 ± 16 [A]
Clean Continental	35 ± 16	53 ± 11 [C]
Polluted Continental	70 ± 25	70 ± 25
Polluted Dust	55 ± 22	55 ± 22 [D]
Clean Marine	20 ± 6	23 ± 5 [E]
Dusty Marine		37 ± 15 [F]

S_a and uncertainty based on CALIPSO constrained retrievals by Liu et al. (2015)

[B] Based HSRL measurements of transported Saharan dust. No λ dependence based on [5]

[C] S_a measured by HSRL for layers classified as clean continental by CALIPSO [6]

[D] Based on the microphysical measurements made during NAMMA. S_a values in agreement with [7] and [8]

[E] S_a measured by HSRL in multiple field campaigns, [8] No λ dependence based on [9]), [10] , [7]

[F] S_a based on mixture of dust and marine aerosol (65/35 by surface area). Uncertainty is larger than either dust or marine alone.

SUMMARY

The aerosol subtyping algorithm in version 4 includes a number of improvements. Specifically, polluted dust is identified in more realistic regions compared to version 3 which showed widespread misclassification. A new dusty marine aerosol type is added to account for dust settling into the marine boundary layer. Smoke layers are no longer misclassified as clean marine aerosol, improving aerosol optical depth estimates for smoke layers in oceanic smoke outflow regions. Aerosol lidar ratios have also been updated in version 4 to reflect current understanding by independent instruments. In addition to the improvements discussed here, stratospheric aerosol types are now identified (volcanic ash, sulfate, smoke, PSC aerosol). Descriptions of aerosol subtyping improvements for version 4 are available on the CALIPSO Data Users Guide [https://www-calipso.larc.nasa.gov/resources/calipso_users_guide/qs/cal_lid_l2_all_v4-10.php].

[1] Vaughan, M., A. Garnier, J. Tackett, M. Avery, S. Young, J. Kar, B. Getzewich, A. Omar, Z. Liu, K.-P. Lee, X. Cai, S. Zeng, Y. Hu, J. Lambeth, R. Ryan, B. Magill, S. Rodier, T. Murray, K. Beaumont, P. Detweiler, P. Lucker, J. Pelon, D. Winker, and C. Trepte, "Introducing CALIPSO's Version 4 Level 2 Lidar Data Products," in *International Laser Radar Conference (ILRC)*, Bucharest, Romania, 2017.

[2] Rogers, R. R., C. A. Hostetler, J. W. Hair, R. A. Ferrare, Z. Liu, M. D. Obland, D. B. Harper, A. L. Cook, K. A. Powell, M. A. Vaughan, and D. M. Winker, "Assessment of the CALIPSO Lidar 532 nm attenuated backscatter calibration using the NASA LaRC airborne High Spectral Resolution Lidar," *Atmos. Chem. Phys.*, vol. 11, pp. 1295-1311, 2011.

[3] Hu, Y., M. Vaughan, Z. Liu, K. Powell, and S. Rodier, "Retrieving Optical Depths and Lidar Ratios for Transparent Layers Above Opaque Water Clouds From CALIPSO Lidar Measurements," *IEEE Geoscience and Remote Sensing Letters*, vol. 4, pp. 523-526, 2007.

[4] Liu, Z., D. Winker, A. Omar, M. Vaughan, J. Kar, C. Trepte, Y. Hu, and G. Schuster, "Evaluation of CALIOP 532 nm AOD over opaque water clouds," *Atmospheric chemistry and physics discussions*, vol. 14, pp. 23583-23637, 2014.

[5] Tesche, M., A. Ansmann, D. Muller, D. Althausen, I. Mattis, B. Heese, V. Freudenthaler, M. Wiegner, M. Esselborn, G. Pisani, and P. Knippertz, "Vertical profiling of Saharan dust with Raman lidars and airborne HSRL in southern Morocco during SAMUM," *Tellus B*, vol. 61, pp. 144-164, Feb 2009.

[6] Rogers, R. R., M. A. Vaughan, C. A. Hostetler, S. P. Burton, R. A. Ferrare, S. A. Young, J. W. Hair, M. D. Obland, D. B. Harper, A. L. Cook, and D. M. Winker, "Looking through the haze: evaluating the CALIPSO level 2 aerosol optical depth using airborne high spectral resolution lidar data," *Atmos. Meas. Tech.*, vol. 7, pp. 4317-4340, 2014.

[7] Papagiannopoulos, N., L. Mona, L. Alados-Arboledas, V. Amiridis, H. Baars, I. Binietoglou, D. Bortoli, G. D'Amico, A. Giunta, J. L. Guerrero-Rascado, A. Schwarz, S. Pereira, N. Spinelli, U. Wandinger, X. Wang, and G. Pappalardo, "CALIPSO climatological products: evaluation and suggestions from EARLINET," *Atmos. Chem. Phys.*, vol. 16, pp. 2341-2357, 2016.

[8] Muller, D., A. Ansmann, I. Mattis, M. Tesche, U. Wandinger, D. Althausen, and G. Pisani, "Aerosol-type-dependent lidar ratios observed with Raman lidar," *J. Geophys. Res.*, vol. 112, Aug 17 2007.

[9] Sayer, A. M., A. Smirnov, N. C. Hsu, and B. N. Holben, "A pure marine aerosol model, for use in remote sensing applications," *Journal of Geophysical Research: Atmospheres*, vol. 117, pp. n/a-n/a, 2012.

[10] Josset, D., J. Pelon, Y. Hu, R. Rogers, Z. Liu, A. Omar, M. Vaughan, P. w. Zhai, and I. Team, "Global scale lidar ratio retrieval over the ocean," presented at the 26th International Laser Radar Conference (ILRC 26), Porto Heli, Greece, 2012.

EPJ Web of Conferences **176**, 02007 (2018)
ILRC 28
https://doi.org/10.1051/epjconf/201817602007

EARTHCARE AEROSOL AND CLOUD LAYER AND COLUMN PRODUCTS

Ulla Wandinger[1]*, Anja Hünerbein[1], Stefan Horn[1], Florian Schneider[1], David Donovan[2], Gerd-Jan van Zadelhoff[2], David Daou[2], Nicole Docter[3], Jürgen Fischer[3], Florian Filipitsch[3]

[1]*Leibniz Institute for Tropospheric Research (TROPOS), Leipzig, Germany, *Email: ulla@tropos.de*

[2] *Royal Netherlands Meteorological Institute (KNMI), De Bilt, The Netherlands*

[3]*Free University of Berlin (FUB), Institute for Space Science, Berlin, Germany*

ABSTRACT

We introduce the development of EarthCARE Level 2 layer products derived from profile measurements of the high-spectral-resolution lidar ATLID and column products obtained from combined information of ATLID and the Multi-Spectral Imager (MSI). Layer products include cloud top height as well as aerosol layer boundaries and mean optical properties along the satellite nadir track. Synergistic column products comprise cloud top height, Ångström exponent, and aerosol type both along-track and across the MSI swath.

1 INTRODUCTION

The Earth Clouds, Aerosol and Radiation Explorer (EarthCARE) is a joint ESA/JAXA mission planned to be ready for launch in 2019 [1]. The platform carries four instruments: a cloud-profiling radar (CPR) operating at 94 GHz, a 355-nm high-spectral-resolution lidar (ATLID), a 7-channel multi-spectral imager (MSI), and a three-view broad-band radiometer (BBR). Current algorithm developments aim at a full chain of atmospheric products, which include stand-alone cloud, aerosol and radiation parameters obtained by the individual instruments (called Level 2a or L2a) as well as synergistic products derived from combined observations (Level 2b or L2b). In this contribution, we focus on a part of the chain dealing with ATLID L2a Layer Products and ATLID-MSI L2b Column Products. ATLID L2a Profile Products are described in an accompanying paper [2].

The high-spectral-resolution lidar ATLID retrievals provide profiles of particle extinction and backscatter coefficients, lidar ratio, and linear depolarization at 355 nm along the track of the satellite [2]. MSI provides cloud cover, phase and type, cloud top height, and cloud optical properties as well as aerosol optical thickness (AOT) at 670 nm (over land and ocean) and 865 nm (over ocean) across a 150-km wide swath.

The EarthCARE mission requirements [3] are based upon the goal to derive the radiative flux at the top of the atmosphere with an accuracy of 10 Wm^{-2} consistent with a 10 km × 10 km snapshot view of the atmosphere. Accordingly, an accuracy of ice and water cloud top height of 300 m is required for the 3-dimensional scene. With respect to aerosols, a detection threshold of 0.05 for the AOT and an accuracy of the vertical layering of 500 m for a horizontal resolution of 10 km have been defined. Such accurate vertical measures can be provided by ATLID, but only for an atmospheric cross section along track. Thus, synergistic cloud and aerosol products by coupling ATLID and MSI observations are proposed to extend the 2-dimensional cross-sections into the 3-dimensional domain. Furthermore, it is required to identify the presence of absorbing or non-absorbing aerosols from natural and anthropogenic sources. For aerosol typing information on the spectral AOT is of interest. This information can be gained by combining ATLID observations at 355 nm with MSI retrievals for wavelengths ≥670 nm.

The 3-dimensional information on clouds and aerosol derived by the coupling of ATLID, CPR and MSI observations will serve as input to broadband radiative-transfer models [4]. In this way, fluxes, heating rates, and radiances can be calculated and top-of-atmosphere (TOA) radiances and fluxes compared to those derived from BBR measurements. If closure of measured and calculated TOA fluxes with the required accuracy can be reached, EarthCARE observations will substantially decrease the uncertainties in our knowledge of global radiative forcing.

2 ATLID LAYER PRODUCTS

ATLID Layer Products are computed by the A-LAY processor. Since they serve as input for the synergistic retrievals described in Sec. 3, they are defined on the EarthCARE Joint Standard Grid (JSG). In order to allow the combination, observations of the EarthCARE instruments are averaged and interpolated to this common grid. It has a resolution of approximately 1000 m in the horizontal and 100 m in the vertical direction and is determined by the actual resolution of the CPR measurements in the horizontal and ATLID in the vertical.

2.1 ATLID Cloud Top Height (A-CTH)

The cloud top height (CTH) is defined as the upper geometrical boundary of the uppermost cloud layer in the atmosphere. The A-CTH product is derived from the ATLID Mie co-polar signal by searching for characteristic signal gradients with a wavelet covariance transform (WCT) algorithm [5, 6] under consideration of configurable threshold settings. Depending on the cloud optical thickness and the signal-to-noise ratio, the boundaries are detected with a horizontal resolution of 1 JSG pixel (thick clouds) or 11 JSG pixels (thin clouds). The product contains a simplified classification of the uppermost cloud as well as quality indicators in terms of level of confidence of the detection and level of consistency with the ATLID Target Classification (A-TC) product [2].

2.2 ATLID Aerosol Layer Descriptor (A-ALD)

The A-ALD product contains geometrical and optical information on aerosol layers. The same WCT algorithm as in the A-CTH retrieval is used to derive aerosol layer boundaries from the ATLID Mie co-polar signal, averaged over 11 JSG pixels horizontally. Appropriate threshold settings allow the discrimination of clouds and aerosol. The A-ALD product provides the upper and lower geometrical boundaries of significant aerosol layers, the optical thickness of each layer (ALOT), the column and the stratospheric AOT at 355 nm. ALOT and AOT are calculated by integrating the ATLID extinction profile taken from the ATLID Extinction, Backscatter and Depolarization (A-EBD) product [2]. In addition, layer-mean values of extinction and backscatter coefficient, lidar ratio and particle linear depolarization ratio are calculated from the A-EBD product. The A-ALD product is defined for cloud-free conditions only. It contains quality indicators in terms of level of confidence for the aerosol layer detection and level of consistency with the A-TC product.

3 ATLID-MSI COLUMN PRODUCTS

The synergistic ATLID-MSI columnar products are derived by the AM-COL processor, which makes use of the A-CTH and A-ALD products described above. It combines this information with MSI column measurements along and across track. The products are defined on the JSG.

3.1 ATLID-MSI Cloud Top Height (AM-CTH)

For the synergistic AM-CTH product, information from A-CTH is used to improve the MSI CTH retrievals in the across-track direction. The MSI CTH retrieval is based on an optimal estimation approach, which infers optical and physical properties of the cloud from measured satellite radiances. The algorithm uses an absorption-free and an absorption solar channel as well as an infrared-window channel at 10.8 µm. Thus, M-CTH represents an infrared effective radiating height. This is a reasonable estimate for the geometrical CTH in case of optically thick clouds, but leads to a significant underestimation of the CTH for multi-layer clouds. The method may also overestimate the CTH of thick clouds in the presence of temperature inversions at cloud top. Thus, the synergistic retrieval is based on the systematic investigation and classification of differences in CTH obtained with ATLID and MSI along track. The CTH difference for the entire swath is extrapolated from the track by use of the cloud classification and different homogeneity criteria. The AM-CTH product contains information on ATLID and MSI CTH differences in the along and across-track directions, together with a quality indicator.

3.2 ATLID-MSI Aerosol Column Descriptor (AM-ACD)

The AM-ACD product provides information on columnar aerosol optical properties. It contains the spectral AOT (355–670 nm over land and 355–670–865 nm over ocean) and the respective Ångström exponents. From the AOT and Ångström exponents the aerosol type is estimated with the help of a look-up table. In the first step, ATLID and MSI collocated data along track are combined. By investigating the horizontal homogeneity of the MSI AOT at 670 nm (identification of aerosol plumes), the product is spread over the swath or parts of it, if possible. The AM-ACD product is defined for cloud-free conditions only. The product also contains a quality indicator, which considers information on multiple aerosol layers provided by the A-ALD product.

4 ALGORITHM VALIDATION BASED ON SIMULATED TEST SCENES

Specific test scenes have been created with the EarthCARE Simulator (ECSIM, [1, 2]) from atmospheric model data to help develop and validate the full chain of EarthCARE processors. In the following, we present a few results obtained with the A-LAY and AM-COL processors for two scenes.

COSMO scene. This small scene was created from model output of COSMO-DE (regional model of the German Meteorological Service) for a domain over Germany, referring to 3 July 2008, 12:00 UTC. It has an extent of 150 km × 100 km and contains partly overlapping ice and water clouds. Boundary-layer aerosol has been artificially added.

MACC scene. This extended scene was created from ECMWF MACC model output and represents a "full frame" case with a length of about 6000 km (EarthCARE products will be provided frame-wise with 8 frames per orbit). The model forecast for 13 January 2015, 12:00 UTC, and a ground track over Africa and Western Europe was selected. The atmospheric scene is dominated by aerosol layers over Northern Africa and by clouds in the tropical region and over Europe as can be seen in the upper panel of Fig. 1, which shows the simulated ATLID attenuated backscatter signal averaged over 11 JSG pixels horizontally.

The middle and lower panels of Fig. 1 show two parts of the scene (see white boxes in upper panel) dominated by clouds (JSG pixels no. 0-1000) and aerosol (JSG pixels no. 2000-3000), respectively. The cloud top height of the A-CTH product and the aerosol layer boundaries of the A-ALD product are overlaid (see legend for details). The color used for the cloud boundary indicates whether the cloud was identified at 1 JSG pixel resolution, which hints to an optically thick cloud, or at 11 JSG pixel resolution, which points to an optically thin cloud. If multiple cloud layers were detected, the cases "thin over thick", "thick over thick" and "thin over thin" are distinguished accordingly. This information is then used in the AM-CTH retrieval scheme to support the cloud classification.

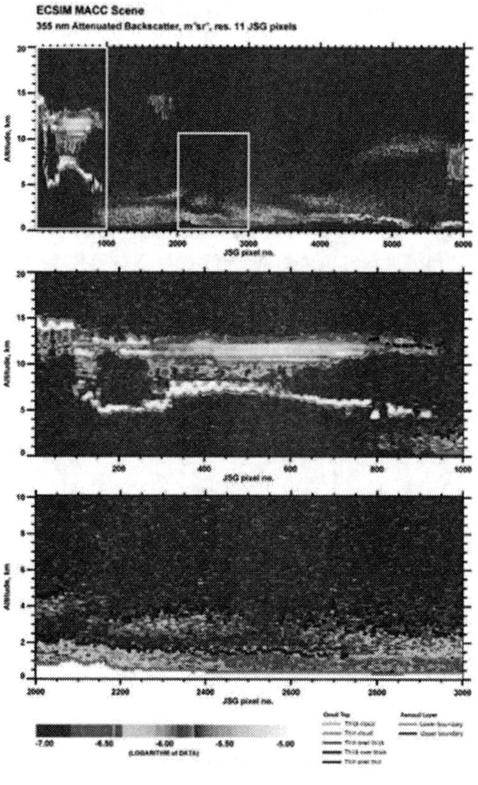

Fig. 1 Time-height contour plot of simulated ATLID attenuated backscatter (upper panel, full frame, JSG pixel no. 0-6000) overlaid with cloud top heights (middle panel, JSG pixel no. 0-1000) and aerosol layer boundaries (lower panel, JSG pixel no. 2000-3000) retrieved for 11 JSG pixels horizontal resolution with the A-LAY processor for the MACC scene. The cloud class in the middle panel is color-coded (see legend).

 https://doi.org/10.1051/epjconf/201817602007

Figure 2 shows the combined CTH retrieval for the COSMO scene. The M-CTH product (upper panel and red symbols in middle panel) indicates a decreasing cloud top height from left to right with a maximum height of 10 km around pixel no. 10. From ATLID an almost constant CTH at 11.5 km is derived over the pixels no. 10 to 55, followed by a sudden decrease to 4.7 km for pixels no. 60 to 70 (black symbols in middle panel). The reason for the discrepancy is the presence of a semi-transparent ice cloud with decreasing optical depth above an opaque water cloud, which results in an effective radiating height in between the two layers from the MSI retrieval. Because the underlying water cloud is broken and thin towards the edge, also its CTH is underestimated by the MSI retrieval due to detected radiation from the surface.

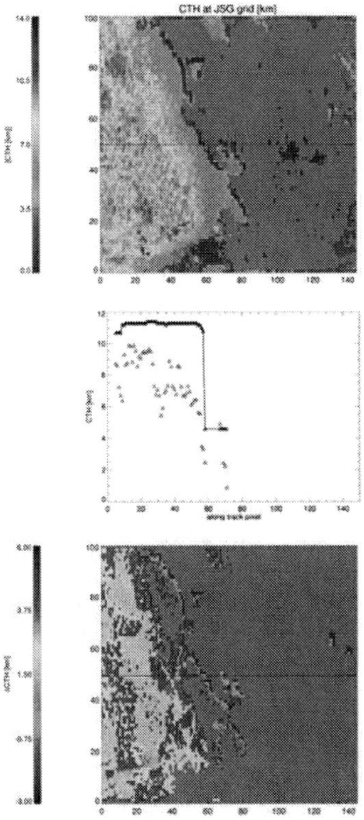

Fig. 2 Cloud top height from MSI with the ATLID track indicated in red (upper panel), MSI CTH (red symbols) and ATLID CTH (black symbols) along track (middle panel), and synergistic CTH difference (lower panel) for the COSMO scene.

Finally, the lower panel of Fig. 2 shows the difference between the estimated true geometrical CTH and the one retrieved from MSI observations over the entire swath based on the combined classification of clouds with MSI and ATLID.

5 CONCLUSIONS

ATLID layer and combined ATLID-MSI column products are part of the complex EarthCARE production scheme, for which several algorithms are currently under development. Altogether they will provide a novel synergistic data set allowing for detailed global investigations of cloud-aerosol-radiation interactions. The algorithms will be further evaluated based on common test scenes before the EarthCARE launch.

ACKNOWLEDGEMENTS

The work presented in this paper is performed within the project Atmospheric Products from Imager and Lidar (APRIL) funded by ESA under contract 4000112018/14/NL/CT.

References

[1] Illingworth, A., et al., 2014: THE EARTHCARE SATELLITE: The next step forward in global measurements of clouds, aerosols, precipitation and radiation. *Bull. Am. Met. Soc.* **96**, 1311–1332.

[2] Donovan, D. P., et al., 2017: EarthCARE ATLID extinction and backscatter retrieval algorithms, this conference.

[3] EarthCARE Mission Requirements Document EOP-SM/1567/TW, issue 5, 2 Nov 2006, available at: http://esamultimedia.esa.int/docs/EarthObservation/EarthCARE_MRD_v5.pdf.

[4] Barker, H. W., et al., 2011: A 3D cloud-construction algorithm for the EarthCARE satellite mission. *Q. J. R. Meteorol. Soc.* **137**, 1042–1058.

[5] Brooks, I., 2003: Finding Boundary Layer Top: Application of a wavelet covariance transform to lidar backscatter profiles, *J. Atmos. Ocean. Tech.* **20**, 1092–1105.

[6] Baars, H., et al., 2008: Continuous monitoring of the boundary-layer top with lidar, *Atmos. Chem. Phys.* **8**, 7281–7296.

SCIENTIFIC MOTIVATION FOR ADM/AEOLUS MISSION

Erland Källén

ECMWF, Shinfield Park, Reading, RG2 9AX, United Kingdom, erland.kallen@ecmwf.int

ABSTRACT

The ADM/Aeolus wind lidar mission will provide a global coverage of atmospheric wind profiles. Atmospheric wind observations are required for initiating weather forecast models and for predicting and monitoring long term climate change. Improved knowledge of the global wind field is widely recognised as fundamental to advancing the understanding and prediction of weather and climate. In particular over tropical areas there is a need for better wind data leading to improved medium range (3-10 days) weather forecasts over the whole globe.

1 INTRODUCTION

Together with temperature, pressure and humidity, wind is one of the basic variables describing the state of the atmosphere. Atmospheric wind observations are required for initiating weather forecast models and for predicting and monitoring long term climate change. Improved knowledge of the global wind field is widely recognised as fundamental to advancing the understanding and prediction of weather and climate [1]. The present in-situ and surface based global observing system for winds has large gaps, in particular over ocean areas and in the tropics. It includes the radiosonde network, commercial aircraft, wind profilers, Doppler radars and surface stations. Winds can also be measured from satellites, cloud track winds from imagers and scatterometers measuring ocean surface winds are the two dominant observation sources. All wind observing systems suffer from a number of limitations, the most important are a lack of global coverage and accuracy limitations. Only the radiosondes and the wind profilers provide vertical wind profiles, most of the other observations are single level or vertically averaged over deep layers. Radiosondes and wind profilers are primarily located over inhabited areas of the northern hemisphere, there are very few observations in the tropics and almost none over the vast ocean areas in the southern hemisphere.

The ADM/Aeolus mission will provide a global coverage of atmospheric wind profiles, urgently needed to enhance the quality of weather forecasts and improve climate analyses [2]. Of particular interest is the potential to improve analysis of tropical weather systems, both in terms of accuracy of the initial state for numerical weather prediction and in terms of tropical circulation monitoring in climate analyses. Improvements in the tropics will lead to improved weather forecasts in mid-latitudes as initial state errors in the tropics propagate to Europe, Asia and North America within a week and thus affect weather forecast quality in the medium range (3-10 days). There are examples of very poor six day forecasts over Europe that can be traced back to initial state errors in the tropics [3]. The wind profiles will also help to better define smaller scale circulation systems in mid-latitudes, examples are atmospheric fronts and polar lows. In the upper troposphere and lower stratosphere (UTLS) the wind profiles will enhance circulation analyses that helps to improve weather prediction and better understand climate change dynamics in the UTLS region.

2 METHODOLOGY

The Aeolus Doppler wind lidar instrument will measure line-of-sight winds in the troposphere and lower stratosphere. Only one horizontal wind component is measured, predominantly in the zonal direction. Aeolus is a polar orbiting satellite flying in a sun-synchronous dawn-dusk orbit at about 350 km height. Wind profile measurements will be provided with a horizontal resolution of about 80 km and a vertical resolution ranging from about 500 meters close to the surface up to about 2 km in the lower stratosphere. To obtain the full, two component, horizontal wind field the Aeolus line-of-sight wind data must be assimilated in a global NWP system. This will be undertaken at ECMWF who will be responsible for providing level 2C wind products using the operational NWP assimilation system. In addition cloud and aerosol information will be provided using information on lidar scattering and backscatter to extinction ratios. Comprehensive information about the Aeolus mission can be found in the Aeolus science report published by ESA [4] as well as in other presentations at the ILRC28 conference.

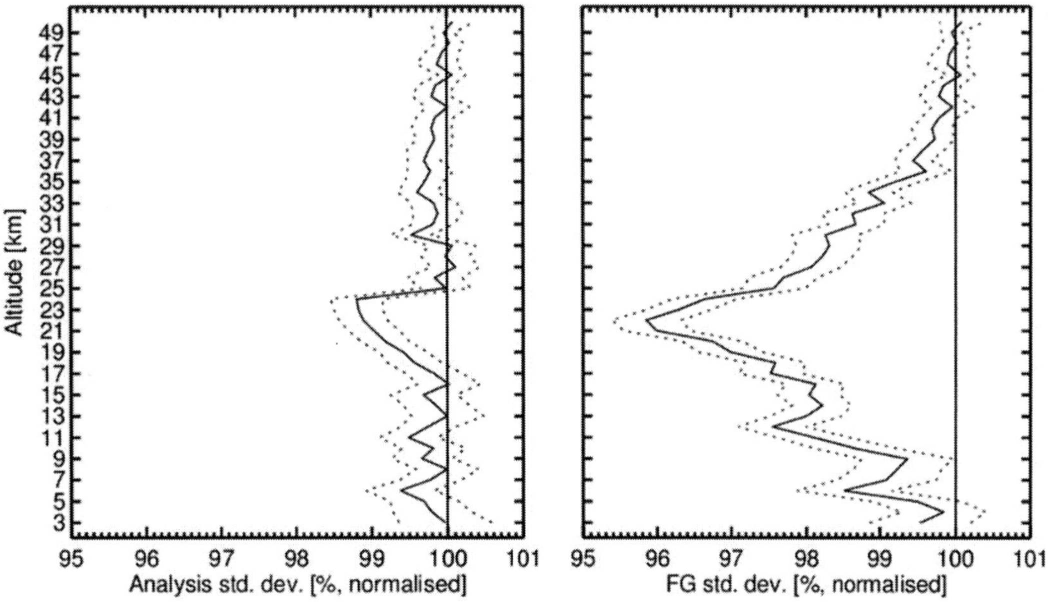

Figure 1 Difference between two assimilation experiments of the fit between analysis (left hand plot) and background (right hand plot) to observed GPSRO bending angles in the tropics.

3 TROPICAL ASSIMILATION

The atmospheric motion and thermodynamic structures in the tropics are quite different from the ones found in midlatitudes. Due to the influence of the rotation of the Earth there is a quasi-stationary balance between temperature, pressure and wind fields in the extra-tropics, i.e. north and south of 30 degrees latitude. Within the tropics the influence of the Earth's rotation on this balance is very different, instead of a quasi-stationary local balance between temperature, pressure and wind we find large scale wave structures that propagate slowly along the equatorial belt. Close to the equator these structures are defined by the wind field, temperature and pressure only exhibit small variations and the winds determine the variations. Further away from the equator temperature, pressure and wind become more balanced but they are linked to the equatorial flow as the large scale wave structures cover several thousands of kilometers [5]. With more wind data we are better able to define tropical wave structures, only mass field data (temperature and humidity) is not sufficient to constrain the wind field. Wind profile data from Aeolus will improve the quality of tropical analyses and reduce initial state errors in numerical weather prediction [6]. Synthetic Aeolus data impact experiments have shown that adding Aeolus data to the ECMWF prediction systems gives about the same positive impact on forecast quality as radiosonde data gives us today [7]. In particular we find cases where medium range forecasts failures over Europe can be traced back to initial state errors in the tropics [3]. Another observation system simulation experiment also illustrates the direct impact of line-of-sight wind observations in reducing short range tropical forecast errors [8].

In addition to providing better defined wind fields in the tropics Aeolus data will also improve the assimilation of mass field data in the tropics. Using the ECMWF assimilation system and performing assimilation of mass field data with and without line-of-sight wind data the fit of analyses and short range forecasts to observed radio occultation (GPSRO) data in the tropics demonstrates the beneficial impact of wind data. In fig 1 we show the relative change of GPSRO bending angle fits to observations comparing one experiment with all conventional wind data removed and the other with line-of-sight wind data included to simulate the upcoming Aeolus wind profiling mission. A reduction of the standard deviation means that the fit improves with more wind data. In figure 1 there is a marked reduction in the UTLS region indicating the importance of wind information in this region.

The Aeolus mission will also provide information about clouds and aerosols from lidar backscatter intensities. This data will be useful for assimilating aerosol information into NWP systems and contribute to improving atmospheric composition monitoring and forecasting.

4 CONCLUSIONS

WMO co-ordinated assessments of observational needs for numerical weather prediction have repeatedly highlighted the primary importance of improved wind observations. The ADM/Aeolus wind profiles will have a significant impact on the quality of weather forecasts and climate analyses. In particular

ACKNOWLEDGEMENTS

Michael Rennie is gratefully acknowledged for providing fig 1.

References

[1] Baker, W. E., Atlas, R., Cardinali, C., Clement, A., Emmitt, G. D., Gentry, B. M., Hardesty, R M., Källén, E., Kavaya, M. J., Langland, R., Ma, Z., Masutani, M., McCarty, M., Pierce, R.B., Pu, Z., Riishojgaard, L. P. R., Ryan, J., Tucker, S., Weissman, M. and Yoe, J. G., 2014: Lidar-measured wind profiles: The missing link in the global observing system. *Bull. Amer. Met. Soc.*, **95**, 543-564.

[2] Stoffelen, A., Pailleux, J., Källén, E., Vaughan, J. M., Isaksen, L., Flamant, P., Wergen, W., Andersson, E., Schyberg, H. and Meynart, R., 2005. The Atmospheric Dynamics Mission for global wind field measurement. *Bulletin of the American Meteorological Society*, **86**, 73-87.

[3] Magnusson, L., 2017: Diagnostic methods for understanding the origin of forecast errors. *ECMWF Technical Memorandum no.* **796**, pp 26.

[4] ESA, 2008: ADM-Aeolus Science report. *ESA report SP-1311*, pp 121.

[5] Žagar, N., N. Gustafsson and E. Källén, 2004: Dynamical response of equatorial waves in four-dimensional variational data assimilation. *Tellus.* **56A**, 29-46.

[6] Žagar, N., 2004. Assimilation of equatorial waves by line-of-sight wind observations. *Journal of the Atmospheric Sciences*, **61**, 1877-1893.

[7] Tan, D. G. and E. Andersson, 2005. Simulation of the yield and accuracy of wind profile measurements from the Atmospheric Dynamics Mission (ADM-Aeolus). *Quarterly Journal of the Royal Meteorological Society*, **131**, 1737-1757.

[8] Horányi, A., Cardinali, C., Rennie, M. and Isaksen, L., 2015. The assimilation of horizontal line-of-sight wind information into the ECMWF data assimilation and forecasting system. Part I: The assessment of wind impact. *Quarterly Journal of the Royal Meteorological Society*, **141**, 1223-1232.

OPERATIONAL USE OF SPACEBORNE LIDAR DATASETS

Franco Marenco*, Gemma Halloran and Mary Forsythe

Satellite Applications, Met Office, United Kingdom
**Email: franco.marenco@metoffice.gov.uk*

ABSTRACT

The Met Office plans to use space lidar datasets from CALIPSO, CATS, Aeolus and EarthCARE operationally in near real time (NRT), for the detection of aerosols. The first step is the development of NRT imagery for nowcasting of volcanic events, air quality, and mineral dust episodes. Model verification and possibly assimilation will be explored. Assimilation trials of Aeolus winds are also planned. Here we will present our first in-house imagery and our operational requirements.

1 INTRODUCTION

Lidar is useful in identifying atmospheric aerosol layers such as volcanic ash, mineral dust, smoke from biomass burning, polluted layers, etc. As it senses the vertical structure of the atmosphere, lidar is also useful in defining the layer height, a quantity that is otherwise weakly constrained using passive satellite imagery.

The CALIPSO mission has been very successful [1], having been in orbit since 2006 and having long exceeded its nominal lifetime. It is difficult to predict when it will fail, but this could happen at very short notice. Moreover, since 2015 the CATS instrument has been installed on the International Space Station (ISS), and it intended to operate in orbit for up to 3 years [2]. A follow-on mission is being considered at the moment.

In the near future, the launch of two missions from ESA are expected: ADM-Aeolus is mainly intended for measuring wind speed, but will provide an aerosol product, and EarthCARE is a mission that specifically targets clouds and aerosols [3,4].

Current lidar imagery from CALIPSO and CATS is available directly on NASA web portals, and it is assumed that in the future ESA will provide it in a similar manner for its missions. These products are probably the best quality that can be achieved. However, level 1 data are available online several days (sometimes several weeks) after their acquisition. Unfortunately these delays make the use of the imagery unlikely for nowcasting or operational data assimilation.

It is however also possible to set up the reception of near real-time (NRT) products. These come as data files, and then imagery has to be produced in-house. The data may also not contain the full spectrum of information, being at a lower data level (e.g. L1 instead of L2).

Coverage is another issue as these sensors have a narrow swath and their repeat cycles vary between 3 and 25 days; moreover the CATS/ISS system only covers the 52S–52N latitude band (see Table 1).

At the Met Office we are currently considering such products, having made contact with the instrument scientists and having started the reception of the NRT products. At the time of writing this abstract, we are in the process of developing in-house imagery. This imagery will be discussed in this presentation, together with a first assessment of the operational requirements.

2 REQUIREMENTS

NRT lidar products could be used in the London VAAC (Volcanic Ash Advisory Centre), based at the Met Office: they can assist in the identification of volcanic ash layers and help assign their height in the atmosphere. This is not possible at the moment using the datasets disseminated online, due to the delay between the acquisition of the measurements and the data availability. The layer height determined by lidar can in principle be used to initialize the Numerical Atmospheric dispersion Modeling Environment (NAME) dispersion model, used for volcanic ash predictions. NRT products are also useful to identify the layer height in air quality and mineral

Table 1. Spaceborne lidar missions considered.

Mission	**Expected Lifetime**	**Coverage**	**Repeat cycle**	**Expected timeliness**	**Product name**
CALIPSO	2006-?	82S-82N	16 d	12-72 h	L1.5 expedited
CATS	2015-2018	52S-52N	3 d	6 h	L1
Aeolus	2017-2020	83S-83N	7 d	3 h	L2A
EarthCARE	2018-2021	83S-83N	25 d	60% data within 6 h	L2A, L2B

dust events.

In the future we could consider testing the data assimilation of lidar datasets to constrain the dust layer height in weather prediction models. The Met Office Unified Model (MetUM) in its Global configuration is run operationally four times a day for the prediction of mineral dust concentration. Currently, MODIS AOD is the only assimilated information on aerosol load, and therefore layer height is fully determined by the model. There will however be limitations with this, as overpasses at a given location are infrequent.

Moreover, lidar signals are noisy, and simulating the measured signal may be tricky as they depend upon all aerosol types at once, and are strongly influenced by the cloud field. Possibly, the verification and assimilation work could be experimented using the layer base and top height information given in the lidar datasets to help constrain the vertical distribution.

3 SYNERGY WITH THE GROUND-BASED LIDAR NETWORK

The Met Office is setting up an operational network with 10 Raymetrics ground-based lidars. The main aim is the detection and monitoring of volcanic events, but the instruments may also help fill other requirements, along the lines discussed above. The imagery from these instruments is currently available internally within 15 minutes of acquisition. Current quicklooks do not show calibrated products or retrieved quantities; thus, they only give a qualitative image of the atmospheric structure. In a successive phase, however, calibrated signals could be delivered and/or retrieved products could be inferred. Such datasets could then be used for model validation and/or data assimilation.

4 DOPPLER WIND LIDAR

ADM Aeolus requires a separate discussion, as this mission's main objective is to provide horizontal line of sight (HLOS) winds for numerical weather prediction (NWP). The Met Office will be exploring the potential of this new instrument.

The Met Office is part of the Aeolus cal/val team and will therefore gain early access to the data products. Monitoring of their quality and coherence with other global wind observations will be put in place and maintained through the lifetime of the mission. For this, NWP model fields will be used as a source of comparison (observations minus background) using established methods for the evaluation of large volumes of observations. The objective here is to identify data quality issues, to produce a reliable observation operator, capable of simulating the Aeolus data from model fields, and to assess biases and uncertainties.

When we have completed this analysis, and as long as the data is of sufficient quality, we intend to trial for assimilation in the Met Office global model and assess where the data provides most benefit and measure the impact of Aeolus on the forecast quality.

Airborne cal/val activities of ADM/Aeolus, using the FAAM BAe-146 research aircraft, are also contemplated, with the aim of

- a direct comparison of measured winds;
- an evaluation of wind variability and issues of scale;
- insight of the effect of heterogeneities in the aerosols and cloud fields on the derived winds;
- studies on scene classification.

Detailed planning of the Aeolus cal/val plans will be established after launch, and pending the availability of sufficient funding.

5 CONCLUDING REMARKS

The project of using spaceborne lidar datasets operationally at the Met Office is in its infancy. This presentation will discuss our requirements, show first imagery products, and describe our plans for the cal/val of Aeolus.

ACKNOWLEDGEMENTS

Thanks to the CALIPSO, CATS, Aeolus and EarthCARE teams at NASA and ESA for the provision of spaceborne lidar datasets and the openness in discussing requirements.

References

Numbered in the order they appear in the text, with brackets. Example of references:

[1] Winker, D.M., J. Pelon, J.A. Coakley, S.A. Ackerman, R.J. Charlson, P.R. Colarco, P. Flamant, Q. Fu, R.M. Hoff, C. Kittaka*, T.L. Kubar, H. Le Treut, M.P. McCormick, G. Mégie*, L. Poole, K. Powell, C. Trepte, M.A. Vaughan, and B.A. Wielicki, 2010: The CALIPSO Mission: A Global 3D View of Aerosols and Clouds, *Bull. Am. Meteorol. Soc.* **91**, 1211-1228.

[2] Yorks, J.E., M.J. McGill, S.P. Palm, D.L. Hlavka, P.A. Selmer, E P. Nowottnick, M.A. Vaughan, S.D. Rodier, and W.D. Hart, 2016: An overview of the CATS level 1 processing algorithms and data products, *Geophys. Res. Lett.* **43**, 4632–4639.

[3] Endemann, M., Dubock, P., Ingmann, P., Wimmer, R., Morancais, D., & Demuth, D., 2004: The Adm-Aeolus Mission - the First Wind-Lidar in Space, *in* Proceedings of the 22nd International Laser Radar Conference (ILRC 2004) held in Matera, Italy, 12-16 July 2004, edited by G. Pappalardo and A. Amodeo, *European Space Agency* ESA SP-561, Paris, 953pp.

[4] A.J. Illingworth, H.W. Barker, A. Beljaars, M. Ceccaldi, H. Chepfer, N. Clerbaux, J. Cole, J. Delanoë, C. Domenech, D.P. Donovan, S. Fukuda, M. Hirakata, R.J. Hogan, A. Huenerbein, P. Kollias, T. Kubota, T. Nakajima, T.Y. Nakajima, T. Nishizawa, Y. Ohno, H. Okamoto, R. Oki, K. Sato, M. Satoh, M.W. Shephard, A. Velázquez-Blázquez, U. Wandinger, T. Wehr, G.-J. van Zadelhoff, 2015: The EarthCARE Satellite: The Next Step Forward in Global Measurements of Clouds, Aerosols, Precipitation, and Radiation, *Bull. Am. Meteorol. Soc.* **96**, 1311-1332.

AEOLUS END-TO-END SIMULATOR AND WIND RETRIEVAL ALGORITHMS UP TO LEVEL 1B

Oliver Reitebuch[1]*, Uwe Marksteiner[1], Marc Rompel[1], Markus Meringer[2], Karsten Schmidt[2], Dorit Huber[3], Ines Nikolaus[4], Alain Dabas[5], Jonathan Marshall[6], Frank de Bruin[7], Thomas Kanitz[7], Anne-Grete Straume[7]

[1]*DLR, Institute of Atmospheric Physics, *oliver.reitebuch@dlr.de*
[2]*DLR, Remote Sensing Technology Institute, Germany*
[3]*DoRIT, Germany*
[4]*University of Applied Sciences Munich, Germany*
[5]*CNRS/Météo-France, CNRM, France*
[6]*Airbus Defence and Space, UK*
[7]*ESA-ESTEC, The Netherlands*

ABSTRACT

The first wind lidar in space ALADIN will be deployed on ESA´s Aeolus mission. In order to assess the performance of ALADIN and to optimize the wind retrieval and calibration algorithms an end-to-end simulator was developed. This allows realistic simulations of data downlinked by Aeolus. Together with operational processors this setup is used to assess random and systematic error sources and perform sensitivity studies about the influence of atmospheric and instrument parameters.

1 INTRODUCTION

The European Space Agency (ESA) is currently implementing a Doppler wind lidar mission named Atmospheric Dynamics Mission ADM-Aeolus [1]. It is considered as a technology demonstrator for future operational wind lidar missions. The launch of the satellite is currently planned for 2018. Aeolus will provide profiles of one component of the horizontal wind vector along the laser line-of-sight (LOS) from ground up to the lower stratosphere (20 km to 30 km) with 250 m to 2 km vertical resolution and a precision of 1 m/s to 3 m/s depending on altitude [2]. A wind profile will be obtained from horizontal averaging over 90 km along track in order to achieve the precision requirement (random error). Systematic errors, which cannot be corrected by a-priori knowledge (e.g., calibration, instrumental corrections) – called an "unknown bias" – are very detrimental for numerical weather prediction (NWP) models. Thus, the requirement for the maximum "unknown bias" of 0.7 m/s is very stringent and is further separated into a wind-independent contribution of 0.5 m/s and a wind-dependent contribution of 0.7 %.

The lidar ALADIN (Atmospheric LAser Doppler INstrument) is based on a direct-detection Doppler wind lidar (DWL) operating at an ultraviolet (UV) wavelength of 354.8 nm [3]. The optical receiver consists of two spectrometers to determine the Doppler shift from the spectrally broad Rayleigh-Brillouin molecular backscatter and the spectrally narrow Mie backscatter from aerosols and cloud particles [4]. Accumulation charge coupled device (ACCD) detectors are used for both the Mie and Rayleigh spectrometer. The minimum vertical resolution of the ALADIN instrument is limited by the ACCD detector to 250 m and only a limited number of 24 atmospheric altitude ranges can be acquired. On the other hand the vertical resolution can be varied from 250 m to 2000 m even within one vertical profile. The number of laser pulse returns accumulated directly on the detector is a settable parameter and determines the minimum along-track horizontal resolution for the raw data, which is typically 3 km for a number of 20 accumulated laser pulses (called "measurement"). Both the Mie and the Rayleigh spectrometer data are stored with this temporal resolution on-board and transmitted to the ground via an X-band downlink every orbit. Further averaging of 600 laser pulse returns (12 s data, 90 km along-track resolution, called "observation") is performed on-ground and is needed to achieve the random error requirement of 1 m/s to 3 m/s.

2 AEOLUS PROCESSING CHAIN

The downlinked data (called Annotated Instrument Source Packets AISP) from ALADIN spectrometers, instrument and platform status (e.g. settings, temperatures) and AOCS (Attitude and Orbit Control System) are processed on-ground to obtain several levels of data products (Level 0, 1A, 1B, 2A, 2B, 2C). For Level 1 data products (processed instrument data) only information from the Aeolus satellite with the ALADIN instrument and the AOCS is used. For Level 2 products additional information from other sources is used, mainly the atmospheric temperature and pressure profiles for correcting the winds derived with the Rayleigh spectrometer from the temperature and pressure dependent signal bandwidth of the Rayleigh-Brillouin scattering [5]. Algorithms for Level 2 data products (derived geophysical wind and aerosol variables) use also a horizontal and vertical grouping of measurements to discriminate clear and cloudy atmospheric scenes [6].

An end-to-end simulator (E2S) was developed for the Aeolus mission with the following objectives:

- testing the ground segment processor chain from Level 0 to Level 2
- assessing the performance of ALADIN with respect to random and systematic errors
- performing sensitivity studies to optimize instrument settings, calibration methods and retrieval algorithms
- validating the Aeolus mission after launch.

Thus, the Aeolus E2S is continuously updated with new functionalities and instrument parameters arising from characterization of the ALADIN instrument on-ground and from experience with the ALADIN airborne demonstrator (A2D). The A2D was developed in order to assess the performance of ALADIN and to optimize the retrieval algorithms with "real" atmospheric signals [7]. Several airborne campaigns with a coherent-detection wind lidar as a reference and the A2D were performed for pre-launch validation of Aeolus from 2007-2009 [8], together with the NASA DC-8 aircraft in 2015 [9] and recently during NAWDEX (North Atlantic Waveguide and Downstream Impact Experiment) in 2016. More than 100 recommendations for testing and operating the ALADIN instrument on-ground and in-orbit and for refinement of the retrieval algorithms and calibration strategy were derived from the experience with the A2D [10] and adapted to the modelling of the ALADIN instrument in the E2S and for the wind retrieval algorithms in the L1B processor.

3 AEOLUS END-TO-END SIMULATOR

The E2S is not only a simulator for the lidar instrument ALADIN interacting with the atmosphere, but also includes modules for the satellite orbit, the AOCS and allows to couple a Digital Elevation Model (DEM) and spatially resolved global information about the surface albedo [11]. DEM and albedo information is important for a correct simulation of the range and intensity of the surface return, which is used for ALADIN calibration and bias correction schemes.

In order to be able to use the E2S for testing the complete chain of processors, the output data are provided in the same format and temporal and spatial resolution as it will be downlinked from the satellite as AISP.

In order to use the E2S for performance assessment, sensitivity studies and optimization of retrieval algorithms it allows for the:

- use of atmospheric vertical profiles (e.g. backscatter and extinction of aerosol and clouds from climatology or earlier lidar missions as LITE or CALIPSO, wind profiles from radiosondes or ECMWF model) with high vertical and temporal resolution, e.g. to simulate atmospheric inhomogeneities equivalent to the measurement data resolution of Aeolus of 3 km; for some sensitivity studies the atmospheric profiles were even varied per laser shot with a pulse repetition rate of 50 Hz
- simulation of spectrally resolved atmospheric backscatter signal of Mie scattering from aerosol and cloud particles and Rayleigh-Brillouin scattering from molecules [12]

- simulation of surface returns using input from a DEM (ACE2 with 9′′ resolution) and global maps of surface reflectance in the UV (ADAM [11] with 0.1° resolution)
- simulation of spectrally broadband solar background signal
- simulation of total instrument optical transmission for transmit and receive paths and spectrally resolved transmission and reflection functions for the Fizeau interferometer and sequential implementation of two Fabry-Pérot interferometers [13]
- simulation of charge transfer and detection on the accumulation CCD including offsets, non-linearity and noise sources (e.g. dark current noise, read-out noise, Poisson detection noise) and subsequent analog-to-digital conversion with 16 bit
- simulation of accumulated signal returns on the Mie and Rayleigh spectrometer ACCD´s from the atmospheric returns and from each transmitted laser pulse, which is used as frequency reference
- simulation of all ALADIN instrument modes including wind measurements with 35° off-nadir pointing, response calibration with 0° nadir pointing, use of the ACCD in imaging mode for telescope focus adjustment, and several internal modes without atmospheric return, e.g. characterization of dark charges, Fizeau and Fabry-Pérot interferometer spectral transmission, or monitoring of laser beam properties.

The simulation of the ALADIN response calibration in nadir mode and the internal modes for characterizing the spectral transmission of the interferometers are especially relevant for assessing the systematic errors. Also several instrument imperfections are included in the E2S, which are relevant for the instrument performance and which were spotted by means of A2D measurements.

4 AEOLUS WIND RETRIEVAL ALGORITHMS

The wind retrieval algorithms are described in detail in the respective Algorithm Theoretical Basis Document (ATBD) [14] and an overview is provided in [6]. The main product from Aeolus will be LOS winds projected to the horizontal plane – called HLOS winds. The L1B algorithms will use only information provided from Aeolus itself (AOCS data, ALADIN wind mode data, ALADIN response calibration mode data) to derive L1B HLOS wind profiles averaged over 1 observation derived from both the Mie and Rayleigh spectrometer data. Thus, the L1B Rayleigh HLOS winds will exhibit a significant bias due to non-correction of the actual atmospheric temperature profile and of the cross-talk of spectrally narrowband Mie returns to the broadband molecular return in case of aerosol and cloud layers. Both effects are corrected for the L2B winds [5, 6].

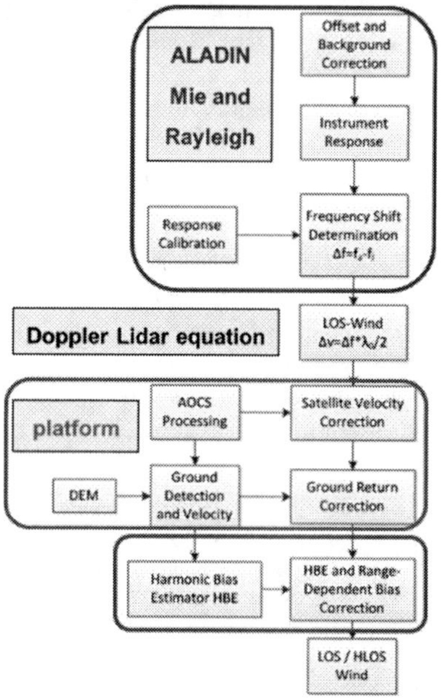

Figure 1 Wind retrieval algorithm up to Level 1B using information from ALADIN Mie and Rayleigh spectrometer data (blue), AOCS and platform data (green) and bias correction (red)

5 CONCLUSIONS

The first wind lidar in space ALADIN will be deployed on ESA´s Aeolus mission and will be operational in 2018. It is not only the first time that the space hardware of a high power UV laser and a Doppler wind lidar was developed, but also the on-ground processors for wind retrieval, calibration and bias correction schemes are novel, i.e. without heritage from earlier space missions. In order to develop, test and optimize the retrieval algorithms before launch, significant effort was undertaken in developing a representative E2S and chain of processors from Level 0 to Level 2 as well as in pre-launch campaigns with the A2D. Nevertheless all algorithm teams and also data product users should be prepared for surprises and serendipity.

ACKNOWLEDGEMENTS

Funding for the development of the E2S and L1B processor was provided by ESA and DLR. The first versions of the E2S and L1B were developed by MDA, Canada.

References

[1] European Space Agency ESA, 2008: ADM-Aeolus Science Report, ESA SP-1311, 121 pp.

[2] European Space Agency ESA, 2016: ADM-Aeolus Mission Requirements Documents, AE-RP-ESA-SY-001, Issue 2, 16/11/2016, 57pp.

[3] Reitebuch, O., 2012: The space-borne wind lidar mission ADM-Aeolus. in Atmospheric Physics – Background, Methods, Trends. U. Schumann (Ed.) *Springer Series on Research Topics in Aerospace,* 815-827.

[4] Reitebuch, O., 2012: Wind lidar for atmospheric research. in Atmospheric Physics – Background, Methods, Trends. U. Schumann (Ed.) *Springer Series on Research Topics in Aerospace*, 487-507.

[5] Dabas, A., Denneulin, M.-L., Flamant, P., Loth, C., Garnier, A., Dolfi-Bouteyre, A., 2008: Correcting winds measured with a Rayleigh Doppler lidar from pressure and temperature effects. *Tellus* **60A**, 206-215.

[6] Tan, D., Andersson., E., de Kloe, J., Marseille, G.-J., Stoffelen, A., Poli, P., Denneulin, M.-L., Dabas, A., Huber, D., Reitebuch, O., Flamant, P., Le Rille, O., Nett, H., 2008: The ADM-Aeolus wind retrieval algorithms. *Tellus* **60A**, 191-205.

[7] Reitebuch, O., Lemmerz, C., Nagel, E., Paffrath, U., Durand, Y., Endemann, M., Fabre, F., Chaloupy, M., 2009: The Airborne Demonstrator for the Direct-Detection Doppler Wind Lidar ALADIN on ADM-Aeolus. Part I: Instrument Design and Comparison to Satellite Instrument. *J. Atmos. Ocean. Tech.* **26**, 2501-2515.

[8] Marksteiner, U., Reitebuch, O., Rahm, S., Nikolaus, I., Lemmerz, C., Witschas, B., 2011: Airborne direct-detection and coherent wind lidar measurements along the east coast of Greenland in 2009 supporting ESA's Aeolus mission, *Proc. of SPIE Europe Int. Symposium Remote Sensing,* **8182,** Prague, Czech Republic, 81820J, 1-8.

[9] Reitebuch, O., Lemmerz, C., Marksteiner, U., Rahm, S., Schäfler, A., Witschas, B., Emmitt, G.D., Greco, S., Kavaya, M.J., Gentry, B., Neely III, R.R., Schüttemeyer, D. (2016): Airborne wind lidar observations in the North Atlantic for preparation of the ADM-Aeolus validation. *Proc. 18th Coherent Laser Radar Conference CLRC*, Boulder, USA.

[10] Reitebuch, O. et al., 2012: Final Report of ADM-Aeolus Campaigns 2004-2012. CD with final report, technical notes, dissertations, peer-reviewed papers, and final meeting handouts. Contract No. 18366/04/NL/MM, 15.06.2012.

[11] Noveltis, 2013: A Surface Reflectance Database for ESA's Earth Observation Missions (ADAM). NOV-3895-NT-12403. Ed. 2, Rev. 1. from http://adam.noveltis.fr/

[12] Witschas, B. 2011: Analytical model for Rayleigh-Brillouin line shapes in air. *Appl. Opt.* **50**, 267-270.

[13] Paffrath, U., Lemmerz, C., Reitebuch, O., Witschas, B., Nikolaus, I., Freudenthaler, V., 2009: The Airborne Demonstrator for the Direct-Detection Doppler Wind Lidar ALADIN on ADM-Aeolus. Part II: Simulations and Rayleigh Receiver Radiometric Performance. *J. Atmos. Ocean. Tech.* **26**, 2516-2530.

[14] Reitebuch, O., Huber, D., Nikolaus, I., (2014). Algorithm Theoretical Basis Document ATBD: ADM-Aeolus Level 1B Products, AE-RP-DLR-L1B-001, V. 4.1, 18.7.2014, 93pp.

AIRBORNE DIRECT-DETECTION AND COHERENT WIND LIDAR MEASUREMENTS OVER THE NORTH ATLANTIC IN 2015 SUPPORTING ESA'S AEOLUS MISSION

Uwe Marksteiner[1*], Oliver Reitebuch[1], Christian Lemmerz[1], Oliver Lux[1], Stephan Rahm[1], Benjamin Witschas[1], Andreas Schäfler[1], Dave Emmitt[2], Steve Greco[2], Michael J. Kavaya[3], Bruce Gentry[4], Ryan R. Neely III[5], Emma Kendall[6], Dirk Schüttemeyer[6]

[1]*DLR, Institute of Atmospheric Physics, Oberpfaffenhofen, Germany, *uwe.marksteiner@dlr.de*
[2]*Simpson Weather Associates SWA, Charlottesville, USA*
[3]*NASA Langley Research Center, Hampton, USA*
[4]*NASA Goddard Space Flight Center, Greenbelt, USA*
[5]*University of Leeds, National Centre for Atmospheric Science, Leeds, United Kingdom*
[6]*ESA-ESTEC, Noordwijk, The Netherlands*

ABSTRACT

The launch of the Aeolus mission by the European Space Agency (ESA) is planned for 2018. The satellite will carry the first wind lidar in space, ALADIN (Atmospheric Laser Doppler INstrument). Its prototype instrument, the ALADIN Airborne Demonstrator (A2D), was deployed during several airborne campaigns aiming at the validation of the measurement principle and optimization of algorithms. In 2015, flights of two aircraft from DLR & NASA provided the chance to compare parallel wind measurements from four airborne wind lidars for the first time.

1 INTRODUCTION

In regular intervals expert teams on Numerical Weather Prediction (NWP) of the World Meteorological Organization (WMO) prioritize the need for vertical wind profiles [1]. Wind information available on a global scale would largely contribute to improvements in NWP and forecasts of climate studies. The European Space Agency implemented the Atmospheric Dynamics Mission ADM-Aeolus [2] and is planning to launch the satellite in 2018. With its unique measurement principle, being applied in the environment of space for the first time, Aeolus is considered to be a technology demonstrator [3]. Along its sun-synchronous orbit the mission will provide wind speeds at heights between ground and the lower stratosphere, i.e. up to about 25 km, with vertical resolutions of 250 m – 2 km depending on altitude and scientific objectives. The strict requirements include for example an altitude dependent precision of 1 m/s to 3 m/s by horizontally averaging the signal over 90 km.

Being the single payload on board the satellite, the direct-detection Doppler Wind Lidar (DWL) instrument ALADIN will measure the wind speed along the laser line-of sight (LOS). A viewing angle of 35° off-nadir in across-track direction enables Aeolus to provide one component of the projected horizontal wind velocity vector. The ALADIN laser transmitter operates in the ultraviolet spectral region at 355 nm at a pulse repetition frequency of 50 Hz. Two spectrometers determine the Doppler shift of the signal backscattered from the atmosphere with respect to the frequency of the emitted laser pulse. Whereas the broadband molecular backscatter signal is analyzed by two sequentially arranged Fabry-Pérot interferometers, a Fizeau interferometer is employed for the narrowband Mie signal emanating from aerosol particles and clouds.

For the purpose of assessing the expectable performance of the space-borne instrument, obtaining measurements from real atmospheric conditions as well as optimizing the retrieval algorithms, an airborne prototype was developed: the ALADIN Airborne Demonstrator [4-6]. As the first airborne direct-detection DWL it has been providing airborne wind measurements since 2005 within the framework of several airborne campaigns. Extensive datasets were gathered particularly from a two-week campaign in 2009 and a three-week campaign in May 2015, both conducted in the North Atlantic Region and over Greenland with Keflavik, Iceland serving as a base station [7]. The latest airborne observations were obtained within the context of NAWDEX (North Atlantic Waveguide and Downstream Experiment) in the same region in September-October 2016. The respective wind measurements are currently being evaluated. During all three campaigns the A2D was deployed together with a coherent 2-µm DWL onboard the DLR Falcon aircraft [8]. In 2015, joint scientific flights were performed with the NASA DC-8 aircraft (Figure 1) which carried the coherent DAWN (Doppler Aerosol Wind) and the direct-detection based TWiLiTE (Tropospheric Wind Lidar Technology Experiment) instrument [9,10]. Thus, for the first time four wind lidars were operated simultaneously during an airborne campaign. This campaign comprised a total number of 13 flights with the A2D out of which 7 were coordinated between Falcon and DC-8.

Figure 1 The participants of the airborne campaign in front of the research aircraft DLR Falcon 20 (left) and NASA DC-8 (right) in Keflavik, Iceland in May 2015.

2 OBJECTIVES AND DEDICATED FLIGHT TRACKS

The objectives of the airborne campaign in 2015 mainly relate to the pre-launch validation of the ADM-Aeolus mission. The primary goals were to:

- confirm and document the technical performance of the A2D and its suitability for the foreseen calibration and validation (Cal/Val) of ADM-Aeolus.
- extend existing datasets on response calibrations over favorable areas for Aeolus calibrations, e.g. regions with high surface albedo.
- extend existing datasets on Rayleigh and Mie wind observations, particularly in highly variable atmospheric conditions.
- rehearse airborne Cal/Val activities after launch of the satellite with focus on airborne flight planning and on coordination with other aircraft and ground validation sites.
- perform co-located measurements with other satellite instruments, e.g. TDS-1 or ASCAT.

From May 11th until May 28th, the DLR Falcon and the NASA DC-8 were operated from Keflavik achieving 35 and 50 flight hours, respectively. Figure 2 shows the flight tracks of the DC-8 and the Falcon (excluding four transfer flights) as well as the dates and corresponding objectives.

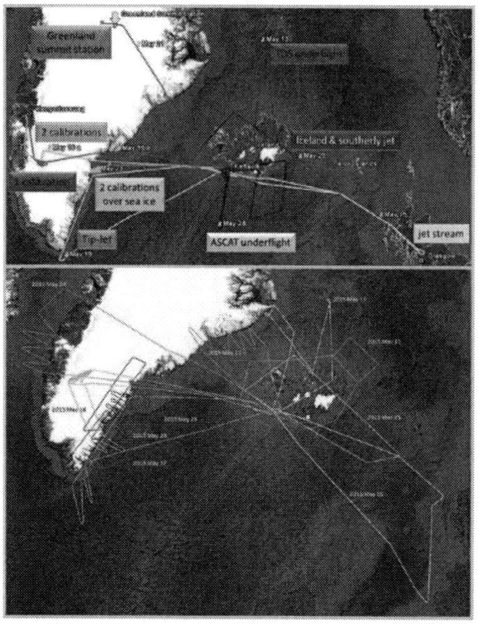

Figure 2 Flight tracks of the DLR Falcon (top) and the NASA DC-8 (bottom) from May 2015. For the Falcon the scientific objectives are given according to the color of the flight track.

Coordinated wind lidar measurements were performed from both aircraft during seven of these flights. At an altitude of about 10 km the Falcon exhibits a maximum endurance of slightly less than 5 hours and a maximum range of about 3500 km. In contrast, an endurance of about 8 hours enables the DC-8 to fly more extended tracks. In addition to the two DWLs the DC-8 deployed a large number of dropsondes [11]. A total number of 101 of these were released during the campaign measuring profiles not only of the horizontal wind vector but also of pressure, temperature and humidity. Therefore, the DC-8 was able to provide valuable information for the A2D calibration mode. One of the flights of the Falcon was dedicated to overpasses of a ground based wind lidar set up at the Greenland summit station [12].

3 AIRBORNE AND GROUND-BASED WIND LIDAR MEASUREMENTS

Due to its high precision of better than 1 m/s, the 2-µm DWL [8] serves as a reference instrument for the A2D. Table 1 lists the main specifications of both lidars.

Table 1 Operating parameters of the 2 DWLs on-board the DLR Falcon aircraft

parameter	A2D	2-µm
detection principle	direct	coherent
wavelength / nm	354.9	2022.5
pulse repetition rate / Hz	50	500
pulse energy / mJ	50 - 60	1 – 2
pulse length / ns	30	400
telescope diameter / m	0.2	0.11
resolution vertical / m	300 - 2400	100
resolution temporal / s	18	40
Line-of-sight	20° off-nadir (fixed)	vertical (fixed) or 20° off-nadir (scanning)

Theoretically, all noise sources for ALADIN are small enough to render shot noise the dominating contributor to the total noise. However, for the A2D we experience higher noise levels, which are primarily related to fluctuations of the co-alignment of transmission and reception paths as well as the internal reference measurements. Figure 3 shows an example of typical wind measurements of the A2D and the 2-µm DWL. This 30-minutes section measured around 2 p.m. on May 19th was obtained during a straight flight leg heading south along the south-eastern coast of Greenland. Only wind measurements outside the telescope overlap region of the A2D (range-gates 8-21 in Figure 3 bottom) are considered for the statistical comparison presented in Figure 4.

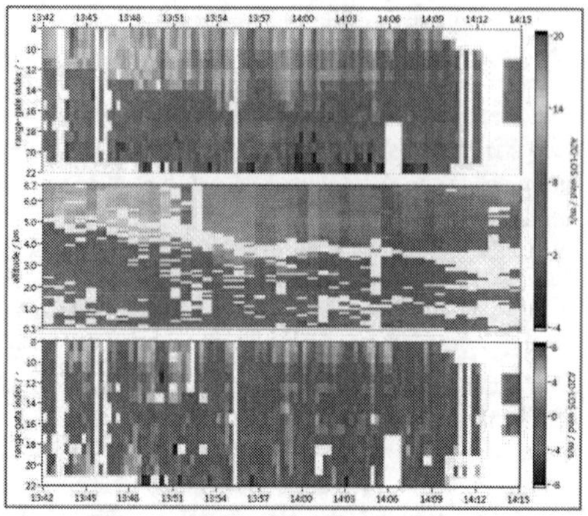

Figure 3 Wind measurements from the A2D Rayleigh channel (top) and the 2-µm DWL (middle) from May 19th in 2015. The wind speed differences after interpolation of 2-µm wind onto the A2D grid (min. coverage 50%) are shown at the bottom.

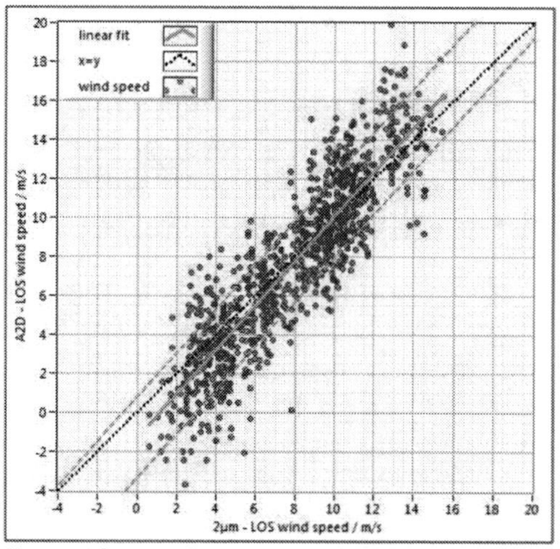

Figure 4 Statistical comparison of A2D Rayleigh winds and winds from the 2-µm DWL (benchmark) for the flight section from 2015/05/19 around 2 p.m. (UTC). Blue dashed lines mark a standard deviation of ±2m/s around the linear fit.

The coarse structure of the A2D and 2-µm wind fields show a remarkable agreement with highest LOS wind speeds on the top left. An edge between high and low winds speeds descending from 5 km to about 3 km seems to come along with a thin layer free of aerosol visible as a horizontal white band in the 2-µm wind field. Generally, white areas represent invalid or quality controlled measurements. As shown in Figure 4 a mean bias of -0.9 m/s, a slope error of 14% and a standard deviation of 2.0 m/s were found from a number of 1021 A2D and 2-µm LOS wind observations.

4 CONCLUSIONS

In 2015, a joint airborne campaign under participation of DLR and NASA was performed over the North Atlantic region, operating four DWLs during coordinated flights of two aircraft for the first time. The current analysis encompasses extensive data sets from the lidar instruments, dropsondes, a ground-based wind lidar on the Greenland Summit Station as well as numerical weather prediction models. Most of the objectives of the airborne campaign could be met and the basis for future coordinated airborne validation campaigns after the launch of ADM-Aeolus has been established.

ACKNOWLEDGEMENTS

The airborne campaign was funded by NASA, ESA and DLR. Funding for the development of the A2D was provided by ESA and DLR.

References

[1] World Meteorological Organisation, 2016: Statements of Guidance for Global NWP. available from *http://www.wmo.int/pages/prog/www/OSY/GOS-RRR.html*.

[2] European Space Agency, 2008: ADM-Aeolus Science Report, ESA SP-1311. .

[3] Reitebuch, O., 2012: The space-borne wind lidar mission ADM-Aeolus, *Atmospheric Physics – Background, Methods, Trends.* in U. Schumann (Ed.) Springer Series on Research Topics in Aerospace, pp. 815-827.

[4] Marksteiner, U., 2013: Airborne wind lidar observations for the validation of the ADM-Aeolus instrument, *PhD Thesis, Technische Universität München, DLR Forschungsbericht 2013-25*, ISSN 1434-8454, 176 p.

[5] Marksteiner, U., Reitebuch, O., Rahm, S., Nikolaus, I., Lemmerz, C., & Witschas, B., 2011: Airborne direct-detection and coherent wind lidar measurements along the east coast of Greenland in 2009 supporting ESA's Aeolus mission, *Proc. of SPIE Europe Int. Symposium Remote Sensing, Prague*, Czech republic, 8182, 81820J, 1-8.

[6] Reitebuch O., Lemmerz, C., Nagel E., Paffrath U., Durand Y., Endemann M., Fabré F., Chaloupy M., 2009: The Airborne Demonstrator for the Direct-Detection Doppler Wind Lidar ALADIN on ADM-Aeolus. Part I: Instrument Design and Comparison to Satellite Instrument. *J. Atmos. Ocean. Tech.* **26**, 2501-2515.

[7] Reitebuch, O., Lemmerz, C., Marksteiner, U., Rahm, S., Schäfler, A., Witschas, B., Emmitt, G.D., Greco, S., Kavaya, M.J., Gentry, B., Neely III, R.R., Schüttemeyer, D., 2016: Airborne wind lidar observations in the North Atlantic for preparation of the ADM-Aeolus validation. *Proc. 18th Coherent Laser Radar Conference CLRC, Boulder, USA, June 27 – July 1, 2016.*

[8] Weissmann M., Busen R., Dörnbrack A., Rahm S., Reitebuch O., 2005: Targeted Observations with an Airborne Wind Lidar, *J. Atmos. Oceanic Technol.* 22, 1706-1719.

[9] Kavaya, M.J., Beyon, J.Y., Koch, G.J., Petros, M., Petzar, P.J., Singh, U.N., Trieu, B.C. & Yu, J., 2014: The Doppler Aerosol Wind (DAWN) Airborne, Wind-Profiling Coherent-Detection Lidar System: Overview and Preliminary Flight Results." *J. Atmos. Oceanic Tech.*, 31, 826-842.

[10] Gentry B., Chen, H., Cervantes, J., Machan, R., Reed, D., Cargo, R., Marx, C. & Jordan, P., 2011: Airborne Testing of the TWiLiTE Direct Detection Doppler Lidar. *Proc. 16th Coherent Laser Radar Conference, Long Beach, California*, USA.

[11] Yankee Environmental System, 2016: http://www.yesinc.com/news/research.html

[12] Shupe M.D., Turner, D.D., Walden, V.P., Bennartz, R., Cadeddu, M.P., Castellani, B.B., Cox, C.J., Hudak, D.R., Kulie, M.S., Miller, N.B., Neely III, R.R., Neff, W.D. & Rowe, P.M., 2013: High and Dry: New Observations of Tropospheric and Cloud Properties above the Greenland Ice Sheet. *Bull. Am. Meteor. Soc.*, 2, 169-186.

UAV-BORNE COHERENT DOPPLER LIDAR FOR MARINE ATMOSPHERIC BOUNDARY LAYER OBSERVATIONS

Songhua Wu[1,2]*, Qichao Wang[1], Bingyi Liu[1], Jintao Liu[1], Kailin Zhang[1], Xiaoquan Song[1]

[1] *Ocean Remote Sensing Institute, College of Information Science and Engineering, Ocean University of China, Qingdao 266100, China , *Email: wush@ouc.edu.cn*

[2] *Laboratory for Regional Oceanography and Numerical Modeling, Qingdao National Laboratory for Marine Science and Technology, Qingdao 266235, China*

ABSTRACT

A compact UAV-borne Coherent Doppler Lidar (UCDL) has been developed at the Ocean University of China for the observation of wind profile and boundary layer structure in Marine Atmospheric Boundary Layer (MABL). The design, specifications and motion-correction methodology of the UCDL are presented. Preliminary results of the first flight campaign in Hailing Island in December 2016 is discussed.

1. INTRODUCTION

Wind profiles and atmospheric boundary layer structure are important for MABL dynamics study. The coherent lidar employing heterodyning technique has been demonstrated to be an effective tool for wind measurements for the marine atmospheric boundary layer where considerable aerosol particles present. The unmanned airborne coherent lidar can help improve the ability of meteorological detection and data assimilation in remote areas, especially for MABL, which compensates for the shortcomings of existing fixed weather stations, buoys and sounding balloons. In order to investigate the capability of automatic observations for coupled ocean-atmosphere dynamics, a compact UAV-borne Coherent Doppler Lidar (UCDL) prototype has been developed by Ocean University of China. Other instruments aboard include a synthetic aperture radar (SAR) for ocean wave measurement and in-situ sensors for temperature and humidity profiling. The UCDL is installed in a small unmanned helicopter which can ben taken off and landed on a research vessel as shown in Figure 1.

Figure 1. Shipborne UAV ocean observation system

2. SYSTEM DESIGN

The UCDL operates at the wavelength of 1.5 μm which is safe for human eyes. A modified Velocity Azimuth Display (VAD) scanning patterns were used, with the maximum measurement range up to 3 km. The Doppler spectral estimates are processed with FPGA and are sent back to the data server in the ship in real time via the UAV datalink up to 100 km. The simplified diagram of UCDL is shown in Figure 2.The system specifications are summarized in

https://doi.org/10.1051/epjconf/201817602012

Table 1. Because of the utilization of the robust semi-conductor single frequency laser and the Erbium doped fiber amplifier, the system works stably even under vibration and temperature variation during flight, more than 20 degree from the sea surface to the height of 3500 m.

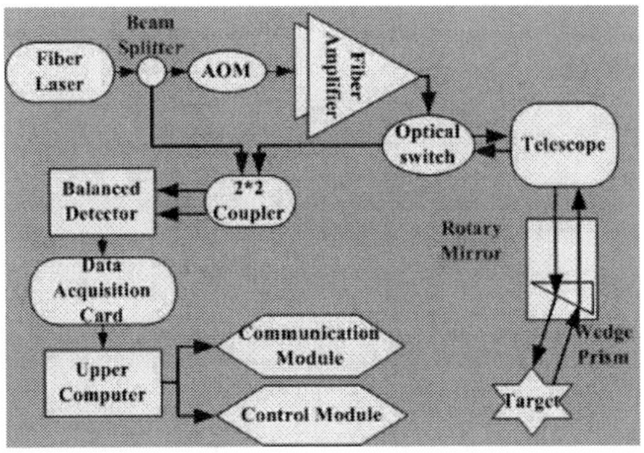

Figure 2. The schematic of UCDL system

Table 1. The specifications of the UCDL

UCDL parameters	
Wavelength	1.5 μm
Pulse Energy	100 μJ
Pulse Duration	200 ns
Pulse Repetition Frequency	10 kHz
Radial Spatial Resolution	30 m
Measurement Range	50 to 3500 m
Wind Profile Average Time	1 min
Velocity Measurement Range	± 60 m/s
Resolution of Velocity	0.3 m/s
Power Consumption	90 w
Weight	23.2 kg
Size	460×370×270 (mm)
Eye Safety Standards	EN60825-1:2007

Before the installation in UAV, all the modules and the lidar as a whole have been tested for the thermal adaptability, the anti-vibration performance, the data link communication, and the electro-magnetic-compatibility test, etc.

3. DATA PROCESSING

The radial velocity is obtained from the peak value of the FFT spectrum data at different range gates. Then the wind speed and direction are determined using VAD fitting method [1]. It is necessary to record and correct the effect of UAV attitude and velocity in real time and calculate the true radial speed. We obtain the UAV attitude and velocity information from the Global Navigation Satellite System (GNSS) and Inertial Measurement Unit (IMU) carried in the UAV, which are used for the retrieval and correction of the UCDL wind speed as listed in table 2.

Table 2. UAV attitude and velocity information

a. Roll angle α (IMU)
b. Pitch angle β (IMU)
c. Yaw angle γ (IMU)
d. Latitude (GNSS)
e. Longitude (GNSS)
f. Height (GNSS)
g. Forward velocity (IMU)
h. Lateral velocity (IMU)
i. Vertical velocity (IMU)

Attitude data (a, b and c in table 2) are used in the calculation of the pointing direction of the UCDL, and the true heading is corresponds to the UAV yaw angle [2]. Longitude and latitude (d and e) provide the geolocation. Amongst others, the height above the sea level (f) is used to determine the location of the UCDL range gates. The velocities (g, h, and i) are the velocity vectors of the UAV. Considering the UAV velocity and the UCDL pointing direction, the components of UAV velocity in each pointing directions could be calculated and corrected, then the real atmospheric wind profile and visualization products is derived based on the algorithm of correction and retrieval, including wind speed and direction information from 50 m above the sea surface to the altitude of the UAV. The height information of MABL could be derived based on the SNR data at the same time.

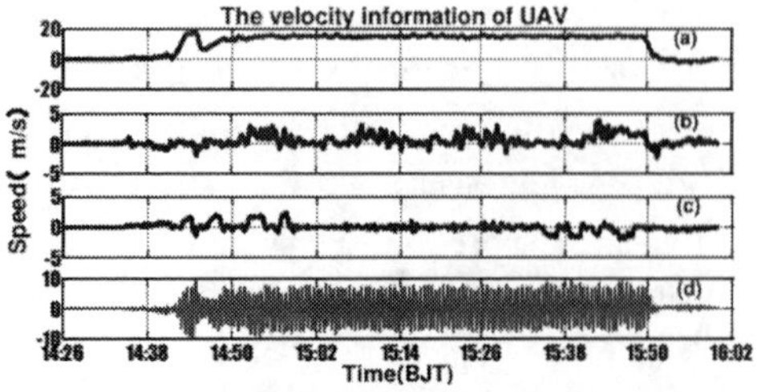

Figure 3. The velocity information of UAV in one flight

Figure 3 is a record of the velocity of UAV by the IMU. Figure 3 (a), (b), and (c) are the forward velocity, lateral velocity, and vertical velocity, respectively. Figure 3 (d) is the projection component of the UAV velocity in the lidar pointing direction, which is used for the velocity correction later on.

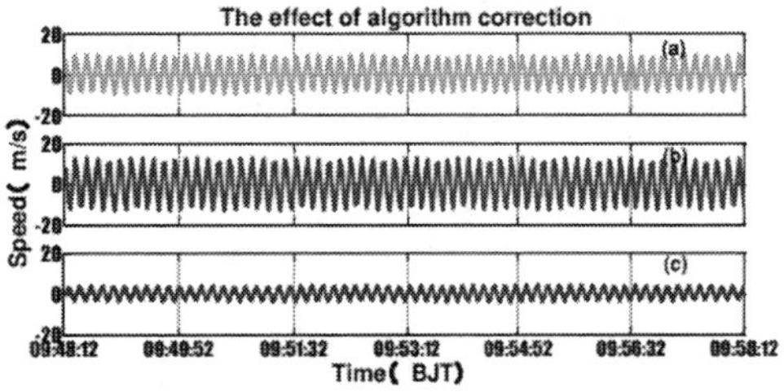

Figure 4. The effect of algorithm correction

Figure 4 shows the measurements and calibration of UCDL. Figure (a) is the raw radial velocity by UCDL, while Figure (b) shows the components of UAV velocity calculated from each pointing direction. The calibrated results of radial wind velocity is shown Figure 4 (c), which is used for wind speed and direction profile retrieval.

4. AIRBORNE EXPERIMENTS

Several field campaigns were conducted at Qingdao and Yangjiang, including the ground measurements, low-altitude flight experiments, and the data link test experiments. The first UAV flight campaign over ocean was carried out in Hailing Island, Yangjiang from November 5 to December 21, 2016. More than ten flights at different heights (800m, 1100m, 1200m, 1500m) from different periods (morning, noon and nightfall) were conducted. Concurrent measurements by in-situ sensors and SAR were performed. Satellite data were also collected as well for further validation analysis.

Figure 5. Airspace and flight routes of the first experiment in Hailing Island

Figure 5 shows three flight plans. The experiments consisted of two phases: UAV was flown from the land (Phase I) and from a barge (Phase II). Route A (Green Line): UAV was taken off / landed at point S1 (Ground) to observe the wind profile of over the Nanpeng Island. Route B (Red Line): UAV was taken off/landed at point S1 (Ground) to monitor both the wind file and MABL structures. Route C (Yellow Line): UAV was taken off at point S2 (Barge), observed the wind profile along the coastal area, and landed at point C6 (Barge). Route A (Green Line) and Route B (Red Line) experiments were performed during Phase I; Route C (Yellow Line) experiment was conducted during Phase II. The UAV was taken off/landed on a barge, while the barge was kept moving from point S2 to C6 during the experiment.

5. RESULTS

During the experiments, the wind profile and

https://doi.org/10.1051/epjconf/201817602012

MABL were observed, some preliminary results from the first flight campaign are presented in Figure 6.

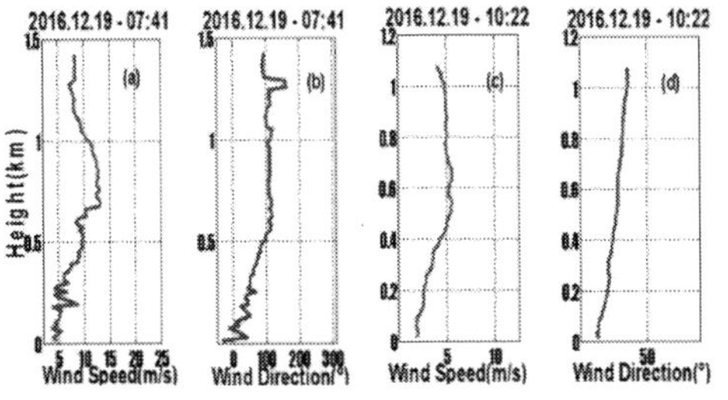

Figure 6 Wind profile measured by UCDL over Hailing Island on Dec 19th, 2016.

Figure 6 (a), (b) are the wind speed and direction profile from 50 m to 1425 m above the sea surface when the UAV was flown at the height of 1500 m at 07:41, December 19, 2016. Figure 6 (c), (d) are the wind speed and direction profile from 50 m to 1125 m above the sea surface when the UAV was flown at the height of 1200 m at 10:22 at the same day. As Figure 6 (b) shows there are wind shear in the direction at the height of 1300 m. During the campaign, a coherent wind lidar, WindPrint S4000 from Seaglet Environmental Technology, conducted ground wind observations for validation.

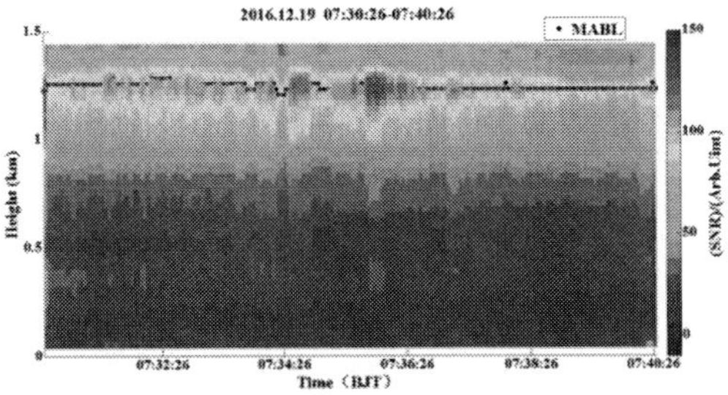

Figure 7 MABL measured by UCDL in Hailing Island on Dec 19th, 2016.

In addition to the wind profile, the MABL structure was derived from the Doppler spectrum of the UCDL. The backscatter signal is retrieved according to the peak height of FFT spectrum and the standard deviation of SNR profile [3]. The MABL height was determined by UCDL from 07:00 to 08:00, December 19, 2016 (as shown in Figure 7). The averaged MABL height is about 1300 m, corresponding to the wind direction shear events shown in Figure 6 (b).

6. CONCLUSION

OUC team developed the UCDL and conducted the first flight campaigns at the coastal area. The wind observation and the velocity correction has been conducted and analyzed. The ground based wind measurements and concurrent atmospheric probing data were obtained for further analysis.

ACKNOWLEGDEMENT

This work is supported by the National High Technology Research and Development Program of China (2014AA09A511) and the National Natural Science Foundation of China (NSFC) under grant 41375016.

References

[1] Browning, K. A., Wexler, R., 1986: The determination of kinetic properties of a wind field using Doppler radar, *Journal of Applied Meteorology* **15**, 1302-1306.

[2] Li Z. G., Liu Z. S., Zhu J. S., Liu B. Y., Reitebuch O., 2014: Wind Retrieval Algorithms For The Wind Products Of The Airborne Coherent Doppler Lidar, Proceeding 'Dragon 3 Mid-Term Results Symposium', Chengdu, China (ESA SP-724, November 2014).

[3] Wang, D. X., Song X. Q., Feng C. Z., Wang X. T., Wu S. H., 2015: Coherent Doppler Lidar Observations of Marine Atmospheric Boundary Layer Height in the Bohai and Yellow Sea, *Acta Optica Sinica* **35**(s1): s101001.

[4] Chouza, F., Reitebuch, O., Benedetti, A., Weinzierl, B., 2016: Saharan dust long-range transport across the Atlantic studied by an

airborneDoppler wind lidar and the MACC model, *Atmospheric Chemistry and Physics* **16**(18), 11581-11600.

A COMPACT LED LIDAR SYSTEM FITTED FOR A MARS ROVER – DESIGN AND GROUND EXPERIMENT

Prane Mariel B. Ong[1,*], Tatsuo Shiina[1], Naohiro Manago[1], Hiroaki Kuze[1], Hiroki Senshu[2], Naohito Otobe[3], George Hashimoto[4], and Yasuhiro Kawabata[5]

[1]*Chiba University, 1-33 Yayoi-cho, Inage-ku, Chiba, 263-8522, Japan, *Email: prane.ong@gmail.com*
[2]*Chiba Institute of Technology, 2-17-1 Tsudanuma, Narashino-shi, 275-0016 Japan*
[3]*Fukuoka University, 8-19-1 Nakakuma, Jonan-ku, Fukuoka, 814-0180 Japan*
[4]*Okayama University, 1-1-1 Tsushima-naka, Kita-ku, Okayama, 700-8530 Japan*
[5]*Meteorological Research Institute, 1-1 Nagamine, Tsukuba, Ibaraki, 305-0052 Japan*

ABSTRACT

A compact LED lidar was constructed and field-tested with the aim to observe the Mars' dust devils. To be able to fit it on the Mars rover, a specialized Cassegrain telescope was designed to be within a 10 cm-cube, with a field of view of 3mrad. The transmitter has 385 nm LED light source with 3 cmϕ opening, 70mrad divergence, 0.75W (7.5nJ/10ns) pulse power, and 500 kHz repetition frequency. The configuration of the optical system is biaxial to easily configure the overlap between their optical axes.

1 INTRODUCTION

Mars explorations and missions have continuously been conducted for several decades now to monitor Mars temperature, biological life, atmosphere, and dust [1]. Although a considerable progress has been made on these, knowledge on the near-ground atmosphere is still lacking due to restricted ground observations [1-4].

One of particular interest is the "dust devil" or strong whirlwind, because of their influence in Martian climate, and how it can affect the rover exploration [5]. Although the motion and size of the observed Martian dust devils can be estimated from images obtained from satellites and rovers, still the quantitative information from these estimates were poorly constrained [6]. Hence, in situ measurements of near-ground atmospheric and dust activities on Mars will be useful in the study of this phenomenon.

Lidar (Light Detection and Ranging) may be able to measure the structure of whirlwinds along the line of sight. Lidar typically uses laser light to quantify the target's distance to the light source. However, a classical laser lidar system is generally too large in volume and in mass to be onboard a Mars rover, and it needs a cooler that requires large amount of power to be operational.

Given these restrictions and the extreme surface temperature condition on Mars, our team have examined the use of LED (light emitting diode) as the light source of the mini lidar system to be installed in a Mars rover. Aside from being energy efficient, heat sink independent, and tough against circumference change, previous researches on LED lidar have proved that the LED lidar can be used in near-ground Earth atmosphere monitoring [7-12]. The beam radiation angle is wide compared to its laser counterpart, but this wide beam divergence of the transmitter will establish a constant overlap in the near-field with the receiver's narrow field of view (F.O.V.). Hence, the system design has been focused on the robustness against tough treatment, transportability (10 cm-cube limit), and capability for near-range monitoring (~<30 m) of dust activities.

2 SYSTEM DESIGN

The goal of the design was to fit the LED lidar within a 10 cm-cube limit. In order to accomplish this, a Cassegrain telescope was chosen as the receiver with specially designed primary mirror to shorten its barrel.

To date, there are two compact LED lidars that were fabricated: version 1, and version 2. Version 2 was the improved version of version 1 specifically in terms of compactness, design, and stability. Both of their receivers have the same optical and electrical parameters. The aperture and focal length were 10 cmϕ and 72.25 mm, respectively. The 2.5 cmϕ diameter secondary mirror has a -25 mm focal length. In both receivers, 45° mirror, iris, interference filter, and

https://doi.org/10.1051/epjconf/201817602013

photomultiplier tube (PMT) detector were incorporated in the rear space of the telescope.

On the other hand, the transmitter for both versions was almost identical except for the optics used. Essentially, the transmitter has a 385 nm wavelength NUV-LED, which is commercially manufactured by Nichia Corporation. It has a 0.75W pulsed power with 10 ns pulse width at 500 kHz pulse-repetition frequency given a beam divergence of about 70 mrad. The transmitter opening is about 3 cmϕ. A high pulse repetition frequency was set to improve the signal-to-noise ratio (SNR) due to the low pulse energy (7.5nJ/10ns). The pulse driver was specially designed to be separated from the transmitter's LED device (with optics unit), to protect it from the varying surface temperature of Mars. The transmitter and receiver parameters are summarized in Table 1.

To be able to follow the high repetition frequency (500 kHz) of the transmitter, the photon counter was fabricated using a high-speed FPGA (Spartan 6, Xilinx) with a time resolution as low as 1 ns or 0.15 m distance resolution equivalent. The power consumption of the current photon counter is 2 W for 1 channel DAQ, but it was verified that it can still be reduced to 1 W.

To be able to perform the ground experiment and the initial performance evaluation, the compact LED lidar transmitter and receiver were housed in a polyvinyl chloride as the frame material (see Figure 1). The transmitter-receiver configuration for both LED lidars was arranged in a biaxial mode with an angle adjuster, so the overlap distance of the transmitter beam with the receiver's optical axis can easily be controlled. Moreover, during experiment, the LED lidar system was mounted in user-controlled tracker to easily adjust the altitude and azimuth directions.

3 SYSTEM PERFORMANCE

Currently, two experiments: wind tunnel experiment, and field experiment, have been done in order to assess the performance of the developed LED lidars. The wind tunnel experiment was conducted using version 1 of the LED lidar, and the field experiment was conducted using version 2 of the LED lidar.

Table 1 Parameters of the transmitter and the receiver.

Transmitter	
Light Source	NUV-LED NCSU034B (Nichia Corp.)
Wavelength	385 nm
Pulse Width	10 ns
Pulsed Power	0.75 W
Pulse-Repetition Frequency	500 kHz
Beam Divergence	70 mrad
Beam Size	3 cmϕ
Cassegrain Receiver	
Barrel	5 cm
F.O.V.	3 mrad
Primary Mirror Aperture	10 cmϕ
Primary Mirror Focal Length	72.25 mm
Secondary Mirror Diameter	2.5 cmϕ
Secondary Mirror Focal Length	-25 mm

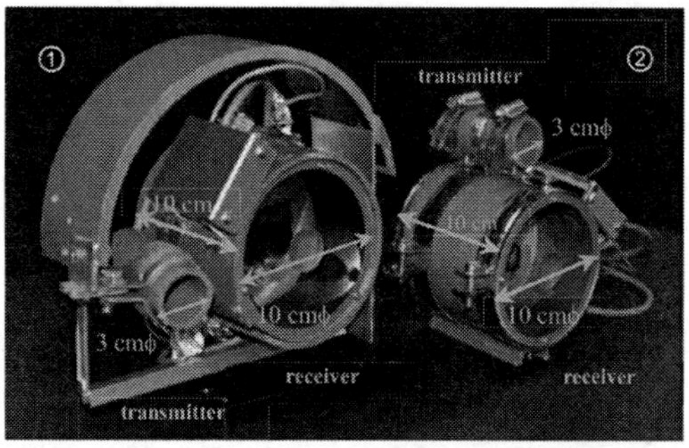

Figure 1 Two versions of the compact LED lidar for near-range ground experiment: ① version 1, and ② version 2.

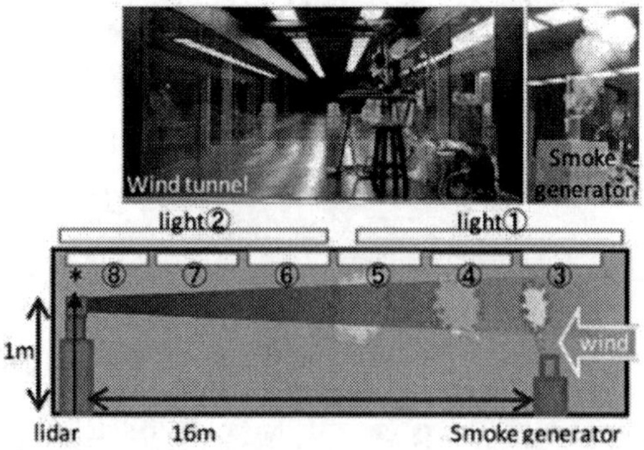

Figure 2 Setup for the wind tunnel experiment in the Japan Meteorological Agency wind tunnel facility.

3.1 Wind Tunnel Experiment

This experiment was conducted in the wind tunnel facility of the Japan Meteorological Agency in Tsukuba, Japan. The LED lidar monitored the smoke's flow, diffusion and convection from a Glycerin smoke generator. The LED lidar and the Glycerin smoke generator arrangement is illustrated in Figure 2. The distance between them was set at 16 m, where the wind direction is downwind with respect to the smoke generator, and the wind speed range is 0-5 m/s. Dust and particle counters were also installed near the smoke generator and were synchronized with the LED lidar measurements. The LED lidar was elevated 1 m from the ground to be vertically centered, since the ceiling height is 2 m.

The backscattered smoke signal coming from the upward-pointing smoke generator with 20 s generation time interval is presented in Figure 3. A distance vs time graph with echo count as the value of the colormap is presented to analyze the smoke's flow over time. Given the smoke interval of 20 s and the wind speed of 1 m/s, it can be seen that at 14-15 m the LED lidar detected a strong backscattered echo from the smoke and gradually decreased overtime, then registered a slight increase after 10 s. This illustrates the diffusion and convection of the smoke. Since we know that the smoke generator was pointing upward, this suggests that it might be the cause why the smoke initially rise for a few seconds before it transitioned to a lower ground. As the smoke diffused overtime in space and rose again in the LED lidar's line of sight due to convection, the measured backscatter smoke signal after 10 s was low compared to the initial reading at 14 m.

We have also measured the backscattered signal from smoke with varying wind speeds: 0.3 m/s, 0.5 m/s, and smoke generator direction (directed towards the LED lidar). And observed that a high wind speed cuts down the smoke convection to a shorter period.

3.2 Field Experiment

In one of the 2 field experiments, the LED lidar was tested first by pointing to a hard target like a building. It was pointed to a building 50 meters away from the source. Figure 3 shows the hard target return from that building. The LED lidar was able to measure the atmospheric echo a few meters away from the system's receiver, because of the overlap between the beam's divergence and receiver's F.O.V., and the hard target return

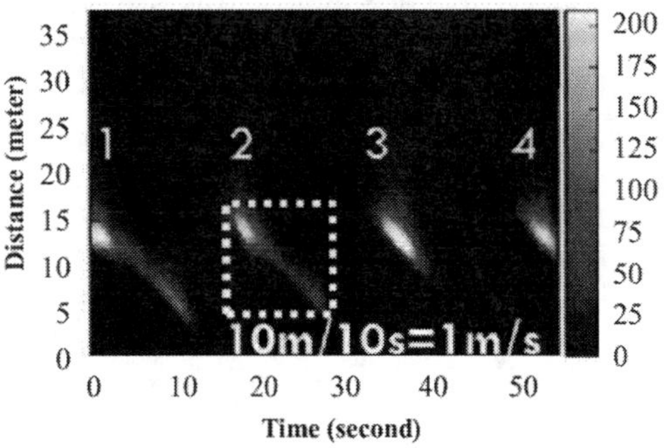

Figure 3 Smoke profile given the wind speed of 1 m/s and smoke interval of 20 s using the LED lidar ver. 1 (Japan Meteorological Agency, Tsukuba, Japan).

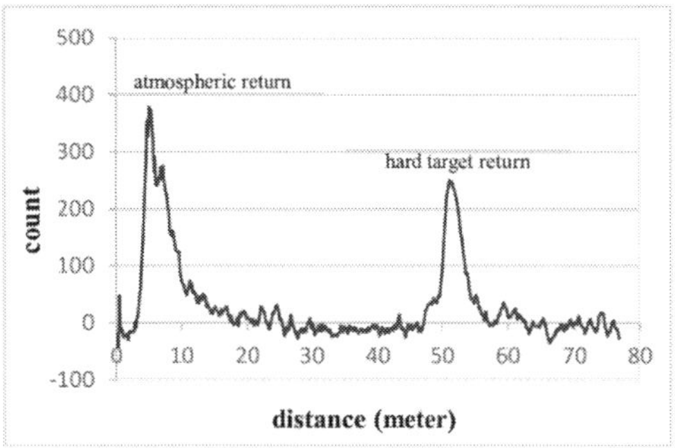

Figure 4 Hard target echo by the LED lidar ver. 2.

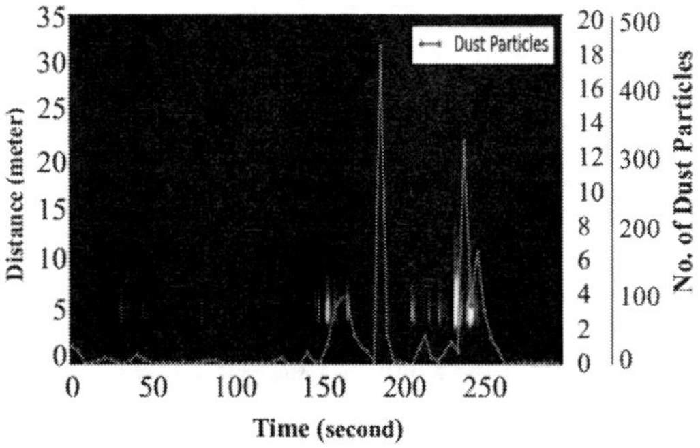

Figure 5 Dust profile with both LED lidar data (ver. 2) and dust counter data measured in Inage-Kaigan field experiment (Chiba, Japan).

(~50m) for an accumulation time of 2 s with distance resolution of 0.15 m or time resolution of 1 ns. The detectable range for the atmospheric return was about 30 m.

The next field experiment was conducted in Inage-Kaigan coast area in Chiba, Japan (35°36'55.6"N, 140°03'48.1"E) on December 6, 2016. The LED lidar was mounted in a tracker to easily configure the elevation and azimuth directions. The system was approximately 30 cm above the ground, with 210° azimuth and 2° elevation angles. The wind direction is downwind with respect to the LED lidar system. At 30 m directly away from the system, we have also installed the anemometer, dust, and particle counters.

The LED lidar was set to a 4 Hz of sampling rate and a distance resolution of 1 ns or 0.15 m distance resolution, giving a total detectable range of 76.8 meters.

Figure 5 presents the distance vs time profile of the dust echo overlayed with dust particles measurement. During this time, the average wind speed is 6 m/s. As can be seen from the figure, there is a 5 s delay in the registered dust counter measurement compared to the LED lidar measurement. This is expected given that the wind direction originates at 340° (NNW), which came from the rear of the LED lidar and towards the dust counter, which is 30 m away. At 245 s, a 10 m dust echo was registered, which was the longest observed dust return, given the observation range of 35 m.

4 SUMMARY

Two compact LED lidars were fabricated to fit the required 10 cm-cube dimension. Version 2 has almost the same parameters (electronics and optics) as version 1, but with more improved design and compactness for stability. They were both tested, although in two different environments. Both have shown promising results in near-range dust echo measurement. Although, the measurement is still restricted to nighttime. The lidar development is still in progress to be able to conduct field experiment during daytime.

ACKNOWLEDGEMENT

This project was supported by the JSPS KAKENHI grant (JP16H03129).

Reference

[1] Garber, S., 2015, April 16: A chronology of Mars exploration, Retrieved from http://history.nasa.gov/marschro.htm.

[2] Guzewich, S.D., Abshire , J. B., Smith, M.D., Riris, H., Sun, X., Gentry, B.M., Yu, A., and Allan,G. R., 2016: MARLI: MARs LIdar for global wind profiles from orbit, *Proceedings of the 47th Lunar and Planetary Science Conference*, 1497pp.

[3] Whiteway, J., Daly, M., Carswell, A., Duck, T., Dickinson, C., Komguem, L., and Cook, C., 2008: Lidar on the Phoenix mission to Mars, J. Geophys. Res. 113, p. E00A08, doi:10.1029/2007/JE003002.

[4] Komguem, L., Whiteway, J. A., Dickinson, C., Daly, M., and Lemmon, M. T., 2013: Phoenix LIDAR measurements of Mars atmospheric dust, Icarus 223, 649–653pp.

[5] Jenner, L., 2007, November 30: Phantoms from the sand: tracking dust devils across Earth and Mars, Retrieved from https://www.nasa.gov/vision/universe/solarsystem/2005_dust_devil.html.

[6] Ferri, F., Smith, P. H., Lemmon, M., and Rennó, N. O., 2003: Dust devils as observed by Mars Pathfinder, *J. Geophys. Res.* **108**, 5133pp. doi:10.1029/2000JE001421.

[7] Koyama, M. and Shiina, T., 2010: Light Source Module for LED Lidar, *Proceedings of 25th International Laser Radar Conference*, S01P-32-1–4.

[8] Grishin M. Eds., 2010: Advances in Solid-State Lasers: Development and Applications, Chapter 8, *INTECH*.

[9] Fukuchi, T., and Shiina, T., Eds., 2012: Industrial Applications of Laser Remote Sensing, *Bentham Science Publishers Ltd.*

[10] Korotcenkov, G., Ed., 2013: Chemical Sensors Volume 4: Chapter 12, *Optical Sensors*.

[11] Shiina, T., Noguchi, K., and Tsuji, K., 2016: Compact and mini Raman lidars for hydrogen gas detection, *Proceeding of ICONE-23, ICONE23-1640*.

[12] Shiina, T., Yamada, S., Senshu, H., Otobe, N., Hashimoto, G. L., and Kawabata, Y., 2016: LED minilidar for Mars rover, *Proc. SPIE 10006*, 10006–15 pp.

APPLICATION OF MULTIPLE-SCATTERING POLARIZATION LIDAR FOR THE EVALUATION OF SPACE-BORNE LIDAR ALGORITHMS

Hajime Okamoto[1]*, Kaori Sato[1], Shuichiro Katagiri[1], Masahiro Fujikawa[1], Tomoaki Nishizawa[2], Nobuo Sugimoto[2], Yoshitaka Jin[2], Atsushi Shimizu[2], Hiroshi Ishimoto[3]

*[1]Research Institute for Applied Mechanics: Kyushu University, Kasuga, Kasuga park 6-1, 816-8580, JAPAN, *Email: okamoto@riam.kyushu-u.ac.jp*

[2]National Institute for Environmental Studies, Tsukuba, Ibaraki 305-8506, JAPAN

[3]Meteorological Research Institute: Japan Meteorological Agency, Tsukuba Ibaraki 305-0052, JAPAN

ABSTRACT

Multi-field of view multiple-scattering polarization lidar (MFMSPL) has been developed to measure enhanced backscattering and depolarization ratio from optically thick clouds. Collocated cloud radar observations and Monte Carlo simulations supported the system performance such as cloud detectability and values of depolarization ratio. The system is used to evaluate/improve algorithms for space-borne lidar such as CALIOP on CALIPSO and ATLID on EarthCARE.

1 INTRODUCTION

Multiple scattering effects in lidar signals have been known to be important to analyze cloud phase and microphysical properties [1]. When optical thickness becomes large, single scattering approximation is not valid and non-zero depolarization ratio appears even for spherical particles. This can be explained as follows; number of backscattered photons affected by multiple scattering events increases as optical thickness increase when other factors are the same and multiple scattering process produces non-zero depolarization ratio. These phenomena have been frequently observed for space-borne lidar observations. Observed depolarization ratio often well exceeded 30% when water clouds were observed as for the case of CALIPSO lidar [2].

Since ice particles often produces large depolarization ratio comparable to that of water clouds in case of space borne lidar observations, there is a fundamental difficulty to use depolarization ratio alone to discriminate cloud phase. To overcome the above situation, it was shown that cloud particle phase could be determined by the combined use of depolarization ratio and the ratio of the attenuated backscattering coefficients for two vertically consecutive layers [3].

We introduced a new ground-based lidar system, Multiple-Field of view Multiple Scattering Polarization Lidar (MFMSPL) that can detect backscattering coefficients and depolarization ratios similar to those observed by CALIPSO lidar [4]. The collocated measurements by 95GHz cloud radar and the MFMSPL showed good agreement in the detection of cloud top boundaries and Monte Carlo simulations verified the performance of the system and accuracy of the observables.

The main objective of the study is to develop and evaluate space-borne lidar algorithms such as cloud mask scheme and cloud particle type algorithms for the analysis of CALIPSO lidar and high spectral resolution lidar, ATLID, on the EarthCARE satellite mission (JAXA-ESA)[5]. The data can be also used to study radiative impacts of clouds and aerosol-cloud interactions.

2 DESCRIPTION OF THE MFMSPL

The MFMSPL has been originally built with eight channels (four parallel and four perpendicular with detectors) and was recently extended to 10-channel system. Five parallel- and perpendicular channel telescopes are inclined with different angles from the vertical direction along a line. CH1, CH3, CH5, CH7 and CH9 detect parallel signals and the tilt angles are 0, 10, 20, 30, and 40mrad, respectively. CH2, CH4, CH6, CH8 and CH10 channels detect perpendicular signals and have tilt angles of 0, 10, 20, 30, and 40mrad, respectively. FOV of each telescope is 10 mrad. N;YAG laser is used and second harmonic wavelength of 532nm is detected. The temporal and vertical resolutions of the data are 10 s and 6m, respectively. The original data are averaged for 5 min and 48m in this study.

3 CLOUD MASK AND CLOUD PARTICLE TYPE FOR MFMSPL

3.1 Cloud detection

We first evaluated the cloud mask algorithm for CALIPSO lidar, KU-mask [6], by using the MFMSPL data. The KU-mask was originally developed based on the cloud mask algorithm for ship-based Mie-type lidar with small FOV[7][8]. It turned out that the KU mask scheme tended to underestimate cloud top regions when water clouds existed.

We developed new cloud mask algorithm that can be applicable for all of the channels of the MFMSPL. The new cloud mask scheme was designed to properly treat the signals from the deeper part of the clouds for off-beam channels where signals became small above cloud bottom regions. The cloud mask result is shown for the total attenuated backscattering coefficient ($\beta_{att,tot}$) estimated by β_{att} for CH5 and β_{att} for CH6(Figure 1).

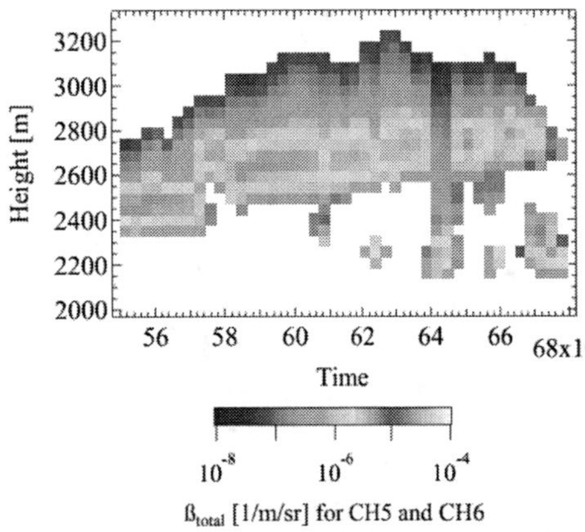

Figure 1 Time-height plot for the total attenuated backscattering coefficient $\beta_{att,tot}$ for CH5 and CH6 of the MFMSPL.

The results for the off-beam channels such as CH3+CH4, CH5+CH6 and CH7+CH8, showed higher cloud top compared with the on-beam channel (CH1+CH2). The algorithm has also a function to identify the fully attenuated pixels.

Then we improved the cloud mask algorithm for space borne lidar on the basis of the algorithms for the MFMSPL. The cloud detection in the deeper bottom part of the clouds was improved by the new cloud mask.

3.2 Analysis of water cloud signatures

Algorithm for cloud particle type (KU-type) that can be applied to CALIPSO lidar, required depolarization ratio and ratio of attenuated backscattering coefficients for the two vertically consecutive layers [3]. The logarithmic expression of the ratio X' was used in the algorithm.

$$X'(R_i) = \log_{10}\left[\frac{\beta_{att}(R_i)}{\beta_{att}(R_{i+1})}\right] \quad (1)$$

, where R_i denotes the distance of the center of the cloud layer i viewed from the satellite. and R_{i+1} denotes the next layer i+1. β_{att} denotes the attenuated backscattering coefficient at layer i. X' was introduced as a proxy of the layer extinction. The two-dimensional diagram on X' and depolarization ratio was created based on the occurrence frequencies of water clouds, randomly oriented ice and horizontally oriented ice particles from CALIPSO lidar signals.

Similarly, X' was estimated using the observed values by the MFMSPL observations with modifications, where R_i is taken to be the cloud height and R_{i+1} corresponded to the next upper layer. Here we examined the depolarization ratio and X' for water clouds observed by the MFMSPL. Time height plot for depolarization ratio and X' was shown in Figures 2a and 2b, respectively.

Analyses of vertical distribution of the depolarization ratio of water clouds showed the values in general monotonically increased from cloud bottom to cloud top. Maximum of depolarization ratio often exceeded 80% near the cloud top regions. On the other hand, X' first increased up to the middle part of the cloud as height increased and turned to slightly decrease or became nearly constant toward cloud top. Large negative values (-1) of X' were found near cloud bottom regions where $\beta_{att,tot}$ increased to some of cloud middle parts.

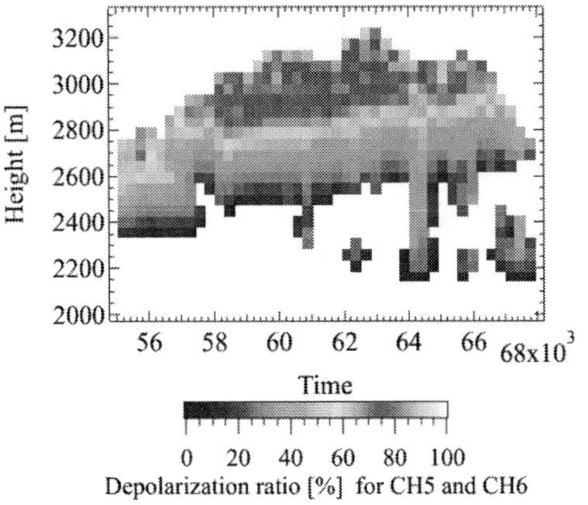

(a)

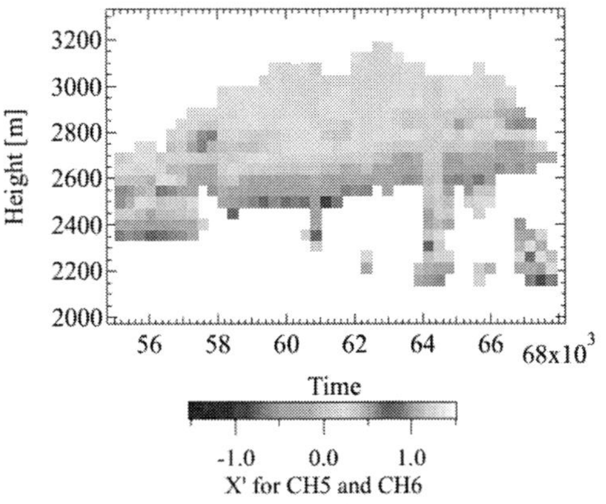

(b)

Figure 2(a) Time-height plot for the depolarization ratio estimated by β_{att} for CH5 and that for CH6 for the same clouds in Figure 1. (b) Same as (a) but for X'.

Next, relationship between depolarization ratio and X' was examined (Figure 3). There are several features found in the relationship. For depolarization ratio smaller than about 40%, as X' increased, depolarization increased for the water clouds. Compared with the diagram used in the KU-type for CALIPSO lidar, the value of depolarization ratio in this analysis were somewhat larger for the same X'. This might be explained as follows; the CALIPSO backscattered signals were results of the integration of signals from both of inner and outer parts. Depolarization ratio from outer part is expected to be larger than that from inner part since signals from outer part are more affected by the multiple scattering contributions compared with the inner parts. Actual observations revealed, i.e., observed depolarization ratio estimated by CH5 and CH6 is larger than that by CH1 and CH2 as expected.

For the depolarization ratio >40%, X' decreased as depolarization ratio increased. These data correspond to the upper cloud layers above about 2800m. These large depolarization ratio with small X' also was found when ice particles appeared so that there is a difficulty to discriminate water from ice particles in these parameter space. Thus it might be necessary to introduce additional discrimination scheme as follows. The cloud phase for the upper regions is considered to be water when the cloud phase of the lower cloud layer connected to the upper layer of interest is identified as water.

Similar modification in the cloud particle type algorithm might be needed for the analysis of cloud phase retrieved from space borne lidar observations.

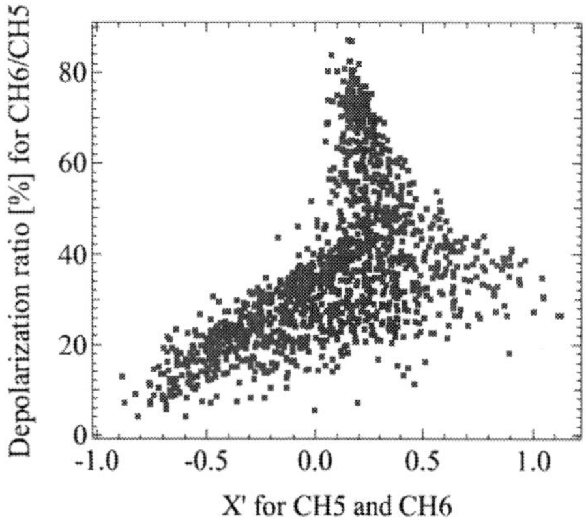

Figure 3 Relationship between the depolarization ratio and X' for CH5 and CH6 for the same clouds in Figure 1.

4 CONCLUSIONS

The MFMSPL data were first used to examine and develop the cloud mask algorithms. The method was also extended to the algorithm for space-borne lidar. The MFMSPL data is especially suitable for the development of the algorithms that were applied to space-borne lidar where multiple scattering often plays a fundamental role.

It was found that our former cloud mask scheme (KU-mask) tended to underestimate bottom part of clouds and new one can overcome this issue.

The relation between depolarization ratio and X' was analyzed by using the MFMSPL data for optically thick water clouds. Similar but somewhat different feature was found for off-beam channels compared with the diagram used in KU-type for the discrimination of cloud particle type. It was suggested that additional treatment might be needed when the discrimination of cloud particle type is conducted in the deeper part of the clouds where multiple scattering contributions are dominant.

ACKNOWLEDGEMENTS

CALIPSO data were obtained from the Atmospheric Sciences Data Center at NASA's Langley Research Center. This study was supported by JSPS KAKENHI Grant Numbers JP17H06139 and JP15K17762, by Green Network of Excellence (GRENE) Program Arctic Climate Change Project and the Arctic Challenge for Sustainability (ArCS), by the Japan Aerospace Exploration Agency for EarthCARE Research Announcement and by the Collaborative Research Program of Research Institute for Applied Mechanics, Kyushu University.

References

[1] Bissonnette, L. R., G. Roy, L. Poutier, S. G. Cober, and G. A. Isaac 2002: Multiple-scattering lidar retrieval method: tests on Monte Carlo simulations and comparisons with in situ measurements, Appl. Opt. 41(30), 6307–6324

[2] Hu, Y., 2007: Depolarization ratio-effective lidar ratio relation: Theoretical basis for space lidar cloud phase discrimination, Geophys. Res. Lett. 34(11), L11812

[3] Yoshida, R., H. Okamoto, Y. Hagihara, and H. Ishimoto 2010: Global analysis of cloud phase and ice crystal orientation from Cloud-Aerosol Lidar and Infrared Pathfinder Satellite Observation (CALIPSO) data using attenuated backscattering and depolarization ratio," J. Geophys. Res. 115, D00H32

[4] Okamoto, H., K. Sato, T. Nishizawa, N. Sugimoto, T. Makino, Y. Jin, and A. Shimizu, T. Takano, M. Fujikawa, 2016: Development of a multiple-field-of-view multiple-scattering polarization lidar: comparison with cloud radar, *Opt. Express*, 24(26), 30053-30067, https://doi.org/10.1364/OE.24.030053

[5] Illingworth A., 2015: THE EARTHCARE SATELLITE: THE NEXT STEP FORWARD IN GLOBAL MEASUREMENTS OF CLOUDS, AEROSOLS, PRECIPITATION AND RADIATION, *Bull. Amer. Meteor. Soc.*, 96, 1311-1332, doi: http://dx.doi.org/10.1175/BAMS-D-12-00227.1

[6] Hagihara, Y., H. Okamoto, and R. Yoshida 2010: Development of combined CloudSat/CALIPSO cloud mask to show global cloud distribution, 2010, *J. Geophys. Res., 115*, D00H33, doi:10.1029/ 2009JD012344

[7] Okamoto, H., Nishizawa T., Takemura, T., Kumagai, H., Kuroiwa, H., Sugimoto, N., Matsui, I., Shimizu, A., Emori, S., Kamei, A., Nakajima, T., 2007: Vertical cloud structure observed from shipborne radar and lidar: Midlatitude case study during the MR01/K02 cruise of the research vessel Mirai, *J. Geophys. Res. Atmos.* **112**, D08216, doi:10.1029/2006JD007628.

[8] Okamoto, H., Nishizawa, T., Takemura, T., Sato, K., Kumagai, H., Ohno, Y., Sugimoto, N., Shimizu, A., Matsui, I., Nakajima, T., 2008, Vertical cloud properties in the tropical western Pacific Ocean: Validation of the CCSR/NIES/FRCGC GCM by shipborne radar and lidar, *J. Geophys. Res. Atmos.* **113**, D24213, doi:10.1029/2008JD009812

EPJ Web of Conferences **176**, 02015 (2018)
ILRC 28
https://doi.org/10.1051/epjconf/201817602015

AN ASSESSMENT OF THE EXPECTED QUALITY OF AEOLUS LEVEL-2B WIND PRODUCTS

Michael P. Rennie*

[1]*European Centre for Medium-Range Weather Forecasts (ECMWF), Reading, UK, *m.rennie@ecmwf.int*

ABSTRACT

The ADM-Aeolus L2B horizontal line-of-sight wind products are discussed, including a brief introduction to the L2B processing methods. Realistic simulation of Aeolus followed by the Ground Segment processing chain is used to assess the expected error statistics of the L2B winds. Examples of the L2B winds simulated from realistic atmospheric cases are discussed. The aim of the paper is to give potential users an impression of the expected characteristics of Aeolus mission's main product.

1 INTRODUCTION

ADM-Aeolus is the fifth satellite to be launched in the Living Planet Programme of the European Space Agency (ESA) [1]. The mission's objective is to provide profiles of high-quality wind observations from the surface to around 30 km altitude, using the Doppler wind lidar instrument ALADIN (Atmospheric LAser Doppler Instrument) in a near-polar sun-synchronous, dawn-dusk orbit. The wind information is the horizontal line-of-sight (HLOS) component, in the off-nadir direction at 35° and perpendicular to the satellite's velocity. The mission is intended to have a minimum lifetime of three years and is expected to be launched in late 2017.

The European Centre for Medium-Range Weather Forecasts (ECMWF) is closely involved in the Aeolus mission because the assimilation of the wind products is expected to significantly improve our weather forecasts, if the mission requirements [3] can be met. Wind profiles are still lacking for global Numerical Weather Prediction (NWP), particularly in the upper troposphere, lower stratosphere and away from populated continents.

ECMWF and KNMI (Royal Netherlands Meteorological Institute) are contracted by ESA to develop the Aeolus Level 2B/C processing software i.e. the retrieval of HLOS wind components from the Level 1B product suitable for use in NWP data assimilation. In the past Météo-France, German Aerospace Center (DLR) and LMD/IPSL (Laboratoire de Météorologie Dynamique) have contributed to the L2B processor [2].

Some features of the L2B products are:

- (H)LOS wind components geolocated with geometric height, latitude, longitude, azimuth angle and time
- Flexible "grouping" algorithm in terms of horizontal averaging length-scale, see Figure 1
- Each observation type (either Mie or Rayleigh) is retrieved from accumulating measurement-level spectrometer counts after classification into clear or cloudy conditions
- Rayleigh winds are corrected for the temperature and pressure sensitivity using a priori information (from NWP model forecasts) and also corrected for Mie cross-talk
- Uncertainty estimates for each wind result and quality flags are derived and provided
- Processing options are easily controlled from a settings file

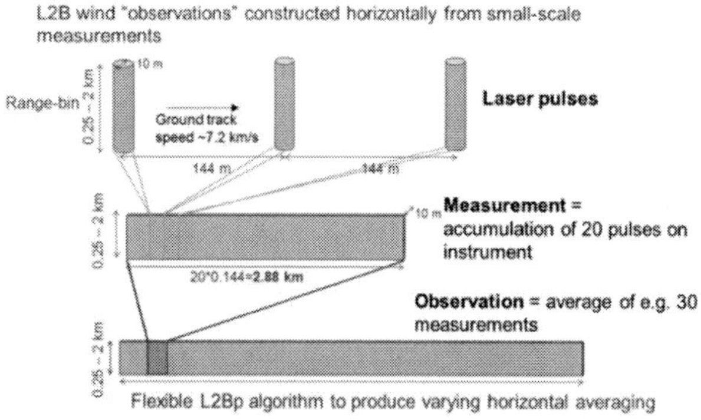

Figure 1. Illustration of how the L2B observation consists of horizontally accumulated "measurement" level data.

 https://doi.org/10.1051/epjconf/201817602015

The L2B processing software is portable, easy to install and is freely available for users to download from: https://software.ecmwf.int/wiki/display/AEOL/ADM-Aeolus+Level-2B+Processor+Package

ECMWF will generate the L2B products during the exploitation phase in near real-time as part of the Ground Segment and will provide the products to ESA for further distribution and archiving. ECMWF will assimilate L2B BUFR product in the global data assimilation analysis, if forecast skill is demonstrated to improve. Related to this is the Level 2C product; which provides vector winds from the ECMWF analysis after the assimilation of L2B winds.

ECMWF has over the years been involved in observation impact studies to assess the potential impact of the Aeolus data. We expect the Aeolus mission to have a positive impact on the analysis and forecast quality, in particular the improvement of winds in the tropics and in the upper troposphere [4]. This work also demonstrated that HLOS winds can provide around 70% the impact of full vector winds.

We now provide an assessment of the Aeolus L2B HLOS wind observation accuracy and precision from realistic simulations and the application of the Ground Segment processing chain.

2 SIMULATION METHODOLOGY

We have employed the Aeolus end-to-end simulator (E2S) and the operational chain-of-processors Level 0, 1A, 1B, CAL suite, and 2B) to investigate L2B wind errors in realistic atmospheric scenarios. The atmospheric inputs to the simulator, i.e. our reference "truth", are derived from ECMWF global model short-range forecasts at resolution TcO1279 (~9km horizontal grid spacing). The E2S atmospheric inputs are:

- Temperature, pressure and molecular backscatter and extinction coefficients derived from these fields
- Particulate backscatter and extinction coefficients derived from model cloud liquid and ice water content using a basic parameterization and assuming a constant lidar ratio of 20; aerosol optical properties are not considered
- HLOS wind derived from model wind components

The atmospheric scenario demonstrated here is a section of an orbit that intersects an equatorial wave with Rossby and Kelvin wave characteristics on 15th March 2014 in the eastern Pacific. Uncertainties in the NWP wind and temperature analysis in this area associated with this wave have been shown to have contributed to a very poor forecast six day forecast over Europe. Therefore this is an interesting case to see what Aeolus observes and how it compares to real wind observations assimilated in the ECMWF operational analysis.

The optical properties and wind inputs to the tropical simulation are shown in Figure 2 and Figure 3.

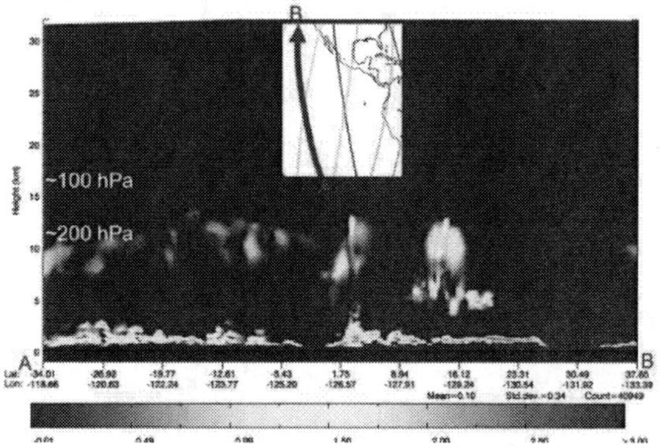

Figure 2. Cross-sectional view of end-to-end simulator input; log10 of the scattering ratio ($1+\frac{\beta_p}{\beta_m}$). *The map (inset) shows the orbit section (A to B).*

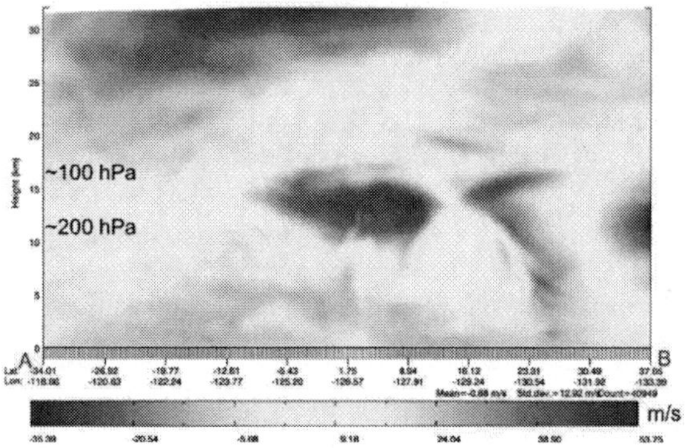

Figure 3. The end-to-end simulator input HLOS wind derived from the ECMWF model forecast i.e. truth to assess L2B winds against. Unit:m/s

The resulting L2B HLOS wind error statistics are examined. Interpreting these L2B error statistics comes with the caveat that the Aeolus simulator is probably not accounting for all sources of instrument noise and hence this will be an underestimate of the real Aeolus error statistics; however there are no known major deficiencies. N.B. this simulation was done with the old higher orbit (408 km) rather than the 320 km which was recently chosen.

3 LEVEL-2B HLOS WIND RESULTS

The resulting L2B Rayleigh-clear and Mie-cloudy HLOS wind observations are shown in Figure 4 and Figure 5 which should be compared to the "truth" HLOS wind of Figure 3 (N.B. the same scales are used for comparison).

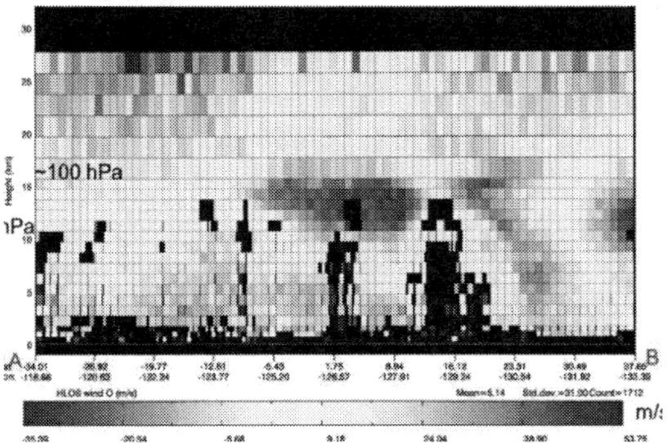

Figure 4. The resultant L2B Rayleigh-clear HLOS wind observations, which can be compared to the "truth" HLOS winds of Figure 3.

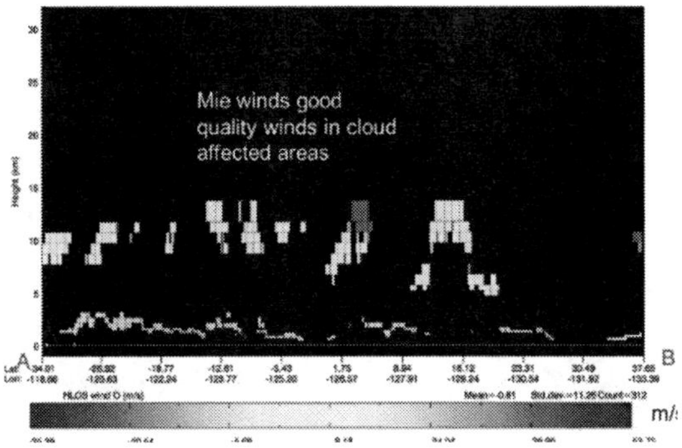

Figure 5. The resultant L2B Mie-cloudy HLOS wind observations, which can be compared to the "truth" HLOS winds of Figure 3.

Aeolus provides a blocky and somewhat noisy view of reality due to the resolution of the observations and the signal-to-noise ratio (SNR). However Aeolus Rayleigh winds are still able to capture the equatorial wave at 100-200 hPa (50 m/s feature in the middle of Figure 3) very well. Also, notice the complementary nature of the Rayleigh and Mie results in terms of coverage.

Aeolus should substantially improve the upper troposphere and lower stratosphere wind observation coverage in the tropics compared to the status quo. The plot of real observations actively assimilated in the ECMWF analysis for this case is shown in Figure 6 (the logarithmic pressure axis and latitude axis is chosen to roughly match the lidar cross-sections for comparison). There are 3700 real observations (mostly AMVs) of u-component wind (in a 20° longitude section) versus ~2000 expected L2B Rayleigh-clear winds (for the Aeolus "curtain"). Only the lower portion of the equatorial wave is sampled with the current wind observing system (and there are gaps around 500-600 hPa). Therefore Aeolus will provide the only direct wind information for the NWP analysis of these upper troposphere/lower stratosphere equatorial waves (apart from a few radiosondes).

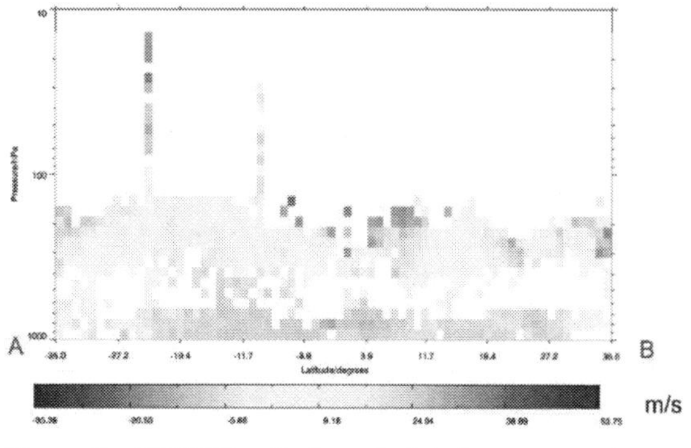

Figure 6. ECMWF actively assimilated u-component wind observations, available in the longitude range 140°W -120°W, for the 12 hour data assimilation cycle at ECMWF on 14th March 2014. Showing that large parts of the equatorial wave feature are not observed.

The expected Aeolus L2B wind error statistics are shown in Figure 7; from a large global sample of realistically simulated data (5.5 orbits); point wind values are considered as "truth". The key features are that L2B Rayleigh-clear error standard deviation is around 3-5 m/s and Mie-cloudy around 1.5-2 m/s, with systematic errors generally less than 0.5 m/s. However note this used "best case" calibration files (worst case systematic errors have been estimated separately to be around 0.5 m/s).

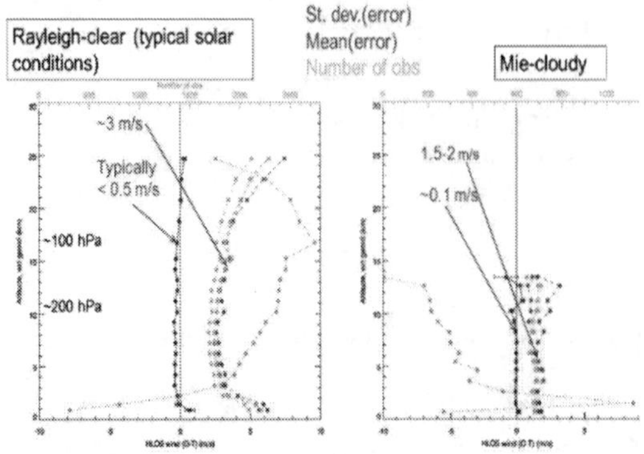

Figure 7. The level of systematic and random errors in the L2B Rayleigh-clear and Mie-cloudy HLOS winds from realistic simulations will all forms of noise "on" in the wind mode simulation. QC was applied to reject Rayleigh winds with L2B error estimates > 8 m/s and Mie winds >3.5 m/s; this removes outliers.

The L2B random error magnitude depends on the size of horizontal accumulation (grouping length) used in the processing; this dependence in shown in Figure 8.

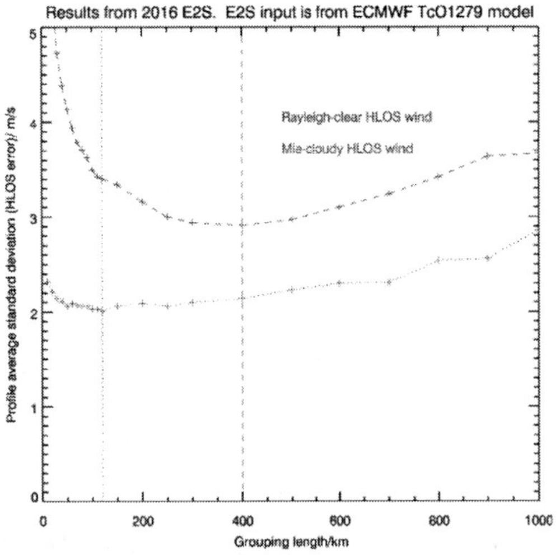

Figure 8. The dependence of the L2B HLOS wind random error (standard deviation) upon the grouping length of the L2B observations.

It is clear that much lower random errors are achievable with the Mie winds compared to the Rayleigh for small horizontal accumulations (e.g. 20 km). The Rayleigh random errors reach a minimum around 400 km accumulation (counteracting the Poisson noise). However not shown here is that the mean errors increase above 100 km averaging due to smoothing the horizontal wind shear across jet streams. In practice at ECMWF we will probably choose around 100 km for the Rayleigh and around 30 km for the Mie grouping length.

4 CONCLUSIONS

The simulation studies show that Aeolus L2B HLOS winds are looking promising in terms of random errors, systematic errors and sampling relative to the current observing system to achieve a significant positive impact in global NWP, particularly in the tropics where upper troposphere/lower stratosphere winds are very much lacking.

ACKNOWLEDGEMENTS

This work is funded by an ESA contract 4200018555/04/NL/MM. Thanks to Lars Isaksen (ECMWF) for advice. Thanks to KNMI (in particular Dr. Jos de Kloe) for setting up the Chain-of-Processors system that allows the end-to-end testing to be run on complex atmospheric data. Thanks to the L1B and L2B teams for general support.

References

[1] European Space Agency ESA, 2008: ADM-Aeolus Science Report, ESA SP-1311, 121 pp.

[2] Tan, D., Andersson., E., de Kloe, J., Marseille, G.-J., Stoffelen, A., Poli, P., Denneulin, M.-L., Dabas, A., Huber, D., Reitebuch, O., Flamant, P., Le Rille, O., Nett, H., 2008: The ADM-Aeolus wind retrieval algorithms. *Tellus* 60A, 191-205.

[3] European Space Agency ESA, 2016: ADM-Aeolus Mission Requirements Documents, AE-RP-ESA-SY-001, Issue 2, 16/11/2016, 57pp.

[4] Horányi, A., Cardinali, C., Rennie, M. and Isaksen, L. (2015), The assimilation of horizontal line-of-sight wind information into the ECMWF data assimilation and forecasting system. Part I: The assessment of wind impact. *Q.J.R. Meteorol. Soc.*, 141: 1223–1232. doi:10.1002/qj.2430

CH4 IPDA LIDAR MISSION DATA SIMULATOR AND PROCESSOR FOR MERLIN: PROTOTYPE DEVELOPMENT AT LMD/CNRS/ECOLE POLYTECHNIQUE

Chomette Olivier[1]*, Raymond Armante[1], Cyril Crevoisier[1], Thibault Delahaye[1], Dimitri Edouart[1], Fabien Gibert[1], Frédéric Nahan[2], Yoann Tellier[1]

[1] *Laboratoire de Météorologie Dynamique (LMD), Ecole Polytechnique, 91128 Palaiseau, FRANCE,*
**chomette@lmd.polytechnique.fr*
[2]*Magellium Paris, 92400 Courbevoie, FRANCE*

ABSTRACT

The MEthane Remote sensing Lidar missioN (MERLIN), currently in phase C, is a joint cooperation between France and Germany on the development of a spatial Integrated Path Differential Absorption (IPDA) LIDAR (LIght Detecting And Ranging) to conduct global observations of atmospheric methane. This presentation will focus on the status of a LIDAR mission data simulator and processor developed at LMD (Laboratoire de Météorologie Dynamique), Ecole Polytechnique, France, for MERLIN to assess the performances in realistic observational situations.

1 INTRODUCTION

The major anthropogenic sources of CH_4 are the emissions from energy production, landfills, waste treatment, cattle, rice fields and incomplete biomass burning. In addition there are significant natural CH_4 sources from wetlands. Due to human activities, its concentration has been multiplied by more than 2 since the pre-industrial period (1750s) and currently reaches approximately 1800 ppb [1].

After carbon dioxide (CO_2), methane (CH_4) is the second anthropogenic greenhouse gas responsible for global warming. Although its global emissions (estimated at around 550 $TgCH_4$ yr-1 [2]) are only 4% of the global CO_2 anthropogenic emissions in units of carbon mass flux, CH_4 has a warming potential that is about 28 times higher than CO_2. Today, surface methane emissions contribute approximately to 30% of the total radiative forcing either directly (20%) [1] or indirectly (10%) via contribution to tropospheric O_3 and H_2O production [1].

Understanding and monitoring the evolution of the mechanisms of sources and sinks of this gas in the atmosphere represents a major challenge to better anticipate the future climate. The goal of MERLIN is to obtain global measurements of the spatial and temporal gradients of atmospheric CH_4 with a precision and accuracy sufficient to constrain methane fluxes significantly better than with the current observations made by passive optical radiometry techniques and limited in precision as well as in spatial and temporal coverage (cloud cover, aerosol interference, nighttime). The main driver for MERLIN are the specification on the CH_4 column-weighted dry-air mixing ratio (XCH_4) split as precision (random error), maximum bias (systematic error) and horizontal sampling as given in Table 1.

Table 1 MERLIN mission requirement for XCH4

	Threshold	Breakthrough	Target
XCH_4 random error	36 ppb (2%)	18 ppb (1%)	8 ppb (0.5%)
XCH_4 systematic error (bias)	3 ppb	2 ppb	1 ppb
Resolution : 50 km horizontal ; total column vertical			

The very low level of systematic error aims at avoiding geographical biases in the XCH_4 fields that could lead to uncertainties in fluxes.

In order to reduce the relative random error due mostly to the signal to noise ratio of the instrument, the data are going to be horizontally averaged over 50km along track.

2 THE MERLIN MISSION

The payload is under the responsibility of the German space agency (DLR), while the platform (MYRIADE Evolutions product line) is developed by the French space agency (CNES). MERLIN is planned to be launch at the end of 2020. The IPDA technique relies on DIAL (Differential Absorption LIDAR) measurements using a pulsed laser emitting at two wavelengths, one wavelength accurately locked on a spectral feature of the methane absorption line, and the other wavelength free from absorption to be used as reference. More details about the instruments can be found in [3].

The MERLIN mission will provide the following data products:

Table 2 MERLIN products

Level 0	For each individual laser shot the range-gated backscatter signal data together with auxiliary data such as laser pulse energy, ranging information, wavelength in a chronological data sequence.
Level 1a	Level 0 data with all calibrations computed and appended.
Level 1b	Integrated signal at echo for off and on wavelengths, range, SNR, slant-path DAOD, geo-location and time.
Level 2	Column-weighted dry-air mixing ratios of methane (XCH_4) shot-by-shot and averaged.
Level 3	Visualizations of level 2 data such as XCH_4 maps.

Level 1 is the responsibility of DLR, Level 2 and 3 are the responsibility of CNES.

We present here the MERLIN measurement principles and the work developed within the MERLIN team at LMD in regards to the LIDAR signals simulation with LIDSIM (LIDar SIMulator) to be able to simulate the Level 0/1a data (see Table 2) and the inversion of this signal done with PROLID (PROcessor LIDar) to obtain the Level 1b and Level 2 product (see Table 2), the column-weighted dry-air mixing ratios of CH_4, referred to as XCH_4, over the satellite sub-track.

3 MEASUREMENT PRINCIPLE

The core element of the MERLIN mission is the nadir-viewing CH4 IPDA LIDAR system. This instrument measures the light back-scattered from the Earth's surface and cloud tops which are illuminated by laser pulses having slightly different wavelengths denoted as λ_{on} and λ_{off} (Fig. 1).

The online wavelength λ_{on} is accurately positioned in the trough of one of the CH4 absorption line multiplets in the 1.64 μm near-infrared spectral region in order to dramatically relax the laser frequency stability and platform pointing requirements.

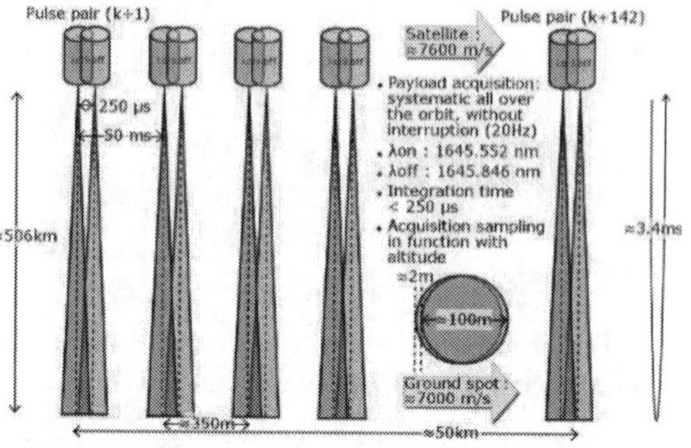

Figure 1 MERLIN measurements concept

The measurement at λ_{off} serves as the reference measurement with negligible absorption by the CH_4 molecules in the path. The on and off wavelengths are spectrally close enough to consider the atmospheric and surface properties to be identical with the exception of CH_4 absorption (and H_2O and CO_2 absorption in a second order). The MERLIN measurement concept is described in Fig. 1.

4 LIDSIM & PROLID

4.1 LIDSIM

The objective of LIDSIM and PROLID is to be used to assess the performance of MERLIN in realistic observational situations (i.e. terrain characteristics, atmospheric conditions).

The inputs of LIDSIM are composed of three modules (Fig. 2)

- The geometry module provides a simulated MERLIN orbit with latitude/longitude of each shot with respect to LIDAR line of sight pointing, satellite altitude, solar zenith angle and time of the shots.

- The scene generation module provides information on the atmospheric conditions and the surface characteristics from different data sources:

o Atmospheric profiles (T, P, H_2O ,O_3), P and T at surface from ECMWF.
o Some information from a Data Elevation Model (DEM) to derive the altitude of each

shot but also the terrain's slope and its roughness. We currently use GMTED2010.

o A global land cover product (GlobCover) to have information about land cover for each shot. This information helps us to create different surface echo linked to land cover classes (all our surface echoes are Gaussian, only the width is modified).

o A database produced from MODIS data to derive the surface reflectivity.

o An inundation climatology to take into account the modifications of surface characteristics brought by inundations.

- The instrument module provides all the instrument characteristics needed to simulate the LIDAR signal in LIDSIM (laser emission, telescope receiver, optics unit, detection and digitalization units).

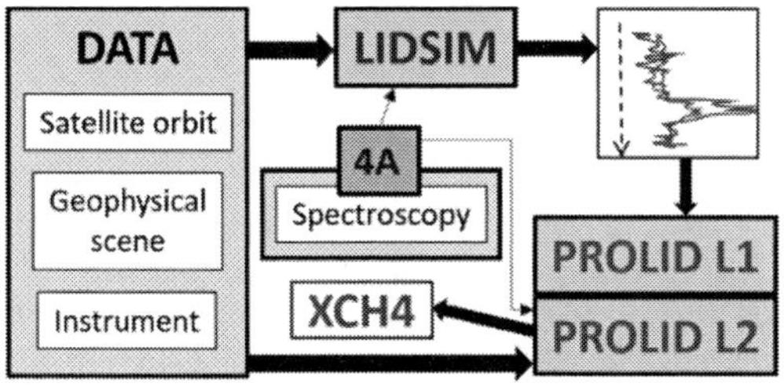

Figure 2 Schematic description and flow diagram of the LIDSIM/PROLID chains

The core of LIDSIM determines the photo-electrons number from the signal (i.e. the photo-electrons detected after being reflected by the ground or a cloud taking into account the absorption and extinction due to the atmosphere) but also from the background (i.e. the photo-electrons generated by the sky background). This step needs a great accuracy on spectroscopic information for the CH_4 multiplet but also for the H_2O and CO_2 in the vicinity of the two MERLIN wavelengths, two gases whose effects must be taken into account in the radiative transfer calculations. For the spectroscopy, GEISA database is used [4] and the Automatized Atmospheric Absorption Atlas (4A) [5] a fast and accurate line-by-line radiative transfer model particularly efficient in the infrared region of the spectrum is used for the radiative computation.

Noise is then added on the signal (laser and background shot noise, laser and background speckle, noise from the detector and from the detection chain). The detection impulse response is also taken into account. The Fig. 3 shows an example of a signal output from LIDSIM.

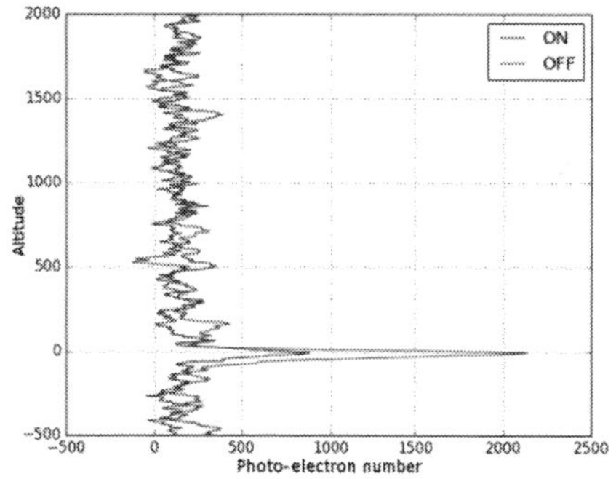

Figure 3 LIDSIM output example. Lidar signal (with noise) for the MERLIN ON and OFF wavelengths

4.2 PROLID

The LIDSIM outputs are directly processed by PROLID. This processor works in two steps. The first one, PROLID-L1, creates MERLIN-L1b like data. Then we obtain the MERLIN-L2 like data (XCH_4) with PROLID-L2. The LIDSIM inputs (Fig. 2) are also needed for PROLID.

PROLID-L1 is still under development at LMD. It is based on the algorithm proposed by DLR. As the impulse response of the detection chain is asymmetrical, the ground echo signal is altered (cf. the asymmetry of the ground echo in Fig. 3). Using a deconvolution process, the asymmetry of the impulse response is rectified and then, the range bias estimation is reduced. PROLID-L1 will also retrieve the scattered energy from the ground echoes and the related SNR for each simulated laser shots.

From the ratio of the LIDAR echoes P_{on}/P_{off}, the Differential Absorption Optical Depth (DAOD) and thus the main data product XCH_4 (defined as a vertical column averaged by a weighting function WF) can be calculated using the following set of equations

$$XCH4 = \frac{\int_0^{Psurf} [CH4]WF(p)dp}{\int_0^{Psurf} WF(p)dp} = \frac{DAOD}{\int_0^{Psurf} WF(p)dp} \quad (1)$$

$$\text{where } DAOD = \frac{1}{2}\ln\left(\frac{P_{on}}{P_{off}}\right) \quad (2)$$

$$\text{and } WF = \frac{\sigma(\lambda_{on}, P, T) - \sigma(\lambda_{off}, P, T)}{g(m_{dry} + m_{H2O}\chi_{H2O})} \quad (3)$$

With P, T the vertical profile of pressure and temperature, σ the methane absorption cross-section (in m^2), m is the molecular mass of a species and χ_{H2O} is the vertical profile of the volume mixing ratio of water vapor with respect to dry air.

To calculate the WF (eq. 3) we need to know the effective absorption cross sections of at least the three atmospheric species CH_4, CO_2 and H_2O. These cross sections depend on atmospheric temperature, pressure and composition and are calculated using 4A. The SSE (Scattering Surface Elevation) needs to be determined in PROLID-L1 in order to know more precisely the surface pressure as the WF is integrated between the top of the atmosphere and the surface.

Once XCH_4 (Eq. 1) is obtained shot by shot, a last module in PROLID-L2 performs the averaging over 50km. The method used is described in [6].

Currently the LIDSIM/PROLID-L2 chain is functional and validated. To validate it, we choose to work with TIGR (Thermodynamic Initial Guess Retrieval) dataset instead of an orbit as input. TIGR is a climatological library of 2311 representative atmospheric situations selected by statistical methods from 80,000 radiosonde reports [7]. Each situation is described, from the surface to the top of the atmosphere, by the values of T, H_2O, O_3 on a given pressure grid.

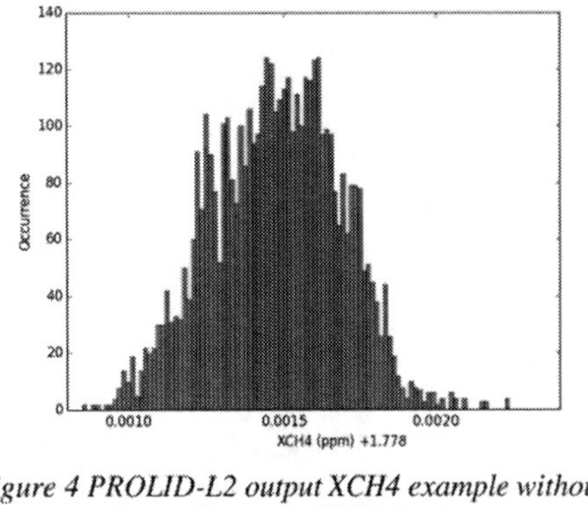

Figure 4 PROLID-L2 output XCH4 example without noise and over the TIGR database

With this database, all atmospheric conditions are taken into account while optimizing time calculation compared to use the orbit data. We also studied the impact of surface reflectivity by adding random value on TIGR. We validate our chain by comparing the XCH_4 calculated by PROLID with the one put as input in LIDSIM without adding noise on the signal. We've found 1.7795±0.0002 ppm (Fig. 4) out of PROLID over the all TIGR database with an amount of CH_4 of 1.78 ppm as input of LIDSIM.

5 PERSPECTIVES

In the future, in addition to complete PROLID-L1, several improvements are planned to be developed in our LIDSIM/PROLID chain such as a better vertical discretization, a more representative surface echo, taking into account the Doppler effect, the aerosols and the clouds.

References

[1] IPCC, 2013: Climate Change 2013: The Physical Science Basis. Contribution of WG I to the Fifth Assessment Report of the Intergovernmental Panel on Climate Change. Cambridge University Press, Cambridge, United Kingdom and New York, NY, USA, 1535 pp.

[2] Kirschke, S. et al..: Three decades of global methane sources and sinks, Nat. Geosci., 6, 813–823, doi:10.1038/ngeo1955, 2013.

[3] Stephan, C. et al.: 2011: MERLIN: a Space-based Methane Monitor, Proceedings of the SPIE, Volume 8159, 815908-15.

[4] Jacquinet-Husson, N.et al.:The 2009 edition of the GEISA spectroscopic database. Journal of Quantitative Spectroscopy & Radiative transfer, 2011, 112, pp.2395-2445.

[5] Cheruy,F. et al.: Contribution to the development of radiative transfer models for high spectral resolution observations in the infrared. Journal of Quantitative Spectroscopy and Radiative Transfer, Volume 53, Issue 6, June 1995, Pages 597-611.

[6] Tellier.Y. et al.: Averaging Bias Correction for Future IPDA Lidar Mission MERLIN, ILRC28 Proceeding (submitted)

[7] Chevallier,F. et al.: TIGR-like atmospheric-profile databases for accurate radiative-flux

 https://doi.org/10.1051/epjconf/201817602016

computation. Quart. J. Roy. Meteorol. Soc. 2000, 126, 777–785.

HIGH SPECTRAL RESOLUTION LIDAR BASED ON QUAD MACH ZEHNDER INTERFEROMETER FOR AEROSOLS AND WIND MEASUREMENTS ON BOARD SPACE MISSIONS

Jean-François Mariscal[1]*, Didier Bruneau[1], Jacques Pelon[1], Mathilde Van Haecke[1], Frédéric Blouzon[2], Franck Montmessin[1], Hélène Chepfer[3]

*[1]LATMOS/IPSL, UVSQ Univ. Versailles St-Quentin, UPMC Univ. Paris 06, CNRS, Guyancourt, France, *jean-francois.mariscal@latmos.ipsl.fr*
[2]CNRS/INSU, Division Technique, Meudon, France
[2]CNRS/LMD, Ecole Polytechnique, Palaison, France

ABSTRACT

We present the measurement principle and the optical design of a Quad Mach Zehnder (QMZ) interferometer as HSRL technique, allowing simultaneous measurements of particle backscattering and wind velocity. Key features of this concept is to operate with a multimodal laser and do not require any frequency stabilization. These features are relevant especially for space applications for which high technical readiness level is required.

1 INTRODUCTION

Remote sensing technique such backscatter lidar has been used to derive aerosol and cloud layer boundaries and extinction. The retrieval of cloud and aerosol optical parameters requires in this case, clear air or use of the hypotheses with respect to the backscatter-to-extinction ratio. As identified in the early phase of lidar development, the high spectral resolution lidar (HSRL) technique allow the differentiation of molecular and particular scattering and more accurate characterization of aerosol or cloud extinction and backscatter coefficients. HSRL has the advantage of not requiring such inputs as molecular and aerosol particle scattering to be separated. Many HSRL systems have been already developed as filtering technique, by use of iodine filters, Michelson or Fabry-Perot interferometers (FPI), to separate the spectrally narrow aerosol scattering from the much broader molecular return. All of these systems operate in direct detection (measurement of the backscattered light power), which has the advantage of relying on both particulate and molecular scattering, allowing extended ranges and capabilities. The direct detection technique has been chosen for space observations of winds or aerosols and clouds using UV-FPI (ESA EarthCare mission). However, whatever the case, all the HSRL techniques developed until now required the emission of a quasi-monochromatic light, that is, the use of a single-mode laser, often injection seeded. Furthermore, most of the systems have also required precise frequency locking of the emitter with regard to the filters. These requirements lead to complex emitter designs that involve delicate servo-loops and can be points of failure in case of space applications.

In contrast, the Quadri-channel Mach Zehnder interferometer (QMZ) operates with a multimode laser. The optical path difference (OPD) of the QMZ matched to the laser free spectral range (FSR) so that the interference state of all the laser modes is exactly the same. The QMZ approach is not to separate the molecular and particulate signals in two distinct channels, but to determine the interference contrast given by the backscattered light that is linked to the backscatter ratio. The contrast is unambiguously determined by the signals given by four detection channels in phase quadrature, whatever the spectral positioning of the laser frequency with regard to the transmission function of the QMZ. In this way, no frequency stabilization is necessary neither for the laser nor the interferometer.

2 METHODOLOGY

2.1 Measurement principle

A schematic of a QMZ is shown in Figure 1. The main distinctive feature compared to a classical Mach Zehnder interferometer (MZI) is the introduction of quarter wave plate (QWP) in on arm of the interferometer, in order to introduce an additional phase difference of π/2 for one direction of polarization with respect to the other. Assuming that no other polarization-dependent path difference occurs, the QMZ provides four output signals in phase quadrature. The two classical MZI output channels are separated by means of polarizers oriented along one axis of the QWP and the detection is made by four non-imaging detectors. In order to be operated with a multimode laser, the OPD, Δ, of the QMZ must match the laser FSR such as

$$\Delta = \frac{c}{FSR}, \quad (1)$$

where c is the velocity of light in vacuum. In this case, the interference state at the QMZ output remains the same regardless the laser modes.

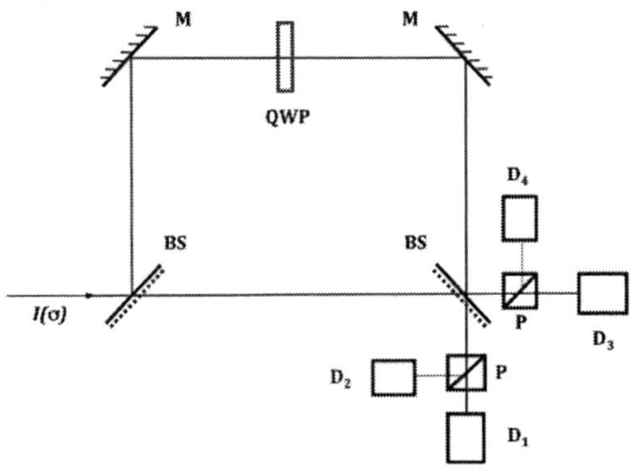

Figure 1 QMZ configuration: M, mirror; BS, beam splitter; QWP, quarter-wave plate; P, polarizers; D1 to D4, detectors

According to [1], with equal fractions of intensity transmitted through the polarizers, the signal delivered by each channel i (from 1 to 4) is written as

$$S_i = \frac{S_t}{4} a_i \left[1 + M_i M_a sin\left(\varphi + (i-1)\frac{\pi}{2}\right)\right], \quad (2)$$

where S_t is the total signal, a_i is the relative photometric sensitivity of channel i, M_i is the instrumental interference contrast of channel i, M_a is the interference contrast given by the atmospheric backscattered signal and φ the interference phase. From Equation 2 we calculate the complex signal Q

$$Q = Q_2 + iQ_1, \quad (3)$$

with

$$Q_1 = \frac{a_3 S_1 - a_1 S_3}{a_3 M_3 S_1 + a_1 M_1 S_3}$$
$$Q_2 = \frac{a_4 S_2 - a_2 S_4}{a_4 M_4 S_2 + a_2 M_2 S_4}. \quad (4)$$

The interference contrast given by the atmospheric backscattered signal is the modulus of the complex signal Q:

$$M_a = |Q|. \quad (5)$$

Always according to [1], the atmospheric backscattered spectrum is the convolution of the emitted spectrum with the spectral broadening caused by the molecules and particles random motion. Therefore, the atmospheric contrast M_a can be expressed as

$$M_a = \frac{\beta_p}{\beta_p+\beta_m} M_p + \frac{\beta_m}{\beta_p+\beta_m} M_m, \quad (6)$$

Where β_p and β_m are the particle and molecule backscatter coefficients, respectively and M_p and M_m are the interference contrast given by particle and molecule coefficients, respectively.

The interference contrast given by a source of spectral full width at half maximum (FWHM), $\delta\nu$, is expressed according to [2] by:

$$M = \exp\left[-\left(\frac{\pi\delta\nu\Delta}{2c\sqrt{ln(2)}}\right)^2\right]. \quad (7)$$

On the one hand, the spectrum backscattered by particles is spectrally broadened by wind turbulence in the probed volume and can be most of time neglected (a radial velocity of $1ms^{-1}$ corresponds to a broadening of 4MHz at 532 nm). The particle spectrum is then, as the laser spectrum, a comb of relatively narrow lines. The response of the laser-matched interferometer to this spectrum is exactly the same as for a single line of typical FWHM $\delta\nu \approx 200$ MHz and leads to an interference modulation $M_p \approx 0.8$ (Equation 7). On the other hand, the spectral broadening of the molecule-backscattered signal is much larger (≈ 1 GHz) due to the wide molecular velocity distribution. The convolution with the emitted frequency comb of a typical 1 GHz FSR produces a nearly constant spectral continuum without interference modulation ($M_m \approx 0$). Thus, the resulting contrast is equal to the instrumental contrast produced by the laser source itself and the retrieved contrast, M_a, is equal to:

$$M_a = R_p M_p, \quad (8)$$

with $R_p = \beta_p/(\beta_p+\beta_m)$ the particle backscatter ratio.

In addition to the contrast, the interference phase of the output signal is obtained by the argument of the complex signal Q

$$\varphi = \arg(Q) = \frac{2\pi\Delta}{\lambda_0}\left(1 + \frac{2V_{LOS}}{c}\right), \quad (9)$$

where λ_0 is the emitted wavelength and V_{LOS} the line-of-sight wind velocity. From the interference phase information we can retrieve the frequency of the backscattered signal and therefore the Doppler shift.

2.2 Calibration and signal processing

The calibration of the QMZ is based on the measurement of the two instrumental parameters: a_i and M_i. These measurements can be done internally by means of a reference channel, consisting in collecting a small amount of light coming from the laser source. During atmospheric measurements, the complex signal is averaged over several laser shots in order to improve the measurement accuracy. As the interference phase may vary during this period, we still record the reference signal in order to perform a phase-compensated summation to avoid any underestimation of the contrast.

For a series of N shots, we have

$$PBR = \frac{\beta_p}{\beta_p+\beta_m} = \frac{\sum_1^N |Q/Q_R|}{N}, \quad (10)$$

and

$$V_{LOS} = \frac{c\lambda_0}{4\pi\Delta}\arg\left(\sum_1^N Q/Q_R\right) \quad (11)$$

where Q_R is the reference complex signal.

3 INSTRUMENT DESIGN

The LATMOS laboratory has been developing for nearly a decade two QMZ interferometers. The first one is integrated to the French airborne lidar LNG, which is a 3 wavelengths and 2 polarizations backscatter lidar, dedicated to the aerosols and clouds studies [3]. The second one is part of the MARBLL lidar breadboard, which was designed with the aim of providing a compact and robust wind lidar for the Mars boundary layer study [2].

The design of the two QMZ devices is based on the same arrangement of fused silica prisms and air gap (Figure 2). In a general way, before entering the prism assembly the atmospheric signal is mixed with the reference signal with an unbalanced coupler (90% for signal and 10% for reference). The atmospheric signal is previously delayed by a 60 m long fiber in order to be temporally separated from the reference. At the output of the coupler a mode scrambler is inserted in order to ensure good mode mixing and good depolarization.

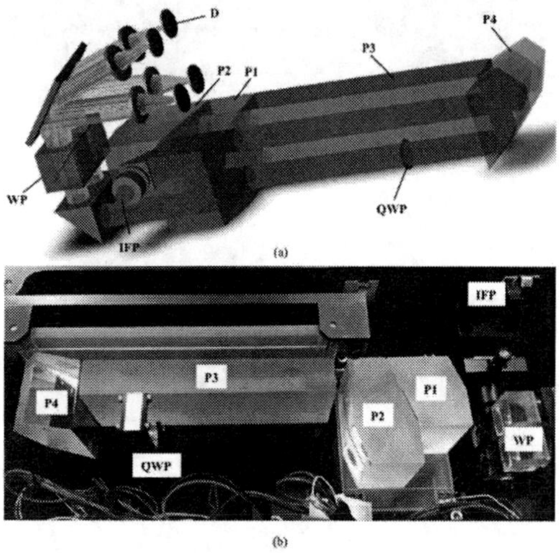

Figure 2 QMZ interferometer: (a) optical design of the QMZ, (b) picture of the MARBLL QMZ. P1, P2, P3, P4, prisms; IFP, input fiber port; QWP, quarter-wave plate; WP, Wollaston polarizers; D, detectors.

At the entrance of the QMZ device, a first lens collimates the light. The first two prisms P1 and P2 enable the splitting and recombination of the two light paths with a nearly nonpolarization-sensitive 50% reflectivity coating. The first arm of the interferometer is made of a thick silica plate P3 and the second arm is an air gap in which the quarter wave plate is inserted. A second parallel pass in each arm is made possible by P4 that reflects the light in a laterally offset path. The two output ports of the interferometer are split in four channels by Wollaston polarizers (WP) and focused on the four detectors (D). The QMZ is field-compensated in order to fit large fiber diameter input without a reduction of the interference contrast.

We can note that the interferometer does not need any tuning element thanks to the high accuracy with which optical components are manufactured (by Winlight Optics, Pertuis, France).

4 CONCLUSION

Technical solutions for space-borne instrument must be the best compromise between scientific performances and technical readiness level (TRL). In terms of technical maturity, monomodal laser does not seem sufficiently high to envisage its deployment onboard space mission. In contrast, the high spectral resolution technique based on QMZ interferometer enable the use of a multimode laser. For instance, the MARBLL interferometer has been designed in order to be used with the same laser as the ChemCam instrument [4] currently in operation on Mars (Curiosity Rover). In addition, the QMZ solution does not require any frequency locking between the laser and the interferometer. In this way, the QMZ concept leads to a simpler and more robust design of the laser and lidar subsystems.

Regarding the measurement performances and when considering particle backscattering lidar, Equation 8 shows that the measurement method is independent of the exact molecular line-width. This avoids the need of precise line-shape modeling based on pressure and temperature profiles, as required for the HSRL vapor-filter and FPI techniques. Note also that the PBR retrieval is not affected by a Doppler shift of the signal, which could be caused by the platform motion and pointing. This last point is a non-negligible advantage for space-borne instrument. Bruneau et al. (2015) [3] describe in details the first results of the 355 nm high spectral resolution airborne lidar LNG and also [5] describe results of the MARBLL first field campaign at the Observatoire de Haute-Provence in France.

In the coming months, thermal vacuum tests will be carried out on the MARBLL QMZ breadboard in order to control the measurement stability over large temperature variations under vacuum (from -20°C to +40°C). The final purpose of these tests is to increase the maturity level of the QMZ concept by moving from TRL 3 to TRL 4.

At last, in preparation of phase 0 of the French-American MESCAL space-borne project, involving the Centre National d'Etudes Spatiales (CNES), the National Aeronautics and Space Administration (NASA) and Airbus Defence and Space Company (ADS), the LATMOS laboratory has been identified to provide a conceptual study and a prototype of the 532 nm HSRL channel based on the QMZ concept. This project, which could take over from the current CALIPSO mission, will be devoted to the study of the climate evolution by measuring clouds and aerosols in the Earth atmosphere.

ACKNOWLEDGEMENTS

We thank the French Space Agency CNES for their continuous support for the development of MARBLL and the phase 0 of MESCAL project.

References

[1] Bruneau, D., Pelon, J., Simultaneous measurements of particle backscattering and extinction coefficients and wind velocity by lidar with a Mach-Zehnder interferometer: principle of operation and performance assessment, *Appl. Opt.* **42**, 1101-1114 (2003)

[2] Bruneau, D., et al., Direct-detection wind lidar operating with a multimode laser, *Appl. Opt.* **52**, 4941-4949 (2013)

[3] Bruneau, D., et al., 355-nm high spectral resolution airborne lidar LNG: system description and first results, *Appl. Opt.* **54**, 8776-8785 (2015)

[4] Maurice, S., Wiens, R.C., Saccoccio, M. et al. Space Sci Rev (2012) 170: 95.doi:10.1007/s11214-012-9912-2

[5] Määttänen, A., et al., Towards a Martian wind lidar, 2017, ILRC 2017, Bucarest, Romania

EPJ Web of Conferences **176**, 02018 (2018) https://doi.org/10.1051/epjconf/201817602018
ILRC 28

TECHNOLOGY ADVANCEMENTS FOR ACTIVE REMOTE SENSING OF CARBON DIOXIDE FROM SPACE USING THE ACTIVE SENSING OF CO_2 EMISSIONS OVER NIGHTS, DAYS, AND SEASONS (ASCENDS) CARBONHAWK EXPERIMENT SIMULATOR

Michael D. Obland[1]*, Joel Campbell[1], Susan Kooi[2], Tai-Fang Fan[2], William Carrion[2], Jonathan Hicks[2], Bing Lin[1], Amin R. Nehrir[1], Edward V. Browell[3], Byron Meadows[1], Kenneth J. Davis[4]

[1]*NASA Langley Research Center, Hampton, Virginia, USA, *michael.d.obland@nasa.gov*
[2]*Science Systems and Applications, Inc., Hampton, Virginia, USA*
[3]*STARSS-III Affiliate, NASA Langley Research Center, Hampton, Virginia, USA*
[4]*Pennsylvania State University, University Park, Pennsylvania, USA*

ABSTRACT

This work describes advances in critical lidar technologies and techniques developed as part of the NASA Active Sensing of CO_2 Emissions over Nights, Days, and Seasons CarbonHawk Experiment Simulator system for measuring atmospheric column carbon dioxide (CO_2) mixing ratios. This work provides an overview of these technologies and results from recent test flights during the NASA Atmospheric Carbon and Transport – America (ACT-America) Earth Venture Suborbital summer 2016 flight campaign.

1 INTRODUCTION

The Active Sensing of CO_2 Emissions over Nights, Days, and Seasons (ASCENDS) CarbonHawk Experiment Simulator (ACES) is a NASA Langley Research Center instrument funded by NASA's Science Mission Directorate. ACES seeks to advance technologies critical to measuring atmospheric column carbon dioxide (CO_2) mixing ratios in support of the NASA ASCENDS mission, which was identified in the U.S. National Research Council's report *Earth Science and Applications from Space: National Imperatives for the Next Decade and Beyond*, as a midterm space mission necessary to better understand global sources and sinks of CO_2 [1]. The ACES design demonstrates advances in: (1) multiple high-efficiency, high-power Erbium-Doped Fiber Amplifiers (EDFAs); (2) enhanced power-aperture product through the use and operation of multiple co-aligned laser transmitters and a multi-aperture telescope design; (3) high-bandwidth, low-noise HgCdTe detector and transimpedence amplifier (TIA) subsystem capable of long-duration operation, and (4) advanced algorithms for cloud and aerosol discrimination. The ACES instrument, an Intensity-Modulated Continuous-Wave (IM-CW) lidar, was designed for high-altitude aircraft operations and can be directly applied to space instrumentation to meet the ASCENDS mission requirements. These advanced technologies are critical for developing an airborne simulator and spaceborne instrument with lower platform consumption of size, mass, and power, and with improved performance.

ACES leverages the measurement techniques and instrument architecture of the Harris, Corp. Multifunctional Fiber Laser Lidar (MFLL) instrument that has flown on multiple measurement campaigns [2]. ACES has 6 times more transmit power, 14 times more power-aperture product, and increased detector subsystem bandwidth compared with the MFLL instrument.

ACES was given the opportunity to fly on the NASA C-130 aircraft as a piggy-back instrument during the summer 2016 Atmospheric Carbon and Transport – America (ACT-America) campaign. ACT-America is part of NASA's Earth Venture Suborbital-2 program and is designed to advance society's ability to predict and manage future climate change by enabling policy-relevant quantification of the carbon cycle. ACT-America will enable and demonstrate a new generation of atmospheric inversion systems for quantifying CO_2 and CH_4 variations via three mission goals: 1) reducing atmospheric transport uncertainties; 2) improving regional-scale estimates of CO_2 and CH_4 fluxes; and 3) evaluating the sensitivity of Orbiting Carbon Observatory-2 (OCO-2) [3, 4] column CO_2 measurements to regional variability in tropospheric CO_2. ACT-America is achieving these goals by deploying airborne and ground-based platforms to obtain data that will be combined with data from existing measurement networks and integrated with an ensemble of atmospheric inversion systems. Two NASA aircraft are used for this experiment: the C-130 (outfitted with both remote and in situ sensors), and the B-200 (outfitted with a complementary suite of in situ sensors). A total of five campaigns, each six weeks in length, are being performed across the eastern half of the United States. The 2016 summer measurement campaign comprised 24 research flights and over 151 hours of science data. Aircraft flight tracks for this campaign are shown in Figure 1.

The ACT-America 2016 summer campaign was the first opportunity for significant flight time for the ACES instrument during a science-focused airborne campaign. During this campaign, ACES collected data over a wide variety of surface reflectivities, terrain, and atmospheric conditions. Simultaneous data collected by in situ instrumentation on the aircraft were used to evaluate the ACES measurements of column-integrated CO_2 mixing ratios.

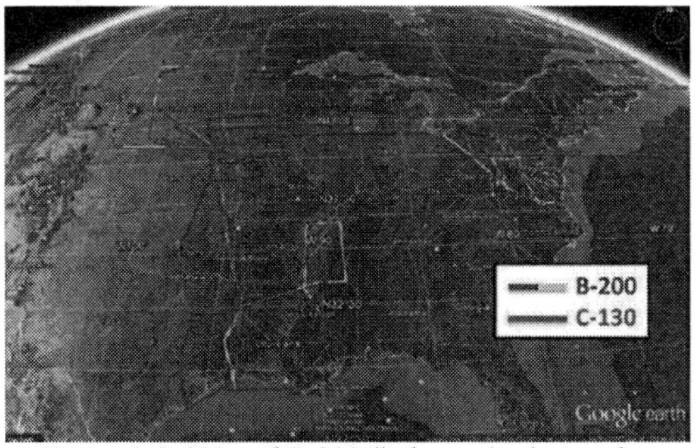

Figure 1 C-130 and B-200 flight tracks during the 2016 ACT-America Summer measurement campaign, which took place July 18 through August 29, 2016.

2 INSTRUMENT

ACES employs the IM-CW measurement technique, in which unique intensity-modulated, range-encoded waveforms are applied to lasers with wavelengths located on and off of a CO_2 absorption line. These lasers are simultaneously transmitted into the atmosphere, and the online and offline signals are then simultaneously received. A measurement of the correlation between the transmitted and received waveforms provides accurate measurements of range to the scattering surface and an integrated optical depth due to absorption by CO_2 within the measurement column. The scattering surface can be ground terrain, water, or clouds [5]. This approach is analogous to mature frequency-modulated continuous wave radar and GPS measurement techniques.

The transmitter, receiver, and detector subsystems are housed in an environmental enclosure originally built for the Global Hawk aircraft. The enclosure is shown in Figure 2 during early testing on the NASA HU-25 aircraft. Data acquisition and control hardware is located in electronic racks that are separate from the instrument enclosure.

Figure 2 The ACES instrument box installed on the NASA HU-25 aircraft for early development tests. The instrument box is present in the foreground. Racks for data acquisition and control systems are located immediately behind the instrument box.

2.1 Laser Transmitter

The ACES laser transmitter includes fabrication of high-efficiency fiber seed laser electronics, modulators, and amplifiers for sensing of CO_2 at 1.57 micron. The transmitter utilizes Master Oscillator Power Amplifier (MOPA) technology meant to resonantly probe the atmosphere using an Integrated-Path Differential Absorption Lidar (IPDA) approach.

ACES simultaneously transmits three beams from commercial Erbium-Doped Fiber Amplifiers

(EDFAs) operating near 1.57 micron. Each EDFA has a transmit power of 10 W, for a total transmit power of 30 W. The EDFAs are seeded by temperature-stabilized seed lasers locked to a CO_2 absorption line. Fine steering and alignment of the transmitted beams is performed by Risley prisms that are housed on an optical bench fed by fiber-coupled collimators from each EDFA. The three laser beams are combined in the far-field and a three-telescope receiver is used to collect return signals from all three beam simultaneously. These signals are then combined on a single detector.

2.2 Detector

The ACES detector is improved from previous detector subsystems by increasing the bandwidth to 4.9 MHz (at a gain of 10^6), reducing overall mass from 18 lb to less than 10 lb, and extending the duration of autonomous, service-free operation from 4 hr to greater than 24 hr. These technology advancements permit higher laser modulation rates, allowing greater flexibility for implementing advanced retrieval algorithms as well as improving range resolution and error reduction.

The baseline ACES detector is an HgCdTe array with 64 diodes wire-bonded together to form a single pixel. The array is designed for integration into a tactical Dewar with an operational temperature range of 60 to 100 K to optimize detector performance. Temperature stability of the detector and Dewar has been shown to be stable to +/- 0.25 K over 1 hr and +/- 1.0 K over 24 hr. The detector NEP is 2.4 $fW/Hz^{1/2}$ with an excess noise factor of about 1.1. While this detector is still in development, a commercially available detector with higher noise and lower responsivity is being used for the ACT-America research flights.

2.3 Receiver

The ACES receiver system consists of a three-telescope design originally built for the constraints of the Global Hawk aircraft. The outgoing laser beams are aligned to the field of view of three fiber-coupled 17.8-cm diameter telescopes. The backscattered light is then collected by the same three telescopes, which are fiber-coupled to the aft optics and a single-detector/TIA subsystem. The receiver design tests the ability of multiple smaller telescopes to provide equal or greater collection efficiency compared with a single larger telescope with a reduced impact on launch mass and cost.

3 RESULTS AND FUTURE PLANS

Some preliminary results show encouraging comparisons with atmospheric model CO_2 optical depth values derived from in situ measurements collected on board the C-130 on spirals under the remote ACES measurements. For example, see Figure 3, where the optical depth differences were found to be less than 2%. Other results show significant deviations from model CO_2 optical depth values due to hardware instabilities. These instabilities are likely due to the use of a commercial off-the-shelf detector in place of the higher quality DRS detector, which is being built and is expected to significantly improve measurement quality. Gain instabilities in the commercial detector likely led to biases in measurements during portions of the flights.

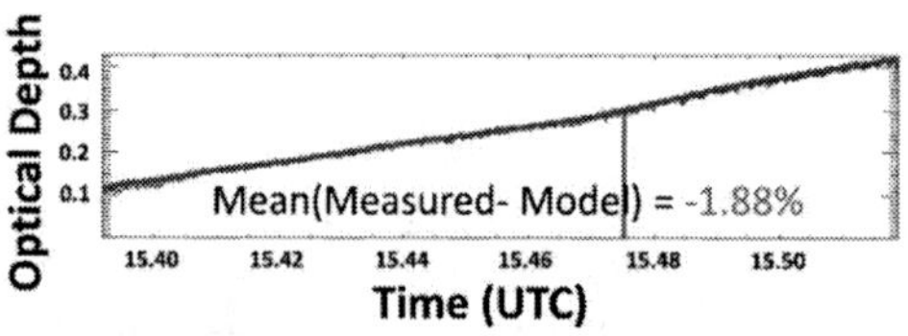

Figure 3 Example measurement data from ACES compared with results from an atmospheric model based on in situ CO_2 and meteorological measurements made on spirals under the ACES measurements.

The IM-CW technique continues to accurately measure range over a variety of surfaces and in the presence of optically thin clouds allowing for retrievals of column CO_2 mixing rations to surface and cloud tops. Advanced correlation techniques have used ACES data to resolve cloud and forest features [6, 7].

Current plans for ACES during the upcoming ACT-America flights are to improve the measurements via improving detector gain stability and data acquisition hardware, and to improve calibration of the instrument through comparisons with other ACT-America measurements of CO_2 column density.

4 CONCLUSIONS

The ACES project is advancing key technologies in support of the NASA ASCENDS satellite mission. These technologies are viewed as critical towards developing an airborne simulator and eventual spaceborne instrument with lower size, mass, and power consumption, and improved performance. The ACES instrument was flown extensively during ACT-America flights on the NASA C-130 aircraft in July and August 2016, and demonstrated encouraging comparisons with in situ measurements. Hardware improvements will continue to increase the quality of measurements in future campaigns, including flights with the next ACT-America campaign in winter, 2017, with the eventual goal of flights on board a high-altitude airborne platform.

ACKNOWLEDGEMENTS

The authors thank the following sources for funding this project: The NASA Earth Science Technology Office (ESTO) Instrument Incubator Program (IIP) and Small Business Innovation Research (SBIR), the NASA Earth System Science Pathfinder Program Office and Earth Venture Program, the NASA Science Mission Directorate, and NASA Langley Research Center.

References

[1] National Research Council, Earth Science and Applications from Space: National Imperatives for the Next Decade and Beyond (National Academies, 2007).

[2] Dobler, J. T., Harrison, F. W., Browell, E. V., Lin, B., McGregor, D., Kooi, S., Choi, Y., Ismail, S., 2013: Atmospheric CO2 column measurements with an airborne intensity-modulated continuous wave 1.57 μm fiber laser lidar, *Appl. Opt.* **52**, 2874-2892.

[3] Crisp D., Atlas, R. M., Breon, F.-M., Brown, L.R., Burrows, J. P., Ciais, P., Connor, B. J., Doney, S. C., Fung, I. Y., Jacob, D. J., Miller, C. E., O'Brien, D., Pawson, S., Randerson, J. T., Rayner, P., Salawitch, R. J., Sander, S. P., Sen, B., Stephens, G. L., Tans, P. P., Toon, G. C., Wennberg, P. O., Wofsy, S. C., Yung, Y. L., Kuang, Z., Chudasama, B., Sprague, G., Weiss, B., Pollock, R., Kenyon, D., Schroll, S., 2004: The Orbiting Carbon Observatory (OCO) Mission, *Advances in Space Research*, **34** (4), 700-709.

[4] Crisp, D., Miller, C. E., DeCola, P. L., 2008: NASA Orbiting Carbon Observatory: measuring the column averaged carbon dioxide mole fraction from space, *J. Appl. Remote Sens.*, **2**, 023508, doi:10.1117/1.2898457.

[5] Lin, B., Nehrir, A., Harrison, F., Browell, E., Ismail, S., Obland, M., Campbell, J., Dobler, J., Meadows, B., Fan, T., Kooi, S., 2015: Atmospheric CO2 column measurements in cloudy conditions using intensity-modulated continuous-wave lidar at 1.57 micron, *Opt. Express*, **23**, A582-A593.

[6] Campbell, J., Lin, B., Nehrir, A., Harrison, F., Obland, M., 2014: Super-resolution technique for CW lidar using Fourier transform reordering and Richardson-Lucy deconvolution, *Opt. Lett.* **39**, 6981-6984.

[7] Campbell, J. F., Lin, B., Nehrir, A. R., Harrison, F. W., Obland, M. D., 2014: High Resolution CW Lidar Altimetry using Repeating Intensity Modulated Waveforms and Fourier Transform Reordering, *Opt. Lett.*, **39**, 6078-6081, http://dx.doi.org/10.1364/OL.39.006078.

EPJ Web of Conferences **176**, 02019 (2018) https://doi.org/10.1051/epjconf/201817602019
ILRC 28

AEROSOL OPTICAL PROPERTIES RETRIEVED FROM THE FUTURE SPACE LIDAR MISSION ADM-AEOLUS

Pauline Martinet[1,*], Thomas Flament[1] and Alain Dabas[1]

*[1]CNRM UMR 3589, Meteo FRance and CNRS, France, *pauline.martinet@meteo.fr*

ABSTRACT

The ADM-Aeolus mission, to be launched by end of 2017, will enable the retrieval of aerosol optical properties (extinction and backscatter coefficients essentially) for different atmospheric conditions. A newly developed feature finder (FF) algorithm enabling the detection of aerosol and cloud targets in the atmospheric scene has been implemented. Retrievals of aerosol properties at a better horizontal resolution based on the feature finder groups have shown an improvement mainly on the backscatter coefficient compared to the common 90 km product.

1 INTRODUCTION

The space-based Doppler mission AEOLUS of the European Space Agency expected to be launched towards end 2017 will be the first High-Spectral Resolution Lidar (HSRL) in space. Operating in the UV the HSRL capability is implemented by directing the backscattered light through 3 successive interferometers. The sensor behind the first one is dedicated to sensing particulate backscatter, its signal is called "Mie signal" and the sum of signals from the other two sensors is called "Rayleigh signal". These signals both contain contributions from molecular and particulate backscatter that are later seperated by computing a "cross talk correction". The system is primarily designed for the measurement of winds, but the HSRL capability enables the measurement of the particulate backscatter and extinction coefficients without any a priori assumption on the aerosol type. The level-2A (L2A) processor has been developed to retrieve vertical profiles of optical properties in 24 bins from the ground up to 30 km with a vertical resolution varying from 250 m to 2 km. The extinction coefficient is then derived through an iterative scheme while the backscatter coefficient is directly derived from the ratio of the pure particulate to molecular signals (more details can be found in [2]).
The ability of the L2A to retrieve clouds and aerosols properties over a 90-km horizontal scale has already been demonstrated. A good retrieval of the overall shape of aerosol and cloud layers (see Fig. 1) with a better accuracy on the backscatter coefficient was shown in [1].

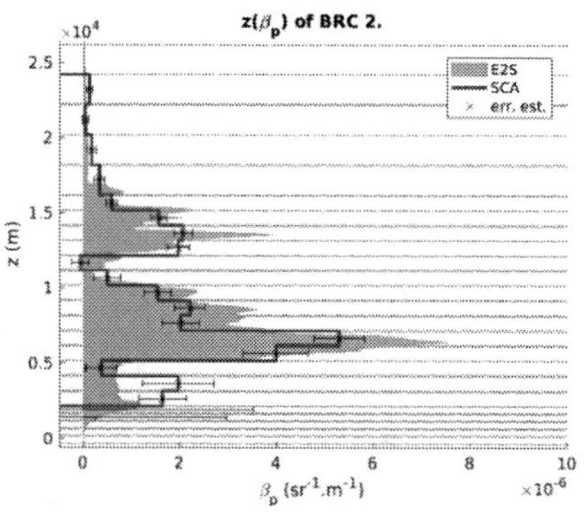

Figure 1: Vertical profile of L2A backscatter coefficients retrieval (red line) compared to the true expected profile (grey) and estimated retrieval errors (horizontal black lines). 90 km horizontal scale product

In order to improve the horizontal resolution of the retrieval and provide an aerosol product at a finer scale than 90 km, a feature finder algorithm has been developed. It identifies homogeneous aerosol and cloud targets on which the lidar signal is accumulated in order to obtain a sufficient signal to noise ratio (SNR).

2 METHODOLOGY

The feature finder developed for ADM-Aeolus is largely inspired by the EarthCare algorithm following the study of [3]. The algorithm is based on the assumption that both clear-sky and cloudy or aerosol loaded Mie signals as well as background noise follow a Gaussian distribution. If we consider a given signal level as a lower threshold, the part of the clear sky signal distribution above the threshold (red area in Fig. 2) is a false-alarm probability, i.e. the probability that the signal level in a clear sky pixel actually exceeds the threshold. The part of the particle-loaded distribution that falls below the threshold (dark green area in Fig. 2) is the probability of missing detection.
The probability of detection on the Mie channel P_{Mie}, i.e the bright green area, can be expressed by:

$$P_{Mie} = 1 - \frac{1}{2}erfc(\frac{S - \delta S}{\sqrt{2}\delta S}) \quad (1)$$

with S the signal level in a particle contaminated bin, δS the noise level, and $erfc$ the complementary error function.
The expression of the SNR is then used to simplify equation 1:

$$SNR = \frac{S}{\delta S} \quad (2)$$

into:

$$P_{Mie} = 1 - \frac{1}{2}erfc(\frac{SNR - 1}{\sqrt{2}}) \quad (3)$$

The complementary $erfc$ function being monotonic, applying a threshold on P_{Mie} is the same as applying a threshold on the Mie SNR. From a database of simulated scenarios, the SNR value enabling a high detection score has been determined.

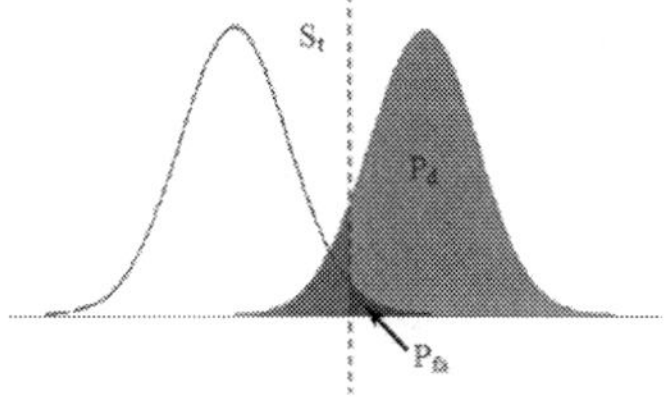

Figure 2: Probability of detection with the distribution for clear sky as a red line and the distribution for the particle-loaded signal as the green area. Figure extracted from the CALIOP ATBD described in [4].

3 RESULTS

3.1 Feature Finder performance

An End-to-End simulator (E2S) developed by ESA to simulate the lidar signals (Rayleigh and Mie channels) received by ADM-Aeolus has been used to simulate simple to more complex atmospheric scenes: standard boundary layer aerosols, additional thin and opaque clouds, real scenes observed during the LITE (Lidar In-space Technology Experiment) experiment. By comparing the aerosol/cloud mask computed by the feature finder to the "true" E2S mask (input particulate backscatter coefficient larger than 10-6 $m^{-1}sr^{-1}$), the expected accuracy of the detection has been evaluated.
Figure 3 shows the feature finder performance on two LITE scenarios: the top panel shows a complex case with thin high clouds and middle level opaque clouds attenuating most of the signal (LITE C)whereas the bottom panel shows a simple case with thin cirrus clouds on top of boundary layer aerosols (LITE A). Particle contaminated pixels well detected by the feature finder are shown in yellow, false alarm in light green, missing aerosol in blue and clear-sky pixels in dark blue. White pixels represent pixels for which a cloud detection is not possible due to the low transmittance of the lidar

signal through the upper clouds (two-way transmittance smaller than 0.1). Taking into account the whole scene, a good detection of 72 % with a false alarm rate of 4 % was obtained for LITE C. The good detection rate is increased up to 83 % if we only consider the atmosphere above the boundary layer in which the lidar signal is probably too noisy to reliably retrieve optical properties at a small scale. For LITE A, a good detection of 45 % is shown for the whole scene but this rate is increased to 86 % when considering the upper part of the atmosphere (above 3 km) corresponding to the cirrus clouds. The developed FF algorithm is thus able to detect high level aerosol and clouds with a very high degree of good detection. However, the detection of boundary layer aerosol is still challenging due to the low noise level.

3.2 Group retrievals

In order to evaluate the capability of the L2A processor to retrieve aerosol properties at a finer horizontal scale compared to the 90 km accumulation length, aerosol retrievals have been performed after averaging lidar signals on small groups identified by the feature finder. These retrievals are then compared to the true E2S properties averaged on the same horizontal features. In order to emphasize the improvement in aerosol properties brought by an increased horizontal resolution, the retrievals at a 90 km horizontal length scale are also compared to the E2S averaged at the same length as used for the feature finder retrievals. We then compute the following ratio:

$$R = log(\frac{abs(aer_{90km} - aer_{E2S,FF})}{abs(aer_{FF} - aer_{E2S,FF})}) \quad (4)$$

where aer_{FF} represents the aerosol retrieval (extinction or backscatter) accumulating the lidar signal on features detected by the algorithm, $aer_{E2S,FF}$ represents the aerosol properties used in the E2S after accumulation on feature detected by the algorithm and aer_{90km} the aerosol retrievals performed by accumulating the lidar signal over 90 km. Figure 4 shows this ratio for the complex LITE C scenario. Values above zero represent an improvement of the retrievals when using a smaller horizontal length, around zero similar performance is found, below zero the 90 km products leads to more consistent results with the E2S inputs. Both for the extinction and the backscatter coefficients, only few pixels show a degradation of the aerosol product if the lidar signal is accumulated on a finer

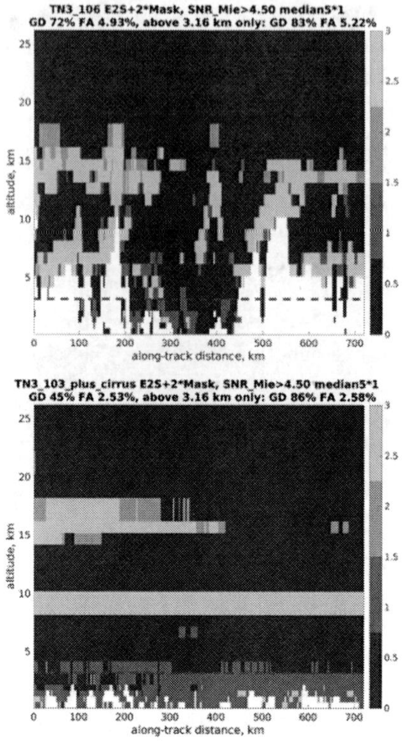

Figure 3: Feature Finder performance of good detection (yellow pixels), false alarm (green pixels), no detection (light blue) for the LITE C scenario with complex heterogeneous clouds (top panel) and the LITE A scenario with homogeneous high thin clouds (bottom panel).

scale. These pixels probably correspond to a strong attenuation of the lidar signal but this needs to be confirmed with a thorough evaluation. This first result demonstrates the robustness of the algorithm used to retrieve aerosol properties. For very long group, the performance is equivalent to the 90 km product as expected but a significant improvement is shown for the backscatter coefficient on small features below targets not attenuating too much the signal.

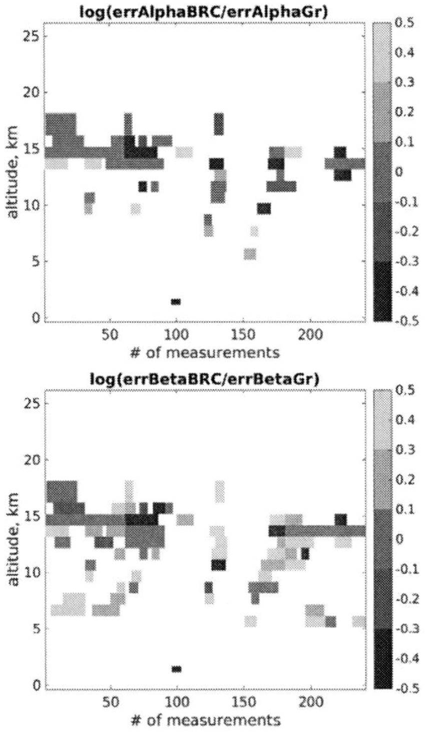

Figure 4: Ratio of the retrieval errors at a small horizontal scale to the retrieval errors at a 90 km accumulation length for the extinction coefficient (top panel) and backscatter coefficient (bottom panel). A logarithm scale is used

.

4 CONCLUSIONS

ADM-Aeolus has demonstrated its capability to fill the gap between the CALIPSO and EarthCare missions by retrieving aerosol and backscatter coefficients with a good accuracy. The newly developed feature finder algorithm has shown encouraging results to provide users with finer scale products in the case of heterogenous targets when the lidar signal is not strongly attenuated. The CAL/VAL activities are now in preparation in the view of the future satellite launch.

ACKNOWLEDGEMENTS

The work described in the paper was funded by the European Space Agency in the frame of the AEOLUS mission. We would like to thank ESA team – namely Frank de Bruin, Anne Straume, Thomas Kanitz – and the other members of the level 1/level 2A development teams for their advice, support and feedbacks.

References

[1] Martinet, P., Dabas, A., Lever, V., Flamant , P., Huber, D., *EPJ Web of Conferences*, **119**, 04001, DOI: http://dx.doi.org/10.1051/epjconf/201611904001

[2] Flamant, P., Cuesta, J., Denneulin, M.-L., Dabas, A. and Huber, D., 2008: ADM-Aeolus retrieval algorithms for aerosol and cloud products. *Tellus*, **60A**, 273–288

[3] Van Zadelhoff, G.-J. and Donovan, D., 2009: A potential FeatureMask algorithm for the EarthCARE lidar. *8th International Symposium on Tropospheric Profiling*, **ISBN 978-90-6960-233-2**

[4] Vaughan, M., Winker,D.M., Powell, K.A, 2005: CALIOP Algorithm Theoretical Baseline Document, Part 2: Feature Detection and Layer Properties Algorithms, PC-SCI-202 Part 2.

 https://doi.org/10.1051/epjconf/201817602020

Averaging Bias Correction for Future IPDA Lidar Mission MERLIN

Yoann Tellier[1]*, Clémence Pierangelo[2], Martin Wirth[3], Fabien Gibert[1]

[1]*Centre National de Recherche Scientifique - Laboratoire de Météorologie Dynamique (CNRS - LMD), France, *yoann.tellier@lmd.polytechnique.fr*
[2]*Centre National d'Etudes Spatiales (CNES), France*
[3]*Deutsches Zentrum für Luft- und Raumfahrt (DLR), Germany*

ABSTRACT

The CNES/DLR MERLIN satellite mission aims at measuring methane dry-air mixing ratio column (XCH_4) and thus improving surface flux estimates. In order to get a 1% precision on XCH_4 measurements, MERLIN signal processing assumes an averaging of data over 50 km. The induced biases due to the non-linear IPDA lidar equation are not compliant with accuracy requirements. This paper analyzes averaging biases issues and suggests correction algorithms tested on realistic simulated scenes.

1 INTRODUCTION

Methane is the second most important anthropogenic greenhouse gas after carbon dioxide **Error! Reference source not found.**. A precise and accurate measurement of atmospheric CH_4 concentration on a global scale is crucial in order to improve the surface flux estimate and thus develop the knowledge of the global methane cycle [2].

The French-German spatial mission MERLIN (Methane Remote Sensing Lidar Mission) aims at measuring the methane dry-air mixing ratio (XCH_4) reaching unprecedented accuracy with targeted systematic error that is less than 0.2% relative error for all latitudes and all seasons.

Integrated Path Differential Absorption (IPDA) lidar measures the laser light (here short pulses) scattered back from the surface in order to retrieve the column content of a specific trace gas along the line of sight. Differential absorption uses the difference in transmission between the on-line signal with a wavelength set about the center of the CH_4 absorption line and the off-line signal which wavelength is significantly less absorbed. From these two signals the Differential Absorption Optical Depth (DAOD) of CH_4 is computed and the corresponding XCH_4 can be derived. The IPDA lidar equation (eq. (1) and (2)) links the backscattered off-line and on-line measurements to the sounded column of XCH_4 [3].

In order to reach the targeted 1% relative random error on XCH_4 measurements, the signal processing of MERLIN requires an horizontal averaging of data over 50 km along track. However, the non linearity of the IPDA lidar equation in combination with the variability of the observed scene (surface elevation, reflectivity, meteorology) along the averaging window induces a bias on the averaged data.

The present paper analyses and compares the biases of several averaging schemes and suggests correction algorithms. A comparative evaluation of the averaging schemes on realistic scenes is a key element to select the best approach (i.e. the least biased) for MERLIN processing.

2 METHODOLOGY

2.1 Principle of IPDA lidar

For every pair of on-line and off-line shots, the DAOD δ can be computed as [4]:

$$\delta = \frac{1}{2} \cdot \ln\left(\frac{P^{\text{off}} \,.\, E^{\text{on}}}{P^{\text{on}} \,.\, E^{\text{off}}}\right) = \frac{1}{2} \cdot \ln\left(\frac{Q^{\text{off}}}{Q^{\text{on}}}\right) \tag{1}$$

where the on and off backscattered signal energies $P^{\text{on/off}}$ are normalized by the ratio of transmitted pulse energies $E^{\text{on/off}}$ ($Q^{\text{on/off}}$ is the normalized signals). Then, we can derive the XCH_4 as:

$$X_{CH_4} = \frac{\delta}{\int_0^{P_{\text{surf}}} WF(p,T) \cdot \mathrm{d}p} = \frac{\delta}{IWF} \tag{2}$$

where P_{surf} is the pressure where the laser beam is backscattered and WF is the weighting function computed from external data describing the measurement sensitivity of XCH_4 along the line of sight.

https://doi.org/10.1051/epjconf/201817602020

These equations are valid for every shot-pairs but averaging over several successive shots is necessary to reach the targeted precision of the measurement.

2.2 Averaging strategies

For the following discussion we will use the triangular bracket notation to denote the arithmetic mean of the quantities Y_i associated to each pulse and the ΔY_i will represent the deviation to this arithmetic mean:

$$\langle Y \rangle = \frac{1}{N}\sum_{i=1}^{N} Y_i \text{ and } \Delta Y_i = Y_i - \langle Y \rangle \quad (3)$$

There are four basic averaging schemes:

1. Averaging of XCH_4 (AVX):

$$\overline{X_{CH_4}}^{avx} = \langle \frac{\delta}{IWF} \rangle \quad (4)$$

2. Averaging of DAOD (AVD):

$$\overline{X_{CH_4}}^{avd} = \frac{\langle \delta \rangle}{\langle IWF \rangle} \quad (5)$$

3. Averaging of quotients (AVQ):

$$\overline{X_{CH_4}}^{avq} = \frac{\frac{1}{2} \cdot \ln \langle \frac{Q^{\text{off}}}{Q^{\text{on}}} \rangle}{\langle IWF \rangle} \quad (6)$$

4. Averaging of signals (AVS):

$$\overline{X_{CH_4}}^{avs} = \frac{\frac{1}{2} \cdot \left(\frac{\langle Q^{\text{off}} \rangle}{\langle Q^{\text{on}} \rangle} \right)}{\langle IWF \rangle} \quad (7)$$

As these equations are not linear and the physical quantities (surface pressure, surface reflectance...) vary for successive shot-pairs, the four averaging schemes are all biased in their own specific way. Correction algorithms must be applied in order to significantly reduce this averaging bias. AVQ (eq. (6)) has not been studied in this paper as the noise distribution leads to the highest bias.

2.3 Bias sources

Bias on DAOD from noise on $Q^{\text{on/off}}$*:*

As the measurements themselves are affected by noise it implies a bias when computing the XCH_4 for every averaging schemes. Indeed, we can write $Q^{\text{on/off}}$ as the sum of their noiseless mean signals $Q_{\text{true}}^{\text{on/off}}$ and a noise term $\Delta Q^{\text{on/off}}$. From this decomposition we can show that for any shot-pair i the true DAOD is biased:

$$\delta_i = \delta_{i,\text{true}} + \ln\left(1 + \frac{\Delta Q_i^{\text{off}}}{Q_{i,\text{true}}^{\text{off}}}\right) - \ln\left(1 + \frac{\Delta Q_i^{\text{on}}}{Q_{i,\text{true}}^{\text{on}}}\right) \quad (8)$$

In the case where the noise follows a normal distribution and has low amplitude (i.e. $\Delta Q_i^{\text{on/off}} / Q_{\text{true}}^{\text{on/off}} \ll 1$), the bias term can be approximated as:

$$\delta_i - \delta_{i,\text{true}} = \frac{1}{4}\left(\frac{1}{(SNR_i^{\text{on}})^2} - \frac{1}{(SNR_i^{\text{off}})^2}\right) \quad (9)$$

Where $SNR^{\text{on/off}}$ are the on-line and off-line signal to noise ratios. Providing high enough SNRs the bias produced by the measurement noise can be estimated.

Bias from shot-by-shot geophysical variations:

Variations of surface reflectivity impacts the AVS. Indeed, this scheme is more sensitive to targets with high reflectivity (i.e. high value of Q_i^{off}) as shown by the following equation:

$$\delta^{avs} = -\frac{1}{2} \cdot \ln\left(\frac{\langle Q^{\text{on}} \rangle}{\langle Q^{\text{off}} \rangle}\right) = -\frac{1}{2} \cdot \ln\left(\sum_{i=1}^{N} \frac{Q_i^{\text{off}}}{N \cdot \langle Q^{\text{off}} \rangle} \cdot e^{-2 \cdot \delta_i}\right) \quad (10)$$

This sum is the weighted average of the shot-by-shot transmissions weighted by the off-signal variations (i.e. proxy of reflectivity variations):

$$w_i = \frac{Q_i^{\text{off}}}{\sum_{k=1}^{N} Q_k^{\text{off}}} \quad (11)$$

So low reflectivity regions contribute less to the average.

Then, when there is a variation of geophysical quantities (e.g. topographic or surface pressure variations) there is a discrepancy between the average DAOD and the single-shot DAOD:

$$\Delta\delta_i = \delta_{i,\text{true}} - \langle \delta \rangle \quad (12)$$

We can derive the bias from equation (10) as:

$$\delta^{avs} - \langle \delta \rangle = -\frac{1}{2} \cdot \ln\left(\sum_{i=1}^{N} w_i \cdot e^{-2 \cdot \Delta\delta_i}\right) \quad (13)$$

Bias when averaging concentrations:

Directly averaging XCH_4 (eq. (4)) is not equivalent to averaging DAOD and IWF separately (eq. (5)). In the first case, we average concentrations and not molecules. So, there is the same contribution from columns with many molecules as from ones with less molecules.

The difference between the two approaches is described by:

$$\overline{X_{CH_4}}^{avd} - \overline{X_{CH_4}}^{avx} = \frac{1}{N}\sum_{i=1}^{N}\frac{\Delta IWF_i}{\langle IWF \rangle} \cdot X_{CH_4,i} \qquad (14)$$

As presented above, several sources of bias appear when computing the average mixing ratio of CH_4. The following section exposes the algorithms that correct for those biases in a practical way.

2.4 Correction algorithms

- For AVX, we need to correct the following sources of bias: bias induced by noise on $Q^{on/off}$ (eq. (9)) and averaging of concentration instead of molecules (eq. (14)).
- For AVD and averaging of IWF, we only need to correct the bias induced by the noise on $Q^{on/off}$ (eq. (9)).
- For AVS and averaging of IWF, several corrections need to be implemented. As AVS is more sensitive to the high reflectance area, the averaging of IWF must be done with consistent weights (eq. (11)):

$$\langle IWF \rangle_w = \sum_{i=0}^{N} w_i \cdot IWF_i \qquad (15)$$

Furthermore the DAOD from averaging signals is biased according to equation (13). It is possible to evaluate this bias by considering as a first order approximation the DAOD computed from average signals pondered by the relative fluctuations of IWF:

$$\Delta\delta_i \approx \delta^{avs} \cdot \frac{\Delta IWF_i}{\langle IWF \rangle_w} \qquad (16)$$

This process could be turned into an iterative correction (beyond the scope of this paper).

Then the last correction to take into account is the bias induced by the equivalent noise on the averaged signals. The resulting SNR can be computed by averaging signals and taking the quadratic mean of the standard deviations in order to apply equation (9).

2.5 Simulation description

The aim of the simulation is to estimate the biases of estimated XCH_4. A global description of the simulation is presented on Figure 1.

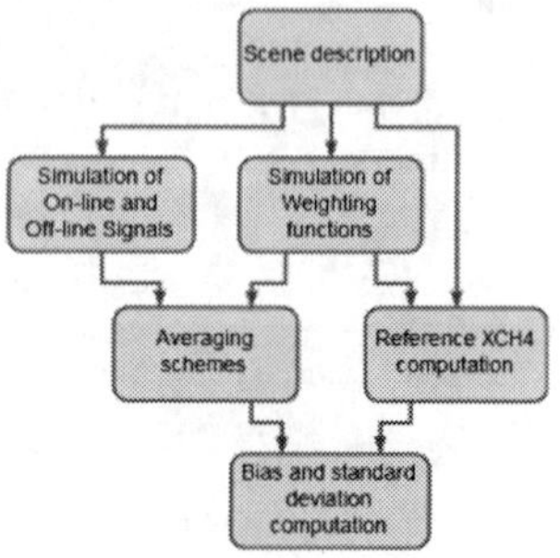

Figure 1 Global description of the simulation

Each simulation considers a typical number of 150 pulses per averaging window, approximately corresponding to 50km. It relies on a description of the CH_4 field, surface pressure and reflectivity description. Then, on-line and off-line signals are computed from surface reflectivity and noise simulation. The weighting functions for each pulse are computed using CH_4 absorption cross-sections from the 4A radiative transfer model [5] and are integrated vertically. Next, we proceed to the computation of averaged XCH_4 on 50km with the different schemes described in section 2.2 and the correction algorithms described in section 2.4.

The number of random samples for noise simulation has to be considerable in order to compute the biases with sufficient accuracy. The simulation described above is repeated 300.000 times to reach an accuracy on the bias of approximately 0.1 ppb for the nominal case (reflectivity of 0.1). This simulation has been tested on realistic scenes.

2.6 Realistic scenes

The averaging schemes have been tested on three realistic scenes located in France (Toulouse, Millau and Chamonix). The physical data and their sources are: Reflectivity from SPOT5, SRTM topography and Interpolated surface pressure from ECMWF.

 https://doi.org/10.1051/epjconf/201817602020

The scenes have been chosen to be representative of the main geophysical variations that can produce bias on XCH_4 estimation. Toulouse, Millau and Chamonix respectively present medium, high and extreme altitude variations (cf. Figure 2) and the same trend for reflectivity variations. The mean reflectivity can be tuned.

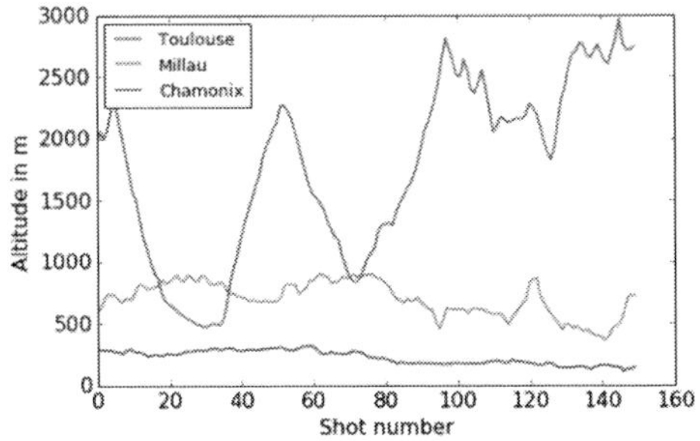

Figure 2 Altitude for the 3 studied scenes

3 RESULTS

The results of the simulation are shown in Figure 3. The averaging scheme that is globally the least biased is AVS. For all reflectivity, the bias stays limited to less than 0.7 ppb. The main cause of bias for this scheme is topographic or meteorological variation.

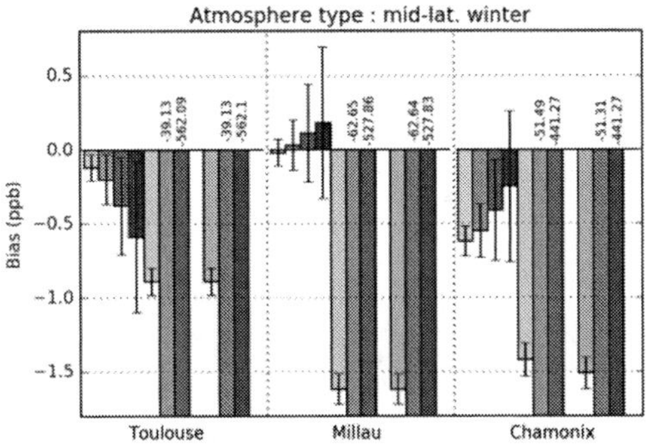

Figure 3 Residual Biases on a 50 km window for the three scenes (Toulouse, Millau, Chamonix) - for the three averaging schemes (blue shades for AVS, green shades for AVD and red shades for AVX) - for four different mean reflectivity (from light to dark colors decreasing reflectivity - 0.1 vegetation ; 0.05 mixed water/vegetation ; 0.025 sea/ocean ; 0.016 ice/snow) - The biases are displayed over the bars that end outside the frame.

AVD and AVX schemes are more affected by the ground reflectivity. The correction given by equation (9) is efficient when reflectivity is high. However, for a lower reflectivity the correction is no longer valid because it comes from a Taylor expansion assuming low amplitude of the noise.

4 CONCLUSIONS

This study shows that it is possible to limit the bias induced by the averaging scheme when an appropriate bias correction is implemented. The AVS scheme is less biased than other schemes in all simulated cases and below the target of 1 ppb allocated for this type of error. An alternative solution would be to take advantage of the two schemes: averaging signals on smaller windows where there is less geophysical variation until we reach a higher SNR that is compliant with the approximation of SNR correction (eq. (9)). Then averaging the resulting columns together on the 50 km window. Current development of this study tends to show that this combined scheme is not better than AVS results. Further studies will check the impact of an imprecise estimation of SNR on the bias correction (eq. (9)).

ACKNOWLEDGEMENTS

The authors wish to express special thanks to Fabien Marnas (Capgemini Technology Services for CNES) for the provision of the interpolated meteorological data used in this study.

References

[1] IPCC, 2013: *Climate Change 2013: The Physical Science Basis. Contribution of Working Group I to the Fifth Assessment Report of the Intergovernmental Panel on Climate Change*. Cambridge University Press, Cambridge, United Kingdom and New York, NY, USA, 1535 pp.

[2] P. Bousquet et al.: Contribution of anthropogenic and natural sources to atmospheric methane variability. Nature 443, 439-443 Available from: 10.1038/nature05132

[3] M. Bode et al., 2014: MERLIN: an integrated path differential absorption (IPDA) lidar for global methane remote sensing, *Proc. of ICSO* 2014, Tenerife, Canary Islands.

[4] C. Pierangelo et al., 2016: MERLIN (methane remote sensing lidar mission): an overview, *Proc. of 27th ILRC*, New York City, USA.

https://doi.org/10.1051/epjconf/201817602020

[5] Scott N.A. et al., 1981: A fast line-by-line method for atmospheric absorption computations: The Automatized Atmospheric Absorption Atlas. J. Appl. Meteor., 20,802-812.

EPJ Web of Conferences **176**, 02021 (2018) https://doi.org/10.1051/epjconf/201817602021
ILRC 28

RETRIEVAL OF ATMOSPHERIC BACKSCATTER AND EXTINCTION PROFILES WITH THE ALADIN AIRBORNE DEMONSTRATOR (A2D)

Alexander Geiss[1,*], Uwe Marksteiner[2], Oliver Lux[2], Christian Lemmerz[2], Oliver Reitebuch[2], Thomas Kanitz[1] and Anne Grete Straume-Lindner[1]

[1]*European Space Agency (ESA-ESTEC), 2200 AG Noordwijk, The Netherlands,*
**alexander.geiss@esa.int*

[2]*German Aerospace Center (DLR), Institute of Atmospheric Physics, Oberpfaffenhofen, 82234 Wessling, Germany*

ABSTRACT

By the end of 2017, the European Space Agency (ESA) will launch the Atmospheric laser Doppler instrument (ALADIN), a direct detection Doppler wind lidar operating at 355 nm. An important tool for the validation and optimization of ALADIN's hardware and data processors for wind retrievals with real atmospheric signals is the ALADIN airborne demonstrator A2D. In order to be able to validate and test aerosol retrieval algorithms from ALADIN, an algorithm for the retrieval of atmospheric backscatter and extinction profiles from A2D is necessary.

The A2D is utilizing a direct detection scheme by using a dual Fabry-Pérot interferometer to measure molecular Rayleigh signals and a Fizeau interferometer to measure aerosol Mie returns. Signals are captured by accumulation charge coupled devices (ACCD). These specifications make different steps in the signal preprocessing necessary.

In this paper, the required steps to retrieve aerosol optical products, i. e. particle backscatter coefficient β_p, particle extinction coefficient α_p and lidar ratio S_p from A2D raw signals are described.

1 INTRODUCTION

The Atmospheric Dynamics Mission Aeolus (ADM-Aeolus) [1, 2] is ESA's second Earth Explorer core mission, with a date of launch by the end of 2017. The direct detection Doppler wind lidar ALADIN carried by ADM-Aeolus emits laser pulses at an ultraviolet wavelength of 355 nm. The mission's first objective is to obtain global observations of horizontally projected line-of-sight wind profiles with regard to testing the capability for future operational wind satellites and the aim to improve numerical weather prediction [3] and climate studies. However, as being the first high spectral resolution lidar (HSRL) in space, ALADIN allows also an accurate vertical profiling of aerosol and cloud optical properties on a global scale as a secondary product. An important tool for the development, testing, validation and optimization of the hardware and data processors of ALADIN, is the ALADIN airborne demonstrator A2D [4, 5]. With the same fundamental concepts and design but slightly different specifications, it can be used to validate and demonstrate the capabilities of ALADIN. The German aerospace centre operated A2D from ground and from an airborne platform during campaigns in 2006, 2007, 2009, 2015 and 2016. Valuable information about the instrument characteristic, calibration and data processing were gained with realistic atmospheric signals and provided important recommendations for ALADIN. However, with focusing on the retrieval of wind profiles, no aerosol optical properties were retrieved with atmospheric measurements. In order to validate and to optimize the aerosol retrieval for ALADIN [6] by using atmospheric measurements, a retrieval of particle backscatter and extinction profiles for A2D must be developed.

https://doi.org/10.1051/epjconf/201817602021

Table 1: Key parameters of A2D.

Platform	Airborne or ground
LOS pointing	20°
Laser	Nd:YAG
Wavelength	354.89 nm
Laser line width	45 Mhz
Pump system	diode
Pulse repetition rate	50 Hz
Pulse duration	25 ns
Pulse energy	50–60 mJ
Laser divergence	90 µrad for $\pm 3\sigma$
Laser beam diameter	0.016 m
Telescope diameter	0.2 m
Configuration	Coaxial
Field of View	100 µrad
Mie receiver	Fizeau
Rayleigh receiver	Fabry-Pérots
ACCD detector	
Imaging zone	16 x 16 pixel
Memory zone	25 x 16 pixel
Altitude range	0–10 km
Vertical resolution	296–2370 m
Min. horizontal resolution	80 m

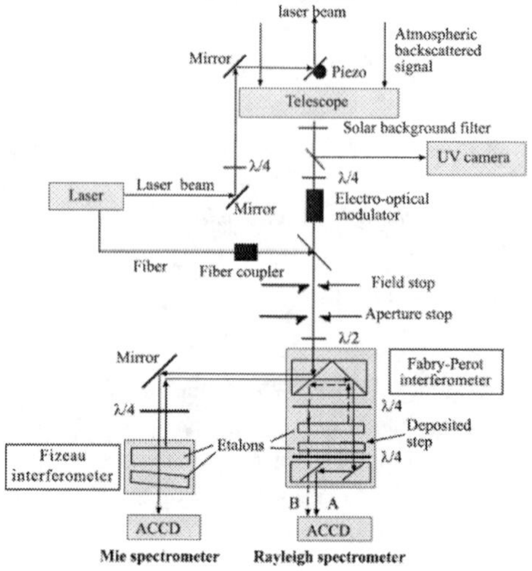

Figure 1: Illustration of the optical configuration of A2D [4].

2 METHODOLOGY

2.1 A2D — principle of operation

The A2D is based on a diode-pumped frequency-tripled Nd:YAG laser emitting pulses with an energy of 50–60 mJ with a repetition frequency of 50 Hz (see Table 1). The outgoing laser light is circularly polarized and the laser beam is steered via a piezo-electrically controlled mirror. Its position is tracked with an UV-camera and a CCD (see Fig. 1). A telescope with 0.2 m diameter collects the backscattered light. The structural design with a secondary mirror and the additional use of an electro-optical modulator to suppress the near-field signal results in a region of incomplete overlap up to a distance of 2000 m from the lidar [5]. Although receiving cross- and parallel-polarized backscattered light only the latter is measured and conducted towards a combination of a Fizeau interferometer to detect the narrowband aerosol return and a dual Fabry-Pérot interferometer to measure the broadband molecular signals. ACCDs are used to capture the signals resulting in 25 available range bins from which 20 are atmospheric range bins. As a consequence, the vertical resolution is limited to 296–2370 m depending on the integration length of the signal. With a measurement summed up over 18 laser pulses and a flight speed of 200 m/s, a minimum horizontal resolution of 80 m is achieved.

2.2 Signal pre-processing

Before an aerosol retrieval can be applied to A2D data, different signal corrections are necessary to obtain pure Rayleigh and Mie signals which can be further processed. At first, a so-called detection chain offset and the background signal must be subtracted from each range bin for both channels. The variable range bin length requires a scaling of the signal intensities of each layer depending on their integration length. The next and fundamental step is a cross-talk correction caused by an incomplete decoupling of both receiver channels. As a consequence, the Mie channel collects Rayleigh

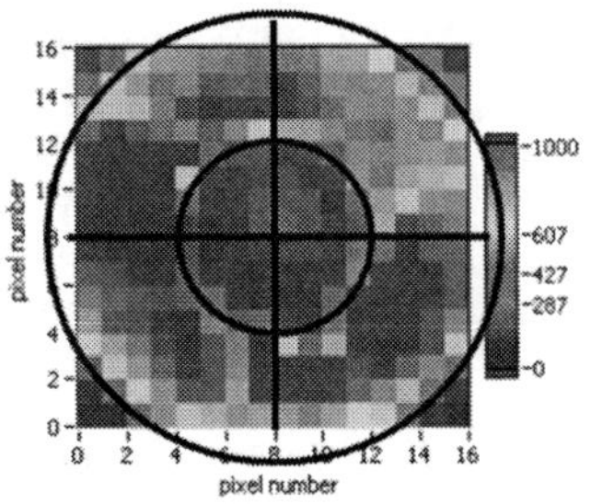

Figure 2: Rayleigh background on the Mie channel in imaging mode. Blue colors in the center are related to the secondary mirror obscuration. Auxiliary lines are marked in black. Courtesy O. Reitebuch (DLR).

signals and the Rayleigh channel is influenced by Mie returns. In the first case, due to the imaging characteristic of the Mie spectrometer (MSP), the optical beam path from the telescope to the MSP is imaged on the Mie ACCD. Thus, an obscuration in the beam path from the secondary mirror and its spider is present on the Mie ACCD as shown in Fig. 2 and requires a so called telescope obscuration (TOBS) correction. For ALADIN the obscuration for every single pixel is derived from an optical model. A constant factor for each pixel is used for the correction. For the A2D, however, a constant TOBS correction is not possible since the illumination of the telescope from the atmospheric signal and the optical path through the A2D is varying. An A2D approach is based on shifting the Mie signal fringe out of the useful spectral range of the Mie receiver (MOUSR) by changing the laser frequency. The remaining telescope image can be monitored and subsequently subtracted from the measurements. The cross-talk contribution from the Mie signal to the Rayleigh channel is corrected according to a method described for ALADIN which is based on using the Mie channel signals [7]. Further corrections for obtaining correct overall Rayleigh intensities must consider the wind induced shift of the Rayleigh-Doppler backscatter spectrum as well as the temperature induced Rayleigh line broadening.

In order to increase the signal to noise ratio, temporal averaging of the pre-processed signal is needed. Because of the coarse resolution of the A2D, a feature mask is required to avoid averaging over different types of particles (e. g. clouds, elevated aerosol, planetary boundary layer). The reliable determination of the ground return for every measured profile is realized by using a digital elevation model (DEM) based on data from the Shuttle Radar Topography Mission (SRTM) [8] with 3 arcsec resolution. To distinguish between different features, i. e. clouds and aerosol, a segmentation algorithm is combined with threshold values in the signal and gradient of the signal.

2.3 Aerosol optical products from A2D

With A2D being a single wavelength high spectral resolution lidar, the independent retrieval of profiles of particle backscatter coefficients β_p, particle extinction coefficients α_p and lidar ratios S_p is possible if an HSRL-retrieval is applied [11, 12]. In case of using only the Mie channel, the Klett-Fernald algorithm [9, 10] allows a determination of β_p only. With data from different flight campaigns being available, especially cases with blown snow at the East coast of Greenland and two flights in the Mediterranean region provide sufficient aerosol backscatter for an aerosol retrieval. Preliminary results of the Mie signal after applying the TOBS correction are shown in Fig. 3 for a flight along the Greenland east coast.

2.4 Validation

In order to verify the accuracy of retrieved aerosol optical products from A2D, a validation with independent measurements of co-located lidar systems is necessary. To quantify the influence of detecting only parallel-polarized light on the aerosol retrieval, a comparison with a depolarization lidar is needed. Particularly, in

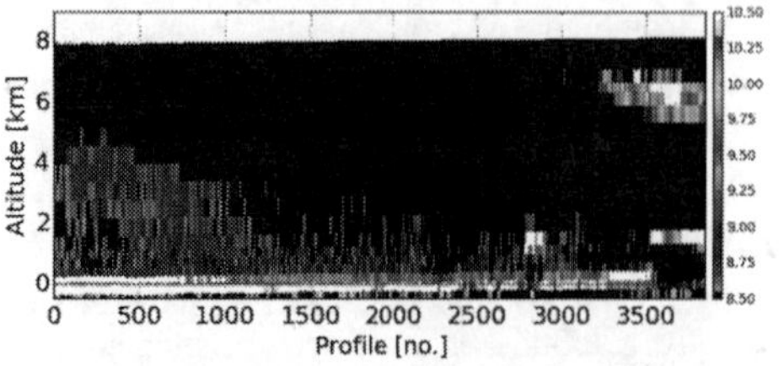

Figure 3: Time height cross section of the A2D Mie signal after TOBS correction (logarithmic scale, a.u.) during flight on 19 May 2015. White colors indicate clouds and violet colors indicate aerosol. The ground is marked as red line.

case of particles with large depolarization ratios (e. g. desert dust, volcanic ash) the determined β_p from A2D can be considerably smaller than the total particle backscatter coefficient. To test the reliability of the feature mask and results from the Klett-Fernald retrieval, data from calibrated ceilometer networks [13] and additional information from e. g. sun photometer measurements can be used.

3 CONCLUSIONS

With A2D having the same fundamental characteristics as ALADIN it is the ideal instrument to perform calibration and validation activities [14] before and after launch of ADM-Aeolus. However, additional steps for a proper aerosol retrieval applicable to A2D and its principle of operation are necessary in order to be able to validate and optimize aerosol products from ALADIN.

References

[1] A. Stoffelen et al., 2005: The Atmospheric Dynamics Mission for global wind field measurements, *Bull. Am. Meteor. Soc.* **86**, 73–87.

[2] European Space Agency, 1999: Atmospheric Dynamics Mission, Report for Mission Selection, *ESA-SP* **1233(4)**, ESTEC, Noordwijk, The Netherlands.

[3] O. Reitebuch, M. Weissmann, 2008: Impact of Airborne and Future Spaceborne Wind Lidar Observations of ADM-Aeolus on Weather Prediction Skills, *24th Int. Laser and Radar Conference (ILRC 2008), Boulder, Colorado*, 23–27.

[4] O. Reitebuch et al., 2009: The Airborne Demonstrator for the Direct-Detection Doppler Wind Lidar ALADIN on ADM-Aeolus. Part I:Instrument Design and Comparison to Satellite Instrument, *J. Atmos. Oceanic Tech.* **26** (12), 2516–2530.

[5] U. Paffrath et al., 2008: The Airborne Demonstrator for the Direct-Detection Doppler Wind Lidar ALADIN on ADM-Aeolus. Part II: Simulations and Rayleigh Receiver Radiometric Performance, *J. Atmos. Oceanic Tech.* **26** (12), 2501–2515.

[6] P. Flamant et al., 2008: ADM-Aeolus retrieval algorithms for aerosol and cloud products, *Tellus* **60A** (2), 273–288.

[7] A. Ansmann et al., 2007: Particle backscatter and extinction profiling with the spaceborne high-spectral-resolution Doppler lidar ALADIN: methodology and simulations, *Applied Optics* **46** (26), 6606–6622.

[8] T. G. Farr et al., 2007: The Shuttle Radar Topography Mission, *Rev. Geophys.* **45**, RG2004.

[9] E. W. Eloranta, 2005: High spectral resolution lidar, *LIDAR, C. Weitkamp* **Springer, 2005** 143–163.

[10] A. Ansmann and D. Müller, 2005: Lidar and atmospheric aerosol particles, *LIDAR, C. Weitkamp* **Springer, 2005** 104–141.

[11] J. D. Klett, 1981: Stable analytical inversion solution for processing lidar returns, *Appl. Opt.* **20** 211–220.

[12] F. G. Fernald, 1984: Analysis of atmospheric lidar observations: some comments *Appl. Opt.* **23** 652–653.

[13] M. Wiegner et al., 2014: What is the benefit of ceilometers for aerosol remote sensing? An answer from EARLINET *Atmos. Meas. Tech.* **7** 1979–2014.

[14] A. G. Straume et al., 2016: ESA's spaceborne lidar mission ADM-Aeolus; recent achievements and preparations for launch, *The 27th International Laser Radar Conference (ILRC 27), New York City, USA*, EPJ Web of Conferences, 119 (2016) 01001.

EPJ Web of Conferences **176**, 02022 (2018) https://doi.org/10.1051/epjconf/201817602022
ILRC 28

EARTHCARE ATLID EXTINCTION AND BACKSCATTER RETRIEVAL ALGORITHMS

David Donovan[1], Gerd-Jan van Zadelhoff[1], David Daou[1], Ulla Wandinger[2]

[1]Royal Netherlands Meteorological Institute (KNMI), De Bilt, The Netherlands,
**Email: donovan@knmi.nl*
[2]Leibniz Institute for Tropospheric Research (TROPOS), Leipzig, Germany

ABSTRACT

ATLID stands for "ATmospheric LIDar" and is the lidar to be flown on the Earth Clouds and Radiation Explorer (EarthCARE) platform in early 2019. ATLID is a High-Spectral Resolution (HSRL) system operating at 355nm. This presentation will introduce the ATLID level-2 retrieval algorithms being implemented in order to derive cloud and aerosol optical properties

1 INTRODUCTION

EarthCARE[1] is an upcoming ESA/JAXA mission expected to fly in 2019. The EarthCARE instruments are:

- A 94 GHz, Doppler Cloud Radar supplied by Japan (CPR)
- An advanced 355 nm High-Spectral Resolution Lidar (ATLID)
- A Multispectral Cloud/Aerosol imager for narrow-band TOA radiances (MSI)
- A 3-view Broad-Band Long- and Short-Wave Radiometer for TOA radiance (BBR)

EarthCARE is a cloud-aerosol-precipitation-radiation process oriented mission for which a comprehensive suite of retrieval algorithms is being developed. As outlined in [1], both (primarily) single-instrument (L2a) and multi-instrument algorithms are being developed. The results of the algorithm chain will be used in an ongoing radiative closure assessment in which the derived aerosol and cloud properties will be used as input to broad-band radiative transfer models which will, in turn, generate TOA flux and radiance estimates which will be compared to the actual BBR measurement. The assessment results will be used to assess the retrieval products, guide scientific and technical improvements and lead to a high-quality set of data suitable for process studies.

ATLID is a polarized three channel lidar operating at 355nm. The lidar delivers profiles of the parallel, so-called, Mie (or particulate) attenuated backscatter, the parallel Rayleigh (molecular) attenuated backscatter and the corresponding total (Mie+Rayleigh) cross-polarized return.

The development efforts are spread over a number of cooperating development teams. In this paper we will describe the ATLID L2a algorithms responsible for target detection, aerosol and cloud optical property retrievals and target classification. Examples of other related developments involving ATLID and MSI can be found in [2].

2 Algorithm overviews

The L2a ATLID algorithms/procedures described here are:

-The Featuremask (A-FM). This algorithm uses a combination of image processing inspired techniques in order to, respectively, detect regions of clouds/aerosols, surface returns, clear air or regions where the lidar return is completely attenuated. At this stage, the valid targets are not classified into different type (e.g. water, ice). This is accomplished in subsequent processing steps.

-The aerosol oriented extinction and backscatter retrieval routine (A-AER). This procedure uses relatively conventional HRSL retrieval methods for determining extinction and backscatter at the 50km+ horizontal scale (e.g. deriving the extinction based on the log derivative of the Rayleigh signal). In order to do this the lidar signals must be appropriately averaged to achieve a target SNR. The averaging mask is based (mainly) on the A-FM produce which is used to avoid averaging intervals which would mix "strong" and "weak" features together which would lead to invalid retrievals.

-The cloud and aerosol extinction, backscatter and depolarization procedure (A-EBD). This routine retrieves the aerosol and cloud extinction and backscatter at the 1-km horizontal scale. At this scale the SNR of the molecular scattering channel return is too low to enable the techniques employed by the A-AER approach. Instead, the method relies on finding the optimal profile of the extinction-to-backscatter ratio (S) that allows one to invert the total lidar signal to produce an extinction profile consistent with the observed Rayleigh channel signal. Multiple-scattering (MS) effects which are important for accurate cloud (and in some cases aerosol) retrievals are included in the procedure which is cast in an optimal-estimation framework [3]. As a-priori information, the S estimates produced by A-AER are used.

-The ATLID Target classification procedure (A-TC). This procedure uses extinction, backscatter and depol. ratio, as well as auxiliary inputs such as temperature and geographic location in order to classify targets into classes such as water or ice cloud or aerosol type. The aerosol typing scheme is based primarily on using the S and depol. ratio to assign a class to the aerosol [4]. The cloud phase determination scheme uses backscatter and depolarization in a manner similar to that employed for the CALIOP retrievals [5].

3 Algorithm Development and Testing

The algorithm development process is being aided by simulated data produced using the EarthCARE Simulator (ECSIM) [1]. One such `scene' has been built using ECMWF MACC data [6] corresponding to a simulated orbit track from sub-Sahara Africa to southern Europe on 2015/01/13, a distance corresponding to a single EarthCARE processing fame (6200 km).

Fig. 1 shows the 355 nm nadir extinction field derived from the MACC data. The corresponding observed CALIOP attenuated backscatter exhibits strong correlation in terms of aerosol and cloud features indicating that the MACC forecast for this time period exhibited a high degree of realism.

The MACC output is at a low resolution (0.5x0.5 Deg.). The cloud and aerosol fields were interpolated to a resolution of 500mx100m. Structure was added by perturbing the mass fields using suitably correlated pseudo-random deviates.

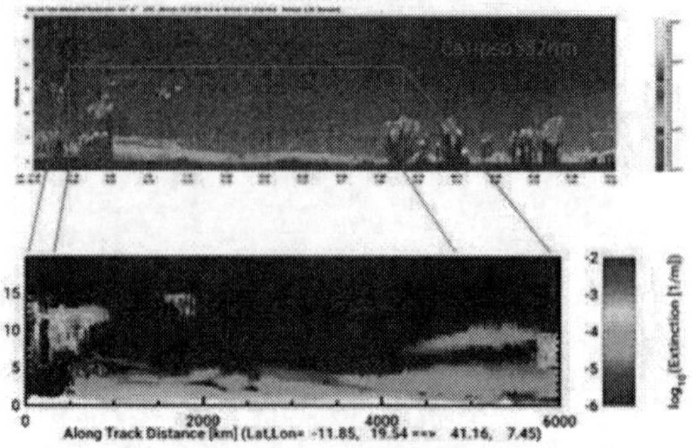

Figure 1: (Bottom Panel) MACC ECSIM scene derivied extinction field. Ice and water clouds as well as aerosol regions (including extensive desert dust) are visable. (Top-Panel) the corresponding observed CALIOP 532 attenuated backscatter image is shown.

3.1 Example Simulations and Retrievals

Example simulated ATLID observations are shown in Fig. 2. Here the observations are plotted with an effective horizontal resolution of 1-km and a vertical resolution of 100m. ECSIM is capable of performing 3-D Monte-Carlo lidar radiative transfer calculations. However, in this case the fast semi-analytical model described in [7] with a physically based parameterization of the depolarized return due to MS in water clouds, was used.

The scene contains regions of ice cloud, water cloud and various types of aerosols. Realistic ice crystal phase functions were used with an adjusted backscatter peak following [8]. The aerosol regions contain mixtures of 4 distinct basic types following [4], namely, Coarse-Dust, Coarse-Marine, Fine-Weakly-absorbing, and Fine-Strongly-absorbing. Mie phase functions were used in the case of the later 3 types while T-matrix spheroids have been used to model the dust aerosols (which are mainly present between 10 and 25 Deg N).

Noise levels corresponding to the expected instrument performance characteristics and background light levels have been included. Cross-talk between the Mie and Ray channels for ATLID is significant and the effect of the cross-talk on the SNR ratio of the cross-talk corrected attenuated backscatters has been accounted for.

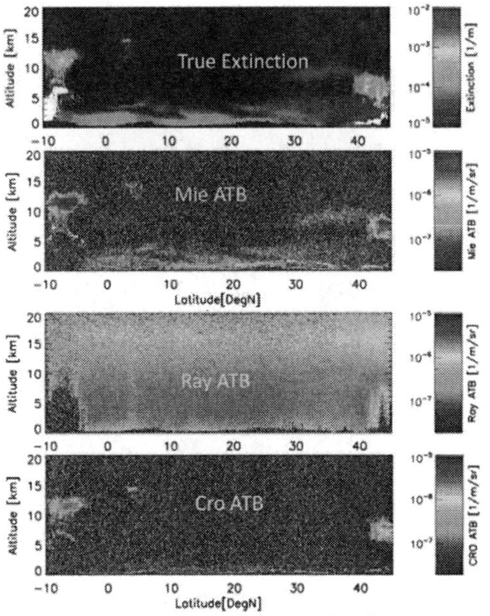

Figure 2: (From Top to bottom) The ECSIM scene extinction and the corresponding ATLID simulated Mie (Particulate), Rayleigh (molecular), and cross polarized attenuated backscatter fields.

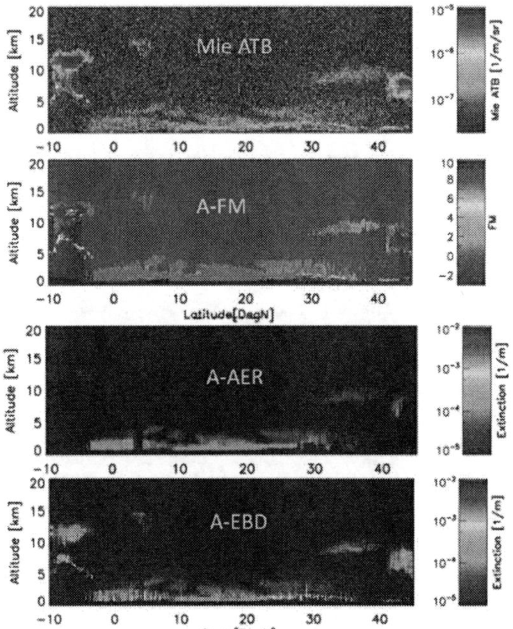

Figure 3: Featuremask, A-AER and A-EBD retrieval results.

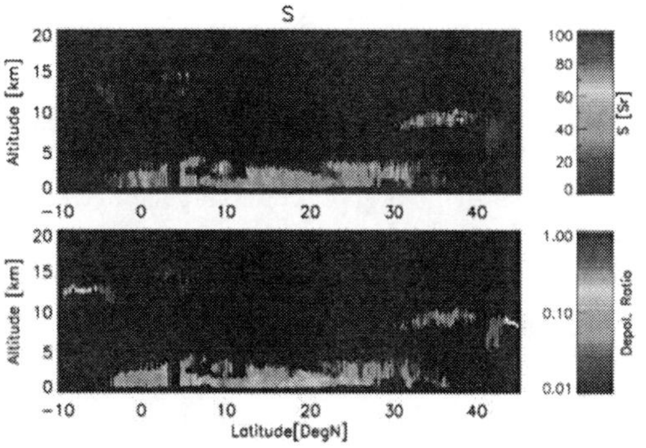

Figure 4: Retrieved A-AER lidar ratio (S) and depolarization ratio.

Sample retrieval results for the Feature mask (A-FM) as well as A-AER extinction and A-EBD extinction are presented in Fig 3. Here it can be seen that the A-FM output (an integer field where higher values correspond to higher detection confidence, negative values correspond to surface returns (-2) and attenuated regions (-1)) does an impressive job of detecting features at high resolution. However, some deficiencies are also noticeable. Namely, there are some issues near the surface in areas of variable surface-height and between about 30 and 40 N between 0 and 5 km in an attenuated region with very low extinctions.

The A-AER and A-EBD extinction fields are also presented in Fig. 3. The A-AER algorithm distinguishes between weak and strong features and horizontally averages the ATB fields avoiding the strong features. Areas below 'thick enough' ice clouds are also avoided due to the possibility of multiple-scattering induced tails [3] extending below the cloud into the aerosol regions. This strategy is reflected in the results which mainly correspond to thin ice cloud regions and aerosol fields. The corresponding higher resolution A-EBD extinction field is seen to be more complete (i.e. Extinction is provided for both cloud and aerosol regions) however, in the case of aerosols, it is considerably noisier than the A-AER product.

The corresponding A-AER extinction-to-backscatter ratio (S) retrievals as well as the depolarization ratio are shown in Fig. 4. Here it can be seen that even though the depolarized signals were not so distinct in the high resolution Cross-ATB presented in Fig. 2., that with appropriate smoothing a usable depolarization ratio can be estimated.

Fig. 5 depicts the aerosol classification for a height of 400 meters above surface based on the data shown in Fig. 4. Here, in general, the aerosol type is reasonably well retrieved.

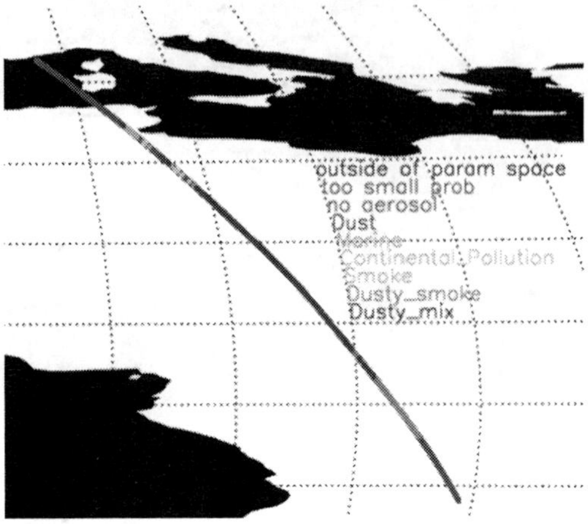

Figure 5: Aerosol type at an altitude of 400 m derived using the S and depolarization data presented in Fig. 4. Here the dark areas correspond to ocean/sea areas. The "no aerosol" regions correspond to areas where high thick clouds prevent lower level aerosols from being detected.

4 CONCLUSIONS

The lidar algorithm development for EarthCARE is well underway. Full-frame data sets have only just recently been generated but already, the use of realistic ECSIM simulations is proving to be an effective means to develop the scientific and technical aspects of the algorithms. This is of particular importance, since for ATLID, suitable proxy sources of actual data are not available (e.g. CALIOP observations cannot be used due to the change in wavelength, different instrument characteristics and the lack of HSRL capability).

In the coming year the algorithms and related codes will continue to be developed with the aim, that when EarthCARE is launched, useful well-understood data products will be able to be rapidly generated and made available to the scientific user community.

ACKNOWLEDGEMENTS

The work presented in this paper is performed under the "Atmospheric Products from Imager and Lidar (APRIL)" project funded by ESA (contract 4000112018/14/NL/CT).

References

[1] Illingworth, A., et al., 2014: THE EARTHCARE SATELLITE: The next step forward in global measurements of clouds, aerosols, precipitation and radiation. *Bull. Am. Met. Soc.* **96**, 1311–1332.

[2] Wandinger et al. 2017, EarthCARE Aerosol and Cloud Layer and Column Products, this conference.

[3] Donovan, D.P., The Expected Impact of Multiple Scattering on ATLID Signals. *EPJ Web of Conferences*, 119 (2016) 01006 doi:http://dx.doi.org/10.1051/epjconf/20161190 1006

[4] Wandinger et al., HETEAC: The Aerosol Classification Model for EarthCARE. *EPJ Web of Conferences*, 119 (2016) 01004 doi:http://dx.doi.org/10.1051/epjconf/20161190 1004

[5] Hu, Y., D. et al., 2009: CALIPSO/CALIOP Cloud Phase Discrimination Algorithm. J. Atmos. Oceanic Technol., 26, 2293–2309, doi: 10.1175/2009JTECHA1280.1.

[6] Morcrette, J.-J et al., Aerosol analysis and forecast in the European Centre for Medium Range Weather Forecasts Integrated Forecast System: Forward modeling. *J. Geophys. Res.*, 114, D06206, doi:10.1029/2008JD011235

[7] Hogan, R., 2008: Fast Lidar and Radar Multiple-Scattering Models. Part I: Small-Angle Scattering Using the Photon Variance–Covariance Method. *J. Atmos. Sci.,* 65, 3621–3635, doi: 10.1175/2008JAS2642.1

[8] Zhou C., and P. Yang, 2015: Backscattering peak of ice cloud particles, Opt. Express, 23, DOI:10.1364/OE.23.011995

VALIDATION OF OPTIMAL ESTIMATION METHOD RETRIEVALS OF MIDDLE ATMOSPHERIC TEMPERATURE

Ali Jalali[1,*], R. J. Sica[1,2] and Alexander Haefele[2,1]

[1]*Department of Physics and Astronomy: University of Western Ontario, Canada,*
**ajalali3@uwo.ca*
[2]*Federal Office of Meteorology and Climatology: MeteoSwiss, Switzerland*

Abstract

OEM (Optimal Estimation Method) retrievals of temperature from lidar measurements are robust and practical (Sica and Haefele, 2015). They offer significant improvements over traditional methods. We will show a climatology of +360 nights of measurements from the Purple Crow Lidar and the improvements offered using an OEM, including the quantitative determination of the top altitude of the retrieval and the evaluation of the various systematic and random uncertainties due to measurement noise.

1 Introduction

Rayleigh scatter measurements from the University of Western Ontario Purple Crow Lidar (PCL) have been used to develop temperature climatologies for the mesosphere and lower thermosphere. Temperature retrievals from Rayleigh-scattering lidar measurements have been performed using the traditional method of Hauchecorne and Chanin [1] since 1980 and the Optimal Estimation Method (OEM) of Sica and Haefele [2] since 2015. The traditional method has some limitations including the removal of the top 10-15 km from the uppermost range of observation, inducing extra uncertainty when merging temperature profiles from different channels in the common range and providing only random uncertainty. An OEM technique was used to overcome these deficiencies and address a solution for each of them. This research investigates the validity of the OEM in comparison with the traditional method.

2 Temperature retrieval methods

The conventional method integrates temperatures downward, and requires the assumption of a "seed" pressure at the highest altitude, taken from a model. The starting point is at the highest altitude of the PCL Rayleigh channel, and the corresponding pressure is obtained from the COSPAR International Reference Atmosphere (CIRA) model. Geophysical variation in the lower thermosphere is sufficiently large to cause temperature retrievals to be unreliable for the top one or two scale heights. Uncertainties due to this pressure assumption make it prudent, without ancillary measurements, to remove the top two scale heights of temperatures from each profile. Unlike the other methods, which include estimates of random uncertainty, the OEM produces a complete uncertainty budget for all parameters in the retrieval procedure. The OEM consists of two elements: a forward model that describes the lidar measurements in the atmosphere, and the *a priori* profiles with corresponding error covariance matrices that constrains solutions to physically meaningful states. The lidar equation and the hydrostatic equilibrium assumption are used as a forward model [2] and the CIRA-86 model as an *a priori*.

The optimum estimate of the state vector, $\hat{x}$, is found by minimizing equation (1), which corresponds to choosing the most likely state from the solution probability density function [3], otherwise known as the cost function. Here $\hat{x}$ is the most probable state, $\mathbf{y}$ is the measurement vector, S_ε is the covariance of the system's state, x_a is the *a priori* vector, and S_a is the *a priori* covariance.

$$Cost = [\mathbf{y} - \mathbf{F}(\hat{\mathbf{x}}, \mathbf{b})]^T \mathbf{S}_{\varepsilon}{}^{-1} [\mathbf{y} - \mathbf{F}(\hat{\mathbf{x}}, \mathbf{b})] + [\hat{\mathbf{x}} - \mathbf{x_a}]^T \mathbf{S_a}{}^{-1} [\hat{\mathbf{x}} - \mathbf{x_a}] \quad (1)$$

The averaging kernel (**A**) represents the sensitivity of the retrieved state ($\hat{x}$) to the true state (x). Sum along the rows of averaging kernel matrix (**Au**, **u** is a vector with unit elements) equal the areas of the averaging kernel matrix at each level. This quantity is shown by **Au** from now on. **Au** represents the amount of information determined from the lidar measurements and how much is due to the *a priori* and is almost unity at accurate retrieval levels [3].

3 Methodology

Measurements from two PCL Rayleigh channels including Low Rayleigh (24-45 km) and High Rayleigh (35-110 km) over 360 nights of measurements from 1999 to 2013 were used to generate the temperature climatologies using the OEM and the traditional method. The OEM was used in order to calculate nightly temperature retrievals, and the temperature climatology procedure is based on the processes presented in Argall and Sica [4]. Three atmospheric models including the US Standard Atmosphere, CIRA and MSIS are considered as an *a priori* in the OEM. An *a priori* should not have a trend during time, in order to avoid the *a priori* inducing a trend in the temperature climatology. The MSIS-00 model's temperature shows a trend due to its complexity and therefore it was not used. The US Standard Atmosphere model provides just yearly temperature profiles which can be used as an *a priori*; however, CIRA-86 provides monthly temperature profiles with smaller temperature variances than those of the US Standard Atmosphere model. Therefore, CIRA-86 was chosen to provide the *a priori* covariance matrix as well as seed pressure at highest altitude for the traditional method. In the case of the traditional method, in order to cut off the temperature to achieve an accurate result, two scale heights (top 15 km) from the highest possible altitude was considered. In the OEM approach, a value of 0.99 for the **Au** was used as a reference for cutoff height instead removing the top 15 km from the temperature retrievals. Below the 0.99 **Au** cutoff, the *a priori* has less than one percent contribution to the temperature retrieval and below this value the results are valid. The retrieved temperatures from the PCL, Low and High Rayleigh channels in the overlap range are not identical using the traditional method. The temperatures should theoretically be identical in the overlapping regions; however, in reality they are frequently different. In the past, in order to create a single temperature profile from different lidar channel measurements, their profiles had to be merged. Therefore, it was necessary to merge two temperature profiles in order to have a temperature profile for the entire PCL profile range. This issue was investigated in Jalali et. al. [5] by using different merging functions in the traditional method. The process involved combining various random uncertainties, choosing overlap ranges, and calculating uncertainties when gluing raw photocounts profiles, all of which were intensive and challenging processes. The OEM does not require merging and provides an easy way to incorporate data from multiple sources, such as combining lidar measurements from different channels.

4 Results

Figures 1 and 2 show two sample PCL temperature profiles using the traditional and the OEM to retrieve temperature from PCL photocounts. In both figures, the altitude of the 15 km cutoff height from the top of profile in the case of the traditional method is close to the 0.99 value in the **Au** for the OEM. Approximately 91 km and 87 km are the 15 km cutoff height in Figures 1 and 2 using the traditional method, respectively. However, 91 km is the cutoff height for the OEM corresponding the value 0.99 for the **Au**. It is noteworthy that there is a large difference in the 15 km cutoff height for the traditional method and the **Au** cutoff height of 0.9 for the OEM case. This difference is about 8.8 km in Figure 1 and 13 km in Figure 2. It can be concluded that the OEM approach allows the use of higher altitudes with known uncertainty in comparison with the removal of the top 2 scale heights using the traditional method. This ability to achieve temperature retrievals at greater heights is one of the main advantages of using the OEM approach. Also, it is necessary to mention that in both figures (1 and 2), the OEM profiles begin at 30 km, while the traditional method's results start from 40 km. This is because the OEM uses both High and Low Rayleigh channel measurements to retrieve a temperature profile, unlike the traditional method which provides two different profiles for each channel and presents just the High Rayleigh channel. Figure 3 shows the difference between the composite years PCL temperature climatology using the OEM and the traditional method between 40 and 90 km. The temperature differences show that there is a good agreement between the two temperature retrieval methods between 40 to 85 km and there is almost no difference between them.

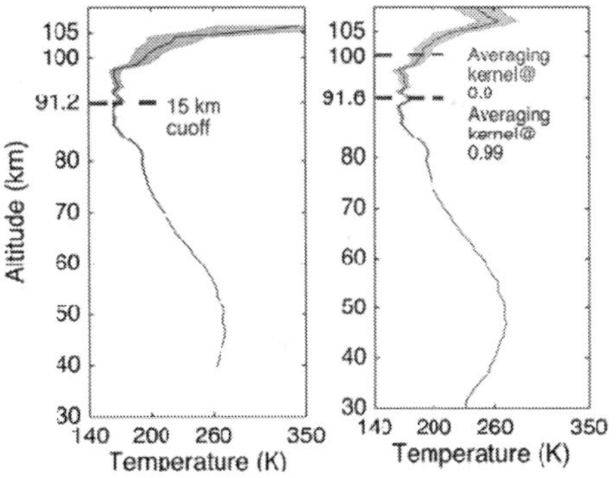

Figure 1: PCL temperature profile for 10 May 1999, integrated over 7.4 hours and with coadded altitude bins to 255 m using the traditional method (left) and the OEM (right). The shaded blue and red profiles are temperature random uncertainty corresponding the traditional method and the OEM respectively. The horizontal dashed lines are the height above which the temperature profile is less than 1%(0.99) and 10 %(0.9) dependent in the choice of an a priori (right panel) and the15 km cutoff height (left panel).

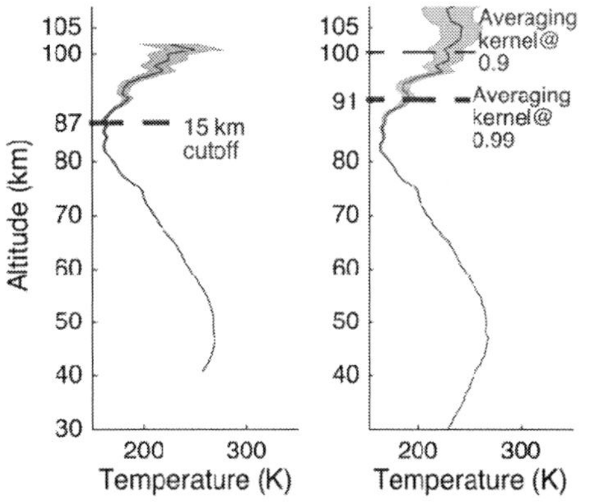

Figure 2: Same as Figure 1 for26 August 2012, integrated over 6.4 hours.

5 Conclusions

Sica and Haefele [2] used a new approach for retreving atmospheric temperature from lidar measurements as an improvement to the traditional method. The temperature climatologies were calculated for the PCL Rayleigh measurements in order to investigate and evaluate this new method. The OEM shows a significant improvement over the traditional method. The OEM provides a mathematical method to determine a cutoff height for temperature retrieval below which the entire temperature is valid. Another advantage of the OEM is the ability to provide one single profile where using measurements from two channels with merging of random uncertainty. The OEM provides a complete uncertainty budget on a per profile basis for all forward model parameters.

https://doi.org/10.1051/epjconf/201817603001

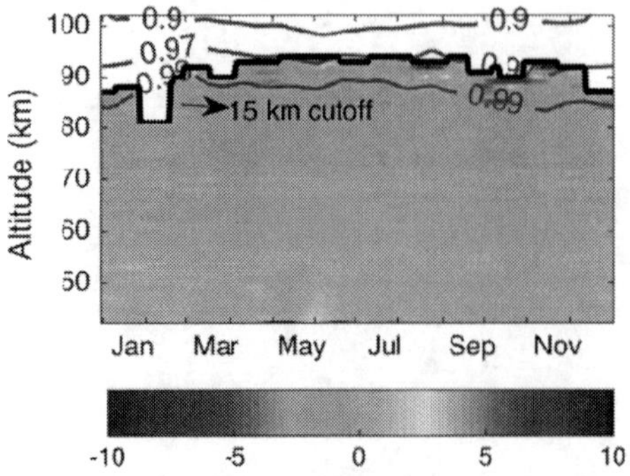

Figure 3: Temperature difference between the OEM and the traditional method. Black line is 15 km cutoff height for the traditional method and blue lines are averaging kernel contours for the OEM. Using OEM the temperature climatology can be extended to almost 100 km with less than 10 % contribution from the a priori.

The 15 km cutoff height from the traditional method corresponds very well with the 0.97 **Au** cutoff. Below these lines the temperature difference is negligible and the two methods are consistent. However, the 0.99 **Au** cutoff (no *a priori* information is in the retrieval) is on average below the 15 km cutoff height. Below the 0.99 cutoff there is almost no difference between the two methods' results except for a few spots close to the 15 km cutoff height. In this paper, only the statistical uncertainty is considered for comparison of the two methods. There are only a few spots with temperature differences larger than 5 K, the causes of which could be due to the systematic uncertainties in different parameters. The advantage of the OEM is shown with the 0.9 **Au** contour line where the *a priori* contribution is only 10 percent. It can be seen that the altitude difference between the 0.9 cutoff and the 15 km cutoff height is roughly 10 km which is an improvement over the traditional method. The uncertainty due to seed pressure is included in the OEM systematic uncertainty budget [2]. Seed pressure uncertainty values for Figures 1 and 2 at the 0.99 and 0.9 **Au** cutoffs are 0.15 K and 2 K, respectevly. In summary, the 15 km cutoff height in the traditional method corresponds approximately **Au** = 0.97 cutoff for the OEM and both methods are consistent up to this cutoff height. However, the OEM determines the percentage of the confidence in the results mathematically and computationally unlike the traditional method, which does not provide a methodology for determining a cutoff height and corresponding uncertainty.

6 Acknowledgements

This project has been supported by the Natural Sciences and Engineering Research Council (NSERC) and the NSERC CREATE Training Program in Arctic Atmospheric Science and the Canadian Network for the Detection of Atmospheric Change (CANDAC).

References

[1] Hauchecorne, A., Chanin, M., 1980: Density and temperature profiles obtained by lidar between 35 and 70 km, *Geophys. Res. Lett.* **7** , 565-568.

[2] Sica, R. J., Haefele, A., 2015: Retrieval of temperature from a multiple-channel Rayleigh-scatter lidar using an optimal estimation method, *Applied Opt.* **54**(8).

[3] Rodgers, C. D: Inverse Methods for Atmospheric Sounding, Theory and Practice, *World Scientific*, 2000.

[4] Argall, P. S., Sica, R. J., 2007: A comparison of Rayleigh and sodium lidar temperature climatologies, *Ann. Geophys.* **25** , 27-35.

[5] Jalali, A., Sica, R. J., Argall, P. S., 2016: Extending and Merging the Purple Crow Lidar Temperature Climatologies Using the Inversion Method, *EPJ Web of Conferences* **119**.

https://doi.org/10.1051/epjconf/201817603002

LIDAR INVESTIGATIONS OF THERMAL REGIME AND AEROSOL STRATIFICATION OF THE STRATOSPHERE OVER TOMSK

Gennady Matvienko, Valeriy Marichev, Dmitry Bochkovsky

V.E. Zuev Institute of Atmospheric Optics, Siberian Branch of the Russian Academy of Sciences, Academician Zuev Square, Tomsk 634055, Russia, E-mail: mgg@iao.ru

ABSTRACT

One of the important applications of lidar techniques is the study of thermal regime and aerosol content of the stratosphere. Such investigations in monitoring mode were started at the Institute of atmospheric optics since 1994 and are continued to date. The main attention is paid for the study of the unexpected disturbances caused by winter stratospheric warming. In this paper we present the results of the study of the vertical distribution of temperature and aerosol over Tomsk of last years.

1 INTRODUCTION

The period of 2015-16, which presents the results was characterized by the absence of volcanic eruptions which could affect the disturbance of the aerosol component of the stratosphere of the northern hemisphere including western Siberian region. So it was a convenient possibility to observe the peculiarities of the temporal variability of the background aerosol loading of the stratosphere over Western Siberia during quite long time interval. In this period, as in previous years [1-5], winter stratospheric warming's were observed. This gave the opportunity to continue the study of the peculiarities of their manifestations.

The represented data array of 161 total signals collected in some nights was used as initial data for analysis in 2015-16.The height interval was from 15 till 50-60 km, the spatial resolution was 192 m. Lidar signals were received in the photopulses counting mode with accumulation of 12×10^4 laser pulses (accumulation time was about 2 hours during a night).

2 METHODOLOGY

Lidar methods of the Raman and elastic aerosol and molecular scattering of light (Raman, Mie and Rayleigh scattering) were used to determine the vertical profile of aerosol characteristics and temperature in the stratosphere. The atmospheric sounding was carried out by laser radiation at a wavelength of 532nm, the reception of the lidar signals at wavelengths 532 and 607 nm.

The optical characteristic R(H) – aerosol scattering ratio (H is the current height) was used as a parameter describing the aerosol vertical stratification. By definition, R(H) is the ratio of the sum of aerosol and molecular backscattering coefficient to the molecular backscattering coefficient. For example, fulfillment of the condition R(H) = 1 means the absence of aerosol at the certain height, and contrary, aerosol appears where R(H) > 1.

The vertical profiles of temperature T(H) were retrieved from the Rayleigh and Raman signals using the formula:

$$T(H)=\frac{P_1(H)P_2(H)}{N(H)H^2}\left[\frac{N(H_m)}{P_1(H_m)P_2(H_m)}T(H_m)+\frac{1}{R*}\int_{Hm}^{H}\frac{N(h)h^2g(h)dh}{P_1(h)P_2(h)}\right]$$

Here *N(H)* - lidar signals, $P_1(H)$ and $P_2(H)$ are the transparencies of the atmosphere from the level of arrangement of the lidar up to the height H at the wavelength of 532 nm (Rayleigh signals) and 532 and 607 nm (Raman signals), respectively; $R*$ is the universal gas constant, $g(h)$ is the gravity acceleration, *Hm* is the maximum height, from which quite reliable for processing signals are detected (so-called calibration height, at which the temperature values *T(Hm)* are set).

3 RESULTS

3.1 Aerosol

The results of measurements of the monthly mean dynamics of the aerosol vertical stratification are shown in Fig. 1 and Fig.2.

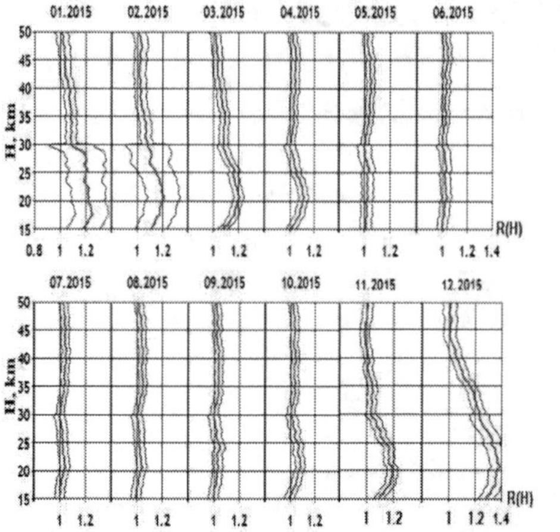

Figure 1 Monthly mean profiles of the aerosol vertical stratification in 2015. Red lines show the mean profiles, black lines show the range of standard deviation

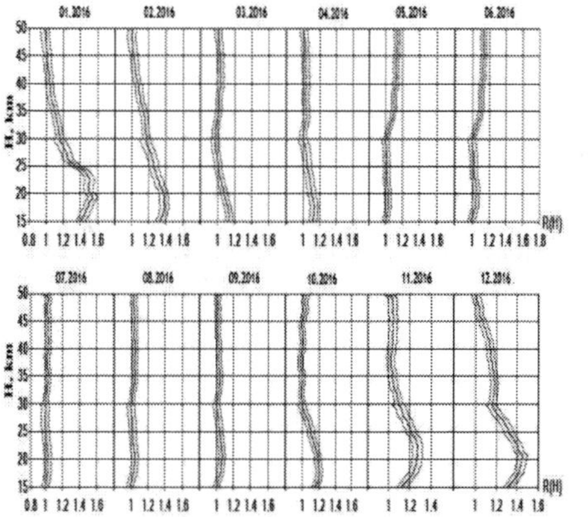

Figure 2 Monthly mean profiles of the aerosol vertical stratification in 2016

The tendency of the maximum aerosol loading of the lower stratosphere in January, decreasing in spring and practical absence of aerosol in May – September is confirmed by comparison with observations in previous years [1, 2]. Aerosol content begins to increase in October and reaches maximum in January.

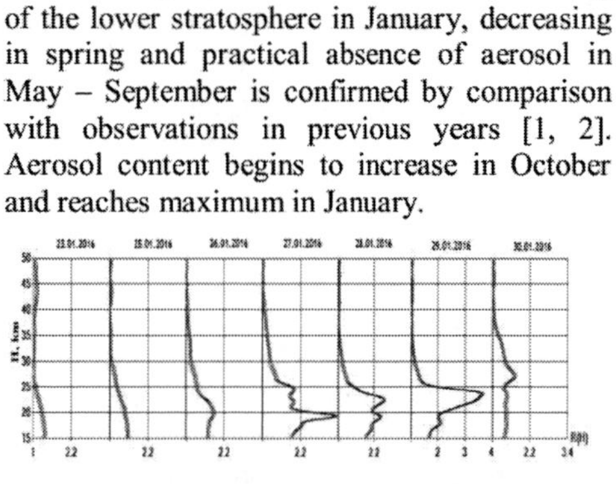

Figure 3 Dynamics of the aerosol vertical stratification in January 2016 (the appearance of polar stratospheric clouds)

The aerosol loading of the stratosphere increased significantly in the first decade of January 2016 and covered the entire stratosphere (Fig. 3). The scattering ratio achieved the maximal level $R = 1.6\text{-}1.8$ at a height of 20 km. Some decrease of the aerosol content took place in the middle of the month. Unusual aerosol stratification was detected in the third decade of the month. It was characterized by the very high aerosol content in the lower stratosphere and aerosol absence in the upper stratosphere for the most days of observation. The layered structure of the aerosol component was observed, which underwent significant vertical variations for the entire period of observations in January 23 to 30. The scattering ratio achieved extremely high values up to $R = 3.5$. The following analysis has shown that distortions of the aerosol component were caused by the appearance of polar stratospheric clouds (PSCs). Our synchronous lidar measurements of temperature discovered the abnormally low temperature values in the lower stratosphere (lower than 190 K), which caused the formation of PSCs.

In March, a sharp decrease in the aerosol loading of the stratosphere was observed: there was no aerosol in the upper stratosphere, and only trace concentrations at heights from 15 to 25 km remained. In the following period of late spring, summer, and early fall, the aerosol component was practically absent in the entire stratospheric layer. This feature revealed, at least, for the Tomsk Region was confirmed by many-year observations.

3.2 Temperature

The study of thermal regime of the stratosphere in 2015 was started in the disturbed period caused by stratospheric warming (SW) [4, 5]. The first manifestations of SW- was observed on 3 December 2014. Figures 4 and 5 show examples of the vertical profiles of temperature obtained during SW 2014-15. Lidar temperature profiles are given in comparison with the profiles of temperature measured from onboard the AURA satellite [6] and taken from the CIRA-86 model [7].

Next event of the SW began at the end of January 2015 and went on until second decade of February (see Fig. 5). Stratospheric warming in winter 2014–2015 was one of the longest SW observed over Tomsk since 1996, which prolonged almost one month. The greatest positive deviations of temperature were observed in the height range 40 up to 45 km and could reach 60K at the height of 40 km (February 5, 2015).

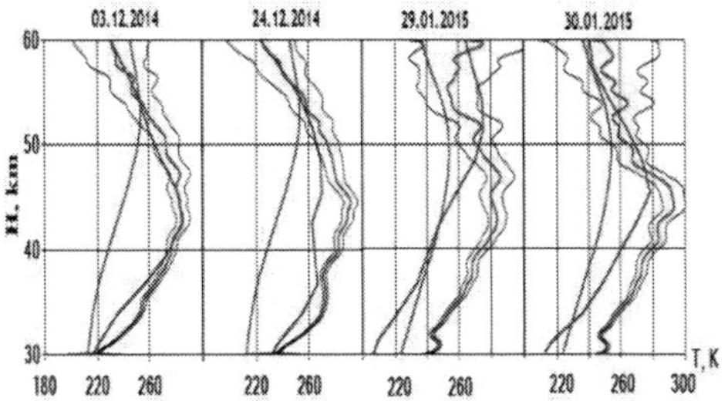

Figure 4 Lidar and satellite observations of the stratospheric warming in December 2014 and January 2015. Red lines show the Rayleigh lidar measurements, black line shows the standard deviation, blue line shows the AURA measurements, and green line shows the monthly mean profile according the CIRA-86 model

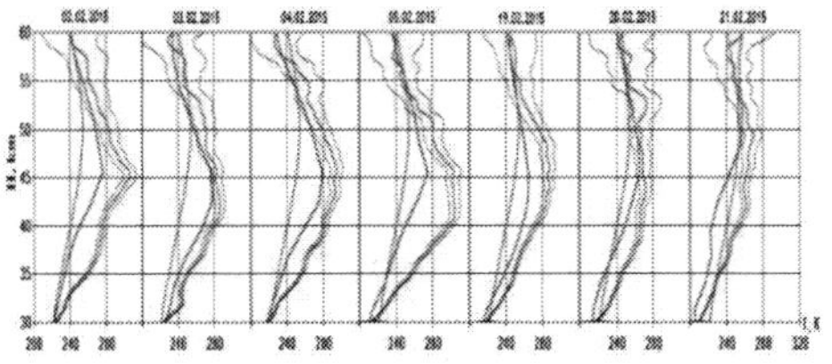

Figure 5 Vertical distribution of temperature in February 2015

The observed SW in 2014–2015 was related to the minor type, when the direction of air mass transfer in the stratosphere did not change from west to east that is confirmed by the data from the web-site [8].

The long-term observations show that the vertical distribution of temperature in Western Siberia in the period March till November is in satisfactory agreement with the CIRA-86 model distribution.

The significant dynamics of the vertical distribution of temperature starts in December (Fig. 6). The positive deviation of temperature from the monthly mean value by approximately 30K appears on December 13 at the height of 50 km. Then, on December 18, the area of "warm" air covers the height range from 37 up to 60 km with the mean value of temperature of about 260K. The maximum warming is observed on December 29, the temperature in the foregoing height range exceeds 260K and reaches 270K at the heights of 40 till 50 km. This event is related with the beginning of the next year stratospheric warming.

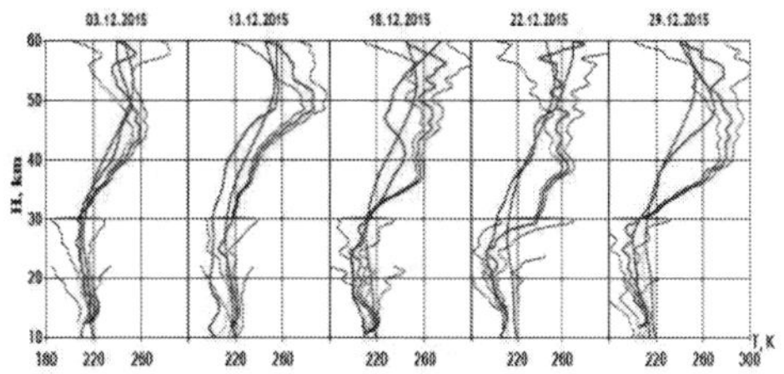

Figure 6 Vertical distribution of temperature in some nights of December 2015

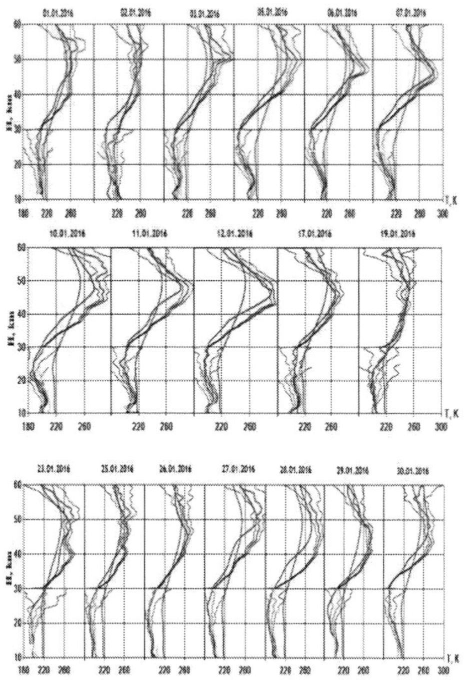

Figure 7 Vertical distribution of temperature in some nights of January 2016. The yellow line shows Raman lidar measurements, lower blue - radiosonde measurements

In January 2016, for the observation period of January 3 to 17, the stable vertical temperature distribution was observed in the form of a wave with a positive deviation from the monthly average value in the height range of 40-50 km and a negative deviation in the lower stratosphere (see Fig. 7). The maximal value of the positive deviation achieved 40 K (January 12, height of 45 km), and that for the negative deviation was 35 K (January 5 and 7, height of about 30 km).In January 19 to 26, some stabilization of the vertical temperature distribution (approaching the model distribution) was observed in the upper stratosphere. However, in the entire layer of the lower atmosphere, the temperature profiles obtained from lidar, balloon, and satellite data remained shifted toward negative values from the model profile.

The last overshoot of warming was observed on February 1 of 2016 in the height range of 40-60 km.

The observed SW in 2015–2016, as the SW in 2014–2015, were related to the minor type [8].

CONCLUSION

The lidar studies of the stratosphere over Tomsk in the period of 2015-2016, as well as earlier years, revealed the following:

1. The loading of the stratosphere with background aerosol occurs in the cold period of a year: start in October, maximum in January, and end in April. In the warm period, there is practically no aerosol in the stratosphere.
2. Every year the winter warming takes place in the stratosphere. It begins in December, is most pronounced in January, and is sometimes extended to February. Weak and, more rarely, strong warmings are observed. The amplitude of positive temperature deviations from the monthly average value can achieve 60 K, and the height of the stratopause can go down to 30 km. In the period of March till November, the vertical temperature distribution is in asatisfactory agreement with the CIRA-86 model distribution.

The work was supported by a grant from the RF President (NS-8199.2016.5) in carrying out monitoring measurements, a Russian Science Foundation grant (no. 14-27-00022) to create a receiver for Raman light scattering and Russian Foundation for Basic Research (project no. 16-05-00901) in the analysis of the lidar sensing data.

References

[1] Marichev V.N. Investigation of variability of the background aerosol vertical structure above Tomsk based on lidar observations in 2010-2011, *Atmos. Ocean Opt.*, 2012, **25**. No. 11. P. 976-984.

[2] Marichev V.N., Bochkosvkii D.A. Studying the variations in background aerosol loading of the stratosphere in 2014, *Proceedings of SPIE*. 2015. **V.9680**. CID: 9680 6U. [9680-209]

[3] Marichev V.N. Lidar investigations of stratospheric warming events above Tomsk in 2008-2010, *Atmos. Ocean Opt*. 2011. **V. 24**. No. 05. pp. 386-391.

[4] Marichev V.N. Investigation into features of manifestation of winter stratospheric warming events over Tomsk from the data of lidar temperature measurements in 2010-2011, *Atmos. Ocean Opt.* 2011. **V. 24.** No. 12. P. 1041-1046.

[5] Marichev V.N. The analysis of the air density and temperature behaviour in the stratosphere above Tomsk in periods of perturbed and quiet states performed according to the results of lidar measurements, *Atmos. Ocean Opt.* 2013. **V. 26.** No. 09. pp. 783-792.

[6] http://mirador.gsfc.nasa.gov

[7] Rees D., Barnett J.J., Labitske K. COSPAR International Reference Atmosphere: 1986. Part II, Middle Atmosphere Models, *Adv. Space Res.* 1990. **V. 10**, N 12. 525 p.

[8] http://users.met.fu-berlin.de/~Aktuell/strat-www/wdiag/eczm.php?alert=8&forecast=all &var= u&lng= eng#fig1

RAYLEIGH LIDAR OBSERVATION OF TROPICAL MESOSPHERIC INVERSION LAYER: A COMPARISON BETWEEN DYNAMICS AND CHEMISTRY

K. Ramesh[1,*], S. Sridharan[1], K. Raghunath[1]
[1] National Atmospheric Research Laboratory (NARL), Gadanki- 517112, A.P., India.
*** Email: karanamram@gmail.com**

ABSTRACT

The Rayleigh lidar at National Atmospheric Research Laboratory, Gadanki (13.5°N, 79.2°E), India operates at 532 nm green laser with ~600 mJ/pulse since 2007. The vertical temperature profiles are derived above ~30 km by assuming the atmosphere is in hydrostatic equilibrium and obeys ideal gas law. A large mesospheric inversion layer (MIL) is observed at ~77.4-84.6 km on the night of 22 March 2007 over Gadanki. Although dynamics and chemistry play vital role, both the mechanisms are compared for the occurrence of the MIL in the present study.

1. INTRODUCTION

LIDAR (LIght Detection And Ranging) is the most efficient technique to capture the nocturnal mesospheric inversion layers with high spatial and temporal resolution. The term Mesospheric Inversion Layer (MIL) refers to the inversion of temperature gradient from negative to positive superposed upon the characteristically decreasing mesospheric thermal structure. However the causative mechanisms of MILs are quite complex, a few suggested mechanisms for their occurrence are gravity wave breaking, gravity wave-tidal interactions, chemical heating, and planetary wave critical level interaction [1,2]. The study of MILs is essential to understand the energetics in the mesosphere and lower thermosphere (MLT) region. Most of the waves which are generated in the lower atmosphere can propagate to the MLT region with increasing amplitude in response to decreasing density. When the waves break, the energy and momentum carried by the wave will be deposited in the background and thereby modulate the thermal and dynamical structures. In addition, the chemical heating by the dominant exothermic reactions among H, O, O_2, O_3, OH, HO_2 enhances the background temperature in the MLT region [3, 4].

In the present study, a large MIL observed on 22 March 2007 over a tropical region, Gadanki (13.5°N, 79.2°E) has been selected to investigate the dominant causative mechanism using mainly the Rayleigh lidar temperature profiles for the duration of ~6 hours.

2. INSTRUMENTATION AND DATA

2.1. NARL LIDAR SYSTEM

The Mie and Rayleigh lidar system installed at National Atmospheric Research Laboratory (NARL), Gadanki (13.5°N, 79.2°E) employs a Q-switched solid state (Nd:YAG) pulsed laser (PL9050) from Continuum, USA. The system operates at 532 nm green laser with the pulse repetition frequency (PRF) of 50 Hz and the energy of ~600 mJ per pulse. The laser beam diameter is 9 mm and it will be expanded into 10 times in order to reduce the beam divergence from ~0.45 mRad (before expansion) to less than 0.1 mRad (after expansion) so that the beam remains within the FOV of the receiver system at all ranges of interest. The Mie (Schmidt-Cassegrain telescope) and Rayleigh receivers (Newtonian telescope) collect the atmospheric back scattered photons which are allowed to fall on photo multiplier tubes (PMTs) of different gains. The PMT output signal will be collected and processed through the MCS-plus (Multi Channel Scalar) PC based photon count system which is working with four data acquisition channels, two from Rayleigh receiver (R, U channels) and two from Mie receiver (P, S channels) for simultaneous photon counting with 12500 laser shots averaged for the time resolution of 250 sec with range resolution (dwell time=2μsec) of 300 m. For example, the typical MCS photon count profiles for P-channel and R-channel are as shown in Figure 1. More details on NARL lidar system can be obtained from *Ramesh and Sridharan*, [2012] [5].

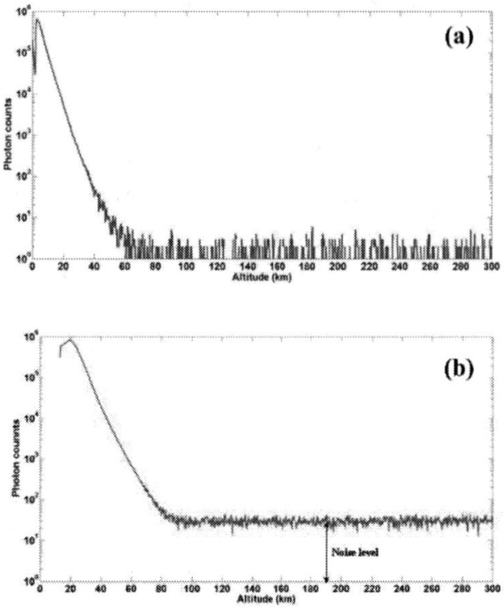

Figure 1 The typical lidar photon count profile for (a) P-channel and (b) R-channel.

2.2 TIMED-SABER DATA

The Sounding of the Atmosphere using Broadband Emission Radiometry (SABER) is a space-borne instrument onboard NASA's Thermosphere-Ionosphere-Mesosphere Energetics and Dynamics (TIMED) satellite. It provides vertical profiles of temperature, ozone volume mixing ratio (O_3 vmr), chemical heating rates due to various exothermic chemical reactions among H, O, O_2, O_3, OH, OH_2, and volume emission rates of different species. More details on SABER instrument and different data sets can be obtained from *Remsberg et al.* [2003] [6], *Mlynczak et al.* [2007] [7], and *Gan et al.* [2012] [8].

3. DATA ANALYSIS

The lidar temperature profiles are determined from the method given by *Hauchecorne and Chanin*, [1980] [9].

In the height range above 30 km where the aerosol contribution is negligible (scattering ratio is greater than one calculated from Mie lidar), the range corrected signal and atmospheric transmission is proportional to the molecular number density. Using the number density taken from an appropriate model (MSIS- 90 or CIRA-86) for the height of 50km where the signal-to-noise ratio is fairly high, the constant of proportionality is evaluated and there by the density profile r(z) is derived. Taking the pressure (P) at the top of the height range (90km) from the model, the pressure profile is computed using the measured density profile, assuming the atmosphere to be in hydrostatic equilibrium. Using the ideal gas relation, the temperature at i^{th} layer $T(z_i)$ is given by

$$T(z_i) = \frac{M g(z_i)\Delta z}{R\,Log(1+X)} \text{ where } X = \frac{\rho(z_i) g(z_i)\Delta z}{P(z_i + \Delta z/2)} \quad (1)$$

and the statistical standard error in the temperature is given by

$$\frac{\delta T(z_i)}{T(z_i)} = \frac{\delta Log|1+X|}{Log|1+X|} = \frac{\delta X}{(1+X)\,Log(1+X)} \quad (2)$$

4. RESULTS

Figure 2a shows the vertical profiles of nightly mean (19:57-02:37 LT) lidar and SABER (12.7°N, 82.6°E, 01:00 LT) temperatures for22-23 March 2007. In addition, the corresponding MSIS-90 (Mass Spectrometer Incoherent Scatter-90) model temperature profile also shown here for comparison. The lidar temperature profile show a large inversion layer at ~77.4-84.6 km with amplitude and thickness of ~34.4 K and 7.2 km respectively. The associated errors in the lidar temperature measurement are 0.1 K–1.0 K from 30 km to 60 km, 1.0 K–5.0 K from 60 km to 80 km and 5.0 K–15.0 K from 80 km to 89 km. However the SABER temperature profile shows the inversion layer at ~73.4-81.3 km with amplitude and thickness of ~31 K and 7.9 km respectively. Here, the upper and lower level temperature and height differences give the inversion amplitude and thickness respectively. Figure 2b shows the SABER ozone (O_3) volume mixing ratio (vmr) profile from which it can be observed that there is no enhancement in the O_3 vmr within the inversion layer.

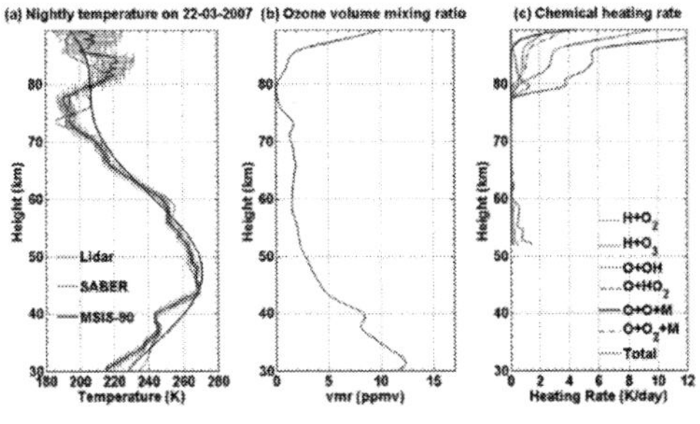

Figure 2 (a) The nightly mean Rayleigh lidar temperature (19:57-02:37 LT) over Gadanki along with TIMED-SABER (12.7°N, 82.6°E, 01:00 LT) and MSIS-90 model temperature profiles for 22-03-2007. The horizontal error bars show the uncertainties in the measurement of lidar temperature. (b), (c) The corresponding SABER O_3 vmr and chemical heating rates.

As shown in Figure 2c, the associated chemical heating rates due to the reactions $O+OH \rightarrow H+O_2$, $H+O_2+M \rightarrow HO_2+M$, $H+O_3 \rightarrow OH+O_2$, $O+HO_2 \rightarrow OH+O_2$, $O+O+M \rightarrow O_2+M$, $O+O_2+M \rightarrow O_3+M$ and their total increases drastically above ~77 km with the dominant reaction between H and O_3 (~3 K/day at ~85 km). However the total heating rate varies as 0.1-5.6 K/day at ~77-85 km which is much smaller than that shown (~30 K/day) by *Ramesh et al.*, [2013] [4].

Figure 3 shows the half-hour integrated lidar temperatures and it can observed from this figure that the inversion layer descends between ~80 and 86 km at the rate of (~1.2 km/hr) diurnal tidal phase speed. Also the maximum inversion temperature decreases from ~286 K (~21:00 hrs) to ~223 K (~02:30 hrs) between ~80 and 85 km during the observation period

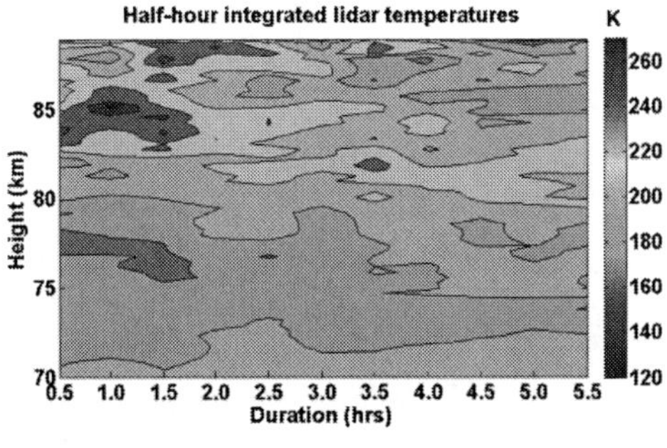

Figure 3 Half-hour integrated Rayleigh lidar temperatures on 22-03-2007 over Gadanki.

Figure 4 shows the 2-dimensional power spectra (FFT) of lidar temperature perturbations and it can be observed from this figure that a small scale gravity wave of T~38 min and λ_z~6.4 km propagates above ~63-82 km with increasing amplitude at the phase speed of c_z~2.8 m/s.

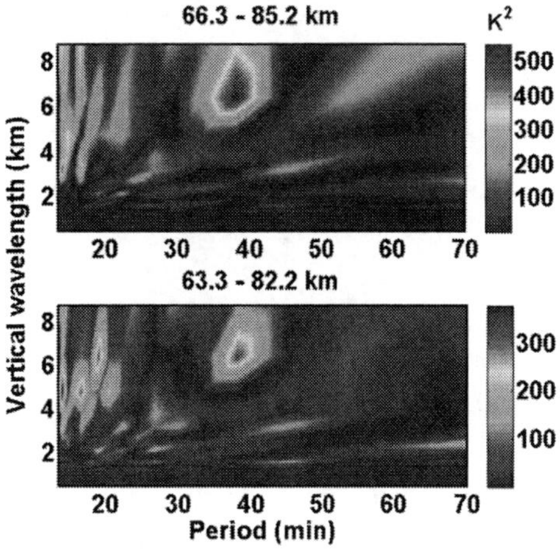

Figure 4 Two dimensional FFT power spectra of lidar temperature perturbations at two different height regions, 63.3-82.2 km and 66.3-85.2 km.

Further Figure 5 shows the vertical profile of lidar nightly mean Brunt–Väisälä frequency square (N^2) calculated from the equation (3).

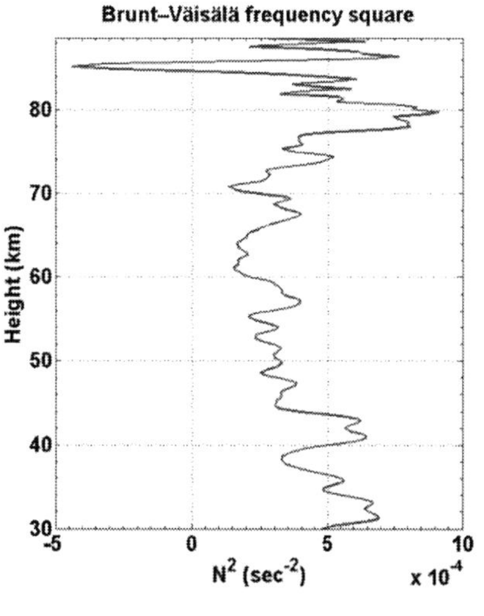

Figure 5 Nightly mean profile of Brunt–Väisälä frequency square for 22-03-2007.

$$N^2(z) = \frac{g(z)}{T(z)}\left[\frac{dT(z)}{dz} + \frac{g(z)}{c_p}\right] \quad (3)$$

Here g(z), T(z) and c_p are acceleration due to gravity, mean temperature and specific heat at constant pressure (1004 J/K/Kg) respectively.

From this figure it can be observed that the condition for convective instability ($N^2<0$) is satisfied at ~85 km which indicates the role of gravity wave breaking for the occurrence of the inversion layer.

5. SUMMARY AND CONCLUSIONS

In the present study, the dominant causative mechanism for the occurrence of a large MIL has been investigated mainly using the Rayleigh lidar observations for 22 March 2007 over Gadanki. A small scale gravity wave (T~38 min, λ_z~6.4) which is propagated from the lower atmosphere attains maximum amplitude at ~ 85 km and breaks at this height region ($N^2<0$). Due to the wave breaking, the heat and momentum flux carried by the wave is deposited in the background and cause the inversion layer. The inversion layer descends at the rate of diurnal phase speed which gives the evidence for the gravity wave tidal interaction during this MIL event. However the smaller values of O_3 vmr and chemical heating rates suggest their negligible role for the occurrence of this MIL.

6. ACKNOWLEDGEMENTS

The authors thank the technical staff for operating and maintaining the Rayleigh lidar system at NARL, Gadanki. They also gratefully acknowledge the NASA's TIMED-SABER team for providing the data used in this study through the website: http://saber.gats-inc.com/.

7. REFERENCES

[1] Meriwether, J.W., Gerrard, A.J., 2004: Mesosphere inversion layers and stratosphere temperature enhancements, *Rev. of Geophys.*, 42, RG3003 http://dx.doi.org/10.1029/2003RG000133.

[2] Meriwether, W.J., Gardner, C.S., 2000: A review of the mesospheric inversion layer phenomenon, *J. Geophys. Res.*, 105, 12405–12416.

[3] Mlynczak, M. G., Solomon, S., 1993: A detailed evaluation of heating efficiency in the middle atmosphere, *J. Geophys. Res.*, 98, 10,517–10,541.

[4] Ramesh, K., Sridharan, S., Vijaya Bhaskara Rao, S., 2013: Dominance of chemical heating over dynamics in causing a few large mesospheric inversion layer events during January-February 2011, *J. Geophys. Res. Space Physics*, 118, 6751–6765, doi:10.1002/jgra.50601.

[5] Ramesh, K., Sridharan, S., 2012: Large mesospheric inversion layer due to breaking of small-scale gravity waves: Evidence from Rayleigh lidar observations over Gadanki (13.5°N, 79.2°E), *J. Atmos. Sol. Terr. Phys.*, 89, 90 – 97.

[6] Mlynczak, M. G., Marshall, B. T., Martin-Torres, F. J., Russell III, J. M., Thompson, R. E., Remsberg, E. E., Gordley, L. L., 2007: Sounding of the Atmosphere using Broadband Emission Radiometry observations of daytime mesospheric $O_2(^1\Delta)$ 1.27 μm emission and derivation of ozone, atomic oxygen, and solar and chemical energy deposition rates, *J. Geophys. Res.*, 112, D15306, doi:10.1029/2006JD008355.

[7] Remsberg, E., Lingenfelser, G., Harvey, V. L., Grose, W., Russell III, J.M., Mlynczak, M., Gordley, L., Marshall, B.T., 2003: On the verification of the quality of SABER temperature, geopotential height, and wind fields by comparison with Met Office assimilated analyses, *J. Geophys. Res.*, 108(D19), 4628, doi:10.1029/2003JD003720.

[8] Gan, Q., Zhang, S. D., Yi, F., 2012: TIMED/SABER observations of lower mesospheric inversion layers at low and middle latitudes, *J. Geophys. Res.*, 117, D07109, doi:10.1029/2012JD017455.

[9] Hauchecorne, A., Chanin, M.L., 1980: Density and temperature profiles obtained by lidar between 35 and 70 km, *Geophys. Res. Lett.*, 7, 565–568.

EXPLORING FINE-SCALE VARIABILITY OF STRATOSPHERIC WIND ABOVE THE TROPICAL LA REUNION ISLAND USING RAYLEIGH-MIE DOPPLER LIDAR

S. M. Khaykin[1]*, A. Hauchecorne[1], J.-P. Cammas[2], N. Marqestaut[2], J.-F. Mariscal[1], F. Posny[2], G. Payen[2], J. Porteneuve[1], P. Keckhut[1],

[1]LATMOS-IPSL, Université Versailles St.-Quenitn, CNRS/INSU, Guyancourt, France,
**Email: sergey.khaykin@latmos.ipsl.fr*
[2]OSUR, UMS3365, Université de la Réunion, Saint-Denis, La Réunion, France

ABSTRACT

A unique Rayleigh-Mie Doppler lidar capable of wind measurements in the 5-50 km altitude range is operated routinely at La Reunion island (21° S, 55° E) since 2015. We evaluate instrument's capacities in capturing fine structures in stratospheric wind profiles and their temporal and spatial variability through comparison with collocated radiosoundings and ECMWF analysis. Perturbations in the wind velocity are used to retrieve gravity wave frequency spectrum.

1. INTRODUCTION

Wind measurements above 30 km are rare. The only information available on a regular basis is zonal mean winds inferred from satellite-borne pressure measurements assuming geostrophic balance. However this does not allow understanding the propagation of gravity waves, whose filtering, dissipation and/or breaking strongly depend on wind speed and direction. Observations of high vertical resolution (~ 1 km) are required for detecting wind shears around the stratopause, which play a crucial role in reflecting or filtering of the waves. Following this requirement, a direct-detection Doppler lidar for wind observations in the middle atmosphere has been developed in France in 1989 [1]. After deployment of the Doppler lidar at Observatory of Haute Provence (OHP) in Southern France (44° N, 6° E, 600 m a.s.l.) in 1993 the technique was shown capable of wind measurements between 5 and 50 km independent of the presence of aerosols [2]. The wind lidar was operated on a regular basis during 1994-1999 and occasionally since then. The observations were used to construct wind climatology in the middle atmosphere and to study gravity wave spectra and small-scale wind fluctuations [3].

In 2012 the Rayleigh-Mie Doppler lidar was deployed at a newly established tropical observatory located on mount Maïdo (21° S, 55° E, 2200 m a.s.l.) at La Reunion island [4]. The Doppler lidar at Maido was operated on a campaign basis since 2013 and routinely, twice a week since September 2015. The wind measurements above La Reunion island represent a substantial value for studying the dynamics of the tropical middle atmosphere – a region where observations are particularly sparse. They are also of big relevance for validation of the upcoming European Space Agency (ESA) ADM-Aeolus satellite mission for global observations of wind. ADM-Aeolus mission will carry a Rayleigh-Mie Doppler lidar, based on the same principle as the French ground-based lidars.

2. INSTRUMENT DESIGN

The horizontal wind components are obtained by measuring Doppler shift between emitted and backscattered light by means of a double-edge Fabry-Perot interferometer with two half-disks areas whose air gap is slightly different, providing two bandpasses. The Doppler shift is due to projection of molecules or particles velocities onto the line of sight of the laser inclined off zenith. The laser beam is sent alternatively to zonal and meridional direction for obtaining both wind components, whereas the vertical pointing is used for measuring the zero Doppler shift reference (in assumption that the vertical wind is negligible).

The Doppler lidar uses a Q-switched, injection-seeded Nd:YAG laser operating at 532 nm in monomode. The pulse repetition rate of the laser is 30 Hz with 24 W mean energy. The 0.3 m^2 telescope of Maido wind lidar is composed of a single rotating mirror, which serves for both the emission and reception.

To extend the altitude range of measurements, there are two pairs of gated photomultipliers (PMT) collecting ~90% (high-gain channels) and ~10% (low-gain channels) of the signals. More detail on the instrument can be found in [5] and references therein.

4. RESULTS

The Doppler wind lidar at Maido observatory is operated routinely since September 2015. In addition to the regular observations, two dedicated measurement campaigns have been held at Maido in September-October 2015 (LIDEOLE-II) and June 2016 (ARISE2). During these campaigns the wind lidar measurements were validated using collocated radiosonde ascents (Modem PTU-wind M-10 GPS radiosondes) conducted simultaneously with lidar acquisitions. Most of the radiosondes at Maido were launched using Totex 3000 gr rubber balloons allowing the payload to reach 40 km altitude for ensuring the largest possible height overlap with the lidar vertical profiles.

4.1 Intercomparison of lidar, radiosondes and operational analysis

The Doppler lidar validation exercise included 9 collocated radiosonde ascents performed within the measurements campaigns since September 2015. Intercomparison of lidar and radiosonde wind profiles showed excellent agreement, with a mean deviation between 0.1 and 2.5 m/s over the 5 - 40 km altitude range and correlation coefficient between 0.85 and 0.99.

Fig. 1 shows an example of comparison of zonal wind profile up to 50 km obtained with lidar (~3 h integration time) and the radiosonde profile. The wind profiles retrieved using low- and high-gain channels of Doppler lidar are fully consistent with those reported by the radiosounding in the entire range of overlap. The maximum air distance between the radiosonde trajectory and lidar sampling area amounted to 45 km at 15 km altitude.

Lidar measurements in the upper stratosphere (above 40 km) were evaluated using ECMWF (European Center for Medium-Range Weather Forecast) T106L137 operational analysis. The model reproduces well the local maximum of easterly wind (-46 m/s) at 37 km seen by lidar and radiosonde and a strong wind shear above, where zonal wind changes to westerly within a ~5 km-thick layer, reaching ~30 m/s at 43 km. We note that in most cases the model tends to smoothen the vertical structures reported by the lidar, which is expectable and can be readily attributed to low vertical resolution of the model at upper stratospheric levels and the assumption on the geostrophic balance. At the same time, the statistical uncertainty (two times standard error) of lidar measurement increases rapidly above 30 km and reaches ~20 m/s at 45 km altitude for a 3 hours acquisition, rendering the lidar retrieval less accurate in the upper stratosphere. This can be remedied by increasing the acquisition interval to more than 4-5 hours.

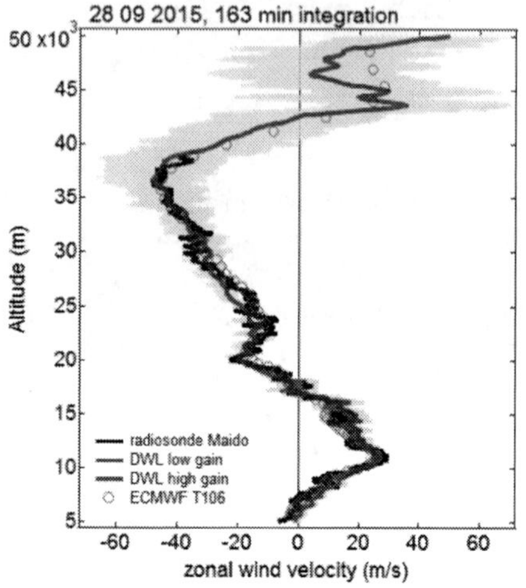

Figure 1. Comparison of zonal wind profiles obtained using Maido Doppler wind lidar on 28.09.2015 with collocated radiosounding and ECMWF operational analysis. Shaded areas indicate the total error of lidar retrieval.

4.2 Capturing fine vertical structures

While the upper stratosphere sampling requires long acquisition time, the lower stratospheric wind profiles can be reliably acquired using sub-hourly integration intervals. Fig. 2 displays an example of lidar – radiosonde intercomparison at 15-23 km altitude range. The lidar wind profile was retrieved by integrating over 50 minutes interval corresponding to the radiosonde's ascent through the stratosphere. The maximum air distance between the radiosonde ascent trajectory and lidar sampling area amounted to 32 km at 18 km altitude.

The lidar profile of zonal wind follows precisely the radiosonde profile, reproducing the vertical structures on a scale of a few hundred meters. The vertical structure of wind profile can be correlated with that of temperature profile taken by the radiosonde and plotted in Fig.2 as a red curve. The cold point tropopause can be found at 17 km altitude. One can see two sub-kilometer-thick layers above the tropopause centered at 17.8 km and 19.2 km altitude and characterized by enhancement in easterly component of wind by about -10 m/s coinciding with local temperature minima. Similar patterns can be seen in meridional wind profiles (not shown). These coincident structures in wind and temperature are indicative of gravity waves manifesting themselves as perturbations in thermodynamical state of the stratosphere.

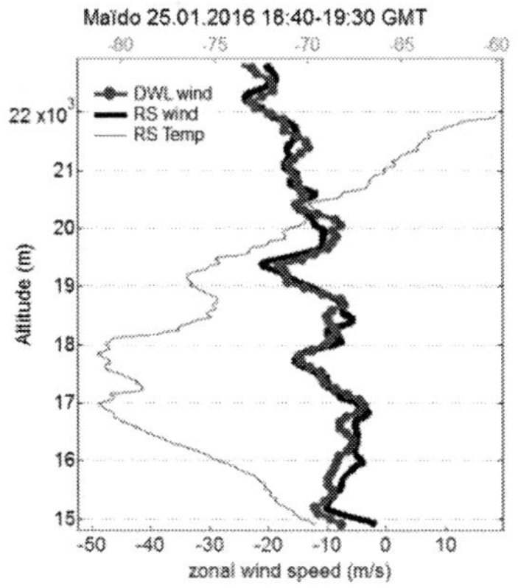

Figure 2. Fine vertical structures in zonal wind profiles (bottom axis) from Maido Doppler wind lidar (blue curve) and collocated radiosounding (black curve). Radiosonde temperature profile (top axis, red curve). The date and time interval of the lidar acquisition/ radiosonde measurement are indicated in the top of the panel.

4.3 Short-scale temporal variability

An important and unique capability of the Doppler lidar technique is the continuous time-resolved measurements of stratospheric wind profile over the course of the night. As opposed to radiosoundings, yielding a snapshot of the atmospheric state, the time-resolved lidar observations are capable of providing information on short-period wind fluctuations due to high-frequency gravity waves.

An experiment carried out within ARISE2 campaign at Maido in June 2016 was aimed at validating the wind temporal variability observed by the lidar using two radiosoundings launched with a 3-hour time interval. The left-hand panels in Fig. 3 shows comparison of radiosonde zonal wind profiles with the time-correspondent lidar measurements (1.5 h non-overlapping integration intervals) and suggest a good agreement for the both soundings. The top right panel compares the two consecutive lidar profiles, demonstrating a remarkable change of the wind profile above 27 km. The bottom right panel quantifies the wind change observed by lidar and compares it with that inferred from two radiosonde ascents. One can see that evolution of the zonal wind profile observed by the lidar is fully confirmed by the radiosounding.

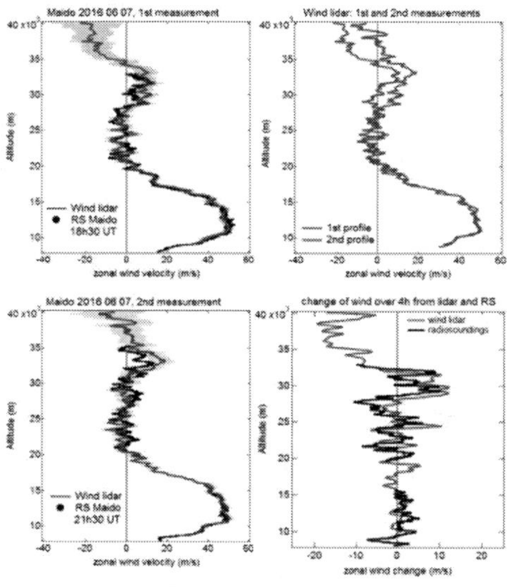

Figure 3. Verification of lidar temporally-resolved wind measurements using two radiosoundings: (top and bottom left) comparison of radiosonde and lidar wind profiles for the two consecutive soundings; (top right) comparison of two independent lidar profiles separated in time by 3 h; (bottom right) wind change over 3 h observed by lidar and radiosondes.

Importantly, in this experiment the radiosonde, carried towards the East by a strong westerly wind in the upper troposphere, was displaced by 100 km with respect to the lidar sampling area at 30 km altitude. This suggests that the temporal variability of wind in the tropical mid-stratosphere (up to 6 m/s per hour) is more important than the meso-scale spatial variability.

4.4 Gravity wave characteristics

After having verified the temporal evolution of wind profile, the measurements on 7th June 2016 were used to derive the frequency spectrum of gravity waves. The continuous lidar measurements of zonal wind spanning 6.4 hours were binned to 10-minute time intervals, allowing to detect waves in the frequency range $0.16 < \omega < 3$ min^{-1}. The frequency power spectrum was computed using Fourier transform for each 1-km layer.

Fig. 4 displays a power spectrum of stratospheric gravity wave frequency as a function of height. Remarkable is the intensification of low-frequency waves ($\omega < 1$ min^{-1}) at 26 km and above, which is consistent with the inference regarding wind change made on the basis of Fig. 3. The higher frequency waves ($\omega > 2$ min^{-1}) manifest themselves mainly above 29 km. The height distribution of gravity wave frequency spectrum obtained from wind measurement is found consistent with that derived from Rayleigh temperature lidar (not shown) which was operated in parallel with the Doppler lidar at Maido.

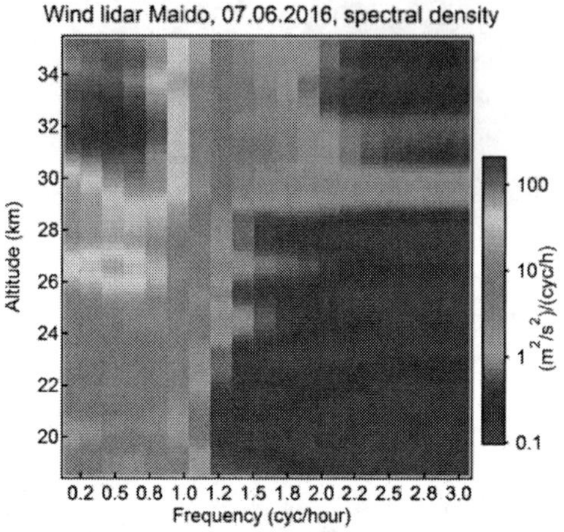

Figure 4. Power spectrum of stratospheric gravity wave frequency as a function of height derived from Maido wind lidar acquisition on 07.06.2016

6. SUMMARY

A Rayleigh-Mie Doppler lidar, capable of wind measurement up to 50 km altitude is operated at a tropical high-altitude Maido observatory since 2013. The results obtained within dedicated validation campaigns allowed evaluating instrument's performance and its capability to capture fine structures in stratospheric wind profiles.

Intercomparison of wind lidars and GPS radiosondes lifted up to 40 km altitude shows excellent agreement. The lidar is shown capable of capturing wind profile structures on a scale of a few hundred meters. Wind lidar measurements in the upper stratosphere are consistent with ECMWF operational analysis, however some discrepancies in the vertical structures are noted.

Temporal and spatial variability of stratospheric wind is explored by combining lidar measurements and collocated radiosoundings. A significant change in wind velocity profile at mid-stratospheric levels observed by lidar over the course of the night is fully confirmed by two successive radiosonde ascents conducted 3 hours apart during the lidar acquisition. We note that the temporal variability of wind in the tropical mid-stratosphere (up to 6 m/s per hour) is more important than the meso-scale spatial variability. This observation has important implications for the validation of ADM-Aeolus satellite mission.

Perturbations in the wind velocity are used to retrieve inertia-gravity wave frequency spectrum in the lower and middle stratosphere. An enhancement of wave activity is found above 26 km where the low-frequency waves ($\omega < 1$ min^{-1}) dominate the spectrum.

ACKWNOLEDGEMENTS

The work was supported by the French National Space Agency (CNES), CNRS-INSU, Université de La Réunion and by the EU H2020 ARISE2 project.

REFERENCES

[1] Chanin, M. L, A. Garnier, A. Hauchecorne, J. Porteneuve (1989): A Doppler lidar for measuring winds in the middle atmosphere, *Geophys. Res. Lett.*, 16(11), 1273–1276.

[2] Souprayen, C., A. Garnier, A. Hertzog, A. Hauchecorne, J. Porteneuve (1999): Rayleigh–Mie Doppler wind lidar for atmospheric measurements. I. Instrumental setup, validation, and first climatological results, *Appl. Opt.*, 38, 12.

[3] Hertzog, A., C. Souprayen and A. Hertzog (2001): Measurements of gravity wave activity in the lower stratosphere by Doppler lidar, *J. Geophys. Res.*, 106, D8, 7879-7890.

[4] Baray, J.-L., Courcoux, Y., Keckhut, P., Portafaix, T., Tulet, P., Cammas, J.-P., Hauchecorne, A. et al. (2013) Maïdo observatory: a new high-altitude station facility at Reunion Island (21° S, 55° E) for long-term atmospheric remote sensing and in situ measurements, *Atmos. Meas. Tech.*, 6, 2865- 2877, doi:10.5194/amt-6-2865-2013.

[5] Khaykin S., Hauchecorne A., Porteneuve J., Mariscal J.-F., D'Almeida E., Cammas J.-P., Payen G., Evan S., Keckhut P. (2016) Ground-based Rayleigh-Mie Doppler liar for wind measurements in the middle atmosphere, EPJ Web of Conferences, EDP Sciences, 2016, The 27th International Laser Radar Conference (ILRC 27), 110, pp.13005.

COMPARISON OF RAYLEIGH-SCATTER AND SODIUM RESONANCE LIDAR TEMPERATURES

Leda Sox[1]*, Vincent B. Wickwar[1], Tao Yuan[1], Neal R. Criddle[1]

[1]Center for Atmospheric and Space Sciences and Physics Department: Utah State University, USA
**leda.sox@gmail.com*

ABSTRACT

We present an unprecedented comparison of temperature measurements from the Rayleigh-scatter (RS) and sodium (Na) lidar techniques. The extension of the RS technique into the lower thermosphere that has been achieved by the group at Utah State University (USU), enables simultaneous, common-volume measurements by the two lidar systems hosted in the Atmospheric Lidar Observatory at USU. The two lidars' nightly averaged temperatures from 80-105 km, based on 19 nights of observations, are explored.

1 INTRODUCTION

Lidar systems remain the most advantageous method for acquiring temperature measurements of the Mesosphere-Lower Thermosphere (MLT) region (~45-115km), in terms of vertical and temporal resolution. Two of the most widely used lidar techniques for MLT studies are RS and Na lidar. RS lidar systems measure elastic backscatter from neutral N_2, O_2, Ar, and O particles in the atmosphere. RS lidar backscatter measurements give relative density profiles, which are then used to calculate absolute temperature profiles [1, 2]. Na lidar measures laser-induced resonant fluorescence from the mesospheric Na layer, which forms in the 80-105 km region of the atmosphere due to meteor ablation. With proper design, Na lidars can measure thermal broadening and Doppler shifts of the laser-induced fluorescent Na spectrum. From this, Na density, temperature, and winds can be deduced [3].

Two climatological studies have compared results from the RS and Na lidar techniques using composite year averages from lidar sites spread around the globe, over different observational periods [4, 5]. In Argall and Sica [4], RS temperatures were found to be 7 K cooler than Na temperatures in the altitude range of 80-95 km. Leblanc et al. [5], found RS temperatures, from 82-88 km, to be 8-14 K colder than the Na temperatures from 80-82 km and 2-6 K colder from 82-88 km. The present study aims to explore these temperature differences using RS and Na lidar measurements made simultaneously, from the same facility, and over the same observational range.

2 METHODOLOGY

Between summer 2014 and summer 2015, there were 19 nights, which spanned all four seasons, when the two USU lidars made simultaneous measurements in the MLT region throughout the night. This dataset is relatively small due to the different observational schedules that are employed by each group. Since the Na lidar can observe over full diurnal cycles, the Na lidar group typically conducts campaigns once a month for three to five days and nights. While the RS lidar cannot currently operate in the daytime, the RS group aims to observe every clear night throughout the year. The difference in observational schedules allows the two lidar systems' datasets to complement each other well, but also means simultaneous measurements are infrequent unless deliberately planned. The RS system is capable of acquiring temperature profiles from 70-115 km, while the Na lidar's temperatures can span 80-110 km. For the most part, however, the two lidar systems measurements overlapped in the 80-105 km range, since these altitude limits can change from night-to-night depending on atmospheric and instrumental conditions for the RS lidar and on the width of the sodium layer for the Na lidar.

Simultaneous, nightly-averaged temperature profiles were calculated for each lidar's dataset to avoid biases induced by atmospheric gravity waves. Each lidar's nightly averages are at least four hours long, and the beginning and end times for both averages were set to be within two minutes of one another before data processing. The temperature profiles from each lidar, along with a profile from the MSISe00 model [6] at the USU location, were investigated for all 19 nights. Figure 1 shows four examples of these lidar comparisons. The error bars plotted with the RS and Na curves were calculated by propagating the measurement error (from photon counting) through each lidar's respective temperature reduction process [2, 3]. A seed temperature is required for the top RS altitude in order to initiate the RS temperature reduction. These were taken from the Na lidar's temperatures [e.g. Figure 1(d)], except on nights when the RS lidar's temperatures went higher than the Na lidar's [e.g. Figure 1(a), (b), and (c)]. On these nights, values from MSISe00 are used.

3 RESULTS

The four temperature curves shown in Figure 1 are from each of the four seasons between summer 2014 and summer 2015. Figure 1 shows that, although the two sets of temperatures agree very well during most of the lidar campaign, differences between the two lidars' temperature measurements are noticeable in some of the 19 nights' observations. For example, in Figure 1(a), very large temperature differences (ΔT up to ~50 K, where $\Delta T = T_{RS} - T_{Na}$) between the two lidars' temperatures are seen above 90 km, while the geomagnetic activity was low. In Figure 1(b), the two temperature curves show the best agreement at all altitudes. Figures 1(c) and (d) exemplify the typical behavior of the whole dataset. This behavior is characterized by the RS lidar temperatures being slightly cooler ($-5\ K \leq \Delta T < 0\ K$) at lower altitudes (80-85 km); the two temperature curves being within a few degrees ($-5\ K \leq \Delta T \leq 5\ K$) of one another in the middle of the altitude range

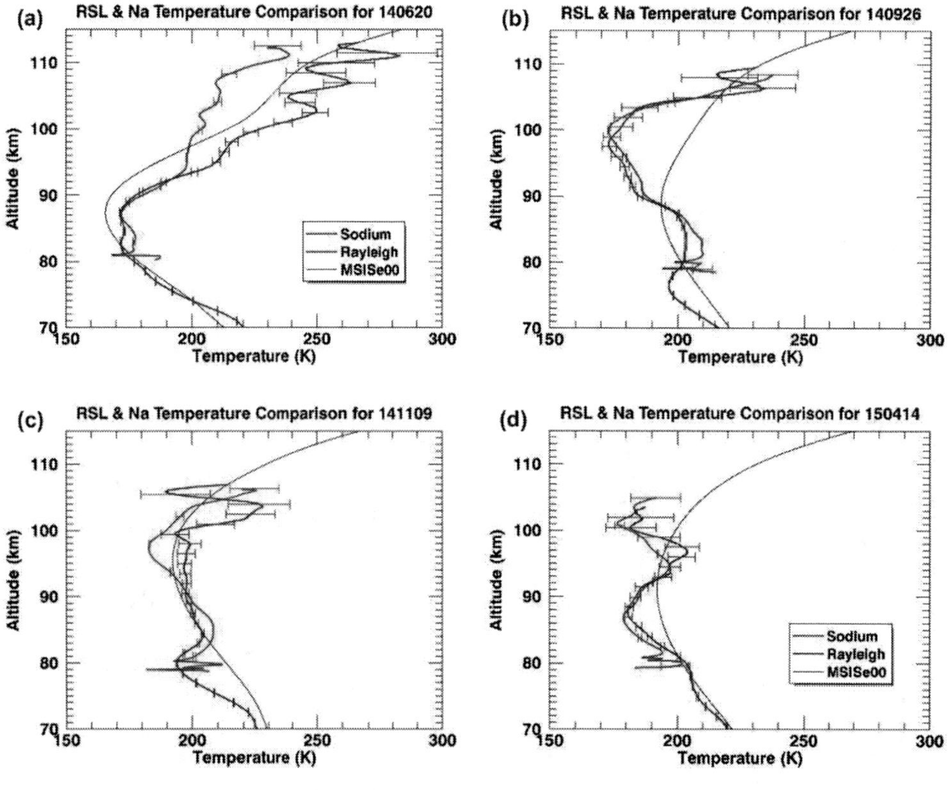

Figure 1 Four representative temperature plots for nightly averages determined using the RS lidar (blue curves) and Na lidar (red curves). MSISe00 model temperatures (black curves) for each date at 6 UT and USU's geographical coordinates, are also given. A profile from each season (summer, fall, winter, and spring) is presented: (a) 20 June 2014, (b) 26 September 2014, (c) 09 November 2014, and (d) 14 April 2015.

(85-95 km); and the RS temperatures being warmer (5 K≤ΔT≤50 K), at higher altitudes (95 km and higher). These differences at altitudes above 95 km are often statistically significant, exceeding both lidars' error bars [e.g. Figure 1(a), (c), and (d)]. This pattern is elucidated in Table 1, which gives the number of nights that ΔT fell into a specific interval at a given height. Here, each lidar's temperatures were averaged over 1 km intervals (e.g. 85 to 86 km, 90 to 91 km, etc.) and then used to calculate ΔT.

On some nights, there is also a difference in the structure of the temperature profile at higher altitudes, where the RS temperatures show stronger and more distinct wave structure than the Na temperatures. Examples of this incongruent wave structure can be seen in Figure 1(a). Similar distinct structures are also observed in 6 of the 19 nights throughout the year (140620, 140723, 140912, 140913, 141029, and 141106, in YYMMDD). These wave features in the RS profiles are puzzling and cannot be explained properly at the moment. The Na lidar has showed that long lasting, large-scale temperature inversion structures exist in the mesopause region due to wave breaking [7]. However, it did not detect any of these structures at these altitudes during these nights. It would be highly difficult for such long-period inertia gravity waves to reach near and beyond 100 km due to the severe atmospheric dissipation they experience, as they slowly propagate upward.

Another interesting observation is that the nights with the best agreement between the two lidars' temperatures, over the entire 80-105 km altitude range, are all close to the fall (140923) and spring (150320) equinoxes, which can be seen in Figure 1(b) and was also observed on 140925 and 150328.

The two lidars' temperatures agree much better with one another than they do with the MSISe00 model temperatures. For the most part, if one lidar's temperature profile is either warmer or colder than the MSISe00 temperatures, then the other lidar's temperatures behave the same way. There are 5 exceptions, however, throughout the year (140620, 140722, 140912, 141108, and 141109), an example of which can be seen in Figure 1(a). In all of these cases, above 90 km, the Na temperatures are colder than MSISe00 temperatures, whereas the RS temperatures are warmer. Two of these nights were also nights when the RS lidar showed the strong wave structure mentioned before. While the overall structure of the lidars' temperature profiles are roughly similar to the MSISe00 structure, there were 7 cases, mostly in the fall and winter, (140912, 140913, 140925, 140926, 141104, 141106, and 141108) where the lidars' mesopauses were at different altitudes than the MSISe00 mesopause, as in Figure 1(b) and (c).

4 CONCLUSIONS

We have presented a unique comparison of simultaneous temperatures acquired by a Rayleigh-scatter lidar and a sodium resonance lidar collocated in the Atmospheric Lidar Observatory on the campus of Utah State University. The two sets of temperature data also overlap in altitude (80-105 km). Our simultaneous, common-volume

Table 1 Number of nights, out of the 19, having a temperature difference interval at specific altitude levels

	80 km	85 km	90 km	95 km	100 km	105 km
20K≤ΔT					3	**2**
15K≤ΔT<20K				1		**2**
10K≤ΔT<15K	1			4	4	1
5K≤ΔT<10K	2		1	**6**	2	1
0K≤ΔT<5K	2	6	6	4	**5**	1
-5K≤ΔT<0K	**5**	**8**	**9**	3	1	2
-10K≤ΔT<-5K	4	3	2	1	3	1
-15K≤ΔT<-10K	2	2	1		1	1
-20K≤ΔT<-15K						
ΔT<-20K						1

RS and Na lidar measurements showed, like previous climatological comparisons in the 80-90 km region [4, 5], that the RS lidar's temperatures were on average colder than the Na lidar's temperatures. However, the magnitude of our differences (-5 K≤ΔT<0 K), was smaller than the previous studies'. Above 95 km, we have shown new results that were not possible in the previous studies, due to the altitude limitations of older RS systems. The results indicate that the RS temperatures are warmer than the Na temperatures in this region (5 K≤ΔT<50 K). The RS lidar temperatures also show stronger and more distinct wave activity than the Na temperatures above about 90 km, which demands more investigation in the future. The best agreement between the two techniques' temperatures, throughout the entire 80-105 km range, occurs on the nights, albeit only three, closest to the equinoxes. The two lidar techniques typically agree with one another better than each does with the MSISe00 model. When they disagree, the RS lidar's temperatures are typically warmer than MSISe00, while the Na lidar's temperatures are colder.

In the future, we aim to take enough simultaneous measurements with the two USU lidars to obtain good coverage throughout all months in order to further explore the day-to-day and seasonal differences between the two systems' deduced temperatures and, especially, to explore the differences above 95 km.

ACKNOWLEDGEMENTS

The original RS lidar was initially upgraded to the much more sensitive configuration with funds from NSF, AFOSR, and USU. The system was further upgraded to bring it on line with funds provided by the Space Dynamics Laboratory Internal Research and Development program, USU, the USU Physics Department, and personal contributions. The Na lidar was supported under NSF AGS grant number 1135882. The RS and Na lidar data presented in this paper were acquired through the dedicated efforts of many USU student operators and engineers.

References

[1] Hauchecorne, A., and Chanin, M.-L., 1980: Density and temperature profiles obtained by lidar between 35 and 70 km, *Geophys. Res. Lett.*, **7**, 565–568, doi:10.1029/GL007i008p00565.

[2] Sox, L., 2016: Rayleigh-Scatter Lidar Measurements of the Mesosphere and Thermosphere and Their Connections to Sudden Stratospheric Warmings, *Utah State Univ.*, Logan, UT, 212pp, http://digitalcommons.usu.edu/etd/5227.

[3] Krueger, D. A., She, C. Y., Yuan, T., 2015: Retrieving mesopause temperature and line-of-sight wind from full-diurnal-cycle Na lidar observations, Appl. Opt., **54**, 9469–9489, doi:10.1364/AO.54.009469.

[4] Argall, P. S., and Sica, R. J., 2007: A comparison of Rayleigh and sodium lidar temperature climatologies, Ann. Geophys., **25**, 27–33, doi:10.5194/angeo-25-27-2007.

[5] Leblanc, T., McDermid, I. S., Keckhut, P., Hauchecorne, A., She, C. Y., Krueger, D. A., 1998: Temperature climatology of the middle atmosphere from long-term lidar measurements at middle and low latitudes, *J. Geophys. Res.*, **103**(D14), 17191–17204, doi:10.1029/98JD01347.

[6] Picone, J. M., Hedin, A. E., Drob, D. P., Aiken A. C., 2002: NRLMSISE-00 empirical model of the atmosphere: Statistical comparisons and scientific issues, *J. Geophys. Res.*, **107**(A12), 1468, doi:10.1029/2002JA009430.

[7] Yuan, T., Pautet, P.-D., Zhao, Y., Cai, X., Criddle, N. R., Taylor, M. J., Pendleton Jr., W. R., 2014: Coordinated investigation of mid-latitude upper mesospheric temperature inversion layers and the associated gravity wave forcing by Na lidar and Advanced Mesospheric Temperature Mapper in Logan, Utah, *J. Geophys. Res. Atmos.*, **119**, 3756–3769, doi:10.1002/2013JD020586.

MINERAL DUST IN CENTRAL ASIA: 18-MONTH LIDAR MEASUREMENTS IN TAJIKISTAN DURING THE CENTRAL ASIAN DUST EXPERIMENT (CADEX)

Julian Hofer[1,*], Dietrich Althausen[1], Sabur F. Abdullaev[2], Abduvosit Makhmudov[2], Bakhron I. Nazarov[2], Georg Schettler[3], K. Wadinga Fomba[1], Konrad Müller[1], Bernd Heinold[1], Holger Baars[1], Ronny Engelmann[1] and Albert Ansmann[1]

*[1]Leibniz Institute for Tropospheric Research (TROPOS), Leipzig, Germany, *hofer@tropos.de*
[2]Physical Technical Institute of the Academy of Sciences of Tajikistan, Dushanbe, Tajikistan
[3]GFZ German Research Centre for Geosciences, Potsdam, Germany

ABSTRACT

Tajikistan is often affected by atmospheric mineral dust. The direct and indirect radiative effects of dust play a sensitive role in the climate system in Central Asia. The Central Asian Dust Experiment (CADEX) provides first lidar measurements in Tajikistan. The autonomous multiwavelength polarization Raman lidar Polly^{XT} was operated for 1.5 years (2015/16) in Dushanbe. In spring, lofted layers of long-range transported dust and in summer/autumn, lower laying dust from local or regional sources with large optical thicknesses occurred.

1 INTRODUCTION

Dust influences weather and climate on earth by direct and indirect effects. Central Asia is strongly affected by climate change which is manifested in Tajikistan in a severe glacier retreat [1]. Glacier melting rates are also influenced by dust due to the darkening of the glacier's surfaces [2]. The glaciers of Tajikistan are the major water resources of the rivers which were originally feeding the Aral Sea, which now became itself a strong dust source [3]. Therefore aerosol observations in Tajikistan are extremely valuable to understand regional and global transport of mineral dust. Only few experiments were performed to characterize the aerosol over Central Asia. During a Soviet-American campaign in 1989 two dust storms were observed [4]. During these dust storms aerosol optical thicknesses (AOT) of up to 3.5 (at 550 nm) were measured and a couple of vertical profiles of the dust distribution were measured by aircraft [5]. Dust transported from the Aral Sea region, from the Middle East and from the Taklamakan desert was reported from more recent lidar measurements conducted in Kyrgyzstan [6, 7]. Nevertheless, the knowledge of the vertical aerosol distribution over Tajikistan and especially the transport of mineral dust over Central Asia is still poor. Therefore the Central Asian Dust Experiment (CADEX) was performed to obtain long-term data on vertical profiles of optical aerosol properties.

2 INSTRUMENTS AND METHODS

The CADEX campaign was conducted at a field site of the Physical Technical Institute of the Academy of Sciences of Tajikistan in Dushanbe. Its coordinates are 38°33'34" N, 68°51'22" E, and its altitude is 864 m a.s.l..
A multiwavelength polarization Raman lidar Polly^{XT} [8] is used for the continuous lidar measurements. All Polly^{XT} systems emit linearly polarized light at 355 nm, 532 nm, and 1064 nm wavelength generated by a Nd:YAG laser and 2nd and 3rd harmonic generators. Vertical profiles of the particle backscatter coefficient are determined for the three emitted wavelengths. The system measures cross-polarized light in two channels to retrieve the particle linear depolarization ratio (PLDR) at 355 nm and 532 nm wavelength. The inelastic channels at 387 nm

and 607 nm allow to determine particle extinction coefficients at 355 nm and 532 nm with the Raman method [9] at night-time. For water vapour measurements a channel at 407 nm is installed. The system's vertical resolution is 7.5 m and the data is stored with a temporal resolution of 30 s.

Additionally, a CIMEL sun photometer from the AERONET station in Dushanbe is collocated with the Polly^{XT} lidar to measure the AOT and retrieve further particle properties.

The dust and non-dust mass concentrations are indirectly calculated with the polarization lidar photometer networking method (POLIPHON) [10, 11].

3 RESULTS

The Polly^{XT} performed measurements from 17 March 2015 until 31 August 2016. During that period, data was acquired for more than 3 hours at 407 days and at 307 days even for more than 20 hours. The quicklooks of all measurements are available at http://polly.tropos.de.

During spring, numerous lofted dust layers up to 10 km above ground were observed. The summer months in Dushanbe are commonly dusty from the ground to certain heights with large AOT. For example, the mean AOT at 500 nm measured with the sun photometer (AERONET level 1.5) in August 2015 was 0.41 and in August 2016 it was 0.28.

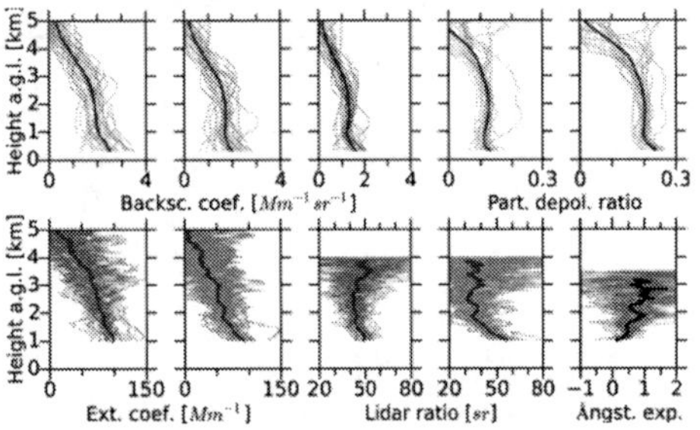

Figure 1: Vertical profiles of Raman-retrieved optical properties (with 563 m vertical smoothing) at each possible night (N=28) during August 2016. Blue for 355 nm, green for 532 nm, red for 1064 nm wavelength, black for Ångström exponent for the 355/532 nm extinction pair. The fat profiles denote the average over the respective profiles.

Figure 1 shows the optical properties measured during August 2016 at each night at which it was possible to retrieve profiles of the optical properties with the Raman method. The extinction coefficients at 355 nm and 532 nm are similar, on average decreasing from about 100 Mm^{-1} to 0 at about 5 km. The average lidar ratio between 1.5 and 3 km is 45 ± 1 sr at 355 nm and 36 ± 3 sr at 532 nm. Note that these cases are not only pure dust cases, but include polluted dust and also pollution events. The PLDR values are large even on average, namely 0.12 ± 0.01 at 355 nm and 0.20 ± 0.01 at 532 nm.

An outstanding profile with larger PLDR values in both wavelengths is measured on 16 August 2016 and will be discussed in detail below. The temporal development of the range corrected signal at 1064 nm at that day as well as the profiles of the optical properties averaged on 16 August 2016, 18:00-19:00 UTC are shown in Fig. 2.

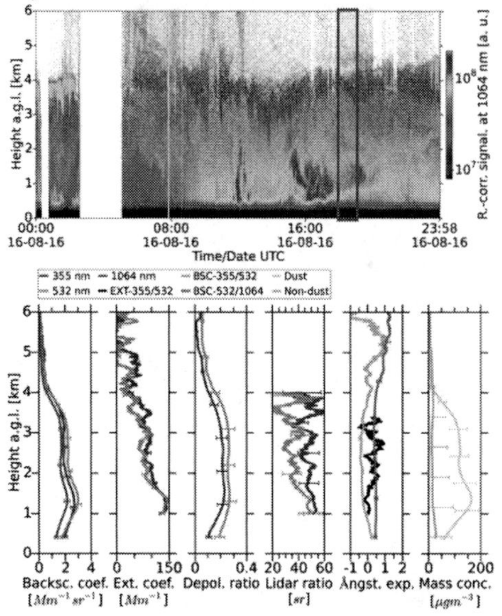

Figure 2: Top: Temporal development of the range-corrected signal at 1064 nm wavelength on 16 August 2016. Bottom: Averaged vertical profiles of the optical properties on 16 August 2016, 18:00–19:10 UTC (pink rectangle on the top).

On this day, there were sporadic clouds and even some rain occurred in the morning. After 12:00 UTC and after 16:00 UTC aerosol plumes were rising up to about 2.5 km height. The backscatter and extinction coefficients at 532 nm wavelength have their maximum at 1.35 km height with about 3 $Mm^{-1}sr^{-1}$ and 150 Mm^{-1} respectively. The PLDR values are large between 1.3 and 3.5 km altitude and indicate dust dominated conditions. The calculated dust mass concentration reaches 164 $\mu g\ m^{-3}$ and the integrated dust mass is 442 mg m^{-2}.
Table 1 summarizes the averaged intensive optical properties and the AOT measured on 16 August 2016, 18:00–19:10 UTC. The PLDR values are lower than known values for pure (Saharan) dust (0.24–0.28 at 355 nm, 0.30–0.32 at 532 nm) [12]. The values at 355 nm are comparable with PLDR values of transported (Saharan) dust (0.21 at 355 nm) [13]. The measured lidar ratios are lower than for Saharan dust (35–70 sr at 335 nm, 45–65 sr at 532 nm) [14] and comparable to Middle Eastern/West-Asian dust (34–39 sr at 532 nm) [15].

Table 1: S_{355},S_{532} lidar ratio, δ^p_{355}, δ^p_{532} PLDR, τ_{lidar} AOT from lidar at 532 nm wavelength, τ_{SPM} AOT from sun photometer at 500 nm wavelength, $\mathring{\alpha}_{355/532}$ extinction-related Ångström exponent.

Date	16 August 2016
Time [UTC]	18:00–19:10
Layer range [km]	0.9–3.8
Averaging [km]	1.7–3.0
S_{355} [sr]	47.7 ± 3.1
S_{532} [sr]	35.5 ± 3.7
δ^p_{355}	0.22 ± 0.01
δ^p_{532}	0.26 ± 0.01
τ_{lidar}	0.42
τ_{SPM} (13:08 UTC)	0.33
$\mathring{\alpha}_{355/532}$	0.38 ± 0.24

4 CONCLUSIONS

The CADEX campaign provided first long-term measurements of vertical profiles of aerosol optical properties over Tajikistan. The presented measurements from August 2016 show dusty summer conditions in the atmosphere over Tajikistan. The aerosol measured on 16 August 2016 is identified as mixed dust with a significant dust contribution.
A comparison of CADEX lidar data with chemical analysis of the aerosol at ground as well as with satellite measurements can be found in another contribution to this conference [16].

ACKNOWLEDGEMENTS

The CADEX project was funded by the German Federal Ministry of Education and Research in the context of "Partnerships for sustainable problem solving in emerging and developing countries" under the grant number 01DK14014. This project has also received funding from the European Union's Horizon 2020 research and innovation programme under grant agreement No 654109. We would like to thank the Academy of Sciences of Tajikistan, Lars Klüser from German Aerospace Center, Emmanouil Proestakis from National Observatory of Athens, Alexandra Chudnovsky from University of Tel Aviv, and Konrad Kandler from TU Darmstadt for their support.

References

[1] Kayumov, A., 2010: Glaciers Resources of Tajikistan in Condition of the Climate Change, *State Agency for Hydrometeorology of Committee for Environmental Protection under the Government of the Republic of Tajikistan*, https://www.wmo.int/pages/prog/www/OSY/Meetings/GCW-IM1/glaciers.pdf.

[2] Gabbi, J., Huss, M., Bauder, A., Cao, F., Schwikowski, M., 2015: The impact of Sa-

haran dust and black carbon on albedo and long-term mass balance of an Alpine glacier, *Cryosphere*, **9** (4), 1385–1400.

[3] Xi, X. and Sokolik, I. N., 2016: Quantifying the anthropogenic dust emission from agricultural land use and desiccation of the Aral Sea in Central Asia, *J. of Geophys. Res. Atmos.*, **121** (20), 12,270–12,281.

[4] Golitsyn, G. and Gillette, D. A., 1993: Introduction: A joint Soviet-American experiment for the study of Asian desert dust and its impact on local meteorological conditions and climate, *Atmos. Environ.*, **27** (16), 2467–2470.

[5] Pachenko, M. V., Terpugova, S. A., Bodhaine, B. A., Isakov, A. A., Sviridenkov, M. A., Sokolik, I. N., Romashova, E. V., Nazarov, B. I., Shukurov, A. K., Chistyakova, E. I., Johnson, T. C., 1993: Optical investigations of dust storms during U.S.S.R.-U.S. experiments in Tadzhikistan, 1989, *Atmos. Environ.*, **27** (16), 2503–2508.

[6] Chen, B. B. and Sverdlik, L. G., 2007: Optical and microphysical characteristics of aerosol structures in Central Asia, *International Conf. on Lasers, Applications, and Technologies 2007*, **6733**, 67330S.

[7] Chen, B. B., Sverdlik, L. G., Imashev, S. A., Solomon, P. A., Lantz, J., Schauer, J. J., Shafer, M. M., Artamonova, M. S., Carmichael, G. R., 2013: Lidar Measurements of the Vertical Distribution of Aerosol Optical and Physical Properties over Central Asia, *Int. J. Atmos. Sci.*, **2013**, 261546.

[8] Engelmann, R., Kanitz, T., Baars, H., Heese, B., Althausen, D., Skupin, A., Wandinger, U., Komppula, M., Stachlewska, I. S., Amiridis, V., Marinou, E., Mattis, I., Linné, H., Ansmann, A., 2016: The automated multiwavelength Raman polarization and water-vapor lidar Polly^{XT}: the neXT generation, *Atmos. Meas. Tech.*, **9** (4), 1767–1784.

[9] Ansmann, A., Riebesell, M. A., Weitkamp, C., 1990: Measurement of atmospheric aerosol extinction profiles with a Raman lidar, *Opt. Lett.*, **15** (13), 746–748.

[10] Ansmann, A., Seifert, P., Tesche, M., Wandinger, U., 2012: Profiling of fine and coarse particle mass: case studies of Saharan dust and Eyjafjallajökull/Grimsvötn volcanic plumes, *Atmos. Chem. Phys.*, **12** (20), 9399–9415.

[11] Mamouri, R. E. and Ansmann, A., 2014: Fine and coarse dust separation with polarization lidar, *Atmos. Meas. Tech.*, **7** (11), 3717–3735.

[12] Freudenthaler, V., Esselborn, M., Wiegner, M., Heese, B., Tesche, M., Ansmann, A., Müller, D., Althausen, D., Wirth, M., Fix, A. Ehret, G., Knippertz, P., Toledano, C., Gasteiger, J., Garhammer, M., Seefeldner, M., 2009: Depolarization ratio profiling at several wavelengths in pure Saharan dust during SAMUM 2006, *Tellus B*, **61** (1), 165–179.

[13] Burton, S. P., Hair, J. W., Kahnert, M., Ferrare, R. A., Hostetler, C. A., Cook, A. L., Harper, D. B., Berkoff, T. A., Seaman, S. T., Collins, J. E., Fenn, M. A., Rogers, R. R.. 2015: Observations of the spectral dependence of linear particle depolarization ratio of aerosols using NASA Langley airborne High Spectral Resolution Lidar, *Atmos. Chem. Phys.*, **15** (23), 13453–13473.

[14] Tesche, M., Ansmann, A., Müller, D., Althausen, D., Mattis, I., Heese, B., Freudenthaler, V., Wiegner, M., Esselborn, M., Pisani, G., Knippertz, P., 2009: Vertical profiling of HSRL in southern Morocco during SAMUM, *Tellus B*, **61** (1), 144–164.

[15] Mamouri, R. E., Ansmann, A., Nisantzi, A., Kokkalis, P., Schwarz, A., Hadjimitsis, D. G., 2013: Low Arabian dust extinction-to-backscatter ratio. *Geophys. Res. Lett.*, **40** (17), 4762–4766.

[16] Althausen, D., Hofer, J., Abdullaev, S. F., Makhmudov, A., Baars, H., Engelmann, R., Fomba, K. W., Müller, K., Schettler, G., Klüser, L., Kandler, K.,: 2017: Combining lidar and other measurements during the Central Asian Dust Experiment (CADEX), *Proceedings of the 28th IRLC*, Bucharest.

ESTIMATE OF RAIN EVAPORATION RATES FROM DUAL-WAVELENGTH LIDAR MEASUREMENTS: COMPARISON AGAINST A MODEL ANALYTICAL SOLUTION

Simone Lolli[1]*, Paolo Di Girolamo[2], Belay Demoz[3], Xiaowen Li[4], Ellsworth J. Welton[5]

[1]*NASA GSFC-JCET, Code 612, Greenbelt, MD, 20771, USA *simone.lolli@nasa.gov*
[2]*Università della Basilicata, Via Sauro 85, 85100 Potenza, Italy*
[3] *UMBC-JCET, 1000 Hilltop Circle, 21250 Baltimore, MD, USA*
[4] *NASA GSFC-GESTAR, Code 612, 20771 Greenbelt, MD, USA*
[5] *NASA GSFC, Code 612, 20771 Greenbelt, MD, USA*

ABSTRACT

Rain evaporation significantly contributes to moisture and heat cloud budgets. In this paper, we illustrate an approach to estimate the median volume raindrop diameter and the rain evaporation rate profiles from dual-wavelength lidar measurements. These observational results are compared with those provided by a model analytical solution. We made use of measurements from the multi-wavelength Raman lidar BASIL.

1. INTRODUCTION

Rain evaporation, while significantly contributing to moisture and heat cloud budgets, is still a poorly understood process with few measurements presently available. In this paper we describe an application of the color ratio technique (*CR*) [1] for the estimation of rain evaporation rates based on the use of a dual-wavelength lidar, which is an active optical device particularly suited to investigate atmospheric constituents, especially aerosols [2], [3], [4] and clouds [5], [6]. Several authors (among others [8]) reported lidar measurements during light precipitation events, i.e. drizzle and virga. In particular in [1], a technique is developed to retrieve the median volume raindrop diameter profile from the measurements performed by two co-located commercial lidars operating at 355 nm and 532 nm. This technique allows retrieving the median volume raindrop diameter profile and the evaporation rate, which in turn it is used to validate a model analytical solution [10]

2. METHODOLOGY

Several approaches to retrieve the median volume raindrop diameter during light precipitation events using two-wavelength lidar backscatter data have been illustrated by different authors [1], [9]. All these approaches take advantage of the sensitivity of the differential backscattering efficiency at the two wavelengths to the rain Drop Size Diameter, DSD (UV_{355nm}-VIS_{532nm} in [1], IR_{905nm}-IR_{1500nm} in [9]). For example, the approach reported in [1] relies on the consideration that the backscattering efficiency at 532 nm is much less sensitive to water sphere diameter with respect to the backscattering efficiency at 355 nm (while the extinction coefficient are wavelength independent). This is depicted in Fig. 1a and 1b, which show how the backscattering efficiencies calculated using an algorithm based on Mie theory for large size parameter values (i.e. for large particles with respect to the sounding wavelength) [1] are dependent on water sphere diameter at 355nm (blue dots) and 532nm (green dots). Integrating those values of the backscattering efficiencies (Fig. 1a) over a modified normalized gamma raindrop size distribution (see Eq. 4 in [1]) for D_0=1-1000 μm, the differences between the backscattering efficiencies in function of the median volume raindrop diameter D_0 become substantial, allowing to define the color ratio (Fig. 1c) as ten times the logarithm of the ratio of the rain backscattering coefficients at the two wavelengths. From the measured color ratio it is possible retrieving the median volume raindrop diameter at each range bin using the curves in Fig. 1c as look-up tables. Accurate information on the method is beyond the scope of this paper and can be found in [1]. Subsequently, the evaporation rate can be estimated as the change of the raindrop volume (shrinking/growing) during the precipitation along the profile for each two adjacent levels as (Eq. 1):

$$E(r) = \frac{(d_{r+1})^3 - d_r{}^3}{(d_{r+1})^3} \quad (1)$$

where the subscripts r and $r+1$ indicate the raindrop diameter d at current range bin r and immediately above ($r+1$). The evaporation rate is closely related to the humidity content of the surrounding atmosphere and to atmospheric processes as breakdown or coalescence. The raindrop diameter and evaporation rate retrieved by lidar with the method described above can be finally compared with the analytical solution model proposed in [10].

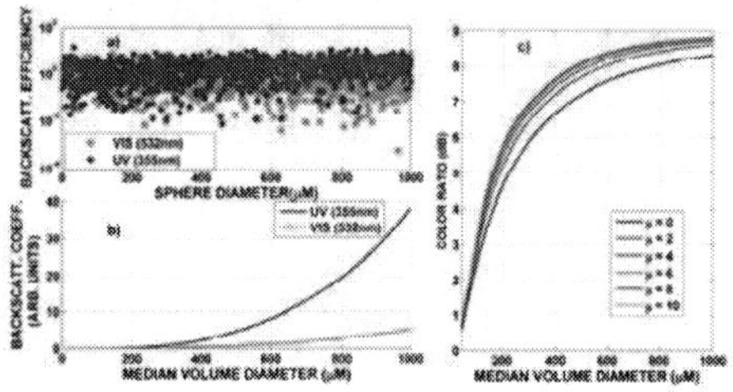

Figure1 a): Backscattering efficiencies obtained with Mie code vs. sphere diameter for 532nm (green dots) and 355nm (blue dots); b) Backscattering coefficient vs. the median volume at two different wavelengths (355 and 532nm, respectively); c) Color ratio obtained from the ratio of the two rain backscattering coefficients in b) for different values of the shape parameter μ of the gamma raindrop distribution

2.1 Analytical Model

The analytical model determines the "evaporation power" of a stationary atmospheric layer, based on the computation of the single raindrop diameter D* (dependent only on temperature and water vapor), defined as the diameter of a single raindrop fully evaporating in the selected layer [10]. The model relies on the assumption of a steady environment with negligible vertical air motion. In [10] these conditions were assumed to be valid for stratiform rain, but in the present paper these conditions are extended to selected virga episodes. Instead of calculating D*, in this study we compute the temporal evolution of each single raindrop diameter present in the precipitation at each different atmospheric layer, from cloud base to complete evaporation [10]. The thickness of each layer is defined from the radiosonde spatial resolution. The temporal evolution of the raindrop diameter only depends on the thermodynamic properties of the considered atmospheric layer (measured by the radiosonde). A raindrop diameter D_1 at cloud base (at a certain altitude h_1) evolves into diameter D_2, at the bottom of the first considered layer (at a certain altitude h_2). Then D_2 becomes the input diameter of the top of the second considered layer and so on, down to the last available layer before complete evaporation. The thermodynamic properties of the atmospheric layer with a thickness defined by h_1-h_2 are determined from the radiosonde.

3. RESULTS

On 23 July 2007 [8], a cyclonic system, developed in the northern Atlantic region, moved toward the COPS measurement site, producing stratiform clouds with consequent stratiform rain occurring in the afternoon. These steady meteorological conditions are optimal to apply the analytical model. The time evolution of the backscattering ratio at 1064 nm for the time interval 13:00-14:35 UT is shown in Fig. 2. The figure clearly reveals the presence of a lidar dark band [8] around 2.8 km, associated with changes in scattering properties of precipitating hydrometeors taking place in the melting layer during the snowflake-to-raindrop transition. This figure also reveals the presence of stratiform clouds, with cloud base at 3.1–3.3 km above ground level (a.g.l.). Ancillary information on the thermodynamic state of the atmosphere was provided by *ad-hoc* radiosondes, launched every three hours during the measurement session [8].

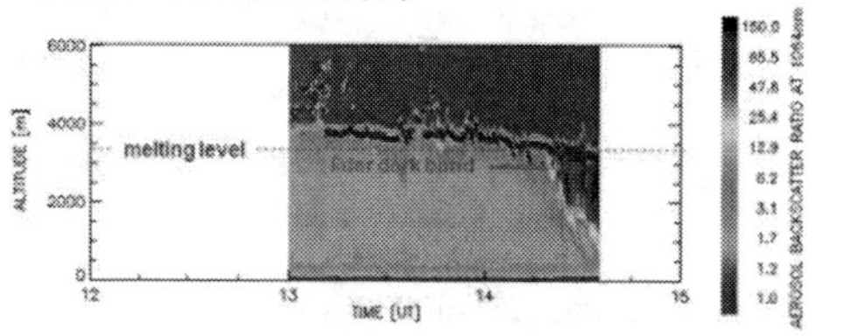

Figure 2 Time evolution of the backscattering ratio at 1064nm for the time interval 13:00-14:35 UT on 23 July 2007 as measured by the lidar system BASIL

The profile of raindrop diameter corresponding to the above estimated *CR* profile is illustrated in Fig. 3 (blue line), with the relative uncertainty that takes into account also the total error on the measured color ratio. This error includes contributions from the signal measurement

https://doi.org/10.1051/epjconf/201817604002

uncertainty, the two-way transmission error, and the atmospheric density uncertainty. The total *CR* error is then related to the uncertainty of the measured range-corrected signal that it is, in turn, related to the Poisson photon-counting statistic at each range level; the uncertainty on the molecular backscattering is related to the molecular density profile and for the error in aerosol transmission is assumed to be constant (8%, [1]) while the error on molecular transmission is related to the again to the molecular density. The median volume raindrop diameter ranges from 100 to 350 μm.

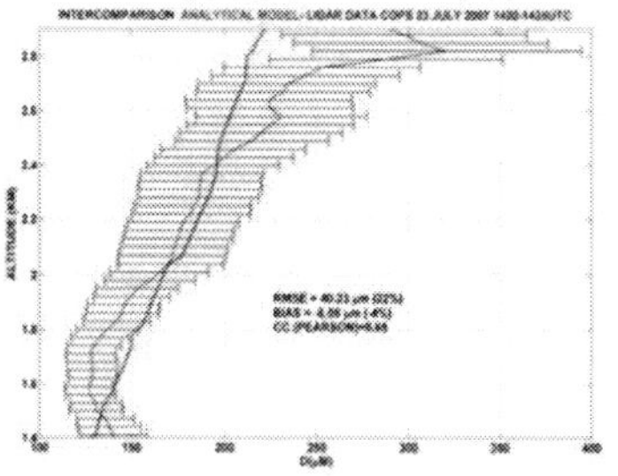

Figure 3: Vertical profile of the median volume raindrop diameter (blue line, with error bar) as determined from the lidar measurements on 23 July 2007 and model analytical solution (red line). The intercomparison shows a 22% Root Mean Square Error and 17% Mean Fractional Error with 5% bias. Regarding the shape, the Pearson Correlation Coefficient shows a positive correlation of 65%.

The model analytical solution for the median volume raindrop diameter D_0 (also reported in Fig. 3, red line) is calculated assuming constant values for water vapor mixing ratio and temperature within each layer, these values taken from the simultaneous and co-located radiosonde profile (launched from the lidar station at 14:06 UTC on 23 July 2007). In this respect, it is to be noticed that, even when launched from the same location of the lidar, the radiosonde is not perfectly co-located with the vertical column illuminated by the lidar, as in fact the former may undergo horizontal drifts due to the wind during its ascent. However, the radiosonde data considered in this case study indicate an overall horizontal drift not exceeding 1.8 km at 3 km; this aspect, together with the verification of the very limited variability of the radar reflectivity profiles at 1.29 and 35.5 GHz (not shown here, but illustrated in Fig. 3 and Fig. 4 of [8]) confirms that the air masses sounded by the lidar and the radiosonde are substantially the same. Fig. 4 compares the vertical profile of the evaporation rate obtained from lidar measurements (blue) and from the analytical model solution (red). Lidar data show that evaporation occurs primarily in the first 400-500 m below the cloud base (down to 2.8 km), with a median volume raindrop diameter reduction in excess of 20 %, possibly associated with mixed phase/melting, while in the remaining portion of precipitation (down to 1.4 km) diameter reduction is smaller. It is to be noticed that lidar measurements catch a much larger variability in the evaporation process than the model. Large negative values in the lidar-based estimate of the evaporation rate profile are found between 2.6 and 2.8 km. We can speculate that these large negative values are partially fictitious and attributable to the progression of the melting process, partially masking evaporation. This speculation is confirmed by the aircraft in-situ measurements, which reveal that melting is still progressing in the first 400-500 m below the cloud base (reported in [8]). We believe that the effect of multiple scattering on these results is negligible. In fact, channels used to measure lidar signals at both 355 and 532 nm are characterized by a very small field-of-view (240 μrad), which minimizes potential signals' contamination by multiple scattering. Additionally, the color ratio is obtained by ratioing these two signals so that, in case of any residual multiple scattering effect on the signals, this cancels out in the color ratio. In order to obtain a proper quantitative assessment of the agreement between the model and the measurements, the root-mean square (RMS) deviation, the bias (BIAS) and the correlation coefficient [3], [11]) of model vs. lidar data have been computed and are found to be 40.23 μm (22%), -6.05 μm (-4%) and 0.65, respectively. Small values for RMS deviation and BIAS and the large value for the correlation coefficient testify the good agreement between model and lidar results.

4. CONCLUSIONS

A comparison between the model analytical solution proposed by [10] and estimates obtained from multi-wavelength lidar data in terms of the median volume raindrop diameter and rain evaporation rate was carried out for two selected case studies: one collected on 23 July 2007 by the Raman lidar system BASIL during the COPS experiment. The inter-comparison reveals good agreement between model and observations, with a correlation between the profiles up to 65%, a root mean square error up to 22% and a bias of 5%. Local higher discrepancies are due to i) the radiosonde soundings different air masses and ii) model assumptions being no longer valid along the profile, as in case of non-steady atmosphere and/or of appearance of collision-coalescence processes.

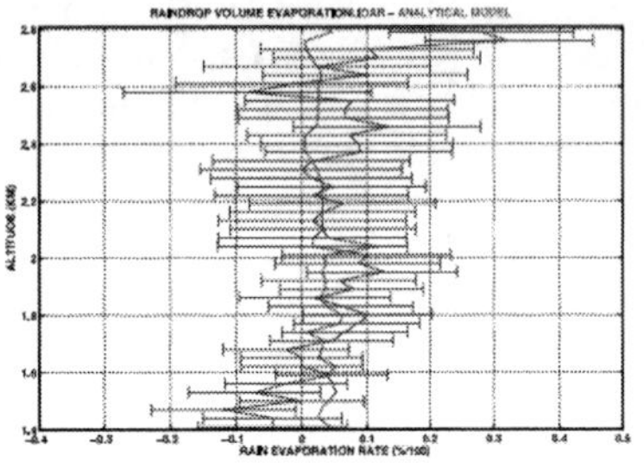

Figure 4: Raindrop evaporation rate (in %/100, blue line) and comparison with the model analytical solution (red line). In the first 200 m below the cloud base the melting process is not completely ended and the rain is mixed with few snowflakes. This determines a strong evaporation (volume shrinkage>20%), which in reality is fictitious. Breakdown is not taken into account by the model. This can explain the difference with the lidar data in the 1.4-1.8 km atmospheric layer. Negative values represent a volume growth of the raindrop.

ACKNOWLEDGEMENTS

This study and the NASA Micro Pulse Lidar Network (MPLNET) are supported by the NASA Radiation Sciences Program (H. Maring).

References

[1] Lolli S. et al, 2013: Evaluating light rain drop size estimates from multiwavelength micropulse lidar network profiling. *J. Atmos. Oceanic Technol.*, **30**, 2798–2807

[2] Di Girolamo, P. et al., 1995: Two wavelength Lidar analysis of stratospheric aerosol size distribution", *J. Aeros. Sci*, **26**, 989-1001, doi: 10.1016/0021-8502(95)00025-8.

[3] Lolli S., et al.,. 2013: 0.355-micrometer direct detection wind lidar under testing during a field campaign in consideration of ESA's ADM-Aeolus Mission, *Atmospheric Measurement Techniques*, **6**, 3349-3358, doi:10.5194/amt-6-3349-2013

[4] Tan, F. et al., 2014: Variation In Daytime Troposphereic Aerosol Via LIDAR And Sunphotometer Measurements In Penang, Malaysia AIP Proceedings 1588, **286**; doi: 10.1063/1.4866962.

[5] J. Campbell, S. Lolli J. Lewis, Y. Gu, E. Welton, 2016: Daytime Cirrus Cloud Top-of-Atmosphere Radiative Forcing Properties at a Midlatitude Site and their Global Consequence. *Journal of Applied Meteorology and Climatology*, http://dx.doi.org/10.1175/JAMC-D-15-0217.1

[6] S. Lolli, J. Lewis, J. R. Campbell, Y. Gu, E. Welton, 2016, "Cirrus Cloud Radiative Characteristics from Continuous MPLNET Profiling at GSFC in 2012. *Óptica pura y aplicada*, **49** (1), 1-6.

[7] Demoz, B. et al., 2000: Raman LIDAR Detection of Cloud Base. *Geophys. Res. Lett.*, **27**, 1899–1902.

[8] Di Girolamo, et al., 2012: Lidar and radar measurements of the melting layer: observations of dark and bright band phenomena, *Atmos. Chem. Phys.*, **12**, 4143–4157.

[9] Westbrook, C. D et al., 2010: Estimating drizzle drop size and precipitation rate using two-colour lidar measurements. *Atmos. Meas. Tech.*, **3**, 671–681, doi:10.5194/amt-3-671-2010

[10] Li, X. and Srivastava, R., 2001: An Analytical Solution for Raindrop Evaporation and Its Application to Radar Rainfall Measurements.", J. of Applied Met., **40** (9), 1607-1616

[11] S. Lolli; P. Di Girolamo. 2015: Principal component analysis approach to evaluate instrument performances in developing a cost-effective reliable instrument network for atmospheric measurements." Journal of Atmospheric and Oceanic Technology, Vol. 32 (9), 1642-1649.

https://doi.org/10.1051/epjconf/201817604003

CHARACTERISTICS OF VOLCANIC STRATOSPHERIC AEROSOL LAYER OBSERVED BY CALIOP AND GROUND BASED LIDAR AT EQUATORIAL ATMOSPHERE RADAR SITE

Makoto Abo*, Yasukuni Shibata, Chikao Nagasawa,

*Tokyo Metropolitan University, Japan; *Email: abo@tmu.ac.jp*

ABSTRACT

We investigated the relation between major tropical volcanic eruptions in the equatorial region and the stratospheric aerosol data, which have been collected by the ground based lidar observations at at Equatorial Atmosphere Radar site between 2004 and 2015 and the CALIOP observations in low latitude between 2006 and 2015. We found characteristic dynamic behavior of volcanic stratospheric aerosol layers over equatorial region.

1. INTRODUCTION

Stratospheric aerosols play important roles in climate regulation and atmospheric chemistry. Based on some satellite data, the stratospheric aerosol optical thickness (AOT) increased after 2000 as the result of a series of moderate but increasingly intense volcanic eruptions [1]. Table 1 shows a list of volcanic eruptions of VEI>=4 in tropical region (Latitude<20) between 2002 and 2015.

Kelut (Kelud) volcano (7.9S, 112.3E) in the Java island of Indonesia erupted on 13 February 2014. The CALIOP observed that the eruption cloud reached 26km above sea level in the tropical stratosphere, but most of the plume remained at 19-20 km [2]. As the plume continued to move west over the Indian Ocean, it thinned and spread out, following local atmospheric circulation. Aerosol index returns show that much of the ash injected into the atmosphere was removed by three days after the eruption. Although most of the SO_2 was injected into the stratosphere, the total SO_2 mass measured was ~0.2 Tg (confirmed by multiple sensors including OMI, AIRS and IASI), making this eruption fairly modest in terms of SO_2 release [3].

In this paper, we report comparisons of lidar and radar observations of stratospheric aerosols and wind at Equatorial Atmosphere Radar (EAR) site in Kototabang (0.2S, 100.3E), Indonesia, and the CALIOP observations data.

Table 1. List of volcanic eruptions of VEI>=4 in tropical region(Latitude<20) between 2002 and 2015.

Volcano	Date	Latitude	Longitude
Ruang	25-Sep-02	2.3N	125E
Reventador	3-Nov-02	0.1S	78W
Manam	28-Jan-05	4.1S	145E
Soufreiere Hills	20-May-06	16.7N	62W
Tavurvur	7-Oct-06	4.3S	152E
Merapi	4-Nov-10	7.5S	110E
Nabro	12-Jun-11	13.4N	42E
Kelut	13-Feb-14	7.9S	112E

2. LIDAR AND RADAR SYSTEM

We had constructed the lidar facility for survey of atmospheric structure from troposphere to lower thermosphere over Kototabang, Indonesia in the equatorial region [4]. The lidar system consists of the Mie and Raman lidars for tropospheric aerosol, water vapor and cloud measurements, the Rayleigh lidar for temperature measurements and the Resonance scattering lidar for metallic species density measurements. The most parts of this lidar system are remotely controlled via the Internet. The full lidar observations started from 2004. The routine observations of clouds and aerosol in the troposphere and stratosphere are continued now.

We have installed 532nm polarization lidar system for stratospheric aerosol monitoring in 2014. We used the nearest operational radiosonde data to calculate the atmospheric molecular density. The lidar backscattered signal was interactively normalized to unity around 27–32 km, where aerosol-free conditions could be assumed. The total linear depolarization ratio d is defined as

$$d = S/(P+S) \times 100(\%)$$

where P and S are the parallel and perpendicular components of the backscattered signals [5,6].

https://doi.org/10.1051/epjconf/201817604003

EAR is a large Doppler radar. It consists of 560 Yagi-antennas in a circular field of 110 m in diameter. It can observe winds and turbulence in the altitude range of 1.5 km to 25 km, and ionospheric irregularities at an altitude above 90 km.

3. RESULTS

Figure 1 shows time-height cross sections of scattering ratio and depolarization ratio observed by 532nm polarization lidar at Kototabang. Over Kototabang, stratospheric aerosol layers were observed on 28 February 2014 about 15 days after the eruption. The peak values of scattering ratio was 4 at 19 km and the values of depolarization ratio was 4% at 19 km. Non-spherical ash particles were probably included in the layers with sulfuric acid particles that were produced from SO2 through chemical reactions. Non-spherical particles were also present in the lower region (13-17km) of the aerosol layer.

On 29 April 2014 (2 months after the eruption), the peak altitude of scattering ratio move upward to 21 km but the peak altitude of depolarization ratio was still at 19 km. Light spherical sulfuric acid particles should be moved upward by the upwelling tropical branch of the Brewer-Dobson circulation but the non-spherical ash should not be moved upward.

Figure 2 shows time-height cross sections of scattering ratio and depolarization ratio based on CALIOP observations over 1S-1N and 80-120E.

Figure 3 shows time-longitude cross sections of maximum scattering ratio based on CALIOP observations over 1S - 1N and 18 - 25km. Stratospheric aerosol cloud was spread in the longitude direction with the lapse of time on February. After March, westward movement is clearly seen. Aerosol should be carried by strong westward equatorial wind generated by QBO.

On 21-26 June 2014 (4 months after the eruption), aerosol transportation from the stratosphere to the troposphere. The peaked stratospheric aerosol layers could not be seen in July 2014. Figure 4 shows time-height cross sections of vertical wind velocity observed by EAR. Vertical resolution is 450m and time resolution is 11 days. In June, we can see down phase structure of vertical wind between 16-18km generated by Kelvin-Helmholtz instability in the tropical tropopause layer[7].

4. CONCLUSION

We have observed 2014 Kelut volcanic stratospheric aerosol from 28 February 2014 at equatorial lidar site located in the Sumatra island of Indonesia. We observed the depolarization maximum to be up to 2km below the backscatter maximum in April 2014. We also observed the vertical transportation process of stratospheric aerosol to troposphere generated by equatorial Kelvin wave instability with lidar observations and radar vertical wind observations.

From CALIOP data, Kelut aerosol cloud movement by strong westward wind generated by QBO was clearly seen. Aerosol layer was spread in the longitude direction with the lapse of time.

As the SO2 mass injected into the stratosphere by Kelut eruption is moderate, these observations are good example of transportation of materials from the equatorial stratosphere to the troposphere.

We have tried trajectory analysis, but the results did not match with CALIOP measurement. The causes is lack of radio sonde data in equatorial region.

There are advantages of high altitude resolution, high temporal resolution and high sensitive depolarization measurement in ground based lidar. On the other hand, satellite based lidar have not enough resolution and sensitivity, but there is advantage of global observation. Using combination of ground based lidar and satellite based lidars data, we can get valuable sample of equatorial stratospheric aerosol vertical and horizontal transportation.

ACKNOWLEDGMENTS

This work was supported by Collaborative Research based on MU Radar and Equatorial Atmosphere Radar. The CALIPSO data were provided by the Atmospheric Science Data Center at NASA Langley Research Center.

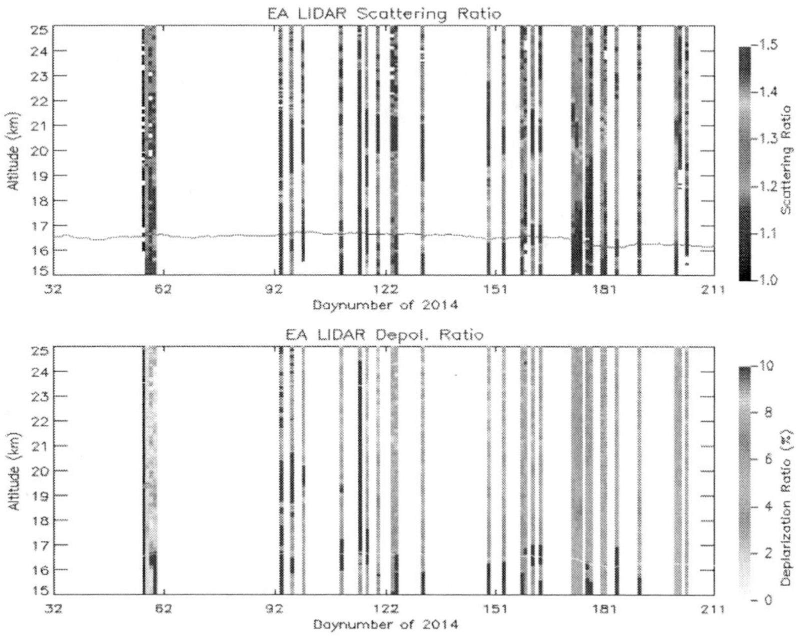

Figure 1 Time-height cross sections of scattering ratio and depolarization ratio observed by 532nm polarization lidar at Kototabang.

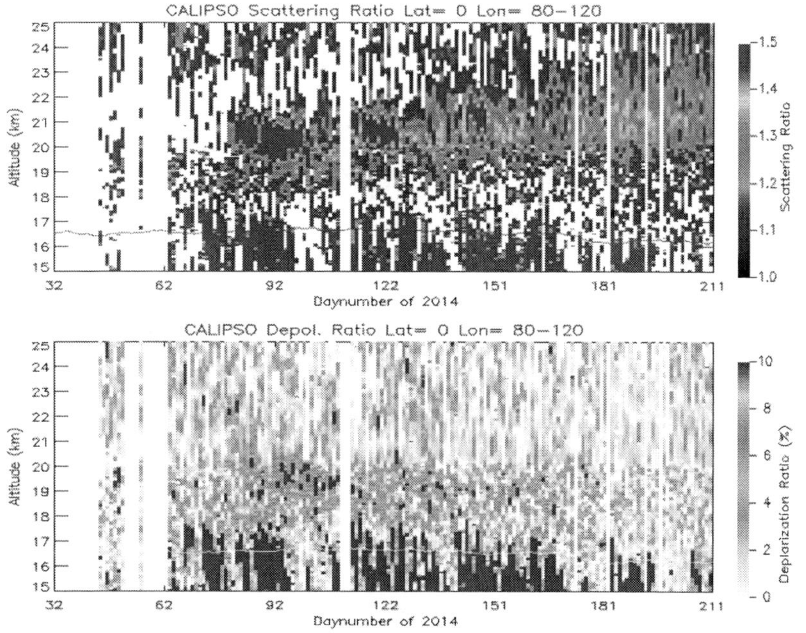

Figure 2 Time-height cross sections of scattering ratio and depolarization ratio based on CALIOP observations over 1S-1N and 80-120E.

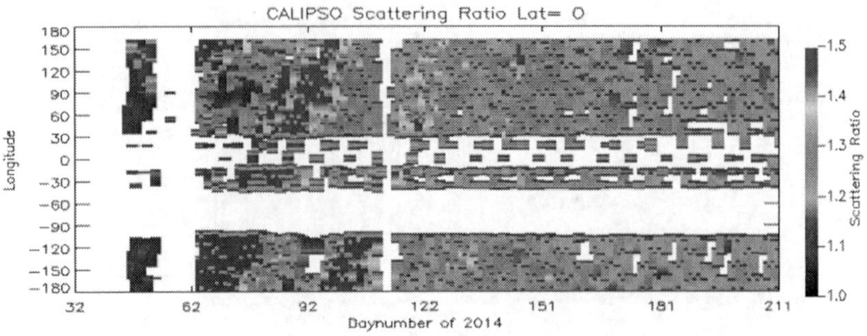

Figure 3 Time-longitude cross sections of maximum scattering ratio based on CALIOP observations over 1S-1N and 18-25km.

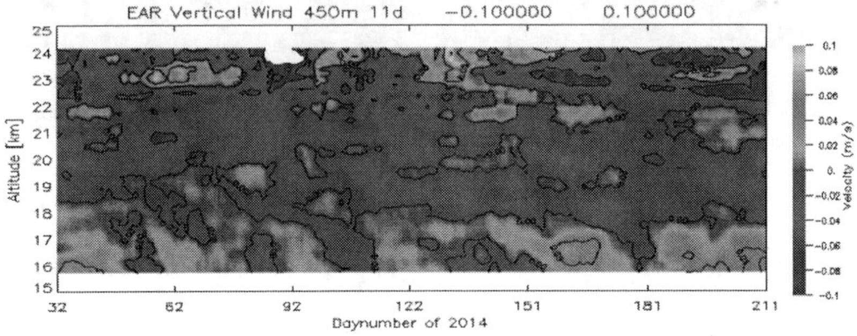

Figure 4 Time-height cross sections of vertical wind velocity observed by EAR. Vertical resolution is 450m and time resolution is 11 days.

REFERENCES

[1] Vernier, J.-P. et al., 2011: Major influence of tropical volcanic eruptions on the stratospheric aerosol layer during the last decade, Geophys. Res. Lett., 38, L12807.

[2] Global Volcanism Program (GVP) 2014: Smithsonian Institution, Kelut Eruption Page. accessed Mar 2015. [Available at http://www.volcano.si.edu/volcano.cfm?vn=263280.]

[3] National Aeronautics and Space Administration (NASA) Goddard Space Flight Center, 2014 (13 February), Kelut (Kelud) Eruption- February 13, 2014 (URL: http://so2.gsfc.nasa.gov/pix/special/2014/kelut/Kelut_summary_Feb14_2014.html)

[4] Nagasawa, C., Abo, M., Shibata, Y., 2006: Lidar System for Observations of Equatorial Lower and Upper Atmosphere, Reviewed and revised papers presented at the 23rd International Laser Radar Conference, No.2O-8, 43-46.

[5] Uchino, O. et al., 2012: On recent (2008–2012) stratospheric aerosols observed by lidar over Japan, Atmos. Chem. Phys., 12, 11975–11984.

[6] Nagai, T., Uchino, O., Fujimoto, T., Sai, Y., Tamashiro, K., Nomura, R., and Sunagawa, T., 1993: Lidar observation of the stratospheric aerosol layer over Okinawa, Japan, after the Mt. Pinatubo volcanic eruption, J. Meteor. Soc. Jpn., 71, 749–755.

[7] Mega, T., Yamamoto, M. K., Luce, H., Tabata, Y., Hashiguchi, H., Yamamoto, M., Yamanaka, M. D., and Fukao, S., 2010: Turbulence generation by Kelvin - Helmholtz instability in the tropical tropopause layer observed with a 47 MHz range imaging radar, J. Geophys. Res., 115, D18115.

 https://doi.org/10.1051/epjconf/201817604004

BOUNDARY LAYER POLLUTION PROFILES FROM A RURAL SITE IN SOUTH KOREA

John Sullivan [1]*, Thomas McGee[1], Anne Thompson[2], Laurence Twigg[3], Grant Sumnicht[3], Ryan Stauffer[1]

[1]*Atmospheric Chemistry and Dynamics Laboratory, NASA Goddard Space Flight Center, Greenbelt, MD 20771 USA, *Email: john.t.sullivan@nasa.gov*

[2]*Earth Sciences Division, NASA Goddard Space Flight Center, Greenbelt, MD 20771 USA*

[3]*Science Systems and Applications Inc., Lanham, MD 20706 USA*

ABSTRACT

During the NASA 2016 KORUS-AQ campaign, the ground based NASA GSFC ozone lidar and balloon borne instrumentation were deployed to the remote Taehwa Forest site (37.3 N, 127.3 E, 151 m AGL) to characterize the transport of pollution downwind of the Seoul metropolitan region. On most days from 02 May to 10 June 2016, continuous hours of lidar profiles of ozone were measured. Select days are shown to represent key ozone events that occurred at the rural site.

1. INTRODUCTION

In order to better assess pollution emission sources and plume evolution in Asia, there has been an international effort to launch the Geostationary Environmental Monitoring Spectrometer (GEMS) to provide hourly measurements of key pollutants (e.g. ozone; nitrogen dioxide (NO_2); sulfur dioxide (SO_2); particulate matter) over the Korean peninsula and the Asia-Pacific region. As a precursor to the Korean GEMS satellite launch and to further investigate the vertical distribution of pollutants impacting South Korea, the United States (U.S.) National Aeronautics and Space Administration (NASA) and the Korean Ministry of the Environment/Korean National Institute of Environmental Research (NIER) conducted an international cooperative field experiment, entitled the Korea-U.S. Air Quality (KORUS-AQ, https://espo.nasa.gov/home/korus-aq/content/KORUS-AQ) study. The campaign was designed to probe pollutants (and chemical gradients) throughout the entire Korean peninsula using specifically designed aircraft flight patterns, ground based measurements (in situ and remotely sensed, and model simulations. The work herein mainly utilizes the synergy of available data to focus on high ozone events at a downwind from Seoul (30 km southeast, Figure 1a) rural site (Taehwa Research Forest, TRF).

Figure 1: Map of South Korea and the Taehwa Forest site (TRF).

2. OBSERVATIONS

2.1 Ozonesondes

An unprecedented O_3-sonde record, which were conditioned and prepared with current community standards [1], was recorded at TRF throughout the KORUS-AQ study from 10 May to 12 June. Afternoon soundings (13:30 to 16:30 KST) of O_3 (top panel, Figure 2) and temperature (bottom panel, Figure 2) illustrate day-to-day variability at TRF. The two case studies (17 May and 09 June) are highlighted with black boxes.

There were large disparities in boundary layer O_3 throughout the campaign period. From 10-16 May, concentrations were mostly between 70-80 ppb, which were associated with cooler temperatures and higher synoptic wind speeds. However by the early afternoon of 17 May, a stagnant high-pressure system located over the Yellow Sea ushered in warmer air, calmer winds, and clearer skies. For the remainder of the campaign until 03 June, a similar meteorological system was preserved, providing favorable conditions for rapid O_3 production of concentrations above 120 ppbv. On 04 and 06 June, mixed precipitation events occurred, limiting O_3 production in the surrounding days. However, by 9-10 June another high-pressure system approached the region, increasing temperatures, stagnating winds, and fostering O_3 production.

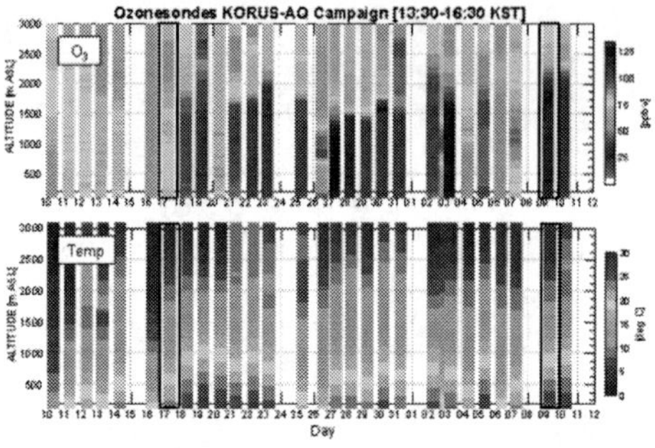

Figure 2: Afternoon soundings (13:30 to 16:30 KST) of O_3 (top panel) and temperature (bottom panel) illustrate day-to-day variability at TRF from 10 May to 12 June.

2.1 GSFC Ozone DIAL

Vertical profiles of O_3 were measured using the NASA Goddard Space Flight Center TROPospheric OZone Differential Absorption Lidar (GSFC TROPOZ DIAL or TROPOZ [2]) in support of the KORUS-AQ study. The TROPOZ system, a charter NASA instrument in the Tropospheric Ozone Lidar Network (TOLNet, http://www-air.larc.nasa.gov/missions/TOLNet/), derives O_3 concentrations to mostly within 10-15% as compared to nearby O_3-sonde profiles [3] and has been previously utilized to characterize O_3 episodes such as stratospheric-tropospheric exchange [4], terrain driven recirculation/transport events [5,6], and regional transport events. During the field study at TRF, the TROPOZ was operated nearly continuously in order to understand sub-hourly (15-min) variability of O_3 during the early morning, afternoon, and evening hours. Figure 3 includes several days where the lidar was operated continuously for 9-13 hrs and illustrates the key O_3 features systematically observed at TRF during the campaign. To add context with auxiliary measurements, each of the TROPOZ measurement days has the 1-min in situ GSFC O_3 monitor (Thermo 49i) data in the lowest bin, a dashed-line solar curve, and coincident ozone soundings previously reported in Figure 2.

3. RESULTS

During every morning at TRF between 07:00-08:00 KST, the surface ozone monitor observed low (<10 ppbv) surface O_3 concentrations. Low early morning values are associated with a stable surface layer, in which surface frictional effects and titration of O_3 occur. Additionally, TROPOZ measurements in six of the eight days in Figure 4 indicate residual layering of O_3 above the stable surface layer with concentrations above 75 ppbv between 08:00 and 10:00 KST. This O_3 resides between 500 and 1500 m ASL, and is related to the extent of vertical mixing during the previous day's planetary boundary growth and decay. An example of residual layer height corresponding to the previous day's ozone mixing height and concentrations is demonstrated in the 9-10 June observations. O_3 with concentrations near 100 ppbv are mixed to near 1700 m ASL by 19:00 KST and result in a residual layer residing at nearly the same height the following morning (as corroborated by the O_3 sonde on 10 June).

As solar radiation and convective mixing increase in the late morning hours (after 11:00 KST), surface concentrations of O_3 generally began to be more representative of concentrations measured aloft from the TROPOZ. During this time, boundary layer growth appears to entrain the residual layer(s) of pollutants helping to increase mid-day O_3 concentrations (e.g. 17-18 May, 9-10 June). Until 14:00 KST, boundary layer O_3 concentrations are mostly between 65-85 ppbv. However, as solar radiation decreases after 14:00 KST most days observe a large (10-40 ppbv) and generally well-mixed contribution of O_3 transported to the site. During all days, O_3 concentrations aloft on each day are above 90

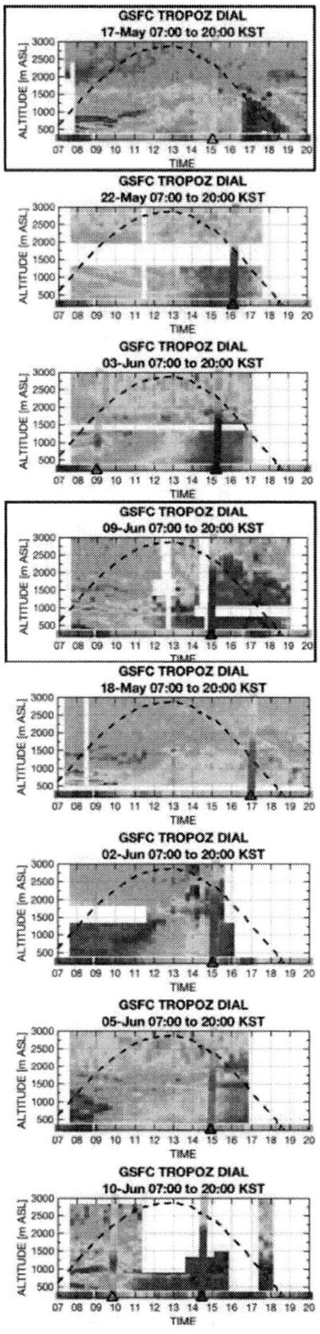

Figure 3: Ozone Lidar Profiles during the KORUS-AQ campaign period.

ppbv after 17:00 KST and several days are higher than 120 ppbv.

As solar radiation diminishes near 19:00 KST, 17 May, 18 May, and 09 June have observations that continue for several hours. During each of these measurement times after 19:00 KST, O_3 concentrations aloft are significantly larger (40-60 ppbv) than the coincident surface concentrations, indicating the rapid depletion and titration of surface ozone from local emissions. This process then results in the polluted aloft residual layer during nighttime hours that will be mixed down on the following morning to enhance the pollution budget.

4. CONCLUSIONS

Several days of nearly continuous ozone lidar observations have yielded a novel assessment of pollution reaching TRF and other rural sites in South Korea. Several case studies indicate rural sites may be experiencing long-term negative effects of O_3 (which are associated with much worse health and agricultural effects) and therefore more stringent domestic regulatory controls on O_3 precursors.

ACKNOWLEDGEMENT

Unless otherwise noted, all ozone lidar data used in this study can be found in the KORUS-AQ data archive (http://www-air.larc.nasa.gov/missions/korus-aq/). This research was supported by appointments to the NASA/USRA Postdoctoral Program at the Goddard Space Flight Center Atmospheric Chemistry and Dynamics Laboratory. The authors gratefully acknowledge support provided by the NASA Tropospheric Chemistry Program and the Tropospheric Ozone Lidar Network (TOLNet).

REFERENCES

[1] Thompson, Anne M., Jacquelyn C. Witte, Herman GJ Smit, Samuel J. Oltmans, Bryan J. Johnson, Volker WJH Kirchhoff, and Francis J. Schmidlin. "Southern Hemisphere Additional O_3sondes

(SHADOZ) 1998–2004 tropical O_3 climatology: 3. Instrumentation, station-to-station variability, and evaluation with simulated flight profiles." *Journal of Geophysical Research: Atmospheres (1984–2012)* 112, no. D3 (2007).

[2] Sullivan, J. T., T. J. McGee, G. K. Sumnicht, L. W. Twigg, and R. M. Hoff. "A mobile differential absorption lidar to measure sub-hourly fluctuation of tropospheric O_3 profiles in the Baltimore–Washington, DC region."*Atmospheric Measurement Techniques* 7, no. 10 (2014): 3529-3548.

[3] Sullivan, John T., Thomas J. McGee, Russell DeYoung, Laurence W. Twigg, Grant K. Sumnicht, Denis Pliutau, Travis Knepp, and William Carrion. "Results from the NASA GSFC and LaRC O_3 lidar intercomparison: New mobile tools for atmospheric research." *Journal of Atmospheric and Oceanic Technology* 32, no. 10 (2015a): 1779-1795.

[4] Sullivan, John T., Thomas J. McGee, Anne M. Thompson, R. Bradley Pierce, Grant K. Sumnicht, Laurence W. Twigg, Edwin Eloranta, and Raymond M. Hoff. "Characterizing the lifetime and occurrence of stratospheric-tropospheric exchange events in the Rocky Mountain region using high-resolution ozone measurements." Journal of Geophysical Research: Atmospheres 120, no. 24 (2015b): 12410-12424.

[5] Sullivan, J. T., et al. (2016), Quantifying the contribution of thermally driven recirculation to a high-ozone event along the Colorado Front Range using lidar, J. Geophys. Res. Atmos., 121, doi:10.1002/2016JD025229.

[6] Sullivan J. T., et al. (2017), Lidar observations revealing transport of O_3 in the presence of a nocturnal low-level jet: Regional implications for "next-day" pollution". Submitted to Atmospheric Environment, 11 Nov 2016.

https://doi.org/10.1051/epjconf/201817604005

RAMAN LIDARS FOR A BETTER UNDERSTANDING OF POLLUTION IN THE ARCTIC SYSTEM (PARCS)

Chazette Patrick[1]*, Raut Jean-Christophe[2], Totems Julien[1], Shang Xiaoxia[1,2], Caudoux Christophe[2], Delanoë Julien[2], and Law Kathy[2]

*[1]IPSL/LSCE, CEA-CNRS-UVSQ, Gif sur Yvette, France, *Patrick.chazette@lsce.ipsl.fr*
[2] LATMOS/IPSL, UPMC Univ. Paris 06 Sorbonne Universités, UVSQ, CNRS, Paris

ABSTRACT

The development of oil and gas drilling and the opening of new shipping routes, in the Barents and Norway seas, poses new challenges for the Arctic environment due to the impact of air pollution emissions on climate and air quality. To improve our knowledge of the interactions between aerosols, water vapor and cloud cover, within the French PARCS (Pollution in the ARCtic System) project, Raman lidar observations were performed from the ground and from an ultra-light aircraft near the North Cape in northern Norway, and coupled with measurements from a 95 GHz ground-based Doppler radar.

1 INTRODUCTION

The French Arctic initiative project Pollution in the ARCtic System (PARCS), aims to improve our knowledge about the impact of local pollution on the Arctic system, including cloud formation, and builds on past efforts that began in the 1980s [1]. The Laboratory of Climate and Environmental Sciences (LSCE) and the Laboratoire de l'Atmosphèere, Milieux et Observations Spatiales (LATMOS) therefore set up an innovative field campaign within the Arctic circle. The experiment took place in May 2016 in the region of Hammerfest (70°39′45″N 23°41′00″E, Norway), 90 km southwest of the North Cape. It is based on the coupling between Raman lidar observations performed from both the ground and an ultralight aircraft (ULA) [2][3]. These observations make it possible to quantify aerosols, water vapor and to identify cloud types via a 95 GHz mini-radar (BASTA) / lidar synergy [4].

The main scientific objectives are to i) sample the aerosol pollution along the coast of northern Norway due to shipping, oil and gas extraction activities, ii) quantify the atmospheric pollution due to local flaring activities and industries, and iii) study the interactions between aerosols and water vapor, cloud liquid droplets, cloud ice crystals.

2 FIELD EXPERIMENT

2.1 Ground-based measurements

The Mobile Atmospheric Station (MAS) was located close to the Hammerfest airport (Figure 1). It was equipped with the 354.7-nm water vapor Raman lidar (WALI) and the 95-GHz mini-BASTA radar [3][4][5]. These instruments carried out continuous measurements between 13 and 26 May 2016 with a final vertical resolution of 15 (1' integration) and 12.5/25/100 m (3" integration) for the lidar and the radar, respectively.

Figure 1. MAS located near Hammerfest airport, equipped with Raman lidar and 95 GHz radar.

The lidar is designed to fulfill eye-safety standards (EN 60825-1). Its emitter is a pulsed Nd:YAG laser (QUANTEL™ Q-Smart). The acquisition system is a PXI (PCI eXtensions for Instrumentation, by National Instruments™) technology with four 12-bit digitizers at 200 MS/s. The wide field-of-view (FOV) ~2.3 mrad allows a full-overlap of the transmission and reception paths beyond ~ 200 m. The acquisition was performed by averaging 1000 laser shots leading to a temporal sampling close to 1 minute. Its pulse energy is ~120 mJ in the UV and the pulse repetition frequency is 20 Hz. The receiver is composed of 3 distinct detection modules using small collector telescopes of 15 cm in diameter. The total number of detection channels is four: the elastic co-polarized and cross polarized channels (with respect to the polarized laser emission), the Raman nitrogen (N_2-channel, ~386.6 nm) and water vapor (H_2O channel, ~407.5 nm) channels.

2.2 Airborne measurements

A Tanarg 912 XS ULA, built by the Air Création Company, with a maximum total payload of approximately 250 kg including the pilot, was used for airborne measurements. Flight duration was 1-2 hours, depending on flight conditions, with a cruise speed around 85-90 km h^{-1}. The 355-nm eye-safe lidar for automatic atmospheric surveys using Raman scattering (LAASURS) was installed on-board the ULA (Figure 2). LAASURS is equipped of the same channels and acquisition system than WALI, except for water vapor. It uses a QUANTEL™ Ultra flash-pumped laser emitter (30 mJ, 20 Hz PRF). A VAISALA™ meteorological probe, type PTU300, was also embedded on the ULA. The ULA location is provided by a Global Positioning System (GPS) and an Attitude and Heading Reference System (AHRS), which are part of the MTi-G system.

Figure 2. N_2-Raman lidar embedded on ULA.

3 ATMOSPHERIC TARGETS

3.1 Aerosols

The measurement of aerosol optical properties in the atmospheric column is very sensitive in the Arctic region since the aerosol concentrations can be very low at this time of year. AERONET measurements give an aerosol optical thickness (AOT) generally lower than 0.1 at 355 nm. These values are confirmed during the campaign where the ground-based lidar (GBL)-derived AOTs remained between 0.05 and 0.13.

We carried out 14 ULA flights in cloudy and windy conditions with a lidar that operated at a temperature down to -17 °C. A flight example is shown in Figure 3. The data are retrieved in synergy with the GBL. Higher aerosol extinction coefficients (AEC) are observed in the northeastern part of the flight. They are probably due to marine sea-salt. During the experiment, aerosol plumes at high altitudes were detected by the GBL, such as the one located at about 7 km above the mean sea level (amsl) in Figure 4. It is highlighted by the aerosol volume depolarization ratio (VDR) and apparent backscatter coefficient (aBC). Its origin remains to be identified more precisely, but it could be a contribution from the major forest fires occurring in Canada during this period.

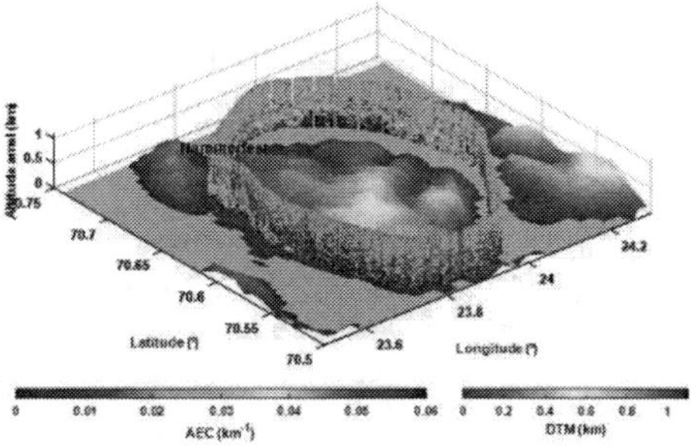

Figure 3. AEC retrieved from the flight around Hammerfest on 16th May 2016.

3.2 Water vapor

Similarly, measuring the water vapor mixing ratio (WVMR) with high relative accuracy is challenging using a Raman lidar, because the air masses are significantly drier and there is always sunlight during summer north of the Arctic Circle. The calibration of the lidar was performed using several vertical ULA flights above the GBL location. The relative standard deviation on the WVMR is assessed to be close to 5% (resp. 25%) within the first km (resp. 3 km), for 200 m/1 h integration (perpetual daytime). A comparison between WALI- and the meteorological probe-derived WVMR profiles is given in Figure 5. Figure 4 also shows the temporal evolution of the WVMR, highlighting a strong variation along time and altitude. Generally, the altitude range of the GBL is limited by the presence of clouds often present during the field-campaign.

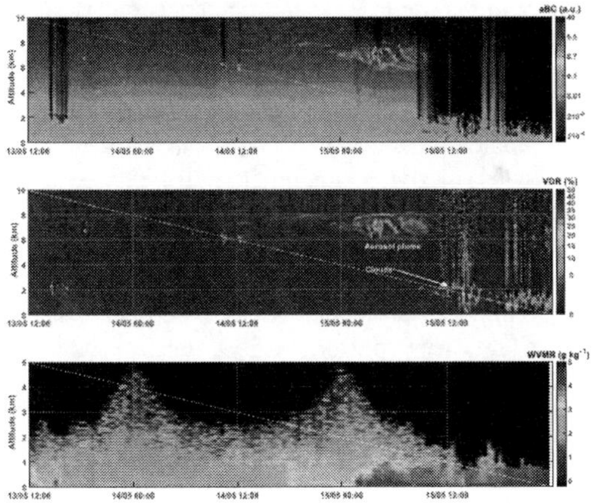

Figure 4. Temporal evolutions, between 13th and 16th May 2016 of the lidar-derived aBC, VDR and WVMR (top, middle and bottom panels, respectively).

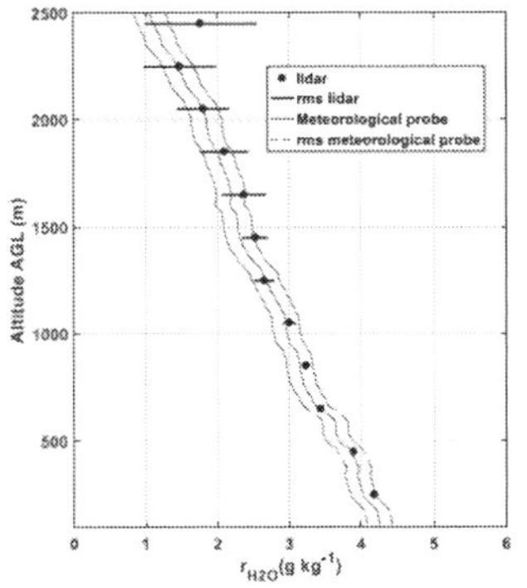

Figure 5. WVMR vertical profiles derived from WALI and the VAISALA ™ probe on 5th May 2016. The red dotted line and the horizontal bars delimit the standard deviation on the meteorological probe and the lidar measurements, respectively.

3.3 Clouds

Simultaneous measurements between WALI and the BASTA mini-radar allow identification of the cloud phase and melting layer (ML) where ice crystals are melting leading to liquid precipitation. The detection of the ML by the lidar/radar synergy has already been described and validated [6]. The lidar/radar synergy is also ideal for characterizing cloud microphysic due their respective sensitivities. Indeed, lidar and radar measurements are roughly proportional to the second and sixth moments of the particle size distribution, respectively [4].

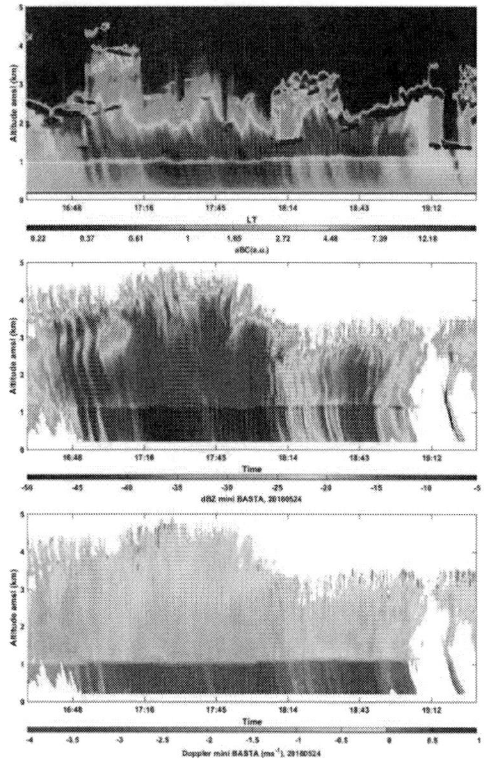

Figure 6. Temporal evolutions on 24th May 2016 of the lidar-derived aBC (top panel), radar reflectivity in dBZ (middle panel), and Doppler radar-derived vertical drop speed (bottom panel).

Figure 6 shows a very rainy situation where the ground temperature was about 6-7 °C. On the upper panel, the temporal evolution of the vertical profile of the aBC is shown. A zone with less scattering is observed close to 1 km amsl (lidar dark band), which corresponds to the ML. The ML is also well identified on radar reflectivity (radar bright band), expressed in terms of dBZ (middle panel), as well as on radar-derived vertical drop speed (bottom panel).

Raindrops are more efficient scatterers on radar signal due to the lower frequency compared to the lidar, and we note significantly higher drop velocities than those of icy crystal particles present in the cloud with a sharp conversion from ice to rain drop at the ML. They have similar signatures on lidar measurements, which are also more sensitive to multiple scattering. The raindrops are less concentrated than liquid cloud drops and therefore allow the signal to penetrate further. Cloud phase detection is one of the advantages of the radar-lidar synergy. For instance, at 18:14 and 1.4 km, the lidar detects a strong return while the radar does not show any modification in its measurements. The very small liquid drops are too small to be identified by the radar (using only reflectivity and Doppler velocity). Such a layer is assumed to be a supercooled layer embedded in the ice layer.

The comparison of the upper and middle panels enables us to classify the water and frozen clouds. The radar reflectivity is indeed much lower in the presence of a cloud water droplets, whereas the lidar signature remains of the same intensity. At the end of the period (after 1900 local time), a precipitating water cloud is observed around 1 km amsl in Figure 6.

4 CONCLUSIONS

During PARCS field-campaign, we collected an original dataset of remote sensing measurements in the Arctic region. Aerosols, water vapor and clouds were simultaneously sampled in the atmospheric column using ground-based and airborne lidars, coupled with a ground-based 95 GHz radar for clouds identification and microphysics. The ML was unmistakably identified by both ground-based lidar and radar. It is a key parameter to constrain and validate the cloud microphysics simulated in mesoscale models.

An important further step will be to simulate cloud cover using the Weather Research and Forecasting (WRF) model and to validate simulations using the field-campaign observations. The water vapor content is also an important element that must be validated in the model because of its influence on cloud formation. These measurements will lead to improvements in our understanding about Arctic clouds and their potential interactions with aerosols.

ACKNOWLEDGEMENTS

This work was supported by the French Institut National de l'Univers (INSU) of the Centre National de la Recherche Scientifique (CNRS) via the French Arctic initiative. The French space agency (CNES) and the Commissariat à l'Energie Atomique et aux énergies alternatives (CEA) are acknowledged for their supports. We also acknowledge valuable help from Tatsuo Onishi (LATMOS/IPSL) in providing tracer forecasts during the experiment. We thank Yoann Chazette, Nathalie Toussaint and Sébastien Blanchon for their help during the field experiment. The ULA flights were performed by Franck Toussaint. The Avinor crew of Hammerfest Airport, represented by Hans-Petter Nergård, and the Air Creation company are acknowledged for their hospitality.

References

[1] Barrie, L. a., 1986: Arctic air pollution: An overview of current knowledge. *Atmos. Environ.* **20**, 643–663. doi:10.1016/0004-6981(86)90180-0

[2] Chazette, P., 2016: Exceptional aerosol pollution plume observed using a new ULA-lidar approach, *Atmos. Environ.,* **141**, 470-480.

[3] Chazette, P., Marnas, F., Totems, J., 2014: The mobile Water vapor Aerosol Raman LIdar and its implication in the framework of the HyMeX and ChArMEx programs: application to a dust transport process, *Atmos. Meas. Tech.,* **7**, 1629–1647.

[4] Delanoë, J., and R. J. Hogan, 2008: A variational scheme for retrieving ice cloud properties from combined radar, lidar, and infrared radiometer, *J. Geophys. Res.*, **113**, D07204, doi:10.1029/2007JD009000

[5] Delanoë J., A. Protat, J.-P. Vinson, W. Brett, C. Caudoux, F. Bertrand, J. Parent du Chatelet, R. Hallali, L. Barthes, M. Haeffelin, and J.-C. Dupont, 2016: BASTA: A 95-GHz FMCW Doppler Radar for Cloud and Fog Studies. J. Atmos. Oceanic Technol. 33, 1023–1038, doi: 10.1175/JTECH-D-15-0104.1.

[6] Sassen, K., Campbell, J.R., Zhu, J., Kollias, P., Shupe, M., Williams, C., 2005: Lidar and Triple-Wavelength Doppler Radar Measurements of the Melting Layer: A Revised Model for Dark- and Brightband Phenomena, *J. Appl. Meteorol.*, **44**, 301–312. doi:10.1175/JAM-2197.1

EPJ Web of Conferences **176**, 04006 (2018) https://doi.org/10.1051/epjconf/201817604006
ILRC 28

OPTICAL PROPERTIES OF AN INDUSTRIAL FIRE OBSERVED WITH A GROUND BASED N_2-RAMAN LIDAR OVER THE PARIS AREA

Xiaoxia Shang[1,2*], Patrick Chazette[2], Julien Totems[2]

*[1]IPSL/LATMOS, Paris, France, *xiaoxia.shang@gmail.com*
[2]IPSL/LSCE, Gif sur Yvette, France

ABSTRACT

This paper presents the first, to our knowledge, lidar measurement of an industrial fire plume, which covered the north of the Paris area on 17th April 2015. The fire started in a textile warehouse and rapidly spread by emitting large quantities of aerosols into the low troposphere. A ground based N_2-Raman lidar performed continuous measurements during this event. Vertical profiles of the aerosol extinction coefficient, depolarization and lidar ratio are derived. A Monte Carlo algorithm was used to assess the uncertainties on the optical parameters, and to evaluate lidar inversion methods.

1 INTRODUCTION

Domestic fires cause casualties and significant property damages. In France, one house fire occurs every 2 minutes, adding up to 263 000 domestic fires each year, causing about 100 deaths and 10 000 injuries (http://iaaifrance.fr/). These fires emit large amounts of gases and aerosols, which are detrimental to human health and degrade visibility. Domestic fires are one of the major sources of aerosols, contributing to both PM_{10} and $PM_{2.5}$.

Weather has a dominant role in fire propagation, especially wind force and direction. In return, the fire influences the dynamics and the chemistry of the atmosphere. Modelling tools are needed to predict regional aerosol emissions from fires and analyze policy options. However, the characterization of fire emissions remains incomplete, mainly due to the difficulties in obtaining smoke samples. Their non-predictability poses an obvious challenge to perform chemical and meteorological measurements.

Lidar is an efficient technique for the detection of various types of particles, such as ash, air pollution, dust, biomass burning aerosols [1][2][3]. Lidar derived parameters can be good constraints for a model [4]. In particular, Raman lidars are becoming well-established tools that are used in the study of numerous areas of importance in the atmospheric sciences [5]. In this study, a ground based N_2-Raman lidar system performed continuous measurements during an exceptionally strong fire in the Paris area on 17th April 2015. The optical properties of the smoke aerosols were characterized using conventional and regularization inversion methods.

Raman lidar data processing usually follows the conventional approach of retrieving profiles of aerosol extinction coefficient (AEC) and aerosol backscatter coefficient, as well as column-averaged values of extinction-to-backscatter ratio, which is also called lidar ratio (LR) [6]. Here, given the large vertical heterogeneity of the atmosphere, regularization methods were applied [7] to estimate a precise LR profile resolving the smoke plume.

2 INSTRUMENTS AND METHODOLOGY

2.1 Fire event

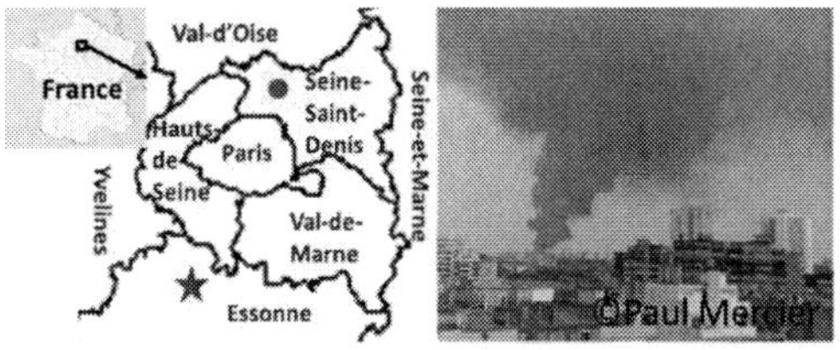

Figure 1. Left: Locations of the textile warehouse on fire (red circle) and of lidar and sun-photometer instruments (purple pentagram). Right: Photo of the smoke plume from far distance.

On 17th April 2015, a violent fire broke out around 2 pm local time (1200 UTC), in a textile warehouse of 12 000 m^2 in La Courneuve, Seine-Saint-Denis, France (48°55'52"N 2°23'52"E, Figure 1). Thick black smoke covered the north area of Paris as shown in Figure 1. With a wind speed of ~22 km/h, the smoke plume rapidly spread from the north to the south of Paris. There was no victim, but damages were assessed around 40 million euros.

Moreover, the traffic was severely disrupted on numerous highways and railways of the Paris area.

2.2 In situ and spaceborne observation

PM_{10} concentrations were observed by the AIRPARIF air quality network at the traffic station "Route nationale 2" (http://www.airparif.asso.fr/), four kilometers downwind of the fire location. The average daytime value on 17th April was ~60 µg/m^3, exceeding the information threshold, but no outlying values were identified during the fire event. The fire mainly injected aerosols into the free troposphere by pyro-convection, just above the planetary boundary layer (PBL) situated close to 1 km above the mean sea level (amsl).

The aerosol optical depth (AOD) and the visible Angstrom exponent are derived from the AERONET sun-photometer station of Palaiseau (http://aeronet.gsfc.nasa.gov/, Figure 1). Elevated AOD values (from 0.6 to 1 at 355 nm) were observed on 17th April, representing the peak value of the whole month (the monthly mean AOD is ~0.2 in cloud-free condition). The visible (440-675 nm) mean Angstrom exponent is found to be ~1.1, representative of carbonaceous particles. Note that the Angstrom exponent for Paris background aerosols is ~1.5.

The AOD values were also derived from the daily MODIS level 2 aerosol product (MYD04_L2), with a spatial resolution of 10 km×10 km. In the 0.4°×0.4° area around Paris, the AOD at 550 nm is 0.55 ±0.09 on 17th April, ~50% larger than the expected one derived from the sun-photometer.

2.3 N_2-Raman lidar

The N_2-Raman lidar LAASURS (Lidar for Automatic Atmospheric Surveys using Raman Scattering, [2]) was put into operation in Palaiseau (48°42'23"N 2°13'22"E, Figure 1), south of Paris, to sample the fire smoke plumes. The straight distance between the locations of the fire and lidar is ~28 km. LAASURS is composed of two reception channels: one dedicated to the measurement of the co- and cross-polarized signals at ~355 nm and one to the inelastic nitrogen Raman backscattered signal at ~387 nm. It enables the retrieval of aerosol optical properties and atmospheric structures with a final resolution of 15 m along the line of sight. A temporal averaging of ~30 min was used in this study to increase the signal-to-noise ratio.

2.4 Lidar inversion methodology

Two inversion methods are applied to retrieve aerosol optical properties (LR, AEC, etc.). The cumulative AOD deduced directly from the N_2-Raman measurements is used as a constraint in the data processing [2].

Inversion of lidar profiles is first performed by applying a conventional approach (one-aerosol-layer approach) [6]. The LR is assumed as range-independent and is the backscatter-weighted mean lidar ratio of the aerosol column (hereafter MLR). The error sources are well known and are mainly due to the vertical heterogeneity of the aerosol layers. The MLR can be poorly representative of the actual LR profile, especially in presence of multiple scattering layers.

Regularization methods may be of advantage if the retrieved backscatter coefficient profile indicates pronounced heterogeneities with altitude. The core principle is to resolve the equation that the cumulative AOD can be expressed as the product of the aerosol backscatter coefficient and the LR. Several regularization methods are available, such as the Tikhonov regularization, which is hereafter applied [8].

3 RESULTS AND DISCUSSION

3.1 Optical characteristics observed by lidar

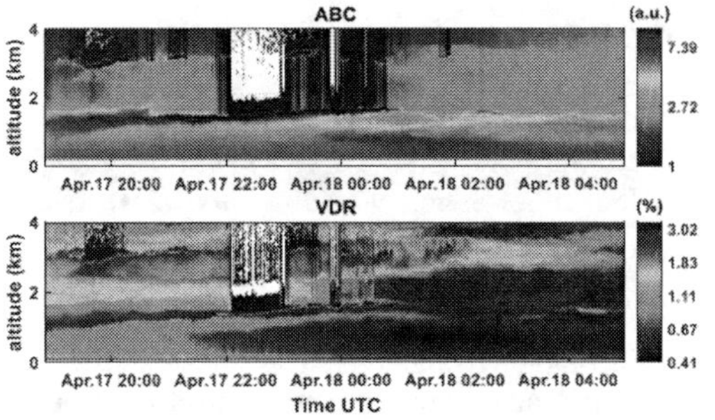

Figure 2. Temporal evaluations, form 1850 UTC on 17th April to 0500 UTC on 18th April 2015, of the attenuated backscattering coefficient (ABC) (top panel), and volume depolarization ratio (VDR) (bottom panel).

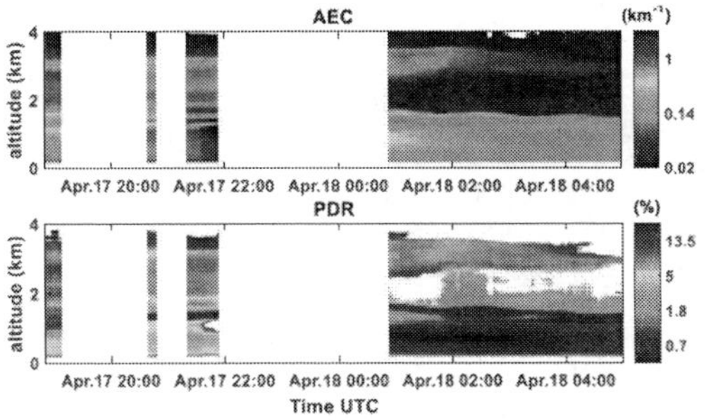

Figure 3. Temporal evaluations, form 1850 UTC on 17th April to 0500 UTC on 18th April 2015, of the aerosol extinction coefficient (AEC), and the particle depolarization ratio (PDR). White gapes indicate cloudy cases which have been removed for lidar inversion. The PDR is only considered for AEC > 0.02 km⁻¹.

The temporal evolutions of both the attenuated backscattering coefficient (ABC) and volume depolarization ratio (VDR) vertical profiles are presented in Figure 2. Three aerosol layers can be easily located in the low troposphere: i) the PBL, under 1.2 km, ii) a thin non-depolarizing layer close to 1.2 km amsl with a high ABC, and iii) a depolarizing layer between 1.8 and ~4 km amsl.

Before the inversion of lidar profiles, the cloudy cases are removed [1]. The vertical profiles of the AEC and the particle depolarization ratio (PDR) are shown in Figure 3. Mean cloud-free vertical profiles are also given in Figure 4 (39 profiles). The standard deviations around the mean values are represented as colored areas.

The aerosol layer located close to 1.2 km amsl is associated with a strong AEC (~0.8 km⁻¹) and a small PDR (~1%). This is the smoke plume coming from the fire accident. The particles appear to be sphere-like. The upper depolarizing aerosol layer presents a PDR~10%, which is probably a mixing of pollution and dust aerosols. The smoke plume has LR ~100 sr, whereas the depolarizing layer has LR ~50 sr. In the PBL, the LR is close to 105 sr, as occasionally observed in the Paris area [2]. Note that the MLR for the same mean vertical profile is ~60 sr. Hence, we note a significant difference, which can lead to an underestimation (overestimation) of the smoke plume (polluted dust layer) AEC.

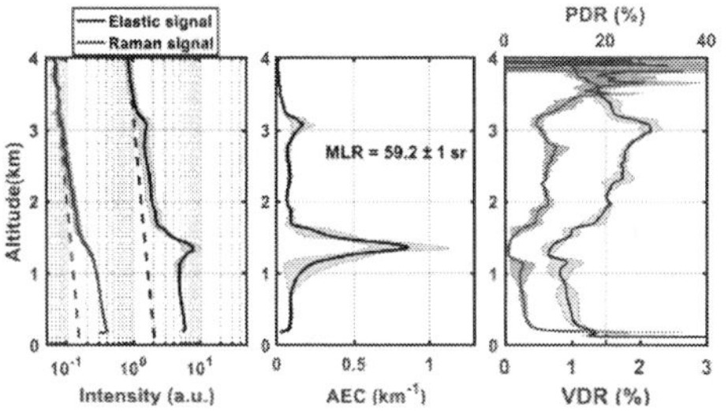

Figure 4. Left: Range-corrected lidar signal of elastic channel (black) and Raman channel (red). Dotted lines indicate theoretical Rayleigh scattering signals. Center: Retrieved AEC (aerosol extinction coefficient) with the used MLR (mean lidar ratio) values. Right: Volume depolarization ratio (VDR, black) and particle depolarization ratio (PDR, red). Gray/Red shaded areas indicate the standard deviations around the mean values, for lidar measurements from 1850 to 2133 UTC on 17th April 2015.

In order to investigate the air mass origins of the polluted-dust layer, back-trajectory analysis in ensemble mode was performed using the NOAA HYSPLIT (HYbrid Single-Particle Lagrangian Integrated Trajectory model, available at http://ready.arl.noaa.gov). Part of the upper aerosol layer seems to originate from North Africa, and may be loaded with dust aerosols. Another strong contribution is found to be recirculated air masses from France, probably containing pollution particles.

3.2 LR: column value or profile

The above results are based on the Klett method, using MLR. Regularization methods can be of advantage for this study because of the pronounced vertical heterogeneities of the aerosol mixing.

A Raman lidar simulator was developed for the sensitivity study of different inversion algorithms. It is an end-to-end simulator containing direct and inverse modules. Predetermined AEC and LR profiles are used as the input of the simulator, and compared with the estimated parameters to assess bias and standard deviations. Lidar system parameters are based on actual measurements. Several regularization methods were simulated under different predetermined atmosphere conditions. Results show that the Tikhonov regularization performs the best estimations, with acceptable uncertainty and a good capacity to reproduce the vertical structure if the SNR is large enough.

As lidar measurements are performed during night-time, the SNR is suitable to use the Tikhonov method for Raman lidar data inversion. It is compared with the Klett inversion results. The comparison shows that LR can be considered as range-independent for most cases (~90%). An example is given in Figure 5 on the left, where the derived AECs using a column LR (in red, Klett method) or LR profile (in blue, Tikhonov method) are similar. However, for the example shown on the right of Figure 5, a difference up to ~50% on the AEC retrieval between 2 methods can be found. The cumulative AOD was better restored with the regularization method. We note however that the regularization algorithm is relatively time consuming.

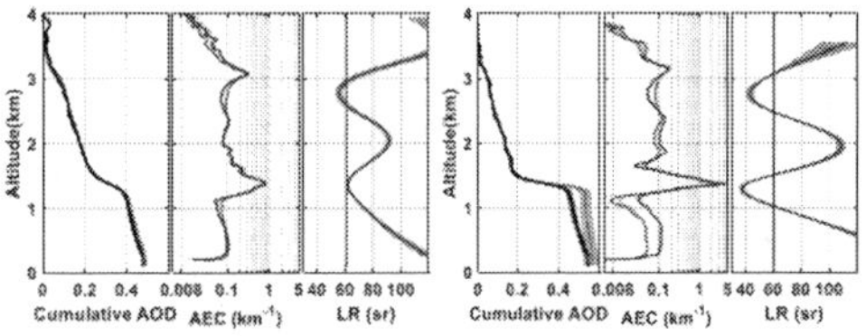

Figure 5. Mean value (lines) and standard deviation (shaded areas) profiles of cumulative aerosol optical depth (AOD), aerosol extinction coefficient (AEC) and Lidar ratio (LR) for ~10 min average centered at 2045 (left) and 2150 (right) UTC of 17th April 2015. Black: AOD derived from Raman channel signal; Red: inversion results using Klett method. Blue: inversion results using Tikhonov method.

4 CONCLUSIONS

A ground based N_2-Raman lidar sampled, for the first time, fire smoke plumes during an exceptional industrial fire event in the Paris area on 17th April 2015. The optical properties of the fire aerosols were characterized using both conventional and regularization inversion methods. This thin smoke plume at 1.2 km has a strong AEC (~0.8 km^{-1}) and a small PDR (~1%), containing sphere-like, absorbent aerosols. The LR of the fire smoke plume was indeed derived as ~100 sr, using the Tikhonov regularization method. The comparison of optical property retrievals between the conventional and regularization inversion methods shows the advantage of a regularization inversion for the heterogeneous atmosphere conditions. The Raman lidar system is proved to be a strong tool to sample the aerosol layers during extreme events, which argues for the existence of lidar networks dedicated to the monitoring of air quality in urbanized area.

ACKNOWLEDGEMENTS

This work was supported by the Commissariat à l'Energie Atomique et aux énergies alternatives (CEA).

References

[1] Chazette, P., Dabas, A., Sanak, J., Lardier, M., Royer, P., 2012: French airborne lidar measurements for Eyjafjallajökull ash plume survey, *Atmos. Chem. Phys.*, **12**, 7059-7072.

[2] Royer, P., Chazette, P., Lardier, M., Sauvage, L., 2011: Aerosol content survey by mini N_2-Raman lidar: Application to local and long-range transport aerosols, *Atmos. Environ.*, 45, 7487-7495.

[3] Chazette, P., Sanak, J., Dulac, F., 2007: New Approach for Aerosol Profiling with a Lidar Onboard an Ultralight Aircraft: Application to the African Monsoon Multidisciplinary Analysis, *Environ. Sci. Technol.*, **41**, 8335-8341.

[4] Wang, Y., Sartelet, K. N., Bocquet, M., Chazette, P., 2014: Modelling and assimilation of lidar signals over Greater Paris during the MEGAPOLI summer campaign, *Atmos. Chem. Phys.*, **14**, 3511-3532.

[5] Whiteman, D.N., 2003: Examination of the traditional Raman lidar technique. I. Evaluating the temperature-dependent lidar equations, *Appl. Opt.*, **42**, 2571-2592.

[6] Klett, J.D., 1981: Stable analytical inversion solution for processing lidar returns, *Appl. Opt.*, **20**, 211-220.

[7] Shcherbakov, V., 2007: Regularized algorithm for Raman lidar data processing, *Appl. Opt.*, **46**, 4879-4889.

[8] Tikhonov, A.E., and Arsenin, V.Y., 1977: Solutions of Ill-posed Problems, Wiley.

ESA'S SPACEBORNE LIDAR MISSION ADM-AEOLUS; PROJECT STATUS AND PREPARATIONS FOR LAUNCH

Anne Grete Straume[1*], Anders Elfving[1], Denny Wernham[1], Frank de Bruin[1], Thomas Kanitz[1], Dirk Schuettemeyer[1], Jonas von Bismarck[2], Fabio Buscaglione[2], O. Lecrenier[3], Phil McGoldrick[4]

[1]*European Space Agency (ESA/ESTEC), PO Box 299, 2200 AG Noordwijk, Netherlands, *Email: anne.straume@esa.int*

[2]*European Space Agency (ESA/ESRIN), Casella Postale 64, 00044 Frascati (Roma), Italy*

[3]*Airbus Defense and Space, 5 rue des Satellites, 31030 Toulouse, France*

[4]*Airbus Defense Defense and Space, Gunnels Wood Road Stevenage, SG1 2AS, United Kingdom*

ABSTRACT

ESA's Doppler Wind lidar mission, the Atmospheric Dynamics Mission (ADM-Aeolus, hereafter abbreviated to Aeolus), was chosen as an Earth Explorer Core mission within the Living Planet Programme in 1999. It shall demonstrate the potential of space-based Doppler Wind lidars for operational measurements of wind profiles and their use in Numerical Weather Prediction (NWP) and climate research. Spin-off products are profiles of cloud and aerosol optical properties. Aeolus carries the novel Doppler Wind lidar instrument ALADIN. The mission prime is Airbus Defence & Space UK (ADS-UK), and the instrument prime is Airbus Defence & Space France (ADS-F).

In April 2016, the ALADIN instrument was successfully tested in ambient conditions by ADS-F in Toulouse. The instrument was then shipped to ADS-UK in Stevenage for integration on the satellite platform. The integration was completed in January 2017 and the satellite was then shipped to Intespace in Toulouse, where it will undergo mechanical testing followed by thermal vacuum testing and the further launch preparations in Belgium and France. The launch is scheduled in December 2017, on board a Vega rocket from French Guiana.

In March 2017, an Aeolus Calibration and Validation (CAL/VAL) Rehearsal Workshop will be held at MeteoFrance in Toulouse, France. The workshop will bringing industry, the user community and ESA together to prepare the Aeolus Commissioning and Operational Phases.

In this paper, the Aeolus mission, its status and launch preparation activities are described.

1. BACKGROUND

The European Space Agency's (ESA's) Living Planet Programme includes two types of complementary user driven missions: the research oriented Earth Explorer missions and the operational service oriented Earth Watch missions. Earth Explorer missions are divided into two types; (i) Core missions being larger missions demonstrating the capabilities of new technologies addressing issues of wide scientific interest, and (ii) Opportunity missions with a similar scope but smaller in terms of cost to ESA. Both types of missions address the research objectives set out in the Living Planet Programme document [1]. All Earth Explorer missions are proposed, defined, evaluated and recommended by the scientific community. Following the recommendations at the User Consultation Workshop in Granada, Spain in 1999, ESA implemented Aeolus as a Core Earth Explorer.

2. ADM-AEOLUS MISSION

2.1 Objectives and motivation

The motivation for the selection of the Aeolus mission was the need for more abundant direct wind profile measurements in the Global Observing System (GOS), which is used by NWP and climate models [2]. In the current GOS, direct wind profile measurements are obtained from radiosondes, commercial aircraft ascends and descends and ground-based wind lidars and radars. The distribution of these measurements is, however, not homogenous, with most observations taken over land in the Northern Hemisphere. Winds can also be inferred from temperature soundings when the flow is in geostrophic balance, which is only the case for large-scale wind fields in the extra-tropics. Air-Motion-Vectors deduced from cloud and aerosol tracking also give valuable wind information at the cloud top or inside optically thin cloud and aerosol layers. However, these observations are limited by the difficulty in performing accurate height-assignments.

The current lack of homogenous sampling of the 3-dimensional wind field in particularly in the tropics and over the oceans leads to difficulties in constraining atmospheric models and the studying of key processes in coupled climate systems. It is therefore expected that the Aeolus mission will largely contribute to the improvement of predictions of small-scales flows and forecasts in observation-sparse regions. Aeolus measurements are furthermore expected to improve short-range forecasts of severe weather. The mission also sets out to demonstrate the potential of space-borne Doppler Wind lidars for operational NWP.

2.2 Instrument concept

Aeolus embarks the direct detection Doppler Wind Lidar instrument ALADIN (Atmospheric LAser Doppler INstrument). ALADIN is a pulsed UV lidar (355 nm, circularly polarized, 50 Hz), measuring continuously along the orbit (Figure 1, left panel). Its high spectral resolution concept allows for the detection of the parallel polarized molecular (Rayleigh) and particle (Mie) backscattered signals in two separate channels, each sampling the wind in 24 vertical height bins (Figure 1, right panel). This makes it possible to deliver winds both in clear and (partly) cloudy conditions down to optically thick clouds. The height of the wind measurement is detected by time-gating. A quasi-global coverage is achieved daily (~ 15 orbits per day) and the orbit repeat cycle is 7 days (111 orbits). The orbit is sun-synchronous with a local equatorial crossing-time of 6 am/pm. A detailed description of the instrument design and operation can be found in [3]. A demonstration of the measurement concept can be found in e.g. [4].

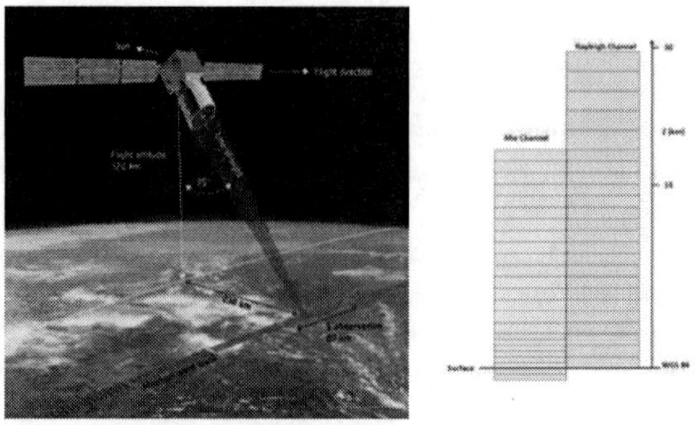

Figure 1: The Aeolus orbit, pointing and sampling characteristics (left) and an example of the vertical measurement binning (right).

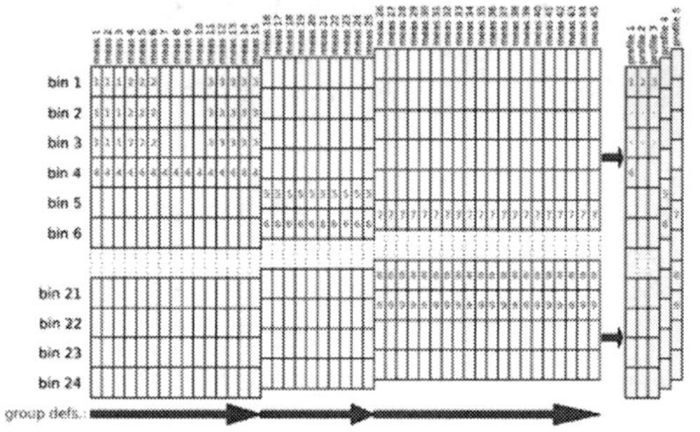

Figure 2: Schematic view of the Aeolus L2b wind observation processing. The number indicate different scene classifications (clouds versus no clouds), resulting in a number of wind profiles for an observation. These are partial or full wind profiles for the Rayleigh (cloud free) and Mie (cloud or aerosol layer winds) channels. Courtesy J. de Kloe (KNMI).

2.3 Products and product requirements

The main product from Aeolus will be horizontally projected line-of-sight (HLOS) wind profile observations (approximately zonally oriented) from the surface up to about 30 km [3]. The atmospheric backscattered signal for the individual laser pulses are averaged on-board to yield ~3 km measurements. These measurements are further averaged on-ground to observations, representing horizontal averaging on a ~87 km scale. The vertical resolution of the layer-average winds vary from 0.25 to 2 km, and can be adapted as a function of the under-laying topography and/or climate zone. The required wind accuracies are 2 m/s in the planetary boundary layer (PBL), 2-3 m/s in the free troposphere, and 3-5 m/s in the stratosphere. The wind profiles are classified into cloudy and cloud free profiles, as illustrated in Figure 2. A detailed description of the Aeolus wind profile retrievals can be found in [5].

Aeolus will also deliver profiles of Mie and Rayleigh co-polarized backscatter and extinction coefficients, scattering ratios and lidar ratios [6] along the lidar line-of-sight. From these parameters it is possible to derive cloud and aerosol information such as layer height, multi-layer cloud and aerosol stratification, cloud and aerosol optical depths (integrated light-extinction profiles), and some information on cloud/aerosol type (lidar ratio).

2.4 ALADIN instrument on-ground characterization in ambient conditions

The ALADIN instrument was successfully tested at ADS-F in Toulouse in April 2016. During this campaign, called the Instrument Full Performance (IFP) test, the instrument was operated at ambient conditions. The main tests included low level tests for instrument characterization and high level tests for instrument performance. The instrument hardware configuration was largely representative for flight, with the exception of the laser cooling. A specially designed Optical Ground Support Equipment (OGSE) was used both in emission and reception to simulate the atmospheric echo (towards the instrument) and to analyse the Aladin emitted beam. The atmospheric echoes were simulated with a dedicated laser, which was modulated in amplitude and frequency to simulate the range of conditions that are expected to be seen by the instrument in orbit. Aladin has been designed to be highly accurate. Therefore the accuracy required to verify its performances is extremely demanding in terms of the OGSE set-up.

Because the IFP test was performed in ambient conditions, both the OGSE and ALADIN were shown to be sensitive to the evolving environment (mainly atmospheric pressure). In the upcoming instrument performance tests in thermal vacuum, the test conditions will greatly improve due to the minimization of ambient related environmental variability. Additional monitoring is also foreseen during the thermal vacuum test to further disentangle the test set-up contribution from the instrument intrinsic quality, and to accurately assess its performances.

Based on the measured transmission in both the Mie and Rayleigh channels and the random error test outcome, the instrument was shown to meet its performance specification. The IFP test hence fulfilled its goal to assess and confirm the instrument performance as far as possible in ambient conditions. The upcoming thermal vacuum tests will enable to consolidate the Aladin instrument performance further in conditions that are much closer to what the instrument will experience in flight.

2.5 Aeolus Mission status

During the Aeolus development phase, a number of technical challenges for the Aeolus system have appeared and have been tackled. These include laser diode performance, optical mounting stability, laser induced damage on the IR and UV coatings, laser-induced contamination in vacuum, and long-term UV energy drift. The transmit lasers and optical bench assembly have been extensively retrofitted and tested, and an in-situ cleaning system permanently purging the instrument with oxygen during its operation has been introduced. The satellite platform was in storage from 2009 until 2015, when it was made ready for the arrival of the instrument in August 2016. The instrument was then integrated onto the spacecraft and the satellite was ready for shipment to the mechanical test facility in Toulouse in January 2017. First light measured by the satellite Rayleigh receivers in 2016, using the on-board software, was a great step forward in the satellite finalization (Figure 3). After the mechanical testing, the satellite will be sent to Liege, Belgium for Thermal Vacuum (TVAC) testing. During the TVAC test, the satellite and the instrument will be functionally tested and verified in space conditions. The mission Payload Data Ground Segment and Flight Operations Segment are currently undergoing the final overall validation testing in preparation for launch. The satellite Final Acceptance Review is planned for October, and the launch from Kourou, French Guyana for December 2017.

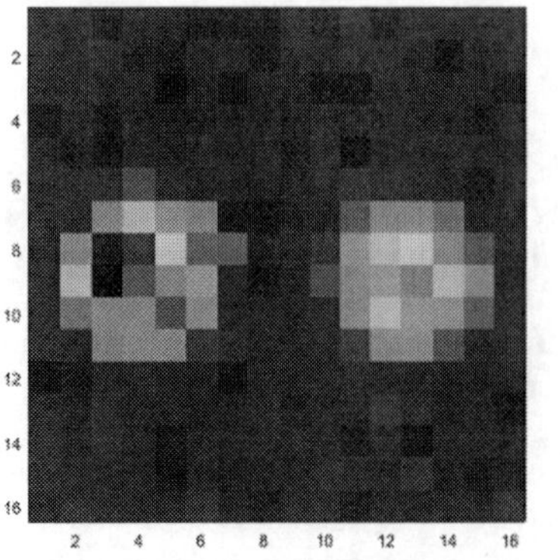

Figure 3: ADM-Aeolus satellite, first light, Rayleigh channel spots imaged on the Aladin detector. Courtesy Airbus Defence & Space.

2.6 Aeolus CAL/VAL preparations

An Announcement of Opportunity (AO) call for CAL/VAL support studies for Aeolus took place in 2007. The AO focussed on the validation of the Aeolus wind and optical properties products during its commissioning and successive 3-year exploitation phase. Sixteen proposals including ground-based and airborne validation experiments were received from teams all around the globe. Due to the delays of the Aeolus programme and the recent changes to the Aeolus operation, a delta-call was released in May 2014. The main objectives of the delta-call was (i) to allow confirmation of the CAL/VAL proposals from the 2007 call and (ii) to call for new proposals. In total 17 proposals were received, whereof 2 were joint national proposals and 3 included a number of institutes from different countries. Most of the proposals concerned validation of both the wind and aerosol/cloud products. Activities of special interest included tropical stratospheric balloon experiments and the inclusion of collocated measurements in polar areas. A Science and CAL/VAL workshop was held in ESA-ESRIN in February 2015 (http://www.aeolus-science-calval-2015.org/). Presentations of the science, instrument and product status, commissioning phase planning and the extensive number of proposals submitted in response to the Aeolus CAL/VAL call in 2014 were presented. The workshop allowed the teams to start activity coordination and exchange of information.

In March 2017, an Aeolus Calibration and Validation (CAL/VAL) Rehearsal Workshop will be held at MeteoFrance in Toulouse, France (http://www.aeolus-calval-2017.org/). The main emphasize of this workshop will be to let users test the ESA data download facilities, to start to work with test datasets, and to test the ESA data reading and exploitation tools. Dedicated panels will further detail the CAL/VAL strategy and coordination.

3. CONCLUSIONS

With the recent completion of the Aeolus satellite, the Mission is nearing its readiness for launch. ESA, industry and the scientific community is getting prepared for Commissioning phase, CAL/VAL and product exploitation.

REFERENCES

[1] European Space Agency, (1998) The Science and Research Elements of ESA's Living Planet Programme. ESA SP-1227, 105 pp.

[2] World Meteorological Organisation (WMO) (2004) Proc. of the 'Third WMO Workshop on the Impact of various observing systems on Numerical Weather Prediction'. WMO/TD No. 1228, 329 pp.

[3] European Space Agency, (2008) ADM-Aeolus Science Report. ESA SP-1311, 121 pp.

[4] Marksteiner, U., Airborne Lidar Observations for the validation of the ADM-Aeolus instrument, PhD Thesis and the Technical University of Munich, http://mediatum.ub.tum.de/doc/1136781/1136781.pdf.

[5] Tan, D. G. H. *et al.*, (2008) The ADM-Aeolus wind retrieval algorithms. Tellus, **60A**, pp. 191-205.

[6] Flamant, P. *et al.*, (2008) ADM-Aeolus retrieval algorithms for aerosol and cloud products. Tellus, **60A**, pp. 273–286.

VALIDATION OF NEW SATELLITE AEROSOL OPTICAL DEPTH RETRIEVAL ALGORITHM USING RAMAN LIDAR OBSERVATIONS AT RADIATIVE TRANSFER LABORATORY IN WARSAW

Olga Zawadzka[1]*, Iwona S. Stachlewska[1], Krzysztof M. Markowicz[1], Anca Nemuc[2], Kerstin Stebel[3]

[1]University of Warsaw, Faculty of Physics, Institute of Geophysics, Warsaw, Poland, zawadzka@igf.fuw.edu.pl
[2]Institute of Optoelectronic Research, Magurele, Romania
[3]Norwegian Institute for Air Research, Kjeller, Norway

ABSTRACT

During an exceptionally warm September of 2016, the unique, stable weather conditions over Poland allowed for an extensive testing of the new algorithm developed to improve the Meteosat Second Generation (MSG) Spinning Enhanced Visible and Infrared Imager (SEVIRI) aerosol optical depth (AOD) retrieval. The development was conducted in the frame of the ESA-ESRIN SAMIRA project. The new AOD algorithm aims at providing the aerosol optical depth maps over the territory of Poland with a high temporal resolution of 15 minutes. It was tested on the data set obtained between 11-16 September 2016, during which a day of relatively clean atmospheric background related to an Arctic air-mass inflow was surrounded by a few days with well increased aerosol load of different origin. On the clean reference day, for estimating surface reflectance the AOD forecast available on-line via the Copernicus Atmosphere Monitoring Service (CAMS) was used. The obtained AOD maps were validated against AODs available within the Poland-AOD and AERONET networks, and with AOD values obtained from the PollyXT-UW lidar. of the University of Warsaw (UW).

1 INTRODUCTION

Despite considerable improvements in the past decades, Europe is still far from achieving levels of air quality that do not pose unacceptable hazards to humans and the environment. Inhabitants in urban areas still can experience significant health problems due to air pollution. Research on the air pollution issue on regional scale, is presently carried out in the frame of ESA-ESRIN founded project entitled SAtellite based Monitoring Initiative for Regional Air quality (SAMIRA). The overall goal of the project is to improve regional and local air quality monitoring through synergistic use of data from present and upcoming satellite missions, traditionally used in-situ air quality monitoring networks, and output from chemical transport models.

Through collaborative efforts of Poland, Romania, the Czech Republic, and Norway, countries confronted with different but pressing air quality problems, the SAMIRA aims at both to support national monitoring obligations and to generate novel research in this area. In SAMIRA, the most relevant pollutants are being addressed; NO_2, SO_2, particulate matter $PM_{2.5}$ and PM_{10}, as well as Aerosol Optical Depth (AOD). The latter can be directly retrieved or indirectly estimated from existing satellites sensors. The satellite retrieval of aerosol properties over land is challenging mainly due to the high surface reflectance, its temporal variability, and spatial inhomogeneity, but also due to anisotropic bi-directional reflectance of land surfaces, which induces higher uncertainties of retrieved parameters. For this paper we choose to exploit the geostationary SEVIRI/MSG sensor [1], as for air quality quasi-continuous, high temporal and spatial resolution monitoring is a necessity. The optimal interpolation approach to retrieve AOD from the SEVIRI data, initially developed by [2], was elaborated for the further improvement. A new version of algorithm was developed and tested on the data set collected between 11-16 September 2016. The obtained AOD values were validated against the next generation PollyXT-UW lidar in Warsaw [3,4], as well as by using the AOD data sets available within the Polish aerosol network Poland-AOD (http://www.polandaod.pl/) and AERONET (https://aeronet.gsfc.nasa.gov/).

2 METHODOLOGY

The modified AOD retrieval algorithm is fed with the high temporal and spatial resolution data from SEVIRI channel 1 (635 nm) and it allows for retrieving AOD at a spatial resolution of about 5×5 km at a frequency of 15 min. The most important parameter for the retrieval process is the surface reflectance. The surface properties are estimated during relatively cloud free reference days (cloud cover < 1/8) with a low AOD (< 0.1 at 500 nm), based on the Top-Of-Atmosphere reflectance measured by the SEVIRI detector. Even if the AOD is low, elimination of influence of atmospheric components, in particular atmospheric aerosols, on the reflectance measured by satellite is still essential. By applying the optimal interpolation method [5] and the additional ground based sources of AOD data the uncertainties related to the aerosol background estimation errors can be reduced to improve the AOD retrieval. For more details of the retrieval scheme refer to [2]. The main changes introduced in the new algorithm include following: I) CAMS AOD forecast data [6] are used as an AOD background information for a reference day; II) all Polish ground-based stations with a continuous AOD measurements are implemented in the code; and III) cloud screening process was improved. Favourable conditions in September 2016 allowed for tests and validation of the new version of algorithm.

Observations and calculations for two 2 days: 11th and 15th September are discussed. On these days, the PollyXT-UW lidar observations were taken at the Radiative Transfer Laboratory (RT-Lab) of the Institute of Geophysics, Faculty of Physics, University of Warsaw (Poland). The RT-Lab site located at central Warsaw (52.21°, N 20.98° E, 112 m a.s.l.) joined the EARLINET [7] in March 2015, and is regularly contributing to the network activities. We use the 12-channel NeXT generation Polly XT system equipped with the photon-counting detection channels at 1064, 532, and 355 nm (elastic), 607 and 387 nm (N2 Raman), 407 nm (H2O Raman), 532 and 355 nm (elastic cross). Since April 2016, lidar is equipped additionally with the Near Range Aerosol Lidar receiver (NARLa) at 532 and 355 nm (elastic), 607 and 387 nm (N2 Raman), which significantly enhances lowermost range of lidar observations.

3 RESULTS AND DISCUSSION

On 11th September, which was a day with well increased aerosol load captured, the back-trajectories calculated using the HYSPLIT model [8], indicated possible occurrence and even accumulation of local pollution over Warsaw due to heat wave conditions (Fig. 1). Results of the NAAPS model simulations [9] suggest that sulphates were significant part of aerosol load on that day (Fig. 2). On the contrary, 15th September was characterized by significantly cleaner conditions, which were related to an Arctic air-mass inflow. Thus, this day was chosen to be used as a reference day for estimating the surface properties. The application of algorithm on SEVIRI data on 11th September provided the AOD maps over Poland with a temporal resolution of 15 minutes. An example of such a map is given in Fig. 3. The obtained AOD-SEVIRI values, (pixel regarded as representative for the Warsaw EARLINET site) were validated against the PollyXT-UW lidar. The lidar-derived AOD at 532nm was obtained from the extinction profiles. Additionally, the SEVIRI AOD data at 635nm were compared with the MFR-7 shadow-band radiometer observations of AOD at 614nm, showing generally a sufficient agreement.

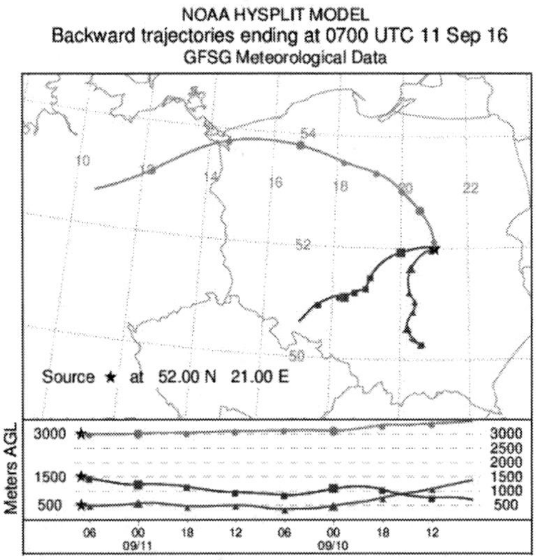

Figure 1 The HYSPLIT back-trajectories obtained for Warsaw at 7:00 UTC for 11th September. The model was run for 48 h with meteorological data from the Global Data Assimilation System (GDAS).

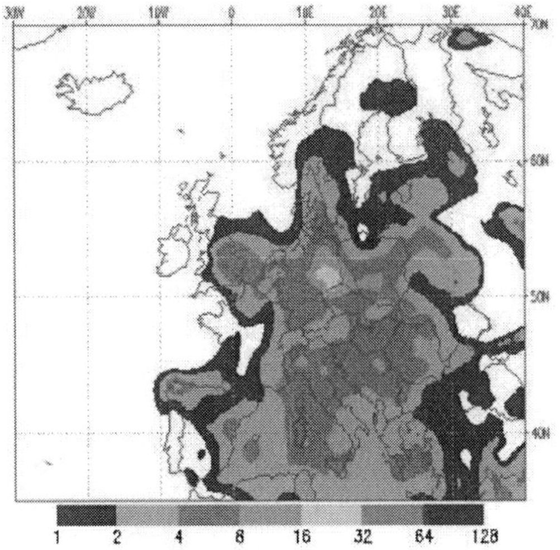

Figure 2 Sulfate surface concentration (µg/m³) on 11 September 2016, obtained from Navy Aerosol Analysis and Prediction System (NAAPS).

The AOD map (Fig. 3) at 635nm was calculated for data acquired at 7:00 UTC on 11th September 2016. Lack of data in northern and western areas of the map are due to the cloud cover. For pixel that covers Warsaw, the AOD SEVIRI for 15 min observations was 0.303. At the same time the AOD measured by the MFR-7 at the RT-Lab in Warsaw equalled 0.391 at 615nm and 0.52 at 500nm. In Figure 4, the daily evaluation of the range and background corrected lidar signal at 1064 nm on 11th September 2016 is shown. The aerosol load is visible up to 3 km, signatures of nocturnal and residual layers are pronounced. At 1-2, 12, and 13-14 UTC some clouds are captured.

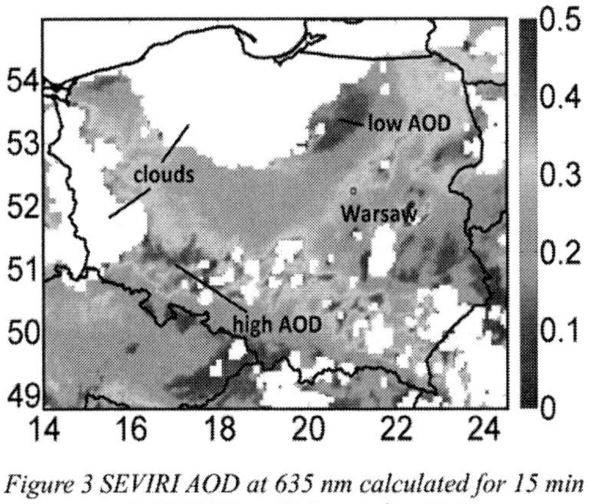

Figure 3 SEVIRI AOD at 635 nm calculated for 15 min at 7:00 UTC on 11 September 2016.

For the data collected at 7:00 UTC, the aerosol extinction and backscatter coefficient profiles at 355nm and 532nm were calculated using the classical Raman approach [10] for the 2h average with 81 points running mean smoothing.

The extinction retrieval was possible down to 400m, as we used the NARLa module. The near- and far-range profiles were calculated separately and combined afterwards. The missing values were interpolated with a constant extinction value at 400m. The maximum value of the particle extinction coefficient at 532 nm of about 0.9 km^{-1} was obtained at about 2 km. From the extinction profile the AOD of 0.113, 0.469, 0.545 was obtained by integration up to 1, 2 and 3 km, respectively. The latter value, was used for the comparison with the AOD-SEVIRI map pixel value for Warsaw.

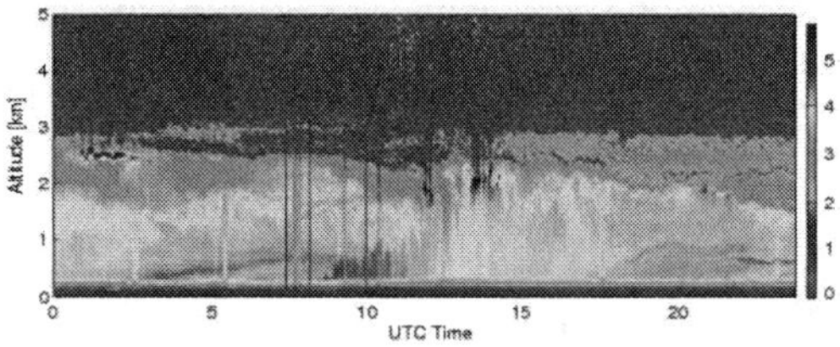

Figure 4 Lidar range corrected signal at 1064 nm on 11 September 2016. Measurements were taken with the PollyXT-UW lidar at the Radiative Transfer Laboratory RT-Lab in Warsaw.

The AOD SEVIRI in Warsaw was lower than the AOD obtained both from lidar and radiometric observations, although we keep in mind that this the data are not directly comparable. First of all, because the difference in the compared ADO-wavelength have to be taken into account. Secondly, because ground-based measurements are representative for a single point, while the SEVIRI data cover area of 5x5 km, and thus the AOD is in fact a mean value for the whole pixel.

In the presented cases the obtained SEVIRI AOD was slightly underestimated in comparison to the lidar and radiometric observations. Such underestimation could be caused by an inaccurate estimation of the surface properties during the reference day (15th September). Analysis of lidar measurements allows to confirm that indeed this day was not a perfectly clean one (not a typical clean background for Warsaw [11]). Although the maximum value of the lidar particle extinction coefficient at 532nm was nearly 10 times lower than the maximum value measured on 11th September, a closer look at the lidar derived intensive and extensive properties indicates an existence of small particles with Ångström Exponent (532/355) strongly oscillating around 2 within the ground and 2.4 km. This is accompanied with a very low linear polarization ratio of 0.018±0.04 at 532nm, lidar ratio of 50±12 sr at 532nm, and the AOD at 532nm of 0.328±0.015 (obtained from the ground up to 3 km). Thus, we interpret the lidar data set obtained on 15th September 2017 as typical for background conditions contaminated with a slight urban anthropogenic origin pollution.

4 CONCLUSIONS

Presented in this paper AOD SEVIRI retrieval algorithm gives opportunity for quasi-continuous measurements providing AOD maps over Poland with high temporary resolution up to 15 min. Regarding the observed AOD differences between various sensors, the valuable information obtained from the lidar allowed for successful validation of the algorithm, as potential areas of problems were identified. Further algorithm adjustments will be undertaken, such as development of an additional procedure for a more appropriate choice of the reference day. The lidar validation of the SEVIRI pixel AOD for Warsaw data will be continued. We will also try to determine any relations of specific weather conditions to aerosol properties [11].

ACKNOWLEDGEMENTS

The scientific work presented in this paper was performed in the frame of the SAtellite based Monitoring Initiative for Regional Air quality (SAMIRA) funded by the ESA-ESRIN Contract 4000117393/16/I-NB and the Technical assistance for Polish Radar and Lidar Mobile Observation System (POLIMOS) funded by the ESA-ESTEC Contract 4000119961/16/NL/FF/mg.

The development of the PollyXT-UW lidar was financed by the Polish Foundation of Science and Technology FNTP-No.519/FNITP/115/2010.

The lidar was developed in collaboration of the University of Warsaw (IGFUW) with the Institute of Tropospheric Research (TROPOS). We would like to acknowledge colleagues from the PollyXT development group of Dr. Dietrich Althausen.

References

[1] Aminou, D.M., 2002: MSG's SEVIRI Instrument. *ESA bulletin* **111**, 15–17.

[2] Zawadzka, O. and K. Markowicz, 2014: Retrieval of Aerosol Optical Depth from Optimal Interpolation Approach Applied to SEVIRI Data, *Remote Sens.* **6**, 7182-7211.

[3] Engelmann R., et al.: The automated multiwavelength Raman polarization and water-vapor lidar PollyXT: the neXT generation, *Atmos. Meas. Tech.* **9**, 1767-1784.

[4] Althausen D., et al.: Polly NET - a network of multiwavelength polarization Raman lidars. Proc. SPIE *Lidar Technologies, Techniques, and Measurements for Atmospheric Remote Sensing IX,* paper no 8894-17, pp.10, SPIE Remote sensing, 23-26 September 2013, Dresden, Germany, (88940I, October 22, 2013).

[5] Rodgers, C., 2000: Inverse Methods for Atmospheric Sounding; Theory and Practice, Series on *Atmospheric, Oceanic and Planetary Physics*; vol. **2**, World Scientific: Singapore.

[6] CAMS - Copernicus Atmosphere Monitoring Service (http://atmosphere.copernicus.eu/).

[7] Pappalardo, G., et al., 2014: EARLINET towards an advanced sustainable European aerosol lidar network, *Atmos. Meas. Tech.* **7**, 2389–2409.

[8] Draxler, R. R., Rolph, G. D., 2010: HYSPLIT (HYbrid Single-Particle Lagrangian Integrated Trajectory). Model access via READY Website (ready.arl.noaa.gov/HYSPLIT.php). NOAA Air Resources Laboratory, *Silver Spring*, MD.

[9] Navy Aerosol Analysis and Prediction System (NAAPS): http://www.nrlmry.navy.mil/aerosol/ (accessed on 9th January 2017).

[10] Baars H., et al.: An overview of the first decade of PollyNET: an emerging network of automated Raman-polarization lidars for continuous aerosol profiling, *Atmos. Chem. Phys.,* **16**, 5111-5137, 2016.

[11] Stachlewska I.S., Zawadzka O., Engelmann R., 2017: Effect of Heat Wave Conditions on Aerosol Optical Properties Derived from Satellite and Ground-Based Remote Sensing over Poland, *Remote Sensing,* vol. **9** (11) pp. art. 1199 10.3390/rs9111199

MINERAL DUST IN CENTRAL ASIA: COMBINING LIDAR AND OTHER MEASUREMENTS DURING THE CENTRAL ASIAN DUST EXPERIMENT (CADEX)

Dietrich Althausen[1,*], Julian Hofer[1], Sabur Abdullaev[2], Abduvosit Makhmudov[2], Holger Baars[1], Ronny Engelmann[1], Khanneh Wadinga Fomba[1], Konrad Müller[1], Georg Schettler[3], Lars Klüser[4] and Konrad Kandler[5]

[1]*Leibniz Institute for Tropospheric Research, Germany, *dietrich@tropos.de*
[2]*Physical Technical Institute of the Academy of Sciences of Tajikistan, Tajikistan*
[3]*GFZ German Research Centre for Geosciences, Germany*
[4]*German Aerospace Center (DLR), German Remote Sensing Data Center, Germany*
[5]*Technical University Darmstadt, Germany*

ABSTRACT

Mineral dust needs to be characterized comprehensively since it contributes to the climate change in Tajikistan / Central Asia. Lidar results from the measurements of mineral dust during CADEX are compared with results of sun photometer measurements, satellite-based measurements, and chemical analysis of ground samples. Although the dust is often advected from far-range sources, it impacts on the local conditions considerably.

1 INTRODUCTION

Central Asia is part of the desert (or dust) belt from the Sahara, over the Saudia Arabian and Iranian deserts, to the East-Asian deserts like the Gobi desert. Within this belt, often mineral dust is lifted up into the atmosphere and advected thousands of kilometers to other locations as for instance to Tajikistan / Central Asia. Such dust events contribute significantly to the climate change in Central Asia.

Chemical measurements at the ground, vertically-resolved lidar measurements, and satellite-based measurements complement each other since the first are much more comprehensively regarding the constituents of the atmospheric particles, while the second relate the atmospheric vertical layers to each other, while the third provide a horizontal prospective. The task of this contribution is to illustrate simultaneous outputs of these three types of mineral dust measurements at Dushanbe.

2 CENTRAL ASIAN DUST EXPERIMENT (CADEX)

Long-lasting vertically resolved aerosol measurements were not available in Tajikistan so far. To fill this gap of knowledge, the Central Asian DUst EXperiment (CADEX) took place at Dushanbe, the capital of Tajikistan, from March 2015 until August 2016. The measurement site was located at the latitude of 38°33'34"N and the longitude of 68°51'22"E, and at the elevation of 864 m a.s.l. on a small hill some hundred meters away from a big road. Within this experiment combined multiwavelengths Raman and polarization lidar and sun photometer measurement were performed to assess the aerosol loading over Dushanbe. Chemical measurements accompanied these measurements to get insights to the chemical constituents of the particles at ground level. Further details to CADEX can be found in [1, 2] and at `http://www.tropos.de/en/research/`

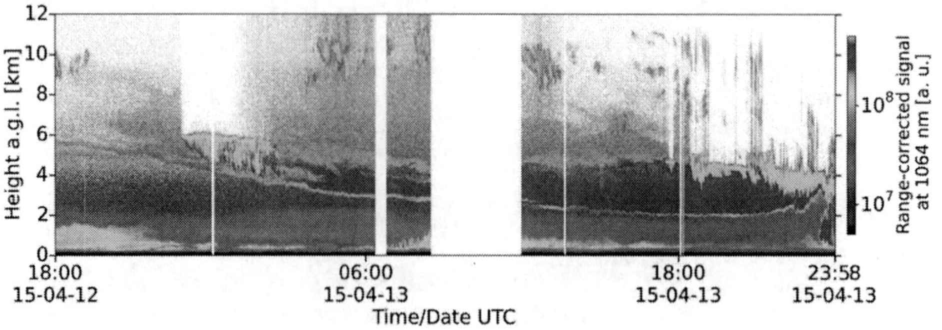

Figure 1: Temporal development of the range-corrected signal at 1064 nm wavelength between 12 April 2015, 18:00 UTC and 13 April 2015, 23:58 UTC.

`projects-infrastructures-technology/research-projects/cadex/`.

3 LIDAR, IASI, AND CHEMICAL MEASUREMENTS

The deployed multiwavelength Raman and polarization (3+2+2) lidar is a Polly^{XT} from TROPOS [3]. This system is part of the Polly^{NET} [4] and its quicklooks are accessible via `http://polly.tropos.de`. An example of a lidar measurement at Dushanbe is shown in Fig. 1. This measurement shows an optically thick lofted dust layer that is decoupled from ground at the beginning and is mixed down at the end of the measurement time. Examples of more profiles are shown in [2]. Other examples are presented in [1, 2]. The identification of aerosol layers containing dust in section 4 is done by using the linear particle depolarization ratio. 10% was set as the minimum value for a dust containing layer. Using this condition, the lower and upper limits of the layers are determined by the maximum gradients of the particle backscatter coefficient profile.

The used sun photometer is made by the company CIMEL and is part of AERONET. Its data (for instance the aerosol optical thickness; AOT) are accessible via `http://aeronet.gsfc.nasa.gov/cgi-bin/type_one_station_opera_v2_new?site=Dushanbe&nachal=2&level=1`.

The Infrared Mineral Aerosol Retrieval Scheme (IMARS) is applied to hyperspectral infrared observations of the Infrared Atmospheric Sounding Interferometer (IASI) on board of the polar orbiting Metop-A satellite (equator crossing time 10:30 LST) in order to retrieve desert dust aerosol optical depth, emission temperature (and thus layer height) and other parameters. A description of the method and its capabilities in full detail can be found in [5, 6].

The collection of PM10 aerosol was carried out on quartz fiber filters (MK 360, MUNKTELL) using a high-volume filter sampler (DHA-80, DIGITEL). These samplings were taken over 48 hours (sampling volume about 1520 m^3) during CADEX. Weighing of filters and all chemical analysis were carried out at TROPOS laboratories according to the techniques described in [7].

4 COMBINATION OF MEASUREMENTS AT GROUND, VERTICAL-RESOLVED, AND SPACE-BASED MEASUREMENTS

Figure 2 summarizes the results from the different measurements for April 2015. Parts a), b), and c) of Fig. 2 show results of the sun photometer, the lidar, and the IASI measurements, but only if lidar and IASI measurements were possible.

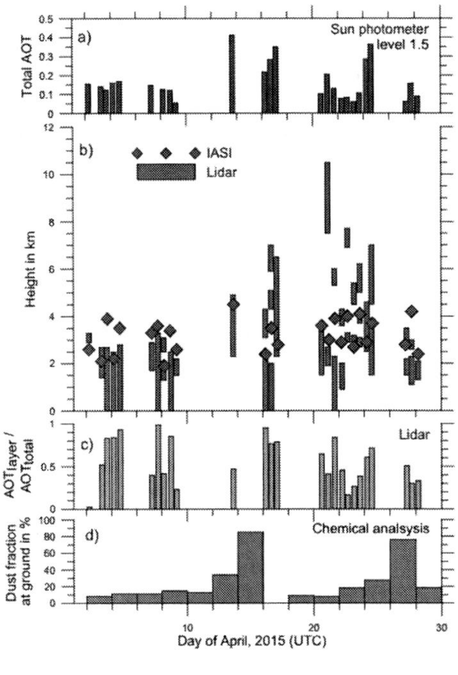

Figure 2: Results from 2 to 30 April 2015. A), b), and c) show only results if lidar AND IASI data are available. a) Aerosol optical thickness (AOT) at 500 nm measured with the sun photometer. b) Dust containing layer heights (red) from lidar measurements. Dust altitude (DALT) from IASI in violet. c) Ratio of aerosol optical thickness of the dust containing layers to the total aerosol optical thickness (from lidar). d) Dust fraction of particles determined by chemical measurements.

The measurements of April 2015 indicate periods with different dust conditions:

- 2 – 9 April: low dust layers, small dust fraction in chemical samples, 6 times: DALT values from IASI > maximum layer heights from lidar, 4 times: DALT values from IASI < maximum layer heights from lidar;
- 9 – 13 April: no IASI measurements, small and moderate dust fraction in chemical samples;
- 13 April: lofted dust layer, large AOT, moderate dust fraction in chemical sample, DALT from IASI < maximum layer height from lidar, down-mixing of dust (compare Fig. 1);
- 14 – 15 April: clouds, rain, very large dust fraction in chemical sample;
- 16 – 17 April: low and lofted dust layers, high dust load (large AOT; large fraction of AOT from layers), DALT values from IASI < maximum layer heights from lidar;
- 17 – 20 April: rainy, low-level clouds, no combined measurements of lidar and IASI;
- 20 – 24 April: often lofted dust layers, moderate dust load and partly large fraction of AOT from layers, DALT values from IASI < maximum layer heights from lidar, at the end of this period large AOT in lofted layers;
- 25 – 26 April: rain and no lidar measurements, very large dust fraction in chemical sample;
- 27 – 28 April: lofted dust layers, low to moderate dust load (AOT), 2 times DALT values from IASI > maximum layer heights from lidar, 1 time: DALT values

from IASI < maximum layer heights from lidar;

- 28 – 30 April: no lidar measurements.

5 DISCUSSION

CADEX provides a unique data set about the aerosol in Central Asia. This contribution describes measurements of April 2015 done with an aerosol lidar, a sun photometer, the space-based IASI, and chemical analysis. Often lofted layers were advected over Tajikistan. 2 periods had been observed with first lofted aerosol layers, then down mixing of dust, rain, and subsequent high dust fractions in the particles at ground. The dust altitudes from IASI were in the range between 1.9 and 4.5 km height, while the lidar-observed heights ranged from ground to 10.5 km height.

ACKNOWLEDGEMENTS

The CADEX project was funded by the German Federal Ministry of Education and Research in the context of "Partnerships for sustainable problem solving in emerging and developing countries" under the grant number 01DK14014. This project has also received funding from the European Union's Horizon 2020 research and innovation programme under grant agreement No 654109.

References

[1] Hofer, J., D. Althausen, et. al., 2017: Mineral Dust in Central Asia: 18-Month Lidar Measurements in Tajikistan during the Central Asian Dust EXperiment (CADEX), this conference.

[2] Hofer, J., D. Althausen, et al., 2016: Central Asian Dust Experiment (CADEX): Multiwavelength Polarization Raman Lidar Observations in Tajikistan, *EPJ Web of Conferences*, **119**, 18006, doi:`dx.doi.org/10.1051/epjconf/201611918006`

[3] Engelmann, R., T. Kanitz, et al., 2016: The automated multiwavelength Raman polarization and water-vapor lidar Polly^{XT}: the neXT generation, *Atmos. Meas. Tech.*, **9**, 1767–1784, doi:`dx.doi.org/10.5194/amt-9-1767-2016`.

[4] Baars, H., T. Kanitz, et al., 2016: An overview of the first decade of Polly^{NET}: an emerging network of automated Raman-polarization lidars for continuous aerosol profiling, *Atmos. Chem. Phys.*, **8**, 5111–5137, doi:`dx.doi.org/10.5194/acp-16-5111-2016`.

[5] Klüser, L., J.R. Banks, et al. 2015: Information content of space-borne hyperspectral infrared observations with respect to mineral dust properties, *Remote Sensing of Environment*, **156**, 204-309, doi:`dx.doi.org/10.1016/j.rse.2014.09.036`.

[6] Klüser, L., C. Di Biagio, et al., 2016: Optical properties of non-spherical desert dust particles in the terrestrial infrared – An asymptotic approximation approach, *J. Quant. Spectrosc. Radiative Transf.*, **178**, 209-223, doi:`dx.doi.org/10.1016/j.jqsrt.2015.11.020`.

[7] Fomba, K. W., Müller, K., et al. 2014: Long-term chemical characterization of tropical and marine aerosols at the Cape Verde Atmospheric Observatory (CVAO) from 2007 to 2011, *Atmos. Chem. Phys.*, **14**, 8883–8904, doi:`dx.doi.org/10.5194/acp-14-8883-2014`.

CHARACTERIZATION OF ATMOSPHERIC THERMODYNAMIC VARIABLES BY RAMAN LIDAR IN THE FRAME OF THE INTERNATIONAL NETWORK FOR THE DETECTION OF ATMOSPHERIC COMPOSITION CHANGE - NDACC

Benedetto De Rosa[1*], Paolo Di Girolamo[1], Donato Summa[1]

[1]Scuola di Ingegneria, Università degli Studi della Basilicata, Italy , [*]Email: benedetto.derosa@unibas.it

ABSTRACT

In November 2012 the Raman Lidar system BASIL, located at the Univ. of Basilicata (Potenza), was approved to enter in NDACC, with the goal of providing accurate routine measurements of the vertical profiles of atmospheric temperature and water vapour mixing ratio. In this presentation we illustrate and discuss water vapour mixing ratio and temperature measurements carried out during these four years and their comparisons with the radiosondes launched from nearby Institute IMAA-CNR (7 km away).

1 INTRODUCTION

Water vapour is a key atmospheric trace gas controlling weather and climate. The effects of water vapour are large in the upper troposphere and lower stratosphere (UTLS), but there are few measurements of water vapour and its long-term variation in this altitude region. Measurements in the UTLS have been traditionally guaranteed by the use of operational radiosondes or balloon-borne frost-point hygrometers, the latter remaining the most reliable source of high quality water vapour measurements in the UTLS region [1], but being yet too expensive to be used on a operational basis.

For the above motivations, in the early 2000s, the international Network for the Detection of Stratospheric Change (NCSC), now Network for the Detection of Atmospheric Composition Change (NDACC), decided to include water vapour Raman lidars among its instruments.This is because, while NDACC remains committed to long-term monitoring of changes in the stratosphere, with an specific emphasis on the evolution of the ozone layer, its priorities have broadened to include issues such as the detection of trends in the overall atmospheric composition and understanding their impacts on the stratosphere and troposphere, and establishing links between climate change and atmospheric composition.

Atmospheric temperature is another parameter of interest for NDACC as in fact changes in climate and atmospheric composition have a significant impact of the thermal structure of the atmosphere.

The Raman Lidar system *BASIL* joined NDACC in November 2012, with the primary goal of providing accurate routine measurements of the vertical profile of both water vapour mixing ratio and temperature, these two parameters being of primary importance for the purposes of NDACC.

2 METHODOLOGY

Measurements discussed in this paper were performed by the Raman Lidar system *BASIL* in Potenza, Italy (40°38'45" N, 15°48'29" E, 730 m above mean sea level). The system is located on the roof of the Scuola di Ingegneria at the Università degli Studi della Basilicata. *BASIL* is capable to perform high-resolution and accurate measurements of atmospheric water vapour and temperature, both in daytime and nighttime, based on the application of the vibrational and rotational Raman lidar techniques, respectively, in the UV [2, 3, 4, 5, 6]. Besides water vapour and temperature, *BASIL*, in its NDACC configuration, also provides measurements of particle backscatter at 355 and 532 nm, particle extinction at 355 nm and particle depolarization at 355 nm. Relative humidity measurements can also be obtained from simultaneous water vapour mixing ratio and temperature measurements [7]. A mobile version of the system has been deployed in a variety of international field campaigns [8, 9]. A valuable contribution to NDACC from BASIL is primarily expected in terms of water vapour mixing ratio and temperature measurements. For this reason, a major emphasis has been put on these parameters, especially in terms of calibration and validation efforts.

In the frame of NDACC, BASIL operates on a routine basis each Thursday, typically from local noon to midnight, covering a large portion of the daily cycle.

2.1 Characterization of atmospheric thermodynamic variables

Raman lidar measurements of the water vapour mixing ratio profile has been extensively discussed in literature (for a review see [2], [10]). The approach is based on the collection of the vibrational Raman backscattered signals from water vapour and nitrogen molecules at the two Raman wavelengths λ_{H_2O} and λ_{N_2}, respectively, i.e.:

$$P_{H_2O}(R) = P_0 \frac{c\Delta t}{2} \frac{A_{tel}}{R^2} O(R) n_{H_2O}(R) \sigma_{H_2O} T_{\lambda_0}(R) T_{\lambda_{H_2O}}(R) \quad (1)$$

$$P_{N_2}(R) = P_0 \frac{c\Delta t}{2} \frac{A_{tel}}{R^2} O(R) n_{N_2}(R) \sigma_{N_2} T_{\lambda_0}(R) T_{\lambda_{N_2}}(R) \quad (2)$$

Water vapour mixing ratio, $x_{H_2O}(R)$, can be obtained from the power ratio of water vapour to molecular nitrogen vibrational Raman signals through the expression:

$$x_{H_2O}(R) = k \frac{P_{H_2O}(R)}{P_{N_2}(R)} \Delta T_{\lambda_{H_2O},\lambda_{N_2}} \quad (3)$$

where k is the lidar system calibration coefficient and $\Delta T_{\lambda_{H_2O},\lambda_{N_2}} = T_{\lambda_{N_2}}(R) / T_{\lambda_{H_2O}}(R)$ is the differential transmission term, which accounts for the different atmospheric transmission at λ_{H_2O} and λ_{N_2}. The calibration coefficient k was determined based on the comparison between simultaneous and co-located water vapour mixing ratio profiles from the Raman lidar and radiosondes for an extended measurement sample. The vertical profile of atmospheric temperature is obtained based on the application of the rotational Raman technique in the UV using the low- and high- quantum number rotational Raman signals $P_{LoJ}(z)$ and $P_{HiJ}(z)$, considering the expressions:

$$R(z) = \frac{P_{HiJ}\ (z[T])}{P_{LoJ}\ (z[T])} \cong \exp\left(\frac{a}{T(z)} + b\right) \quad (4)$$

$$T(z) = \frac{a}{\ln R(z) - b} \quad (5)$$

with a and b being two calibration constants, which can be determined through the comparison of BASIL measurements with simultaneous and co-located temperature measurements from a different profiling sensor (for example a radiosonde).

3 RESULTS

As an example of the measurements carried out in the frame of NDACC, figure 1 illustrates the time evolution of the water vapour mixing ratio over an approx. 4 h period of from 16:43 UTC to 21:00 UTC on 27 November 2014. Figure 1 is plotted as a succession of 5 min averaged consecutive profiles. To reduce signal statistical fluctuations, vertical smoothing is applied to the data to achieve an overall vertical resolution of 150 m.

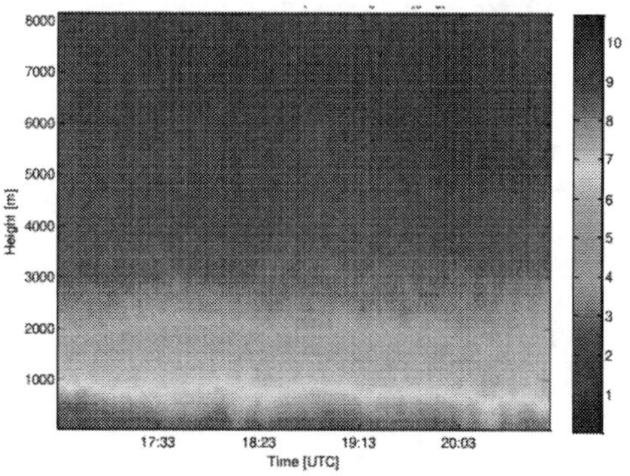

Figure 1 Time evolution of the water vapour mixing ratio over an approx. 4h period of from 16:43 UTC to 21:00 UTC on 27 November 2014.

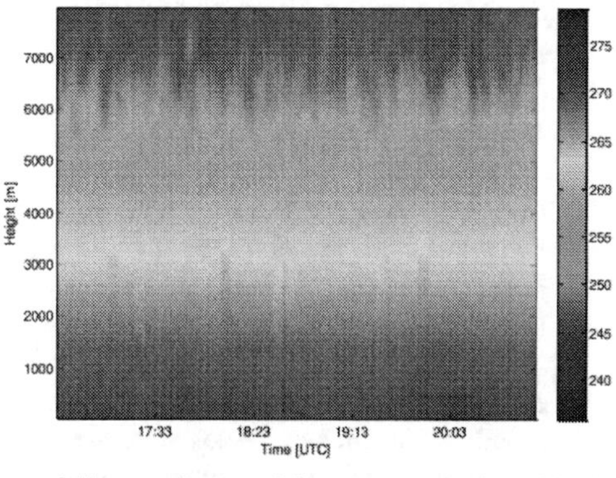

Figure 2 Time evolution of the temperature from 16:43 UTC to 21:00 UTC on 27 November 2014.

Figure 2 shows the evolution of atmospheric temperature in the same time interval of figure 1. In this case, in order to obtain a sufficiently high signal statistics, it was necessary to consider a vertical resolution of 300 m. Figure 1 reveals the presence of a humid layer extending up to 1.5 km, identifying the convective boundary-layer. The top of the convective boundary layer is clearly visible also in figure 2, being identified by the strong temperature gradient at 1.5 km. Comparisons between the Raman lidar at Scuola di Ingegneria and the radiosondes launched from IMAA-CNR (7 km away) were carried out in the period 9 October 2014- 7 May 2015. An overall number of 11 comparisons were possible. The sample is limited because before 20 November 2014 radiosondes were not routinely launched from IMAA-CNR (routine launches are carried out on Thursday, i.e. on the same day when BASIL is operated in the frame of NDACC). Figure 3 illustrates the vertical profiles of water vapour mixing ratio (panel a) and temperature (panel b) on 27 November 2014, considering an integration time of approx. 4 h between 16:43 and 21:00 UTC. The vertical resolution is 150 m for both water vapour and temperature measurements. For comparison, the almost simultaneous water vapour mixing ratio and temperature profiles measured by a radiosonde launched from of Tito Scalo at 18:00 UTC on 27 November 2014 are also included in the figure. The agreement between BASIL and the radiosonde is good for both parameters, especially when considering the distance between the two sites. The water vapour mixing ratio profile from BASIL is found to reach approx. 12 km, with a sensitivity level of 0.05 g/kg. The capability to reach 12 km and a sensitivity level of 0.03-0.05 g/kg have been observed in most of the 2 h water vapour profiles measured in clear sky conditions by BASIL in the frame of NDACC. When considering measurements integrated over the entire night, the lidar profile is found to extend up to approx. 18-20 km. More results from this measurement effort, also focusing on the statistical analysis on the collected data, will be illustrated and discussed at the Conference. Temperature profiles are typically measured by BASIL with a vertical resolution of 300 m. Up to 8 km the agreement with the radiosonde is very good. Considering the 11 possible comparisons between BASIL and the radiosondes in the period 9 October 2014- 7 May 2015, the water vapour mixing ratio mean relative bias was estimated (figure 4).

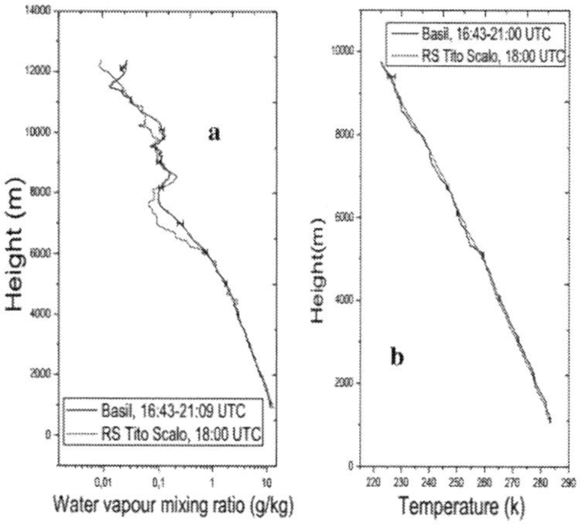

Figure 3 Vertical profiles of water vapour mixing ratio (panel a) and temperature (panel b) as obtained considering a time integration between 16:43 and 21:00 UTC on 27 November 2014, together with the simultaneous profile as measured by a radiosonde launched from the nearby site of Tito Scalo at 18:00 UTC on 27 November 2014.

Results in the figure reveal that, with the only exception of a couple of points, bias values are within ± 10% up to 14 km. These results demonstrate the capability of BASIL to comply with the NDACC observational requirements.

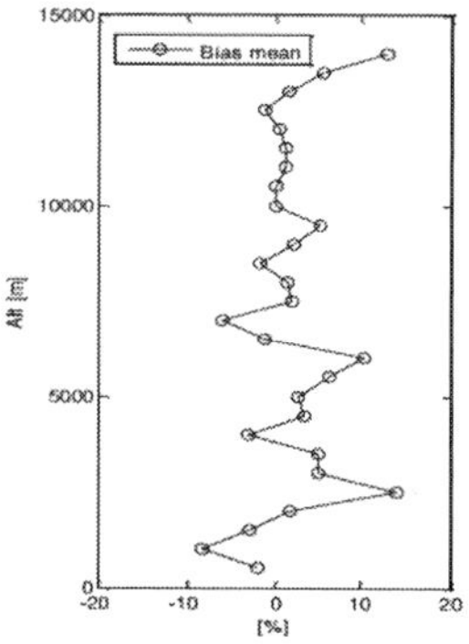

Figure 4 Vertical profile of the vapour mixing ratio mean bias (expressed in %) between BASIL and the radiosondes

Figure 5 illustrates the vertical profiles of the mean relative and absolute bias and RMS deviation of BASIL vs. the radiosondes for temperature measurements as obtained considering the above mentioned 11 days. Above the planetary boundary layer up to 9.5 km the value are low, in particular the bias absolute are within ± 1 K.

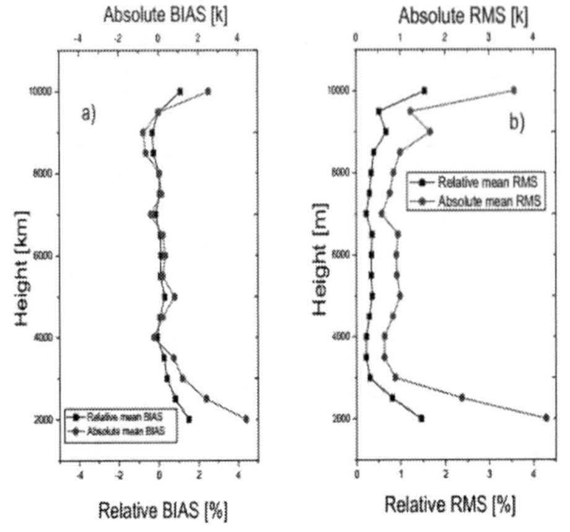

Figure 5 Vertical profiles of the relative and absolute mean bias (left panel) and RMS deviation (right panel) for the temperature measurements

ACKNOWLEDGEMENTS

We wish to warmly thank Istituto di Metodologie per l'Analisi Ambientale, Consiglio Nazionale delle Ricerche, in Tito Scalo, for the provision of the radiosnde data.

References

[1] Vomel, H., H. Selkirk, L. Miloshevich, J. Valverde-Canossa, J. Valdes, E. Kyro, R. Kivi, W. Stolz, G. Peng, and J. A. Diaz,2007: Radiation dry bias of the Vaisala RS92 humidity sensor. J. Atmos. *Oceanic Technol.*, **24**, 953-963.

[2] Whiteman, D. N., 2003a: Examination of the traditional Raman lidar technique. I. . Evaluating the temperature-dependent lidar equations, *Appl.Optics*,**42**,2571-2592.

[3] Behrendt, A. and Reichardt, J.: Atmospheric temperature profiling in the presence of clouds with a pure rotational Raman lidar by use of an interference-filter-based polychromator, *Appl Optics*,**39**,1372-1378,2000.

[4] Di Girolamo, P., R. Marchese, D. N. Whiteman, B. B. Demoz, 2004: Rotational Raman Lidar measurements of atmospheric temperature in the UV. *Geophys. Res. Lett.* **31**,doi:10.1029/2003GL018342.

[5] Di Girolamo, P., A. Behrendt, and V. Wulfmeyer, 2006. Spaceborne profiling of atmospheric temperature and particle extinction with pure rotational Raman lidar and of relative humidity in combination with differential absorption lidar: performance simulations, *Appl. Opt.* **45**, 2474-2494, doi:10.1364/AO.45.002474.

[6] Bhawar, R., P. Di Girolamo, D. Summa, C. Flamant, D. Althausen, A. Behrendt, C. Kiemle, P. Bosser, M. Cacciani, C. Champollion, T. Di Iorio, R. Engelmann, C. Herold, Müller, D., S. Pal, M. Wirth, V. Wulfmeyer, 2011: The Water Vapour Intercomparison Effort in the Framework of the Convective and Orographically-Induced Precipitation Study: Airborne-to-Ground-based and airborne-to-airborne Lidar Systems, *Quarterly Journal of the Royal Meterological Society* **137**,325-348.

[7] Di Girolamo P., D. Summa, R.-F Lin, T. Maestri, R. Rizzi, G. Masiello, 2009: UV Raman Lidar measurements of relative humidity for the characterization of cirrus cloud microphysical properties, *Atmos. Chem. Phys.* **9**, 8799-8811, doi: 10.5194/acp-9-8799-2009.

[8] Di Girolamo, P., C. Flamant, M. Cacciani, E. Richard, V. Ducrocq, D. Summa, D. Stelitano, N. Fourrié and F. Saïd, 2016: Observation of low-level wind reversals in the Gulf of Lion area and their impact on the water vapour variability, *Q. J. Roy. Meteor. Soc.* **142 (Suppl 1)**, 153–172, doi: 10.1002/qj.2767.

[9] P. Di Girolamo, D. Summa, B. De Rosa, M. Cacciani, A. Scoccione, A. Behrendt, V. Wulfmeyer, 2017: Characterization of Boundary Layer Turbulent Processes by Raman Lidar: Demonstration of the Measurement Capabilities of the Raman Lidar System BASIL, *Atmos. Chem. Phys.* **17**, 745-767, doi:10.5194/acp-17-745-2017.

[10] Whiteman, D. N., 2003b Examination of the traditional Raman lidar technique. II. . Evaluating the ratios for water vaour and aerosols.temperature-dependent lidar equations, *Appl.Optics*,**42**,2593-2608.

MONITORING THE INDUSTRIAL SOURCES OF AEROSOL IN CUBATAO, BRAZIL, USING A SCANNING ELASTIC LIDAR AND A LIDAR DOPPLER

Renata F. da Costa[1]*, Marcia T. A. Marques[2], Fernanda de M Macedo[2], Izabel da Silva Andrade[2], Elaine Cristina Araujo[2], Thais Correa[2], Maria Helena Goncalves de Andrade Salani[2], Daniel Silveira Lopes[3], Maria Lucia Goncalves Guardani[3], Eduardo Landulfo[2], Roberto Guardani[1]

*[1] University of Sao Paulo, Chemical Engineering Department, Brazil, *renata.facundes.costa@usp.br*
[2] IPEN – Energy and Nuclear Power Research Institute, Brazil
[3] CETESB – Sao Paulo State Environmental Agency, Brazil

ABSTRACT

Field campaigns with a scanning multiwavelength elastic lidar coupled with a Doppler system to monitor industrial atmospheric aerosol emissions were carried out, with the objective of monitoring aerosol emission sources and plume dispersion. Since the technique provides information on the spatial and temporal distribution of aerosol concentration, the implementation of a systematic monitoring procedure is proposed as a valuable tool in air quality monitoring applied to regions of interest.

1 INTRODUCTION

The area around the Brazilian city of Cubatao is known for the highly concentrated industrial activities, which, combined with the specific geographical situation, results in a series of problems related with air quality [1].

Cubatao is located in a narrow coastal plain surrounded by a steep mountain range to the north, west, and east, and by the sea to the south. In the region, there is a large oil refinery, a steel production site, a cement plant, plus a number of petrochemical and fertilizer plants, and the largest harbor in South America. At ca. 1 km west and northwest a 600-1000 m high mountain shell retains air circulation. Due to the local topography and proximity of the ocean, wind direction and velocity show daily changes that affect air quality, and frequent events of high pollutants concentration in the industrial area are recorded. A significant number of events of high levels of PM2.5, PM10, O_3, NO_2 and SO_2, are recorded by the local authority (CETESB), which operates 5 air quality monitoring stations in the region [2].

2 SYSTEM SETUP

For the study presented here two lidar systems were used. A three-wavelength elastic backscatter lidar system and a Doppler lidar.

2.1 Elastic lidar

The three-wavelength elastic backscatter system operates with a Nd:YAG laser (CFR 450, Quantel S.A.) at 355 nm, 532 nm, and 1064 nm, transmitting pulses of 7±2 ns in duration at a fixed repetition rate of 20 Hz and a divergence of less than 0.3 mrad. The receiver used to collect the backscattered laser light is a 150 mm diameter Dall-Kirkham telescope with an effective focal length of 1000 mm.

2.2 Doppler Lidar

The Doppler lidar is a pulsed lidar model Windcube V2 from Leosphere that detects the frequency shift from the backscattered light due the particles movement, the frequency shift is proportional to the line-of-sight velocity and the wind velocity and the wind direction can be inferred. The lidar was set up to measure vertical profile of wind velocities and wind direction in 12 levels from 40 to 290 m with a vertical resolution of 20 to 30 m, recording 10 minutes averaging with a temporal resolution of approximately 1 Hz. The Doppler lidar was co-located to the Elastic lidar during the measurement campaign.

3 METHODOLOGY

The campaigns for data collection were carried out in Cubatao during August 2016, under dry weather conditions, to monitor the spatial and temporal distribution of aerosols in the region. The lidar systems were installed in the center of an industrial complex where most fertilizer plants and a steel industry are located.

3.1 Data acquisition

For data acquisition, an elevation of 12 degrees and a window of 40 degrees in azimuthal angle were adopted. Three strategic sectors were scanned with these elevation and azimuthal angles in 1 degree steps. Each sector was scanned three times with an integration time of 5 seconds, totalizing 15 seconds for each primary data set. The first sector was facing the north direction (center in 12° N), the second was facing 50° NE, and the third and last sector was facing 100° E, as azimuthal angle reference.

Information about wind velocity and wind direction were acquired at the same site and time with the Doppler lidar for 12 levels of measurements, 40, 60, 80 ,100 ,120, 140, 160, 180, 200, 230, 260 and 290 m.

3.2 Data analysis

In order to analyze the data, plan position indicator (PPI) plots of the range-corrected lidar backscattered signal were created for each of the three wavelengths, by interpolating the data between azimuthal angle steps.

The range-corrected lidar backscattered signal at a given distance from the system is expressed as:

$$\mathrm{R.C.S.}(\lambda) = \mathrm{K}\,\mathrm{O}(r)\beta(\lambda)e^{-2\int_{r_0}^{r}\alpha(\zeta,\lambda)d\zeta} \quad (1)$$

where R.C.S.(λ) is the range corrected lidar signal for dark-current, zero-bin, background and range-squared distance at wavelength λ [3], K is an instrumental constant accounting for the effective telescope area, pulse length and optical efficiency, O(r) is the overlap function, which refers to the geometrical overlap between the laser beam and the telescope field of view depending on the distance r from the lidar system, $\beta(\lambda)$ is the atmospheric backscatter coefficient for both molecules and particle contributions. $\alpha(\zeta, \lambda)$ is the atmospheric extinction coefficient (molecule plus particle contribution) which depends on the wavelength and the traveling distance ζ.

To improve the analysis, a study of the ratio between the ln R.C.S. signal was made using 1064 and 355 nm, as describe in equation below:

$$\mathrm{A}_{\mathrm{mod}} = -\frac{\ln\left(\mathrm{R.C.S.}(1064)\big/\mathrm{R.C.S.}(355)\right)}{\ln\left(1064\big/355\right)} \quad (2)$$

4 RESULTS

Figures 1, 2 and 3 show the R.C.S for 355, 532, 1064 nm signals, for sectors 1, 2 and 3 respectively, in the form of PPI plots of the three sectors, overlaid to the region map. The plots show an intense signal in the third sector compared to the other two. The observed difference in signal strength for different wavelengths is an indication that there is a higher concentration of larger particles in this sector, due to the higher scatter intensity at 1064 nm.

The interpretation of the scattered lidar data can only be carried out in combination with the Doppler lidar data for the wind field. A summary of the wind field measurements at two altitudes from ground level is shown in Figure 4 indicating that the wind was in mainly in the N to S direction. Thus, the observed aerosol plume is probably being generated in the north sector (first sector of the measurements), where the fertilizer plants are located, and transported by the wind to the south sector. The lidar system was able to detect the spatial distribution of the plume over the measurement region, as shown in Figures 1 to 3, for 1064 nm. A similar interpretation of these data is valid for the other wavelengths, too.

Figure 5 shows the results of the Angstrom exponent modified. This variable provides an indirect relationship with the particle size. The larger the exponent, the smaller the particle size, thus the higher points of concentration of large particles become evidenced.

It can be observed from figure 6 relative humidity, measured during the acquisition period, did not change significantly, thus its effects were neglected in this analysis. These data were provided by the harbor of Santos and the instrumentations were about 1 km of distance from lidar.

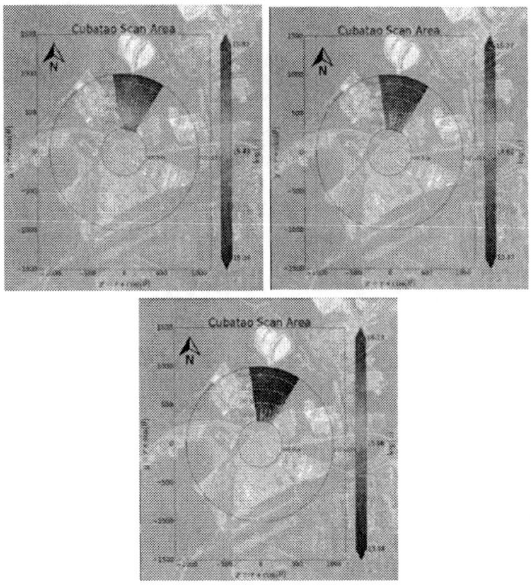

Figure 1 First sector PPI plot of 355 nm (left), 532 nm (right) and 1064 nm (bottom) R.C.S. Source: Google earth V 7.1.8.3036. (January 26, 2017). Cubatao, Brazil. 23°50'57.71"S, 46°23'19.80"W, Eye alt 6.24 km. Digital Globe 2016. http://www.earth.google.com [January 17, 2017].

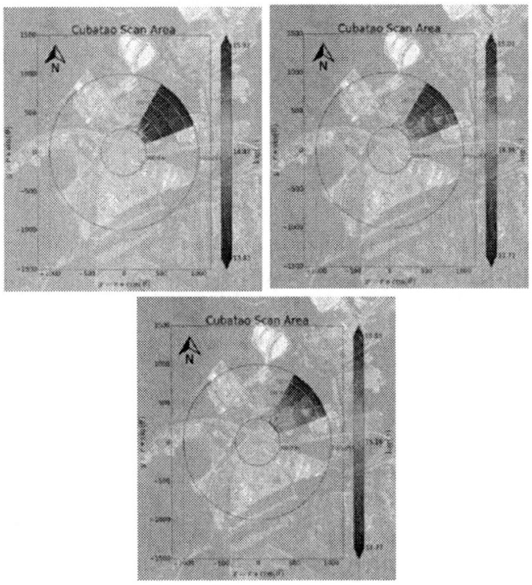

Figure 2 Second sector PPI plot of 355 nm (left), 532 nm (right) and 1064 nm (bottom) R.C.S. Source: Google earth V 7.1.8.3036. (January 26, 2017). Cubatao, Brazil. 23°50'57.71"S, 46°23'19.80"W, Eye alt 6.24 km. Digital Globe 2016. http://www.earth.google.com [January 17, 2017].

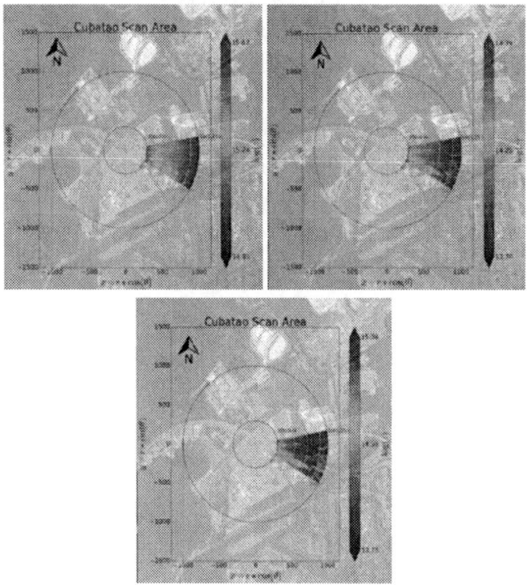

Figure 3 Third sector PPI plot of 355 nm (left), 532 nm (right) and 1064 nm (bottom) R.C.S. Source: Google earth V 7.1.8.3036. (January 26, 2017). Cubatao, Brazil. 23°50'57.71"S, 46°23'19.80"W, Eye alt 6.24 km. Digital Globe 2016. http://www.earth.google.com [January 17, 2017].

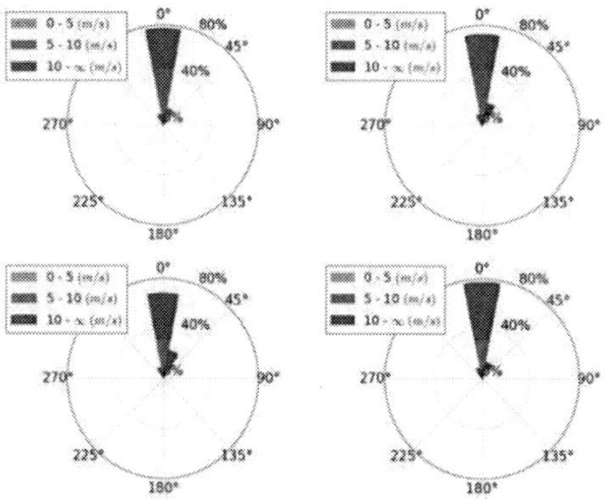

Figure 4 Wind roses indicating wind direction and wind velocity for (left to right) 80, 120, 160 and 200 m

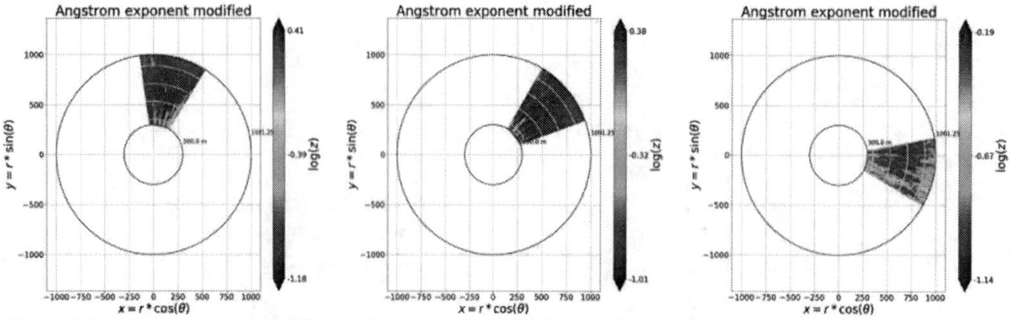

Figure 5 Angstrom exponent modified to first, second and third sector from left to right.

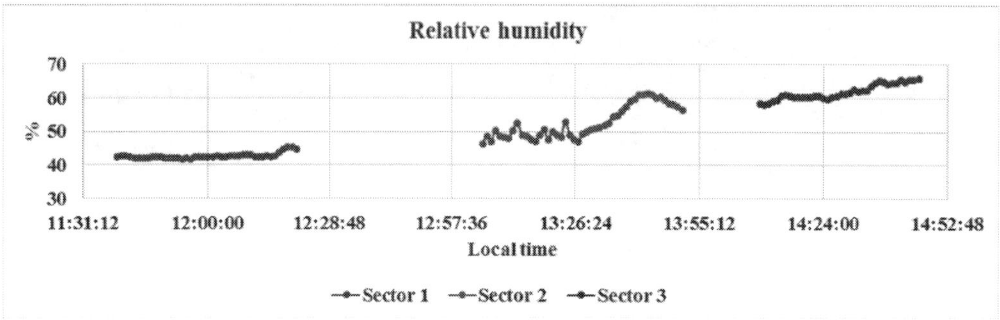

Figure 6 Relative humidity during data acquisition period.

5 CONCLUSIONS

Although only a small set of monitoring results is shown in this study, the results serve to illustrate the potential of the lidar technique to provide information not only on the aerosol concentration distribution over a region of interest, but also to detect specific locations where aerosols are emitted to the atmosphere, as, for instance, industrial stacks. Thus, a backscatter lidar combined with a Doppler lidar can provide real time information on the spatial distribution of aerosol concentration for different particle size ranges, enabling to identify point emission sources (stacks), or distributed sources, like resuspension, or traffic related emissions.

ACKNOWLEDGEMENTS

The authors would like to thank the supporting agencies Coordenação de Aperfeiçoamento de Pessoal de Nível Superior (CAPES), Conselho Nacional de Desenvolvimento e Pesquisa (CNPq), Fundação de Amparo à Pesquisa do Estado de São Paulo (FAPESP) and Comissão Nacional de Energia Nuclear (CNEN) for financing the research.

References

[1] Steffens J., Guardani R., Landulfo E., Moreira P. F.M. Jr. and daCosta R. F., 2011: Study on correlations between Lidar scattered light signal and air quality data in an industrial area, *Proc. Environmental Sciences* **4**, 95-102.

[2] Da Costa, R. F., Marques, M. T. A., Lopes, D. S., Guardani, M. L. G., Macedo, F. d. M., Landulfo, E., Guardani, R., 2016: Monitoring the environmental impact of aerosol loading and dispersion from distinct industrial sources in Cubatao, Brazil, using a scanning lidar, Proc. of SPIE **10006**, 1000608.

[3] Guerrero-Rascado, J. L., da Costa R. F., Bedoya, A. E., Guardani R., Alados-Arboledas, L., Bastidas, A. E., Landulfo E., 2014: Multispectral elastic scanning lidar for industrial flare research: characterizing the electronic subsystem and application. *Optics Express* **22** (25), 31063–31077.

EPJ Web of Conferences **176**, 04012 (2018) https://doi.org/10.1051/epjconf/201817604012
ILRC 28

BIOMASS BURNING AEROSOL OVER ROMANIA USING DISPERSION MODEL AND CALIPSO DATA

Victor Nicolae[1*], Alexandru Dandocsi[1,2], Luminita Marmureanu[1], Camelia Talianu[1,3]

[1]*National Institute of R&D for Optoelectronics, *Email: victor.nicolae@inoe.ro*
[2]*University POLITEHNICA of Bucharest, Faculty of Applied Sciences, Bucharest, Romania*
[3]*Institute of Meteorology, University of Natural Resources and Life Sciences, Vienna, Austria*

ABSTRACT

The purpose of the study is to analyze the seasonal variability, for the hot and cold seasons, of biomass burning aerosol observed over Romania using forward dispersion calculations based on FLEXPART model. The model was set up to use as input the MODIS fire data with a degree of confidence over 25% after transforming the emitted power in emission rate. The modelled aerosols in this setup was black carbon coated by organics. Distribution in the upper layers were compared to Calipso retrieval.

1 INTRODUCTION

One of the driving forces of EARTH's radiative budget is played by the aerosols particles. The origin of the particles is related with anthropogenic or natural sources, such as industry, volcanic emissions, or wild fires. Aerosols emitted by fires contain mainly two components that have different optical behavior: black carbon (BC) which absorbs solar radiation, thus having a direct positive impact on the radiative budget and second organic carbon (OC), that mainly scatter solar radiation[1]. Pollution resulted from biomass burning combined with urban pollution is due to anthropogenic activities and represents the main source of primary aerosols, having a significant impact on human health and climate[2]. Forty percent of the total emission of black carbon is related to biomass burning. Understanding the evolution, the regional distribution, and behavior of BC is important for climate modeling, but also for regulation related to emission control. The high uncertainty related to aerosols[1]represents a major challenge to quantify and to anticipate the effects on air quality, climate and meteorology, at local, regional and global level.

Different characteristics related with the type of vegetation, moisture, aging processes and the degree of mixing with other aerosols, have an important impact on aerosol optical properties.

In this study we analyze the biomass burning aerosols distribution over Romania using FLEXPART model, during the hot and cold season, by evaluating the output of the model when Calipso data were available.

2 METHODOLOGY

This study has been performed for Romania during summer 2016 and winter 2016-2017 based on the FLEXPART aerosol dispersion model (version 9.3) and MODIS Fire Radiative Power (FRP) observations. Assessment of the results using the CALIPSO satellite measurements over Romania was also performed.

2.1 Flexpart model

FLEXPART ("FLEXible PARTicle dispersion model") is a Lagrangian model designed for computing the long-range transport, diffusion, dry and wet deposition, and radioactive decay of air pollutants from point sources for a large number of particles (assimilated with infinitesimal small air areas)[3][4]. The particle dispersion model calculations can be done assuming two ways of transport in the atmosphere: passive transport without involving removal processes and a mode assuming aerosol-like removal by dry and wet deposition. In this analysis, the contribution of the different biomass burning aerosol sources were calculated with FLEXPART forward simulations for 20000 particles based on 6-hourly Global Forecast System (GFS) real-time analysis and 1-hour forecast products and with $0.5^0x0.5^0$resolution [3].

To cover as much biomass burning aerosol types, we defined four anthropogenic sources based on the Selected Nomenclature for Air Pollution (SNAP): Industrial (SNAP 34, SNAP 5, SNAP6),

Urban (SNAP 1, SNAP 2, SNAP 6), Agriculture (SNAP 10), Traffic (SNAP 7, SNAP 8).The SNAP source categories used in this study are shown in Table 1.

Table 1: SNAP source categories

SNAP	Sector name
1	Energy industries
2	Non-Industrial combustion
34	Industry (combustion and processes)
5	Extraction and distribution of fossil fuels
6	Product use
7	Road transport
8	Non-road transport and other mobile sources
9	Waste treatment
10	Agriculture

The biomass burning aerosol (BBA) generated for this analysis was considered as BC coated by organics, with a mono mod lognormal size distribution (accumulation mode) and the size range between 0.1μm - 1 μm. The BBA composition was considered taking account of the TNO MACC III anthropogenic emission inventory for time period 2000 - 2011 [4]. We considered three species: BC, Primary organic matter (POM), and Sulphate (SO_4), which could generate the BBA defined in this study. The correlation between sources and species mass contribution to aerosols is presented in Fig. 1.

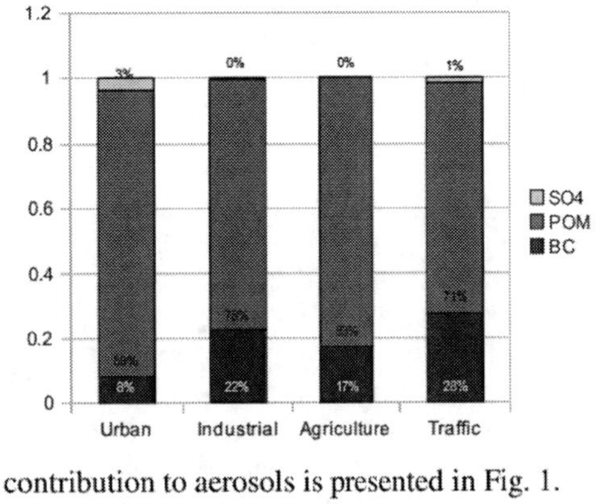

Figure 1 Biomass burning tracer generated for 4 anthropogenic sources

In the model, the species are run in forward mode without involving removal processes and tracer particles are dropped from the simulation after 3 days. This aerosol tracer was added in model with the emission algorithm using MODIS fire detections, landuse information, and emission factors.

2.2 MODIS fire detections

The assimilated data in the model is based on AQUA and TERRA satellite retrieval Fire Radiative Power (FRP). FRP represents a measure of radiative energy emitted per time unit in case of a fire incident. The FRP value is directly proportional to the fuel burned in the fire event. The evaluation of thermal anomalies detected by MODIS sensors are based on brightness temperature difference at 4 μm [5].The concentration was retrieved from FLEXPART based on emission rate derived from FRP and emission coefficient established by Ichokuet al., 2014 [7]. The fire period was estimated for 1 day, including in the model the smouldering period.

2.3 Calipso measurements

The Caliop sensor from the Calipso satellite it is a depolarisation LIDAR that provide vertical profiles of high resolution of aerosols and clouds. CALIOP has 2 lasers Nd:YAG, but uses just one for every measurements. Every laser produce simultaneously pulses at 1064 and 532nm. The laser footprint at the earth surface is 70m at every 333m. The instrument provides continuous measurements during day and night[8]. The three receptions channels of the CALIOP are focused on backscattering at 1064nm and two channels that measure the polarized components of the backscatter signal at 532nm. Calipso data are sent once per day to the ground segment, being processed and archived. Level 2 data from Calipso were used in order to retrieve the biomass burning aerosol layers. Defining a minimum and a maximum longitude range (20.25E, 29.68E respectively) and a minimum and a maximum latitude range (43.61N, 48.25N respectively) a perimeter in the Romanian region was defined. The data selected in this perimeter have characteristics of smoke aerosols which means a lidar ratio of 70 sr in the visible (VIS) wavelength and 40 sr in the infrared (IR) wavelength. The *Cloud – Aerosol Discrimination (CAD)* score of this data had been constrained to be lower than -25 and higher than -100. Negative values means aerosol products and the numeric value represents the level of confidence. We selected the days with more than 85 points in the perimeter described above, and we used the level 2 aerosol profile products from Calipso which provides profiles of backscatter coefficient, extinction coefficient, and the particle depolarization ratio from the VIS region and relative humidity profiles which were used to provide the mass concentration profiles. Using the OPAC simulation for the mass-extinction efficiencies and the methodology described by [9] and [10] based on:

$$\beta_{high}^{part} = \beta_{total}^{part} \cdot \frac{\left(\delta_{total}^{part} - \delta_{d}^{part}\right)}{\left(\delta_{high}^{part} - \delta_{low}^{part}\right)} \cdot \frac{\left(1 + \delta_{high}^{part}\right)}{\left(1 + \delta_{total}^{part}\right)}, (1)$$

where δ_{total}^{part} is the particle depolarization ratio and β_{total}^{part} is the total particle backscatter coefficient.

3 RESULTS AND DISCUSSIONS

A comparison of observed and model-estimated BBA concentration for 2 cases during the summer 2016 is shown in Fig. 2,3 underlining the sensitivity on BBA presence also from Caliop measurements and from FLEXPART. Another two cases, when Calipso data were (01.04.2016 and 15.09.2016) available, were analyzed confirming the BBA presence (not shown).

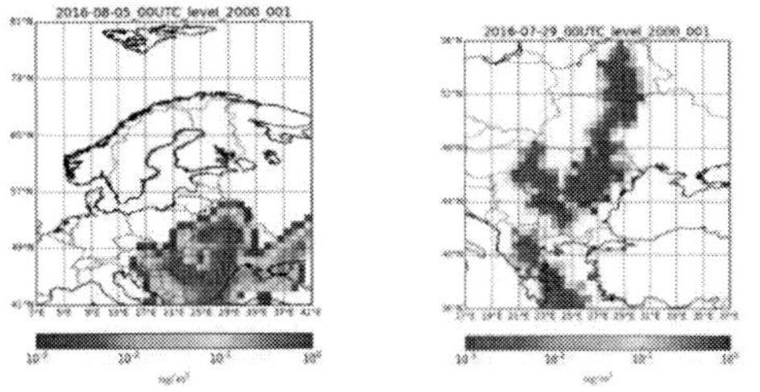

Figure 2BBA distribution over Romania on 05 August respectively 29 July 2016

A comparison profile to profile from Calipso and FLEXPART are still too early to be done due to some uncertainties related to meteorological simulations, approximations in the PBL differences between grid estimation and punctual measurement, altitude of smoke injection in to the atmosphere, time of smoldering and sub-grid effect. Previous studies demonstrate also that these limitations can influence in a significant way the comparisons at small scale. [11]

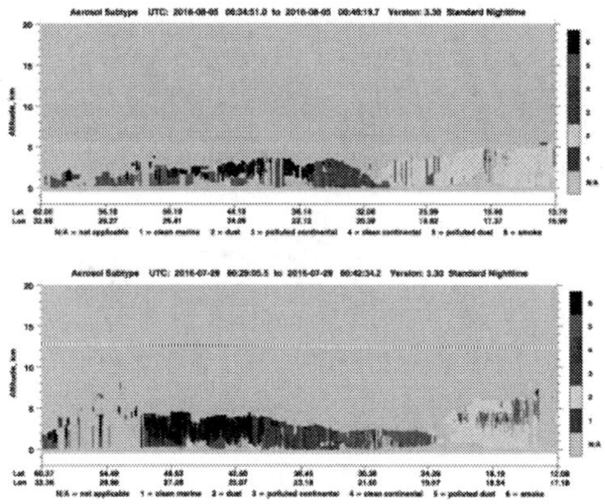

Figure 3Position (altitude, latitude and longitude) and type of the aerosol layers detected by CALIPSO for 5.08.2016, and 29.07.2016 black color indicating smoke aerosol particles.

The model used in this study is suitable tool to identify the plumes at different altitudes, affected areas by high concentrations in BBA, and to asses from the qualitatively point of view the plume evolution. In this context we analyzed the most affected areas by BBA over Romania, generating sensitivity maps (model based) on two seasons. The summer average BBA distribution (computed from first of July to end of August) over Romania show important influences from surrounding area. (Fig.4).

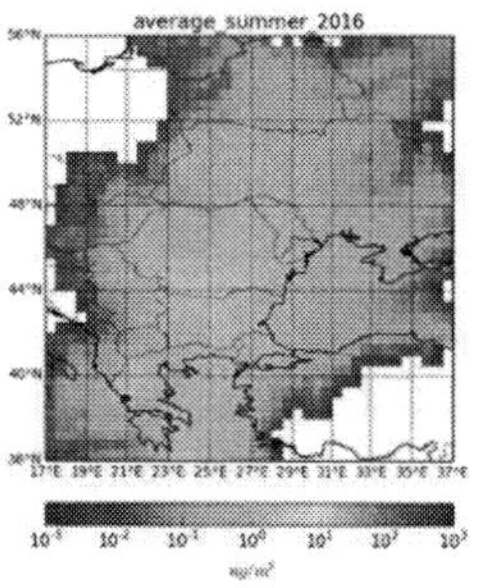

Figure 4 Average summer distribution of BBA mass concentration over Romania

The presence of vegetation fires in the South of Europe influences the aerosol concentrations over Romania, higher in the South part and lowest in center of the country, the Carpathian mountains acting as a natural barrier.

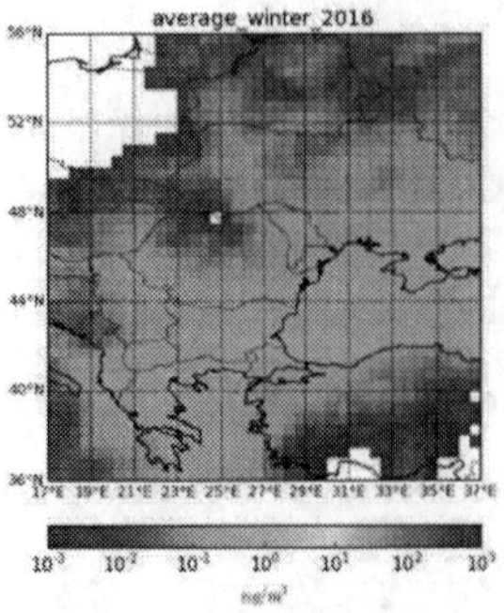

Figure 5 Average winter distribution of BBA mass concentration over Romania

During the cold season considered, from November to January, lower influences of BBA aerosol were observed over Romania (Fig.5). It must be mentioned that the data assimilated from the satellite MODIS sensors, during winter, is not taking into account the residential heating emission, that represents one of the main sources of BBA during cold season. The important uncertainties during winter are related also to the higher cloud cover during winter[12], that underestimate the thermal hot spot presence and also the BBA occurrence in to the atmosphere.

4 CONCLUSIONS

Seasonal variability shows a clear variability of BBA over Romania. Low concentrations of BBA produced by wild fires are recorded during winter.

From this preliminary results the model shows a clear underestimation of the BBA mass concentration especially during winter. This approach is suitable especially for the hot season when cloud cover is lower and thermal hot spot are visible from satellite.

This first comparison indicate that models correctly represent, on average, the BBA distribution, but their performance for simulating vertical profiles structure and the exact amount of BBA should be further explored.

ACKNOWLEDGEMENTS

The research leading to these results has received funding from the European Union's Horizon 2020 Research and Innovation Programme, under Grant Agreement no 692014-ECARS and from the National Core Program- PN 16 40 01 01.

References

[1] IPCC (2013), Contribution of Working Group I to the Fifth Assessment Report of the Intergovermental Panel on Climate Change in Summary for Policymakers in Climate Change edited by T. F. Stocker, D. Qin, G. K. Plattner, M. Tignor, S. Allen, J. Boschung, A. Nauels, Y. Xia, V. Bex, P. Midgley, Cambridge University Press.

[2] Bond, T. C. et al. 2013: Bounding the role of black carbon in the climate system: A scientific assessment, *J. Geophys. Res.*, 118, 5380–5552, doi:10.1002/jgrd.50171.

[3] Saha, S., et al. 2011, updated daily. NCEP Climate Forecast System Version 2 (CFSv2) 6-hourly Products. Research Data Archive at the National Center for Atmospheric Research, Computational and Information Systems Laboratory.

[4] Stohl, A., et al., G.,2005: Technical note: The Lagrangian particle dispersion model FLEXPART version 6.2, *Atmos. Chem. Phys.*, 5, 2461–2474,doi:10.5194/acp-5-2461-2005,.

[5] Kuenen J. J. P., et al. 2014, TNO-MACC_II emission inventory; a multi-year (2003–2009) consistent high-resolution European emission inventory for air quality modelling, *Atmos. Chem. Phys.*, 14, 10963–10976,

[6] Davies, D. K., et al. Fire Information for Resource Management System: Archiving and Distributing MODIS Active Fire Data, *Geos. & Remote Sens.*,**47**, 72–79, doi:10.1109/tgrs.2008.2002076,2009.

[7] Ichoku, C., and Ellison, L., 2014: Global top-down smoke-aerosol emissions estimation using satellite fire radiative power measurements, *Atmos. Chem. Phys.*, **14**(13), 6643–6667.

[8] Winker D., et al. 2007, Initial performance assessment of CALIOP, *Geophysical Res. Lett.*, **34**, 19, L19803.

[9] Tesche, M., et al. (2011). Optical and microphysical properties of smoke over Cape Verde inferred from multiwavelengthlidar measurements. *TellusB, 63*, 677–694

[10] Nemuc A., et al. Assessment of aerosol's mass concentrations from measured linear particle depolarization ratio (vertically resolved) and simulations, *Atmos. Meas. Tech.*, 6, 3243–3255, 2013, doi:10.5194/amt-6-3243-2013

[11] Foy de B., S. P. et al., Aerosol plume transport and transformation in high spectral resolution lidar measurements and WRF-Flexpart simulations during the MILAGRO Field Campaign,*Atmos. Chem. Phys.*, 11, 3543–3563, 2011, doi:10.5194/acp-11-3543-2011

[12] National Meteorological Administration, 2008: Climate of Romania, *Editura Academiei Române*, Bucharest (in Romanian)

MONITORING AND QUANTIFYING PARTICLES EMISSIONS AROUND INDUSTRIAL SITES WITH SCANNING DOPPLER LIDAR

L. Thobois[1*], P. Royer[1], R. Parmentier[1], M. Brooks[2], A. Knoepfle[2], J Alexander[2], P.Stidwell[2], R. Kumar[2]

*[1]Leosphere: Lidar Environmental Observations, Orsay, France, *lthobois@leosphere.com*
[2]ECOTECH, Australia

ABSTRACT

Scanning Coherent Doppler Lidars have been used over the last decade for measuring wind for applications in wind energy [1], meteorology [2] and aviation [3]. They allow for accurate measurements of wind speeds up to a distance of 10 km based on the Doppler shift effect of aerosols. The signal reflectivity (CNR or Carrier-to-Noise Ratio) profiles can also be retrieved from the strength of the Lidar signal. In this study, we will present the developments of algorithm for retrieving aerosol optical properties like the relative attenuated backscatter coefficient and the mass concentration of particles. The use of these algorithms during one operational trial in Point Samson, Western Australia to monitor fugitive emissions over a mine will be presented. This project has been initiated by the Australian Department of Environment Regulations to better determine the impact of the Port on the neighboring town. During the trial in Summer, the strong impact of turbulence refractive index on Lidar performances has been observed. Multiple methodologies have been applied to reduce this impact with more or less success. At the end, a dedicated setup and configuration have been established that allow to properly observe the plumes of the mine with the scanning Lidar. The Lidar data has also been coupled to beta attenuation in-situ sensors for retrieving mass concentration maps. A few case of dispersion of plumes will be presented showing the necessity to combine both the wind and aerosol data.

1 INTRODUCTION

Mining, quarrying and cement manufacturing plants, as with other heavy industrial activities involve a series of operations which can produce significant quantities of fugitive particles like dust or soot, with potentially adverse effects on the local and neighboring environments. Effective control of their occurrence is necessary in order to comply with environmental regulations involving the maximum level of particle emissions, and also in order to optimize the operational processes of these industries. At some locations, industrial emissions can also be limited by the background level of particles as in desert areas, where dust concentration can be close to the maximum level of particles permitted. In addition to conventional in-situ particles counters located at specific points, the monitoring in 2D of particles around industrial plants allows for better characterization and thus management of particles emissions.

2 EXPERIMENTAL SETUP

A WINDCUBE200S Scanning Doppler Lidar [4] was installed for a 3-month campaign at an Iron Ore Port facility on the West coast of Australia (see on figure 1). The main objective of this campaign is to quantify the emissions from the port Iron Ore stockpiles and ship loading operations (see orange area on figure 1) and to determine its impact on the nearby town (see blue area on figure 1).

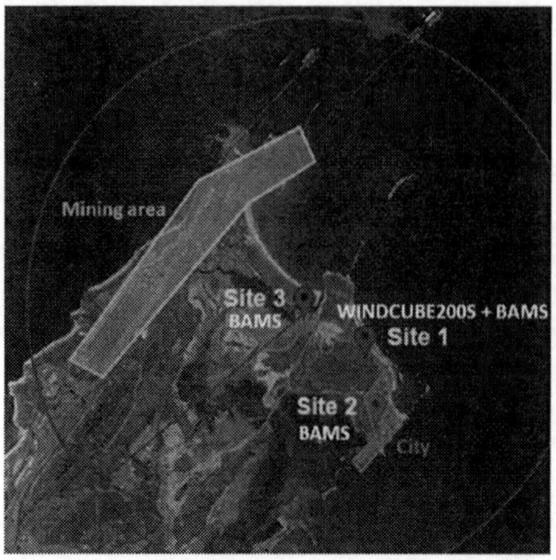

Figure 1 Experimental setup

The scanning Lidar has been lifted 9.5m above the ground level in order to reduce the impact of the refractive index turbulence on the Lidar instrumental function.

Figure 2 WINDCUBE200S and BAMS installation on site 1

Lidar measurements are performed with a range resolution of 50m, an accumulation time of 1 sec and a scan rate of 1°/sec. The scenarios of measurements are detailed in Table 1 and consist in a 10min sequence with one full PPI (Plan Position Indicator), 2 RHI cross-sections (Range Height Indicator) and 2 min of vertical LOS (Line-Of-Sight).

	Azimuth		Elevation		Duration
	Start	End	Start	End	
PPI	0	360	0.25°		360 sec
RHI 1	167°		-2	45	47 sec
RHI 2	308°		-2	45	47 sec
Vertical LOS	0°		90°		120 sec

Table 1 Measurement scenarios

The relative attenuated backscatter coefficient ($\beta_{rel.att.}$) is computed by removing the instrumental function (F) to the Carrier-to-Noise Ratio (CNR):

$$\beta_{rel.att.}(r)=\frac{CNR(r)}{F(r)} \qquad (1)$$

The instrumental function is determined by fitting a lorentzian shape function on horizontal measurements in homogeneous atmospheric conditions.

The optical-to-mass conversion factor A is determined by correlation between in-situ BAMS (Backscatter Moessbauer Spectrometer) 1-hour averaged mass concentration (C_{BAM}) and the 1-hour averaged relative attenuated backscatter coefficient ($\beta_{rel.att.}$) from lidar measurements at the BAMS location:

$$A(t)=\frac{\overline{\beta_{rel.att.(t)}}}{C_{BAM(t)}} \qquad (2)$$

This conversion factor is then applied to Lidar data to give a real-time estimate of the mass concentration. This method has been used since several comparisons between Lidar and BAMS data have shown that these two quantities are highly correlated in the same kind of environment.

The correlation between measured BAMS and lidar-derived concentrations is given on figure 3.

 https://doi.org/10.1051/epjconf/201817604013

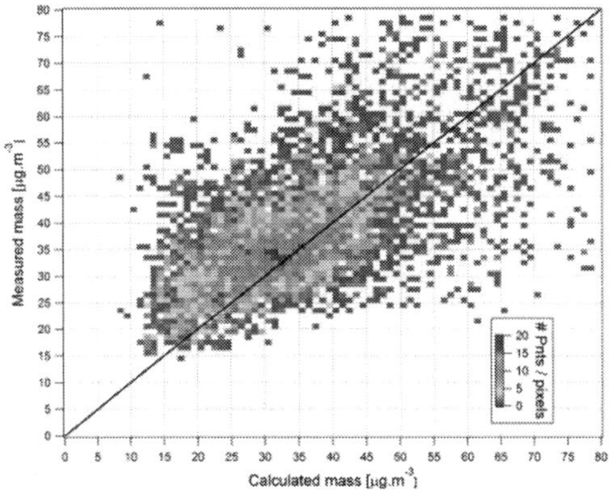

Figure 3 Correlation between measured BAMS mass concentrations and calculated WINDCUBE mass concentrations

3 RESULTS

BAMS measurements and real-time estimated aerosol concentrations from the Lidar are available online http://airodis.ecotech.com.au/der-Lidar/index.html.

Two examples of mass concentrations observations are given on figure 4. Concentrations are divided into 6 categories:

- Level 0 (in transparent) for concentrations < 40 µg/m^3
- Level 1 (in yellow) for concentrations > 40 µg/m^3
- Level 2 (in orange) for concentrations > 60 µg/m^3
- Level 3 (in red) for concentrations > 100 µg/m^3
- Level 4 (in purple) for concentrations > 350 µg/m^3
- Solid (in grey) for concentrations > 500 µg/m^3

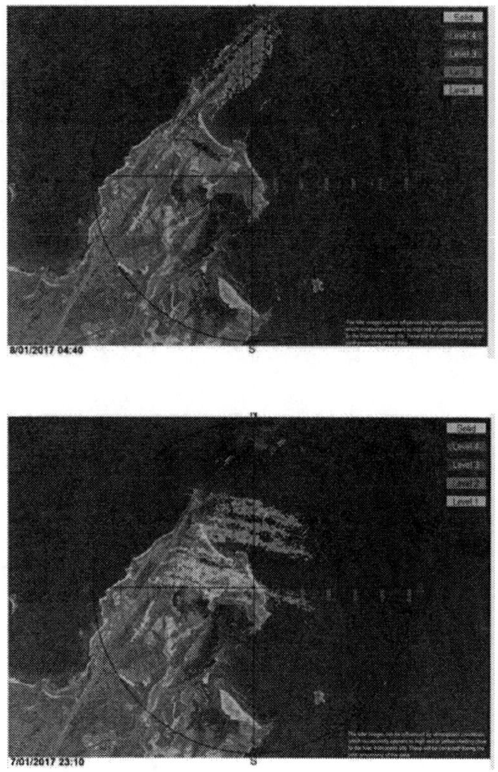

Figure 4 Mass concentration observations over the mining site

The plumes dust plumes are clearly visible on the two examples with an extension of more than 4 km. In the first case the concentration are higher and level 3 is reached in the center of the plume. The Scanning Doppler Lidar also enables to localize the emission source.

4 CONCLUSIONS

We present here an example of use of scanning Doppler Lidar for the monitoring and the quantification of particle emissions over an industrial mining site.

Doppler Lidar relative attenuated backscatter measurements were correlated to BAMS in-situ concentrations in order to derive the optical-to-mass conversion factor. This conversion is then applied to the Lidar measurement in order to give a real-time estimate of the mass concentration. This real-time information is particularly important for mining industry to regulate and reduce the dust emissions with water sprays.

As perspectives, the wind and mass concentration data will be combined and analyzed to better determine the dispersion of the aerosol with wind and turbulence. Besides, these studies show that Coherent Doppler Lidars can provide aerosol optical properties with a good accuracy and sensitivity like direct detection Lidars if the phase of the coherent signal is not affected by the refractive turbulence index.

[1] Krishnamurthy et al., 2012: Coherent Doppler lidar for wind farm characterization, Wind Energy, DOI: 10.1002/we.539, 16, 189-206.

[2] Frehlich et al., 1998: Coherent Doppler lidar measurements of wind field statistics, Boundary-Layer Meteorology, 86, 233-256.

[3] Shun and Chan, 2008: Applications of an Infrared Doppler LIDAR in Detection of Windshear, Journal of Atmospheric and Oceanic Technology, DOI: http://dx.doi.org/10.1175/2007JTECHA1057.1, 25, 637-655

[4] Thobois et al., Sept. 2016: New York State Mesonet, Meteorological Technology International, 62-65.

EPJ Web of Conferences **176**, 04014 (2018)
ILRC 28
https://doi.org/10.1051/epjconf/201817604014

SCANNING ELASTIC LIDAR OBSERVATIONS OF AEROSOL TRANSPORT IN NEW YORK CITY

Adrian Diaz*, Victor Dominguez, Selma Dobryansky, Yonghua Wu, Mark Arend, Daniela Viviana Vladutescu, Barry Gross, Fred Moshary

NOAA CREST Center and City College of New York, New York, NY 10031, USA,
**adiaz001@citymail.cuny.edu*

ABSTRACT

In this study, spatial distribution of aerosols in New York City is observed using a scanning eye-safe 532 nm elastic-backscatter micro-pulse lidar system. Observations show dynamics of the boundary layer and inhomogeneous distribution and transport of aerosols. The data acquired are complemented with simultaneous measurements of particulate matter and wind speed and direction. Furthermore, the system observations are validated by comparing them with a co-located multi-wavelength lidar.

1. INTRODUCTION

Lidar imaging of the spatial distribution of aerosols can shed light on atmospheric dynamics, aerosol transport and particulate pollution. Due to the harmful effects pollution has on human health, it is of special interest to study the aerosol distribution within the boundary layer of highly populated urban areas like New York City. Currently, there are ground-based lidars that measure vertical distribution of aerosols above New York City. However, vertically pointing instruments are unable to obtain multi-dimensional views of the atmosphere, hence scanning lidars are needed instead.

Previous implementations of scanning lidars at other locations have combined transmitter sources in different regions of the spectrum with different lidar techniques, some of them include elastic-backscatter lidar in the UV [1], differential absorption lidar in the IR [2] and Raman-shifted and polarization lidar in the IR [3,4]. The requirements for a scanning lidar are rapid scan and moderate range. These can be achieved with a high pulse energy transmitter that would allow for moderate range measurements with reduced averaging time. Nevertheless, these tend not to be eye-safe and their operation must follow strict guidelines set forth by the US Federal Aviation Administration, and American National Standards for Safe Use of Lasers Outdoors [5].

On the other hand, micro-pulse lidars (MPL) meet the necessary eye-safety requirement for scanning and are simple and low cost. The idea of a MPL was developed by NASA in 1992 as a result of technological improvements on solid states devices accomplished at the time. It consists of a transmitter source with pulse energy in the order of micro Joules and a repetition rate of thousands of Hertz. When appropriately expanded, the energy transmitted by this source becomes eye-safe at all ranges. Moreover, the receiver employs a compact avalanche photo diode (APD) detector for effective photon counting acquisition [6]. Currently, MPL's most known use is in the NASA Micro-Pulse Lidar Network (MPLNET), which continuously measures aerosol and cloud vertical structure, as well as boundary layer heights from multiple sites around the world [7].

In this study, a commercial scanning eye-safe MPL is used to measure multi-dimensional distributions of aerosols in New York City. These measurements, combined with simultaneous measurements of particulate matter and wind speed and direction, provide information necessary to understand boundary layer dynamics and aerosol transport.

2. METHODOLOGY

The instrument used, the Mini-MPL, is a scanning eye-safe elastic-backscatter MPL located at the City College of New York (CCNY) and manufactured by Sigma Space with the specifications shown on table 1.

Spatial distributions of aerosols were measured in New York City by scanning along the elevation (or polar) angle while pointing towards Bronx and Manhattan, respectively at 73° and 173°

 https://doi.org/10.1051/epjconf/201817604014

azimuth angles measured with respect to the north (figure 1). The scans consisted of sweeps from 0° to 90° elevation angles with 1° or 2° increments. At every angle, measurements were taken with 15 seconds average and 75 m range resolution. Using these scanning parameters, two-dimensional cross-sections were obtained with sufficient SNR within the first 5 km horizontally and up to 3 km vertically.

Table 1 Mini-MPL specifications.

Transmitter	
Wavelength	532 nm
Repetition rate	2500 Hz
Pulse Energy	4 μJ
Receiver	
Diameter	8 cm
Field of view	220 μrad
Detector	APD photon counting
Range resolution	5/15/30/75 m

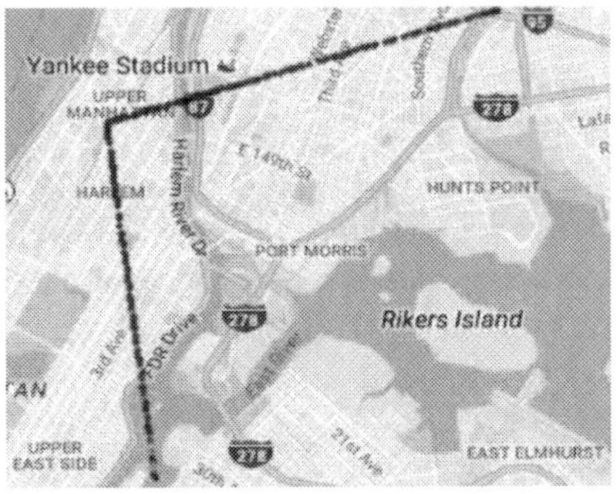

Figure 1 Map showing the directions of the scans.

3. RESULTS

Multiple scans were carried out during afternoon hours on July 28th 2016 (figure 2). Measurements from the first two scans show a few clouds (depicted in red). Moreover, an increase in the overall amount of aerosols can be observed throughout the day.

A co-located air monitoring station managed by the New York State Department of Environmental Conservation includes a continuous fine particulate matter (PM 2.5) sampler. It uses a 1405 TEOM™ to provide hourly PM 2.5 surface concentration measurements. Data retrieved from this station shows an initial low concentration followed by a considerable increase towards the end of the day (figure 3). This is in agreement with the overall amount of aerosols observed from the scanning measurements.

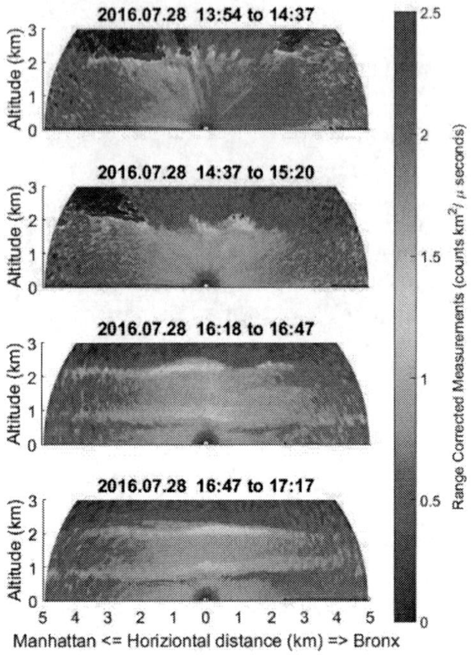

Figure 2 Aerosol distributions in New York City at different times on July 28th 2016.

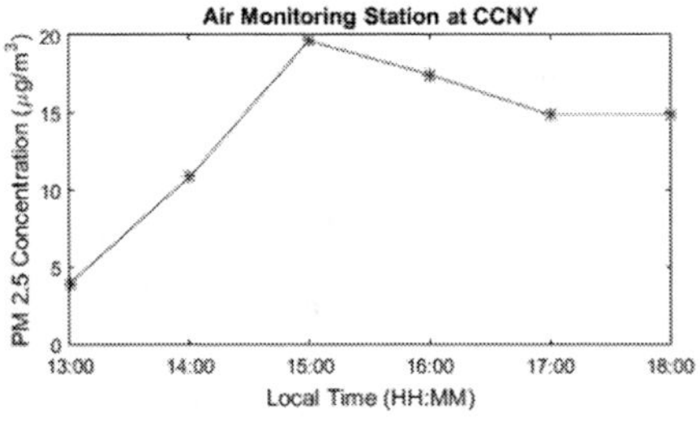

Figure 3 PM 2.5 concentration measurements from an air monitoring station at CCNY.

The last two scans show two aerosol layers at 2 km and between 700 m and 1.2 km. It appears that the aerosol layers move to the north from Manhattan to Bronx. To further investigate this, measurements of wind speed and direction at three different locations were retrieved from the NYCMetNet network. One of them includes a radar wind profiler (Scintec LAP3000) reporting winds up to 2 km height and is located in Jersey City approx. 15 km from CCNY (figure 4). The other two weather stations report surface winds and are located at Yankee Stadium and Central Park West approx. 2 and 4.5 km from CCNY, respectively (figure 5).

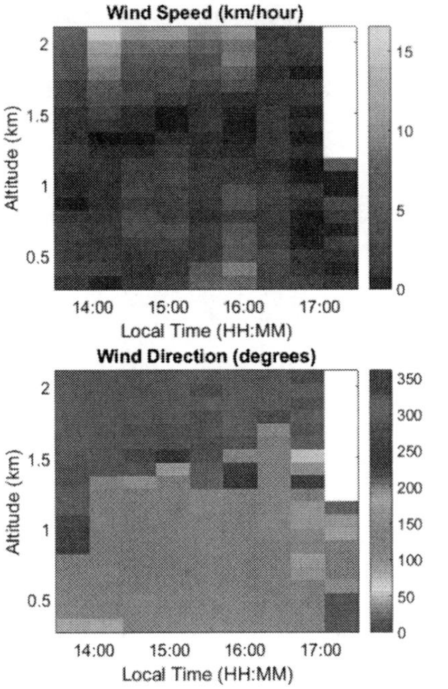

Figure 4 Wind speed and direction measurements from radar wind profiler.

The data from the three different stations consistently show winds originating from the south (around 180°) with speeds between 4 and 10 km/hour. Moreover, measurements from the wind profiler indicate that the southerly winds extend up to 1.4 km height. This confirms that the lower aerosol layer observed on the scanning measurements is indeed being transported from south to north.

The increase in the amount of aerosols throughout the day, and the appearance of the aerosol layers were also observed on simultaneous measurements taken with a co-located Vaisala CL 51 ceilometer (figure 6).

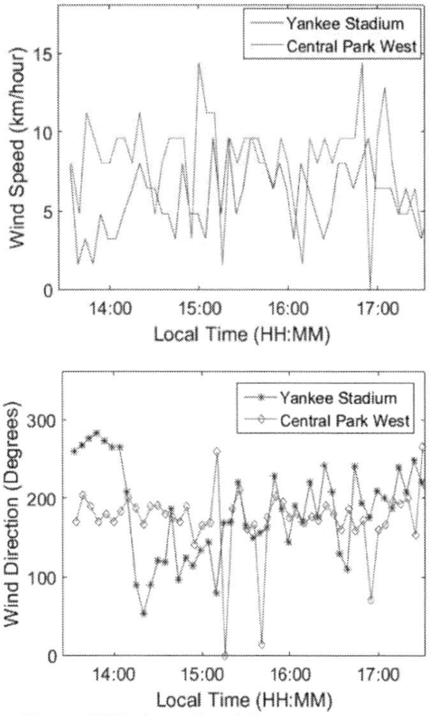

Figure 5 Wind speed and direction measurements from weather stations at Yankee Stadium and Central Park West.

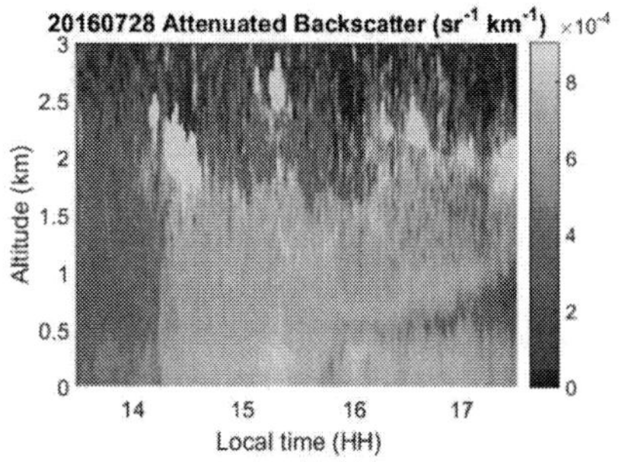

Figure 6 Simultaneous ceilometer measurements.

On July 20th, measurements were taken using the MPL fixed at 90° elevation with 30 meters resolution and 1 minute averages from 11 to 15:30 local time. Measurements using the 532nm channel of a co-located multi-wavelength lidar were taken during the same period of time for comparison purposes. It can be observed that both measurements are consistent and show fairly similar distributions (figure 7).

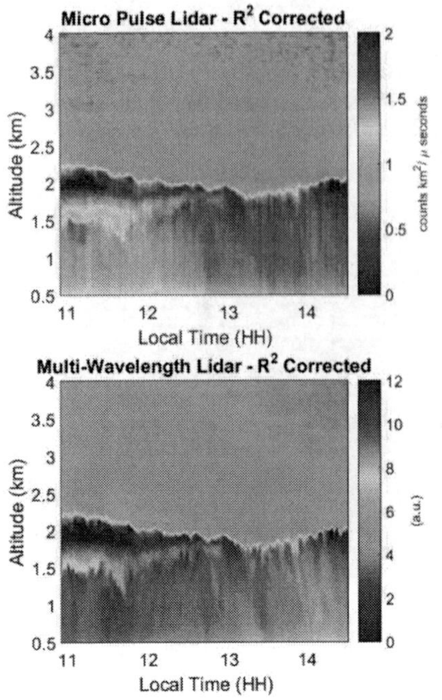

Figure 7 Measurements from (Top) MPL and (Bottom) multi-wavelength lidar.

4. CONCLUSIONS

Boundary layer dynamics in New York City have been observed by acquiring multi-dimensional distributions of aerosols with a scanning micro pulse lidar. Obtained measurements are supported by simultaneous PM 2.5 concentration measurements and the use of a ceilometer. Simultaneous wind speed and direction measurements were used to better understand dynamics and evolution of aerosol distribution. Additionally, measurements from this instrument are in agreement with the ones obtained with a co-located multi-wavelength lidar. In future studies, the effects of changes that might occur during the duration of a scan will be taken into account.

ACKNOWLEDGEMENTS

This study was supported by National Oceanic and Atmospheric Administration (NOAA) under Grant #NA11SEC4810004 and by New York State Energy Research and Development Authority (NYSERDA).

References

[1] Behrendt, A., Pal, S., Wulfmeyer, V. and Lammel, G., 2011. A novel approach for the characterization of transport and optical properties of aerosol particles near sources–Part I: Measurement of particle backscatter coefficient maps with a scanning UV lidar. *Atmospheric environment*, 45(16), pp.2795-2802.

[2] Späth, F., Behrendt, A., Muppa, S.K., Metzendorf, S., Riede, A. and Wulfmeyer, V., 2016. 3-D water vapor field in the atmospheric boundary layer observed with scanning differential absorption lidar. *Atmospheric Measurement Techniques*, 9(4), pp.1701-1720.

[3] De Wekker, S.F. and Mayor, S.D., 2009. Observations of atmospheric structure and dynamics in the Owens Valley of California with a ground-based, eye-safe, scanning aerosol lidar. *Journal of Applied Meteorology and Climatology*, 48(7), pp.1483-1499.

[4] Mayor, S.D., Spuler, S.M., Morley, B.M. and Loew, E., 2007. Polarization lidar at 1.54 μm and observations of plumes from aerosol generators. *Optical Engineering*, 46(9), pp.096201-096201.

[5] American National Standards Institute, Z136.6, 2015. American National Standard for Safe Use of Lasers Outdoors, *American National Standards Institute*, New York.

[6] Spinhirne, J.D., 1993. Micro pulse lidar. *IEEE Transactions on Geoscience and Remote Sensing*, 31(1), pp.48-55.

[7] Welton, E.J., Campbell, J.R., Spinhirne, J.D. and Scott III, V.S., 2001, February. Global monitoring of clouds and aerosols using a network of micropulse lidar systems. In *Second International Asia-Pacific Symposium on Remote Sensing of the Atmosphere, Environment, and Space* (pp. 151-158). International Society for Optics and Photonics.

MEASUREMENTS OF WATER VAPOR PROFILES WITH COMPACT DIAL IN THE TOKYO METROPOLITAN AREA

Makoto Abo[1*], Tetsu Sakai[2], Phong Pham Le Hoai[1], Yasukuni Shibata[1], Chikao Nagasawa[1]

[1]*Tokyo Metropolitan University, Japan, *Email: abo@tmu.ac.jp*
[2]*MRI, Japan*

ABSTRACT

In recent years, the frequency of occurrence of locally heavy rainfall that can cause extensive damages, has been increasing in Japan. For early prediction of heavy rainfall, it is useful to measure the water vapor vertical distribution upwind cumulus convection beforehand. For that purpose, we have been developing compact water vapor differential absorption lidar (DIAL).

We show the results of the measurements with lidar in summer when the local heavy rainfall frequently occurs in Japan. We also show the preliminary result of the assimilation of the lidar data to the numerical model and impact on the heavy rainfall prediction.

1 INTRODUCTION

Locally heavy rainfall is a severe weather hazards to affect urban areas in Japan. For early prediction of the heavy rainfall, numerical weather model is employed using the conventional meteorological station data. However, the lead time and accuracy of the prediction are limited because of the coarse spatial and temporal resolutions of the data. To improve them, it is useful to measure the water vapor distribution upwind of cumulus convection beforehand and assimilate the data into the model. Figure 1 shows schematic of this study.

For that purpose, we have developed compact and mobile diode-laser-based differential absorption lidar (DIAL) that can measure the vertical distribution of water vapor in the lower troposphere.

2 DIAL SYSTEM

The DIAL employs two distributed Bragg reflector (DBR) lasers operating at 829.054 nm for the online wavelength and 829.124 nm for the offline wavelength with tapered semiconductor optical amplifier (TSOA) in a master oscillator power amplifier (MOPA) configuration. The lidar measure the vertical distribution of the water vapor in the lower troposphere with a vertical resolution of a few tenths of a meter and temporal resolution of a few tenths of a minute.

Table 1 Specifications of DIAL system.

Transmitter :	
Laser	DBR diode laser x2 (seed laser), 80 mW+TSOA
Wavelength	829.124 nm (Off-line) 829.054 nm (On-line)
Pulse Energy	4 uJ
Repetition	10 kHz
Receiver :	
Telescope	Cassegrain
Diameter	35 cm
Field of View	0.25~2.8 mrad (adjustable)
Detector	PMT (Hamamatsu H9422P-50)
Operation	Photon Counting
Interference Filter	
Center wavelength	829.11 nm
FWHM	0.275 nm
Transmission	47%

3 MEASUREMENT RESULTS

We show the results of the 24H continuous measurements with lidar in summer (26 August 2016) when the local heavy rainfall frequently occurs in Japan in Figure 2.

Figure 3 shows the results of comparison with JMA mesoscale analysis (MSM) and radiosonde at Tsukuba, 80 km northeast from the lidar observation site Hino.

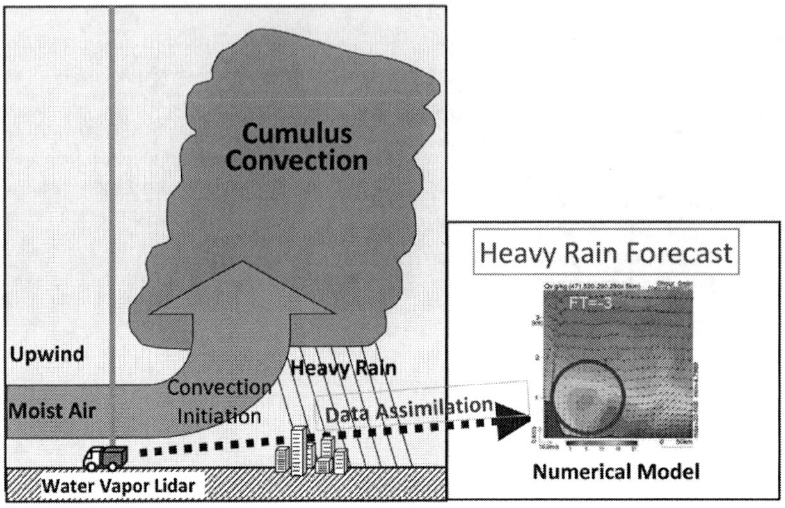

Figure 1. Schematic of this study.

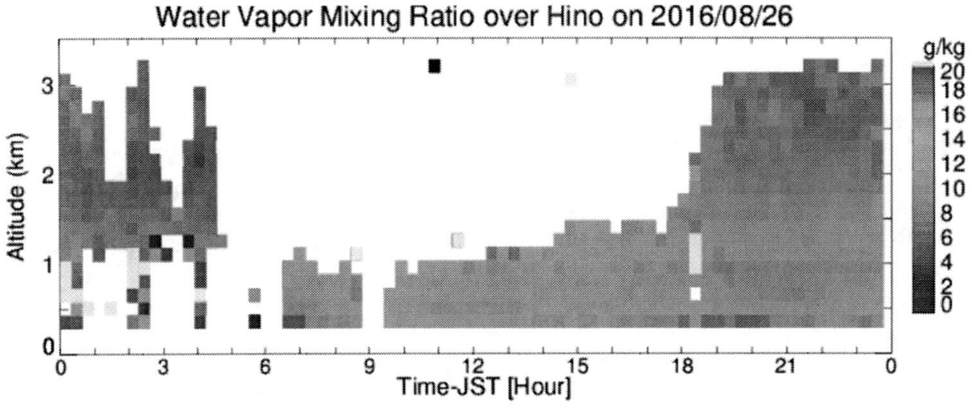

Figure 2. Temporal and height cross section of water vapor mixing ratio obtained with compact DIAL over Hino (35.7°N, 139.4°E) on 26 August 2016.

https://doi.org/10.1051/epjconf/201817604015

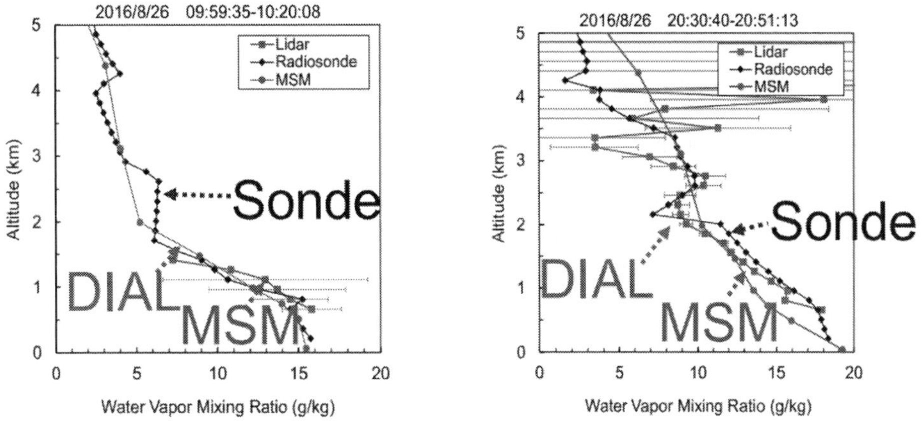

Figure 3 The results of comparison with JMA mesoscale analysis (MSM) and radiosonde at Tsukuba (80 km northeast).

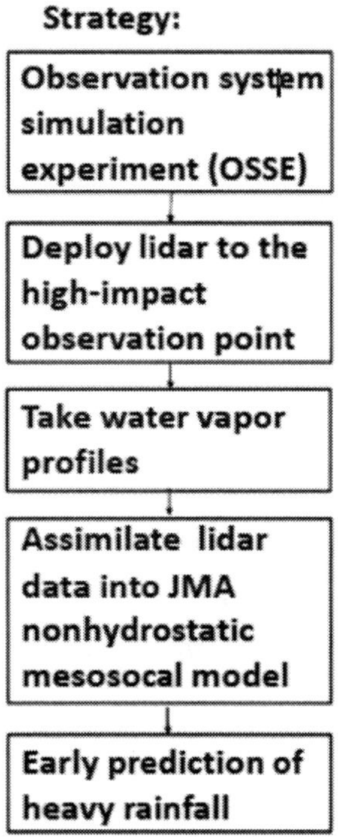

Figure 4. Strategy of early prediction of the heavy rainfall with numerical weather model and water vapor profiles observed by lidar.

4 CONCLUSIONS

We have been developing water vapor lidars for local heavy rainfalls forecast. Now measurement range is 0.5−3 km in night time and 1 km in daytime with 150 m vertical and 20 min time resolution with uncertainty within 30%.

We are charanging to increase daytime measurement range to 3 km for DIAL using etalon and increasing laser power.

Next step, we are planning to conduct field experiment in Tokyo in the summer of 2017 and to assimilate the lidar data to mesoscale model and study the impact on it.

ACKNOWLEDGEMENTS

This work was supported by JSPS KAKENHI Grant Number JP26282115.

References

[1] Le Hoai, P.P., M. Abo, T. Sakai (2016): Development of field-deployable Diode-laser-based water vapor DIAL, EPJ Web of Conferences 119 05011, doi: 10.1051/epjconf/20161190501.

[2] Repasky, K., D. Moen,, S. Spuler, A. Nehrir, and J. Carlsten (2013): Progress towards an autonomous field deployable diode laser-based differential absorption lidar (DIAL) for profiling water vapor in the lower troposphere, Remote Sens., 5, 6241–6259, doi:10.3390/rs5126241.

[3] Spuler, S. M., Repasky, K. S., Morley, B., Moen, D., Hayman, M., and Nehrir, A. R.(2015): Field-deployable diode-laser-based differential absorption lidar (DIAL) for profiling water vapor, Atmos. Meas. Tech., 8, 1073-1087, doi:10.5194/amt-8-1073-2015.

Airborne Two-Micron Double-Pulse IPDA Lidar Validation for Carbon Dioxide Measurements over Land

Tamer F. Refaat[1]*, Upendra N. Singh[1], Jirong Yu[1], Mulugeta Petros[1], Ruben Remus[1] & Syed Ismail[2]

[1]*NASA Langley Research Center, Hampton, VA 23681, USA,*
**Email: tamer.f.refaat@nasa.gov*
[2]*Analytical Services and Materials, Inc., Hampton, VA 23666, USA*

ABSTRACT

An airborne double-pulse 2-μm Integrated Path Differential Absorption (IPDA) lidar has been developed at NASA LaRC for measuring atmospheric CO_2. IPDA was validated using NASA B-200 aircraft over land and ocean under different conditions. IPDA evaluation for land vegetation returns, during full day background conditions, are presented. IPDA CO_2 measurements compare well with model results driven from on-board *in-situ* sensor data. These results also indicate that CO_2 measurement bias is consistent with that from ocean surface returns.

1 INTRODUCTION

Atmospheric carbon dioxide (CO_2) plays a key role in Earth's environment and climate and it influences processes in the atmosphere, biosphere, and hydrosphere. Large uncertainties in quantifying CO_2 fluxes arise due to data quality and insufficient spatial and temporal coverage of the gas distributions. Active optical remote sensing has been recommended to improve the understanding of CO_2 fluxes including sources and sinks. For more than 20 years, NASA Langley Research Center (LaRC) has been involved in maturing 2-μm technologies, including pulsed laser transmitters, for lidar systems that are focused on meeting the science objectives for CO_2 measurements. Recently, an airborne CO_2 double-pulse 2-μm integrated path differential absorption (IPDA) lidar was developed at LaRC [1-3].

Airborne field experiments were conducted using NASA B-200 aircraft to test and evaluate the CO_2 measurement capabilities of the 2-μm IPDA. The IPDA was tuned to the CO_2 R30 strong absorption line at 2050.9670 nm. This line is optimum for lower tropospheric weighted column CO_2 sensing. Flights were conducted over land and ocean under different conditions. Initial IPDA validation focused on low reflectivity oceanic surface returns during full day background conditions [3]. On April 5, 2014 IPDA CO_2 measurements were compared to airborne flask air-sampling CO_2 measurements conducted by NOAA over the Atlantic Ocean. IPDA performance modeling was conducted to evaluate measurement sensitivity and bias errors. IPDA signals compare well with predicted model results including altitude dependence. Off-off-line testing was conducted to evaluate the IPDA systematic and random errors. Results over the ocean showed altitude-independent differential optical depth offset of 0.0769. Measured CO_2 differential optical depth random error of 0.0918 compared well with the predicted value of 0.0761. IPDA CO_2 column measurement compared well with model-driven, concurrent different altitude air-sampling from NOAA. CO_2 differential optical depth of 1.0054 ±0.0103 was retrieved from 6-km altitude, 4-GHz offset on-line operation and 10 s average. IPDA ranging resulted in less than 3 m uncertainty [3].

In this paper a separate B-200 airborne validation experiment for the 2-μm IPDA lidar will be presented. On April 10, 2014, the IPDA was evaluated over East Virginia vegetation targets, from different altitudes. On board aircraft sensors, *in-situ* sensor (LiCor) and balloon sonde were used to obtain CO_2 and meteorological profiles. These profiles were applied to model and compare with CO_2 column IPDA measurements over land. Careful analysis allows reduction or elimination of range biases with the pulsed IPDA.

2 METEOROLOGY

Meteorological data are required for IPDA lidar modeling and retrievals. These data are acquired from other instruments and can limit the accuracy of IPDA measurements. In this analysis, temperature and pressure profiles were obtained from balloon sonde. Dry-air H_2O and CO_2 mixing ratios, x_{wv} and x_{cd}, were obtained from aircraft and *in-situ* sensors. US standard atmospheric (USSA) model was included as a reference. Figure 1 shows the NASA-B200 GPS altitude, R_A, and line-of-sight distance, R_L, obtained using the roll and pitch angles. Ground elevation, R_G, was obtained from Google Maps Application Programming Interface web-service [4].

Figure 2 shows altitude profiles of x_{wv} obtained from sonde and aircraft and *in-situ* sensors compared with USSA model. The figure also shows x_{cd} profiles obtained from *in-situ* compared to USSA with updated surface level of 411 ppm. A 2 km boundary layer (BL) height could be estimated from x_{wv} profiles. This is indicated from fixed amount within BL followed by sharp drop in lower troposphere. Near-surface high x_{wv} deviation between sonde and *in-situ* profiles is observed in BL with closer match at higher altitudes. CO_2 profile shows lower near-surface amounts due to photosynthesis followed by gradual increase toward BL top. Sharp decline in lower troposphere indicates CO_2 trapping. Data clusters in both x_{cd} and x_{wv} profiles are due to fixed altitudes flight track, as listed in Table 1. Statistical analysis conducted for these clusters to obtain mean and standard deviation. Linear interpolation was used to join the means and linear extrapolation to extend profiles down to surface and up to 4.5 km altitude.

3 IPDA MODELING

The 2-μm IPDA measured double-path optical depth mainly results from CO_2 and H_2O absorption and aerosol optical depths. The CO_2 double-path differential optical depth, $\Delta\tau_{cd}$, can be modeled as

$$\Delta \tau_{cd} = 2 \cdot \int_{R_A}^{R_G} \Delta \sigma_{cd} \cdot N_{cd} \cdot dr \qquad (1)$$

for a nadir airborne IPDA at an altitude R_A operating over ground elevation R_G; $\Delta\sigma_{cd}$ is the CO_2 differential absorption cross section obtained at on- and off-line wavelengths, λ_{on} and λ_{off}, respectively; N_{cd} is the CO_2 number density (in m^{-3}); and r is the range. In this equation, aerosols optical depth cancel due to the small on- and off-line wavelength separation in $\Delta\sigma_{cd}$ measurements.

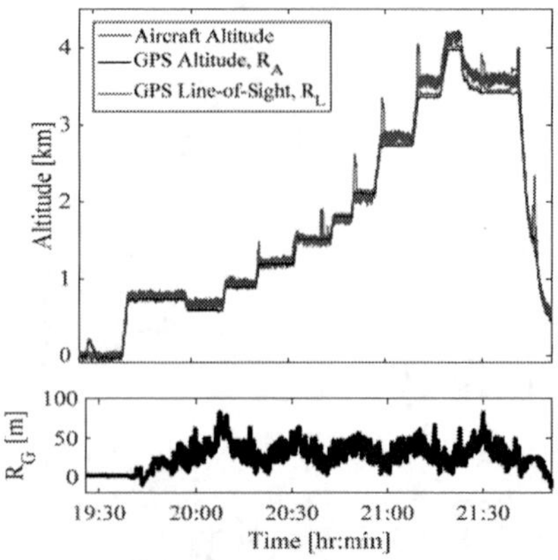

Figure 1 April 10, 2014 NASA B-200 flight altitude, obtained from aircraft sensors and GPS, the derived line-of-sight distance (top) and the corresponding ground elevation, R_G (bottom).

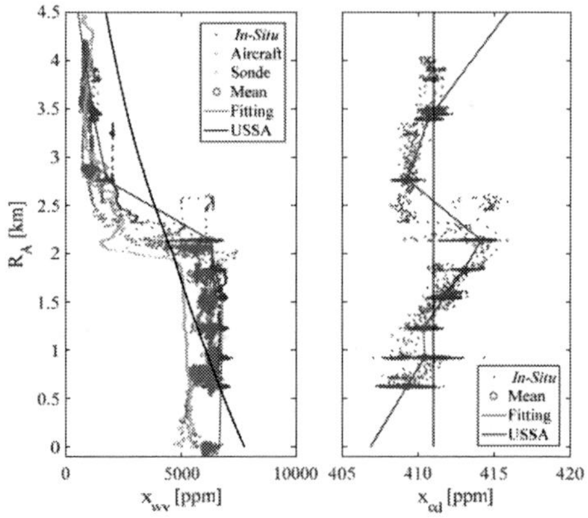

Figure 2 Comparison between x_{wv} (left) and x_{cd} (right) profiles obtained from different sensors and fittings.

H_2O interference is not included due to relatively lower differential optical depth. For example, modeling indicated the highest H_2O interference error on CO_2 measurement is 0.09% at 624.2 m altitude the nearest to the surface.

Figure 3 shows the predicted CO_2 differential optical depth using both *in-situ* and USSA model for this flight. On-line wavelengths were set at 3 and 4 GHz offsets from line center. Agreement between *in-situ* and USSA model in $\Delta\tau_{cd}$ estimates is attributed to R30 line properties, characterized by low temperature dependence and low H_2O interference. The figure also shows the IPDA measurements and modeling over ocean surface for comparison with land vegetation targets [3].

4 IPDA MEASUREMENTS

During this IPDA validation, the instrument was operated almost continuously with on-line set to 3 GHz. Similar to the ocean validation, analysis is focused on high-signal detection channel. Trans-impedance amplifier (TIA) gain was set to 10^3 V/A for lowest altitudes (up to 1.3 km) and 10^4 V/A for higher altitudes, with digitizer full-scale range of 1 and 2 V, respectively. This was done to avoid saturation and increase signal-to-noise ratio (SNR). Figure 4 shows on- and off-line sample return signals from different altitudes.

4.1 Ranging

IPDA column length, R_C, was obtained by converting the time delay between the near-field residual scattering and return pulse peaks into distance using the speed of light. Table 1 lists ranging results corresponding to the averaged data cluster timings. Figure 5 compares R_C to R_L at three altitudes. Less than 3 m IPDA range accuracy was demonstrated over ocean and calibrated targets [2-3]. Larger error in IPDA column length measurement ($\varepsilon_R = \delta(R_C - R_L)$) was observed over land that is a characteristic of range variability due to vegetation (trees).

4.2 CO_2 differential optical depth

IPDA measured $\Delta\tau_{cd}$ is obtained by ratioing the return power, P, return pulse width, t, and transmitted laser energy, E, for the on- and off-line wavelengths, according to

$$\Delta\tau_{cd} = \ln\left[\frac{P_{off}\cdot t_{off}}{E_{off}} \bigg/ \frac{P_{on}\cdot t_{on}}{E_{on}}\right] \qquad (2)$$

Table 2 lists statistical results of the IPDA $\Delta\tau_{cd}$ measurement mean, $\Delta\tau_{cd,g}$, as compared to simulation prediction, $\Delta\tau_{cd,c}$. A comparison of these results that are similar to those from ocean target are included in Figure 3 [3]. The observed offset or bias at each altitude, $\Delta(\Delta\tau_{cd})$, is the difference between the *in-situ* driven model and Gaussian mean (i.e., $\Delta(\Delta\tau_{cd}) = \Delta\tau_{cd,g} - \Delta\tau_{cd,c}$). Similar to ocean surface observations, IPDA bias trends higher at higher altitudes and via versa. These biases are consistent in spite of IPDA operation at 3 and 4 GHz over ocean and only at 3 GHz over land.

Table 1 Comparison between flight altitude, ground elevation, line-of-sight, IPDA column measurement and column measurement uncertainty (in m).

R_A	R_G	R_L	R_C	ε_R
3394.1	30.2	3363.9	3366.0	6.8
2759.9	30.4	2729.5	2729.1	7.0
2140.3	34.1	2106.2	2111.3	7.5
1838.6	34.0	1804.6	1807.8	7.4
1544.5	27.2	1517.3	1527.6	8.0
1228.7	29.7	1199.0	1201.1	6.6
920.8	31.4	889.4	894.9	8.2
624.2	43.9	580.3	585.5	7.6

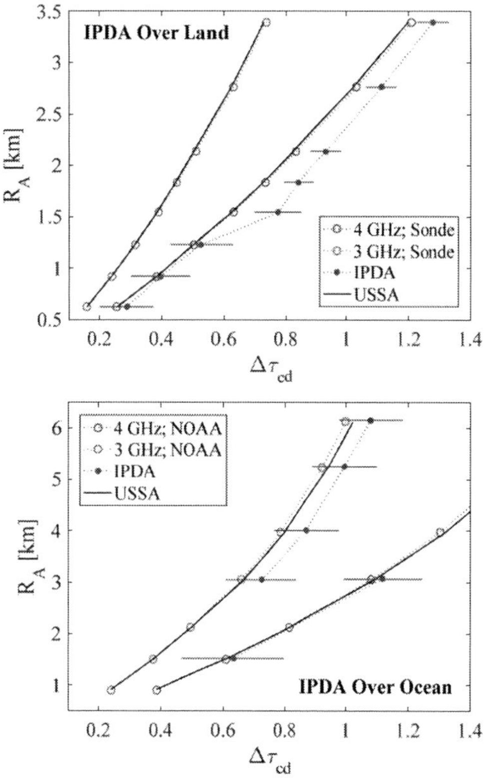

Figure 3 Comparison between IPDA CO_2 differential optical depth early afternoon measurements and simulations over land vegetation (top) and ocean (bottom). Land and ocean simulations were obtained from in-situ and NOAA air-sampling, respectively [3].

Characterization of the higher altitude bias through off-off line testing over ocean revealed consistent results [3]. Although these biases were originally attributed to the TIA, different TIA gain settings were used in this flight as compared to the ocean flight. This indicates that the digitizer is more likely the reason for such offsets. At low altitudes digitizer input was set to lower full-scale range than higher altitudes to accommodate TIA gain settings. Digitizer different input ranges resulted in different biases. This is confirmed for both IPDA validations over ocean and land.

5 CONCLUSIONS

Airborne double-pulse 2-μm IPDA lidar capability for CO_2 remote sensing was demonstrated over land and ocean. As an active remote sensor, this IPDA is capable of enhancing spatial and temporal resolution of CO_2 measurement over different target conditions during day and night. IPDA operation at the strong CO_2 R30 line at 2050.9670 nm allows optimum lower tropospheric weighted column measurements with low temperature dependence and H_2O interference errors. IPDA modeling was conducted for nadir operation targeting land vegetation at variable elevations using USSA model and meteorological data collected by on-board LiCor *in-situ* sensor and balloon sonde. CO_2 differential optical depth measurement biases are consistent for different flights, over land and ocean, and correlates with digitizer settings, which can be characterized by off-off line IPDA testing.

ACKNOWLEDGEMENTS

This work was funded and supported by NASA Earth Science Technology Office.

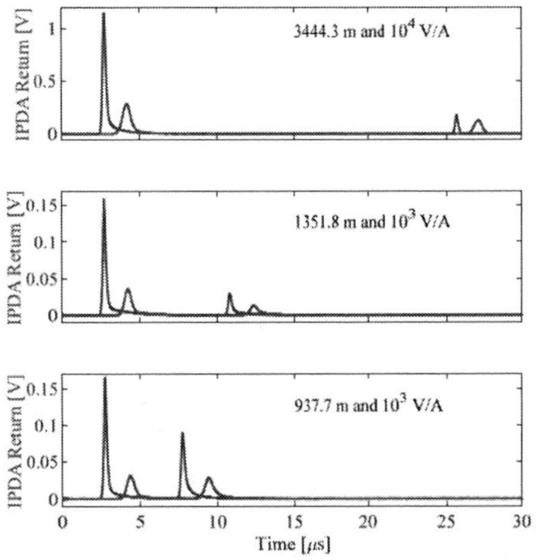

Figure 4 100-shot (10 s) averaged on-line (blue) and off-line (red) return signals from different altitudes.

Table 2 Statistical comparison of simulated and measured 2-μm IPDA CO_2 double-path differential optical depth.

R_A	$\Delta\tau_{cd,c}$	$\Delta\tau_{cd,g}$	$\delta(\Delta\tau_{cd,g})$	$\Delta(\Delta\tau_{cd})$
3394.1	1.2088	1.2778	0.0515	0.0690
2759.9	1.0325	1.1122	0.0500	0.0797
2140.3	0.8355	0.9331	0.0489	0.0976
1838.6	0.7319	0.8433	0.0516	0.1114
1544.5	0.6288	0.7752	0.0789	0.1463
1228.7	0.5078	0.5280	0.1007	0.0202
920.8	0.3851	0.3963	0.0970	0.0112
624.2	0.2560	0.2862	0.0867	0.0302

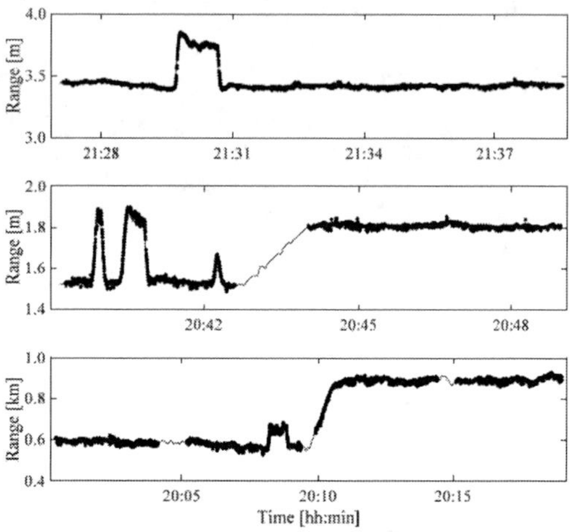

Figure 5 Comparison between IPDA lidar column length measurements (per-shot black dots) and line-of-sight (red profiles) at different altitudes.

References

[1] U. Singh, B. Walsh, J. Yu, M. Petros, M. Kavaya, T. Refaat, N. Barnes, 2015: Twenty years of Tm:Ho:YLF and LuLiF laser development for global wind and carbon dioxide active remote sensing, *Opt. Mater. Exp.* **5**, 827-837.

[2] T. Refaat, U. Singh, M. Petros, R. Remus, J. Yu, 2015: Self-calibration and laser energy monitor validations for a double-pulsed 2-μm CO_2 integrated path differential absorption lidar application, *Appl. Opt.* **54**, 7240-7251.

[3] T. Refaat, U. Singh, J. Yu, M. Petros, R. Remus, S. Ismail, 2015: Double-pulse 2-μm integrated path differential absorption lidar airborne validation for atmospheric carbon dioxide measurement, *Appl. Opt.* **55**, 4232-4246.

[4] *https://developers.google.com/maps/documentation/elevation*

TWOMEY EFFECT IN SUBTROPICAL STRATOCUMULUS CLOUDS FROM UV DEPOLARIZATION LIDAR

Martin de Graaf[1,2,*], Jessica Brown[3] and David Donovan[1]

[1]*Royal Netherlands Meteorological Institute (KNMI), Utrechtseweg 297, 3731 GA De Bilt, The Netherlands* **martin.de.graaf@knmi.nl*
[2]*TU Delft, Geoscience and Remote Sensing (GRS), Stevinweg 1, 2628 BN Delft, The Netherlands* *m.degraaf@tudelft.nl*
[3]*Wageningen University & Research, Droevendaalsesteeg 4, 6708 PB Wageningen, The Netherlands*

ABSTRACT

Marine stratocumulus clouds are important climate regulators, reflecting sunlight over a dark ocean background. A UV-depolarization lidar on Ascension, a small remote island in the south Atlantic, measured cloud droplet sizes and number concentration using an inversion method based on Monte Carlo (MC) modelling of multiple scattering in idealised semi-adiabatic clouds. The droplet size and number concentration were modulated due to smoke from the African continent, measured by the same instrument.

1 INTRODUCTION

The interaction between clouds and aerosols, and the resulting impact on radiative forcing is one of the most uncertain mechanisms in the global climate system. To gain a better understanding of these important interactions a number of international campaigns have taken place in the southeastern Atlantic region in 2016. This region is characterised by a consistent layer of marine boundary layer stratocumulus clouds, capped by a temperature inversion created by widespread subsidence. Large biomass burning events in the southern African dry season lead to high concentrations of aerosols being advected across and above these cloud layers. This region has often been described as a 'natural laboratory' for examining cloud and aerosol interactions. Here we present the first results from the Ascension Island Initiative (ASCII) campaign. The aim of the ASCII campaign was to identify whether the microphysical properties of marine stratocumulus clouds altered with the presence of aerosols. To achieve this, a depolarisation UV lidar was set up on Ascension in September 2016. This month was selected as the burning season in southern Africa. It was also selected to coincide with a period of intensive observations from a number of other international campaigns including LASIC, CLARIFY and NASA-ORACLES, based in Walvis Bay, all of which also focused on understanding and investigating the interaction of biomass burning smoke and clouds in the southeast Atlantic region.

Figure 1: Aerosol Direct Radiative Effect in Wm^{-2} over stratocumulus clouds in the Atlantic Ocean on 24 September 2016, derived from colocated OMI and MODIS reflectance measurements. The pink triangles indicate the locations of Ascension Island (left) and St. Helena (right). The yellow star indicates Walvis Bay in Namibia.

Ascension Island is a remote, volcanic island located at 8°S; 14°W, 1600 km from Africa and 2250 km off the coast of Brazil, see Figure 1. The typical clouds found in subtropical marine regions, such as around Ascension Island, are low lying bands of stratocumulus capping the boundary layer and are typically found between 1-1.5 km. These marine stratocumulus clouds are very important for the global climate, as they have a high albedo compared to the dark ocean over which they occur and reflect around 30% of the incoming solar radiation[1]. An estimated 4% increase in their cover could offset the warming due to a doubling of CO_2[2]. There are many ways in which aerosols interact with clouds. In this paper we focus on the first aerosol indirect effect (Twomey effect).

2 METHODOLOGY

The ASCII campaign was conducted on Ascension Island from 1–30 September 2016, with the aim to investigate the effect of biomass burning aerosols on cloud microphysical properties using a ground-based UV depolarisation lidar. Observations from the lidar were processed using an inversion method which produces estimates of cloud microphysical properties such as the effective radius and cloud droplet number density. This new inversion method is based on Monte Carlo modelling of multiple scattering within idealised semi-adiabatic clouds[3]. The method utilizes the propensity of light to become depolarized when it undergoes multiple scattering within a liquid water cloud. The backscattered radiation from spherically symmetrical and uniform elements such as liquid cloud droplets retains the polarization of the incident light under single scattering. However, the returning light will be partially depolarized due to multiple scattering within liquid water clouds. The inversion method used here is based on a simple cloud-base representation, with a linear liquid water content lapse rate Γ_l as well as a constant cloud droplet number density N_c. As the liquid water content (LWC) increases with height while the number density remains constant, the effective radius R_{eff} is increased with height. The extinction coefficient α also increases with height. The droplet size distribution used is a single-mode modified-gamma distribution. This simple but representative liquid water cloud model enables the computation of multiple scattering inside the cloud using Monte Carlo modelling. The Earth Clouds and Aerosols Radiation Explorer (EarthCARE) simulator (ECSIM) lidar-specific MC forward model was used to compute LookUp Tables (LUTs) for various values of cloud-base height, lidar field of view (FOV), R_{eff} and Γ_l, see Table 1 for the values used.

Table 1: Range of parameters used in the MC simulations

Parameters	Values
Cloud base [km]	0.5, 1.0, 2.0, 4.0
FOV [mrad]	0.5, 1.0, 2.0, 4.0
R_{eff}	2.0,2.6,3.3,4.3,5.6, 7.2,9.3,12.0
Γ_l	0.1,0.2,0.4,0.6,0.8,1.0 1.2,1.4,1.6,1.8,2.0

3 RESULTS

During the campaign, overcast skies with light drizzle was common. The average surface temperature was 21.32°C, with an average precipitation rate of 0.11 mm hr^{-1} and an east-southeasterly wind direction. Generally a well defined stratocumulus deck was present at around 1 to 1.5 km, as well as a persistent temperature inversion. Due to high solar radiance and the high position of the sun during the day there was a large amount of solar background noise, meaning aerosol signals were generally only visible during the night. Aerosols were visible on 19 out of the 25 days observed and

 https://doi.org/10.1051/epjconf/201817605002

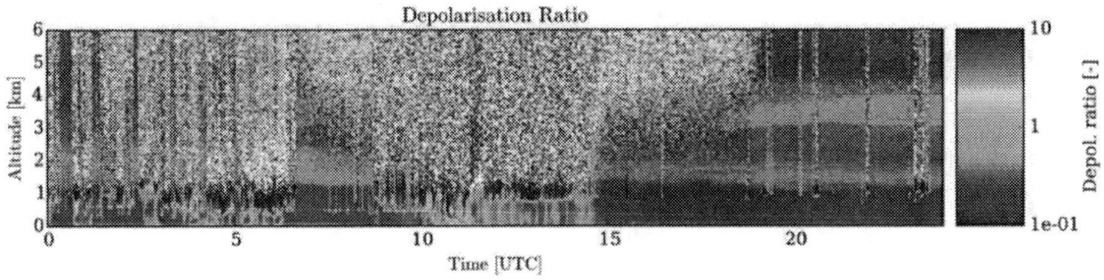

Figure 2: Depolaristion measurement from the UV depol lidar on Ascension Island on 20 September 2016, showing broken clouds at the top of the boundary layer, with two aerosol layers, one near the cloud bottom and one in the free troposphere (around 4 km).

occurred between heights of 1.5 to 4.5 km.

Figure 2 shows an example measurement from the UV depolarisation lidar on 20 September 2016. The depolarisation signal shows broken clouds during the morning and afternoon, and clear skies in the evening and night. In the clear sky parts, several aerosol layer are visible.

During overcast periods, cloud parameters were retrieved from the depolarisation measurements. An example is shown in Figure 3 for 22 September 2016. The cloud droplet number density and effective radius were retrieved at a height of 100 m above the cloud base. To identify the effects of smoke on cloud properties, each day was classified as either a clean, mixed

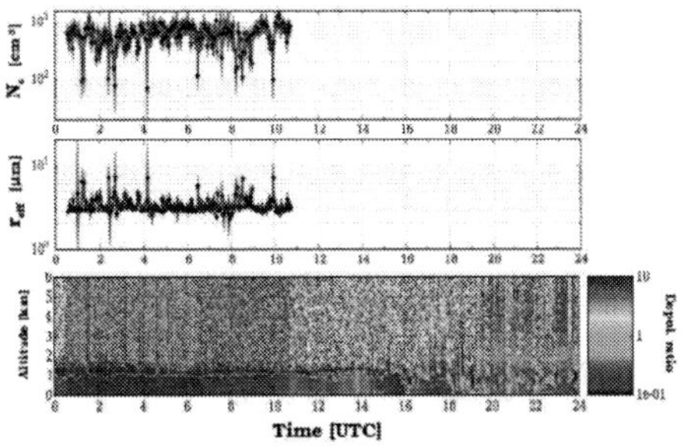

Figure 3: Inversion results for the 22nd of September. This is an example of an ideal period to invert, with a very well-defined ATB peak and an absence of below cloud aerosols or drizzle.

or separated case. Clean cases indicates days when no aerosols are visible while mixed cases were when aerosols are visible directly above the peak attenuated backscatter. This indicates that aerosols may have been mixing with the cloud layer. Separated cases were days when aerosols are seen in a layer which is vertically separated from the cloud layer below it. The majority of days fall in the mixed category, with dramatically fewer days falling under the clean and separated cases. The cloud droplet number density and effective radii were averaged for each case. Generally there is a clear differentiation in the mean values of each inversion product between each of the three cases. The $R_{eff}(100)$ ranged between 3.4 to 4.2 μm, with the highest value occurring for the clean case, see Figure 4. Both the mixed and separated cases had similar mean values for $R_{eff}(100)$. The N_c peaked in the mixed case at 550 cm^{-3} with the smallest value occurring in the clean case.

4 CONCLUSIONS

During the ASCII campaign the majority of days fell under the mixed case with aerosol layers being visible directly about the peak attenuated backscatter. This suggests that the aerosols may be mixed into the cloud layer. The presence of aerosols at heights of less than 2 km supports the idea that the widespread

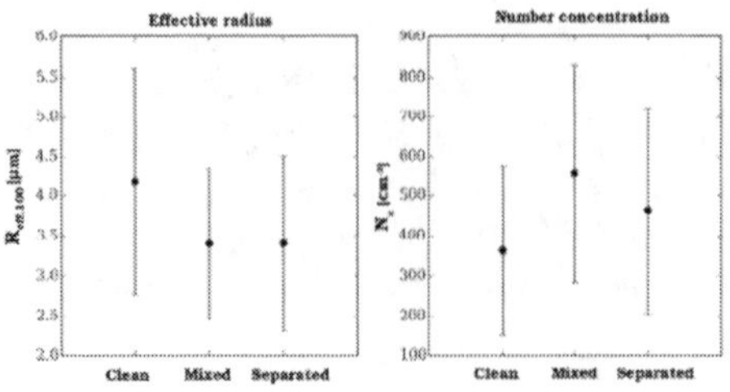

Figure 4: Average cloud droplet effective radii and number concentrations at 100 m above cloud base for the different identified cases: clean *(no smoke),* mixed *(aerosols around cloud base) and* separated *(aerosols at high altitude).*

subsidence occurring across the southeastern Atlantic region mixes smoke downwards as it moves away from the African coast at high altitudes. The mixed cases show an increased cloud droplet number concentration as well as a smaller effective radius at 100 m above cloud base. This suggests a cloud aerosol indirect effect, with aerosol particles acting as CCN. The N_c increased by 200 cm^{-3} from clean cases while the $R_{eff}(100)$ dropped by almost $1\mu m$. The separated cases sit in between the clean and mixed for $R_{eff}(100)$ and N_c. This may be explained by slight mixing of smoke from the separated layer into the clouds, which is not clearly seen in the lidar. In a previous study over the southeastern Atlantic a drop in cloud droplet effective radius of 30% was found [4], similar to what is observed here (a decrease by 20%). A key difference here is that in that study satellite observations (cloud top) were examined, while the inversion method here focuses on cloud properties near the cloud base. In addition, other factors which may have lead to this difference between cases, such as varying meteorological conditions, have not been examined in this study. Possible meteorological changes are currently being investigated to extract a definite conclusion about the Twomey affect over Ascension Island due to smoke incursions.

ACKNOWLEDGEMENTS

This campaign was funded by the Koninklijke Hollandsche Maatschappij der Wetenschappen (KHMW). Karolina Sarna is gratefully acknowledged for her initial involvement in the project.

References

[1] Bennartz, R., 2007, Global assessment of marine boundary layer cloud droplet number concentration from satellite. *Journal of Geophysical Research 112* (D2), D02201.

[2] Albrecht, B. A., D. A. Randall, S. Nicholls, B. A. Albrecht, D. A. Randall, and S. Nicholls,1988, Observations of Marine Stratocumulus Clouds During FIRE. *Bulletin of the American Meteorological Society 69* (6), 618–626.

[3] Donovan, D. P., H. Klein Baltink, J. S. Henzing, S. R. de Roode, and A. P. Siebesma, 2015, A depolarisation lidar-based method for the determination of liquid-cloud microphysical properties. *Atmospheric Measurement Techniques 8* (1), 237–266.

[4] Costantino, L. and F.-M. Bréon, 2013, Aerosol indirect effect on warm clouds over South-East Atlantic, from co-located MODIS and CALIPSO observations. *Atmospheric Chemistry and Physics 13 (1)*, 69–88.

LIDAR DERIVED PROPERTIES OF AIR-MASSES ADVECTED FROM UKRAINE, SAHARA AND CARPATHIAN MOUNTAINS TO WARSAW, POLAND ON 9 - 11 AUGUST 2015

Lucja Janicka[1], Dominika Szczepanik[1], Karolina Borek[1], Birgit Heese[2], Iwona S. Stachlewska[1]

[1]*University of Warsaw, Faculty of Physics, Institute of Geophysics, Warsaw, Poland, iwona.stachlewska@igf.fuw.edu.pl*

[2]*Leibnitz Institute for Tropospheric Research, Leipzig, Germany*

ABSTRACT

The aerosol layers of different origin, suspended in the atmosphere on 9-11 August 2015 were observed with the PollyXT-UW lidar in Warsaw, Poland. The HYSPLIT ensemble backward trajectories indicate that the observed air-masses attribute to a few different sources, among others, possible transport paths from Ukraine, Slovakia, and Africa. In this paper, we attempt to analyse and discuss the properties of aerosol particles of different origin that were suspended over Warsaw during this event.

1 INTRODUCTION

At the end of July a heat-wave span over Europe, causing in some countries absolute temperature records, e.g. 40.3°C record in Germany. In Poland, the temperatures were slightly lower, still from August 2nd, we had to face one of the biggest heat waves on record; until August 15[th,] the temperatures very often exceeded 35°C. At the beginning of the second half of the month, it became significantly cooler, still high temperatures were observed untill the end of the month. In general, August 2015 proved to be an exceptionally warm month, with the positive temperature anomaly of 4.71K as calculated in relation to the period 1961-1990 and of 3.73K to the period 1981-2010. It was also the warmest August in Poland after 1807.During the first two weeks of August 2015, extremely dynamical weather conditions were caused by a stationary front, which was located in the high pressure system that was formed over Europe. Such special weather conditions may contribute to an advection of multiple layers of several air-masses. These can be observed only by the means of the quasi-continuous ground-based or satellite lidar observations. For such multi-layer cases, interpretation of the passive remote sensors that provide information on the spectrally dependent columnar aerosol optical depth is practically impossible.

Here we discuss lidar derived particle properties characteristic for different aerosol layers that were suspended in the atmosphere over Warsaw from 9th to 11th August 2015. The detected layers are characterized by distinctively different properties and origin. We measured biomass burning particles from wild fires in Ukraine, the weak layer of depolarizing particles transported form Africa, and the air-masses that few days earlier circulated over Western Carpathians in Slovakia. Moreover, other layers with uncertain origin were observed, indicating possible mixtures of the above mentioned particle types.

2 METHODOLOGY

The Radiative Transfer Laboratory (RT-Lab) in Warsaw, Poland (52.21°N : 20.98°E, 112 m asl) is equipped with the 12-channel PollyXT-UW lidar [1] built in a scientific collaboration of the UW and TROPOS. The lidar was designed as an aerosol-depolarization-Raman lidar (thus at the UW it is alternatively called the ADR Lidar), and comprises two independent detection modules, which can be applied for simultaneous data collection in the near- and far-range regime from the lidar. The main, integral unit of the lidar is the far-range unit, so-called $3\beta+2\alpha+2\delta$+WV detection module, which provides quasi-automated measurements at the following 8 channels: 1064, 355 and 532nm (for elastic total scattering), 532 and 355nm (for elastic cross-scattering), 607 and

387nm (Raman scattering on N2) and 407nm (Raman scattering on H2O). The compact near-range aerosol-Raman lidar receiver (NARLa) was built to enhance the lidar lower detection limit. The NARLa (so-called $2\beta+2\alpha$) is designed for 355 and 532nm (elastic), and 607 and 387nm (Raman N2) detection. It was installed for regular observations in April 2016 and thus the near-range data are not used here.

The PollyXT-UW lidar provides regular high-quality measurements within the lidar networks: since July 2013 to the world-wide PollyNET [2] (http://polly.tropos.de) and since March 2015 to the EARLINET [3] (https://www.earlinet.org). It also delivers the data to the Polish aerosol network PolandAOD (http://www.polandaod.pl).

The lidar optical properties that are discussed in this paper were calculated using the TROPOS data evaluation software (Verlauf5GN); details on the applied evaluation scheme are given in [4]. The day and night-time optical properties profiles are derived from lidar signals using the classical approaches: particle extinction and backscatter coefficients retrieval is done with Raman channels (for all calculations Ångström exponent of 1 is assumed) [4]; the ±45 method is used for linear depolarization retrieval [5]; the water vapour mixing ratio and relative humidity profiles are derived as in [6] (obtained from dusk to dawn, as we measure with 407nm channel during night-time only). Thus, for evaluation of aerosol optical properties and moisture solely lidar data are used. Those are however calibrated with use of (p_0,T_0) at the RT-Lab in Warsaw and the radiosonde data for (p,T)-profiling available at 00UTC, 12UTC at WMO12374 site in Legionowo (25km North).

3 RESULTS AND DISCUSSION

The multiple aerosol layers observed over Warsaw on 10th August 2015 are shown in Fig.1.

The upper sub-figure depicts the daily evolution of the 1064nm elastic signals (in arbitrary units) up to 5km altitude. Note that in the lowermost range (< 600m), the geometrical overlap between laser beam and full field of view of the telescope is incomplete. The signals measured above 5km are not shown, as above that height neither any significant aerosol signatures nor cloud systems are revealed.

The aerosol structures and layering are depicted in lower sub-figure of Fig.1. In each case, they were discerned manually based on the trajectory analyzes that was combined with the optical properties derived from the lidar measurements.

The backward-trajectories were calculated using the HYSPLIT model [7], beginning with all simulations at Warsaw location at different hours and different altitudes. Each time the model was run for 4 to 9 days using the meteorological data from the Global Data Assimilation System (GDAS). The plots of the trajectories are not shown here for brevity.

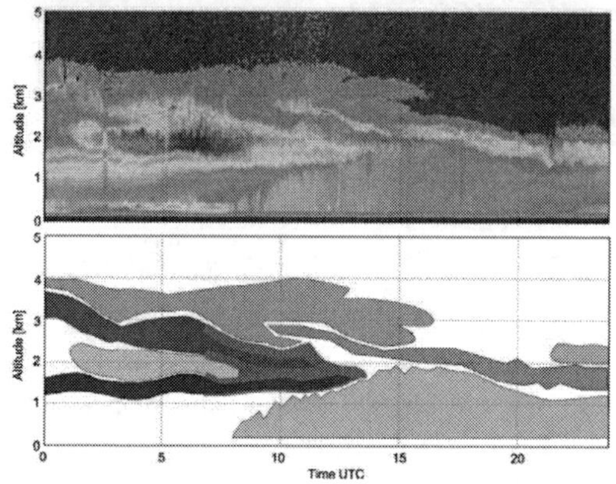

Figure 1 Lidar range corrected signal at 1064 nm on 10 August 2015 measured at the Radiative Transfer Laboratory RTLab in Warsaw (upper sub-figure, red colours indicate high aerosol load, blue - low) along with attributed aerosol layering based on the back-trajectory and optical properties analyses (lower sub-figure, each colour defines layer described in text).

In Fig.2, the results obtained on August 10th at about 00:00UTC and 22:00UTC from lidar data averaged over 45min are shown as examples. They depict the particle backscatter coefficient profiles plotted along with the profiles of volume linear depolarization ratios. We chose these two times, as they depict what it seems to be a distinctly different composition of aerosol air-masses that were observed on that day: at 00:00 UTC profiles in Fig.2 are corresponding to the dark blue, red and green air-masses defined in the lower sub-figure of Fig.1; at 22:00 UTC profiles in Fig.2 are corresponding to the pink, blue and grey air-masses in the lower sub-figure of Fig.1.

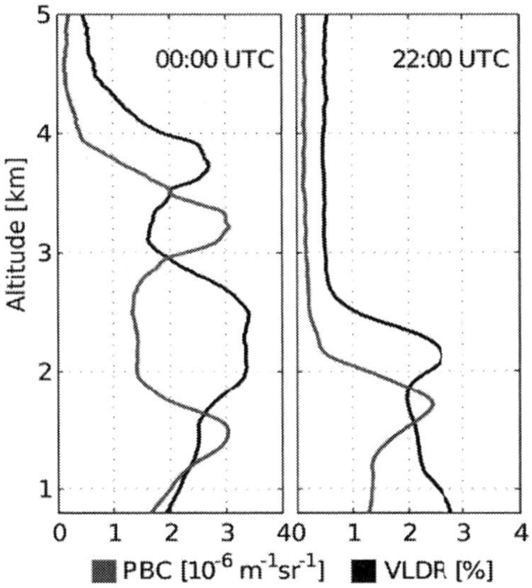

Figure 2 Profiles of the particle backscatter coefficient (in green) plotted along with the profiles of volume linear depolarization ratios (in black) obtained form the PollyXT-UW lidar in Warsaw on 10 August 2015.

As mentioned above, the HYSPLIT backward trajectory analysis and their interpretation against lidar derived aerosol optical properties and water vapor profiles (if available) was performed. This analysis clearly indicate, that the defined in Fig.1 yellow and red layers, can be attributed to an air-mass that likely contained the biomass burning particles from the wild-fires originating in Eastern Europe (mostly Ukraine: in red; and likely Belarus: in yellow). Indeed, one of the standard MODIS products which is shown in Fig.3, depicts the accumulated locations of the fires detected by MODIS on board of the Terra and Aqua satellites passing over Europe and Western Asia. The composite shown in Fig. 3 was obtained over a 10-day period from 9th to 18th August 2015, which most closely coincide with the analyzed period. Also the MODIS 10-days fire composite obtained before August 9th, indicated fire activity in this area; it is not shown here for brevity. Note that, each colored dot indicates a location where MODIS detected at least one fire during the composition period. Moreover, the plume of smoke from the peat-lands fires that occurred in vicinity of the Chernobyl nuclear power plant exclusion zone was captured at the Terra MODIS image of 10th August 2015 (not shown here for brevity). In Fig.1, in green, the air-masses from the West, and probably also from the Western Sahara seem to have brought the pollution that was likely mixed with the mineral dust particles. For this layer, the lidar derived linear depolarization was higher and the lidar derived relative humidity was low. However, this layer is difficult to define. In Fig.1, in violet, we have likely a mixture of aerosols that were transported from the East and the South of Europe, remaining very close to the ground on the entire transport pathway. The interpretation of the optical properties of this data is challenging. According to the backward trajectories, the dark blue air-masses in Fig.1 seem to have circulated very close to the surface over the southern border of Poland, The Czech Republic and Slovakia about 3 to 9 days before they were arriving in Warsaw. Also here the interpretation of the data is not a straight forward one.

Figure 3 The MODIS global fire map for 09-18 August 2015, color ranges from red where the fire count is low to yellow where number of fires is large.

In the same figure, in the area that is depicted in light-blueish and gray color, we discern layers with the origin found very hard to be assessed. For both of these layers, the ensemble backward-trajectories point either into the western Europe or to Ukraine. In both layers, similar optical properties and low values of relative humidity were derived from the lidar data. However, we treat the two layers separately, as the light-blueish one seems to be very weekly defined with a little higher particle linear depolarization ratios, than in the gray layer. Finally, in Fig.1 in pink, there is an aerosol that was suspended within the boundary layer, that was ascending by the convection during the day, and reaching at the maximum at around 2km – typical height for summer-time well mixed boundary layer in Warsaw. Thus it is likely that within the pink layer a typical polluted anthropogenic urban aerosol was captured.

4 CONCLUSIONS

For selected times on August 10th, from the lidar signals approximately averaged over 45min for every consecutive hour over the day, the particle extinction and backscatter coefficients profiles, as well as the particle linear depolarization ratio profiles were obtained (not shown for brevity). For these profiles, also the related lidar ratios at 355nm and 532nm, and the extinction Ångström exponents at both wavelengths were calculated.

In general, in the main layers, the obtained results reveal different features than typical biomass burning values observed over Warsaw [6, 8, 9]. Here, high lidar ratios in the range of 84-98 sr at 355nm and 72-88 sr at 532nm. The extinction-related Ångström exponent was from 1.1 to 1.7. A relatively low volume linear depolarization of 0.9-2.0% at 355nm and 1.9-2.7% at 532nm was obtained. In some cases, e.g. as in Fig.2, left at 3-4km and right around 2.5km, the depolarizing layer (values up to in this layer 4.6%) was clearly observed only above the backscattering layer. Such a shift is not unusual for the layers in which the biomass burning particles are observed along with the mineral dust particles, which likely seem to occurred on that day.

4 OUTLOOK

In the near future, we are planning to extend this study, mainly for calculating the sets of optical properties for data obtained on other days during the event. The manual aerosol typing was found to hinder the analysis, as time consuming. Thus, use of advanced automated approach for accurate definition of existing layers is planned. Finally, retrieval of particles micro-physical parameters for specific layers of interest that will be defined based on holistic view on the event shall follow.

ACKNOWLEDGEMENTS

The development of the PollyXT-UW lidar was financed by the Polish Foundation of Science and Technology FNTP-No.519/FNITP/115/2010.

This work was partly financed by ESA-ESTEC Contract 4000119961/16/NL/FF/mg (POLIMOS).

References

[1] Engelmann, R., et al.: The automated multiwavelength Raman polarization and water-vapor lidar PollyXT: the neXT generation, Atmos. Meas. Tech., 9, 1767-1784, doi:10.5194/amt-9-1767-2016, 2016.

[2] Althausen, D., et al.: Polly NET - a network of multiwavelength polarization Raman lidars, Proc. SPIE, Lidar Technologies, Techniques, and Measurements for Atmospheric Remote Sensing IX, Dresden, 23-26 Sept 2013, 88940I paper no.17, 2013.

[3] Pappalardo, G., et al.: EARLINET towards an advanced sustainable European aerosol lidar network, Atmos. Meas. Tech., 7, 2389–2409, 2014.

[4] Baars, H., et al.: An overview of the first decade of PollyNET: an emerging network of automated Raman-polarization lidars for continuous aerosol profiling, Atmos. Chem. Phys., 16, 5111-5137, 2016.

[5] Freudenthaler, V., et al.: Depolarization ratio profiling at several wavelengths in pure Saharan dust during SAMUM 2006, Tellus B, 61(1), 165-179, 2009.

[6] Stachlewska I.S., Costa-Suros M., Althausen D.: Raman lidar water vapor profiling over Warsaw, Poland, Atmospheric Research, vol. 194, 15 September 2017, pp. 258–267, DOI 10.1016/j.atmosres.2017.05.004, 2017.

[7] Draxler, R. R., Rolph, G. D., 2010: HYSPLIT accessed via NOAA ARL READY Website (http://ready.arl.noaa.gov/HYSPLIT.php)

[8] Janicka L., Stachlewska I.S., Veselovskii I., Baars H.: Temporal variations in optical and microphysical properties of mineral dust and biomass burning aerosol derived from daytime Raman lidar observations over Warsaw, Poland, Atmospheric Environment, vol. 169, pp. 162-174, doi:10.1016/j.atmosenv.2017.09.022, 2017

[9] Ortiz-Amezcua P., Guerrero-Rascado L.L., Granados-Muñoz M.J., Benavent-Oltra J.A., Böckmann Ch., Samaras S., Stachlewska I.S., et al., Microphysical characterization of long-range transported biomass burning particles from North America at three EARLINET stations, Atmos. Chem. Phys., 17, 5931-5946, doi:10.5194/acp-17-5931-2017, 2017.

STUDY OF MIXED PHASE CLOUDS OVER WEST AFRICA: ICE-CRYSTAL CORNER REFLECTION EFFECTS OBSERVED WITH A TWO-WAVELENGTH POLARIZATION LIDAR

Igor Veselovskii[1,2], Philippe Goloub[3], Thierry Podvin[3], Didier Tanre[3], Albert Ansmann[4], Michail Korenskiy[1], Anatoli Borovoi[5], Qiaoyun Hu[3], Valentin Bovchaliuk[3] David N. Whiteman[6]

[1]Physics Instrumentation Center, Moscow, Russia
[2]Joint Center for Earth Systems Technology, UMBC, Baltimore, USA
[3]Univ. Lille, Laboratoire d'Optique Atmosphérique, Lille, France
[4]Leibniz Institute for Tropospheric Research, Leipzig, Germany
[5]Institute of Atmospheric Optics, Tomsk, Russia
[6]NASA GSFC, Greenbelt, USA

ABSTRACT

Lidar sounding is used for the analysis of possible contribution of the corner reflection (CR) effect to the total backscattering in case of ice crystals. Our study is based on observations of mixed phase clouds performed during the SHADOW campaign in Senegal. Mie-Raman lidar allows measurements at 355 nm and 532 nm at 43 dg. off-zenith angle, so the extinction and backscattering Ångström exponents can be evaluated. In some measurements we observed the positive values of backscattering Ångström exponent, which can be attributed to the corner reflection by horizontally oriented ice plates.

1 INTRODUCTION

One of intriguing questions in the study of scattering phase function of ice crystals is the possible contribution of corner reflection (CR) effect to the total backscattering [1]. This so-called corner reflection may occur in case of ice crystals of simple shape, such as hexagonal plates and columns [2,3]. For example, if two faces of a crystal are perpendicular, they operate as a 2D corner reflector in backscattering directions. In contrast with specular reflection from horizontally oriented plates which occurs only for zenith pointing sounding geometry, the corner reflection from randomly oriented crystals may be predominant at any measurement angles [4].

The effects of diffraction and interference can be accounted for in the frame of physical-optics approximation [4]. Application of this model to the simplest case of randomly oriented hexagonal columns demonstrates that, in contrast with extinction coefficient which is spectrally independent, the account of Fraunhofer diffraction, associated with backscattered plane-parallel beams, results in spectrally dependent backscattering coefficient.

Corner reflection may occur also for oriented plates when they are probed by tilted lidars. Corresponding computations demonstrate that the effect is maximal when lidar is tilted at 30-40 dg of zenith [5]. We should also keep in mind, that the corner reflection may be sensitive to the crystal deformation [5,6]. The surface roughness of crystal faces can also suppress the CR effect [1]. Thus there is a need of experimental data for better estimating the CR contribution to the backscatter of real ice clouds and, as previously mentioned, we are going to use the spectral dependence of the backscattering coefficient to identify the relevant cases.

In 2015-2016 the field campaign SHADOW was organized in Senegal in the framework of the CaPPA (Chemical and Physical Processed in the Atmosphere) project (http://www.labex-cappa.fr/). The LILAS - Raman lidar used during the campaign, allows measurements of particle extinction, backscattering coefficient and depolarization ratio at 355 nm and 532 nm. In our presentation we analyze their spectral dependence to evaluate the possible contribution of CR effect.

2 EXPERIMENTAL SETUP

The instrumentation site is located at the IRD (Institute for Research and Development) Center, Mbour, Senegal (14^{0}N, 17^{0}W). The observations were performed during March-April 2015 and December 2015-January 2016. Detailed information regarding the site and the instruments is reported in a recent publication [7]. The LILAS is based on a tripled Nd:YAG laser with a 20 Hz repetition rate, and pulse energy of 90/100/100 mJ at 355/532/1064 nm. The aperture of the receiving telescope is 400 mm. The lidar allows evaluation of three particle backscattering (355 nm, 532 nm, 1064 nm) and two extinction coefficients (355 nm, 532 nm). To improve the system performance at 532 nm, the rotational Raman channel was used instead of the vibrational one. The sounding was performed through an open window at a fixed off-zenith angle of 43 dg., which was the minimal possible angle, so the effect of specular reflection was excluded.

Depolarization measurements were performed at 532 nm only in March-April 2015 while depolarization data at 532 nm and 355 nm were both available for the December 2015 – January 2016 period. The relative uncertainty of depolarization measurements is estimated as ±15% [7].

3 RESULTS OF THE MEASUREMENTS

Pure ice clouds occur mainly at altitudes where the atmospheric temperature is below the temperature of homogeneous water freezing (about -40^{0}C), while at lower altitudes the ice crystals may coexist with super-cooled liquid water (SLW) droplets. During the SHADOW experiment we acquired about 170 hours of clouds observations in 5 km – 14 km height range and, in approximately 25% of this time, SLW layers occurred. The liquid water layers were confined by the 10 km height.

A typical example of a mixed phase cloud observed during the 16-17 December 2015 night is shown in Fig.1. We provide the height-tempo distributions of range corrected lidar signal as well as the particle depolarization ratio at 532 nm. Between 22:00 and 23:40 UTC the backscattering coefficient at 532 nm presents strong enhancement inside the layer centered at approximately 9 km altitude. The depolarization ratio in this layer decreases, indicating the presence of SLW. The atmospheric temperature, measured by the radio-sonde in Dakar varies from -28C at 8500 m to -35C at 9500 m, which are the heights corresponding to the bottom and to the top of the liquid layer observed in M'Bour. The SLW layer is observed for 1.5 hours and then dissipates. Below the SLW layer the particle depolarization ratio increases up to ~40%, indicating the presence of ice crystals. The optical depth of ice phase is significantly lower comparing to that of liquid one, because conversion of SLW droplets to ice crystals is associated to a size increase and a decrease of droplet number density.

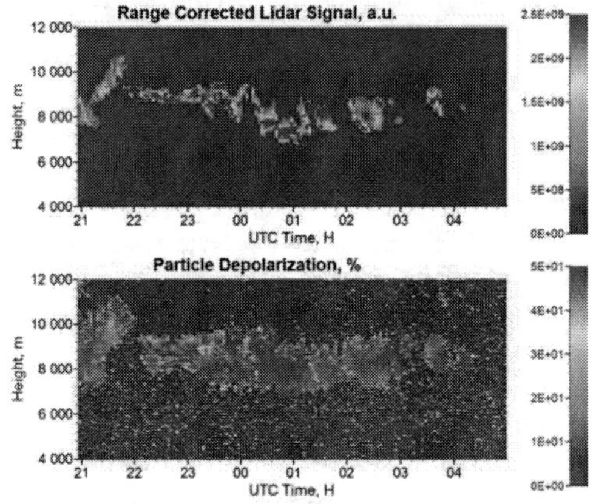

Figure 1. Range corrected lidar signal and (b) particle depolarization ratio at 532 nm measured in the 16-17 December 2015 night.

The vertical distributions of cloud backscattering coefficients and depolarization ratios at 355 nm and 532 nm are reported in Fig.2 for three time slots: 21:00-21:45 (pure ice cloud), 22:10-23:10 (liquid layer), and 00:00-01:00 UTC (ice cloud with residual SLW). The values of optical depth (OD) at 532 nm derived from Raman signals for these 3 periods are 1.49, 1.41 and 0.8 respectively. For the 21:00-21:45 time slot both particle depolarization ratios δ_{355} and δ_{532} are about 45±7% and do not present height dependence. The height averaged lidar ratio at both wavelengths is 22±4.5 sr. Depolarization ratio in the SLW layer (second time slot 22:10-23:10 UTC) is low, δ_{532} is 8±1.5% at 8700 m (near the liquid layer base), still it may indicate the presence of ice crystals in liquid water surroundings. Above 9200 m the depolarization ratio increases, reaching 17% at 9400 m altitude. This increase of the depolarization ratio may be partly related to the contribution of multiple scattering. The averaged lidar ratios in liquid layer at both wavelengths are 15.5±3 sr, which is lower than in ice cloud. For the 00:00-01:00 time slot the depolarization ratio varies within the 25%-45% range due to residual SLW. The lidar ratio at both wavelengths is 17±3.5 sr, which corresponds to an intermediate value between ice crystals and liquid droplets.

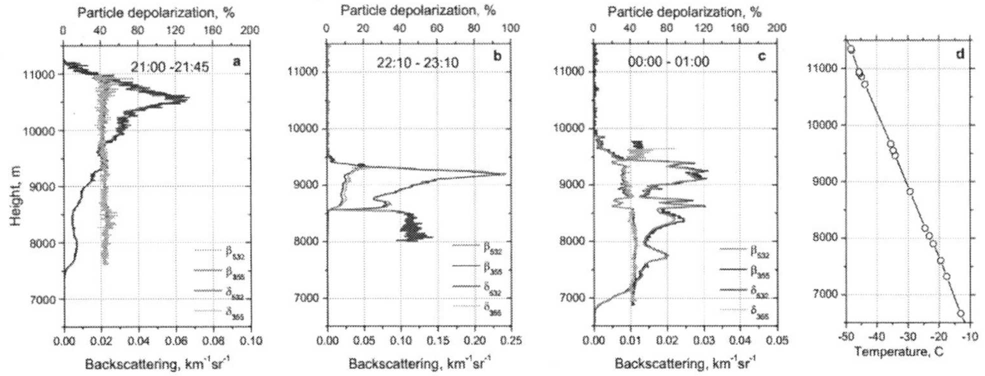

Figure 2. Vertical profiles of backscattering coefficients and particle depolarization ratios at 355 and 532 nm. The measurements were performed in the night 16-17 December 2015 and results are averaged over the temporal interval (a) 21:00 – 21:45 UTC, (b) 22:10-23:10 UTC, (c) 00:00 – 01:00 UTC. Plot (d) shows temperature profile obtained from radio-sonde measurements in Dakar on 15 December at 12:00 UTC.

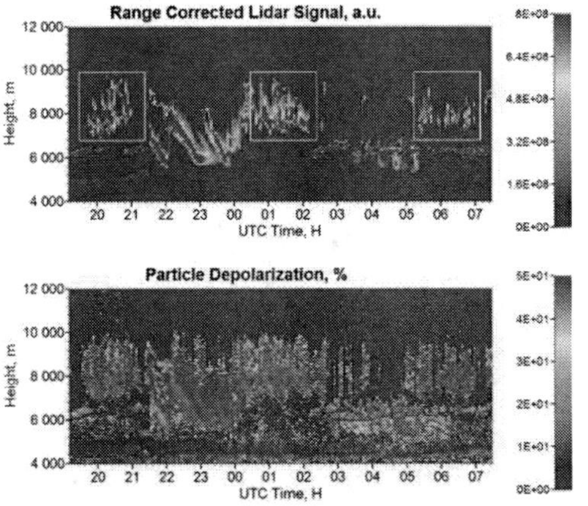

Figure 3. Range corrected lidar signal and particle depolarization ratio at 532 nm measured on 12-13 January 2016. White rectangulars mark the regions where spectral dependence of backscattering coefficient occurred.

For the three time slots, backscattering, extinction coefficients and depolarization ratios at 355 nm and 532 nm are close. No deviations of the backscattering Ångström exponent (BAE) from the zero value exceeding the measurement uncertainty (which we estimate to be below ±0.2), that could be attributed to the crystals corner reflection, are observed.

We have analyzed 15 representative measurement periods where ice clouds occurred and spectral dependence of backscattering coefficient was observed only on 12-13 January 2016. The height-temporal distributions of range corrected lidar signal and depolarization ratio for that period are shown in Fig.3. The positive values of BAE were observed for time intervals 20:00-21:00, 00:30 – 02:30 and 5:30-7:00 UTC. For these intervals the lidar signals inside the 7000 m – 9000 m height range are typical for ice precipitation, so presence of oriented ice crystals is possible.

Fig.4 shows the vertical profiles of backscattering coefficients (β) and particle depolarization ratios (δ) at 355 nm and 532 nm averaged over 00:30 – 02:30 time period. Inside the height range 7000 - 9000 m β_{355} exceeds β_{532} and corresponding BAE reaches a maximal value of 0.6 at 8100 m. Within the same height range the ratio $\delta_{355}/\delta_{532}$ decreases

 https://doi.org/10.1051/epjconf/201817605004

from 1.0 to 0.8. The observed spectral dependence of backscattering coefficient and depolarization ratio can probably be explained by the corner reflection of oriented plates. The polarization planes of output laser radiation at 532 nm and 355 nm are orthogonal (532 nm is vertical while 355 nm horizontal), so depolarization ratios of oriented plates in the case of corner reflection become different [5]. Similar spectral dependence of backscattering and depolarization was observed also for 20:00-21:00 period. The extinction coefficients α_{532} and α_{355} are close for all the considered height-temporal intervals. Thus the observed spectral dependence is attributed only to backscattering coefficients and depolarization ratios.

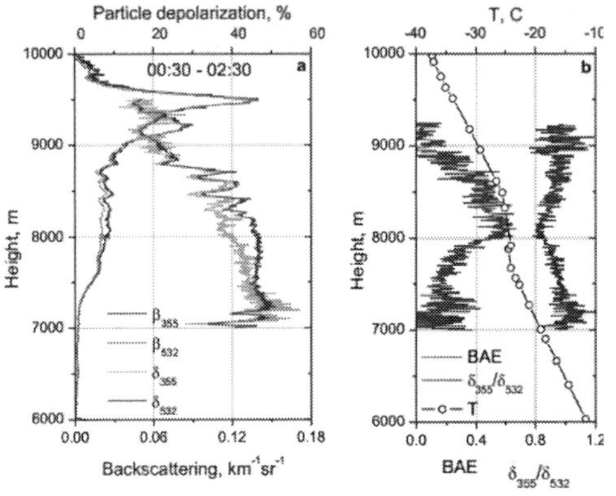

Figure 4. Vertical profiles of (a) backscattering coefficients and particle depolarization ratios at 355 nm, 532 nm, and (b) BAE together with ratio $\delta_{355}/\delta_{532}$ on 12-13 January 2016. Results of lidar measurements are averaged over 00:30 – 02:30 UTC period. Open symbols show temperature profile measured with radio-sounding in Dakar on 13 January at 00:00 UTC.

4 CONCLUSIONS

Most of the time, the contribution of the corner reflection to the total backscattering was not significant, but based on our knowledge, it is the first time that spectral dependence of backscattering coefficient, that can be related to the corner reflection of oriented ice plates, is observed. This low contribution of CR can be the result of crystals deformation or surface roughness. Another possible explanation can be the relative small amount of crystals of simple shape in the cases we selected.

ACKNOWLEDGEMENTS

Development of the software for lidar data analysis was supported by the Russian Science Foundation (project no.16-17-10241). The CaPPA project (Chemical and Physical Properties of the Atmosphere) is funded by the French National Research Agency (ANR) through the PIA (Programme d'Investissement d'Avenir) under contract "ANR-11-LABX-0005-01" and by the Regional Council " Nord-Pas de Calais - FEDER".

References

[1] Zhou, C., and Yang, P: Backscattering peak of ice cloud particles, Opt. Expr. 23, 11995-12003, 2015.

[2] Borovoi, A., Grishin, I., Naats, E., and Oppel, U.: Backscattering peak of hexagonal ice columns and plates, Opt. Lett. 25, 1388-1390, 2000.

[3] Del Guasta, M.: Simulation of lidar returns from ideal and deformed hexagonal ice prisms in cold cirrus by means of face tracing," J. Geophys. Res. **106,** D12, 12,589–12,602, 2001.

[4] Borovoi, A., Konoshonkin, A., and Kustova, N.: Backscattering by hexagonal ice crystals of cirrus clouds, Opt. Lett. 38, 2881-2884, 2013.

[5] Del Guasta, M., Vallar, E., Riviere, O., Castagnoli, F., Venturi, V., Morandi, M.: Use of polarimetric lidar for the study of oriented ice plates in clouds, Appl. Opt. 45, 4878-4887, 2006.

[6] Borovoi, A., Kustova, N., and Konoshonkin, A.: Interference phenomena at backscattering by ice crystals of cirrus clouds, Opt. Exp. 23, 24557-24571, 2015.

[7] Veselovskii, I., Goloub, P., Podvin, T., Bovchaliuk, V., Derimian, Y., Augustin, P., Fourmentin, M., Tanre, D., Korenskiy, M., Whiteman, D., Diallo, A., Ndiaye, T., Kolgotin, A., Dubovik, O.: Study of African dust with multi-wavelength Raman lidar during the "SHADOW" campaign in Senegal, Atm. Chem. Phys. 16, 7013–7028, 2016.

STRENGTHS AND LIMITATIONS OF THE NATALI CODE FOR AEROSOL TYPING FROM MULTIWAVELENGTH RAMAN LIDAR OBSERVATIONS

Doina Nicolae[1]*, Camelia Talianu[1,2], Jeni Vasilescu[1], Victor Nicolae[1], Iwona S. Stachlewska[3]

*[1]National Institute of R&D for Optoelectronics, Romania, *nnicol@inoe.ro*
[2]University of Natural Resources and Life Science, Institute of Meteorology, Austria
[3]University of Warsaw, Faculty of Physics, Institute of Geophysics, Poland

ABSTRACT

A Python code was developed to automatically retrieve the aerosol type (and its predominant component in the mixture) from EARLINET's 3 backscatter and 2 extinction data. The typing relies on Artificial Neural Networks which are trained to identify the most probable aerosol type from a set of mean-layer intensive optical parameters. This paper presents the use and limitations of the code with respect to the quality of the inputed lidar profiles, as well as with the assumptions made in the aerosol model.

1 INTRODUCTION

Global and local properties of atmospheric aerosols have been extensively observed using both space-borne and ground-based instruments, especially during the last decade. Researchers are exploiting satellite remote sensing observations to characterize aerosol layers and to both, assess and propose parameterizations in regional and global models. Although detailed studies of aerosol optical properties have been made, mixed aerosols are still difficult to characterize.

An important advance in the remote sensing of aerosols is the development of continental-scale ground based lidar networks, which provide quality assured optical profiles on a large temporal and spatial scale. EARLINET, the European Aerosol Research Lidar Network data (http://ww.earlinet.org) is relevant for climatology, but also for special events, with strong aerosol influence: Saharan dust outbreaks, forest-fire smoke plumes, photochemical smog or volcano eruptions [1].

In the frame of the ESA funded NATALi project, we developed an algorithm based on a combination of Artificial Neural Networks, to estimate the most probable aerosol type from a set of multispectral data. The algorithm has been implemented in a Python code and adjusted for running on the EARLINET profiles (3 backscatters, 2 extinctions and one linear particle depolarization ratio).

The algorithm relies on the ability of specialized Artificial Neural Networks (ANN) to resolve the overlapping values of the intensive optical parameters calculated for each identified layer in the multiwavelength Raman lidar profiles [2]. The ANNs were trained using synthetic data, for which a new aerosol model was developed. Aerosols were considered spheroids and built up using OPAC-defined internal mixtures, with the associated microphysical properties picked up from GADS. The intensive optical properties obtained from this model were compared to literature and found to be consistent with observations. Variability of the optical properties was achieved by considering different number mixing ratios and values of relative humidity. In addition, we included the uncertainty of the observations as a prerequisite hypothesis in order to match real lidar data. These requirements have added to the complexity of the ANNs selected to make the retrieval, because of the significant overlap of the input values for the intensive optical parameters.

Two parallel typing schemes were implemented in order to accommodate data sets containing or not the measured linear particle depolarization ratios (LPDR): a) identification of mixtures from 14 aerosol mixtures (high resolution typing) if the LPDR value is available in the input data files; b) identification of 5 predominant aerosol types (low resolution typing) if the LPDR is not provided. For each scheme, three ANNs are run simultaneously, and a voting procedure selects the most probable answer.

2 METHODOLOGY

The whole algorithm has been integrated in a Python code. NATALI software is build on three

 https://doi.org/10.1051/epjconf/201817605005

modules: (a) Input module: to convert the input files into the specific format read by the ANNs; (b) Typing module: to run the ANNs and decide on the most probable aerosol type; (c) Output module: to save the results and logs.

The input module reads the lidar files in EARLINET NetCDF format, checks for the availability of all required parameters (β1064, β532, β355, α532, α355, and optionally δ532), identifies the layer geometrical boundaries, calculates the intensive optical parameters within each layer, their mean value and associated uncertainty.

Layer boundaries are calculated by applying the gradient method on the 1064 nm backscatter coefficient profile [3]. The inflexion points of the second derivative of the profile data (smoothed with the Savistky-Golay filter) give the top and bottom of each identified layer. Low or fine structure of the aerosol layers is revealed by a respectively higher or lower value of the smoothing parameter (adjustable) *FINESSE*. Only layers with a thickness larger than 300 m are considered relevant, for the reason of significant signal-to-noise ratio.

The intensive optical parameters and their associated uncertainties are computed for the middle part of each layer for which the signal-to-noise ratio is highest (no less than 200m mid-layer), to exclude the margins which are affected by the smoothing.

The linear particle depolarization ratio is picked-up directly from the EARLINET b532 file, if existing. For each layer, and for all of the above arrays, the module calculates averages and associated uncertainties.

Several filters are applied on the data, and only layers which pass the following criteria are further considered for typing:

- availability of all necessary intensive optical parameters
- values of the intensive optical parameters between acceptable limits [4]
- the relative error of each intensive optical parameter is lower than 50%

For each layer and for each intensive optical parameter, the module generates a number of values (N, adjustable) between [average – uncertainty] and [average + uncertainty]. Data are then scrambled, considering that any combination has a similar probability to describe the reality. The cluster of possible combinations of intensive optical parameters is prepared for the ANN input.

The typing module runs in parallel the ANNs for each data set representing a layer, and applies the voting procedure to identify the most probable aerosol type.

In the case that the depolarization is available, the module runs in parallel 6 ANNs: 3 for high resolution (A1H, A2H, A3H) and 3 for low resolution typing (A1L, A2L, A3L). The probable aerosol type is provided by the high resolution ANNs, while the predominant type is provided by the low resolution ANNs. As such, if typing in high resolution fails (for reasons of insufficient quality of the input data), the user has still access to information in low resolution.

If the depolarization is not available, the module runs in parallel 3 ANNs (B1L, B2L, B3L), and returns the most probable predominant aerosol type. By comparison to the low resolution typing when depolarization is available, in this case the ANNs cannot distinguish the Volcanic type, as it overlaps completely (in all existing parameters) with Dust or Continental polluted types. As such, with 3ANNs only 5 predominant types are retrieved, namely: Continental, Continental polluted, Dust, Marine, and Smoke.

A voting procedure selects the most probable answer out of the three (possibly different) individual returns. Selection is done based on the confidence level of the ANN outputs, and the stability over the uncertainty range (percentage of agreement for values between error limits). The output module prepares and saves the files in 2 formats: csv and human-readable (telegrams), and writes the log.

The software's code structure resembles the three module approach described earlier: (a) the input module: nt_input.py; (b) the typing module: nt_typing.py; (c) the output module: nt_output.py. These three modules are orchestrated by the natali.py script, which contains the high-level algorithm and calls the required module methods.

Apart from its main purpose, NATALI algorithm proved to have side-applications. In this paper, we show the capability of the software to test the quality of the optical data and identify incorrect calibration or incorrect cloud screening.

3 RESULTS

Two data sets with continuous Raman measurements at Bucharest and Warsaw were analyzed with NATALI and presented in this paper. No additional filtering was applied to the data. Examples of the optical profiles used as inputs for the NATALI code are presented in Figs. 1 and 3 respectively. The result of the aerosol typing in low resolution (in the absence of the linear particle depolarization ratio, i.e. retrieving only the predominant component in the mixture) is presented in Figs. 2 and 4 respectively. In each of the time series, the black rectangle marks the data set for which the optical profiles are shown.

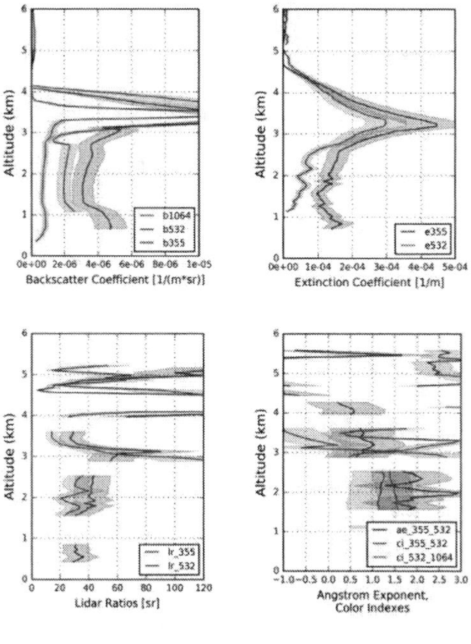

Figure 1 Aerosol optical profiles for Bucharest observations, 09/07/2012: backscatter coefficients (upper left panel), extinction coefficients (upper right panel), lidar ratios (bottom left panel), Angstrom coefficients (bottom right panel)

It can be seen from the examples at Bucharest site that NATALI cannot perform the typing if the uncertainty of the input mean-layer intensive optical parameters is higher than 100%. In this case, NATALI returns "Unknown" and flags the output. This situation generally appears for high altitude layers, where there are few particles and the signal-to-noise ratio of the lidar, especially of the Raman channels, is low.

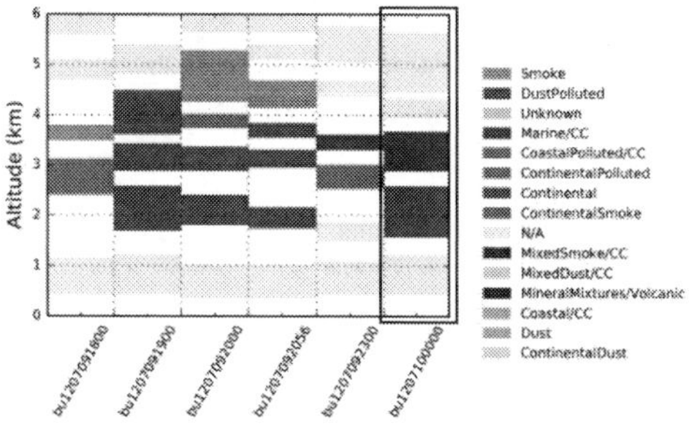

Figure 2 Results of the aerosol typing for Bucharest observations, 09/07/2012

In the incomplete overlap region, the signal-to-noise ratio is high but the values of intensive optical parameters are outside the acceptable ranges, therefore NATALI again returns "Unknown". This is due either to an incorrect overlap correction of the raw lidar signals, or to missing information (different cuts of the backscatter or extinction profiles. For the layers with significant load and above the full overlap of the lidar NATALI is able to estimate the aerosol type even if the calibration is not perfect and/or some input parameters are not within the typical range. One particular advantage of NATALI is that it can be used to identify regions where clouds were not removed completely. For example, the marked profile in Fig. 2 shows the presence of marine particles at 3-4 km altitude in a continental site. Looking at the corresponding optical profiles (Fig. 1), it is clear that clouds are present in the lidar products. From the optical point of view, cloud particles are close to marine aerosols, large and with a low lidar ratio.

Analysis of the Warsaw observations show the difficulty of NATALI to distinguish between smoke particles and continental polluted. The extinction and backscatter-related Angstrom coefficients are very similar for these aerosol types, with smoke having a slightly higher values of the lidar ratios. However, when considering the uncertainty of the measurements, this differences in the lidar ratio are no longer visible.

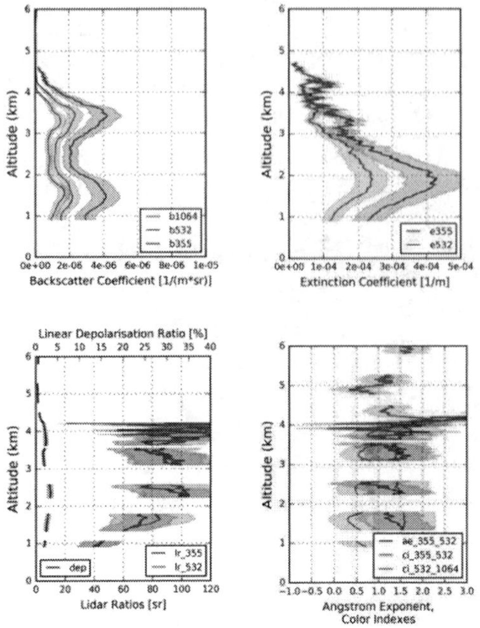

Figure 3 Aerosol optical profiles for Warsaw observations, 10/08/2015: backscatter coefficients (upper left panel), extinction coefficients (upper right panel), lidar ratios (bottom left panel), Angstrom coefficients (bottom right panel)

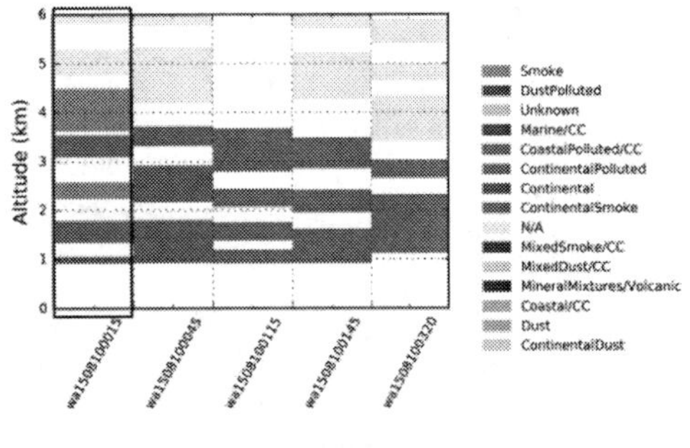

Figure 4 Results of the aerosol typing for Warsaw observations, 10/08/2015

Nevertheless, the output of the code seems fairly stable with time, considering the sensitivity of the retrieval with respect to the calibration of the input optical data.

4 CONCLUSIONS

The aerosol typing algorithm NATALI was embedded into a Python code that allows fast analysis of the multispectral lidar data.

Optical profiles are displayed in parallel with the return of the ANNs regarding the aerosol type (if depolarization information is available) or the predominant component in the mixture (otherwise). The retrieval is stable over a large range of data uncertainty, however it may fail if the signal-to-noise ratio is too low or the optical profiles are not calibrated correctly. The neural network is able to recognize the pattern of noisy data, such pattern has to be corrected, otherwise the retrieval results will be misleading. The sensitivity of the retrieval to the quality of the optical data makes NATALI a useful tool for quick check-up of the lidar products..

ACKNOWLEDGEMENTS

The work presented in this paper was performed in the frame of the project Neural network Aerosol Typing Algorithm based on LIdar data (NATALI) funded by ESA under contract 4000110671/14/I-LG. Also, this project has received funding from the European Union's Horizon 2020 research and innovation programme under grant agreement No. 654109 ACTRIS-2, and grant agreement No. 692014 ECARS. The development and evaluation of Warsaw lidar was done in the frame of the Polish National Science Center project NCN 2012/05/E/ST10/01578

References

[1] Tesche, M., et al., Optical and microphysical properties of smoke over Cape Verde inferred from multiwavelength lidar measurements, *Tellus B*, 63, 677–694, doi:10.1111/j.1600-0889.2011.00549.x, 2011

[2] Nicolae, D., et al., Independent retrieval of aerosol type from lidar, *EPJ Web of Conferences* **119**, 18002, DOI: 10.1051/epjconf/201611918002, 2016

[3] Belegante, L., et al., Retrieval of the boundary layer height from active and passive remote sensors. Comparison with a NWP model, *Acta Geophysica*, 62(2), 276–289, doi:10.2478/s11600-013-0167-4, 2014

[4] Wandinger U. et al, HETEAC: The Aerosol Classification Model for EarthCARE, *EPJ Web of Conferences* **119**, 01004, DOI:10.1051/epjconf/201611901004, 2016

MODELLING OF POLLEN DISPERSION IN THE ATMOSPHERE: EVALUATION WITH A CONTINUOUS 1β+1δ LIDAR

Michaël Sicard*[1,2], Rebeca Izquierdo[3,4,5], Oriol Jorba[6], Marta Alarcón[3], Jordina Belmonte[4,5], Adolfo Comerón[1], Concepción De Linares[4,5], José Maria Baldasano[7]

[1]*Remote Sensing Laboratory, Universitat Politècnica de Catalunya, Barcelona, Spain, *Email: msicard@tsc.upc.edu*
[2]*Ciències i Tecnologies de l'Espai - Centre de Recerca de l'Aeronàutica i de l'Espai / Institut d'Estudis Espacials de Catalunya (CTE-CRAE / IEEC), Universitat Politècnica de Catalunya, Barcelona, Spain*
[3]*Departament de Física, Universitat Politècnica de Catalunya, Barcelona, Spain*
[4]*Departament de Biologia Animal, Biologia Vegetal i Ecologia, Universitat Autònoma de Barcelona, Bellaterra, Spain*
[5]*Institut de Ciencia i Tecnología Ambientals (ICTA), Universitat Autònoma de Barcelona, Bellaterra, Spain*
[6]*Earth Sciences Department, Barcelona Supercomputing Center – Centro Nacional de Supercomputación, Barcelona, Spain*
[7]*Environmental Modeling Laboratory, Technical University of Catalonia, Barcelona, Spain*

ABSTRACT

Pollen allergenicity plays an important role on human health and wellness. It is thus of large public interest to increase our knowledge of pollen grain behavior in the atmosphere (source, emission, processes involved during their transport, etc.) at fine temporal and spatial scales. First simulations with the Barcelona Supercomputing Center NMMB/BSC-CTM model of *Platanus* and *Pinus* dispersion in the atmosphere were performed during a 5-day pollination event observed in Barcelona, Spain, between 27 – 31 March, 2015. The simulations are compared to vertical profiles measured with the continuous Barcelona Micro Pulse Lidar system. First results show that the vertical distribution is well reproduced by the model in shape, but not in intensity, the model largely underestimating in the afternoon. Guidelines are proposed to improve the dispersion of airborne pollen by numerical prediction models.

1 INTRODUCTION

Respirable allergenic materials such as pollen grains are a serious public health concern worldwide with the most prevalent impacts among children and adolescents [1][2]. Allergic airway diseases may increase further in the future due to intensive human activities that perturb the environment and change land management practices, which could modify the pollen amount, pollen allergenicity, duration of pollen season, and pollen spatial distributions [3][4].

Several numerical models have been used to simulate the dispersion of pollen in the atmosphere in Europe [5][6] and in the United States [7]. Most of these simulations are confined to short-term pollination events affecting a rather small region and restricted to just a few pollen types. While the evaluation of these simulations is almost always performed against in-situ pollen concentration measurements, nearly no information is known about their performance in the atmospheric column. In Europe, only the SILAM (System for Integrated modeLling of Atmospheric composition [6]) model from the Finnish Meteorological Institute is providing forecast of pollen grain number concentration at continental scale, but without performing routine evaluation.

This work presents the very first simulations with the Nonhydrostatic Multiscale Meteorological Model on the B grid at the Barcelona Supercomputing Center (BSC) with a newly developed Chemical Transport Model (NMMB/BSC-CTM) of the dispersion of *Platanus* and *Pinus* pollen in the atmosphere which occurred during a 5-day pollination event in Barcelona, Spain, between 27 – 31 March, 2015. The model evaluation in the atmospheric

 https://doi.org/10.1051/epjconf/201817605006

column is performed against continuous lidar measurements. The results are discussed in terms of the pollen parametrization and the schemes used in the model.

2 INSTRUMENTS AND MODELLING

2.1 Pollen measurements

During the event considered here, the daily and hourly concentration of 93 types of pollen was measured from pollen samples obtained using volumetric suction pollen trap based on the impact principle, the standardized method in European aerobiological networks. Further details on the instrument and method can be found in [8][9].

Measurements of the vertical distribution of the particle backscatter coefficient and of the volume and particle depolarization ratios were made with the Barcelona Micro Pulse Lidar (MPL) system, Sigma Space model MPL-4B, part of the MPLNET (Micro Pulse Lidar Network, http://mplnet.gsfc.nasa.gov/) network. The Barcelona MPL optical layout uses an actively controlled liquid crystal retarder which makes the system capable to conduct polarization-sensitive measurements by alternating between two retardation states [10]. The signals, called "co-polar", P_{co}, and "cross-polar", P_{cr}, are used to calculate the total signal, P, as $P = P_{co} + 2P_{cr}$, which is inverted using the two-component elastic lidar inversion and a constant lidar ratio, LR, of 50 sr to obtain the particle backscatter coefficient. P_{co} and P_{cr} are also used to calculate the volume depolarization ratio, δ^V, as:

$$\delta^V(z) = \frac{P_{cr}(z)}{P_{co}(z) + P_{cr}(z)} \quad (1)$$

Pollen has formerly been distinguished from other particle types thanks to its depolarization capabilities (see references in [8]). For this reason the particle depolarization ratio is also calculated using Eq. (4) of [8]. To quantify the pollen contribution to the particle backscatter ratio the method of [11] is used assuming that the external mixing of particles observed is composed only of depolarizing (pollen) and much less depolarizing (everything but pollen) particles. The pollen backscatter coefficient, β_{pol}^{obs}, is then calculated using Eqs. (7) and (8) of [8]. The vertical height, h_{pol}, up to which the pollen plume extends is calculated with a simple threshold method [8].

2.2 Pollen modelling

The NMMB/BSC-CTM model is the online air quality forecast system developed at BSC [12][13]. Meteorology, transport schemes and the output configuration are the same as in [9].

The two pollen types largely predominant during the pollination event were *Pinus* and *Platanus* [8]. *Pinus* is an important genus of the Spanish landscapes; *Platanus* is an ornamental tree very common in the street of metropolitan areas. They are selected in this study due to their high airborne pollen concentrations. For this reason the aerosol scheme is set to bulk *Pinus* and *Platanus* aerosols. The main parameters, reported in Table 1, are taken from the literature. Geographical distributions are obtained from the Cartography of habitats of Catalonia for *Pinus* and from the Barcelona's City Council Open Data Service for *Platanus*. *Pinus* tree density, 650 tree/ha, is from the Forest Inventory of Catalonia while for *Platanus* the number of tree per pixel is directly entered in the model.

Table 1 Pollen parametrization of the aerosol scheme

	Pinus	*Platanus*
Emission factor	81 g/day/tree [14]	2.48 g/day/tree [15]
Grain diameter	59 µm [16]	19 µm [7]
Grain density	560 kg/m^3 [16]	920 kg/m^3 [7]

3 EVALUATION OF THE FIRST SIMULATIONS

The pollen simulations are performed at fine temporal (1 hour) and spatial (1 x 1 km^2) scales. The pollen concentration, C_{pol}^{mod} in kg/m^3, is calculated at the coordinates of the measurement through a bilinear interpolation and converted to pollen backscatter coefficient as:

$$\beta_{pol}^{mod} = \frac{C_{pol}^{mod} \times 10^3 \times \sigma^*}{LR} * 10^6 \left[Mm^{-1} sr^{-1} \right] \quad (2)$$

where σ^* is the specific extinction cross-section at 532 nm in m^2/g. In a first approximation a value of 0.6 m^2/g, taken from [17] and measured for Saharan dust particles, is used. For the 2 days of the event with no (or low) nocturnal pollen near-surface activity (27 and 29 March, respectively noted 27M and 29M, see [8]), Figure 1 shows the comparison model-observation of the daily averaged pollen backscatter coefficient, the hourly difference $\beta_{pol}^{\mathrm{mod}} - \beta_{pol}^{obs}$ and the score of the model in terms of fractional bias and correlation coefficient calculated from the minimum height up to h_{pol}.

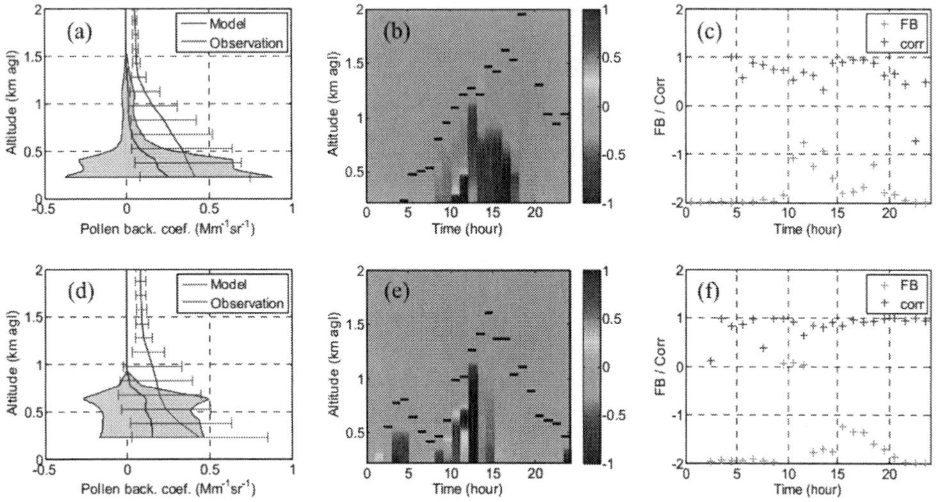

Figure 1 Daily averaged pollen backscatter coefficient (± standard deviation) on (a) 27M and (d) 29M; Hourly difference $\beta_{pol}^{\mathrm{mod}} - \beta_{pol}^{obs}$ ($Mm^{-1}sr^{-1}$) on (b) 27M and (e) 29M (h_{pol} is reported by horizontal black dashes); Hourly fraction bias (FB) and correlation coefficient ($corr$) on (c) 27M and (f) 29M

A first result is the good performance of the model to reproduce the shape of vertical profiles ($corr > 0.5$ in general). Although the daily averaged $\beta_{pol}^{\mathrm{mod}}$ falls in both cases in the standard deviation of β_{pol}^{obs}, it appears largely underestimated at all heights. The temporal evolution of their hourly difference shows that the strongest underestimations occur in the afternoon. On both days some overestimations are also visible around 10:00 and below 0.75 km. On 29M this overestimation is compensated ($FB \simeq 0$) by the underestimation of the model above 0.75 km. During the night on 29M (before 06:00) the pollen layers observed < 0.8 km are not reproduced at all by the model. Several research lines are contemplated to improve these first simulations: (i) complete the tree distribution, (ii) improve the emission scheme and make it time-dependent, (iii) study the sensitivity of the model to pollen size, (iv) identify the mechanisms responsible for the pollen transport during nighttime, (v) study the impact of the changes sea-land ↔ land-sea breeze regimes on the concentration at the ground and in the atmospheric column.

ACKNOWLEDGEMENTS

Lidar data analysis were supported by the ACTRIS (Aerosols, Clouds, and Trace Gases Research Infrastructure Network) Research Infrastructure Project funded by the European Union's Horizon 2020 research and innovation programme under grant agreement n. 654169; by the Spanish Ministry of Economy and Competitivity (project TEC2015-63832-P) and of Science and Innovation (project UNPC10-4E-442) and EFRD (European Fund for Regional Development); by the Department of Economy and Knowledge of the Catalan autonomous government (grant 2014 SGR 583). This work is contributing to the ICTA 'Unit of Excellence' (Spanish Ministry of Economy and Competitivity, project MDM2015-0552).

References

[1] Nathan, R., Meltzer, E., Selner, J., and Storms, W.: Prevalence of allergenic rhinitis in the United States, J. Allergy Clin. Imm., 99, 808–814, 1997.

[2] WHO (World Health Organization), Phenology and human health: allergic disorders, Copenhagen, WHO Regional Office for Europe, 55 pp., 2003.

[3] Beggs, P. J.: Impacts of climate change on aeroallergens: past and future, Clin. Exp. Allergy, 34, 1507–1513, 2004.

[4] D'Amato, G., Cecchi, L., Bonini, S., Nunes, C., Annesi-Maesano, I., Behrendt, H., Liccardi, G., Popov, T., and van Cauwenber, P.: Allergenic pollen and pollen allergy in Europe, Allergy, 62, 976–990, 2007.

[5] Veriankaitė, L., Siljamo, P., Sofiev, M., Šauliene, I., and Kukkonen, J.: Modeling analysis of source regions of long-range transported birch pollen that influences allergenic seasons in Lithuania, Aerobiologia, 26, 47–62, 2010.

[6] Sofiev, M., Siljamo, P., Ranta, H., Linkosalo, T., Jaeger, S., Rasmussen, A., Rantio-Lehtimäki, A., Severova, E., and Kukkonen, J.: A numerical model of birch pollen emission and dispersion in the atmosphere. Description of the emission module, Int. J. Biometeorol., 57, 45–58, doi: 10.1007/s00484-012-0532-z, 2013.

[7] Zhang, R., Duhl, T., Salam, M.T., House, J.M., Flagan, R. C., Avol, E.L., Gilliland, F.D., Guenther, A., Chung, S. H., Lamb, B.K., and VanReken, T.M., Development of a regional-scale pollen emission and transport modeling framework for investigating the impact of climate change on allergic airway disease, Biogeosciences 11, 1461–1478, doi:10.5194/bg-11-1461-2014, 2014.

[8] M. Sicard, R. Izquierdo, M. Alarcón, J. Belmonte, A. Comerón, J. M. Baldasano, Near-surface and columnar measurements with a micro pulse lidar of atmospheric pollen in Barcelona, Spain, Atmospheric Chemistry and Physics, vol. 16, issue 11, pp. 6805-6821, doi: 10.5194/acp-16-6805-2016, 2016.

[9] M. Sicard, R. Izquierdo, O. Jorba, M. Alarcón, J. Belmonte, A. Comerón, J. M. Baldasano, Atmospheric dispersion of airborne pollen evidenced by near-surface and columnar measurements in Barcelona, Spain, Proc. SPIE 10001, 100010L, A. Comerón, E. I. Kassianov, K. Schäfer, J. W. Jack, R. H. Picard, K. Weber (Ed.), SPIE, Washington (EE.UU.), doi: 10.1117/12.2244517, Edinburgh, United Kingdom, 26 – 29 September 2016.

[10] Flynn, C. J., Mendoza, A., Zheng, Y., and Mathur, S., "Novel polarization-sensitive micropulse lidar measurement technique", Optics Express 15(6), 2785-2790 (2007).

[11] Shimizu, A., Sugimoto, N., Matsui, I., Arao, K., Uno, I., Murayama, T., Kagawa, N., Aoki, K., Uchiyama, A., and Yamazaki, A.: Continuous observations of Asian dust and other aerosols by polarization lidars in China and Japan during ACE-Asia, J. Geophys. Res., 109, D19S17, doi: 10.1029/2002JD003253, 2004.

[12] Pérez, C., Haustein, K., Janjic, Z., Jorba, O., Huneeus, N., Baldasano, J. M., Black, T., Basart, S., Nickovic, S., and Miller, R.L., "Atmospheric dust modeling from meso to global scales with the online NMMB/BSC-Dust model – Part 1: Model description, annual simulations and evaluation", Atmospheric Chemistry and Physics 11, 13001–13027 (2011).

[13] Jorba, O., Dabdub, D., Blaszczak-Boxe, C., Pérez, C., Janjic, Z., Baldasano, J. M., Spada, M., Badia, A., and Gonçalves, M., "Potential significance of photoexcited NO2 on global air quality with the NMMB/BSC chemical transport model", Journal of Geophysical Research 117(D13301), 1-16 (2012).

[14] Williams, C.G., "Conifer reproductive biology. Springer", New York, 167 pp. (2009).

[15] Bricchi, E., Frenguelli, G., Mincigrucci, G., "Experimental results about *Platanus* pollen deposition", Aerobiologia 16, 347–352 (2000).

[16] Jackson, S.T., and Lyford, M.E., "Pollen Dispersal Models in Quaternary Plant Ecology: Assumptions, Parameters, and Prescriptions", The Botanical Review 65(1), 39-75 (1999).

[17] Pérez, C., Nickovic, S., Baldasano, J. M., Sicard, M., Rocadenbosch, F., and Cachorro, V. E.: A long Saharan dust event over the western Mediterranean: Lidar, sun photometer observations, and regional dust modeling, J. Geophys. Res., 111, D15214, doi:10.1029/2005JD006579, 2006.

PROPERTIES OF HORIZONTALLY ORIENTED ICE CRYSTALS OBSERVED BY POLARIZATION LIDAR OVER SUMMIT, GREENLAND

RYAN R. NEELY III[1,2,*], ROBERT A. STILLWELL[3], SEBASTIAN COLE[1], JEFFREY P. THAYER[3]
MATTHEW SHUPE[2,4], MARISSA GOERKE[5], SAMUEL DORSI[5], JOSEPH ULANOWSKI[6]
[1]*School of Earth and Environment and the National Centre for Atmospheric Science, University of Leeds,*
**Presenting Author*
[2]*Cooperative Institute for Research in Environmental Science, University of Colorado*
[3]*Aerospace Engineering Sciences, University of Colorado,*
[4]*NOAA/Earth System Research Laboratory, Boulder, Colorado*
[5]*Polar Field Services, Littleton, Colorado*
[6]*University of Hertfordshire*

ABSTRACT

A source of error in microphysical retrievals and model simulations is the assumption that clouds are composed of only randomly oriented ice crystals. This assumption is frequently not true, as evidenced by optical phenomena such as parhelia. Here, observations from the Cloud, Aerosol and Polarization Backscatter Lidar at Summit, Greenland are utilized along with other sensors and beam imaging to examine the properties of horizontally oriented ice crystals and the environment conditions in which they occur.

1 INTRODUCTION

The microphysical properties of clouds play a significant role in determining their radiative effect; one of these properties is the orientation of ice crystals. A source of error in current microphysical retrievals and model simulations is the assumption that clouds are composed of only randomly oriented ice crystals (ROIC). This assumption is frequently not true, as evidenced by optical phenomena such as parhelia (commonly referred to as sundogs). Here, observations from the Cloud, Aerosol and Polarization Backscatter Lidar (CAPABL) at Summit, Greenland are utilized along with instruments that are part of the Integrated Characterization of Energy, Clouds, Atmospheric state and Precipitation at Summit (ICECAPS) project in order to determine when, where and under what conditions horizontally oriented ice crystals (HOIC) occur at Summit, Greenland.

2 METHOD

Summit, Greenland is located 3.2km above mean sea level on the Greenland Ice Sheet (GrIS) at 72.6°N, 38.5°W [1,2,3]. In 2010, the ICECAPS project established an intensive cloud observatory at the station [1]. ICECAPS includes a wide range of instrumentation with the ability to capture turbulent and thermodynamic properties of the atmosphere and clouds. Specific instrumentation included in the ICECAPS suite be used as part of this study includes the Cloud Aerosol Polarization and Backscatter Lidar (CAPABL), millimeter-wave cloud radar (MMCR) and twice-daily radiosonde launches [1].

In this work we build upon the initial results of [1], [2] and [3] to examine the occurrence of HOIC over Summit, Greenland using CAPABL and supporting observations. HOIC exhibit both non-zero values of depolarisation and diattenuation when observed in a non-zenith/non-nadir orientation (non normal to the crystal face) while ROIC will only exhibit non-zero values of depolarisation. CAPABL was developed using the lidar polarisation theory of Hayman and Thayer (2012) [4]. To identify HOIC, CAPABL makes observations of diattenuation and depolarisation [2, 3]. CAPABL accomplishes this by observing multiple planes of linear polarization, and photon counting and analog detection schemes, to self-evaluate, correct, and optimize signal combinations to improve cloud classification. For more information on how CAPABL implements the theory of Hayman and Thayer (2012) to make observations of HOIC please see [2, 3]. Using the measures of diattenuation and depolarisation ratio, CAPABL is the only instrument with the ability to unambiguously identify HOIC on an operational basis [2]. In addition to the novel polarimetric observations we also show results from opportunistic observations of images projected onto the snow-covered surface near CAPABL during the Arctic night that originate from scattering in CAPABL's transmitted beam. This provides a unique validation of CAPABL's primary observations.

2.1 HOIC Occurrence Categorisation

The main study period considers a 335-day period starting on July 2nd, 2015 and ending on May 31st, 2016. Throughout this period CAPABL sampled at a temporal resolution of 5s per polarisation and a spatial resolution of 7.5m. For an overall improved signal-to-noise ratio we have integrated this raw data to have a resolution of 30m and 20s (80 seconds for a complete 4 polarisation retrieval of diattenuation). Approximately 5 minutes of data per day was missed at midnight UTC to allow for system diagnostics [3]. Observations from lower than 0.2 km above ground level (AGL) were disregarded in order to minimize the influence of HOIC in fog/diamond dust and blowing snow as well a systemic effects due to the overlap of CAPABL's laser and telescope [3]. An additional day (December 6th, 2016) of HOIC observations will also be considered in the discussion as a unique phenomenon was captured using CAPABL's laser and digital photography that provides additional information about the crystal habits during HOIC events.
Having established the temporal and spatial domain of the study, criteria were set to identify the thermodynamic phase and orientation of the observed ice using the method outlined by [3]. Depolarisation values set for the identification of liquid and ice is consistent with previous studies of Arctic cloud by [5], [6] and [3]. Values of combined diattenuation were computed by multiplying CAPABL's two diattenaution observations together, which have opposite sensitivity to systematic effects, in order to remove false positives within the data as described in [3]. The threshold value of 0.005 used in the diattentuation criteria here is slightly more stringent (i.e. √0.005>.05) than the value used by [2].

2.2 Identification of Atmospheric Conditions

Values of temperature, relative humidity and horizontal wind speed were obtained from twice-daily radiosonde launches at Summit. Data for all variables from radiosondes were linearly interpolated to correspond to the range and time resolutions of measurements made by CAPABL. From this, observations were linearly interpolated with time for each day, creating a grid with the same spatial and temporal domain as the CAPABL observations but containing radiosonde observations.

2.3 Analysis of Temporal Variability

The frequencies of HOIC occurrence were calculated using all available days for each month of the study period. Frequencies of occurrence (F) of each atmospheric condition were calculated for each month using Equation 1 where N denotes the number of detections identified as a result of the masking process for HOIC, ice and liquid, and x is either HOIC, ice or liquid interchangeably.

$$F_x = \frac{N_x}{N_{HOIC} + N_{ROIC} + N_{Liq}} \times 100 \quad (1)$$

The number of detections of each data type identified as a result of the masking process was summed in order to establish the total number of quality-controlled measurements per month.

All days containing HOIC were examined and the nature of the occurrence was classified as cirrus, stratiform or falling/precipitating. These classifications were based on CAPABL's observation plus co-incident observations of reflectivity, vertical fall speeds and spectral width provided by the MMCR [1,7]. The role of liquid was not considered in the analysis of cirrus cloud as they are composed solely of ice crystals. Liquid was also not considered in the analysis of the falling/precipitating classification because liquid droplets within Arctic clouds are too small to be precipitated [8]. Having classified the occurrences of HOIC, the number of occurrences per classification was totalled. A day containing more than one classification was counted in each of the classifications within the day.

3 RESULTS OF POLARIMETRIC OBERVATIONS

HOIC were positively identified in cirrus cloud, stratiform cloud or precipitating from cloud on 86 days of the 335-day study period. Figure 1 shows an example product of the masking process and applied to CAPABL's data as well as the ancillary observations from the MMCR on April 22, 2016 from 00:00 to 24:00 UTC.
HOIC were observed on 3 occasions during this period. These occurrences were identified from the masking procedure (results shown in the

middle panel of Figure 1) and inspection of the corresponding radar observations in Figure 1. At ~16:00 there is substantial HOIC present and this has been separated into 3 distinct occurrences using the combination of CAPABL and NOAA MMCR observations shown in Figure 1: 1) at 4km a large pocket of HOIC is identified in cirris, 2) in small deck of liquid bearing stratiform cloud at 1.5km and 3) in precipitating ice near the surface. Following these occurences HOIC was also observed at 18:00 in precipitating/falling ice that is connected to the higher clouds. Finally, at 20:00 and into the next day, the cirrus clouds at 4 km show a strong HOIC signal that is typical of HOIC in cirrus found on other days. This method is used throughout the study to broadly classify the type of HOIC occurrence.

Combining the HOIC mask with the information provided by interpolated radiosonde profiles bounding the observaiton period (for this example for profiles from 12Z on April 22, 2016 and 00Z on April 23, 2016 were used) we create a pixel by pixel comparison of the conditions in which HOIC compared to ROIC occurred througout the day. For this example, we have not broken this into the classifications of cirrus, striatiform or precipitating ice due to the limited sample size. From the boxplots of Figure 2 it can be said the HOIC are occurring over a much smaller range of temperatures than ROIC, and a slightly higher temperature of -25.4°C as opposed to the -26.9°C of ROIC. The range of relative humidity (RH) over which HOIC are occurring is much wider than that of ROIC. HOIC also has a higher median value of 57.2% compared to that of the ROIC, 51.2%. HOIC also have a lower median horizontal wind speed of 18.5ms^{-1} as opposed to 19.6 ms^{-1} of ROIC. Further differences can be seen in the height ranges of both ROIC and HOIC. ROIC occurred over a larger vertical range than HOIC, and typically occurred higher in the atmosphere with a median height of 1.45 km, compared to the 0.96 km of HOIC.

The overall results of the occurrence and classification analysis are summarized by month in Figure 3. Examining the entire data set in a manner similar to Figure 2, we find that the temperature of HOIC in all cases is significantly higher than that of ROIC. We also find that the mean RHs are higher for HOIC than they are for ROIC in all cases. The mean horizontal wind speeds that HOIC occur under are always significantly higher than those of ROIC. Finally, the mean heights of HOIC in all cases were always significantly lower than that of ROIC.

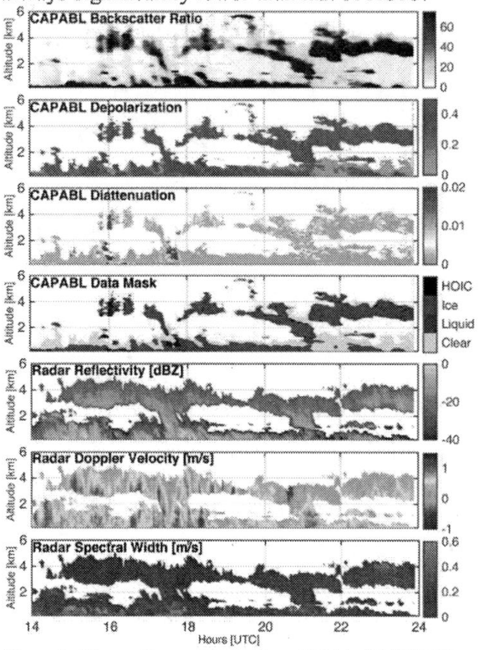

Figure 1: Observations made on from 14:00 to 24:00 UTC on April 22, 2016 from CAPABL and NOAA MMCR. Positive values in the Doppler velocity indicate a downward direction. The red colouring in the diattenuation plot exceed the 0.005 threshold for HOIC. HOIC occurs at 16:00 in cirrus at 4km, in stratiform cloud at 1.5km and in precipitating ice near the surface, at 18:00 in precipitating/falling ice and at 20:00 onwards in cirrus cloud at 4km.

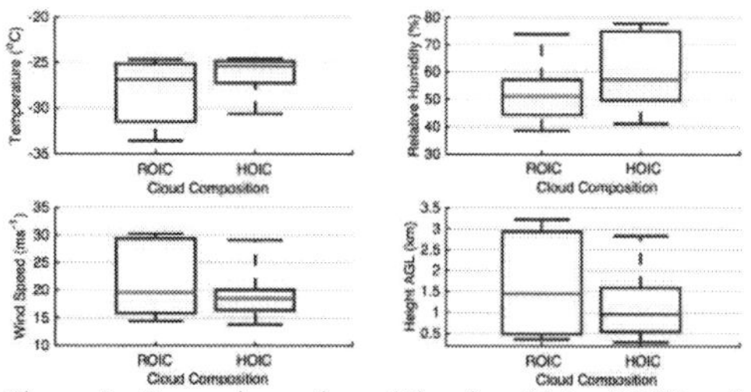

Figure 2: Comparison of condition in which ROIC and HOIC occurred in Figure 1 using CAPABL's masking from Figure 1 and interpolated radiosonde profiles launched 12Z on April 22, 2016 and 00Z on April 23, 2016. Boxes include the 25th to 75th percentiles and whiskers extend to the 10th and 90th percentiles.

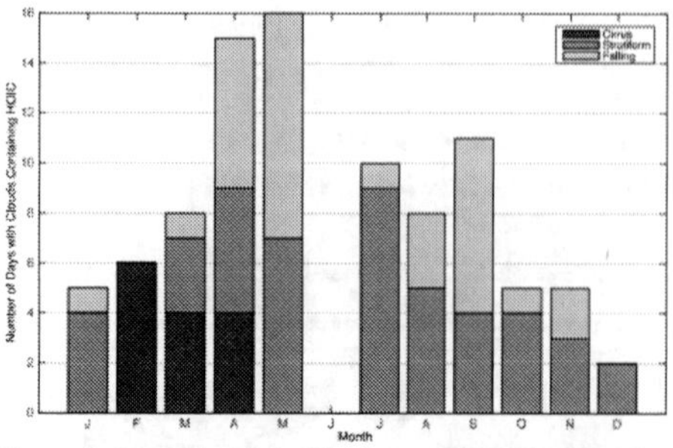

Figure 3: Monthly totals of days containing HOIC. Monthly totals are split into the 3 categories of cirrus, stratiform of falling/precipitated HOIC. June is not considered and therefore has no data.

4 UNIQUE VALIDATION OF CAPABL'S HOIC CLASSIFICATION

Ongoing work to understand HOIC at Summit includes recent observations of a phenomenon manifesting itself as succinct flashes of scattered light originating from CAPABL's transmit beam on the snow surface near the building housing the instrument (Figure 4). When photographed, the flashes reveal 2-D pattern similar to the forward scattering used to characterize cloud and aerosol particles. Using this information, theoretical 2-D patterns were computed using the ADDA discrete dipole approximation [9]. The simulations reveal that the flashes are originating from HOIC with large "flutter" angles (5°-8°). The modelling also reveals that the population of ice crystals consists of 100-300μm pristine hexagonal plates, as well as much larger (1-4mm) more irregular plates and columns. These observations qualitatively agree with CAPABL's and the MMCR's observations. More work will be done to make a quantitative comparison.

5 DISCUSSION AND CONCLUSIONS

The observations presented here provide the first documentation of the conditions under which HOIC occur at Summit, Greenland and are the first steps in the development of a climatology. This study has shed light on when and where HOIC are occurring but is subject to limitations such as poor resolution of the temperature and humidity observations used; these limitations however can be overcome with more and new observations. Future work will investigate the role of turbulence in the formation of HOIC by utilizing the observations from the MMCR. Sampling of temperature, water vapour and horizontal wind at a higher temporal resolution would also be beneficial for future research. Analysis of temporal variabilities in future work will extend and the study period used here and be used to explore the radiative properties of HOIC, especially their long wave effect.

Figure 4: Scattering from the CAPABL's transmit beam onto snow from HOIC. This image consisits of 30 superimposed images each at 1.3 s exposure time. Recorded on 6 December 2016. CAPABL is behind the camera.

ACKNOWLEDGEMENT

This material is based upon work supported by the National Science Foundation Graduate Research Fellowship Program under Grant No. DGE 1144083, National Science Foundation Grants PLR-1303864, PLR-1303879, PLR-1314156, and ATM-0454999. All the authors would like to especially thank the staff and science technicians at Summit Station as well as the entire team at Polar Field Services for their support and dedication to help maintain the instrumentation and collect the data presented in this work.

REFERENCES

[1] Shupe, M. D., et al.: BAMS, 94, 169–186, 2013.
[2] Neely, R. R., et al.: JTECH, 1635-1655, 2013.
[3] Stillwell, R. A., et al.: Atmos. Meas. Tech. Discuss., in review, 2016.
[4] Hayman and Thayer: JOSAA, 400-409, 2012.
[5] Intrieri, J., et al.: J. Geo. Res., 2002.
[6] Shupe, M. D., and Uttal, T.: J. App. Met., 1544-1562, 2005.
[7] Moran, K.P., et al.: BAMS, 79, 443-455, 1998.
[8] Morrison, H., et al.: Nat. Geo., 5, 11–17, 2012.
[9] Yurkin, M. A. and Hoekstra, A. G.: J. Quantit. Spectr. Rad. Transf. **112**, 2234-2247 (2011).

A NOVEL AUTOMATIC CALIBRATION SYSTEM FOR WATER VAPOR RAMAN LIDAR

Giovanni Martucci[1]*, Jordan Voirin[1,2], Valentin Simeonov[2], Ludovic Renaud[1], and Alexander Haefele[1]

[1] *Federal Office of Meteorology and Climatology, MeteoSwiss, Payerne, Switzerland,*
**Giovanni.martucci@meteoswiss.ch*
[2] *Ecole Polytechnique Fédérale de Lausanne - EPFL, CH-1015 Lausanne, Switzerland*

ABSTRACT

RALMO is operated at MeteoSwiss and provides continuous measurements of water vapor since 2008. Water vapor needs calibration to be used for climatological and meteorological studies. Primarily, RALMO is calibrated against a reference radiosonde. A secondary new calibration system based on an internal LED has been installed to perform automatic relative calibrations correcting for instrumental drifts. This study characterizes the secondary calibration system and validates it against the primary.

1 INTRODUCTION

The RAman Lidar for Meteorological Observations (RALMO) is operated at the Aerological station at Payerne (46°480 N, 6°560 E, 491ma.s.l.) by Federal Office of Meteorology and Climatology MeteoSwiss in collaboration with the École Polytechnique Fédérale de Lausanne (EPFL) and provides continuous measurements of water vapor, temperature and aerosols almost uninterruptedly since 2008 [1],[2]. This long time series needs careful calibration to be used for climatological studies and operational meteorology. RALMO detects the Raman scattered light from atmospheric water vapor (H_2O) and nitrogen (N_2), which allows to derive the water vapor mixing ratio (ω). The profiles of ω obtained by RALMO are calibrated using collocated radiosounding. The calibration factor obtained by scaling the ω profile obtained by RALMO to the ω profile from the radiosounding is corrected automatically by a system that accounts for the uneven aging (f_{aging}) of the N_2 and H_2O photomultipliers. The expression for ω can be obtained rearranging the terms in the LIDAR equation at each altitude z for H_2O and N_2:

$$\omega(z) = C\frac{E_{H2O}(z)}{E_{N2}(z)}\frac{T_{H2O}(z,\lambda_{H2O})}{T_{N2}(z,\lambda_{N2})} \quad (1)$$

Where E_x is the received energy, λ_x is the Raman-shifted wavelength and T_x is the atmospheric transmission at the Raman-shifted wavelengths. The factor C is the system calibration factor and it depends on:

$$C = \frac{\xi(\lambda_{N2})\gamma(\lambda_{N2})\sigma_{N2}}{\xi(\lambda_{H2O})\gamma(\lambda_{H2O})\sigma_{H2O}} \quad (2)$$

Where ξ is the acquisition system's optical efficiency at the wavelength λ, $\gamma(\lambda_x)$ is the detectors' sensitivity and σ_x is the Raman-backscattering cross section. The ratio of transmissions $T_{H2O}(z,\lambda_{H2O})/T_{N2}(z,\lambda_{N2})$ can be approximated to a constant value in the troposphere accounting for a departure of 5-10% from unity due essentially to the unknown aerosol distribution [3].

2 INSTRUMENT DESCRIPTION

RALMO has been designed based on the following specifications: a narrow receiver's field of view; a narrowband grating polychromator that isolates the ro-vibrational Q-branches and high laser energy allowing to minimize the contamination by solar background and to maximize the signal-to-noise ratio; a near real-time data availability; long-term data consistency and system stability; autonomous operation; minimal maintenance and eye safety to ensure the possibility to measure without interfering with the surroundings. Such a system ensures high accuracy and precision of the measured parameters, the traceability of the measurement and high data availability since 2008. Table 1 summarizes the system parameters as it is described more in detail by Dinoev et al., 2013 [1]. RALMO laser's source emits at 1064 nm its fundamental harmonic, it then uses the third harmonic at 355 nm for measurements. A beam expander reduces the beam divergence at the instrument's output and the expanded beam is transmitted into the atmosphere. The light backscattered by the atmosphere is collected by the LIDAR's telescope (4 mirrors of 30 cm diameter) and transmitted by optical fibers separately to the H_2O and Aerosol/Temperature polychromators for the wavelength separation. The photomultipliers detect the optical signals and convert them into electrical signals, which are acquired and stored for the data analysis.

Table 1 RALMO system parameters

Transmitter	Receiver
Nd:YAG laser – third harmonic: Wavelength: 354.7 nm Spectral line width: 0.7 cm−1 Rep. rate: 30 Hz Pulse energy: 300 mJ Pulse duration: 8 ns *Beam expander – Galilean type:* Expansion ratio 15× *Transmitted beam:* Divergence: calculated 0.06 mrad measured 0.09 ± 0.02 mrad Diameter: 140 mm	*Four fiber-coupled parabolic mirrors:* Focal length: 1m Diameter: 0.3m Axial displacement: 235mm FOV: 3 x 0.20 and 1 x 0.22 mrad *Diffraction grating:* Bandwidth (FWHM): 0.33 nm Central wavelength H2O 407.45 nm, N2 386.7 nm Efficiency: 33% (peak at H2O) *Photodetectors (Hamamatsu)* H2O R7600U-200 N2 and O2H6780

3 METHODOLOGY

We define two types of calibration: the primary calibration ("external") made by collocated radiosounding and the secondary ("internal") made by an internal stabilized UV Light Emitting Diode (LED) source shining onto the N_2 and H_2O photomultipliers (PMTs). As a result of the external calibration, RALMO is scaled to the radiosounding $\boldsymbol{\omega}$ profile and the calibration factor C (eq.2) is obtained. The internal calibration system allows performing automatic relative corrections of C that account for the differential aging f_{aging} of the N_2 and H_2O PMTs. A single LED source is installed into the polychromator unit, shone onto the N_2 and H_2O PMTs and acquired by Licel acquisition system at 100 Hz. The two PMTs are Hamamatsu photomultipliers: H6780-04 for the N_2 signal and a R7600U-200 for the H_2O. The LED illuminates the same regions of the PMT's photocathodes that are illuminated by the atmospheric signal during normal measurements. This is essential for the accuracy of the calibration correction because of the high spatial inhomogeneity of the PMTs sensitivity [4]. The original design of the internal calibration procedure is provided by Simeonov et al., 2014, [5], here we provide a description of its latest development along with the main equations on which the method is based. The differential aging, f_{aging}, causes a drift in the calibration factor in eq. 2. Due to f_{aging}, the true signal, S_{true}, of the atmospheric compound X, is not equal to the measured signal $S_{X,\,obs}$. The true signal can then be expressed as,

$$S_{X,true}(t) = f_{aging,X}(t)\, S_{X,obs}(t) \tag{3}$$

The equation 1 can be rewritten in the following way,

$$\boldsymbol{\omega}(t) = C\frac{S_{H_2O,true}(t)}{S_{N_2,true}(t)} = C\frac{f_{H_2O}(t)}{f_{N_2}(t)}\frac{S_{H_2O,obs}(t)}{S_{N_2,obs}(t)} = C_{ext}(t)\frac{S_{H_2O,obs}(t)}{S_{N_2,obs}(t)} \tag{4}$$

C_{ext} is the calibration factor obtained by comparison with the radiosounding. If the two PMTs were aging at the same rate ($f_{aging} = 1$) then the ratio of their sensitivity would not change in time. The true LED signal ratio of H_2O/N_2 must then be constant and can be written as,

$$A_{LED} = \frac{S_{H_2O,LED,true}(t)}{S_{N_2,LED,true}(t)} = \frac{f_{H_2O}(t)}{f_{N_2}(t)}\frac{S_{H_2O,LED,obs}(t)}{S_{N_2,LED,obs}(t)} \tag{5}$$

By inversion of (5) the differential aging correction ratio is,

$$\frac{f_{H_2O}(t)}{f_{N_2}(t)} = A_{LED}\frac{S_{N_2,LED,obs}(t)}{S_{H_2O,LED,obs}(t)} \tag{6}$$

We can then rewrite the water vapor mixing ratio:

$$\boldsymbol{\omega}(t) = C\,A_{LED}\frac{S_{N_2,LED,obs}(t)}{S_{H_2O,LED,obs}(t)}\frac{S_{H_2O,obs}(t)}{S_{N_2,obs}(t)} \tag{7}$$

$$= C'\frac{S_{N_2,LED,obs}(t)}{S_{H_2O,LED,obs}(t)}\frac{S_{H_2O,obs}(t)}{S_{N_2,obs}(t)}$$

$$= C' r_{aging}(t)\frac{S_{H_2O,obs}(t)}{S_{N_2,obs}(t)}$$

Comparing equations (4) and (7), at time zero, t_0,

$$C' = \frac{C_{ext}(t_0)}{r_{aging}(t_0)} \tag{8}$$

Once the calibration factor is determined, we can calculate the water vapor mixing ratio at any time $t \geq t_0$,

$$\omega(t) = C_{ext}(t_0)\frac{r_{aging}(t)}{r_{aging}(t_0)}\frac{S_{H_2O,obs}(t)}{S_{N_2,obs}(t)} \quad (9)$$

The corrected calibration factor, C^*, is then:

$$C^*(t) = \frac{r_{aging}(t)}{r_{aging}(t_0)} C_{ext}(t_0) \quad (10)$$

4 RESULTS

After having defined the corrected calibration factor, C^*, we validate the time series of $C^*(t)$ against the external calibration method. The validation consists of comparing eq.(10) with the external calibration factors $C_{ext}(t_i)$ at specific times $t_i > t_0$. When the method based on eq. (10) is used operationally, the LIDAR profiles of ω are calibrated automatically. It is then important that, during the assessment phase, the time series $C^*(t)$ matches the reference values $C_{ext}(t_i)$. Figure 1 shows the comparison between the time series $C^*(t)$ (yellow) and the reference $C_{ext}(t_i)$ values since July 2014. The $C_{ext}(t_i)$ values are obtained by calibration with different radiosondes: the operational SRS-C34 and SRS-C50 (black), the RS92 (red), the RS41 (blue) and the SnowWhite (green).

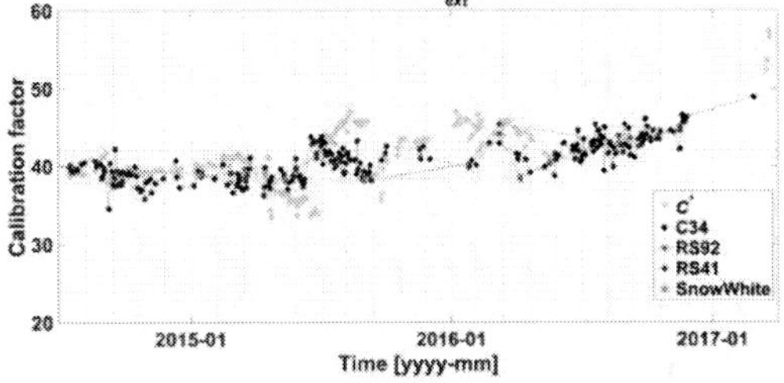

Figure 1 timeseries of the corrected calibration factor $C^*(t)$ (yellow) vs radiosounding $C_{ext}(t_i)$ calibrations based on LED ratio.

The timeseries of $C^*(t)$ does not match perfectly the timeseries of the $C_{ext}(t_i)$ points. Especially during the year 2015 differences up to 20% can be observed between $C^*(t)$ and $C_{ext}(t_i)$. In order to be a reference for $C^*(t)$, the $C_{ext}(t_i)$ points must be determined accurately. The $C_{ext}(t_i)$ factors are obtained by selecting automatically the optimal vertical layers along the ω profile where the agreement between RALMO and the radiosounding has a correlation coefficient > 0.6. The selected optimal region (composed of multiple layers) is also a function of the distance of the radiosonde from RALMO and of the homogeneity of the atmosphere between the two. Figure 2 shows an example of automatic calibration with the selection of the optimal ranges in blue dots.

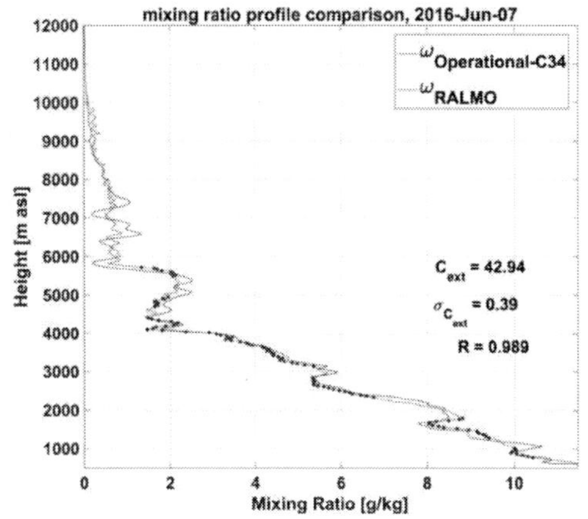

Figure 2 automatic selection of optimal region of RALMO vs RS.

All points respecting the condition R > 0.6 are linearly fitted and the slope of the fit gives the value of C_{ext}. For the case shown in figure C_{ext} = 42.94, the correlation coefficient is R = 0.99 and the fit error is 1%. Similar procedures to perform an automatic calibration based on external ω profiles are described by [6],[7], [8]. Analyzing the $C^*(t)$ and $C_{ext}(t_i)$ data in Figure 1, we cannot be satisfied with the reproducibility of the $C_{ext}(t_i)$ points by the $C^*(t)$ timeseries. Different reasons can make the use of the internal calibration difficult, first of all the strong sensitivity of the internal calibration system to temperature changes along the optical path. It appears that small changes in the room temperature translate into changes of the incident angle of the LED light shining onto the PMT's leading to the variability of the r_{aging} term. One way to avoid temperature dependencies of the automatic calibration system is to use the ratio of the solar background at the N_2 and H_2O wavelengths instead. The sun photons that travel towards RALMO are measured as a reference (same atmospheric path through the year) when the sun Zenith angle is at 20°, i.e. the maximum elevation angle on the 21st of

December at Payerne. The solar background (BKG) ratio replaces the LED ratio in the above equations and the new r_{aging} factor becomes:

$$r_{aging} = \frac{S_{N_2,BKG,obs}(t)}{S_{H_2O,BKG,obs}(t)} \qquad (11)$$

Using the BKG it allows to go back to the beginning of the data timeseries in 2008, while the LED data are available only since 2014. The new timeseries of $C^*(t)$ based on BKG is shown in Figure 3.

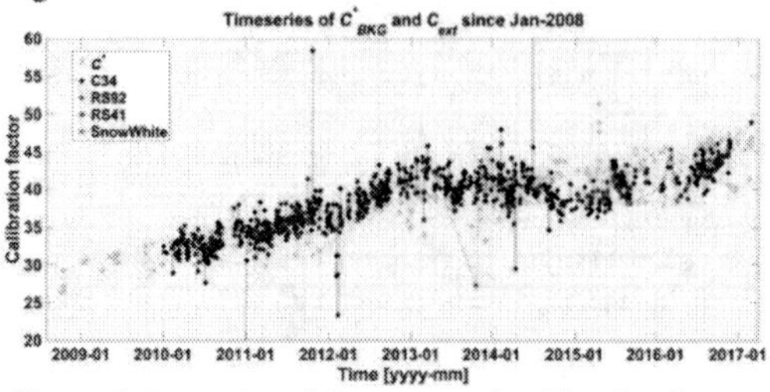

Figure 3 timeseries of the corrected calibration factor $C^*(t)$ (yellow) vs radiosounding $C_{ext}(t_i)$ calibrations based on BKG ratio.

5 CONCLUSIONS

An automatic procedure to calibrate the water vapor mixing ratio (**ω**) retrieved from RALMO has been developed and tested at the Aerological station of MeteoSwiss, Payerne. The method consists of the automatic correction of the calibration factor C obtained by external calibration of RALMO **ω** profile by comparison with radiosounding. The correction expressed by eq.(10) is based on an automatic monitoring of the differential aging of the N_2 and H_2O PMTs. Two different sources have been tested to calculate the differential aging of the PMTs: a LED (eq.6) and the solar background (eq.11). The C^* values obtained using the LED showed a limited reproducibility of the C_{ext} factors, essentially due to temperature dependence of the calibration system. The C^* values obtained using the BKG showed a better agreement with the C_{ext} factors and provided the possibility to extend the procedure over the entire RALMO datasets.

ACKNOWLEDGEMENTS

Dr. Shlomo Festig from Soreq Nuclear Research Center, Israel, has significantly contributed to the design and creation of the new internal calibration procedure.

References

[1] Dinoev, T., Simeonov, V., Arshinov, Y., Bobrovnikov, S., Ristori, P., Calpini, B., Parlange, M., and van den Bergh, H.: Raman Lidar for Meteorological Observations, RALMO – Part 1: Instrument description, *Atmos. Meas. Tech.*, **6**, 1329–1346, doi:10.5194/amt-6-1329-2013, 2013

[2] Brocard, E., Philipona, R., Haefele, A., Romanens, G., Ruffieux, D., Simeonov, V., and Calpini, B.: Raman Lidar for Meteorological Observations, RALMO – Part 2: Validation of water vapor measurements, *Atmos. Meas. Tech.*,**6**, 1347–1358, doi:10.5194/amt-6-1347-2013, 2013.

[3] Whiteman, D. N.: Raman lidar system for the measurement of water vapor and aerosols in the Earth's atmosphere, *Appl. Opt.*, **31**, 3068–3082, 1992.

[4] Simeonov V., Larcheveque G., Quaglia P., van den Bergh H., and Calpini B., "The influence of the PMT spatial uniformity on lidar signal," Appl.Opt, 38, 5186-5190 (1999).

[5] Simeonov V., Fastig S., Haefele A., and Calpini, B.: Instrumental correction of the uneven PMT aging effect on the calibration constant of a water vapor Raman lidar, *Proc. SPIE* **9246**, Lidar Technologies, Techniques, and Measurements for Atmospheric Remote Sensing X, 92460N (October 20, 2014); http://dx.doi.org/10.1117/12.2066802

[6] Dionisi, D., Congeduti, F., Liberti, G. L., and Cardillo, F.: Calibration of a Multichannel Water Vapor Raman Lidar through Noncollocated Operational Soundings: Optimization and Characterization of Accuracy and Variability, *J. Atmos. Ocean. Tech.*, **27**, 108–121, doi:10.1175/2009JTECHA1327.1, 2010.

[7] Foth, A., Baars, H., Di Girolamo, P., and Pospichal, B.: Water vapour profiles from Raman lidar automatically calibrated by microwave radiometer data during HOPE, *Atmos. Chem. Phys.*, **15**, 7753-7763, doi:10.5194/acp-15-7753-2015, 2015.

[8] Hicks, S., Sica, J. R., and Martucci, G.: Calibration of the MeteoSwiss RAman Lidar for Meteorological Observations (RALMO) Water Vapour Mixing Ratio Measurements using a Radiosonde Trajectory Method, contribution to ILRC28, Bucharest, 2017

LIDAR NETWORK OBSERVATION OF DUST LAYER EVOLUTION OVER THE GOBI DESERT IN MAY 2013

Kei Kawai[1]*, Kenji Kai[1], Yoshitaka Jin[2], Nobuo Sugimoto[2], Dashdondog Batdorj[3]

[1]*Nagoya University, Japan, *kawai.kei@e.mbox.nagoya-u.ac.jp*
[2]*National Institute for Environmental Studies, Japan*
[3]*National Agency for Meteorology and Environmental Monitoring, Mongolia*

ABSTRACT

A lidar network captured the evolution of a dust layer in the Gobi Desert on 22–23 May 2013. The lidar network consists of a ceilometer and two AD-Net lidars in Mongolia. The dust layer was generated by a strong wind due to a cold front and elevated over the surface of the cold front by an updraft of the warm air in the cold-front system. It was evolving from the atmospheric boundary layer to the free troposphere while moving 600 km through the desert with the cold front.

1 INTRODUCTION

Asian dust originates in arid and semi-arid regions of East Asia such as the Gobi Desert and the Taklimakan Desert [1, 2]. A strong wind raises the dust from the ground to the atmospheric boundary layer (ABL), which is the bottom layer of the atmosphere. If the dust reaches the free troposphere (FT) over the ABL, it can be transported over a long range by westerlies [3, 4, 5]. However, the capping inversion between the ABL and the FT prevents the dust in the ABL from reaching the FT. The mechanism to elevate the dust from the ABL to the FT is not clear due to little observational evidence.

Some studies have suggested that an updraft of a cold-frontal system elevates dust from the ABL to the FT by using a space-borne lidar and numerical models [6, 7]. Moreover, our previous study has shown this fact by using a ceilometer installed in the central part of the Gobi Desert [8]. In addition to the ceilometer, two AD-Net lidars have been operated in the eastern part of the desert [9]. The ceilometer and the two lidars constitute a lidar network in the desert. The present study takes advantage of the lidar network and reveals the process that the cold-frontal system elevates dust from the ABL to the FT.

2 OBSERVATIONS AND DATA

A lidar network in the Gobi Desert consists of a ceilometer and two AD-Net lidars. The ceilometer is installed at Dalanzadgad (43.58°N, 104.42°E, 1470 m above sea level (ASL); D in Fig. 1), which is in the central part of the desert. The AD-Net lidars are placed at Sainshand (44.87°N, 110.12°E, 937 m ASL; S in Fig. 1) and Zamyn-Uud (43.72°N, 111.90°E, 962 m ASL; Z in Fig. 1), which are in the eastern part of the desert. The distances among three sites are 480 km (D–S), 600 km (D–Z), and 190 km (S–Z).

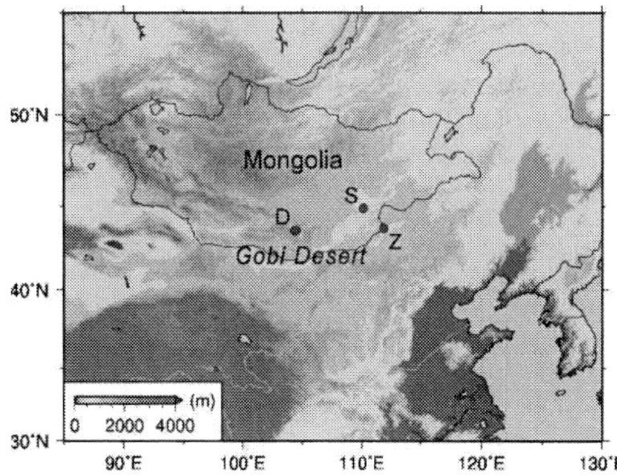

Figure 1 Topographic map around Mongolia. The red-filled circles with the characters of D, S, and Z indicate the locations of Dalanzadgad, Sainshand, and Zamyn-Uud, respectively.

The ceilometer at Dalanzadgad is a Vaisala CL51 ceilometer, which employs a diode laser of a 910-nm wavelength as the light source. It outputs a vertical profile of attenuated backscatter coefficient with a height resolution of 10 m every 6 seconds. The present study used the observation data averaged in 15 minutes and 30 m. The ceilometer was installed in Tsukuba, Japan, before moved to Dalanzadgad and corrected by a dual-wavelength lidar [10].

AD-Net is a lidar observation network for Asian dust and other aerosols in East Asia [9]. The AD-Net lidars at Sainshand and Zamyn-Uud obtain vertical profiles of attenuated backscatter coefficients at wavelengths of 532 nm and 1064 nm and volume depolarization ratio at a wavelength of 532 nm. The time and height resolutions of the parameters are 15 minutes and 30 m, respectively.

The present study used surface weather charts produced by Japan Meteorological Agency and surface meteorological data provided from the SYNOP reports. The present weather codes of 06–09, 30–35, and 98 are regarded as dust events which consist of dust storm, floating dust, and blowing dust [11].

3 RESULTS AND DISCUSSION

A dust event occurred in the Gobi Desert on 22–23 May 2013. Vertical distribution of dust was successfully observed by the lidar network in the desert during the dust event. The local standard time (LST) in Mongolia is 8 hours ahead of the coordinated universal time (UTC).

3.1 Pressure pattern

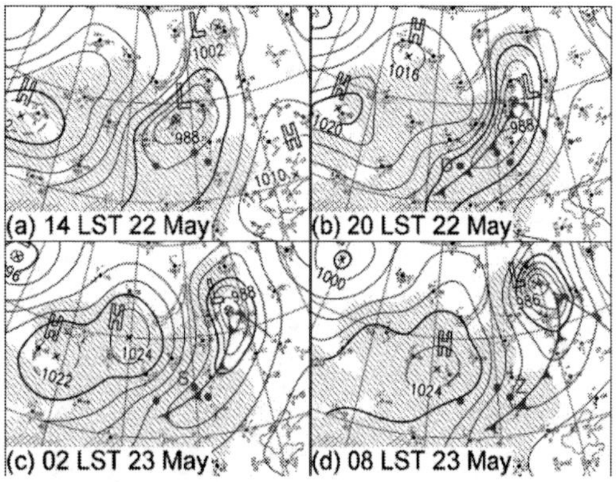

Figure 2 Surface weather charts around Mongolia on 22–23 May 2013 (Japan Meteorological Agency). The red dots with the characters of D, S, and Z indicate the locations of Dalanzadgad, Sainshand, and Zamyn-Uud, respectively.

Figure 2 presents surface weather charts around Mongolia on 22–23 May 2013. A low pressure was in northern Mongolia at 14 LST on 22 and moved northeastward until 08 LST on 23. The low pressure was accompanied with cold and warm fronts from 20 LST on 22. The fronts were generated between 14 LST and 20 LST on 22. The cold front moved southeastward across the Gobi Desert with a speed of 30 km/h. It passed Dalanzadgad between 14 LST and 20 LST on 22, Sainshand between 20 LST on 22 and 02 LST on 23, and Zamyn-Uud between 02 LST and 08 LST on 23.

3.2 Meteorological field

Figure 3 indicates surface winds and dust phenomena over Mongolia at 20 LST on 22 May 2013. Strong northwest winds were observed in the central part of the Gobi Desert. The wind speed ranged from 6 m/s to 16 m/s. The strong winds were located behind and along the cold front. In addition, dust phenomena were observed in the region of the strong winds. In summary, the dust outbreak during this dust event was caused by the strong winds which were generated by the cold front.

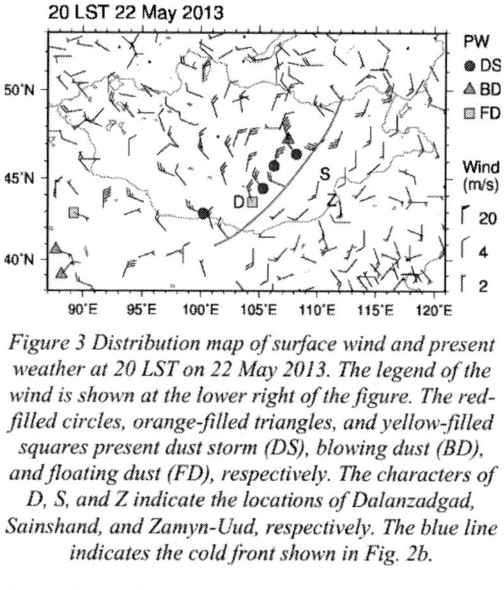

Figure 3 Distribution map of surface wind and present weather at 20 LST on 22 May 2013. The legend of the wind is shown at the lower right of the figure. The red-filled circles, orange-filled triangles, and yellow-filled squares present dust storm (DS), blowing dust (BD), and floating dust (FD), respectively. The characters of D, S, and Z indicate the locations of Dalanzadgad, Sainshand, and Zamyn-Uud, respectively. The blue line indicates the cold front shown in Fig. 2b.

3.3 Dust elevation mechanism

Figure 4 presents the observation results at Dalanzadgad, Sainshand, and Zamyn-Uud from 14 LST on 22 to 18 LST on 23 in May 2013. At Dalanzadgad, a dust layer is shown by large attenuated backscatter coefficient over the ground between 14 LST on 22 and 02 LST on 23 (Fig. 4a).

Judging from each vertical profile (not shown), the top height of the dust layer reached 1.8 km at 20 LST on 22. The dust layer was the densest at 21 LST on 22. During the dust event, the wind speed ranged from 6 m/s to 10 m/s, and the wind direction was between the west and the northwest. This strong wind raised the dust from the ground. The cold front passed between 17 LST and 20 LST on 22 because the wind speed increased though the wind-direction change was not clear. Cold air advected over the ground below a height of 1.8 km after the cold front passed. The cold air included little dust from 2230 LST on 22. The leading edge of the cold air was not clear because the dust layer extended from the warm air to the cold air. The top of the cold air corresponds to the surface of the cold front. Some dust floated at a height range of 0.8–1.6 km between 22 LST on 22 and 00 LST on 23. It was located over and along the surface of the cold front. Our previous study has shown that the dust was elevated by an updraft of the warm air in the cold-frontal system [8].

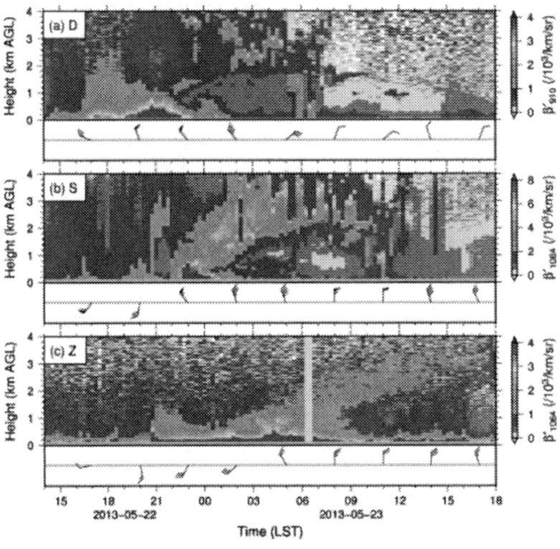

Figure 4 Time-height cross sections of attenuated backscatter coefficient (each upper) and time series of surface wind (each lower) on 22–23 May 2013 at (a) Dalanzadgad (D), (b) Sainshand (S), and (c) Zamyn-Uud (Z). A half and full barb and a pennant indicate the wind speeds of 1, 2, and 10 m/s, respectively.

3.4 Extension of dust distribution

At Sainshand, the cold front passed between 20 LST and 23 LST on 22 May 2013 as shown by the wind-direction change from the south to the northwest (Fig. 4b). The wind speed increased from 6 m/s to 11 m/s during the wind-direction change. A dust layer was observed near the ground from 22 LST on 22 to 02 LST on 23. After that, cold air advected over the ground with little dust. The depth of the cold air gradually increased from 1.0 km to 2.2 km until 07 LST on 23. As with Dalanzadgad, the dust was elevated by the updraft of the warm air along the surface of the cold front. The dust formed a layer with a vertical thickness of about 2 km. The distribution area of the dust was much larger than that at Dalanzadgad. The dust layer reached the free troposphere and mixed with clouds.

At Zamyn-Uud, the cold front passed between 02 LST and 05 LST on 23 as shown by the wind-direction change from the southwest to the north (Fig. 4c). As with Sainshand, a dust layer was elevated by the updraft of the warm air along the surface of the cold front. The vertical thickness of the dust layer was about 1 km and smaller than that at Sainshand. The dust layer also reached the free troposphere and mixed with clouds.

Figure 5 illustrates an observational model of dust elevation and extension during the dust event. The dust distribution along the surface of the cold front was larger at Sainshand and Zamyn-Uud than that at Dalanzadgad. The dust layer was evolving from the atmospheric boundary layer to the free troposphere while moving across the desert with the cold front. This extension of the dust layer was caused by the updraft of the warm air in the cold-frontal system.

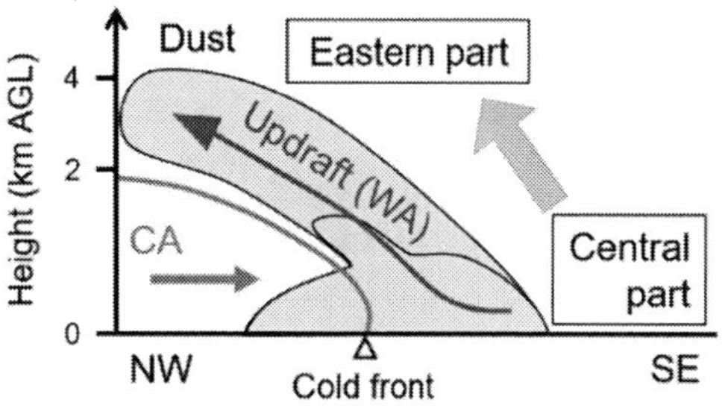

Figure 5 Observational model of dust elevation and extension in the Gobi Desert on 22–23 May 2013. CA and WA mean cold air and warm air, respectively.

4 CONCLUSIONS

The Gobi Desert lidar network captured the evolution of a dust layer over the source region on 22–23 May 2013. The lidar network consists of a ceilometer and two AD-Net lidars in the central and eastern parts of the desert, respectively. The AD-Net lidars are 480 km and 600 km away from the ceilometer. During the dust event, a cold front associated with a low pressure moved southeastward across the desert with a speed of 30 km/h. The cold front generated strong northwest winds (6–16 m/s), which raised dust from the ground and formed a dust layer in the atmospheric boundary layer. The dust layer was elevated along the surface of the cold front by an updraft of the warm air in the cold-frontal system. The updraft extended the dust layer to the free troposphere while the cold-frontal system was moving across the desert.

ACKNOWLEDGEMENTS

The present study is supported by Grant-in-Aid for Scientific Research from Japan Society for the Promotion of Science (JSPS) (No. 16H02703) and by JSPS Core-to-Core Program (B. Asia-Africa Science Platform).

REFERENCES

[1] Sun, J., Zhang, M., Liu, T., 2001: Spatial and temporal characteristics of dust storms in China and its surrounding regions, 1960-1999: Relations to source area and climate, *J. Geophys. Res.*, **106**, 10325-10333.

[2] Wu, J., Kurosaki, Y., Shinoda, M., Kai, K., 2016: Regional characteristics of recent dust occurrence and its controlling factors in East Asia, *Sci. Online Lett. Atmos.*, **12**, 187-191.

[3] Kai, K., Okada, Y., Uchino, O., Tabata, I., Nakamura, H., Takasugi, T., Nikaidou, Y., 1988: Lidar observation and numerical simulation of a Kosa (Asian Dust) over Tsukuba, Japan during the Spring of 1986, *J. Meteor. Soc. Japan*, **66**, 457-472.

[4] Uno, I., Eguchi, K., Yumimoto, K., Takemura, T., Shimizu, A., Uematsu, M., Liu, Z., Wang, Z., Hara, Y., Sugimoto, N., 2009: Asian dust transported one full circuit around the globe, *Nature Geosci.*, **2**, 557-560.

[5] Yumimoto, K., Eguchi, K., Uno, I., Takemura, T., Liu, Z., Shimizu, A., Sugimoto, N., 2009: An elevated large-scale dust veil from the Taklimakan Desert: Intercontinental transport and three-dimensional structure as captured by CALIPSO and regional and global models, *Atmos. Chem. Phys.*, **9**, 8545-8558.

[6] Hara, Y., Yumimoto, K., Uno, I., Shimizu, A., Sugimoto, N., Liu, Z., Winker, D.M., 2009: Asian dust outflow in the PBL and free atmosphere retrieved by NASA CALIPSO and an assimilated dust transport model, *Atmos. Chem. Phys.*, **9**, 1227-1239.

[7] Alizadeh, C.O., Zawar-Reza, P., Sturman, A., 2012: Atmospheric forcing of the three-dimensional distribution of dust particles over Australia: A case study, *J. Geophys. Res.*, **117**, D11206.

[8] Kawai, K., Kai, K., Jin, Y., Sugimoto, N., Batdorj, D., 2015: Dust event in the Gobi Desert on 22–23 May 2013: Transport of dust from the atmospheric boundary layer to the free troposphere by a cold front, *Sci. Online Lett. Atmos.*, **11**, 156–159.

[9] Sugimoto, N., Matsui, I., Shimizu, A., Nishizawa, T., Hara, Y., Xie, C., Uno, I., Yumimoto, K., Wang, Z., Yoon, S.C., 2008: Lidar network observations of tropospheric aerosols, *Proc. SPIE*, **7153**, 71530A.

[10] Jin, Y., Kai, K., Kawai, K., Nagai, T., Sakai, T., Yamazaki, A., Uchiyama, A., Batdorj, D., Sugimoto, N., Nishizawa, N., 2015: Ceilometer calibration for retrieval of aerosol optical properties, *J. Quant. Spectrosc. Radiat. Transfer*, **153**, 49–56.

[11] World Meteorological Organization (WMO), 1995: Manual on Codes, International Codes, Vol. I.1 (Annex II to WMO Technical Regulations) Part A–Alphanumeric Codes, *WMO*, Geneva, Switzerland, 492pp.

EPJ Web of Conferences 176, 05010 (2018) https://doi.org/10.1051/epjconf/201817605010
ILRC 28

FLUORESCING AEROSOLS AND CLOUDS: INVESTIGATIONS OF CO-EXISTENCE

Jens Reichardt[1,*], Ronny Leinweber[1] and Anne Schwebe[1]

[1]*Richard-Aßmann-Observatorium, Deutscher Wetterdienst, Am Observatorium 12, 15848 Lindenberg, Germany, *jens.reichardt@dwd.de*

ABSTRACT

RAMSES of the Lindenberg Meteorological Observatory, Germany, is the first multipurpose lidar to routinely measure the fluorescence spectra of atmospheric aerosols. Combined with the other measurement parameters (cloud water content and optical properties, moisture and temperature), this capability allows one to study the co-existence of clouds and fluorescing aerosols for the first time. The fluorescence receiver is briefly described, and measurement examples are presented and discussed.

1 INTRODUCTION

Its scientific relevance notwithstanding, lidar studies of cloud-aerosol co-existence and, prospectively, interaction have not been attempted so far because conventional elastic-backscatter or Raman lidars cannot distinguish between scattering from different particles, and so the aerosol signal is masked in the presence of clouds. With its new fluorescence spectrometer, the spectroscopic Raman lidar RAMSES [1, 2] is the first instrument to make such a distinction: Pure condensed water does not fluoresce while experience shows aerosols in the vast majority of cases do. So fluorescence observed within clouds unveils the occurrence of aerosol particles. In this contribution the integration of the fluorescence spectrometer into the RAMSES instrument is outlined. Furthermore, aerosol observations are discussed, particularly with regard to the interdependence between aerosol type, spectral characteristics and source region; and with regard to the co-existence of aerosols and clouds.

2 FLUORESCENCE SPECTROMETER

Similarly to the water spectrometer [2], the fluorescence spectrometer has been added to the RAMSES far-range receiver. Like the former, it consists of four main components, (1) the fiber coupler which images the atmospheric return signals transmitted by a dichroic beamsplitter used to suppress visible light in the vibrational-rotational Raman detection channels (BS7, see Fig. 4(b) of [1]) onto the circular front face (2.3-mm diameter) of a, (2) quartz/quartz fiber bundle (100-μm cores, 0.12 numerical aperture). The fiber bundle, rectangularly reshaped, is coupled into a, (3) 1/8-m grating spectrograph (600 grooves per millimeter, 500-nm blaze). In its focal plane a, (4) 32-channel multianode photomultiplier single-photon-counting detection system is mounted (SP32-20, Licel GmbH, Berlin, Germany), yielding a spectral resolution of about 12.8 nm per detection channel. The wavelength range of the spectrometer covers the near UV and the visible light spectrum, however, because of the spectral characteristics of the above-mentioned beamsplitter, fluorescence spectra are only attainable > 440 nm. Fortunately, the transmitted signal of the 387-nm Raman band of molecular nitrogen is sufficiently strong to be used for spectrum calibration. As a first step, relative calibration of the spectrum measurement is obtained by inserting a light source of known spectral characteristics into the telescope hall and illuminating the far-range receiver field stop. An absolute calibration is then achieved by comparing measured and calculated (from spectroscopic and concurrent radiosonde data) molecular-nitrogen Raman signals. Light extinction by particles and molecules is taken into account.

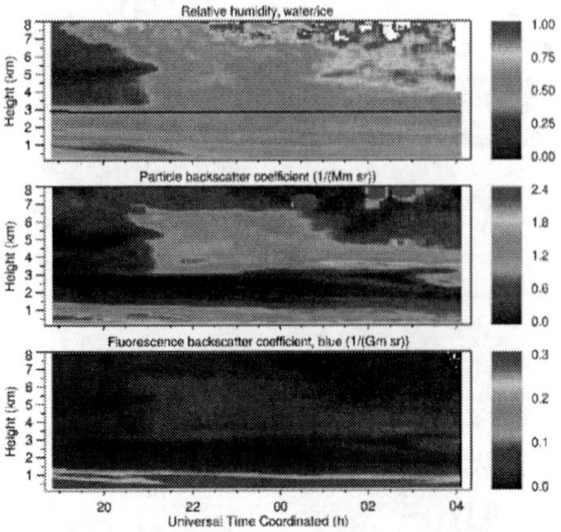

Figure 1: Temporal evolution of relative humidity (with respect to liquid water and ice above and below 0°C, respectively; 0°C-isotherm indicated by black curve), particle backscatter coefficient, and fluorescence backscatter coefficient (455–530 nm spectral range) as measured with RAMSES during the night of 2–3 April 2016. Whitened stretches indicate times and heights where data did not pass quality control.

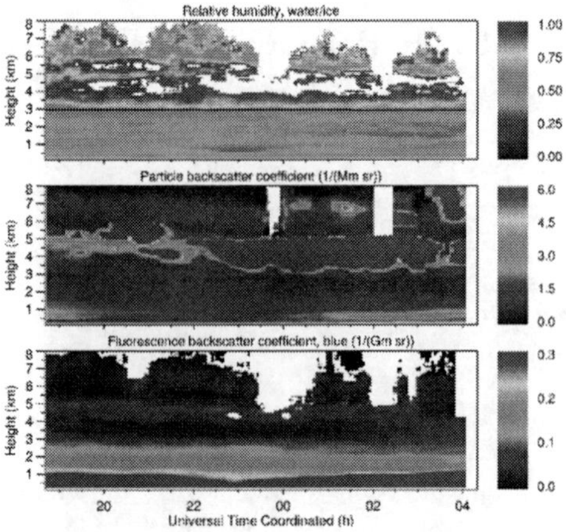

Figure 2: Similar to Fig. 1 but for the night of 3–4 April 2016.

3 MEASUREMENT EXAMPLES

3.1 Saharan dust

As a first measurement example, the Saharan dust event in early April 2016 is presented in Figs. 1–3. RAMSES was operating continuously, but only nighttime measurements are discussed when data from the fluorescence spectrometer are available. Figure 1 depicts the advent of the dust plume which originated from Algeria and Mali according to simulations with the FLEXPART Lagrangian particle dispersion model [3]. Fluorescence is low, and, with a spectral maximum missing > 440 nm, the spectrum resembles those usually observed for boundary-layer aerosols under unperturbed conditions (see Fig. 7). Optical properties are homogeneous. Particle depolarization ratios, δ_{par}, of about 30–35%, and lidar ratios, S_{par}, between 60 and 80 sr (not shown) indicate particle diameters of about 0.8 μm, according to finite-difference time-domain scattering computations [4] for quartz-rich irregular aerosol particles with an aspect ratio of 1. A short-lived change in particle properties to values typical of ice clouds ($\delta_{\mathrm{par}} = 15\%$, $S_{\mathrm{par}} = 25$ sr) is observed at the leading edge of the aerosol plume at 20:45 UTC (Fig. 1), but the elastic backscatter coefficient, β_{par}, is rather small ($\sim 3.5\,\mathrm{Mm^{-1}sr^{-1}}$), and the relative humidity with respect to ice, RHi, low (about 60%). Similar events occur at later times and slightly higher altitudes at much higher RHi values. Note that sporadic cirrus clouds were present above 9 km. The strong fluorescence below 1.5 km is indicative of a polluted boundary layer.

The next day, a mostly frozen cloud appeared in a moistened layer between 3.5 and 5.5 km. Interestingly, obscured by elastic cloud scattering otherwise, the fluorescence measurement reveals the presence of the weakly-fluorescing dust aerosol in the cloud-top layer (Fig. 2). Origin and spectral characteristics of the aerosol are very similar to what was found the day before. As an example of a combined measurement of elastic and fluorescence cloud and aerosol properties, Fig. 3 shows the RAMSES observation at 22 UTC. RHi hardly reaches ice saturation within the cloud, elastic-backscatter

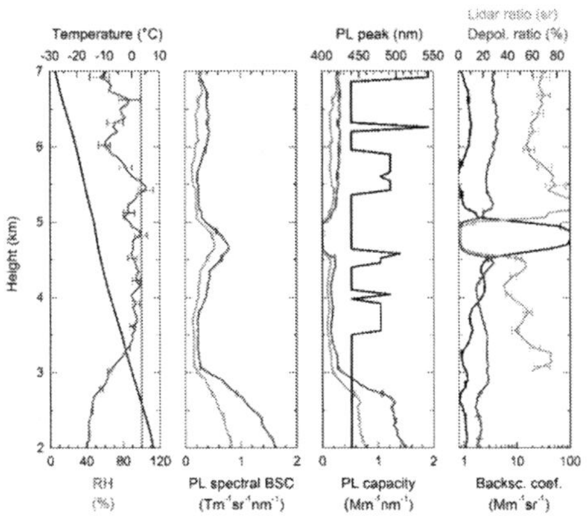

Figure 3: Relative humidity with respect to ice, and temperature (interpolated; from 6-hourly radiosonde data); mean green (540–615 nm spectral range) and blue (455–530 nm) fluorescence spectral backscatter coefficients; green and blue fluorescence capacities (mean fluorescence spectral backscatter coefficient divided by particle backscatter coefficient), and wavelength of spectral maximum; and particle backscatter coefficient, and depolarization and lidar ratios as measured on 3 April 2016 at 22 UTC. Bars indicate statistical errors of the lidar measurement. Note the logarithmic scale of the elastic backscatter coefficient.

peak and fluorescence spectral backscatter coefficient maxima roughly coincide in height. Cloud optical properties are characteristic of horizontally aligned ice particles. Note that fluorescence capacity is several times larger below 3 km, which suggests differences in aerosol composition even though the fluorescence spectra (not shown) in the wavelength range currently accessible with RAMSES is similar.

3.2 Biomass-burning aerosol (BBA)

On 9 September 2016 aerosols from Africa were measured in the lower free troposphere (2–3.7 km; ($\delta_{\text{par}} < 10\%$, $S_{\text{par}} = 60\,\text{sr}$)) again (Fig. 4), but the source region was slightly more to the southwest. Interestingly, the fluorescence properties differ vastly from those of the

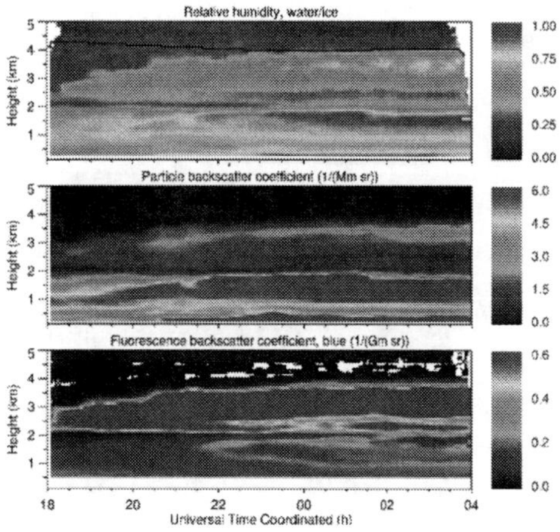

Figure 4: Similar to Fig. 1 but for the night of 9–10 September 2016.

Apr 2016 cases: The aerosol fluoresces much more strongly (e.g., blue fluorescence capacity of about 5–7 $\text{Mm}^{-1}\text{nm}^{-1}$ as compared to $\leq 1\,\text{Mm}^{-1}\text{nm}^{-1}$), and the spectrum exhibits a distinct local maximum at 490–500 nm (Fig. 7). This may indicate the presence of BBA because fluorescence spectra of local Easter bonfires show a similar peak, albeit at longer wavelengths (Fig. 7). Relative humidity with respect to liquid water, RHw, is between 35% and 80%, β_{par} correlates with RHw, but humidity is insufficient for cloud formation.

Hardly visible in Fig. 4, but yet confirmed by a Lindenberg in-situ optical backscatter sonde launched at 23 UTC, is another BBA layer at heights around 5 km. It is remnant of a larger BBA event that could be observed for several days starting on 6 September 2016 (Fig. 5). While β_{par} is near the detection limit, the BBA shows a brilliant fluorescence display. The spectrum closely resembles the spectrum measured near the ground during the bonfires. The air masses of the vertically well-defined layer were advected from North America, very dry ($< 20\%$ RHi) and thus void of clouds. Incidentally, a thin cirrus was observed aloft ($> 9\,\text{km}$) during the whole night.

Finally, Fig. 6 illustrates a BBA incident dur-

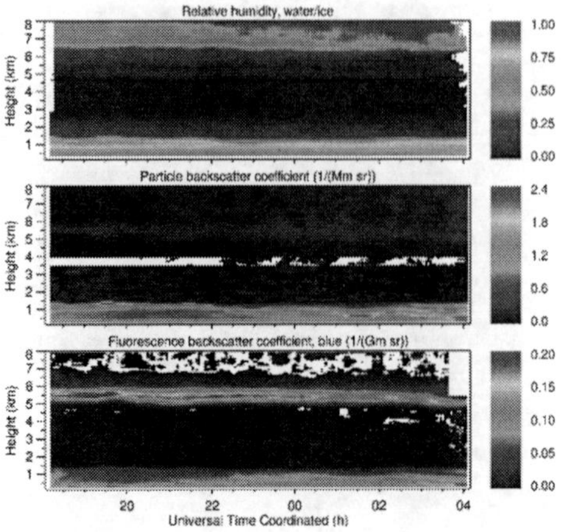

Figure 5: Similar to Fig. 1 but for the night of 6–7 September 2016.

ing which cirrus clouds occurred. The measurement was taken in the night of 7–8 September 2015, note the fine fluorescence streaks. A remarkable observation is that the descending cloud-top layer is strongly correlated with a BBA filament. Unfortunately, data quality is insufficient to determine RHi within the cirrus.

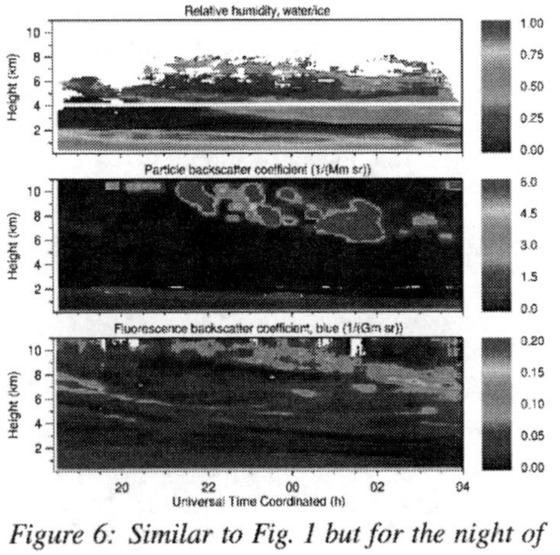

Figure 6: Similar to Fig. 1 but for the night of 7–8 September 2015.

In conclusion, the fluorescence spectrometer has proven to be a promising tool to study aerosol-cloud co-existence. The information extracted from the fluorescence spectra is sufficient to allow for a rough aerosol classification, which is an important result per se because backtrajectory calculations are often inconclusive and not meaningful enough to determine aerosol type. Experimental work is underway to extend measurements of the fluorescence spectrum to wavelengths < 440 nm.

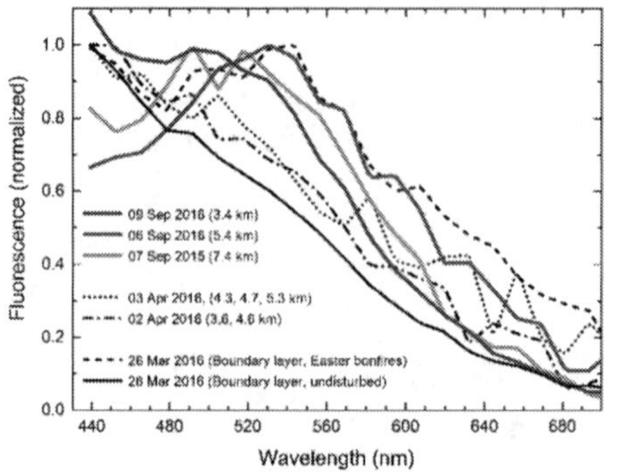

Figure 7: Normalized fluorescence spectra of free-troposphere aerosol. Dates and approximate heights are indicated in the legend. The typical fluorescence spectrum of boundary-layer aerosol as well as the fluorescence spectrum of air laden with biomass-burning aerosol near the ground are shown for comparison.

References

[1] Reichardt, J., 2014: Cloud and aerosol spectroscopy with Raman lidar, *J. Atmos. Ocean. Technol.* **39** (9), 1946-1963.

[2] Reichardt, J., Wandinger, U., Klein, V., Mattis, I., Hilber, B., Begbie, R., 2012: RAMSES: German Meteorological Service autonomous Raman lidar for water vapor, temperature, aerosol, and cloud measurements, *Appl. Opt.* **51** (34), 8111-8131.

[3] Stohl, A., Forster, C., Frank, A., Seibert, P., Wotawa, G., 2005: Technical note: The Lagrangian particle dispersion model FLEXPART version 6.2, *Atmos. Chem. Phys.* **5** (9), 2461-2474.

[4] Yang, P., Liou, K. N., Mishchenko, M. I., Gao, B.-C., 2000: Efficient finite-difference time-domain scheme for light scattering by dielectric particles: application to aerosols, *Appl. Opt.* **39** (21), 3727-3737.

EPJ Web of Conferences **176**, 05011 (2018) https://doi.org/10.1051/epjconf/201817605011
ILRC 28

EXTINCTION MEASUREMENTS WITH LOW-POWER HSRL SYSTEMS—ERROR LIMITS

Ed Eloranta

*Affiliation: University of Wisconsin-Madison, US, *eloranta@ssec.wisc.edu*

ABSTRACT

HSRL measurements of extinction are more difficult than backscatter measurements. This is particularly true for low-power, eye-safe systems. This paper looks at error sources that currently provide an error limit of ~10^{-5} m^{-1} for boundary layer extinction measurements made with University of Wisconsin HSRL systems. These eye-safe systems typically use ~300mW transmitters and 40 cm diameter receivers with a 10^{-4} radian field-of-view.

1 INTRODUCTION

The HSRL provides both backscatter and extinction cross-sections measurements. The backscatter cross-section measurement is derived from a ratio of signals; calibration is easy and robust. The extinction cross-section measurement is more difficult. It can be derived by taking the log of the HSRL molecular signal as a function of range, r, and then differentiating with respect to r:

$$\beta_e(r) = \frac{1}{2} \cdot \frac{d}{dr}[log(\frac{\eta(r)\cdot\beta_m(r)}{r^2\cdot S_m(r)})] \quad (1)$$

This provides the total extinction cross-section, β_e as function of the HSRL recovered molecular profile, S_m, the overlap correction, η and the molecular scattering cross-section, β_m. βm is computed from an independent sounding. The extinction due to particulates alone can be derived by simply subtracting the molecular extinction cross-section.

It is instructive to write the equation as follows:

$$\beta_e(r) = \frac{1}{2\eta}\frac{d\eta}{dr} + \frac{1}{2\beta_r}\frac{d\beta_m}{dr} - \frac{1}{r} - \frac{1}{2S_m}\frac{dS_m}{dr} \quad (2)$$

The molecular extinction cross-sections in the boundary layer is ~ 10^{-5} m^{-1}. Aerosol extinction on clear days is often less than the molecular. Thus, it would desirable to keep uncertainty of each term of the equation to less than $10^{-6}m^{-1}$. This paper will consider extinction cross-section errors segregated by the terms of equation 2.

2 OVERLAP CORRECTION

Figure 1 plots the overlap correction term of equation 2 for one of our lidars. Our low-power, ~300mW, eye-safe systems employ a 10^{-4} radian field-of-view to reduce sky noise. The small field-of-view coupled with a 40 cm diameter receiver requires overlap corrections for altitudes below 6km.

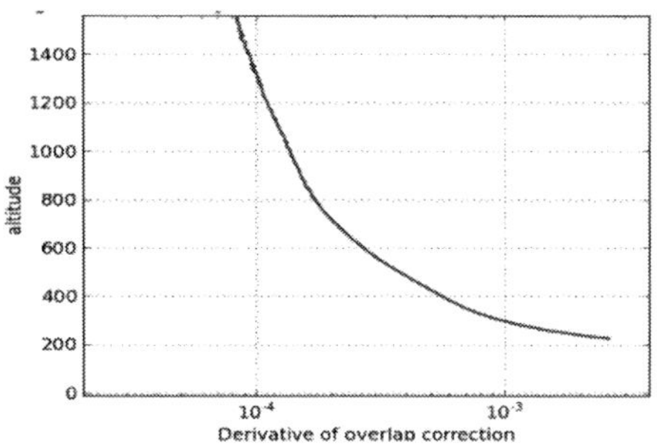

Figure 1. First term of equation 2 for a UW HSRL with a 100urad field-of-view.

Notice that in order to maintain an overlap correction error in the extinction below 10^{-6} m^{-1} at 300 m, the overlap correction must be known at the 0.1% level.

Recent UW HSRL systems include a wide-field-of-view (WFOV) channel to determine $\eta(r)$. This consists of a 2.5-cm diameter telescope mounted behind the secondary mirror of the main telescope. It is coupled via optical fiber to a second molecular HSRL channel consisting of a interference filter, an etalon to block sky background light, and a molecular iodine filter to block aerosol scattering. This channel duplicates the molecular of the HSRL except that it views a 1 mr angular field-of-view rather that the 0.1 mr viewed by the main molecular channel. The WFOV channel provides complete overlap, with $\eta(r)$ equal to 1, at ~500 m from the lidar. Figure 2 shows the WFOV-to-molecular channel ratio after it is normalized to one above 6 km. It also shows $\eta(r)$. This is obtained from the ratio of the signals in the WFOV and molecular channels at ranges between 500 m and 6km. Generation of the correction factor below 500 m relies on cases with a well-mixed boundary that extends above 500 m where we can measure both the extinction and the backscatter cross-section accurately. Below 500m η(r) is determined from the ratio of the expected return from the expected molecular to the measured molecular return. The expected molecular return is computed from the density profile corrected by the extinction computed from the measured backscatter profile and the backscatter phase function measured above 500m.

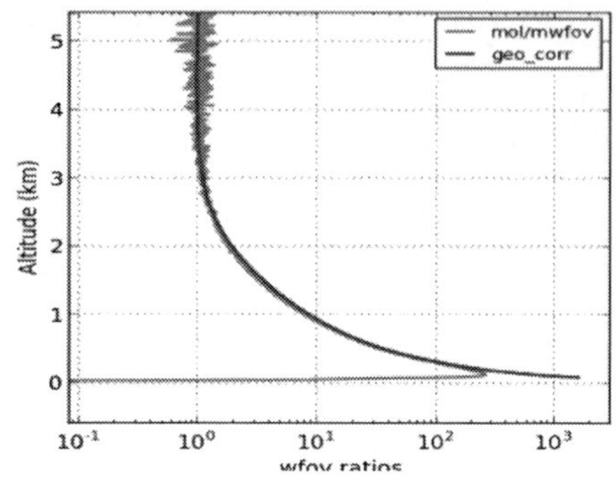

Figure 2. The ratio between the WFOV and the standard molecular channel signals scaled to one above 6km (cyan) and η(r) (blue). The WFOV/molecular ratio correction is smoothed to generate η(r) values at altitudes with full overlap of the WFOV channel. Below full WFOV overlap, η(r) is computed using the measured molecular return and the expected atmospheric return. The atmospheric return is computed from the density profile corrected for extinction. Extinction is estimated from the backscatter cross-section and backscatter phase function measured in the mixed layer.

The WFOV channel views a solid angle 100 times as large as the main channel using an objective with 0.0039 times the area. Because it has much poorer noise performance than the molecular channel, overlap corrections must be computed from time averages of the WFOV signal. These overlap corrections provide good correction when generated from, and applied, to night-time data.

Even though the HSRL telescope uses an athermal design, employing carbon fiber construction, and Zerodur optical elements, the system is unable to maintain the necessary stability during mid-day. Overlap variations appear to result from some combination of thermal changes in the transceiver-telescope and beam spread caused by atmospheric turbulence. A time-varying overlap correction is required. This is done by observing the ratio of WFOV-to-molecular channel ratio as a function of time and altitude. The overlap correction changes in direct proportion to measured changes in this ratio. These are used to dynamically adjust the overlap correction. In order to achieve adequate signal-to-noise, the ratio is fit with a first- or second-order Savitzky-Golay filter inside a rolling window centered on the observation time. Typical window sizes range from 1 to 2 hours in time and 100-300 meters in altitude. Figure 3 shows a typical example of the backscatter along with the corresponding extinction and the dynamic overlap correction. Plots of extinction and extinction corrections are shown three different altitudes in figure 4. At mid-day (4 UT) the correction was $-8x10^{-5}$ 1/m.

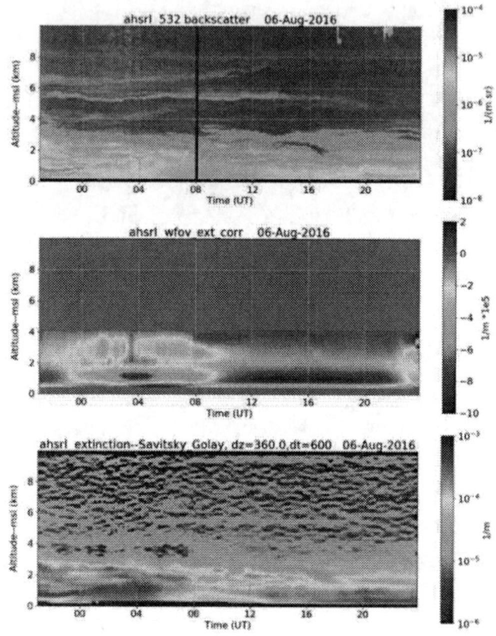

Figure 3. Aerosol Backscatter cross-section (top), dynamic extinction cross-section correction (middle), and extinction cross section (bottom) derived for HSRL data acquired on 6-Aug-16.

3 TEMPERATURE PROFILE ERRORS

The second term on the right-hand side of Equation 2 is evaluated from an independently measured temperature profile. The molecular extinction cross-section is directly proportional to the atmospheric density so that for a vertically pointing lidar:

$$\frac{1}{\beta_m}\frac{d\beta_m}{dz} = \frac{1}{\rho}\frac{d\rho}{dz} = -\frac{1}{T}\left(\frac{dT}{dz} - \frac{g}{R}\right) \quad (3)$$

Where: ρ = density, g = acceleration of gravity, z = altitude, and R = gas constant.

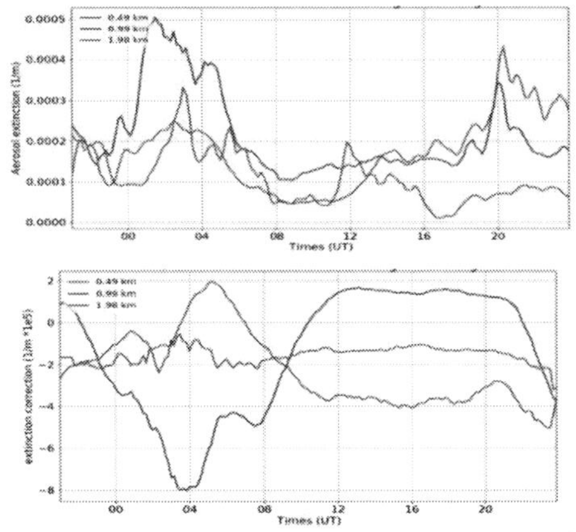

Figure 4. Derived extinction cross-section(upper) and corrections due variations in first term of equation 3 for the data shown in figure 3. Curves are shown for altitudes of 490 (red), 990(blue) and 1990(green) meters.

The temperature resolution to achieve a given β_e error can be obtained by differentiating Equation 3.

$$\Delta\beta_e = \frac{1}{T}\cdot\left(\frac{dT}{dz}+\frac{g}{R}\right)\cdot\Delta T + \frac{1}{T}\cdot\Delta\frac{dT}{dz} \quad (4)$$

A 2 K temperature error at 300 K, and a lapse rate of 6 K/km produces a only ~10^{-6} m^{-1} error from the first term of equation 4. Lapse rate errors are more significant. Near surface temperature profiles transition from inversion to adiabatic lapse rates on most summer mornings providing extinction variations:

$$\frac{1}{T}\Delta\frac{dT}{dz} \sim \frac{1}{300}\cdot\frac{10}{1000} \sim 3\cdot 10^{-5}m^{-1} \quad (5)$$

Figure 5 shows lapse rates derived from soundings taken at 6-hour intervals in Seoul, South Korea. Notice the low altitude transitions between inversion and lapse conditions. Their effect is clearly seen in the 0.53 km time history of HSRL extinction. At 16:00 UT the extinction suddenly decreases by nearly $10^{-5}m^{-1}$as the lapse rate transitions from lapse to inversion and then increases abruptly by nearly $2x10^{-5}$ m^{-1} at ~1:00 UT when the inversion conditions switch back to lapse. Between these transitions the derived extinction goes negative. This most likely due to the complex terrain of Seoul reflecting a difference between atmospheric conditions at the radiosonde site and the lidar site.

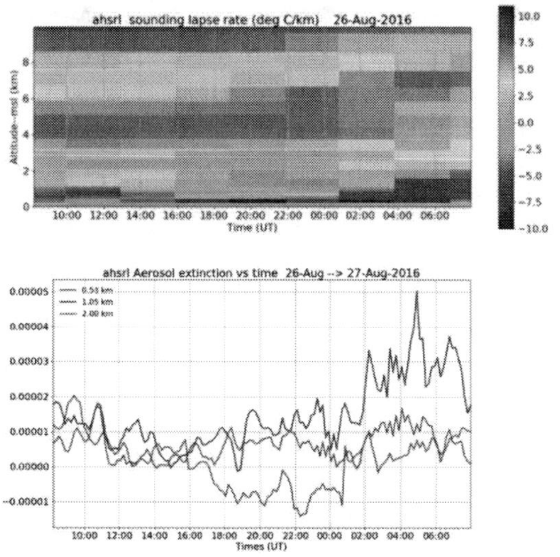

Figure 5. Lapse rate vs time and altitude (top) and derived extinction cross-sections at altitudes of 0.53km (red),1.05km (blue) and 2.0km (green). Notice the jumps in extinction at 0.53km occurring at 16:00 and 1:00 UT. These correspond to changes from inversion to lapse in the lapse rate plot.

4 PHOTON COUNTING STATISTICS

We now consider the final term of Equation 2. The molecular profile, S_m, is derived from the number of photons detected in the HSRL molecular, N_m, and combined, N_c, channels of the instrument:

$$S_m(r) = k(r)\cdot(N_m(r) - C_{am}N_c(r)) \quad (6)$$

Outside of clouds N_m and N_c are of similar magnitude, while $C_{am} < 0.0005$, the leakage of aerosol signal into the molecular channel, can be neglected outside of clouds. The last term of Equation 2 can then be written as:

$$\frac{1}{S_m(r)}\frac{dS_m}{dr} = \frac{1}{k(r)}\frac{dk(r)}{dr} + \frac{1}{N_m(r)}\frac{d(N_m(r)+N_b)}{dr} \quad (7)$$

Errors in the logarithmic range derivative of the molecular signal are dependent on how accurately the HSRL calibration separates molecular and particulate signals as well as statistical photon counting errors. Notice that the range derivative now also includes the background counts, N_b, from sky and detector noise. Although the average number of background counts does not vary with range, counting errors will appear as range variations. The first term on the right hand side includes the range variation of the calibration coefficients used to separate molecular and particulate photons. These coefficients are determined through calibration of the spectral bandpass of the receiver channels and knowledge of the atmospheric temperature profile. They are quite well measured and are almost always dominated by photon counting error.

Photon counting errors are described via Poisson statistics where the signal variance is equal to the number of photon counts. The problem posed by the N_m term can be seen by approximating the error for best case where the molecular photon counting rate is near the point of detector saturation. The Geiger-mode avalanche photodiode detectors used in our HSRL systems are unable to count more than ~1 photon per 7.5 m range bin. With the system operating at a 4 kHz repetition rate and a 2.5 second averaging time, a single range bin can count a maximum of 10^4 photons in one sample. Approximating the derivative from the counting error for 2.5-second profiles and a separation of one range bin shows that even in the case of strong signals, photon counting statistics can provide a large source of error.

$$\frac{1}{N_m(r)}\frac{dN_m(r)}{dr} \sim \frac{1}{10^4}\frac{\sqrt{2}\cdot\sqrt{10^4}}{7.5} \sim 2\cdot10^{-3}\,\mathrm{m}^{-1} \quad (8)$$

This error is approximately three orders of magnitude larger than the $10^{-6}\,\mathrm{m}^{-1}$ extinction cross-section we might like to measure. Clearly, it will be necessary to average over longer time and space intervals to reduce errors near those that are desired.

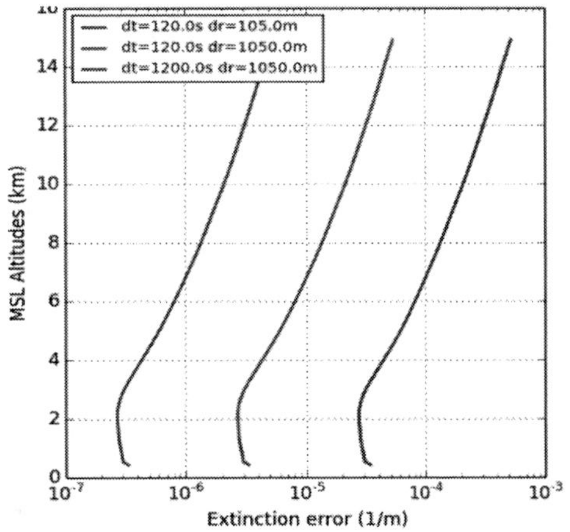

Figure 6. Extinction errors due to photon counting statistics for different time and range averages estimated from the last term on the left-hand side of Equation 2. These results computed from clear-air returns on 2-June-2012.

Figure 6 shows that photon counting errors in the extinction measurement can be reduced below $10^{-6}\,\mathrm{m}^{-1}$ below 7 km using 20-minute time and 1 km range averaging. This amount of averaging may be practical in stable cloud free conditions, but may obscure the temporal and spatial scales of the atmospheric structures in more variable conditions. Typical time averages ~20 minutes and altitude averages of ~200 m provide a photon counting error limit below $10^{-5}\,\mathrm{m}^{-1}$ with our current systems.

5 CONCLUSIONS

Although extinction cross-section is a more useful quantity than backscatter cross-section, it is more difficult to measure. With current hardware and algorithms boundary layer measurement errors of ~$10^{-5}\,\mathrm{m}^{-1}$ can often be obtained with current UW HSRL system using ~300mW transmitters and 10^{-4} mr field-of-view receivers. Limits are imposed by photon counting noise along with imprecise knowledge of the overlap function and the atmospheric density profile. We are working to derive temperature profiles from lidar data and thus reduce uncertainties due to temperature.

ACKNOWLEDGEMENTS

This research was supported by US Department of Energy contract 258032.

References

[1] Eloranta, E.W., "High Spectral Resolution Lidar", in *Lidar: Range-Resolved Optical Remote Sensing of the Atmosphere,* K. Weitkamp editor, Springer-Verlag, New York, 2005.

[2] University of Wisconsin lidar group web site: lidar.ssec.wisc.edu

[3] Savitzky, A; Golay, M.J.E. (1964), "Smoothing and Differentiation of Data by Simplified Least Squares procedure", *Analytical Chemistry,* **36**, (8), 1627-163

Micro-Pulse, Differential Absorption Lidar (DIAL) Network for Measuring the Spatial and Temporal Distribution of Water Vapor in the Lower Atmosphere

Scott Spuler*[1], Kevin Repasky[2], Matt Hayman[1], Amin Nehrir[3]

[1]*National Center for Atmospheric Research, Earth Observing Lab, Boulder, CO 80307, USA, *spuler@ucar.edu*
[2]*Montana State University, Electrical and Computer Engineering, Bozeman, MT 59717, USA*
[3]*NASA Langley Research Center, Hampton, VA 23681, USA*

ABSTRACT

The National Center for Atmospheric Research (NCAR) and Montana State Univeristy (MSU) are developing a test network of five micro-pulse differential absorption lidars to continuously measure high-vertical-resolution water vapor in the lower atmosphere. The instruments are accurate, yet low-cost; operate unattended, and eye-safe – all key features to enable the larger network needed to characterize atmospheric moisture variability which influences important processes related to weather and climate.

1. INTRODUCTION

Water vapor is one of the fundamental thermodynamic variables that define the state of the atmosphere. It is highly variable in space and time and influences many important processes related to weather and climate. Three recent National Research Council (NRC) reports highlight the need for improved water vapor measurements as a necessary step toward improving mesoscale numerical weather prediction and quantitative precipitation forecasting skills [1, 2, 3]. A recent review paper detailing the state of the art for thermodynamic profiling,[4] also stresses the importance of water vapor profiling in the lower troposphere and suggests low-power water vapor differential absorption lidar (DIAL) as a promising candidate for a compact remote sensing instrument that can be deployed in networks.

Developments in semiconductor laser and amplifier technology have enabled low cost, compact lasers with narrow linewidths and long lifetimes. Semiconductor lasers are well suited to lidar systems designed for network deployment where the instrument needs to be capable of unattended operation for long periods without maintenance and should not pose a risk to aircraft or bystanders. Technology using a narrow-line semiconductor laser, such as external cavity laser diodes (ECDL), to seed a tapered semiconductor optical amplifier (TSOA) were developed and demonstrated for the purpose of water vapor profiling using the differential absorption lidar technique at 828 nm [5, 6, 7]. Following this seminal work, a field-capable micro-pulse DIAL (MPD) instrument was collaboratively designed, built and tested by MSU and NCAR [8, 9]. This eye-safe and reliable design has been shown to deliver accurate retrievals of water vapor in the lower troposphere and produce scientifically significant data [10].

NCAR and MSU are now developing a five unit water vapor profiling testbed – necessary to capture nonlinear mesoscale moisture variability caused by land surface effects, such as terrain variations, lakes, vegetation and irrigation. Since water vapor is highly variable in time and three-dimensional space, five stations allow for measuring mean water vapor values within four triangular areas and could quantify non-linearities in gradients across the network domain. The addition of wind profile measurements collocated with each of these stations would provide for information on moisture advection and moisture divergence.

2. INSTRUMENT DESIGN

The instrument, shown schematically in Figure 1, utilizes a diode-laser-based master oscillator power amplifier (MOPA) configured transmitter, and a shared telescope transmitter and receiver to achieve opto-mechanically stable, and eye-safe, operation. A multistage filtered receiver is implemented to allow continuous measurement in high background conditions (e.g., bright daytime clouds).

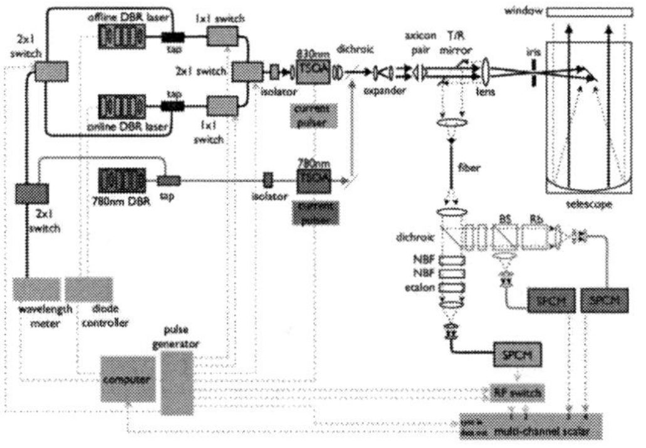

Figure 1. Schematic of the current MPD system combined with a new channel (shown in red) – a diode-laser-based (DLB) High Spectral Resolution Lidar (HSRL). BS = beam-splitter; T/R = transmit receive; Rb = Rubidium vapor cell; SPCM = Single Photon Counting Module.

A pair of distributed Bragg reflector (DBR) semiconduc-

 https://doi.org/10.1051/epjconf/201817605012

tor lasers nominally centered at 828 nm are interleaved at 100 Hz and used to seed a TSOA. Current pulses to the TOSA amplify the laser and produce approximately 5μJ over 900 ns duration. The beam is shaped into an annular ring with an axicon pair and then expanded to fill the inner part of the telescope with a radius of 10 cm. The laser repetition rate, set to 7 kHz, limits the maximum altitude of the lidar system to 21 km. Direct exposure to the laser exiting the telescope is within eye-safe limits. The backscattered light received is directed into a 105μm core multimode fiber in the receiver system. The light from the fiber is collimated and passes through a pair of interference filters and solid etalon, with a combined passband of 0.014 nm for daylight rejection. The filtered light is then directed into a fiber coupled single photon counting module (SCPM), which utilizes a silicon avalanche photodiode operated in so-called 'Geiger-mode'.

To date, the NCAR water vapor MPD has been deployed at two field campaigns – FRAPPE (Colorado, USA) in the summer of 2014, and PECAN (Kansas, USA) in the summer of 2015 within a 6.1m long intermodal shipping container equipped as mobile lidar laboratory. The system was run unattended in all weather and provided near-continuous profiles of water vapor – including periods of bright daytime clouds – from 300 m above ground level to 4 km (or cloud base, whichever comes first) with 150 m nominal vertical resolution and 1-5 minute temporal resolution. The MPD was greater than 95% operational in both field projects resulting in a high continuity of water vapor profile time series as shown in Fig. 2. Well-established water vapor observing systems, including radiosondes, infrared and microwave radiometers, and ground-based global positioning system receivers were collocated with the MPD at these field campaigns. Data comparisons showed excellent agreement between the MPD and these systems, with the MPD providing distinct operational and resolution advantages [10]. An example comparison with a series of radiosondes is shown as Fig. 3. The continuous water vapor measured by MPD agrees well with the 11 radiosonde launched during this 12 hr time period. Notably, the MPD was the only one of three collocated continuously profiling instruments that detected the two elevated moist layers from 1 to 3km AGL during this time period. [10]

Following the PECAN deployment, a more portable field enclosure was constructed to simplify fielding the water vapor MPD. A picture of the field prototype from Feb 2016 is shown as Fig. 4. The new enclosure was designed to 1.) be reasonably small and lightweight so that it could be moved with a forklift and not require a crane for lifting on/off transport vehicles, 2.) provide a weatherproof and thermally stable environment for the instrument, and 3.) provide instrument vibration isolation, to minimize the amount of disassembly and packing needed for shipment to field projects. The enclosure has external dimensions of 1 m depth, 2 m width, and 2 m height and was designed to have extra space around the instrument to allow easy alignment access and allow for future changes and upgrades.

NCAR has already demonstrated one potential upgrade technology – a diode-laser-based (DLB) high spectral resolution lidar (HSRL) channel based on the same low-cost lidar architecture [11, 12]. This HSRL channel, is added to the MPD as shown in Figure 1. The upgraded instrument combines water vapor with quantitative measurements of aerosols and could significantly contribute to studies of aerosol-cloud interaction. There is additional synergy as HSRL directly measures the aerosol-to-molecular backscatter ratio and would provide accurate Rayleigh Doppler (RD) corrections for the water vapor spectroscopy (although this correction term is small, since the MPD instrument was designed to be insensitive to RD errors [10]). In addition, HSRL do not require a correction for the geometric overlap function to obtain the aerosol backscatter coefficient and are well suited to quantitative retrievals at comparatively low altitudes. As part of the network deployment, the DLB-HSRL offers opportunities for higher resolution measurements of mixing layer height for model and satellite comparison studies as well as continuous monitoring of a key observational parameter for weather and climate.

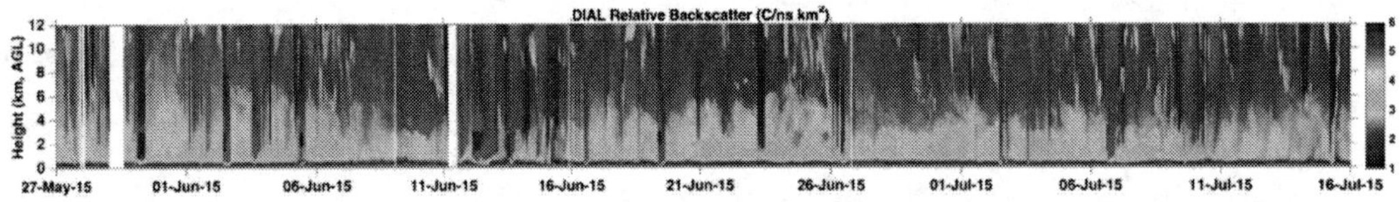

Figure 2. Time series of decimated 5-min/75-m MPD profiles showing (top) range-corrected relative backscatter and (bottom) absolute humidity (gm^{-3}). Vertical white bars signify missing data due to MPD downtime or attenuated signal above optically thick clouds.

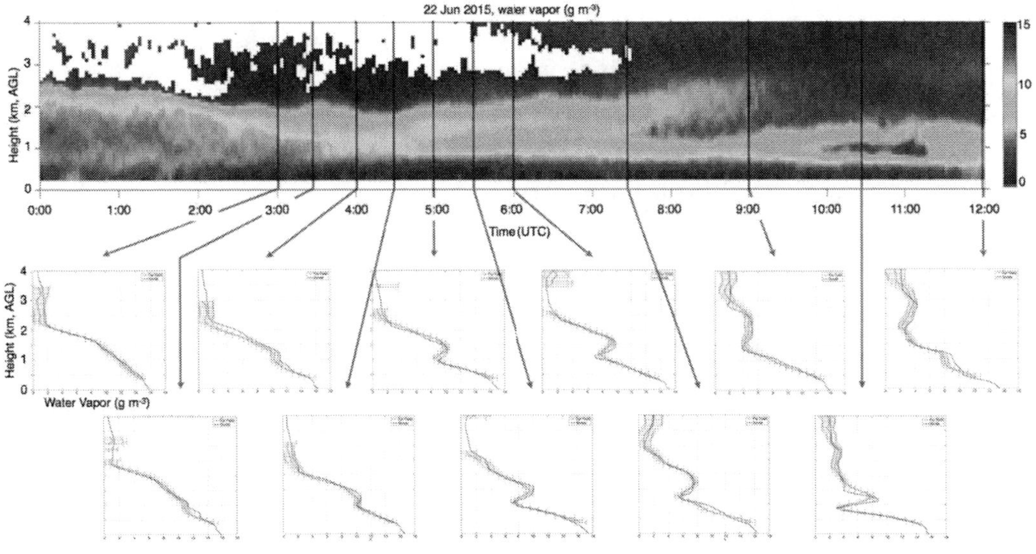

Figure 3. Water vapor profile time series measured with the MPD on 22-Jun-2015. The profiles show elevated layers of water vapor during the 12 hr time span. The gray lines indicate sonde launch times and point to the corresponding comparison profiles. Water vapor profiles measured by the sonde and MPD are shown in blue and red; respectively. The horizontal red lines represent the MPD measurement error at each 75-m height interval.

MSU and NCAR have also developed a performance model for a low-cost temperature channel to integrate into the MPD architecture. For the water vapor DIAL, a temperature insensitive absorption feature was chosen, and the backscatter signal from the online and offline wavelengths are used to retrieve the number density profile as a function of range. However, if a temperature sensitive absorption feature is chosen for a molecule with a well know atmospheric mixing ratio such as oxygen (O_2), temperature as a function of range can be retrieved via the DIAL technique. For low error DIAL temperature retrievals (e.g., <1C), the aerosol-to-molecular backscatter ratio must be known [14]. This value can be obtained from the DLB-HSRL which is close in wavelength. The estimated performance of the proposed oxygen MPD temperature channel is comparable to a rotational Raman lidar [15].

3. NETWORK CONSTRUCTION PLANS

At the time of this writing, two fully functional water vapor MPD instruments exist (only one unit has been packaged into its field environmental enclosure). These two units are being modified to match one another and after the design is finalized, three cloned units will be built. All of the units will include design modifications to improve reliability in the field, e.g., one potential upgrade is a compact seed laser package with integrated optics to provide a fiber coupled and optically isolated output. This upgrade effort is part of a NASA funded SBIR [13] and would reduce alignment time at the start of new field deployments. The instrument environmental enclosure, based on the current design, will allow for long term deployments and include temperature control and remote internet access. Additionally, the instrument control will be modified to improve reliability for the

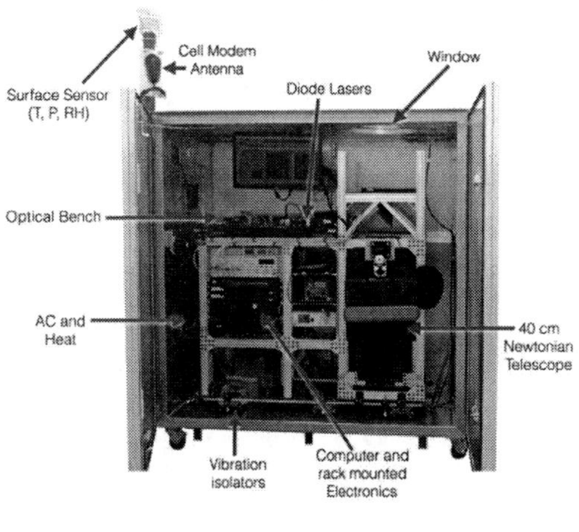

Figure 4. Picture of the MPD system within its portable field enclosure. The unit (with a size of 1m x 2m x 2m) is the first of the small network. The next units to be built will be similar in design although slightly smaller

network by incorporating instrument monitoring, alarm, and control features. Processing routines have been automated and will provide the data in an appropriate format for ease of data assimilation. When the construction phase is complete, each MPD instrument will be tested using radiosondes for validation. The MPD testbed is anticipated to be ready for service in 2019, and will be made available to the larger research community for field campaigns and deployment through NCAR.

4. SUMMARY

A small testbed network of validated micro-pulse water vapor DIAL instruments is under construction. The diode-laser-based transmitter is inherently low-maintenance, low-cost, and designed to be eye-safe at the exit port. Each node provides continuous profiles of water vapor with complete daytime coverage – including periods of bright clouds – from 300 m above ground level to 4 km (or cloud base, whichever comes first) with 150 m nominal vertical resolution and 1 to 5 minute temporal resolution. A low-cost HSRL channel can readily be integrated into the instrument which would significantly contribute to studies of aerosol-cloud interaction. The five unit testbed network has a planned completion date in 2019. The testbed will be a significant step towards improving our understanding of the distribution of atmospheric water vapor, and provide the opportunity to measure the impact potential of a national scale network.

ACKNOWLEDGMENTS

NCAR is sponsored by the National Science Foundation. The authors would like to acknowledge the support of the National Science Foundation grant number 1624736.

REFERENCES

[1] National Research Council, 2009: *Observing Weather and Climate from the Ground Up: A Nationwide Network of Networks*. The National Academies Press, Washington, DC.

[2] National Research Council, 2010: *When Weather Matters: Science and Service to Meet Critical Societal Needs*. The National Academies Press, Washington, DC.

[3] National Research Council, 2012: *Weather Services for the Nation: Becoming Second to None*. The National Academies Press, Washington, DC.

[4] Wulfmeyer, V., R. M. Hardesty, D. D. Turner, A. Behrendt, M. P. Cadeddu, P. Di Girolamo, P. Schlssel, J. Van Baelen, and F. Zus, 2015: A review of the remote sensing of lower tropospheric thermodynamic profiles and its indispensable role for the understanding and the simulation of water and energy cycles *Rev. Geophys.*, **53**, 819–895.

[5] Nehrir, A. R., Repasky, K. S., Carlsten, J. L., Obland, M. D., and Shaw, J. A., 2009: Water Vapor Profiling Using a Widely Tunable, Amplified Diode-Laser-Based Differential Absorption Lidar (DIAL). *Journal of Atmospheric and Oceanic Technology*, **26(4)**, 733–745.

[6] Nehrir, A. R., Repasky, K. S., and Carlsten, J. L., 2011: Eye-Safe Diode-Laser-Based Micropulse Differential Absorption Lidar (DIAL) for Water Vapor Profiling in the Lower Troposphere. *Journal of Atmospheric and Oceanic Technology*, **28(2)**, 131–147.

[7] Nehrir, A. R., Repasky, K. S., and Carlsten, J. L., 2012: Micropulse water vapor differential absorption lidar: transmitter design and performance. *Optics express*, **20(22)**, 25137–51.

[8] Repasky, K., Moen, D., Spuler, S., Nehrir, A., and Carlsten, J., 2013: Progress towards an Autonomous Field Deployable Diode- Laser-Based Differential Absorption Lidar (DIAL) for Profiling Water Vapor in the Lower Troposphere. *Remote Sens.*, **5**, 6241–6259.

[9] Spuler, S. M., Repasky, K. S., Morley, B., Moen, D., Hayman, M., and Nehrir, A. R., 2015: Field deployable diode-laser-based differential absorption lidar (DIAL) for profiling water vapor. *Atmospheric Measurement Techniques*, **8**, 1073–1087.

[10] Weckwerth, T. M., K. Weber, D. D. Turner, and S. M. Spuler, 2016: Validation of a Water Vapor Micropulse Differential Absorption Lidar (DIAL). *J. Atmospheric and Oceanic Technology*, **33**, 2353–2372.

[11] Hayman, M., S. Spuler, B. Morley, and E.W. Eloranta, 2015: Design of a low cost diode-laser-based high spectral resolution lidar. 27th International Laser Radar Conference, New York, NY, USA.

[12] Hayman, M., and S. Spuler, 2017: Demonstration of a low cost diode-laser-based high spectral resolution lidar (HSRL). 28th International Laser Radar Conference, Bucharest, Romania.

[13] Overton G, 2016: Lidar: Differential absorption lidar demands high-power, narrowband DBR sources *Laser Focus World*, **52(9)**, 14–15.

[14] Bösenberg, J., 1998: Ground-based differential absorption lidar for water-vapor and temperature profiling: methodology. *Applied Optics*, **37(18)**, 3845–3860.

[15] Repasky, K., Spuler, S., and Hayman, M., 2017: Modeling of a Semiconductor Based Differential Absorption Lidar for Temperature Profiling in the Lower Troposphere *AMS Annual Meeting*, Seattle WA, USA 24-Jan-2017 .

EPJ Web of Conferences **176**, 05013 (2018) https://doi.org/10.1051/epjconf/201817605013
ILRC 28

GREENLITE™: A YEAR OF CARBON DIOXIDE MONITORING OVER PARIS, FRANCE, AND RECENT PROGRESS IN MONITORING METHANE

Jeremy Dobler[1]*, T. Scott Zaccheo[2], Timothy Pernini[2], Nathan Blume[1], Michael Braun[1]

*[1]Harris Corporation: Environmental Solutions, Fort Wayne, IN, USA, *Jeremy.Dobler@harris.com*
[2]Atmospheric and Environmental Research, Lexington, MA, USA

ABSTRACT

GreenLITE™ is a ground-based laser absorption spectroscopy system capable of measuring and mapping CO_2 concentrations over areas up to 25 km^2. The system was deployed for COP21 as a demonstration and has now completed a year of CO_2 measurements over the city of Paris, France. We will discuss lessons learned and relevant data from the year-long deployment. Recently, the system has demonstrated the same measurement capability for CH_4, and results from preliminary testing are presented.

1 INTRODUCTION

It has long been recognized that prediction of human induced effects on climate change requires a deeper understanding of the overall carbon cycle. The two primary greenhouse gases humans are contributing which can impact global climate change are carbon dioxide (CO_2) and methane (CH_4), and although there has been significant progress in understanding emissions of these gases, there are still many gaps in our accounting. Furthermore, the primary sources of the human generation of these gases are connected to urban areas, and not surprisingly this is dominated by large metropolitan areas. With this factor in mind, and a growing global desire to limit the human impact on climate change, the 21st Conference of Parties (COP21) met in Paris, France, in December 2015 to discuss global commitments for reduction of emissions. Uncertainties in the emissions of greenhouse gases from cities can be large, and point measurements within the city structure experience large local variability and are ill suited for inversions, leaving upwind and downwind gradients as the best source of urban concentration data for inversions [1].

Although these methods have been shown to significantly reduce uncertainties between bottom-up and top-down estimates of urban emissions, they provide little information on the spatial distribution within the city boundaries, and further improvement is needed to accurately evaluate mitigation results and further reduce emissions. To this end, an integrated-path differential absorption laser-based system was deployed to Paris in November 2015. The system provided spatially averaged concentration measurements over a 25 km^2 area within the heart of the city, along with estimates of the time varying 2-D spatial distribution. The following sections describe the measurement approach and review results from a full year of CO_2 measurements and recent local experiments using a similar system for measuring CH_4.

2 METHODOLOGY

The Greenhouse gas Laser Imaging Tomography Experiment (GreenLITE™) utilizes a unique continuous wave laser absorption spectroscopy approach to enable long-path atmospheric concentration measurements originally developed for an airborne instrument described in reference [2].

Figure 1: Top left – A GreenLITE™ transceiver. Top right – An installed GreenLITE™ reflector. Bottom – System layout as installed in Paris with 15 reflectors and two transceivers covering ~25 km^2.

In addition to the long-path concentration measurements, GreenLITE™ utilizes a configuration that results in multiple intersecting paths, which can be used to provide estimates of the 2-D spatial distribution of the concentration within the field of regard.

2.1 Measurement Description

The GreenLITE™ transceivers deployed in Paris, France, each use two fiber-coupled distributed feedback lasers (DFBs) to generate laser light at a wavelength that is absorbed by the gas of interest (1571.112 nm) and another nearby wavelength that has significantly lower absorption (1571.052 nm). Both wavelengths are subjected to unique low-frequency amplitude modulations and combined in fiber before being simultaneously transmitted via an F/2 25.4 mm off-axis parabola (OAP) to a 127 mm hollow retroreflector, which returns the light to be collected by a custom 152.4 mm F/4 coaxial OAP-to-fiber receiver telescope and sent to a room-temperature 1.5 mm InGaAs detector. The electrical signal is then amplified via a transimpedance amplifier and digitized. The return signal amplitude and phase from the individual wavelengths are then reconstructed using a digital lock-in amplifier approach. Prior to transmission, a small portion of the beam is sent to a reference InGaAs detector in order to monitor the relative transmitted signal levels for normalizing the return. A simple block diagram is given in Figure 2.

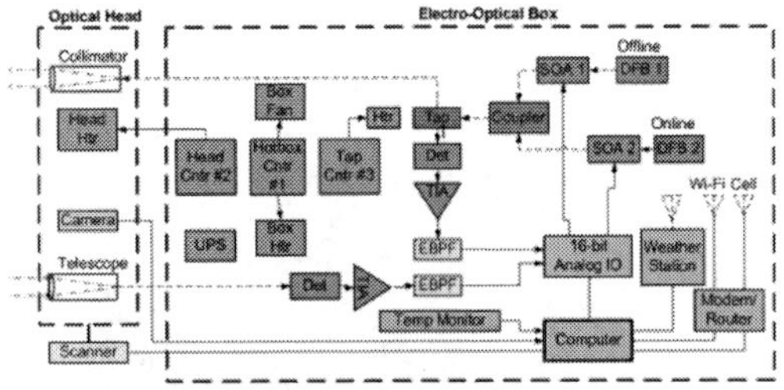

Figure 2: Basic block diagram of the GreenLITE™ transceiver. Acronyms in the figure are as follows: Htr – heater, Cntr – controller, UPS – uninterruptible power supply, Det – detector, TIA – transimpedance amplifier, EBPF – electronic bandpass filter, SOA – semiconductor optical amplifier, DFB – distributed feedback laser, and IO – input/output.

The resulting signals are used to calculate the differential absorption between the two wavelengths over the path-length determined by the transmitted channel phase delay relative to the reference phase. The transceivers are mechanically scanned to each of the 15 retroreflectors in sequence, and require 10 seconds at each to retrieve the differential transmission along the path defined by the transceiver and reflector locations as illustrated in Figure 1. The differential absorption information is sent, along with atmospheric state data (temperature, pressure, relative humidity, wind speed and direction), to a cloud-based custom signal processing software package. The information is used in conjunction with a radiative transfer model (LBLRTM [3]) to convert the observed data into a dry-air mole-fraction value, in parts per million by volume (ppmv), for each 10 second measurement. The combination of measurements from the overlapping paths is used in a sparsely-sampled tomographic-like reconstruction algorithm to produce the estimate of the 2-D spatial distribution. Additional information describing the system can be found in references [4, 5]. The data products from the processed system data are continually streamed near-real-time via a user web interface.

3 RESULTS

The GreenLITE™ system operated continuously over the city of Paris, France, from November 04, 2015, to November 14, 2016. An example of a typical web-based output at a given point in time is provided in Figure 3.

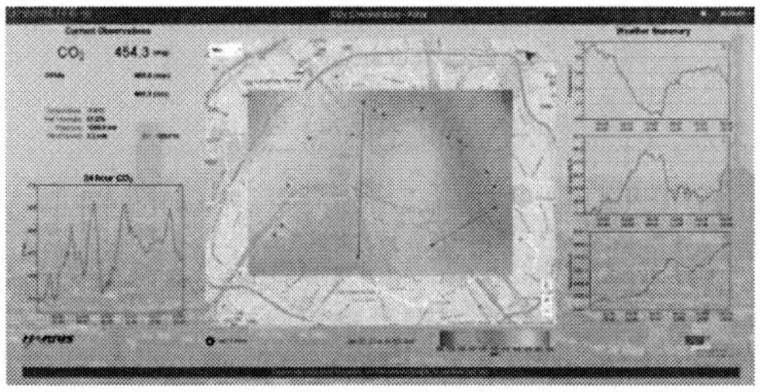

Figure 3: GreenLITE™ web interface output for 1/20/2016 at 20:50 GMT. The left-hand side shows the current spatially averaged CO_2 concentration and the concentration as a function of time (plotted in lower left corner). Also on the left, the maximum and minimum CO_2 concentrations across the field are shown with the current atmospheric state data. The center shows the 2-D estimate of the spatial distribution. The right-hand side shows the measured weather parameters.

Over the one-year period, 3.57 million (M) raw differential transmission samples were collected, 1.91 M samples passed quality control with dry air mole fraction retrieved, and 64 thousand tomographic 2-D reconstructions were generated. Out of the 376 days the instrument was installed, valid measurements were collected on all but 6 days, resulting in 370 days of operation.

There were two primary instruments that had data available during the deployment of the GreenLITE™ system. A Picarro cavity ring-down spectrometer (CRDS) and a Thermo Fisher Delta Ray had coincident measurements available during the Paris deployment; comparisons are shown in Figure 4.

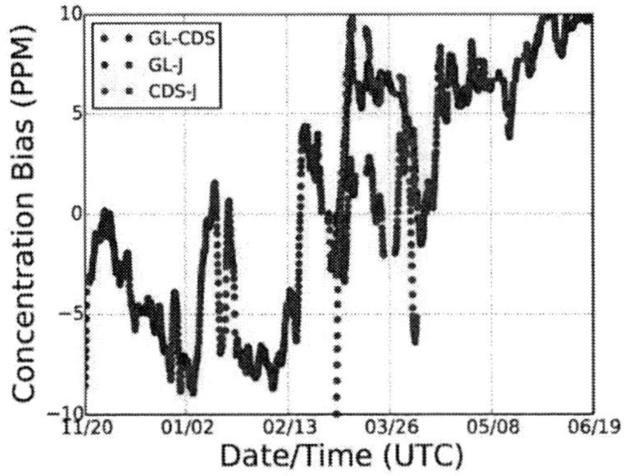

Figure 4: Comparison of GreenLITE™ (GL) to a Picarro CRDS (CDS) located outside of the GL footprint and the GL versus the Thermo Fisher Delta Ray (J), as well as the CDS versus J.

In general, the in situ instruments agreed well with GreenLITE™, especially the Thermo Fisher Delta Ray, which was located near one of the GreenLITE™ transceivers for a 5-week period in March - April 2016. The only long-term record was from the Picarro at the CDS site, which showed an offset of about 5.2 ppm, on average, over the year, and showed a seasonal variation of about 5 ppm, as shown in Figure 5.

An example of other interesting observations over the course of the experiment includes information on daily cycles that are clearly driven by human activity. Figure 6 shows an example of this, comparing weekday diurnal cycles to weekend diurnal cycles. Other things that were seen are clear correlations between CO_2 and nitrogen dioxide (NO_2), and an anti-correlation between CO_2 and ozone (O_3), using GreenLITE™ and AirParif data.

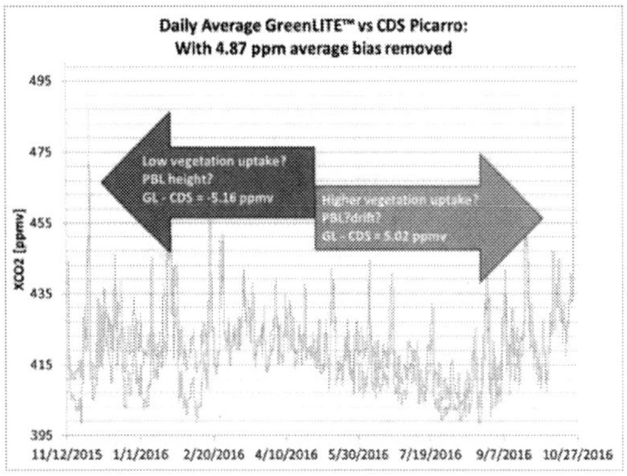

Figure 5: Full-year comparison of hourly averages between GreenLITE™ and the CDS Picarro instrument with the average bias removed. Residual error is nearly equal but in opposite directions from Nov - April and May - Oct. Further experiments are required to understand these differences, but they appear to be related to the biogenic signature seen by the Picarro located at the CDS site, which is in an urban park.

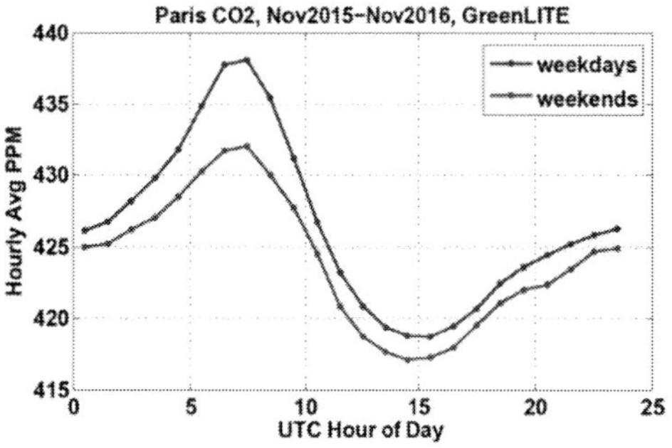

Figure 6: An example of GreenLITE™ observations clearly capturing differences from weekdays to weekends driven by human activity.

Recently, GreenLITE™ has undergone experiments at the Harris Farm Test Site in New Haven, Indiana, to demonstrate the same capability for CH_4 monitoring as that demonstrated in Paris for CO_2. This included a controlled release, which was clearly identified above the background, as seen in Figure 7. The left-hand plots in Figure 7 represent the time just after the start of the release near the center of the measurement grid at 19:15 GMT. The righthand plots represent the full strength of the release at

19:20 GMT, including the windswept plume due to a 1.9 m/s wind out of the southwest.

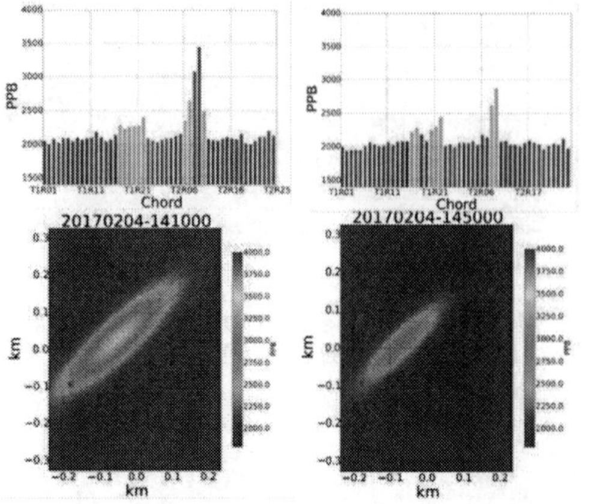

Figure 7: A GreenLITE™ collection of CH_4 during a controlled release at the Harris Farm Test Site (Top ppb), along with corresponding 2-D reconstructions (bottom ~0.4 km W X 0.6 km H).

4 CONCLUSIONS

The GreenLITE™ instrument has demonstrated reliable autonomous operation, measuring CO_2 over the city of Paris, France, for a one-year period. The differences in the GreenLITE™ measurements and the independent in situ measurements are comparable to comparisons between the in situ instruments themselves, which is promising but not sufficient for validation. Further examination of the data set, including combinations with additional information such as boundary layer height and 3-D wind fields, or direct insertion into existing inversion schemes, are planned for the coming year.

Recent experiments have demonstrated the ability of the GreenLITE™ measurement concept to be applied to CH_4 measurements by clearly capturing a controlled release at the Harris Farm Test Site.

We are currently seeking opportunities to participate in validation efforts focused on determining the absolute accuracy of the open path measurements and the 2-D reconstructions for both CO_2 and CH_4.

ACKNOWLEDGEMENTS

The authors acknowledge Laboratoire des Sciences du Climat et de l'Environnement, LSCE/IPSL, CEA-CNRS-UVSQ, and Université Paris-Saclay for their support in getting GreenLITE™ installed in Paris and providing access to the in situ data. We also acknowledge the support of Enviroearth for the installation and removal of the hardware and Harris management for funding the Paris demonstration.

References

[1] F. Bréon, G. Broquet, V. Puygrenier, F. Chevallier, I. Xueref-Rémy, M. Ramonet, E. Dieudonné, M. Lopez, M. Schmidt, O. Perrussel and P. Ciais, "An attempt at estimating Paris area CO2 emissions from atmospheric concentration measurements," *Atmos. Chem. Phys.,* vol. 15, pp. 1707-1724, 2015.

[2] J. T. Dobler, F. W. Harrison, E. V. Browell, B. Lin, D. P. McGregor and S. Kooi, "Atmospheric CO2 column measurements with an airborne intensity-modulated continuous wave 1.57um fiber laser lidar," *Applied Optics,* vol. 52, no. 12, pp. 2874-2892, 2013.

[3] J. T. Dobler, N. Blume, M. Braun, T. S. Zaccheo, T. G. Pernini and C. Botos, "Greenhouse Gas Laser Imaging Tomography," in *27th Inernational Laser Radar Conference*, New York, 2015.

[4] S. A. Clough, M. W. Shephard, E. J. Mlawer, J. S. Delamere, M. J. Iacono, K. Cady-Pereira, S. Boukabara and P. D. Brown, "Atmospheric radiative transfer modeling: a summary of the AER codes, Short Communication," *J. Quant. Spectrosc. Radiat. Transfer,* vol. 91, pp. 233-244, 2005.

[5] J. T. Dobler, T. S. Zaccheo, N. Blume, M. Braun, C. Botos and T. G. Pernini, "Spatial mapping of greenhouse gases using laser absorption spectrometers at local scales of interest," in *Lidar Technologies, Techniques, and Measurements for Atmospheric Remote Sensing XI, Upendra N. Singh; Doina N. Nicolae, Editors, Proceedings of SPIE Vol. 9645*, Tolouse, France, 2015.

TRIPLE-WAVELENGTH LIDAR OBSERVATIONS OF THE LINEAR DEPOLARIZATION RATIO OF DRIED MARINE PARTICLES

Moritz Haarig [1,*], Albert Ansmann [1], Holger Baars [1], Ronny Engelmann [1], Dietrich Althausen [1], Stephanie Bohlmann [1,a], Josef Gasteiger[2] and David Farrell [3]

[1]*Leibniz Institute for Tropospheric Research (TROPOS), Germany, *haarig@tropos.de*
[2]*Faculty of Physics, University of Vienna, Austria*
[3]*Caribbean Institute for Meteorology and Hydrology (CIMH), Barbados*
[a]*now at: Finnish Meteorological Institute, Kuopio, Finland*

ABSTRACT

For aerosol typing with lidar, sea salt particles are usually assumed to be spherical with a consequently low depolarization ratio. Evidence of dried marine particles at the top of the humid marine aerosol layer with a depolarization ratio up to 0.1 has been found at predominately maritime locations on Barbados and in the Southern Atlantic. The depolarization ratio for these probably cubic sea salt particles has been measured at three wavelengths (355, 532 and 1064 nm) simultaneously for the first time and compared to model simulations.

1 INTRODUCTION

The oceans are a large source of aerosol particles. The emitted sea salt is a widely spread aerosol type and serves as cloud condensation nuclei and ice-nucleating particle over the ocean [1]. The highly hygroscopic sea salt particles are usually of spherical shape in humid marine environments, resulting in a low particle linear depolarization ratio (PLDR = 0.03 ± 0.01 at 532 nm [2]).

Under dry conditions, sea salt particles have a cubic-like shape [3]. With increasing relative humidity sea salt particles grow by water-uptake. Once a relative humidity of 70-74% is reached, sea salt particles are spherical [4].

In a laboratory study, Sakai [5] measured the PLDR of spherical sea salt particles (0.01) and crystalline sea salt particles (0.08) at 532 nm.

In lidar field measurements, first evidence of an enhanced depolarization ratio was reported by Murayama [6] for dry marine particles based on measurements in Tokyo in 1999, but a clear separation from a potential dust influence was not possible.

Recent lidar measurements have been performed in the trade wind zone in the Eastern Caribbean and in the Southern Atlantic environments to study pure marine aerosols.

Pure marine conditions prevailed during the winter campaign (February-March 2014) of the Saharan Aerosol Long-Range Transport and Aerosol-Cloud-Interaction Experiment (SALTRACE) [7]. The campaign was conducted at the Caribbean Institute for Meteorology and Hydrology (CIMH), located at the west coast of Barbados (13°N, 59°W). From 23 till 25 February 2014 enhanced depolarization was continuously observed at the top of the marine aerosol layer (MAL) because of dried marine particles. The free troposphere was absolutely dust-free so that no disturbance by non-marine aerosols was possible. For the first time, the particle linear depolarization ratio at three wavelengths (355, 532 and 1064 nm) of non-spherical sea salt layers has been measured. A case study of the dry marine particle depolarization ratio at three wavelengths will be discussed in section 3.

2 LIDAR SYSTEM

The multi-wavelength polarization Raman lidar BERTHA (Backscatter Extinction lidar-Ratio

Temperature Humidity profiling Apparatus) of the Leibniz Institue for Tropospheric Research (TROPOS) is a container-based mobile lidar system. It enables the measurement of the depolarization ratio at three wavelengths simultaneously. Currently it operates as a 3+2+3 lidar system (3 backscatter coefficients, 2 extinction coefficients and 3 depolarization ratios). A more detailed description has been given at the 27th ILRC by Haarig [8] and in [9].

The signals are detected with a range resolution of 7.5 m and a time resolution of 10 s. To ensure the good quality of the depolarization measurements, a Δ90° calibration [10] was performed regularly. Vaisalla RS92 radiosondes were launched and the temperature and pressure profiles were used to compute the molecular contribution to the measured lidar signals.

3 MEASUREMENT EXAMPLE

A very pronounced case of a thin layer of dried marine particles was observed over Barbados on 24 February 2014. A well mixed marine aerosol layer reached up to 2 km (Fig. 1). The feature of interest is the enhanced signal in the upper 200 m of the aerosol layer measured in the 1064 nm cross polarized channel. This feature is not visible in the color plot of the 1064 nm range-corrected total signal (upper panel in Fig. 1).

The radiosonde launched at 23:07 UTC shows a strong temperature inversion of 4 K at 2 km height (Fig. 2). In the MAL, RH was about 70% to 80%. RH decreased to values of about 5% within about 200 m just above MAL top.

Marine particles lost their water shell and their spherical shape at the top of the MBL and this caused enhanced depolarization. Figure 2 shows the relative humidity profile from the radiosonde and the 1064 nm volume depolarization. It can be seen that the volume depolarization partly reaches 0.06 till 0.08 at

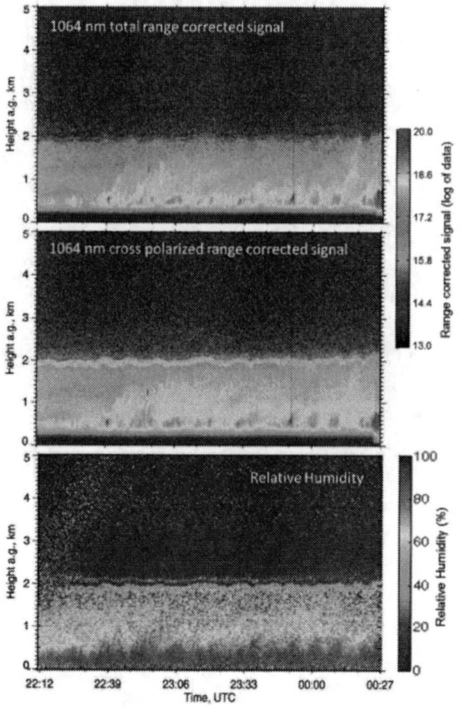

Figure 1: Range-corrected lidar signals for the 1064 nm total channel (upper panel) and 1064 nm cross polarized channel (central panel), measured at Barbados on 24 February 2014, 22:12 – 00:27 UTC. RH (lower panel) is derived from the water vapor channel (407 nm) and the temperature profile of the radiosonde (23:07 UTC).

the top of the MAL.

The particle backscatter coefficient and the PLDR for the three wavelengths (355, 532 and 1064 nm) of the BERTHA lidar system are shown in Fig. 3. In the 2 hour profiles, a maximum PLDR of 0.055 (355 nm), 0.068 (532 nm), and 0.038 (1064 nm) is reached in the thin layer of dried marine aerosol. Below in the humid MAL, the PLDR is below 0.02. The laboratory measurements by Sakai [5] for dried sea salt particles (PLDR = 0.08 at 532 nm) are in good agreement with our finding. The

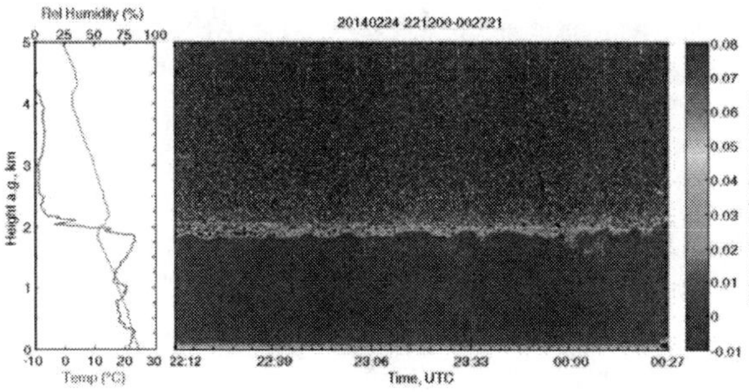

Figure 2: The temperature and relative humidity profile of the radiosonde launched at 24 February 2014 23:07 UTC (left) and the height-time display of the 1064 nm volume linear depolarization ratio (right).

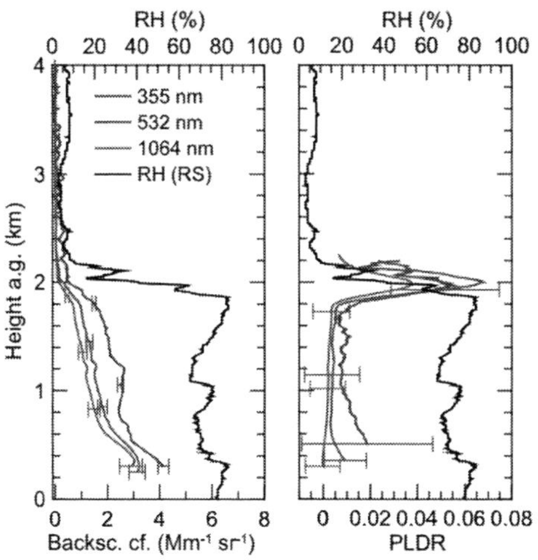

Figure 3: Vertical profiles of particle optical properties observed on 24 February 2014, 22:12 - 24:00 UTC. Shown are the particle backscatter coefficient (left) and the PLDR (right) at 355 nm (blue), 532 nm (green) and 1064 nm (red). The profile of RH measured by the radiosonde is shown in both plots. Note that the drastic change in RH from 80% to less than 10% occurs within 200 m.

profile of RH was added in Fig. 3 to show the coincidence of the decrease of RH with the increase of PLDR.

Model calculations predict a depolarization ratio of up to 0.15 at all three wavelengths for perfectly cubic sea salt particles for realistic size distributions of dried marine particles. However the dried sea salt particles cannot be of perfectly cubic shape, but have much smoother edges. Thus, the measured values for the enhanced depolarization ratio have to be smaller than the model predictions. Furthermore, mixing of spherical and cubic particles may occur at MAL top, what also decreases the depolarization ratio.

Similar results have been found on board of a research vessel over the Southern Atlantic. An extended Portable Lidar System (Polly^{XT}) [11] of TROPOS was deployed on board of the German research vessel *Polarstern* in the framework of OCEANET. The PLDR at 355 and 532 nm of dried sea salt particles was measured.

4 CONCLUSIONS

By means of a triple-wavelength polarization lidar the spectral dependence of the depolarization ratio of dried marine particles has been unambiguously measured for the first time. The observed enhanced depolarization ratio is in good agreement with laboratory studies. Discrepancies to model calculations indicate that the dried marine particles have a cubic-like shape with smoothed edges.

References

[1] DeMott, P. J., Hill, T. C. J., McCluskey, C. S., Prather, K. A., Collins, D. B., Sullivan, R. C., Ruppel, M. J., Mason, R. H., Irish, V. E., Lee, T., Hwang, C. Y., Rhee, T. S., Snider, J. R., McMeeking, G. R., Dhaniyala, S., Lewis, E. R., Wentzell, J. J. B., Abbatt, J., Lee, C., Sultana, C. M., Ault, A. P., Axson, J. L., Diaz Martinez, M., Venero, I., Santos-Figueroa, G., Stokes, M. D., Deane, G. B., Mayol-Bracero, O. L., Grassian, V. H., Bertram, T. H., Bertram, A. K., Moffett, B. F., Franc, G. D., 2016: Sea spray aerosol as a unique source of ice nucleating particles, *Proceedings of the National Academy of Sciences*, **113**, 5797-5803

[2] Groß, S., Esselborn, M., Weinzierl, B., Wirth, M., Fix, A., Petzold, A., 2013: Aerosol classification by airborne high spectral resolution lidar observations, *Atmospheric Chemistry and Physics*, **13**, 2487-2505

[3] Chamaillard, K., Jennings, S., Kleefeld, C., Ceburnis, D., Yoon, Y., 2003: Light backscattering and scattering by nonspherical sea-salt aerosols, *Journal of Quantitative Spectroscopy and Radiative Transfer*, **79-80**, 577 - 597

[4] Tang, I., Tridico, A. and Fung, K., 1997: Thermodynamic and Optical Properties of sea salt aerosols, *Journal of Geophysical Research* **102**, 23269-23275

[5] Sakai, T., Nagai, T., Zaizen, Y., Mano, Y., 2010: Backscattering linear depolarization ratio measurements of mineral, sea-salt, and ammonium sulfate particles simulated in a laboratory chamber, *Appl. Opt.* **49**, 4441-4449

[6] Murayama, T., Okamoto, H., Kaneyasu, N., Kamataki, H. and Miura, K., 1999: Application of lidar depolarization measurement in the atmospheric boundary layer: Effects of dust and sea-salt particles, *Journal of Geophysical Research* **104**, 31781-31792

[7] Weinzierl, B., Ansmann, A., Prospero, J. M., Althausen, D., Benker, N., Chouza, F., Dollner, M., Farrell, D., Fomba, W. K., Freudenthaler, V., Gasteiger, J., Groß, S., Haarig, M., Heinold, B., Kandler, K., Kristensen, T. B., Mayol-Bracero, O. L., Müller, T., Reitebuch, O., Sauer, D., Schäfler, A., Schepanski, K., Spanu, A., Tegen, I., Toledano, C., Walser, A., 2017: The Saharan Aerosol Long-range Transport and Aerosol-Cloud-Interaction Experiment (SALTRACE): overview and selected highlights, *Bulletin of the American Meteorological Society*, **98**, 1427–1451

[8] Haarig, M., Althausen, D., Ansmann, A., Klepel, A., Baars, H., Engelmann, R., Groß, S., Freudenthaler, V., 2016: Measurement of the Linear Depolarization Ratio of Aged Dust at Three Wavelengths (355, 532 and 1064 nm) Simultaneously over Barbados, *EPJ Web of Conferences*, **119**, 18009

[9] Haarig, M., Ansmann, A., Althausen, D., Klepel, A., Groß, S., Freudenthaler, V., Toledano, C., Mamouri, R.-E., Farrell, D. A., Prescod, D. A., Marinou, E., Burton, S. P., Gasteiger, J., Engelmann, R., and Baars, H., 2017: Triple-wavelength depolarization-ratio profiling of Saharan dust over Barbados during SALTRACE in 2013 and 2014, *Atmos. Chem. Phys.*, **17**, 10767–10794

[10] Freudenthaler, V., Esselborn, M., Wiegner, M., Heese, B., Tesche, M., Ansmann, A., Müller, D., Althausen, D., Wirth, M., Fix, A., Ehret, G., Knippertz, P., Toledano, C., Gasteiger, J., Garhammer, M., Seefeldner, M., 2009: Depolarization ratio profiling at several wavelengths in pure Saharan dust during SAMUM 2006, *Tellus B*, **61**, 165-179

[11] Engelmann, R., Kanitz, T., Baars, H., Heese, B., Althausen, D., Skupin, A., Wandinger, U., Komppula, M., Stachlewska, I. S., Amiridis, V., Marinou, E., Mattis, I., Linné, H., Ansmann, A., 2016: The automated multiwavelength Raman polarization and water-vapor lidar PollyXT: the neXT generation, *Atmos. Meas. Tech.*, **9**, 1767-1784

EPJ Web of Conferences **176**, 05015 (2018) https://doi.org/10.1051/epjconf/201817605015
ILRC 28

TWO-YEAR OPERATION OF THE LIDAR1200: FROM FINE-SCALE TROPOSPHERIC STRUCTURES TO LOWER STRATOSPHERIC WATER VAPOR DETECTION

Hélène Vérèmes[(1,2)]*, Guillaume Payen[2], Philippe Keckhut[3], Valentin Duflot[(1,2)], Jean-Luc Baray[4], Jean-Pierre Cammas[(1,2)], Jimmy Leclair de Bellevue[1], Françoise Posny[1], Stéphanie Evan[1], Jean-Marc Metzger[2], Nicolas Marquestaut[2], Franck Gabarrot[2], Susanne Meier[5], Holger Vömel[6] and Ruud Dirksen[5]

[1]Laboratoire de l'Atmosphère et des Cyclones, UMR8105, Saint-Denis de La Réunion, France,
**Email: helene.veremes@univ-reunion.fr*
[2]Observatoire des Sciences de l'Univers de La Réunion, UMS3365, Saint-Denis de la Réunion, France,
[3]Laboratoire ATmosphères, Milieux, Observations Spatiales-IPSL, UMR8190 CNRS, UVSQ, UPMC, Guyancourt, France,
[4]Laboratoire de Météorologie Physique, UMR6016, Observatoire de Physique du Globe de Clermont-Ferrand, Clermont-Ferrand, France,
[5]Deutscher Wetterdienst, Meteorological Observatory Lindenberg, Lindenberg, Germany,
[6]National Center for Atmospheric Research, Boulder, CO, USA

ABSTRACT

The 2-year lidar water vapor database (November 2013 - October 2015) of the Maïdo Observatory (Reunion Island / 21°S,55.5°E) is now processed. The performances of the lidar in providing accurate vertical structures are shown to be good. The ability to measure quantities of a few ppmv in the lower stratosphere is demonstrated (based on Cryogenic Frost point Hygrometer sonde/lidar profiles comparisons) for a 48-hour integration time period, up to 22 km (with a vertical resolution of 1.3 km).

1 INTRODUCTION

Since the inauguration of the Maïdo Observatory in October 2012 [1], a Raman lidar system (called "Lidar1200") has been monitoring, several nights a week, water vapor and temperature, from the ground up to the lower stratosphere and in the stratosphere up to the mesosphere, respectively [2].

The calibration methodology of the water vapor Lidar1200 dataset is based on the detection of quasi-stationary periods of the hourly calibration coefficient derived from GNSS observations. The GNSS technique has been chosen in order to improve the reliability of the calibration and the robustness of the water vapor measurement on a long-term perspective. This methodology has been successfully used in the 2013-2015 dataset [3]. The average standard uncertainty in the transfer of the calibration is around 3 %. The 2-year database has been processed and a first evaluation of the mean performances has been done, results are presented in the next section.

The Lidar1200 was involved in 5 measurement campaigns at the Maïdo Observatory: MALICCA-1 (MAïdo LIdar Calibration CAmpaign) (April 2013) [2][4], MALICCA-2 (November 2013), LIDEOLE-1 (November 2014), MORGANE (Maïdo ObservatoRy Gas and Aerosol NDACC Experiment) (May 2015) and LIDEOLE-2 (September 2015). The field campaigns play an important role in the evaluation of the performances of the Lidar1200. The results of the instrumental comparisons led during the MORGANE campaign illustrate the quality of the Lidar1200 data. The comparison between water vapor lidar profiles and CFH sonde's data are detailed in the next section.

With an average of 1 to 2 measurement night(s) per week and the intensive measurement campaigns, the Lidar1200 will be able to supply regular measurements. The Maïdo Observatory is located in the Southern Hemisphere having tropical and subtropical characteristics (depending on the season). Its data could be used to document various dynamical processes. The analysis of the first 2 years of data of the Lidar1200, detailed in the last sections, led to the characterization of structures that could be detected on the lidar water vapor profiles: from fine-scale structures in the troposphere to seasonal variability up to the Upper Troposphere / Lower Stratosphere (UT/LS).

2 PERFORMANCES OF THE LIDAR1200 IN WATER VAPOR MONITORING

2.1 Mean performances

Mean profiles have been calculated and differentiated into three classes characterized by their integration time: 10, 40 and 240 minutes [3]. The measurement sessions have been averaged and processed with the algorithm used on a routine basis. The mean total uncertainty for the three profiles is lower than 15 % below 9 km asl regardless of the integration time. The threshold of the total uncertainty of 20 % is reached at 11, 13 and 16 km asl for 10-, 40- and 240-min profiles, respectively [3]. Those uncertainties are representative of a mean profile for each integration time, those values should be evaluated case-by-case depending on the measurement conditions (cloud fraction, moon phase...). To reach the tropopause with a weaker uncertainty, it appears important to integrate data on a longer integration time period than 4 hours (which represents a typical night of measurement at the Maïdo Observatory under good meteorological conditions).

2.2 Performances in the UT/LS

A 9-hour session has been operated in September 2015. The profiles reached 16 km with an uncertainty of ~35 % [3]. To cross the tropopause with an uncertainty lower than 20 %, several nights of measurement are necessary. The profiles of each campaign (integrating 12 to 48 hours) demonstrated that at least 10 hours of integration are necessary to reach the lower stratosphere [3]. To increase the upper limit of the profiles of only 1 km, it requires twice or three times that value. A threshold of 48 hours (MORGANE campaign) has been identified to find a significant improvement of the performances; the water vapor mixing ratio is retrieved up to 22 km with an uncertainty lower than 0.8 ppmv / 22 % and with a vertical resolution of 1.3 km [3].

2.3 Validation: comparison with CFH sonde's data

5 CFH sondes were launched simultaneously with Lidar1200 measurements. The water vapor mixing ratio profiles from both techniques were compared [3]. When taking the CFH measurements as a reference, the results led to the validation of the Lidar1200 profiles between 2.2 and 22 km asl and to the establishment of integration time criteria (depending on the altitude and the thickness of the observed water vapor structures). The average biases between the CFH and the Lidar1200 data in the lower and middle troposphere, in the upper troposphere and in the lower stratosphere (Fig. 1) were lower than 12 % [3]. The CFH and the Lidar1200 are in agreement up to the lower stratosphere (22 km asl), with respect to their uncertainties. This agreement validates the data processing scheme of the Lidar1200. The first results between the CFH and the Lidar1200 profiles (MORGANE campaign) shows that the instruments disagree above 22 km asl [3]. It will be further investigated through future measurement campaigns.

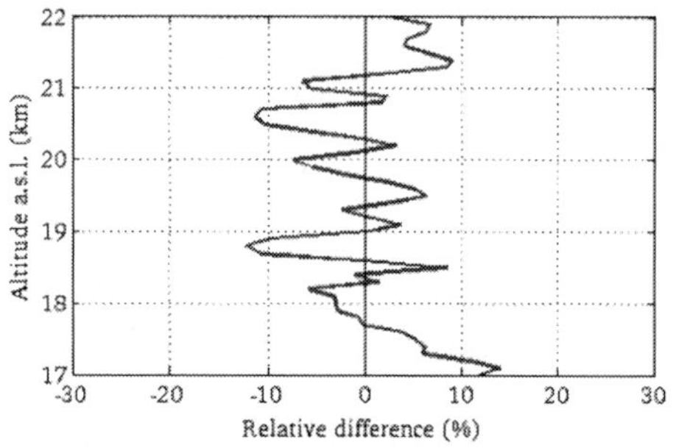

Figure 1 Difference between the lidar (integration of all the data) and the mean CFH water vapor mixing ratio profiles of the MORGANE campaign between 17 and 22 km asl (CFH data are taken as reference)

3 DETECTION OF FINE-SCALE STRUCTURES IN THE TROPOSPHERE

Comparisons with the CFH radiosoundings of the field campaigns show that the Lidar1200 was able to detect a structure of 100-m thick with a time step of 5 minutes at 9 km asl [3]. The Lidar1200 is able to detect and monitor tropospheric water vapor structures of a few hundred meters and a small lifetime up to the middle troposphere.

 https://doi.org/10.1051/epjconf/201817605015

4 PRELIMINARY CHARACTERIZATION OF SEASONAL VARIATIONS OF WATER VAPOR

4.1 Seasonal cycle over Reunion Island

The analysis of monthly profiles extracted from the Lidar1200 dataset (between November 2013 and October 2015) illustrates the ability of the lidar to characterize the seasonal variability of the water vapor [3]. The monthly profile corresponds to the integration of the data of one month with the same calibration coefficient, which represents 8 to 48 hours of measurements. The temporal series shows a clear seasonal cycle from the ground up to the middle troposphere. Some variability appears in the upper troposphere and in the lower stratosphere but no seasonal cycle could be characterized. The vertical range of this seasonal cycle and the amplitude of the water vapor variability between summer and winter from those preliminary results [3] are different from the first water vapor climatology established in 2012 on a previous configuration of the system [5]. The improvement of the system and the data processing and the specific calibration methodology benefited the characterization of the seasonal cycle of water vapor in the South-West Indian Ocean.

4.2 Signatures of water vapor transport

Seasonal profiles were calculated from the monthly profiles: the 3 monthly profiles of each season were averaged. Their analysis illustrates the ability of the lidar to detect signatures of water vapor transport. In the lower troposphere, the vertical structure of the humidity field associated with the trade winds inversion (relative maximum at the basis and relative minimum at the top of the inversion layer) is detailed, as its thickening during the austral summer, in Figure 2. It documents the role of the cumulus clouds in the moistening of the upper part of the trade winds layer. The results show the detection in altitude of the humid air flux in the middle troposphere, convected by the general Hadley circulation, around 10-11 km in austral summer and 8-9 km in austral autumn (Fig. 2).

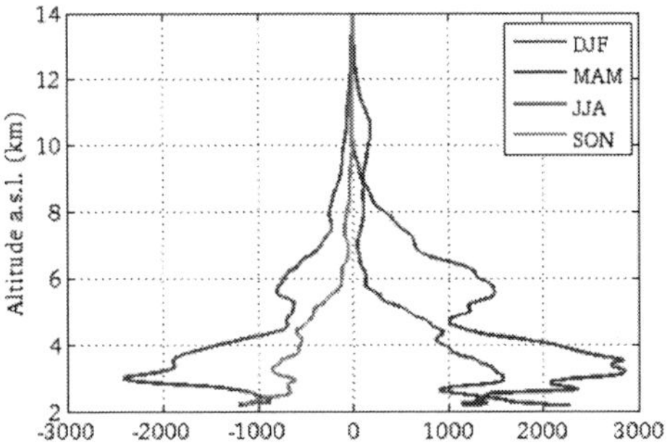

Figure 2 Difference between seasonal and the total mean water vapor mixing ratio profiles (summer in blue, autumn in green, winter in red and spring in cyan) [data: quasi-stationary periods / Lidar1200]

5 CONCLUSIONS

The Lidar1200 is able to provide high-resolution (vertical and temporal) water vapor profiles in the troposphere on a routine basis. This system is able to reach the lower stratosphere and to detect water vapor mixing ratios of only a few ppmv. A rigorous work on the design of the system and on the evaluation of the data has been carried out on the instrument in order to meet the definition of « a reference observation » [6]. This lidar is dedicated to provide regular water vapor measurements to international networks such as NDACC (Network for the Detection of Atmospheric Composition Change) and GRUAN (GCOS Reference Upper Air Network).

A user's guide for the time of integration of the data has been defined empirically thanks to the results of the instrumental comparisons [3] and is based on a compromise between the vertical resolution and the uncertainty. The detection of structures of a few hundred meters can be realized with 10-min integration files (with a vertical resolution lower than 100 m). 40 minutes are necessary to realize instrumental comparisons in the troposphere. To reach the upper troposphere, an integration time of 4 hours should be covered. Around 10 hours of integration are required to cross the tropopause with a vertical resolution of about 1.3 km.

The exploitation of the Lidar1200 database on its own and with others measurements of the Maïdo Observatory will bring new results on the characterization of the tropospheric water vapor variability, the monitoring of the water vapor in lower stratosphere, the study of stratosphere-troposphere exchanges [7] and the validation of satellite measurements.

ACKNOWLEDGEMENTS

The lidar data used in this publication were obtained as part of the Network for the Detection of Atmospheric Composition Change (NDACC) and a level 2 product as daily vertical water vapor profiles will be publicly available through the NDACC portal and the French atmospheric data portal. The raw data and other products will be available on demand and available at the OPAR web portal. We acknowledge the ACTRIS project and the support of the European Community (Research Infrastructure Action under the FP7 "Capacities" specific program for Integrating Activities, ACTRIS Grant Agreement no. 262254). Acknowledgements are addressed to the European Commission (FEDER program), Région Réunion and CNRS for their strong support in the building of the Maïdo facility. Acknowledgements are also addressed to the Université de La Réunion and CNRS for their strong support of the OPAR station and the OSU-R activities. The research leading to these results was supported by the French LEFE INSU-CNRS Program (VAPEURDO and VEGA projects). The authors gratefully acknowledge E. Golubic, P. Hernandez and L. Mottet, Y. Courcoux, J. Porteneuve, A. Hauchecorne and D. Dionisi.

References

[1] Baray et al., 2013: Maïdo observatory: a new high-altitude station facility at Reunion Island (21° S, 55° E) for long-term atmospheric remote sensing and in-situ measurements, *Atmos. Meas. Tech.*, **6**, 2865-2877.

[2] Dionisi et al., 2015: Water vapor observations up to the lower stratosphere through the Raman lidar during the Maïdo Lidar Calibration Campaign, *Atmos. Meas. Tech.*, **8**, 1425-1445.

[3] Vérèmes et al., 2017: A Raman lidar at Maïdo Observatory (Reunion Island) to measure the water vapor in the troposphere and in the lower stratosphere: calibration and validation, *Atmos. Meas. Tech. Discuss.*, in review.

[4] Keckhut et al., 2015: Introduction to the MALICCA campaign dedicated to the validation of Upper Air Meteorological parameters, *J. Appl. Remote Sens*, **9**(1), 094099.

[5] Hoareau et al., 2012: A Raman lidar at La Reunion (20.8°S, 55.5°E) for monitoring water vapour and cirrus distributions in the subtropical upper troposphere : preliminary analyses and description of the future system, *Atmos. Meas. Tech.*, **5**(6), 1333-1348.

[6] Immler et al., 2010: Reference Quality Upper-Air Measurements: guidance for developing GRUAN data products, *Atmos. Meas. Tech.*, **3** (5), 1217–1231.

[7] Vérèmes et al., 2016: Multiple subtropical stratospheric intrusions over Reunion Island: Observational, Lagrangian, and Eulerian numerical modeling approaches, *J. Geophys. Res. Atmos.*, **121**, 14,414–14,432.

OPO LIDAR SOUNDING OF TRACE ATMOSPHERIC GASES IN THE 3 – 4 µM SPECTRAL RANGE

Oleg A. Romanovskii[1,2]*, Sergey A. Sadovnikov[1], Olga V. Kharchenko[1], Semen V. Yakovlev[1,2]

[1] *V.E. Zuev Institute of Atmospheric Optics, Tomsk, Russia, *roa@iao.ru*
[2] *National Research Tomsk State University, Tomsk, Russia*

ABSTRACT

The applicability of a KTA crystal-based laser system with optical parametric oscillators (OPO) generation to lidar sounding of the atmosphere in the spectral range 3–4 µm is studied in this work. A technique developed for lidar sounding of trace atmospheric gases (TAG) is based on differential absorption lidar (DIAL) method and differential optical absorption spectroscopy (DOAS). The DIAL-DOAS technique is tested to estimate its efficiency for lidar sounding of atmospheric trace gases. The numerical simulation performed shows that a KTA-based OPO laser is a promising source of radiation for remote DIAL-DOAS sounding of the TAGs under study along surface tropospheric paths. A possibility of using a PD38-03-PR photodiode for the DIAL gas analysis of the atmosphere is shown.

1 INTRODUCTION

To overlap the near- and middle IR region, radiation from optical parametric oscillators based on nonlinear crystals is used [1]. However, laser radiation is required to be monochromatic in the standard DIAL technique. Finally, the problem of standard DIAL measurements at only two radiation wavelengths means that the disturbing absorption is ignored. The result of this is error introduced by a priori uncertain absorption coefficient. These disadvantages do not affect differential optical absorption spectroscopy (DOAS) [2]. However, DOAS capabilities for vertical profiling are restricted; only path-average measurements have been carried out until now. A technique which combines the advantages of both methods—spatial resolution of the DIAL technique and identification of gases in DOAS—could be a promising approach to a solution of the problem. The aim of this work is the development of a technique for lidar sounding of trace atmospheric gases, which combines DIAL and DOAS, and its validation in a numerical experiment for estimating the capabilities of lidar sounding of the gas composition of the atmosphere in the 3–4 µm spectral range with an OPO-based laser system.

2 DIAL AND DOAS TECHNIQUE FOR LIDAR SOUNDING

The P_R signal from a scattering layer of Δz in thickness can be represented as

$$P_R(z,\lambda) = P_0(\lambda)\frac{A_D}{z^2}\eta(\lambda)O(z)\Delta z\beta(z,\lambda)e^{-2\tau(z,\lambda)} \quad (1),$$

where $P_0(\lambda)$ is the laser radiation power, A_D is the instrumental function width, $O(z)$ is the overlapping region of the laser beam and detector's field-of view, $\beta(z,\lambda)$ is the mass coefficient of backscattering radiation, $\eta(\lambda)$ is the efficiency of the receiving-transmitting system, Δz is the spatial resolution along the sounding path, and $\tau(z,\lambda)$ is the attenuation coefficient. The concentration of a gas under study is calculated by the DIAL equation

$$n(z) = \frac{1}{2\Delta\sigma_{abs}\Delta z}\ln\left(\frac{P_R(z,\lambda_{on})P_R(z+\Delta z,\lambda_{off})}{P_R(z,\lambda_{off})P_R(z+\Delta z,\lambda_{on})}\right) \quad (2),$$

where σ_{abs} is the absorption cross-section.

Disadvantages of the DIAL technique are caused by uncertainties of a priori absorption coefficients at two wavelengths. DOAS allows these disadvantages to be avoided, by monitoring the transmission in the UV, visible and IR regions in a broad band (20–100 nm). The first step in DOAS is the calculation of the ratio of the spectrum observed (I_{OBS}) to the reference spectrum (I_{REF}), which is found from the laser spectrum (I_0) measured with the same detector. An atmospheric spectrum at a known content of absorbing gases can be used as I_{REF}. Calculating the ratio (I_{OBS}/I_{REF}) and taking its natural logarithm, the optical depth can be found:

$$-\ln\left(\frac{I_{OBS}}{I_{REF}}\right) = \sum_i \sigma^i_{abs}([n_i(z)]_{OBS}\Delta z_{OBS} - [n_i(z)]_{REF}\Delta z_{REF}) + Y(\beta) \quad (3)$$

 https://doi.org/10.1051/epjconf/201817605016

Equation (3) includes the wavelength-dependent absorption (the first term in Eq. (3)) and the scattering Y(β), which is mainly determined by the backscattering coefficient β. The ratio is independent of the laser radiation spectrum or the spectral dependence of the receiving optics, spectrometer, and detector, which is an important advantage of DOAS. The so measured magnitude is equal to the difference in the concentration of absorbing gases in the reference case and in the case of real atmosphere. A differential spectrum is usually retrieved by several hundreds of points; the number of fitting parameters is no more than six. Thus, equation (2) becomes overdetermined and is solved with the least-square method. Under this approach, the fitting coefficients are varied for the fitting spectrum to match the best with the spectrum observed. In this case, if the absorption coefficients are known, the integral content of each gas can be found.

Simultaneous determination of the concentrations of several gases is an important advantage of DOAS as compared to the DIAL technique.

3 OPO LASER SYSTEM FOR REMOTE SOUNDING OF THE ATMOSPHERE

In this work, we consider a laser system (designed by SOLAR LS Company, Minsk) which is a part of a OPO DIAL lidar and ensures the tunable generation of nanosecond radiation pulses in the 3–4 μm range. Tables 1 and 2 present main specifications of the pumping laser and radiation converter.

Table 1 Specifications of pumping laser

Pulse frequency	10 Hz
Output energy: at 1064 nm	350 mJ
Pulse length at 1064 nm, FWHM	10...13 ns
Beam diameter at 1064 nm	≤ 6 mm
Beam divergence at 1064 nm	~ 1.5 mrad
Stability of pulse energy at 1064 nm	± 2.5 %

Table 2 Specifications of radiation converter

Wavelength tuning range	3…4 μm
Radiation line width	< 5 cm^{-1}
Pulse energy, in the tuning curve peak	> 5
Pulse frequency	10 Hz
Radiation divergence	</= 2 mrad
Wavelength tuning control	with 3 SMs

The DIAL-DOAS technique developed for TAG measurements was validated for estimation of the lidar signal levels, using the specifications of the above described KTA-based OPO laser system.

4 SIMULATION OF TAG LIDAR MEASUREMENTS

Sounding of some atmospheric gases (HCl, HBr, NO_2, and N_2O) along surface atmospheric paths (in the range 1–5 km) has been numerically simulated. The standard midlatitude summer model was used in the simulation. The absorption of all main atmospheric gases was considered; the HCl, HBr, and NO_2 concentrations were taken equal to 1 ppm, and the N_2O concentration, to 1.5 ppm. Table 3 represents the informative wavelengths used in the numerical simulation and appropriate for DIAL-DOAS sounding of the TAGs under study. Spatially and spectrally resolved lidar echo signals in the TAG wavelength range calculated for surface tropospheric paths are shown in Figs. 1a and 1b for N_2O and NO_2, respectively, and in Figs. 2a and 2b, for HBr and HCl, respectively.

Table 3 Informative wavelengths

Gas	λ_{abs}, μm (in air)	ν_{abs}, cm^{-1} (in air)	T_{gas}	T_{dist}
HBr	3.98730	2507.962	0.79	0.99
HCl	3.57191	2799.622	0.60	0.93
N_2O	3.87664	2579.55	0.85	0.90
NO_2	3.42577	2919.051	0.73	0.89

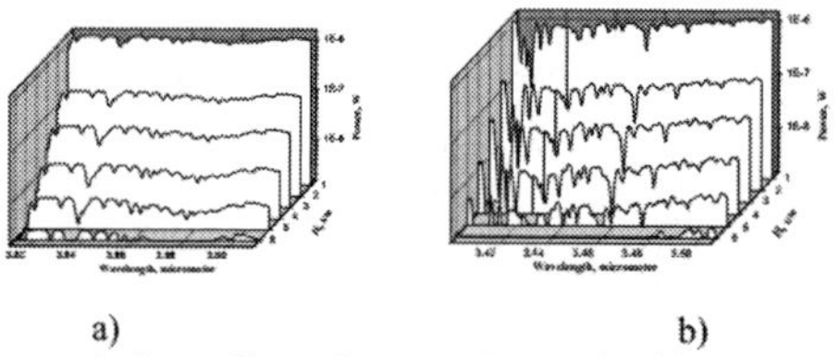

a) b)

Figure 1. Spatially and spectrally resolved lidar echo signals of (a) N_2O and (b) NO_2 sounding

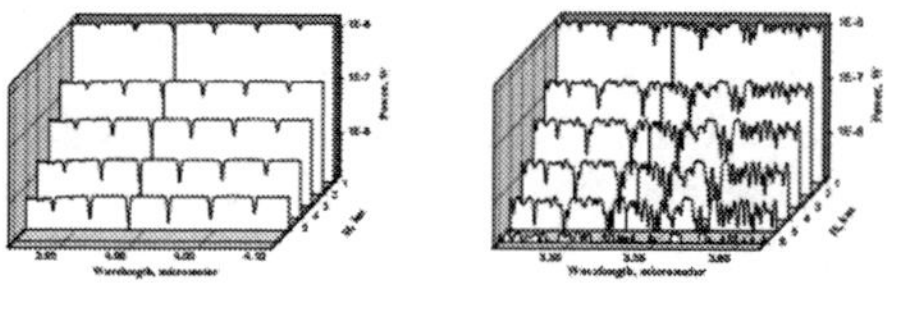

a) b)

Figure 2. Spatially and spectrally resolved lidar echo signals of (a) HBr and (b) HCl sounding

 https://doi.org/10.1051/epjconf/201817605016

Figures 1 and 2 show that the level of lidar echo signals exceed the level of photodetector noise equivalent power (NEP = 10^{-9} W) in the whole altitude range under study (0–5 km).

The numerical simulation results show a possibility of retrieving lidar signals along paths up to 5 km long during sounding of N_2O and NO_2, as well as HCl and HBr, with KTA-based OPO radiation in the 3–4 μm range.

5 MEASUREMENT RESULTS

Energy parameters of a laser system have been measured in experiments with an OPO-based setup for estimation of optimal lidar output parameters. Two improved schemes for measuring output laser parameter are used.

In the first scheme, OPHIR PE10-C, PE25-C, and PE25-BB energy meters (EM) are used as the receiving unit.

A pumping wavelength of 1.064 nm is transformed by means of an optical parametric oscillator (OPO) based on nonlinear KTP and KTA crystals. Turning the crystal allows the radiation wavelength to be tuned in the 3–4 mm spectral range. A gold-coated end mirror allows the laser radiation to be directed to the EM detection area.

The measurements result in the wavelength dependence of the radiation pulse energy. In the second scheme, a gold-coated end mirror directs the laser radiation to a topographical target, which is represented by a diffusing reflector with an albedo of 0.8. The main specifications of the setup are given in Table 4.

A sounding path is 20 m long. Backscattered radiation is gathered by a receiving Cassegrain telescope (main mirror diameter is 30mm and focal length is 120 mm) and focused to the PD38-03-PR photodiode detection area. An aperture and BS-15 light filter are included in the scheme to prevent scattered radiation from saturating the photodiode.

The level of signals with the use of the above optical elements allows a suggestion about possibilities of sounding the atmosphere along a surface tropospheric path up to 1 km long.

The experimental geometry is shown in Figure 3. The time dynamics of a single radiation pulse (curve 1) and the shape of a sync pulse (curve 2) are shown in Fig. 4.

Table 4 Specification of the experimental setup

Laser	
Wavelength turning range, μm	3–4
Maximal pulse energy, mJ	9
Pulse length, ns	10–13
Pulse frequency, Hz	10
Beam divergence, mrad	2
Optical receiving system	
Main mirror diameter, mm	300
Focal length, mm	120
Analog-to-digital converter of signals	
Number of analog outputs	2
Transmission band, MHZ	200
Resolution, bit	8
Sampling frequency over all channels, GS/s	1
Photodetector	
Spectral range, μm	1–4
Diameter of photodiode detection area, μm	300
Input rise time, ns	50

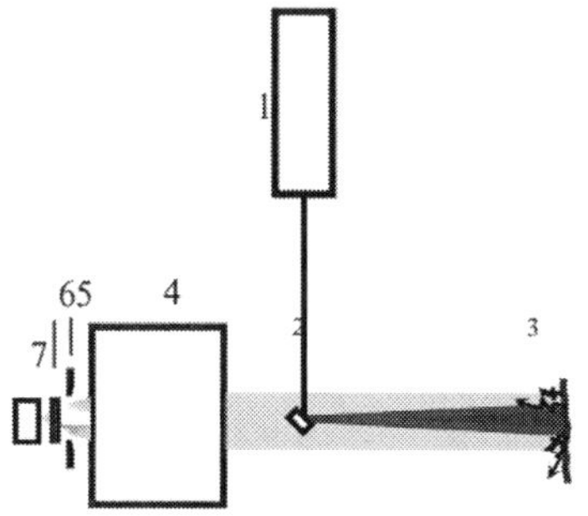

Figure 3 Geometry of the experiments: OPO-laser (1), mirror (2), diffusing reflector (3), receiving telescope (4), aperture (5), light filter (6), and photodiode (7)

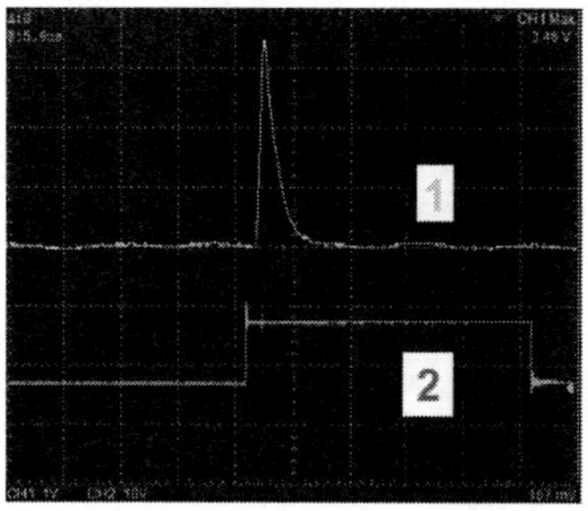

Figure 4 Time dynamics of a pulse of laser radiation at a wavelength of 3.335 μm (1) and sync pulse shape (2)

Figure 5 – 6 shows energy parameters of the experimental setup measured in the first scheme with the use of EM and PD38-03-PR photodiode for path measurements with the receiving telescope. To check the correct operation of the laser throughout the wavelength tuning range, the pulse energy was measured with different EMs.

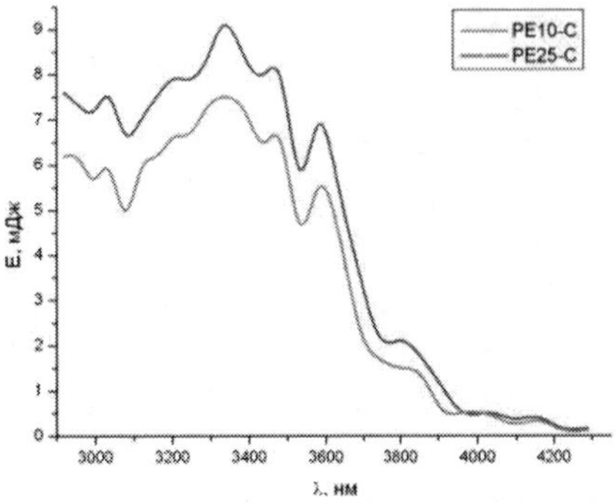

Figure 5 Pulse energy as a function of the radiation wavelength when using PE10-C and PE25-C EMs as radiation detectors

Figures 5 – 6 shows the effect of water vapors on the lasing lines in the 4-μm spectral region without nitrogen purging of the OPO cavity, i.e., the effect of strong water vapor absorption lines is obvious.

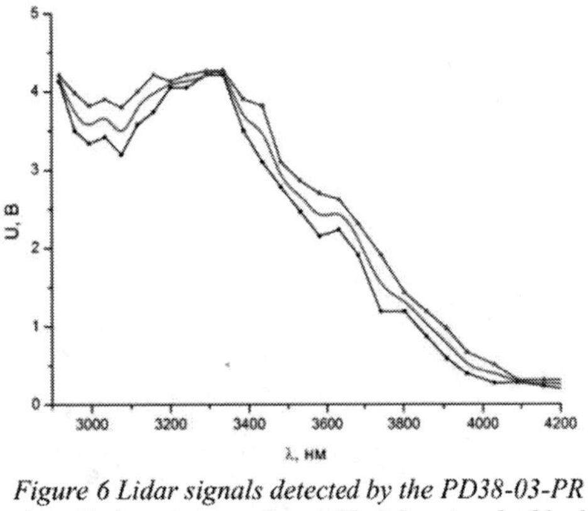

Figure 6 Lidar signals detected by the PD38-03-PR photodiode: minimum detectable lidar signals (black curve), maximum detectable lidar signals (blue curve), result of averaging a statistical sample of lidar signals (red curve)

6 CONCLUSIONS

The technique developed for lidar sounding of TAGs, which combines DIAL and DOAS, and its validation in numerical experiments confirm prospects of the use of the selected informative wavelengths. The numerical simulation performed shows that OPO laser is a promising source of radiation for remote DIAL-DOAS sounding of the TAGs under study along surface tropospheric paths. A possibility of using a PD38-03-PR photodiode for the DIAL gas analysis of the atmosphere is shown.

ACKNOWLEDGEMENTS

The work was supported by the President of the Russian Federation (grant no. MK-1367.2017.5 for the support of young Russian scientists).

References

[1] Mitev V., Babichenko S., Bennes J., 2013: Mid-IR DIAL for high-resolution mapping of explosive precursors, *Proceeding of SPIE*, 2013, **8894**, 88940S.

[2] Platt U., Perner D., and Patz H. W., 1979: Simultaneous measurement of atmospheric CH_2O, O_3, and NO_2 by differential optical-absorption, *J. Geophys. Res*, **84**, pp. 6329–6335.

Potential Retrieval of Aerosol Microphysics From Multistatic Space-Borne Lidar

Nathaniel Levitan[1], Barry Gross[1,2*] Fred Moshary [1,2], Yonghua Wu [1,2]

[1]*Dept. Elec. Engr., City College of New York, USA, *gross@ccny.cuny.edu*
[2]*NOAA-CREST, New York, USA*

ABSTRACT

HSRL lidars are being considered for deployment to space to retrieve aerosol microphysics. The literature is mostly focused on the monostatic configuration; but, in this paper, we explore whether additional information for the retrieval of microphysics can be obtained by adding a second detector in a bistatic configuration. The information gained from the additional measurements can under certain conditions reduce the ill-posed nature of aerosol microphysics retrieval and reducing the uncertainty in the retrievals.

1 INTRODUCTION

Aerosol microphysics is an important part of characterizing the atmosphere and better understanding the effects of aerosols on earth's energy balance. Conventional multispectral backscatter lidars simply do not have the information content needed. To address this limitation, a novel space-born multistatic lidar system has been proposed in the literature [1] to better characterize the vertical distribution of aerosol microphysics on a global scale. The proposed multistatic configuration allows for the retrieval of more intensive aerosol properties because it measures information about the scattering matrix elements at angles other than those obtained from pure backscattering (180°). While the sensitivity of the difference in the scattering matrix elements from the off-nadir and on-nadir angles was demonstrated in previous works [1], we are not aware of any effort to assess the improvement in the potential retrieval of realistic aerosol size distributions. In this paper, we explore the potential to improve the retrieval of the size distribution parameters and the refractive indices through the addition of the off-nadir measurements.

We compare the bistatic case to the traditional monostatic 3β + 2α High Spectral Resolution Lidar (HSRL) measurement configuration, being considered for deployment to space by NASA [2]. Many ground-based studies [2] have indicated that this is the minimum measurement dataset needed to retrieve information about the aerosol microphysics and an air-borne monostatic 3β + 2α HSRL system has been deployed by NASA Langley in preparation for a possible space-born mission. The ground-based studies have also concluded that HSRL must be used because of its ability to separate measurements of extinction and backscatter at the physical layer without an estimation of the lidar ratio (The variability of the lidar ratio causes a large increase in the uncertainty if it is estimated and not measured). In the multistatic configuration, the concept of β is extended to the multi-angular domain and it reflects the signal scattered back to the sensor at scattering angle θ (rather than simply backscattering at 180°). Specifically, we define this quantity as

$$\beta_{dbs}(\theta) = \frac{\omega_0 \beta_e F(\theta)}{4\pi} \quad (1)$$

where $F(\theta)$ is an element of the scattering matrix. As a result, in the bistatic configuration, there are 6 measurements of β_{dbs} per polarization in the 3β + 2α configuration (3 at each angle).

The concept of backscattering can be expanded to the polarization domain by considering the entire scattering matrix. For example, in the case of a lidar emitting circularly polarized radiation, the measured Stokes vector can be related to the atmospheric aerosol distribution as

https://doi.org/10.1051/epjconf/201817605017

$$\begin{bmatrix} I(z) \\ Q(z) \\ U(z) \\ V(z) \end{bmatrix} = \frac{1}{4\pi} \begin{bmatrix} C_I(z)\beta_s F_{11}(\theta) \\ C_Q(z)\beta_s F_{12}(\theta) \\ C_U(z)\beta_s F_{34}(\theta) \\ C_V(z)\beta_s F_{44}(\theta) \end{bmatrix} e^{(sec\theta-1)\int_0^r \beta_e(z')dz'} \quad (2)$$

The inclusion of an off-nadir angle in the bistatic configuration allows for access to the polarization measurements for retrieval of the microphysics of spherically symmetric particles. At 180°, information about spherically symmetric particles is contained in F_{11} and the other polarization channels can only be used to determine the particle's nonsphericity. In contrast, at off-nadir angles, the other elements of the scattering matrix also contain information about spherically symmetric particles and this information can be exploited to improve the retrievals of the microphysics [1].

For a particle size distribution, the directional backscatter coefficient has a simple expansion as

$$\beta_{dbs}(\theta) = \frac{1}{4\pi}\int_{-\infty}^{\infty} \frac{dV(r)}{dlnr}\frac{3}{4}\pi r^{-3}\beta_s(r)F(\theta,r)dlnr \quad (3)$$

From this expression, we can define the particle size distribution backscatter weighting function as

$$W_{dbs}(r,\eta) = \frac{3}{4}\pi r^{-3}\beta_s(r)P(\theta,r) \quad (4)$$

For every refractive index and axial ratio distribution, the weighting function in equation 4 is known from either Mie theory (if the axial ratio distribution states that the particles are spherical) or the T-matrix method [3]. Because of the weighting functions are known, the inversion of the size distribution from measurements of β_{dbs} takes the form of the solution of the Fredholm integral problem in equation 5.

$$\beta_{dbs}(\theta) = \frac{1}{4\pi}\int_{-\infty}^{\infty} \frac{dV(r)}{dlnr} W_{dbs}(r,\theta)dlnr \quad (5)$$

Measurements of the extinction coefficient are also related to the size distribution through a similar Fredholm integral problem listed in equations 6 and 7.

$$W_e(r,\eta) = \frac{3}{4}\pi r^{-3}\beta_e(r) \quad (6)$$

$$\beta_e = \frac{1}{4\pi}\int_{-\infty}^{\infty} \frac{dV(r)}{dlnr} W_e(r)dlnr \quad (7)$$

It has been extensively documented [4,5] that the problem of inverting equations 5 and 7 to retrieve the size distribution and refractive index is ill-posed, even when the number of unknown microphysics parameters is greater than the number of measurements. This is because, especially when the refractive index is unknown and the best weighting function has to be selected, the inversion is not unique. To address the ill-posedness, the retrieval algorithm needs regularization to select only those solutions that are likely to occur in nature. In other words, some information about the typical distributions atmospheric aerosol needs to be incorporated into the algorithm.

2 METHODOLOGY

This paper seeks to explore the benefits of adding a second receiver at an off-nadir angle for a space born 3β + 2α HRSL system. To do so, we simulated measurements by using aerosol microphysics retrievals from AERONET [6]. Specifically, we use the six bimodal lognormal size distribution parameters and the complex refractive indices retrieved by AERONET to simulate the measurements. We used 10,000 randomly selected AERONET retrievals to generate a database of 10,000 sets of simulated multistatic lidar measurements. These AERONET retrievals and simulated measurements form our characterization of typical atmospheres that can be used by a retrieval algorithm to regularize the ill-posed problem of microphysics inversion from lidar data. We consider both monostatic 3β + 2α measurement set and the bistatic 3β + 2α measurement set. The monostatic measurement set consists of measurements of the extinction coefficient (α) at 355 and 532 nm and measurements of the backscatter coefficient (β) at 355, 532, and 1064 nm. The bistatic measurement set we consider augments the monostatic measurements with six measurements of β_{dbs} at 170° (same three wavelengths and two polarization bands sensitive to F_{11} and F_{12}, respectively). We chose 170° to perform the analysis because the error caused by horizontal aerosol inhomogeneity is minimal at this angle [1].

We perform a statistical analysis to determine the anticipated error in microphysics retrieval for the monostatic and bistatic configuration. We add Gaussian error to each of the measurements in the retrieval subset with a standard deviation/measurement ratio of 0.1 for β measurements and 0.15 for α measurements (which corresponds to 10% and 15% error, respectively). Then, for every retrieval measurement, we computed a weight for every database measurement in the whole database based on the joint Gaussian PDF, the standard deviation/measurement ratios above, and an assumption that the errors in the individual measurements are uncorrelated. These weights allow us to get an estimated retrieval for each measurement in the retrieval subset by taking the weighted average of the database aerosol distribution parameters (bimodal lognormal size distribution parameters and refractive indices). Then, across all the retrievals in the retrieval subset, we find the mean retrieval, mean retrieval error, and the standard deviation of the retrieval error for all the aerosol parameters. The standard deviation of the retrieval error and the mean retrieval error as a percentage of the mean retrieval are shown in the results. In this simlation study, we focus only on non absorning aerosols and limit the imaginary ref index < .01. This was needed since without the constraint, significant errors in this method can result. However, if we make use of simultaneous measurements from OMI and filter out high Absorbing Indices, the results of this study can be used to extract useful vertical PSD information.

3 RESULTS

The results of the statistical analysis we described in the methodology section are presented in Tables 1 and 2.

First, we note that the low mean retrieval errors in Table 2 (as compared to the standard deviations in Table 1) indicates that the bias of our retrievals is low and the error is mostly driven by random retrieval scatter, which is seen in the standard deviation. The standard deviations in Table 1 clearly demonstrate the information gain for retrieval of aerosol microphysics by adding a detector at 170° in that the retrieval error standard deviations are reduced by more than 50%. Specifically, the average error in the retrieval of all the aerosol parameters is 19.5% in monostatic case and 7.86% in the bistatic case.

Table 1: Standard deviation of aerosol parameters error as percentage of mean retrieval

	Standard deviation as percentage of mean retrieval (Monostatic 180°)	Standard deviation as percentage of mean retrieval (Bistatic 170° and 180°)
Fine Mode Volume Concentration	29.06%	14.07%
Fine Mode Volume Median Radius	19.16%	5.47%
Fine Mode Lognormal Std. Dev	14.01%	5.93%
Coarse Mode Volume Concentration	60.26%	19.81%
Coarse Mode Volume Median Radius	17.08%	8.73%
Coarse Mode Lognormal Std. Dev	10.22%	5.81%
Refractive Index Real Part 1022 nm	3.28%	1.56%
Refractive Index Imaginary Part 1022 nm	2.98%	1.55%

CONCLUSIONS

In this paper, we explored the ability of a space-born 3β + 2α HRSL multistatic lidar to perform retrievals of aerosol microphysics from space. We found that the bistatic configuration does work to reduce the ill-posedness of the inverse problem as it significantly reduces the percentage error in retrieval space compared to the monostatic configuration when the complex refractive index is sufficiently low. The non-uniqueness of the solution is the primary difficulty in the retrieval of aerosol microphysics from lidar

	Mean retrieval error as percentage of mean retrieval (Monostatic 180°)	Mean retrieval error as percentage of mean retrieval (Bistatic 170° and 180°)
Fine Mode Volume Concentration	3.44%	2.33%
Fine Mode Volume Median Radius	0.68%	0.37%
Fine Mode Lognormal Std. Dev	0.71%	0.1%
Coarse Mode Volume Concentration	0.14%	0.13%
Coarse Mode Volume Median Radius	0.75%	0.25%
Coarse Mode Lognormal Std. Dev	0.3%	0.001%

Table 2: Mean retrieval error of aerosol parameters as percentage of mean retrieval

measurements and this reduction in the ambiguity is of great aid to inversion algorithms. Further climatological constraints on the aerosol microphysics parameters as well as Ssuitable filtering of absobing cases using OMI should be incorporated in a future retrieval algorithm for the multistatic configuration that would further increase the accuracy by reducing the size of space of typical atmospheric aerosol parameters.

ACKNOWLEDGMENTS

We would also like to thank all the groups that maintained AERONET sites used in this paper. This work is partially supported by NOAA under the grant CREST agreement #NA17AE1625.

References

[1] Mishchenko, M., Alexandrov, M., Cairns, B., & Travis, L. (2016). Multistatic aerosol–cloud lidar in space: A theoretical perspective. *J. Quant. Spect. Radiat. Transfer*, *184*, 180-192.

[2] Burton, S., Chemyakin, E., Liu, X., Knobelspiesse, K., Stamnes, S., & Sawamura, P. et al. (2016). Information content and sensitivity of the 3β+ 2α lidar measurement system for aerosol microphysical retrievals. *Atmospheric Measurement Techniques*, *9*(11), 5555-5574.

[3] Dubovik, O., Holben, B., Lapyonok, T., Sinyuk, A., Mishchenko, M., Yang, P., & Slutsker, I. (2002). Non-spherical aerosol retrieval method employing light scattering by spheroids. *Geophysical Research Letters*, *29*(10), 54-1-54-4.

[4] Veselovskii, I., Kolgotin, A., Griaznov, V., Müller, D., Wandinger, U., & Whiteman, D. (2002). Inversion with regularization for the retrieval of tropospheric aerosol parameters from multiwavelength lidar sounding. Applied Optics, 41(18), 3685.

[5] Chemyakin, E., Burton, S., Kolgotin, A., Müller, D., Hostetler, C., & Ferrare, R. (2016). Retrieval of aerosol parameters from multiwavelength lidar: investigation of the underlying inverse mathematical problem. *Applied Optics*, 55(9), 2188.

[6] Dubovik, O. & King, M. (2000). A flexible inversion algorithm for retrieval of aerosol optical properties from Sun and sky radiance measurements. *J. Of Geophys. Res.: Atmos.*, *105*(D16), 20670

LIDAR ICE NUCLEI ESTIMATES AND HOW THEY RELATE WITH AIRBORNE IN-SITU MEASUREMENTS

Eleni Marinou[1,2,*], Vassilis Amiridis[1], Albert Ansmann[3], Athanasios Nenes[4,5,6], Dimitris Balis[2], Jann Schrod[7], Ioannis Binietoglou[1,8], Stavros Solomos[1], Dimitra Mamali[9], Ronny Engelmann[3], Holger Baars[3], Michael Kottas[1], Alexandra Tsekeri[1], Emmanouil Proestakis[1,10], Panagiotis Kokkalis[1], Philippe Goloub[11], Bojan Cvetkovic[12], Slobodan Nichovic[12], Rodanthi Mamouri[13,14], Michael Pikridas[14], Iasonas Stavroulas[14], Christos Keleshis[14], Jean Sciare[14]

*[1]IAASARS, National Observatory of Athens, Athens, 15236, Greece, *elmarinou@noa.gr*
[2]Department of Physics, Aristotle University of Thessaloniki, Thessaloniki, 54124, Greece
[3]Leibniz Institute for Tropospheric Research (TROPOS), Leipzig, 04318, Germany
[4]Schools of Chemical and Biomolecular Engineering and Earth and Atmospheric Sciences, Georgia Institute of Technology, Atlanta, GA 30332, USA
[5]Institute of Chemical Engineering Sciences, FORTH, Patras, GR-26504, Greece
[6]Institute of Environmental Research and Sustainable Development, National Observatory of Athens, Palea Penteli, GR-15236, Greece
[7]Institute for Atmospheric and Environmental Sciences, Goethe University Frankfurt, 60438 Frankfurt am Main, Germany
[8]National Institute of R&D for Optoelectronics, Magurele, Romania
[9]TU Delft, Department of Geoscience and Remote Sensing, the Netherlands
[10]Laboratory of Atmospheric Physics, Department of Physics, University of Patras, 26500, Greece
[11]Laboratoire d'Optique Atmospherique, University of Lille, France
[12]Republic Hydrometeorological Service of Serbia, 11000 Belgrade, Serbia
[13]Cyprus University of Technology, Dep. of Civil Engineering and Geomatics, Limassol, Cyprus
[14]Energy, Environment and Water Research Center, The Cyprus Institute, Nicosia, 2121 Aglantzia, Cyprus

ABSTRACT

By means of available ice nucleating particle (INP) parameterization schemes we compute profiles of dust INP number concentration utilizing Polly-XT and CALIPSO lidar observations during the INUIT-BACCHUS-ACTRIS 2016 campaign. The polarization-lidar photometer networking (POLIPHON) method is used to separate dust and non-dust aerosol backscatter, extinction, mass concentration, particle number concentration (for particles with radius > 250 nm) and surface area concentration. The INP final products are compared with aerosol samples collected from unmanned aircraft systems (UAS) and analyzed using the ice nucleus counter FRIDGE.

1 INTRODUCTION

Despite their low concentration in the atmosphere, ice nucleating particles (INP) are crucial for the evolution of ice in clouds, as they act as seed-surfaces for water in order to enable the development and growth of ice. Mineral dust is one of the most important IN aerosol types and has been identified in liquid-water, mixed phase and cirrus clouds in field and laboratory studies [1]. Nevertheless, there are several questions on ice nucleation processes such as the abundance and vertical profile of INP as a function of cloud nucleation conditions (temperature and supersaturation) and the aerosol particle spectrum. Furthermore, field data of INP are limited and the number of observations is still relatively small. Gaps in our knowledge of the detailed impact of aerosols on the evolution of liquid-water, mixed phase and cirrus clouds drive atmospheric research towards field studies of aerosol-cloud-dynamics interaction.

During April 2016, a joint field campaign of INUIT (Ice Nuclei Research Unit), BACCHUS (Impact of Biogenic versus Anthropogenic emissions on Clouds and Climate: towards a Holistic UnderStanding) and ACTRIS (Aerosols, Clouds, and Trace gases Research InfraStructure) projects took place at Cyprus, focusing on aerosol, clouds and ice nucleation in the Eastern

Mediterranean. We use the data from this field campaign to compare remote sensing with airborne in-situ retrievals of INP concentrations (INPC).

2 DATA AND METHODOLOGY

2.1 CAMPAIGN MEASUREMENTS

During the campaign, continuous lidar observations were performed at Nicosia (35°08'26"N, 33°22'52"E, 181m a.s.l.) with the Polly-XT lidar of National Observatory of Athens. The system is a 12 channel system with near-range capabilities (3 elastic, 2 Raman, 2 depolarization, 1 water-vapor, 2 near-range elastic, and 2 near-range Raman signals) [2]. Since the middle of the campaign, lidar observations were complemented by a SUN/SKY/LUNAR AERONET photometer, used to evaluate the horizontal variability of the atmosphere between Nicosia and 2 additional monitoring sites located 30 km westerly, where additional in-situ INP measurements were performed. UAS, operating from the Cyprus institute airfield (Orounda, 35°05'42"N, 33°04'53"E, 327 m a.s.l.) performed observations with the aim to measure the abundance of ice nucleating particles in the lower troposphere [3]. Seven km further west than the airfield, at the Cyprus Atmospheric Observatory (Agia Marina Xyliatou, 35°02'19"N, 33°03'28"E, 532m a.s.l.), an AERONET sun photometer, along with a suite of aerosol and INP in-situ instruments were located.

2.2 POLLY-XT OBSERVATIONS

A first snapshot of the aerosol time-height variability is given in Figure 1 (30/3/2016 - 27/4/2016). During that period, the Eastern Mediterranean was affected by air mass advection from Europe, northern Africa and the Atlantic as indicated by the backward trajectories and DREAM8b dust model [4]. During April 4 to 7 several elevated structures were observed reaching heights up to 10km. The sun photometer observed aerosol optical depths (AOD, 550 nm) values up to 0.3 and Angström exponents (AE, 380 - 870 nm) between 0.5 – 1.0. Atmospheric conditions favored the transport of dust from the west parts of North Africa over the Continental Europe and towards Cyprus. Following this period, we observed five Saharan dust events. The first event (April 8 to 11) was very intense with AOD values of 0.6 to 1.0 and AE less than 0.2. The events of April 15 - 16, 21, and 26 reach AOD values of 0.5 and AE of 0.1 - 0.4. On April 24 minor dust event reached AODs of 0.3 and AE of 0.3 – 0.5. During four out of the five events, ice nucleation was observed in the top of the dust plums, and we observed volume depolarization ratio values (at 532 nm) greater than 32%.

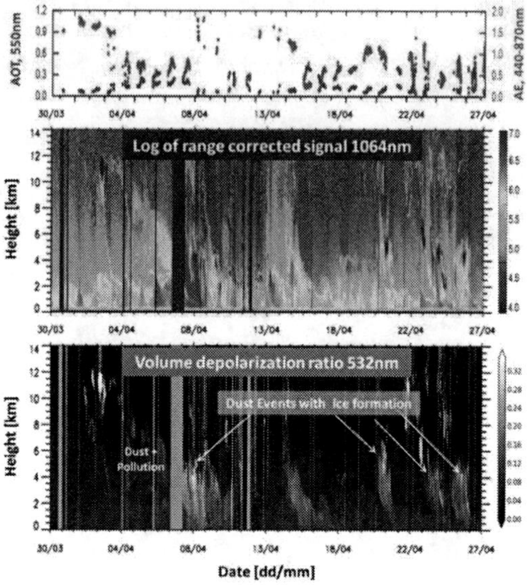

Figure 1: AERONET optical thickness and Ångström exponent, Range-corrected lidar signal and volume linear depolarization ratio measured with Polly-XT at Nicosia from 30 March -27 April 2016.

In order to retrieve INPC profiles from the lidar measurements we follow the methodology described in [5]. In brief, we separate the backscatter profile in dust and non-dust component using POLYPHON technique assuming particle depolarization ratio for dust of 0.31. Then, we calculate the extinction coefficient for the dust and non-dust components from which we estimate mass concentration, particle number concentration (for particles with radius >250nm) and surface area concentration profiles using AERONET parameterizations. Finally, we estimate INP concentrations using the well-known parameterizations of DeMott et al. 2015 (D15), Niemand et al. 2012 (N12) and Steinke et al. 2015 (S15) [5]. Figure 2 provides an example of the estimated products during 21/4/2016 at 1-2 UTC.

INP concentrations start above 5 km a.g.l., where the temperature is less than -10 °C (N12) and above 6 km a.g.l. where the temperature is less than -20 °C (D15, S15). There are a number of distinct pathways of dust ice nucleation: contact freezing, immersion freezing, condensation freezing and deposition nucleation. N12 and D15 quantifies the immersion freezing at or above water saturation.

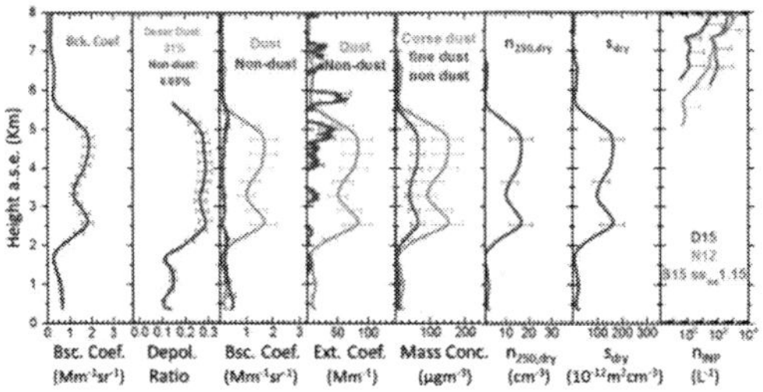

Figure 2: From left to right Polly-XT profiles of: (i) backscatter coefficient, (ii) particle depolarization ratio, (iii) separated backscatter coefficient of dust and non-dust aerosols, (iv) extinction coefficient of dust and non-dust aerosol, (v) mass concentration of coarse dust, fine dust and non-dust aerosol, (vi) large particle number concentration (vii) surface area concentration, (viii) INPC computed with different parameterization schemes. Case of 21/4/2016, 1-2 UTC.

2.3 UAS MEASUREMENTS

Two different types of UAS were used for INP sampling, namely Cruiser and Skywalker X8, able to carry a payload of up to 10 kg and 2 kg respectively. 42 flights were performed in total, which generated a total of 52 samples over 19 days. INPC were measured by electrostatic precipitation of aerosol particles onto silicon wafers onboard the UAS, followed by laboratory analysis of the samples in the ice nucleus counter FRIDGE [2]. FRIDGE usually addresses the deposition / condensation freezing mode(s). During the campaign, the UAS reached maximum altitudes of 2.5 km a.g.l. (2.85 km a.s.l), making these measurements ideal for comparison with the lidar data.

2.4 CALIPSO

In this work we use CALIPSO (The Cloud-Aerosol Lidar and Infrared Pathfinder Satellite Observation) [8] for INPC monitoring. During the campaign period, CALIPSO satellite overpassed Nicosia site within 5 km distance two times at 5 and 21 April 2016. From these two cases, we use the case of the 21st of April that coincided with one of the dust events, and CALIPSO L2 product gave quality assured retrievals. The atmospheric curtain of CALIPSO is shown in Figure 3. In this plot, the green dotted line highlights Nicosia station and the box shows the part we use for the INPC retrievals (±80 km for the station). The curtain shows the elevated dust plume and cloud formation at 80 km away from the station, on the top of the dust plume and in heights greater than 5km.

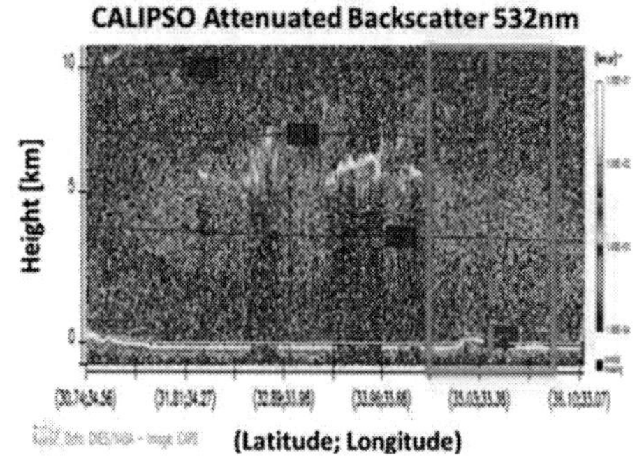

Figure 3: CALIPSO Attenuated Backscatter at 532nm above Nicosia on 21/4/2016, 10:34 UTC. Green dotted line: Polly-XT location. Green box: ±80 km track used for the comparison.

3 CASE OF 21 APRIL 2016

On that day, atmospheric conditions favored the transport of dust from both Sahara Desert and the Arabian Peninsula. The two plums merged over Cyprus during the time of the event, with the major contribution coming from Sahara (DREAM8b dust model). Figure 4 (top) shows the 24hr Polly-XT measurements of that day. Above the station there was formation of ice clouds between 2 and 10 UTC. We were able to derive level 2 cloud-free optical profiles and estimate INPC close to the cloud. These profiles are shown in Figure 4 (bottom). From left to right: Raman retrievals with Polly-XT during 0-1 UTC (fig 4i) and 1-2 UTC (fig 4ii), CALIPSO retrievals at 10:34 UTC (fig 4ii) and Klett retrievals with Polly-XT between 14-15 and 15-16 UTC. INPC estimates are ~ 0.4 L^{-1} at 5 km (-10 °C) and ~ 4 L^{-1} at 6 km (-20 °C) before the cloud formation (Polly-XT) and right next to the clouds (CALIPSO). After 14 UTC, and while the dust plume was descending, INPC values lower than $0.2\ L^{-1}$ were observed at heights between 5 and 6 km. The results show a very good agreement between the Polly-XT and CALIPSO INPC retrievals.

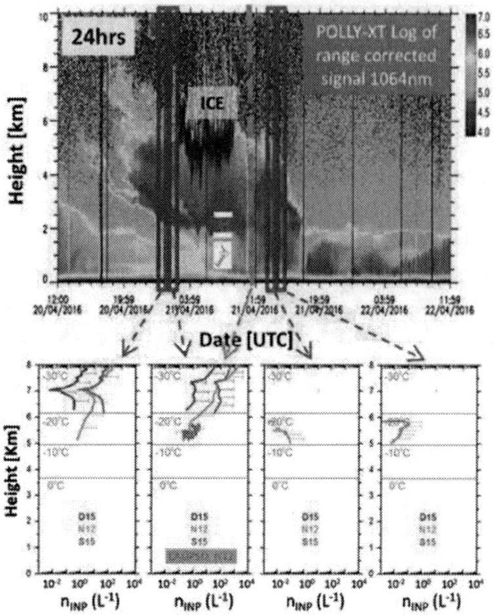

Figure 4: Top: Polly-XT 24hr range-corrected signal at Nicosia during the event of 20-21 April. Purple boxes periods used for INP conc .estimations. Green line: CALIPSO overpass time. White lines: UAS flights period and altitudes. Bottom: INPC for the time periods (from left to right):0-1 UTC, 1-2UTC (Polly-XT) & 10:34 UTC (CALIPSO), 14-15, 15-16UTC.

4 CONCLUSIONS

In Figure 5 the in-situ (UAS) and lidar INPC estimations are compared for three different temperatures (-20°C, -25°C and -30°C). When we apply the N12 and D15 parameterizations for the same height as the UAS flight (2.55 km a.g.l) we get between one to two orders of magnitude higher values compared to the FRIDGE measurements. When we compare the UAS measurements with the lidar INPC values in the actual height of -20°C, -25°C and -30°C (6 km, 6.5 km and 7 km a.g.l.), the N12 parameterization provide INP concentration values closer to the measurements while D15 gives one order of magnitude lower values. We will investigate this topic more intensively in the next time.

ACKNOWLEDGEMENTS

The authors acknowledge ACTRIS2 project (grant agreement No 654109).

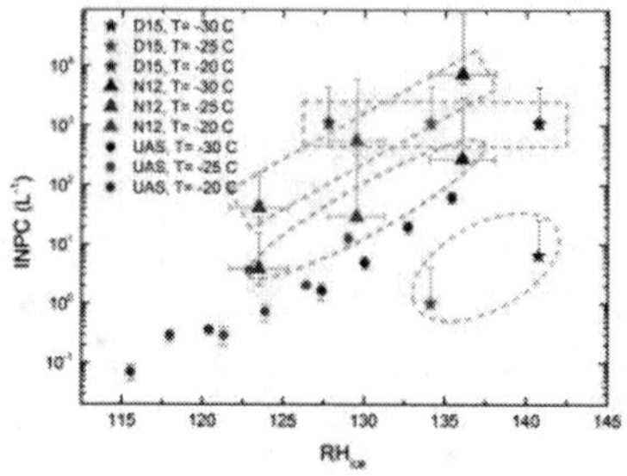

Figure 5: INP concentrations during the event. Dots: UAS measurements. Triangles and stars: lidar estimations based on N12 and D15 respectively. Peach dotted squares correspond to N12 and D15 INPC values for the air mass at the height of UAS flight (2.55 km a.g.l.). Blue dotted circles correspond to INP estimates at the actual heights of Temps -20, -25 and -30 °C during the time window 1-2 UTC.

References

[1] Knippertz, P., Stuut, J.-B.W.: Mineral dust: A key player in the earth system, doi:10.1007/978-94-017-8978-3, 2014.

[2] Engelmann, R., et al.: The automated multi-wavelength Raman polarization and water-vapor lidar PollyXT: the neXT generation, Atmos. Meas. Tech., 9, 1767-1784, doi:10.5194/amt-9-1767-2016, 2016.

[3] Schrod, J., et al.: Ice nucleating particles over the Eastern Mediterranean measured by unmanned aircraft systems, Atmos. Chem. Phys. Discuss., doi:10.5194/acp-2016-1098, in review, 2016.

[4] Nickovic, S., et al.: A model for prediciton of desert dust cycle in the atmosphere, J. Geophys. Res., 106, 18113–18129, doi:10.1029/2000JD900794, 2001.

[5] Mamouri, R.-E. and Ansmann, A.: Potential of polarization lidar to provide profiles of CCN- and INP-relevant aerosol parameters, Atmos. Chem. Phys., 16, 5905-5931, doi:10.5194/acp-16-5905-2016, 2016.

https://doi.org/10.1051/epjconf/201817605018

[6] Winker, D. M., et al.: Overview of the CALIPSO Mission and CALIOP Data Processing Algorithms, J. Atmos. Oceanic Technol., vol 26, pp. 2310–2323, doi: 10.1175/2009JTECHA1281.1, 2009.

VALIDATION OF THE TOLNET LIDARS DURING SCOOP (SOUTHERN CALIFORNIA OZONE OBSERVATION PROJECT)

Thierry Leblanc[1*], Maria-Jose Granados-Munoz[1], Kevin Strawbridge[2], Chris Senff[3], Andy Langford[3], Tim Berkoff[4], Guillaume Gronoff[4], Russel DeYoung[4], Bill Carion[4], G. Chen[4], John Sullivan[5], Tom McGee[5], M. Jonhson[6], S. Kuang[7] and Mike Newchurch[7]

*[1]California Institute of Technology, Jet propulsion Laboratory, USA, *thierry.leblanc@jpl,nasa.gov*
[2]Environment and Climate Change Canada, Canada
[3]NOAA-Earth System Laboratory, USA
[4]NASA-Langley research Center, USA
[5]NASA-Goddard Space Flight Center, USA
[6]NASA-Ames Research Center, USA
[7]University of Alabama in Huntsville, USA

ABSTRACT

Five TOLNet lidars participated to a validation campaign at the JPL-Table Mountain Facility, CA in August 2016. All lidars agreed within ±10% of each other and within ±7% of the ozonesondes. Centralized data processing was used to compare the uncertainty budgets. The results highlight the TOLNet potential to address science questions ranging from boundary layer processes to long range transport. TOLNet can now be seen as a robust network for use in field campaigns and long term monitoring.

1 INTRODUCTION

The Tropospheric Ozone Lidar Network (TOLNet) is a North-America-based network of six ozone differential absorption lidars dedicated to the study of tropospheric ozone variability at timescales ranging from a few minutes to several decades, and to satellite validation and model evaluation. The four mobile systems comprise AMOLITE from Environment Canada, LMOL from NASA Langley Research [1], TOPAZ from NOAA-Earth System Research Laboratory [2], and TROPOZ from NASA-Goddard Space flight Center [3]. The two systems built at fixed locations are operated by the University of Alabama in Huntsville, and the JPL-Table Mountain Facility [4].

In August 2016, leveraging from their nearby deployment for field campaigns, it was decided to deploy the TOPAZ and TROPOZ lidars at the Table Mountain Facility for an intercomparison campaign before returning to their home base. The other two TOLNet mobile lidars joined the campaign efforts, making a unique opportunity for five of the six TOLNet lidars to be inter-compared and validated. In support of the validation, the JPL lidar group launched 17 ozonesondes.

In this extended abstract, we show the preliminary results obtained from "Level 2" SCOOP data produced for the blind phase of the validation exercise. The focus is on the strictly-coincident lidar-ozonesonde profiles, starting at launch time and extending 30-minutes after launch. In addition to lidar-lidar and lidar-sonde comparisons, the lidars uncertainty budgets will be compared using a newly developed centralized data processing algorithm.

2 DEPLOYMENT

Three lidars (TOPAZ, TROPOZ and TMT) were co-located at the TMF main facility (elev. 2285 m above mean sea level-a.s.l.), while the two others (AMOLITE, LMOL) were located less than a quarter mile away (elev. 2270 m), making all five systems nearly co-located. The campaign officially started on August 10, 2016 and ended on August 16, one day too early due to a local wildfire mandatory evacuation order. Each lidar acquired between 2 and 20 hours of data per day. The automated, autonomous system AMOLITE acquired data continuously during and after the campaign, making unprecedented 31 days uninterrupted tropospheric ozone measurements above the site. The TOPAZ instrument is equipped with a two-axis scanner, allowing pointing of the laser beam in any direction at various elevation angles besides the zenith. All lidars use more or less advanced aerosol measurement capability.

For ozone measurements, all lidars use the same differential absorption technique, but at different wavelength pairs constrained by hardware configurations [1][2][3][4]. Three instruments use quadrupled Nd:YAG lasers to produce 266 nm beams. AMOLITE then uses a CO_2 Raman cell to shift the wavelength to 287 nm and 299 nm, while the JPL-TMF (later referred to as "TMT") and TROPOZ systems use a pair of D_2 and H_2 Raman cells to shift to 289 nm and 299 nm. The other two lidars (TOPAZ and LMOL) use similar Ce:LiCAF lasers pumped by a quadrupled Nd:YLF laser to produce two wavelengths, tuned to approximately 287 nm and 294 nm during the SCOOP campaign.

The receiver configurations and detectors vary between the instruments, with telescope sizes ranging between 5 cm and 91 cm depending on instrument and intensity range considered. For ozone, all systems use photomultiplier tubes, and Licel transient recorders with analog detection and photon counters. Ozone profiles are retrieved either using both types of data acquisition systems (AMOLITE, LMOL, TOPAZ) or using photon-counting mode only (TROPOZ and TMT). In any case, the combination of multiple analog and photon counting pairs allows coverage of the entire lower and middle troposphere (3-10 km) by all five instruments and coverage of the upper troposphere (10-16 km) by three of them (AMOLITE, TROPOZ, and TMT).

During the campaign, the JPL lidar group launched ozonesondes at all possible times of the day, specifically, 2 sondes before sunrise, 4 sondes in the daytime morning hours, 4 sondes around midday, 3 sondes in the late afternoon hours, and 4 sondes after dark. Several surface ozone monitors, a ceilometer, water vapor Raman lidar, and stratospheric ozone and temperature lidar complemented the ozonesonde and tropospheric ozone lidar measurements.

3 DATA PROCESSING

For the blind phase of this validation exercise, all lidar groups were required to retrieve profiles with prescribed time and vertical resolutions. One data version included profiles retrieved at 10-minutes time resolution in order to produce curtain plots of ozone and allow altitude and time-dependent ozone variability investigations.

Another data version consisted of averaging the lidar raw data acquired during the first 30 minutes of each ozonesonde flight. For this dataset, a common vertical effective resolution scheme was used for all lidars. Averaging kernels were computed for each lidar separately, and for the ozone profiles, in order to obtain an effective resolution increasing linearly from 100 m near the surface to 1500 m at 8 km a.s.l., and then constant resolution of 1500 m at all altitudes above 8 km. The effective resolution definition follows the NDACC (Network for the Detection of Atmospheric Composition Change) -standardized "Impulse Response"-based definition [5].

The differences between two coincident 30-min lidar profiles and between a 30-min lidar and ozonesonde profile were computed. Because of technical or operational constraints, all lidars were fully operational during only 7 of the 17 launches. In order to increase the comparison statistics, the lidar profiles were compared one-by-one to the ozonesondes, resulting in about 13 to 16 sonde-coincident profiles for each lidar. The ozonesonde profiles were then used as the reference transfer to compare the lidars.

4 RESULTS

The campaign mean of the relative differences between the JPL-TMF tropospheric ozone lidar (TMT) and ozonesonde is shown in Figure 1. The number of coincidences (right plot) decreases with altitude, due to the decrease of the lidar signal-to-noise ratio and resulting lower profile cut-off from nighttime to daytime, as well as to the presence of optically thick clouds at 5-6 km on certain days. The grey-shaded area indicates the combined ozonesonde and lidar total standard uncertainties.

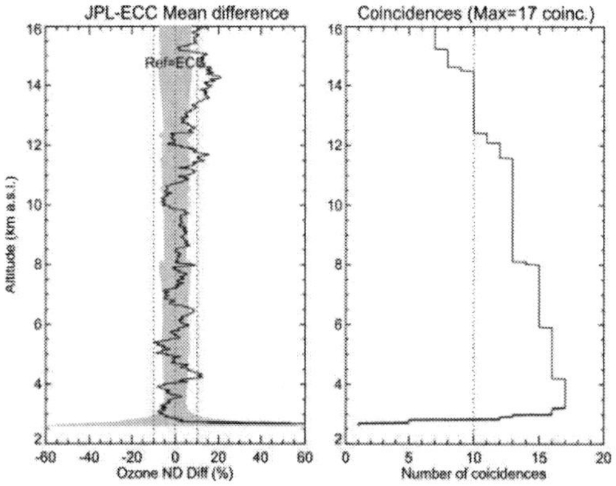

Figure 1 Left: Campaign-mean of the ozone relative differences (%) between TMT lidar profiles and ozonesonde; Right: Number of coincidences as a function of altitude

Overall, the TMT lidar and sonde profiles agree within ±7%, within the reported combined uncertainties. One exception is above 12 km where the difference can locally reach 15% (lidar higher than sonde). This bias is partially explained by possible loss of sonde-lidar co-location, but also by the fact that the TMT far-field (narrow-fov) was occasionally impacted by slight misalignment. The TMT lidar-sonde difference also exceeds the combined uncertainty estimates at 4 and 5 km, due to residual, uncorrected cloud contamination that occurred during the Aug. 13 and Aug. 16 mid-day measurements.

The mean differences shown in figure 1 for TMT were computed for all five lidars and composited together in Figure 2. The absolute differences (ppbv) are shown on the left, and the relative differences (%) are shown on the right. The maximum number of coincidences with the ozonesondes varies between 13 and 17, depending on the lidar instrument considered. Below about 10 km, all lidars agree with the ozonesondes within 4 ppbv or 7%. Larger differences can be observed above 10 km, partly explained by the increase in detection noise in the lidar data, as well as a possible loss of co-location between the sondes and the lidars.

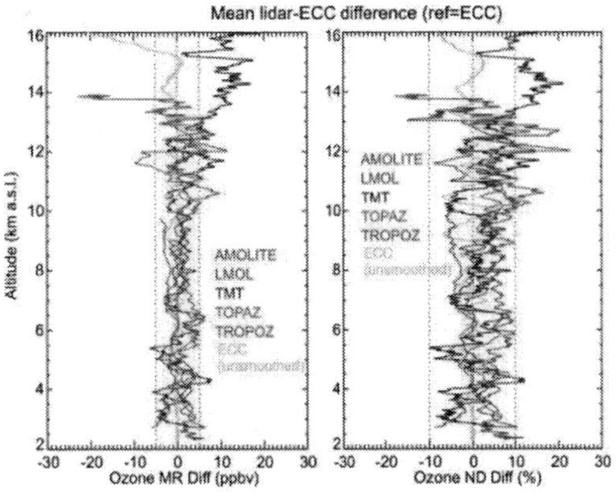

Figure 2 Campaign-mean of the differences (left: ppbv; right: %) between all coincident 30-min average lidar and ozonesonde profiles

In order to further evaluate the capability of the TOLNet lidars, the consistency of their uncertainty budgets was investigated. A centralized data processing algorithm was recently developed at JPL, with the objective to produce profiles of quality similar to that of the individual in-house data processing algorithms, but with the added capability to produce profiles with standardized vertical resolutions and uncertainty budgets.

Figure 3 shows the ozone uncertainty budget computed by the centralized data processing algorithm for a 30-min nighttime ozone profile measured by TOPAZ on August 16, 2016. The colored curves show the individual uncertainty components, while the dash, black curves show the total uncertainty (ppbv on the left, % on the right).

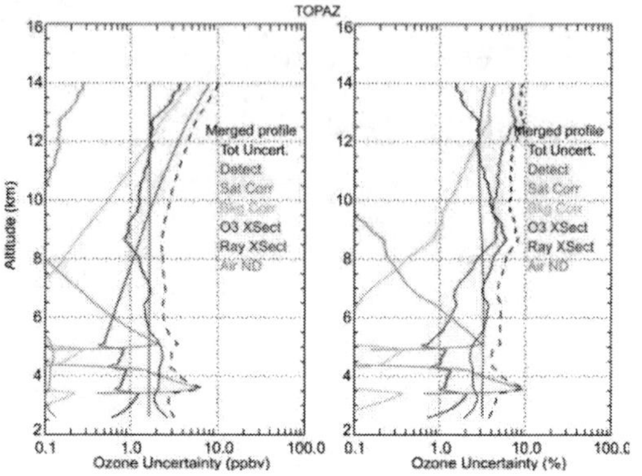

Figure 3 Example of standardized uncertainty budget, in this case for TOPAZ (30-min profile, nighttime)

Interestingly, in this example, at least four components play a significant role in the total budget. Below 3 km, the ozone absorption cross-section uncertainty dominates (3%). At 3.5 km the saturation correction uncertainty peaks at 9%. Above 4 km, uncertainty owed to the ozone absorption cross-section and Rayleigh extinction cross-sections uncertainties is of the same order of magnitude (3%). Finally, above 10 km, the detection noise uncertainty dominates (4% increasing to 8%). This example of uncertainty budget is quite different from that computed for TMT, where saturation correction uncertainty and detection noise play a more significant role throughout the profile (not shown).

The ozone total uncertainty computed by the centralized data processing algorithm for all five lidars and for the same 30-min averaged profile is plotted in Figure 4. The figure shows that total uncertainty can vary significantly from one lidar instrument to another. Larger uncertainty for TMT below 5.5 km is due to a combination of low STNR and inadequate common SCOOP vertical resolution scheme used in the near-field channels. Larger uncertainty at 8 km is associated with a large saturation correction uncertainty for the far-field channels. The TMT instrument was designed and optimized for nighttime, upper tropospheric and lower stratospheric measurements, which explains its atypical uncertainty budget compared to the other four systems optimized for air quality studies.

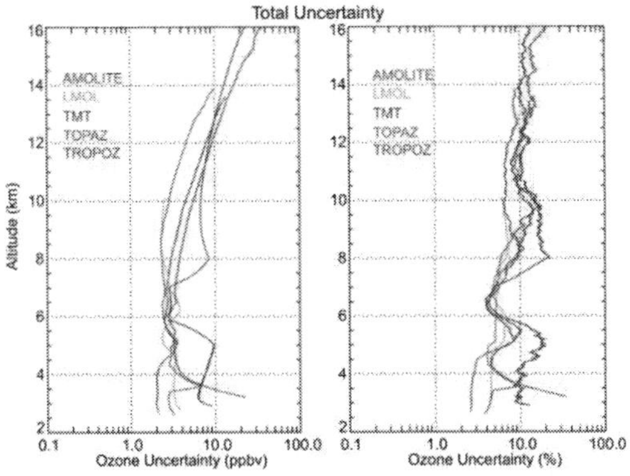

Figure 4 Total uncertainty computed by the centralized data processing algorithm for all 5 lidars (30-min profile, nighttime)

5 CONCLUSIONS

In August 2016, five (out of six) TOLNet lidars were deployed at the JPL-Table Mountain Facility, CA for a 7-day-long intercomparison campaign. More than 50 hours of lidar measurements were acquired for each of the lidar systems, and 18 ozonesondes were launched from the site (1 to 6 launches per day).

All lidars were found to agree within ±10% of each other and within ±7% of the ozonesonde throughout the altitude range 3-10 km a.s.l., which is typically within the reported uncertainties.

A centralized data processing algorithm was used to further compare the lidars, specifically allowing a quantitative intercomparison of the uncertainty budgets using standardized definitions. Inter-instrument differences in the magnitude of individual components such as detection noise uncertainty or saturation correction uncertainty revealed that dominant uncertainty components can vary significantly from one lidar to another, thus highlighting the diversity of the TOLNet "fleet" and its potential for science applications ranging from small-scale boundary layer processes to long range transport in the upper troposphere.

Yet the present results from the SCOOP Level 2 data remain preliminary. A fully-validated version (known as "SCOOP Level 3" data) will be released during the spring of 2017 and publicly available for download from the TOLNet website https://www-air.larc.nasa.gov/missions/TOLNet. The lidar-lidar and lidar-sonde differences, as well as ozone total uncertainty are expected to decrease with this new release.

The SCOOP campaign results showed that the TOLNet lidars are now mature enough to constitute a robust reference network for use in field campaigns as well as long term monitoring.

ACKNOWLEDGEMENTS

This work is supported by the TOLNet program developed by NASA's Science Mission Directorate. Part of this work was carried out at the Jet Propulsion Laboratory, California Institute of Technology, under agreements with the National Aeronautics and Space Administration. The authors would like to thank the entire JPL-TMF Staff and Lidar Group for their sustained hard work throughout the campaign.

References:

[1] De Young, R., et al., 2017: Langley mobile ozone lidar: ozone and aerosol atmospheric profiling for air quality research, *Appl. Opt.* **56**(3), 721-730, 2017.

[2] Alvarez, R. J., et al., 2012: Reconfiguration of the NOAA TOPAZ Lidar for Ground-based Measurement of Ozone and Aerosol Backscatter, *Proceedings of the 26th ILRC*, Porto Heli, Greece, 25-29 June 2012, pp. 249 – 252, 2012.

[3] Sullivan, J. T., et al., 2014: A mobile differential absorption lidar to measure sub-hourly fluctuation of tropospheric ozone profiles in the Baltimore–Washington, D.C. region, *Atmos. Meas. Tech.*, **7**, 3529-3548, 2014.

[4] McDermid, I. S., et al., 2002: Redesign and improved performance of the tropospheric ozone lidar at the Jet Propulsion Laboratory Table Mountain Facility, *Appl. Opt.*, **41**, 7550-7555, 2002.

[5] Leblanc, T., et al., 2016: Proposed standardized definitions for vertical resolution and uncertainty in the NDACC lidar ozone and temperature algorithms – Part 1: Vertical resolution, *Atmos. Meas. Tech.*, **9**, 4029-4049, 2016.

https://doi.org/10.1051/epjconf/201817605020

MONITORING OF DISPERSED SMOKE-PLUME LAYERS BY DETERMINING LOCATIONS OF THE DATA-POINT CLUSTERS

Vladimir Kovalev*, Cyle Wold, Alexander Petkov, and Wei Min Hao

*Missoula Fire Sciences Laboratory, MT 59808, USA, *E-mail: vladimirakovalev@fs.fed.us*

ABSTRACT

A modified data-processing technique of the signals recorded by zenith-directed lidar, which operates in smoke-polluted atmosphere, is discussed. The technique is based on simple transformations of the lidar backscatter signal and the determination of the spatial location of the data point clusters.

The technique allows more reliable detection of the location of dispersed smoke layering. Examples of typical results obtained with lidar in a smoke-polluted atmosphere are presented.

1 INTRODUCTION

The worldwide increase of the number of wildfires is considered as a result of global warming. On the other hand, wildfires further influence climate change. Therefore, monitoring wildfires and predicting their possible distribution is an imperative task of atmospheric science. However, the determination of the exact location of the smoke plume is a significant challenge, especially when monitoring dispersed smoke layers and plumes is made in the areas where the dispersion process creates an extended transition zone between regions of polluted and clear air.

Lidar is a good tool for monitoring smoke layer behavior. In principle, it can easily distinguish regions with high levels of backscatter from regions of clear atmosphere and detect the boundary between different atmospheric layers. When determining the boundaries of the aerosol formations with lidar, generally, some relative characteristics of the backscatter signal from areas of interest are used [1–3]. In particular, when using the gradient method, one can find the smoke-plume boundary location as the range where the derivative of the square-range-corrected lidar signal either is maximum or decreases from the maximum down to a fixed, user-defined level [1]. Unfortunately, there is no way to establish an optimal value for this level, which would be acceptable for different atmospheric situations. The use of other techniques, including the wavelet technique [2], also requires the selection of concrete parameters, and might be a significant issue.

The situation becomes especially challenging when one should define the upper height of the area of increased backscattering, created by dispersed smoke. The boundary of such layering is difficult to define as the conventional criteria generally assume a sharp boundary between the clear sky and the area of increased backscattering. The data-processing method which allows determining the upper height of the region of increased backscattering with poorly defined boundaries is considered in [3].

In our current study, the method in [3] is modified and used together with the application of the cluster principle, which allows more reliable determination of the locations of dispersed smoke boundaries. This methodology can be used both for one-directional and scanning lidar; however, below this method is considered only for profiling the atmosphere with a zenith directed lidar.

2 METHODOLOGY

In the study [3], the recorded lidar signal, $P_\Sigma(r)$, which is the sum of the variable backscatter signal, $P(r)$ at the range r and the range-independent offset, B, is transformed in the auxiliary function $Y(x)$. For the zenith-directed lidar, the latter is defined as the function of independent x in the form,

$$Y(x) = P_\Sigma(x)x = [P(x) + B]x, \qquad (1)$$

where $x = h^2$, and h is the height. The sliding derivative of this function, dY/dx, is calculated and the intercept point of each local slope fit of the function with the vertical axis is found as the difference of two terms,

$$Y_0^*(x) = Y(x) - \frac{dY}{dx}x. \quad (2)$$

By using the intercept function instead the derivative of the square-range-corrected backscatter signal, the determination of the systematic offset, *B*, can be avoided. The retrieval technique is based on the normalized intercept function, which is defined as,

$$Y_0(x) = \frac{Y_0^*(x)}{x} = \frac{Y(x)}{x} - \frac{dY}{dx}. \quad (3)$$

For determining the upper boundary of the dispersed smoke plume of interest, the absolute value of this difference is calculated,

$$|Y_0(x)| = \left|\frac{Y(x)}{x} - \frac{dY(x)}{dx}\right|, \quad (4)$$

and then transformed into the function of height, $|Y_0(h)|$. The maximum value of this function,

$$Y_{0,\max} = \max|Y_0(h)|, \quad (5)$$

within the total height interval from h_{min} to h_{max} is found, and used for calculating the data points of interest,

$$Y_{0,\chi}(h) = |Y_0(h)| - \chi Y_{0,\max}; \quad (6)$$

here the independent variable χ can be chosen in the range $0 \leq \chi < 1$, so that $Y_{0,\chi}(h) \geq 0$.

The same as in [3], the consecutive values of χ with the fixed step $\Delta\chi = 0.05$ are used in this study; that is, $\chi_{\min} = 0$, $\chi_1 = 0.05$, $\chi_2 = 0.1$, etc. For each discrete χ, the corresponding height of a smoke layer, $h_{max}(\chi)$ was determined; the latter was found as the maximum height where $Y_{0,\chi}(h)$ was a non-zero value.

The main challenge is the selection of the optimal level, χ_{opt}, and the corresponding height of interest, $h_{max}(\chi_{opt})$. In the previous study [3], such a selection was based on the calculation of the differences between the adjacent heights $h_{max}(\chi_{k-1})$, $h_{max}(\chi_k)$, and $h_{max}(\chi_{k+1})$. The value of χ_{opt} was determined using the following two conditions. First, the difference $h_{max}(\chi_{opt} - \Delta\chi) - h_{max}(\chi_{opt})$ should be maximal. Second, the next consecutive increase of χ, that is, selection of χ equal to $\chi_{opt} + \Delta\chi$, then $\chi_{opt} + 2\Delta\chi$, etc., should result in an invariable, or at most, slowly decreasing height $h_{max}(\chi)$.

As mentioned, the method in [3] works properly when the boundary between the layer with increased backscattering and clear air can be properly defined. The situation becomes much more challenging when lidar data are noisy and are obtained in smoke-polluted layers where the dispersion processes create an extended transition zone between the regions of polluted and clear air. In such situations, the use of the modified data-processing technique based on the analysis of the cluster locations of the data points, $h_{max}(\chi)$, and their temporal changes is extremely effective.

3 ANALYSIS OF EXPERIMENTAL DATA

To clarify the modified data processing technique, let us consider Figs. 1 and 2, in which real square-range-corrected lidar signal and corresponding dependence of $h_{max}(\chi)$ on χ are respectively shown. Here and below we will consider data, obtained from the lidar backscatter signals measured close to the zenith direction (80°); these signals were obtained on August 4, 2016 near Missoula, Montana, during lidar profiling of the atmosphere influenced by three wildfires: Roaring Lion Fire [4], Cedar Fire [5], and Observation Fire [6].

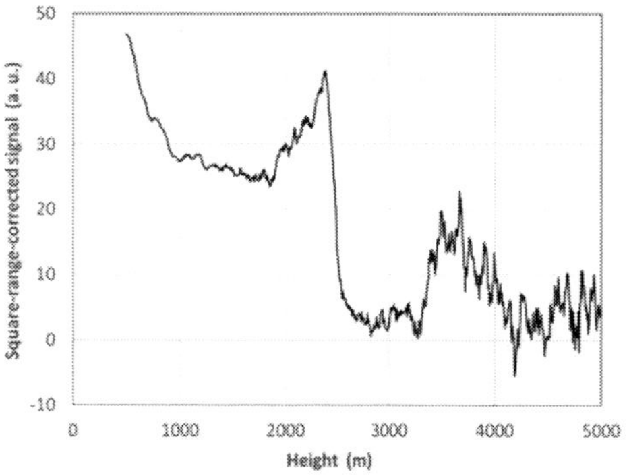

Fig. 1. The square-range corrected signal measured in the atmosphere influenced by three closely located wildfires.

The square-range corrected lidar signal shown in Fig. 1 is not extremely noisy, so that the use of gradient methods for determining the height of smoke layering could yield acceptable results within the height up to ~3000 m. However, it could be difficult to discriminate the useful spikes caused

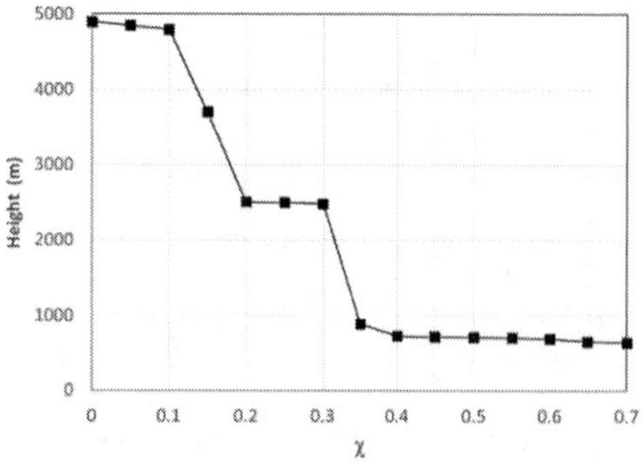

Fig. 2. Dependence of the height $h_{max}(\chi)$ on χ extracted from the signal in Fig. 1. The data points close to the height 5000 m are originated by noise.

by the actual change in the backscattering signal from the spikes originated by random noise over the larger heights. Using the criteria given in [3] and analyzing the corresponding dependence $h_{max}(\chi)$ on χ in Fig. 2, one can overcome this issue, but only for heights up to ~ 3000 m. In particular, one can easily distinguish two distinct layers, which upper heights are located at ~700 – 800 m and ~ 2500 m. However, the origin of the data point at $h \approx 3700$ m at $\chi = 0.15$ is unclear. Moreover as mentioned, the dependencies of $h_{max}(\chi)$ on χ that obey the criteria in [3] are not generally obtained when the smoke

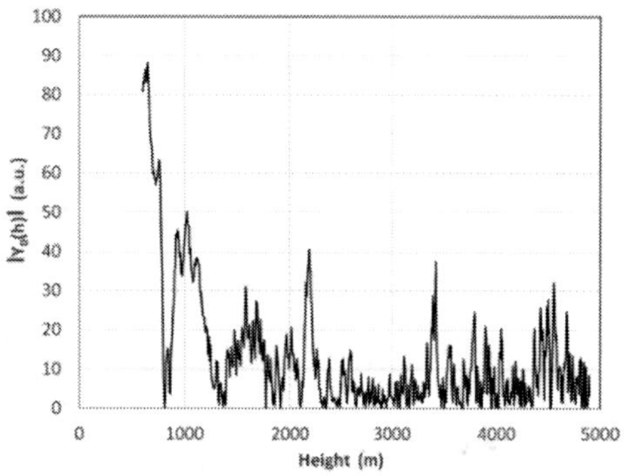

Fig. 3. The typical function $|Y_0(h)|$ versus height obtained from a noisy signal.

plume is strongly dispersed and the backscatter signals from that area are too noisy. Such a very typical case is shown in Figs. 3 and 4, where the

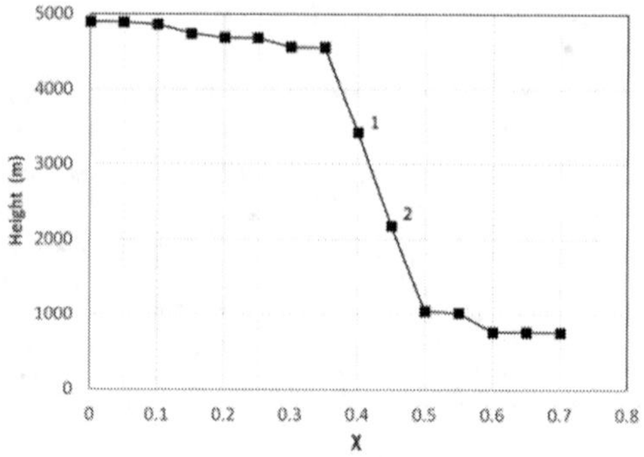

Fig. 4. Dependence of the height $h_{max}(\chi)$ on χ extracted from the function $|Y_0(h)|$ in Fig. 3.

function, $|Y_0(h)|$ and the corresponding dependence $h_{max}(\chi)$ on χ, obtained on the same day, August 4, 2016, are presented. Unlike Fig. 2, in which one can easily discriminate two separated layering heights, the same criteria do not allow reliable discrimination of smoke-layer heights in Fig. 4. In particular, it is unclear whether points 1 and 2 which correspond to $\chi = 0.4$ and $\chi = 0.45$ are originated by the noise or by the backscatter signal. Note that eight points, located in the vicinity of maximum height, $h_{max} \approx 5000$ m, are originated by increased noise level at the far end of lidar range. This noise is more intense than the noise in Fig. 2, where the noise data points are obtained within the range of χ, from 0 to 0.15, whereas in Fig. 3, this range extends from 0 to 0.35.

Thus, in a common case, the data points in the dependencies $h_{max}(\chi)$ on χ can be originated from three sources: (1) the data points originated from the random noise on the signal far end; these data points are located at or close to the maximum height, h_{max}; (2) the data points originated from the change in the backscatter signal on the upper boundary of smoke area, and (3) the data points, who's source is unclear; they can originate from either the first or the second source.

To solve the problem of discriminating the backscatter and noise data points, the auxiliary cluster principle can be used. To clarify this principle, let us initially consider dependencies $h_{max}(\chi)$ on χ, obtained for three consecutive time periods, at local times 15:30, 15:31, and 15:34 (Fig. 5). In all these curves, the boundary of the increased

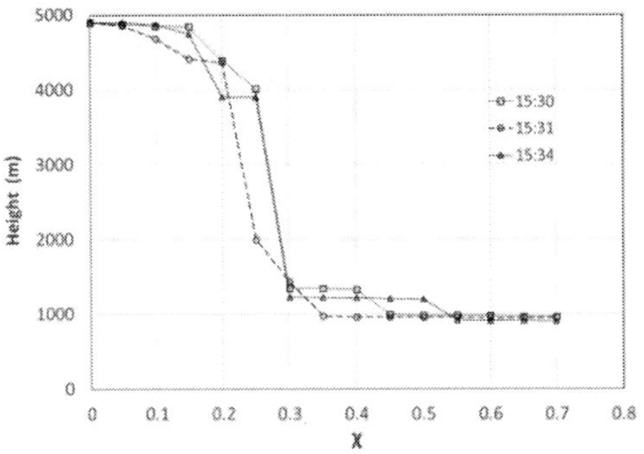

Fig. 5. The dependence of the height $h_{max}(\chi)$ on χ obtained from the signals measured at 15:30, 15:31, and 15:34 local time (MST).

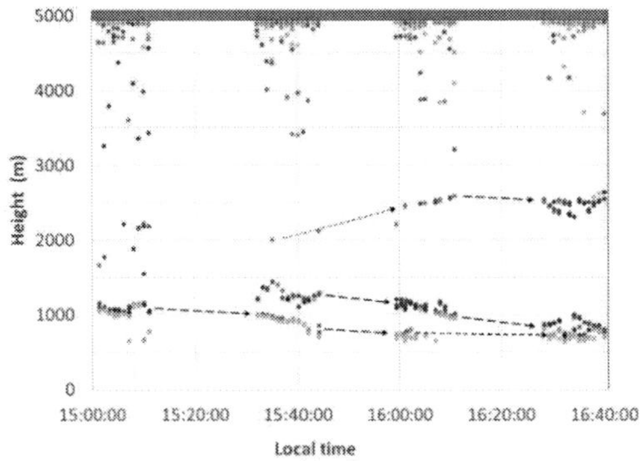

Fig. 6. Data points $h_{max}(\chi)$ versus time for the period from 15:00 to 16:15 of local time.

backscatter layer within the height interval ~1000-1300 m is clearly seen. The same as in the previous figures, the far-end data create noise data points, located close to maximum height, h_{max} = 5000 m. The data point at the height 2000 m can be seen only at 15:31; therefore, this point also most likely originated from random noise. The origin of the three data points close to ~ 4000 m is unclear and cannot be clarified without additional analysis of the data points obtained before 15:30 and after 15:34. This principle of separation between the casual and systematically appearing data points in the two-dimensional dependence of data points $h_{max}(\chi)$ on time and locating clusters of such data points is the basis of the cluster method.

To illustrate this method, in Fig. 6 the set of data points $h_{max}(\chi)$ versus time for the extended period from 15:00 to 16:15 of local time is given. One can discriminate three separate layers in this figure: (1) the lower layer, which height decreases with time from 1000 m to ~600 m; (2) the layer, which during the period 15:32 – 15:44 is located at the heights, ~ (1200 - 1400 m) and then goes down to ~700 m; and (3) the upper layer which appears at 15:30 at the height ~2000 m and then rises. After 16:30 and until 16:40 (the termination of our measurements), this layer at the height ~2500 m was reliably fixed. The scattered data points at the heights larger than 3000 m originated from signal random noise.

4 SUMMARY

The modified lidar data-processing methodology is presented which allows more reliable determination of the upper heights of significantly dispersed smoke layering in presence of increased signal noise.

References

[1] L. Menut, C. Flamant, J. Pelon, and P. H. Flamant, 1999: Urban boundary-layer height determination from lidar measurements over the Paris area, *Appl. Opt.* **38**, 945-954.

[2] I. M. Brooks, 2003: Finding boundary layer top: Application of a wavelet covariance transform to lidar backscatter profiles, *J. Atmos. and Oceanic Technol.* **20**, 1092-1105.

[3] V. Kovalev, A. Petkov, C. Wold, and W. M. Hao, 2011: Lidar monitoring of the regions of intense backscatter with poorly defined boundaries", *Appl. Opt.* **50**, 103-109.

[4] http://inciweb.nwcg.gov/incident/4913/.

[5] http://inciweb.nwcg.gov/incident/4874/ .

[6] http://inciweb.nwcg.gov/incident/4819/ .

CLASSIFYING STAGES OF CIRRUS LIFE-CYCLE EVOLUTION

Benedikt Urbanek[1]*, Silke Groß[1], Andreas Schäfler[1], Martin Wirth[1]

*[1]Deutsches Zentrum für Luft- und Raumfahrt, Institut für Physik der Atmosphäre, Oberpfaffenhofen, Germany, *benedikt.urbanek@dlr.de*

ABSTRACT

Airborne lidar backscatter data is used to determine in- and out-of-cloud regions. Lidar measurements of water vapor together with model temperature fields are used to calculate relative humidity over ice (RH_i). Based on temperature and RH_i we identify different stages of cirrus evolution: homogeneous and heterogeneous freezing, depositional growth, ice sublimation and sedimentation. We will present our classification scheme and first applications on mid-latitude cirrus clouds.

1 INTRODUCTION

Cirrus clouds are an important constituent of the Earth-atmosphere system. Their radiative properties depend both on macroscopic and microphysical parameters. As knowledge on cirrus clouds is still incomplete, they impose high uncertainties on climate prediction [1]. For instance it is unclear how cirrus properties change, as the cloud evolves in time.

To study possible dependencies on cirrus evolution, we developed a method that identifies regions inside a measured cloud corresponding to different thermodynamic states with regards to cirrus formation, particle growth and sublimation. This is done by evaluating temperature and RH_i, as those two are main parameters determining the thermodynamic state.

In order that ice can form in the atmosphere, RH_i must reach or surpass 100 %. Such areas are called ice-supersaturated regions (ISSR). The existence of an ISSR does not imply the existence or formation of a cirrus cloud. For the homogeneous freezing (HOM) of solution droplets at cirrus temperature (< 235 K), high supersaturations in the order of 140 % are necessary [2]. Additionally, solid aerosol particles can act as ice nuclei and thus lead to freezing under a much broader range of conditions (heterogeneous freezing, HET).

Once ice particles are present, remaining supersaturation is depleted by deposition of water vapor onto existing crystals (DEP). Also regions of subsaturation with respect to ice (SUB) can emerge, when heavy ice particles sediment out of the ISSR or RH_i is reduced by warming, leading to sublimation of ice crystals.

Our method can help to put in-situ and remote sensing data in perspective to cirrus evolution and to investigate the respective optical, microphysical and radiative properties of cirrus clouds.

We present its application in a case study of a lee wave influenced cirrus cloud, obtained during the ML-CIRRUS experiment 2014 and first results from a statistical analysis of cloud optical properties. We also aim to combine our classification with trajectory based methods to reveal information on the temporal development of classified evolution stages

2 METHODOLOGY

2.1 Differential Absorption Lidar WALES

WALES is an airborne lidar that measures the tropospheric water vapor concentration below the aircraft by simultaneously emitting laser pulses at three online and one offline wavelength in the water vapor absorption band around 935 nm [3]. Additionally, WALES features a channel at 1064 nm and one HSRL channel at 532 nm; both, polarization sensitive. An analysis of its accuracy can be found in [4]. WALES provides collocated measurements of humidity, backscatter ratio (BSR) and aerosol depolarization ratio (ADEP). We calculate RH_i from WALES humidity data and ECMWF analysis and forecast temperature data, that we interpolate both temporally and spatially onto the lidar measurement cross-section (see [5], their Sec. 2).

2.2 ML-CIRRUS campaign

In March and April 2014, the Mid-Latitude Cirrus mission ML-CIRRUS was carried out, designed to investigate natural and traffic induced cirrus clouds. Therefore the German research aircraft HALO was equipped with a combined in-situ and remote sensing payload and performed 16 research flights above Central Europe and the North Atlantic, probing mid-latitude cirrus clouds originating from e.g. air traffic, warm conveyor belts, jet streams, or mountain waves [6].

2.3 Classification scheme

Our classification scheme is outlined in Fig. 1. First a threshold for BSR of 2 is used to determine in- and out-of-cloud regions. Then we employ temperature dependent RH_i thresholds for HOM and HET and a RH_i value of 100 % to distinguish DEP/ISSR and SUB regions. Freezing onset parameterization ($RH_{i,HOM}$, $RH_{i,HET}$) is taken from [2,7].

Outside of clouds, $ISSR_{out}$, HET_{out} and HOM_{out} represent regions of ever more favorable ice nucleation. Inside, HET_{in} and HOM_{in} regions show active ice nucleation. DEP and SUB regions are in a later stage of evolution dominated by crystal growth and ice sublimation.

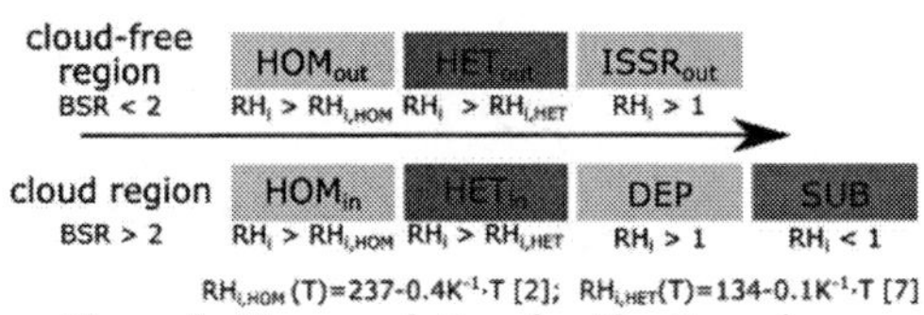

Figure 1: Cirrus evolution classification scheme

3 RESULTS

3.1 Case study ML-CIRRUS 2014-03-29

We apply our classification scheme to a cirrus case that was obtained during ML-CIRRUS on 29 March 2014. A trough, extending from west Ireland to the Iberian Peninsula and further to the western part of North Africa (Fig. 2), leads to high southerly winds with speeds of up to 35 ms^{-1} over Southern France and Spain at 300 hPa.

The flight path (Fig. 2, red) was chosen to run against the main wind direction over Southern France, intersecting a cirrus cloud influenced by the gravity lee wave of the Pyrenees. Our case study focuses on this southward flight leg (Fig. 2 white).

In Fig. 3 a), BSR along the chosen flight leg is plotted. On the lee side, north of the Pyrenees (14:19-13:34 UTC), a high cirrus cloud is observed, with BSR values up to 200, that extends from a height of 7 km to 11 km. ADEP of more than 30 % and temperatures below 240 K (not shown) clearly indicate a pure ice cloud.

Furthermore, in the region between the two clouds (14:34-14:43 UTC) gravity lee waves can be spotted at about 10 km and also in the lower aerosol layer.

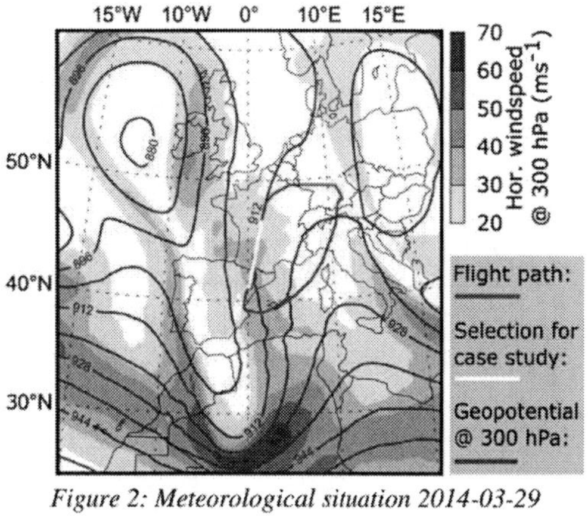

Figure 2: Meteorological situation 2014-03-29

https://doi.org/10.1051/epjconf/201817605021

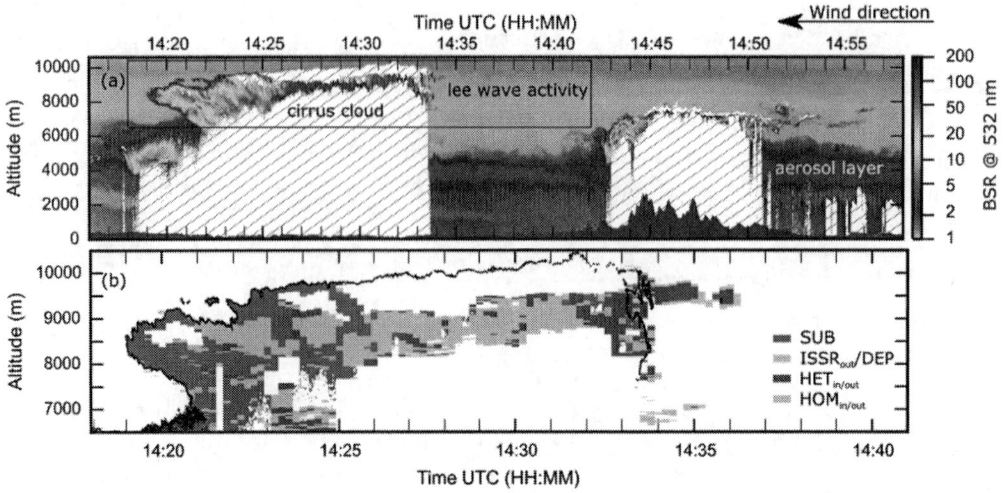

Figure 3: (a) Backscatter ratio at 532 nm cross-section measured by WALES. (b) Visualization of evolution classification, cloud border: black contour line (BSR=2)

Fig. 3 b) shows classified evolution stages. Cloud-free regions and cloud regions can be distinguished by the black contour line for a BSR value of 2.

A humid layer, at about 9.5 km, reaches ice supersaturation (ISSR) in the two crests of the gravity lee wave to the south of the cloud (14:34-14:36 UTC). Here values of RHi are even higher than the threshold for HET freezing. At the cloud edge, also the HOM freezing threshold is surpassed (14:33 UTC). The southern section of the cirrus is dominated by ice nucleation and represents the youngest part of the cloud.

In the middle (14:26-14:32 UTC), a section of moderate supersaturation (DEP) is located. This is an already well developed part of the cirrus that is dominated by depositional growth of ice crystals. After an initial ascent (14:32-14:34 UTC), the cloud top level slopes from over 10 km down to under 9 km at the northern edge. This indicates a large-scale decent reducing supersaturation and evoking the intermediate DEP region as well as large connected areas of subsaturation (SUB) in the northern part of the cloud. Here it is starting to break up, as ice particles are sublimating.

3.2 Statistical Analysis

This classification can also be applied to the complete lidar data set to study optical cirrus properties in a more statistical approach. Fig. 4 shows the distribution of ADEP occurrence during ML-CIRRUS for in-cloud evolution stages. They feature a maximum around 40 % with a tendency towards higher values going from HET to DEP to SUB. SUB also shows a pronounced secondary maximum at 52 %.

Our ongoing research aims to resolve such multimode features by taking other factors

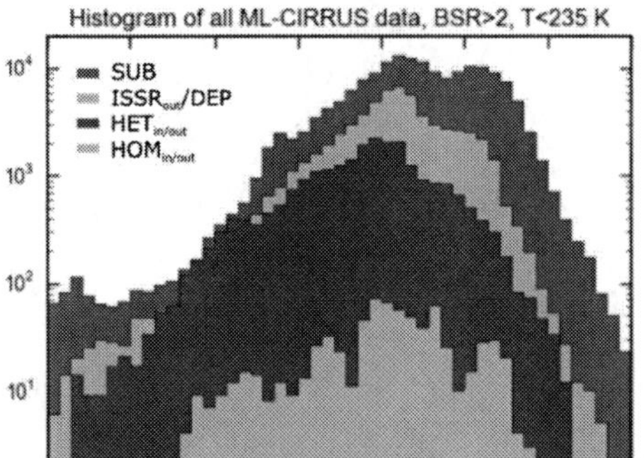

Figure 4: Distribution of Linear Depolarization

into account like cloud type, formation process and meteorological context. Therefore it seems to be promising to also combine this method with results from trajectory based classifications that identify in-situ and liquid-origin cirrus types.

4 CONCLUSIONS

Airborne DIAL water vapor measurements are used together with model temperature data and parameterizations for freezing onset conditions to identify all relevant stages of cirrus cloud evolution. In cloud-free air (BSR < 2), ice supersaturated regions ($ISSR_{out}$) as well as regions of homogeneous (HOM_{out}) and heterogeneous freezing (HET_{out}) are determined. These indicate favorable areas of cirrus cloud formation. Inside the cloud, ice nucleation (HOM_{in}, HET_{in}), depositional growth (DEP) and sublimation regions (SUB) are distinguished, representing the formation, growing and break up phases of a cirrus cloud.

Our method is used to investigate the large airborne lidar data set obtained during ML-CIRRUS. It facilitates the study of the spatial distribution of evolution stages and helps to set in-situ and other remote sensing data in perspective to cirrus evolution. By bringing together those data sources, the specific optical and microphysical properties of different cirrus stages can be explored. Combining our method with trajectory based classifications promises to also reveal details of the temporal succession of evolution stages.

ACKNOWLEDGEMENTS

This work is funded by DLR VO-R young investigator group. The authors like to thank the staff members of the HALO aircraft for preparing and performing the measurement flights, Christiane Voigt, Andreas Minikin and everybody contributing to the successful planning and execution of ML-CIRRUS, and ECMWF for providing model data.

References

[1] Baran, A. J., 2012: From the single-scattering properties of ice crystals to climate prediction: A way forward, *Atmos. Res.,* **112**, 45-69.

[2] Koop, T., B. Lou, A. Tsias, and T. Peter, 2000: Water Activity as the Determinant for Homogeneous Ice Nucleation in Aqueous Solutions, *Nature,* **406**, 61-614.

[3] Wirth, M, A. Fix, P. Mahnke, H. Schwarzer, F. Schrandt, and G. Ehret, 2009: The airborne multi-wavelength water vapor differential absorption lidar WALES: System design and performance, *Appl. Phys. B*, **96**, 201-213.

[4] Kiemle, C. et al., 2008; First airborne water vapor lidar measurements in the tropical upper troposphere and mid- latitudes lower stratosphere: accuracy evaluation and inter-comparisons with other instruments, *Atmos. Chem. Phys.*, **8**, 5245-5261.

[5] Urbanek, B., S. Groß, A. Schäfler, and M. Wirth, 2016: Determining stages of cirrus life-cycle evolution: a cloud classification scheme, *Atmos. Meas. Tech. Dis.,* doi:10.5194/amt-2016-332, in review.

[6] Voigt, C., et al., 2016: ML-CIRRUS – The airborne experiment on natural cirrus and contrail cirrus with high-altitude long-range research aircraft HALO, *B. Am. Meteorol. Soc.*, doi:10.1175/BAMS-D-15-00213.1.

[7] Krämer, M., et al., 2016: A microphysics guide to cirrus clouds – Part 1: Cirrus types, *Atmos. Chem. Phys.*, **16**, 3463-3483.

AEROSOL OPTICAL PROPERTIES VARIABILITY DURING BIOMASS BURNING EVENTS OBSERVED BY THE EOLE-AIAS DEPOLARIZATION LIDARS OVER ATHENS, GREECE (2007-2016)

Maria Mylonaki[1], Alexandros Papayannis[1*], Rodanthi Mamouri[2], Athina Argyrouli[1], Panagiotis Kokkalis[1], Georgios Tsaknakis[1] and Ourania Soupiona[1]

[1]*National Technical University of Athens, Laser Remote Sensing Unit, Physics Department, Zografou, Greece, *apdlidar@central.ntua.gr*

[2]*Cyprus University of Technology, Dep. of Civil Engineering and Geomatics, Limassol, Cyprus*

ABSTRACT

The EOLE multi-wavelength aerosol Raman-depolarization lidar, and the AIAS depolarization lidar, in synergy with a sun photometer (CIMEL), were used, in the period 2007-2016, to provide the vertical profiles of the aerosol optical properties over Athens, Greece. More than 30 biomass burning events (fresh and aged smoke particles) were observed, with smoke layers between 1.5 up to 4-5 km height, while their duration ranged from 1-3 days. Lidar ratio (LR) values ranged from 40-105 sr (at 355 nm) and from 40-100 sr (at 532 nm), while the linear particle depolarization ratio (LPDR) at both 355 and 532 nm, remained <7%. The extinction-related Ångström exponent (AE_a) at 355 nm/532 nm) ranged from 0.3 to 2.1. Additionally, a case of a near-range transport of biomass burning aerosols arriving over Athens up to 4 km height, between 27 and 28 June 2016, was studied. For this case, we found LRs of the order of 70±5 sr (355 nm) and 65±15 sr (532 nm) and AE_a(355 nm/532 nm) around 1.

1. INTRODUCTION

Wildfires play a substantial role in terrestrial ecosystems and have severe atmospheric and climatic effects (e.g. emission/production of global warming gases/aerosols: CO_2, O_3, aerosol particles, etc.) due to the emission of huge quantities of gases. Recent estimations of the global annual burned areas are of the order of more than 350 Mha per year [1-2], while the annual emissions of CO_2 are estimated to be of the order of 2.1 Pg C per year [1]. For the period 1979-2013 large wildfires caused a doubling of global burnable area strongly affected by the prolonged fire weather seasons in conjunction with an increased frequency of long fire weather seasons after 1996 [2].

In Europe, during the summer and early autumn period's large wildfires and agricultural biomass burning (BB) events, for several days, occur quite frequently, mainly in Southern Russia, Ukraine, Southern Europe and the Balkans, thus affecting large areas of Eastern-South-Eastern Europe through long-range transport of air masses [2-6].

Moreover, during the recent years (2010-today) of economic recession, in many European countries (Greece, Poland, UK, Spain, Ukraine, etc.), during the winter period, people have switched their domestic heating from oil to uncontrolled wood and biomass combustion [7]. This has caused a detrimental air quality deterioration by augmenting the ambient particulate matter (PM), exceeding by far the EU daily limit values, especially in Athens, Greece [8]. For instance, actually Athens is considered as the 4th most air polluted city in the European Union due to wood burning activities during winter time [7].

Wildfire and wood burning atmospheric emissions contain/produce greenhouse gases like CO_2, CH_4, NOx, NMHCs, N_2O, O_3, and other toxic ones, like CO, VOCs, Hg, PAHs, as well as biomass burning aerosols, which strongly affect air quality up to several thousands of kilometers [3]. As these aerosols play a substantial role in Earth's climate (e.g. direct and indirect effect) [3] and human health, it becomes very important to measure their optical, microphysical and chemical properties in a global scale on a long-term basis.

Although intensive campaigns have been performed worldwide to measure and characterize the biomass burning aerosols, long-term measurements are still missing, especially in areas like the Mediterranean region, which is largely affected by long-range transport of BB particles from Southern Russia and Ukraine [5], and by numerous forest fires and biomass burning events occur on an annual basis [6, 8-10].

Raman lidars have proven to be an appropriate tool in aerosol characterization of biomass burning aerosols, since they can provide the vertical profiles of the aerosol optical (β_{aer}, α_{aer}, LR, LPDR) and microphysical properties with high spatial and temporal resolution during many BB events observed mainly over Europe [10-16]. Besides, their chemical properties can be

 https://doi.org/10.1051/epjconf/201817605022

estimated aloft, using thermodynamic models and relative humidity observations [17].

2. METHODOLOGY

A synergy of instruments (Raman and elastic depolarization lidars and sun photometer) and tools (HYSPLIT model and satellite data from MODIS sensor) were used to derive the aerosol optical properties aloft, during biomass burning events observed over Athens, for a 10-year period (2007-2016). The Athens Raman depolarization lidar system (EOLE) was run in parallel with the AIAS depolarization lidar, in the frame of ARIADNE, the Greek lidar network [18]. Both systems are based in the Laser Remote Sensing Unit (LRSU) of the National Technical University of Athens (NTUA) (37.9°N, 23.6°E, 200 m a.s.l.).

EOLE (Fig. 1-left) is able to perform independent and simultaneous measurements of the vertical profiles of extensive aerosol parameters, β_{aer} (at 355, 532 and 1064 nm), α_{aer} (at 355 and 532 nm), as well as of the water vapor mixing ratio in the troposphere (using the H_2O Raman channel at 407 nm). AIAS (Fig. 1-right) provides the aerosol backscatter coefficient and the linear particle depolarization ratio at 532 nm. The AERONET CIMEL sun photometer (Fig. 1-middle) is operating at the National Observatory of Athens (NOA) (www.aeronet.gsfc.nasa.gov).

Figure 1. EOLE lidar system (left), NOA CIMEL sun photometer (middle), AIAS mobile aerosol depolarization lidar system (right).

In addition, HYSPLIT model and MODIS sensor data were used for the characterization of the air masses arriving over our site.

Moreover, the EOLE multi-wavelength Raman lidar data were used to derive the vertical profiles of the aerosol intensive properties: LR (at 355 and 532 nm), AE_a and backscatter-related Ångström exponent (AE_b) of the aerosols at 355-532-1064 nm, and LPDR at 355 nm. Due to the weak Raman signals, the complete set of aerosol products can be derived for tropospheric layers only during nighttime. The data processing was based on the Single Calculus Chain (SCC) developed within EARLINET [19] and on manual retrievals. Sun photometry data were provided by the AERONET CIMEL sun photometer.

For this study, based on the combined data from the EOLE and AIAS lidars, we compared our results (LR and LPDR at 355 and 532 nm, AEa at 355 nm/532 nm), with those reported in the literature [12-16].

3. RESULTS

In the period under study we sampled more than 30 clear biomass burning events, with fresh and aged smoke particles over Athens, Greece with smoke layers between 1.5 up to 4-5 km height, and duration of 1-3 days. Fig. 2 shows LR versus LDR (left), LR versus AE_a (middle), as well as LPDR versus AE_a (right), all at 532 nm, except for AE_a(355 nm/532 nm), for fresh (≤ 3 days) and aged (> 3 days) BB aerosols. The data are taken from [10-16] (denoted with squares) and from our measurements (denoted with stars). We can see that our data are, mostly, in good accordance with literature data obtained in Europe.

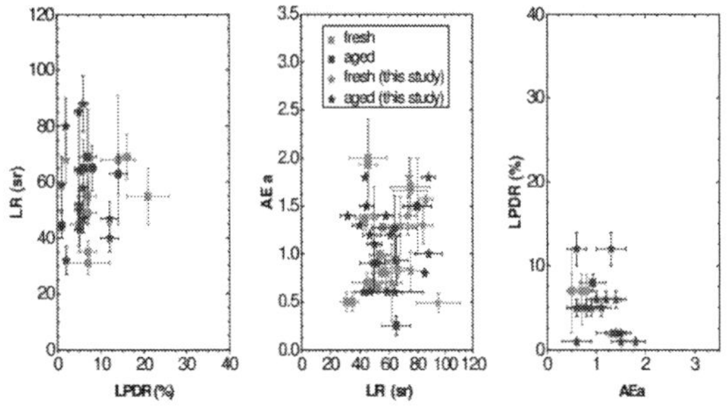

Figure 2. LR (532 nm) versus LPDR (532 nm) (left), LR (532 nm) versus AE_a (middle), LPDR (532 nm) versus AE_a (right) (355nm/532 nm) for fresh and aged BB particles [literature data (squares) and current study data (stars)].

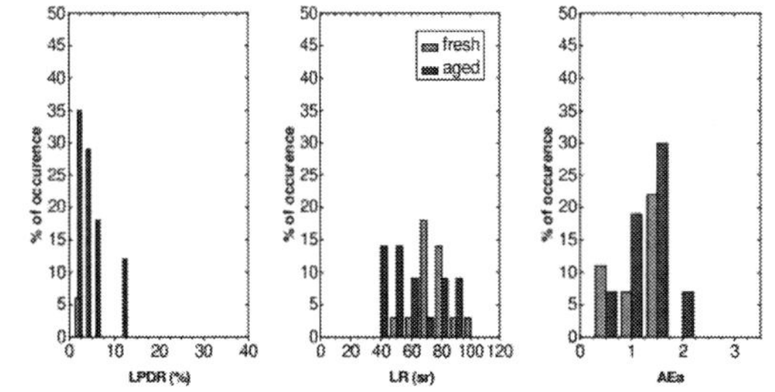

Figure 3. Frequency of occurrence (%) of LPDR (532 nm), LR (532 nm) and AE_a(355nm/532nm) for fresh and aged BB particles (our study).

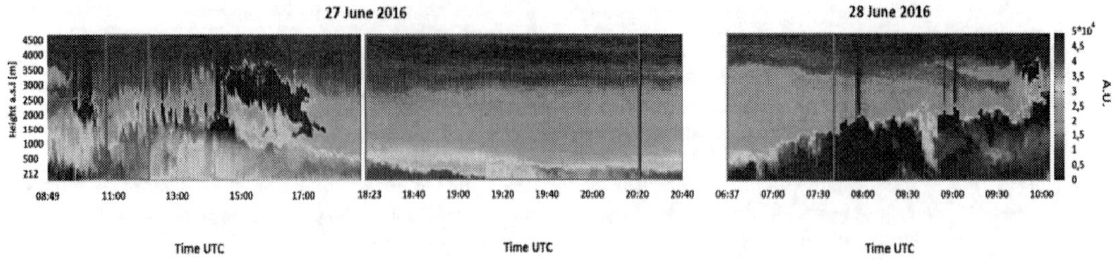

Figure 4. Spatio-temporal evolution of the range-corrected (RCS) backscattered lidar signals at 1064 nm obtained by EOLE over Athens for 27 June (day and night) and 28 June (day) 2016.

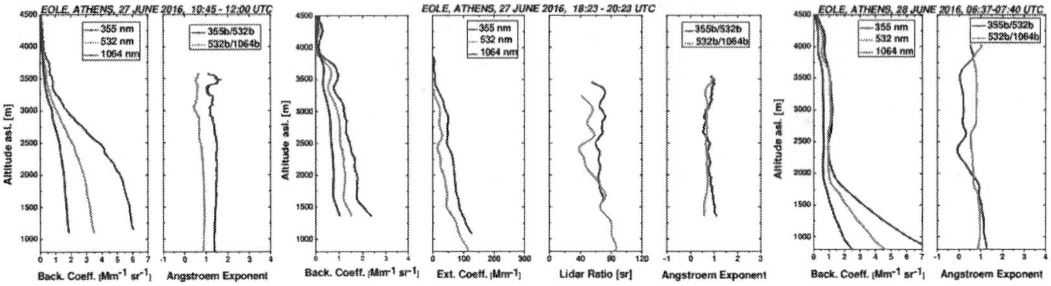

Figure 5. Vertical profiles of β_{aer}, α_{aer}, LR, AE_b at 355, 532, 1064 nm, obtained by the EOLE system on 27 June (day and night) and 28 June (day) 2016.

In Fig. 3 we present the frequency of occurrence (%) of LPDR, LR and AE_a(355 nm/532 nm) for fresh and aged BB particles at 532 nm, as measured over our station. We see that in more than 80% of the cases we studied, LPDR<7% (aged), 40 sr<LR<105 sr (fresh and aged), while 1.2<AEa<1.6 (aged) and 0.4<AEa<1.4 (fresh). Based on observations we found that LR(355nm)= 73±13sr (fresh) and 60±22 sr (aged); LR(532nm) =71±15sr (fresh) and 60±18 sr (aged). These values are in good agreement with relevant published ones [10-16].

4. CASE STUDY

A case study of fresh BB particles from a wildfire occurred about 20 km from NTUA, in Parnis (Attika) mountain, between 27-28 June 2016 will be analyzed, regarding the aerosol optical properties: β_{aer} (355, 532, 1064 nm), α_{aer}(355, 532 nm), LR(355, 532 nm), AE_a(355nm/532nm) and AE_b(355nm/532nm) and AE_b(532nm/1064nm).

In Fig. 4 we present the spatio-temporal evolution of the range-corrected (RCS) lidar signal at 1064 nm obtained by EOLE for 27 June (08:49-20:40 UTC, day and night) and 28 June (06:37-10.00 UTC, day) 2016. Smoke layers are seen between 2.5 and 4.5 km height (in light blue-green color) over the locally produced particles within the PBL (from ground up to 2-2.5 km height, in yellow-red color). Scattered clouds are formed at the top of the PBL or above it at 1.5-3.5 km height (i.e., between 14:30-17:00 UTC), on 27 June 2016.

In Fig. 5 we show the vertical profiles of β_{aer}, α_{aer}, LR, AE_b obtained by the EOLE system at 355, 532, 1064 nm, on 27 (10:45-12:00 UTC & 18:23-20:23 UTC) and 28 June (06:37-07:40 UTC) 2016. We found LRs of the order of 70±5 sr (355 nm) and 65±15 sr (532 nm) and AE_a(355 nm/532 nm) around 1.

5. CONCLUSIONS

Multi-wavelength aerosol Raman lidar and sun photometer data were used, over a 10-year period (2007-2016), to provide the vertical profiles of the aerosol optical properties over Athens, Greece. We studied more than 30 intense biomass burning events originating from forest and agricultural fires mainly over Europe (Russian Federation, Ukraine, Romania, Italy, Greece, etc.). These events comprised fresh and aged smoke particles, with layers between 1.5 up to 4-5 km height and of duration of ~1-3 days. The statistical analysis of the aerosol optical properties derived by EOLE and AIAS [LR(355-532 nm, LPDR(532 nm), AE_a(355 nm/532nm)] showed that: 40 sr<LR<105 sr (355 nm) and 40 sr<LR<100 sr (532 nm), LPDR<7%, in most cases and 0.3<AE_a(355 nm/532 nm)<2.1. Our observations, when compared with those reported in the literature concerning various biomass burning events (fresh and aged) observed by Raman lidars, showed good agreement. A case study (27-28 June 2016) of near-range transport of biomass burning aerosols over

Athens showed smoke layers up to 4-4.5 km a.s.l., with LR≈70±5 sr (355 nm), LR≈65±15 sr (532 nm) and AE_a (355 nm/532 nm)≈1.

ACKNOWLEDGMENTS

The research leading to these results has received funding from the European Union 6th & 7th Framework Programmes (FP6 and FP7 under grant agreement RICA-025991 and no 262254, as well as from Horizon 2020 ACTRIS research/innovation programme under grant agreement n° 654109. The work of AP and AA has received funding from the European Union Seventh Framework Programme (FP7/2007-2013): People, ITN Marie Curie Actions Programme (2012-2016) in the frame of ITaRS, under grant agreement n° 289923. The authors gratefully acknowledge the NOAA Air Resources Laboratory for the provision of the HYSPLIT transport and dispersion model and READY website (http://www.ready.noaa.gov) used in this publication. MODIS-FIRMS data were provided by EARTHDATA (EOSDIS-NASA). The AERONET sun photometer data were provided by the National Observatory of Athens [PI Dr. V. Amiridis]. The Biomedical Research Foundation of the Academy of Athens is acknowledged for the provision of the lidar mobile platform.

References

[1] Chuvieco, E., et al., 2016: A new global burned area product for climate assessment of fire impacts, *Global Ecology and Biogeography*, **25**, pp. 619-629.

[2] Jolly, W. M., et al., 2015: Climate-induced variations in global wildfire danger from 1979-2013, *Nature Communications*, **6**:7537, doi: 10.1038/ncomms8537.

[3] IPCC, 2014: Intergovernmental Panel on Climate Change: Climate Change 2013: *Synthesis Report*.

[4] Wotawa, G., 2001: Interannual variability of summertime CO concentrations in the Northern Hemisphere explained by boreal forest fires in North America and Russia, *Geophys. Res. Lett.,* **28**, 4575-4578.

[5] Amiridis, V., et al., 2009: Optical characteristics of biomass burning aerosols over Southeastern Europe determined from UV-Raman lidar measurements, *Atmos. Chem. Phys*., **9**, 2431-2440.

[6] Bougiatioti, A., et al., 2014: Processing of biomass-burning aerosol in the eastern Mediterranean during summertime, *Atmos. Chem. Phys*., **14**, 4793-4807.

[7] European Environment Agency (EEA), Air Quality in Europe, 2016 Report, *EEA Report* No **28**/2016, 2016.

[8] Paraskevopoulou, D. et al., 2016: Long term characterization of organic and elemental carbon in the $PM_{2.5}$ fraction: the case of Athens, Greece, *Atmos. Chem. Phys.* **14**, 13313-13325.

[9] Bougiatioti, A., et al., 2016: Biomass-burning impact on CCN number, hygroscopicity and cloud formation during summertime in the eastern Mediterranean, *Atmos. Chem. Phys*., **16**, 7389-7409.

[10] Nicolae, D., et al., 2013: Characterization of fresh and aged biomass burning events using multi-wavelength Raman lidar and mass spectrometry, *J. Geophys. Res*., **118**, 2956-2965.

[11] Wandinger, U., et al., 2002: Optical and microphysical characterization of biomass burning and industrial-pollution aerosols from multi-wavelength lidar and aircraft measurements, *J. Geophys. Res.*, **107**, D21, 8125.

[12] Balis, d., et al., 2003: Raman lidar and sun photometric measurements of aerosol optical properties over Thessaloniki, Greece, during a biomass burning episode, *Atmos. Environ*., **37**, 4529-4538.

[13] Müller, D., et al., 2005: Raman lidar observations of aged Siberian and Canadian forest fire smoke in the free troposphere over Germany, in 2003: microphysical particle characterization, *J. Geophys. Res.*, **110**, D17201.

[14] Gross, S., et al., 2013: Aerosol classification by airborne high spectral resolution lidar observations, *Atmos. Chem. Phys.*, **13**, 2487-2505.

[15] Giannakaki, E., et al., 2016: Optical and microphysical characterization of aerosol layers over S. Africa by means of multi-wavelength depolarization and Raman lidar measurements, *Atmos. Chem. Phys.*, **16**, 8109-8123.

[16] Müller, D., et al., 2007: Aerosol-type-dependent lidar ratios observed with Raman lidar, *J. Geophys. Res.*, **112**, D16202.

[17] Mamouri et al., 2012: Multi-wavelength Raman lidar, sun photometric and aircraft measurements in combination with inversion models for the estimation of the aerosol optical and physico-chemical properties over Athens, Greece, *Atmos. Meas. Tech*., **5**, 1793-1808.

[18] Papayannis, A., et al., 2012: ARIADNE: The Greek Lidar Network, *Proc. 26th International Laser Radar Conference*, Porto Heli, Greece, 349-352.

[19] d'Amico, G., et al., 2012: EARLINET Single Calculus Chain for automatic lidar data processing: First tests on optical products, *Proc. 26th International Laser Radar Conference*, Porto Heli, Greece, 331-334.

EPJ Web of Conferences **176**, 05023 (2018) https://doi.org/10.1051/epjconf/201817605023
ILRC 28

ARCTIC BIOMASS BURNING AEROSOL EVENT–MICROPHYSICAL PROPERTY RETRIEVAL

Böckmann, Christine[1,*], Ritter, Christoph[2] and Ortiz-Amezcua, Pablo[3]

[1] *University of Potsdam, Institute of Mathematics, Germany, *bockmann@uni-potsdam.de*
[2] *Alfred Wegener Institute for Polar and Marine Research, Germany*
[3] *University of Granada, Department of Applied Physics, Spain*

ABSTRACT

An intense biomass-burning (BB) event from North America in July 2015 was observed over Ny-Ålesund (Spitsbergen, European Arctic). An extreme air pollution took place and aerosol optical depth (AOD) of more than 1 at 500nm occurs in middle and lower troposphere. We analyse data from the multi-wavelength Raman-lidar KARL of Alfred Wegener Institute to derive microphysical properties of the aerosol of one interesting layer from 3186 to 3306 m via regularization. We found credible and confidential microphysical parameters.

1 INTRODUCTION

The data for this work was obtained in Ny-Ålesund, Spitsbergen, in the European Arctic on 10 July 2015. An intensive event of BB aerosol originating from the boreal North America was observed for several days around that period at different Arctic sites (Markowicz et al. [1]) and produced an AOD(500) >1. Profiles of extinction and backscatter were obtained by the "3+2" Raman lidar KARL according to the method of Ansmann [2] with 10min / 30m resolution.

This event of BB aerosol and the KARL lidar are described also in [3]. The extinction and backscatter coefficient profiles are shown in Fig 1. For the inversion of the microphysics we selected an altitude range from 3186m to 3306m because a contemporaneous radiosonde showed a humidity about 80-85% and, hence, we expected a larger effective radius and a lower refractive index (RI) as was reported for dry BB aerosol in literature. Wandinger et al [4] derived for example effective radii of 0.25μm and RI of 1.56–1.66 for the real (Re) and 0.05i–0.07i for the imaginary part (Im).

2 METHODOLOGY

The model relating the optical parameters $\Gamma(\lambda)$ with the volume size distribution $v(r)$ is described by the action of a Fredholm integral operator of the 1st kind with the kernel function $K(r,\lambda;m) = \frac{3}{4r}Q(r,\lambda;m)$ and

$$\Gamma(\lambda) = \int_{r_{\min}}^{r_{\max}} K(r,\lambda;m)v(r)\mathrm{d}r, \qquad (1)$$

where λ is the wavelength, r is the radius, $r_{\min}$, $r_{\max}$ are suitable lower and upper radius bounds, m is the complex RI, $\Gamma(\lambda)$ denotes either the extinction or backscatter coefficients, and Q stands for either the extinction or the backscatter (dimensionless) Mie efficiencies respectively. The wavelength in our measurement cases can only take three discrete values 355, 532, and 1064 nm, since all the measurements were performed with the Raman lidar forming data sets of 3 backscatter coefficients in all three wavelengths and 2 extinction coefficients in the first two. Identifying $\Gamma(\lambda)$ as our measurement data and $v(r)$ as the unknown volume distribution, the problem reduces to the inversion of Eq. (1). Knowing the volume distribution, we can then extract the following microphysical parameters

- total surface-area concentration (μm^{2}cm^{-3}) $s_t = 3\int \frac{v(r)}{r}\mathrm{d}r$
- total volume concentration (μm^{3}cm^{-3}) $v_t = \int v(r)\mathrm{d}r$
- total number concentration (cm^{-3}) $n_t = 3/4\pi \int \frac{v(r)}{r^3}\mathrm{d}r$
- effective radius (μm) $r_{\text{eff}} = 3\frac{v_t}{s_t}$.

In addition, the complex RI and the single scattering albedo (SSA) in 355 nm and 532 nm are retrieved. Note, that in this work the com-

Table 1: Retrieved RI and SSA using both methods.

	TSVD best point	TSVD mean	TSVD deviation	ITER
RI Re	1.45	1.48	±0.02	1.49
RI Im	0.001	0.007	±0.005	0.01
SSA (355)	0.981	0.919	±0.030	—
SSA (532)	0.987	0.942	±0.017	—

mon assumption of wavelength-independent RI is made, as a member of the predefined grid introduced, see e.g. [5, 6, 7]. Solving Eq. (1) requires discretization, regularization and a parameter choice rule, see e.g. [5, 8].

2.1 TSVD regularization

We discretized Eq. (1) with spline collocation. The volume distribution $v(r)$ is approximated with respect to the B-spline functions ϕ_j by $v_n = \sum_{j=1}^{n} b_j\phi_j$, reducing the problem to the determination of the coefficients b_j. The continuous problem (Eq. (1)) is now replaced by a discrete one $Ab = \Gamma$, where the matrix elements of the linear system are

$$A_{ij} = \int_{r_{\min}}^{r_{\max}} K(\lambda_i, r; m)\phi_j(r)\mathrm{d}r \qquad (2)$$

and Γ is a vector now. By doing so, we have already projected our problem to a finite n-dimensional space. Clearly, the decision about the projected dimension (n) and order (d) of the base functions ϕ_j is critical, since an appropriate representation of our solution strongly depends on it. This is not done at once; on the contrary, the algorithm constructs a linear system ($5 \times n$) for each value of n and d specified and every RI within our predefined grid. For example, we first fix the refractive index, and calculate the kernel functions, then decide for candidates for n and d, e.g., $n = 3,\ldots,8$ and $d = 3,4$ which define a B-spline function and finally calculate the matrix elements from Eq. (2). For more particular details on the B-spline basis in the frame of a non-negative size distribution, we are looking for, see [5].

Each linear system is solved by first expanding the matrix using Singular Value Decomposition (SVD). Potential noise in our matrix will be magnified as a result of the singular values clustering to zero. We would like to prevent this behavior by including only a part of the SVD, i.e defining a certain cut-off level k, above which the most noisy solution coefficients are filtered out. This regularization procedure is known as Truncated SVD (TSVD), see [5].The parameter choice rule consists in selecting an appropriate triple (n,d,k) heuristically.

2.2 Iterative regularization

A particular iterative regularization (ITER) was used additionally to retrieve microphysical properties of this event. This method is based on Runge-Kutta regularization, see [8, 9]. Since the noise level of the backscatter and extinction profile is often not known or only a roughly estimation is available, a heuristical parameter choice rule is used here too, the well-known L-curve method. This method was developed to retrieve the fine mode only until 1.25μm.

3 RESULTS

The retrieval by TSVD was first done with a refractive index grid with a resolution of 40x40 points and a range for the Re of RI from 1.3-1.8 and for the Im from 0-0.05. We found that the best possible RI's lay on a diagonal pattern what is an indication of precise measurements in agreement with investigated simulations, see [10]. The best points are situated in the range 1.4-1.55 for Re and 0-0.025 for Im. Therefore, the retrieval was done a second time on those ranges with a grid resolution of 20x20, see Fig. 2 (top). We selected the 12 points with a relative residual norm below 8% and a similar shape of the volume distribution. The best point (7.22%) was found for $m = 1.45 + 0.00132i$. Averaging all 12 selected RI points results in $m = 1.48 + 0.00713i$, see Table 1. Additionally, the second algorithm ITER found a similar value $m = 1.49 + 0.01i$ which is in accor-

https://doi.org/10.1051/epjconf/201817605023

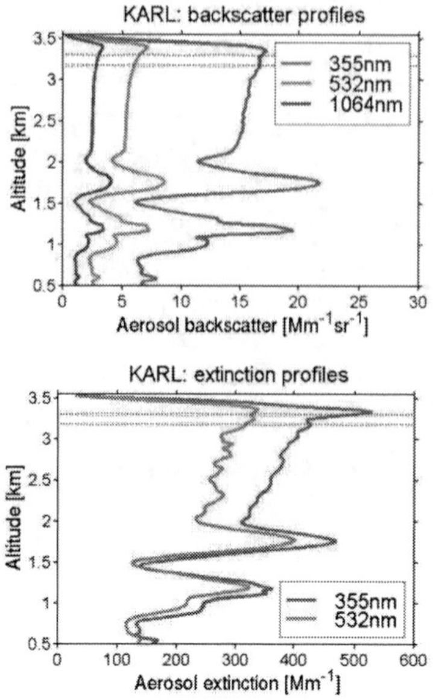

Figure 1: Optical backscatter and extinction coefficient profiles from Raman lidar KARL and aerosol layer of interest (dotted horizontal lines).

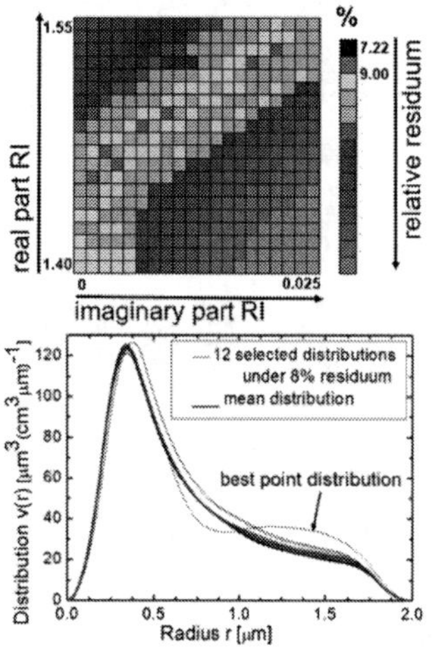

Figure 2: Top: Retrieved complex refractive index grid. Bottom: 12 retrieved volume distributions (thin black lines) and mean distribution (thick blue line).

dance with the deviation level from the first algorithm, see Table 1. Comparing both algorithms for the used B-spline bases yields: Both

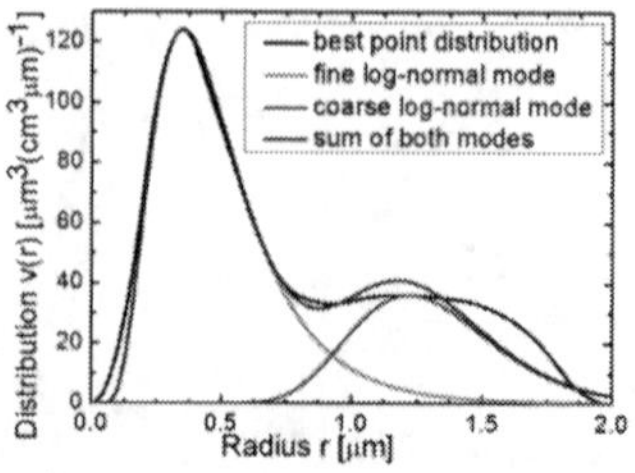

Figure 3: Retrieved volume distribution of the best point, see Fig. 2 with TSVD and two log-normal modes for fine and coarse mode (green lines).

Table 2: Retrieved microphysical properties of the inverted and fitted distributions. Values in brackets stand for: (.) best point and {.} used r_{min}=0.125 μm (regular used r_{min}=0.001 μm).

	mean TSVD	fine mode	coarse mode
r_{med}	—	0.19 (0.22)	1.60 (1.11)
σ	—	1.71 (1.63)	1.08 (1.24)
r_{eff}	0.45 ±0.01 (0.46)	0.40 — (0.39)	1.63 — (1.24)
v_t	83.7 ±2.24 (87.0)	66.3 — (59.2)	6.1 — (24.02)
s_t	561 ±2.5 (564)	501 — (454)	11.25 — (58.2)
n_t	{477} ({479})	602 (478)	0.35 (3.5)

determined for all selected distributions $n = 8$ (only two outliers $n = 7, 9$) and $d = 3$ for the number and degree of the used splines, respectively. Moreover, with respect to TSVD no additional cut of any singular value was necessary. The retrieved mean volume distribution, see Fig. 2 bottom (blue line), and in particular the best point distribution shows obviously a bi-modal volume distribution, one for the fine mode and one for the coarse mode particles. It is commonly assumed that the number distribution is a log-normal distribution, this property is also true for the surface-area and the volume distribution with the same geometric standard deviation as well. Therefore, we separated the distribution into two modes by using two log-normal distributions such that the sum of both modes (brown line) fits the retrieved volume distribution well, see Fig. 3. Finally, we summarized all microphysical parameters in Table 2 for the whole distribution as well as for the two modes separately. We remark, that for the total number concentration of the complete mean distribution the Aitken-mode (first part of the fine mode) was excluded since even very small uncertainties in the volume distribution by dividing by r^3 with very small radii lead to a huge amplification.

4 CONCLUSIONS

We found for the effective radius of the complete mean distribution $r_{\mathrm{eff}} = 0.45 \pm 0.01\,\mu\mathrm{m}$ as well as for the fine and coarse mode 0.40μm and 1.24μm, respectively. Our expectation was approved. Because of the large humidity 80-85% in the regarded particle layer, the effective radius of the BB-particles is larger as usual in the literature, e.g. 0.25μm in [4]. Furthermore, in agreement the RI is indeed lower in both parts for such wet particles, compare the Introduction last part and Table 1. Analyzing the whole profile is an ongoing work but some of the particle properties between 2km and 3.3km altitude are also discussed in [3].

ACKNOWLEDGEMENTS

The work has been supported partially by the European Union (EU) Seventh Framework Program for research, technological development and demonstration under grant agreement No. 289923 - ITaRS and EU's Horizon 2020 research and innovation programme under grant agreement No. 654109 (ACTRIS-2). This work was also partially funded by the Spanish Ministry of Education, Culture and Sports through grant FPU14/03684 with the support of the Erasmus+ programme of the EU.

References

[1] Markowicz, K.M., Pakszys, P., Ritter, C., Zielinski, T., Udisti, R., Cappelletti, D., Mazzola, M., Shiobara, M., Lynch, P., Zawadzka, O., Lisok, J., Petelski, T., Makuch, P., Karasinski, G.: 2016, Impact of North American intensive fires on aerosol optical properties measured over the European Arctic in July 2015, accepted at JGR, doi: 10.1002/2016JD025310.

[2] Ansmann, A., Wandinger, U., Riebesell, M., Weitkamp, C., and Michaelis, W. 1992: Independent measurement of extinction and backscatter profiles in cirrus clouds by using a combined Raman elastic-backscatter lidar, Appl. Opt. 31, 7113-7113.

[3] Ritter, C. and Böckmann, C.: Observation of an intensive Biomass Burning Event over Spitsbergen (this issue).

[4] Wandinger, U., Müller, D., Böckmann, C., Althausen, D., Matthias, V., Bösenberg, J., ... and Ansmann, A. (2002). Special Section: Lindenberg Aerosol Chracterization Experiments (LACE): LAC 7 Optical and microphysical characterization of biomass-burning and industrial-pollution aerosols from multiwavelength lidar. Journal of Geophysical Research-Part D-Atmospheres, 107(21), doi: 10.1029/2000JD000202.

[5] Böckmann, C., 2001: Hybrid regularization method for the ill-posed inversion of multiwavelength lidar data to determine aerosol size distributions, *Appl. Opt.* **40**, 1329–1342.

[6] Kirsche, A., 2003: Entwicklung der Runge-Kutta Iteration und Anwendung als Regularisierungsverfahren, *Diploma thesis University of Potsdam*.

[7] Samaras, S., Nicolae, D., Böckmann, C., Vasilescu, J., Binietoglou, J., Labzovskii, L., Toanca, F., and Papayannis, A., 2015: Using Raman-lidar-based regularized microphysical retrievals and Aerosol Mass Spectrometer measurements for the characterization of biomass burning aerosols, *J. Comput. Phys.* **299**, 156–174.

[8] Böckmann, C. and Kirsche, A., 2006: Iterative regularization method for lidar remote sensing, *Comput. Phys. Commun.* **174**, 607–615.

[9] Osterloh, L., Pérez, C., Böhme, D., Baldasano, J. M., Böckmann, C., Schneidenbach, L., and Vicente, D., 2009: Parallel software for retrieval of aerosol distribution from LIDAR data in the framework of EARLINET-ASOS, *Comput. Phys. Commun.* **180**, 2095–2102.

[10] Müller, D., Böckmann, C., Kolgotin, A., Schneidenbach, L., Chemyakin, E., Rosemann, J., Znak, P., and Romanov, A., 2016: Microphysical particle properties derived from inversion algorithms developed in the framework of EARLINET, *Atmos. Meas. Tech.* **9**, 5007–5035.

PROPERTIES OF ARCTIC HAZE AEROSOL FROM LIDAR OBSERVATIONS DURING IAREA 2015 CAMPAIGN ON SPITSBERGEN

Iwona S. Stachlewska[1], Christoph Ritter[2], Christine Böckmann[3], Ronny Engelmann[4],

[1]*University of Warsaw, Faculty of Physics, Institute of Geophysics, IGFUW, Warsaw, Poland,*
iwona.stachlewska@igf.fuw.edu.pl
[2]*Alfred-Wegener-Institute for Polar and Marine Research, AWI, Potsdam, Germany*
[3]*University of Potsdam, Potsdam, Germany*
[4]*Leibnitz Institute of Tropospheric Research, TROPOS, Leipzig, Germany*

ABSTRACT

Arctic Haze event was observed on 5-8 April 2015 using simultaneously Near-range Aerosol Raman Lidar of IGFUW and Koldewey Aerosol Raman Lidar of AWI, both based at AWIPEV German-French station in Ny-Ålesund, Spitsbergen. The alterations in particle abundance and altitude of the aerosol load observed on following days of the event is analyzed. The daytime profiles of particle optical properties were obtained for both lidars, and then served as input for microphysical parameters inversion. The results indicate aerosol composition typical for the Arctic Haze. However, in some layers, a likely abundance of aqueous aerosol or black carbon originating in biomass burning over Siberia, changes measurably the Arctic Haze properties.

1 INTRODUCTION

During the transition form early spring to summer, the Arctic front expands over a large fraction of Northern Hemisphere and moves towards lower latitudes. This causes strong intrusions at different altitudes in the troposphere of polluted air-masses originating from industrial areas in Europe, Siberia, and/or North America. Due to the stable atmospheric conditions, low cloud cover, and low precipitation during late winter and early spring in the Arctic, this long-range transported aerosol can be trapped in the Arctic troposphere even for a month. The Arctic Haze particles are of a sub-micron size (Ångström exponent 1-1.5), well-aged (accumulation mode), with substantial fraction of sulfate (cooling effect) and soot (warming effect). Presence of absorbing particles in the troposphere causes a local warming at the altitude of their existence and, at the same time, a cooling below it at the ground. Measurements have shown that black carbon found in Arctic Haze events over Spitsbergen seem to be a minor but an important component, due to its absorbing properties [1]. In the frame of the iAREA Project (Impact of Absorbing aerosols on Radiative forcing in the European Arctic), a field campaign was performed from 26 March to 7 May 2015 at Ny-Ålesund. One of the aims of this campaign was at profiling of absorbing particles during haze season by using the iCAP comprising particle counter and micro-aethalometer installed under a tethered balloon. These in-situ data profiles were combined with ground-based observations of advanced Raman lidar and allowed to obtain a new methodology for absorption coefficient profiling [2]. Another aim, discussed in this paper, was at exploring solely the Raman lidar data for complex retrieval of optical and microphysical properties of Arctic Haze.

2 INSTRUMENTS

The Near-range Aerosol Raman lidar receiver (NARLa) [3], was developed to serve along with the advanced 8-channel ($2\alpha+3\beta+2\delta$+WV) aerosol-depolarization-Raman lidar (ADR lidar) of the NeXT generationn Polly^{XT}-type [4] which is used for regular EARLINET measurements at IGFUW Radiative Transfer Laboratory (RTLab) in Warsaw. NARLa development was dedicated to improve Raman detection range within boundary layer at 4-channels ($2\alpha+2\beta$). The NARLa is based on a Newtonian type telescope of a 5cm diameter and a 20cm focal length, which is combined with a fiber core of 0,4mm, resulting in FOV of 2mrad.

The elastic signals are detected at 355, 532nm and Nitrogen Raman channels at 387, 607nm. For all channels signals are acquired in a photon counting mode with a low overall dead-time of about 2ns. The first 256 signal bins are allocated for a pre-trigger. The signal is obtained up to 40km with a 7.5m spatial resolution. For experiment discussed in this paper, NARLa was installed next to Koldewey Aerosol Raman Lidar (KARL) [5] at the AWIPEV station in Ny-Ålesund from 27 March to 6 May 2015. The KARL's laser was used as a radiation source for NARLa. KARL is a coaxial Nd:Yag based system emitting IR, VIS, and UV pulses of ~10W/color (beam divergence is 0.6mrad). Detection is done at 9-channels ($2\alpha+3\beta+2\delta+2WV$) using a 70cm diameter receiving mirror. An aperture stop is adjustable (position and diameter) and allows for measurements in the troposphere (overlap range <700m) and stratosphere. KARL is member of the NOAA-NDACC network.

3 METHODOLOGY

Pre-selection of KARL and NARLa signals was done with a purpose to obtain a data set of particle optical properties, as appropriate for performing inversion of particle microphysical parameters at various layers. Selection was based on inspecting the range and background corrected signal plots at 532nm, then all signals were evaluated using the VerlaufNG5t.vi software approved for standard EARLINET retrieval [6]. The daytime profiles of aerosol extinction coefficient were obtained for both lidars using the classical Raman technique with Ansmann approach; then the particle backscatter coefficient profiles were calculated (Fig.1). All NARLa profiles were obtained with 7.5m/20min resolution with the same smoothing applied prior to the optical retrieval. KARL data were evaluated with 30m/10min resolution without smoothing. Calculations were done with use of RS92-Vaisala sonde launched everyday at 11/23UTC in Ny-Ålesund. From retrieved optical profiles, profiles of lidar backscatter-to-extinction ratio and Ångström exponent were obtained(Fig.1). The lidar derived aerosol optical depth was obtained by integrating the extinction profiles, interpolated with the extinction coefficient value derived at the completed overlap range height.

The microphysical retrieval algorithm is based on a concept of using the truncated singular value decomposition as a regularization method. This method was adapted to work for the retrieval of the particle size distribution (PSD) and is called the hybrid regularization technique [8,9,10], since it uses a triple of regularization parameters to invert a mono-modal or bimodal PSDs. Inversion of an ill-posed problem (such as PDS retrieval) is a challenging task, as small measurement errors and tiny rounding errors often will be hugely amplified during the solution process, unless an appropriate regularization method with carefully determined regularization parameters is used. Therefore, we use two regularization techniques in parallel, for comparison purposes. The iterative regularization method is based on Páde iteration. Here, the well-known L-curve method serves as a regularization parameter choice rule. The approximated PSD is a linear combination of B-splines with appropriately determined weights. The B-splines of order d are polynomials of degree $d-1$. Since the distribution of the B-spline nodes also plays a critical role, we modified the first algorithm such that a non-equidistant node grid can be used. Later, the iterative method was equipped with an adaptive non-equidistant node grid. One more challenge was considered; as the complex refractive index (CRI) is actually also unknown, a grid of viable alternatives for the CRI (all combinations of real parts of CRI and imaginary parts of CRI) is assumed [7, 8, 9].

3 RESULTS AND DISCUSSION

Analysis of the NARLa and KARL data, obtained between 5-8 April 2015 at Ny-Ålesund shows enhanced aerosol loads with max. up to 3.5km, in comparison to e.g. 3 April when only background aerosol was observed [3]. During the event, the obtained mean particle backscatter coefficient values, calculated for a 100m thick layer at an altitude corresponding to the max. aerosol load, were relatively low: 0.78-1.76 Mm-1 sr-1 at 355nm, 0.51-0.93 Mm-1 sr-1 at 532nm, 0.32-0.67 Mm-1 sr-1 at 1064nm. However, aerosol optical depth (AOD) calculated from the ground to the upper age of aerosol layer, was high of 0.1-0.15 at 532nm and of 0.13-0.19 at 355nm. The Ångström exponent (AE >1.6) indicated small particles and lidar ratios (LR) of 58-82sr at 532nm, 46-74sr at 355nm indicated polluted arctic conditions. The color ratio of lidar ratios at 532/355 was changing gradually form 1.1 to 1.4 during the event. The linear particle depolarization ratios (KARL measurements; not shown, for brevity), were a background one or very small. These values confirm likeliness of the Arctic Haze occurrence.

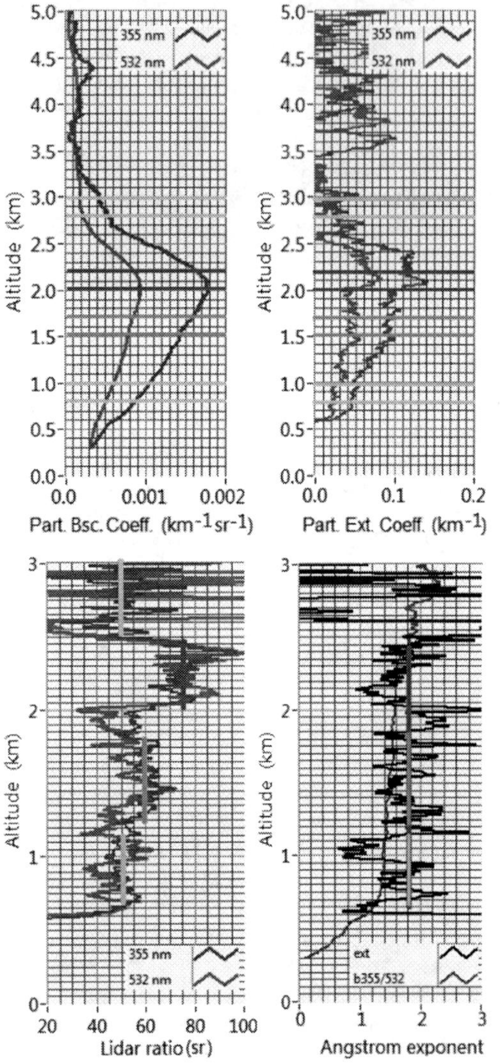

Figure 1 Profiles of optical properties retrieved from NARLa data at 22:40UTC on 6 April 2015.

The microphysical inversion was performed based on NARLa 2α+2β at 355, 532nm supported with KARL β at 1064nm. In general, results obtained for the entire event at different layers and different days show following features: A monomodal PSD with an accumulation mode of particles with an effective radius (*Radii*) of 407-501nm, the real part of complex refractive index (*Re* CRI) of 1.42-1.51 significantly higher than for water particles, and the imaginary part (*Im* CRI) of 0.007-0.015 indicating likely absorption. Particle concentration *N* was in the range of 60-362 part/cm3. Single scattering albedo was very low (SSA <0.92 at 355nm and <0.94 at 532nm), again in favor of absorbing particles.

A careful analysis of the inversion data (in temporal and vertical extent) indicate considerable changes in the aerosol composition during this event. For the highest aerosol abundance of the event, observed on April 6 (Fig.1, upper panels), at selected ranges (colored boxes) the optical properties are visibly different. Herein, we interpret differences that can be seen in the weakest aerosol load at an altitude of circa 2.8km (pink) and in the strongest aerosol load at circa 2.3km (red). Especially, we attempt to explain the origin of a high LR (~75sr, Fig.1, lower panels) obtained in the latter layer.

In free-troposphere, in uppermost layer at 2.8km, (Figs.1,2 in pink), the *Re* CRI of 1.51, *Im* CRI of 0.04, monomodal PSD with *Radii* of 500nm and *N* of 60part/cm3 was derived. This values can be interpreted along with the radiosounding data, as dry absorbing particles (Fig.3, RH<65%). For this layer, ensemble 10-days back-trajectories indicate likely biomass burning (black carbon) transport from over Siberia.

In the layer at 2.3km (Figs.1,2 in red), the *Re* CRI of 1.48 and *Im* CRI of 0.03, with bi-modal PSD, which is dominated by particles with a fine mode *Radii* of 210nm and very high *N* of 501part/cm3. In the second larger mode with *Radii* of 1130nm, a very low N of 0.46part/cm3 was retrieved. These results indicate presence of only slightly less absorbing particles, which in this case were wet and accumulated above an inversion layer (Fig.3, RH >85%).

https://doi.org/10.1051/epjconf/201817605024

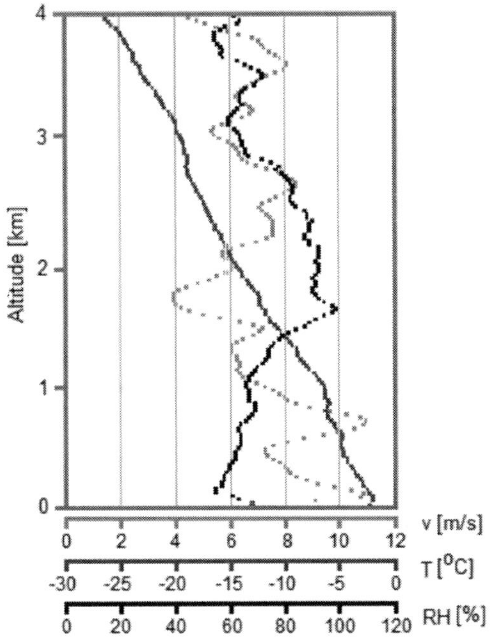

Figure 2 Termodynamical profiles obtained by radiosonding in Ny-Ålesund 23UTC on 6 April 2015.

4 CONCLUSIONS

A characteristics of the optical and microphysical particle properties during the Arctic Haze event of 5-8 April 2015, in Ny-Ålesund, Spitzbergen was obtained and discussed. Evidence for existence of a humid layer and its influence on the properties of Arctic Haze is given. This results complement studies of humid layers, as characterized by extremely high lidar ratios, that are occurring regardless of polluted or background Arctic conditions and were reported during several airborne campaigns of AWI [e.g. 10].

ACKNOWLEDGEMENTS

The NARLa was developed in collaboration of IGFUW and TROPOS in the frame of SONATA-BIS Project: *Integrated study of climate processes involving absorbing aerosols* funded by National Science Centre of Poland (2012/05/E/ST10/01578).

The iAREA campaign was financed within the project: *Impact of absorbing aerosols on radiative forcing in the European Arctic* funded by Polish-Norwegian Research Programme (Pol-Nor/196911/38/2013).

This research work was partly supported by the EUFP7 for research, technological development and demonstration (Grant No.289923 ItaRS), and the EU's Horizon 2020 research and innovation programme (Grant No. 654109 ACTRIS-2).

References

[1] Yamanouchi T., et al. 2005: Arctic Study of Tropospheric Aerosol and Radiation (ASTAR) 2000: Arctic haze case study, *Tellus* 57B, 141–152

[2] Markowicz, K.M, C. Ritter, J. Lisok, P. Makuch, I.S. Stachlewska, D. Cappelletti, M. Mazzola, M.T. Chilinski, 2017: Vertical variability of aerosol single-scattering albedo and black carbon concentration based on in-situ and remote sensing techniques during iAREA campaigns in Ny-Ålesund, *Atmos. Environ.*, vol. 164, pp. 431-447, 10.1016/j.atmosenv.2017.06.014

[3] Stachlewska I.S., et al. 2016: Near-range receiver unit of NeXT generation PollyXT used with Koldeway Aerosol Raman Lidar in Arctic, *ILRC27, New York*

[4] Engelmann R., et al. 2016: The automated multi-wavelength Raman polarization and water-vapor lidar PollyXT: the neXT generation, *Atmos. Meas. Tech.* 9, 1767-1784

[5] Ritter C., Neuber R., Schulz A., Markowicz K.M., Stachlewska I.S., Lisok J., Makuch P., Pakszys P., Markuszewski P., Rozwadowska A., Petelski T., Zielinski T., Becaglie S., Traversie R., Udistie R., Gausaf M., 2016: 2014 iAREA campaign on aerosol in Spitsbergen – Part 2: Optical properties from Raman-lidar and in-situ observations at Ny-Ålesund, Atmos. Environ. Vol.141 1-19 10.1016/j.atmosenv.2016.05.053

[6] Baars H., et al. 2016: An overview of the first decade of PollyNET: an emerging network of automated Raman-polarization lidars for continuous aerosol profiling, *Atmos. Chem. Phys.*, 16, 5111-5137.

[7] Böckmann C., et al. 2017: Arctic biomass burning aerosol event microphysical property retrieval, *ILRC28, Bucharest,* (this issue).

[8] Müller D., et al. 2016: Microphysical particle properties derived from inversion algorithms developed in EARLINET, *Atmos. Meas. Tech.* 9, 5007-5035.

[9] Samaras S., et al. 2015: Using Raman-lidar-based regularized microphysical retrievals and aerosol mass spectrometer for characterization of biomass burning aerosols, *J.Comput.Phys.* 299,156-174.

[10] Stachlewska I.S. and Ritter C. 2010: On retrieval of lidar extinction profiles using Two-Stream and Raman techniques, *Atmos.Chem.Phys.*, 10, 2813-2824

LOWER-FREE TROPOSPHERIC OZONE DIAL MEASUREMENTS OVER ATHENS, GREECE

Michail Mytilinaios[1], Alexandros Papayannis[1]*, and Georgios Tsaknakis[1]

[1]*National Technical University of Athens, Laser Remote Sensing Unit, Phys. Dept., Zografou, Greece,*
**apdlidar@central.ntua.gr*

ABSTRACT

A compact ozone differential absorption lidar (DIAL) was implemented at the Laboratory of Laser Remote Sensing of the National Technical University of Athens (NTUA), in Athens, Greece. The DIAL system is based on a Nd:YAG laser emitting at 266 nm. A high-pressure Raman cell, filled with D_2, was used to generate the λ_{ON} and λ_{OFF} laser wavelength pairs (i.e., 266-289 nm and 289-316 nm, respectively) based on the Stimulated Raman Scattering (SRS) effect. The system was run during daytime and nighttime conditions to obtain the vertical profile of tropospheric ozone in the Planetary Boundary Layer (PBL) and the adjacent free troposphere.

1 INTRODUCTION-TROPOSPHERIC OZONE

Ozone plays a significant role in the tropospheric chemistry as the major precursor of the hydroxyl radicals (OH), which in turn is the primary removal agent for most of the atmospheric pollutants [1]. In addition, ozone as a greenhouse gas influences the radiative balance of the troposphere and consequently the climate change [2].

Moreover, high ozone levels over urban areas contribute to the formation of photochemical smog while can cause health problems to humans and plants [3].

The ozone budget in the troposphere and its concentration changes are controlled by both chemical and dynamical processes: photochemical production in the Planetary Boundary Layer (PBL), downward transport from the stratosphere and horizontal transport from a source region [4].

Ozone is produced by photochemical reactions involving primary pollutants such as carbon monoxide (CO), nitrogen oxides (NO_X) and hydrocarbons (HC), which accelerate the process [1,3]. The transport of ozone from the stratosphere occurs during stratospheric-tropospheric exchanges, which in the extratropics are associated mainly with tropopause folds and cutoff lows events [5-11].

Furthermore, ozone in the free troposphere can be easily transported from a source region to another, due to its relatively longer lifetime and the higher wind velocities, as detected over the Eastern Mediterranean [8-10].

2 METHODOLOGY-EXPERIMENTAL SETUP

2.1 Methodology

By applying the well-known lidar equation [12] at the two wavelengths (λ_{ON} and λ_{OFF}), we get the ozone concentration n_{O3} [4]:

$$n_{O_3}(z) = \frac{1}{2\Delta\sigma_i}\left[\frac{d}{dz}\ln\frac{P(\lambda_{OFF},z)}{P(\lambda_{ON},z)}\right] \quad (1)$$

where, P_i = number of photons received at channel i (i=*ON* or *OFF* wavelength), $\Delta\sigma_i = \sigma(\lambda_{ON})-\sigma(\lambda_{OFF})$ for O_3 and $\sigma(\lambda)$ = absorption cross section of ozone (cm^2) at wavelength λ.

The corresponding *statistical error* (ε_1) if given by the following equation [4]:

$$\varepsilon_1(z) = \frac{1}{2n_{O_3}(z)\Delta\sigma_i\Delta z[N(z)]^{1/2}}\sqrt{\frac{2}{SNR(z)_1^2}+\frac{2}{SNR(z)_2^2}} \quad (2)$$

where, $$SNR(z)_i = \frac{P_i(z)}{(P(z)_i+P_{bi}(z))^{1/2}} \quad (3)$$

and $P_{bi}(z)$ is the atmospheric and electronic background noise at channel i, Δz is the range resolution of the measurement, and N(z) is the number of laser shots (related to the integration time).

The corresponding *systematic error* (ε_2), if we omit the laser beam absorption by other gases, is given by the following equation [4]:

$$\varepsilon_2(z) = \frac{1}{n_{O_3}(z)\Delta\sigma_i}\left[\frac{1}{2}\cdot\frac{d}{dr}\ln\frac{\beta(\lambda_{ON},z)}{\beta(\lambda_{OFF},z)} - \Delta\alpha_{RAY}(\lambda,z) - \Delta\alpha_{aer}(z,\lambda)\right] \quad (4)$$

where, $\beta(\lambda_i,z)$ is the atmospheric backscatter coefficient at wavelength λ_i (i=*ON* or *OFF*) at range z, $\Delta\alpha_{RAY}(\lambda,z) = \alpha_{RAY}(\lambda_{ON},z)-\alpha_{RAY}(\lambda_{OFF},z)$ and $\Delta\alpha_{aer}(\lambda,z)=\alpha_{aer}(\lambda_{ON},z)-\alpha_{aer}(\lambda_{OFF},z)$, where $\alpha_{RAY}(\lambda_i,z)$ and $\alpha_{aer}(\lambda_i,z)$ are the extinction coefficients of

 https://doi.org/10.1051/epjconf/201817605025

atmospheric molecules and particles, respectively, at wavelength λ_i (*i=ON* or *OFF*) at range z.

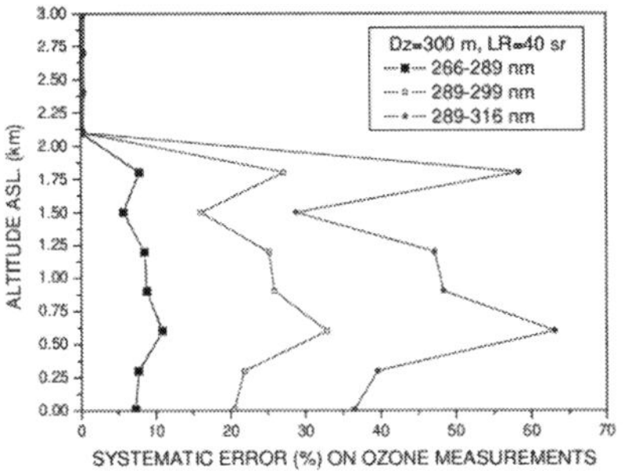

Figure 1. *Systematic error (ε_2) in % for three different wavelength pairs: 266-289 nm, 289-299 nm and 289-316 nm.*

In Fig. 1 we present simulations of the systematic error ε_2, in a polluted urban atmosphere (with an intense presence of aerosols), for three wavelength pairs (266-289 nm, 289-299 nm, 289-316 nm), a range resolution of 300 m and a typical lidar ratio (LR), $LR(\lambda,z)=\alpha_{aer}(\lambda,z)/\beta_{aer}(\lambda,z)$, of 40 sr. From this figure we see that we can derive the ozone profiles within the first 2 km (typically within the Planetary Boundary Layer) with a systematic error $\varepsilon_2<10\%$, if we use the 266-289 nm wavelength pair.

2.2 Experimental setup

To produce the ON and OFF wavelength we use the Stimulated Raman Scattering (SRS) effect by optically pumping, with the 4th harmonic of a Nd:YAG laser (17-20 mJ/pulse at 266 nm), a 1-m long high-pressure Raman cell filled with a Raman active gas (D_2 at 14 atm).

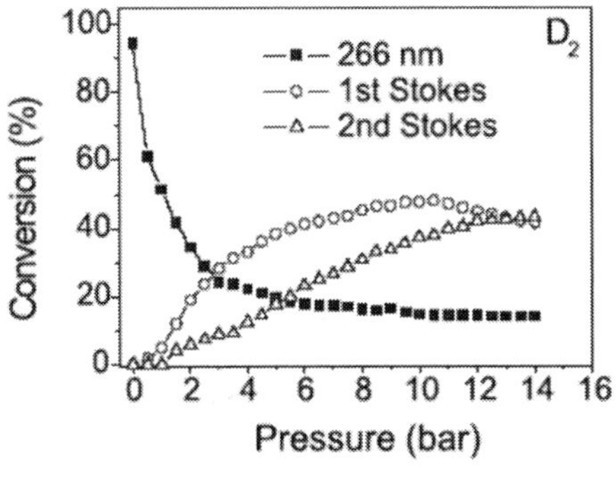

Figure 2. *Conversion efficiency (%) at the exit of a 1-m long Raman cell, at both the 1st and 2nd Stokes wavelengths (289 and 316 nm, respectively) and 266 nm [13].*

Previous experiments [13] have shown that at a pressure of 14 bar (D_2) we can achieve an energy conversion efficiency at the exit of the Raman cell of the order of 40% at both the 1st and 2nd Stokes wavelength (289 and 316 nm, respectively), and of ~18% at the fundamental wavelength 266 nm (Fig. 2). By adding H_2 in the Raman cell, the wavelength of 299 nm, can also be generated [13].

The Differential Absorption Lidar (DIAL) system for ozone profiling in the lower troposphere of the Laser Remote Sensing Unit (LRSU) of NTUA (EOLE and DIAL lidar systems) is based in Athens (37.96°N, 23.78°E, 220 m a.s.l.). The system is based on the 4th harmonic of a pulsed Nd:YAG laser, emitting 17-20 mJ/pulse at 266 nm (10 Hz repetition rate). It is used to optically pump a 1-m long stainless steel Raman cell filled with D_2 (14 atm), through the SRS effect. The laser pulses produced at 266 nm, 289 nm and 316 nm, have energies of the order of 3 mJ (266 nm), and 6 mJ (both at 289 and 316 nm).

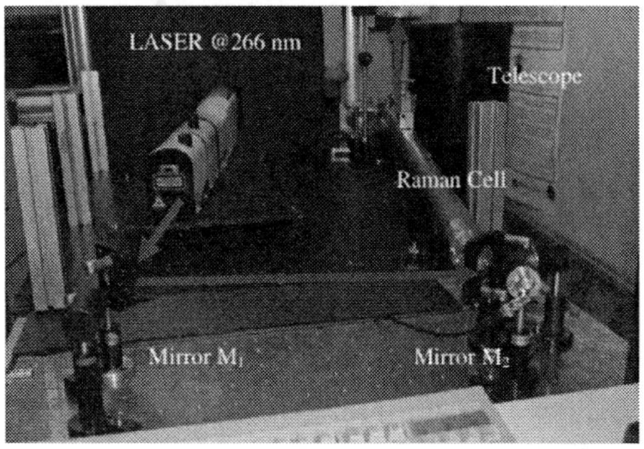

Figure 3. *The experimental setup of the LRSU ozone DIAL system (emitting and receiving units). The ON and OFF DIAL wavelength are produced from a high-pressure Raman call filled with D_2 (14 atm).*

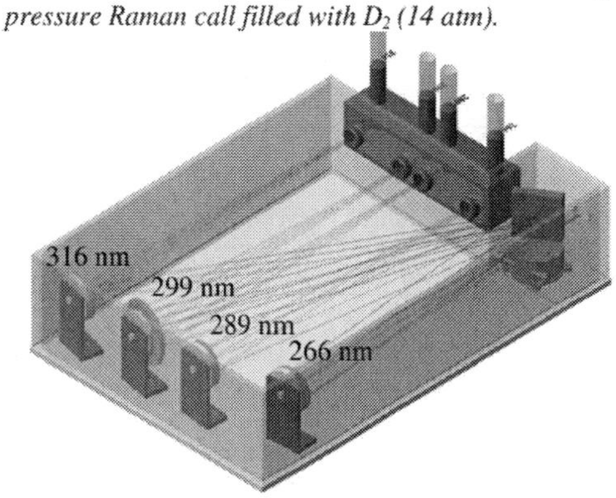

Figure 4. *The multi-wavelength ozone spectrometer (auto CAD simulations) for the spectral separation of the ON and OFF lidar signals.*

A receiving Newtonian telescope of 250 mm diameter and 1125 m focal length (f/4.5) is used to collect the backscatter laser beams at 266, 289 and 316 nm (Fig. 3). This telescope is optically coupled to a discriminating spectrometer through a pure SiO_2 optical fiber [14]. The backscattered lidar signals between 266-316 nm are spectrally separated by a Czerny-Turner grating spectrometer (4960 lines/mm) with a spectral resolution better than 0.5 nm (Fig. 4). Four concave mirrors are used to focus the lidar signals at the photocathode of compact photo-multiplier tubes (PMTs) at four respective wavelengths: 266, 289, 299 and 316 nm. High performance transient recorders (Licel GmbH) are used to digitize the received lidar signals (12 bits - 40 MHz - 8192 signal bins) at 266, 289 and 316 nm. The typical number of averaged lidar signals is 4000, corresponding to 6 min time resolution.

3 EXPERIMENTAL RESULTS

In Fig. 5, we present the vertical profile of the ozone number density (moles/cm^3) obtained during daytime (10:15-14:21 UTC) over Athens, on 02 December 2016, based on a combination of two-wavelength pairs: 266-289 nm (inside the PBL, from ground up to 2 km height) and 289-316 nm (in the free troposphere). This profile has been corrected for the atmospheric backscatter and for the molecular and particulate differential extinction (cf. equation 4). The latter one has been based on aerosol measurements obtained by our DIAL system at 316 nm and at 355 nm by the collocated multi-wavelength Raman lidar of NTUA (EOLE system).

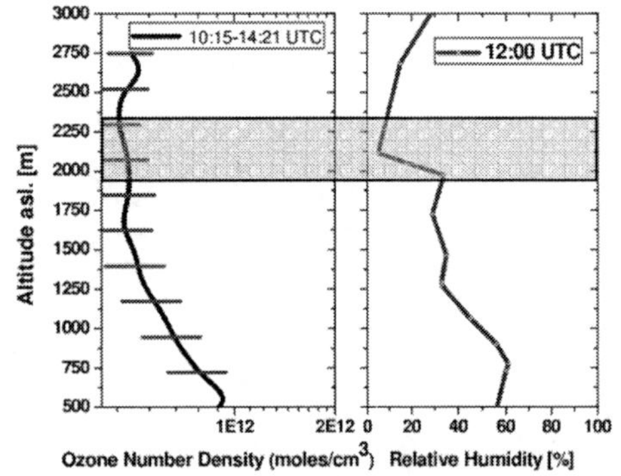

Figure 5. *left: Vertical profile of the ozone number density (moles/cm^3) and right: relative humidity, obtained over Athens during daytime by the NTUA DIAL system (wavelength pairs 266-289 nm and 289-316 nm) and a radiosonde, respectively.*

The corresponding vertical profile of the relative humidity (RH) obtained by radiosonde (cf. Fig. 5-right) launched by the Hellenic National Meteorological Service (HNMS), shows a very low value of RH (6%) observed around 2.1 km height. This very low value of RH may be related to a recent stratospheric-tropospheric transport (STT) event, occurred over Athens. This can be corroborated by three factors:

a) High values of absolute vorticity and winds field over Athens (observed a few hours before) at 500 hPa surface (cf. Fig. 6),

b) Total ozone data showing a cut-off low (denoted by a brown circle) event [5-7, 9-10] with high ozone values over Greece (cf. Fig. 7).

c) Strong downward winds around 2 km height (skew-T diagram not shown).

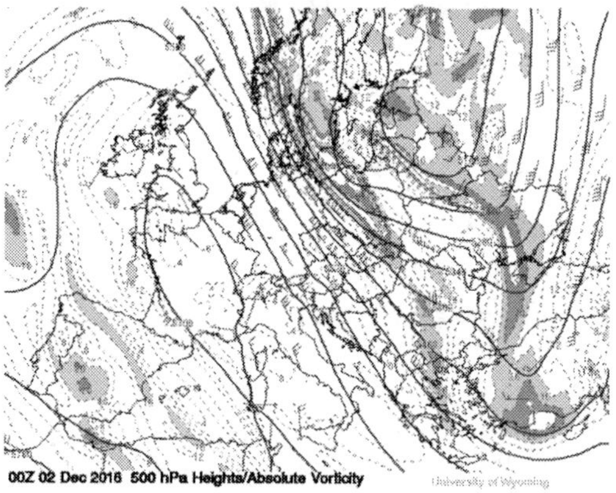

Figure 6. *Absolute vorticity map and winds field at 500 hPa surface, on 02 December 2016 (00:00 UTC) [image credit: University of Wyoming, USA].*

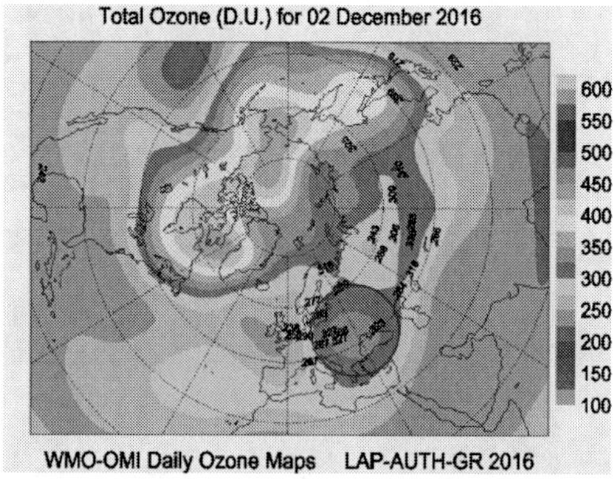

Figure 7. *Total ozone map obtained by OMI-Aura satellite on 02 December 2016.*

4 CONCLUSIONS

In this paper we presented a typical vertical profile of the ozone number density obtained over Athens, using a two-wavelength pair DIAL system (266-289 nm and 289-316 nm), which coincided with a deep STT event [9-11], observed down to 2.1 km height.

Systematic ozone measurements are to be performed over Athens, starting on February 2017, during daytime and nighttime.

ACKNOWLEDGMENTS

The AutoCAD simulation of Figure 4 has been kindly provided by Dr. Jacques Porteneuve (IPSL, CNRS, France). Total ozone map has been provided by the WMO Ozone mapping center (LAP-AUTH, Greece). Radiosonde data has been obtained by the Hellenic National Meteorological Service. Absolute vorticity map and winds field, at 500 hPa surface, have been provided by the University of Wyoming (USA).

References

[1] Brasseur G., Orlando J. and Tyndall J., (Eds.), *Atmospheric Chemistry and Global Change*, *Oxford University Press*, pp. 654, 1999.

[2] Stevenson, D. S., et al., 2013: Tropospheric ozone changes, radiative forcing and attribution to emissions in the Atmospheric Chemistry and Climate Model Intercomparison Project (ACCMIP), *Atmos. Chem. Phys.*, **13**, 3063-3085.

[3] Finlayson-Pitts, B. J., and Pitts, J. N., Jr., 1999: Chemistry of the Upper and Lower Atmosphere, *Academic Press*, USA, pp. 969.

[4] Papayannis, A., Ancellet, G., Pelon, J., and Mégie, G., 1990: Multiwavelength lidar for ozone measurements in the troposphere and the lower stratosphere, *Appl. Opt.*, **29**, 467-476.

[5] Ancellet, G., Beekmann, M., Papayannis, A., 1994: Impact of a cut-off low development on downward transport of ozone in the troposphere, *J. Geophys. Res.*, **99**(D2), 3451-3468.

[6] Galani, E., Balis, D., Zanis, P., Zerefos, C., Papayannis, A., Wernli, H., and Gerasopoulos, E., 2003: Observations of stratosphere-totroposphere transport events over the eastern Mediterranean using a ground-based lidar system, *J. Geophys. Res.*, **108**(D12), 8527, doi:10.1029/2002JD002596.

[7] Stohl, A., et al., 2003: Stratospheric-tropospheric exchange. A review and what we have learned from STACCATO, *J. Geophys. Res.*, **108**(D12), 8516, doi:10.1029/2002JD002490.

[8] Kourtidis, K., et al., 2002: Regional levels of ozone in the troposphere over the eastern Mediterranean, *J. Geophys. Res.*, **107**(D18), 8140, doi:10.1029/ 2000JD000140, 2002.

[9] Papayannis, A., Balis D., Zanis P., Galani E., Wernli H., Zerefos C., Stohl A., Eckhardt S., Amiridis V., 2005: Sampling of an STT event over the Eastern Mediterranean region by lidar and electrochemical sonde. *Ann. Geophys.*, **23**, 2039-2050.

[10] Akritidis, D., Pozzer1, A., Zanis, P., Tyrlis, E., Škerlak, B., Sprenger, M., and Lelieveld, J., 2016: On the role of tropopause folds in summertime tropospheric ozone over the eastern Mediterranean and the Middle East, *Atmos. Chem. Phys.*, **16**, 14025-14039.

[11] Trickl, T., et al., 2016: How deep are stratospheric intrusions LUAMI 2008, *Atmos. Chem. Phys.*, **16**, 8791-8815.

[12] Measures. R., 1992: Laser Remote Sensing. Fundamentals and Applications, Krieger Publ., pp. 510.

[13] Tzortzakis, S., Tsaknakis G, Papayannis A, Serafetinides A. A., 2004: Investigation of the spatial profile of stimulated Raman scattering beams in D_2 and H_2 gases using a pulsed Nd:YAG laser at 266 nm, *Appl. Phys.*, **B79**, 71-75.

[14] Kalabokas, P., Papayannis, A., Tsaknakis, G. and Ziomas, I., 2012: A study on the atmospheric concentrations of primary and secondary air pollutants in the Athens basin performed by DOAS and DIAL measuring techniques, *Science Total Environ.*, **414**, 556-563.

Investigating potential wet bias in the Purple Crow Lidar water vapor measurements

Jeffrey VanKerkhove[1,*], R. J. Sica[1], R. Wing[2] and P. S. Argall[3,1]

[1]*Department of Physics and Astronomy, University of Western Ontario, London, Canada, *Email: jvankerk@uwo.ca*
[2]*CNRS/INSU, LATMOS-IPSL, Paris, France*
[3]*School of Applied Science and Technology, Fanshawe College, London, Canada*

ABSTRACT

The Purple Crow Lidar is a large aperture lidar, capable of retrieving water vapor into the stratosphere. A comparison with the ALVICE lidar in 2012 showed water vapor measurements were consistently larger than those of ALVICE in the lower stratosphere, prompting an investigation of the system. Processing approaches and additional instrumental corrections are considered.

1 INTRODUCTION

Water vapor plays a fundamental role in Earth's radiative budget, as well as in atmospheric chemistry and circulation. Its role is paramount in the coupling of the Upper Troposphere and Lower Stratosphere (UTLS), where small changes in water content can have a great effect on global-scale processes.

Located near London, Canada, the Purple Crow Lidar (PCL) measures atmospheric temperatures and water vapor using a frequency doubled Nd:YAG laser at 532 nm in 1000 mJ pulses at a 30 Hz repetition rate. Water vapor content is calculated using Raman backscatter returns from molecular nitrogen (at 607 nm) and water vapor (at 660 nm). Nitrogen counts are detected with a Licel transient recorder which is capable of analog and photon counting (PC) detection modes, while water vapor uses a Stanford Research Systems multi-channel scaler/averager and only uses photon counting. The incorporation of a liquid Mercury mirror allows for a large aperture telescope (2.65 m diameter), which can detect water vapor into the lower stratosphere.

A validation campaign with the ALVICE lidar was carried out in 2012, showing a relative wet bias in PCL's water vapor measurements in the lower stratosphere [1]. This investigation intends to determine if the wet bias is due to environmental, instrumental, or algorithmic issues.

Water vapor measurements from lidar are calculated via the volume mixing ratio, which can be written as [2]:

$$w_v = \frac{M_W}{M_{air}} \frac{n_N}{n_{air}} \frac{\sigma_N}{\sigma_W} \frac{\xi_N}{\xi_W} \frac{\Gamma_N}{\Gamma_W} \frac{N_W}{N_N} \tag{1}$$

where M is molecular weight, n is number density, σ is the backscatter cross section, ξ is instrument efficiency, Γ is the atmospheric transmission function, and N are the processed counts. The subscripts N and W indicate quantities related to the molecular nitrogen and water vapor channels respectively. Calculation of the processed counts includes applying a dead time correction to the PC counts [3], subtracting the background, and then merging with the analog counts [4].

2 METHODOLOGY

In this section, we discuss additional corrections considered in the calculation of processed counts and the water vapor mixing ratio.

2.1 Altitude offset

Differences between the analog and PC data acquisition cause an altitude offset in the raw count data. The offset is determined by sending a function generator pulse through the detectors and shifting the analog and PC count arrays so that the pulses line up. This test was carried out by summer research student Sean Hartery, finding a 68.1 m nitrogen channel offset.

2.1.1 Warm up correction

Identified in [5], the Licel transient recorder underestimates count rates until the system has warmed up. For the PCL nitrogen channel, this warm up time is about 40 minutes (Figure 1). If data acquisition has begun before the unit has warmed up, a correction can be applied.

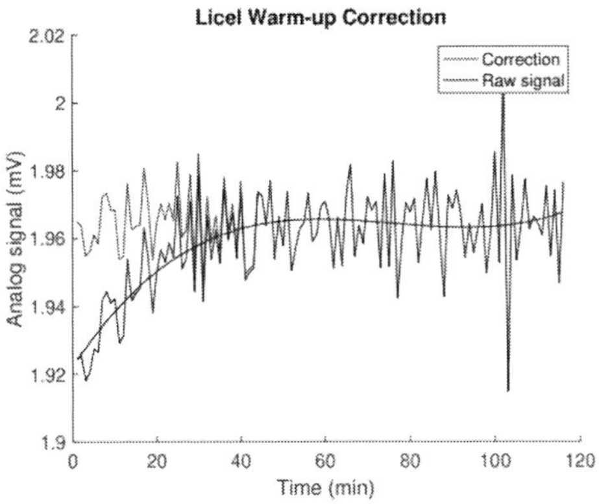

Figure 1: Warm up correction for the nitrogen analog counts.

2.2 Temperature dependence correction

The Raman cross section spectrum varies with temperatures found at different altitudes, resulting in signal loss if narrowband filters are used. The correction factor is calculated using an approach similar to the method discussed in [7], which requires knowledge of the temperature-dependent Raman cross sections and interference filters for nitrogen and water vapor returns (Figure 2).

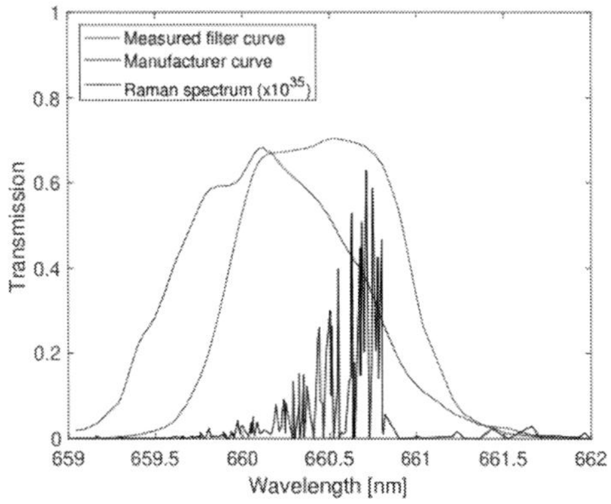

Figure 2: Interference filter for the water vapor channel and the water vapor Raman cross section at 20°C. Note how the measured filter transmission varies from the manufacturer's specifications.

3 RESULTS

Figure 3 shows how the additional corrections mentioned in the Methodology section affect the overall water vapor mixing ratio. The 'warm up' correction has little overall effect, increasing the mixing ratio by only 0.03% when applied. The altitude offset correction is dominant in the lowermost region of the retrieval, where the analog counts are important, but is not a major consideration in the PC-dominated UTLS region. The temperature dependence correction has the largest contribution to the mixing ratio in the UTLS region, increasing the value by 10.9% at 14 km.

4 CONCLUSIONS

Although the corrections discussed here do affect the retrieval to varying degrees, none account for the lower stratospheric wet bias observed during the validation with the ALVICE lidar. A thorough characterization of the back-

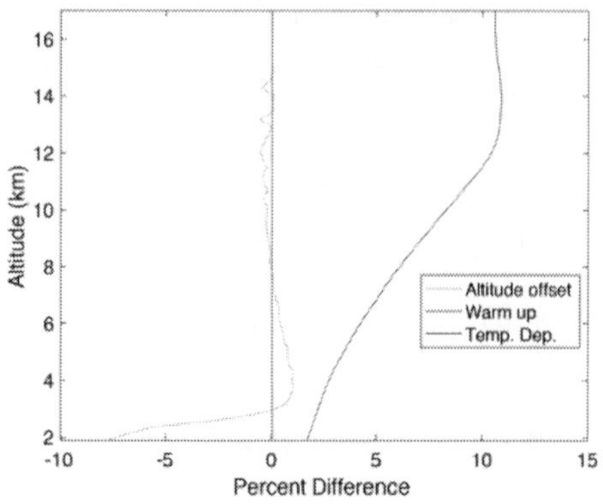

Figure 3: Percent differences between the original and corrected water vapor volume mixing ratio profiles.

ground is still required, as well as consideration of additional parameters, including differential overlap. Instrumental or environmental fluorescence may also contribute the observed wet bias, and should be tested for. However, since PCL operates in the visible region, it is not as susceptible to fluorescence as UV lidars.

ACKNOWLEDGEMENTS

The authors would like to thank the NSERC CREATE Training Program in Arctic Atmospheric Science and the Canadian Network for the Detection of Atmospheric Change (CANDAC) for their support. We would also like to acknowledge the support of the Canadian National Science and Engineering Research Council.

References

[1] R. Wing. Multi-Sensor Calibration and Validation of the UWO-PCL Water Vapour Lidar. Master's thesis, University of Western Ontario, 2012.

[2] D. Whiteman, S. Melfi, and R. Ferrare. Raman lidar system for the measurement of water vapor and aerosols in the earth's atmosphere. *Appl. Opt.*, 31:3068, 1992.

[3] D. Donovan, J. Whiteway, and A. Carswell. Correction for nonlinear photon-counting effects inlidar systems. *Appl. Opt.*, 32:6742, 1993.

[4] R. Newsom, D. Turner, and B. Mielke. Simultaneous analog and photon counting detection for raman lidar. *Appl. Opt.*, 48:20, 2009.

[5] M. Walker, D. Venable, and D. Whiteman. Gluing for raman lidar systems using the lamp mapping technique. *Appl. Opt.*, 53:8535, 2014.

[6] R. Sica, S. Sargoytchev, and S. Argall. Lidar measurements taken with a large-aperture liquid mirror. 1: Rayleigh-scatter system. *Appl. Opt.*, 34(30):6925, 1995.

[7] Whiteman, D. N., Examination of the Traditional Raman lidar technique. I. Evaluating the temperature- dependent lidar equations, *Appl. Opt.*, 42, 15, 2003.

[8] R. Sica and A. Haefele. Retrieval of water vapor mixing ratio from a multiple channel raman-scatter lidar using an optimal estimation method. *Appl. Opt.*, 55(4):763, 2016.

CALIPSO RECORDINGS AND MONITORING DUST STORMS OVER THE OPEN SEAS IN SOUTH OF THE IRAN PLATEAU

Hamid R. Khalesifard[1]*, Farizeh Bayat[1]

[1] *Department of Physics and Center for Research in Climate Change and Global warming, Institute for Advanced Studies in Basic Sciences, No. 444 Prof. Sobouti Blvd., Zanjan 4513766731, Iran, *khalesi@iasbs.ac.ir*

ABSTRACT

Open seas in the south of the Iran plateau are under the influence of heavy dust storms which are originating either from the Tigris and Euphrates basin, the Arabian Peninsula or Hamoun lake. We have used the recordings of the CALIPSO satellite to investigate the seasonal variations as well as the origins of the dust storms over the region. CALIPSO data set shows dust activities are frequent during May to September in the interested region and the Hamoun lake has considerable impacts on it.

1 INTRODUCTION

Iran, a country on the dust belt, is surrounded by several sources such as: Tigris and Euphrates basin in west, the region between the Caspian and Aral seas in north, and the Arabian Peninsula in south [1–3]. The well-known northwesterly Shamal wind, transfer dust plumes from the Mesopotamian region toward the Iran plateau and the Persian Gulf area [1, 2, 4]. Such dust activities may be intensified by other dust sources (like those inside the Arabian Peninsula) and influence on a quite wide region which extends up to the Arabian sea, Indian Ocean and Indian subcontinent [5]. Dust sources in the Arabian Peninsula mostly are located in the south and east sides of the Peninsula, where they are quit active during April to September [1, 4]. These sources mostly affect the Oman and Arabian seas, and even the Indian subcontinent but don't have considerable impacts on the Iran plateau [2, 4–6].

On the other hand, advected dust from the Hamoun lake in East Iran also affects the atmosphere over the Oman and the Arab seas and sometimes even the Indian Ocean. Such dust plums may also influence southwest Asian regions like Southeast Iran and West Pakistan [2, 7–9]. East and southeast of Iran are under the influence of northerly wind known as Levar or "Sistan 120-day wind". Levar mostly is blowing during mid-May to mid-September and can also lead to some changes in the climate of the Oman sea [4, 10, 11].

This work is an effort on monitoring dust storms over the Persian Gulf and the Oman sea by using CALIPSO recordings during 2006 to 2015. For this purpose, columnar averaged particulate depolarization ratio (PDR), column integrated attenuated backscatter (IAB) and columnar aerosol optical depth (AOD) have been extracted from the CALIPSO level 2 data products. In the next part of this manuscript, after introducing the data set and calculations in section 2, results and discussions are presented in section 3 and finally the work is concluded in section 4.

2 DATA SET AND CALCULATIONS

Four colored rectangles in Figure 1 depict the investigation regions. The regions are chosen in a way to cover almost all of the atmosphere over the open waters in south of the Iran plateau but prevent the land. The figure also depicts the 16-day, ground tracks of CALIPSO over the whole area. To investigate dust activities in the specified regions, AOD, PDR and IAB at 532 nm, have been extracted from the CALIPSO level 2/ 5km aerosol profile V3, during June 2006 to December 2015. To assign a single value to each of the mentioned parameters, their averaged value over each ground track which is just inscribed inside one of the rectangles has been calculated. In order to finding the AOD, PDR and IAB single value for each ground tracks, all features which are not aerosols screened. Note that all these parameters have been obtained by considering data uncertainty and their valid range. After computing these attributed

 https://doi.org/10.1051/epjconf/201817605027

parameters to each ground tracks, their corresponding monthly and annual averaged values for each rectangle have been calculated.

One of the most essential factors for monitoring aerosols transport is wind speed and direction evaluations. In order to achieve this purpose, we have used daily vector wind NCEP/NCAR reanalysis with 2.5° × 2.5° spatial resolution.

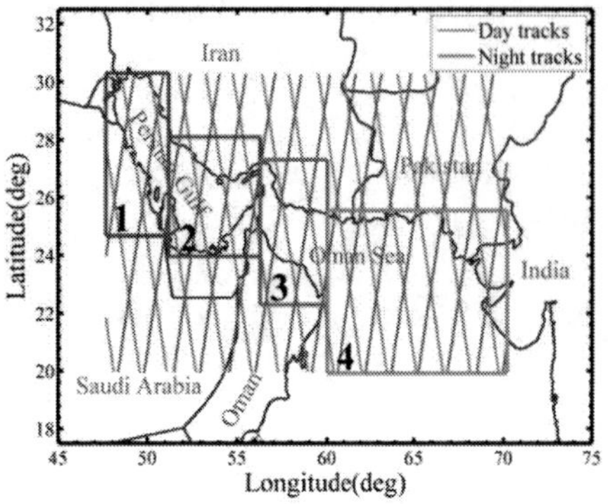

Figure 1. 16-day CALIPSO ground tracks, regions 1 to 4 are covering the investigation area.

3 RESULTS AND DISCUSSIONS

Figure 2 shows monthly variations of IAB, averaged over the whole data set. Comparing Figure 2d with Figures 2a-2c, one can find, region 4 shows different behavior especially during June to September. PDR more or less shows the same behavior for all regions (not shown here) but again monthly values of AOD in region 4 are different respect to other regions (not shown here).

These differences will be more clear when one looks to variations of AOD versus IAB (Figure 3). Figures 3a-3c show the variations of 4-month averaged values of AOD versus IAB for February to May (FMAM), June to September (JJAS), and October to January (ONDJ). In Figure 3 all obtained AOD and IAB values for each 4-month, in any region, have been averaged and just a rectangle has been assigned to them. The rectangle is centered on the averaged value of AOD and IAB and its dimensions in the x and y directions are two times of the standard deviations of IAB and AOD data points. Figure 3b very clearly shows that in JJAS, both the AOD and IAB have considerable increases in region 4. This is in temporal coincidence with the 120-day Sistan, Levar wind and the dry annual period for the Hamoun lake [12].

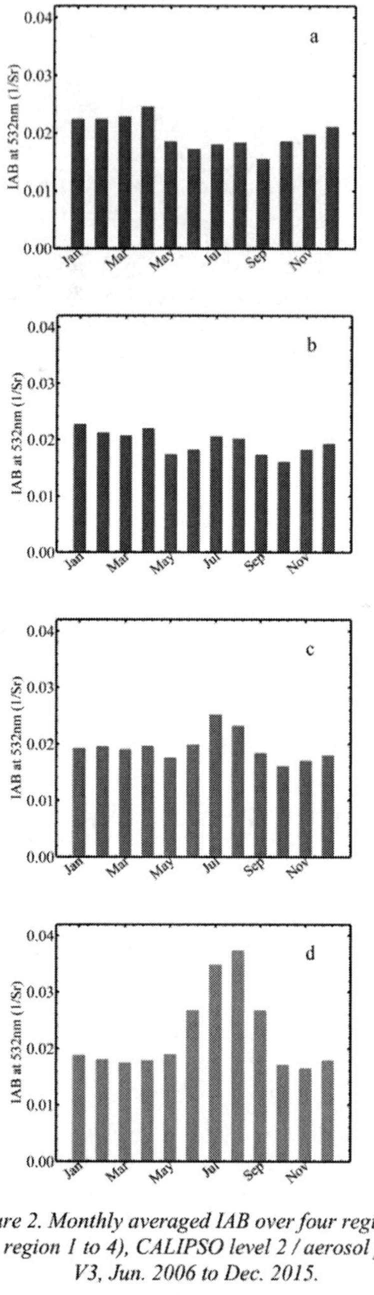

Figure 2. Monthly averaged IAB over four regions (a to d, region 1 to 4), CALIPSO level 2 / aerosol profile V3, Jun. 2006 to Dec. 2015.

 https://doi.org/10.1051/epjconf/201817605027

Four-month averaged 700 hPa wind field over the region are shown in Figures 4a-4c. The wind patterns are calculated using NCEP/NCAR reanalysis for years 2006-2015. Similar behavior also can be observed when one looks to the variations of PDR versus IAB (not shown here).

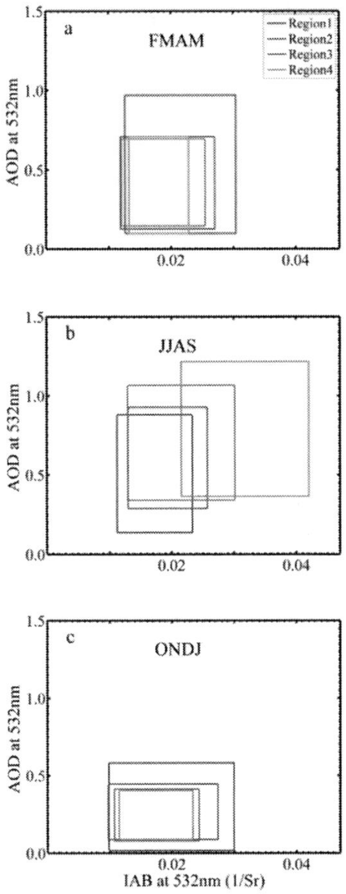

Figure 3. 4-month averaged AOD Vs. IAB, a- February to May, b- June to September, c- October to January, CALIPSO level 2 / aerosol profile V3, Jun. 2006 to Dec. 2015.

4 CONCLUSIONS

The obtained results which parts of them are shown in Figure 2 and Figure 3 indicate that the atmosphere over the Persian Gulf, Oman and Arab seas experiences frequent dust outbreaks especially during spring and summer times. As we divided the whole area into four sub regions (Figure 1), it has been found that in region 1, the occurrence rate of dust storms has its maximum value in May. But for

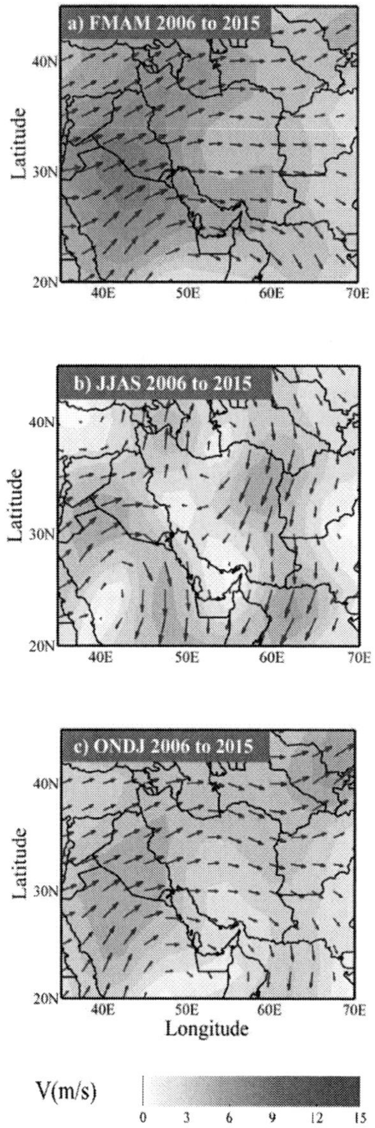

Figure 4. 4-month averaged wind field over the Iran plateau during 2006 to 2015 at 700 hPa, a- February to May, b- June to September, c- October to January, NCEP/NCAR reanalysis.

the other three regions, this is happening in July. Looking to the wind pattern over the region [12] one can find that region 1 is quite affected by dust storms originated from the Tigris and Euphrates basin but regions 2 and 3 are mostly under influence of dust plumes advected from the Arabian Peninsula. The interesting case is with region 4. Looking to Figure 3c one can clearly find that during June to September, dust storms are very frequent and intense in this region. This is in exact temporal coincidence with the annual cycle of 120-day Sistan Levar wind (Figure 4b).

In other words, this northerly wind takes the dust from the bed of seasonal Hamoun lake and transport it to Southeast Iran, West Pakistan, Oman and Arab seas and sometimes it even may affect the Indian ocean. It is interesting to see that CALIPSO recordings especially on IAB can distinguish between the dust that advected from the Arabian Peninsula and those from the Hamoun lake. Therefore, IAB can be introduced as a powerful tool on characterization of atmospheric particles, especially when its variations respect to AOD and PDR have been considered.

ACKNOWLEDGEMENTS

Authors are grateful to NASA Langley Research Center for providing the CALIPSO data (http://eosweb.larc.nasa.gov/). The NCEP/NCAR reanalysis data is prepared through the NOAA Climate Diagnostics Center/National Oceanic and Atmospheric Administration, at Boulder, Colorado (http://www.esrl.noaa.gov/). They also thank Ruhollah Moradhaseli for valuable discussions.

References

[1] Prospero, J. M., Ginoux, P., Torres, O., Nicholson, S. E. and Gill, T. E., 2002: Environmental characterization of global sources of atmospheric soil dust identified with the Nimbus 7 Total Ozone Mapping Spectrometer (TOMS) absorbing aerosol product, *Rev. Geophys*, **40** (1), 1002pp.

[2] Abdi Vishkaee, F., Flamant, C., Cuesta, J., Flamant, P. and Khalesifard, H.R., 2011: Multiplatform observations of dust vertical distribution during transport over northwest Iran in the summertime, *Journal of Geophysical Research: Atmospheres*, **116** (D5).

[3] Ginoux, P., Prospero, J. M., Gill, T. E., Hsu, N. C. and Zhao, M., 2012: Global-scale attribution of anthropogenic and natural dust sources and their emission rates based on MODIS Deep Blue aerosol products, *Reviews of Geophysics*, **50** (3).

[4] Goudie, A. and Middleton, N. J., 2006: Desert dust in the global system, *Springer Science & Business Media*.

[5] Badarinath, K. V. S., Kharol, S.K., Kaskaoutis, D. G., Sharma, A. R., Ramaswamy, V. and Kambezidis, H. D., 2010: Long-range transport of dust aerosols over the Arabian Sea and Indian region— A case study using satellite data and ground-based measurements, *Global and Planetary Change*, **72** (3), 164-181.

[6] Kaskaoutis, D. G., Gautam, R., Singh, R. P., Houssos, E. E., Goto, D., Singh, S., Bartzokas, A., Kosmopoulos, P. G., Sharma, M., Hsu, N. C. and Holben, B. N., 2012: Influence of anomalous dry conditions on aerosols over India: transport, distribution and properties, *Journal of Geophysical Research: Atmospheres*, **117** (D9).

[7] Abdi Vishkaee, F., Flamant, C., Cuesta, J., Oolman, L., Flamant, P. and Khalesifard, H.R., 2012: Dust transport over Iraq and northwest Iran associated with winter Shamal: A case study, *Journal of Geophysical Research: Atmospheres*, **117** (D3).

[8] Rashki, A., Kaskaoutis, D. G., Eriksson, P. G., Qiang, M. and Gupta, P., 2012: Dust storms and their horizontal dust loading in the Sistan region, Iran. *Aeolian Research*, *5*, 51-62.

[9] Rashki, A., Kaskaoutis, D. G., Goudie, A. S. and Kahn, R. A., 2013: Dryness of ephemeral lakes and consequences for dust activity: the case of the Hamoun drainage basin, southeastern Iran, *Science of the Total Environment*, 463, 552-564.

[10] Middleton, N. J., 1986: A geography of dust storms in South-west Asia, *Journal of Climatology*, **6** (2), 183-196.

[11] McMahon, H., 1906: Recent survey and exploration in Seistan. *The Geographical Journal*, **28** (3), 209-228.

[12] Alizadeh-Choobari, O., Zawar-Reza, P. and Sturman, A., 2014: The "wind of 120days" and dust storm activity over the Sistan Basin. *Atmospheric Research*, 143, 328-341.

CLEAR-AIR LIDAR DARK BAND

Paolo Di Girolamo[1], Andrea Scoccione[2], Marco Cacciani[2], Donato Summa[1], Jan H. Schween[3]

[1]*Scuola di Ingegneria, Università degli Studi della Basilicata, Italy , Email:* digirolamo@unibas.it

[2]*Dipartimento di Fisica, Università di Roma "La Sapienza", Italy*

[3]*Institut fuer Geophysik und Meteorologie, Universität zu Köln, Germany*

ABSTRACT

This paper illustrates measurements carried out by the Raman lidar BASIL in the frame of HOPE, revealing the presence of a clear-air dark band phenomenon (i.e. the appearance of a minimum in lidar backscatter echoes) in the upper portion of the convective boundary layer. The phenomenon is clearly distinguishable in the lidar backscatter echoes at 1064 nm. This phenomenon is attributed to the presence of lignite aerosol particles advected from the surrounding open pit mines in the vicinity of the measuring site..

1 INTRODUCTION

In the frame of the HD(CP)2 Observational Prototype Experiment (HOPE), the Raman lidar system BASIL was deployed and operated over the period April-May 2013 in the Atmospheric Supersite JOYCE, located within the Jülich Research Centre (Central Germany, Lat.: 50°54' N; Long.: 6°24' E, Elev. 105 m). This site is approximately 3 km West of the *Tagebau-Hambach* open-pit lignite mine, which represents the largest operational lignite mine on Earth with a maximum depth of ~200m (Figure 1). The dump of this mine forms the hill *Sophienhöhe* which reaches 302 m and is partially re-cultivated with forest. A second open-pit lignite mine, named *Tagebau Inden*, is located approximately. 3 km South-West of the Supersite.

Both mines lie in sectors of 29-114° (red shaded area in figure 1) and 180-239° (blue shaded area in figure 1) relative to the location of the Raman lidar. If wind is blowing from these directions, lignite particles from the two open-pit mines or the surrounding hill are lifted up from the ground and transported over the lidar site, with appreciable effects on the measurements.

Evidence of this particle transportation was found in lidar elastic backscatter echoes on a variety of case studies during HOPE, with the appearance of a specific odd feature in the upper portion of the convective boundary layer (CBL). Specifically, a minimum in lidar backscatter echoes at 1064 nm, with a backscatter reduction of approximately 1 dB, is observed. This feature is found to have a vertical extent of approximately 100 m and persist over a period of several hours, with an alternation of intensifications and attenuations of the phenomenon. A similar feature with a comparable temporal duration and backscatter reduction had been first reported by [1] in the presence of light precipitation events; this phenomenon, referred to as lidar dark band, had been demonstrated to be ascribable to the changes in scattering properties of precipitating particles taking place during the snowflake-to-raindrop transition in the proximity of the melting level.

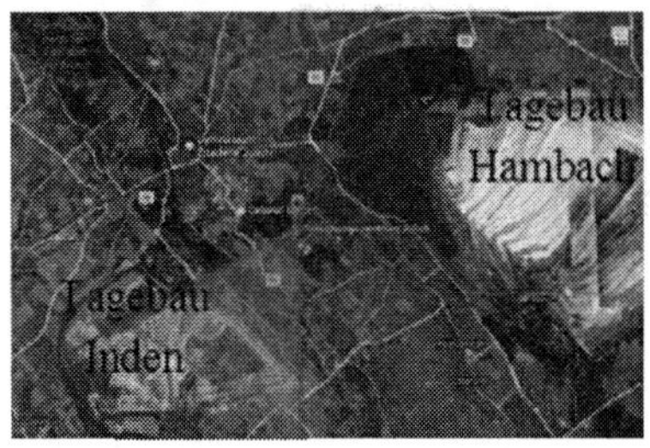

Figure 1: Locations of the Raman lidar system BASIL within the Jülich Research Centre (pink dot), the two open-pit lignite mines Tagebau Hambach and Tagebau Inden, and the hill Sophienhöhe.

Instead, the phenomenon reported in the present research effort appears in clear-air conditions and in the presence of strong convective activity in the boundary layer: we will refer to it in the following as to the clear-air dark band phenomenon or the convective dark band phenomenon. In this paper we provide experimental evidence and a possible interpretation for the occurrence of this phenomenon.

2 RESULTS

Figure 2 illustrates the time–height cross-section of the range-corrected backscatter signal at 1064 nm, RCS_{1064}, as measured by BASIL (upper panel), and the vertical wind speed, as measured by the Univ. of Cologne wind lidar (lower panel), in the time interval 12:00 - 13:00 UTC on 18 April 2013. For the purpose of these measurements, the two lidars were located within a distance of 80 m. The upper panel of the figure clearly reveals the presence of a significant aerosol loading within the boundary layer, which is tracing the presence of a well-mixed and quasi-stationary CBL at this time of the day, extending up to approximately 2000 m. The figure also reveals the presence of alternating updrafts and downdrafts. The largest variability of RCS_{1064} is observed in the interfacial layer, as a result of the penetration of aerosol-rich air rising from the ground and the entrainment of aerosol-poor air sinking from the free troposphere.

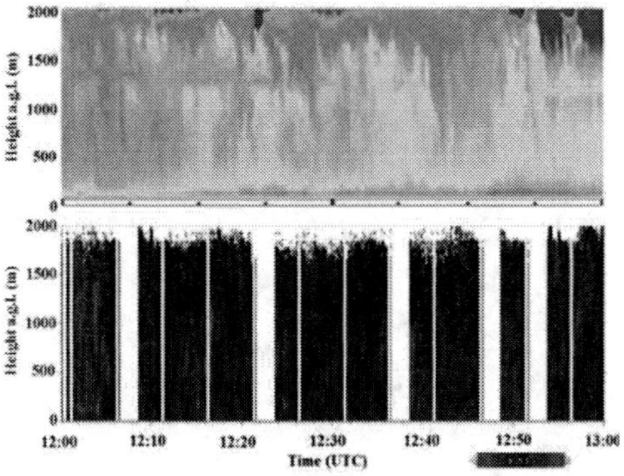

Figure 2: Time–height cross-section of RCS_{1064} (upper panel, with the black dashed line highlighting the presence of the backscatter reduction) and the vertical wind speed (lower panel) in the time interval 12:00 - 13:00 UTC on 18 April 2013. RCS_{1064} is expressed in arbitrary units with low to mid to high values represented with blue to green to red.

A persistent minimum in lidar backscatter is observed around 1200 m, with alternating intensity fluctuations. This minimum persist albeit clear presence of up-drafts (orange eddies, with positive vertical wind speed values) and down-drafts (blue eddies, with negative vertical wind speed values), and thus cannot be related to a layered structure of aerosol in the mixing layer. Note that the backscatter minimum occurs preferably during up-drafts, but not during down-drafts. This behavior is clearly highlighted in figure 2, where the black dashed lines indicating the lidar backscatter minima only appear in temporal coincidence with the vigorous updrafts, testified by the positive vertical speed values (up to 5 m s^{-1}) measured by wind lidar, but do not appear in coincidence with the downdrafts (negative vertical speed values down to -5 m s^{-1}).

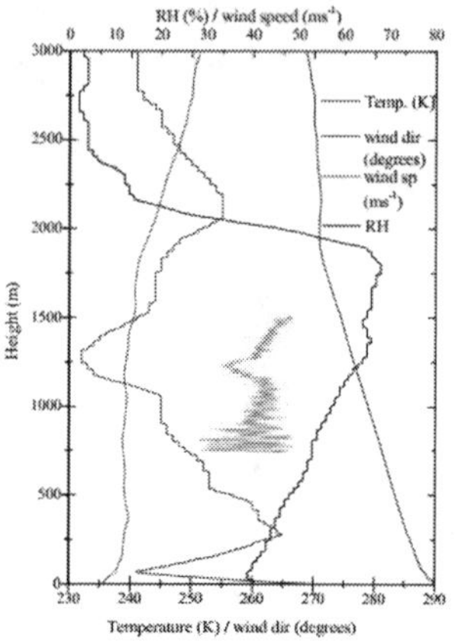

Figure 3: Vertical profile of RCS_{1064}, at 13:00 UTC on 18 April 2013 (1 min average, green line), together with the vertical profiles of temperature (blue line), RH (black line) and wind direction (red line) and speed (purple line) as measured by the radiosonde launched from the near-by station of Hambach (4 km E-SE).

Figure 3 illustrates the vertical profile of RCS_{1064} at 13:00 UTC on 18 April 2013 (1 min average, green line – for the purpose of clarity only shown in the height interval 750-1500 m), together with the vertical profiles of temperature, relative humidity (RH) and wind direction and speed, as measured by the radiosonde launched from the near-by station of Hambach (4 km E-SE). The clear-air dark band is found to extend from 1150 to 1275 m, with a vertical extend of 125 m and a minimum in lidar backscatter at 1225 m (backscatter reduction = 17%, corresponding to 0.8 dB). This band takes place few hundred meters below both the Lifting Condensation Level (LCL, at 1725 m or 814 mbar)) and the freezing level (at 1630 m or 823.2 mbar). The figure reveals that wind is blowing from directions in the interval from 265° at surface to 232°. More specifically, the lidar backscatter reduction is located in the same height region (1125-1450 m) where wind is found to blow from directions in the interval 232-239°, i.e. from the directions where the Tagebau Inden open-pit lignite mine is located. CBL wind direction measurements from radiosondes may be difficult to interpret as they may reflect rotations taking place within the convective plumes. However, similar values (236-242°) in the same height interval are also present in the 1 h (12:00 - 13:00 UTC) average profile measured by the wind lidar (not shown here), with values throughout the whole profile from 270° (at surface) to 230° (at 1600 m).

Figure 3 also reveals that the air at this height is characterized by RH values in the range 60-62 %. Lignite particles advected by the wind on the lidar site are captured and ingested within the updrafts and downdrafts associated with the intensive convective activity present at the lidar site. As a result of the adiabatic cooling, air-parcels undergo a sudden RH increase from values in the range 60-62 % (this being the environmental RH value at the base of the dark band) to values in the range 75-80 % (these latter are the values reached within the lifting air-parcel assuming an ideal adiabatic cooling with no air entrainment into the convective plumes or external air within the lifting air-parcel). This sudden increase of RH has important effects on the size growth of the uplifted lignite aerosols.

Most clear-air dark band observations we collected in the absence of a cloud topped CBL. However, clouds in the upper portion of the CBL are observed in figure 2 at 12:53-13:00 UTC (orange-brown features in the upper panel).

3 HYGROSCOPIC AND SCATTERING BEHAVIOR OF LIGNITE PARTICLES

Dried lignite particles produced in open pit lignite mines have a very marked hygroscopic behavior [2,3], and, as a result of this behavior, effectively absorb moisture from the atmosphere. Measurements of particle size distribution of lignite particles escaped from heavy industrialized areas (mining and power stations operations), in the form of fly ash or fugitive dust, have been reported by several authors (among others, [4]; [5]). Specifically, [4] were able to measure the particle size distribution of fly ash injected into the atmosphere from elevated stacks in power stations, with a prominent mode at approximately 8 µm and with about 80% of the particles smaller than 10 µm. Civiš and Hovorka (2010) [5] reported size distribution measurements for brown coal with an average particle size of 1.84 µm. All these authors revealed a limited degree of poly-dispersion for atmospheric lignite particles. Considering a log-normal size distribution, the degree of poly-dispersion or width of the particle distribution is expressed in terms the percentage standard deviation of the log of the distribution. Specifically, in [4] a percentage standard deviation σ of 16 % is reported. A narrower size distributions has been reported by [5], with σ equal to 5 %.

The solution effect typically dominates hygroscopic particles' growth when the radius is small (smaller than the critical radius r_c), which results in small solution droplets being in equilibrium with water vapour at RH values less than 100 %. At this stage, small increases in RH determine small droplet size growths until equilibrium is newly reached. This mechanism is possibly responsible for the lignite particle growth below the LCL, ultimately leading to the appearance of a minimum in lidar backscatter echoes (i.e. the above mentioned clear-air dark band phenomenon). The above derived RH change from 60-62 to 75-80% would accordingly result in a particle growth by ~ 15-30 %. In this study, we are considering an initial size for the dry lignite particles of 1.84 and 8 µm, as reported by Civiš and Hovorka (2010) [4] and Triantafyllou et al. (2006) [5], respectively, based on measurements of dust samples from surface lignite mines (see more details in section 3).

Scattering properties of lignite particles have been simulated based on the application of a light scattering code for spheres using Mie theory (http://philiplaven.com/mieplot.htm). In order to properly simulate these scattering processes, accurate information on particle refractive index are required, besides those on particle size distribution already provided above. Accurate measurements of lignite refractive index have been reported by Lohi et al. (1992) [6], who observed values of the real and imaginary part of the complex refractive index of 1.70 and $1x10^{-6}$, respectively. Simulations of the scattering properties of lignite particles are illustrated in figure 4. Results are expressed in terms of backscattering efficiency Q_{back} as a function of particle radius. These simulations are obtained considering a log-normal size distribution, with a value of σ of 5 %. The simulation in the upper panel of the figure considers a minimum radius of 1.84 µm, as given by [4], while the simulation in the lower panel considers a minimum radius of 8 µm, as given by [5]. Both simulations consider a sounding wavelength of 1.064 µm, which is the laser wavelength used for the dark band lidar measurements illustrated in figure 2. Results illustrated in this figure 4 reveal that, for specific minimum radius values (for example, 2.4 µm, 3.8 µm, 16.5 µm, 28 µm, 40 µm), a reduction in backscatter efficiency of approximately 1 dB is observed for a particle size growth by 15-30 %, as the one experienced by lignite particles during their adiabatic ascent. Thus, we believe that the observed dark band is related to the oscillations of the backscatter efficiency, ultimately leading to Mie back-scattered signal intensity fluctuations.

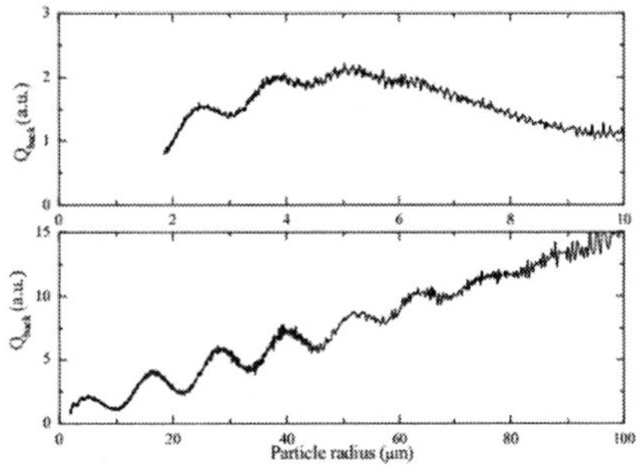

Figure 4: Simulations of the backscattering efficiency as a function of particle radius for lignite particles.

The growth of particles to equilibrium size associated with increasing RH can be continued up to a RH of 100 % and slightly beyond. Cloud formation at the top of the CBL can take place above the LCL if critical saturation ratio, corresponding to the peak of the Koehler curve, is reached. The critical saturation ratio is typically reached for super-saturation values in the range 0.05-1%, depending on the composition (and consequently the level of hygroscopicity) and size of the aerosol particle acting as condensation nuclei. In the case of lignite particles, typical values of critical radius, r_c, and saturation ratio, S_c, are in the range 0.1-10 µm and 0.05-1%, respectively. Up to this point RH has to be increased for the droplet to grow. However, if RH slightly exceeds the critical saturation ratio, the particle is enabled to grow beyond the critical radius and its critical saturation ratio falls below S_c. As a consequence, the vapour condensates on the droplet, which will continue to grow without the need for further increase in the saturation ratio. When this occurs, clouds can form on the top of the CBL. These processes are responsible for the clouds observed in the upper portion of the CBL at 12:53-13:00 UTC (orange-brown features in the upper panel of figure 2). In the clouds, the droplet growth process does not continue indefinitely as many droplets are present and all of them compete for the same available water vapour.

4 DISCUSSION

Dark bands are observed in lidar backscatter echoes in the upper portion of the CBL. These bands appear in clear sky conditions characterized by strong convective activity and are mostly confined to updrafts. Adiabatic cooling within the updrafts leads to an RH increase and a consequent particles' growth, especially of hygroscopic particles. If we assume that most of the particles we observe are hygroscopic lignite particles from the surrounding lignite open-pit mines and their size distribution is mono-disperse or very narrow, we must conclude that the observed dark band is related to the oscillations of the backscatter efficiency as described by Mie-theory, ultimately leading to intensity fluctuations of the Mie back-scattered radiation. If the particle size distribution would be less narrow the backscatter oscillations would smear out if not disappear.

The fact that the dark band is confined to the up-drafts could be explained in two ways: either the adiabatically warming and thus decrease in RH in down-drafts does not lead to an inversion of the particle growth, i.e. there is a hysteresis and humidified particles do not evaporate the water they took up during their way up. Or down-drafts transport other particles than the up-drafts. These particles might be less hygroscopic and thus change their size less with RH. Alternatively their size probability distribution function might be not as narrow as in the updrafts resulting in a smear out of the oscillations of the backscatter efficiency.

References

[1] Sassen, K. and Chen, T., 1995: The lidar dark band: An oddity of the radar bright band, *Geophys. Res. Lett.* **22**, 3505-3508.

[2] Lignites of North America, Volume 23

[3] Krawczykowska, A., Marciniak-Kowalska, j., 2012: AGH Journal of Mining and Geoengineering, vol. 36, No. 4, 57-65.

[4] Triantafyllou, A. G., Zoras, S., Evagelopoulos V., 2006: Particulate matter over a seven year period in urban and rural areas within, proximal and far from mining and power station operations in Greece, *Environ. Monitor. and Assess.* **122**, 41-60.

[5] Civiš, M., and Jan Hovorka, J., 2010: Characterization of main dust samples from a surface lignite mine using a resuspension chamber, *J. Res. Applic. Prof. Security*.

[6] Lohi, A., Wynnyckyj, J. R., Rhodes, E., 1992: Spectral measurement of the complex refractive index of fly ashes of canadian lignite and sub-bituminous coals, *Can. J. Chem. Eng.* **70**, 751-758.

SEVENTEEN-YEAR SYSTEMATIC MEASUREMENTS OF DUST AEROSOL OPTICAL PROPERTIES USING THE EOLE NTUA LIDAR SYSTEM (2000-2016)

Ourania Soupiona[1], Maria Mylonaki[1], Alexandros Papayannis[1], Athina Argyrouli[1], Panayotis Kokkalis[1] and Georgios Tsaknakis[1]

[1]*National Technical University of Athens, Laser Remote Sensing Unit, Physics Department, Zografou, Greece,* apdlidar@central.ntua.gr

Abstract

A comprehensive analysis of the seasonal variability of the optical properties of Saharan dust aerosols over Athens, Greece, is presented for a 17-year time period (2000-2016), as derived from multi-wavelength Raman lidar measurements (57 dust events with more than 80 hours of measurements). The profiles of the derived aerosol optical properties (aerosol backscatter and extinction coefficients, lidar ratio and aerosol Ångström exponent) at 355 nm are presented. For these dust events we found a mean value of the lidar ratio of ~52±13 sr at 355 nm and of ~58±8 sr (not shown) at 532 nm (2-4 km a.s.l. height). For our statistical analysis, presented here, we used monthly-mean values and time periods under cloud-free conditions. The number of dust events was greatest in late spring, summer, and early autumn periods. In this paper we also present a selected case study (04 April 2016) of desert dust long-range transport from the Saharan desert.

1. INTRODUCTION

Dust events occur in the Eastern Mediterranean region mostly from early spring to early autumn, while during winter months dust transport over the area remains quite low as previously shown [1-5]. Aerosol lidar measurements are carried out in Athens since the year 2000. Each year Saharan dust events occurs that heavily influence the atmospheric conditions over the city of Athens. The optical properties of the resulting dust aerosol for the time-period 2000-2016, are presented in this work.

2. INSTRUMENTATION

The Raman lidar system (EOLE) of the National Technical University of Athens (NTUA), in the frame of the European Aerosol Research Lidar network (EARLINET) was used to derive the vertical profiles of aerosols over Athens, Greece. The lidar system is located at the Laser Remote Sensing Unit (LRSU) of NTUA (37.9°N, 23.6°E, 200 m a.s.l.). LRSU is equipped with an advanced 6-wavelength Raman lidar system, with a very recently added (2016), depolarization channel at 355 nm (Fig. 1), able to perform independent and simultaneous measurements of the vertical profiles of the aerosol backscatter (at 355, 532 and 1064 nm) and extinction coefficient (at 355 and 532 nm). The data processing was based on the Single Calculus Chain (SCC) developed within EARLINET [6] and on manual retrievals. Air-mass back-trajectory analysis from the HYSPLIT and the BSC DREAM models was used to verify the lidar measurements of dust events.

Figure 1. The LRSU multiwavelength Raman lidar (EOLE) system.

3. METHODOLOGY

The NTUA aerosol lidar measurements during Saharan dust outbreaks [1,4] are based on early warnings, using forecast model data, provided by the DREAM model [7]. However, the dust lidar measurements are performed only under cloud-free weather conditions. Also, due to the weak Raman signals, the complete set of aerosol products (3 backscatter and 2 extinction coefficient profiles) can be derived for tropospheric layers only during nighttime. Thus, the number of the observed dust days is always fewer than the forecasted ones. In Fig. 2 we present the monthly and seasonal distribution of the dust cases percentage of occurrences observed by EOLE, for the period 2000-2016.

In this study the main emphasis has been given on the mean aerosol properties at 355 nm, as the 1st Stokes of the laser radiation at 355 nm shifted by N_2 (at 387 nm) was the first Raman channel to be implemented by EOLE in 2003, some years before the implementation of the 2nd Raman channel (at 607 nm). Thus, the longest database on the aerosol properties over Athens refers to the 355 nm channel. We note here, that according to

Groß et al. [3], during the Saharan Aerosol Long-range Transport and Aerosol-Cloud interaction Experiment (SALTRACE) in Barbados in June and July 2013, the lidar ratio (LR) within the pure dust layer was found to be wavelength independent, with mean values of 53±5 sr at 355 nm.

According to Groß et al., 2015 [3], during the Saharan Aerosol Long-range Transport and Aerosol-Cloud interaction Experiment in Barbados in June and July 2013, the lidar ratio (LR) within the pure dust layer was found to be wavelength independent with mean values of 53±5 sr at 355 nm.

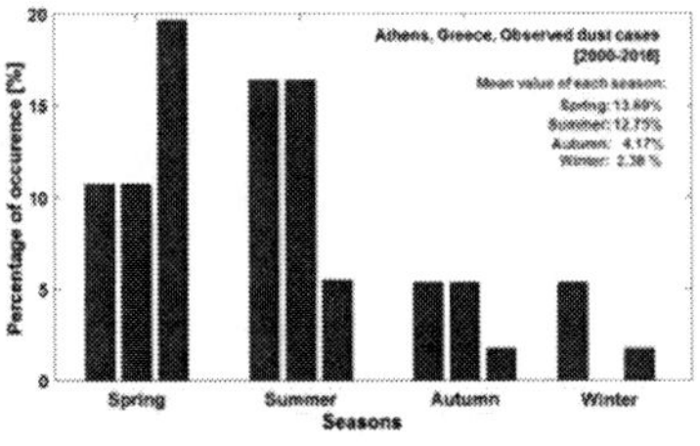

Figure 2. Percentage of occurrence (%) of the monthly (blue bars) evolution of the mean number of dust cases observed by EOLE over Athens, for the period 2000-2016. Mean values for each season are also mentioned.

4. STATISTICAL ANALYSIS

In Fig. 3 (upper and middle graphs) we present the seasonal evolution (from spring to winter) of the vertical profiles of the aerosol optical properties (backscatter-β_{aer} and extinction-α_{aer} coefficients) at 355 nm, as well as their mean values (red lines) and the corresponding statistical error (red error bars). The seasonal variation of the lidar ratios and their mean values (red lines) are presented in Fig. 3 (lower graph) for the height range between 2 and 4 km, where usually most of the Saharan dust layers are observed [4].

Fig. 4 shows the frequency of occurrence (%) of the lidar ratio (57 dust cases) values measured at 355 nm (2-4 km height), where LRs varied from 22 and 77 sr; the corresponding statistical mean of LR was 52±13 sr. However, 73% of the dust cases fall within 40sr<LR<70sr, with the most frequent LR value being between 40-50 sr (~37% of the dust cases).

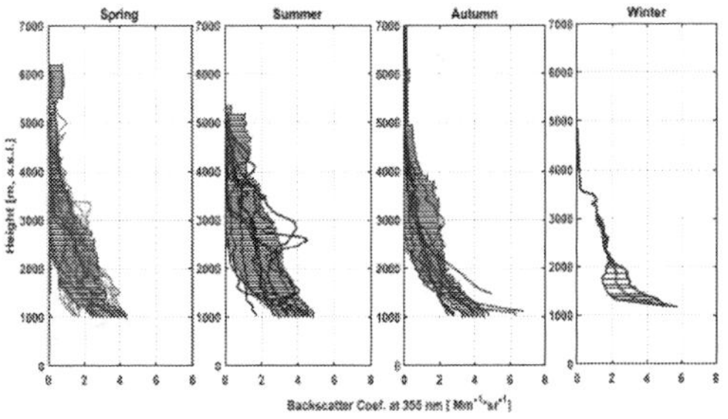

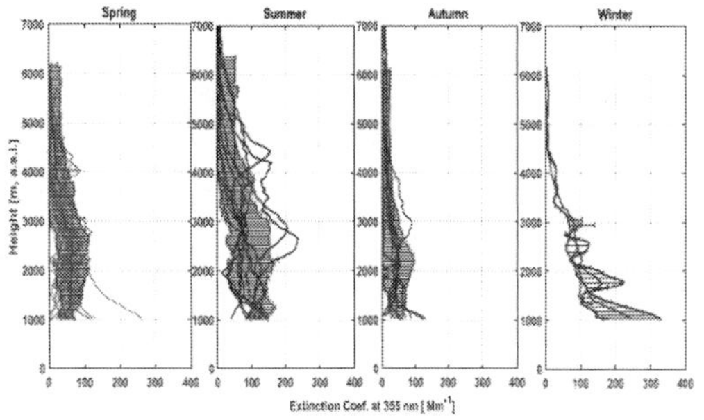

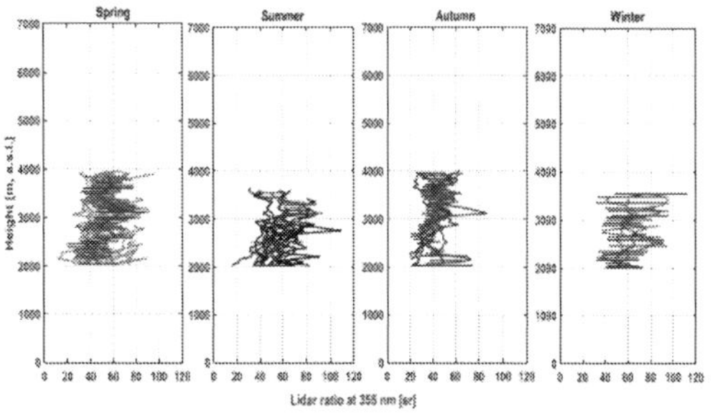

Figure 3. Seasonal variation of the vertical profiles of the aerosol optical properties: backscatter (upper graph), extinction (middle graph) coefficients and lidar ratio (lower graph) at 355 nm, as well as their mean values (red lines) and the corresponding statistical error (red error bars) for 57 dust cases during the period 2000-2016.

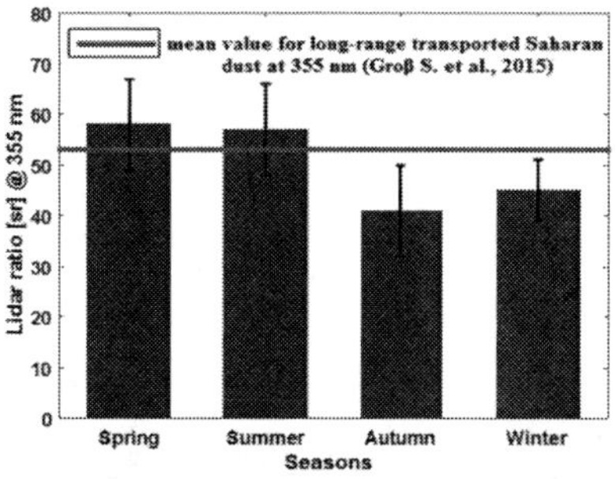

Figure 4. Frequency distribution of the lidar ratio at 355 nm between 2-4 km a.s.l., as retrieved from 57 observations of dust cases over Athens.

5. CASE STUDY OF 04 APRIL 2016

The dust event of 04 April 2016 observed over Athens was forecasted by the BSC-DREAM model for 12:00 and 18:00 UTC (Fig. 5a). The dusty air masses are distinguished by their green-yellow color, spreading from N-NW-W Saharan towards Italy and Greece. The corresponding dust load over Athens was of the order of 0.2 g/m^2. According to the BSC-DREAM forecast, the vertical profile of the dust concentration [shown in

Fig. 5b, for 12:00 UTC (left) and 18:00 UTC (right)] shows the presence of dust particles from near ground up to 4-4.5 km height (also between 4.5-6 km at 18:00 UTC), with maximum mass concentration of ~25 μg/m^3, around 1.5 km height, indicating an episode of rather low intensity. Indeed, EOLE detected dust layers in the range between 1 and 3 km and 5.5-7 km height.

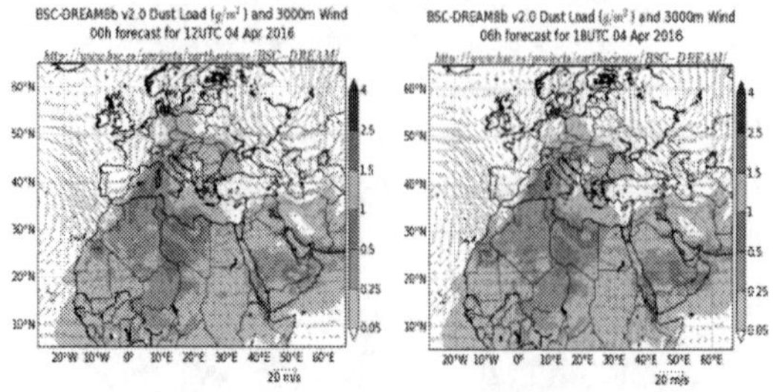

Figure 5a. Dust forecast for 04 April 2016 provided by BSC-DREAM model: dust load in g/m^2 for 12:00 UTC (left) and 18:00 UTC (right).

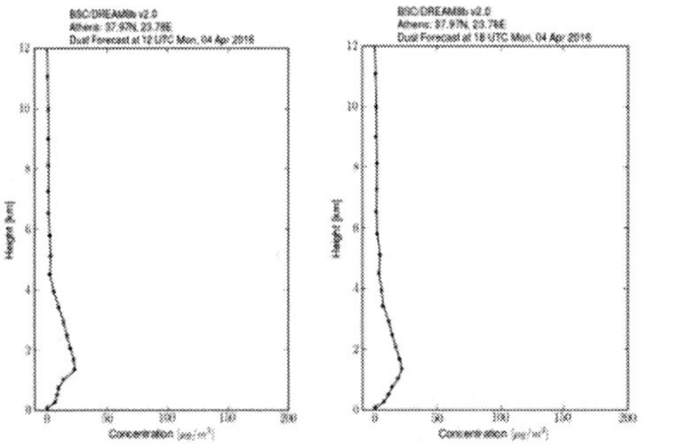

Figure 5b. Dust forecast for 04 April 2016 provided by BSC-DREAM model: dust concentration profile in μg/m^3 for 12:00 UTC (left) and 18:00 UTC (right).

Figure 6. Spatio-temporal evolution of the range-corrected lidar signal at 1064 nm obtained over Athens on 04 April 2016 (11:45-19:21 UTC).

In Fig. 6 we present the spatio-temporal evolution of the range-corrected lidar signal at 1064 nm obtained over Athens on 04 April 2016 (11:45-19:21 UTC). The aerosol layers are clearly visible, mainly from 1-3 km and around 5.5-6.5 km height.

Indeed, the aerosol optical properties (β_{aer} and the backscatter-related Ångström exponent AE_b) obtained by EOLE from 11:45-13:07 UTC (cf. Fig. 7-upper graph) show a thick layer from 1.2 to 3 km height, with AE_b (at 532-1064 nm) around 1.1. Similar properties show the aerosol profiles obtained during nighttime from 17:45-19:21 UTC (cf. Fig. 7-lower graph) where the lower aerosol layer re-mains within about the same height range (1.7-3.5 km) and the AE_b(532-1064 nm) is again below1, from 1 to 3.5 km height (typical for the presence of Saharan dust) [1-3,8].

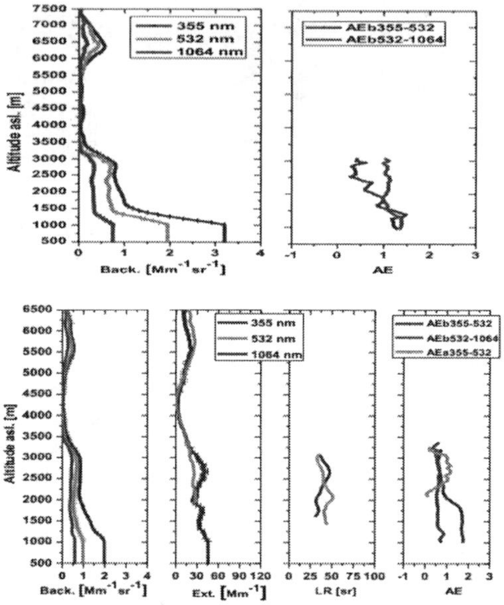

Figure 7. Aerosol optical properties obtained by EOLE at 355, 532, 1064 nm, on 04 April 2016; *upper graph*: profiles of β_{aer} and AE_b(355-532 nm and 532-1064 nm) (11:45-13:07 UTC); *lower graph*: profiles of β_{aer}, α_{aer}, LR, AE_b,(355-532 nm and 532-1064 nm) and AE_a (355-532 nm) (17:45-19:21 UTC).

Furthermore, the 6-day air mass back trajectories arriving over Athens on 04 April 2016, between ~1.8-3 km, are shown in Fig. 8 [for 12:00 UTC (left) and 18:00 UTC (right)] show that the source region of these aerosol layers is the Saharan desert (mainly from Algeria-Tynisia-Libya).

Indeed, SeaWiFS detected the arrival of Saharan dust over Greece on 04 April 2016 (Fig. 9).

Moreover, the AERONET sun photometer of the National Observatory of Athens detected the arrival of

 https://doi.org/10.1051/epjconf/201817605029

Saharan dust, showing aerosol optical thickness at 440 nm of ~0.33 (09:00-15:00 UTC - not shown here).

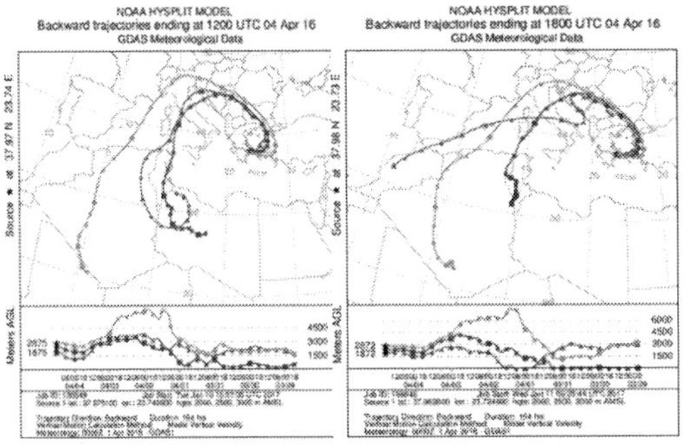

Figure 8. Air mass backward trajectories arriving over Athens on 04 April 2016 at 12:00 UTC (left) and 18:00 UTC (right), based on the NOAA Hysplit model.

Figure 9. SeaWiFS true color image on 04 April 2016.

Indeed, the true color image provided by the Sea-viewing Wide Field-of-view Sensor-SeaWiFS (Fig. 10) on 04 April 2016, clearly shows the desert dust sweeping Greece from SW to the NE (Athens is marked by asterisk).

5. CONCLUSIONS

Multi-wavelength aerosol Raman lidar data (57 dust events) have been analyzed for the period 2000-2016, regarding the main aerosol optical properties (β_{aer}, α_{aer}, LR, AE_a and AE_b) of Saharan dust particles over the city of Athens. Mean values of 52±13 sr and 58±8 sr, were found for LR at 355 and 532 nm, respectively.

The height range, where usually most of the layers related to Saharan dust intrusions were observed, was found between 2 and 4 km, while the occurrence of the dust events was greater during spring and summer months.

ACKNOWLEDGMENTS

The LRSU-NTUA multi-wavelength Raman lidar is founding member of the EARLINET network (www.earlinet.org). The research leading to these results has received funding from the European Union 6th and 7th Framework Programmes (FP6 and FP7 under grant agreement RICA-025991 and no 262254, as well as from the Horizon 2020 ACTRIS research/innovation programme under grant agreement n° 654109. AP and AA have received funding from the European Union Seventh Framework Programme (FP7/2007-2013): People, ITN Marie Curie Actions Programme (2012-2016) in the frame of ITaRS, under grant agreement n° 289923. The authors gratefully acknowledge the NOAA Air Resources Laboratory for the provision of the HYSPLIT transport and dispersion model and READY website (www.ready.noaa.gov) used in this publication. SeaWiFS image has been provided by NASA GSFC, Ocean Biology Processing Group. The dust forecast and concentration profiles have been provided by BSC-DREAM model (www.bsc.es/ESS/).

REFERENCES

[1] Papayannis, A., et al., 2005: Measurements of Saharan dust aerosols over the Eastern Mediterranean using elastic backscatter-Raman lidar, spectro-photometric and satellite observations in the frame of the EARLINET project, Atmos. Chem. Phys., 5, 2065–2079.

[2] Balis, D., V. Amiridis, S. Nickovic, et al., 2004: Optical properties of Saharan dust layers as detected by a Raman lidar over Thessaloniki, Greece, Geophys. Res. Lett., 31, L13104, doi:10.1029 /2004GL019881.

[3] Groß, S., et al., 2015: Optical properties of long-range transported Saharan dust over Barbados as measured by dual-wavelength depolarization Raman lidar measurements, Atmos. Chem. Phys., 15, 11067–11080, 2015.

[4] Papayannis, A., et al., 2008: Systematic lidar observations of Saharan dust over Europe in the frame of EARLINET (2000–2002), J. Geophys. Res., 113, D10204, doi:10.1029/2007JD009028.

[5] Amiridis, V., et al., 2005: Four-year aerosol observations with a Raman lidar at Thessaloniki, Greece, in the framework of EARLINET, J. Geophys. Res., 110, D21203, doi: 10.1029/ 2005JD06190.

[6] d'Amico, G., et al.: EARLINET Single Calculus Chain for automatic lidar data processing: First tests on optical products, 2012: Proc. 26th International Laser Radar Conference, Porto Heli, Greece, 331-334.

[7] Basart, S., et al., 2012: Development and evaluation of the BSC-DREAM8b dust regional model over Northern Africa, the Mediterranean and the Middle East, Tellus, B64, 18539, http://dx.doi.org/10.3402/ tellusb.v64i0.18539.

[8] Papayannis, A., et al., 2012; Optical-microphysical properties of Saharan dust aerosols and composition relationship using a multi-wavelength Raman lidar, in situ sensors and modelling: a case study analysis, Atmos. Chem. Phys., 12, 4011-4032.

EPJ Web of Conferences **176**, 05030 (2018) https://doi.org/10.1051/epjconf/201817605030
ILRC 28

OBSERVATION OF AN INTENSE BIOMASS BURNING EVENT OVER SPITSBERGEN

Christoph Ritter[1], Christine Böckmann[2]

[1]*Alfred-Wegener-Institute for Polar and Marine Research, Research Department Potsdam, D-14473 Potsdam, Germany, critter@awi-potsdam.de*
[2]*University of Potsdam, Institute of Mathematics, Germany*

ABSTRACT

During July 2015, an extremely strong event of biomass burning (BB) originating from the boreal America was observed over Spitsbergen with a "3+2" Raman Lidar. In this work we show some results from the lidar and a contemporaneous launched radiosonde to demonstrate that this aerosol was only hygroscopic at humidities above 80%.

1. INTRODUCTION

The Arctic is climatologically a very sensitive region, where the temperature fluctuations are approximately twice as large as in the global mean, an effect, which is called Arctic Amplification [1]. Especially absorbing aerosol as soot might be important when it acts via radiative forcing in the atmosphere ("dimming") or reducing the albedo via deposition on ground ("darkening").
During summer time the minimal aerosol load is measured in the Arctic [2] with AOD_{500} values around 0.05. In Spitzbergen small biogenic particles from the Nordic ocean dominate in in-situ measurements during that season [3], while in remote sensing instruments sporadically BB events can be seen aloft. Due to its location at the edge of the Gulf stream Spitsbergen is a location where generally Arctic and Marine influences mix.

The data of this work was obtained at the research village of Ny-Ålesund, 78.9°N, 11.9°E, Spitsbergen in the European Arctic. The Alfred-Wegener-Institute for Polar- and Marine research operates several instruments at that site for many years, among others the "3+2+2" Raman lidar KARL ("Koldewey Aerosol Raman Lidar") and daily launches of Vaisala RS-92 radiosondes around 11 UT.

On the afternoon of 9 July 2015 the AOD started to increase dramatically, indicating the beginning of the event. Our lidar observations have been performed on 10 July, the day with maximal pollution. On that day the AOD_{500} reached values > 1. The event lasted for a couple of days in the troposphere and was also recorded by other stations in the European Arctic [4].

2. DESCRIPTION OF THE EVENT

The data for this contribution was recorded with the "3+2+2+2" Nd:YAG Raman Lidar KARL. This system was described in [5] and consists of a 70cm mirror with approx. 1.5mrad field of view. It uses a spectra 290-50 Pro laser with about 10W in each color (355nm, 532 and 1064nm). Apart from the elastic wavelengths (355 and 532nm also in cross polarization) the N2 Raman channels of 387nm and 607nm and water vapor (407nm and 660nm) were recorded.

The aerosol event was evaluated with a resolution of 30m/10min according to Ansmann [6] using the radiosonde data from our site to correct for the molecular scattering. As the stratosphere was free of aerosol, a backscatter ratio of 1.05 at 532nm in 13km – 15km altitude and an Angström exponent of backscatter of -1 was assumed as a boundary condition. (This is our long-term experience for this site to obtain reasonable color ratios for cirrus clouds and aerosol.) For the following discussion it must be noted that no smoothing was applied to the lidar profiles. Hence it is assumed that each data point with the given resolution has independent information content. Due to the very high extinction and backscatter values the error component due to noise in the data is less than 1% for the backscatter and about 7% for the extinction.

The backscatter (at 532nm) and color ratio (355nm / 532nm) of the event is depicted in Figs 1 and 2. It can be seen that the aerosol is mainly confined below 3.6km altitude. The structures above 10km altitude around UT 12 and in 5km altitude in the afternoon are clouds. Times and heights where the backscatter was influenced by multiple scattering have been removed from the figures (white stripes).

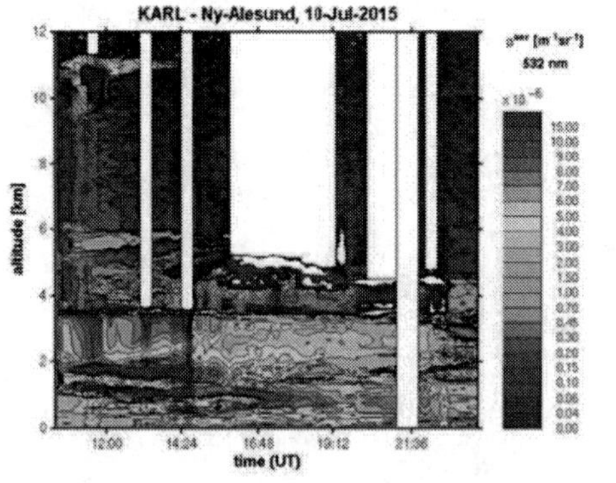

Fig 1: Overview of the backscatter coefficient of the aerosol event.

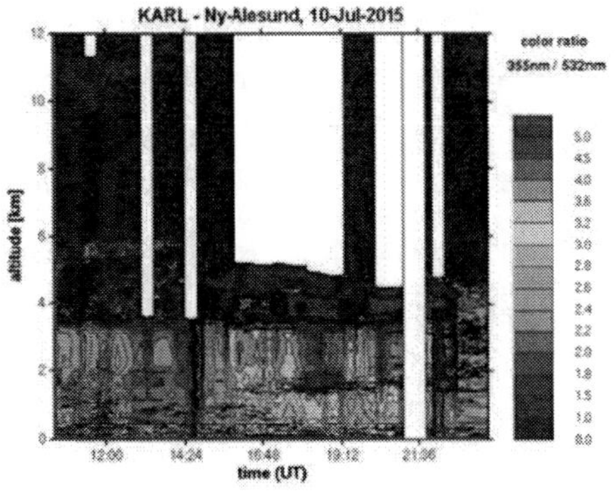

Fig 2: Overview of the color ratio (defined as the ratio of aerosol backscatter at 355nm to 532nm) of the aerosol event.

Generally the aerosol load declines over the day and some clear structures and variability is seen in the contour plots.

The origin of the BB event is indicated in Fig. 3 from hysplit trajectories using the NCEP reanalysis data. Very strong forest fires have been reported in Western Canada for the whole month of June [4]. In early July a plume of polluted air was located over Alaska from which the air moved via the North Pole to Spitsbergen within 5 days. Note that due to the sparseness of meteorological data longer trajectories cannot be calculated in the Arctic.

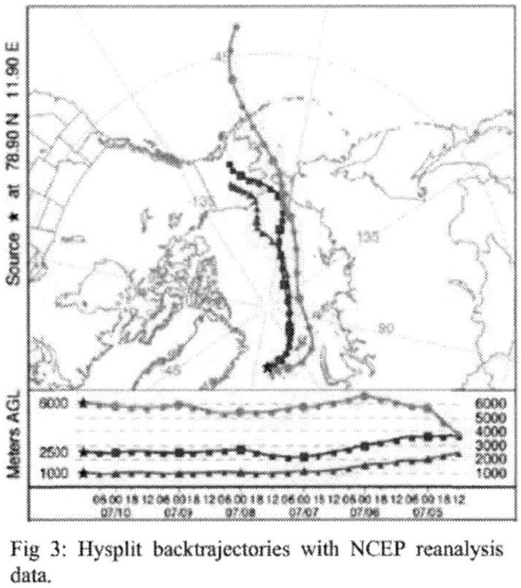

Fig 3: Hysplit backtrajectories with NCEP reanalysis data.

The profile of temperature and contemporary lidar backscatter are plotted in Fig 4. The potential temperature and the relative humidity are depicted in Fig. 5. It can be seen that the stratification is stable in the lowest 2.1km altitude (with a very low inversion layer), while between 2.1km and 3.5km the atmosphere is close to neutral stratified. Nevertheless pronounced gradients both in the humidity and the backscatter exist. For this reason a mixing did not happen as this would have destroyed the gradients. Therefore, by analyzing the aerosol event in detail in altitudes between 2.1km and 3.5km we can learn something on the hygroscopic properties of this event, as the rel. humidity increased from 50% to more than 90%. An elevated inversion at 3.5km altitude caps the aerosol and the atmosphere becomes much clearer.

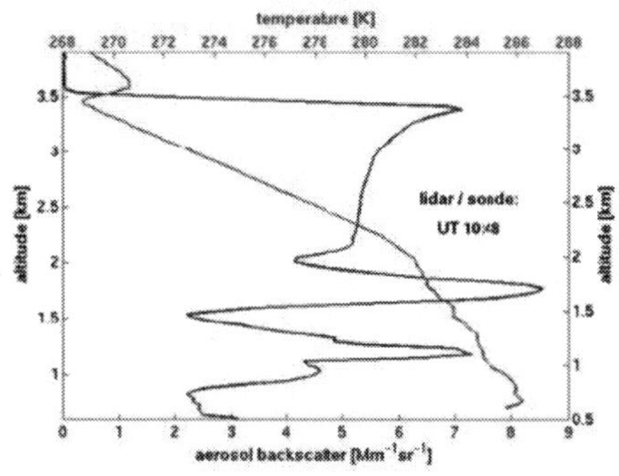

Fig 4: Contemporary profiles of temperature and aerosol backscatter from radiosonde and lidar

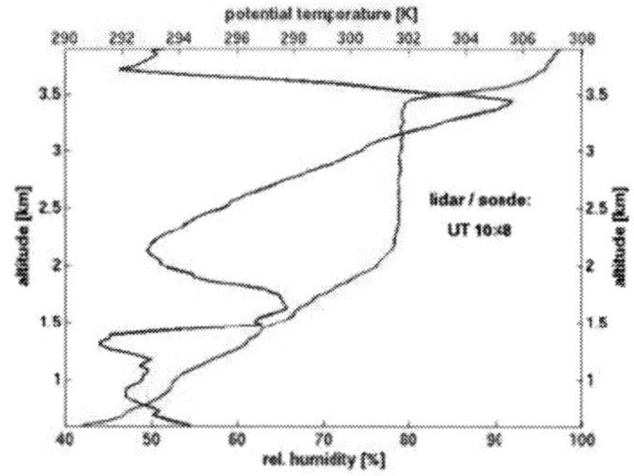

Fig 5: Profiles of potential temperature and relative humidity from the radiosonde

By inspection of the backscatter coefficient in the interval between 2.1 and 3.4 km altitude (Fig. 6), which corresponds to an increase of relative humidity from 50% to 90%, (back-) scattering enhancement factors can be calculated [7], as we assume that the aerosol in this height interval has the same origin. Quite low backscatter enhancements have been found of 1.4 for 532nm and 1064nm and only 1.2 for 355nm.

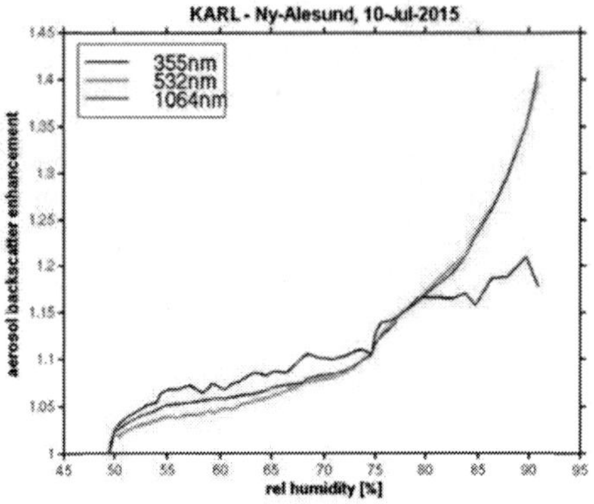

Fig 6: Increase of aerosol backscatter as function of the rel. humidity

To analyze the hygroscopicity of this event further, in Fig. 7 and 8 scatter plots are presented which show the lidar ratios and the color ratios as a function of the relative humidity. In this work we define the color ratio simply by the ratio between the aerosol backscatter of the shorter wavelength divided by the aerosol backscatter of the longer wavelength. Due to the error estimation given above relative changes of the color ratio of 2% and 8% for the lidar ratio should be significant.

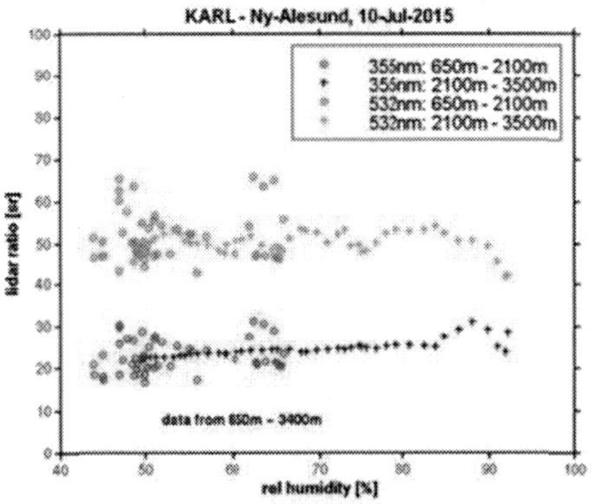

Fig 7: Lidar ratios as function of the rel. humidity

It can be seen from Fig. 7 that the lidar ratios are higher for the visible branch and lower for the UV. The values in both colors are similar to what is found at the site during the Arctic Haze season in spring [8]. Moreover it can be seen that below 2.1 km altitude the lidar ratios are more variable. In Ny-Ålesund a complex orography of nearby mountains produces generally a complicated wind shear pattern. Only in about 2km altitude the wind turns into its synoptic direction, as was analysed be inspection of radiosonde data at the site [9]. Therefore we assume that the aerosol arriving at above 2.1km have been in the free troposphere for at least several days, which apparently produced a quite uniform particle distribution. Increasing the relative humidity should lead to water uptake of the aerosol which increases the size and changes the shape and refractive index and, hence, the lidar ratio. However, only a weak influence at humidities > 80% can be seen, especially at the visible wavelength.

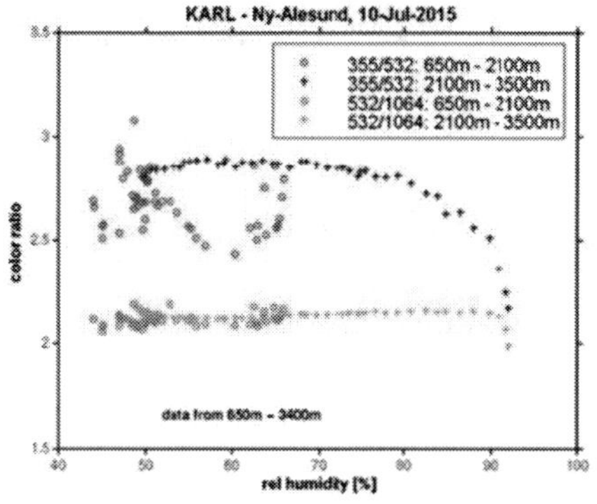

Fig 8: Color ratios as function of the rel. humidity

Fig. 8 shows the dependence of the color ratio on the relative humidity. Again, a higher variability was found in the wind shear zone, whereas in the free troposphere the color ratio is quite uniform up to about 80% humidity. The color ratio between 532nm and 1064nm is given in green and cyan. Here a possible hygroscopic growth of the particles is more obvious than in the color ratio of the shorter wavelengths (355nm to 532nm). In total the hygroscopicity of this BB aerosol seems to be quite low.

Additionally we analysed the linear aerosol depolarisation. The signal induced error for this depolarisation is in the order of 1‰. We clearly found higher depolarising particles close to the ground and more spherical aerosol aloft. Apparently the aerosol with complicated geometry seems to sediment down faster and hence is seen closer to the surface in an aged plume. Fig. 8 shows the situation. Also in black the humidity is given for comparison. Above 2 km there is a subtle trend that higher humidity decreases the depolarisation slightly. Nevertheless the depolarisation of this event is mainly a function of the altitude and not of the relative humidity.

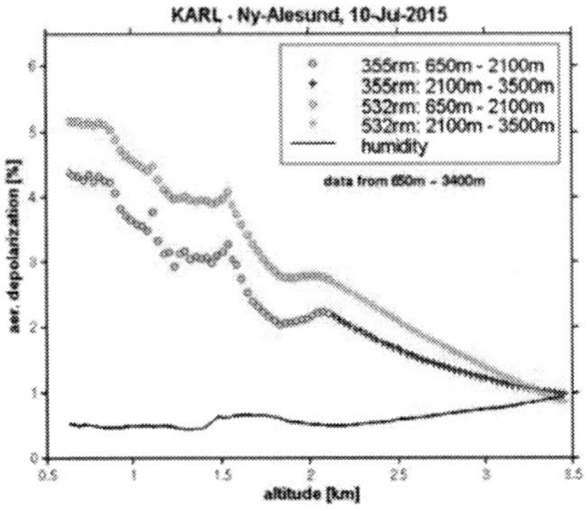

Fig 9: Aerosol depolarization as function of the altitude

3. SUMMARY AND CONCLUSIONS

A case study for a BB event in the inner Arctic, which produced an extremely high aerosol load of an aerosol backscatter at 532nm > 8 $Mm^{-1}sr^{-1}$, was presented. The intensive quantities lidar ratio, color ratio and aerosol depolarisation were analysed as a function of humidity measured by a contemporary radio sonde. It was found that the BB event was quite hydrophobic. Only at relative humidities > 80% a decrease of the color ratio indicated some hydroscopic growth while the influence of the humidity on the lidar ratio was only subtle. The aerosol depolarisation could not be analysed in this way as it clearly increased towards the surface. However, above 1500m the depolarisation decreases such that an inversion of the microphysical properties using Mie theory seems to be justified. This will be done in a separate work [10], this conference. Based on the results presented here a refractive index lower than for pure BB aerosol and an increased particle diameter compared to dry conditions is expected at 3.4km altitude where the relative humidity excels 90%.

4. REFERENCES

[1] Serrece, M.C. and Francis, J.A.: (2006) The Arctic Amplification Debate, Climate Change, **76**, 241 - 264, doi: 10.1007/s10584-005-9017-y

[2] Toledano, C., Cachorro, V.E., Gausa, M., Stebel, K., Aaltonen, V., Berjón, A., Ortiz de Galisteo, J.P., Frutos, A.M., de Bennouna, Y., Blindheim, S., Myhre, C.L., Zibordi, G., Wehrli, C., Kratzer, S., Hakansson, B., Carlund, T., Leeuw, G. de, Herber, A. and Torres, B.: (2012) Overview of sun photometer measurements of aerosol properties in Scandinavia and Svalbard, Atmospheric Environment, **52**, 18-28, doi:10.1016/j.atmosenv.2011.10.022

[3] Tunved, P., Ström, J. and Krejci, R.: (2013) Arctic aerosol life cycle: linking aerosol size distributions observed between 2000 and 2010 with air mass transport and precipitation at Zeppelin station, Ny-Ålesund, Svalbard, Atmos. Chem. Phys. **13**, 3643 – 3660, doi:10.5194/acp-13-3643-2013

[4] Markowicz, K.M., Pakszys, P., Ritter, C., Zielinski, T., Udisti, R., Cappelletti, D., Mazzola, M., Shiobara, M., Lynch, P., Zawadzka, O., Lisok, J., Petelski, T., Makuch, P., Karasinski, G.: (2016) J. Grophys. Res. doi: 10.1002/2016JD025310

[5] Hoffmann, A., Ritter, C., Neuber, R., Beninga, I., Schmid, J.: 2010: A redesigned Raman Lidar for Cloud and Aerosol Profiling in the, *ILRC 25*, St. Petersburg

[6] Ansmann, A., Wandinger, U., Riebesell, M., Weitkamp, C., and Michaelis, W. 1992: Independent measurement of extinction and backscatter profiles in cirrus clouds by using a combined Raman elastic-backscatter lidar, *Appl. Opt.* **31**, 7113-7113

[7] Granados-Muñoz, M. J., Navas-Guzmán, F., Bravo-Aranda, J. A., Guerrero-Rascado, J. L., Lyamani, H., Valenzuela, A., Titos, G., Fernández-Gálvez, J., and Alados-Arboledas, L. 2015: Hygroscopic growth of atmospheric aerosol particles based on active remote sensing and radiosounding measurements: selected cases in southeastern Spain, *Atmos. Meas. Tech.*, **8**, 705-718, doi:10.5194/amt-8-705-2015

[8] Ritter, C. , Neuber, R. , Schulz, A., Markowicz, K. M., Stachlewska, I., Lisok, J., Makuch, P., Pakszys, P., Markuszewski, P., Rozwadowska, A., Petelski, T., Zielinski, T., Becagli, S., Traversi, R., Udisti, R. and Gausa, M. (2016): 2014 iAREA campaign on aerosol in Spitsbergen – Part 2: Optical properties from Raman-lidar and in-situ observations at Ny-Ålesund, Atmospheric Environment, **141**, pp. 1-19. doi: 10.1016/j.atmosenv.2016.05.053

[9] Schulz, A. (2012) Die arktische Grenzschichthöhe auf der Basis von Sondierungen am Atmosphärenobservatorium in Ny-Ålesund und im ECMWF-Modell, diploma thesis, Univ. Potsdam, https://epic.awi.de/32026/1/Diplomarbeit_Schulz_PBL_2012.pdf

[10] Böckmann, C. and Ritter, C.: Arctic Biomass Burning Aerosol Event - Microphysical property retrieval (this issue)

FIRST RESULTS OF CIRRUS CLOUDS PROPERTIES BY MEANS OF A POLLY^{XT} RAMAN LIDAR AT TWO MEASUREMENT SITES

Kalliopi – Artemis Voudouri[1], Elina Giannakaki[2,3], Mika Komppula[2] and Dimitris Balis[1]

[1]Laboratory of Atmospheric Physics, Physics Department, Aristotle University of Thessaloniki, Greece
kavoudou@auth.gr
[2]Finnish Meteorological Institute, P.O.Box 1627, FI-70211, Kuopio, Finland
[3]Department of Environmental Physics and Meteorology, Faculty of Physics, University of Athens, Greece

ABSTRACT

Geometrical and optical characteristics of cirrus clouds using Raman lidar Polly^{XT} measurements at different locations are presented. The Polly^{XT} has been participated in two long-term experimental campaigns, one close to New Delhi in India and one at Elandsfontein in South Africa, providing continuous measurements and covering a wide range of cloud types. First results of cirrus cloud properties at different latitudes, as well as their temporal distributions are presented in this study. An automatic cirrus clouds detection algorithm is applied based on the wavelet covariance transform. The measurements at New Delhi performed from March 2008 to February 2009, while at Elandsfontein measurements were performed from December 2009 to January 2011.

1 INTRODUCTION

Cirrus clouds that are classified as high clouds have been characterized as having a significant double role in the Earth's climate system, by cooling or heating the Earth [1]. Thus systematic monitoring and detailed characterization of their geometrical and optical properties at different geographical locations is important in global climate analysis.

Active remote sensing as lidar instruments are a useful tool for providing measurements of vertical distribution of the optical properties of cirrus clouds. The most obvious advantage of lidar in cloud research is the opportunity of continuously monitoring of the cloud evolution and the high spatial resolution. Elastic backscatter and Raman lidars have been used for the retrieval of the geometrical and the optical properties of cirrus clouds already mentioned in literature [2]. From the reflected and transmitted polarized components of the backscattered light the profile of the depolarization ratio can be obtained as well, providing significant information about the sphericity of the retrieved particles and thus their liquid or solid phase. However, measurements can be restricted to those days with the presence of low clouds and rain.

Long term studies of cirrus clouds and comparisons between different measurement sites are limited. The aim of this work is to present lidar derived statistical results on geometrical and optical properties of cirrus clouds in two subtropical areas, estimating their dependence on geographical locations.

2 METHODOLOGY

The processing of the measurements and the detection of cirrus clouds geometrical boundaries was made by an automatic algorithm. This algorithm is based on a wavelet covariance transform (WCT) that is used to detect discontinuities in lidar signal as the base and the top of the cloud, with respect to the signal to noise ratio. This method is based on seeking high correlation between lidar signal and the Haar wavelet [3]. The WCT is performed to the range corrected signal, after subtracting the zero and the background measurement and after normalization with a maximum value below 1km. The normalization guarantees the applicability of the method at all atmospheric conditions. The dilation was chosen to be 225m, proportional to the cloud geometry. A threshold of -0.1 was used as a detection value [3]. The base of the cloud is defined one height bin below the altitude at which the wavelet is lower than the chosen threshold and the top is defined in the upper symmetric area.

Lidar profiles were checked before analyzed based on signal to noise ratio, specifically SNR was chosen to be greater than 2 for the cirrus detection.

Once the cirrus boundaries are detected, the cloud optical depth can be determined by integrating the particle extinction coefficient within the cloud boundaries. The effect of multiple scattering which is a function of the cloud height, the receiver field of view (FoV), the size and shapes of ice-crystals [4] will also be calculated in next steps of our analysis.

The WCT derived profile on 25th of February 2010 at Elandsfontein station is presented. The cloud base was calculated at 7.15km from the raw signal, while the base identified by WCT method was 7.03km. The cloud boundaries identified by WCT method is in good agreement with the cloud boundaries observed by the range corrected signal at 1064nm as shown in Fig.1 (left).

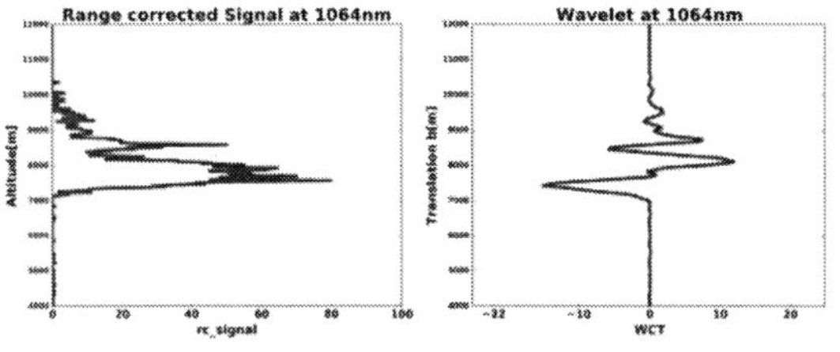

Figure 1: (left) Normalized range-corrected lidar signal at channel 1064nm and (right) resulting covariance transform as a function of translation b at dilation a=225 m. The specific WCT profile feature is used to identify cloud layers.

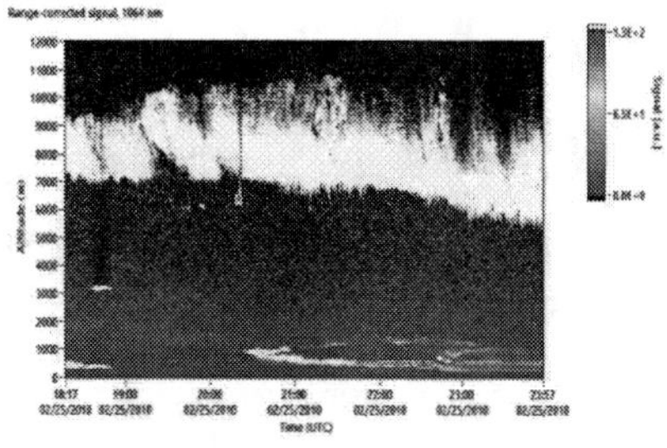

Figure 2: Range corrected backscatter signal at 1064nm observed with Polly^{XT} on 25th of February 2010 at Elandsfontein station.

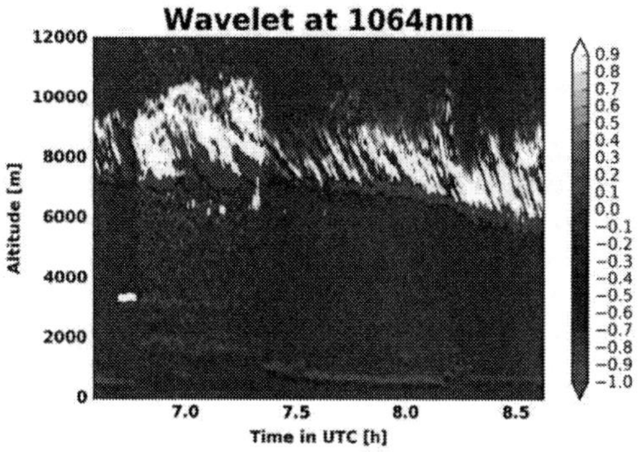

Figure 3: Cirrus cloud evolution as determined from the Polly^{XT} observations by applying the WTC on 25th of February 2010 at Elandsfontein station.

3 LIDAR SYSTEM AND MEASUREMENT SITE

The transportable Raman lidar Polly^{XT} supplied by the Finnish Meteorological Institute (FMI) has been operated to perform continuous measurements of different kind of clouds.

Full description of the system is provided in literature [5] and [6]. Polly^{XT} is equipped with a Nd:YAG laser emitting at 1064nm, which after doubling and tripling the frequency, emits at the wavelengths of 532 and 355nm respectively. The vertical resolution of the signal profiles is 30m and all measurements are available online at http://polly.tropos.de. The system was operating automatic and almost full time during the period 2008-2011.

The Polly^{XT} has participated in two campaigns in different sites over the word, within the framework of the EUCAARI project [7], covering a wide range of clouds. Measurements have been performed in Gual Pahari (28°260N, 77°090E, 243 m a.s.l) in India from March 2008 to March 2009, and in Elandsfontein (26°15′S, 29°26′E, 1745 m a.s.l) about 150km from Johannesburg in South Africa from December 2009 to January 2011. The experimental sites of different latitudes provide a better analysis of the different cirrus properties. In this study only nighttime data are considered.

4 RESULTS

During the period March 2008 to January 2011 65 night-time measurements of cirrus clouds were performed and analyzed. Only night-time data were analyzed in order to retrieve the optical properties with Raman method. The occurrence of cirrus clouds is about 16% of the total observation (only nighttime measurements were consider) and cirrus are located between 8km and 11km height typically. Missing data are attributed to low SNR. The geometrical parameters analyzed were obtained as mean values for the cloud. Fig. 4 presents an overview of the cirrus measurements during the analyzed period. As it can be seen no seasonal pattern, can be concluded.

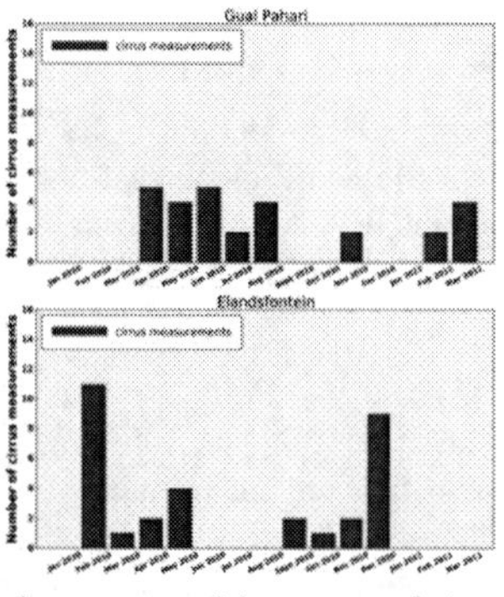

Figure 4: Occurrence of the measured cirrus clouds over Gual Pahari during the period 2008-2009 (left) and Elandsfontein during the period 2009-2010(right).

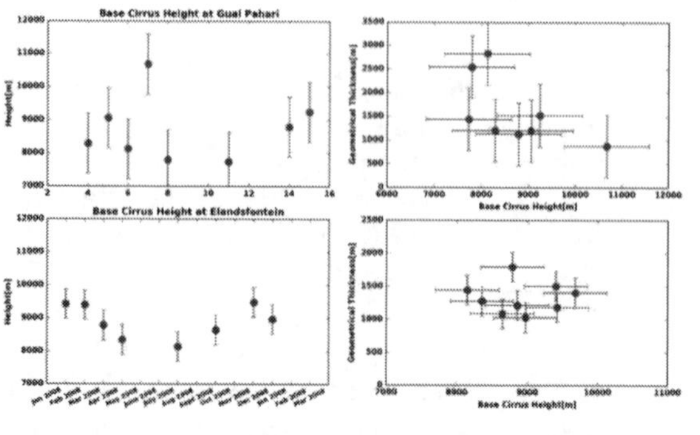

Figure 5: Variations of monthly mean base altitudes of Cirrus clouds at the two measurement sites (right). Dependence of cirrus cloud thickness on altitude cirrus base (left). The standard deviation from the mean value is also presented.

Fig. 5 presents the defined monthly cirrus base and its correlation with the cirrus thickness. Base heights showed a broad (7.5-11km) distribution. Results indicate that geometrical thickness spreads from 0.8 to 3km, with an average value of 1.5km. Results derived from other studies show that mean thickness can vary at different latitudes. In Thessaloniki, Greece, cloud thickness range from 1 to 5 km with mean value of 2.7 ±0.9 km [8], while in China, calculated thickness confined to a range of 0.8–4 km [9]. The dependence of cirrus cloud thickness on the altitude cirrus base is also presented in Fig. 5, showing greater variation for lower cirrus base at Gual Pahari. The geometrical thickness of cirrus clouds depends on the formation mechanism, the cloud altitude and the cloud temperature and should be further examined. Mean values of cirrus base and geometrical thickness are presented in Table 1.

Table 1: Mean values and standard deviation of cirrus base and thickness for Gual Pahari and Elandsfontein.

	Gual Pahari (28°26N, 77°09E)	Elandsfontein (26°15′S, 29°26′E)
Base, km	*8.7±0.9*	*8.9±0.45*
Thickness, km	*1.6±0.6*	*1.3±0.2*

5 CONCLUSIONS

Lidar derived data at two different latitude experimental sites (Gual Pahari and Elandsfontein) were processed in order to evaluate the geometrical and optical properties of cirrus clouds. An automated cloud detection algorithm was applied to determine the cloud base and top height and the cloud thickness and showed a good efficiency in detecting clouds. Cirrus clouds were detected at 16% of the nighttime measurements that were analyzed with an average mid-cloud height of 8.5km. Similar geometrical mean values were derived for both subtropical areas, while properties showed greater distribution at Gual Pahari. The dependency of cirrus clouds thickness on temperature will be further examined.

Optical properties of cirrus clouds i.e. backscatter and extinction coefficient profiles and depolarization ratio will be retrieved in next steps of our research.

ACKNOWLEDGEMENTS

The author would like to thank Finnish Meteorological Institute of Atmospheric Research Centre of Eastern Finland for providing the lidar dataset for analysis.

Elina Giannakaki acknowledges the support of the Academy of Finland (project no. 270108).

References

[1] Liou K.N., 1986: The influence of Cirrus on Weather and Climate Process: a Global Perspective, *Monthly Weather Review*, **114**, pp.1167-1199

[2] Ansmann, A., Wandinger U., Riebesell M., Weitkamp C., Michaelis W., 1992:Independent measurement of extinction and backscatter profiles in cirrus clouds by using a combined Raman elastic backscatter lidar, *Appl. Opt.*, **31**, 7113–7131

[3] Baars H., Ansmann A., Engelmann R., Althausen D., 2008: Continuous monitoring of the boundary-layer top with lidar, *Atmos. Chem. Phys.*,**8**, 7281-7296

[4] Eloranta, E., 1998: Practical model for the calculation of multiply scattered lidar returns, *Appl. Opt.*, *37*, 2464–2472, doi:10.1364/ao.37.002464.

[5] Althausen D., Engelmann R., Baars H., Heese B., Ansmann A., Muller D., Komppula M., 2009: Portable Raman lidar PollyXT for automated profiling of aerosol backscatter, extinction, and depolarization. *J.Atmos Ocean Tech* **26**:2366-2378. Doi10.1175/2009JTECHA1304.1

[6] Engelmann R., Kanitz T., Baars H., Heese B., Althausen D., Skupin A., Wandinger U., Komppula M., Stachlewska IS, Amiridis V., Marinou E., Mattis I., Linne H., Ansmann A., 2015: EARLINET Raman lidar PollyXT: the next generation. Atmos Meas Tech Discuss **8**:7737-7780. doi10.5194/amtd-8-7737-2015

[7] Kulmala, M., Asmi, A.,Lappalainen, H.K., et al.: General overview: European Integrated project on Aerosol Cloud Climate and Air Quality interaction (EUCAARI) – integrating aerosol research from nano to global scales, 2011: *Atmos. Chem. Phys.*, **11**, 13061-13143, doi:10.5194/acp-11-13061-2011

[8] Giannakaki, E., Balis, D.S., Amiridis, V., Kazadzis, S., 2007. Optical and geometrical characteristics of cirrus clouds over a Southern European lidar station. *Atmos. Chem. Phys.* **7**, 5519–5530.

[9] Wang, Jin, Zhang, Lei, Huang, Jianping, Cao, Xianjie, Liu, Ruijin, Zhou, Bi, Wang, Hongbin, 2013. Macrophysical and optical properties of mid-latitude cirrus clouds over a semi-arid area observed by micro-pulse lidar. *J. Quant. Spectrosc. Radiat. Transf.* **122**, 3–12.

FEASIBILITY STUDY TO MEASURE HDO/H_2O ATMOSPHERIC PROFILES THROUGH A RAMAN LIDAR

Gian Luigi Liberti[1], Davide Dionisi[1]*, Frederique Cheruy[2], Camille Risi[2]

*[1]Institute of Atmospheric Sciences and Climate, Italy, *d.dionisi@isac.cnr.it*
[2]Institut Pierre-Simon Laplace, Laboratoire de Météorologie Dynamique, France

ABSTRACT

The Raman lidar technique as currently applied for the retrieval of WV mixing ratio profiles allows, in theory, to estimate HDO/H_2O atmospheric profiles. The objective of this study is to develop a lidar simulator and to study the feasibility of the ground-based Raman lidar HDO measurement. Thus, the characteristics of a realistic lidar system for the estimation of HDO/H_2O atmospheric profiles are investigated through a set of numerical simulations.

1 INTRODUCTION

Improving the climate models' treatment of the complex water cycle is one of the key goals set by the Intergovernmental Panel on Climate Change. In this context, the isotopologue ratios of water (e.g. $HD^{16}O/H_2{}^{16}O$) contain valuable information about the history of the water content and are a powerful tool for investigating the water cycle (e.g. moistening and dehydration in the tropical troposphere [1], precipitation evaporation in the lower troposphere [2], etc...).

Remote-sensing measurements of HDO are a recent development and there are several satellites providing global HDO datasets [3]. However, absolute measurement calibration is dependent on accurate spectroscopy, while retrieval validation requires in situ profiling capability [3]. In principle, the Raman lidar technique allows estimating HDO/H_2O atmospheric profiles. Expected advantages of this technique, that could be used to validate the existing measurements, are the temporal and vertical sampling (depending on the required integration) but, more important, independent retrieval for each layer.

The objective of this study is to present a set of numerical tools developed to investigate the feasibility of a Raman lidar system to estimate the HDO/H_2O profiles. In particular:

- to define the optimal position and width of the filter for the measurement of the Raman backscattering from HDO.
- to define the signal integration over the vertical and over the time of the signal needed to reach a useful SNR
- to assess the capability to estimate variability of the HDO/H_2O ratio.

The above objectives were obtained developing a radiative transfer model able to simulate a Raman lidar (Sect.2) and using a set of profiles of species of interests, as produced by a Climate Model (Sect.3), as input to produce a statistical dataset of estimated the measurements.

2 RAMAN LIDAR SIMULATOR

The model used to simulate the received lidar signal (*P*) for a given wavelength can be schematized as follow:

$$P = INS \cdot EXT \cdot BKS + BKG + NOISE \quad (1)$$

INS: contains all terms describing the characteristics of the lidar system. The current version includes the following variables (in parenthesis the values adopted for the simulation discussed in this study):

- laser power (0.8 J), frequency (10 Hz) and wavelength (354.71333 nm) assumed as monochromatic.
- lidar altitude (0 aslm)
- overall transfer efficiency, accounting for transmission of optical components and quantum efficiency of the photomultiplier. Current version assumes a spectrally independent value (0.5).
- telescope radius (1 m).

Full overlap is assumed for the whole atmosphere.

EXT: contains all terms needed to account for the two wavelengths extinction. The following species have been included:

-***Atmospheric gases.*** O_3[4], NO_2 [5] and SO_2 [6] including temperature and pressure dependence when available in the original data base of absorption cross sections.

- ***Molecular scattering.*** Molecular scattering extinction is estimated according to [7].
- ***Aerosols extinction.*** In the version used to produce the results discussed, the profile of aerosols spectral extinction for each wavelength is computed from a highly parametrized representation based on 3 parameters: AOT@380 nm (0.2), Angstrom exponent (1.5) and a scale height (2 km) having assumed the aerosols profile decreasing exponentially.

BKG: this term account for the contribution from background assumed as constant (0.1 photons/s) and spectrally independent.

NOISE: noise is assumed as the square root of the signal including the background.

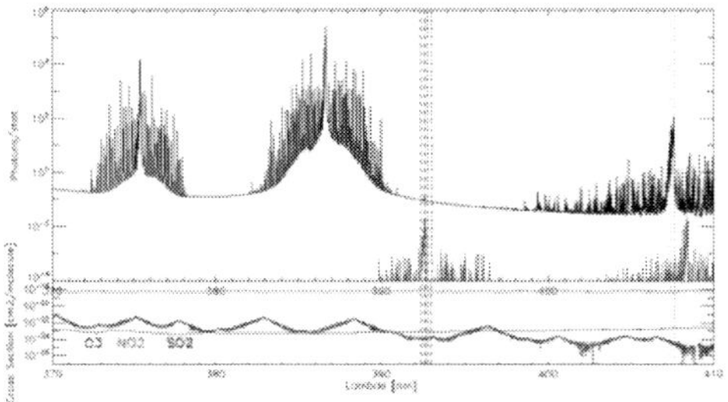

Figure 1.Example of spectrum of simulated numbers of photons per shot received by the system at z=75m for the modeled roto-vibrational Raman spectra of N_2, O_2, H_2O and HDO (black curve) and, as a reference, from HDO (purple curve). Black vertical dashed lines indicate the filter width of the HDO, N_2+ and H_2O channels selected through the lidar simulator. Absorbing cross section of O_3, NO_2 and SO_2 are also depicted (lower panel).

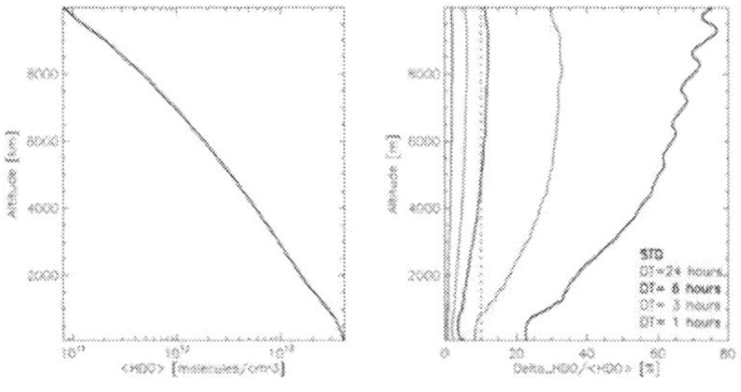

Figure 2. left: Mean profile of the HDO concentration in the outputs used; right: mean profiles of relative [%] HDO temporal variability for different time intervals.

BKS: Raman backscattering cross sections for: N_2, O_2, H_2O and HDO. Bi-atomic molecules, roto-vibrational Raman spectra are computed according to [8]. Cross sections for the H_2O and HDO Raman lines are computed according to [9] that provide tables for parallel and perpendicular polarizability coefficients and other relevant information for H_2O and HDO. Partition function is estimated with a 2nd order fit of the values tabled in [10] within the temperature range 175-325 K.
Single wavelength Raman cross sections contribution is computed summing contributions of each single line whose line broadening is estimated assuming collisional broadening as dominant process being the troposphere the region of interest.

The atmosphere is divided in 75 m thick layers from 0 to 10 km. Spectrally the model is limited by the availability of precomputed radiative properties: 370-410 nm with a resolution of 0.01 nm. Figure 1 shows an example of simulated spectrum.

3 LMDZ5A simulated data sets

The General Circulation Model (GCM) LMDZ5A (Laboratoire de Météorologie Dynamique- Zoom) is the atmospheric component of the Institut Pierre-Simon Laplace coupled model (IPSL-CM5A). Simulations were performed with forced sea surface conditions following the AMIP protocol [11] with the isotopic version of LMDZ [12] for the 1979-present period. Horizontal wind fields are nudged by ECMWF reanalyses [13] to ensure that simulated meteorological conditions are realistic on a day-to-day basis. Hourly outputs for the year 2007 at Rome Tor Vergata (41.8°N, 12.6°E) are used. Simulated profiles of H_2O, HDO, O_3, T as well as surface pressure are used as input for the lidar simulator. Profiles of SO_2, NO_2 and aerosols are assumed as constant.
Figure 2 shows the average HDO profile and its relative temporal variability for the outputs used in this study.

4 PRELIMINARY RESULTS

Once produced a set of simulated spectra from the whole time series of model derived profiles, the first step has been the estimation of the optimal position for a channel to measure the HDO Raman scattering. This has been done by searching, within

 https://doi.org/10.1051/epjconf/201817605032

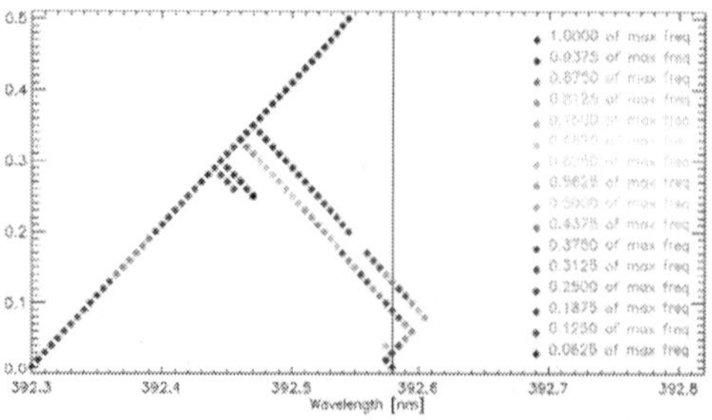

Figure 3 Central wavelength at which the maximum signal occurs (color=frequency of occurrence) as a function of the channel width. Only levels below 3 km are used to select the position of the HDO channel.

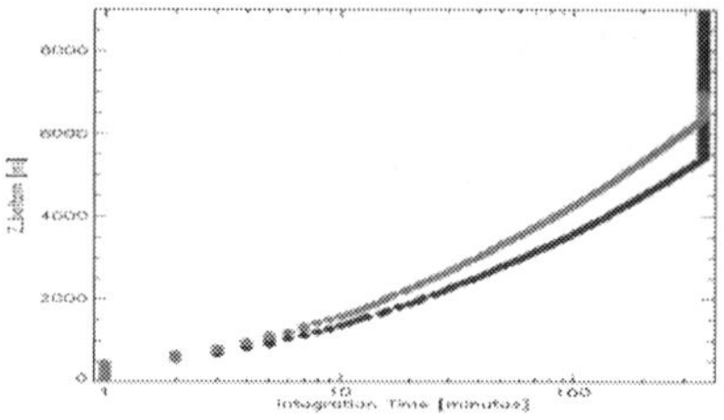

Figure 4 Integration time required to obtain for 95% of the simulated profiles SNR≥10 as a function of height. Signal is integrated over 1/3 (Red/Green) km thick layers estimated every 75 m

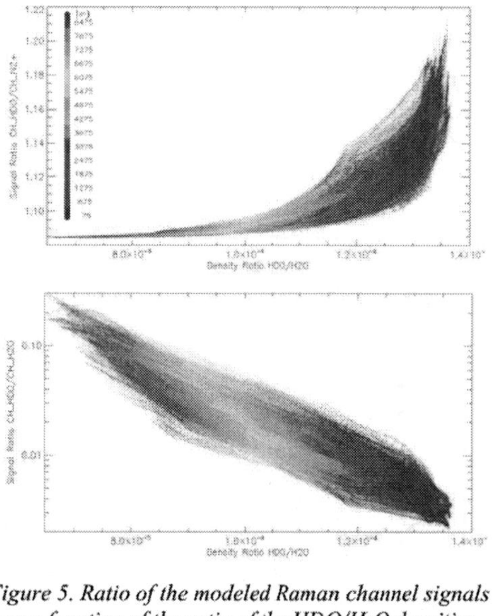

Figure 5. Ratio of the modeled Raman channel signals as a function of the ratio of the HDO/H₂O densities (color scale indicates the signal height). Upper: HDO/H₂O. Lower. HDO/N₂+.

a spectral interval including the largest HDO Raman signal, the position of an ideal car-box like spectral response channel maximizing the SNR as estimated from the whole set of simulated profiles. The search was done varying the channel width from 0.01-0.5 nm. Figure 3 shows the results of the channel definition statistics. The following results are relative to a 0.25 nm width channel centered at 392.5 nm.

The second step has been the estimation of expected temporal and vertical resolution. In order to do this, for the HDO channel, as defined in the previous step, the SNR was estimated by integrating over time from 1' to 360' a 1' step for different vertical integration layers. Fig.4 shows the profiles of integration time required to reach a single channel SNR≥10 for 95% of the analyzed cases for 1 and 3 km layer thickness.

A third step consisted in investigating the sensitivity to the variability of HDO/H_2O ratio of simulated signal ratios for different combinations of acquisition channels. Fig.5 shows the scatterplot of signal ratio for two combinations of Raman channels (see Fig.1) as a function of the HDO/H_2O model derived density ratio.

5 CONCLUSIONS

Preliminary results show that, as expected, retrieving HDO/H_2O profiles with Raman technique is a challenging issue. With a highly performing and somehow ideal system, useful information can be obtained from nighttime measurements on variability day-to-day variability of HDO vertical distribution in the boundary layer and lower free troposphere. However, in the remaining part of the troposphere, the uncertainty of the estimated ratio is of the order of magnitude of the expected dynamic of the variable of interest. In particular:

- from the simulation analyses it appears a strong contamination of the N_2 wings Raman signal in the selected HDO spectral window (Fig.1). This may be partially overestimated due to the assumption on the functional form of the line broadening as uniquely due to pressure broadening as well as by the number of rotational levels (60) considered to produce the N_2 Raman cross section spectra.
- The ratio of the HDO and H_2O channels (Fig.5 upper panel) shows an inverse dependence from

the ratio of density. In fact, the HDO channel is sensitive to the air density through the contamination of N_2 wings and a moister atmosphere corresponds to lower pressure. A possible solution to remove this dependence is measuring the residual N_2 contribution with a channel (N_2+) close to the HDO one (see Fig.1 and Fig.5 lower panel)

- The results of the simulations indicate that, ideally, integrating the measurements to obtain a signal to noise ratio higher than 10 allows the estimation of the HDO day-by-day variability in the boundary layer.

Further activities will include:

1) simulation of more favorable scenarios: more powerful system (doubling the laser emission), site located at higher altitude
2) simulation for different regimes as for example sub-tropical regions;
3) inclusion of more realistic features in the simulator: spectral response of the system, spectrally dependent background, Raman backscattering from other potentially interfering species (e.g. CO_2);
4) estimation of uncertainties due to the effects of variability of other atmospheric components (gas, aerosols) responsible for the extinction.
5) definition of a processing algorithm to retrieve HDO/H2O profiles and estimation of associated final uncertainty budget.

Finally, the developed code can be adapted to simulate other lidar systems including instrument proposed for candidate satellite missions.

ACKNOWLEDGEMENTS

LMDZ simulations used the HPC resources of IDRIS under the allocation 292 made by GENCI.

References

[1] Galewsky, J., and J. V. Hurley, An advection-condensation model for subtropical water vapor isotopic ratios, J. Geophys. Res., 115, D16116, doi:10.1029/2009JD013651, 2010.

[2] Worden, J., D. Noone, and K. Bowman, Importance of rain evapo- ration and continental convection in the tropical water cycle, Nature, 445, 528–532, 2007.

[3] Risi, C., et al., Process-evaluation of tropospheric humidity simulated by general circulation models using water vapor isotopologues: 1. Comparison between models and observations, J. Geophys. Res., 117, D05303, doi:10.1029/2011JD016621, 2012.

[4] Brion, J., et al.: Absorption Spectra Measurements for the Ozone Molecule in the 350–830 nm Region, J. Atmos. Chem., 30, 291-299, 1998.

[5] Voigt, S., J.Orphal, J.P.Burrows. The temperature and pressure dependence of the absorption cross-sections of NO2 in the 250–800 nm region measured by Fourier-transform spectroscopy. J. Photochem. Photobiol. A 149, 1–7. 2002.

[6] Rothman et al.: The HITRAN 2012 molecular spectroscopic database. J. Quant. Spectrosc. Rad. Transf. 130, 4–50, 2013.

[7] Eberhard, W. L.: Correct equations and common approximations for calculating Rayleigh scatter in pure gases and mixtures and evaluation of differences, Appl. Opt., 49, 1116-1130, 2010.

[8] Wandinger, U. Raman Lidar. In Lidar Range-Resolved Optical Remote Sensing of the Atmosphere. Ed. C.Weitkamp. Springer S. in Optical Sciences, Vol. 102, pp.240-271, ISBN 978-0-387-40075-4, 2005.

[9] Avila G. et al.: The Raman spectra and cross-section of H_2O, D_2O and HDO in the OH/OD stretching regions. J. Mol. Spectrosc. 228, 38-65, 2004.

[10] Avila G. et al: The rotational Raman spectra and cross sections of H_2O, D_2O, and HDO, J. Mol. Spectrosc. 220, 259-275, 2003.

[11] Gates, W.L.: AMIP: The Atmospheric Model Intercomparison Project. Bull. Am. Meteorol. Soc. 73, 1962-1970. 1992.

[12] Risi C. et al, Water stable isotopes in the LMDZ4 general circulation model: Model evaluation for present-day and past climates and applications to climate interpretations of tropical reocrds. J. Geophys. Res. 115, D12118. 2010

[13] Uppala et al. The ERA-40 re-analysis. Q. J. R. Meteorol. Soc. 131, 2961-3012. 2005

LONG TERM LIDAR MEASUREMENTS OF AEROSOL PROPERTIES OVER THESSALONIKI

Nikolaos Siomos[1,*], Kalliopi A. Voudouri[1], Eleni Giannakaki[2,5], Vasilis Amiridis[3], Maria Filioglou[1,2], Alexandros Papayannis[4] and Dimitris S. Balis[1]

[1]*Laboratory of atmospheric physics, Physics Department, Aristotle University of Thessaloniki, Greece, *Email: nsiomos@physics.auth.gr*
[2]*Finnish Meteorological Institute, Atmospheric Research Centre of Eastern Finland, Kuopio, Finland*
[3]*Institute for Astronomy, Astrophysics, Space Applications and Remote Sensing, National Observatory of Athens, Athens, Greece*
[4]*National Technical University of Athens, Physics Department, Laser Remote Sensing Laboratory, Athens, Greece*
[5]*Department of Environmental Physics and Meteorology, Faculty of Physics, University of Athens, Greece*

ABSTRACT

In this study we present some first results on the potential of 15 years of lidar measurements over the lidar station of Thessaloniki to compile a climatology of the aerosol properties for the period 2001-2015. This is examined on a monthly, seasonal and annual basis. Both the profile structure and the columnar properties of the aerosol extinction and backscatter products are examined. The results are compared for consistency against co-located sunphotometer measurements.

1 INTRODUCTION

The Thessaloniki Lidar System (THELISYS) operates as a UV Raman and elastic lidar since 2000. The initial setup included a Raman channel at 387nm and two elastic channels at 355nm and 532nm. In 2008 a second Raman channel at 607nm was added and in 2012 a third elastic channel at 1064nm was included as well. The station participates in the European Lidar Network and measurements are conducted and processed systematically according to the network standards and schedule [1]. The measurements from the period 2001-2008 were analyzed and by [2] and [3] and focus mostly on the origin and typing of the observed aerosol layers. Here we analyze the long term variability of the aerosol backscatter and extinction coefficient at 355nm during the period 2001-2015. The seasonality is analyzed for both the vertical profiles and the columnar optical depth (AOD). The AOD seasonality is also compared with sunphotometer data for consistency. The data processing procedure is described in detail in section 2.

2 METHODOLOGY

The analysis is based on the aerosol extinction and backscatter profiles at 355nm that were produced after processing with our operational algorithms that are following the EARLINET quality assurance procedures for the period 2001-2015. The extinction profiles are retrieved with the Raman inversion [4] and correspond solely to night time measurements. The elastic backscatter profiles that we present here are retrieved by applying the Klett inversion [5], that requires the assumption of a constant lidar ratio that is chosen from climatological values. The elastic backscatter profiles correspond to daytime measurements. All profiles are publicly available in the EARLINET database (`doi:10.1594/WDCC/`

EN_all_measurements_2000-2016). Due to hardware limitations in the lidar system, the profiles typically cover a variable altitude range for each measurement. To avoid this behavior we extent each profile down to the ground level, that is 60m for Thessaloniki, assuming that the extinction and backscatter coefficients remain constant below the lower retrieved value. The monthly averages are calculated first and the seasonal averages are derived for the monthly averages. The lidar data availability, however, is neither constant nor always high for the analyzed periods. In order to examine if the averages are representative of the corresponding time period, data with higher temporal availability are required to be used as an independent reference. For this purpose, aerosol optical depth data from a co-located CIMEL sunphotometer are compared with the corresponding lidar values. The instrument is part of the Aerosol Robotic Network AERONET and the level 1.5 daily averages that are produced by the AERONET algorithms are used here. In order to achieve representative seasonal AOD values, averages of less than 10 days per month and less than 2 months per season were excluded from the analysis. Since the sunphotometer doesn't operate in the same wavelengths as the lidar, it is necessary to convert the AOD values using an angstrom exponent. For the 355nm the sunphotometer AOD at 440nm was converted using the angstrom 440-675nm in order to have the longest possible timeseries. The lidar ratio value that was used to retrieve each backscatter profile is applied in order to convert the integrated backscatter to AOD, mention later in the paper as backscatter based AOD. It is important to mention that unlike the Raman extinction profiles where the overlap of the lidar system is calculated and applied individually per case, the elastic backscatter product is not overlap corrected. The profiles are restricted above the full overlap height. This could affect the AOD values. The results of the analysis are presented in section 3.

3 RESULTS

Time height cross sections of the Raman extinction and elastic backscatter profiles at 355nm are displayed in figure 1. Most of the aerosols are usually located below 5km for the whole period. Higher values appear at the years 2001 to 2005 possibly affected by issues related to the higher overlap of the system during that period. A seasonal behavior is observed here. Higher values appear usually in the summer and lower values in the winter. Layers at higher altitudes are more frequent during summer. The

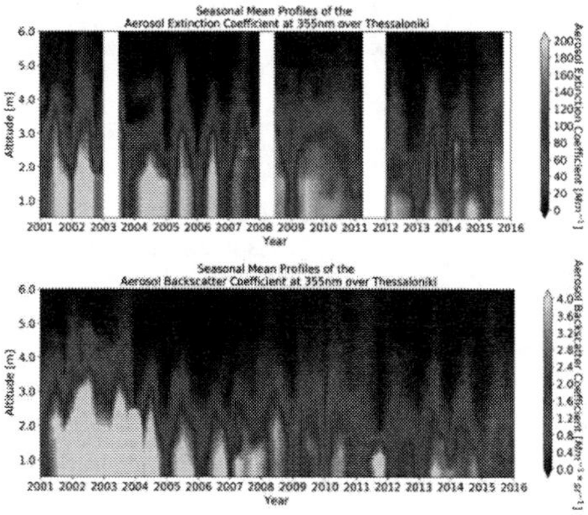

Figure 1: Seasonal mean time-height cross section of the aerosol extinction coefficient (upper panel) and the aerosol backscatter coefficient (lower panel) at 355nm for the period winter 2001 to autumn 2015.

annual cycle from the whole period and the corresponding AOD values are presented in figures 2 and 3 respectively. Only months with more than 5 years of available data are included in the figures. The maximum values appear in the summer months, that is, August for the extinction profile and July for the backscatter profile. A secondary maximums appear in October. The extinction annual cycle also shows high values in March and April while a third maximum appears in February of the backscatter annual cycle.

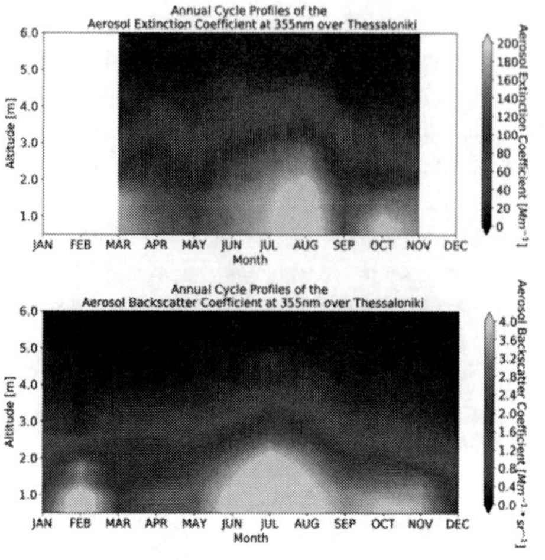

Figure 2: Mean annual cycle of the aerosol extinction coefficient (upper panel) and the aerosol backscatter coefficient (lower panel) at 355nm over Thessaloniki.

Differences could be attributed to the fact that the Raman extinction dataset contains only nighttime measurements while the elastic backscatter dataset contains mainly daytime measurements (see section 2). When compared with the sunphotometer annual cycle the agreement is best for April, May, October and November, as far as the lidar extinction AOD is considered. The maximum of the sunphotometer AOD also occurs during July but the lidar value is overestimated. Overall, the monthly variations of the annual cycle are reproduced from the lidar extinction AOD with the exception of September. As far as the backscatter based AOD is considered, February, May, Jun, August and October are the months that show the best agreement with the sunphotometer. The variability, however, of the February average backscatter based AOD is high. January, September and December are quite underestimated. Moreover, the secondary maximum

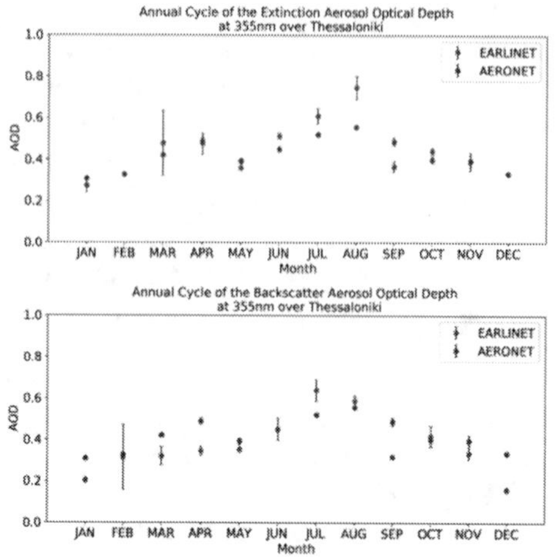

Figure 3: Comparison of the annual cycle calculated with the sunphotometer AOD at 355nm and with the extinction (upper panel) and backscatter (lower panel) AOD at 355nm.

in April isn't reproduced. The lack of an overlap correction in the elastic backscatter profiles (see section 2) could explain the differences between the two lidar products.

Figure 4 shows the monthly (upper panel) and seasonal (lower panel) mean AOD timeseries from the lidar and the sunphotometer for the Raman extinction and the elastic backscatter based AOD at 355nm. We have chosen the period Autumn 2003 - Autumn 2015 in order to have comparable lidar and sunphotometer datasets. The least square fit line is calculated for both the AOD values. The sunphotometer trend is considered more accurate than the lidar trend since it is produced by higher temporal frequency data (see section 2). The trend based on the monthly means is $-11.2 \cdot 10^{-3}$ per year against $-9.6 \cdot 10^{-3}$ per year for the sunphotometer while the trend based on the seasonal means is $-12.4 \cdot 10^{-3}$ per year against $-10.1 \cdot 10^{-3}$ per year respectively. Despite the lower data availability of the lidar extinction dataset, the trends are well reproduced.

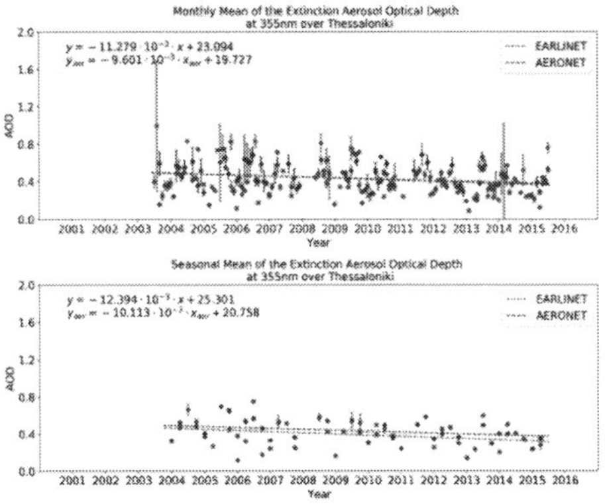

Figure 4: Comparison of the sunphotometer AOD at 355nm and with the lidar extinction AOD at 355nm based on the monthly mean (upper panel) and the seasonal mean (lower panel) values.

4 CONCLUSIONS

The analysis of the long term lidar measurements of Thessaloniki shows that the variable and usually limited availability and the different sampling of the climatological EARLINET lidar data provide consistent results with reference data. The comparison between the sunphotometer AOD and the lidar extinction AOD at 355nm resulted in similar monthly and seasonal trends during the period 2003-1015. As far as the annual cycle is considered, the monthly variability of the lidar extinction is compatible with the sunphotometer one. Higher mean values appear during summer and the maximum AOD occurs in August. The agreement is best for April, May, October and November. By including more lidar products in the analysis, it is possible to investigate the trends of additional optical products such as the lidar ratio using the Raman backscatter product at 355nm or the angstrom exponent 355nm-532nm using the elastic backscatter at 532nm. The clustering of the lidar measurements by their aerosol type using wind backward trajectories is also going to be implemented in the future.

ACKNOWLEDGEMENTS

This work has been conducted in the framework of EARLINET (EVR1 CT1999-40003), EARLINET-ASOS (RICA-025991) and ACTRIS-2 funded by the European Commission. This project has received funding from the European Union's Horizon 2020 research and innovation programme under grant agreement No 654109.

References

[1] G. Pappalardo, A. Amodeo, A. Apituley, A. Comeron, V. Freudenthaler, H. Linné, A. Ansmann, J. Bösenberg, G. D'Amico, I. Mattis, L. Mona, U. Wandinger, V. Amiridis, L. Alados-Arboledas, D. Nicolae, and M. Wiegner. Earlinet: towards an advanced sustainable european aerosol lidar network. *Atmospheric Measurement Techniques*, 7(8):2389–2409, 2014.

[2] V. Amiridis, D. S. Balis, S. Kazadzis, A. Bais, E. Giannakaki, A. Papayannis, and C. Zerefos. Four-year aerosol observations with a raman lidar at thessaloniki, greece, in the framework of european aerosol research lidar network (earlinet). *Journal of Geophysical Research: Atmospheres*, 110(D21):n/a–n/a, 2005. D21203.

[3] E. Giannakaki, D. S. Balis, V. Amiridis, and C. Zerefos. Optical properties of different aerosol types: seven years of combined raman-elastic backscatter lidar measurements in thessaloniki, greece. *Atmospheric Measurement Techniques*, 3(3):569–578, 2010.

[4] Albert Ansmann, Ulla Wandinger, Maren Riebesell, Claus Weitkamp, and Walfried Michaelis. Independent measurement of extinction and backscatter profiles in cirrus clouds by using a combined raman elastic-backscatter lidar. *Appl. Opt.*, 31(33):7113–7131, Nov 1992.

[5] James D. Klett. Stable analytical inversion solution for processing lidar returns. *Appl. Opt.*, 20(2):211–220, Jan 1981.

EPJ Web of Conferences **176**, 05034 (2018) https://doi.org/10.1051/epjconf/201817605034
ILRC 28

ANALYSIS OF DIFFERENT CLOUD SIGNATURES USING MULTIWAVELENGTH RAMAN LIDAR RETRIEVALS

Simona Andrei[1], Flori Ţoancă[1], Anca Nemuc[1], Alex Dandocsi[1, 2], Livio Belegante[1], Doina Nicolae[1]

[1] *National Institute of R&D for Optoelectronics, Bucharest-Măgurele, 077125, Romania, simona.andrei@inoe.ro*

[2] *University Politehnica of Bucharest, Bucharest, 060042, Romania*

ABSTRACT

The present study is focused on microphysics analysis of different cloud types observed by a multi-wavelength Raman lidar. Particle's content from a wide variety of cloud types, produced in various atmospheric environments, were investigated using data provided by lidar and validated using ECMWF's ERA Interim reanalysis. The results emphasized the capability of lidar instruments to detect multiple cloud layers, and to discriminate between ice particles and water droplets found within the studied clouds.

1 INTRODUCTION

Clouds represent one of the most important components of weather and climate system, having an extensive influence on Earth's radiation budget. Together with aerosols, they contribute with the largest uncertainty in estimating and interpretations of the Earth's changing energy budget [1]. The evolution of a cloud is determined by the balance between dynamical, radiative and microphysical processes that define the manner in which the atmospheric radiation is modulated [2]. Clouds and cloud systems are organized by larger-scale circulations into different regimes such as deep convection, or midlatitude storm tracks guided by the tropospheric westerly jets [1]. Depending on the atmospheric environment features, clouds are composed of water droplets, ice crystals or a mixture of both [3]. The phase partition of condensed water into liquid droplets and ice crystals is temperature depending, but there is much uncertainty regarding the extent to which ice and liquid water coexist, and how to represent the ice-to-liquid water ratio below 0^0C. Observations have emphasized the presence of liquid water at temperatures as cold as - 30°C to - 40°C [4]. Remote sensing observations were used in different campaigns to study cloud types, and their features such as liquid/ice water content or ice crystals [5]. Lidars are able to probe few tens of centimeters to a few tens of meters above cloud base where the maximum cloud supersaturation occurs [6]. Lidar depolarization was used to distinguish between liquid and solid phases of water [7]. Ice crystals, considered non-spherical particles can exhibit depolarization ratios greater than 20%, and can be easily distinguished from cloud droplets or super cooled droplets with lower depolarization ratios [8].

This study aims to correlate the information from lidar observations and ERA Interim reanalysis in order to assess and validate the cloud particle types measured in different atmospheric environments.

2 METHODOLOGY

The observations of clouds have been performed by a multiwavelenth depolarization Raman lidar [9] located at Bucharest-Măgurele (44.35^{0}N, 26.03^{0}E, 93m ASL). Volume linear depolarization ratio (VDR) was used to discern between quasi-spherical particles with low depolarization ratios and non-spherical particles with higher depolarization ratios [10]. High temporal and vertical resolution of this parameter can provide information on cloud dynamics and structure. For water droplets the VDR is low, since they do not produce depolarization, while for ice crystals the VDR can go well above 40% [11]. By using lidar VDR, a first estimation of ice-to-liquid water ratio had been extracted: 0% for water droplets to 100% for ice crystals.

ECMWF - ERA Interim reanalysis data [12] were used to investigate the meteorological context and the atmospheric behavior with respect to the cloud particles (ice, super cooled water droplets and water droplets). The analysis has been focused on profiles of temperature (T), specific cloud ice water content (CIWC) and specific cloud liquid water content (CLWC).

3 RESULTS

Three cases of different cloud types were investigated in order to establish the correlations between lidar's optical parameters and meteorological parameters.

3.1 Study case 1 (May 16th 2014)

Lidar measurements performed on May 16th 2014 spotted the presence of a cloud with a gradual vertical expansion (Figure 1). The cloud base had a fast decreasing from 5 to 2 km altitude and, at the end of measurements lidar has detected the initiation of precipitation. Time series of VDR show values ranging from 6 to 18% corresponding to water droplets up to 35% corresponding to ice crystals.

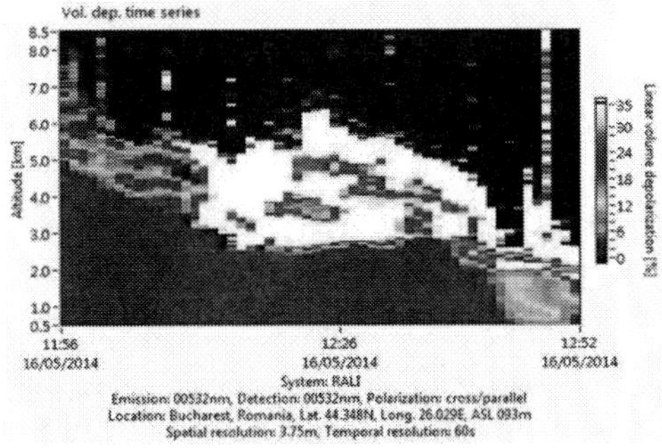

Figure 1 Lidar time series of VDR for May 16th 2014

Meteorological analysis had revealed the presence of a strong cyclonic cut-off with direct influence on Romanian territory. Air circulation had enhanced the warm and humid air mass advection over Bucharest-Măgurele (not shown here) which enabled the development of a Cumulonimbus cloud.

The typical shape of the cloud development with updrafts and downdrafts passing the measurement site is emphasized in Figure 2a. Ice-to-liquid water ratio indicates the presence of water droplets (0-20%) within the anvil cloud and precipitation regions, while a mixture of water droplets, super cooled water and ice crystals (20-100%) has been observed between 12:11 and 12:41 UTC. Thermal profile for Bucharest-Măgurele (Figure 2b) shows a strong decreasing with the altitude reaching -49^{0}C on 300hPa. The profiles of specific ice and liquid water content (Figure 2c) shows a maximum of CLWC at 900hPa, while CIWC profile shows a remarkable increasing between 775 and 250hPa with a maximum value at 350hPa.

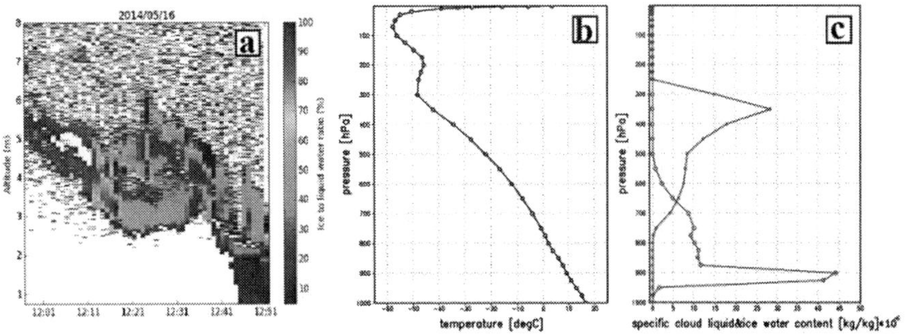

Figure 2 Time series of ice to liquid water ratio from lidar measurements (a), profile of T^{0}C (b) and CIWC (blue line) & CLWC (red line) (c) from ECMWF ERA - Interim reanalysis for May 16, 2014 [12TC]

3.2 Study case 2 (May 14th, 2015)

Lidar measurements performed during the evening hours of May 14th, 2015 (Figure 3) had noticed the passing of a cloud with a decreasing base from 10 to 2 km. Time series of VDR show high values between 18:00 and 19:00 UTC time interval.

Time series of ice-to-liquid water ratio (Figure 4a) reveals the presence of a pre-frontal cumuliform development, with values ranging from 0 to 60% corresponding to the presence of warm and cooled water droplets.

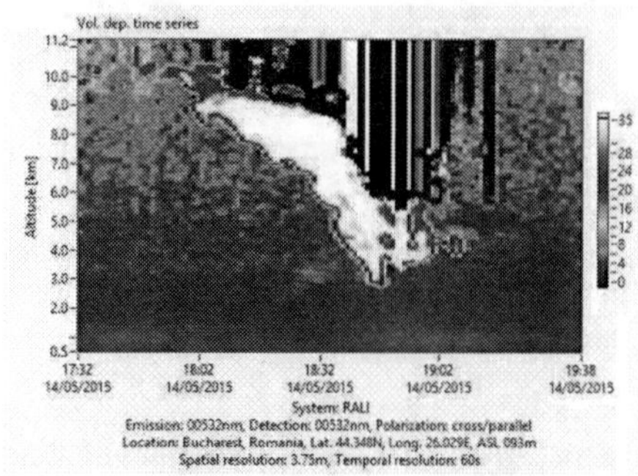

Figure 3 Lidar time series of VDR for May 14th 2015

Thermal profile (Figure 4b) emphasizes the presence of warm dry air within the lower troposphere (0 to 800hPa). The thermal decreasing with the altitude reached -60^{0}C at the top of troposphere. Vertical distribution of specific ice and liquid water content (Figure 4c) indicates two layers with increasing values: one, between 800 and 500hPa, and the other one, between 400 and 200hPa. Within the first layer CLWC presents a maximum value at 650hPa and a maximum value for CIWC at 600hPa. A weaker peak of CIWC is indicated within the second layer, between 300 and 25hPa.

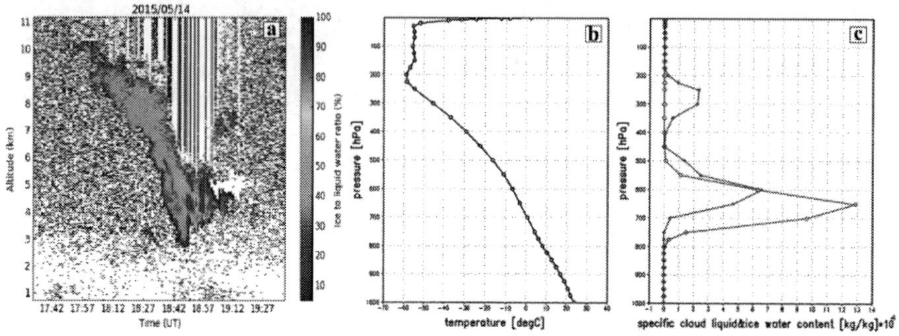

Figure 4 Time series of ice to liquid water ratio from lidar measurements (a) profile of T^{0}C (b) and CIWC (blue line) & CLWC (red line) (c) from ECMWF ERA - Interim reanalysis for May 14, 2015 [18TC]

3.3 Study case 3 (May 19th, 2016)

Measurements performed on May 19th, 2016 emphasize the presence of Cirrus cloud having its base at around 8 km at the beginning of the measurement and decreasing at around 6.5 km during the last hour. The VDR (Figure 5) shows low values (4 to 14%) indicating the presence of spherical particles (water droplets). At the level of 4 km a Cumulus cloud can be observed (red circle) developed in the middle of an aerosol layer present between 2.5 and 4 km (not shown here).

Ice to liquid water ratio (Figure 6a) shows low values (0 to 20%) indicating the dominance of cold water particles. Thermal profile (Figure 6b) emphasizes the presence of warm dry air within the lower troposphere (0 to 800hPa) and the decreasing with the altitude reached -56^{0}C at the tropopause level.

Figure 5 Lidar time series of VDR for May 19th 2016

Vertical distribution of atmospheric humidity (Figure 6c) indicates increasing values within two layers, first one - between 850 and 550hPa, and the second one – between 400 and 250hPa. Within the first layer, CIWC presents a maximum value at 600hPa and a maximum value of CLWC at 650hPa. Within the second layer, discreet peak of CIWC content presents a maximum at 350hPa

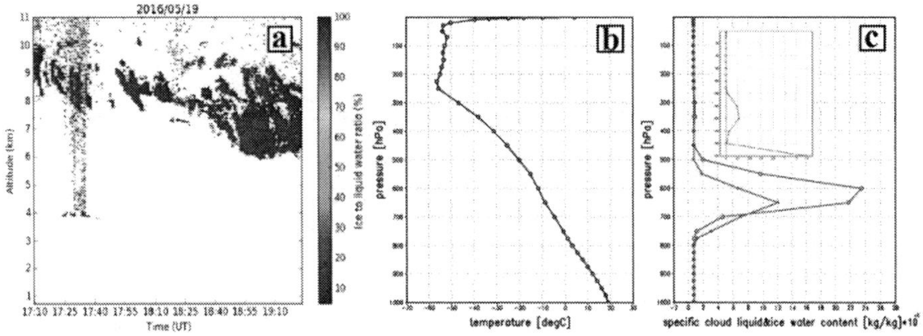

Figure 6 Time series of ice to liquid water ratio from lidar measurements (a profile of T⁰C (b) and CIWC (blue line) & CLWC (red line) (c) from ECMWF ERA - Interim reanalysisfor May 19th, 2016 [18TC]

4 CONCLUSIONS

The profiles presented in this study shows the capability of lidar instruments to detect different cloud signatures. Based on the VDR the lidar instruments are able to distinguish between spherical (water droplets) and non-spherical particles (ice particles). The ice to liquid water ratio profiles from lidar measurements and specific cloud ice and liquid water content from reanalysis data are in good agreement. Synoptic meteorological contexts explain well the cloud types detected by lidar. Vertical distributions of cloud particles for the analyzed cases are consistent with those mentioned by theory [3].

ACKNOWLEDGEMENTS

The research leading to these results has received funding from ANCS Core program, project no. 16.40.01.01 and the E.U. Horizon2020 Programme (GA 692014) - ECARS.

References

[1] Intergovernmental Panel on Climate Change, 2013: Chapter 7 of the 5th assessment report of the Intergovernmental Panel on Climate Change.

[2] Shuppe, M.D., et al., 2008: A focus on mixed-phase clouds: The status of ground-based observational methods, *Bull. Amer. Meteor. Soc.*, **89**, 1549–1562.

[3] Pruppacher, H. R., Klett, J. D., 1997: Microphysics of clouds and precipitation (2nd ed.), Springer.

[4] Hogan, R. J., et al., 2003: Characteristics of mixed-phase clouds. I: Lidar, radar and aircraft observations from CLARE'98, *Q. J. R. Meteorol. Soc.*, **129** (592), 2089–2116.

[5] Myagkov, A., et al., 2016: Relationship between temperature and apparent shape of pristine ice crystals derived from polarimetric cloud radar observations during the ACCEPT campaign, *Atmos. Meas. Tech.*, **9**, 3739–3754.

[6] Bühl, J., et al., 2013: Towards a quantitative characterization of heterogeneous ice formation with lidar/radar: Comparison of CALIPSO/CloudSat with ground-based observations, Geophys. Res. Lett., 40, 4404–4408, doi:10.1002/grl.50792.

[7] Sassen, K., 1991: The polarization lidar technique for cloud research: a review and current assessment, *Bull. Am. Meteor. Soc.*, **72**, 1848-1866.

[8] Pal, S. R., and Carswell, A. I., 1973: Polarization properties of lidar backscattering from clouds, *Appl. Optics*, **12** (7), 1530-1535.

[9] Klett, J. D., 1985: Lidar inversion with variable backscatter/extinction ratios, *Appl. Optics*, **24**, 1638–1643.

[10] Freudenthaler, V. 2016: About the effects of polarising optics on lidar signals and the Δ90 calibration, *Atmos. Meas. Tech.*, **9**, 4181-4255.

[11] Groβ, S. et al.., 2013: Aerosol classification by airborne high spectral resolution lidar observations, *Atmos. Chem. Phys.*, **13**(5), 2487–2505.

[12] Dee D.P., et al., 2011: The ERA-Interim reanalysis: configuration and performance of the data assimilation system. *Q. J. R. Meteorol. Soc.* **137**: 553–597.

DIFFUSE SUNLIGHT BASED CALIBRATION OF THE WATER VAPOR CHANNEL IN THE UPC RAMAN LIDAR

Constantino Muñoz-Porcar[1]*, Adolfo Comeron[1], Michaël Sicard[1], Ruben Barragan[1], David Garcia-Vizcaino[1], Alejandro Rodríguez-Gómez[1], Francesc Rocadenbosch[1]

*[1]Universitat Politècnica de Catalunya, Barcelona, Spain, *constan@tsc.upc.edu*
[2]Ciències i Tecnologies de l'Espai - Centre de Recerca de l'Aeronàutica i de l'Espai / Institut d'Estudis Espacials de Catalunya (CTE-CRAE / IEEC), Universitat Politècnica de Catalunya, Barcelona, Spain

ABSTRACT

A method for determining the calibration factor of the water vapor channel of a Raman lidar, based on zenith measurements of diffuse sunlight and on assumptions regarding some system parameters and Raman scattering models, has been applied to the lidar system of Universitat Politècnica de Catalunya (UPC; Technical University of Catalonia, Spain). Results will be analyzed in terms of stability and comparison with typical methods relying on simultaneous radiosonde measurements.

1 WATER VAPOR MIXING RATIO

The water vapor mixing ratio (*WVMR*) in the atmosphere is defined as the ratio between the water vapor mass in a given volume and the mass of the rest of the volume. As the water vapor mass is always going to be much lesser than the mass of the rest of molecules in the volume, this latter mass can be approximated by the mass of the dry air. Taking into account, moreover, that the molecular weight of water molecules is 18, the mean molecular weight of the air is 28.96 (page 4 of ref. [1]) and that the nitrogen (N_2) volume proportion in the air (hence the proportion of its molecule number concentration) is 78.08%, the mixing ratio can be expressed as

$$\omega = \frac{0.7808 \times 18 \times N_W}{28.96 \times N_N}, \qquad (1)$$

where N_W is the number concentration of water vapor molecules and N_N the number concentration of nitrogen molecules.

2 WATER VAPOR MIXING RATIO FROM RAMAN LIDAR SIGNALS

The detected signal voltage (or the count rate per bin) from a given range R in the nitrogen and water vapor Raman channels of a lidar instrument is given by the lidar equation:

$$S_X(R) = \frac{E\, A_r\, c\, K_X O_X(R) N_X(R) \dfrac{\partial \sigma_{X\,eff}}{\partial \Omega}}{2R^2} \exp\left\{ -\int_0^R \left[\alpha(r,\lambda_0) + \alpha(r,\lambda_X) \right] dr \right\}, \qquad (2)$$

where X is either W or N (referring respectively to water vapor and nitrogen channels); E is the emitted pulse energy at λ_0 (355 nm in our case); A_r is the effective receiving area; c is the speed of light; K_X is, for each detected wavelength, a system constant taking into account the end-to-end transmittance of the receiving optics, the photodetector responsivity and a transducer constant; $O_X(R)$ is the overlap function profile; the exponential term takes into account the extinction along the propagation path of both the emitted pulse (λ_0) and the scattered Raman-shifted radiation (λ_X); $\partial \sigma_{X\,eff} / \partial \Omega$ is the effective differential Raman backscatter cross-section of the species, which takes into account the frequency selection produced in the narrowband interference filters (see Eq. (6)) [2]. No temperature dependence of Raman scattering has been considered in this formulation.

From Eqs. (1) and (2):

$$\omega(R) = 0.4853 \times \frac{K_N}{K_W} \frac{O_N(R)}{O_W(R)} \frac{\dfrac{\partial \sigma_{N\,eff}}{\partial \Omega}}{\dfrac{\partial \sigma_{W\,eff}}{\partial \Omega}} \times \frac{S_W(R)}{S_N(R)} \times \exp\left\{ \int_0^R \left[\alpha(r,\lambda_W) - \alpha(r,\lambda_N) \right] dr \right\}. \qquad (3)$$

Differences in transmission at the Raman frequencies accounted for in the exponential term can be neglected in common situations: Rayleigh contribution, computed using atmospheric models, results in differences below 3% for altitudes where water vapor content is meaningful, while differences in aerosol extinction would only produce significant errors in heavily aerosol loaded atmospheres ($AOD > 2$) [3]. If, moreover, identical overlap functions are assumed, the *WVMR* can be obtained multiplying the ratio between the signal profiles corresponding to water vapor and nitrogen Raman channels by a calibration factor that has to be obtained for each lidar instrument. This factor depends on both the effective Raman backscatter cross-sections and the system constants in each channel:

$$\omega(R) = F_{CAL} \frac{S_W(R)}{S_N(R)}, \tag{4}$$

with

$$F_{CAL} = 0.4853 \times \frac{\dfrac{\partial \sigma_{N\,eff}}{\partial \Omega}}{\dfrac{\partial \sigma_{W\,eff}}{\partial \Omega}} \frac{K_N}{K_W}. \tag{5}$$

The effective Raman backscatter cross-section can be calculated in each channel as a summation of the spectral lines weighed by the normalized transmission of the corresponding interference filter at each frequency [2]:

$$\frac{\partial \sigma_{X\,eff}}{\partial \Omega} = \sum_i \frac{\partial \sigma_X(\lambda_i)}{\partial \Omega} t_X(\lambda_i), \tag{6}$$

where λ_i is the wavelength corresponding to each line of the Raman spectrum, $\partial \sigma_X(\lambda_i)/\partial \Omega$ and $t_X(\lambda_i)$ being respectively the corresponding differential backscatter cross-section and the normalized transmission function of the interference filter for the specie X.

Filter transmission functions have been provided by the filter manufacturer, while differential Raman cross sections for each spectral line have been computed following [4] in the case of water vapor and [5] in the case of Nitrogen. As an example, Figure 1 shows the relative distribution of the spectral lines for nitrogen at 275K and the transmission curve of the corresponding interference filter used in the UPC system. In our case, the effective differential Raman backscattered cross-sections have resulted 2.41×10^{-34} m²/sr for nitrogen (computed between 384 and 390 nm) and 7.03×10^{-34} m²/sr for water vapor (computed between 405 and 410 nm). The ratio between both effective cross-sections appearing in Eq. (5) is thus 0.34.

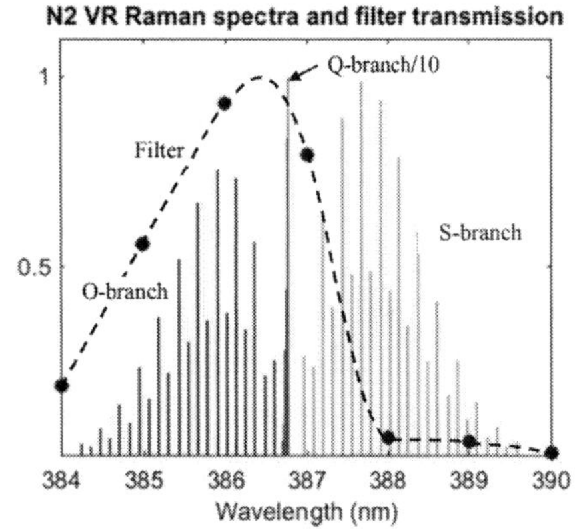

Figure 1 N_2 Raman spectra and interference filter transmission curve.

3 CALIBRATION FACTOR USING DIFFUSE SUNLIGHT RADIATION

The ratio between system constants K_X needed in Eq. (5) to find the calibration factor can be derived from the measurement of diffuse sunlight radiation in N_2 and water vapor channels of the lidar instrument [6]. Once the "dark" offset –due to electronics and photodetector dark current in the case of an analog channel, or to dark counts for a photoncounting channel– has been removed from the detected signal, we obtain a background-radiation induced DC voltage S_{BX} that is

$$S_{BX} = K_X A_r \Omega_X B_X L_{\lambda X}, \tag{7}$$

where X stands again for either N_2 or H_2O, Ω_X is the receiver field of view for either the N_2 or the H_2O Raman channel, B_X is the interference filter effective bandwidth, and $L_{\lambda X}$ is the spectral radiance at the corresponding Raman-shifted wavelength. Then we have

$$\frac{S_{BN}}{S_{BW}} = \frac{K_N}{K_W} \frac{\Omega_N}{\Omega_W} \frac{B_N}{B_W} \frac{L_{\lambda N}}{L_{\lambda W}}. \tag{8}$$

Eq. (8) shows that if the ratios between Ω_X, B_X and $L_{\lambda X}$ are known, the ratio between the system constants K_X, required in Eq. (5) for calculating the calibration factor F_{CAL}, can be obtained by computing the ratio of the measured background-radiation induced offsets S_{BX}:

$$\frac{K_N}{K_W} = \frac{\Omega_W}{\Omega_N}\frac{B_W}{B_N}\frac{L_{\lambda W}}{L_{\lambda N}}\frac{S_{BN}}{S_{BW}} \quad (9)$$

We have assumed as an approximation that the field of view in both channels is the same. Regarding the data provided by the manufacturer of the filters it has been also assumed that the effective bandwidths ratio B_W/B_N is 0.91. The ratio between spectral radiances, which depends on the day of the year, the time and the aerosol loading conditions, has been estimated for each individual calibration using a radiative transfer model (GAME [7]) and auxiliary measurements (lidar, sun photometer, radiosondes, etc.). Finally, the background-induced signals S_{BN} and S_{BW} can be obtained either directly from diffuse sunlight radiation measurements or from lidar signals at 25-30 km, where backscatter contribution can be assumed to be negligible.

A set of calibration measurements during 2016 has been used to obtain the ratio of background photo-counting mode signals in the lidar instrument. Simulations to obtain in each case the ratio of spectral radiances have been also performed. With these results and with the rest of estimated parameters, the calibration factor F_{CAL} has been calculated in each case using Eqs. (5) and (9). Figure 2 shows the resulting values for a set of 18 calibrations performed in February 23th between 8:00 UTC and 17:30 UTC. The relative detected illumination intensity in each measurement is plotted too. In this case background offsets where obtained from lidar signals between 25 and 30 km. Table 1 Calibrations between Feb. and June, 2016.

(N: Number of calibrations; SD: standard deviation)
**Background values from lidar signals at 25 - 30 km*

Date	*N*	F_{CAL}	*SD*
22/02/2016*	13	0.230	2.56%
23/02/2016*	18	0.225	2.10%
31/03/2016*	13	0.206	3.31%
11/04/2016	1	0.208	-
13/04/2016	1	0.207	-
18/04/2016	1	0.205	-
03/05/2016	1	0.207	-
20/06/2016	1	0.209	-
23/06/2016	9	0.209	1.47%

4 COMPARISON WITH CALIBRATIONS USING RADIOSONDE DATA

shows the results corresponding to other calibrations between February and June 2016.

All sets of multiple calibrations performed in the same day from dawn to sunset (22/02, 23/02, 31/03 and 20/06) show no significant dependence on illumination intensity or solar zenith angle. Results show in general good short-term stability: The relative standard deviation during the 7 dates of calibrations between 31/03 and 23/06 is only 0.64%. On the other hand, between calibrations made in February and all the rest, starting in 31/03 and ending in 23/06, a significant decrease of the calibration factor can be observed (~10%). This change, produced by at the moment unidentified changes in the instrument conditions, indicate that periodic calibrations should be performed.

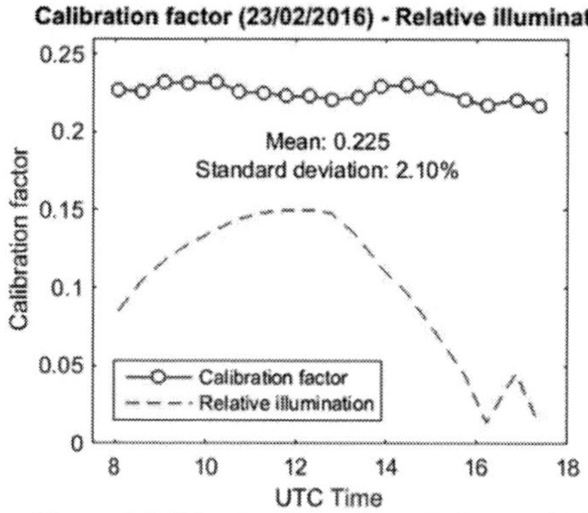

Figure 2 Calibration factor F_{CAL} (February 23th, 2016). Relative illumination is also plotted.

Table 1 Calibrations between Feb. and June, 2016.
(N: Number of calibrations; SD: standard deviation)
**Background values from lidar signals at 25 - 30 km*

Date	*N*	F_{CAL}	*SD*
22/02/2016*	13	0.230	2.56%
23/02/2016*	18	0.225	2.10%
31/03/2016*	13	0.206	3.31%
11/04/2016	1	0.208	-
13/04/2016	1	0.207	-
18/04/2016	1	0.205	-
03/05/2016	1	0.207	-
20/06/2016	1	0.209	-
23/06/2016	9	0.209	1.47%

5 COMPARISON WITH CALIBRATIONS USING RADIOSONDE DATA

The calibration factor F_{CAL} can be also obtained by comparison with simultaneous measurements from a reference instrument [3]. In our case, 150 minutes, night-time lidar measurements and almost collocated, simultaneous radiosondes data have been used to estimate the calibration factor of the lidar instrument in dates close to cases in section 3. The estimated calibration factor has been obtained from the Raman signals $S_N(R)$, $S_W(R)$ and the *WVMR* values provided by the radiosonde $\omega_{RS}(R)$, which is taken as reference. From Eq. (4):

$$F_{\mathrm{CAL}} = \frac{\omega_{RS}(R) S_N(R)}{S_W(R)} \quad (10)$$

Two cases have been studied: The first one in January 28th, 2016, which is relatively close to February background-based calibrations, and the second one in June 20th, which is coincident with another one. Figure 3 shows, in the latter case, the calibration factor as a function of the altitude where the estimation has been computed and its mean value and standard deviation within a useful calibration range. The upper limit of this useful range is determined by the region where water vapor signal is strong enough for good estimations. Lidar and radiosonde profiles are usually not coincident either in the lowest part of the atmosphere, which obliges to limit also the minimum height of the calibration range. Table 2 summarizes the results of both tests, which show good agreement with the ones obtained using diffuse sunlight calibrations.

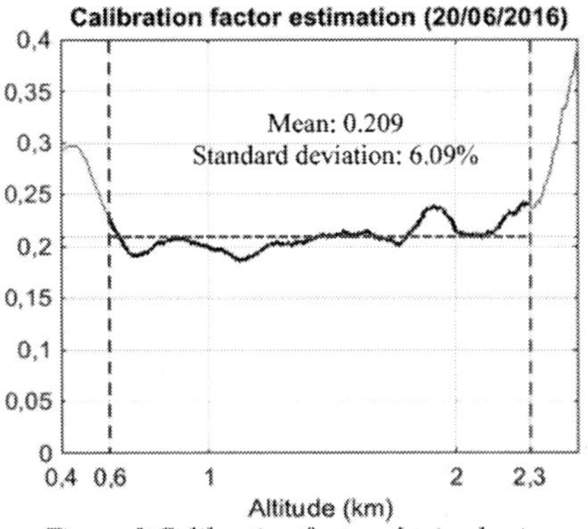

Figure 3 Calibration factor obtained using a radiosonde as reference instrument (20/06/2016).

Table 2 Comparison between calibrations.

Radiosonde			Diffuse sunlight	
Date	**Mean**	***SD***	**Date**	F_{CAL}
28/01/2016	0.226	8.32%	22/02/2016	0.230
20/06/2016	0.209	6.09%	20/06/2016	0.209

6 CONCLUSIONS

The calibration factor required for water vapor mixing ratio measurements has been estimated from direct measurements of diffuse sunlight radiation. Results show good short-term stability, no significant dependence on illumination intensity or solar zenith angle and good agreement with the calibration factor obtained with the classical method based on the use of simultaneous radiosonde measurements. Mid-term changes of the calibration factor reveal the need of performing periodic calibrations of the instrument to assure accurate water vapor measurements. The method can be therefore used as an easy and reliable alternative to the typical calibration relying on radiosondes.

ACKNOWLEDGEMENTS

This work has received funding from the European Union H2020 programme under ACTRIS-2 project (GA 654109), the Spanish Ministry of Economy and Competitiveness – European Regional Development Funds under TEC2015-63832-P project, and from Generalitat de Catalunya (Grup de Recerca Consolidat) 2014-SGR-583.

References

[1] R. M. Measures, Laser remote sensing: fundamentals and applications. Malabar, Fla.: Krieger, 1992.

[2] D. D. Venable et al., Lamp mapping technique for independent determination of the water vapor mixing ratio calibration factor for a Raman lidar system., Appl. Opt., 50 (23), pp. 4622–4632, 2011.

[3] I. Mattis et al., Relative-humidity profiling in the troposphere with a Raman lidar., Appl. Opt., 41(30), pp. 6451–62, 2002.

[4] G. Avila et al., The Raman spectra and cross-sections of H2O, D2O, and HDO in the OH/OD stretching regions, J. Mol. Spectrosc., 228(1), pp. 38–65, 2004.

[5] F. Liu and F. Yi, Lidar-measured atmospheric N_2 vibrational-rotational Raman spectra and consequent temperature retrieval, Opt. Express, 22(23), p. 27833, 2014.

[6] V. Sherlock et al., Methodology for the independent calibration of Raman backscatter water-vapor lidar systems, Appl. Opt., 38(27), pp. 5816–5837, 1999.

[7] P. Dubuisson et al., Water vapor retrieval over ocean using near-infrared radiometry, J. Geophys. Res., 109(D19), Oct. 2004.

DETECTION OF SAHARAN MINERAL DUST AEROSOL TRANSPORT OVER BRAZILIAN NORTHEAST THROUGH A DEPOLARIZATION LIDAR

Anderson G. Guedes[1]*, Eduardo Landulfo[2], Elena Montilla-Rosero[3,5], Fábio J.S. Lopes[2], Judith J. Hoelzemann[3], José Henrique Fernandez[3], Marcos P. A. Silva[1], Renata S. S. Santos[3], Juan L. Guerrero-Rascado[4], Lucas Alados-Arboledas[4]

[1] *Graduate Program in Climate Sciences, Federal University of Rio Grande do Norte, Brazil, andersongguedes@yahoo.com.br*
[2] *Institute of Energy and Nuclear Research, Brazil*
[3] *Department of Atmospheric and Climate Sciences, Federal University of Rio Grande do Norte, Brazil*
[4] *Andalusian Institute for Earth System Research, University of Granada, Spain*
[5] *School of Science, Physical Sciences Department, EAFIT University, Colombia*

ABSTRACT

In this study we present results of linear volume depolarization ratio profiles obtained by a depolarization lidar in operation in Natal, Brazil. The DUSTER system has 4 channels, namely: 1064, 532 s/p and 355 nm. This system is calibrated with a half-wave plate using the $\Delta 90^0$ methodology. The data obtained from this system is correlated with AERONET sunphotometer data, and, when available, CALIPSO satellite data. In addition a trajectory model (HYSPLIT) is used to calculate backward trajectories to assess the origin of the dust polluted air parcels. The objective is to create a transport database of Saharan dust.

1 INTRODUCTION

Long range transport of dust is a well-known phenomenon important to the atmospheric radiative budget with global impact on climate, air quality and human health [1-5]. It is estimated that about 3 billion tons of dust from the Sahara are transported each year and there are considerable suspicions about the influence of these aerosols on the Earth's radiative balance. Thus, creating an observational network for such an environmental occurrence is a challenge which requires a lot of resources. The existing observation networks [6] are concentrated regionally and there still are some gaps to be filled in order to obtain full global coverage.

In South America LALINET is one of these networks with the goal of measuring biomass burning plumes, volcanic ashes and dust transport into and out of the continent and more recently a lidar system (DUSTER) was developed at Natal-RN (5°50′29′′S, 35°11′57′′W, sea level), in Brazil's northeastern region. DUSTER is a depolarization lidar system originally designed to detect and identify aerosol dust layers from Saharan Desert which are transported across the Atlantic Ocean. These observations are in coordination with an AERONET sunphotometer installed in Natal and also with CALIPSO's observation of these transport events.

The city of Natal is the capital of the state of Rio Grande do Norte (RN), located at the coast in Northeastern Brazil. The city and surroundings comprise a large diversity of different ecosystems, such as dune landscape, mangrove and Atlantic forests. This place has an area of approximately 167 km^2 and an increasing population of about 877,662 inhabitants, according to the Brazilian Institute of Geography and Statistics (IBGE) in 2016. The climate of Natal is considered tropical wet with dry summer (austral) and an annual mean temperature of 26 ^{0}C. The temperature may drop to 24 ^{0}C in winter and reach about 30 ^{0}C during summer, according to the National Institute for Meteorology (INMET). Solar irradiation yields elevated indices of up to 3000 hours of insolation per year, the average precipitation is 1465 mm/year (INMET) and occurs primarily from March through June with a maximum in April (265 mm).
The so called window of opportunity to observe dust transported into the continent at these latitudes is strongly influenced by the ITCZ (Intertropical Convergence Zone) migration

https://doi.org/10.1051/epjconf/201817605036

putting this region under a variety of circulation patterns similar to those observed in the Caribbean counterpart around Barbados [5]. Due to these mechanisms the dust transport into this region occurs most likely during December, January and February (DJF) at ≈ 5°N and the less likely chance to observe this phenomenon should be in June, July and August (JJA) at ≈ 20° N. This migration is strongly affected major oscillations such as ENSO and El Niño. The predominant observed transport takes place westwards of the African continent under strong influence of trade winds with some portions towards north over the Mediterranean sea. This pattern of transport follows a trend towards the Caribbean in JJA months and towards South America (SA) in DJF. This trend was tracked and confirmed by CALIPSO [7] and a seasonal dependence was also observed in the measured plume altitude, which was higher in JJA and slightly above the Planetary Boundary Layer (PBL) in DJF. In SA the transport over the Amazon basin is a known process [8].

2 METHODOLOGY

The DUSTER has 4 channels at 1064 nm, 532 nm (p), 532 nm (s) and 355 nm coupled to a 300 mm Cassegranian Carbon-Plastic telescope. Each channel has interference filters to block sunlight background with bandwidths of (2 nm FWHM at each channel). The laser source emits at 1064 nm, 532 nm and 355 nm, with energy per pulse of about 300 mJ, 150 mJ and 90 mJ, respectively. The laser repetition rate is 20 Hz and pulse duration is (7±2) ns. The detection box has an Avalanche Photo-diode (APD) for 1064 nm and 3 photomultipliers, for both polarizations at 532 nm and 355 nm. Their signal is digitized by a transient recorder operating in analog and photoncounting modes (Licel model TR20-80, bandwidth□10 - 250 MHz) (Figure 1).

The depolarization measurement capability of DUSTER let to distinguish and to identify unambiguously the volume linear depolarization ratio parameters. This parameter is defined as the ratio of the cross polarized lidar return signal to the parallel-polarized backscatter signal, the signals are selected in the optical reception box by means of a polarizing beam splitter cube (PBC).

To retrieve the linear volume depolarization ratio $\delta_v$the$\Delta 90^0$calibration [9] method was implemented using a$\lambda/2$plate (HWP) for retrieving the amplification factor η^*. The HWP was located in front of laser emitter and several measurements were done under clear sky conditions to get the amplification due to polarization changes.

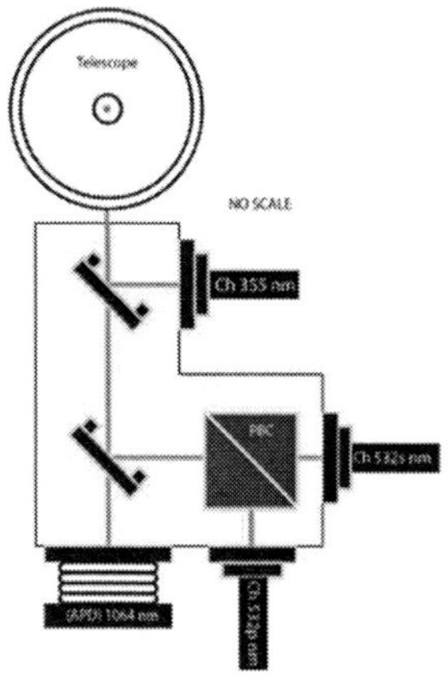

Figure 1. Sketch of the receiver unit of lidar system DUSTER with the depolarization channel 532 nm, labeled.

2.2 Sunphotometer

The AErosol RObotic NETwork (AERONET) is an international network of ground-based sun photometers that provides automatic sun and sky scanning measurements. The sunphotometer (model CIMEL CE318 radiometer included in AERONET) located at Natal is setup about 200 m away from the DUSTER Lidar system. This performs radiances measurements in order to retrieve the Aerosol Optical Thickness for aerosols at several wavelengths. Besides the AOD, it is possible to obtain the aerosol size distribution (AE), the phase function, single scattering albedo and extinction-to-backscatter ratio. This procedure assures measurement accuracy to within 13%.

3 RESULTS

DUSTER started its operation in February 2016 following LALINET's measurement protocol, with regular measurements on most days of week. Initially, the measurements were carried out to check system performance and depolarization calibration procedure. By$\Delta 90^0$calibration a η^* value of 0.35 was retrieved using laser line filters to 532 nm with FWHM bandwidth of 2 nm. On March 11 a distinctive layer was observed at 2.5 to ≈ 3 km which was identified as mostly consisting of dust aerosols, due to the DUSTER measured a volume linear depolarization ratio in the dust layer (2.5 to 3 km height) of the 0.17±0.01 for 532 nm (Figure 2).

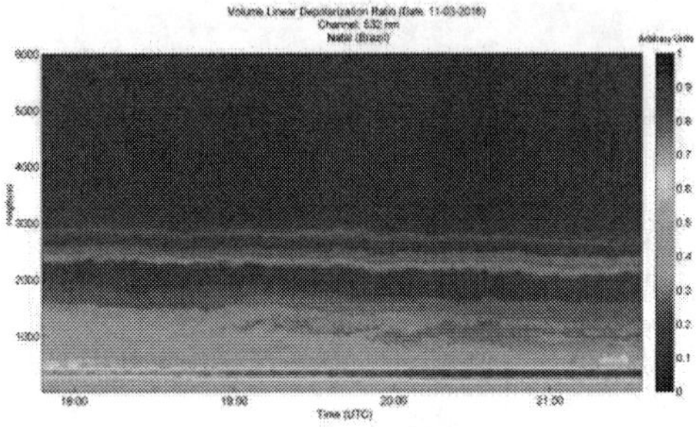

Figure 2. Volumetric Depolarization Ratio obtained with DUSTER on March 11 (17:48 UTC-21:35UTC) analyzing the upper plume with a strong possibility of being dust transported over the ocean.

One local aspect worth mentioning is the fact that observing these plumes becomes a challenge as they move away due to a steady wind field present in the Natal area, certainly an aspect easily recognizable by the presence of trade Winds. The presence of low clouds also gives an additional difficulty to the observing site.

Synoptically, according to the automatic INMET meteorological station in Natal, this day was a clear day without precipitation, with an average temperature of 29°C, ranging from 27°C in the early morning to 29°C at 21 UTC and a relative humidity of 63-88%. The predominant wind direction were from Southeast and its velocity 3.7 m/s (18 UTC) - 3.1 m/s (22 UTC). Solar radiation were 2476 kJ/m^2 (18 UTC) - 78.41 kJ/m^2 (21 UTC).

The Natal AERONET sunphotometer station data allow to obtain diurnal Angström Exponent (AE) (note: days where data were collected) during the first semester of 2016. Some periods in February, mid March and early June are corresponding to dust aerosol (Figure 3). In April it was not possible to obtain data with the instrument due to technical problems. The low AE values suggest the presence of an aerosol compatible with desert aerosols.

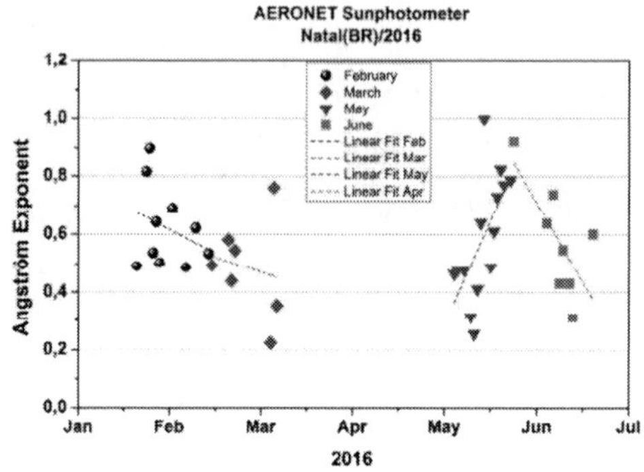

Figure 3. Variation of AE for the months of February to June.

In order to validate the source of these dust plumes a HYSPLIT trajectory model simulation was carried out to provide back trajectories which ended on March 11 and described the air masses from 10 days back predominantly to the Northern African West Coast as shown in Figure 4.

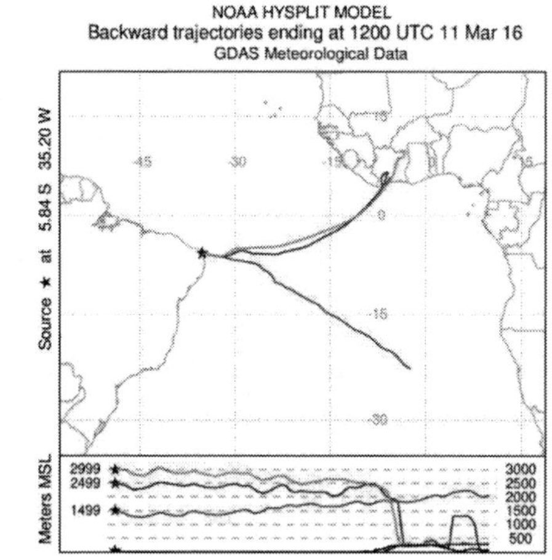

Figure 4. Dust track direction and trajectory from Saharan region towards Natal obtained with HYSPLIT Model.

The shortest distance from the CALYPSO satellite in relation to the DUSTER site for the day of March 11 was 194 km, a good data correlation was not recommendable.

4 CONCLUSIONS

The data obtained with DUSTER in the first halfof 2016 have shown episodes where the aerosols travelling over the city of Natal are compatible with dust. The 2017 campaign (Dec 2016, Jan 2017 and Feb 2017) and improvements in the depolarization calibration process will contribute to more accurate analysis with the equipment.

ACKNOWLEDGEMENTS

We acknowledge the *Conselho Nacional de Desenvolvimento Científico e Tecnológico* CNPq/MCTIC for the research grants under process number 400430/2014-2 and the research project CNPq 477713/2013-0. Also, we thank the *Coordenação de Aperfeiçoamento de Pessoal de Nível Superior* CAPES/MEC for their support via a doutorado-sanduiche research grant number 88881.133708/2016-01.

References

[1] Mahowald, N.M., Ballantine, J.A., Feddema, J., and Ramankutty, N., "Global trends in visibility for dust sources", Atmospheric Chemistry and Physics 7 (12), 3309-3339 (2007).

[2] Antón, M., Valenzuela, A., Román, R., Lyamani, H., Krotkov, N., Arola, A., Olmo, F. J., and Alados- Arboledas, L., "Influence of desert dust intrusions on ground-based and satellite-derived ultraviolet irradiance In Southeastern Spain", journal of Geophysical Research: Atmospheres 117 (D19), (2012).

[3] DeMott, P.J., Prenni, A.J., Liu, X., Kreidenweis, S.M., Petters, M.D., Twohy, C.H., Richardson, M. S., Eidhammer, T., and Rogers, D. C., "Predicting global atmospheric ice nuclei distributions and their impacts On climate", Proceedings of the National Academy of Sciences 107 (25), 11217-11222 (2010).

[4] Longueville, F. D., Hountondji, Y. C., Henry, S., and Ozer, P., "What do we know about e effects of desert Dust on air quality and human health in west Africa compared to other regions", Science of The Total Environment 409 (1), 1-8 (2010).

[5] Prospero, J. M., Collard, F.-X., Molinié J., and Jeannot, A., "Characterizing the annual cycle of african Dust transport to the caribbean basin and south america and its impact on the environment and air quality", Global Biogeochemical Cycles 28 (7), 757-773 (2014)

[6] Guerrero-Rascado, J. L., Landulfo, E., Antuña, J. C., de Melo Jorge Barbosa, H., Barja, B., Bastidas, A. E., Bedoya, A. E., da Costa, R.F., Estevan, R., Forno, R., Gouveia, D.A., Jiménez, C., Larroza, E. G., da Silva Lopes, F.J., Montilla-Rosero, E., Arruda Moreira, G., Nakaema, W. M., Nisperuza,D., D., M., Otero, L., Papandrea, S., Pallota, J. V., Pawelko, E., Quel, E. J., Ristori, P., Rodrigues, P. F., Salvador, J., S. Anchez, M. F., and Silva, A., "Latin American network (LALINET) For aerosol research: Diagnosis on network instrumentation", Journal of Atmospheric and Solar-Terrestrial Physics 138-139, 112-120 (2016).

[7] Z. Liu, A. Omar, M. Vaughan, J. Hair, C. Kittaka, Y. Hu, K. Powell, C. Trepte, D. Winker, C. Hostetler, R. Ferrare, and R. Pierce, Calipso lidar observations of the optical properties of saharan dust: a case study of long-range transport, J. Geophys. Res. 113 (2008), D07207.

[8] R. Swap, M. Garstang, S. Greco, R. Talbot, and P. Kallberg, Saharan dust in the amazon basin, Tellus 44B (1992), 133–149.

[9] Freudenthaler, V., "On the e effects of polarizing optics on lidar signals and the Δ90 calibration", Atmospheric Measurement Techniques 9 (9), 4181-4255 (2016).

EPJ Web of Conferences **176**, 05037 (2018) https://doi.org/10.1051/epjconf/201817605037
ILRC 28

APPLICATION OF A MULTIPLE SCATTERING MODEL TO ESTIMATE OPTICAL DEPTH, LIDAR RATIO AND ICE CRYSTAL EFFECTIVE RADIUS OF CIRRUS CLOUDS OBSERVED WITH LIDAR.

Diego Gouveia[1,*], Holger Baars[2], Patric Seifert [2], Ulla Wandinger[2], Henrique Barbosa[1], Boris Barja[3], Paulo Artaxo[1], Fabio Lopes[4], Eduardo Landulfo[4] and Albert Ansmann[2]

[1]*Applied Physics Department. Institute of Physics, University of Sao Paulo, Sao Paulo, SP, Brazil, *diego.gouveia@usp.br*
[2]*Leibniz Institute for Tropospheric Research (TROPOS), Leipzig, Germany*
[3]*Atmospheric Research Laboratory, University of Magallanes, Punta Arenas, Chile*
[4]*Laser Environmental Application Laboratory, Instituto de Pesquisas Energeticas e Nucleares, Av. Prof. Lineu Prestes 2242, Sao Paulo, Brazil 01258-010, Brazil*

ABSTRACT

Lidar measurements of cirrus clouds are highly influenced by multiple scattering (MS). We therefore developed an iterative approach to correct elastic backscatter lidar signals for multiple scattering to obtain best estimates of single-scattering cloud optical depth and lidar ratio as well as of the ice crystal effective radius. The approach is based on the exploration of the effect of MS on the molecular backscatter signal returned from above cloud top.

1 INTRODUCTION

Lidar systems are an indispensable tool in remote sensing of macro and microphysical properties of cirrus clouds, especially considering optically thin clouds (COD <0.3) that are not usually detected by Radar or other passive instruments [1] and correspond to about 80% of the total cirrus clouds occurrences in the tropics [2] and middle latitudes, where they were estimated to account for about half of the net cirrus radiative forcing [3].

However, it is well known that measurements with lidar are highly influenced by multiple scattering effects, where a substantial fraction of the photons scattered in the forward direction remains close to the receiver field of view (RFOV) where they are subject to subsequent scattering events until they are detected by the system [4]. Thus, the amount of photons detected that come from multiple scattering depends primarily on the width of the RFOV, the volume covered by it and the width and intensity of the diffraction peak. Therefore, the effects of multiple scattering are particularly intense in cloud measurements, which are optically thick media and have particles large in relation to the wavelength of the laser. This generates a narrow diffraction peak that remains within the RFOV, and can change the molecular signal measured by the lidar far above the top of the cloud [5].

By not considering the effects of multiple scattering on the retrieval of the cloud properties leads to a considerable underestimation of the extinction coefficient and the extinction-to-backsatter (lidar-) ratio (LR). Fortunately, we have available accurate forward models [6, 7] that allow us to calculate the intensity of multiply scattered photons by clouds in lidar signals.

In this work, we present an iterative approach to obtain the optical depth and average lidar ratio of cirrus clouds corrected for the effects of multiple scattering using the elastic backscatter signal, for the case that there is information about the size of the ice crystals as well as errors associated. As done by [5], a possible way to estimate the effective radius of ice crystals is studied by taking advantage of the distortions caused by multiple scattering effects

https://doi.org/10.1051/epjconf/201817605037

on the molecular signal above the top.

2 Multiple Scattering In Cirrus

There are several ways that one can obtain the optical depth and an average value for the lidar ratio (LR) of cirrus clouds by elastic backscattering lidar measurements. A simple way is to find the value of the LR in which the cloud optical depth calculated by the integration of the extinction coefficient profile computed by the Klett method (COD_{Klett}) equals the value of the optical depth calculated by the transmittance method (COD_{trans}). An equivalent way would be to find the value of the LR in which the backscatter ratio below and above the cloud equals 1. In any case, both the backscatter signal from within the cloud and the molecular above the top are affected by the multiple scattering effects, leading to smaller values of COD and LR than the real one.

To visualize the effects of multiple scattering on the lidar measurements, we used a model developed by Hogan [7] to simulate an attenuated elastic backscatter profile with a cirrus cloud between 13 and 15 km when measured by a UV lidar system (355nm) with laser divergence of 0.36 mrad and RFOV of 1.75 mrad (both full angles). We used extinction coefficients of 0.3 and 0.15 km^{-1}, with optical depth of 0.45, 30 sr lidar ratio and effective radius of 20 μm. Figure 1 shows the resulting profiles of attenuated backscatter coefficient for the single scattering case (SS) and for the case including the effect of multiple scattering (MS). It is possible to observe that above the cloud base the photons that were forward scattered and that remained in the RFOV generate an extra signal in the backscatter profile, influencing the signal even above the top of the simulated cloud (15 km). The ratio of the single and multiple scattering signal shows that for this case the single scattering backscattered photons may be as small as 70% of the actual signal measured by the lidar. We can also see that the SS/MS ratio has a dependence on the distance above the cloud, which occurs because the forward scattered photons have an angular distribution around the emission axis, gradually leaving the RFOV and thus distorting the molecular scattering expected by a purely molecular atmosphere, which is used to calibrate the lidar signal through the Rayleigh scattering theory.

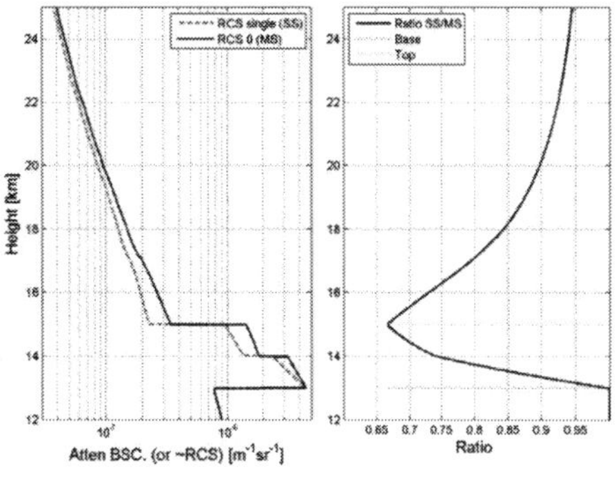

Figure 1: Simulated lidar signal with and without the effects of multiple scattering (left). Single to multiply scattered ratio (right).

3 Correction for Multiple Scattering

The abovementioned non-linear distortions in the molecular signal, which may be greater than 10% even at heights of 10 km above the top of the cloud, demonstrate that only with a MS correction for the whole lidar profile (not only within the cloud) it is possible to derive the optical properties through elastic lidar techniques. This can be done using the Hogan model [7] that is fast enough to be used in an iterative solution, using as the first input of the model an estimate of the extinction profile measured without any correction (iteration 0) and thus calculate a first approximation of the ratio SS/MS which is used to correct the lidar profile. After the first correction the new extinction profile is used to improve the estimation of the correction factor and so on in the following iterations (iteration 1,2,3 ...). Also, ice crystal size information from a lookup table, e.g., from [1], is needed to run the model. Figure 2 shows the extinction coefficient profile for the uncorrected signal (0) and the correction results after some iterations. The square shape of the extinction profile helps us to see how the extinction profile is deformed by the effects of the multiple scattering and that the calculated value for the lidar ratio is very different from the calculated value in the profile without the effects of multiple scattering. We also see that few iterations are sufficient to reasonably correct the extinction profile.

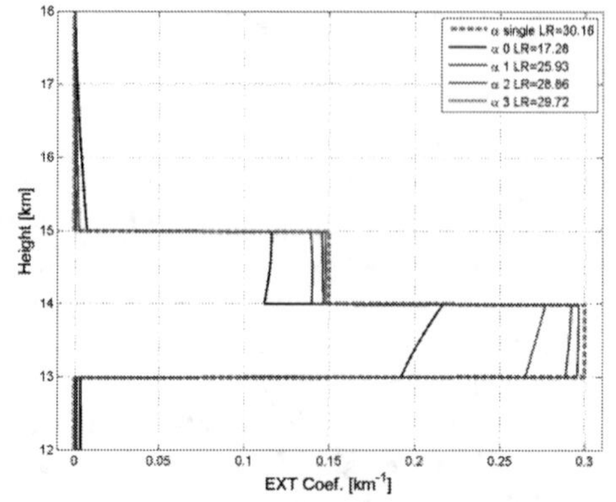

Figure 2: Extinction coefficient profile obtained from the simulated lidar signal with no effects of multiple scattering (single) and also from the MS signal for each of the iterations (from 0 to 3)

Figure 3 shows the calculated values of COD and LR in each of the iterations. We can see that the iterative method accurate corrects the values of the optical depth and the lidar ratio after a few iterations, converging to values very close to the simulated true values. The differences probably come from numeric errors of rounding and discretizing the model, but even so the relative error is quite small. We see that the correction depends heavily on the value of the effective radius chosen to run the model, with the correction being as good as the accuracy of the effective radius. Thus, for real applications it is necessary to have a good parameterization of the effective radius of the ice crystals, or the result can generate more error than the non-corrected case.

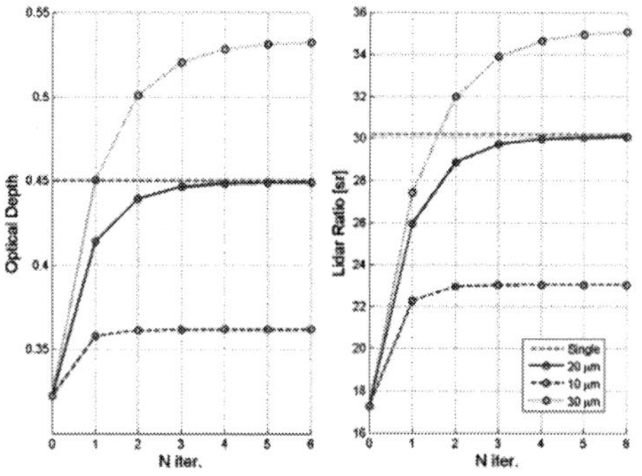

Figure 3: Evolution of the retrieved optical depth and the lidar ratio when corrected by the iterative method, when calculated with the correct value of the effective radius of the simulated ice crystal (20μm) and when it is ± 10 μm.

4 Effective Radius

We saw that the correction of the multiple scattering effects in the COD and LR retrieval can be done very easily as long as accurate information of the effective radius of the ice crystals are available. Otherwise, the problem becomes undetermined, since both the extinction coefficient and the particle size contribute greatly to the intensity of the multiple scattering. However, due to the geometric nature of the multiple scattering effects, the influence of ice crystal size can be isolated by the detailed analysis of the behavior of the MS effect on the molecular signal above the cloud top: relatively large crystals will produce much more narrower frontal scattering which leaves the telescope field of view with a different rate than for smaller ice crystals. Thus, we can use the expected value for the molecular scattering given by the Rayleigh theory (which involves an independent measurement of the state of the atmosphere) to track the specific effect of a given size of ice crystals. To investigate this, we simulate a cloud with an effective radius of 30 μm and calculate the correction for the MS for several values of extinction coefficient and effective radius (see figure 3). By calculating the RMSE of the linear fit between the corrected signal above the top of the cloud and the purely molecular signal given by the theory, we saw that there is a global minimum that is reached when the correct effective radius value is considered. At the moment, sensitivity of this method is being tested on real measurements to evaluate its applicability.

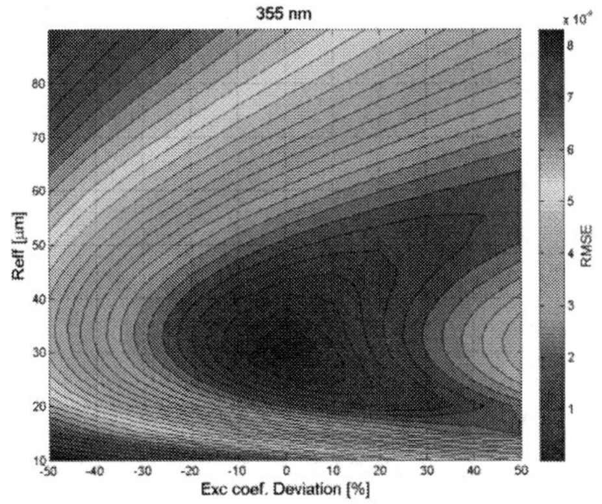

Figure 4: RMSE from the linear fit between the molecular signal above the cloud top and the expected from the Rayleigh theory.

5 CONCLUSIONS

In this paper we presented an efficient iterative way to retrieve cirrus optical depth and lidar ratio from elastic backscattering lidar systems, correcting for multiple scattering effects. Also, we discussed a possible way to retrieve the effective radius by comparing the molecular signal from above the cloud top with the one expected by the Rayleigh theory. The next steps are going to study the sensibility of this technique for the application to lidar measurements and to investigate the dependence on the laser wavelength.

ACKNOWLEDGEMENTS

We thank Professor Robert Hogan for provision of the MS model (http://www.met.reading.ac.uk/clouds/multiscatter/).

References

[1] Wang, Z., Sassen, K., 2002: Cirrus Cloud Microphysical Property Retrieval Using Lidar and Radar Measurements. Part II: Midlatitude Cirrus Microphysical and Radiative Properties, *J. Atmos. Sci.*, 59, 2291–2302.

[2] Gouveia, D. A., Barja, B., Barbosa, H. M. J., Seifert, P., Baars, H., Pauliquevis, T., Artaxo, P., 2017: Optical and geometrical properties of cirrus clouds in Amazonia derived from 1 year of ground-based lidar measurements, *Atmos. Chem. Phys.*, 17, 3619-3636.

[3] Kienast-Sjögren, E., Rolf, C., Seifert, P., Krieger, U. K., Luo, B. P., Krämer, M., Peter, T., 2016: Climatological and radiative properties of midlatitude cirrus clouds derived by automatic evaluation of lidar measurements, *Atmos. Chem. Phys.*, 16, 7605-7621.

[4] Wandinger, U., 1998: Multiple-scattering influence on extinction- and backscatter-coefficient measurements with Raman and high- spectral-resolution lidars, *Appl. Opt.*, 37, 417–427.

[5] Reichardt, J., Reichardt, S., 2006: Determination of cloud effective particle size from the multiple-scattering effect on lidar integration-method temperature measurements, *Appl. Opt.*, 45, 2796-2804.

[6] Eloranta, E. W., 1998: Practical model for the calculation of multiply scattered lidar returns, *Appl. Opt.*, 37, 2464-2472.

[7] Hogan, R., 2006: Fast approximate calculation of multiply scattered lidar returns, *Appl. Opt.*, 45, 5984-5992.

AEROSOLS OBSERVATIONS WITH A NEW LIDAR STATION IN PUNTA ARENAS, CHILE

Boris Barja[1]*, Felix Zamorano[1], Pablo Ristori[2], Lidia Otero[2], Eduardo Quel[2], Nobuo Sugimoto[3] Atsushi Shimizu[3] and Jorge Santana[4].

[1]*Atmospheric Research Laboratory (LIA), University of Magallanes, Punta Arenas, Chile.*
**email:bbarja@gmail.com*
[2]*Division Lidar, CEILAP (UNIDEF-CONICET). Buenos Aires, Argentina.*
[3]*National Institute for Environmental Studies (NIES). Tsukuba. Ibaraki, Japan.*
[4]*Institute of Astronomy, Geophysics and Atmospheric Sciences (IAG). University of São Paulo, Brazil.*

ABSTRACT

A tropospheric lidar system was installed in Punta Arenas, Chile (53.13°S, 70.88°W) in September 2016 under the collaboration project SAVERNET (Chile, Japan and Argentina) to monitor the atmosphere. Statistical analyses of the clouds and aerosols behavior and some cases of dust detected with lidar, at these high southern latitude and cold environment regions during three months (austral spring) are discussed using information from satellite, modelling and solar radiation ground measurements.

1 INTRODUCTION

Atmospheric particles as aerosols and clouds have a great influence on the climate, weather and daily human live. Aerosols have important influences on human health, air quality and climate changes. Their concentrations in the atmosphere have significantly increased since the preindustrial times [e.g. 1]. Their capacity to scatter and absorb solar radiation and work as cloud condensation nuclei, allows atmospheric aerosols to affect atmospheric radiation and dynamics [e.g. 2]. Also, atmospheric aerosols can be a major component of haze pollution, reducing atmospheric visibility and being dangerous to human health [3]. Another type of aerosol recognized as an important component of the climatic system is the dust. Dust can travel large distances across the earth surface and affects all of Earth's climatic zones from the tropics to the poles [4]. Specifically, high latitude dust produced particularly by glacial and periglacial processes in the high and cold regions have attracted little research attention despite increasing evidence that local and regional high-latitude dust emissions can affect the climatic system [e.g. 5].

On the other hand, clouds a component of the atmosphere that affect the solar and terrestrial radiation and the energy balance in the climatic system. Clouds are a very important issue around the Antarctic continent. Clouds around and in Antarctica region are amongst the least well observed due to the remote environment [6].

From the last statements the importance to conduct measurements of aerosols and clouds with ground based instruments around high latitudes regions is evident.

With the objective to long term monitoring the aerosols and clouds in the atmosphere a tropospheric lidar system was installed in the Atmospheric Research Laboratory belonging to University of Magallanes in the city of Punta Arenas, Chile (53.13°S, 70.88°W) during September 2016 under the collaboration project Savernet - South American Enviroment Risk Network (Chile, Japan and Argentina) [7].

In the following, the characteristics of lidar system are reported. Preliminary results of three first months of measurements are presented in Sect. 3, from September to December of 2016 (austral spring) with some cases of dust detected with lidar. Results and discussions using information from solar radiation ground measurements, data HYSPLIT back trajectories and Navy Aerosol Analysis and Prediction System (NAAPS) Global Aerosol Model results are presented in Sect. 4 to evaluate the source of the measured dust, followed by conclusions in Sect. 5.

2 METHODOLOGY

The lidar system installed in the measurement site of Atmospheric Research Laboratory, University of Magallanes (53.13°S, 70.88°W) is a Multi-wavelength Raman Depolarization Aerosol lidar.

https://doi.org/10.1051/epjconf/201817605038

It was installed through the Savernet Project [7]. Details about the lidar system are summarized in Table 1 and detailed information is given in [7]. A laser (Continuum Surelite II – 10Hz), was used as the energy source, the acquisition system from Licel is operating at 20 MSPS and controlled by a computer. The emission – reception configuration is coaxial. The beams are not expanded since low divergence is not desired for this instrument with the tropospheric measurements objective. The light is oriented by a fused silica prism. Three wavelengths are sensed in the spectrometric box: 1064 nm, 532 nm and 355 nm; polarization is available in the UV and visible channels. Also Raman signals are measured at 384 nm, 408 nm and 607 nm. The observation protocol consists in acquiring 3 minutes of lidar signals every 15 minutes on a 24 h/day basis. The data is actually being sent for analysis and fast visualization to CEILAP and NIES.

Table 1 Summarized characteristics of the Multi-wavelength Raman Depolarization Aerosol lidar installed in Punta Arenas.

Emission	Continuum Surelite II – 10Hz 3 wavelength (1064, 532, and 355 nm)
Reception	Telescope 20 cm diameter. 1064 nm, Polarization (532 and 355 nm) Raman (384 nm, 408 nm and 607 nm)
Acquisition	Licel system operating at 20 MSPS

In the present work we are using only information from the 1064 and 532 nm channels, the signals from these channels were used to obtain the attenuated backscatter coefficient at 1064 and 532 nm, the volume depolarization ratio and the dust and sphere extinction coefficients for the 532 nm channel. These quantities are processed with the used algorithm in AD-Net [8, 9]. The lidar ratio used during the processing was 50 sr, the value used for the AD-Net network for dust studies.

Methodology for discriminating between dust and spherical aerosols is based in the depolarization of the lidar signal [8, 9]. So we can determine the extinction coefficient component and make a preliminary analysis of the presence and predominance of the dust particles, spherical particles or mixture of both particles types. Spherical particles are related with aerosols from contamination and sea salt aerosols.

For the analysis of some cases of dust and spherical particles we used information of observation and modelling from the National Research Laboratory (NRL) aerosol page. Modelling was the NAAPS (Navy Aerosol Analysis and Prediction System) Global Aerosol Model. Also we used the NOAA's HYSPLIT atmospheric transport and dispersion modeling system to show the origin of the air mass that arrives to the site during the days of studied measurements [10].

3 RESULTS

We can see in Figure 1, five panels showing the time series of attenuated backscatter coefficient for 532 nm (first box), the volume depolarization ratio for 532 nm (second box), the attenuated backscatter coefficient for 1064 nm (third box), these three first boxes show the altitude until 18 km. The fourth and fifth boxes show the dust and spherical extinction coefficients for 532 nm until an altitude of 9 km. In these last two boxes we can see the color gray to show no measurements. These graphics correspond for the September month of 2016. Similar graphics (not showed) were obtained for the remaining months of the spring season of 2016 (October, November and December).

These graphics show the high frequency of clouds during the measurement period. It was also possible to determine the presence of dust and spherical aerosols over Punta Arenas. We can see in Figure 1, in the fourth box the presence of dust particles during the days of 23 and 24 September 2016, with values of the dust extinction coefficient around 0.1 km^{-1} almost up to 2 km of altitude. During these days there are lower values of the spherical extinction coefficient. So the contribution of the dust aerosol to the site is significantly higher than that of the spherical aerosols.

In Figure 1, we can see also a higher contribution of spherical particle aerosols with lower contribution from dust particles during the days 16 and 17 September 2016. We are looking in more detail in these days.

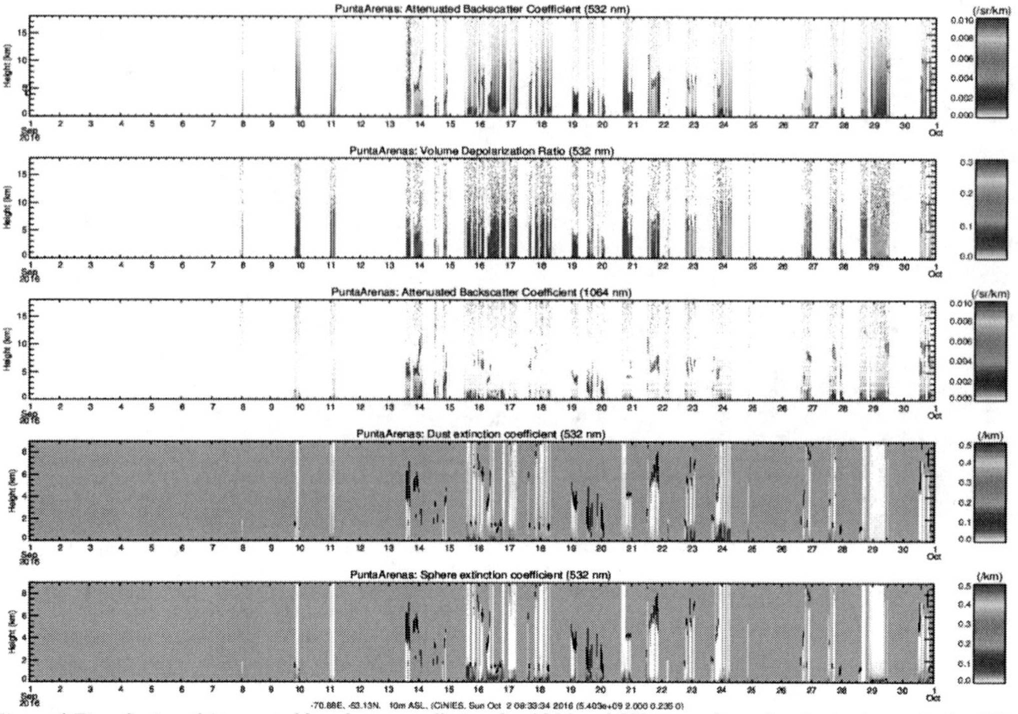

Figure 1 Time Series of Attenuated bacskatter coefficient for 1064 and 532 nm, volume depolarization ratio for 532 nm and dust and spherical extinction coefficient at 532 nm (first, second, third and fourth panel respectively), during the month of 2016.

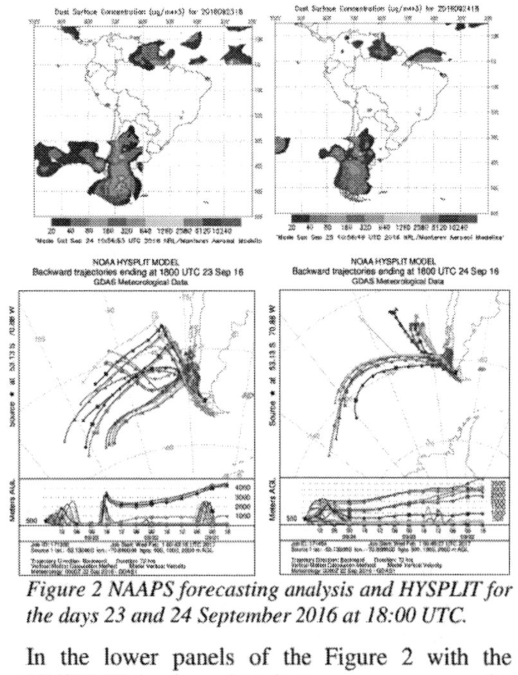

Figure 2 NAAPS forecasting analysis and HYSPLIT for the days 23 and 24 September 2016 at 18:00 UTC.

In the lower panels of the Figure 2 with the HYSPLIT backward trajectory we can see the higher input from the continent in the lower altitudes of the atmosphere, denoting the possible contribution of the regional dust to our site of measurements. The high dust concentration at our site is noted in the upper panels of the Figure 2 with values between 40 $\mu g/m^3$ and 80 $\mu g/m^3$. These values of concentration were obtained from the NAAPS forecasting analysis for the September 23 and 24, 2016.

The NAAPS model for the days 15 and 16 September shows no dust. The HYSPLIT back trajectories indicate the origin of the air masses measured at the site to be over the Ocean at all altitudes. So there is a difference between these two days with different origins of the air masses showing different contribution to the aerosols measured in the site.

We select another day from the NAAPS forecasting model with presence of high levels of concentration of dust in surface near to 160 micrograms per meter cubic. This situation occurs during the day of October 7, 2016, and similar conditions stay during the next two days. Figure 3 show the NAAPS analysis for this day in the left

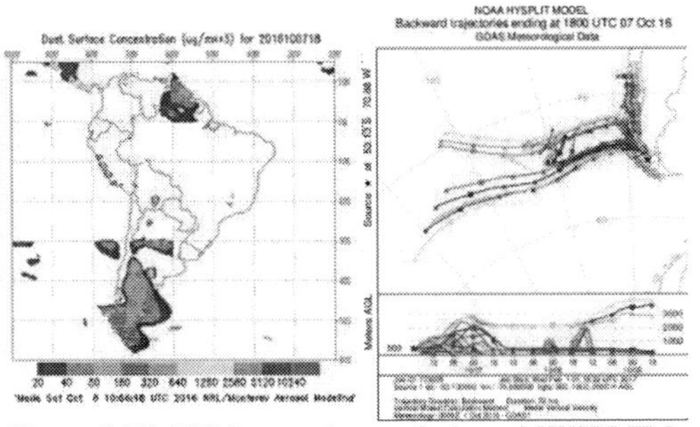

Figure 3 NAAPS forecasting analysis and HYSPLIT for the days 07 October 2016 at 18:00 UTC.

panel and the right panel shows the HYSPLIT back trajectories for the same date.

There is evidence of the presence of dust in our site from the modeling results but lidar signals show low contribution of dust during this day.

4 CONCLUSIONS

We show the first results of measuremnts with lidar in Punta Arenas during 2016. The analysis of the Mie channels shows that there is possible to determine the dust and spherical aerosols in agreement with other information in surface. But there are other days that there is no coincidence between the two sources of information.

So there is needed to make future improvement to the analysis of lidar signals in our site, tuning the method from the AD-Net to our location.

ACKNOWLEDGEMENTS

This work was supported by the project SAVERNet financing by JICA, JST through SATREPS. Also we acknowledge the NAAPS and HYSPLIT teams.

References

[1] Forster, P., Ramaswamy, V., Artaxo, P., et al., 2007. Radiative Forcing of Climate Change, in Climate Change 2007: The Physical Science Basis. Contribution of Working Group I to the Fourth Assessment Report of the Intergovernmental Panel on Climate Change, edited by S. Solomon, D. Qin, M. Manning, Z. Chen, M. Marquis, K. B. Averyt, M. Tignor, and H. L. Miller, pp. 129–234, Cambridge Univ. Press, Cambridge, United Kingdom and New York, NY, USA.

[2] Guo, L., A. Turner, and E. Highwood, 2016: Local and Remote Impacts of Aerosol Species on Indian Summer Monsoon Rainfall in a GCM. J. Climate, 29, 6937–6955, doi: 10.1175/JCLI-D-15-0728.1.

[3] Chameides, W. L. and M. Bergin, 2002. Soot Takes Center Stage. Science. Vol. 297, Issue 5590, pp. 2214-2215 DOI: 10.1126/science.1076866.

[4] Goudie, A.S. and Middleton, N.J., 2006. Desert dust in the global system, Berlin Heidelburg, New York, Springer, 287 p.

[5] Bullard, J. E., et al., 2016, High-latitude dust in the Earth system, Rev. Geophys., 54, 447–485, doi:10.1002/2016RG000518.

[6] Listowski, C. and Lachlan-Cope, T.: The Microphysics of Clouds over the Antarctic Peninsula – Part 2: modelling aspects within Polar WRF, Atmos. Chem. Phys. Discuss., doi:10.5194/acp-2016-1135, in review, 2017.

[7] Quel, E., Sugimoto N., Otero L., Jin, Y., Ristori, P., Nishizawa, T., González, F., Papandrea, S., Shimizu, A., Mizuno, A., 2016, Aerosols monitoring network to create a volcanic ash risk management system in argentina and chile. EPJ Web of Conferences 119, 19006, DOI: 10.1051/epjconf/201611919006.

[8] Shimizu, A., N. Sugimoto, I. Matsui, K. Arao, I. Uno, T. Murayama, N. Kagawa, K. Aoki, A. Uchiyama, and A. Yamazaki 2004, Continuous observations of Asian dust and other aerosols by polarization lidars in China and Japan during ACE-Asia, J. Geophys. Res., 109, D19S17, doi:10.1029/2002JD003253.

[9] Atsushi Shimizu, Tomoaki Nishizawa, Yoshitaka Jin, Sang-Woo Kim, Zifa Wang, Dashdondog Batdorj, and Nobuo Sugimoto, Evolution of a lidar network for tropospheric aerosol detection in East Asia, Optical Engineering 56(3), 031219.

[10]Stein, A.F., Draxler, R.R, Rolph, G.D., Stunder, B.J.B., Cohen, M.D., and Ngan, F., 2015. NOAA's HYSPLIT atmospheric transport and dispersion modeling system, Bull. Amer. Meteor. Soc., 96, 2059-2077, http://dx.doi.org/10.1175/BAMS-D-14-00110.1

DEVELOPMENT OF WAVELENGTH LOCKING CIRCUIT FOR 1.53 MICRON WATER VAPOR MONITORING COHERENT DIFFERENTIAL ABSORPTION LIDAR

Masaharu Imaki[1], Ryota Kojima[1], and Shumpei Kameyama[1]

[1]Mitsubishi Electric Corporation, 5-1-1 Ofuna, Kamakura, Kanagawa, JAPAN, 247-8501,
**Imaki.Masaharu@dc.MitsubishiElectric.co.jp*

ABSTRACT

We have studied a ground based coherent differential absorption LIDAR (DIAL) for vertical profiling of water vapor density using a 1.5µm laser wavelength. A coherent LIDAR has an advantage in daytime measurement compared with incoherent LIDAR because the influence of background light is greatly suppressed. In addition, the LIDAR can simultaneously measure wind speed and water vapor density.

We had developed a wavelength locking circuit using the phase modulation technique and offset locking technique, and wavelength stabilities of 0.123 pm which corresponds to 16 MHz are realized. In this paper, we report the wavelength locking circuits for the 1.5 um wavelength.

1 INTRODUCTION

Water vapor in the atmosphere affects growing cumulonimbus which cause localized heavy rainfall, therefore the measurement of water vapor density profile is required for the early prediction of this rainfall. Height profiling of water vapor measurement using incoherent DIAL technique was reported in Ref. 1 and 2. The continuous height profile measurement was demonstrated, however, the measurement accuracy was influenced by solar background light. On the other hand, coherent DIAL technique can overcome this issue and was reported in Ref. 3 and 4. Good agreement result on water vapor density in comparison with that of radiosonde soundings was shown.

Since late 1990s, we have been developing a wind sensing coherent Doppler LIDARs with 1.5 µm wavelength. The LIDAR with 1.5 µm wavelength has advantages such as its compactness and reliability by using telecom products. Here, we consider to apply this technology to DIAL, and perform preliminary study on a ground based coherent DIAL for vertical profiling of water vapor density using a 1.53 µm laser wavelength. In the followings, we discuss the concept, configuration, and simulation results of the LIDAR.

2 WAVELENGTH SELECTION

The main requirements for the ON wavelength selection in the 1.5 µm wavelength region are (i) strong line intensity and high weighting function near the ground in the region of the band of optical amplifier, and (ii) low temperature dependence of absorption. In the band of optical amplifier, the strong line intensity is in shorter wavelength for 1520 nm - 1570 nm range, and the value is 3.066×10^{-24} (cm/molecule) at 1531.374 nm wavelength.

The height dependence of the absorption coefficient k is expressed by

$$k(\nu,P,T)=S_T(T)\cdot\frac{1}{\pi}\cdot\frac{\gamma_{self}(P,T)}{\gamma_{self}(P,T)^2+\left(\nu-\nu_{shift}(P)\right)^2} \quad (1)$$

where S_T is the line intensity, γ_{self} is the self-broadened harfwith at half maximum, and ν is the wavenumber in vaccume. Figure 1 shows the height dependence of the absorption coeffient at

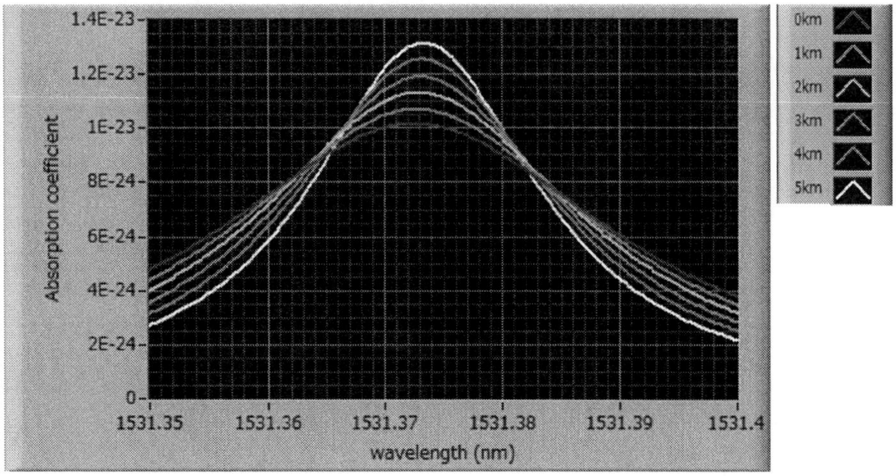

Figure 1. Height dependence absorption coeffient at 1531.374 nm wavelength.

1531.374 nm wavelength using the U.S. standard atmosphere 1976 model. The center of the absorption line has a large dependence on temperature and the pressure. In the absorption line wing, there is low dependence on the temperature and the pressure region exist. From these results, we have selected the absorption wavelength at 1531.3828 nm [5].

4 WAVELENGTH LOCKING CIRCUIT

Figure 2 shows the wavelength locking circuit. For the ON wavelength, we utilize two CW lasers because the absorption coefficient of water vapor in the 1.5 μm wavelength region is relatively low. One of the CW lasers is locked to the HCN R18 absorption line (1531.276 nm) by the phase modulation technique and the other CW laser is locked to the water vapor absorption line by the offset locking technique.

Figure 3 (a) schematically shows the error signal of the phase modulation technique. If the laser wavelength controlled to be at the center of the absorption line, the side-band signals which is positive and negative amplitude are canceled, and then, the amplitude of the error signal is zero. If the laser wavelength has an offset from the absorption line center, a difference appears between the positive and negative amplitude of the side-band signals, and then, the error signal has some amplitude. Therefore, the wavelength can be stabilized by controlling the laser so as to make the amplitude of the error signal to be zero.

Figure 3 (b) is the error functions of the offset locking technique. The difference of the wavelength between the HCN R18 absorption line and water vapor absorption line is 0.1068 nm which corresponds to the 13.663 GHz frequency difference. Therefore, the beat frequency of HCN locked laser and CW laser for water vapor absorption is controlled to 13.663 GHz by using the filter edge method, by stabilizing the amplitude of the beat frequency to be constant.

We had developed this wavelength locking circuit. Figure 4 shows the results of wavelength stabilities of phase modulation technique and offset locking technique. By using this locking system, the wavelength stabilities for HCN R17 absorption line attain 0.07 pm which corresponds to 9.05 MHz, and the wavelength stabililities for water vapor absorption line attain 0.13 pm which corresponds to 16.2 MHz.

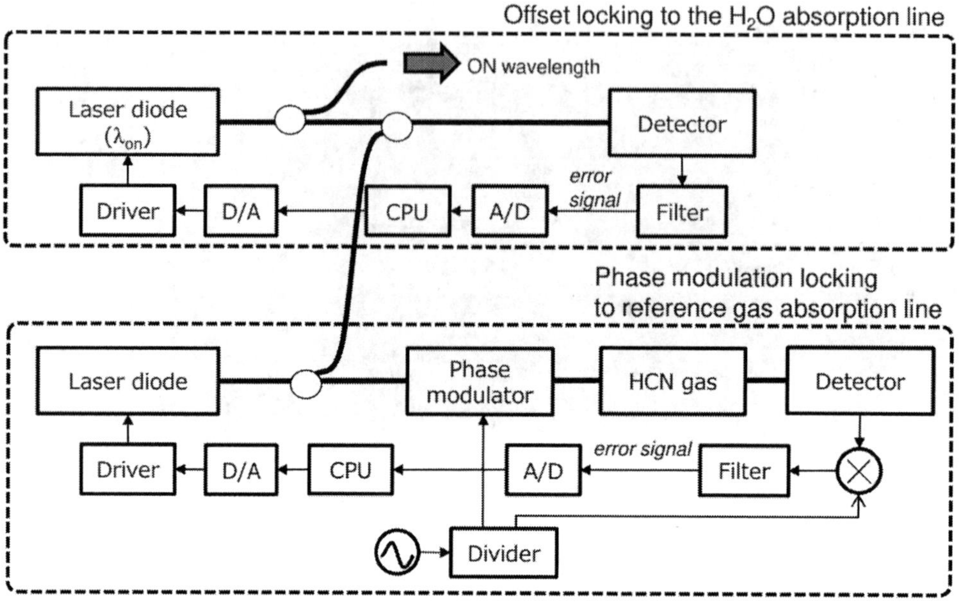

Figure 2. Schematic diagram of coherent differential absorption LIDAR.

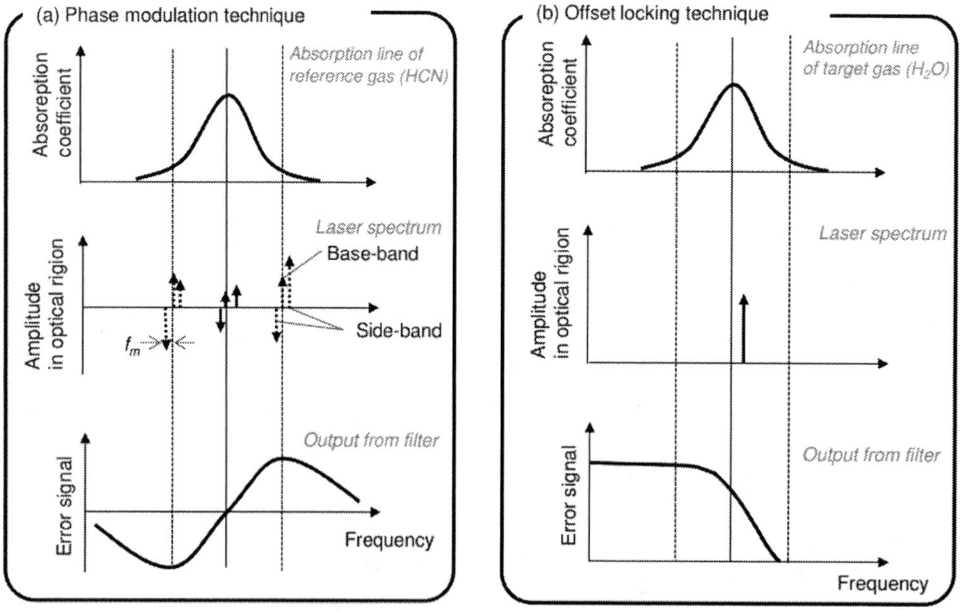

Figure 3. Error signal of the wavelength locking circuits.

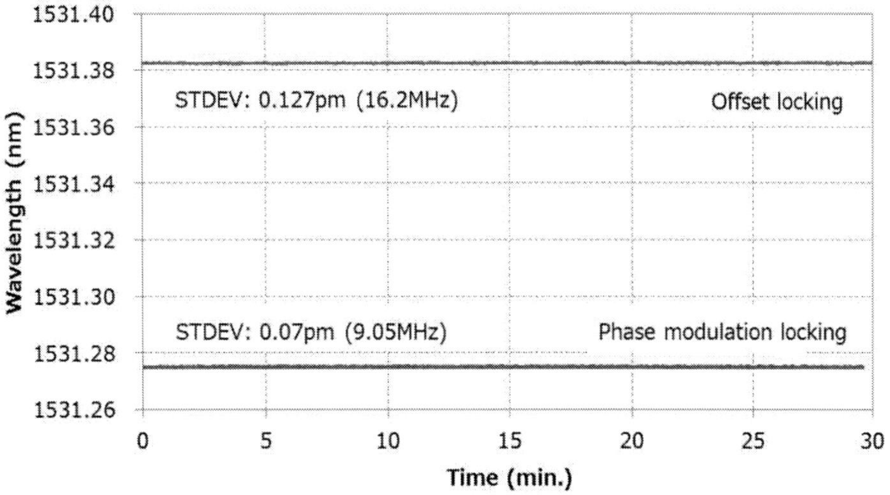

Figure 4. The data of wavelenght locking.

5 CONCLUSION

In conclusions, we reported on the result of the wavelength locking circuit for ground based coherent DIAL for vertical profiling of water vapor density using 1.5 μm laser wavelength.

Future work is the development of a benchtop model of coherent DIAL system, and demonstration of water vapor measurements.

References

[1] R. Nehrir and K. S. Repasky, "Eye-Safe Diode-Laser-Based Micropulse Differential Absorption LIDAR (DIAL) for Water Vapor Profiling in the Lower Troposphere," Journal of Atmospheric and Oceanic Technology **28** (2011) 131-147.

[2] S. M. Spuler, K. S. Repasky, D. Morley, D. Moen, M. Hayman, and A. R. Nehrir, "Field-deployable diode-laser-based differential absorption LIDAR (DIAL) for profiling water vapor," Atmospheric Measurement Techniques **8** (2015) 1073-1087.

[3] M. J. Kavaya, S. W. Henderson, E. C. Russell, R. M. Huffaker, and R. G. Frehlich, "Monte Carlo computer simulations of ground-based and space-based coherent DIAL water vapor profiling," Appl. Opt. **28** (1989) 840-851.

[4] R. M. Hardesty, "Coherent DIAL measurement of range-resolved water vapor concentration," Appl. Opt. **23** (1984) 2545-2553.

[5] M. Imaki, R. Kojima, T. Yanagisawa, and S. Kameyama, "Preliminary study on ground based coherent differential absorption LIDAR for vertical profiling of water vapor density using 1.53 μm wavelength," 18th Coherent Laser Radar Conference (2016).

EPJ Web of Conferences **176**, 05040 (2018) https://doi.org/10.1051/epjconf/201817605040
ILRC 28

STATISTICS OF OPTICAL AND GEOMETRICAL PROPERTIES OF CIRRUS CLOUD OVER TIBETAN PLATEAU MEASURED BY LIDAR AND RADIOSONDE

Guangyao Dai[1,2*], Songhua Wu[1,3], Xiaoquan Song[1,3], Xiaochun Zhai[1]

[1]*Ocean University of China: Ocean Remote Sensing Institute, Qingdao, China, *dai@tropos.de*
[2]*Leibniz Institute for Tropospheric Research, Leipzig, Germany*
[3]*Laboratory for Regional Oceanography and Numerical Modeling, Qingdao National Laboratory for Marine Science and Technology, Qingdao, China*

ABSTRACT

Cirrus clouds affect the energy budget and hydrological cycle of the earth's atmosphere. The Tibetan Plateau (TP) plays a significant role in the global and regional climate. Optical and geometrical properties of cirrus clouds in the TP were measured in July-August 2014 by lidar and radiosonde. The statistics and temperature dependences of the corresponding properties are analyzed. The cirrus cloud formations are discussed with respect to temperature deviation and dynamic processes.

1 INTRODUCTION

Cirrus clouds play a significant role in the energy budget and the hydrological cycle of the earth atmosphere. Cirrus affects the climate via two opposite effects; an infrared greenhouse effect and a solar albedo effect [1]. Since the significant role the cirrus plays, a clear understanding of their optical and geometrical properties is highly essential for climate modeling studies.

The TP is a vast elevated plateau in the middle of the Eurasian continent with averaged elevation 4.5 km above mean sea level (a.m.s.l.), and it has an important role in the global and regional climate system. The TP lies at a critical and sensitive junction of four climatic systems: the Westerlies, the East Asian monsoon, the Siberian cold polar airflow and the Indian monsoon [2]. During July and August 2014, atmospheric measurements were carried out in the TP for the first time to our knowledge using a ground-based high power depolarization and Raman lidar.

2 INSTRUMENT AND EXPERIMENT

A three-wavelength combined elastic-backscatter Raman lidar, water vapor, cloud and aerosol lidar (WACAL) is used to perform continuous measurements of cirrus clouds. The vertical resolution of the signal is equal to 3.75 m and the temporal resolution is 30s. In our study, the radiosonde of GTS1 type is utilized. The radiosonde provides temperature with accuracy of $\pm 0.2\,°C$, relative humidity accuracy of $\pm 5\%$ and pressure accuracy of ± 1 hPa.

During July and August 2014, we conducted the third TP Experiment of Atmospheric Sciences (TIPEX III) by using WACAL in Naqu (31.48 °N, 92.06°E, 4508 m). To investigate the temperature dependence of the optical and geometrical properties of cirrus, simultaneous observations by WACAL and radiosonde are combined. Consequently, only the concurrent measurements by WACAL and radiosonde at nighttime were taken into account and the optical and geometrical characteristics as well as the temperature of cirrus clouds were analyzed.

3 METHODOLOGY

We determined the linear volume depolarization ratios (LVDR) using the "±45°-calibration" [3]. The extinction coefficients are calculated by the Fernald method but they are corrected by removing the multiple-scattering effect. Also, the WACAL is capable of detecting the cloud base height and even the cloud top height if the clouds can be penetrated by laser. To determine the cloud height, an algorithm that combines "differential zero-crossing algorithm" and "threshold algorithm" is used.

4 RESULTS AND DISCUSSION

During the TIPEX III, 21 measurements of cirrus clouds were performed. The occurrence frequency of the cirrus clouds is about 20.75%.

4.1 Geometrical characteristics

In Figure 1, we present the cirrus structure and the vertical distribution of the occurrence frequency of cirrus base/top heights (CBH/CTH) measured during the TIPEX III. The maximum frequencies of the CBH and CTH are 33.33% at a height of 6-7 km above ground level (AGL) and 23.80% at a height of 7-8 km AGL, respectively. The CBH ranges from 4 to 12 km AGL, with a mean value of 7.4 ± 2.0 km. The CTH ranges from 5 to 13 km AGL, with a mean value of 8.5 ± 2.0 km.

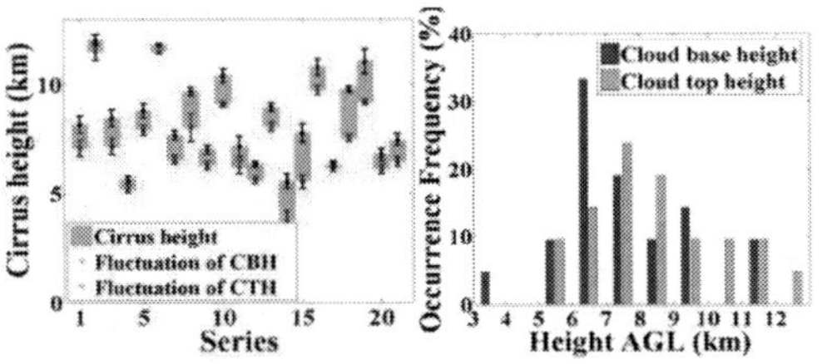

Figure 1: (a) Cirrus structure and (b) the histogram of cloud base/top heights.

Histograms of cirrus cloud middle height and mean temperature are presented in Fig. 2. Cirrus clouds are generally observed in the TP from 4 to 12 km AGL, with mid-cloud temperatures ranging from -79.7 to -26.0 °C, and have a mean temperature of -46.1 ± 15.5 °C. There were two peaks of occurrence frequencies of the cirrus middle height, including 28.57% at 6-7 km and 19.05% at 8-9 km AGL, respectively. From Fig. 2 (b), the maximum frequency of the cirrus middle temperature is 38.10% between -40 and -30 °C.

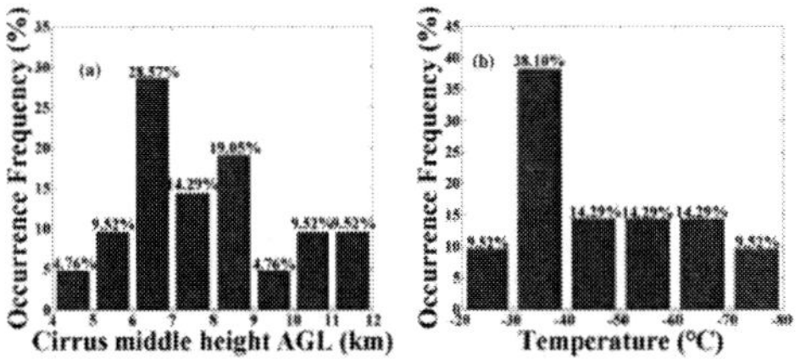

Figure 2: (a) Cirrus middle height occurrence frequency and (b) temperature occurrence frequency.

From Fig. 3, the cloud thickness ranges from 0.12 to 2.26 km, with 81% of the case studies have thickness between 0.5 and 2 km. The cirrus thickness distribution has a skewed normal distribution. The maximum frequency of the cirrus thickness is 38.10% between 0.5 and 1 km while the minimum frequency is 4.76% between 2 and 2.5 km.

https://doi.org/10.1051/epjconf/201817605040

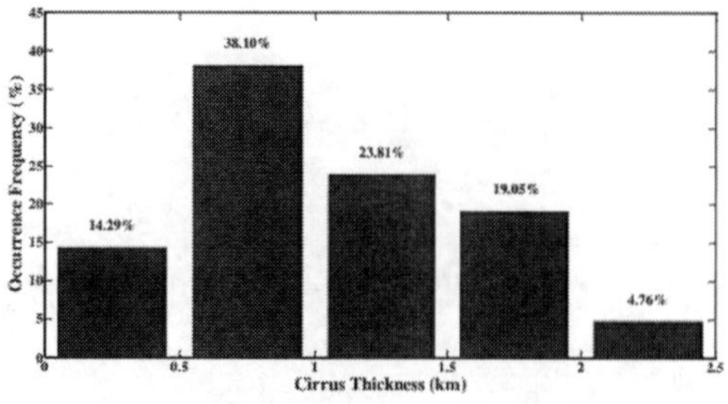

Figure 3: Cirrus cloud thickness

4.2 Optical properties

In Fig. 4, the LVDR, the extinction coefficient and the corresponding statistical uncertainties are presented. From this figure, the LVDR ranges from 0.3 to 0.55, while the extinction coefficient ranges from 10^{-5} m^{-1} to 0.005 m^{-1}. For further investigation, we present the distribution and histogram of the cirrus clouds optical depth in Fig. 5. The optical depth of all the cirrus clouds is between 0.01 and 3. It is found that 14.29% of cirrus clouds are sub-visible (optical depth<0.03), 66.67% are optically thin (0.03<optical depth<0.3) and 19.04% are opaque (optical depth>0.3).

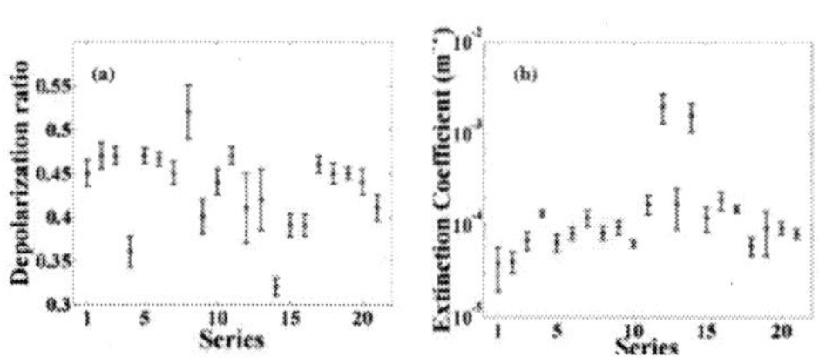

Figure 4: (a) LVDR and (b) Extinction coefficient and statistical uncertainties

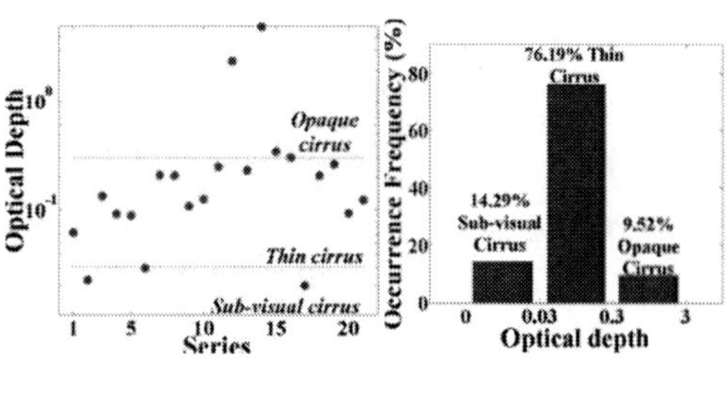

Figure 5: (a) Distribution of optical depth and the classification of all the cirrus clouds, (b) Occurrence frequency of different types of cirrus clouds.

4.3 Temperature dependence of optical properties

Figure 6 shows the LVDR dependences on temperature and cirrus middle height. The LVDR of cirrus varies from 0.3 to 0.55. With the increase of the temperature, the values of the LVDR are gradually reduced. We propose that with the decrease of the temperature and the increase of cirrus middle height, the irregular ice types with larger size and greater distortion are formed, which result in the increase of LVDR of the cirrus clouds. According to Fig. 6, it can be seen that the LVDR with higher temperature and lower altitude is much more sensitive to the ambient temperature.

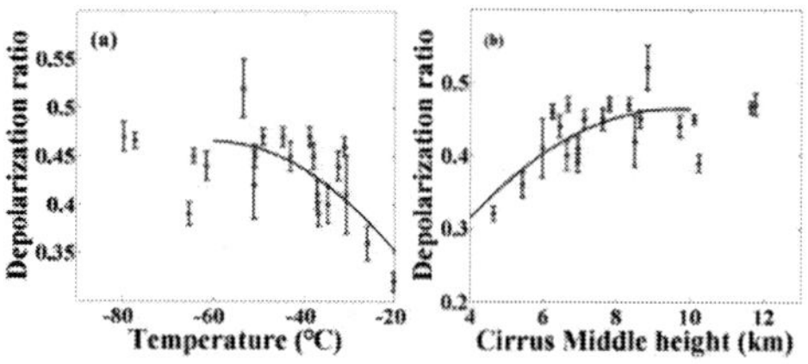

Figure 6: (a) Scatter diagram of LVDR and temperature, (b) Scatter diagram of LVDR and cirrus middle height.

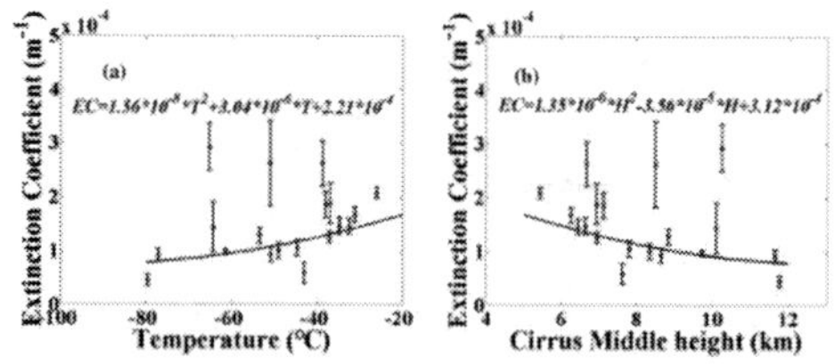

Figure 7: (a) Scatter diagram of extinction coefficient and temperature, (b) Scatter diagram of extinction coefficient and cirrus middle height.

Figure 7 shows the extinction coefficient

dependences on temperature and cirrus middle height. From Fig. 7, the mean extinction coefficient decreases with decreasing temperature. Although the cirrus radiative properties depend on cloud microphysics, it is the ambient temperature that governs the radiative transfer since it dominates the cloud microphysics. Due to the stability of the larger cirrus particle radius in the extremely low temperature, the temperature sensitivity of the extinction coefficient is found gradually weak with the decreasing temperature. Consequently, temperature has a bigger impact on the extinction coefficient for warmer and lower cirrus.

Figure 8 presents the temperature anomalies during July and August 2014 over Naqu. The warm color corresponds to the positive deviation while the cool color corresponds to negative deviation. Moreover, the CBH and CTH are also plotted in Fig. 8. The lidar observations can be separated in six stages: 10 to 11 July, 18 to 22 July and 31 July to 08 August for the cirrus appearances stages and 12 to 17 July, 23 to 30 July and 09 to 16 August for the cirrus blank stages. During the 3 cirrus appearances stages, the temperatures were higher than the average value, which indicates that the formation of the cirrus clouds may be related to the local ground heating of TP. The cold perturbations were found at a height of about 12 km AGL. During the 3 cirrus blank stages, warmer air mass existed at a height of above 12 km AGL. The tropospheric temperatures with the presence of cirrus were about 5 °C higher than those when no cirrus clouds existed. Consequently, the Rossby waves can best be recognized at a height of 16.5 km a.m.s.l. in the tropospheric layer and with a duration cycle of about 10 days. To conclude, the formation of cirrus clouds over TP is dominated by the Rossby wave and deep convective activity caused by local ground heating.

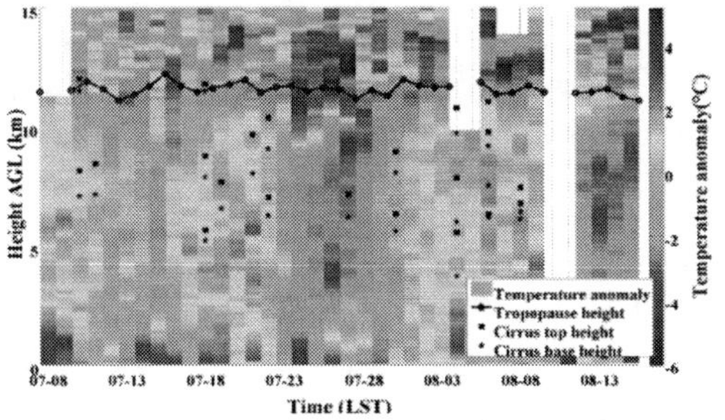

Figure 8: Temperature anomalies during July and August 2014.

5 SUMMARY AND CONCLUSION

This study provides an analysis of the geometrical and temperature characteristics as well as the optical properties of cirrus clouds over TP. The results show that cirrus clouds are occurring between 8.5 to 16.5 km a.m.s.l., with temperatures ranging from -79.7 to -26.0 °C . The cloud thickness ranges from 0.12 to 2.26 km with a mean thickness of 1.07 ± 0.56 km. The LVDR differs from 0.3 to 0.55, with a mean value of 0.43 ± 0.045.

The cirrus clouds are classified as sub-visible cirrus (14.29%), optically thin cirrus (66.67%) and opaque cirrus (19.04%) judging by the optical depth.

From the temperature anomaly from July to August in 2014, the formation of cirrus clouds has apparent relationship with the dynamic processes of Rossby wave and deep convective activity over TP.

ACKNOWLEDGEMENTS

This work was supported by the National Natural Science Foundation of China (NSFC) under grant 41375016 and 91337103, by the China Special Fund for Research in the Public Interest under grant GYHY201406001.

References

[1] Giannakaki, E., Balis, D., Amiridis, V., and Kazadzis, S., 2007: Optical and geometrical characteristics of cirrus clouds over a Southern European lidar station, *Atmos. Chem. and Phys.*, **7**(21), 5519-5530.

[2] Wu, S., Dai, G., Song, X., Liu, B., and Liu, L., 2016: Observations of water vapor mixing ratio profile and flux in the TP based on the lidar technique, *Atmos. Meas. Tech.*, **9**(3), 1399-1413.

[3] Freudenthaler, V., Esselborn, M., Wiegner, M., Heese, B., Tesche, M., Ansmann, A., Müller, D., Althausen, D., Wirth, M., and Fix, A., 2009: linear volume depolarization ratios profiling at several wavelengths in pure Saharan dust during SAMUM 2006, *Tellus B*, **61**(1), 165-179.

 https://doi.org/10.1051/epjconf/201817605041

VERTICAL SEPARATION OF THE ATMOSPHERIC AEROSOL COMPONENTS BY USING POLIPHON RETRIEVAL IN POLARIZED MICRO PULSE LIDAR (P-MPL) MEASUREMENTS: CASE STUDIES OF SPECIFIC CLIMATE-RELEVANT AEROSOL TYPES

Carmen Córdoba-Jabonero[1*], Michaël Sicard[2,3], Albert Ansmann[4], Ana del Águila[1], and Holger Baars[4]

[1]*Instituto Nacional de Técnica Aeroespacial (INTA), Atmospheric Research and Instrumentation Branch, Torrejón de Ardoz (Madrid), Spain,* *cordobajc@inta.es

[2]*Remote Sensing Laboratory, Universitat Politècnica de Catalunya (UPC), Barcelona, Spain*

[3]*Ciències i Tecnologies de l'Espai - Centre de Recerca de l'Aeronàutica i de l'Espai/Institut d'Estudis Espacials de Catalunya (CTE-CRAE/IEEC), Universitat Politècnica de Catalunya, Barcelona, Spain*

[4]*Leibniz Institute for Tropospheric Research (TROPOS), Leipzig, Germany*

ABSTRACT

POLIPHON (POlarization-LIdar PHOtometer Networking) retrieval consists in the vertical separation of two/three particle components in aerosol mixtures, highlighting their relative contributions in terms of the optical properties and mass concentrations. This method is based on the specific particle linear depolarization ratio given for different types of aerosols, and is applied to the new polarized Micro-Pulse Lidar (P-MPL). Case studies of specific climate-relevant aerosols (dust particles, fire smoke, and pollen aerosols, including a clean case as reference) observed over Barcelona (Spain) are presented in order to evaluate firstly the potential of P-MPLs measurements in combination with POLIPHON for retrieving the vertical separation of those particle components forming aerosol mixtures and their properties.

1 INTRODUCTION

It is widely known that aerosols contribute to the climate change due to their effects in the radiative balance of the atmosphere. However, particular types of aerosols (dust, marine, pollution, smoke, volcanic ash, pollen, ...) present specific properties, contributing in a different way to climate implications [1].

Lidar depolarization measurements together with the lidar ratio (LR) estimations make possible the classification of aerosols according to their type. Besides, polarized Raman (inelastic) lidar systems can separate independently the extinction from the scattering, fact that, in addition to their depolarization elements, can also give more information on the properties of the aerosols. In our case, a polarized Micro-Pulse Lidar (P-MPL), the standard system within NASA/MPLNET (Micro Pulse Lidar NETwork, mplnet.gsfc.nasa.gov), is routinely in operation at two Spanish aerosol lidar stations: UPC/Barcelona-BCN (NE Iberian Peninsula, IP) and INTA/El Arenosillo-ARN (Huelva, SW IP). The P-MPL is an elastic (no Raman) and monochromatic (532 nm) system which includes also a depolarization-sensitive channel. The system operates in an automatic and continuous 24/7 mode.

BCN and ARN sites are located, respectively, in a coastal urban and coastal rural environment. Hence, the 'aerosol mixing' is relatively high at both stations, and then the inversion of the aerosol properties from lidar measurements becomes crucial, requiring the application of determined methods. In particular, the retrieval of their optical properties with an elastic and monochromatic lidar, as the P-MPL, is carried out using the well-known Klett-Fernald inversion algorithm (KF method, [2, 3]). This retrieval method is applied to the vertical profiles of the lidar backscattered signal in synergy with simultaneous sun-photometer measurements that provide ancillary data of the Aerosol Optical Depth (AOD), that is the constraint condition of the KF inversion convergence (i.e., [4]).

Regarding the microphysical properties, retrieval procedures like LIRIC, GARRLiC, GRASP, etc. are more demanding on the lidar performance requirements and cannot be applied to P-MPL measurements. Hence, we use the POLIPHON (POlarization-LIdar PHOtometer Networking) method [5, 6], developed in the Leibniz Institute for Tropospheric Research (TROPOS, www.tropos.de). POLIPHON retrieval consists in the vertical separation of a maximum of three components in aerosol mixtures of aerosol types with clearly different particle depolarization ratios (see more details in [5]).

The aim is to make use of the potential of P-MPLs measurements in combination with POLIPHON in order to: 1) provide a vertical component discrimination of the aerosol mixtures; 2) retrieve their properties; and 3) evaluate their radiative forcing. The systematic application of POLIPHON to P-MPL measurements could yield an interesting climatology in view of those aerosols affecting the climate change. In addition, our findings open perspectives for CALIPSO/CALIOP (Cloud-Aerosol Lidar with Orthogonal Polarization) observations, since it is also an elastic polarized lidar.

Case studies of specific climate-relevant aerosols, dust, fire smoke, pollen and clean conditions (reference case), observed over BCN are examined. Results will be extrapolated to the aerosol observations to be performed over ARN, where the P-MPL system was recently deployed.

2 METHODOLOGY

2.1 Measurement station and lidar system

Barcelona (BCN) station is located at the North East Iberian Peninsula (41.4ºN, 2.1ºE, 115 m a.s.l.), by the coast of the Mediterranean Sea, in the North campus of the Universitat Politècnica de Catalunya (UPC). A P-MPL is in routine operation since 2014. The system acquires vertical aerosol profiles with a relatively high frequency (2500 Hz) using a low-energy (~ 7 µJ) Nd:YLF laser at 532 nm. Polarization capabilities rely on the signal registered at both the co- and cross channels. The P-MPL signal follows the NASA/MPLNET requirements of 1-min integrating time and 15 m vertical resolution. BCN is also a NASA/AERONET site.

Parallel and perpendicular P-MPL range-corrected signals (p-RCS and s-RCS, respectively) are obtained from [7]. The volume linear depolarization ratio (δ^V) is obtained from the ratio of s-RCS to p-RCS. The particle linear depolarization ratio (δ_p) is calculated by the procedure shown in [8]. The total particle backscattering coefficient (β_{total}) is obtained by applying the KF algorithm [2, 3] to the total RCS (=p-RCS+s-RCS), constrained to the AERONET AOD. In this study, AERONET data (AOD and the Ångström exponent, AEx) have been hourly averaged in order to coincide with the hourly averaging applied to P-MPL measurements.

2.2 POLIPHON method

POLIPHON procedure can be applied in 1 or 2 steps, retrieving the separation of two or three components in aerosol mixtures, respectively. In particular, the 2-step version is only used for the dust case in order to obtain the separation of three components: dust coarse, dust fine and non-dusty aerosols; otherwise, the discrimination is only performed for two components. POLIPHON details can be found in [5, 6]. Table 1 shows the POLIPHON version (1-step or 2-step) used in each aerosol mixture case, in addition to the particular δ_i values assumed for each specific aerosol component (i=1, 2, 3, depending on the number of separated components, regarded as 'pure' aerosols) [5, 6, 9, 10] presented in this work: dust, fire smoke and pollen (a clean case is also included for reference).

Table 1. Aerosol cases, date of the event (time shown in Figures 1 and 2), AERONET data at that time, and parameters used in the POLIPHON retrieval algorithm.

			AERONET data		POLIPHON retrieval (*)	Linear depolarization ratio for each aerosol component			
Aerosol case	Date	Time (UTC)	AOD	AEx		δ_1	δ_2	δ_3	Ref.
DUST	05 July 2016	03:00 15:00	0.33 0.29	0.52 1.78	2-step	0.39	0.16	0.05	[5, 6]
SMOKE	23 May 2016	06:00	0.14	1.32	1-step	0.15	0.05	---	[9]
POLLEN	23 March 2016	15:00	0.10	1.74	1-step	0.40	0.05	---	[10]
CLEAN	08 November 2016	15:00	0.04	1.33	1-step	0.05	0.01	---	[5, 6]

(*) 1-step: Separation of two aerosol components.
2-step: Separation of three aerosol components.

The contribution of each aerosol component is expressed in terms of the relative ratio between the specific height-integrated backscattering coefficient (β_i) for each aerosol component and β_{total}, i.e., the relative β_i/β_{total} ratio (in %), and calculated along the day.

3 RESULTS

3.1 Dust case

A dusty event occurred over BCN station on 5 July 2016. Dust incidence is mostly observed for the first half of the day with a high loading of large particles. They seem to be almost completely removed for the second part of the day, and only smaller dust particles and non-dusty aerosols are present.

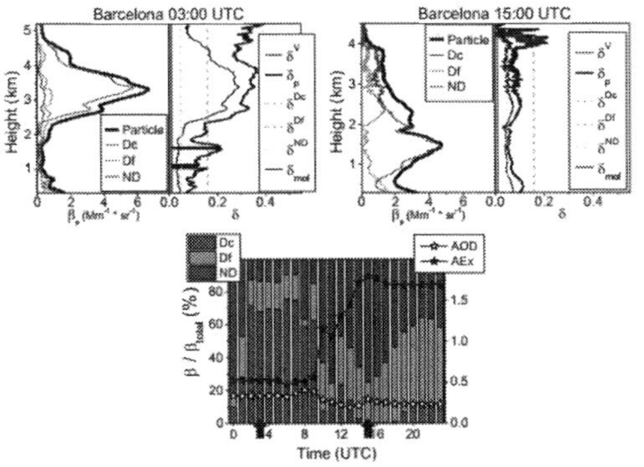

Figure 1. Dust case. (Up) Both vertical backscattering coefficient (β_p) (total particle and for each specific component, left panels), and linear depolarization ratio (volume, δ^V, and particle, δ_p, right panels) at 03:00 UTC (left) and 15:00 UTC (right). (Down) Evolution of the relative contribution of each aerosol component, β/β_{total} (%), along the day. (Note: Dc - dust coarse; Df – dust fine, and ND – non-dusty aerosols). The time of those profiles shown in the upper panels are marked by black arrows. Daily AERONET/AOD and AEx values are also included (white and black stars, respectively).

These results are shown in Figure 1, where the daily evolution of the three separated aerosol components (in relative percentage respect to the total backscattering coefficient) (Fig. 1 down) is shown. In addition, the vertical profiles of both the total and the three separated backscattering coefficients (Dc and Df denote, respectively, dust coarse and fine particles, and ND is for non-dusty aerosols) (left panels) together with both the volume and particle linear depolarization ratios (δ^V and δ_p, respectively) (right panels) are also sown in Figure 1 at two moment of the day: 03:00 UTC with high coarse dust presence (Fig. 1, up-left), and 15:00 UTC (no dust coarse) (Fig. 1, up-right).

AERONET/AOD and AEx values reported along the day (night-time data are assumed equal to the first and last daytime values in each case) also confirm these results; in particular, AEx is close to 0.5 (coarse particles predominance) and higher than 1.5 (fine particles predominance), respectively, in the first and second part of day. AERONET values are also included in Table 1.

3.2 Smoke, pollen and clean cases

Smoke, pollen and clean episodes were observed over BCN station on 23 May 2016, 23 March 2016 and 8 November 2016, respectively. In all these cases, a 1-step POLIPHON retrieval is used to separate the aerosol mixtures in two components. Particle linear depolarization ratio values for each 'pure' aerosol case are shown in Table 1. As for the dusty case, the evolution of the relative percentage between aerosol components in terms of the backscattering coefficient is shown in Figure 2 (right) (the same as illustrated in Figure 1), together to the vertical profiles of the backscattering coefficients and linear depolarization ratios at a given moment of the day (Fig. 2-left).

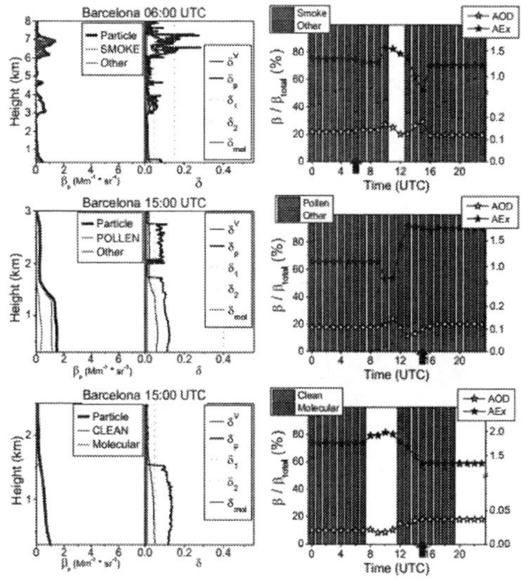

Figure 2. The same as Fig. 1, but for all other cases (from top to bottom panels): fires smoke, pollen and clean conditions.

In the smoke case, a high proportion of smoke particles are detected in a layer between 6 and 7.5 km height. They are observed almost during all the day, corresponding to a 50%, approximately, out of total aerosol loading. In addition, results reflect that the AEx parameter decreases just when the smoke contribution also decreases from around 13:00 to 16:00 UTC (see Fig. 2-up).

Pollen particles (Fig. 2-center) are found no higher than 1.5 km height over the surface and from around 10:00 UTC on, but with a contribution no higher than 20%. The morning increase of AEx might be due to the formation of local, urban aerosols which are much smaller particles compared to pollen grains. Although at the same time β/β_{total} also increases, this result suggests that local aerosols (and not pollen) dominate the columnar-averaged optical properties.

The clean case (see Fig. 2-down) shows that local aerosols are responsible of the background clean conditions in the BCN surroundings (AOD < 0.05). Indeed, the presence of these local aerosols is mostly observed for the daytime period.

4 CONCLUSIONS

Four different aerosol mixture cases have been examined: dust, smoke and pollen, and clean conditions as reference. Good results were achieved in the component separation of aerosol mixtures by applying POLIPHON retrieval to P-MPL measurements. The evolution of the relative aerosol contribution retrieved along the day for each aerosol component is also plausible in comparison with AERONET data. However, assuming a given value for the linear depolarization ratio for 'pure' aerosols is crucial in that retrieval. Further cases of aerosol mixtures are on-going studied.

ACKNOWLEDGEMENTS

This work is supported by the Spanish Ministerio de Economía y Competitividad (MINECO) under grant CGL2014-55230-R. Lidar measurements in Barcelona were supported by the ACTRIS (Aerosols, Clouds, and Trace Gases Research Infrastructure Network) Research Infrastructure Project funded by the European Union's Horizon 2020 research and innovation programme (grant agreement n. 654169); by the Spanish MINECO (project TEC2015-63832-P) and the Spanish Ministry of Science and Innovation (project UNPC10-4E-442) and EFRD (European Fund for Regional Development); by the Department of Economy and Knowledge of the Catalan autonomous government (grant 2014 SGR 583).

References

[1] Myhre et al, 2013: Anthropogenic and natural radiative forcing. In: Stocker et al. (Eds.), Climate Change 2013: The Physical Science Basis. Contribution of Working Group I to the Fifth Assessment Report of the Intergovernmental Panel on Climate Change. Cambridge University Press, Cambridge, NY, USA.

[2] Fernald, 1984: Analysis of atmospheric lidar observations: some comments, Appl. Optics, 23, 652–653.

[3] Klett, 1985: Lidar inversion with variable backscatter/extinction ratios, Appl. Optics, 24, 1638–1643.

[4] Córdoba-Jabonero et al., 2014: Lidar Ratio discrimination retrieval in a two-layer aerosol system from elastic lidar measurements in synergy with sun-photometry data. In: Proceedings of the International Conference in Atmospheric Dust, ProScience, 1, 243-248.

[5] Tesche et al., 2011: Profiling of Saharan dust and biomass-burning smoke with multiwavelength polarization Raman lidar at Cape Verde, Tellus B, 63 (4), 649-676.

[6] Ansmann et al., 2012: Profiling of fine and coarse particle mass: case studies of Saharan dust and Eyjafjallajokull/Grimsvotn volcanic plumes, Atmos. Chem. Phys., 12, 9399-9415.

[7] Flynn et al., 2007: Novel polarization-sensitive micropulse lidar measurement technique, Optics Express, 15 (6), 2785-2790.

[8] Cairo et al., 1999. Comparison of various depolarization parameters measured by lidar, Appl. Optics, 38, 4425–4432.

[9] Burton et al., 2012: Aerosol classification using airborne High Spectral Resolution Lidar measurements - methodology and examples, Atmos. Meas. Tech., 5, 73-98.

[10] Sicard et al., 2016: Near-surface and columnar measurements with a micro pulse lidar of atmospheric pollen in Barcelona, Spain, Atmos. Chem. Phys., 16, 6805-6821.

CIRRUS CLOUDS PROPERTIES DERIVED FROM POLARIZED MICRO PULSE LIDAR (P-MPL) OBSERVATIONS AT THE ATMOSPHERIC OBSERVATORY 'EL ARENOSILLO' (SW IBERIAN PENINSULA): A CASE STUDY FOR RADIATIVE IMPLICATIONS

Ana del Águila[1*], Laura Gómez[1], José Manuel Vilaplana[1], Mar Sorribas[1], and Carmen Córdoba-Jabonero[1]

[1]*Instituto Nacional de Técnica Aeroespacial (INTA), Atmospheric Research and Instrumentation Branch, Atmospheric Observatory 'El Arenosillo' (Huelva), Spain,* *aguilapa@inta.es

ABSTRACT

Cirrus (Ci) clouds are involved in Climate Change concerns since they affect the radiative balance of the atmosphere. Recently, a polarized Micro Pulse Lidar (P-MPL), standard system within NASA/MPLNET has been deployed at the INTA/Atmospheric Observatory 'El Arenosillo' (ARN), located in the SW Iberian Peninsula. Hence, the INTA/P-MPL system is used for Ci detection over that station for the first time. Radiative effects of a Ci case observed over ARN are examined, as reference for future long-term Ci observations. Optical and macrophysical properties are retrieved, and used for radiative transfer simulations. Data are compared to the measured surface radiation levels and all-sky images simultaneously performed at the ARN station.

1 INTRODUCTION

Cirrus (Ci) clouds play an important role in Earth radiative forcing and Climate Change [1]. The degree of understanding of their climate implication is still low; hence, Cirrus features are widely required in data assimilation by global climatic models. In particular, the contribution of Ci-contrails formed from air traffic is also poorly known. In addition, Ci cloud properties, both macrophysical and optical, affect the solar radiation levels reaching the ground, enhancing or negating the global warming effect. Thereby, a better understanding of Ci cloud properties is needed in order to explain their influence in the radiative balance.

Active remote sensing instruments, as lidar systems, are generally used to detect Ci clouds heights. In particular, a polarized Micro-Pulse Lidar (P-MPL), standard NASA/Micro-Pulse NETwork (MPLNET, mplnet.gsfc.nasa.gov) system, was recently deployed at the INTA/Atmospheric Observatory 'El Arenosillo' (ARN), where flight routes frequently cross over this area as well. This P-MPL has been used for Ci detection over ARN for the first time; a Ci cloud case observed over ARN on 28 October 2016 is presented in this work, as reference for future long-term Ci observations. The aim is to compare the retrieved Ci optical and macrophysical properties to the surface radiation levels measured at ARN station. Radiative transfer simulations are also performed to study the impact on surface of the Ci clouds. Results can contribute to validation purposes of the next ESA's EarthCARE mission, whose scientific goal is related to the radiation-aerosol-cloud interactions.

2 METHODOLOGY

2.1 Measurement station and lidar system

The Atmospheric observatory 'El Arenosillo' (ARN) is located at the Southwest Iberian Peninsula (37.1°N 6.7°W, 40 m a.s.l.), inside a natural protected environment near Doñana National Park and less than 1 km from the Atlantic coastline. ARN is managed by the Spanish Institute for Aerospace Technology (INTA), and also is a NASA/AERONET site.

The new polarized Micro Pulse Lidar (P-MPL) model 4B (Sigma Space Corp., USA) was deployed at ARN station at the end of September 2016. The P-MPL system is an elastic lidar with a highly-pulsed (2500 Hz) and low-energy (~ 7 μJ) Nd:YLF laser at 532 nm, operational in full-time continuous mode (24h/7d). Polarization parameters are determined from the co- and cross-channels, obtaining both the parallel and perpendicular range-corrected signals (p-RCS and s-RCS, respectively) by applying [2]. P-MPL routine measurements follow the NASA/MPLNET requirements of 1-min integrating time and 75-m vertical resolution. Due to the variability of the Cirrus clouds, registered 1-min lidar profiles are 10-min averaged in order to obtain a relatively good signal-to-noise ratio without neglecting potential Ci variability.

2.2 Retrieval of the Ci properties

Macrophysical and optical properties of Ci clouds observed on 28 October 2016 (unfortunately, no data are available after noon) have been retrieved by using the procedure described in [3]. In particular, both top and base heights, the thickness, the backscattering ratio (BSR, total-to-molecular backscattering coefficient ratio), the volume linear depolarization ratio (δ^V), the Ci Cloud Optical Depth (CCOD), and the Lidar Ratio (LR) are derived. Ci clouds are also classified according to their CCOD: sub-visual (svCi, CCOD < 0.03), semi-transparent (stCi, CCOD: 0.03-0.3) and opaque (opCi, CCOD > 0.3).

2.3 Solar radiation measurements

Solar radiation data were obtained by a set of radiometers installed on a Kipp&Zonen 2AP solar tracker. The Global (G) and Diffuse (F) radiations were measured by using two Kipp & Zonen CMP21 pyranometer (spectral range: 285 - 2800 nm), shadowing the one measuring the diffuse component. A Kipp & Zonen SHP1 pyrheliometer (200 - 4000 nm) and a YES UV.B-1 pyranometer (270 - 400 nm) were used to measure, respectively, the Direct (D) and Erythemal effective UV (UV-Ery) radiations. Data were also 10-min averaged, as for the lidar profiles. Those instruments are regularly calibrated with traceability to the World Radiation Reference.

2.4 Radiative transfer simulations

Radiative transfer simulations were performed to calculate the (spectrally-integrated) global (G), direct (D) and diffuse (F) irradiances on the surface within the spectral range from 285 to 2800 nm, as well as the Erythemally-weighted UV irradiance. The radiative transfer model (RTM) used for these calculations was the pseudo-spherical SDISORT code [4], using the US standard atmosphere as input for the model. In order to determine the impact of the Cirrus clouds on the surface radiation levels, two scenarios were considered: (1) No Ci clouds occurrence; and (2) the presence of an homogenous layer of Ci clouds (until noon), whose lidar-derived altitude and optical depth were input parameters into the RTM.

3 RESULTS

3.1 Optical and macrophysical properties of Ci clouds

A layer of Ci clouds was observed over ARN on 28 October 2016 (unfortunately, no data were available from 11:45 UTC on). The vertical BSR (R), volume linear depolarization ratio (δ^V), Ci Cloud Optical Depth (CCOD), and Lidar Ratio (LR) were derived. Their temporal evolution is shown in Figures 1 (R and δ^V) and 2 (CCOD and LR, together to their standard deviation).

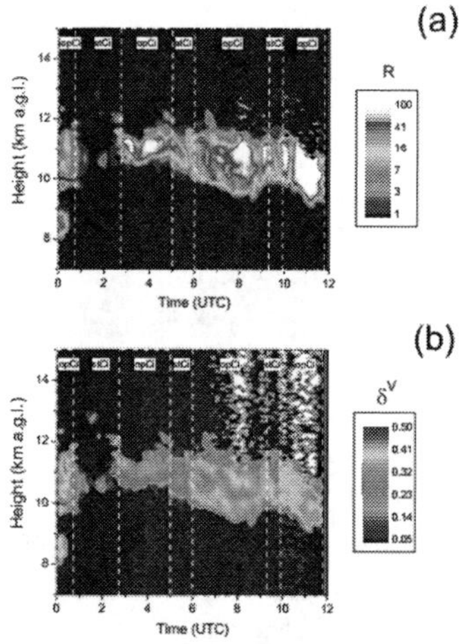

Figure 1 Temporal evolution of a Ci case observed over ARN on 28 October 2016 in terms of: (a) Backscattering ratio (R), and (b) Linear Volume Depolarization Ratio (δ^V). Ci CCOD-type is also marked at the top.

On the top of all figures, the Ci predominant CCOD-type category for consecutive Ci layers is labelled. Seven Ci groups are found, alternating between stCi and opCi categories; but no svCi were observed. During the overall observational period (almost 12 h, first half of the day), opCi clouds were more frequent (68% / ~ 8 h) than stCi (32% / ~ 4 h) (see Fig. 1). In particular, CCOD values higher than 1 are retrieved in the daytime intervals: 07:50-08:20 UTC, and 10:30-11:30 UTC, where the highest R and δ^V values are found

 https://doi.org/10.1051/epjconf/201817605042

(see Figs. 1 and 2-up). Regarding LR, lower values are found for opCi (around 20-40 sr) than for stCi (around 30-50 sr) (see Fig. 2-center). In the overall period, the mean values of CCOD and LR are 0.8 ± 0.7 and 28 ± 6 sr, respectively, being 1.2 ± 0.8 and 25 ± 5 sr, respectively, for daytime.

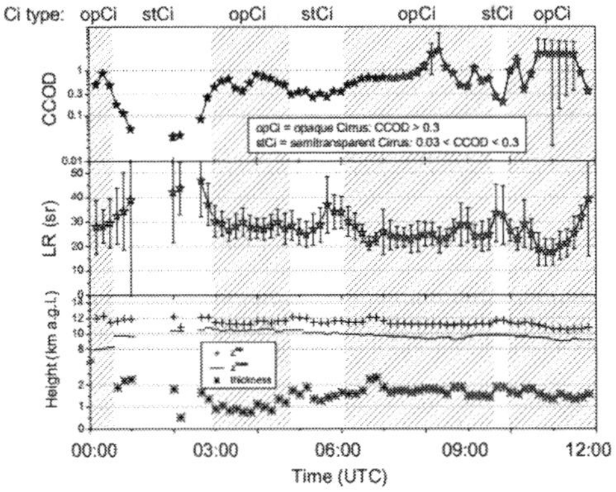

Figure 2. Optical and macrophysical properties: (UP) Cirrus Cloud Optical Depth (CCOD, full stars); (CENTER) Lidar Ratio (LR, open stars); and (DOWN) both top and base heights ('+' and '-' symbols, respectively) and thickness (stars) of the Ci cloud layers. Ci CCOD-type is labelled at the top and marked by white (stCi) and shaded (opCi) bands. Daytime interval is shown in red.

Ci clouds are found just below the tropopause height-level (around 13 km height), showing the top and base boundaries of the Ci layers (z^{top} and z^{base}, respectively) together with their thickness along the day until noon in Figure 2-down. z^{base} ranges between 8.0 and 10.8 km above ground level (a.g.l.), and z^{top} ranges between 10.7 and 12.3 km a.g.l. Their thickness varies between 0.5 and 2.3 km. Temperatures at the Ci top levels (data not shown) are lower than that threshold determined for ice cloud formation (-38°C).

3.2 Surface radiation levels

Surface solar radiation measurements and all sky images (not shown) show a good agreement with the lidar retrievals. Global (G), diffuse (F), and direct (D) components of the total integrated solar spectrum (including the Erythematically-weighted radiation, UV-Ery) reaching the ground are examined. The discussion will be addressed for the first half of the daytime since lidar measurements are only available until noon (see Figure 3, shaded area).

The solar radiation intensity increases (decreases) with the solar elevation (zenithal angle) reaching the maximum at noon. Short wavelengths are rather less affected by Ci clouds than those in the visible range because the diffuse/global ratio is larger in the shortwave range and consequently less sensitive to the presence of these high Ci clouds in the sun path [5]. Indeed, the effect in the Erythemal radiation (UV-Ery in Fig. 3) is only significant when the CCOD is large enough and Ci clouds are in the sun path. This is in agreement with lidar results, since a dense opCi occurrence with CCOD values around 2 is observed in the interval 10:30-11:30 UTC, affecting clearly all the radiation components, and especially evident for the UV-Ery radiation (see Figs. 2-up and 3).

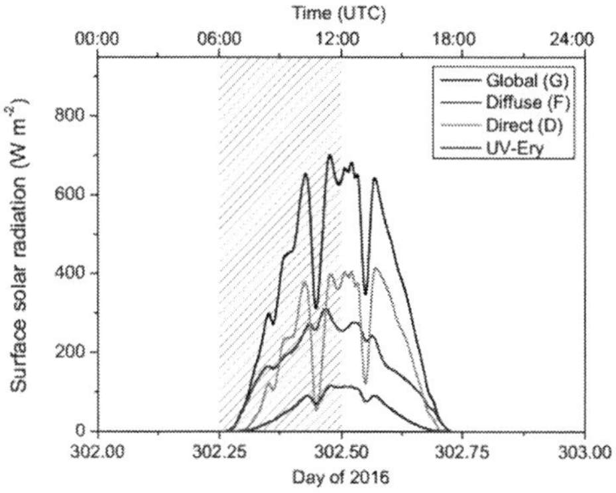

Figure 3. Daily evolution of the measured Global (G), Diffuse (F), Direct (D) and Erythemal effective UV (UV-Ery) radiations on the surface on 28 October 2016. Shaded area corresponds to the simultaneous lidar measurement period.

However, the most sensitive to these Ci clouds is the direct component, where larger variations are also observed, even with CCOD < 2 (see Figs. 2-up and 3). This different behaviour between the Erythemal radiation (UV-Ery) and the direct total solar component (D) suggests the presence of Cirrus instead of lower clouds attenuating both diffuse and direct components of the radiation. The diffuse/global ratio, as expected, increases under Ci cloudy skies [5].

The fraction of overcast sky and the density of the Ci clouds changes from morning to noon and, consequently, the evolution of the solar radiation (as the solar zenith angle decreases) departs from the typical 'bell' that would be expected in case of clear sky conditions (see Fig. 4). In addition, it is observed that this solar affected by Ci clouds 'bell' is different between measurements and simulations. However, the daytime strongest attenuation measured at around 11:00 UTC coincides with a quite dense Ci cloud (CCOD ~ 2), as before exposed, detected by the lidar in front of the sun, and this sun occultation strongly affected both UV and total radiation budget.

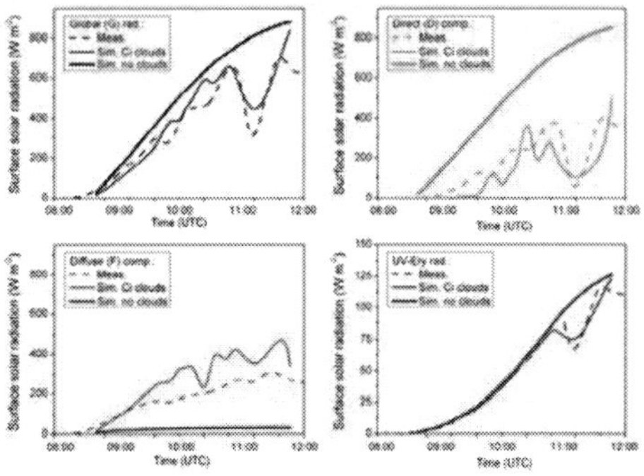

Figure 4. (From left to right, and up to down): Surface Global (G, black), Direct (D, green), Diffuse (F, red), and Erythemal effective UV (UV-Ery, blue; note the different solar intensity-axis scale) radiations during the first half part of the daytime on 28 October 2016, as obtained from: measurements (shaded line), and simulations under both no clouds (thicker solid lines) and Ci clouds presence (thinner solid lines).

Therefore, differences between measurements (registered under a given cloud cover state of the sky) and simulations (performed using the lidar optical products derived for Ci detected just above the lidar sky viewing) observed from morning to noon are associated to the fact that the Ci clouds responsible of sun occultation (strong solar attenuation) can be out from the field of view of the lidar system at that given moment.

4 CONCLUSIONS

A case study of Ci clouds observed over the Atmospheric Observatory 'El Arenosillo' (ARN) along the day (unfortunately, only lidar data until noon) has been examined. In general, LR and CCOD present an apparent opposite relationship: higher CCOD values correspond to lower LRs. The direct solar component is the most affected radiation by Ci presence. A good agreement is achieved between lidar and radiation data: dense opCi occurrence with a CCOD ~ 2 is observed in the interval 10:30-11:30 UTC when radiation components are strongly affected. Differences found between measurements and simulations of the surface solar radiation can be related to the particular sun occultation situation and the cloud cover portion as seen by the solar instrumentation and the field of view of the lidar system. Further Ci cases are to be examined in a future work.

ACKNOWLEDGEMENTS

This work was supported by the Spanish Ministerio de Economía y Competitividad (MINECO) under grants CGL2014-55230-R (AVATAR) and CGL2014-56255-C2-2-R (ICARO).

References

[1] Myhre et al., 2013: Anthropogenic and natural radiative forcing. In: Stocker, T.F., Qin, D., Plattner, G.-K., Tignor, M., Allen, S.K., Boschung, J., Nauels, A., Xia, Y., Bex, V., Midgley, P.M. (Eds.), Climate Change 2013: The Physical Science Basis. Contribution of Working Group I to the Fifth Assessment Report of the Intergovernmental Panel on Climate Change. *Cambridge University Press*, Cambridge, NY, USA.

[2] Flynn et al., 2007: Novel polarization-sensitive micropulse lidar measurement technique, *Optics Express*, **15** (6), 2785-2790.

[3] Córdoba-Jabonero et al., 2017: Diversity on subtropical and polar cirrus clouds properties as derived from both ground-based lidars and CALIPSO/CALIOP measurements, *Atmos. Res.*, **183**, 151-165.

[4] Dahlback and Stammes, 1991: A new spherical model for computing the radiation field available for photolysis and heating at twilight, *Planet. Space Sci.*, **39**, 671-683.

[5] Costa et al., 2014: Surface UV Radiation in the South of Portugal: Monitoring and Assessment of Cloud Effects, *AIP Conference Proceedings*, **04**, 1531, 852.

REFINEMENT OF THE CALIOP CLOUD MASK ALGORITHM

Shuichiro Katagiri[1]*, Kaori Sato[1], Kohei Ohta[1], Hajime Okamoto[1]
[1]*Research Institute for Applied Mechanics, Kyushu University, Kasuga, Kasuga koen 6-1, 816-8580, JAPAN, *Email: katagiri@riam.kyushu-u.ac.jp*

ABSTRACT

A modified cloud mask algorithm was applied to the CALIOP data to have more ability to detect the clouds in the lower atmosphere. In this algorithm, we also adopt the fully attenuation discrimination and the remain noise estimation using the data obtained at an altitude of 40 km to avoid contamination of stratospheric aerosols. The new cloud mask shows an increase in the lower cloud fraction. Comparison of the results to the data observed with a PML ground observation was also made.

1 INTRODUCTION

In the Earth climate system, clouds play a major role in the energy budget by means of atmospheric radiation processes [1] [2] and water cycles through distribution of rain on the ground [3], and besides transport water into the higher atmosphere. These important influences on the earth's climate are so complicated that a lot of observations cannot draw a convergent result. As for the atmospheric radiation, clouds have two competing effects: they partly reflect the solar radiation and therefore cool the atmosphere, while they partly interrupt the infrared radiation from the earth and thereby warm the atmosphere. The data sets of vertical cloud distribution are accordingly needed for the evaluation of the Earth climate system, especially for general circulation models (GCMs). Cesana *et al.* [4] evaluated cloud phase products from the Cloud-Aerosol Lidar and Infrared Pathfinder Satellite Observations (CALIPSO) [5] which include our product, and showed inconsistency among the products. Consequently, our algorithm using the CALIPSO data to make the cloud mask is to be refined. This algorithm is summarized in [6][7], and we improve it.

2 METHODOLOGY

We use the attenuated total backscattering coefficient β at 0.532 μm observed by the Cloud Aerosol Lidar with Orthogonal Polarization (CALIOP) onboard CALIPSO to discriminate the cloudy pixels. When the pixels with larger β than the threshold value β_{th} then the ones are recognized as the cloudy pixels. Mainly our algorithm is based on the threshold value β_{th} derived from the results of shipborne observations in the mid-latitudes [8] and in the tropical western Pacific [9] which is described as

$$\beta_{th}(z,R) = \frac{\beta_{th,aerosol} + \beta_{th,noise}(z,R)}{2} - \frac{\beta_{th,aerosol} - \beta_{th,noise}(z,R)}{2} tanh(z-5), \quad (1)$$

where

$$\beta_{th,aerosol} = 10^{-5.25}\ [1/\mathrm{m/sr}], \quad (2)$$

and

$$\beta_{th,noise}(z,R) = \{P_m(z,R) + P_n + \sigma_n\}R^2 \quad (3)$$

where z and R denote altitude of the target pixel and the distance between CALIPSO and the surface of the earth, respectively, P_n and σ_n are the remain noises mean and standard deviation, respectively, after the calibration which is estimated with the data obtained from the high altitude to reduce the effects of clouds, aerosols, and polar stratospheric clouds (PSCs), and P_m is described as

$$P_m(z,r) = \beta_{mol}(z)/R^2. \quad (4)$$

Here $\beta_{mol}(z)$ is the volume molecular backscattering coefficient at an altitude of z calculated with the reanalysis data according to Hostetler *et al.* [10]. In this refinement of the algorithm, we changed the altitude where $\beta_{th,noise}$ is determined from 20 km to 40 km to avoid the stratospheric aerosol contamination. And if the pixels below the higher cloud layer have dominant signals, then the ones also are masked to be cloudy ones. In addition, we mask out the pixels on which the lidar signals are fully attenuated to discard undistinguishable pixels.

 https://doi.org/10.1051/epjconf/201817605043

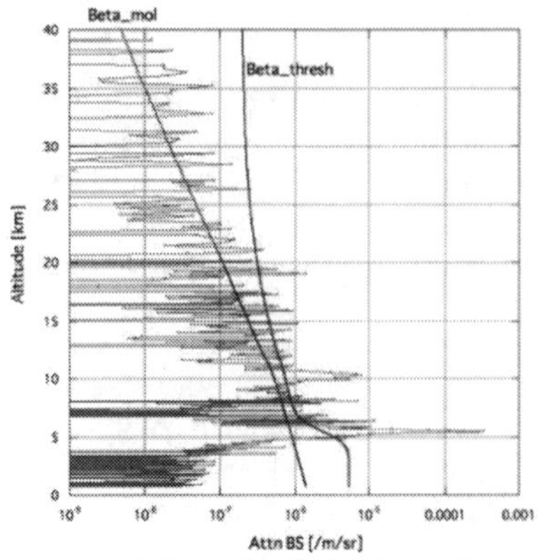

Figure 1 Example of the relationship, β (red line), β_{mol} (green) and β_{th} (blue).

Figure 1 illustrates the example of the raw data, which is the attenuated total backscattering coefficient β, received by CALIPSO (red lines), the calculated volume molecular backscattering coefficient β_{mol} (green), and the threshold value β_{th} (blue).

3 RESULTS

Figure 2 shows the raw signal β received by CALIOP (top panel) and the fully attenuated mask (bottom, blue colored) which is decided by referring to whether the surface return can be detectable or not. This fully attenuated mask roughly seems to be good, although sometimes noises near the surface disturb masking.

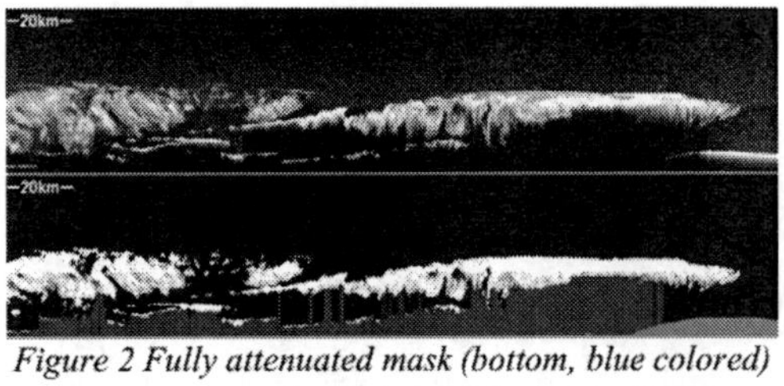
Figure 2 Fully attenuated mask (bottom, blue colored) and raw signal β *(top)*

The area colored blue is recognized as the fully attenuated pixels, therefore when the cloud fraction is estimated, they are not counted. Hence this improves the statistical cloud fraction. Figure 3 shows the cloud mask over Ny-Ålesund, Norway, where the left panel shows the original cloud mask which omits the detailed signals from clouds, the center panel shows the refined mask which deals with smaller signals than the original cloud mask, and the right panel shows the cloud mask using Micro pulse lidar observation from the ground. At this moment, there exists very optically thick low cloud below the higher cloud. These figures indicate that refined mask is extended around the top of the low optically thick cloud (below the red area in the center panel) compared to the original one, that is, the refined algorithm can mask deeper into the optically thick cloud than original one. The MPL laser beam cannot be transmitted through this thick low cloud, on the other hand, the laser beam from CALIPSO also is fully attenuated at the top of the cloud. Both lidar beam signals suggest the low cloud top is optically very thick, and therefore we should take into account the fully attenuated pixels in case that there is so optically thick cloud that the lidar cannot detect anything below the cloud.

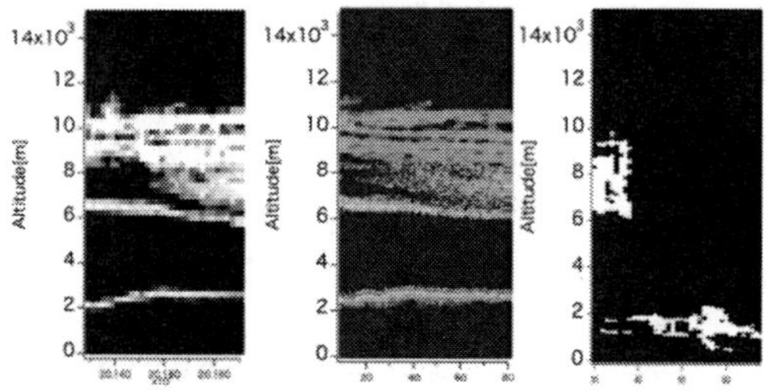

Figure 3 Cloud mask over Ny-Ålesund, original (left), refined (center), and Micro pulse lidar (MPL) (right)

Figure 4 shows the latitudinal cross section of the ice cloud fraction in January, 2009.

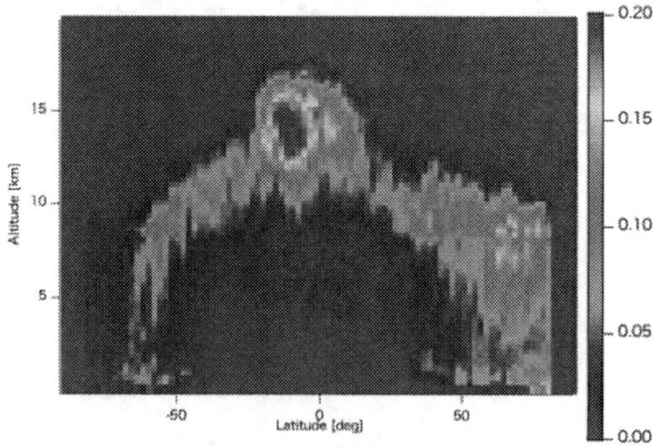

Figure 4 Cloud fraction (ice cloud)

Figure 5 shows the latitudinal cross section of the water cloud fraction. These two averaged cloud fractions have the vertical resolution of 240 m, and the horizontal resolution of 1.1 km. We use the cloud phase discrimination algorithm made by Yoshida [11], which uses the relation between the depolarization ratio σ at 0.532 μm and the differential of attenuated backscattering coefficients x' at 0.532 μm to distinguish ice and water particles, where x' is defined as

$$x' = log_{10}\left[\frac{\partial\beta}{\partial z}\right] . \quad (5)$$

Water clouds generally have larger extinction coefficients than ice clouds at 0.532 μm, therefore the ratio of water clouds between x' and σ getting larger more slowly than ice clouds. This property is used in the cloud phase discrimination algorithm.

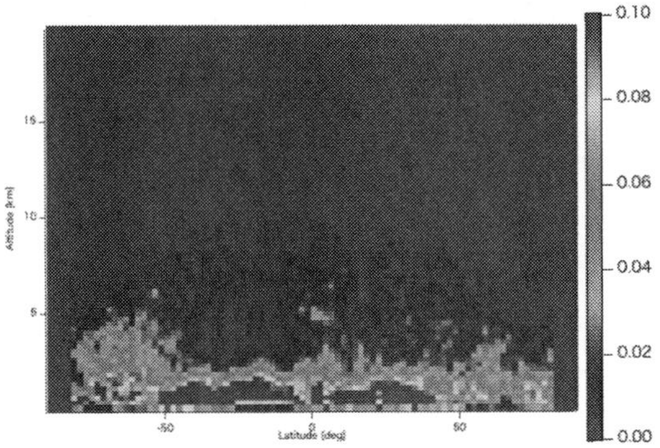

Figure 5 Cloud fraction (water cloud)

The refined algorithm result of the ice cloud mask is almost the same as the original one. However, the result of the water cloud takes slightly larger values than original algorithm. This indicate that the discrimination of the fully attenuated pixel is efficient for the statistical treatment of the lower clouds.

4 CONCLUSIONS

Our algorithm of the cloud mask was refined to have the fully attenuation discrimination, therefore there can be seen larger cloud fraction especially in the water cloud cases. This reduces the uncertainty of the statistical cloud fraction. The comparison to the ground observation by MPL indicates this refinement spread the ability of masking cloud pixels in lower optically thick clouds.

ACKNOWLEDGEMENTS

CALIPSO data were obtained from the Atmospheric Sciences Data Center at NASA's Langley Research Center. MPL is conducted by National Institute of Polar Research (NIPR). This study was supported by JSPS KAKENHI Grant Numbers JP17H06139, JP15K17762, JP25247078 and JP15K17762, by Green Network of Excellence (GRENE) Program Arctic Climate Change Project and the Arctic Challenge for Sustainability (ArCS), by the Japan Aerospace Exploration Agency for EarthCARE Research Announcement and by the Collaborative Research Program of Research Institute for Applied Mechanics, Kyushu University.

References

[1] Charlock, T. P., Ramanathan, V., 1985: The albedo field and cloud radiative forcing produced by a general circulation model with internally generated cloud optics, *J. Atmos. Sci.* **42**, 1408–1429.

[2] Ramanathan, V., Cess, R. D., Harrison, E. F., Minnis, P., Barkstrom, B. R., Ahmad, E., Hartmann, D., 1989: Cloud-radiative forcing and climate: Results from the Earth Radiation Budget Experiment, *Science* **243**, 57–63.

[3] Trenberth, K. E., Fasullo, J. T., Mackaro, J., 2011: Atmospheric moisture transports from ocean to land and global energy flows in reanalyses, *J. Climate* **24**, 4907–4924, doi:10.1175/2011JCLI4171.1.

[4] Cesana, G., Chepfer, H., Winker, D., Getzewich, B., Cai, X., Jourdan, O., Mioche, G., Okamoto, O., Hagihara, Y., Noel, V., Reverdy, M., 2016: Using in situ airborne measurements to evaluate three cloud phase products derived from CALIPSO, *J. Geophys. Res. Atmos.* **121**, 5788–5808, doi:10.1002/ 2015JD024334.

[5] Winker, D. M., Vaughan, M. A., Omar, A. H., Hu, Y., Powell, K. A., Liu, Z., Hunt, W. H., Young, S. A., 2009: Overview of the CALIPSO mission and CALIOP data processing algorithms, *J. Atmos. Oceanic Technol.* **26**, 2310–2323, doi:10.1175/2009JTECHA1281.1.

[6] Hagihara, Y., Okamoto, H., Yoshida R., 2010: Development of a combined CloudSat-CALIPSO cloud mask to show global cloud distribution, *J. Geophys. Res. Atmos.* **115**, D00H33, doi:10.1029/2009JD012344.

[7] Hagihara, Y., Okamoto, H., Luo, Z. J., 2014, Joint analysis of cloud top heights from CloudSat and CALIPSO: New insights into cloud top microphysics, *J. Geophys. Res. Atmos.* **119**, 4087–4106, doi:10.1002/2013JD020919.

[8] Okamoto, H., Nishizawa T., Takemura, T., Kumagai, H., Kuroiwa, H., Sugimoto, N., Matsui, I., Shimizu, A., Emori, S., Kamei, A., Nakajima, T., 2007: Vertical cloud structure observed from shipborne radar and lidar: Midlatitude case study during the MR01/K02 cruise of the research vessel Mirai, *J. Geophys. Res. Atmos.* **112**, D08216, doi:10.1029/2006JD007628.

[9] Okamoto, H., Nishizawa, T., Takemura, T., Sato, K., Kumagai, H., Ohno, Y., Sugimoto, N., Shimizu, A., Matsui, I., Nakajima, T., 2008, Vertical cloud properties in the tropical western Pacific Ocean: Validation of the CCSR/NIES/FRCGC GCM by shipborne radar and lidar, *J. Geophys. Res. Atmos.* **113**, D24213, doi:10.1029/2008JD009812.

[10] Hostetler, C. A., *et al.*, 2006: CALIOP algorithm theoretical basis document, calibration and level 1 data products, Rep. PC-SCI-201, NASA Langley Res. Cent., Hampton, Va.

[11] Yoshida, R., Okamoto, H., Hagihara, Y., Ishimoto H., 2010, Global analysis of cloud phase and ice crystal orientation from Cloud-Aerosol Lidar and Infrared Pathfinder Satellite Observation (CALIPSO) data using attenuated backscattering and depolarization ratio, *J. Geophys. Res. Atmos.* **115**, D00H32, doi:10.1029/2009JD012334.

WILDFIRE SMOKE TRANSPORT AND IMPACT ON AIR QUALITY OBSERVED BY A MULLTI-WAVELENGTH ELASTIC-RAMAN LIDAR AND CEILOMETER IN NEW YORK CITY

Yonghua Wu, Wilson Peña, Barry. Gross, Fred Moshary*

NOAA-CREST and the City College of New York, NY, NY 10031, USA. *moshary@ccny.cuny.edu

ABSTRACT

The intense wildfires from the western Canada in May 2016 injected large amount of smoke into the atmosphere. This paper presents integrated observation of the event by a lidar, ceilometer, and satellite together with models and an assessment of smoke plume impacts on local air quality in New York City (NYC) area. A dense aloft plume on May 20 and a boundary layer plume on May 25 are analyzed. The smoke mixing into planetary-boundary-layer (PBL) and strong diurnal variation of PBL-top are shown. For the 2ndcase, the ground $PM_{2.5}$ measurements show a significant increase in both the urban and upwind non-urban areas of NYC. The smoke sources and transport paths are further verified by the satellite observations and HYSPLIT model data.

1 INTRODUCTION

An intense wildfire began on May 1, 2016, roared into Fort McMurray, Alberta, Canada, forcing the largest wildfire evacuation in the province's history https://en.wikipedia.org/wiki/Fort_McMurray).
The wildfire spread across approximately 1,500,000 acres across northern Alberta and into Saskatchewan before it was declared to be under control on July 5, 2016. The smoke from the wild fires travelled across the US to the Gulf coast, even to Europe according to the satellite imageries (https://weather.com/science/environment/news/fort-mcmurray-wildfire-smoke-travels-europe). The emissions from the wild fires include aerosols and gaseous compounds such as carbon monoxide (CO), carbon dioxide (CO_2), nitrogen oxide (NOx), etc., and can adversely affect the downwind air quality and climate radiation in the regional-continental scale through long-range transport [1].

In addition, the wildfire emissions have acute or long-term health implications on the exposed populations, according to the various health agencies. Among the major components of wildfire smoke, fine particles ($PM_{2.5}$) affect ambient air quality and have various effects on human health [2]. The study by Le et al. [3] suggests that rapid increases in $PM_{2.5}$ concentrations resulting from wildfire smoke can affect the health of elderly populations with increased rate of hospitalization for respiratory and cardiovascular diagnoses in the downwind area thousands of kilometers away from the fires. Weber et al. [4] show that high $PM_{2.5}$ exposures is associated with increased risk of asthma, myocardial infarction and heart failure in NYC metropolitan and surrounding areas.

In this study, the observations of Canadian smoke plume transport to NYC are analyzed with a multi-wavelength Elastic-Raman lidar, a ceilometer (Vaisala CL-51), an AERONET sun/sky radiometer, satellite observations and the model output products on May 20 and 25, 2016. Ground $PM_{2.5}$ in the urban and upwind non-urban area of NYC are also investigated to assess the smoke influences on the local air quality, as well as its speciation such as organic carbon (OC), elemental carbon (EC) and sulfate. Transport paths and origins of smoke plumes are investigated by the NOAA-HMS (Hazard Mapping System) fire and smoke product and HYSPLIT backward trajectory analysis.

2 METHODOLOGY

At CCNY, a suite of ground-based remote sensing instruments are deployed on a building roof, near a standard surface air quality monitoring station by the New York State Department of Environment Conservation. The instruments include a 3-wavelength Elastic-Raman lidar, a ceilometer, and an AERONET Cimel sun/sky radiometer. The lidar transmits three wavelengths (1064-, 532- and 355-nm) with a flash lamp-pumped Nd: YAG laser at a repetition rate of 30 Hz (Spectra-physics Quanta-Ray PRO-320) [5]. Three elastic-scattering and two Raman-scattering returns by nitrogen and water vapor molecules excited by 355-nm laser output are collected by a receiver telescope (Ø50-

cm). The signals are acquired by LICEL transient recorders (TR40-160) and recorded with 1-min average, and full return signals are detected with the starting altitude 0.5 km. The ceilometer measures laser backscatter at a wavelength of 910 nm with a range resolution of 10 m [6]. It makes fully automatic 24-hr/7-day observations in all-weather condition. It provides attenuated backscatter and calibrated range-corrected backscatter coefficient products, which allows monitoring of the PBL process and evolution for air pollution studies. The aerosol contribution in the lidar overlap region (<0.5-km) can be corrected by using the ceilometer-measured backscatter profile near the surface or by extending a constant of aerosol extinction at 0.5 km to the surface. In addition, AERONET products from a Cimel sun/sky radiometer provide column aerosol optical depth (AOD), Angstrom exponent (AE), microphysical parameters (volume size distribution and refractive index and single-scattering albedo [7]. The AOD can be used to constrain the lidar-ratio in the lidar retrieval of aerosol extinction profile under clear sky.

3 RESULTS and DISCUSSION

3.1 Case-1: Aloft plumes on May 20, 2016

Figure 1(a) shows the aloft plumes at 5 ~ 7 km altitude observed by the lidar on May 20, 2016. The aerosol extinction coefficients in the aloft layer are much larger than the ones in the PBL, and some thin layers are seen below 5-km. The Angstrom exponents in Fig.1(b) are around 1.8 in the aloft layer, which indicate a fine-mode dominant particle size distribution. They are likely associated with the smoke. The total AOD attains 0.4 at 532-nm and are mostly contributed from the aloft plume as shown in Fig.1(c). Meanwhile, the satellite NOAA-GASP (GOES Aerosol/Smoke Product) AOD map clearly shows a high-AOD belt (red color in Fig.1d) from Canada to NYC area, which indicates the smoke transport path. Furthermore, Fig.1(e) shows the wildfire point sources (red +) and smoke area (yellow) from the NOAA-HMS product, and the air backward trajectories with the end point at CCNY at 5- and 7-km altitude, 60-hour duration from the NOAA-HYSPLIT model. The results indicate that the main wildfire points were located in the west Canada and the border of US-Canada, and an area of light to moderately dense residual smoke was moving east across a large

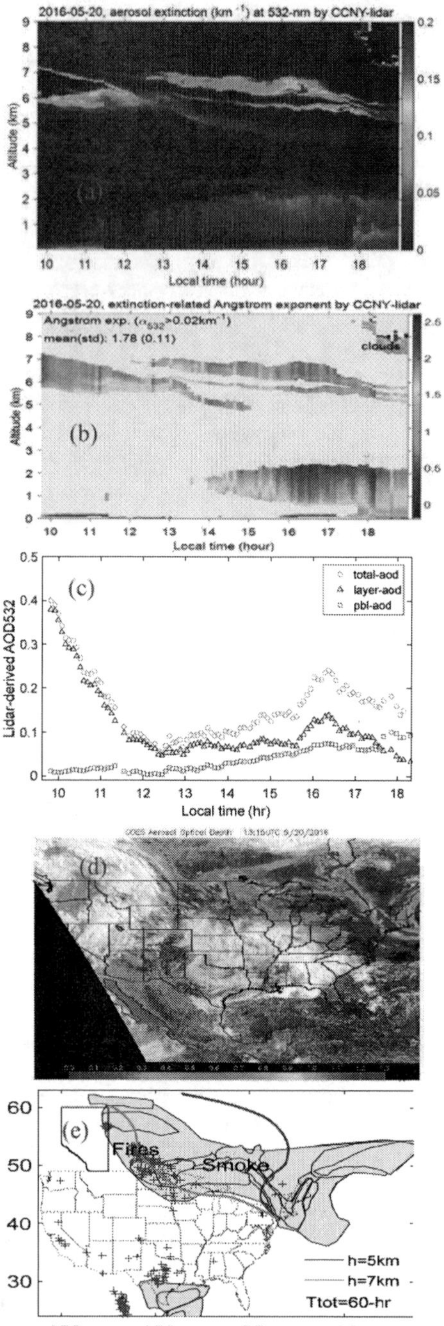

Fig.1 (a)-(c) Aerosol extinction, Angstrom exponent and AOD from the CCNY-lidar, (d) AOD spatial distribution from NOAA-GASP, (e) Fire points (red+) and smoke (yellow) and trajectories on May 20, 2016.

Section of the northern US/southern/eastern Canada.

3.2 Case-2: PBL-plumes on May 25, 2016

Figure 2 shows the time-height distributions of aerosols measured by the ceilometer and lidar on May 25, 2016. Fig.2(a) indicates two aerosol layers in the PBL: one below 1-km starting from 6:00 am and other one at ~3 km at noon. In particular, the PBL-tops showed a dramatic increase that reached up to 3 km in the afternoon, which indicate strong convective or turbulent process that resulted in plumes mixing well within the PBL. In Fig.2(b) from the lidar daytime measurements, the high PBL-tops over the convective daytime period are consistent with the ceilometer data while there are some aloft aerosols observed above the PBL. The aerosol extinction coefficients are derived from the CCNY-lidar and are shown in Fig. 2c. On May 25, the extinction of aerosols was as high as 0.2 km^{-1} at 532-nm with the Angstrom exponents of 1.5~1.7. Meanwhile, the consistent aerosol extinctions at 355-nm between the two independent retrievals from the Raman- and Mie-scattering signals, indicate the reliability of retrievals. In addition, the lidar-derived aerosol optical depths vary from 0.2 to 0.4 at 532-nm. The mean of Raman-lidar derived lidar ratio in the PBL is 62.7 +/- 11 sr at 355-nm.

Figure 3 shows the fires sources (red +), smoke areas (yellow) and air back-trajectories with the ending point at CCNY (urban), Queens College(urban) and Newburgh (non-urban) sites (12:00 UTC, 72-hr long, at the altitudes of 0.5-km, 1-km and 2-km). Interestingly, the similar trajectories of transport are indicated among the three sites and at the different altitudes. There were numerous wildfires at Fort McMurray in Alberta, Canada and some wildfires in the southwest US. The trajectories indicate that the aloft smokes over NYC was transported from Canada.

Figure 4 gives the temporal variation of ground $PM_{2.5}$ in the urban (CCNY) and nonurban (Newburgh) or upwind areas of NYC on May 24-29, 2016. A significant increase can be seen starting on May 25; and the $PM_{2.5}$ concentrations show consistent temporal variation at these sites. The $PM_{2.5}$ concentrations at CCNY and Newburgh were well correlated (R=0.84). Such good correlation probably indicates that the $PM_{2.5}$ at the urban and upwind non-urban sites were from the similar sources, which is consistent

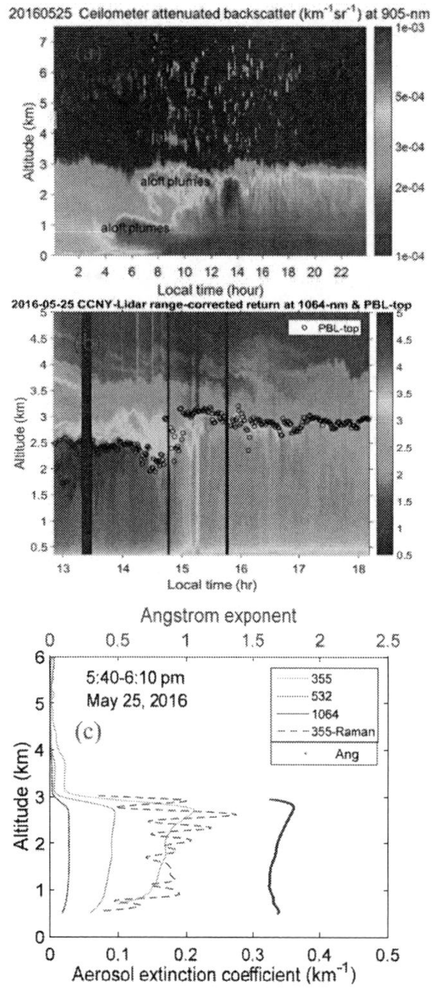

Fig.2 Attenuated backscatter, extinction and Angstrom exponent from ceilometer and lidar on May 25, 2016

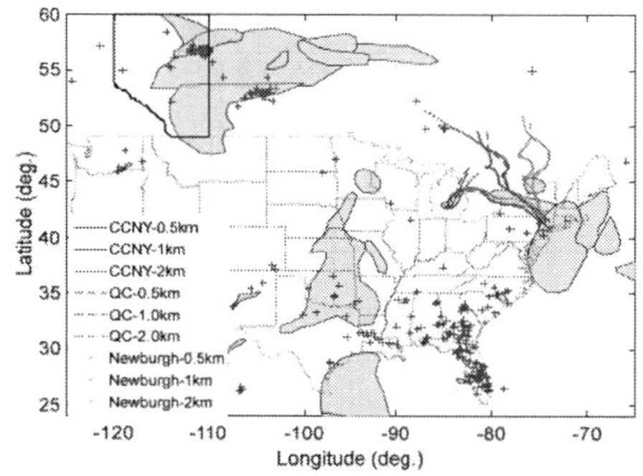

Fig.3 Fire sources (red+), smoke (yellow), backward trajectories (green-black lines) by NOAA-HMS and HYSPLIT model on May 25

https://doi.org/10.1051/epjconf/201817605044

with the long-range transport of aerosols.

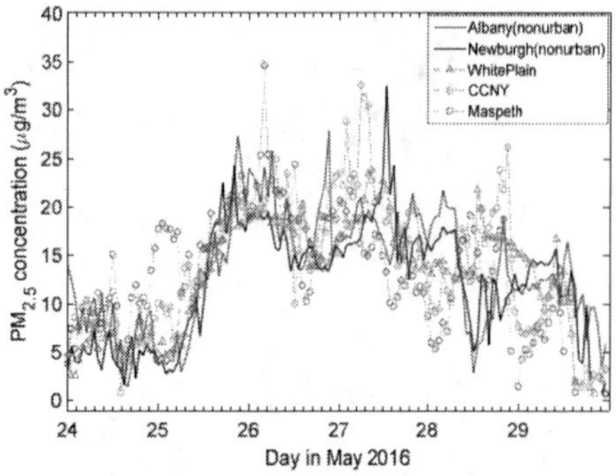

Fig.4 Ground $PM_{2.5}$ concentrations in the urban and upwind non-urban area of NYC.

We analyzed the temporal variation of the OC, EC, sulfate of $PM_{2.5}$ and trace gases (CO and O_3) at Queen College site in NYC. Both the OC and EC showed consistent increase for this event. For instance, the OC increased from 2 μg/m^3 to 8 μg/m^3 while the EC increased up to 4 μg/m^3. Meanwhile, the sulfate also indicated an increase trend. These values are much larger than the monthly average of OC (1.0~4 μg/m^3), EC (0.5~1.4 μg/m^3) and sulfate (1~3 μg/m^3) in NYC [8, 9]. Thus, the results indicate the smoke-like plume transport and mixture with the industrial aerosols (sulfate) in the upwind region along the transport paths. Furthermore, the peak values of hourly ozone concentrations on May 25-26 were greater than the NAAQS value of 70 ppb.

4 CONCLUSIONS

This paper analyzes two episodes of the wildfire smoke transport from Canada to NYC on May 20 and 25, 2016. The optical properties of aerosols and their influences on the local air quality are assessed. Aloft plumes at 6~7 km on May 20, contributed to the total AOD. The PBL-plumes below 3-km on May 25 are well mixed and the strong diurnal variation of PBL-top are demonstrated from the profiling observations of ceilometer and lidar, along with and strong convection process and high PBL-top at ~3 km in the daytime. Importantly, the ground $PM_{2.5}$ concentrations show consistent increase from 5 to 30 μg/m^3 in the urban and upwind non-urban area of NYC, and a good correlation (R=0.8) indicates similar sources that for the regional transport event. The increases of OC, EC and sulfate indicate the smoke plumes transport and mixture with the industrial emissions in the upwind area of the region studies. The sources and transport paths from the western Canada to the site are clearly demonstrated by the NOAA-HMS and HYSPLIT model.

ACKNOWLEDGEMENTS. This study was supported by National Oceanic and Atmospheric Administration (NOAA) under the CREST Grant # NA11SEC4810004.

References

[1] Liu, Y., and et al., 2014: Wildland fire emissions, carbon, and climate: Wildfire–climate interactions. Forest Ecol. Manage., 317, 80–96.

[2] Fann, N., et al., 2016: Ch.3: Air Quality Impacts. The Impacts of Climate Change on Human Health in the United States: A Scientific Assessment. U.S. Global Change Research Program, Washington, DC, 69–98.

[3] Le, G. E., et al., 2014: Canadian Forest Fires and the Effects of Long-Range Transboundary Air Pollution on Hospitalizations among the Elderly, *ISPRS Int. J. Geo-Inf., 3*, 713-731.

[4] Weber, S. A., et al., 2016: Assessing the impact of fine particulate matter (PM2.5) on respiratory-cardiovascular chronic diseases in the New York City Metropolitan area using Hierarchical Bayesian Model estimates, Environ. Res., 151, 399-409.

[5] Wu, Y., S. Chaw, B. Gross, et al., 2009: Low and optically thin cloud measurements using a Raman-Mie lidar, *Appl. Opt.*, **48**, 1218-1227.

[6] Münkel, C., et al., 2004: New optical concept for commercial lidar ceilometers scanning the boundary layer. Proc. SPIE, 5571, 364-374.

[7] Holben, B.N., and et al., 1998: AERONET A federated instrument network and data archive for aerosol characterization, *Rem. Sens. Environ.*, 66, 1-16.

[8] Rattigan, O. V., et al., 2010: Multi-year hourly PM2.5 carbon measurements in New York: Diurnal, day of week and seasonal patterns, Atmos. Environ., 44 (16), 2043-2053.

[9] Rattigan, O. V., et al., 2016: Long Term Trends in New York: $PM_{2.5}$ Mass and Particle Components, Aerosol and Air Quality Research, 16(5), 1191-1205.

ATMOSPHERIC CO_2 MEASUREMENTS WITH A 2-μm DIAL INSTRUMENT

Erwan Cadiou[1,2], Jean-Baptiste Dherbecourt[1]*, Guillaume Gorju[1], Jean-Michel Melkonian[1], Antoine Godard[1], Jacques Pelon[3], Myriam Raybaut[1]

[1]ONERA, The French Aerospace Lab, Chemin de la Hunière, F-91761 Palaiseau cedex, France
[2]CNES, centre de Toulouse, 18 avenue Edouard Belin, F-31401 Toulouse cedex 9, France
[3]LATMOS, Université Pierre et Marie Curie, 4 Place Jussieu Paris, France

ABSTRACT

We report on ground-based atmospheric concentration measurements of carbon dioxide, using a pulsed direct detection differential absorption lidar operating at 2051 nm. The transmitter is based on a tunable parametric source emitting 10-mJ energy, 10-ns duration Fourier-limited pulses. Range resolved concentration measurements have been carried out on the aerosol back-scattered signal. Cloud signals have been used to get long range integrated-path measurements.

1 INTRODUCTION

Space born differential absorption lidar (DIAL) measurements give some new opportunities to better understand the cycles of the three main greenhouse gases: CO_2, CH_4 and water vapor. Such instruments allow monitoring these gases during day and night and at high latitude which current passive missions cannot do. The spatially resolved data provided would better constraint the greenhouse gases transport models and help identify their sources and sink [1,2]. Then several DIAL missions are in preparation: Merlin (CNES/DLR) to measure CH_4 and ASCENDS (NASA) for CO_2. One of the challenges of these missions is to get robust laser sources emitting at least two wavelengths (inside and outside an absorption line of the gas) to perform the DIAL measurement.

In the 2 µm spectral region, the gases of interest present particularly well-suited lines for space-borne applications which can be efficiently addressed by solid state laser technologies [3,4]. In this paper we present long-range absorption measurements on the R30 2-µm absorption line of the CO_2. The lidar transmitter is based on a nested cavity oscillator (NesCOPO) whose wide tunability in the 1.98 - 2.3 µm range allows targeting the three species of interest [5,6].

2 LASER TRANSMITTER

The experimental set-up is depicted in Fig. 2. The whole system is pumped by a single frequency Nd:YAG laser delivering 100 mJ-energy, 12 ns pulses at a 30 Hz repetition rate. In order to reach a high energy level while maintaining a nearly diffraction limited beam quality, a MOPA architecture has been implemented.

The master oscillator is a NesCOPO with a 10 mm long type II periodically poled lithium niobate (PPLN) nonlinear crystal used close to degeneracy and composed of several gratings with different poling periods. Single-longitudinal-emission is ensured by a dissociation of the two-cavity length in the 10 % range, leading to a single coincidence of signal and idler modes within the parametric gain bandwidth. The two mirrors of the signal cavity are mounted on piezo-electric transducers allowing shifting the signal and idler mode combs. In this way, the coincidence between the two combs can be adjusted and the emitted wavelength tuned with a very high precision.

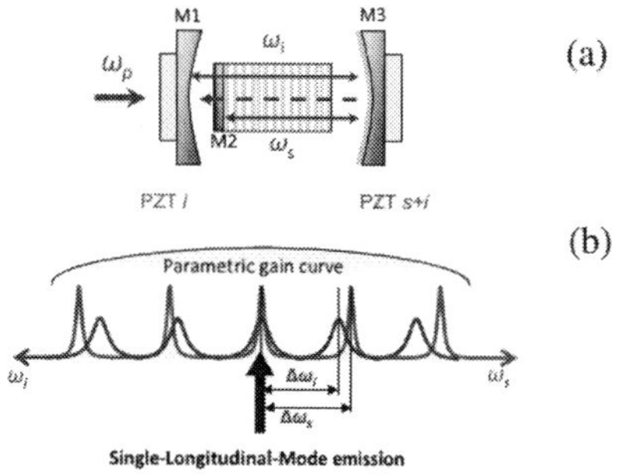

Figure 1 (a) NesCOPO device. Signal and idler cavities can be adjusted independently by the mean of piezoelectric transducers. (b) Signal and Idler modes overlap: with adequate dissociation, a single frequency is emitted.

 https://doi.org/10.1051/epjconf/201817605045

The NesCOPO is followed by parametric amplifier stages including a type 0, 12 mm long PPLN crystal and four 20 mm long bulk KTP (Potassium titanyl phosphate) crystals. The oscillation build-up time of the NesCOPO is compensated by a delay line inserted in the pump beam path before the amplifiers. To maintain the beam quality at the output the KTP crystals are oriented in a walk-off compensation configuration. A dichroic mirror is inserted between the second and the third crystal to mitigate back-conversion defects, following the approach from [7].

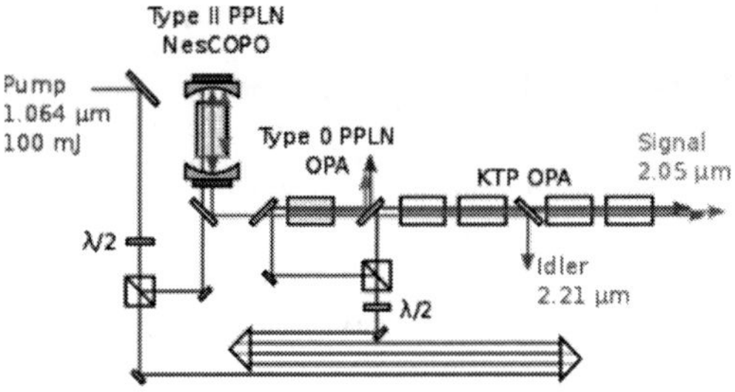

Figure 2 Schematic diagram of the MOPA overall design

The emitter delivers 10 ns pulses tunable between 1980 nm and 2300 nm. In the following lidar experiments, the available energy at 2051 nm was in the 5 - 10 mJ range. The signal beam factor was measured from the 16 % - 84 % knife edge widths method, and estimated to be $M^2 < 1.1$ for both vertical and horizontal directions.

The signal frequency is measured shot by shot by a WSU 10 High Finesse wavemeter. In order to match with the wavemeter spectral sensitivity, the signal frequency is doubled with a KTP crystal. The half maximum linewidth of the emitter was characterized to be narrower than 60 MHz through heterodyne beat-note analysis with a reference laser diode [5].

3 SPECTRAL PURITY

An important property that characterizes the ability of a laser source to perform precise DIAL measurements is its spectral purity. It is defined as the ratio of the energy contained in a narrow spectral bandwidth to the total output energy. The purity requirements for spaceborne greenhouse gases remote sensing are extremely high with a ratio better than 99.9 % [3]. The spectral purity of the parametric MOPA source can be monitored with the system depicted in Fig. 2 (a) following the approach from [8]. The measurement consists in tuning the source at the center of a spectroscopic line in order to strongly attenuate the main longitudinal mode of the signal through a multipass cell and detect the side modes transmission. In this purpose the MOPA output signal is centered on the 2051 nm CO_2 line and coupled to a CO_2 30 m multi-pass cell. The pressure of the cell is chosen at 50 mBar, which is sufficiently high to get a transmission inferior to 10^{-4} on a 600 MHz spectral bandwidth around the center of the line and sufficiently low so that the side modes of the OPO experience a small attenuation (lower than 10 %) as depicted on Fig. 2.

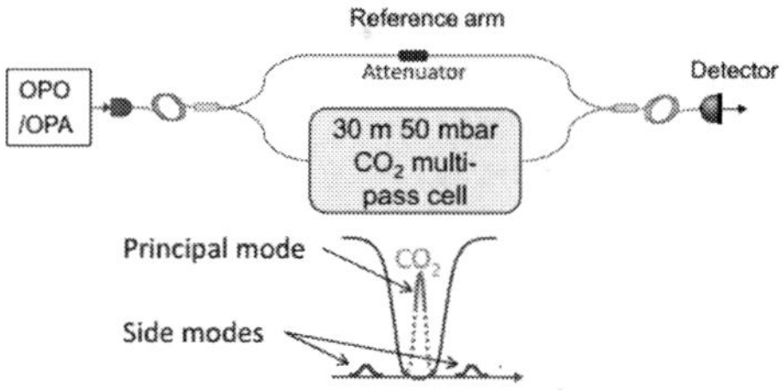

Figure 2 Schematic diagram of the line-center spectral purity measurement experimental set-up. Reference and cell transmitted signals are detected with a 100 ns delay on the same detector.

The transmission of the cell τ_{Cell} is deduced from the ratio between the energy at the output of the cell and the energy of a reference arm. In our experiment both energies were measured on the same detector. This is made possible by the 100 ns delay separating the pulses of the two arms introduced by the multi-pass-cell. An attenuator is placed on the reference arm to put the reference signal in the same amplitude range as the cell signal. The energy at the output of the cell being superior to the side mode energy, the spectral purity S can be limited by:

$$S = \frac{E_{primary\ mode}}{E_{total}} \geq 1 - \tau_{Cell} \qquad (1)$$

The cell transmission is measured on a shot by shot basis (at 30 Hz). Fig. 3 shows this transmission measurement for 400 consecutive

https://doi.org/10.1051/epjconf/201817605045

shots. The purity appears to be better than 99.98% for 95% of the shots.

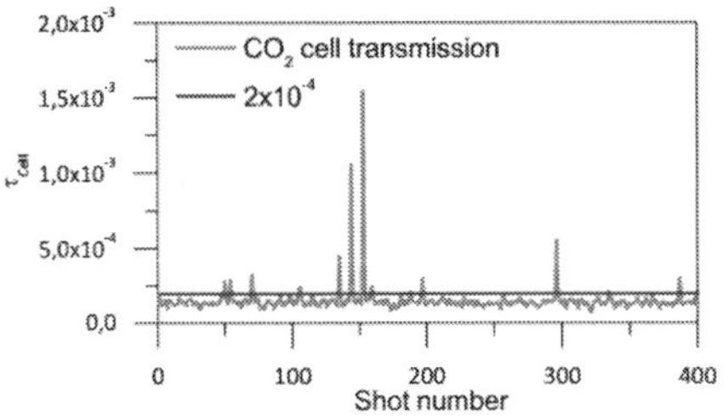

Figure 3 CO_2 cell transmission measured for 400 shots.

4 LIDAR EXPERIMENTS AND RESULTS

To implement atmospheric measurements the emitter output is coupled with a receiver collecting backscattered light. The receiver configuration is shown in Fig. 4. A Newtonian reflector focuses the backscattered light on a 500 µm pin-hole conjugated with a 30 MHz bandwidth uncooled extended InGaAs photodiode. The resulting field of view of 650 µrad is adapted to the transmitted laser beam. Emissions and reception axis overlap to get signal on a maximum range. The lidar axis is oriented at a 45° elevation angle which enables to reach clouds at the top of the boundary layer. The resulting signal is acquired with a 8-bits, 500 MHz digitizer.

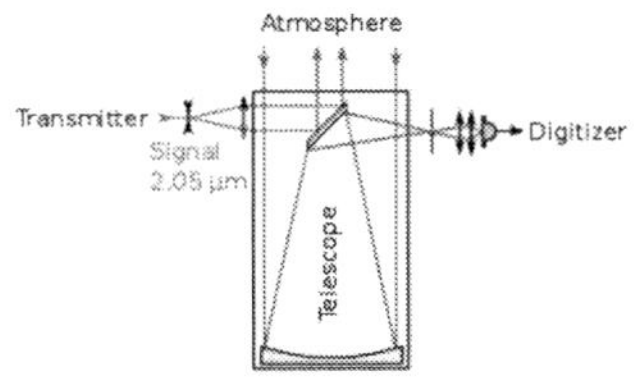

Figure 4 Schematic diagram of the receiver

Table 1 Experimental parameters

Transmitter		Receiver	
λ_{on}	2050.9689 nm	Aperture	25.4 cm
λ_{off}	2050.8008 nm	Detector type	InGaAs
Energy	5-10 mJ	NEP	3 pW/√Hz
Linewidth	60 MHz	Bandwidth	30 MHz
Divergence	500 µrad	FOV	650 µrad

First concentration measurements have been carried out on the CO_2 R30 absorption line at 2 µm. Experimental conditions are described in table 1. In order to have on-signal and off-signal backscatter coefficients as close as possible, the emitted wavelength alternates between on-line and off-line at each shot. Following the acquisition, signals are processed according to the following steps:

1. Background subtraction of blind records;
2. Subtraction of the pretrigger offset;
3. Traces are normalized in energy. For the data used in this paper, the energy is measured by the lidar detector on the light scattered on the transmitter optics.
4. Average at each wavelength;
5. Smoothing by a 50-m hann window.

The logarithm of the ratio of the resulting traces P_{on} and P_{off} yields the single path differential optical depth:

$$DAOD(r) = \frac{1}{2}\ln\left(\frac{P_{off}(r)}{P_{on}(r)}\right) \quad (2)$$

Local dry air volume mixing ratio X_{CO_2} is retrieved by evaluating the slope of the single path differential optical depth:

$$X_{CO_2}(r) = \frac{1}{\left(1 - X_{H2O}(r)\right)WF(r)}\frac{dDAOD}{dr} \quad (3)$$

where WF is the weight function defined as :

$$WF(r) = \left(\sigma_{ON}(r) - \sigma_{OFF}(r)\right)\rho_{air}(r) \quad (4)$$

with ρ_{air} the air density, and σ_{ON} and σ_{OFF} the cross sections calculated according to the Voigt profile and the Hitran 2012 data base.

Typical on-line and off-line signals are depicted on Fig. 4 (a) for a total accumulation of 8000 shots. The Fig. 4 (b) shows X_{CO_2} obtained for a $\Delta r = 200\ m$ resolution.

On each range gate Δr, the slope of the differential optical depth is evaluated by a linear least square regression. Standard deviation errors are calculated from the noise measured on the pretrigger signal. In the first 600 m, the retrieved concentration lies between 350 and 400 ppm with a precision varying from 5 to 15% depending of the signal to noise ratio (SNR) which decreases with the measurement range. Some biases coming from the digitizer are expected to introduce additional errors up to 25% of the retrieved CO_2 concentration for low SNR beyond 750 m.

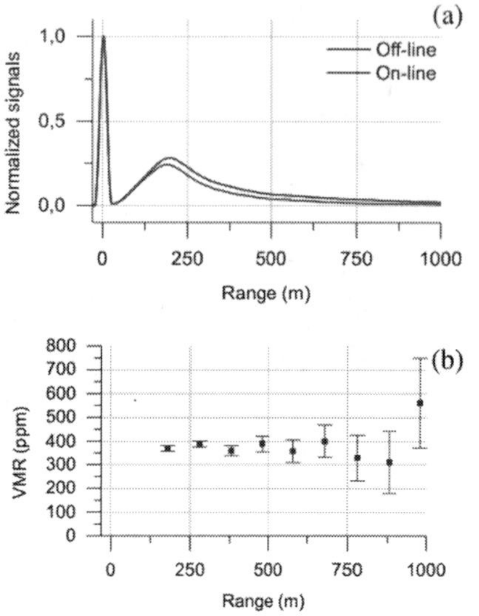

Figure 4 (a) OFF-line and ON-line return signals for a total of 8 000 shots (10 minutes) (b) Atmospheric dry air CO_2 mixing ratio retrieved from the signal with a 200 m spatial resolution. Error bars represents the RMS measurement errors due to the detector noise.

5 CONCLUSIONS

We presented CO_2 DIAL measurements using a NesCOPO allowing targeting the three main green-house gases of interest. Preliminary range-resolved measurements on CO_2 have been presented. Further improvements will consist in improving the performances of the instrument by limiting biases and optimize the transmitter and receiver units, and carrying out measurement on other species, namely CH_4 and H_2O.

ACKNOWLEDGEMENTS

This work is partially supported by the CNES Research and Technology program.

References

[1] D. M. Hammerling, S. R. Kawa, K. Schaefer, S. Doney, and A. M. Michalak, 2015: Detectability of CO2 flux signals by a space-based lidar mission, *J. Geophys. Res. Atmos.* **120**, 1794–1807.

[2] C. Weaver, C. Kiemle, S. R. Kawa, T. Aalto, J. Necki, M. Steinbacher, J. Arduini, F. Apadula, H. Berkhout, and J. Hatakka, 2014: Retrieval of methane source strengths in Europe using a simple modeling approach to assess the potential of spaceborne lidar observations, *Atmos. Chem. Phys.* **14**, 2625–2637.

[3] G. Ehret, C. Kiemle, M. Wirth, A. Amediek, A. Fix, and S. Houweling, 2008: Space-borne remote sensing of CO2, CH4, and N2O by integrated path differential absorption lidar: a sensitivity analysis, *Appl. Phys. B* **90**, 593–608.

[4] J. Caron and Y. Durand, 2009: Operating wavelengths optimization for a spaceborne lidar measuring atmospheric CO2., *Appl. Opt.* **48**, 5413–22.

[5] J. Barrientos Barria, A. Dobroc, H. Coudert-Alteirac, M. Raybaut, N. Cézard, J.-B. Dherbecourt, T. Schmid, B. Faure, G. Souhaité, J. Pelon, J.-M. Melkonian, A. Godard, and M. Lefebvre, 2014: Simultaneous remote monitoring of atmospheric methane and water vapor using an integrated path DIAL instrument based on a widely tunable optical parametric source, *Appl. Phys. B* **117**, 509–518.

[6] J. Barrientos Barria, D. Mammez, E. Cadiou, J.-B. Dherbecourt, M. Raybaut, T. Schmid, A. Bresson, J.-M. Melkonian, A. Godard, J. Pelon, and M. Lefebvre, 2014: Multispecies high-energy emitter for CO_2, CH_4, and H_2O monitoring in the 2 μm range," *Opt. Lett.* **39**, 6719–22.

[7] G. Arisholm, O. Nordseth, and G. Rustad, 2004: Optical parametric master oscillator and power amplifier for efficient conversion of high-energy pulses with high beam quality, *Opt. Express* **12**, 4189.

[8] G. Ehret, A. Fix, V. Weiss, G. Poberaj, and T. Baumert, 1998: Diode-laser-seeded optical parametric oscillator for airborne water vapor DIAL application in the upper troposphere and lower stratosphere, *Appl. Phys. B Lasers Opt.* **67**, 427–431

LIDAR STAND-ALONE RETRIEVAL OF ATMOSPHERIC AEROSOL MICROPHYSICAL PROPERTIES DURING SLOPE

Pablo Ortiz-Amezcua[1,2]*, Stefanos Samaras[3], Christine Böckmann[3], Jose Antonio Benavent-Oltra[1,2], Juan Luis Guerrero-Rascado[1,2], Roberto Román[1,2], Lucas Alados-Arboledas[1,2]

[1]*Andalusian Institute for Earth System Research (IISTA-CEAMA), Granada, Spain, *portizamezcua@ugr.es*
[2]*Department of Applied Physics, University of Granada, Spain*
[3]*Institute for Mathematics, University of Potsdam, Germany*

ABSTRACT

Two cases from SLOPE campaign at Granada are analyzed in terms of particle microphysical properties using novel software developed at Potsdam University. Multiwavelength Raman lidar measurements of particle extinction and backscatter coefficients as well as linear particle depolarization ratios are used as input for the software. The result of the retrieval is a 2-dimensional particle volume distribution as a function of radius and aspect ratio, from which the particle microphysical properties are obtained.

1 INTRODUCTION

Aerosol particles play an essential role on atmosphere and climate, and it is thus important to have vertically resolved information about their microphysical properties. Multiwavelength Raman lidar measurements allow for stand-alone microphysical retrievals, i. e., the calculation of particle microphysical properties using only Raman lidar data and applying regularization methods [1].

Most of the regularization methods use Mie theory to model aerosol particles as an ensemble of spheres, what usually provides fairly good results for small particles like biomass burning particles. However, using depolarization measurements makes possible to generalize the models to account for spheroidal particles and therefore, the retrievals include also information about the particle shape.

In this work, new software developed at the University of Potsdam [2] using a 2-dimensional approach is tested applying it to measurement cases from ACTRIS-2 experimental campaign SLOPE (Sierra Nevada Lidar AerOsol Profiling Experiment). That campaign was carried out in Granada from May to August 2016, and was designed for gathering data useful for testing the retrieval schemes through inversion of remote sensing observations. It combined active and passive remote sensing together with in-situ measurements at several levels in the northern slope of Sierra Nevada mountain range.

2 METHODOLOGY

The present study has been carried out using data corresponding to measurements taken at the Andalusian Institute for Earth System Research (IISTA-CEAMA) in Granada (37.16° N, 3.61°W, 680 m a.s.l.). This is a medium-size city in Southeastern Spain, located in a natural basin delimited by mountains reaching 3000 m a.s.l.; the air masses are mainly coming from the Atlantic Ocean, Europe, North Africa and the Mediterranean Basin [3]. The station is included in several research networks like EARLINET (European Aerosol Research Lidar Network) [4], and AERONET (Aerosol Robotic Network) [5].

The vertical profiles of particle optical properties were obtained using signals from the multiwavelength Raman lidar system MULHACEN (LR331D400, Raymetrics Inc.), configured in a monostatic biaxial alignment pointing vertically to the zenith. The light source is a pulsed Nd:YAG laser with emission at 355, 532 and 1064 nm wavelength. The radiation is collected by a 40 cm-diameter Cassegranian telescope, and it is then split into seven channels allowing the detection of elastic signals at 355 nm, 532 nm (in parallel and perpendicular polarizations) and 1064 nm and three Raman channels at 387 and 607 nm (nitrogen Raman shifted signal from 355 and 532 nm, respectively) and 408 nm (water vapor Raman shifted signal

 https://doi.org/10.1051/epjconf/201817605046

from 355 nm). This instrument has an overlap limitation due to its intrinsic setup, but it will not be taken into account in this work, since our study focuses in heights above 2 km a.s.l. [6]. The vertical profiles of particle backscatter (β) and extinction (α) coefficients were retrieved from night-time lidar signals using the Raman algorithm [7]. The ratio between the perpendicular and parallel components of the particle backscatter coefficient at 532 nm, known as linear particle depolarization ratio (δ_P) was also obtained following the calculation and calibration methods described in [8].

The set of 3 β (at 355, 532 and 1064 nm), 2 α (at 355 and 532 nm) and 1 δ_P (at 532 nm) was employed as input for the regularization software to obtain particle microphysical properties. This software [2] is based on a generalization of Mie model in two dimensions accounting for non-spherical particles. In this way, an ensemble of spheroids characterized by their volume-equivalent radius (r) and aspect ratio (a) is considered, and the aerosol volume "shape-size" distribution $v(r,a)$ is obtained by regularization [9, 10]. A precalculated database is used for the calculation of the kernel functions produced by Mieschka tool based on T-matrix method [11]. The best solution is chosen from a grid of wavelength-independent real (RRI) and imaginary (IRI) parts of refractive index by means of least residual error. The microphysical properties are then obtained from the retrieved distribution: volume concentration (v_c), effective radius (r_{eff}), effective aspect ratio (a_{eff}) and single scattering albedo (SSA).

Columnar microphysical data retrieved from a CIMEL CE-138 sun-photometer, belonging to AERONET and collocated with the lidar system, were used to compare with the lidar stand-alone regularization results. The algorithm used by AERONET retrieves microphysical properties such as the particle size distribution, complex refractive index, and effective radius, from almucantar spectral sky radiance measurements performed by ground-based sun photometers [5].

3 RESULTS

Two measurement cases were picked from the database created during SLOPE. In order to assess the performance of the regularization algorithm in different scenarios, cases with different aerosol types were selected, namely an aged smoke plume probably mixed with local pollution on 18th May, and a mineral dust event from Sahara desert on 9th June. The analyses of the cases are presented separately in the following subsections.

3.1 Case I: 18th May 2016

Several aerosol plumes coming from North America (according to backward trajectories, not shown here) were detected in May 2016. Fire detection maps from satellite measurements indicated wide regions of active forest fires that seemed to be the source of the aerosol arriving to Granada.

In Fig. 1, optical profiles from Raman lidar measurements are depicted for the case on 18th May 2016 at 22:00-22:30 UTC. Particle backscatter coefficients showed values up to 2 $(Mm \cdot sr)^{-1}$ at 532 nm in an aerosol layer centered around 2.5 km a.s.l. Low δ_P at 532 nm (less than 8%) and strong spectral dependence of β, with β-related Angström Exponents (AE_β) around 2 indicated that the measured plume contained smoke particles from the North American Forest Fires, and thus small spherical particles were expected [12].

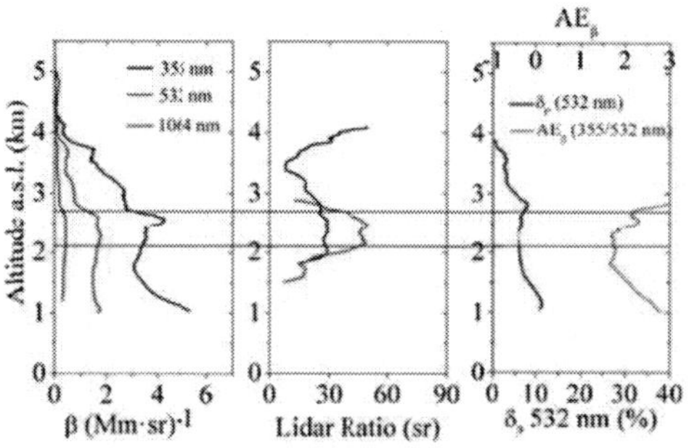

Figure 1: particle backscatter coefficient, lidar ratio, linear particle depolarization ratio and Angström exponent profiles for 18th May 2016 at 22:00-22:30 UTC. Brown lines indicate the region selected for microphysical analysis

A region from 2.1 and 2.7 km a.s.l. was selected (Fig. 1) and its averaged optical properties were used as input for the regularization algorithm. The retrieved 2-dimensional particle volume distribution (not shown here) presented a single fine mode with microphysical properties included

https://doi.org/10.1051/epjconf/201817605046

in Table 1, together with the results of the retrieval from the closest sun-photometer measurement. The results agree with typical values for aged smoke [12], and also include the information about shape: a_{eff} 1.126 and the mode width of the a distribution, a_{width} 0.048 suggest spherical dominance and a hint of prolate particles.

Table 1: particle microphysical properties retrieved from lidar and from sun-photometer for 18th May 2016

SSA is averaged for all wavelengths

	Lidar 22:00-22:30 UTC	**AERONET 16:49 UTC**
RRI	1.42	1.511
IRI	0.007	$5\cdot10^{-5}$
r_{eff} (μm)	0.26	0.295
a_{eff}	1.126	-
a_{width}	0.048	-
SSA	0.96	0.992

3.2 Case II: 9th June 2016

The second case corresponds to a mineral dust plume coming from Sahara Desert, according to backward trajectories analysis (not shown here). The optical profiles for 9th July 2016 at 01:00-02:00 UTC (Fig. 2) confirm the decoupled strong dust layer between 2.5 and 5 km a.s.l., with β more than 5 $(Mm\cdot sr)^{-1}$ at 532 nm. Particle lidar ratios around 50 sr for both 355 and 532 nm channels, AE_β less than 0.5 and δ_P almost 30 % reveal the presence of pure mineral dust particles in accordance to [13].

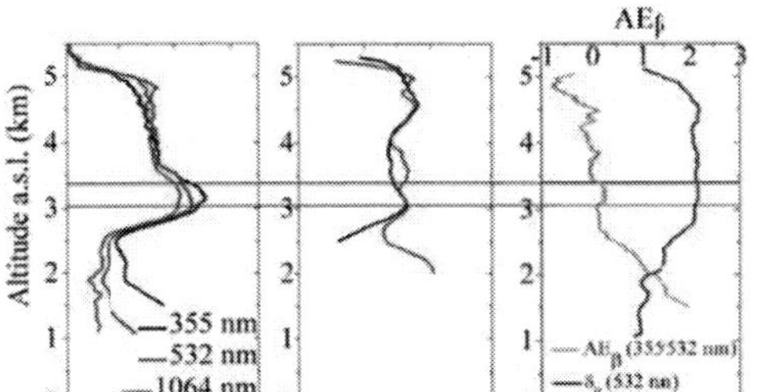

Figure 2: particle backscatter coefficient, lidar ratio, linear particle depolarization ratio and Angström exponent profiles for 9th June 2016 at 01:00-02:00 UTC. Brown lines indicate the region selected for microphysical analysis

For the microphysical retrieval, the optical properties were averaged in the region from 3.05 to 3.4 km a.s.l. The results are shown in Table 2, compared with the AERONET retrievals some hours later. The results suggest large prolate spheroidal particles, with r_{eff} 0.38 μm and a_{eff} 1.14 (slightly higher than in Case I). There is a noteworthy similarity with the values from the sun-photometer retrieval, which can be an indication that the properties of the studied particles fairly represent the effective properties of the whole atmospheric column for this case. It has to be noted that the columnar r_{eff} was calculated only from the part of the size distribution exactly coinciding with the range that can be analyzed with the lidar regularization algorithm, i. e., from 0 to 2.2 μm.

Table 2: particle microphysical properties retrieved from lidar and from sun-photometer for 9th June 2016

SSA is averaged for all wavelengths

** r_{eff} calculated from size distribution until 2.2 μm*

	Lidar 01:00-02:00 UTC	**AERONET 06:20 UTC**
RRI	1.38	1.45
IRI	0.003	0.002
r_{eff} (μm)	0.38	0.4*
a_{eff}	1.14	-
a_{width}	0.047	-
SSA	0.98	0.96

4 CONCLUSIONS

This study presents the microphysical analysis of two different aerosol plumes measured with Raman lidar during the experimental campaign SLOPE at Granada in 2016.

The addition of depolarization measurements to the typical input setup for the regularization algorithm, making possible to use a more advanced 2-dimensional model for the particle volume distribution, was shown to be a successful option to obtain more information about size and shape of the analyzed particles. The effective radius was found 0.26 μm for the aged smoke case, and 0.38 μm for the mineral dust case. The effective aspect ratios indicated in both cases the predominance of prolate particles, although this feature was slightly more pronounced for the dust case.

ACKNOWLEDGEMENTS

This work was supported by the Andalusia Regional Government through project P12-RNM-2409, by the Spanish Ministry of Economy and Competitiveness through projects CGL2013-45410-R and CGL2016-81092-R, and grant FPU14/03684, and by the Erasmus+ Programme of the European Union. The financial support for EARLINET in the ACTRIS Research Infrastructure Project by the European Union's Horizon 2020 research and innovation program under grant agreement n. 654169 is gratefully acknowledged. FEDER program is acknowledged for the instrumentation used in this work. This work was also partially funded by the University of Granada through "Plan Propio. Programa 9. Convocatoria 2013".

References

[1] Müller, D., Böckmann, C., Kolgotin, A., Schneidenbach, L., Chemyakin, E., Rosemann, J., Znak, P., Romanov, A., 2016: Microphysical particle properties derived from inversion algorithms developed in the framework of EARLINET, *Atmos. Meas. Tech.* **9**, 5007-5035.

[2] Samaras, S., Böckmann, C., Nicolae, D., 2016: Combined sphere-spheroid particle model for the retrieval of the microphysical aerosol parameters via regularized inversion of lidar data, *EPJ Web of Conferences* **119**, 23022.

[3] Alados-Arboledas, L., Alcántara, A., Olmo, F. J., Martínez-Lozano, J. A., Estellés, V., Cachorro, V., Silva, A. M., Horvath, H., Gangl, M., Díaz, A., Pujadas, M., Lorente, J., Labajo, A., Sorribas, M., Pavese, G., 2008: Aerosol columnar properties retrieved from CIMEL radiometers during VELETA 2002, *Atmos. Env.* **42** (11), 2654-2667.

[4] Pappalardo, G., Amodeo, A., Apituley, A., Comeron, A., Freudenthaler, V., Linné, H., Ansmann, A., Bösenberg, J., D'Amico, G., Mattis, I., Mona, L., Wandinger, U., Amiridis, V., AladosArboledas, L., Nicolae, D., Wiegner, M., 2014: EARLINET: towards an advanced sustainable European aerosol lidar network, *Atmos. Meas. Tech.* **7**, 2389-2409.

[5] Holben, B., N., Eck, T. F., Slutsker, I., Tanré, D., Buis, J. P., Setzer, A., Vermote, E., Reagan, J. A., Kaufman, Y. J., Nakajima, T., Lavenu, F., Jakowiak, I., Smirnov, A., 1998: AERONET – A Federated Instrument Network and Data Archive for Aerosol Characterization, *Remote Sens. Environ.* **66**(1), 1-16.

[6] Navas-Guzmán, F., Guerrero-Rascado, J. L., Aldados-Arboledas, L., 2011: Retrieval of the lidar overlap function using Raman signals, *Óptica Pura y Aplicada* **44**(1), 71-75.

[7] Ansmann, A., Wandinger, U., Riebesell, M., Weitkamp, C., Michaelis, W., 1992: Independent measurement of extinction and backscatter profiles in cirrus clouds by using a combined Raman elastic-backscatter lidar, *Appl. Opt.* **31**, 7113-7131.

[8] Bravo-Aranda, J. A., Navas-Guzmán, F., Guerrero-Rascado, J. L., Pérez-Ramírez, D., GranadosMuñoz, M. J., Alados-Arboledas, L., 2013: Analysis of lidar depolarization calibration procedure and application to the atmospheric aerosol characterization, *Int. J. Remote Sens.* **34**, 3543-3560.

[9] Böckmann, C., 2001: Hybrid regularization method for the ill-posed inversion of multiwavelength lidar data in the retrieval of aerosol size distributions, *Appl. Opt.* **40**(9), 1329-1342.

[10] Böckmann, C., Osterloh, L., 2014: Runge-Kutta type regularization method for inversion of spheroidal particle distribution from limited optical data, *Inverse Probl. Sci. En.* **22**, 150-165.

[11] Mishchenko, M., I., Travis, L., D., Mackowski, D. W., 1996: T-matrix computations of light scattering by nonspherical particles: A review, *J. Quant. Spectrosc. Ra.* **55**(5), 535-575.

[12] Ortiz-Amezcua, P., Guerrero-Rascado, J. L., Granados-Muñoz, M. J., Benavent-Oltra, J. A., Böckmann, C., Samaras, S., Stachlewska, I. S., Janicka, Ł., Baars, H., Bohlmann, S., Alados-Arboledas, L., 2016: Microphysical characterization of long-range transported biomass burning particles from North America at three EARLINET stations, *Atmos. Chem. Phys. Discuss.*, in review.

[13] Guerrero-Rascado, J. L., Ruiz, B., Alados-Arboledas, L., 2008: Multi-spectral Lidar characterization of the vertical structure of Saharan dust aerosol over Southern Spain, *Atmos. Env.* **22**, 2668-2681.

A LIDAR AT CLERMONT-FERRAND—FRANCE TO DESCRIBE THE BOUNDARY LAYER DYNAMICS, AEROSOLS, CIRRUS AND TROPOSPHERIC WATER VAPOR

J.L. Baray[1, 2]*, P. Fréville[2], N. Montoux[1], A. Chauvigné[1], D. Hadad[1] & K. Sellegri[1]

*[1]LaMP, Université Clermont Auvergne, CNRS, France, *J.L.Baray@opgc.fr*
[2]OPGC Université Clermont Auvergne, CNRS, France

ABSTRACT

A Rayleigh-Mie-Raman LIDAR provides vertical profiles of tropospheric variables at Clermont-Ferrand (France) since 2008, in order to describe the boundary layer dynamics, tropospheric aerosols, cirrus and water vapor. It is included in the EARLINET network. We performed hardware/software developments in order to upgrade the quality, calibration and improve automation. We present an overview of the system and some examples of measurements and a preliminary geophysical analysis of the data.

1 INTRODUCTION

Long-term observations of water vapor, aerosols and cirrus clouds are of crucial importance to monitor the atmospheric composition and to understand the weather climate system. Satellites observations may be completed by in situ and remote sensing systems from ground based stations organized in networks such as GAW (Global Atmospheric Watch), ACTRIS (Aerosol, Clouds and Trace gases Research Infrastructure) and its LIDAR component EARLINET (European Aerosol Research LIDAR Network) [1], GRUAN (GCOS Reference Upper Air Network) or NDACC (Network for the Detection of Atmospheric Composition Change).

A Rayleigh-Mie-Raman LIDAR system is in operation since 2008 at Clermont-Ferrand and provides regular tropospheric profiles of water vapor, aerosols and cirrus clouds. It is also used to estimate the aerosol boundary layer height and indicate if the PUY global GAW station is in the mixing layer or in the free troposphere. Here we present a short technical overview of the LIDAR system operating at Clermont-Ferrand, work of calibration and data quality assurance mainly performed in the framework of EARLINET and examples of measurements in their geophysical context.

2 DESCRIPTION OF THE LIDAR SYSTEM

The system was designed by Gordien Strato and built by Raymetrics. It uses a Quantel CFR400 laser. The optical part is deployed outside, on the roof over the room hosting the computer and electronic device (Figure 1). The main technical characteristics are given in the Table 1.

The optical part containing the laser, the beam expender, the telescope and the optical box for the receiving wavelength separation using beamsplitters is thermoregulated, and protected by a glass. This glass allows operation in any weather condition, without modifying the polarization of emitted and received light. An automatic labview acquisition software based on the signals discriminates several situations such as "weak signal", "clouds", or "normal conditions of measurements". The system can work 24 hours a day, 7 days a week, if the conditions are good, and manage stop/restart automatically in case of bad weather conditions. Details on the algorithm and data processing are given in [2].

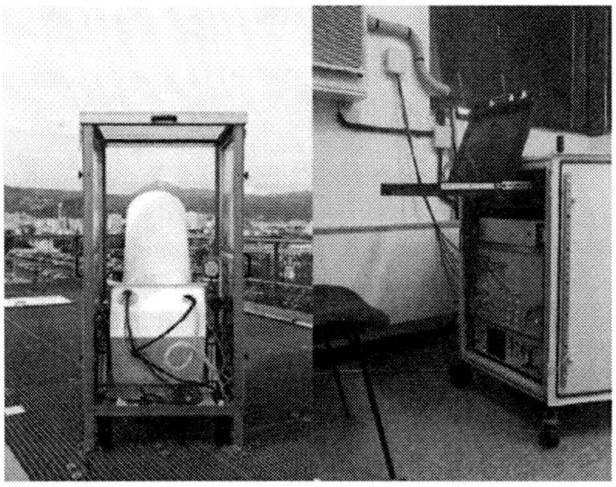

Figure 1: Optical (left, © Cyril Fresillon) and electronical (right © Jean-Luc Baray) parts of the LIDAR system.

https://doi.org/10.1051/epjconf/201817605047

Table 1 Main technical characteristics of the LIDAR

Laser	Nd:Yag
Nominal energy	60 mJ per pulse at 355 nm
Frequency	10 Hz
Beam diameter after expander	50 mm
Beam divergence	0.14 mrad
Reception channels	Elastic ⊥ : 354.7 nm Elastic // : 354.7 nm Raman N_2 : 386.6 nm Raman H_2O : 407.5 nm
Telescope diameter	400 mm
Field of view	0.25 to 1 mrad
Partial / Full Overlap	500 m / 1100 m
Detectors	Hamamatsu R7400 Photomultipliers
Vertical Sampling	7.5 m

3 NETWORKS

At the national French level, the LIDAR system has been involved in the SOERE ROSEA, ORAURE, ATMOS, and now in the research infrastructure ACTRIS-FR. Data (range corrected signals, level 1 profiles) are transmitted in real time to the French AERIS- ICARE data center, in the framework of ACTRIS-FR. At the European level, we participate to the EARLINET network and are involved in the ACTRIS-2 WP2. In this framework, our action consists in:

- Providing regularly aerosols backscattering coefficient profiles to different categories of the EARLINET database (climatology, cirrus, calipso overpasses…);
- Carrying out regular quality assurance tests on the instrument. Those tests (dark measurements, telecover tests, depolarisation calibrations and Rayleigh fit checks) are used (i) to optimize the alignment procedure, (ii) to check regularly the LIDAR signals quality and (iii) to determine some instrumental correction factors for the processing. They are submitted at least once a year to LIDAR calibration center (lical) in Bucharest.
- Participating to network measurement campaigns such as the SCC (single calculus chain) exercise from 9 to 12 July 2012, during a Saharian dust intrusion event over Europe [3].

4 EXAMPLES OF MEASUREMENTS

4.1 WATER VAPOR PROFILES AND STRATOSPHERE-TROPOSPHERE EXCHANGE

The water vapor retrieval is based on the fact that in the troposphere, N_2 is in a constant proportion of dry air. The mixing ratio of water vapor is then derived from the ratio between the H_2O and the N_2 Raman backscattered signals. Independent water vapor measurement or model profile has to be used to calibrate the LIDAR mixing ratio profile. The use of GPS columns at the top of the puy de Dôme station is currently evaluated but the GPS station of puy de Dôme is distant of 12 km from the LIDAR. Unfortunately, no collocated radiosonde measurement is routinely performed at this site. Consequently, the calibration procedure has been performed normalizing by water vapor mixing ratios from the ECMWF ERA-Interim reanalysis between 3 and 5 km height.

Since 2013, more than 100 night-time profiles per year can reach more than 4 km high. This number increases from year to year as the system improves and reaches more than 195 night-time profiles in 2015. An interesting example of water vapor measurement was during the night between 29 and 30 January 2013. Figure 2 is the time series of one minute water vapor profiles.

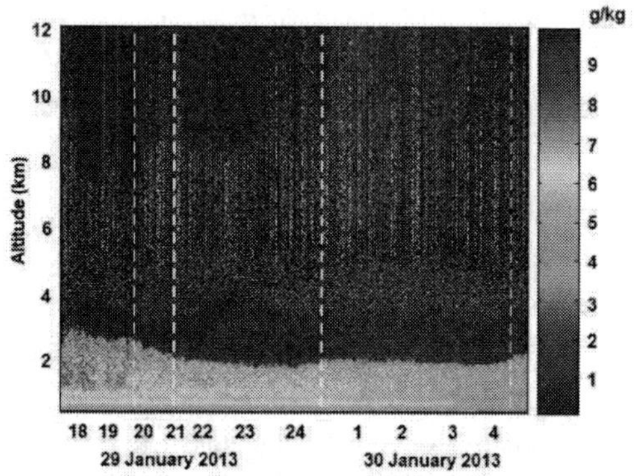

Figure 2: Time series of one minute water vapor profiles during the night between 29 and 30 January 2013. The time scale is not regular because some cloudy files have been removed.

It shows a humid layer near the ground, capped by a dry layer. The limit between the two layers varies during the night from 2 to 3 km. The upper limit of the profiles is around 5-6 km.

By integrating all the files obtained during this night, the upper limit is extended to 8 km (Figure 3). This allows a dry layer to be detected between 5 and 7 km. Four-day backtrajectories suggest a polar and stratospheric origin for a major part of the air masses arriving over Clermont-Ferrand during this case study (Figure 4). The dynamical analysis based on ECMWF ERA-Interim, and Reverse Domain Filling calculations confirm that this stratosphere-troposphere exchange event was associated to the polar jet stream [4].

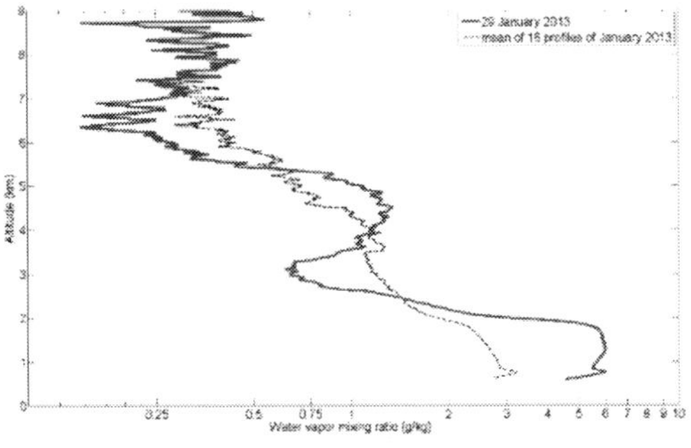

Figure 3: Profile calculated by integrating the whole night between 29 and 30 January 2013 (in red), compared to the mean of 16 profiles of January 2013 (black dashed line)

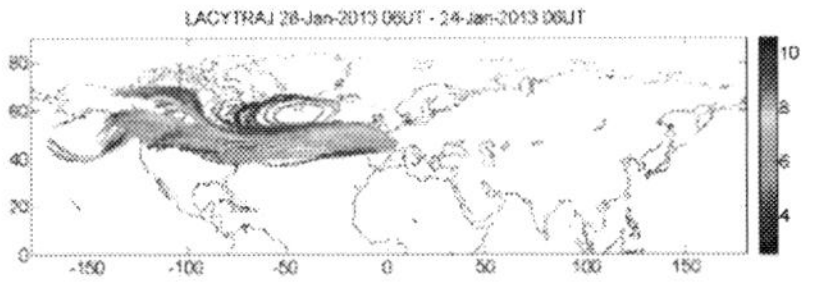

Figure 4: Four-day backtrajectories arriving over Clermont-Ferrand during the case study, calculated with the LACYTRAJ model [5]. Colors indicate the altitude in km.

4.2 AEROSOLS PROFILES, CIRRUS AND MIXING LAYER HEIGHT

Rayleigh-Mie channels allow to calculate backscattering coefficient profiles by applying Fernald-Klett inversion on range corrected signals (Pr^2) using specific LIDAR ratios. This methodology allowed to characterize the Eyjafjallajökull volcanic plume in April–May 2010, in combination with in situ measurements at the microphysical station at the top of the puy de Dôme [6].

Natural and anthropogenic aerosols are most concentrated in the atmospheric mixing layer, where they are removed by rain, particle coagulation, or sedimentation. In the free troposphere, aerosol particles are subjected to fewer removal processes increasing their lifetime and impact on the climate. It is then important to survey the evolution of the limit between the mixing layer and the free troposphere. One way to do this is to use the wavelet covariance transform technique on backscattered LIDAR profiles. This methodology shows that the mixing layer height may vary a lot during one day (Figure 5). In the example of 4 May 2016, variations during the night are due to the difficulty for the algorithm to distinguish the nighttime residual layer and the mixing layer. During daytime, the mixing layer height is better retrieved and increases due to convective uplift.

In combination with is-situ observations at the top of the puy de Dôme station, the lidar has also allowed to evidence feeding processes of the free troposphere with aerosol particles from the mixing layer during a cold period in February 2012 [7].

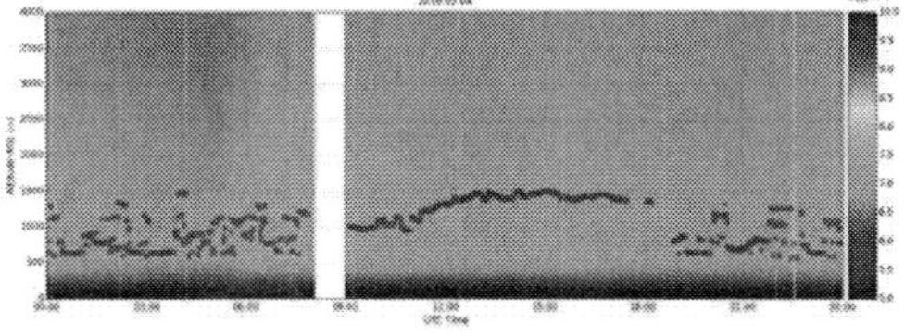

Figure 5: Example of mixing layer height estimation, on 4 May 2016. Extracted from the ICARE on-line data archive (http://www.icare.univ-lille1.fr/).

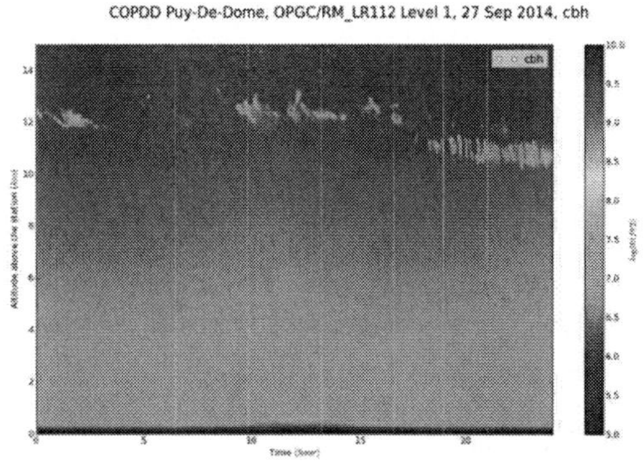

Figure 6: Time series of $log(Pr^2)$ during the 27 September 2014. Extracted from the ICARE on-line data archive.

A special day presenting aerosols near the ground and cirrus cloud near the tropopause on 27 September 2014 is presented on Figure 6.

The aerosol layer is observed below 4-5 km with high values of Pr^2. A cirrus event appears also between 11 and 13 km, in addition to the aerosol event.

The origin of the aerosol layer could be a mix of volcanic emissions from the Bardabunga (64°N, 17°W) eruption, and biomass burning in North America.

5 CONCLUSIONS AND FUTURE PLANES

We provided a technical description of the Rayleigh-Mie-Raman LIDAR in operation at Clermont-Ferrand (France) since 2008, and examples of measurements of tropospheric aerosols, cirrus and water vapor profiles. The continuation of routine measurements in the long term is very important for climate issue and we plan to ensure it.

In operation since now 9 years, the laser does not currently provide the nominal power despite regular maintenance operations. We plan to change it as soon as the budget will be completed. It could be the first step of a multi-year investment program whose objective is an increasing evolution of our LIDAR system, with 3 wavelengths in emission and an additional near-range channel to better document the boundary layer dynamics.

ACKNOWLEDGEMENTS

This work is dedicated to Vincent Giraud (1965-2009), who was involved in LIDAR activities at Clermont-Ferrand, at the beginning. The LIDAR system is operated by OPGC observatory and LaMP research laboratory. It has been supported by Clermont-Auvergne, CNRS, CNES, ROSEA-ORAURE and now ACTRIS-FR, and regional founds (FEDER, CPER, Region Auvergne). We acknowledge the ICARE service center for furnishing the BASIC/STRAT figures. We thank all the people involved in CO-PDD instrumented site, and in the LIDAR data processing (M. Hervo, G. Payen, D. Dionisi).

References

[1] Pappalardo, G.; Amodeo, A.; Apituley, A.; Comeron, A.; Freudenthaler, V.; Linné, H.; Ansmann, A.; Bösenberg, J.; D'Amico, G.; Mattis, I.; et al. EARLINET: Towards an advanced sustainable European aerosol LIDAR network. Atmos. Meas. Tech., 7, 2389–2409, 2014.

[2] Fréville P., N. Montoux, J.-L. Baray, A. Chauvigné, F. Réveret, M. Hervo, D. Dionisi, G. Payen, K. Sellegri, Lidar developments at Clermont-Ferrand – France for atmospheric observation, Sensors, 15, 3041-3069; doi:10.3390/s150203041, 2015.

[3] Sicard, M., et al.: EARLINET: potential operationality of a research network, Atmos. Meas. Tech., 8, 4587-4613, doi:10.5194/amt-8-4587-2015, 2015.

[4] Baray J.-L. , Y. Pointin, J. Van Baelen, M. Lothon, B. Campistron, J.-P. Cammas, Olivier Masson, D. Hadad, A. Colomb, C. Hervier, Y. Bezombes, C. Duroure, and F. Tridon, Case study and climatological analysis of upper tropospheric jet stream and stratosphere-troposphere exchanges using VHF profilers and radionuclide measurements in France, submitted, Journal of Applied Meteorology and Climatology, 2016.

[5] Clain G., J.-L. Baray, R. Delmas, P. Keckhut, J.-P. Cammas, A lagrangian approach to analyse the tropospheric ozone climatology in the tropics: Climatology of Stratosphere - Troposphere exchange at Reunion Island, Atmospheric Environment, 44, 968-975, 2010.

[6] Hervo, M., et al. Physical and optical properties of 2010 Eyjafjallajökull volcanic eruption aerosol: Ground-Based, LIDAR and airborne measurements in France. Atmos. Chem. Phys., 12, 1721–1736, 2012.

[7] Freney E., K. Sellegri, E. Asmi, C. Rose, A. Chauvigne, J.-L. Baray, A. Colomb, M. Hervo, N. Montoux, L. Bouvier, D. Picard, Experimental evidence of the feeding of the free troposphere with aerosol particles from the mixing layer, Aerosol and Air Quality Research, 16 : 702-716, doi: 10.4209/aaqr.2015.03.0164, 2016.

A METHODOLOGY FOR CLOUD MASKING UNCALIBRATED LIDAR SIGNALS

Ioannis Binietoglou[1,2]*, Giuseppe D'Amico[3], Holger Baars[4], Livio Belegante[1], Eleni Marinou[2]

[1]*National Institute for R&D for Optoelectronics, Romania, *ioannis@inoe.ro*
[2]*National Observatory of Athens, Greece*
[3]*Consiglio Nazionale delle Ricerche - Istituto di Metodologie per l'Analisi Ambientale, Italy*
[4] *Leibniz Institute for Tropospheric Research (TROPOS), Germany*

ABSTRACT

Most lidar processing algorithms, such as those included in EARLINET's Single Calculus Chain, can be applied only to cloud-free atmospheric scenes. In this paper, we present a methodology for masking clouds in uncalibrated lidar signals. First, we construct a reference dataset based on manual inspection and then train a classifier to separate clouds and cloud-free regions. Here we present details of this approach together with an example cloud masks from an EARLINET station.

1 INTRODUCTION

Lidar systems have evolved into key observing tools of atmospheric aerosols, providing measurements of aerosol optical properties with high spatial and temporal resolution. In Europe, the European aerosol research lidar network (EARLINET) is allowing the study of aerosol vertical structure at a continental scale. Two major concerns, however, of such a distributed lidar networks is the homogeneity of the processed results and the ability to provide measurements in real-time. To tackle both issues, EARLINET is developing the Single Calculus Chain (SCC), a tool for automatic processing signals from heterogenous lidar systems [1].

Most aerosol processing algorithms in the SCC can be only applied in cloud-free conditions. This is a crucial step, as even a small misinterpretation of clouds as aerosols, could have strong effect in the retrieved lidar products and hinder any further use of the data. In EARLINET, the process of identifying cloud-free regions is done separately at each lidar station before submitting data for processing in the SCC, but this procedure is manpower-intensive and unfit for real-time analysis. Therefore, an accurate, automatic cloud-masking procedure is a necessary first step for constructing a fully automatic processing chain.

In this paper, we outline a methodology for cloud-masking based on uncalibrated lidar systems. We follow a supervised learning approach, trying to find the optimal discrimination parameters based on a reference cloud-masked dataset. The cloud masking procedure is developed for the SCC using data from several EARLINET lidar systems. In this paper, we present an example using the Polly^{XT} lidar systems of the National Observatory of Athens [2]. In section 2 we outline a general supervised learning methodology for cloud-masking and give an example application that is being tested for the SCC, while in section 3 we present specific results of this procedure.

2 METHODOLOGY

The aim of the methodology is to assign a cloud label to each lidar bin. The algorithm takes as input uncalibrated pre-processed lidar signals. We treat each lidar scene separately, assuming no knowledge of the instrument constant of the measuring system. This is a realistic assumption as the algorithm is designed for research lidar systems, that their parameters can change often. Being based on uncalibrated lidar signals, the results of the algorithm cannot be completely objective, and so they are not designed to be used for studying clouds. The aim instead is just to identify cloud-free regions suitable for the retrieval of aerosol properties.

2.1 Supervised learning

Supervised learning is separated in two phases (Fig. 1) [3]. In the *training* phase, a flexible classification algorithm is tuned to the specific problem of cloud masking based on a pre-classified reference dataset. The aim of this step is to find the optimal parameters to separate cloudy and cloud-free bins. In the second phase, the trained classifier is used to classify new data i.e. create a cloud mask for new lidar observations.

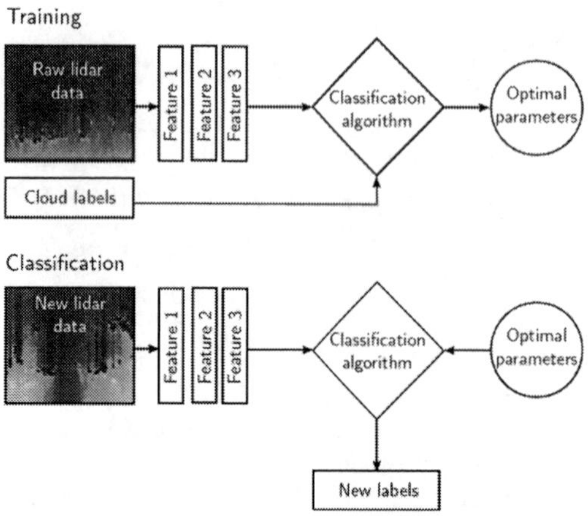

Figure 1. Overview of the two phases of the classification procedure.

Specifically, the first step of the training phase is *feature extraction*: instead of performing the classification using actual lidar data, each lidar bin is described by a set of parameter (features) that are selected to maximize the difference of aerosol and cloud regions. Such features could be, for example, the value of the backscattered signal, the slope in adjacent pixels, the variability of the signal in the region around the bin, etc. Second, we construct a reference dataset that contains a set of lidar scenes and a known cloud mask. The reference cloud mask can be created e.g. through manual inspection of lidar signals, or using ancillary collocated instruments like a cloud radar. Third, we use the reference dataset to tune (train) a classification algorithm to assign to each possible combination of features to the cloudy or non-cloudy categories. Seen in another way, the aim of this step is to separate the feature space in two sub-spaces, one assigned as cloudy and one as cloud-free. The surface that separates the two spaces is called the decision boundary of the classification. There are several well documented algorithms to find optimal decisions boundaries, depending on the number of features and the required stability [4]. Some of these algorithms can also give a probability that each new point belongs to one of the two classes. Finally, in practice the reference dataset is split in three parts: The first part, called the training dataset, is used to find the optimal decision boundary; the second part, called the validation dataset is used to evaluate the performance of each classification methods. Finally, a last part of the references dataset is used to quantify the performance of the chosen classifier. The end results of this process is a trained classifier i.e. a fixed procedure to assign each new set of features to one of the two classes, cloudy or non-cloudy.

In the classification phase (Fig. 2, bottom), the algorithm is used to classify lidar data not involved in the training procedure. The feature extraction procedure is performed in an identical way as the training phase. The output of this phase is the new cloud mask and, depending on the used classification algorithm, the probability that each bin belongs to the cloud or non-cloud categories.

2.2 Reference dataset

We constructed a reference dataset using 8 atmospheric scenes measured by Polly^{XT} lidars. Polly^{XT}s are autonomous, portable, multi-wavelength lidar systems developed by the TROPOS institute in Leipzig, Germany. They are typically operated with a temporal resolution of 30s and a vertical resolution of 7.5m. For constructing the reference dataset, we use measurements from Leipzig (Germany), Athens (Greece), and Finokalia (Greece). The 6-hour-long scenes were selected to include a wide range of cloud types, from low-level water clouds to optically thin cirrus, and aerosol burdens, including desert dust intrusions. In this way, the classifier can be trained in a wide range of atmospheric scenarios, covering most situations encounter in European measurements sites. For each scene, the reference cloud mask was constructed using manually selected thresholds in the values of range-corrected signal and edge detection value.

2.3 Features extraction

The goal of feature extraction is to convert the uncalibrated, pre-processed, lidar signals to a set of features, appropriately selected to assist the classification procedure. To make the mask as less restrictive as possible, we used only the 1064nm elastic channel. Regions with low signal-to-noise ratio are excluded from further analysis. The range-corrected signals are first normalized, aiming to homogenize the data from different lidar systems. For the normalization, we select data between full overlap range and 12km, and use their median as a normalization factor. To exclude extreme values

 https://doi.org/10.1051/epjconf/201817605048

from this region, we apply an iterative 2-sigma clipping procedure, i.e. rejecting data that are more than 2 standard deviations away from the mean value [5]. In this way, we roughly exclude clouds from the normalization dataset and make the normalization procedure more robust.

For this demonstration, we perform the cloud masking using three features. First, we use the Sobel operator as a 2D edge detection filter. The Sobel operator is a very efficient filter for calculating the discrete gradient of the signal in both the time and vertical axis. To each bin, we assign the norm of the 2D gradient vector. Second, we estimate the standard deviation of the normalized signal for a box 5x5 bins centered around the bin, as cloudy regions are expected to have much larger variability than aerosol regions. Third, we estimate the ratio of minimum to maximum normalized value for a box 5x15 centered around the bin. For cloudy regions, this ratio will have values close to zero, while in aerosol regions the value will be closer to 1. These three features were found helpful to discriminate cloudy from non-cloudy regions, but other features are also evaluated to further improve the procedure.

2.4 Classification algorithm

For the classification, we use a simple logistic regression classifier, which outputs a linear decision boundary and provides the probability that each bin belongs to one of the two classes. To prevent overfitting, we apply a L_2 regularization constraint, i.e. we try to find the optimal decision boundary including an extra penalty factor for the square of the boundary coefficients [3]. The optimization problem is solved using stochastic gradient descent that is a very efficient algorithm for fitting linear classifiers [4].

As a post-processing step, profiles classified as cloud free and having no signals at the full-overlap altitude are marked as "fog/low clouds", that is used as a general category to mark cases were the laser beam is completely extinguished and a consequence not properly observed.

3 RESULTS

In this section, we present an example of the cloud masking procedure, applied to a complex lidar scene obtained in Cyprus using the Polly^{XT}-NOA system. The top panel of Fig. 2 shows the logarithm of the range-corrected (RC) signal at 1064nm, while the bottom panel shows the constructed cloud mask. The scene covers the period from 6:00 to 12:00 UTC during which the planetary boundary layer (PBL) is rising and around 10:00 UTC clouds start forming on its top. A lofted aerosol layer is locater around 3km a.g.l, while several cloud layers are observed initially above 8km but then down to 4km.

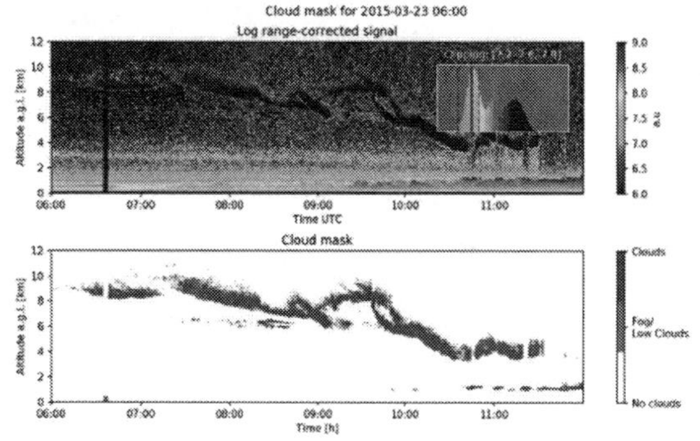

Figure 2. Example cloud mask applied to Polly^{XT}-NOA measurements on 23d of March, 2015. (top panel) Log RC signal at 1064nm. (inset figure) Histogram of log RC signal; the vertical lines shown the min, median, and max values of the clipping procedure. (bottom panel) Produced cloud mask.

The inset figure of Fig. 2 shows the histogram of log-RC values from 500m to 12km. The vertical dashed lines indicate the limits of the 2-sigma clipping procedure: only points between these values were used to calculate the median value used for normalization. Most cloud values have been excluded from the normalization making the procedure more robust.

As seen at the bottom panel of Fig.2, the cloud-masking algorithm can successfully detect most cloudy bins and separate the cloudy structures from the rest of the atmosphere. In the cases of low clouds, the regions marked as clouds are wider than the ones observed in the range-corrected signal. This is expected, as all features used in the classification take into account a region around each bin. Also, note that several bins at the edges of cirrus clouds are not classified consistently. This is also expected from a bin-based method, as such bins typically have characteristics similar to the ones observed within aerosol layers.

 https://doi.org/10.1051/epjconf/201817605048

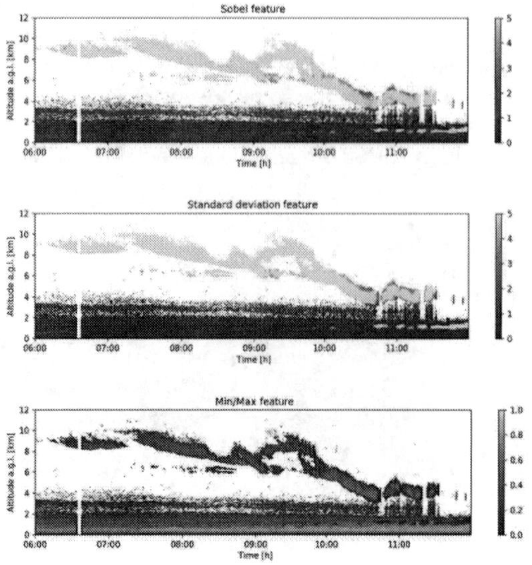

Figure 3. Time-height plots for the features used for the classification. (top panel): Sobel operator feature. (middle panel) standard deviation feature. (bottom panel): min/max feature.

Fig. 3 presents time-height plots of the three features uses in the example. In the specific case, both the Sobel operator and the standard deviation feature seems to clearly highlight cloudy and non-cloudy regions. The use of a combination of features, however, proves to make the classification more robust, and provide reliable cloud mask even in cases were simple thresholds in one parameter would fail.

Figure 4. Log-probability that each bin is cloudy.

Finally, Fig. 4 presents the predicted log probability that a pixel is cloudy. Using this information, the strictness of the algorithm output can be easily tuned, by changing the probability threshold of what is considered cloud.

4 CONCLUSIONS

We have presented an approach for assigning a cloud mask to uncalibrated lidar signals using a supervised training algorithm. We have also presented a set of features that seem to be effective in the classification procedure. However, several challenges remain to be solved. First, we need to study the effect of different spatial and temporal resolutions of the input lidar data in the quality of the cloud-mask. Seconds, we need to expand the reference dataset to include a broader set of atmospheric scenes observed by different lidar systems. Finally, we need to extend the approach to other wavelengths, making the use of the developed algorithm as broad as possible.

ACKNOWLEDGEMENTS

The research leading to these results has received funding from the European Union's Horizon 2020 Research and Innovation program under grant agreements No. 654169 (ACTRIS) and No. 602014 (ECARS).

References

[1] D'Amico, G., Amodeo, A., Baars, H., Binietoglou, I., Freudenthaler, V., Mattis, I., Wandinger, U., and Pappalardo, G., 2015: EARLINET Single Calculus Chain – overview on methodology and strategy, *Atmos. Meas. Tech.*, 8, 4891-4916.

[2] Engelmann, R., Kanitz, T., Baars, H., Heese, B., Althausen, D., Skupin, A., Wandinger, U., Komppula, M., Stachlewska, I. S., Amiridis, V., Marinou, E., Mattis, I., Linné, H., and Ansmann, A., 2016: The automated multiwavelength Raman polarization and water-vapor lidar PollyXT: the neXT generation, *Atmos. Meas. Tech.*, 9, 1767-1784.

[3] Duda, R.O., Hart, P.E. and Stork, D.G., 2012. *Pattern classification*. John Wiley & Sons.

[4] Pedregosa F., Varoquaux G., Gramfort A., Michel V., Thirion B., Grisel O., Blondel M., Prettenhofer P., Weiss R., Dubourg V., Vanderplas J., 2011: Scikit-learn: Machine learning in Python. *J. Mach. Learn. Res.* 12(Oct), 2825-2830.

[5] Pergola, N., Pietrapertosa, C., Laçava, T. and Tramutoli, V., 2001: Robust satellite techniques for monitoring volcanic eruptions. *Ann. Geophys.*, 44(2), 167-177.

https://doi.org/10.1051/epjconf/201817605049

WILD FIRE AEROSOL OPTICAL PROPERTIES MEASURED BY LIDAR AT HAIFA, ISRAEL

Birgit Heese[1,*], Julian Hofer[1], Holger Baars[1], Ronny Engelmann[1], Dietrich Althausen[1] and Yoav Y. Schechner[2]

[1]*Leibniz Institute for Tropospheric Research - TROPOS, Germany, *heese@tropos.de*
[2]*Technion - Israel Institute of Technology, Israel*

ABSTRACT

Optical properties of fresh biomass burning aerosol were measured by lidar during the wild fires in Israel in November 2016. A single-wavelength lidar Polly was operated at the Technion Campus at Haifa. The detector with originally two channels at 532 and 607 nm was recently upgraded with a cross- and a co-polarised channel at 532 nm, and a rotational Raman channel at 530.2 nm. Preliminary results show high particle depolarisation ratios probably caused by soil dust and large fly-ash particles.

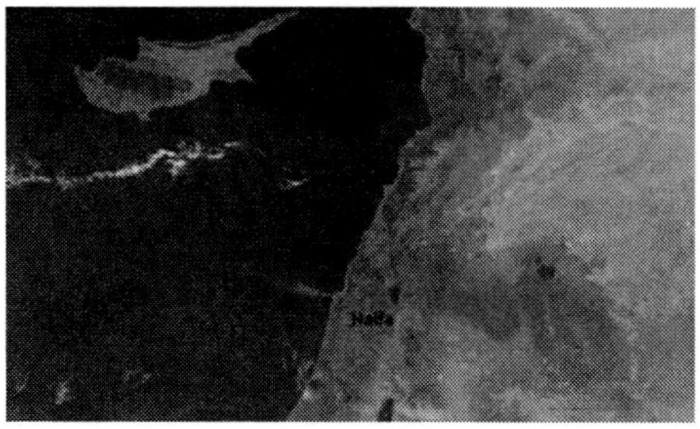

Figure 1: MODIS image on 24 November 2016 at 11:06 UTC (https://earthdata.nasa.gov).

1 INTRODUCTION

Wild fires in Israel started on 22 November 2016 and lasted until 25 November 2016. They were reported at several spots around the country, but the largest ones spread across Haifa. The view from the satellites perspective (Fig. 1) clearly shows the smoke plume from the fires stretching out from Haifa onto the Mediterranean. The fires had coincided with strong and dry easterly winds. The wind direction and speed during the wild fires measured at Haifa University are shown in Fig. 2. Easterly winds set in on 19 November and the wind speed increased significantly during that week. On 24 November, maximum wind gusts up to 70 km/h were measured. This was the day with the strongest fires in Haifa. This incident offers the opportunity to study fresh wild fire optical properties with the recently installed Polly lidar at Haifa.

2 METHODOLOGY

The lidar measurements at Haifa, Israel are performed by TROPOS since late July 2016 in collaboration with the Viterby Faculty of Electrical Engineering at the Technion. The deployed lidar is the first single-wavelength Polly built in 2003. The lidar was recently upgraded. Originally it had two channels at 532 and 607 nm. After the upgrade it now has one cross- and one co-polarised channel at 532 nm, and one rotational Raman channel at 530.2 nm. Thus, it is now capable to measure the linear particle depolarisation ratio profile in addition to the particle backscatter and extinction coefficient profiles. Using these profiles and collocated sun photometer measurements, we are able to apply the polarization-lidar photometer networking (POLIPHON) method [1] to separate coarse and fine mode particles in the vertical profile.

https://doi.org/10.1051/epjconf/201817605049

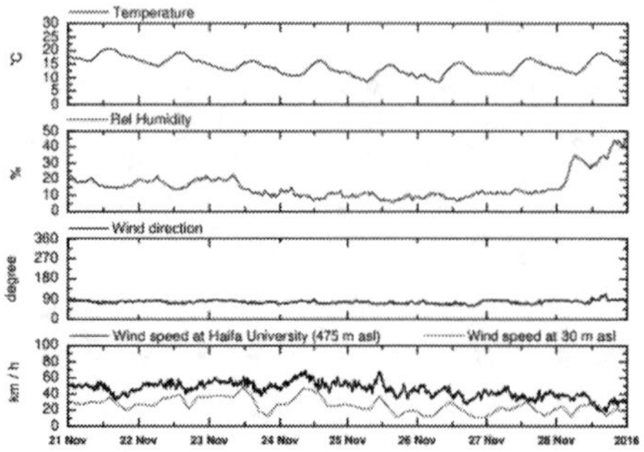

Figure 2: Temperature, relative humidity, wind speed, and wind direction measured at Haifa University on 21 – 28 November 2016. (Israel Meteorological Service)

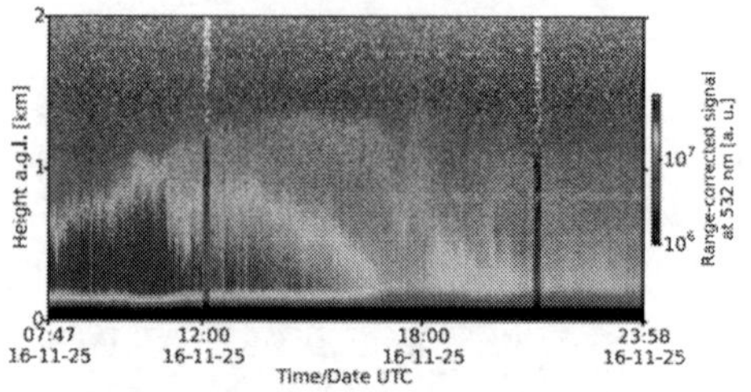

Figure 3: Range corrected lidar signal on 25 November 2016 measured at Haifa. The vertical low-signal bars at 12 and 21 UTC are due to calibration periods. Horizontal stripes are artefacts.

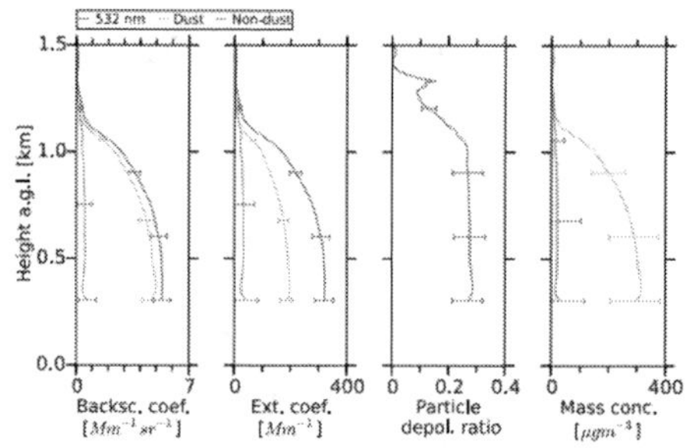

Figure 4: Profiles of particle backscatter and extinction coefficient, linear particle depolarization ratio, and mass concentration (from left to right) at 532 nm, separated into dust and non-dust fractions on 25 November 2016, 10:00-11:00 UTC.

During daytime, the particle backscatter and extinction coefficient profiles are calculated using the Klett method [2] (lidar ratio = 60 sr, vertical smoothing: 82.5 m). During night-time the Raman method [3] was applied (vertical smoothing: 307.5 m). For molecular corrections we used radio soundings from the station Bet Dagan of the Israel Meteorological Service (IMS), which is close to Tel Aviv airport. The linear particle depolarisation ratio is derived using regular, internal calibration measurements following the EARLINET quality assurance standards [4].

3 RESULTS

On 25 November 2016, one day after the major fires in Haifa, high values of the range-corrected lidar signal were observed (see Fig. 3). The measurements started at 7:47 UTC, when the evolution of the boundary layer was already in progress. High backscatter signals can be seen in the lower boundary layer up to about 1 km height over the measurement site - especially before 11 UTC. Also the respective profiles of particle backscatter coefficient (Fig. 4) show comparatively high values of up to 5.5 $Mm^{-1}sr^{-1}$. This is twice as much as observed on the days before and the following days. Then the values of the particle backscatter coefficient were about 2-2.5 $Mm^{-1}sr^{-1}$. The particle extinction coefficient values were calculated using a lidar ratio of 60 sr, assuming a mixture of pollution, burning, and dust particles, which leads to a value of about 300 Mm^{-1} on that day. On the days before and after the fire period, the particle extinction coefficients were around 150 Mm^{-1}, including the ones derived from Raman measurements. As a preliminary step to identify different particle types, we applied the POLIPHON method to separate the particle backscatter coefficient profile into dust and non-dust contributions.

This method uses the measured particle linear depolarisation ratio, which was as high as 0.26-0.27 in the lower boundary layer on this day. For the separation into dust and non-dust particle backscatter coefficients, a linear particle depolarisation ratio of 0.31 for dust particles and 0.05 for non-dust particles was assumed. These values are derived from several previous measurement campaigns. The conversion into mass concentration is then realized by applying particle specific extinction-to-mass conversion factors based on prior long-term sun photometers observations and AERONET data analysis (see [5] and references therein for details).

These preliminary results lead to the conclusion, that a considerable amount of large particles must have been present in the atmosphere during the wild fires. These large particles are possibly soil dust and flying ash lifted into the atmosphere by the observed strong easterly winds and by turbulence associated with the fires. Dust from the desert regions east of Israel may also have been transported by the winds towards the coast regions.

HYSPLIT 24 h back-trajectory calculations for arrival time of 11:00 UTC (Fig. 5) show that the air masses at the lower altitudes of 500 m, 1000 m and 2000 m height are coming from the regions east and north-east of Haifa.

An example of a lidar measurement after the fire period is shown in Fig. 6. On 28 November during daytime, the boundary layer starts to develop around 06:00 UTC and is reaching up to about 2 km in the evening. The upper part of the boundary layer persists as the residual nighttime layer, while the lower part is decaying slightly towards midnight. Fig. 7 shows the corresponding lidar profiles for the measurement period from 18:00-20:00 UTC. The values of the particle backscatter coefficient were

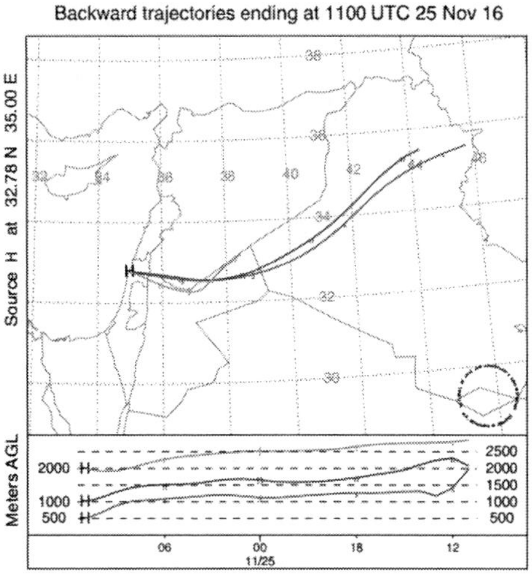

Figure 5: HYSPLIT backward trajectories ending at Haifa on 25 Nov. 2016 at 11:00 UTC.

between 2 $Mm^{-1}sr^{-1}$ and 3 $Mm^{-1}sr^{-1}$ in the lower boundary layer and drop slowly below 2 $Mm^{-1}sr^{-1}$ above 1 km height.

The particle extinction coefficients are around 150 Mm^{-1} below and 100 Mm^{-1} above 1 km height. Because it was evening, the lidar ratio profile could be calculated using the Raman method. The lidar ratio is rising from 50 sr to 60 sr in the lower layer and is 40 sr in the residual boundary layer. The linear particle depolarisation ratio increases from values of 0.15 in the lower boundary layer up to 0.2 at the top at 2 km height. These high values indicate a non-negligible part of coarse mode particles present above the measurement site. However, the conversion into dust and non-dust particle contributions show that the dust mass concentration has decreased from about 300 μgcm^{-3} on 25 November to 60 μgcm^{-3} on 28 November. This is less than a third of the amount of large particles than during the wild fires.

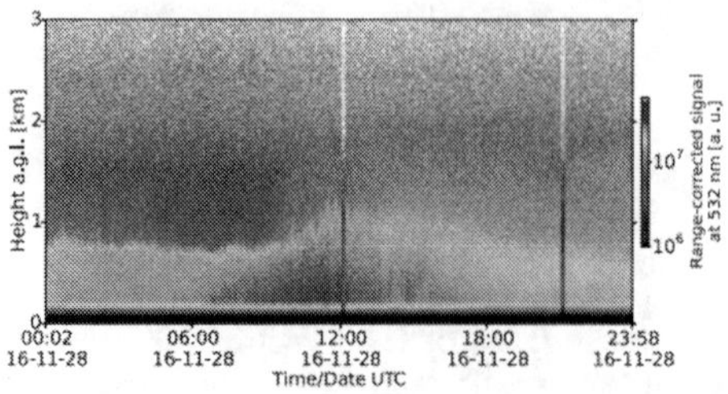

Figure 6: Range-corrected lidar signal on 28 November 2016 measured at Haifa. The vertical low-signal bars at 12:00 and 21:00 UTC are due to calibration periods. Note that the height scale ranges to 3 km in this graph.

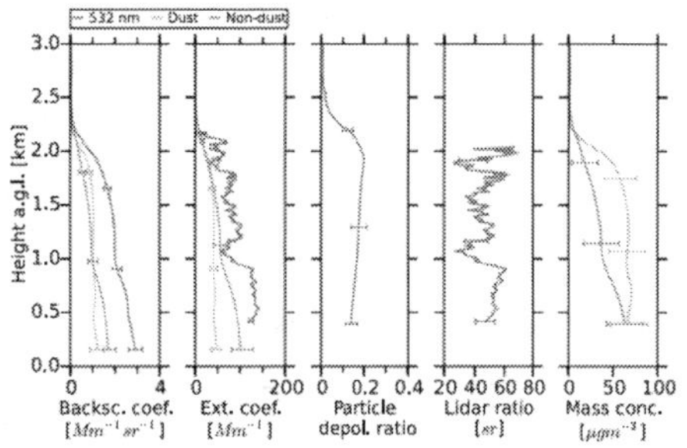

Figure 7: Profiles of particle backscatter and extinction coefficient, linear particle depolarization ratio, lidar ratio, and mass concentration (from left to right) at 532 nm, separated into dust and non-dust fractions on 28 November 2016, 18:00-20:00 UTC. These profiles were calculated using the Raman method.

4 CONCLUSIONS

These analyses are still preliminary. Separation into different particle contributions to the lidar properties will be extended, including further separation into fine and coarse mode dust. In March 2017 the one wavelength Polly at the Technion site in Haifa was replaced by a multi-wavelength Polly^{XT} lidar of the new generation [6]. This will allow for more comprehensive lidar data retrievals and a larger set of optical particle properties, including lidar ratios and depolarisation ratios at two wavelengths as well as the respective Ångström exponents.

ACKNOWLEDGEMENTS

We are grateful to Rotem Zamir, Vadim Holodovsky and Amit Aides for assistance in the lidar operation in Haifa. This project is funded by the German-Israeli Foundation (GIF grant I-1262-401.10/2014) with support of the Norman and Helen Asher Fund. The work of Yoav Y. Schechner is conducted in the Ollendorff Minerva Center. Minerva is funded through the BMBF (German federal ministry for education and research). TROPOS has also received funding from the European Union's Horizon 2020 research and innovation programme under grant agreement No 654109.

References

[1] Mamouri, R.E. and A. Ansmann, 2014: Fine and coarse dust separation with polarization lidar, *Atmos. Meas. Tech.* **7**, 3717–3735.

[2] Klett, J. D., 1981: Stable analytical inversion solution for processing lidar returns, *Appl. Optics* **20**, 211–220.

[3] Ansmann A., et al., 1992: Independent measurement of extinction and backscatter profiles in cirrus clouds by using a combined Raman elastic-backscatter lidar, *Appl. Opt.* **31**, 7113–7131.

[4] Freudenthaler V., et al., 2009: Depolarization ratio profiling at several wavelengths in pure Saharan dust during SAMUM 2006. *Tellus* **61B**, 165-179.

[5] Mamouri, R. E. and A. Ansmann, 2017: Potential of polarization/Raman lidar to separate fine dust, coarse dust, maritime, and anthropogenic aerosol profiles, in preparation.

[6] Engelmann, R., et al., 2016: The automated multiwavelength Raman polarization and water-vapor lidar PollyXT: the neXT generation, *Atmos. Meas. Tech.* **9**, 1767-1784, doi:10.5194/amt-9-1767-2016.

WATER VAPOR AND AEROSOL LIDAR MEASUREMENTS WITHIN AN ATMOSPHERIC INSTRUMENTAL SUPER SITE TO STUDY THE AEROSOLS AND THE TROPOSPHERIC TRACE GASES IN ROME

D. Dionisi[1]*, A.M. Iannarelli[2], A. Scoccione[3], G.L. Liberti[1], M. Cacciani[3], S. Argentini[1], L. Baldini[1], F. Barnaba[1], M. Campanelli[1], G. Casasanta[1], H. Diémoz [4], L. Di Liberto[1], G.P. Gobbi[1], I. Petenko[1,5], A.M. Siani[3], J. Von Bismarck[6], S. Casadio[2]

*[1]Institute of Atmospheric Sciences and Climate, Italy, *d.dionisi@isac.cnr.it*
[2]Serco Spa, Italy
[3]Physics Department, "Sapienza" University of Rome, Italy
[4]ARPA Valle d'Aosta
[5]A.M. Obukhov Institute of Atmospheric Physics, Russia
[6]ESA-ESRIN EOP-GMQ

ABSTRACT

A joint instrumental Super Site, combining observation in urban ("Sapienza" University) and semi-rural (ESA-ESRIN and CNR-ISAC) environment, for atmospheric studies and satellites Cal/Val activities, has been set-up in the Rome area (Italy). Ground based active and passive remote sensing instruments located in both sites are operating in synergy, offering information for a wide range of atmospheric parameters. In this work, a comparison of aerosol and water vapor measurements derived by the Rayleigh-Mie-Raman (RMR) lidars, operating simultaneously in both experimental sites, is presented.

1 INTRODUCTION

The characterization of the planetary boundary layer (PBL) is important in a variety of fields such as air quality studies, weather forecasting, climate change modeling, and Earth Observation satellite Cal/Val activities. Aerosols have a significant impact on the earth radiative budget and on air pollution. Furthermore, aerosols trapped within the PBL, can be used as tracers to study the boundary layer vertical structure and its temporal variability. The water vapor is an important atmospheric variable, its mixing ratio is useful as a tracer of air parcels and in understanding energy transport within the atmosphere.

In this context, "Sapienza" University, ESA-ESRIN and CNR-ISAC decided to cooperate and to exploit the remote sensing instrumentation, located in their experimental sites, in a coordinated effort (see Figure 1):

- BAQUNIN (Boundary-layer Air Quality Using Network of INstruments) Super-Site instruments are located at "Sapienza" University, in the city centre (41.90°N, 12.51°E and 75m ASL).
- CIRAS observatory (CNR-ISAC Rome Atmospheric Supersite,) is located in the semi-rural foot area of Tor Vergata (41.88°N, 12.68°E, 107 m ASL, http://www.isac.cnr.it/en/infrastructures/ciras-cnr-isac-rome-atmospheric-supersite).

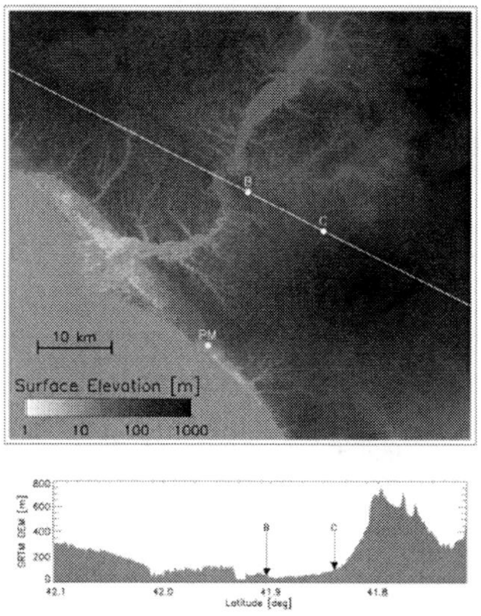

Figure 1. Upper panel: surface elevation a.s.l. of Tiber valley of the study area. Lower panel: elevation transect along with the location of two sites. B indicates the BAQUNIN site, C the CIRAS site and PM the position of the radio-sounding site of Pratica di Mare.

In figure 1, the upper panel depicts the study area, the locations and the surface elevation of the two sites. The lower panel shows the elevations transect along with the locations of BAQUNIN (B), CIRAS (C) and Pratica di Mare WMO station (PM).

This synergy of sites gathers a unique ensemble of advanced instrumentation for atmospheric remote-sensing and for the validation of atmospheric composition satellite data. The list of available instruments at both sites is reported in table 1.

Table 1. Instrument suite

Instrument type	Products	BAQUNIN	CIRAS
Elastic and Raman lidars	Aerosol backscattering and extinction , H2O profiles, AOD, cloud boundaries	✓	✓
Ceilometer	Aerosol backscattering profile, cloud boundaries		✓
SODAR	Thermal turbulence, wind speed and direction profiles	✓	✓
RADAR	Precipitation		✓
MFRSR	Total column of AOD, O3, H2O	✓	
POM 01 L (PREDE)	AOD, SSA, AE, Refr. Indx, PF, VSD	✓	
Brewer	Total Column of O3, NO2 and SO2, UV Dose and UV Index	✓	
Pandora	Total Column of NO2, O3, HCHO, others	✓	✓
Meteorological sensors	Air temperature, RH, pressure, wind, SW and LW radiation	✓	✓
CIMEL	Total column of AOD and H2O, SAE, AAE, SSA, Refr. Indx, VSD	✓	✓

From the instruments in Table 1, among others, the following products are listed: Aerosol backscattering and extinction profiles, Aerosol Optical Depth (AOD), Single Scattering Albedo (SSA), Angstrom Exponent (AE), Scattering Ångström Exponent (SAE), Absorption Ångström Exponent (AAE), Real and imaginary part of Refractive Index (Refr. Indx), Phase Function (PF), and Volume size distribution (VSD).

The simultaneous employment of the advanced lidars located in both sites to study the variability of aerosol and water vapor is presented in this work.

2 Rayleigh-Mie-Raman Lidar systems

Lidar is a well-established technique for measuring both aerosols and water vapor vertical profiles [1] along the time. Furthermore having two lidars located in different environments, urban and semi-rural, gives the possibility to observe the differences of aerosol and water vapor distributions. In the following paragraphs a short description of the two Rayleigh-Mie-Raman (RMR) lidars is reported.

2.1 RMR Lidar at BAQUNIN Super Site

The transmitter is a Nd:Yag laser with 3 wavelengths 1064, 532 and 355 nm and pulse frequencies 30Hz.

The receiver is composed by 4 telescopes to detect the backscattered radiation at 5 wavelengths. A Cassegrain telescope (the diameter is 500 mm) is used to detect the elastic signal at 355 nm and to the Raman signals at 387 nm (N_2) and 407 nm (H_2O). The radiation is separated via an optical system composed by interferential filters and dichroic mirrors. The detection altitude range is 0.5 - 10 km. The elastic radiation at 532 nm is collected by two telescopes [2], the first one is reflective (60 mm diameter) and dedicated to the lowest portion of the troposphere (0.05 – 5 km), the second is a Cassegrain telescope (160 mm diameter) covering the 0.5 – 10 km altitude range. This signal is further split into two channels for each telescope in order to collect the total backscatter and backscatter-polarized parallel to the emitted radiation. For the elastic radiation at 1064 nm, a Cassegrain telescope (160 mm) is used, and the range of the altitude is 0.125-10 km.

For all channels, the vertical resolution is 7.5 m and the temporal resolution is 10 seconds which corresponds to an integration of 300 laser pulses. This system operates in both day and night-time conditions.

2.2 RMR Lidar at CIRAS Super Site

The transmitter is based on an Nd:YAG laser with 2nd (532 nm) and 3rd (355 nm) harmonic generators.

Backscattered radiation is collected and analyzed at 4 wavelengths of interest: 355 and 532 nm for the elastic backscatter and 386.7 and 407.5 nm for Raman scattering of N_2 and H_2O molecules, respectively. The receiver is in a multiple-telescope configuration [3, 4], allowing the sensing of a wide-altitude atmospheric interval. For the Raman and elastic backscatter, two different collection channels (CHs) are employed: one for the lower range (0.1 – 5 km altitude, using

https://doi.org/10.1051/epjconf/201817605050

a 30 cm telescope and an optical fiber) and one for the upper range (2–13 km altitude, using an array of nine 50 cm aperture telescopes and an optical fiber bundle). An additional 15 cm telescope is used to collect the 532 nm elastic backscatter from the lower atmosphere (0.5–8 km).

The acquisition vertical resolution is 75 m and the signals are integrated over 60 second (600 laser pulses) and recorded. The RMR system is operated in manual mode and it provides 40 – 60 nighttime measurements per year. The system operates in both day and night-time conditions.

3 PRELIMINARY RESULTS

An example of simultaneous lidar RMR measurements is presented in this section. Figure 2 shows the logarithm of the Range Corrected Signal (RCS) at the wavelength of 532 nm acquired by the BAQUNIN Lidar (upper panel) and CIRAS Lidar (lover panel) on October 10, 2015 between 16.30 and 18.00 UT.

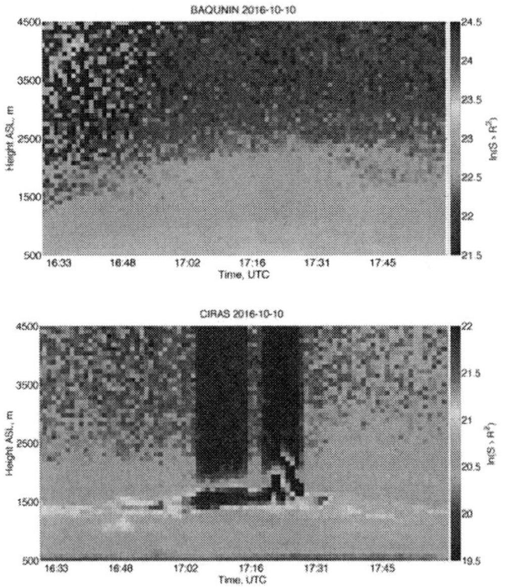

Figure 2. Time-height cross-section of the logarithm of the range corrected signal at 532 nm derived by the Baqunin and Ciras RMR lidar measurements (upper and lower panel, respectively) acquired on October 10, 2015 between 16:30 and 18:00 UT.

Both measurements have been re-gridded on the same temporal and spatial resolution (1 min and 75 m, respectively). Taking into account the different characteristics of two instruments, the ranges of the value scales of the plots are optimized for each system, in order to better appreciate the atmospheric structures. At both sites, aerosol layers are present up to 2000-2300 m, but the CIRAS lidar shows also a high backscattering layer around 1300-1500 m, which correspond to the passage, in temporal sequence, of aerosols, cloud and, again, aerosol. In terms of the aerosol load, the AOD values measured at the two sites seems to agree quite well. The values of simultaneous measurements are reported in table 2.

Table 2. AOD values derived by measurements at the two site, using Lidars at 532 nm, MultiFilter Rotating Shadowband Radiometer at 496 nm (BAQUNIN) and Cimel sun photometer at 500 nm (CIRAS)

	Lidar (17:00 UTC)	**Passive sensor (15:30 UTC)**
BAQUNIN	0.14±0.02	0.12±0.01
CIRAS	0.14±0.05	0.11±0.01

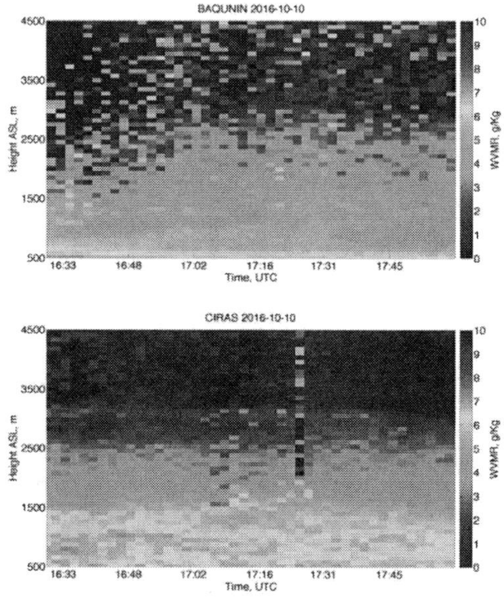

Figure 3. Time-height cross-section of the water vapor mixing ratio (WVMR) derived by the Baqunin and Ciras RMR lidar measurements (upper and lower panel, respectively) acquired on October 10, 2015 between 16:30 and 18:00 UT.

Figure 3 shows the time-height cross sections of the water vapor mixing ratio (WVMR) derived by both lidars. In this case, the temporal and spatial resolution of both measurements is 2 min and 75 m, respectively. The WVMR lidar data have been calibrated using, as reference, the radiosounding launched at the WMO station # 16245 (Pratica di Mare, see figure 1) by the Italian Meteorological Service of Military Aeronautics on October 10, 2015 at 12 UTC.

The comparison between the plots of figure 3 highlights, over the CIRAS site, the presence of a moist layer between 500 and 1500 m, which seems bounded by the observed aerosol/cloud layer. This humid layer is not observed over the BAQUNIN site. This feature is well seen in Figure 4, where the temporal evolution of the WVMR lidar profiles integrated for 30 minutes (blue and red lines for BAQUNIN and CIRAS lidar measurements, respectively) are plotted together with the radiosonde WVMR profile (green line). Below 1500 m, the atmosphere is moister at the CIRAS site of (the mean difference is around 0.78 g/Kg). An opposed behavior seems to occur above 1500 m, where the BAQUNIN lidar measures slightly higher values of WVMR compared to CIRAS lidar (the mean difference is around 0.12 g/Kg).

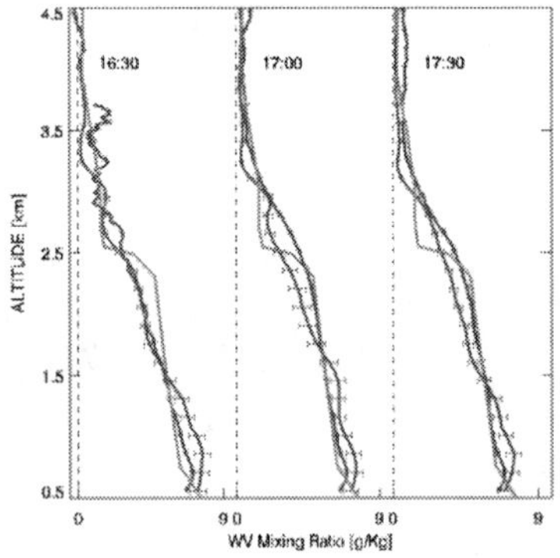

Figure 4. Temporal evolution of the 30-min integrated WVMR profiles derived by BAQUNIN and CIRAS lidars (blue and red lines, respectively) for the October 10, 2015. As reference, the radiosonde WVMR profile is also plotted (green line).

4 CONCLUSIONS

With the objective of providing independent and high-quality atmospheric composition ground-based data, which are required to accurately validate the upcoming European Earth Observation (EO) missions and payloads (e.g. S3, S5p, S5, S4, ADM-Aeolus, EarthCARE, METOP-SG, MTG), a coordinated instrumental Super site located in both urban (BAQUNIN) and semi-rural (CIRAS) environment of Rome has been setup. In this work, an example of the employment of RMR lidars located at both sites to study the temporal and spatial variability of aerosol and water vapor is presented. In the near future, the development of synergistic and coordinated ground-based measurement techniques and algorithms is planned to provide validation products that combine different atmospheric parameters (e.g. the combination of atmospheric chemistry and aerosol profiles). Within this frame, the Neural network Aerosol Typing Algorithm based on LIdar data (NATALI, http://natali.inoe.ro/) will be applied to both lidar datasets to classify aerosols.

ACKNOWLEDGEMENTS

Part of this activity has been funded by ESA Sensor Performance, Product and Algorithms Maintenance and Operations of the Earth Observation Payload Data System (IDEAS+ Contract NO.: VEGA/AG/14/01713 with SERCO S.P.A., WP-3360).

References

[1] Casadio, S., et al., 1996: Convective characteristics of nocturnal urban boundary layer as observed with Doppler sodar and Raman lidar. Boundary-Layer Meteorol., 79, 375-391.

[2] Di Girolamo, P., et al., 2017: Characterisation of boundary layer turbulent processes by the Raman lidar BASIL in the frame of $HD(CP)^2$ Observational Prototype Experiment. Atmos. Chem. Phys, 17, 745-767.

[3] Congeduti, F., et al., 1999: The multiple-mirror lidar '9-eyes.' *J. Opt.*, **1A**, 185–191.

[4] Dionisi D., F. Congeduti, G.L. Liberti, F. Cardillo, 2010: Calibration of a Multichannel Water Vapor Raman Lidar through Noncollocated Operational Soundings: Optimization and Characterization of Accuracy and Variability. J. Atm. Ocean. Tech., **27**, 108–121.

RETRIEVALS OF ICE-WATER CONTENT FROM AN AIRBORNE ELASTIC LIDAR IN TROPICAL CONVECTIVE CLOUDS

Konstantin Baibakov[1]*, Mengistu Wolde[1], Cuong Nguyen[1], Alexei Korolev[2], Ivan Heckman[2]

[1]*Flight Research Laboratory, National Research Council (NRC), Ottawa, Canada,*
**Konstantin.Baibakov@nrc-cnrc.gc.ca*
[2]*Environment and Climate Change Canada (ECCC), Toronto, Canada*

ABSTRACT

Ice water content (IWC) is one of the critical parameters in determining the cloud radiative impact. In this work lidar-based IWC retrievals obtained in tropical mesoscale convective systems are evaluated in the context of an extensive in-situ and remote sensing instrumentation suite. Based on a test case of May 27, 2015 lidar-derived IWC values at 50 m above the aircraft were on average within 25% of the in-situ IWC measurements obtained using an isokinetic probe.

1 INTRODUCTION

Ice and mixed-phase clouds cover roughly 25% of the Earth's surface and can have a significant and variable effect on the Earth's radiation budget [1]. The ice water content (IWC), is one of the critical parameters in determining the cloud radiative impact and hydrological cycle. In addition, high IWC conditions are recognized by aviation authorities as potentially dangerous for operation of commercial aviation [2]. A substantial effort has been expended to estimate IWC from various in-situ and remote sensing measurements. In particular, several works have previously shown the feasibility of IWC retrievals from lidar extinction profiles, especially in relation to spaceborne lidars. However, in many cases this technique is applied to mid-latitude clouds with IWC values on the order of mg/m^3. We have found no works that evaluate the potential of lidar IWC retrievals in the high ice water content (HIWC) environment, with IWC values reaching as high as 2-3 g/m^3. Furthermore, the quality of the remotely sensed IWC measurements is often evaluated through comparisons with in-situ data. This can turn out to be problematic in HIWC, because the standard measurement techniques of estimating IWC, either from 2D particle imagery or hot-wire probes, can suffer from numerous instrumental issues [3]. Recently developed isokinetic probes (IKP) [4] were designed to solve most of these issues and can provide accurate measurements at high IWC concentrations (up to 15 g/m^3) and air speeds (up to 200 m/s) [idem]. In May 2015, the NRC Convair-580 aircraft participated in the HAIC (High Altitude Ice Crystals)-HIWC project from an operational base in Cayenne (French Guiana, 5°N 52°W). The project was aimed at better understanding of the HIWC environment with an eye towards reducing the dangers of flying in ice crystals icing (ICI) conditions. The Convair was jointly equipped by NRC and ECCC and included state of the art in-situ and remote sensing instrumentation. In particular, two quasi-identical Airborne Elastic Cloud Lidar systems (AECL-zenith and AECL-nadir) were installed onboard the Convair and acquired close to 40 hours of data. An IKP probe was also used during the campaign. The purpose of this paper is to briefly describe the AECL lidar system, discuss the AECL performance and evaluate AECL capabilities in supplying IWC information as compared to in-situ measurements. Section 2 describes the instrumentation and data processing with a focus on AECL. Section 3 presents the main results of the AECL-based IWC retrievals based on a representative test case of May 27, 2015. Finally, Section 4 summarizes our preliminary findings.

2 METHODS AND INSTRUMENTATION

2.1 AECL lidar description and data processing

Airborne Elastic Cloud Lidar (AECL) built by Alpenglow Instruments, LLC. is a compact airborne elastic lidar operating at 355nm. The lidar design is shown in Figure 1. AECL features a rotatable beam bender section that allows the instrument to be configured in either zenith or nadir orientation. The change in configuration, however, can not be performed while in-flight.

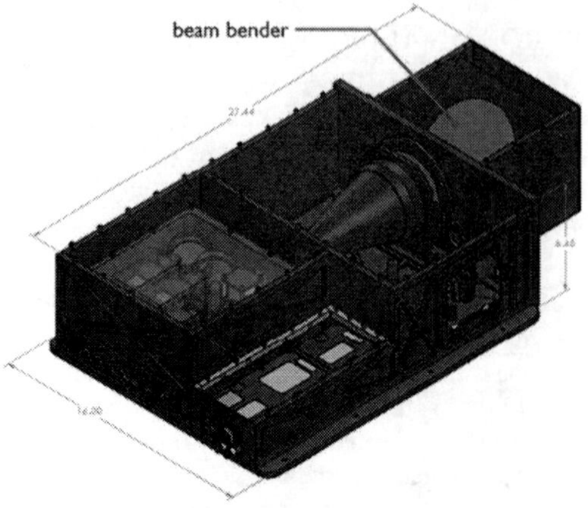

Figure 1. AECL design with a rotatable beam bender

AECL can estimate a degree of particle-induced depolarization of the emitted laser pulse by separately measuring parallel and perpendicular polarization components of the return signal, $P_{\parallel}$ and $P_{\perp}$. The (linear) depolarization ratio, LDR = $P_{\perp}/ P_{\parallel}$, can often provide an insight into the types of particles encountered by the laser beam with lower LDR values often associated with spherical (liquid) drops.

Table 1 Technical specifications of the AECL transmitter, receiver and data acquisition system

Transmitter	
Laser Wavelength	355nm YAG
Pulse Repetition Frequency	20 Hz
Pulse Width	≈8 ns
Pulse Energy	12mJ
Beam Divergence	$0.4x10^{-3}$ radians
Receiver	
Diameter	≈100 mm
Field of View	≈1000 μrad
Data System	
Number of Channels	Four: Low and high gain parallel and perpendicular signals
Detector	PMT
Range Resolution	1.5m, 3.0m, 6.0m, 12m (programmable)

The essential AECL technical specifications are listed in Table 1 while the lidar data processing chain is briefly described below. First, the return signal is background- and range-corrected to compensate for parasite (internal) signal sources and the R^2 signal drop off with range, respectively. Then the incomplete overlap correction is also applied to compensate for the effect of the laser beam not being entirely within the telescope FOV at the lowest ranges. We estimate that the full overlap is achieved at 150 m. Comparing the AECL results in the few lowest range bins against the in-situ (flight-level) data requires careful consideration of the incomplete overlap problem. In this work, we have used the method of [5] to correct the signal below 150 m by using calibration data in clear air. Despite the correction, we acknowledge that evaluating lidar data in the overlap region can be problematic and associated with a source of error. Once all the basic corrections have been applied, the lidar vertically resolved extinction coefficients are derived using a Klett technique with a boundary condition in the far range. This method requires an assumption of the particulate extinction-to-backscatter (or lidar) ratio, LR. In this work we used a constant vertically averaged value of LR=20 Sr consistent with climatological values from satellite statistics and other works. Only the lidar profiles in zenith direction were used for Klett inversions. Finally, once the extinction coefficient (κ) profiles have been computed, IWC profiles can be calculated using a power law parametrization:

$$IWC = a\kappa^b \tag{1}$$

where we used the values of a=527 and b=1.32 as per [1].

2.2 In-situ instrumentation

During the campaign, Convair-580 carried a multitude of in-situ instruments designed for cloud studies. Among them were particle imaging probes (including SPEC 2D-S and DMT PIP) as well as IKP for bulk IWC measurements. The imaging probes provide a visual information on particle size and shape. Particle probes' data was also used to derive cloud extinction coefficient and particle median mass size and concentration. IWC information can also be derived from imaging probes using empirical size-to-mass parameterization. The IKP provides reliable measurements of IWC by measuring a change in absolute humidity from continuously aspirating and evaporating all encountered particles and ensuring that the air speed at the probe inlet matches the free stream air speed [3].

2.3 NAWX radar

The NAWX radar system [6] has polarimetric and Doppler capabilities at both frequencies (X-band: 9.41 GHz [3.2 cm], W-band: 94.05 GHz [3.2 mm]) and can switch electronically between zenith, nadir and side looking antennas. The equivalent radar reflectivity (Z_e) is used to characterize cloud structures in ice environments.

3 RESULTS

Figure 2 shows a dataset obtained on May 27, 2015 between 11:08 and 11:17 during the HAIC-HIWC flight campaign. Listed from top to bottom are: 1) lidar backscatter power (arbitrary units), 2) lidar depolarization (arbitrary units), 3) lidar extinction coefficient profiles obtained from Klett inversion (km^{-1}, log-scale), 4) radar W-band reflectivity (dBZ), 5) computed particle concentration, L^{-1}, 6) computed particle median diameter, µm, 7) extinction coefficient from in-situ data and lidar extinction at 50 m above aircraft (km^{-1}), 8) IWC measured by IKP and lidar-based IWC taken at 50 m above aircraft (g/m^3). White line in panes 1 and 4 signifies the last bin below which lidar returns are considered too weak to be used for data analysis. Figure 2 is a typical measurement scene from the HAIC-HIWC campaign and contains several features of interest relevant to AECL performance. The aircraft stayed at an altitude of ~5.5 km between 11:08 and 11:10 and then gradually ascended to 6 km by 11:17. Before 11:12:30 AECL penetration range is relatively high which allows the detection of multiple cloud layers up to 1.5 km above the aircraft. Lidar data indicates the presence of liquid droplets (high backscatter and low depolarization ratio) above 6 km with layers of glaciated particles (low backscatter and high depolarization ratio) immediately below. The time after 11:12:30 is associated with significantly reduced lidar penetration range and can be subdivided into the following relevant time periods (TP): TP1 (11:12:00-11:12:30), TP2 (11:13:05-11:13:45) and TP3 (11:13:45-11:16:30).

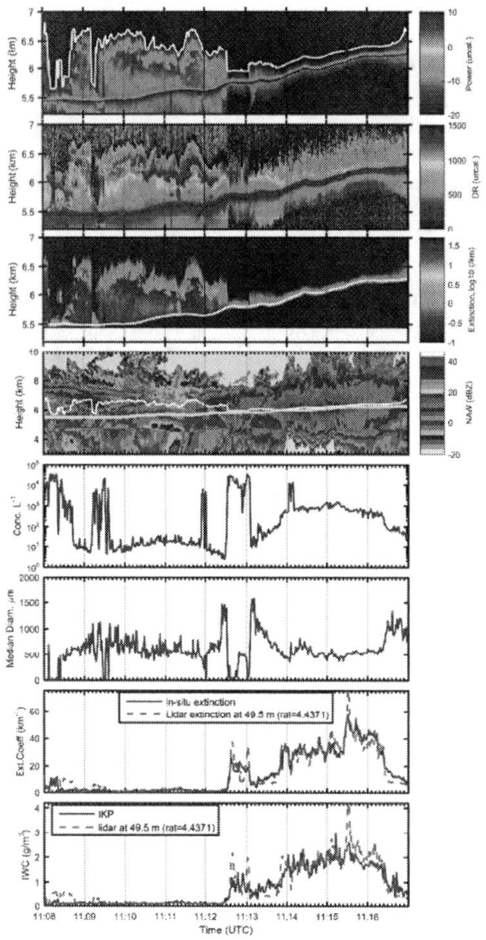

Figure 2 Lidar, radar and in-situ responses during the HAIC-HIWC on May 27, 2015. See text for details.

While TP1 and TP3 are both associated with relatively high lidar extinction values (reaching values >20 km^{-1}), in-situ measurements show that the extinction is caused by different types of particles. Referring to panes 5 and 6 of Figure 2, it can be seen that particle concentrations are an order of magnitude higher during TP1 than during TP3 (10^3 vs 10^4 L^{-1}). At the same time, the particle median mass diameter, MMD, defined as the size that splits the mass size distribution into two parts equal by mass, gets as low as 50-60 µm during TP1 but stays at an average value of ~500 µm during TP3. These observations imply that TP1 is dominated by small liquid cloud droplets, while TP3 consists of larger glaciated particles.

The Cloud Imaging Probe (CPI) imagery (not shown) corroborates this analysis, showing small regular particles during TP1 and predominantly unrimed columns during TP3. TP3 is associated with the High Ice Water Content environment – a primary subject of investigation in the campaign. Previous results show (e.g. [7]) that MMD in HIWC varies between 200 and 500 μm and decreases with increasing TWC and/or decreasing temperature. These statistics are consistent with the results of Figure 2. During TP3, lidar penetration range is significantly reduced from ~1.2 km at 11:10:35 to ~200m at 11:15:00. The occurrence of HIWC is also characterized by strong W-band radar reflectivity of up to 17 dBZ. The extinction pane of Figure 2 presents the in-situ extinction estimated from the particle size distribution as well as Klett-derived extinction taken at ~50 m above the aircraft and scaled by 4.44. Throughout the campaign Klett-derived extinction profiles had to be generally scaled by a factor of ~4 to match the in-situ extinction. While the source of this discrepancy is currently being investigated, low lidar extinction values can be explained, in part, by insufficient overlap and multiple scattering corrections as well as errors in the initial parameters used in Klett inversions, especially the cloud lidar ratio. It's clear nevertheless that relative variations in lidar extinction match well those from the in-situ based retrievals with R^2=0.86 for the entire time period between 11:12:30 and 11:17:00. Lidar extinction values were then used in Eq. (1) to derive IWC concentrations. Pane 8 shows a comparison between the IWC time-series derived from the IKP probe (IWC_{IKP}) and AECL-based IWC values taken at 50 m above aircraft (IWC_{lid}). During TP3 IWC_{lid} values were on average within 25% of the IWC_{IKP} and generally matched well the IWC_{IKP} in terms of relative variations.

4 CONCLUSIONS

In May 2015 AECL sampled regions of HIWC near Cayenne (French Guiana) with IWC values as high as 2-3 g/m^3. Lidar IWC values derived from extinction using a power law parametrization were on average within 25% of the IKP measurements. The relatively good agreement is important for a number of reasons. First, a consistency between fundamentally different instruments yields more credence to both the absolute values and the relative variations of IWC in the HIWC environment. Second, it shows that the parametrization developed by [1] produces results consistent with in-situ measurements in HIWC environment. Third, it shows a potential for lidar-only IWC retrievals in the future once the inherent uncertainties in lidar data processing are accounted for.

ACKNOWLEDGEMENTS

Funding for the NRC Convair-580 participation in the HAIC-HIWC field project was provided by NRC and ECCC. We would like to acknowledge the NRC and ECCC project crew that supported the field project. We also acknowledge the staff at Alpenglow Inc. for their technical support.

References

[1] Heymsfield, A. et al, 2014, Relationships between Ice Water Content and Volume Extinction Coefficient from In Situ Observations for Temperatures from 0° to –86°C: Implications for Spaceborne Lidar Retrievals J. Appl. Meteor. Climatol., 53, 479–505, doi: 10.1175/JAMC-D-13-087.1.

[2] M. L. Grzych and J. G. Mason, Conditions Associated with Jet Engine Power Loss and Damage Due to Ingestion of Ice Particles: What We've Learned Through 2009 14th Conf. Aviat. Range, Aerosp. Meteorol., 2010.

[3] Wendisch M., J.L. Brenguier, 2013: Airborne Measurements for Environmental Research: Methods and Instruments. Willey, 641p.

[4] Davison, C. R. et al, 2012, Naturally Aspirating Isokinetic Total Water Content Probe: Pre-flight Wind Tunnel Testing and Design Modifications AIAA Paper 2012-3040, 4th Atmospheric and Space Environments Conference, 25 - 28 June 2012, New Orleans, Louisiana.

[5] Sasano, Y. et al, 1979, Geometrical form factor in the laser radar equation: an experimental determination Appl. Opt. 18, 3908-3910.

[6] Wolde, M. and Pazmany, A., 2005, NRC Dual-Frequency Airborne Radar For Atmospheric Research, Proc. 32nd Conf. on Radar Meteorology, Albuquerque, USA.

[7] Leroy, D. et al., 2016, HAIC/HIWC field campaigns - Specific findings on ice crystals characteristics in high ice water content cloud regions", 8th AIAA Atm. and Space Env. Conf., AIAA 2016-4056.

POLARIZATION LIDAR FOR ATMOSPHERIC MONITORING

Qiaojun Liu[1]*, Chengxuan Wu[1], Andrew Yuk Sun Cheng[1], Zhangjun Wang[1], Xiangqian Meng[1], Chao Chen[1], Xianxin Li[1], Xingtao Liu[1], Hao Zhang[1], Fangyi Zong[1]

*[1]Shandong Academy of Sciences Institute of Oceanographic Instrumentation, Qingdao266001, China, *Email: qiaojunliu@163.com*

ABSTRACT

Aerosol plays an important role in global climate and weather changes. Polarization lidar captures parallel and perpendicular signals from atmosphere to research aerosols. The lidar system we used has three emission wavelengths and could obtain the atmospheric aerosol extinction coefficient, backscattering coefficient and depolarization ratio. In this paper, the design of the lidar is described. The methods of data acquisition and inversion are given. Some recent results are presented.

1 INTRODUCTION

Lidar is a powerful tool for atmospheric aerosols monitoring with high temporal resolution [1-4]. Aerosol plays an important role in global climate and weather changes. The traditional Mie scattering lidar technique assumes the aerosol particles are spherical and homogeneous [5]. But it is not consistent with the actual atmospheric environment. Non-spherical or inhomogeneous particles will introduce a depolarized component into the backscattering. Lidar polarization technique greatly expands the capabilities of atmospheric monitoring [6-10]. Polarization lidar transmits a linearly polarized laser pulse and uses a beam splitter to separate the perpendicular and parallel signals of the backscattered light. The ratio of these two signals can be used to analyze the physical and optical characteristics of the aerosols and clouds.

2 METHODOLOGY

The lidar system we established has three emission wavelengths and six receiving channels. It could obtain the atmospheric aerosol extinction coefficient, backscattering coefficient at 355nm, 532nm and 1064nm, and depolarization ratio at 532nm. The system layout is depicted below the figure 1.

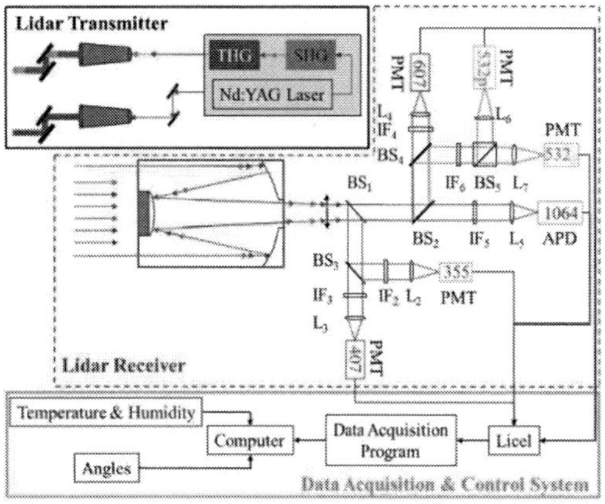

Figure 1 Design of lidar system

The system is based on a Nd:YAG laser followed by a second- and a third-harmonic generator, the whole emitting at the 355 nm, 532 nm and 1064 nm wavelengths. The receiver telescope is a Schmidt-Cassegrain telescope with 300mm diameter. The system has six receiving channels. They collect elastic scattering signals at 355nm, 1064nm, Raman scattering signals at 407nm, 607nm, and the parallel and perpendicular signals at 532nm. Signals are detected by photomultiplier tubes and APD as indicated in fig.1. This system can detect the atmospheric aerosol extinction coefficient, backscattering coefficient and depolarization ratio. System parameters and photo are showed in table 1 and figure 2, respectively.

Table 1 Lidar system parameters

Laser	Nd:YAG
Wavelength	355, 532 and 1064 nm
Single Pulse Energy	80 mJ@355 nm, 50mJ@532 nm,

 https://doi.org/10.1051/epjconf/201817605052

	100mJ@1064 nm
Repetition Rate	20 Hz
Divergence Angle	0.15 mrad
Telescope type	Schmidt-Cassegrain
Telescope diamater	300mm
Telescope FOV	0.3mrad

Figure 2 Photo of lidar system

The range(r)-resolved volume linear depolarization ratio δ is defined as

$$\delta(r)=\frac{p_{rs}(r)/k_s}{p_{rp}(r)/k_p}=\frac{\beta_s(r)}{\beta_p(r)}\exp\{\int_0^r[\alpha_p(r')-\alpha_s(r')]dr'\} \quad (1)$$

where $p_{rs}(r)$ and $p_{rp}(r)$ are the lidar signals with polarization perpendicular and parallel to the polarization of the transmitted laser light, respectively. k_s and k_p are the system constants of perpendicular and parallel channels. $\beta_p(r)$ and $\beta_s(r)$ are the backscatter coefficients for scattering parallel and perpendicular relative to the polarization of the transmitted laser beam. $\alpha_p(r)$ and $\alpha_s(r)$ are parallel and perpendicular components of extinction coefficient at the distance r. Normally, $\alpha_p(r)=\alpha_s(r)$ [11], so

$$\delta(r)=\frac{\beta_s(r)}{\beta_p(r)}=k\frac{P_{rs}(r)}{P_{rp}(r)} \quad (2)$$

Where $k=k_p/k_s$, is a calibration factor.

3 RESULTS

Examples of the measurement results are depicted in figure 3, figure 4 and figure 5. They all show 24-hour continuous monitoring of volume depolarization ratio on different days in September, 2015. Clouds exist between 6 and 9 kilometers in figure 3, between 8 and 10 kilometers in figure 4. In figure 5, clouds exist at above 10 kilometers and dropped to about 7 kilometers gradually. Figure 6 shows aerosol extinction coefficient profile at 3:01am on September 18th. It seems that extinction coefficient increases between 10 and 12km at 3:01am, which has the same trend with figure 5.

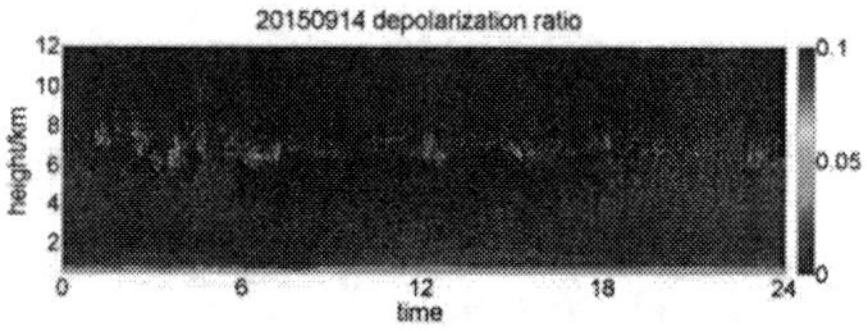

Figure 3 The depolarization ratio on September 14th,2015

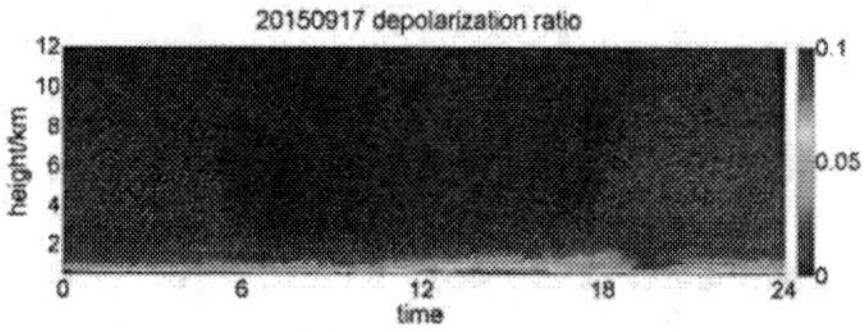

Figure 4 The depolarization ratio on September 17th,2015

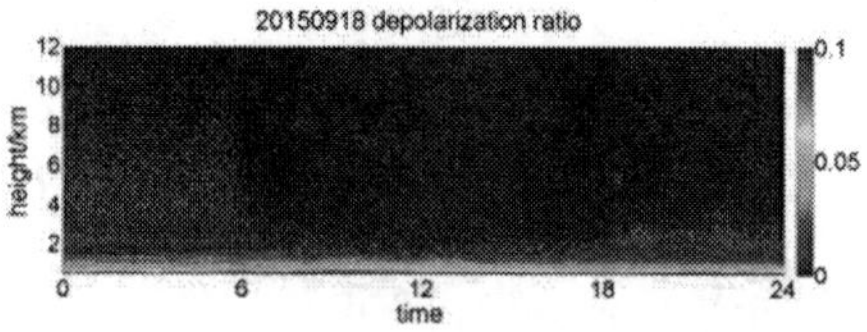

Figure 5 The depolarization ratio on September 18th,2015

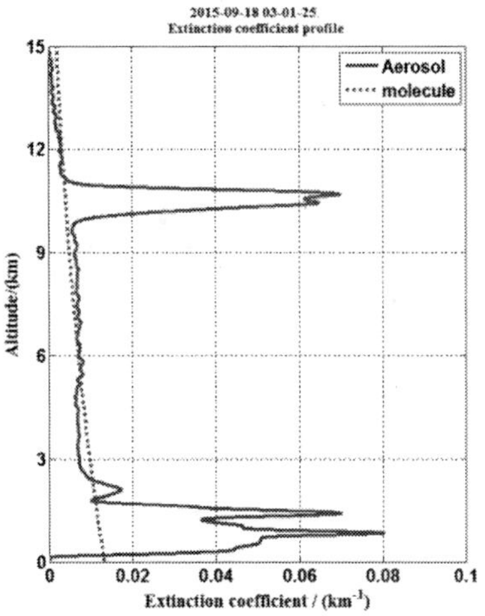

Figure 6 The aerosol extinction coefficient profile at 3:01am on September 18th

4 CONCLUSIONS

A polarization lidar system has been developed for atmospheric monitoring. Designs of the lidar system and the inversion method used to obtain depolarization ratio are discussed. Some recent results measured are also presented. Further research and analysis should to be done to get more understanding about the non-spherical particles and clouds in the atmosphere.

ACKNOWLEDGEMENTS

This work is supported by Shandong province excellent young and middle-aged scientists research award fund with grant No. BS2014HZ009, Shandong Academy of Sciences Youth Foundation with grant No. 2015QN019, Qingdao applied basic research program with grant No. 15-9-1-91-JCH and No. 15-9-1-102-JCH, Progress Program of Science and Technology of Shandong with grant No. 2012GHY11519.

References

[1] Measures, R.M., 1984. Laser Remote Sensing, Fundamentals and Applications, pp. 259-269, John Wiley and Sons, New York.

[2] Bates T., Huebert B., Gras J., Griffiths F., Durkee P., 1998. International global atmospheric chemistry (IGAC) projects's first aerosol characterization experiment (ACE 1): overview, Journal of Geophysical Research, 103(D13), pp. 16297-16318.

[3] H. Takahashi, H. Naoe, Y. Igarashi, Y. Inomata and N. Sugimoto, 2010. Aerosol concentrations observed at Mt. Haruna, Japan, in relation to long-range transport of Asian mineral dust aerosols, Atmospheric Environment, 44(36), pp. 4638-4644.

[4] J. N. PORTER, B. R. LIENERT et al, 2002: A Small Portable Mie–Rayleigh Lidar System to Measure Aerosol Optical and Spatial Properties , Journal of Atmospheric and oceanic Technology, VOL 19, 1873

[5] Klett J D, 1981: Stable analytical inversion solution for processing lidar returns, Appl Opt., 20(2), 211

[6] Allen, R.J., and C.M.R. Platt, 1977: Lidar for multiple backscattering and depolarization observations. Appl. Opt., 16,3193

[7] G.P.Gobbi: Polarization Lidar return from aerosols and thin clouds: a framework for the analysis. Applied Optics, 1998, vol.37, No.24, pp5505-5508.

[8] O.Uchino, I.Tabata, K.Kai and Y.Okada: Polarization Properties of Middle and High Lever Clouds Observed by Lidar. J. Meteor. Soc. Japan, 1988, vol.66, pp607-616

[9] R.M.Schotland, K.Sassen and R.Stone: Observation by Lidar of Linear Depolarization Ratios for Hydrometeors. J. Appl. Meteor. 1971, Vol.10, pp1011-1017

[10] Pal, S.R., and A.I. Carswell, 1973: Polarization properties of lidar backscattering from clouds. Appl. Opt.,12,1530.

[11] K.Sassen: Lidar Backscatter Depolarization Technique for Cloud and Aerosol Research. pp.393-416.

IDENTIFICATION OF LONG-RANGE TRANSPORT OF AEROSOLS OVER AUSTRIA USING EARLINET LIDAR MEASUREMENTS

Talianu Camelia[1,2*]

[1]*Institute of Meteorology, University of Natural Resources and Life Sciences, Vienna, Austria,*
[2]*National Institute or R&D for Optoelectronics, Magurele, Romania *camelia.talianu@boku.ac.at*

ABSTRACT

The aims of the study is to identify the paths of the long-range transported aerosols over Austria and their potential origin, and to estimate their properties, using lidar measurements from EARLINET stations closest to Austria from Germany and Romania and aerosol transport models. As of now, there is no lidar station in Austria. The study is part of a project to estimate the usefulness of a lidar station located in Vienna, Austria.

1 INTRODUCTION

Aerosols are produced locally, with given properties; they are then transported over short-to long-distances, usually in turbulent movements. Transport and dispersion of aerosols depend on aerosols' properties, on meteorological conditions, and on surface properties. During transport, aerosols can also mix with other atmospheric, so more sources contribute to the aerosol budget in a given place.

Lidar systems are useful tools to determine the properties of long-range transported aerosols and their vertical distribution [1-3]. EARLINET [4], a network of lidar systems with stations distributed over Europe, provides comprehensive ground-based lidar data for aerosol vertical distribution as well as aerosols' optical and microphysics properties, allowing a detailed study of long-range transported aerosols in Europe.

To identify the source regions of aerosols, a statistical analysis of back-trajectories can be performed. The back-trajectories analysis relates the aerosol mass loading changes at a receptor location to spatially-fixed sources, identifying the sources by a source – receptor matrix calculation [5].

Due to its geographical position in Central Europe, Austria is affected mainly by local and long-range transport of aerosols from variable sources. Marine aerosols with sea-salt particles content are rarely observed above Austria, as the nearest sea, Adriatic Sea, is located at a distance of few hundred kilometers, so aerosols measured in Austria are predominantly continental aerosol.

The purpose of this analysis is to study the long-range transport of aerosols over Austria, assuming a receptor site at Vienna. The study is based on cluster analysis, using measurements from EARLINET lidar stations closest to Austria, MACC reanalysis data [6] and the FLEXPART aerosol dispersion model [7].

2 METHODOLOGY

The analysis has been performed for Vienna (48.21°N, 16.36°E) as receptor site, using data from the period March - May 2014. This period has been reported of Austria as "seventh warmest spring (March - May) in its 247-year period of record, at 1.5°C higher than the 1981–2010 average" [8]

The vertical distributions of aerosols over Vienna were obtained from MACC reanalysis data related to dust aerosol and smoke. Here, smoke was considered a mixture of organic mater, sulphate and black carbon.

For the cluster analysis, clusters with trajectories traversing over four EARLINET stations around Austria: Garmisch-Partenkirchen (47.47°N, 11.06°E), Munich-Maisach (48.20°N, 11.45°E), Leipzig (51.35°N, 12.43°E) (all stations located in Germany), and Bucharest (44.35°N, 26.03°E, Romania), were selected.

The contribution of the different aerosol sources was evaluated comparing the spatial distribution of the layers as determined from the lidar measurements with the FLEXPART forward and backward simulations for each observed aerosol layers. From the lidar measurements, the aerosol layers were calculated using a wavelet analyses applied on the derivative of back-scattering coefficients. The transport of aerosols (dust or smoke) and the source-receptor sensitivity were calculated with the FLEXPART model using 6-hourly GFS meteorological data interleaved with operational forecasts every 3 hours. The model was run for 7 days using a backward simulation with a tracer released within a volume estimated from lidar measurements and MACC observations, where no lidar measurements were available (at receptor site). The tracer release period ranges were taken of the order of two hours when estimated from lidar measurements, and from one to three days when estimated from MACC observations.

The types of the aerosols in Vienna were estimated from the analysis of the aerosol layer optical properties obtained from lidar measurements, correlating the layers at the lidar stations with the FLEXPART transport layers in the cluster analysis.

The regions of the potential aerosol sources were taken from NASA Fire Information for Resource Management System (FIRMS) maps [9] for fire investigation, and Barcelona Supercomputer Center regional dust forecast model - BSC-DREAM8b [10], for dust investigation over Europe.

3 EVENT SELECTED

Figure 1 shows the mixed ratio of dust and smoke over Vienna for Mar - May 2014, extracted from MACC reanalysis data.

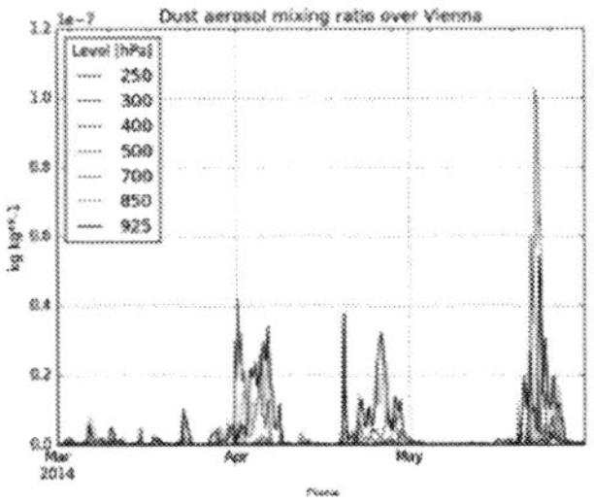

Figure 1 Dust aerosol over Austria, 1 Mar - 1 Jun 2014

Three events of dust aerosol over Austria are shown in Fig. 1: 30 Mar – 7 Apr 2014, 20 Apr – 29 Apr 2014 and 20 May – 26 May 2014. The data selected from these periods are correlated with lidar measurements from lidar stations.

4 RESULTS

For all cases, the cluster analysis was performed for two levels: 850 hPa (aprox. 1200 m) and 750 hPa (aprox. 3000 m)

Case 1: 30 Mar – 7 Apr 2014

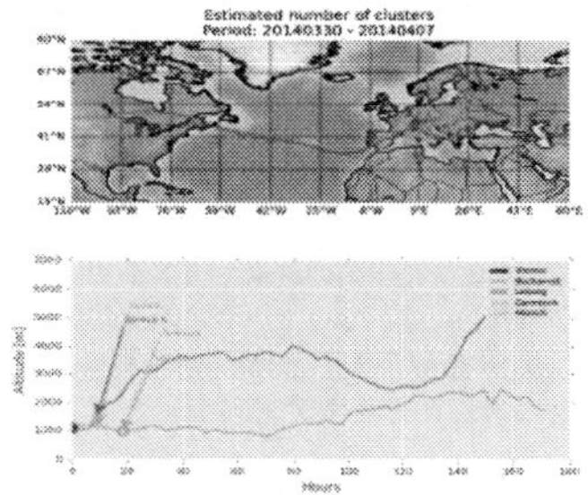

Figure 2 Cluster analysis applied on backtrajectories starting at 00:00 UTC, 850 hPa

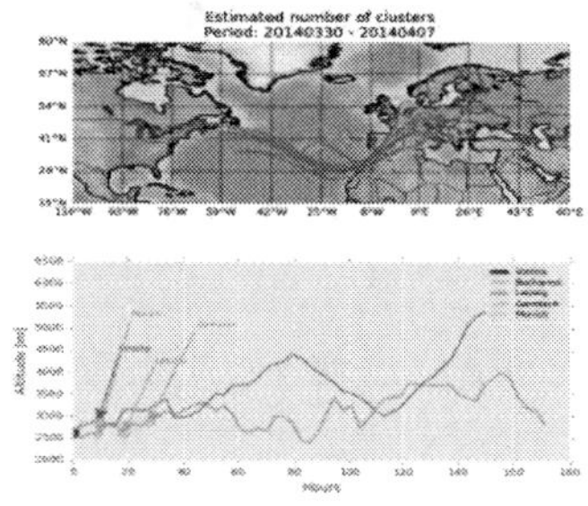

Figure 3 Cluster analysis applied on backtrajectories starting at 00:00 UTC, 700 hPa

There are two clusters with trajectories over German lidar stations started at 850 hPa (Fig. 2) and two clusters for trajectories started at 700 hPa (Fig. 3)

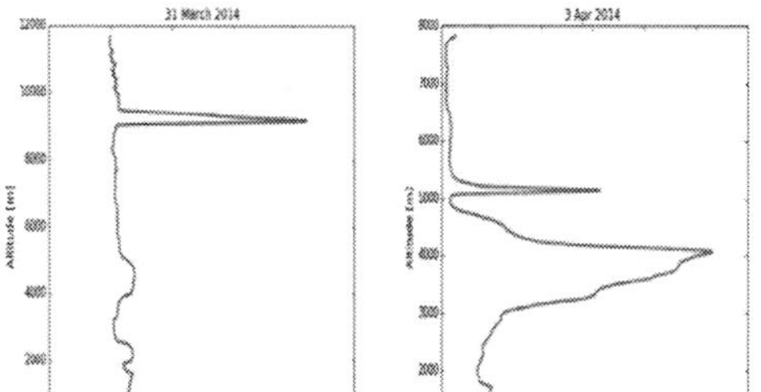

Figure 4 Garmisch lidar station: Backscatter coefficients at 313 nm, 31 March 2014, 11:30 UTC (left) and 3 Apr 2014 11:30 UTC (right)

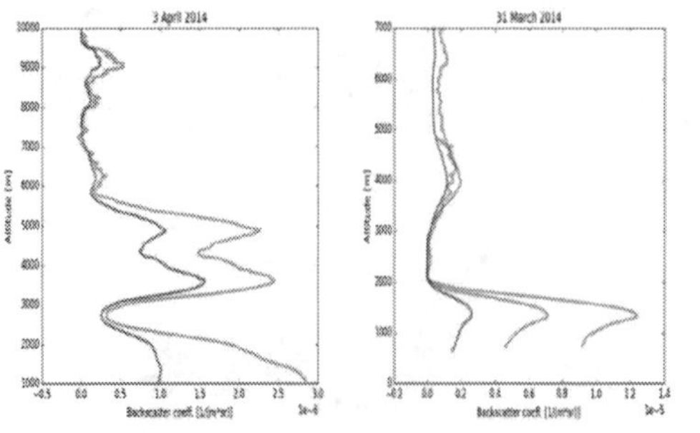

Figure 5 Leipzig lidar station: Backscatter coefficients at 1064 nm, 532 nm, 355 nm, 31 March 2014, 11:30 UTC (left) and 3 Apr 2014 18:30 (right), 532 nm and 355 nm

The dates selected and the altitudes for Vienna and the considered lidar stations measurements for this case are shown in Table 1. For the lidar stations, the first altitude is obtained from FLEXPART at the station coordinates, while the second altitude is obtained from the lidar measurements.

Table 1

Vienna		Lidar Station		
Date	Alt. (m)		Date	Alt. (m) model/lidar
1 Apr	2615	Leipzig	31 Mar	2950 / 3100
1 Apr	1115	Garmisch	31 Mar	1800 / 1900
1 Apr	1115	Leipzig	31 Mar	1250 / 1300
4 Apr	1115	Garmisch	3 Apr	1100 / 1200
4 Apr	2615	Leipzig	3 Apr	2800 / 3000
4 Apr	2615	Garmisch	3 Apr	2980 / 3100

Case 2: 20 Apr – 29 Apr 2014

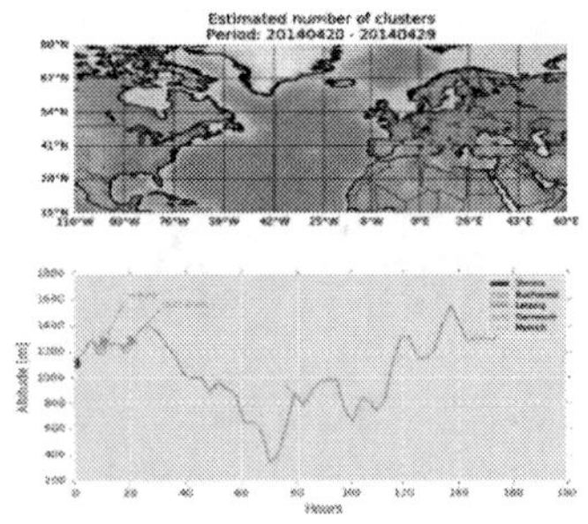

Figure 6 Cluster analysis applied on backtrajectories starting at 00:00 UTC, 850 hPa

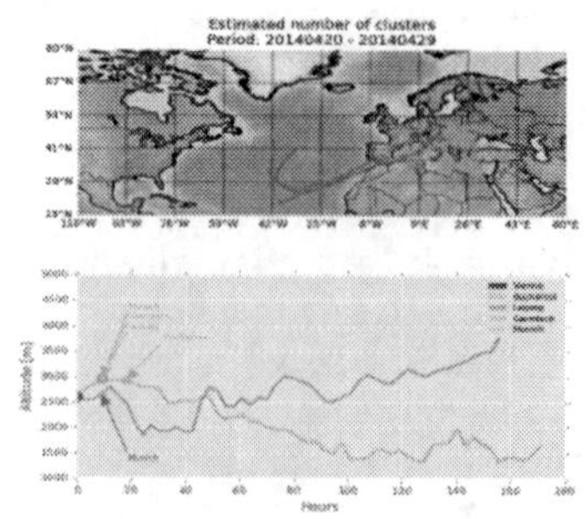

Figure 7 Cluster analysis applied on backtrajectories starting at 00:00 UTC, 700 hPa

There are only one cluster with trajectories started at 850 hPa (Fig. 5) and two clusters over all Earlinet lidar stations, trajectories started at 700 hPa (Fig. 6)

Table 2 shows, similar to Table 1, the results obtained for this case. The lidar measurements are performed at: 4:00 UTC Garmisch, 17:00 UTC Bucharest, 16:00 UTC Leipzig, 20:00 UTC Munich.

Table 2

Vienna		Lidar Station		
Date	Alt. (m)		Date	Alt. (m) model/lidar
27 Apr	2615	Bucharest	26 Apr	3100 / 3200
27 Apr	1115	Bucharest	26 Apr	1200 / 1000
25 Apr	2615	Leipzig	24 Apr	3000 / 3200
25 Apr	2615	Garmisch	24 Apr	2680 / 2800
25 Apr	2615	Munich	24 Apr	2800 / 2850
25 Apr	1115	Leipzig	24 Apr	1300 / 1200

Case 3: 20 May – 26 May 2014

For this case, a single cluster was identified for each of the two levels, 850 hPa and 700 hPa passing over Leipzig station. Table 3 shows the results obtained for this case.

Table 3

Vienna		Lidar Station		
Date	Alt.(m)		Date	Alt. (m) model/lidar
23 May	2615	Leipzig	22 May	2715 / 2600
23 May	1115	Leipzig	22 May	1120 / 1300

From the optical properties obtained from the lidar measurements, it results that the Case 1 is a mixture between dust and smoke particles. This mixture could be possible due to the same transport pathway for both sources, from North America over West Sahara. The three-dimensional structure of the aerosol distribution revealed by the lidar systems, combined with MACC observations and the three-dimensional structure of the aerosol plume from FLEXPART aerosol transport model (see Figs. 8 and 9) confirm the presence of the two different components, smoke and dust coming from the two aerosol sources : smoke from North America (see Fig 10) and dust from Sahara (see Fig. 11). Case 2 and case 3 are Saharan dust intrusions over Europe.

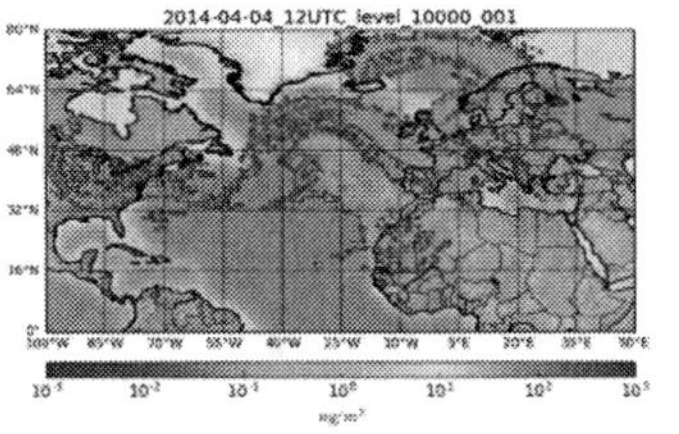

Figure 8 Integrated column , 4 Apr 2014

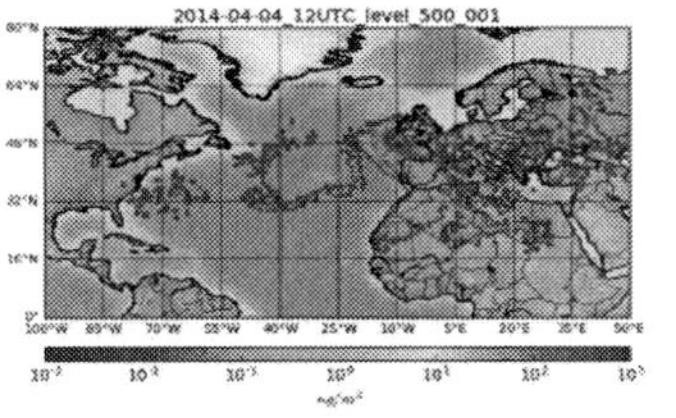

Figure 9 Footprint (500 m) , 4 Apr 2014

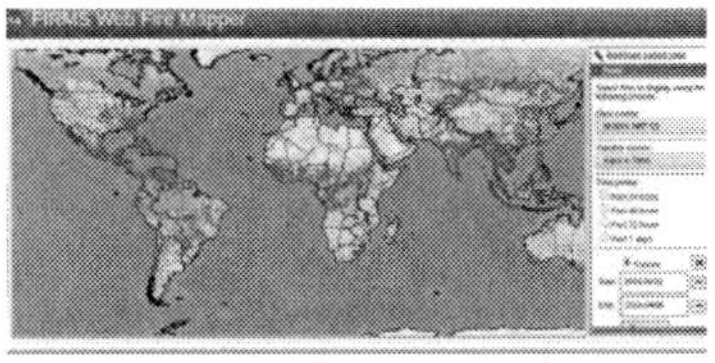

Figure 10 FIRMS , 1 Apr - 5 Apr 2014

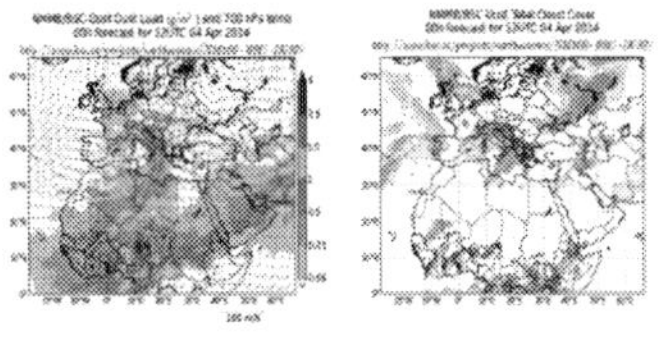

Figure 11 Dust aerosol over Europe, 4 Apr 2014

5 CONCLUSIONS

This study of aerosols over Austria, based on a cluster analysis of aerosol layers determined from lidar measurements correlated with layers from the FLEXPART model revealed an influence of long-range transport (combined or alternated) of Saharan dust and aerosols emissions of wildfires from North America. Two cases of dust and one case of mixture of dust from Sahara and smoke from biomass burning from American wildfire were identified for the period analyzed, Spring 2014.

ACKNOWLEDGEMENTS

This work is supported by Austrian Science Fund (FWF): project number M 2031 Meitner-Programm.

References

[1] Tesche, M. et al.: J. Geophys. Res.-Atmos., 114, D13202, doi:10.1029/2009JD011862, 2009.

[2] Müller, D. et al.: Geophys. Res. Lett., 34, L05803, doi:10.1029/2006GL027936, 2007.

[3] A. Nemuc, J. Vasilescu, C. Talianu, L. Belegante, and D. Nicolae, Atmos. Meas. Tech., 6, 3243–3255, 2013.

[4] Pappalardo, G. et al.: Atmos. Meas. Tech., 7, 2389–2409, doi:10.5194/amt-7-2389-2014, 2014.

[5] P. Seibert and A. Frank: Atmos. Chem. Phys., 4, 51–63, 2004.

[6] MACC Project, http://www.gmes-atmosphere.eu/

[7] Stohl, A. et al: Atmos. Chem. Phys., 5, 2461–2474, 2005.

[8] NOAA National Centers for Environmental Information, State of the Climate: Global Analysis for Annual 2014

[9] https://earthdata.nasa.gov/earth-observation-data/near-real-time

[10] DREAM Forecast Model, http://www.bsc.es/ESS/nmmb_bsc-dust/NA-ME-E/NA-ME-E

LIDAR INVESTIGATIONS OF OZONE IN THE UPPER TROPOSPHERE – LOWER STRATOSPHERE: TECHNIQUE AND RESULTS OF MEASUREMENTS

Oleg A. Romanovskii[1, 2]*, Alexey A. Nevzorov[1], Alexey V. Nevzorov[1], Olga V. Kharchenko[1]

*[1] V.E. Zuev Institute of Atmospheric Optics, Tomsk, Russia, *roa@iao.ru*
[2] National Research Tomsk State University, Tomsk, Russia

ABSTRACT

The main aim of the research is to develop the technique for laser remote ozone sensing in the upper troposphere – lower stratosphere by differential absorption method for temperature and aerosol correction and analysis of measurement results. The authors have determined wavelengths, promising to measure ozone profiles in the upper troposphere – lower stratosphere. We present the results of DIAL measurements of the vertical ozone distribution at the Siberian lidar station in Tomsk. The recovered ozone profiles were compared with IASI satellite data and Kruger model.

1 INTRODUCTION

Among techniques for measuring the spatial distribution of the concentration of an atmospheric gas the most sensitive is the differential absorption method (DIAL). Laser sounding of the ozonosphere became routing at some observatories from the second half of 1980s [1-2]. It allows receiving data on vertical distribution of ozone (VDO), which successfully supplement similar data received by contact methods with the use of ozonesondes, rockets, and satellites (TOMS, SAGE-II, TERRA, MetOp, etc.). Laser sounding of the vertical distribution of stratospheric ozone has been carried out at the Siberian Lidar Station (SLS) of Institute of Atmospheric Optics SB RAS (Tomsk, 56.5° N, 85.0° E) since 1989. A long period of lidar observations of stratospheric ozone has shown that a part of the ozonosphere the most important for the study is located in the lower stratosphere, where ozone is affected by the dynamic factor. The aims of this work are selection of VDO sounding wavelengths, design of ozone lidar, development of algorithm and program for VDO retrieval with temperature and aerosol correction, and analysis of the measurement results, including the comparison between the ozone vertical profiles retrieved from the ground-based DIAL lidar system data and the profiles retrieved from IASI/MetOp data.

2 SELECTION OF WAVELENGTHS

Lidar measurements of VDO are carried by DIAL in the UV spectral range 200–370 nm (Hartley–Huggins band). In this spectral range, the absorption cross-section σ is several times larger than that for wavelengths used in stratospheric measurements (e.g., $\sigma_{299} = 4.4 \cdot 10^{-19}$ cm^2 for the wavelength λ_{on} = 299 nm and $\sigma_{308} = 1.4 \cdot 10^{-19}$ cm^2 for λ_{on} = 308 nm). KrF laser radiation (248 nm) or the fourth harmonic of the Nd:YAG laser (266 nm) are usually used for measurements of tropospheric ozone, in combination with a technique based on the stimulated Raman scattering (SRS) in H_2, D_2, CO_2, and other gases. The most common are hydrogen and deuterium. Table 1 represents possible sets of wavelengths that correspond to the 1st, 2nd, and 3rd Stokes (*C*) frequencies of SRS conversion in H_2, D_2, and CO_2.

Table 1. Sets of wavelengths that correspond to Stokes (C) frequencies of SRS conversion in H_2, D_2, and CO_2

Pumping radiation	Wavelengths (nm) corresponding to the Stokes frequencies (*C*) of the SRS conversion		
	in H_2 C1 C2	In D_2 C1 C2 C3	in CO_2 C2 C3
Nd:YAG, 266 nm KrF, 248 nm	299 341 277 313	289 316 268 291 319	287 299

Different wavelength combinations are used in practice in different altitude ranges in the troposphere and lower stratosphere [2]. We have estimated possibilities of vertical ozone profile sounding in the upper troposphere – lower stratosphere at the wavelength pair 299/341 nm. The calculations have shown that these wavelengths allow the sounding up to about 22 km and ozone measurements in the troposphere. The sounding altitude maximum is determined, first, by the range of signal detection at λ_{on}, which is always shorter than the range of signal detection at λ_{off} due to stronger ozone absorption. In view of this, $\lambda_{on} = 299$ nm is preferable to 277 or 292 nm. In addition, wavelengths of 299 and 341 nm are implemented in one sounding beam, in contrast to, e.g., the 292/313 nm pair 2 (see Table 2). Thus, the wavelength pair 299/341 nm is of higher information content for VDO measurements in the troposphere – lower stratosphere (5–22 km altitude range).

3 LIDAR FOR SENSING THE VDO IN THE UPPER TROPOSPHERE – LOWER STRATOSPHERE

The sounding wavelengths selected (299/341 nm) were used in the ozone lidar designed. Its block diagram is shown in Fig. 1. The 4th harmonic (266 nm) of the fundamental frequency of a Nd:YAG laser (LS-2134UT laser, LOTIS TII company, Minsk) is used as a laser radiation source, which is then SRS converted in hydrogen in the 1st (299 nm) and 2nd (341 nm) Stokes components.

The recording channel of the lidar is equipped with PMTs (R7207-01) and HAMAMATSU amplifier/discriminators (C3866). The pumping pulse energy at a wavelength of 266 nm is 60 mJ.

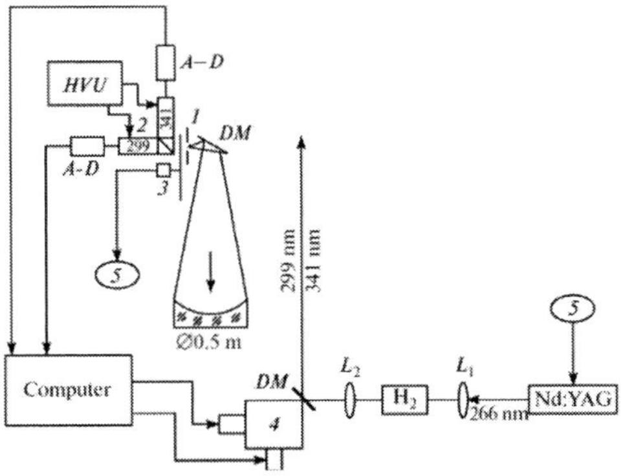

Figure 1 Block-diagram of the lidar: (1) field diaphragm, (2) cell for spectral selection with a PMT, (3) mechanical shutter, (4) automated adjustment unit of the rotating output mirror, (5) system for synchronizing the shutter operation time and the moment of emission of laser pulses, (DM) deflecting mirrors, (Nd:YAG) solid-state laser, (H2) SRS conversion cell with H2, (A–D) amplifiers– discriminators, (HVU) high-voltage power supply units for the PMT, and (L1, L2) lenses

4 TECNIQUE FOR VDO RETRIEVAL

Initial equations for calculation of ozone concentration during DIAL lidar sounding of the atmosphere have the form

$$N_{on}(H) = c \cdot \left[\beta^a_{on}(H) + \beta^m_{on}(H)\right] \cdot \exp\left[-2\int_0^H \alpha^a_{on}(H) + \alpha^m_{on} + k_{on} \cdot n(H)\right] \quad (1)$$

$$N_{off}(H) = c \cdot \left[\beta^a_{ff}(H) + \beta^m_{off}(H)\right] \cdot \exp\left[-2\int_0^H \alpha^a_{off}(H) + \alpha^m_{off} + k_{off} \cdot n(H)\right]$$

where *N(H)* is the echo-signal recorded at corresponding wavelengths, *C* is the instrumental constant, α^a is the aerosol extinction coefficient, β^a is the aerosol backscattering coefficient, *k* is the absorption coefficient, *n(H)* is the ozone concentration. The final equation is derived with the use of mathematical transformations

$$n(H) = \underbrace{\frac{1}{k_{on}(H,T) - k_{off}(H,T)}}_{A} \cdot \left\{ \underbrace{\frac{d}{dH}\ln\left[\frac{N_{off}(H)}{N_{on}(H)}\right]}_{B} - \underbrace{\frac{d}{dH}\ln\left[\frac{\beta^a_{off}(H) + \beta^m_{off}(H)}{\beta^a_{on}(H) + \beta^m_{on}(H)}\right]}_{C} - \underbrace{2 \cdot \left[\alpha^a_{off}(H) - \alpha^a_{on}(H)\right]}_{D} - \underbrace{2 \cdot \left[\alpha^m_{off}(H) - \alpha^m_{on}(H)\right]}_{F} \right\}$$

(2)

Actual variations in the air temperature can cause variations in the ozone absorption coefficient, which results in systematic errors in VDO retrieval. Therefore, it is reasonable to carry out a correction to the temperature dependence in the technique and algorithm for VDO retrieval. The model of behavior of ozone absorption cross-sections [3, 4], was used in the technique. Transforming the data from [3, 4] in a polynomial, we derive equations for calculation of the part A in Eq. (2):

$$K_{299}(H,T) - K_{341}(H,T) = 5.8815E\text{-}16 - 1.1538E\text{-}17 \cdot (T(H)-273) + 9.0281E\text{-}20 \cdot (T(H)-273)^2 - 3.5194E\text{-}22 \cdot (T(H)-273)^3 + 6.8356E\text{-}25 \cdot (T(H)-273)^4 - 5.2918E\text{-}8 \cdot (T(H)-273)^5 \quad (3)$$

Aerosol scattering exceeds the molecular one by several times at high atmospheric aerosol content, which significantly distorts ozone profiles retrieved under unconsidered scattering and attenuating properties of the atmosphere. In the considered algorithm for VDO retrieval, the aerosol correction is considered in the equations *C* and *D* by means of introduction of a real distribution of the scattering ratio $R_{off}(H)$, while VDO in the usual, undisturbed atmosphere can be calculated at $R_{off}(H)=1$. Mathematically transformed equations *C*, *D*, and *F* are:

$$C=\frac{d}{dH}\left\{\ln\left[\left(\frac{\lambda_{off}}{\lambda_{on}}\right)^{x}\cdot\left[1-\frac{1}{R_{off}(H)}\right]+\frac{1}{R_{off}(H)}\cdot\left(\frac{\lambda_{off}}{\lambda_{on}}\right)^{4}\right]\right\}$$

$$D=2\cdot 0.04\cdot\left\{\beta^{a}_{off}(H)\cdot\left[1-\left(\frac{\lambda_{off}}{\lambda_{on}}\right)^{x}\right]\right\}=2\cdot 0.04\cdot\left\{\left[R_{off}(H)-1\right]\cdot\beta^{m}_{off}(H)\cdot\left[1-\left(\frac{\lambda_{off}}{\lambda_{on}}\right)^{x}\right]\right\}$$

$$F=2\cdot 0.119\cdot\beta^{m}_{off}(H)\cdot\left[1-\left(\frac{\lambda_{off}}{\lambda_{on}}\right)^{4}\right], \quad (4)$$

where, at corresponding wavelengths λ (*on* at an absorption line and *off* out of the absorption line), $R_{off}(H)$ is the real distribution of the scattering ratio, x is the parameter that characterizes the particle size, $\beta^{a}_{off}(H)$ is the aerosol backscattering coefficient, $\beta^{m}_{off}(H)$ is the molecular backscattering coefficient. The typical errors of the retrieved O_3 profiles are in the range from 10 to 25%. To decrease the retrieval errors, the temperature correction of ozone absorption coefficients is used in the software. A high aerosol concentration in the 0–20 km altitude range should be considered during lidar signal retrieval when sounding at 272/289 nm and 299/341 nm wavelengths; therefore, aerosol correction is considered in the software. The technique and software developed were used for VDO retrieval in the upper stratosphere – lower stratosphere at 299/341 nm wavelengths.

5 MEASUREMENT RESULTS

VDO profiles retrieved from lidar sounding data are compared with the IASI/MetOp measured profiles. The IASI is mounted onboard the European Space Agency meteorological satellite MetOp. IASI provides for spectra of high radiometric quality with a resolution of 0.5 cm $^{-1}$ in the range from 625 to 2760 cm^{-1} [5]. The ozone profiles were retrieved from satellite sounding data in the range 1025–1075 cm $^{-1}$. The range of lidar signal detection was from 5 to 18 km. The ozone vertical profile error was from 6 to 18%. The coordinates and time of the lidar and satellite sounding are given in Table 2.

Figure 2 exemplifies the comparison of lidar and satellite vertical profiles of ozone measured on April 26, 2015; the VDO profiles are compared with the Krueger model [6]. All ozone measurements are shown by three curves: the retrieved lidar profile, IASI profile, and Krueger model. Figure 2 shows that the profiles retrieved from lidar and satellite data agree quite well throughout the layer of measurements. The lidar profile mainly repeats the form of the IASI curve, even with overestimated lidar data near 9 and 13 km. Above 7 km height, the measured profiles differ from the model profile, which is natural for the ozone dynamics at certain days of measurements.

Table 2. Coordinates and time of satellite and lidar sounding

Date	Siberian Lidar Station		MetOp (IASI) satellite	
	Greenwich time	Coordinates (56.5 N, 85.0 E)	Greenwich time	Coordinates
13.01.2015	11:53–13:45		14:17	56.472 N, 85.387 E
19.02.2015	12:39–14:13		14:53	56.681 N, 85.164 E
05.03.2015	13:05–14:56		15:02	56.472 N, 85.118 E
20.03.2015	13:32–15:24		14:53	56.691 N, 85.124 E
06.04.2015	14:25–16:17		15:41	56.254 N, 84.935 E
26.04.2015	15:11–17:03		15:26	56.585 N, 84.594 E

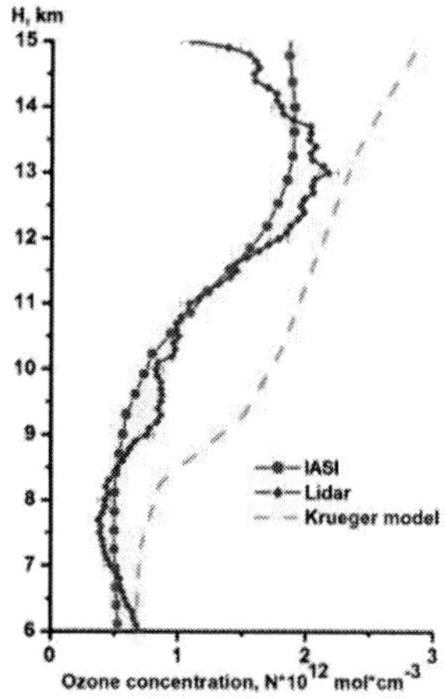

Figure 2 Comparison of vertical profiles of ozone retrieved from lidar and satellite data for April 26, 2015

It should be noted that the lidar operates in the photon counting mode at a spatial resolution of 100 m, while the IASI profile has a more dynamic step of 150±100 m, which connected with the ozone profile retrieval algorithm of this interferometer [7]. Figure 3 shows all the considered cases of comparison of ozone profiles measured at the SLS and retrieved from IASI data (see Table 2). The measurements are reduced to the total altitude range 6–15 km for convenience. Increased ozone concentrations measured at the SLS as compared to IASI are seen in the range from 8.5 to 12.5 km throughout the whole observation period.

https://doi.org/10.1051/epjconf/201817605054

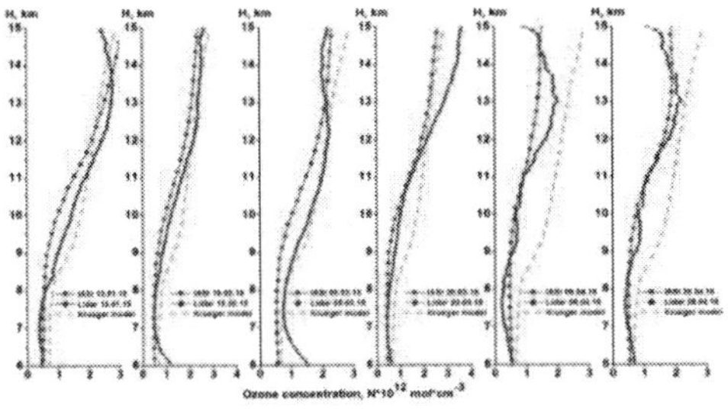

Figure 3 Comparison of vertical profiles of ozone

The following conclusions can be drawn from the analysis of the ozone profiles average over the period under study (Fig. 4). The difference between the lidar and IASI profiles of the ozone concentration growth with altitude. Ozone shows pronounced annual variations; therefore, to find relative errors of its measurements with IASI, the difference in the concentrations was normalized to the lidar data: (Lidar – IASI)/Lidar. These data are shown in Fig. 4c; it is seen that the mean relative difference is positive in the altitude ranges 6–6.4 and 8.4–15 km and attains 80.07% at an altitude of 13.7 km.

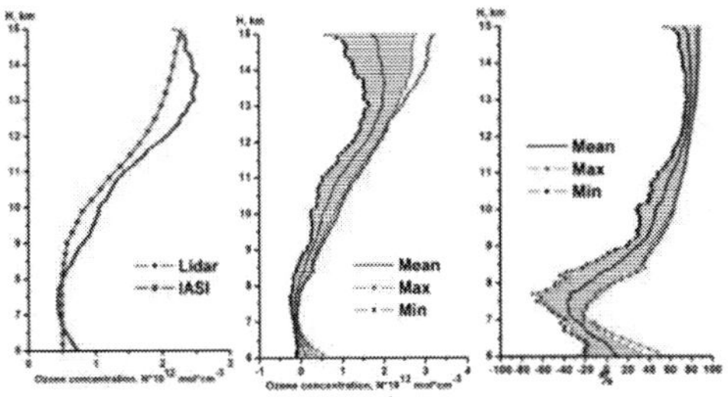

a *b* *c*

Figure 4 Mean vertical profiles of ozone (a), their difference (Lidar – IASI) in abs. units (b), and the relative difference (Lidar – IASI)/lidar (c)

The comparison performed shows that the absolute differences in the lidar and IASI measured ozone concentrations can change from –0.25 to $3.21*10^{12}$ mol. cm^{-3}.

6 CONCLUSIONS

The results of use of the developed techniques for VDO DIAL retrieval with temperature and aerosol correction confirm prospects of the wavelengths chosen (299 and 341 nm) for ozone lidar sounding. Results of lidar measurements at 299 and 341 nm agree with model estimates, which point out to acceptable accuracy of ozone sounding in altitudes near 6–18 km.

ACKNOWLEDGEMENTS

The work was supported by the Russian Science Foundation (Agreement No. 15-17-10001).

References

[1] Bukreev V.S., Vartapetov S.K., Veselovskii I.A., Shablin Yu.S., 1996: Excimer-laser-based lidar system for stratospheric and tropospheric ozone measurements, *Quantum Electron.*, **26**(6), 546-551.

[2] Eisele H., Scheel H.E., Sladkovic R., Trickl T., 1999: High resolution lidar measurements of stratosphere - troposphere exchange, *J. Atmos. Sci.*, **56**(3), 319-330.

[3] Malicet J., Daumont D., Charbonnier J., Parisse C., Chakir A., Brion J., 1995; Ozone UV spectroscopy. II. Absorption cross-sections and temperature dependence, *J. Atmos. Chem.*, **21**(3), 263-273.

[4] Zhu H., Qu Z.W., Grebenshchikov S.Y., Schinke R., Malicet J., Brion J., Daumont D., 2005; Huggins band of ozone: Assignment of hot bands, *J. Chem. Physics.*, **122**(2), Article Number 024310.

[5] Clerbaux C., Boynard A., Clarisse L., George M., Hadji-Lazaro J., Herbin H., Hurtmans D., Pommier M., Razavi A., Turquety S., Wespes C., and Coheur P.-F., 2009: Monitoring of atmospheric composition using the thermal infrared IASI/MetOp sounder, *Atmos. Chem. Phys.*, **9**(16), 6041-6054.

[6] Krueger A.J., Minzner R.A., 1976: Mid-latitude ozone model for the 1976 U.S. Standard Atmosphere, *J. Geophys. Res.*, **81**(D24) 4477-4488.

[7] August T., Klaes D., Schlüssel P., Hultberg T., Crapeau M., Arriaga A., O'Carroll A., Coppens D., Munro R., Calbet X., 2012: IASI on Metop-A: Operational Level 2 retrievals after five years in orbit, *J. Quant. Spectrosc. and Radiat. Transfer.* **113**(11), 1340–1371.

RETRIEVAL OF THE TROPOSPHERIC AEROSOL MICROPHYSICAL CHARACTERISTICS FROM THE DATA OF MULTIFREQUENCY LIDAR SENSING

S. Samoilova, [1] M. Sviridenkov, [2] I. Penner, [1,*] G. Kokhanenko, [1] and Yu. Balin [1]

1V.E. Zuev Institute of Atmospheric Optics, Tomsk 634055, Russia
2A.M. Obukhov Institute of Atmospheric Physics, Moscow 119017, Russia
**Corresponding author: penner@iao.ru*

ABSTRACT

Regular lidar measurements of the vertical distribution of aerosol optical parameters are carried out in Tomsk (56^{0}N, 85^{0}E) since April, 2011. We present the results of retrieval of microphysical characteristics from the data of measurements by means of Raman lidar in 2013. Section 2 is devoted to the theoretical aspects of retrieving the particle size distribution function $U(r)$ (SDF) assuming a known complex refractive index m (CRI). It is shown that the coarse fraction cannot be retrieved unambiguously. When estimating $U(r)$ and m together (section 3), the retrieved refractive index is non-linearly related to the optical coefficients and the distribution function, which leads to appearance of different, including false values of m. The corresponding $U(r)$ differs only slightly, so the inaccuracy in m does not essentially affect the retrieval of the distribution function.

1. INTRODUCTION

At present, lidar systems are applied for routine observations in the European Aerosol Research LIdar NETwork (EARLINET) [1], the Asian Dust NETwork (AD-Net) [2], and the lidar network in CIS countries (CIS-LiNet) [3]. At night, they allow three backscattering coefficients $\beta(\lambda_i)$ to be evaluated for λ_i = 355, 532, and 1064 nm and two extinction coefficients $\sigma(\lambda_k)$ for λ_k = 355 and 532 nm, further referred to as $(3\beta+2\sigma)$. The algorithm used by us allows the errors to be decreased down to ~ 5% (σ) and ~ 2% (β), including the free troposphere. The spectral set of the optical coefficients and the errors of their estimation allow us to proceed to reconstruction of the vertical distribution of the microphysical aerosol characteristics.

2. RETRIEVAL OF THE SIZE DISTRIBUTION FUNCTION UNDER THE ASSUMPTION OF KNOWN REFRACTIVE INDEX

The determination of the particle size distribution function $U(r)$ (SDF) for the known complex refractive index m (CRI) is reduced to the inversion of a system of linear algebraic equations. There is a problem for the lidar data: the parameters of the fine fraction are retrieved well for any regularization matrix, but the parameters of the coarse fraction are estimated ambiguously. The stabilizer order in fact specifies the range of the correct retrieval of the coarse aerosol mode parameters. This in turn leads to biased estimates of U_{coarse}. This clearly expressed dependence of the solution obtained on the restrictions on its smoothness is primarily caused by insufficient information, since for the coarse aerosol mode the spectral dependence of the optical coefficient becomes weak, and they contain in fact only two independent values.

From the mathematical point of view, the aerosol optical and microphysical characteristics are related with each other by the system of Fredholm's integral equations of the first kind:

$$\int_{r_{\min}}^{r_{\max}} K_j(m,r)U(r)d\ln r+\varepsilon_j^{meas}=g_j^{meas}, \quad (1)$$

where g_j^{meas} are the extinction and backscattering coefficients measured at the corresponding wavelengths, ε_j^{meas} are the measurement errors, $K_j(m,r)$ are the kernels of the equations depending on the refractive index and particle size. After the expansion, system of integral equations (1) is reduced to the system of linear algebraic equations for the vector of the weighting coefficients $\vec{u}=u_k$:

$$A\vec{u} + \vec{\varepsilon} = \vec{g}\,, \qquad (2)$$

where $\vec{\varepsilon}$ is the sum of experimental ε_j^{meas} and mathematical ε_j^{math} errors. Depending on the preliminary assumption about the expected properties of u_k, the regularized solution of Eq. (2) can be found in the following form:

$$\vec{u} = \left(A^T A + \alpha Q\right)^{-1} A^T \vec{g}\,, \qquad (3)$$

where α is the regularization parameter, and Q is the smoothing matrix.

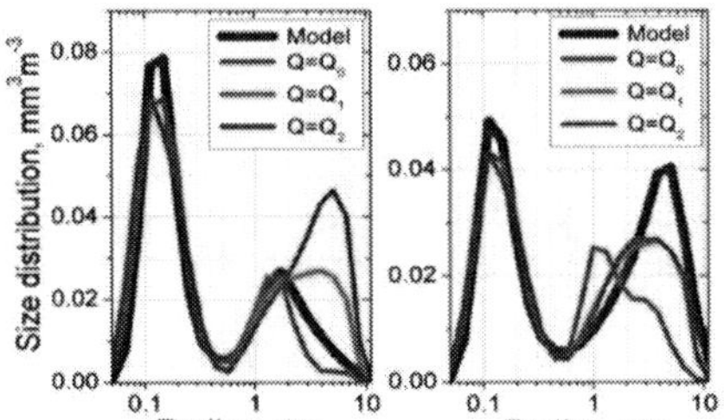

Figure 1. Examples of the SDFs reconstructed for two models; stabilizers of the zero ($Q = Q_0$), first ($Q = Q_1$) and second orders ($Q = Q_2$) were used.

Figure 1 demonstrates the results of estimating the size distribution function by the Tikhonov method (3). The biased estimates of U_{coarse} are predetermined by the *a priori* choice of the stabilizer order that in turn can lead to different values both on the abscissa and the ordinate, in particular, it fixes the geometrical mean radius of coarse particles R_{coarse} (model R^{true}_{coarse} and estimated R^{est}_{coarse}, see [4]). At the same time, the size distribution function of fine particles is retrieved well.

The statistical regularization method gives more accurate results:

$$\vec{u} = \left(A^T A + W_\varepsilon W_u^{-1}\right)^{-1} A^T \vec{g}\,, \qquad (4)$$

where $\vec{u}$ is the sought vector of weight coefficients, and W_ε, W_u are the covariance matrices of the errors, respectively. The matrix W_u^{-1} is non-diagonal, and the preliminary estimation W_u can make it possible to consider the data on the occurrence of the two aerosol modes (in fact, on the presence of several extrema), thereby cutting off a considerable number of physically unjustified solutions in the regularization stage.

Let us approximate the coarse u_k by the polynomial $b_0 + b_1 \ln r + b_2 (\ln r)^2$ for estimation W_u . Then Eq. (2) is rewritten in the form

$$b_0\vec{l}_0 + b_1\vec{l}_1 + b_2\vec{l}_2 = \vec{g}_{coarse}, \qquad (5)$$

where $l_{0k} = \sum A_{jk}$, $l_{1k} = \sum A_{jk} \ln r_k$, $l_{2k} = \sum A_{jk} (\ln r_k)^2$, $g_{coarse,j} = g_j - g_{fine,j}$. The solution of (5) for the coefficients b_i is reduced to a solution of the system of three linear equations $B\vec{b} = \vec{d}$, where the components of the matrix B and the vector $\vec{d}$ have the following forms: $B_{ik} = \sum l_{ij} l_{kj}$, $d_k = \sum g_{coarse,j} l_{kj}$. For the known coefficients b_i, the mean radius is estimated from the formula $R^{est}_{coarse} \approx \exp(-0.5 b_1 / b_2)$.

The accuracy of the estimation of u_k is ambiguous: the estimate is correct up to $R^{true}_{coarse} = R^{est}_{coarse} \le 3\,um$. It displaces toward higher values with increasing particle radius R^{est}_{coarse}, increases uncontrollably, and fails to be observed at $R^{true}_{coarse} \approx 4um$; then its displacement toward smaller values R^{est}_{coarse} is observed. The instability of the estimate for $R^{true}_{coarse} > 3um$ is also observed for other estimations of u_k; however, the boundaries of the range 3 and 4um are stable. Thus, $R^{true}_{coarse} = 3um$ can be considered as an upper boundary of the correct estimation of the mean radius of the coarse fraction (and of the SDF as a whole and $U(r) = \sum_{fine+coarse} B_k(r) u_k$) from lidar measurements. However, the statistical regularization method allows application of the *plausible* $U(r)$ (plausible here implies the SDF retrieved previously from lidar or other measurements) for estimation of W_u. The results are presented in more details for 462 empirical; models obtained at the Zvenigorod AERONET site in 2011–2012 [4].

3. SIMULTANEOUS RETRIEVAL OF THE SIZE DISTRIBUTION AND REFRACTIVE INDEX

Although in practice m and $U(r)$ are determined in parallel, there is a principle difference: CRI is a parameter of the kernel functions of the system of Fredholm's equations (1) and is not linearly related neither with the optical coefficients nor the distribution function. It makes the problem more complicated, because there are no standard methods for solving the non-linear inverse problems. It is not easy even to answer the question: is it always possible to estimate m from lidar $(3\beta+2\sigma)$ measurements? The study of possible errors does not completely answer this question, because the errors can be caused both by the quality of the inversion algorithms and by efficiency of the used a priori data, and so on.

In general form, the problem is reduced to minimization of the discrepancy functional:

$$\Phi(m) = \frac{1}{5}\sum_{j=1}^{5}\frac{\left|g_j^{meas} - g_j^{calc}(m)\right|}{g_j^{meas}} * 100\% . \qquad (6)$$

On the one hand, it depends on the measured optical coefficients g_j^{meas} (right part of Eq. (1)) and on the other hand on the same coefficients $g_j^{calc}(m)$, simulated for the all possible values of the real m_{real} and imaginary m_{image} parts of the refractive index. Application of iterative procedures like in [6] is unfounded, because a consequence of the undetermined data used is the solution dependence on the initial approximation. It is worthwhile to use the relative error in Eq. (6), because the coefficients σ and β can differ by two orders of magnitude.

$\Phi(m)$ values generate a complicated gully surface on the (m_{real}, m_{image}) plane, and the low values areas are situated at the "bottom" of the gully at $\Phi(m) \le \delta(\varepsilon^{meas}, \varepsilon^{math}) \approx 10\%$. At $\varepsilon \neq 0$, the boundaries of the area lead to the appearance of the false local minima. Thus, in the presence of errors, multiple values of the refractive index on the (m_{real}, m_{image}) plane correspond to the unique set of the values of the optical coefficients and distribution function too, and the problem of retrieving m is undetermined even for a known SDF.

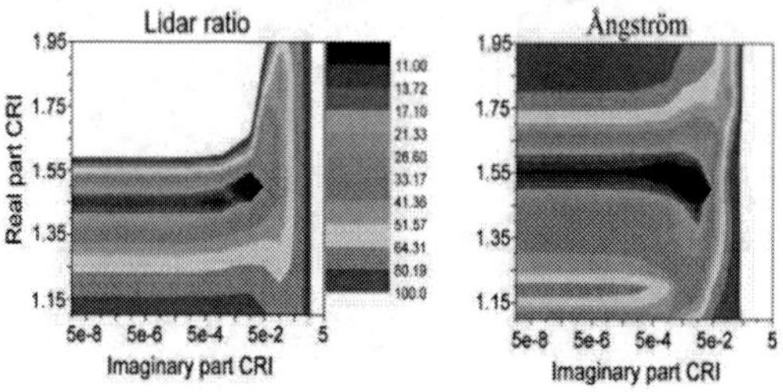

Figure 2. Different discrepancy functionals at the known SDF.

The calculation of $\Phi(m)$ for an already retrieved $U(r)$ is simpler. Two average discrepancy functionals are shown in Fig. 2: for the lidar ratio (left panel) and for the backscattering Angstrom exponent (right panel) obtained from measurements on September 24, 2013 (see details below). The first functional is analogous to Eq. (6) and leads to simultaneous increase of the real and imaginary parts of the refractive index for the minima of $\Phi(m)$. Increase of the imaginary part for decreasing real part is observed for the second functional, and the area of their crossing provides a unique value of CRI.

The results of the retrieval of the optical and microphysical characteristics from lidar data are shown in Fig. 3. Aerosol is transported in the boundary layer (from 0.5 up to 1.2 km) and the middle troposphere (from 4.0 up to 5.3 km) from Northern Europe, Atlantic Ocean, Canada with a north-westerly direction. Variations of the total to Rayleigh scattering ratio are shown in Fig. 3a. Parameters of large particles are not known exactly, since in the lower layer $R_{coarse}^{est} = 9.08um$, and in the upper layer $R_{coarse}^{est} = 4.79um$. So we assume that $R_{coarse}^{true} \in [3 \div 4um]$ and use the regularization matrix $Q = Q_1$ to obtain W_u for

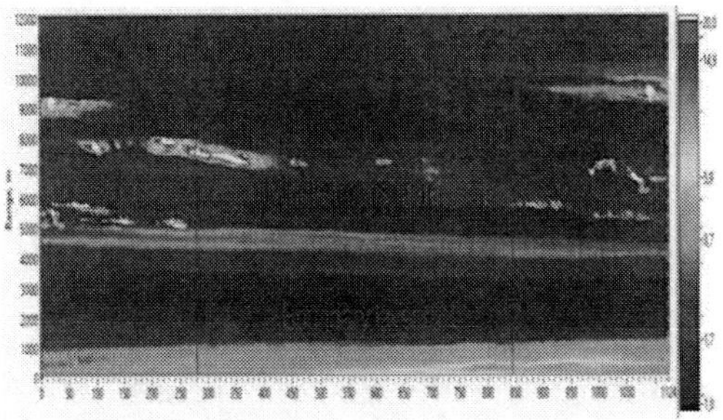

Figure 3a. Spatial-temporal variation of the scattering ratio at 1064 nm, April 03, 2013.

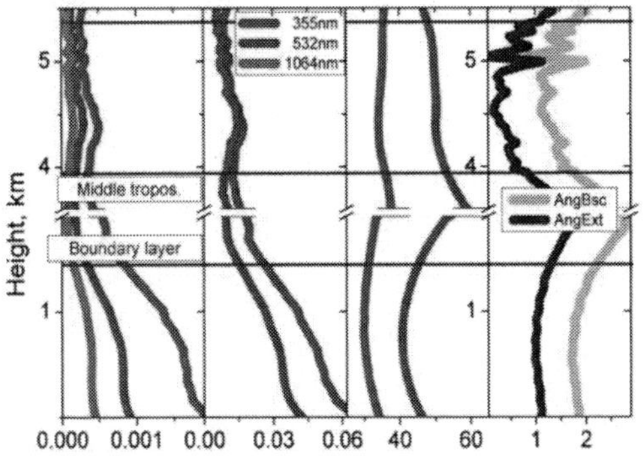

Figure 3b. Retrieval of the optical characteristics: backscatter (left panel), extinction (central left panel), lidar ratio (right central panel), Ångström exponent for 355/532 (right panel).

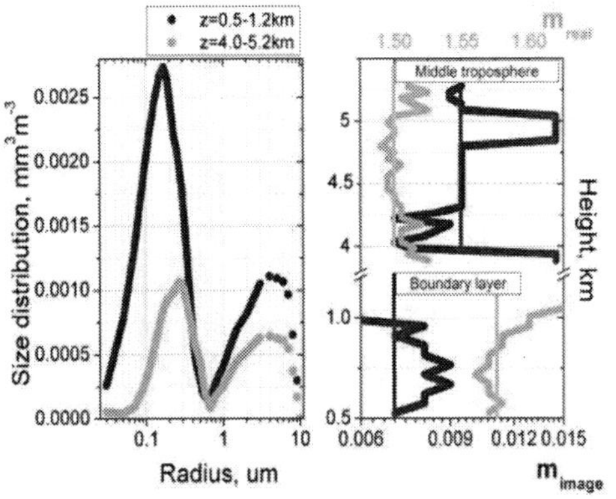

Figure 3c. Retrieval of SDF (left panel) and CRI (right panel) from lidar measurements on April 03, 2013.

both layers. Average characteristics for the boundary layer are $m^{mean} = 1.58 + i0.0065, R_{fine} = 0.19um, R_{coarse} = 4.0um,$ and for the free troposphere $m^{mean} = 1.50 + i0.009$, $R_{fine} = 0.25um$, $R_{coarse} = 3.5um$. However, possible ambiguity in the determination of the coarse particle parameters and possible false refractive index estimation have to be taken into account.

ACKNOWLEDGEMENT

This work was supported in part by the Russian Foundation for Basic Research, under grant N 16-48-700307.

REFERENCES

[1] Pappalardo G., Amodeo A., Apituley A., et. al, 2014: EARLINET: towards an advanced sustainable European aerosol lidar network, *Atmos. Meas. Tech.,* **7**, 2389-2409, doi:10.5194/amt-7-2389-2014.

[2] Murayama T., Sugimoto N., Uno I., et al, 2001: Ground-based network observation of Asian dust events of April 1998 in east Asia, *J. Geophys. Res.,* **106(D16)**, 18345–18359.

[3] Chaikovsky A.P., Ivanov A.P., Balin Yu.S., et al, 2006: CIS–LINET – Lidar Network for Monitoring Aerosol and Ozone in CIS Regions, *Reviewed and Revised Papers Presented at the 23d ILRC,* Nagasava C. and Sugimoto N. , eds. (Nara, Japan), 671–672.

[4] Samoilova S.V., Sviridenkov M.A., and Penner I.E., 2016: Retrieval of the particle size distribution function from the data of lidar sensing under the assumption of known refractive index, *Appl. Opt.,* **55(28),** 8022-8029.

[5] Samoilova S.V., 2014: Retrieval of complex refractive index from lidar measurements: possibilities and limitations, *Atmos. and Oceanic Optics,* **27(3)**, 197-206.

[6] Dubovik O., M.D. King, 2000: A flexible inversion algorithm for retrieval of aerosol optical properties from Sun and sky radiance measurements, *J. Geophys. Res.,* **105(D16),** 20673 –20696.

https://doi.org/10.1051/epjconf/201817605056

REFURBISHMENT OF DURBAN FIXED UKZN LIDAR FOR ATMOSPHERIC STUDIES – CURRENT STATUS

Venkataraman Sivakumar

School of Chemistry and Physics, Discipline of Physics, University of KwaZulu Natal, Westville, Durban 4001, South Africa

venkataramans@ukzn.ac.za *or* svsk74@gmail.com

ABSTRACT

The fixed LIDAR system at University of KwaZulu-Natal (UKZN) in Durban was installed in 1999 and operated until 2004. In 2004, the system was relocated and operation closed due to various technical and instrument problems. The restructuring of the LIDAR system was initiated in 2013 and it is now used to measure vertical aerosol profiles in the height range 03-25 km. Here, we describe the present system in detail, including technical specifications and results obtained from a recent LIDAR calibration campaign.

1 INTRODUCTION

Besides their unquestionable importance in atmospheric physics, significant gaps in the scientific knowledge of aerosols (including vertical distributions) especially in the southern hemisphere still exist. LIDAR systems are unique in providing vertical profiles of aerosol characteristics, in contrast to any other instruments. Although LIDAR systems for atmospheric studies exist in many developed/developing countries, yet found to be a state-of-the-art for South Africa and African countries. There are currently three different LIDAR systems available in South Africa, one is located in Pretoria and a further two in Durban. Both the LIDARs (fixed and portable) in Durban are operated at University of KwaZulu-Natal (UKZN). The fixed LIDAR system was installed in 1999 and was operational from 1999 to 2004 under a French-South Africa bi-lateral agreement between the Reunion University and the Service d'Aéronomie (CNRS, IPSL,and Paris) for atmospheric studies. The system had three different receiving channels and used for measuring both aerosols in the upper troposphere and lower stratosphere (i.e.10-35 km) together with middle atmosphere temperature in the height range 30 to 70 km [1]. The system was operated during specific campaigns for several weeks each year from 1999 to 2004. After 2004, the system was relocated from one of the university campuses (Howard's College) to another (Westville) and no successful operations were possible due to various technical and instrument problems. Due to financial and human resource limitations, the restructuring of the LIDAR was not initiated until 2013 with the aim to make it re-operational. The refurbishment of the system is still on going and currently it is able to perform vertical aerosol profile measurements in the upper troposphere and lower stratosphere (i.e.,10-35 km). Figure-1 illustrates the current LIDAR system during operation. System details and technical specifications are described below as well as results obtained during a recent LIDAR calibration campaign.

Figure-1: *Pictorial representation of UKZN fixed LIDAR system.*

2 SYSTEM DESCRIPTION

The UKZN-Durban fixed LIDAR is capable of measuring temperature profiles and aerosol backscatter coefficients. It employs a pulsed Nd:YAG laser operating at 30 Hz and a fundamental wavelength of 1064 nm. A frequency doubled laser emitting in the green, at 532-nm with approximately 150 mJ per pulse is used for transmission. The emitted laser beam after 10 times of expansion is directed vertically upward. The backscattered signals from the atmosphere are received by three different telescopes, simultaneously. The telescopes consist of parabolic mirrors with diameters of 445 mm (two) and 200 mm (one) and are designed to cover a wide dynamic range. The larger telescope (445 mm) was initially used for investigation of upper atmospheric levels while the smaller one (200 mm) covers the lower atmosphere. However, in the present reconfiguration, we have used both two large telescopes for addressing the laser backscatter signals from the lower atmosphere. We have also used the optical fiber combiner to reduce the number of data acquisitions into one. The received photons are collected at the focus point of the telescope and then transmitted to data-acquisition after passing through narrow-band filter (± 1 nm) and photo-multiplier tubes (PMT). The present data-acquisition (Licel) systems provide range-resolved measurements for every 7.5 m. The signal is then processed in an off-line system to retrieve the aerosol profiles. The future plan is to upgrade the system for measuring temperature in the altitude range 10- 70 km by combining Rayleigh and the Raman-N_2 channels using the large telescopes while aerosol profiles will be retrieved in the 4-38 km range based on data provided by the smaller telescope [2]. Important specifications of the current LIDAR system are provided in the table below.

Transmitter	
Laser Source	Nd:YAG – Spectra Physics
Operating Wavelength	1064, 532 nm
Energy per pulse	120 mJ
Beam Expander	10 x
Pulse width	8-9 n sec
Pulse repetition rate	30 Hz
Beam Divergence	0.5 mRad
Receiver	
Telescope	Newtonian
Diameter	445 mm and 200 mm
Field of View	1 mrad
IF Filter (BW)	1 nm
PhotoMultiplier Tube	
Type	Hamamatsu R7400-U20
Optical Fibre	Multimode, 600 µm core
Filter bandwidth	1 nm
Data Processing	
Basic data structure	Analog & Photon count
Range Resolution	7.5 m
Data	
Aerosol/Cloud	4 km to 25 km (current)
Temperature	30 km to 80 km (year-2)
Water vapour	4 km to 25 km (year-3)

3 OBSERVATIONS AND CALIBRATION

After refurbishment of most of the components of the UKZN fixed LIDAR we describe recent observations to calibrate the system for the first time. The LIDAR calibration campaign was performed from 05th to 09th September 2016 using the Council for Scientific and Industrial Research (CSIR)-National Laser Centre (NLC) mobile LIDAR and UKZN portable LIDAR. All the three LIDAR systems were operated simultaneously and at the same site (UKZN Westville campus). Since, the UKZN fixed LIDAR system is not portable, the CSIR-NLC Mobile and Portable LIDAR were placed close to the fixed LIDAR location (see figure 2 and 3 below).

It is worth to note here the following points for comparison when addressing the calibration of UKZN fixed LIDAR system.

- UKZN fixed LIDAR operates with Nd:YAG Laser (Spectra Physics) with ~150 mJ at 30 Hz for 532 nm. Two large telescopes (each has 445 mm in diameter) are employed in this system.
- CSIR-NLC LIDAR operates with Nd:YAG Laser (Continuum) with ~165 mJ at 20 Hz for 532 nm. The receiving telescope has a diameter of 400mm.
- UKZN portable LIDAR operates with Nd:YAG Laser (Quantel) with ~40 mJ at 10 Hz for 355 nm, there is no 532 nm employed in this system. The receiving telescope has almost half the diameter of CSIR-NLC mobile LIDAR, i.e., 200 mm.

Apart from the above significant changes in specifications, more details about the CSIR-NLC mobile LIDAR one may refer to Sivakumar et al., 2009 [3] and UKZN portable LIDAR into Sivakumar et al., 2015 [4].

Figure-2: *Pictorial representation of LIDAR calibration set-up location site at UKZN (source from Google maps). Distance between the two location is ~220 m.*

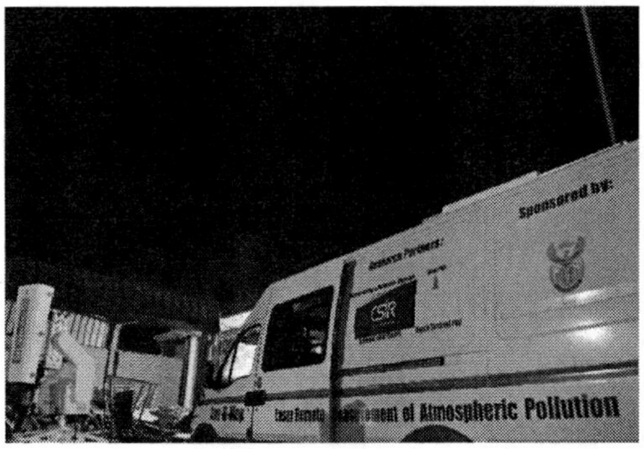

Figure-3: *Pictorial representation of UKZN portable LIDAR (on left) and NLC mobile LIDAR(right) during calibration campaign at UKZN.*

A sample height-time color map of backscatter signal returns obtained for the night 06 September 2016 are shown in figure 4(a-c). The backscattered signal returns for the fixed LIDAR show a general feature of lower boundary layer aerosol structure and passage of clouds over the site. It is obvious from the figure 4(a) that a clear and strong backscattered signal returns from UKZN fixed LIDAR system in comparison to other two LIDAR system. The height of occurrence of passing clouds is unambiguously detected by all the LIDAR which indirectly gives evidence of good system performance. It is evident from the figure that the detection of thin cloud is in the height range 5 to 6 km and around 7 km. There was also a passage of high thick cloud during the middle of observations. Apart from the above cloud passage, the backscatter signal illustrates the detection of boundary layer around 2 km and the significant presence of aerosols.

Though all the three systems detect unambiguously the height of cloud occurrence, there are differences in backscatter signal strength. This is due to the fact that the system has different transmission and receiving capability, as indicated earlier. Thus, the results obtained indicate the system with high reception may provide a significant higher backscatter signal in comparison to the one with high transmission with similar area of reception (CSIR-NLC Mobile LIDAR). Furthermore, the UKZN portable LIDAR system has low transmission and smaller area of reception and does not illustrate most features of the cloud due to low backscatter signal (see. figure-4c). The less conspicuous features of the cloud in fig. 4c could also be due, at least partially; to the lesser contrast between the backscatter of clouds and Rayleigh backscatter at 355 nm (other two LIDAR's are operated at 532 nm) besides the smaller area of reception with low energy of transmission.

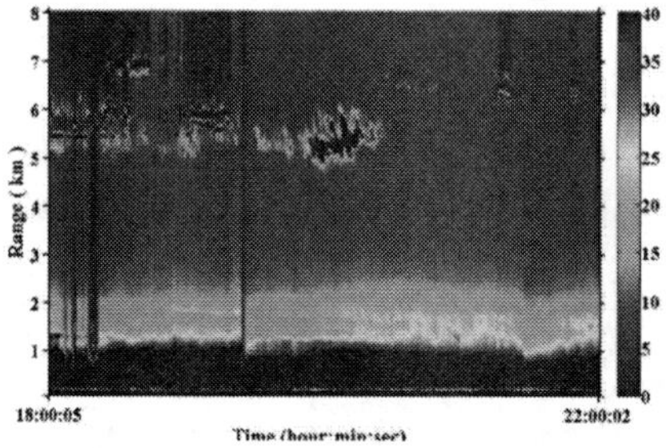

Figure 4a. *Height-time (GMT) color map of UKZN Durban Fixed LIDAR backscatter signal returns for the night of 06 September 2016.*

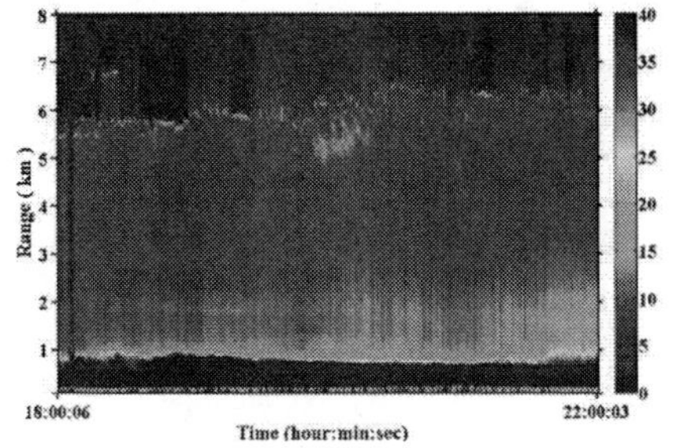

Figure 4b. *Height-time (GMT) color map of CSIR-NLC mobile LIDAR backscatter signal returns for the night of 06 September 2016.*

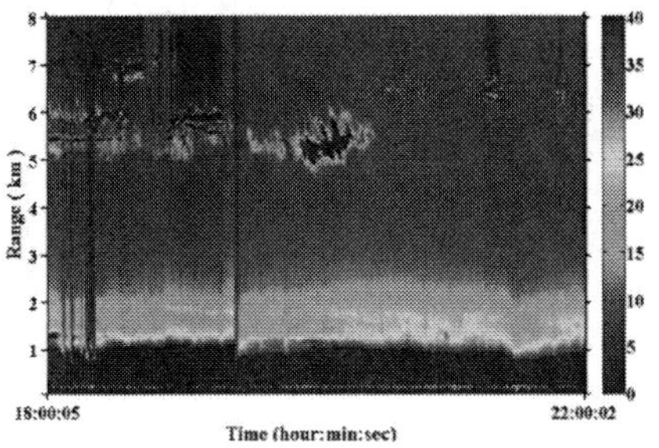

Figure 4c. *Height-time (GMT) color map of UKZN Portable Durban LIDAR backscatter signal returns for the night of 06 September 2016.*

4. FUTURE PERSPECTIVES

The current plan is to restructure the complete system and optimize operation.. The system will then be employed for long-term data acquisition in order to investigate temperature and aerosol variability over Durban. Simultaneous measurements are envisaged together with the CSIR-NLC mobile LIDAR systems which is currently planned to be installed at University of Zululand (170 km from UKZN) and LIDAR observations at Maido Observatory at Reunion University (ReunionIsland).

ACKNOWLEDGEMENTS

We are thankful to the different South Africa funding agency, National Research Foundation (NRF) (Grant no: 78682 and 78663), African Laser Centre (ALC), Centre National de la Recherché Scientifique (CNRS) (France) and Directorate General for Scientific Research and Technological Development (DG-RSDT) (Algeria).

REFERENCES

[1] Moorgawa, A, Bencherif H, Michaelis M M, Porteneuve J, Malinga S, 2007: The Durban atmospheric LIDAR, *Optics & Laser Technology*, 39, 306-312.

[2] Mbatha N M, Sivakumar V, Bencherif H, Malinga S B, Pillay S R, Moorgawa A and Michaelis M M, 2012: Analyses of the Middle atmosphere thermal structure over Durban using a ground-based Rayleigh LIDAR system and satellite experiments (HALOE and SABER), *S. Afr. J. Sci.*, 108, 612.

[3] Sivakumar, V., Tesfaye, M., Alemu, W., Moema, D., Sharma, A., Bollig, C., and Mengistu, G., 2011: CSIR South Africa mobile LIDAR—First scientific results: comparison with satellite, sun photometer and model simulations, *S. Afr. J. Sci.* 105.11-12.

[4] Sivakumar, V., Ogunnuyi, J., Adesina, J., Singh P., and Shikwambana, L., 2015: A new 2-D scan portable Durban LIDAR for atmospheric studies – system description and first scientific result, *Proc. of 31st Annual conference of South African society for atmosphere science, ISBN 978-0-620-67825-4.*

MULTIANGLE LIDAR OBSERVATIONS OF THE ATMOSPHERE

Pawar Lalitkumar Prakash[1]*, Yogesh Kumar Choukiker[2], K Raghunath[3]

[1]*VIT University, Vellore, India 632014, *lalitpwr61@gmail.com*
[2]*VIT University, Vellore, India 632014,*
[3]*National Atmospheric Research Laboratory, Gadanki, India 517112,*

ABSTRACT

Atmospheric Lidars are used extensively to get aerosol parameters like backscatter coefficient, backscatter ratio etc. National Atmospheric Research Laboratory, Gadanki (13°N, 79°E), India has a powerful lidar which has alt-azimuth capability. Inversion method is applied to data from observations of lidar system at different azimuth and elevation angles. Data Analysis is described and Observations in 2D and 3D format are discussed. Presence of Cloud and the variation of backscatter parameters are seen in an interesting manner.

1 INTRODUCTION

Study of Atmospheric aerosols is of general interest. Aerosol parameters like backscatter coefficient, extinction coefficient, backscatter ratio and optical depth are important for these studies. Observations from a vertical looking lidar system are plenty around the world. However observations from a scanning lidar are very limited and aerosol studies from such a lidar opens up new frontiers of research fields. Here a lidar system which is capable of movement in both azimuth and elevation angles is employed to observe atmosphere in a conical volume. National Atmospheric Research Laboratory (NARL), Gadanki, India has a powerful backscatter lidar that is capable of taking observations in the above said mode. The data is subjected to Inversion technique to derive backscatter ratio which gives an estimate of aerosols in the atmosphere.

1.1 System Description

The lidar system consist of monostatic biaxial system where a Nd:YAG laser with a 2nd harmonic generator transmitting a beam of 10-mm width at 532-nm wavelength is used as transmitter which generates the laser beam of width 10mm. The beam is expanded to a beam diameter width of 80mm. The repetition frequency of output pulse is 50Hz and average power is 30 W. Newtonian telescope with diameter of 750mm is used as receiver. The data collected by telescope is given to PMT and discriminator combination through an optical fiber cable having core diameter of 1.5mm and numerical aperture of 0.37. To collect data from optical cable very high gain photomultiplier tube is used which have gain of order 10^7. In receiver system in detector unit, interference filter is used which having 50% peak transmission. To remove the noise signal from collected signal discriminator is used, which is driven by high speed negative pulses generated by PMT. After the noise removal, photon counting is done with the use of multichannel scalar card.

Important parameters of the system are given in Table 1.

Table 1 Specifications of Lidar System

Parameter	Specification
Laser	Nd:YAG
Laser power	30W @50Hz repetition rate
Wavelength of operation	532nm
Receiver	Newtonian type of telescope
Detector	PMT
Type of Feed	Optical fiber
Region of operation	5-35km
Signal Recovery	Photon counting method
Time integration	4 minutes
Height resolution	300m

2 METHODOLOGY

To determine the important parameters like backscatter coefficient, extinction coefficient, backscatter ratio and optical depth, methods as proposed by Fernald [1], Klett [2], Kano [3] and Hamilton [4].

In the Lidar used here, telescope and laser beam are ganged together such that axis of telescope and laser beam is parallel all time. So when telescope is moved in a particular azimuth angle and elevation angle, laser beam is also tilted in the same direction. To do this kind of movement telescope controller software is used. In user interface of software, values for tilts in the North-South (NS) and East-West (EW) or Azimuth (AZ) and Elevation (EL) are given.

If any alignment is disturbed because of any reasons, proper alignments are made by giving NS offset and EW offset so that overlap takes place at required height.

Here time resolution of 250 seconds and height resolution of 300m is taken. So it takes 4 minutes to generate each raw data profile of photon counts.

2.1 Scanning with constant elevation angle:

Fernald [1] method is used to calculate the backscatter coefficient for range of 5-35km. Backscatter ratio is calculated from the received power (photon counts) that is due to aerosols in atmosphere, the lidar equation which relates the transmitted and received power is given in Eq 1. Backscatter parameters are calculated by solving the equation.

$$P_r(z) = K\beta(Z)\, T_1^2(Z)\, T_2^2(Z) / Z^2 \quad \ldots (1)$$

Where,

K = System constant which includes the optical losses at both transmitting and receiving sides and receiving telescope aperture.

$\beta(Z)$ = sum of $\beta_1(Z)$ aerosol and $\beta_2(Z)$ molecular backscattering coefficient.

$= [\beta_1(Z) + \beta_2(Z)]$

$T_1(Z)$ = Transmittance by aerosols

$= \exp[-\int_0^z \sigma_1(z)\, dz]$

$T_2(Z)$ = Transmittance by molecules

$= \exp[-\int_0^z \sigma_2(z)\, dz]$

Where $\sigma_1(z)$ and $\sigma_2(z)$ are extinction coefficients of aerosol and molecular respectively.

$$\beta_1(z\text{-}1) = \frac{X(z-1)}{\frac{X(z)}{\beta 1(Z)} + S1[X(z)+X(z-1)]\Delta z} \quad \ldots (2)$$

$$\sigma_1(z\text{-}1) = \frac{X(z-1)}{\frac{X(z)}{\sigma 1(z)} + [X(z)+X(z-1)]\Delta z} \quad \ldots (3)$$

Where,

$S1 = \frac{\sigma 1(z)}{\beta 1(Z)}$ which is extinction to backscatter ratio. [5]

The above equations are for turbid atmosphere and not applicable for all types of atmosphere conditions. In above equations X (z) is a range normalized signal which is given by

$$X(z) = P(z)\, z^2 \quad \ldots (4)$$

Δz = Range resolution

After calculating the backscatter coefficient due to aerosol and extinction coefficient, backscatter ratio can be computed as follows,

$$SR(z) = 1 + \frac{\beta 1(z)}{\beta 2(Z)} \quad \ldots (5)$$

Backscatter ratio is used to determine the amount of aerosols in the total observation volume. It gives the information regarding how aerosols concentration is varying with range.

The lidar system is operated in the following mode: System is scanned from 0-360° in azimuth direction with step resolution angle of 5° at fixed off-zenith angle of 40°.

3 RESULTS

3.1 Results with fixed elevation angle:

The lidar system is installed at Gadanki, India which is a rural site. The lidar location is around 375 m above from the sea level.

Experiment was carried out in clear sky and cirrus cloud conditions. The experiment started at 1905 hrs on 23/01/2017 and ended at 0305 hrs on 24/01/2017.

The system was operated in the above mode for several nights. In the azimuth angle of 50° and 55° in north-east direction because of obstruction from tree, data was not collected and hence a gap in the plot. In Figure 1 and 2, range squared intensity is plotted against azimuth angle and range. (It is not converted to height).

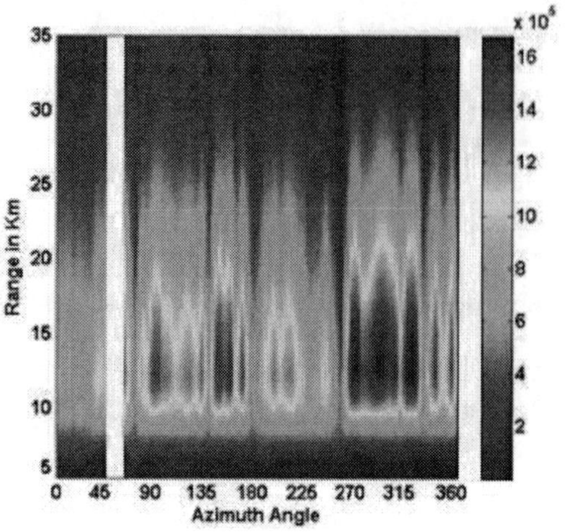

Figure 1 Raw Photon Count Profile on 23/1/2017 which is range corrected for range of 5-35km.

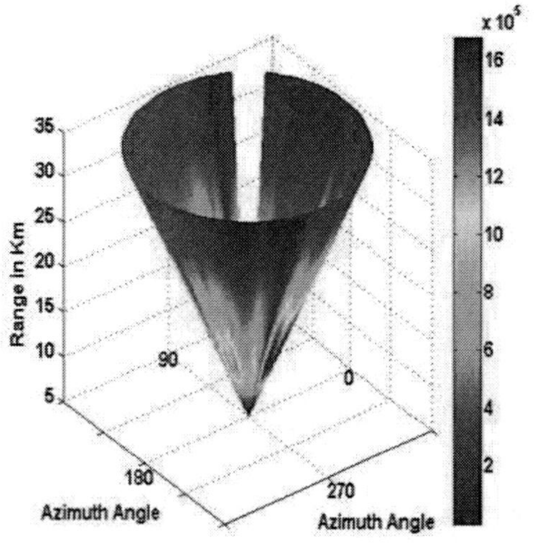

Figure 2 Raw Photon Count Profile on 23/1/2017 which is range corrected for range of 5-35km.

Backscatter ratio is calculated from Eq. 2-5 assuming turbid atmosphere. Molecular backscatter values are taken from CIRA model. Backscatter ratio profiles are shown in Figure 3 and 4.

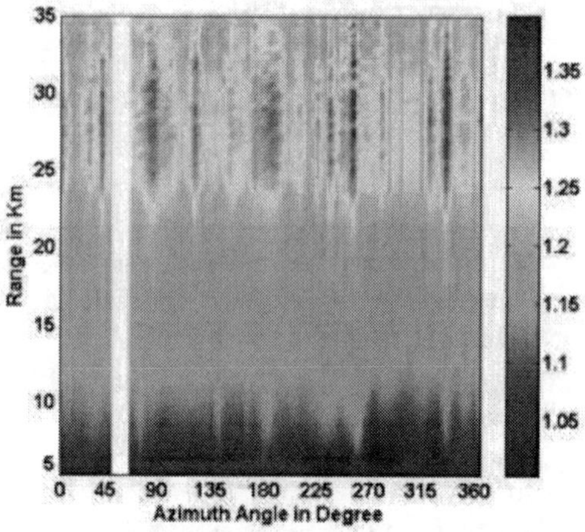

Figure 3 Backscatter Ratio Profile on 23/1/2017 for range of 5-35km.

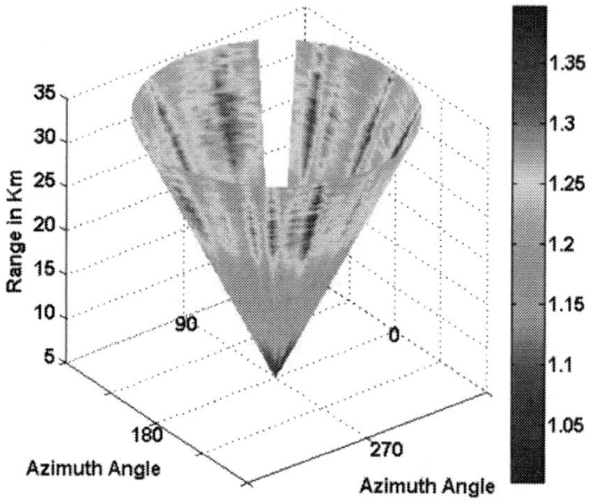

Figure 4 Backscatter Ratio Profile on 23/1/2017 for range of 5-35km.

Lidar data was operated with 2 minute time integration on 24/01/2017. The experiment started at 1905 hrs on 24/01/2017 and ended at 0004 hrs on 25/01/2017. The observations are at 40 degree off-zenith angle and steps of 5 degrees in azimuth angle and is shown in Figure 5. The backscatter ratio profile has shown cirrus cloud presence. Backscatter ratio values of 1-1.25 on 23/1/2017 in Figure 3 & 4 and upto 7 was observed on 24/1/2017 in Figure 5. It may be noted that turbid atmosphere was assumed in backscatter ratio value computation. The values for generalized atmosphere will be calculated in due course using appropriate solution. The high values of backscatter ratio beyond range of 15km are being looked into. Employing Fernald method for slant angle observations is also being studied.

Observations with varied elevation angles can be explored for further cloud evolution studies.

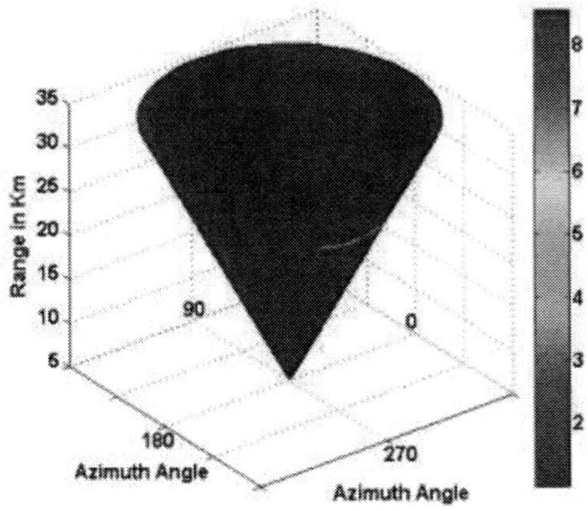

Figure 5 Backscatter Ratio Profile on 24/1/2017 for range of 5-35km.

4 CONCLUSIONS

The observations from multi angle lidar is presented and visualized. The backscatter ratio plots in a conical atmosphere volume give an idea of the presence of aerosols. It also gives the presence of clouds at appropriate heights hitherto not possible with vertical looking systems. Computation of other aerosol parameters like backscatter ratio and optical depth will give more insight into these studies.

ACKNOWLEDGEMENTS

First author would like to thank National Atmospheric Research Laboratory Gadanki, India, for giving opportunity to carry out his Master's internship in the lidar studies.

References

[1] Frederick G. Fernald 1984: Analysis of atmospheric lidar observations: Some comments, *Appl. optics,* **23**(5), 652-653.

[2] J. D. Klett 1981: Stable analytical inversion solution for processing lidar returns, *Appl. optics*, **24**, 211-220.

[3] M. Kano 1968: On the determination of Backscattering and Extinction Coefficient of the Atmosphere by Using a Laser Radar, *Pap. Meteorol. Geophys.* **19**, 121-129.

[4] P. M. Hamilton 1969: Lidar measurement of backscatter and attenuation of atmospheric aerosol, *Atmos. Environ.* **3**, 221-223

[5] K Raghunath, Y Bhavani Kumar, V Siva Kumar, P B Rao, Kohei Mizutani, Tetsuo Aoki, Motoaki Yasui & Toshikazu Itabe, 2000:Indo-Japanese lidar observations of aerosols over a tropical latitude, *Indian Journal of radio & Space Physics*,**29**, 239-244.

DUST MASS CONCENTRATIONS FROM THE UK VOLCANIC ASH LIDAR NETWORK COMPARED WITH IN-SITU AIRCRAFT MEASUREMENTS

Martin Osborne*[1], Franco Marenco[2], Mariana Adam[2], Joelle Buxmann[2], Jim Haywood[2]

[1]*Univeristy of Exeter, UK, *martin.osborne@metoffice.gov.uk*
[2]*Met Office, UK*

ABSTRACT

The Met Office has recently established a series of 10 lidar / sun-photometer installations across the UK, consolidating their ash / aerosol remote sensing capabilities [1]. In addition to this network, the Met Office have acquired the Civil Contingency Aircraft (MOCCA) which allows airborne in-situ measurements of ash / aerosol scattering and size-distributions. Two case studies are presented in which mass concentrations of Saharan dust are obtained remotely using lidar returns, and are then compared with those obtained in-situ. A thorough analysis of the mass concentration uncertainty will be provided at the conference.

1 INTRODUCTION

The UK Met Office acts as the London Volcanic Ash Advisory Centre (VAAC) and is responsible for issuing forecasts and safety information in the event of a volcanic eruption in the North Atlantic region. Data from the new lidar network (fig 1) and the instrumentation carried on MOCCA will be used by expert forecasters to supplement modelling and satellite data. This could be particularly useful where cloud is obscuring satellite products.

This study aims to provide an initial assessment of the estimates of aerosol mass concentration retrieved using the new lidar network by using Saharan dust as a proxy for volcanic ash. The assessment is made by comparing remotely retrieved estimates of mass concentrations with in-situ airborne measurements made using MOCCA.

2 LIDAR NETWORK AND MOCCA

The new lidar network consists of nine fixed locations and one mobile facility (fig 2). The lidars (Raymetrics LR111-300) operate at 355nm and have N2 Raman and depolarization (polar and co-polar) channels. Data is transmitted to the Met Office HQ and can be accessed and visualized in near real time

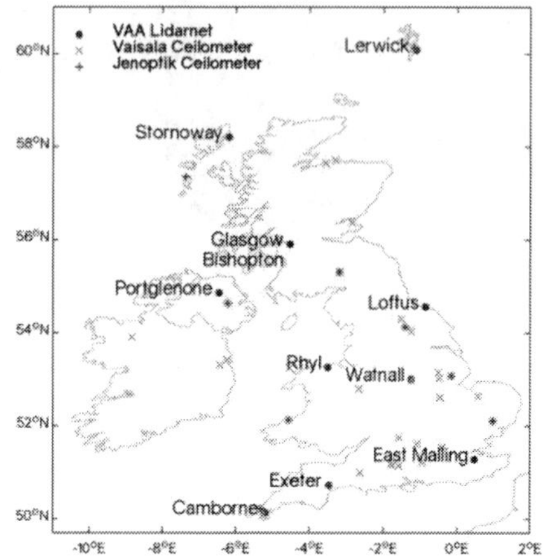

Figure 1.The Met Office Lidar-sunphotometer operational network. Also shown are the location of the operational ceilometers.

MOCCA is a Cessna 421 light aircraft equipped with a three wavelength nephelometer and an optical particle counter (Cloud and Aerosol Spectrometer (CAS)). MOCCA also carries a Leosphere ALS450 backscatter / depolarization lidar operating at 355nm. This can be positioned to point upwards or downwards before takeoff.

3 METHODOLOGY

3.1 In – situ mass estimate

Figure 2 upper and lower show output from the Met Office operational numerical weather prediction (NWP) model for dust aerosol optical depth (AOD) for the 1st and 25th November 2016 respectively. On both occasions AODs of around 0.1 – 0.2 were forecast. The output for the model also suggested that the skies over Camborne in the South West of the UK would be relatively cloud free. On each of these days, MOCCA was directed to make flights over the Camborne lidar site (fig 3) in order to perform in-situ measurements with the nephelometer and the CAS. During both

 https://doi.org/10.1051/epjconf/201817605058

flights, the dust layers were first mapped using the on board Leosphere lidar, and from this data altitudes were chosen for in-situ sampling runs. The data from the CAS were processed using an assumed refractive index for desert dust. The resulting volume size distributions were multiplied by a nominal density for desert dust of 2.6 μg cm^{-3} to arrive at an estimated mass concentration.

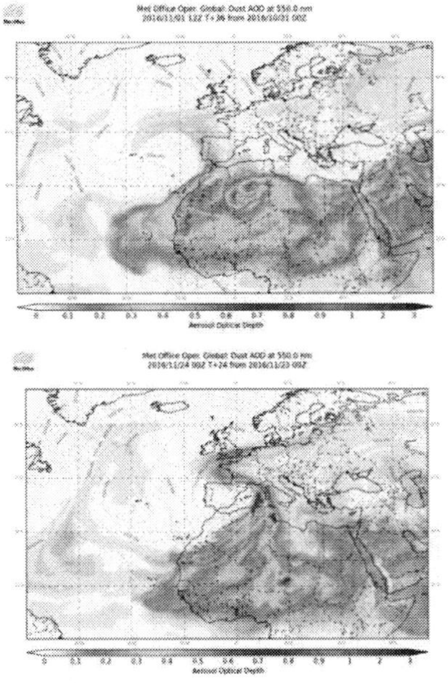

Figure 2. Dust AOD forecasts from the Met Office NWP model. Upper 1/11/16 and lower 25/11/16

3.2 Lidar mass estimate

Using the average of ~ 3hours of Raman channel data from the night before each flight, the methods described in [2] & [3] were applied to data from the lidar at the Camborne installation. The resulting profiles for extinction and backscatter were used to calculate mean lidar ratios (LR) both in and out of the dust layers. The LR uncertainty was determined as the STD over the 3h of measurements during night time. In order to obtain aerosol extinction & backscattering profiles from the elastic channel during day light hours, this lidar ratio was then used in Fernald / Klett retrievals [3,4]. It was assumed that the dust layer during day time had the same optical properties as during the night and thus the same LR. We feel this is justified as the dust had been transported from the source over 4 to 5 days, and there is no reason to assume that it had modified significantly in the ~ 12 hrs between the night time retrievals and those during the day. We did not consider the Raman channel during day time due to the high level of noise. The particle depolarization ratio (PDR) was then calculated using the resulting backscatter profiles and the depolarization channel.

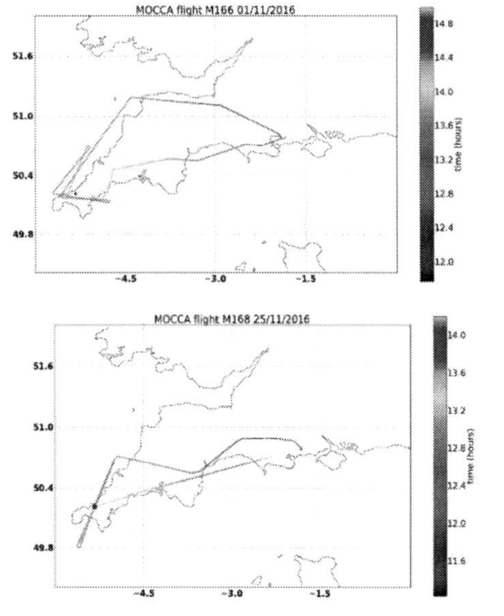

Figure 3. MOCCA flight tracks for (a) 1/11/16 and (b) 25/11/16. Camborne is shown in blue

Table1: Parameters used in lidar retrievals

Parameter	Value
Lidar ration in dust layer	01/11= 45 sr ± 6.9
	25/11= 65 sr ± 18.0
Lidar ratio below dust	01/11 = 20 sr ± 3
	25/11= 20 sr ± 6.5
Dust density	2.6 μg cm^{-3}
Non-dust density	1.5 μg cm^{-3}
Specific extinction dust	0.6 m^{2}g^{-1}
Specific extinction non-dust	0.16 m^{2}g^{-1}
Particle depolarization ratio dust	0.2
Particle depolarization ratio non-dust	0.001

Finally, using the methods described in [5] mass concentration estimates were obtained for dust, and non-dust aerosols. The values of parameters used in these retrievals are summarized in table 1. Molecular scattering was calculated using temperature and pressure profiles from the Met Office UKV 1.5km NWP model.

4 RESULTS

Figure 4 upper and lower show ~ 24 hours of data from the elastic and depolarization channels of the Camborne lidar, recorded on 1st and 25th November 2016 respectively.

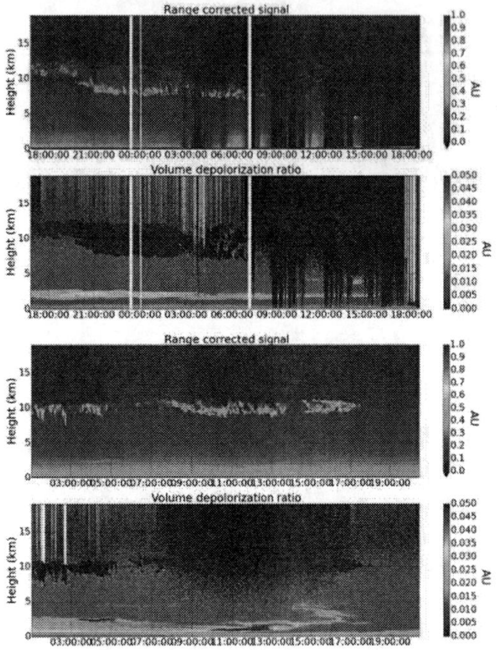

Figure 4.Range corrected signals for Camborne lidar elastic channel and volume depolarization ratio. Upper 1st November 2016 and lower 25th November 2016 .

Clear depolarizing layers can be seen in the depolarization channel on both days at between ~ 1 and ~2 km. Given this clear depolarization, and the prediction of Saharan dust made by the Met Office NWP model, we identify this layer as desert dust.

4.1 Lidar mass concentrations

Figure 5 upper and lower show the mass estimates arrived at using the process outlined above. The PDR in the dust layer on the 1st November is lower than expected. This is potentially due to the dust being polluted and becoming hydrated. The relative humidity recorded by MOCCA was ~ 80% in the dust layer.

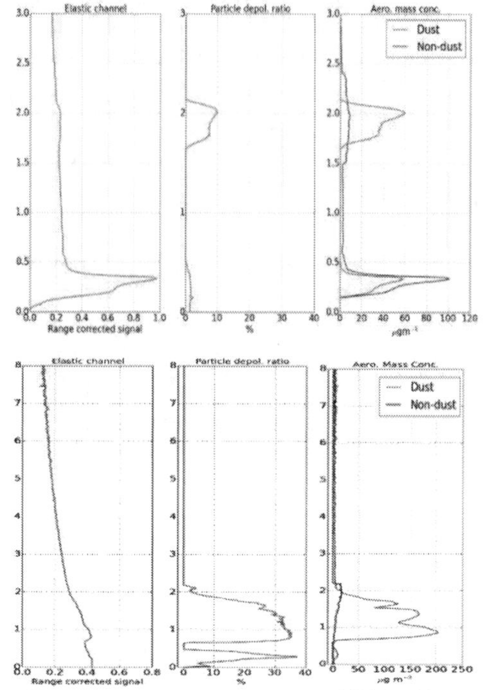

Figure 5. Lidar mass concentrations. Upper 1/11/16 and lower 25/11/16

The relative humidity recorded by MOCCA was ~ 80% in the dust layer. The PDR in the dust layer on the 25th November is more in line with values reported in the literature for desert dust. The relative humidity in this layer was around 40%.

4.2 In-situ mass concentrations.

Figure 6 upper and lower show the in-situ mass concentrations obtained using the CAS on the 1st and 25th of November 2016 respectively. In Fig 6 upper, the two peaks in the mass concentration correspond to MOCCA flying into and out of a highly in-homogeneous dust layer. In the dust layer the CAS mass concentrations are often between 40 and 70 μgm^{-3}, showing some agreement with the 40 to 60 μgm^{-3} obtained in the lidar retrievals. Figure 6 lower shows CAS mass concentrations between 100 and 200 μgm^{-3}, again, showing some agreement with the lidar retrievals. Based on estimates in [7], we expect a factor of two uncertainty for the mass derived from CAS.

Following [6], we expect an uncertainty of around 40% for lidar derived mass concentrations.

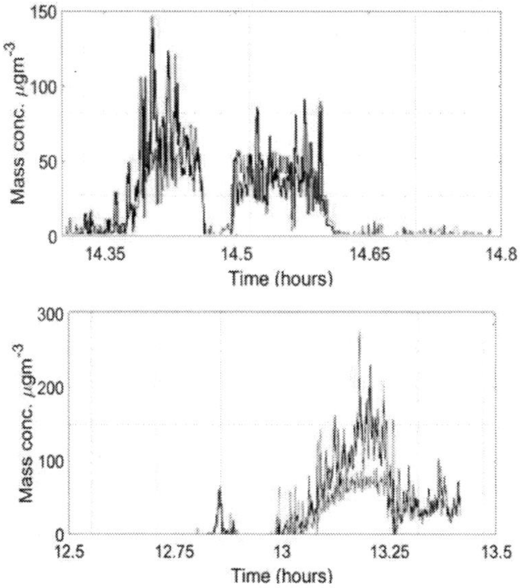

Figure 6. In-situ mass concentrations recorded by the CAS on board MOCCA (Black lines). Upper 1/11/16 and lower 25/11/16.

5 CONCLUSIONS

Estimates of mass concentration of Saharan dust were obtained remotely using data from a new volcanic ash lidar network. A height resolved lidar ratio was derived using N2 Raman channel data, and separate masses for dust and non-dust aerosols were obtained using data from elastic and depolarization channels. The results were compared with those obtained in-situ using the MOCCA aircraft and found to show some agreement. Such estimates for ash mass concentration are of interest to many users (including VAAC and ICAO) [8], and given that model uncertainty is estimated at a factor of 1.5 [9], even uncertain estimates would be invaluable to VAAC forecasters. Detailed analysis of the two case studies described above (including a thorough analysis of the mass concentration uncertainties), and further case studies in which Sun-photometer data and data from the aircraft nephelometer are utilized, will be presented at ILRC.

ACKNOWLEDGEMENTS

We would like to thank all the Met Office teams involved in the VA Lidar-sunphotometer project and also in the operation and flying of MOCCA. We are grateful to the Civil Aviation Authority and Department for Transport for funding both the lidar / sunphotometer project and MOCCA.

References

[1] Adam, M., Buxmann, J., Freeman, N., Horseman, A., Salmon, C., Sugier, J, Bennett, R,: 2017: The UK Lidar-Sunphotometer operational volcanic ash network, 28 *ILRC*

[2] Ansmann, A., Riebesell, M., Weitkamp, C., 1990: Measurement of atmospheric aerosol extinction profiles with a Raman lidar, *Opt. Lett.* **15**, pp. 746-748.

[3] Ansmann, A., Muller, D., Lidar and atmospheric aerosol particles. In: Weitkamp, C., 2006: Lidar: Range-Resolved Optical Remote Sensing of the Atmosphere, 1st edition. Singapore: Springer Science & Business, p. 105 – 138.

[4] Fernald, F. G., 1984: Analysis of atmospheric lidar observations: some comments, *Appl. Optics*, **23**, pp. 652–653.

[5] Klett, J. D., 1985: Lidar inversion with variable backscatter/extinction ratios, *Appl. Opt., 24*, pp 1638–1643.

[6] Ansmann, A., Tesche, M., Seifert, P., Gross, S., Freudenthaler, V., Apituley, A., Wilson, K. M., Serikov, I., Linne, H., Heinold, B., Hiebsch, A., Schnell, F., Schmidt, J., Mattis, I., Wandinger, U., Wiegner, M., 2011: Ash and fine-mode particle mass profiles from EARLINET-AERONET observations over central Europe after the eruptions of the Eyjafjallajokull volcano in 2010, *Journal of Geophysical Research*, **116** pp. 1-15.

[7] Johnson, B., et al. (2012), In situ observations of volcanic ash clouds from the FAAM aircraft during the eruption of Eyjafjallajökull in 2010, J. Geophys. Res., 117, D00U24.

[8] Martucci G. and A. Haefele, 2014: Review of Requirements for a European network of Automatic LIDARs and Ceilometers for cloud, aerosol and volcanic ash monitoring (http://eumetnet.eu/).

[9] Webster, H. N., et al. (2012), Operational prediction of ash concentrations in the distal volcanic cloud from the 2010 Eyjafjallajökull eruption,J. Geophys. Res., 117 D00U08.

INVESTIGATING THE EFFECT OF HYGROSCOPICITY OF AEROSOLS ON OPTICAL PROFILES OF PBL OBSERVED BY DUAL-WAVELENGTH LIDAR

Wei-Nai Chen[1], Yung-Chang Chen[2], Hui-Ming Hung[2], M. Roja Raman[3]

[1]*Research Center for Environmental Changes, Academia Sinica, Taipei, Taiwan.*
E-mail: wnchen@rcec.sinica.edu.tw
[2]*Department of Atmospheric Science, National Taiwan University, Taipei, Taiwan.*
E-mail: hmhung@ntu.edu.tw
[3]*UGC - SVU Centre for MST Radar Applications, S V University, Tirupati, Andhra Pradesh, India.*
E-mail: ramansvu@gmail.com

ABSTRACT

The light scattering and radiation properties of aerosols are mainly dominated by hygroscopicity. In this study, the relationship between the wavelength dependent ratio of lidar scattering signals (color ratio) and relative humidity and the application of the color ratio to identify the cloud base is examined.

1. INTRODUCTION

Changes in aerosol scattering properties can highly influence the Earth's climate by affecting the Earth's radiation budget. The knowledge of the direct and indirect interaction of atmospheric aerosols on climate is still quite low compared to other atmospheric constituents, not only due to high spatial and temporal variability of aerosols but also due to hygroscopicity. In the ambient atmosphere, the aerosol particles may absorb moisture from the surrounding air and increase their size causing significant change in their size distribution and their associated optical and microphysical properties[1].

In previous studies we already noticed that the inter-relationship between the aerosol Ångström exponent and particle depolarization can reveal information on particle size and shape [2]. In that study, it was reported that the aerosol hygroscopicity might be the reason for the observed distortion in the linear correlation between Ångström exponent and particle depolarization. In this study, we try to combine Mie scattering theory and Köhler theory to study the effect of aerosol hygroscopicity on PBL profiles. We also try to use the lidar color ratio (**CR**) as indicator of particle size to investigate the inter-relationship between color ratio, hygroscopicity (κ), and relative humidity.

2. METHOD

2.1. Color Ratio

The particle size can be inferred from size-dependent light scattering properties such as the Ångström exponent and color ratio [2; 3; 4] (and references therein). The Ångström exponent had been wildly used to understand the particle size. However, the Ångström exponent might be highly uncertain or unavailable because the value of the Ångström exponent highly depends on the accuracy of lidar inversion. Therefore, the ratio of lidar backscatter at different wavelengths (color ratio) might be a simple indicator and more suitable than the Ångström exponent for Planetary Boundary Layer (PBL) studies.

In this study, vertical profiles of lidar backscatter measured by Research Centre for Environmental Changes / Academia Sinica and National Taiwan University (RCEC/ASNTU) Lidar installed in National Taiwan University, Taipei, Taiwan is used for a case study. The RCEC/ASNTU Lidar is a dual wavelength depolarization lidar operated at 355 nm and 532 nm designed to measure aerosols and clouds up to 10 km during day and 20 km at night[2]. The particulate backscatter color ratio **CR** is defined as

$$CR = \frac{P_{532}}{P_{355}} \tag{1}$$

where P_{532} and P_{355} are particulate backscatter at 532 and 355nm respectively.

2.2. Hygroscopicity

The saturation ratio of a liquid aerosol droplet in aerosol in its volume equivalent diameter, S_{eq}[5], over an aqueous solution droplet can be calculated from

$$S_{eq} = a_w \times e^{A/D} \tag{2}$$

$$a_w = \frac{n_w}{n_w + in_s} \tag{3}$$

$$A = \frac{4\sigma_w M_w}{\rho_w R_v T} \tag{4}$$

where a_w is the activity of water in solution, ρ_w is the density of water, M_w is the molecular weight of water, σ_w is the surface tension of the solution/air interface, R is the universal gas constant, T is temperature, and D is the diameter of the droplet, D_d is the dry diameter.

For a multicomponent system (multiple solutes + water) at equilibrium, Eq. 2 yields the equation defining κ–Köhler theory

$$S_{eq}(D) = \frac{D^3 - D_d^3}{D^3 - D_d^3(1-\kappa)} \times \exp\left(\frac{4\sigma_w M_w}{\rho_w R_v T D}\right) \tag{5}$$

Where κ can then be understood as expressing the volume (or mass or moles, with appropriate unit conversions) of water that is associated with a unit volume of dry particle.[5] The overall value for κ is given by the simple mixing rule

$$\kappa = \sum_i \varepsilon_i \kappa_i$$

The wavelength dependent backscattering coefficient of grown particles can be calculated using

$$\beta(\lambda) = \frac{\pi D^2}{4} N_d Q_{back}(\lambda) \quad (6)$$

where N_d is volume number density of particle and Q_{back} is the backscattering efficiency calculated from Mie theory with complex refractive index. In this study, $m = 1.44 + 0.43i$ is used for most of the simulation.

In general, the value of κ for soluble inorganic aerosols is in the range 0.4-1.4; for organic aerosols it is in the range of 0.01-0.5 and for black carbon is less than 0.01.

Our previous studies found κ is around 0.2 for aerosols in Taipei City and 0.1- 0.2 for aerosols in Kaohsiung City [6].

3. RESULTS

3.1. Hygroscopic Growth

Figure 1 shows the typical particle size distribution (solid black) observed in Taipei. The hygroscopic growth size distributions from *dry* to RH=98% for aerosols with κ=0.2 (solid blue) and 0.61 (dashed black) are shown for comparison. Particles with κ=0.61 are about two times larger than particle with κ=0.2 when RH=98%.

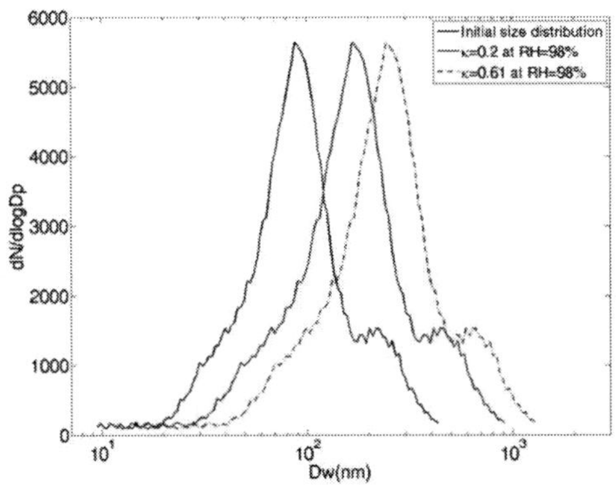

Figure 1: Dry particle size distribution (solid black) and hygroscopic particles growth from "dry" to RH=98% for particles with κ=0.2 (solid blue) and 0.61 (dashed blue).

The corresponding backscattering coefficient as a function of RH for aerosols with κ=0.2 is shown in Fig. 2. The complex refractive index of aerosol is set to $m = 1.44 + 0.043i$. Here the Particle Backscatter Coefficient (PBC) or β_{par} increases slowly with increase in RH up to RH<80%. After RH is higher than 80%, the β_{par} increases rapidly with increase in RH. At RH ≈98% the backscatter is about 10 times larger than the backscatter at RH=40%.

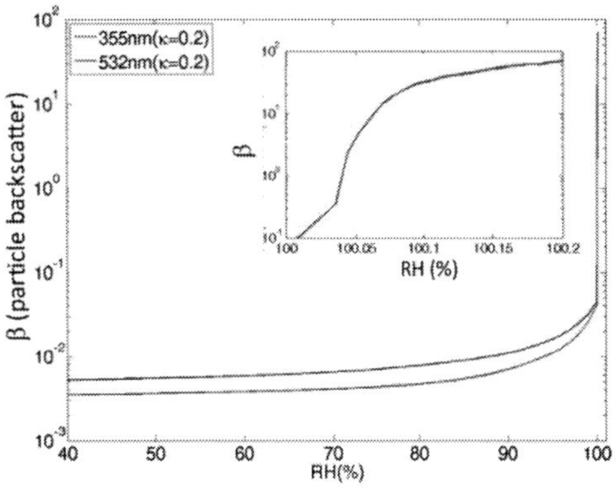

Figure 2: Hygroscopic growth of backscatter with varying RH for aerosols with m=1.44+0.043i and κ=0.2.

3.2. Color Ratio vs. RH

The calculated color ratio **CR** for different RH is shown in Fig. 3. It can be noticed here that the color ratio shows opposite tendency at lower and higher RHs, respectively. In general, the value of the color ratio is about 0.5 for dry particles. The color ratio slightly decreases with increase in RH from "dry" condition to a critical RH_T about 50% (κ=0.2), 60% (κ=0.4), or 75% (κ=0.61). If RH is larger than RH_T, the color ratio turns to increase rapidly with increase in RH and reaches a maximum *CR* of ~0.7-0.8 at RH≈95%. This result implies a κ-dependent turning point RH_T might be useful as an indicator of κ.

3.3. Case Study

A case observed on 13 January 2009 is selected to simulate and investigate the impact of hygroscopic growth. Fig. 4 shows the time-height evolution of range corrected signal ($P \times z^2$) for 532 nm observed by RCEC/ASNTU lidar. The height of boundary layer is about 1.5 km throughout the day. The height of mixed layer was determined using gradient algorithm [7] and is around 1 km from 00Z - 06Z and decreased to 700m - 500m after 06Z.

A radiosonde was launched on 13 January 2009 at 00Z to measure the vertical profile of atmospheric parameters. The profile of measured RH is shown in Fig. 5. Based on this RH profile, the relative backscattering coefficients for 355 nm and 532 nm are calculated and shown in Fig. 6. Aerosols are assumed being well mixed and hygroscopic growth follow the κ-Köhler theory. The initial size distribution is shown in Fig. 1. The complex refractive index of dry aerosol is set to $m = 1.44 + 0.043i$ and κ is 0.2 (the typical value of κ observed in Taipei [6]).

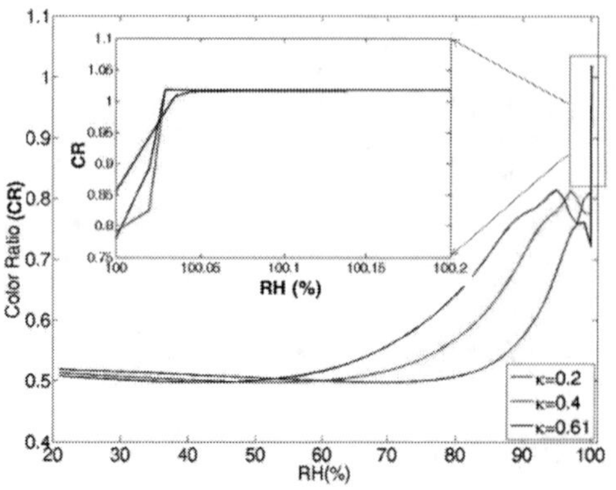

Figure 3: Calculated Color Ratio for different RH for aerosols with κ=0.2 (black), 0.4 (red), 0.61 (blue) and complex reflectivity m=1.44+0.043i.

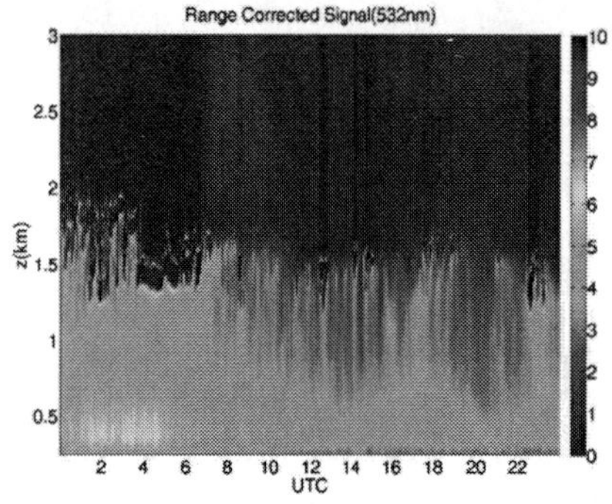

Figure 4: Time-height variation of RCEC/ASNTU Lidar range corrected signal (RCS) at 532 nm observed on 13 January 2009 at Taipei

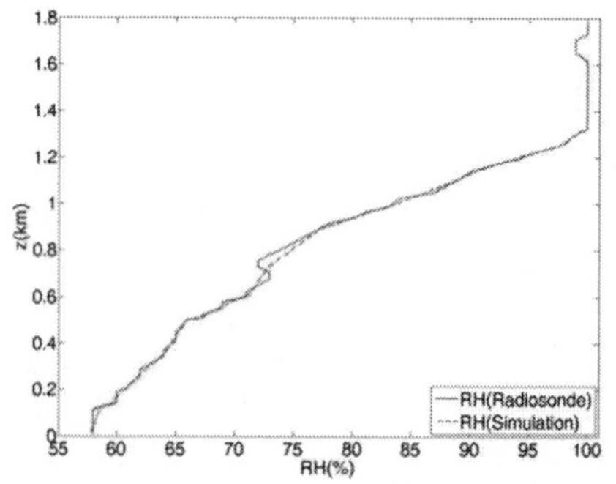

Figure 5: Radiosonde measured (solid green) RH profile on 13 January 2009 at 00Z. Smoothed RH profile (dashed green) used for hygroscopic simulation.

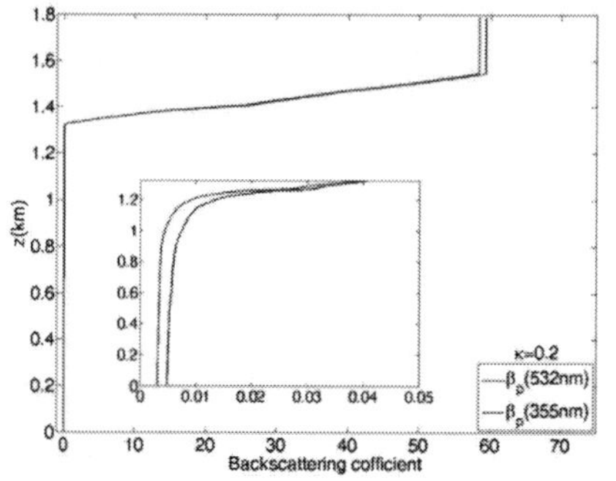

Figure 6: Simulated lidar backscattering coefficients for 355 nm (blue line) and 532 nm (red line) with κ=0.2 and m=1.44+0.043i. The initial size distribution is shown in Fig. 1 and RH profile is shown in Fig. 5.

As shown in Fig. 5, RH was about 55% at ground level. RH gradually increased with height and reached RH=98% at height $z\approx$1.25 km. A small inversion in RH was found at $z\approx$0.8 km, which has been smoothed (dashed green) for hygroscopic simulation. The simulated lidar backscattering coefficients for two wave lengths are shown Fig. 6. Backscattering coefficients for 355 nm (β_{355}) and 532 nm (β_{532}) at ground level (z=0) are about 0.005 and 0.003 $m^{-1}sr^{-1}$, respectively. β_{355} and β_{532} slightly increase with height up to $z\approx$1 km or RH$\approx$90%. β_{355} and β_{532} turn to rapidly increase to 0.04 $m^{-1}sr^{-1}$ when RH$\approx$98%. Beyond RH=98%, aerosols are activated as cloud droplets and the corresponding β_{355} and β_{532} increase to 55 at $z\approx$1.4 km, which are about 1000 times higher than β_{355} and β_{532} at surface level.

The height distribution of observed (solid red) and calculated (solid blue) color ratios are shown in Fig. 7. The observed **CR** decreases from 0.6 to 0.4 with height increasing from surface level to 1.3 km. **CR** turns to rapidly increases from 0.4 to 1.4 with height (or RH) increasing from 1.3 km to 1.4 km. The calculated **CR** slightly increases from 0.42 to 0.6 with height increasing from surface level to 1.2 km. The calculated **CR** turns rapidly increases from 0.6 to 1 with height increasing from 1.2 km to 1.3 km.

The similarities and discrepancies between observation and simulation can be noticed here. The calculated **CR** starts to rapidly increase at lower RH than observed **CR**. The vertical structures of observed and calculated **CR** are much different between 1.3 km and 1.5 km. This discrepancies might be owing to incorrect κ value or RH. Fig. 8 shows the calculated **CR** profiles based on the RH profile (dashed blue) modified from radiosonde measurement (solid green) to match the observed **CR**. This result shows, a small uncertainty would cause large difference from reality. Another implication from Fig. 7 and Fig. 8 is κ might be over estimated. It can be noticed that the simulated **CR** profile with κ=0.1 is more closer to lidar observation.

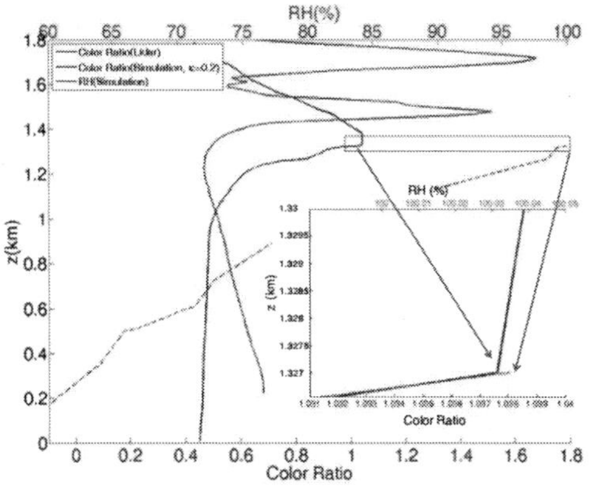

Figure 7: Observed (red) and simulated (greed) vertical profile of Color Ratio (**CR**) with κ=0.2 and RH profile (blue) observed by radiosonde.

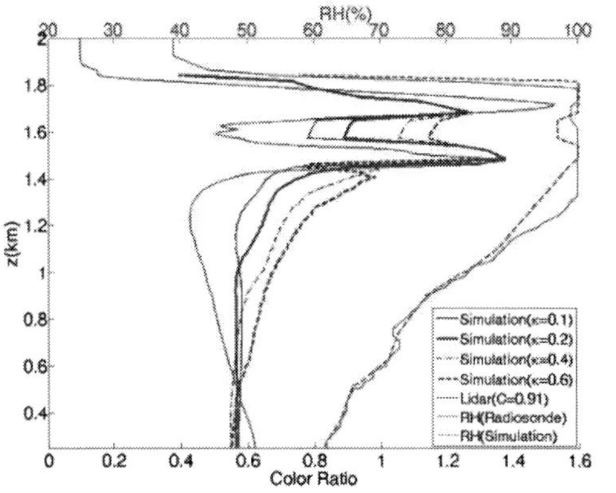

Figure 8: Same as Fig. 7 but with various κ and modified RH profile (blue).

4. SUMMARY

- In general, **CR** of "dry" particle slightly decreases with increase in RH from 0% to a critical RH_T about 60% (κ=0.2), 70% (κ=0.4), or 80% (κ=0.61). Color ratio turns to increase rapidly with increase in RH when RH is larger than RH_T.
- The discrepancies might be owing to incorrect κ value or RH. Measured RH is found highly uncertain when RH higher than 95%.
- Considering the value of κ about 0.1 can make the simulation and observation more closer but it is lower than typical value of κ observed by CCN counter. More study is needed to reveal the possible reasons. Full investigation will be presented in the conference.

REFERENCES

1. Fernández, A. J., Apituley, A., Veselovskii, I., Suvorina, A., Henzing, J., Pujadas, M., and Artíñano, B. Study of aerosol hygroscopic events over the Cabauw experimental site for atmospheric research (CESAR) using the multi-wavelength Raman lidar Caeli. *Atmospheric Environment*, 120(C):484–498, November 2015.
2. Chen, W.-N., Tsai, F., Chou, C., Chang, S., Chen, Y., and Chen, J. Optical properties of Asian dusts in the free atmosphere measured by Raman lidar at Taipei, Taiwan. *Atmospheric Environment*, 41(36):7698–7714, 2007.
3. Ansmann, A., Bösenberg, J., Brogniez, G., Elouragini, S., Flamant, P. H., Klapheck, K., Linn, H., Menenger, L., Michaelis, W., Riebesell, M., Senff, C., Thro, P.-Y., Wandinger, U., and Weitkamp, C. Lidar Network Observations of Cirrus Morphological and Scattering Properties during the International Cirrus Experiment 1989: The 18 October 1989 Case Study and Statistical Analysis. *Journal of Applied Meteorology*, 32(1):1608–1622, October 1993.
4. Liu, Z., Vaughan, M., Winker, D., Kittaka, C., Getzewich, B., Kuehn, R., Omar, A., Powell, K., Trepte, C., and Hostetler, C. The CALIPSOLidar Cloud and Aerosol Discrimination: Version 2 Algorithm and Initial Assessment of Performance. *JOURNAL OF ATMOSPHERIC AND OCEANIC TECHNOLOGY*, 26(7):1198–1213, July 2009.
5. Petters, M. D. and Kreidenweis, S. M. A single parameter representation of hygroscopic growth and cloud condensation nucleus activity. *Atmospheric Chemistry and Physics*, 7(8):1961–1971, April 2007.
6. Hung, H.-M., Lu, W.-J., Chen, W.-N., Chang, C.-C., Chou, C. C.-K., and Lin, P.-H. Enhancement of the hygroscopicity parameter kappa of rural aerosols in northern Taiwan by anthropogenic emissions. *Atmospheric Environment*, 84(C):78–87, February 2014.
7. Flamant, C., Pelon, J., Flamant, P. H., and Durand, P. Lidar determination of the entrainment zone thickness at the top of the unstable marine atmospheric boundary layer. *Boundary-Layer Meteorology*, 83(2):247–284, 1997.

ACKNOWLEDGMENTS

This study was supported by Taiwan National Science Council (105-2119-M-001-025 and 103-2923-M-001-003-MY3). We also thank Academia Sinica, Taiwan for providing Post-doctoral fellowship funded by National Science Council, Taiwan.

EPJ Web of Conferences **176**, 06001 (2018)
ILRC 28

https://doi.org/10.1051/epjconf/201817606001

CHARACTERIZATION OF AIRCRAFT DYNAMIC WAKE VORTICES AND ATMOSPHERIC TURBULENCE BY COHERENT DOPPLER LIDAR

Songhua Wu[1,2]*, Xiaochun Zhai[1], Bingyi Liu[1], Jintao Liu [1]

*[1]Ocean Remote Sensing Institute, Ocean University of China, China, *wush@ouc.edu.cn*
[2]Laboratory for Regional Oceanography and Numerical Modeling, Qingdao National Laboratory for Marine Science and Technology, China

ABSTRACT

Field observations for the wake vortices by Coherent Doppler Lidar (CDL) have been carried out at the Beijing Capital International Airport (BCIA) and Tianjin Binhai International Airport (TBIA) to investigate the wake vortices evolution characteristics and the near-ground effect. This paper introduces the dynamic wake vortices and atmospheric turbulence monitoring technique, successfully demonstrating that the CDL can capture the key characteristics of wake vortices in real-time, including wake vortices intensity, spatial-temporal evolution and so forth.

1 INTRODUCTION

The formation of aerodynamic lift force as a result of aircraft motion in airspace is always accompanied by the formation of a pair of vortices. Strong vortices generated by heavy aircraft present a potential hazard to other aircraft following close behind. In times of increasing air traffic and aircraft dimensions, the wake vortices issue is becoming more important for flight safety, airport capacity, and aircraft design. Several research programs based on simulations, model calculations, wind tunnel and so forth have been carried out [1-3]. For experimental investigations of wake vortices in the atmospheric environment, the Coherent Doppler Lidar (CDL) has been proved to be the most effective and flexible tool along with the developed measuring strategies and methods for processing of raw Lidar data [4-9].

In order to investigate the wake vortices evolution characteristics and the near ground effect, field observation of wake vortices utilizing CDL has been carried out at BCIA and TBIA since 2014. This paper gives a brief analysis of the dynamic wake vortices and atmospheric turbulence characteristics monitoring technique.

2 METHODOLOGY

Range Height Indicator (RHI) scanning mode is used with a fixed azimuth angle and varying elevation during the field experiment. As can be seen in Figure 1(a), the color indicates the averaged tangential speed in each detection volume. The center of the pair of positive and negative radial velocity can roughly illustrate the core position of each wake vortex at height of 200 *m*. The Doppler spectrum broadening due to wake vortices is provided in Figure 1(b). To derive wake vortices characterization in real time, the range bins of wake vortices are firstly determined based on Doppler spectrum width.

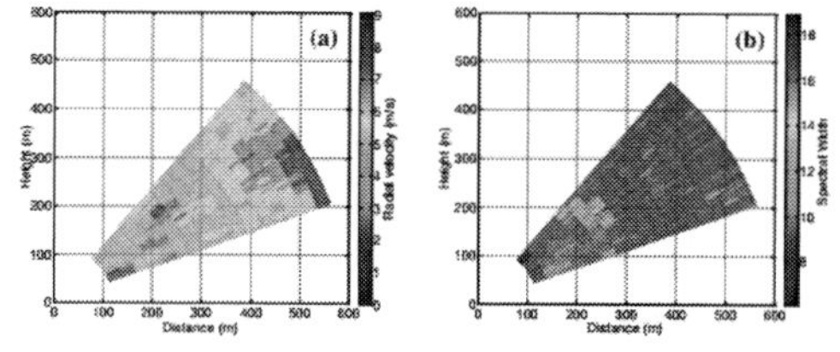

*Figure 1 **(a)** Distribution of the mean of positive and negative velocity along the laser beam **(b)** Distribution of the spectrum width along the laser line*

After the determination of wake vortices range bin, the wake vortices exact location (i.e. the elevation angle of wake vortex) is obtained using the specific distribution of the Doppler spectrum. Figure 2(a) shows an example of the Doppler spectrum versus the velocity and elevation angle at a fixed range which equals to the distance from the Lidar to the left vortex core. The spectrum shown in Figure 2(b) are taken from the data in Figure 2(a) at different elevation angles, labeled with 1, 2, and 3. Graph 1 shows a spectrum measured outside the vortex area, and graph 2 and 3 show the measurement below and above of the left wake vortex core, respectively. According to the result in Figure 2(b), one can select an appropriate threshold to find two points of the intersection with the spectrum curve to the right and to the left of the main spectrum peak, which consists of the velocity envelopes for exact core position determination.

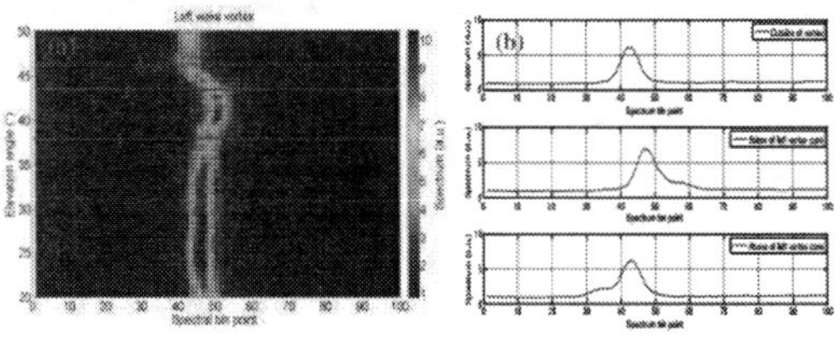

*Figure 2 **(a)** Doppler spectrum measured at different elevation angles and fixed range bin **(b)** spectrum of **(top)** outside, **(middle)** below and **(bottom)** above the left vortex core.*

3 RESULTS

3.1 Wake vortices characteristics evolution

The velocity envelope distribution with regard to transverse distance near the core areas from descending scan is shown in Figure 3 marked with dotted line. For each radius, a pair of positive and negative velocities can be obtained. The mean value of the velocity pair stands for the velocity along the circle with this radius. In this case, the circulations of these two wake vortices are calculated from the mean value within a distance of 5 *m* and 15 *m* from the vortices core position, which are 336 m^2/s and 318 m^2/s, respectively. It is noted that the envelopes include the radial wind velocity and it needs to be derived and then subtracted before the calculation of wake vortices circulation. Figure 4 shows the B744 wake vortices spatial-temporal evolution process on 16:38:07-16:35:51 LT 5 March 2014 at BCIA. Obviously, the wake vortices move away from Lidar because of the crosswind.

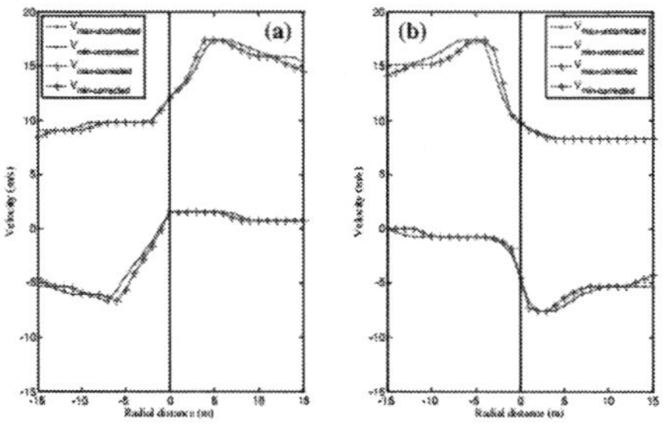

*Figure 3 Velocity envelope distribution around the core area before (dotted line) and after correction (plus line) for descending scan. **(a)** left wake vortex **(b)** right wake vortex.*

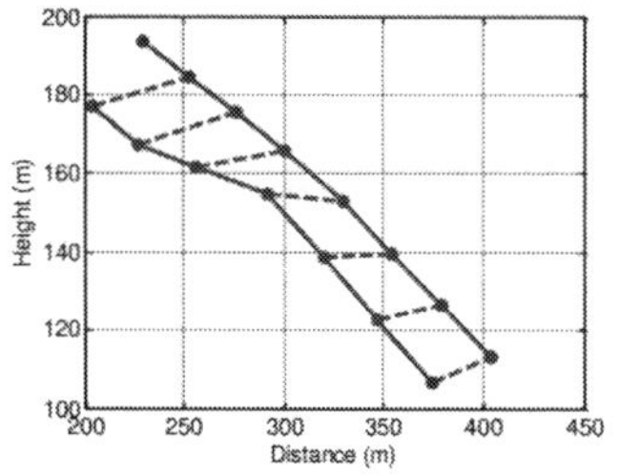

Figure 4 Wake vortex time-spatial evolution process after a B744 flew over during 16:38:07-16:38:51 LT 5 March 2014 at BCIA.

In order to capture the wake vortices with high updating rate, both ascending and descending scans are used alternately to make a reciprocating scanning. However, as the wake vortices usually move downwards, the descending (ascending) Lidar scan results in an expanded (compressed) spatial scale of wake vortices, as a consequence, the value of vortex circulation is overestimated (underestimated) as shown in Figure 5(a). Therefore, the "real" dimension of wake vortices should be firstly corrected to calculate the circulation precisely [5]. Basically, the trajectory and speed of vortices obtained from a series of descending and ascending scans can be estimated to correct the expanded (compressed) wake vortices to obtain its exact dimension and circulation. The results marked with plus line in Figure 3 is the corrected distribution of vortices velocity envelope with transverse distance for descending scan. It can be seen that after the scale of x-axis and interpolation, the corresponding velocity envelope values are less than the uncorrected ones, correcting the overestimate for the expanded wake vortices.

Figure 5(b) shows the wake vortices circulation evolution after correction. Compared to Figure 5(a), the fluctuations of wake vortex circulation value, due to the relative motions of the vortices and laser probing beam, are corrected. Expect for the obvious diminishing trend, it is found that the circulation of left wake vortex (upstream vortex in this case) is always greater than the right wake vortex's (downstream vortex), which is consistent with the fact that the lifetime of upstream vortex is longer than downstream vortex's in the crosswind affection.

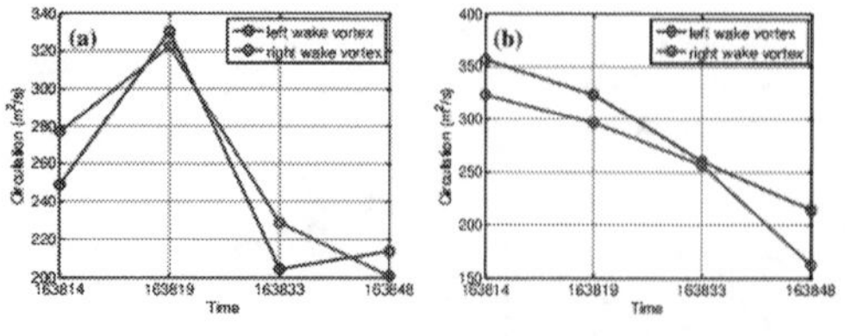

*Figure 5. Wake vortex circulation evolution **(a)** before correction and **(a)** after correction during 16:38:07-16:38:51 LT 5 March 2014 at BCIA.*

3.2 Atmospheric turbulence retrieval using RHI scanning mode

The spatial statistics of the velocity field are described by the structure function. In many cases, the structure function has a simple description in terms of two parameters: the integral length scale L_i and the turbulence energy dissipation rate (TEDR) ε. Before each aircraft overflight, the line-of-sight (LOS) velocity components measured by CDL can be used to calculate the vertical profiles of the crosswind and ε, important parameters for understanding the effect of the atmospheric turbulence on the decay of aircraft wake vortices. The algorithm for extracting high resolution turbulence parameters using longitudinal structure function based on RHI mode has been described in detail by Smalikho et al. [6]

Generally, the values of crosswind velocities at different altitudes h should be calculated as follows,

$$V_T(h) = \frac{1}{N_a}\sum_n\sum_l V_D(h_{nl}) / \cos\varphi_n, \qquad (1)$$

where the summation is performed satisfying the condition $h-\delta/2 < h_{nl} < h+\delta/2$, and N_a is the number of the values h_{nl} in the averaging layer.

Then the 2D distribution of the mean radial velocity can be estimated as

$$< V_D(R_l,\varphi_n) >= V_T(h_{nl})\cos\varphi_n, \qquad (2)$$

which is shown in Figure 6(b). The fluctuation of the mean radial velocity is shown in Figure 6(c), according to the equation as

$$V_D^{'}(R_l,\varphi_n) = V_D(R_l,\varphi_n) - < V_D(R_l,\varphi_n) >, \qquad (3)$$

An estimate of the ε can be obtained from the longitudinal structure function of the wind radial speed as follows,

$$\hat{D}_v(r_i,h) = (N_s L)^{-1}\sum_{n=1}^{N_s}\sum_{h-\delta<h_{kli}<h+\delta}^{L} D_v(r_i,h_{kli},n), \qquad (4)$$

$$\hat{D}_e(r_i,h) = (N_s L)^{-1}\sum_{n=1}^{N_s}\sum_{h-\delta<h_{kli}<h+\delta}^{L} D_e(r_i,h_{kli},n), \qquad (5)$$

$$D_v(r_i,h_{kli},n) = [V_D^{'}(R_k+r_i,\varphi_l,n) - V_D^{'}(R_k,\varphi_l,n)]^2, \qquad (6)$$

$$D_e(r_i,h_{kli},n) = [e(R_k+r_i,\varphi_l,n) - e(R_k,\varphi_l,n)]^2, \qquad (7)$$

where $r_i = i\Delta R, i = 1,2,...,16$, $h_{kli} = (R_k + r_i/2)\sin\varphi_l$, $R_k = R_0 + k\Delta R$.

To investigate the influence of turbulence on the aircraft wake vortices, we retrieve the height profiles of ε from Lidar data measured during 20 scans. In accordance with the results of numerical simulation, this measurement period of 5 min should be optimal. Figure 7 shows the structure function estimates of turbulence using 20 RHI scans on 17:35:34-17:56:18 18 August 2016 at TBIA with an azimuth angle $\varphi = 30°$ and height $H = 210m$. The ε, standard deviation of the transverse radial velocity σ and the integral length scale L_i is 2.02×10^5 m^2/s^3, 0.21 *m/s* and 281.11 *m*, respectively. Figure 8 shows the height profiles of L_i, σ and ε at corresponding time period.

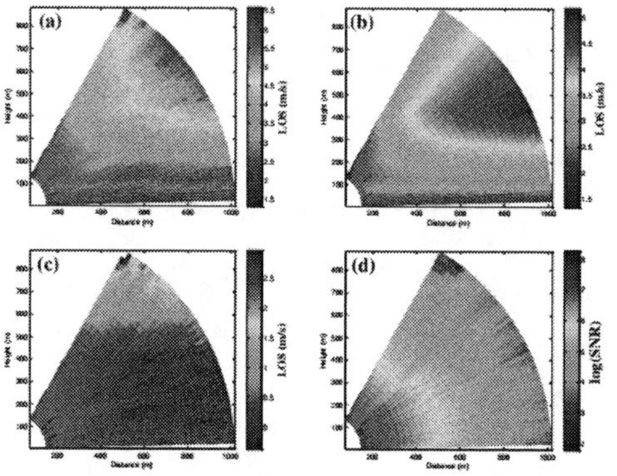

*Figure 6 The 2D distribution of **(a)** radial velocity **(b)** mean radial velocity **(c)** radial velocity fluctuation **(d)** Signal-to-Noise Ratio (SNR) results using RHI scan mode during 17:35:34-17:56:18 LT August 18 2016 at Tianjin airport*

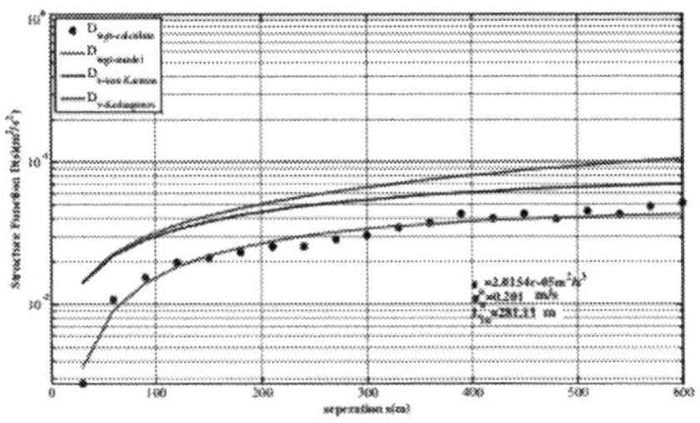

Figure 7. Structure function estimates of turbulence using 20 RHI scans on 17:35:34-17:56:18 LT August 18 2016 at Tianjin airport with an azimuth angle $\varphi = 30^{\circ}$ height $H = 210m$.

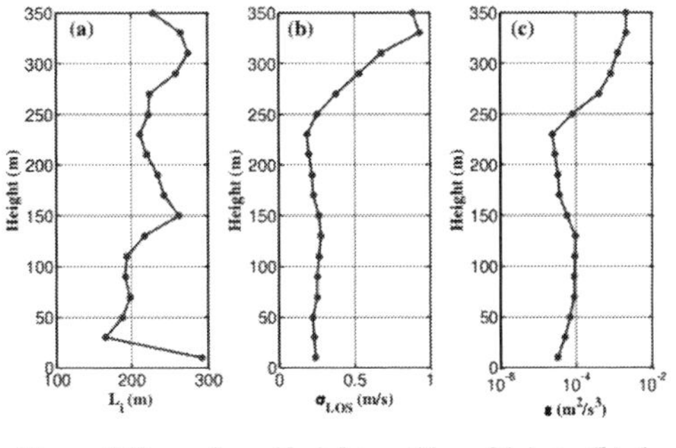

*Figure 8 Examples of height profiles of **(a)** ε, **(b)** the standard deviation of LOS σ_{LOS}, **(c)** the integral length scale L_i retrieved by velocity structure function method.*

4 CONCLUSIONS

In this paper, dynamic wake vortices monitoring algorithm based on the spectral width estimation is introduced. The atmospheric turbulence characteristics are also retrieved using Lidar structure function in RHI mode. Further comparisons and validations are planned in the future. The BCIA and TBIA wake vortices campaigns successfully demonstrate that the concurrent atmospheric turbulence and wake vortices monitoring utilizing coherent Doppler Lidar can describe the key characteristics of wake vortices in real-time, including the wake vortices intensity, spatial-temporal evolution and the interaction with the atmospheric wind and turbulence. The influence of the dynamic state of the atmosphere on the aircraft wake vortices and the near ground effect of wake vortices will be further studied in detail.

ACKNOWLEDGEMENTS

This work is supported by a joint research program between Boeing-COMAC Sustainable Aviation Technology Center and OUC.

References

[1] Holz-aring, F., and pfel. 2003: Probabilistic two-phase wake vortex decay and transport model. *J. Aircr.*, **40** (2), 323-331.

[2] Proctor, F. H., Hamilton, D. W., and Han, J. 2000: Wake vortex transport and decay in ground effect: vortex linking with the ground.

[3] Gerz, T., Holzäpfel, F., and Darracq, D. 2002: Commercial aircraft wake vortices. Progress *Aerosp. Sci.*, **38** (3), 181-208.

[4] Köpp, F., Rahm, S., and Smalikho, I. (2004). Characterization of Aircraft Wake Vortices by 2-μ m Pulsed Doppler Lidar. *J. Atmos. Oceanic. Technol.*, **21** (2), 194-206.

[5] Wu, S., Liu, B., and Liu, J. 2016: Aircraft Wake Vortex Measurement with Coherent Doppler Lidar. In EPJ Web of Conferences (Vol. 119, p. 14008). EDP Sciences.

[6] Smalikho, I. N., and Rahm, S. 2010: Lidar investigations of the effects of wind and atmospheric turbulence on an aircraft wake vortex. *Atmos. Oceanic Opt.*, **23** (2), 137-146.

[7] Holzäpfel, F., Holzäpfel, F., Stephan, A., Stephan, A., Heel, T., Heel, T., ... and Körner, S. 2016: Enhanced wake vortex decay in ground proximity triggered by plate lines. *Aircr. Eng. Aerosp. Technol.*, **88** (2), 206-214.

[8] Smalikho, I. N., and Banakh, V. A. 2015: Estimation of aircraft wake vortex parameters from data measured with a 1.5-μm coherent Doppler lidar. *Opt. Lett.*, **40** (14), 3408-3411.

[9] Smalikho, I. N., Banakh, V. A., Holzäpfel, F., and Rahm, S. 2015: Method of radial velocities for the estimation of aircraft wake vortex parameters from data measured by coherent Doppler lidar. *Opt. Exp.*, **23** (19), A1194-A1207.

EPJ Web of Conferences **176**, 06002 (2018) https://doi.org/10.1051/epjconf/201817606002
ILRC 28

COMPARISON AND VALIDATION OF WAKE VORTEX CHARACTERISTICS COLLECTED AT DIFFERENT AIRPORTS BY DIFFERENT SCANNING LIDAR SENSORS

Ludovic Thobois[1]*, Jean-Pierre Cariou[1], Valerio Cappellazzo[2], Christian Musson[2], Vincent Treve[2]

[1]*Leosphere, France, *lthobois@leosphere.com*
[2]*EUROCONTROL, Belgium*

ABSTRACT

Today, the demand for increasing airport capacity is high, in particular for increasing runway throughput from an ATM perspective. Runway capacity is often directly linked with the minima longitudinal separation between aircraft on approach phase or between aircraft on departure. The separation minima are based on surveillance capabilities and on wake turbulence (WT) in order to mitigate respectively collision risk and WT-induced accidents, therefore WT hazard becomes a major concern for ATM. For ten years, many research LIDAR systems have been used for better understanding wake vortices behaviors in the operational environment within large range of wind and turbulence conditions. All these studies[1][2] helped to design new concepts of wake separations between aircrafts thanks to the proven capabilities of LIDAR systems to assess the risks of wake vortex (WV) encounters through the circulation retrievals. The re-categorization project, called RECAT [8], has been launched by a joint EUROCONTROL – FAA initiative in order to renew and optimize the out-of-date currently applied ICAO regulations on distance separation. Nowadays, the first phase of regional RECAT projects, which consists in defining new distance separation matrices composed of six/seven static aircraft categories instead of three, entered the operational phase and is deployed in several airports in United States and Europe. In addition, other concepts like Time-Based Separation have also been studied and deployed in London Heathrow. The airports where these solutions have been deployed obtained significant benefits as increased runway throughput and improved resilience to disruptions. For implementing such new WT solutions at an airport, a local safety assessment before the implementation and a risk monitoring after are usually needed. Before implementation, it may be required to determine for the targeted airport the relative variations of risk of wake vortex encounters, given the local ATM rules, the traffic mix, the weather conditions and their impact on the wake vortex decay. After implementation, the risk monitoring might perform in-depth analysis of wake vortex encounter reported by pilots. For all the mentioned steps, the use of scanning Doppler LIDARs is the only experimental sensor capable of measuring the localization and the circulation of the wake vortices and to provide ground truth wake vortex measurements. Next generation operational LIDARs need to be developed to address in a cost effective way these operational needs. Furthermore, a specific configuration and methodology need to be developed to ensure the accuracy of the wake vortex data. Such a LIDAR based wake vortex solution has been tested at Paris Charles De Gaulle which implemented the RECAT-EU wake separation scheme. The wake vortex circulation, initial spacing and decay measured have been compared to the data collected in London Heathrow by a different LIDAR sensor. The results indicated that the initial circulation, the time to demise, the decay curve evolution and the vortex spacing are very coherent between the two databases.

1 INTRODUCTION

Wake vortices (WV) are a consequence of the pressure differential between the lower and upper surface of the wing moving through the air. The trailing vortices rolls-up into a pair of coherent, counter-rotating vortices that can persist for a considerable length of time after the aircraft has flown by and could pose a potential hazard to the safe flight of an aircraft that is following the lead aircraft and encountering these vortices. The trailing vortices' rate and time to dissipate (decay) depend on the weight, size, wing configuration and speed of the aircraft as well as atmospheric conditions. Specifically, the roll-up of two-vortex system (2VS) formed behind an aircraft is composed of two counter-rotating vortices of initial total circulation Γ_0 and initially spaced by a distance b_0 with s the spacing factor that depends on the combined effects of wing and horizontal tail plane (HTP) loadings. The initial total circulation is related to the aircraft weight, W_l, flight speed V_l, wing span b_l and wing and horizontal tail plane loadings:

$$\Gamma_0 = \frac{W_l}{\rho V_l b_l s} \quad (1)$$

Where ρ is the air density, the sub index l refers to leader aircraft and s is the spacing factor defined as the ratio between the initial lateral spacing between the vortices b_0 and the aircraft span b_l. The vortex total circulation Γ_{tot} decreases in time, starting from Γ_0 and following a two-phase decay, with a decay rate that depends on the atmospheric turbulence, characterized by the Eddy Dissipation Rate (EDR), ϵ, and thermal stratification, characterized by the Brunt-Väisälä frequency[4], N. To measure properly WV, scanning coherent Doppler Lidar (CDL) are used when configured with high resolution (<10m) and high update rate (10s). Commonly, the experimental setup used is to perform a vertical slice perpendicular to the runway in order to characterize WV in a monitoring area of 500m width and at an altitude of one wingspan. For optimizing the distance separations to be applied between aircrafts, huge wake vortex databases are required to be collected by CDLs. In this study, two types of CDLs have been used: the first CDL is a WindTracer with which a database has been collected a few years ago at Heathrow (EGLL-1) airport and the second CDL is a WIDCUBE200S which has been deployed at Charles De Gaulle (CDG) airport 1 year ago. The objective of this work is now to compare the two WV databases collected by two sensors with different WV characterization algorithms and at two airports with different atmospheric environments. Therefore it is of interest to assess the wake vortex characteristics recorded by both sensors in order to identify similarities and differences of the two campaigns. The comparison is focused on Γ_0 and b_0 measurements as they are important parameters that drive the entire WV physics as described above.

2 METHODOLOGY

2.1 Wake vortex algorithms for CDLs

Many studies show the capabilities to track WV within the Doppler spectra and also within the radial wind data [5][6][7]. In the past two years, WV algorithms have been developed based on the fitting of a WV theoretical model on radial wind speeds [7]. This method is used in the WINDCUBE WV characterization algorithm which detects, localizes and computes the circulation of WV. This last step is performed by a least mean square optimization of a Hallock-Burnham WV theoretical model on the radial wind speeds, as proposed by Smalikho [7]. This algorithm has been validated on real wake vortex data collected at Paris-Charles-De-Gaulle airport in 2011 and at Toulouse-Blagnac airport in 2014 [9].

2.2 Two-Phase Decay curve

All the measured data from the CDG and EGLL-1 databases have been used to reconstruct a generic (per aircraft type) two-phase decay curve that represents the WV physics. Each phase is analytically described by a decreasing exponential function characterized by coefficients α_1 and α_2 and with Γ_0 as initial circulation. The transition between the two phases occurs at $t = t_c$ which can be viewed as a measured time-to-demise (dashed lines in Figure 1 and Figure 2). The mathematical formalization of the fit $\Gamma^*(t)$ can be written as:

$$\Gamma^*(t) = \begin{cases} \Gamma_0 \exp(-\alpha_1 t) & for\ t \le t_c \\ \Gamma_0 \exp(\alpha_2 t_c - t) - \alpha_1 t_c & for\ t > t_c \end{cases} \quad (2)$$

Four parameters have to be fixed: Γ_0, α_1, α_2 and t_c. They are determined by minimizing residues between the fit $\Gamma^*(t)$ and the data $\Gamma(t)$. Residues are defined by the least squares:

$$res = \sum_{i=1}^{N} (\Gamma^*(\alpha_1, \alpha_2, t_c, \Gamma_0; t_i) - \Gamma(t_i))^2 \quad (3)$$

With the use of numerical computing software, a (local) solution to this multi-variables minimization problem without constraint is found. In the fit no extrapolation of the lateral WV position has been considered; it conservatively assumed that the WV lateral positions after the last measured point remain constant.

3 RESULTS

 https://doi.org/10.1051/epjconf/201817606002

The wake vortex algorithm has been applied to collect wake vortex data at Paris-Charles de Gaulle airport. The analyses are conducted for the CDG Database containing about 26000 tracks from 19th November 2015 to 22nd September 2016 with the Leosphere LiDAR and compared with the EGLL-1 Database containing about 55000 tracks collected from October 2008 until December 2010 where a different LIDAR has been used.

3.1 Wake Vortex Circulation and Decay

In Figure 1 and Figure 2 the comparison between the CDG Database (blue curve), and the EGLL-1 Database (red curve) with the two-phase decay curve are presented. The resulting plot is the mean of all the decay curves, in every plot there is the indication of total number of tracks in the database, the number of tracks taken in account by the method (between brackets) and the number of tracks that contains NaN values. As we can see in the results the initial vortex circulation Γ_0, the time to demise and the decay curve are very coherent between the two databases. This is valid for all the aircraft types where a sufficient number of tracks for the comparison are collected.

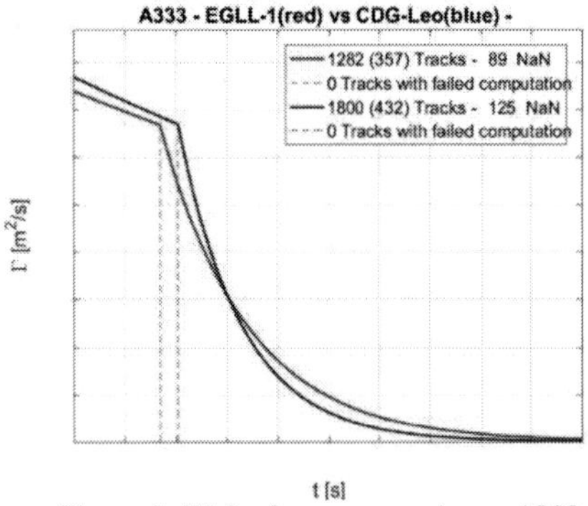

Figure 1: Wake decay comparison - A333

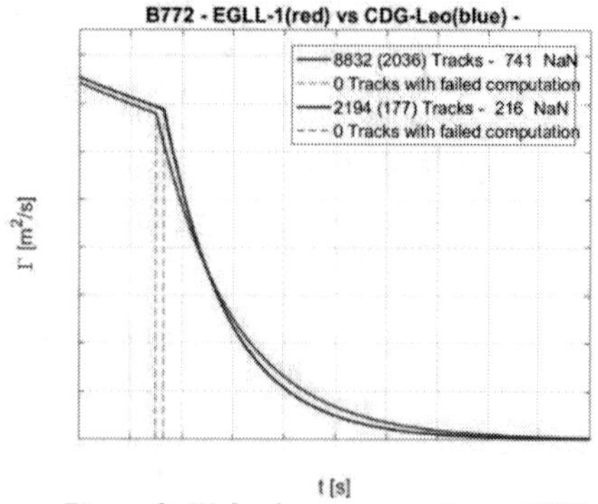

Figure 2: Wake decay comparison - B772

3.2 Initial vortex spacing

The data collected have also been analyzed to compare the initial vortex spacing (b_0). The b_0 are obtained following the 2 steps methodology presented in the RECAT-EU Safety Case [8]. For step 1 of the b_0 determination process, the b_0 measured in London Heathrow (EGLL-1 database) are used as altitude threshold for the CDG measurement selection (avoiding an iterative process). The Step 2 is used for better characterization of the b_0 for all the aircraft that do not have the complete roll-up of the vortex at one wingspan altitude. The mean and the median values of b_0 are calculated using all selected tracks for each aircraft type. The CDG results are compared in *Table 1* with the results of EUROCONTROL analysis of the EGLL-1 database.

The vortex spacing factor $s = b_0/b$ measured in CDG datasets are found to be, on average, between 2% and 14% bigger than those of the EGLL-1 dataset, however for many aircraft types the number of tracks collected that satisfies the altitude and time conditions identified by the methodology is limited. Some high differences like in the case of B752 could be justified by the different operating conditions at the two airports. The differences observed for the Boeing 777 family can be associated to a bad identification of the different versions B772 and B77W leading to an error of 7-11%. Some measurements of b_0 are also available for the A388-800 (A388), the obtained values are found consistent (only 2.15% deviation) with the A388 EGLL-1 measurements. Then, only the aircraft for which about 100 tracks (or more) were collected and with no problems in the identification and operations were taken into account. Applying this methodology, the computed ratio between the b0 measured in EGLL-1 and the median b0 measured in CDG shown a systematic error on the b0 measurement of about 2% (see *Table 2*).

AC	rel % to EGLL	# ac
A332	0.38%	336
A333	2.02%	282
A343	-2.46%	102
A388	-2.15%	107
B744	-1.88%	6
B752	-14.74%	529
B763	-6.19%	250
B772	-11.79%	529
B77W	-8.64%	654

Table 1 - b0 analysis of CDG Dataset

AC	b0 Ratio (%)	Wingspan	Aircraft Type
A332	98	60.3	H
A333	97	60.3	H
A343	100	60.3	H
A388	96	79.75	J
B744	102	64.44	H
B752	111	38.05	M
B763	101	47.57	H
B772	111	60.93	H
B77W	107	64.8	H

Table 2 - b0 Ratio

4 CONCLUSIONS

To sum up, the comparison of the two datasets show good agreements in terms of wake vortex decay, initial circulation, time to demise, and vortex initial spacing. For this last parameter, a mean error of about 2% has been obtained with a given methodology consisting in the sufficient data sampling for each aircraft type and in the filtering of the aircrafts that were badly identified or not operated in the same way at the two airports. Even if these data collections have been performed at different airports, by different Lidar sensors and despite the fact they have been analyzed by different wake vortex characterization algorithms, there is already a reliable degree of consistency between the two wake vortex datasets, which create a solid basis for the future work. The final goal is to have a robust methodology to build consistent and universal wake vortex datasets with different scanning LIDAR sensors. This way, all those datasets can be used in support of safety assessment necessary to deploy wake solutions that can tackle the airports capacity issues.

ACKNOWLEDGEMENTS

This paper is based on the LIDAR data campaign jointly conducted with NATS (in Heathrow) and DSNA (in Paris-Charles-de-Gaulle).

References

[1] CREDOS: Crosswind-Reduced Separations for Departure Operations, EU Project Program: FP6-AEROSPACE, Ref.: 30837, Record Number: 81468.

[2] FAR WAKE Fundamental Research on Aircraft Wake Phenomena, EU project Funded under: FP6-AEROSPACE Ref.: 12238, Start date: 2005-02-01, End date: 2008-05-31.

[3] Greene, G. C., "An Approximate model of vortex decay in atmosphere," J. Aircraft, Vol. 23, July 1986, pp. 566–573.

[4] De Visscher, I. et al., "Aircraft vortices in stably stratified and weakly turbulent atmo-spheres: simulation and modeling," AIAA J., Vol. 51, No. 3, March 2013, pp. 551–566.

[5] Köpp, F., et al. "Characterization of aircraft wake vortices by 2-micro pulsed doppler lidar", Journal of Atmospheric and Oceanic Technology, vol. 21, no. 2, pp. 194–206, 2004.

[6] Dolfi-Bouteyre, A., et al., "1.5 m all fiber pulsed lidar for wake vortex monitoring, Conference on Lasers and Electro-Optics", Optical Society of America, 2008.

[7] Smalikho, I.N.; Banakh, V.A.; Holzäpfel, F.; Rahm, S. Method of radial velocities for the estimation of aircraft wake vortex parameters from data measured by coherent Doppler lidar. 2015. Optics Express 23(19) A1194-A1207.

[8] EUROCONTROL, European Proposal for revised Wake Turbulence Categorisation and Separation Minima on Approach and Departure, "RECAT - EU", Safety Case Report, 2013.

[9] Thobois, L., et al. "Deployment of a next generation and operational LIDAR solution for monitoring wake vortices for supporting new wake turbulence regulations (Invited)", 8th AIAA Atmospheric and Space Environments Conference, AIAA Aviation, (AIAA 2016-3273)

https://doi.org/10.1051/epjconf/201817606003

APPLICATION OF LIDAR AND OTHER PROFILING TECHNIQUES TO STUDY THE IMPACT THAT SEVERE WEATHER HAZARDS HAVE ON THE NEW YORK CITY BUILT ENVIRONMENT

Mark Arend[1]*, Mark Campmier[1], Aris Fernandez[1], Fred Moshary[1]

[1]NOAA CREST Center and City College of New York, New York, NY 10031, USA,
**marend@ccny.cuny.edu*

ABSTRACT

The complexity of urban boundary layer dynamics poses challenges to those responsible for the design and regulation of buildings and structures in the urban environment. Lidar systems in the New York City Metropolitan region have been used extensively to study urban boundary layer dynamics. These systems, in conjunction with other sensing platforms can provide an observatory to perform research and analysis of turbulent and inclement weather patterns of interest to developers and agencies.

1 INTRODUCTION

Global urbanization inspires the construction of new highly elevated skyscrapers to unprecedented heights while existing buildings age. The concurrence of these newer high rises (including the presence of construction sites) along with the older buildings, causes significant engineering challenges for builders and regulatory agencies who are responsible for considering the impact of these buildings by extreme weather events (e.g. high winds and continued erosion induced by tropical and extra-tropical storms). Characterizing the interaction between urban weather systems, the urban boundary layer and the built environment in a large metropolitan region like New York City can benefit from ground based remote sensing observations (such as lidar) especially when these observations are taken in to consideration with other observing platforms in order to provide a more complete description of the potential risks that are inherent within the complex built environment. The early stages of instrumenting the New York City region with such beneficial assets will be presented and a plan for enhancing the observatory suite to better serve stakeholders will be discussed. Ongoing studies that will involve planning for the optimal enhancement of the observatory will be given.

2 METHODOLOGY

The beginnings of a ground based observatory based in the NYC Metro area has been evolving over the years that consists of a Doppler wind lidar system, other ground based wind profilers (including a Radar Wind Profiler), a ceilometer for characterizing cloud and boundary layer heights and a Microwave Radiometer for profiling temperature, water vapor, liquid water content. Fig. 1 below shows the locations of these instruments in the NYC metro area.

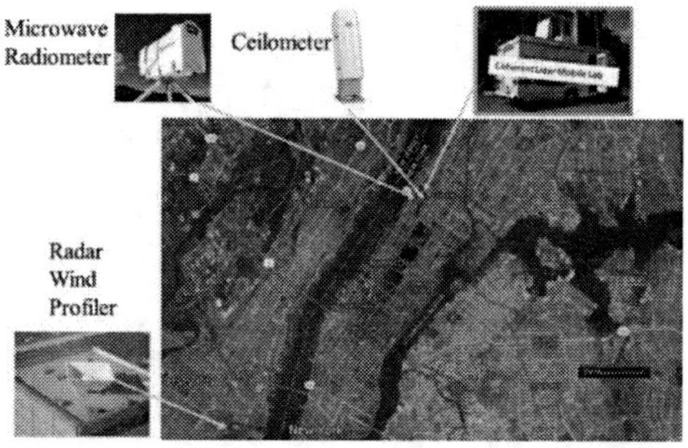

Figure 1 Locations of some of the Optical Remote Sensing Laboratory instruments in the New York Metropolitan area (coherent Doppler lidar, Microwave radiometer, Ceilometer and Radar Wind Profiler)

There have been a number of previous reports that describe the instruments shown in Fig. 1 as well as other atmospheric boundary layer instruments and studies including the construction and evaluation of the Coherent Doppler Lidar (CDL) [1-4]. In addition, a number of previous reports incorporate these instruments (including additional instrumentation) to study the dynamics of the urban atmospheric boundary layer in the NYC metropolitan region with applications to air quality [5,6], wind hazards [7], turbulent boundary layer and stability studies [8-11].

3 RESULTS

To expand these studies a focus is made on a few case studies namely (1) an event during the summer of 2015 in which a number of the instruments were cooperating and for which the evolution of the boundary layer stability is observed by pointing the CDL vertically and observing the fluctuations of the vertical velocity along with the diurnal cycle of the solar forcing and (2) an extreme wind event during the landfall of Hurricane Sandy on October 29-30, 2012.

3.1 Observations of diurnal forcings

Fig. 2 shows a time series of the profiles from the microwave radiometer, CDL and Ceilometer over the period of two days on July 2nd and July 3rd, 2015. The Microwave Radiometer is picking up a clear diurnal signal of the temperature profile during this episode. It should be noted that the synoptic conditions for this event were not necessarily extreme conditions, the temperatures were in the lower 70s (morning) to mid 80s (afternoon), and there were weak winds from the NW as a mild cold front passed through and there was low humidity over the few days.

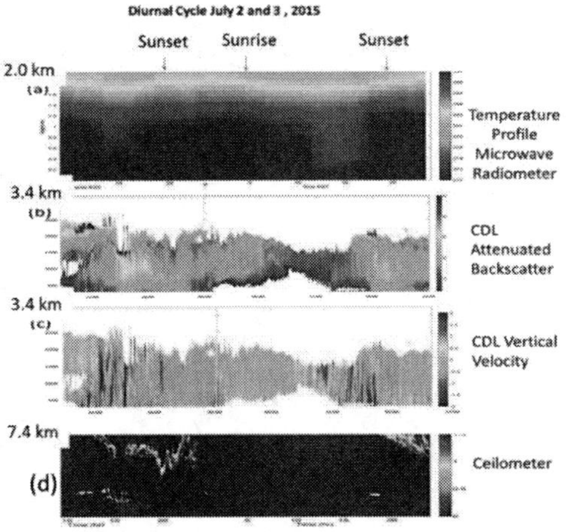

Figure 2. Time series of vertical profiles from the (a) Microwave radiometer, (b and c) Coherent Doppler Lidar, and (d) Ceilometer during July 2 and July 3, 2015. The time of Sunset, Sunrise and then Sunset are indicated with arrows above the top plot.

Although the ceilometer is showing the existence of some high level clouds in the evening of July 2nd, the diurnal forcing observed in the temperature profiles of the radiometer before sunset on July 2nd appears again on July 3rd. These forcings are coincident with a clear signal of vertical fluctuations observed by the CDL before sunset as well as the apparent formation of low level clouds. These observations illustrate the utility of co-operating Doppler lidar systems with other lidars and profilers in order to study the evolution of atmospheric stability. Such studies of the evolution of atmospheric stability become even more important when the synoptic forcing becomes so strong that the winds become hazards to structures as shown in the next example.

3.2 Wind hazards

Wind hazards can be studied using ground based remote sensing profilers in order to asses the exposure hazards depending on the type urban landscape. This can provide added value to the commonly accepted standards utilized by Architects, Civil and Construction Engineers as s shown in Fig. 3 below taken from the SEI/ASCE (American Society of Civil Engineers Structural Engineering Institute) handbook [12]. The "three second peak gust wind speed" recorded at NWS stations is typically used as a design guideline when assessing design loads. "Exposure coefficients" are then applied to adjust for height and terrain dependent wind exposure as summarized in Fig. 3.

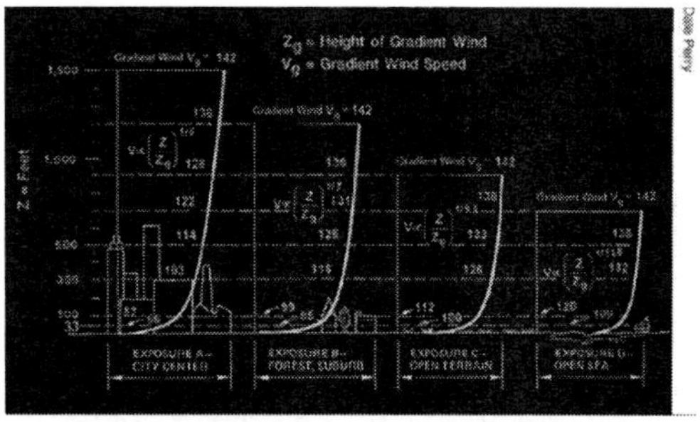

Figure 3. Guidelines from the American Society of Civil Engineers Structural Engineering Institute for assessing the exposure to the built environment for various terrains and heights based on simple assumptions related to surface observations of wind.

As an example on how wind profilers might be useful to better asses the exposure in built environments, a case study during the landfall of Hurricane sandy is studied to point out the utility of profiling winds using ground based remote sensing. In this particular example, as illustrated in Fig. 4, a project in which a crane was at risk during the storm presented concerns. The crane collapsed at 2:30 pm EDT (13:30 EST). Also shown in Fig. 4 are the wind barbs from the Radar Wind Profiler. In addition, an anemometer located on the top of a nearby sky scraper recorded the wind speed and direction throughout the period that Hurricane Sandy approached land fell on land. The coincident observations from various profilers as well as surface stations of events like those illustrated in Fig. 4

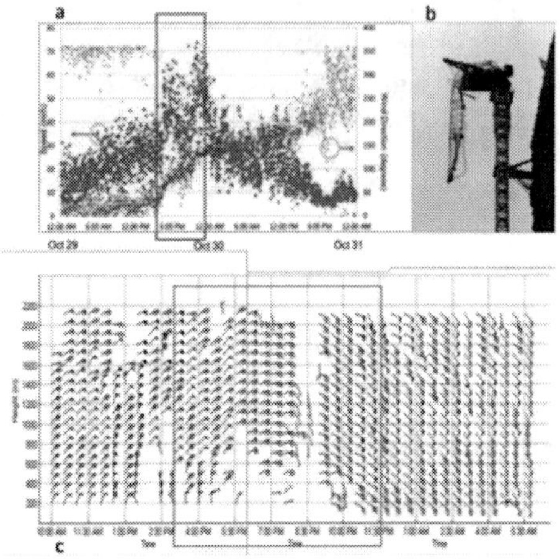

Figure 4. Observations made during the landfall of landfall of Hurricane Sandy. (a) Wind speed (blue) and direction (red) from an anemometer on a weather tower located on top of a building very near to the location of where the crane collapsed (b). (c) Wind barbs from the Radar Wind Profiler (as shown in Fig. 1) are presented indicating the same episode (square boxes indicate the similar time periods.

Another source of observations that can be used in conjunction with the vertical profilers during such extreme events are a collection of weather stations in the NYC metro area. Fig. 5 shows an illustration of the wind speeds over the NY/NJ metropolitan area during the landfall of Hurricane Sandy. Taking all of the observations in to consideration allows for more advanced assessment of the risks that such an extreme event can have not only at the surface but at many vertical levels. Studies such as these can also provide better guidance for entities (both commercial and governmental) who are concerned about setting standards.

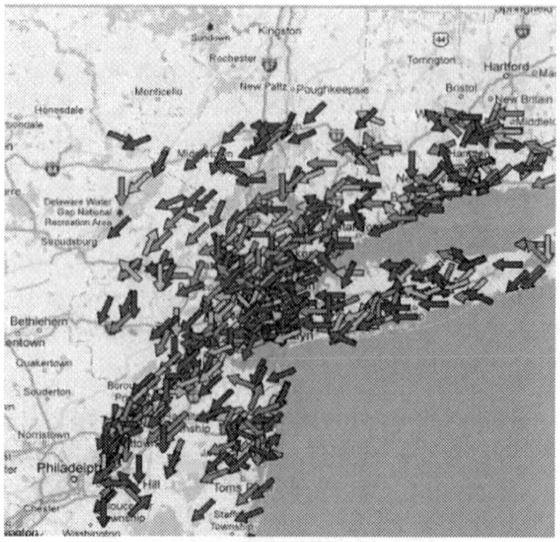

Figure 5. Wind speed and directions as observed by surface weather stations in the NY/NJ Metropolitan region during the landfall of Hurricane Sandy (Oct 29 17:30 EST). Color codes represent wind speed in mph, purple 0-5, blue 5-10, green 10-15, yellow 15-20, red greater than 20.

4 CONCLUSIONS

Lidar and other profiling techniques have been used to study the impact that severe weather hazards have in relation to the built environment with a focus on wind hazards and turbulent conditions. In addition, results of an ongoing study funded by the NYC Department of Buildings to evaluate and suggest updates to amend the administrative code of the City of New York, in relation to the effects of wind on certain buildings will be presented.

ACKNOWLEDGEMENTS

Partial support for this work comes from NOAA-CREST, the CUNY CREST Institute and the CUNY High Performance Computing Facility

"NOAA-CREST program is funded by NOAA/EPP Grant # NA11SEC4810004"

"The CUNY HPCC is operated by the College of Staten Island and funded, in part, by grants from the City of New York, State of New York, CUNY Research Foundation, and National Science Foundation Grants CNS-0958379, CNS-0855217 and ACI 1126113."

Any opinions, findings, conclusions, or recommendations expressed in this publication are those of the author(s) and do not necessarily reflect the view of the U.S. Department of Commerce National Oceanic and Atmospheric Administration.

References

[1] Arend, M ; Abdelazim, S., Lopez, M., Moshary, M., 2013: A comparison of two embedded programming techniques for high rep rate coherent Doppler lidars, *Proc. SPIE 8731, Laser Radar Technology and Applications XVIII, 87310E (May 28, 2013); doi:10.1117/12.2016018.*

[2] Abdelazim, S., Santoro, D., Arend, M. F., Moshary, F., Ahmed, S., 2015: Development and Operational Analysis of an All-Fiber Coherent Doppler Lidar System for Wind Sensing and Aerosol Profiling, *Geoscience and Remote Sensing, IEEE Transactions on* **53** (12), 6495-6506.

[3] Abdelazim, S., Santoro, D., Arend, M., Moshary, F., Ahmed, S., 2016: Signal to Noise Ratio Characterization of Coherent Doppler Lidar Backscattered Signals, *EPJ Web of Conferences* **119** 17014, DOI: 10.1051/epjconf/201611917014

[4] Abdelazim, S., Santoro, D., Arend, M., Moshary, F., & Ahmed, S. 2015: The challenges of implementing and testing two signal processing algorithms for high rep-rate coherent Doppler lidar for wind sensing, *Proc. SPIE 9474, Signal Processing, Sensor/Information Fusion, and Target Recognition XXIV, 94741B,* DOI:10.1117/12.2176897.

[5] Gan, C., Y. Wu, B. Madhavan, B. Gross, F. Moshary, 2011: Application of active optical sensors to probe the vertical structure of the urban boundary layer and assess anomalies in air quality model PM2.5 forecasts, *Atmos. Environ.*, 45 (37), 6613-6621.

[6] Vladutescu, V., Wu, Y., Gross, B., Moshary, F., Ahmed, S.,Blake, R., Razani, M., 2012: Remote Sensing Instruments used for Measurement and Model Validation of Optical Parameters of Atmospheric Aerosols *IEEE Transactions on Instrumentation & Measurement 61, 1733-1746 (2012)*

[7] Arend, M. F., Gutiérrez, E. J., Gonzalez, J. E., Moshary, F., Bornstein, B., Martilli, A., 2014, Vertical Profiles and Horizontal Distributions of Observed and Modeled Wind and Pressure Fields Over the NYC Metro Area During Superstorm Sandy as a Tool for Understanding Impacts and Risks for a Vertically Built Environment, *AMS Annual Meeting, Atlanta GA, 2014*

[8] Arend, M., Valerio, I., Neufeld, S., Bishir, B., Wu, Y., Moshary, F., Melecio-Vazquez, D., and Gonzalez, J., 2016: Exploring the Turbulent Urban Boundary by Use of Lidars and Microwave Radiometers, *EPJ Web of Conferences* **119** 15005, DOI: 10.1051/epjconf/201611915005

[9] Melecio-Vázquez, D., González-Cruz, J.E., Arend, M., Han, Z., Dempsey, M., Booth, J., Gutierrez, E. and Tongue, J., 2015: New York Metro-Area Boundary Layer Catalogue: Boundary Layer Height and Stability Conditions from Long-Term Observations. *9th International Conference on the Urban Climate jointly with 12th Symposium on the Urban Environment. July 20-24, 2015, Centre de Congres Pierre Baudis, Toulouse, France.* URL:http://www.meteo.fr/icuc9/LongAbstracts /nomtm9-2-2991314_a.pdf

[10] Melecio D., and J. Gonzalez, P. Ramamurthy, and M. Arend, 2016: The Structure and Evolution of the Urban Boundary Layer During an Extreme Heat Event, *22nd Symposium on Boundary Layers and Turbulence. June 20-24, 2016, Sheraton Salt Lake City Hotel, Salt Lake City, UT. Conference Presentation.* https://ams.confex.com/ams/32AgF22BLT3BG /webprogram/Paper295973.html

[11] Melecio D., J. Gonzalez, P. Ramamurthy, and M. Arend, 2017: Modeling and Observations of the Structure and Evolution of the Urban Boundary Layer of New York City, *97th American Meteorological Society Annual Meeting: 13th Symposium of the Urban Environment. January 22-27, 2017, Washington State Convention Center, Seattle, WA. Conference Presentation.*

[12] ASCE. 2010. Minimum Design Loads for Buildings and Other Structures. *ASCE/SEI Standard 7-10*

BOUNDARY LAYER TEMPERATURE MEASUREMENTS OF A NOCTUAL URBAN BOUNDARY LAYER

Simon Holloway[1], Hugo Ricketts[1,2,*] and Geraint Vaughan[1,2]

[1]*Centre for Atmospheric Science, University of Manchester, Manchester, M13 9PL, United Kingdom* *h.ricketts@manchester.ac.uk
[2]*National Centre for Atmospheric Science, University of Manchester, Manchester, M13 9PL, United Kingdom*

ABSTRACT

A low-power lidar system based in Manchester, United Kingdom has been developed to measure temperature profiles in the nocturnal urban boundary layer. The lidar transmitter uses a $355nm$ diode-pumped solid state Nd:YAG laser and two narrow-band interference filters in the receiver filter out rotational Raman lines that are dependent on temperature. The spectral response of the lidar is calibrated using a monochromator. Temperature profiles measured by the system are calibrated by comparison to co-located radiosondes.

1 INTRODUCTION

Pollution transport within and out of the atmospheric boundary layer (ABL) is a topic of widespread interest in atmospheric science. The development of the ABL during the day is well understood, but at night the dynamics is more complex, and recreating the evolution of the nocturnal boundary layer (NBL) with a computer model can be difficult. To help understand the dynamics of the NBL, a reliable way to measure temperature is required.

A temperature lidar has been developed and calibrated to measure temperature in the NBL over Manchester, United Kingdom. The lidar utilises a low-powered diode-pumped solid state (DPSS) Nd:YAG laser which is cheaper to run and more reliable than flashlamp-pumped lasers. In the receiver, the light is collected using photomultiplier tubes (PMTs) after passing through narrowband interference filters to isolate the rotational Raman scattering signal from the elastic scattering.

To calibrate the lidar, detailed measurements were carried out using a high resolution monochromator as well as launching radiosondes.

2 METHODOLOGY

2.1 Theory

The lidar system used in this study relies on measuring rotational Raman scattering of the transmitted laser beam. The lidar uses a low power laser so it can only be operated at night, as the solar background obscures the signal during the day. Figure 1a shows the rotational Raman spectra for nitrogen and oxygen at $270K$. As the temperature increases, the distribution in intensity of the Rotational Raman lines shifts further away from the transmitted wavelength. By selecting the wavelengths shown by the filter profiles it is possible to relate the ratio of the light passing through the "near" and "far" filters to the temperature of the atmosphere [1]. This ratio is shown in Figure 1b.

2.2 Transmitter

The lidar transmitter is a diode-pumped solid state Nd:YAG WAVE-Y-1000-355 laser built by Elforlight Ltd which is tripled to produce a wavelength of $355nm$. The laser has a pulse energy of $240\mu J$ and pulse repetition frequency (PRF) of $1000Hz$. Using a low powered laser with high PRF allows measurements in the

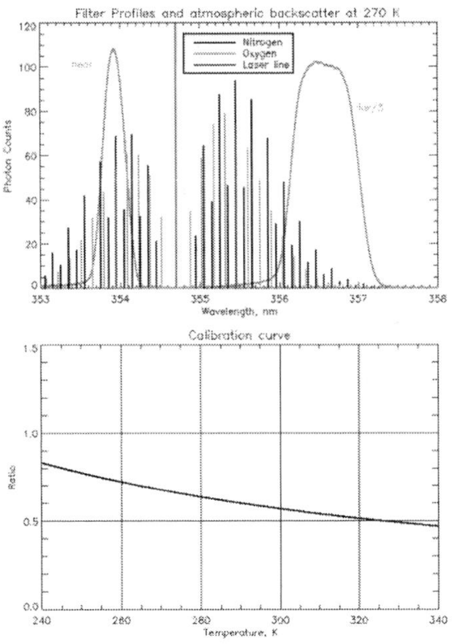

Figure 1: (a) Nitrogen and Oxygen rotational Raman spectral lines at 270K. The spectral responses of the two Raman receiver channels are overlaid. (b) Lookup table to convert the ratio of the Raman channels to temperature.

boundary layer without saturation of the detectors. The transmission optics expand the beam by a factor of $10x$ to minimise divergence and to comply with eye-safety.

2.3 Receiver

The backscattered light is collected using a $60cm$ diameter Nasmyth-Cassegrain type telescope and diverted into a beam splitter which splits the light into three interference filters: a "near" filter (a narrowband filter to detect rotational Raman spectra close to the transmission wavelength), a "far" filter (another narrowband filter to detect rotational Raman spectra further away from the transmission wavelength); the remaining light is passed though a broader $355nm$ filter to obtain the elastic backscatter. The light from each of these filters is passed through a spatial filter to minimise background noise and then measured using a PMT.

Figure 1a shows the narrowband filters which were chosen after modelling which wavelengths would provide the most sensitivity to the change in temperature. Initially, both near and far filters were chosen on the Stokes side of the elastic backscatter line. However, the near filter showed signs of degradation and was replaced by a filter on the anti-Stokes side, so any future degradation would not lead to breakthrough of the elastic signal.

3 RESULTS

3.1 Calibration

To calibrate the lidar system, a halogen lamp was focussed on the main aperture of the telescope. This allowed the whole system to be calibrated simultaneously, including the beam splitter and collimators. The resulting receiver response channels are shown in Figure 1a. The near and far filters have a central wavelength of $353.988nm$ and $356.485nm$ with a full width at half maximum value (FWHM) of $0.315nm$ and $0.783nm$, respectively.

Figure 1a shows the receiver calibration using unpolarised light. In fact, because the Rotational Raman scattering is 57% polarised parallel to the laser it was necessary to calibrate the receiver separately for polarisations parallel and perpendicular to the laser and combine these in the proportion $4:3$ to obtain the appropriate calibration curve.

For accurate temperature measurements (better than $1K$) the spectral calibration needed to be supplemented by an adjustment derived from a co-located radiosonde, using a Windsond system which provided a vertical profile of pressure, temperature, humidity and wind. This

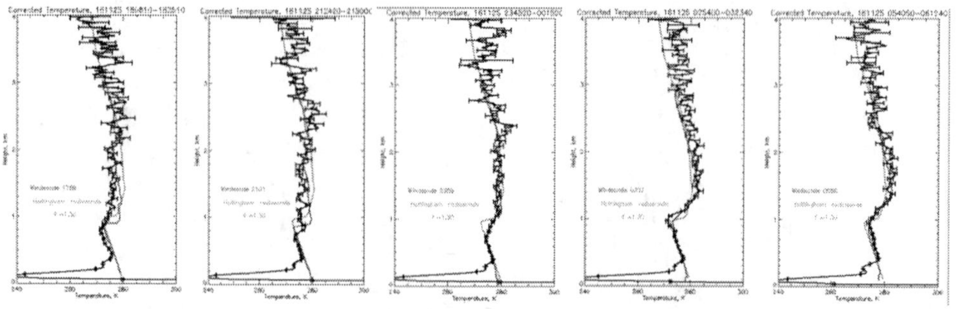

Figure 2: The lidar-derived temperature coinciding with the five Windsond ascents on the night of 24th/25th November 2016 [Times in UTC]. The data has been overlap-corrected. The lidar data is in black with error bars. The red and green lines are the co-located Windsond and the Nottingham routine Met Office radiosonde ascents, respectively.

also allowed an overlap correction to be derived, which was needed with this system below $1km$. A total of 17 sondes were launched in the winter of 2016/17. This enabled a direct comparison of temperatures calculated from the lidar data and the sondes, after using one of the sondes to adjust the calibration.

On the night of 24th/25th November 2016 five sondes were launched throughout the night to check whether the lidar experienced any drift overnight. Figure 2 shows the five sonde ascents and their corresponding lidar profiles. No significant drift is visible between ascents. The overlap correction for the lidar data was calculated using the 00UT sonde ascent and demonstrates that temperature can be retrieved between the heights of $0.5km$ and $2-3km$, after the overlap correction.

3.2 Preliminary lidar observations

Data presented in Figures 3 and 4 were collected using $15m$ range gates, with a time resolution of $30s$. These were vertically smoothed using a moving boxcar average of $75m$ and temporally smoothed with a boxcar of width 5 minutes. The aerosol indicator plotted in Figure 4 was produced by taking the ratio between the corrected count rates of the elastic and far-field channel signals. Below $400m$ there is an error in data caused by saturation of the elastic channel. Data uncertainty was determined by taking the Poisson noise arising from the PMTs from each of the channels.

Figures 3 and 4 show temperature and aerosol measurements from the lidar on the night of 24th/25th November 2016, respectively. In both figures some low-level intermittent cloud was obscuring the lidar beam at about 900 metres between 18UT and 24UT. When cloud is not present, Figure 3 clearly shows the inversion at about 800 metres at the beginning of the night. The inversion then increases to about $1km$ by the end of the night. This is corroborated by the increase in the height of the aerosol in the residual layer observed in Figure 4.

A more quantitive analysis of the temperature data in Figure 3 shows an increase in temperature above the inversion of about $7-8K$, which is also observed in the sonde launches in Figure 2. The data also show a slight increase in temperature of about $2K$ at about 500 metres. This is not observed in the sonde data and is under further investigation.

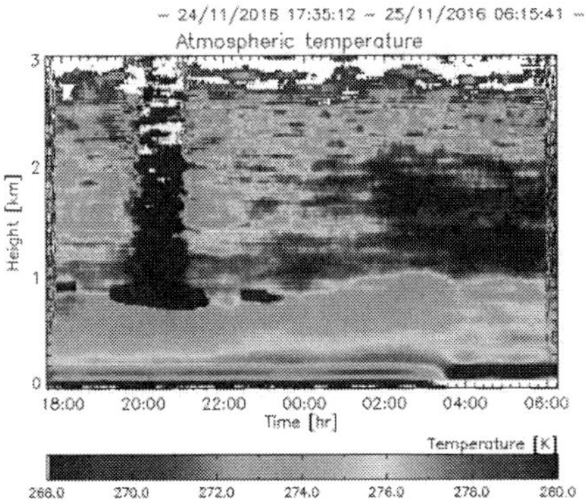

Figure 3: Plot of temperature above the lidar for 24th/25th November 2016. Some low-level cloud is visible between 18UT and 24UT at about 900 metres as well as the after 04UT at 200 metres.

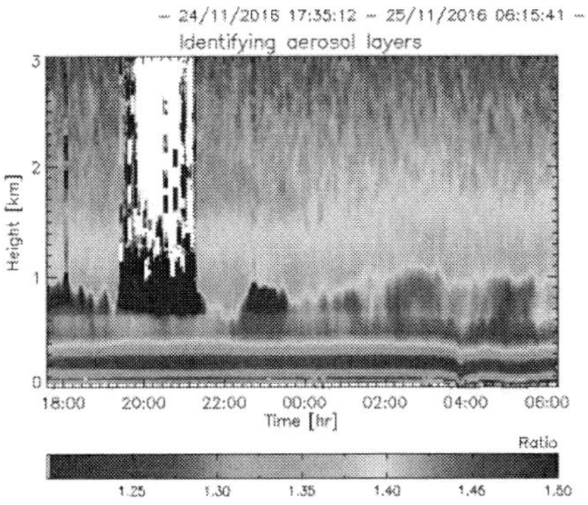

Figure 4: Plot of the ratio of elastic to far channel backscatter above the lidar for 24th/25th November 2016. This functions as an indicator of aerosol and shows the lifting of the inversion overnight.

4 CONCLUSIONS

We have shown that a boundary-layer temperature lidar based on a DPSS laser is capable of measuring the evolution of the boundary-layer inversion over the course of the night over the city of Manchester in the United Kingdom, with the lidar temperatures comparing well with a series of co-located radiosondes. The evolution is also consistent with the aerosol backscatter profiles as measured by the elastic channel. Further refinements to the lidar will enable the measurements to extend lower in the atmosphere and follow the evolution of residual layers within the boundary layer.

ACKNOWLEDGEMENTS

Simon Holloway wold like to thank the Natural Environment Research Council of the United Kingdom for his PhD funding (NERC Grant: NE/L501591/1). Professor Geraint Vaughan and Dr Hugo Ricketts would like to thank the National Centre for Atmospheric Science (NCAS) of the United Kingdom.

References

[1] Vaughan, G., Wareing, D. P., Peplar, S. J., Thomas, L., Mitev, V., 1993: Atmospheric Temperature Measurements made by Rotational Raman-scattering, *Appl. Opt.* **32**, 2758-2764.

LIDAR STUDY OF WIND TURBULENCE, LOW LEVEL JET STREAMS, AND ATMOSPHERIC INTERNAL WAVES IN THE BOUNDARY LAYER OF ATMOSPHERE

Viktor Banakh[1]*, Igor Smalikho[2]

[1,2]*Institute of Atmospheric Optics of Siberian Branch of the Russian Academy of Sciences, Russia,*
**banakh@iao.ru*

ABSTRACT

The results of lidar study of wind turbulence, low level jet streams, and internal atmospheric waves in the stable boundary layer of atmosphere on the coast of Lake Baikal are presented. Few events of the atmospheric internal waves (AIWs) were registered during the experimental campaign. All the registered AIWs were observed in the presence of low level jet streams. Two dimensional time–height patterns of the wind turbulence dissipation rate during AIW events were obtained as well.

1 INTRODUCTION

Wind turbulence, low level jet streams, and atmospheric internal waves (AIWs) were studied on the coast of the Lake Baikal in summer campaigns of 2015 and 2016 with the use of a Stream Line lidar (Halo Photonics). The lidar was mounted 340 m away from the coast of the lake, on the territory of the Baikal Astrophysical Observatory of Institute of Solar-Terrestrial Physics, Siberian Branch, Russian Academy of Sciences (51°50'47.17"N, 104°53'31.21"E); the lidar altitude above the lake level was 180 m. The analysis of the lidar data processing results shows several AIW events occurred in the presence of jet streams. Two jet streams were observed simultaneously during some AIW events. Using the initial data of these measurements (arrays of radial velocities and signal-to-noise ratios (SNR)) we also retrieved the time–height patterns of the turbulence kinetic energy dissipation rate and calculated the relative error of assessment of the dissipation rate from the lidar data.

2 METHODOLOGY

Stream Line lidar measurements were carried out under the following parameters: elevation angle $\varphi = 60°$; number of pulses used for data accumulation $N_a = 3000$; angular speed of probing beam scanning $\omega_c = 10°/\text{s}$; measurement time per scanning of 36 s; resolution in the azimuth angle $\Delta\theta = 2°$, and the number of rays per scanning $M = 180$.

To retrieve the vertical profiles of the wind speed U, the wind direction angle θ_V, and the vertical component of the wind velocity V_z, the method of filtered sine wave fitting was used [1, 2]. This method allows estimation of the mean wind velocity with acceptable accuracy from very noisy arrays of radial velocities $\hat{V}_r(R_k, \theta_m, n)$ containing up to 50% of bad estimates $\hat{V}_r$. Here $R_k = R_0 + k\,\delta R$ is the distance, R_0 is the minimal sounding range, $k = 0, 1, 2, ..., K$, $\delta R = 3$ m, $\theta_m = m\Delta\theta$ is the azimuth angle, $m = 1, 2, 3, ..., M$, n is the number of current scan. As shown in [3] for N_a=3000, the minimal signal to noise ratio is 0.005 (–23 dB), under which the wind speed estimation error does not exceed 0.1 m/s.

The turbulence kinetic energy dissipation rate ε is one of key parameters of the wind turbulence. The turbulence structure in its inertial interval is completely determined by this parameter and obeys the Kolmogorov law [4]. To assess the turbulence energy dissipation rate ε, we used the method of transverse (azimuth) structure function of the radial velocity measured by the lidar during conical scanning [5]. Therefore, a high resolution in the azimuth angle was important during the measurements. At the same time, the tangential speed of the volume sounded movement $V_t = \omega_c R_k \cos\varphi$ should be much more than the mean wind speed at the measurement altitude. For correct estimate of the dissipation rate ε from the array $\hat{V}_r(R_k,\theta_m,n)$, the probability of bad estimation of the radial velocity P_b should be equal to zero. The minimal SNR, at which the lidar estimate of the dissipation rate by this method still satisfactorily agrees with the assessments of ε from acoustic anemometer data, is equal to 0.05 [5].

The time–height patterns of the turbulence energy dissipation rate $\varepsilon(h,t)$ were calculated with a height step of 26 m and a time step of 36 s from the array of radial velocities $\hat{V}_r(R_k,\theta_m,n)$ measured with the use of data received for $N = 25$ scannings (measurement time is 15 min) for each vertical profile of ε. For each height *h*, the SNR values from the array of $\hat{\mathrm{SNR}}(R_k,\theta_m,n)$ estimates were averaged over all the azimuth angles θ_m and over $N = 25$ scannings. Then the relative error of the radial velocity σ_e was calculated by the equation $\sigma_e = \left(1+\frac{1}{\mathrm{SNR}}\right)\frac{\Delta v}{\sqrt{N_a}}$, where $\Delta v = 0.4$ m/s; it was used for calculation of the relative error of the dissipation rate E_ε.

3 RESULTS

The Stream Line lidar measurements on August 23, 2015, showed an AIW with a period of 9 min and longitudinal and vertical components of the velocity of about 1 m/s and 0.3 m/s, respectively, which was propagating through the region of experiment for almost 6 h in the presence of two jet streams. The initial data measured by the lidar from 11:00 to 17:00 on August 23, 2015, satisfied the condition SNR > 0.05 up to 500 m. We limited to this altitude in the retrieval of the vertical profile of the dissipation rate, while the profiles of the wind velocity components were retrieved up to height of about 900 m.

Figure 1 shows the time–height patterns of the wind speed $U(h,t)$, vertical component of the wind velocity $V_z(h,t)$, $\varepsilon(h,t)$, $\mathrm{SNR}(h,t)$, $\sigma_e(h,t)$, and $E_\varepsilon(h,t)$. The SNR is below a threshold of 0.005 in the black regions.

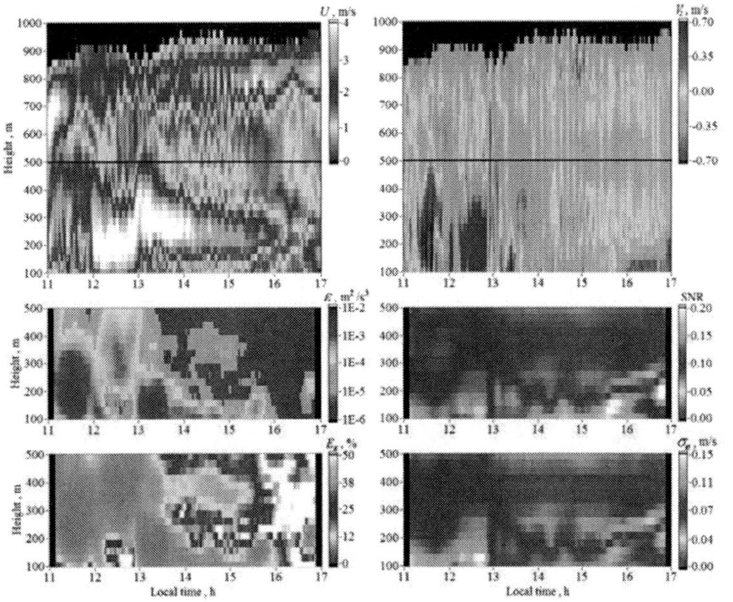

Figure 1 Time–height patterns of the wind speed U, vertical component of the wind velocity V_z, turbulence energy dissipation rate ε, SNR, relative error of the dissipation rate E_ε, and the radial velocity error σ_e retrieved from the Stream Line lidar measurements on the coast of the Lake Baikal on August 23, 2015

SNR varied from 0.05 go 0.2 in the 100–500 m layer during that measurement. The radial velocity error does not exceed 0.15 m/s. The data on the wind speed in Fig. 1 clearly show the presence of two jet streams with speed maxima at altitudes of about 250 and 750 m, respectively (hereinafter, the altitudes are given relative to the lidar location point). The directions of these jet streams are almost perpendicular to each other (the bottom jet stream was directed from north to south through mountains, and the top jet stream was directed from east to west, i.e., from the lake side) [3]. The atmospheric conditions promoted generation of an AIW. Near-harmonic oscillations of $U(h,t)$ and $V_z(h,t)$ are clearly seen in Fig. 1 from 14:20 to 15:00. The data on the turbulence energy dissipation rate in Fig. 1 show that $\varepsilon(h,t)$ takes values from a wide range, attaining a maximum of about 0.006 m^2/s^3 before the generation of a jet stream. Relatively strong wind turbulence is also observed under the jet stream. The turbulence is quite weak inside the jet stream, including during an AIW event.

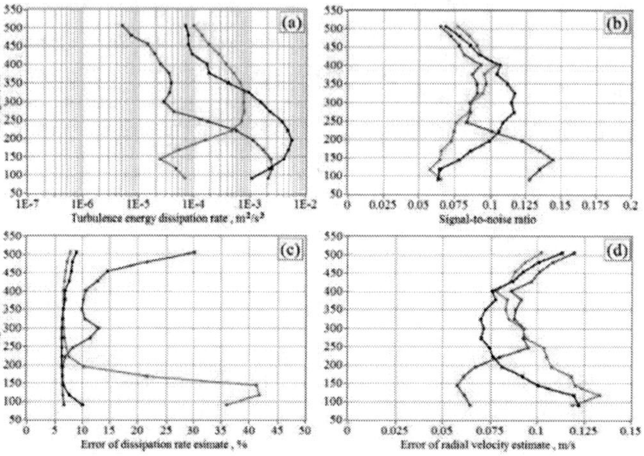

Figure 2 Vertical profiles of the turbulence energy dissipation rate (a), SNR (b), relative error of the dissipation rate estimate (c), and the radial velocity error (d) retrieved from the lidar measurements at 11:30 (black curve), 12:30 (red curve), and 13:30 (blue curve). Data are taken from Figure 1

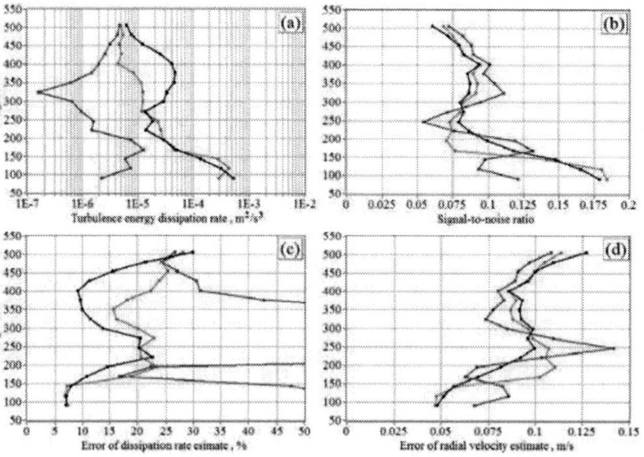

Figure 3 Vertical profiles of the turbulence energy dissipation rate (a), SNR (b), relative error of the dissipation rate estimate (c), and the radial velocity error (d) retrieved from the lidar measurements at 14:30 (black curve), 15:30 (red curve), and 16:30 (blue curve). Data are taken from Figure 1

Figures 2 and 3 show the vertical profiles of ε, SNR, E_ε, and σ_e retrieved from Stream Line lidar measurements with an interval of 1 h (the data are taken from Fig. 1).The turbulence energy dissipation rate obviously can be retrieved from lidar measurements with an error less than 30% at $\varepsilon \sim 5\cdot10^{-6}$ m^2/s^3. The SNR should be no less than 0.075 in this case. The turbulence spatial structure should include an inertial interval with the upper boundary of no less than a half of the longitudinal size of the sounded volume Δz ($\Delta z = 30$ m for the Stream Line lidar).

4 CONCLUSIONS

The results of lidar study of wind turbulence, low level jet streams, and internal atmospheric waves in the boundary layer of atmosphere on the coast of Lake Baikal are presented.

For estimation of the mean wind velocity the modified method of filtered sine wave fitting [2, 3] was used. The method modified allows a 1.5-fold increase in the altitude of retrieval of the mean wind velocity profiles from lidar data as compared to the traditional sine wave fitting.

The method of azimuth structure function [5] was applied for estimation of the turbulent energy dissipation rate from radial velocities measured by the lidar with the use of conical scanning by probing beam around the vertical axis. The validity of this method depends on the magnitude of the dissipation rate measured and the lidar signal to noise ratio. For minimal possible values of the dissipation rate, the method is valid if the signal to noise ratio of lidar echo signal exceeds 0.24.

Few events of the internal atmospheric waves were registered during the experimental campaigns. All the registered AIWs were observed in the presence of low level jet streams. The periods and amplitudes of wave variations of wind velocity during AIWs and parameters of jet streams were determined.

Two dimensional time–height patterns of the wind turbulence dissipation rate during AIW events were obtained as well. It is shown that the wind turbulence in the area occupied by jet streams is very weak. Possibly, it is absent at all, and wind velocity variations in the area of jet streams are caused by mesoscale atmospheric processes. In the process of dissipation of AIWs the wind turbulence strength increases.

ACKNOWLEDGEMENTS

This study was supported by the Russian Foundation for Basic Research, Project No. 16-42-700072.

References

[1] Banakh, V.A., and Smalikho, I.N., 2013: Coherent Doppler Wind Lidars in a Turbulent Atmosphere, *Artech House*, Boston and London, 248 pp.

[2] Banakh, V.A., Smalikho, I.N., 2016: Wind sensing in an atmospheric boundary layer by means of micropulse coherent Doppler lidars, Opt. Spectrosc. **121** (1), 1140–1147.

[3] Banakh, V.A., Smalikho, I.N., 2016: Lidar observations of atmospheric internal waves in the boundary layer of atmosphere on the coast of Lake Baikal, *Atm. Meas. Tech.* **9** (10), 5239–5248.

[4] Kolmogorov, A.N., 1941: Local structure of turbulence in incompressible viscose fluid at very large Reynolds numbers, *Doklady AN SSSR* **30** (4), 299-303 [in Russian].

[5] Smalikho, I.N., Banakh, V.A., Falits, A.V., Rudi, Yu.A., 2015: Estimation of the turbulent energy dissipation rate from the data measured by the "StreamLine" lidar in the surface layer of atmosphere, *Atmos. Ocean. Opt.* **28** (10), 901–905 [in Russian].

THE MARBLL EXPERIMENT: TOWARDS A MARTIAN WIND LIDAR

Anni Määttänen[1]*, François Ravetta[2], Franck Montmessin[1], Didier Bruneau[1], Jean-François Mariscal[1], Mathilde Van Haecke[1], Guillaume Fayolle[1], Christophe Montaron[1], David Coscia[1]

*[1]LATMOS/IPSL, UVSQ Université Paris-Saclay, UPMC Univ. Paris 06, CNRS, Guyancourt, France, *anni.maattanen@latmos.ipsl.fr*
[2]LATMOS/IPSL, UPMC Univ. Paris 06 Sorbonne Universités, UVSQ, CNRS, Paris, France

ABSTRACT

Operating a lidar on Mars would fulfill the need of accessing wind and aerosol profiles in the atmospheric boundary layer. This is the purpose of the MARs Boundary Layer Lidar (MARBLL) instrument. We report recent developments of this compact direct-detection wind lidar designed to operate from the surface of Mars. A new laser source has been developed and an azimuthal scanning capability has been added. Preliminary results of a field campaign are presented.

1 INTRODUCTION

The LATMOS laboratory has been developing since 2008 a compact and robust direct-detection wind lidar based on a Mach-Zehnder interferometer concept and designed to operate from the surface of Mars. MARBLL uses a mature state-of-the-art wind lidar technology, including a Nd:Yag laser emitting at 1064 nm. The Mach-Zehnder interferometer is used as a spectral analyzer. The wind measurement, as for all Doppler wind lidars (DWLs), is based on measuring the Doppler shift induced by the radial velocity of aerosols backscattering laser photons to the telescope. Traditionally the DWLs use methods based on quasi-monomode laser emission and require precise frequency locking between the emitter and the analyzer. Such methods have not yet been demonstrated during a space mission. The original MARBLL design included a laser inherited from the ChemCam instrument [1] on the Mars Science Laboratory Curiosity rover. Lidars are perfectly well suited to the dusty conditions encountered on Mars, with suspended dust particles highly reflective at the laser wavelength (ω_0=0.98). Bruneau et al. (2013) [2] describe the instrument, the optical design and the results of the first field campaign at the OHP site (Observatoire de Haute-Provence, France).

2 METHODOLOGY

2.1 Description of the campaign

MARBLL was set up for a week at the Gérard Mégie station of the OHP site in southeastern France in September 2016. The station performs continuous geophysical measurements including atmospheric balloon radio soundings (p, T, z, **V**) and provides a platform for instrumental campaigns, such as ours. The OHP site (5°43'E, 44°56'N) hosts also a mast (ICOS project, www.icos-ri.eu) that provides, for example, wind measurements at three heights (10, 50 and 100 meters) and a compact automated aerosol lidar (CE376, www.cimel.fr) working at 532 nanometers. These measurements can be compared with MARBLL winds and particle backscatter profiles.

The original Mach-Zehnder interferometer (MZI) was designed specifically to match the optical requirements of the ChemCam laser EDU 1 (the optical path of the prisms inside the MZI is aligned with the laser cavity). The new set-up of the instrument uses another laser source (for all laser parameters see Table 1). The new laser, E3M, has been conversely designed in-house to fit the parameters required by the heritage interferometer. The mechanical design includes the possibility for a motorized scan over a desired range of azimuth angles and a manual zenithal angle variation. The electronics design was modified to include a correction enabling the scanning sequence.

The campaign was focused on validating the portability of the instrument and the consequent optical mounting and re-alignment, the mechanical design and the electronics design that pilots and synchronizes the scan and measurement sequences.

Table 1 E3M and EDU1 laser specifications

Laser	E3M (in-house)	EDU1 (MSL heritage)
Crystal	Nd:YAG	Nd:KGW
Wavelength	1064 nm	1067 nm
Energy	12,5 mJ	>24 mJ
Pulse duration	4.4 ns	<8 ns
Repetition rate	100 Hz	1-10 Hz

2.2 Performed measurement sequences

We performed three types of measurements:

1. Simple zenithal or oblique measurements;
2. VAD (Velocity-Azimuth Display) scans of different durations at 45° zenithal angle with different azimuthal angle steps within the 180° total scan range allowed at the site;
3. DBS (Doppler Beam Swinging) measurements (3-5 azimuthal directions with different zenithal angles).

The laser was operated at 100 Hz frequency and the measurement integration times were varied. Post-processing tests were made to find the optimal averaging of data: in the results presented here a 10s averaging has been used (1000 laser shots were averaged for one profile) and data with signal-to-noise ratio SNR<10 were discarded.

3 RESULTS

We present here the MARBLL measurements made with static pointing and also during a VAD sequence. Results are compared with other concomitant measurements (see below). The DBS sequences are yet to be analyzed and are therefore not presented here.

3.1 Wind profile comparisons

The first simple oblique measurements were performed at the beginning of the campaign (September 14th). Measurements were made at a 45° zenith angle during 15 minutes towards North and then East. This gave us the opportunity for a first quick comparison of the MARBLL wind measurements with a radiosonde launched simultaneously and with the wind sensors of the ICOS mast. The results for the North wind component measured by MARBLL simultaneously with the radio sounding are shown in Figure 1. The comparison shows a fair agreement within the ABL where sufficient amounts of aerosols allow MARBLL to collect enough signal exceeding the SNR 10 threshold.

Figure 2 shows a similar measurement for the Eastward component from a measurement sequence performed two days later (September 16th, see section 3.2). The wind profiles are analyzed from measurements performed with the lidar pointing towards West at 45° zenith angle for one minute. A radio sounding was launched at around the same time and allowed further comparison in addition to the ICOS data. The comparison is good overall, but the ICOS mast indicated much stronger winds at 100m than both the radio sounding and MARBLL, whereas around 50m the agreement between the three measurements seems to be better.

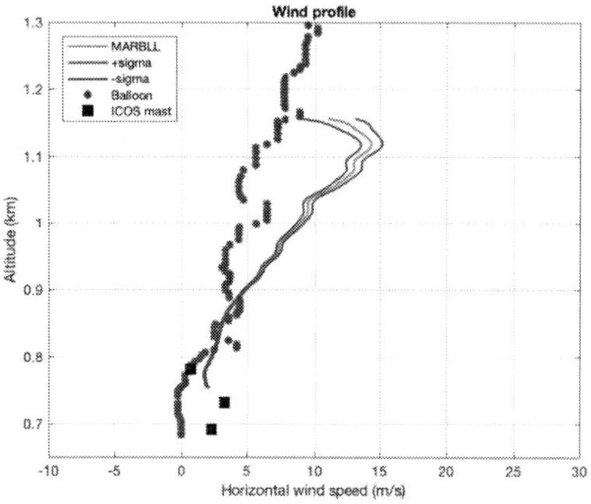

Figure 1: A comparison of the northward wind component profile measured on September 14, 2016 by MARBLL (green line) with simultaneous radio soundings (blue asterisks, smoothed over 15 points) and the ICOS anemometers (black squares). Altitude is the altitude above sea level (OHP is at 681 m asl). MARBLL measurements with SNR<10 and radial wind standard deviation>5m/s are not displayed.

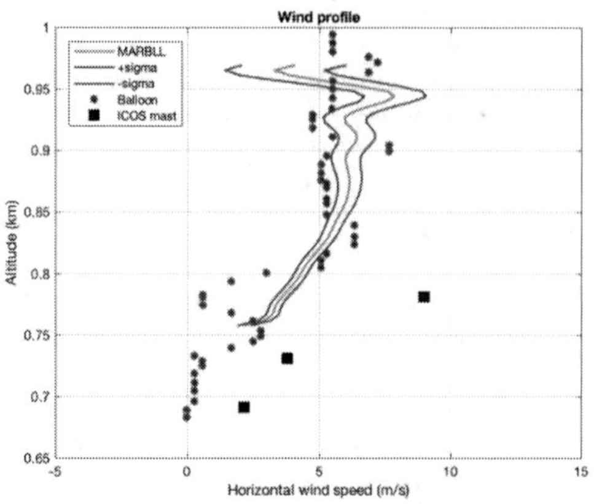

Figure 2: Same as Figure 1, but for the eastward wind component on September 16th.

Figure 3 compares the MARBLL measurements obtained 100 m above ground level, the same elevation as the ICOS wind sensors. The agreement is particularly convincing. Note that this plot also shows the variability of the wind field in the boundary layer. In such turbulent conditions, no perfect match can be expected when comparing MARBLL to radio soundings.

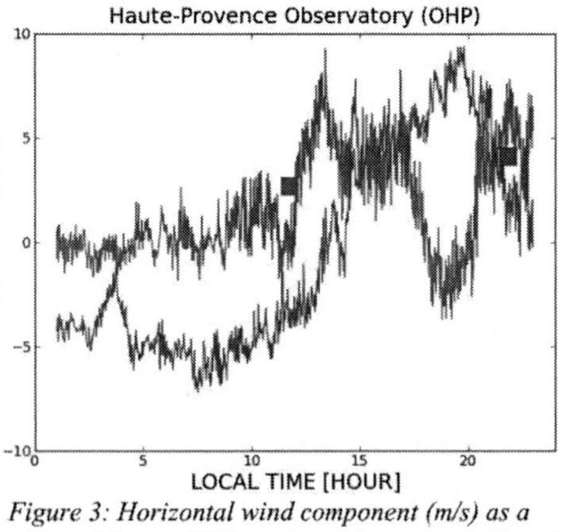

Figure 3: Horizontal wind component (m/s) as a function of time at OHP on September 14 (meridional wind in red) and on September 16 (zonal wind in green). Continuous lines are the measurements from the ICOS mast (100 m above ground level), squares are MARBLL measurement at the same altitude.

3.2 180° VAD sequences

The first validated, full 180° VAD sequences were performed on September 16th with two adjacent radio soundings. During the first sequence, MARBLL scanned the atmosphere at 45° zenithal angle from West (through North) to East with 3° intervals with 60s integration time per direction. We fit the data to extract the expected sinusoidal behavior of the wind and thus access both the wind speed and direction along the vertical [3].

The VAD analysis was performed only in the very shallow aerosol-laden ABL between about 70 m to 520 m (0.75-1.2 km asl). At higher altitudes, the error on individual wind measurement gets significant, but the sinusoidal behavior can still be fitted (not shown).

Wind speed and direction are shown in Figure 4 together with the radio soundings. The profiles reveal similar tendencies, but in the lowest 200 meters the balloon measurements seem perturbed by the oscillating motion of the sounder (not shown). Above about 1 km the MARBLL errors increase significantly. Some assumptions regarding the properties of the background wind field (both stationary and homogeneous) are intrinsically made in the VAD analysis. MARBLL VAD thus provides an average view of the wind field from which radio soundings can depart substantially given the localized nature in both space and time of their measurements.

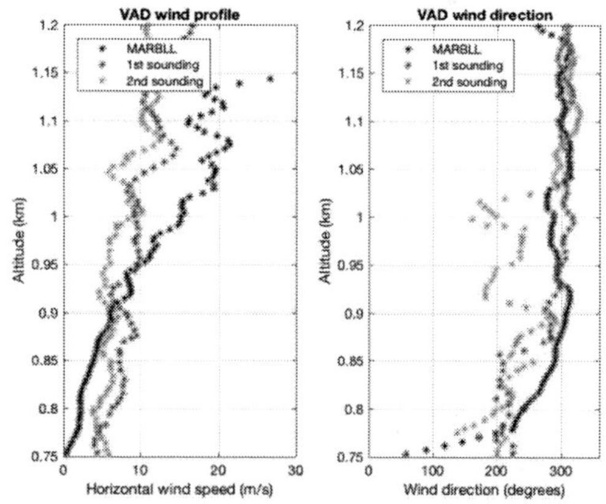

Figure 4: Profiles of wind speed (left) and direction (right) extracted from the VAD analysis and two radio soundings (1st in the beginning of the MARBLL sequence, 2nd 30 minutes after the end of the sequence). MARBLL errors increase significantly above 0.95 km.

3.3 Aerosol layer and cloud detection

The Phoenix lidar has been so far the only lidar on the Martian surface. Phoenix lidar made pioneering observations of precipitating water ice clouds in the polar atmosphere [4]. Since one of the goals on MARBLL on Mars will be cloud detection and aerosol profile measurements from the surface, the last morning of the campaign was devoted to performing a long static zenithal measurement sequence that was started slightly before sunrise and was continued during three hours to monitor the development of the boundary layer. Unfortunately, the campaign week was very rainy, which reduced greatly the aerosol concentrations. Thus, direct comparison of aerosol profile is not possible. However, clouds were detected during this sequence. Comparison of the partly precipitating cloud structures, as seen by both lidars (MARBLL on Figure 5, ICOS on Figure 6), shows great consistency.

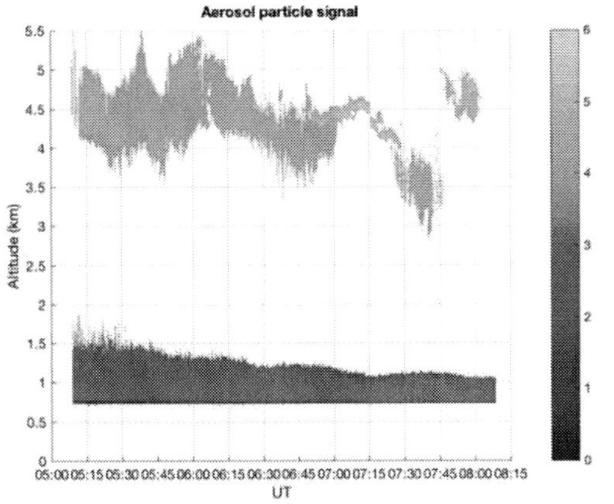

Figure 5: MARBLL range corrected signal @ 1064 nm (arbitrary units, SNR > 10) as a function of altitude from 5 UT to 8 UT (sunrise at 5:20 UT).

4 CONCLUSIONS

The OHP MARBLL validation campaign suffered from the aerosol-poor conditions at the OHP site during the rainy campaign week, limiting the signal to a very shallow layer near the ground. Nevertheless, several key measurements (wind speed and direction, cloud geometry) were made and compared to radio soundings, wind measurements on a mast and aerosol lidar data. Further comparison of MARBLL with another wind profiler (radar or lidar) is foreseen. The OHP campaign allowed us to validate several key aspects of the instrument in "real" (although terrestrial) conditions: in particular, the VAD scan capability has been validated, in line with expectations made for possible Mars operations. The OHP campaign set the major milestones of the MARBLL development roadmap for 2017 and beyond. A new measurement campaign is planned in dustier conditions to better assess the instrument performances in Mars-like conditions.

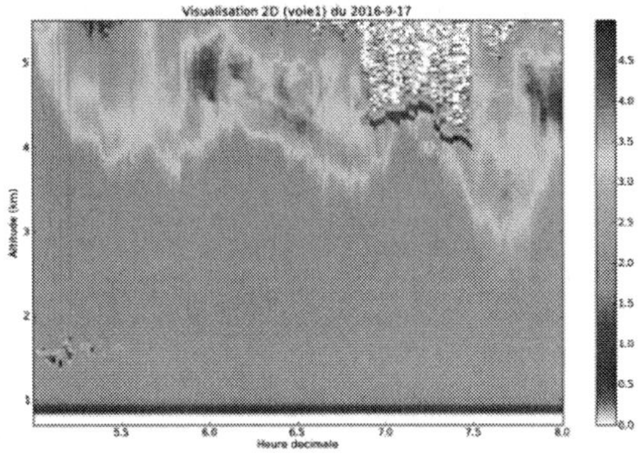

Figure 6: ICOS lidar range-corrected signal (arbitrary units) @532 nm vs. altitude from 5 to 8 UT.

ACKNOWLEDGEMENTS

We are grateful to Pierre-Eric Blanc and Ilja Reiter for providing access to the ICOS data, and to the entire staff of the Gérard Mégie station for their support. We also thank the FR 3098 ECCOREV and the OSU PYTHEAS. We thank the French Space Agency CNES for their support to the development of MARBLL.

References

[1] Maurice, S., Wiens, R.C., Saccoccio, M. et al. Space Sci Rev (2012) 170: 95. doi:10.1007/s11214-012-9912-2

[2] Bruneau D., et al., Direct-detection wind lidar operating with a multimode laser, Appl. Opt. 52, 4941-4949 (2013).

[3] Werner, C., chapter 12 in "Lidar", Vol. 102, Springer Series in Optical Sciences, pp 325-354

[4] Whiteway, J. A. et al., Mars Water-Ice Clouds and Precipitation, Science 325 (5936), pp. 68-70, DOI: 10.1126/science.1172344.

EPJ Web of Conferences **176**, 06007 (2018) https://doi.org/10.1051/epjconf/201817606007
ILRC 28

PLANETARY BOUNDARY LAYER HEIGHT VARIABILITY OVER ATHENS, GREECE, BASED ON THE SYNERGY OF RAMAN LIDAR AND RADIOSONDE DATA: APPLICATION OF THE KALMAN FILTER AND OTHER TECHNIQUES (2011-2016)

Dimitrios Alexiou[1], Panagiotis Kokkalis[1], Alexandros Papayannis[1]*, Francesc Rocadenbosch[2,3], Athina Argyrouli[1], Georgios Tsaknakis[1], and Chris G.Tzanis[4]

[1]*National Technical University of Athens, Laser Remote Sensing Unit, Physics Department, Zografou, Greece, Email: apdlidar@central.ntua.gr*
[2]*Remote Sensing Laboratory (RSLab), Department of Signal Theory and Communications, Universitat Politècnica de Catalunya, E-08034 Barcelona, Spain*
[3]*Institute for Space Studies of Catalonia–Aeronautics and Space Research Center, Universitat Politècnica de Catalunya, E-08860 Barcelona, Spain*
[4]*Division of Environmental Physics and Meteorology, Department of Physics, National and Kapodistrian University of Athens, Athens, Greece*

ABSTRACT

In this paper we studied the temporal evolution of the Planetary Boundary Layer height (PBLH) over the basin of Athens, Greece during a 5-year period (2011-2016) using data from the EOLE Raman lidar system. The lidar data (range-corrected lidar signals-RCS) were selected around 12:00 UTC and 00:00 UTC for a total of 332 cases: 165 days and 167 nights. Extended Kalman filtering techniques were used for the determination of the PBLH. Moreover, several well established techniques for the PBLH estimation based on lidar data were also tested for a total of 35 cases. Comparisons with the PBLH values derived from radiosonde data were also performed. The mean PBLH over Athens was found to be of the order of 1617±324 m at 12:00 UTC and of 892±130 m at 00:00 UTC, for the period examined. The mean PBLH growth rate was found to be about 170±64 m h^{-1} and 90±17 m h^{-1}, during daytime and nighttime, respectively.

1. INTRODUCTION

The Planetary Boundary Layer (PBL) is the lowest part of the troposphere that is strongly influenced directly by the presence of the Earth's surface and responds to surface forcing with a timescale of about 1 hour or less [1]. The knowledge of the PBLH is very important because air pollutants are trapped within the PBL, thus affecting human health, and atmospheric modeling needs this information to provide air pollution forecasts.

Atmospheric aerosols are present mostly in the lower troposphere where they play a crucial role in the Earth's climate [2]. In the lidar technique aerosols can be used as tracers for the atmospheric motion and for the study of the PBL structure [1].

2. METHODOLOGY

Several criteria are currently used to retrieve the PBL height from radiosonde and lidar data, or even from sodar data [3-5]. In this study we will calculate the PBLH (or mixing height) over Athens, Greece, using the extended Kalman filter (KF) technique.

2.1. THE EXTENDED KALMAN FILTER APPROACH

The Kalman filter is an adaptive filter inherited from classic control theory [6] that enables the state vector of a dynamic linear system to be estimated and tracked with time (e.g., position coordinates of an aircraft). This filter can also be applied to non-linear systems - as is the case here - via linearization, which gives rise to the Extended Kalman Filter (EKF) technique. The filter operates by minimizing the error between the estimated and the true state vector in a mean-square error sense over time. Because the filter makes use of not only present information (the measurements), but also past estimates, as well as related covariance statistics, it provides an optimal solution over time. Recently, Lange et al., [7], [8], departing from previous works of [9-10], has successfully applied the EKF to estimate the daytime PBLH from tropospheric backscatter lidar and radar signals (returns), respectively. In what follows, we stick to the algorithm and notation given in [7-8]:

In lidar applications, the EKF uses range-corrected backscatter returns, $U(R)$, at successive discrete times, t_k (in what follows, the "observables", $\mathbf{z}_k$, k a reminder of discrete time; formally, $\mathbf{z}_k(R)$, R omitted for brevity) as a proxy of the total backscatter coefficient and, in turn, of the atmospheric (aerosol) load. Central to the method is the assumption of an abrupt mixing-layer (ML)-to-free-troposphere (FT) transition in $U(R)$, which is modeled by an erf-like function (Fig. 1). The morphology of the erf function of Fig. 1 gives rise to the state vector,

$$\boldsymbol{x_k} = \left[R_{bl,k}, a_k, A_k, c_k\right]^t \qquad \text{Eq.(1)}$$

to be estimated at each recursive loop of the filter.

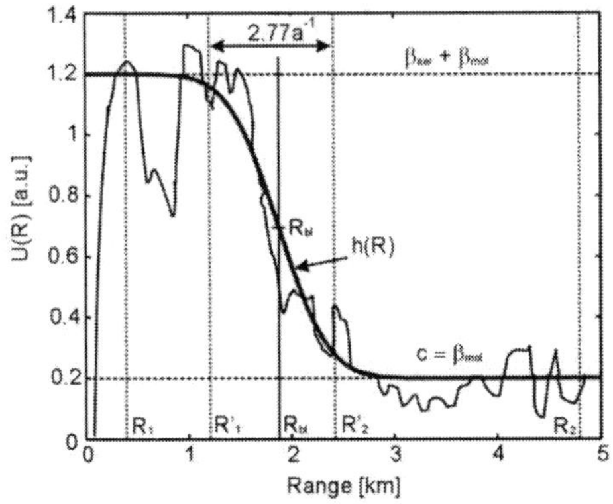

Figure 1. The erf-like transition model. h(R) stands for the total backscatter coefficient, R_{bl} is the PBLH, a is the form factor related to the entrainment-zone (EZ) thickness ($2.77a^{-1}$), A is the ML-to-FT transition amplitude, and c is the FT molecular background (adapted from [8]).

The filter requires two models to operate:

(i) *The state-vector model,* in which *t*he dynamics of the state-vector from time t_k to time t_{k+1} are modeled by a Gauss-Markov transition model,

$$\boldsymbol{x_{k+1}} = \boldsymbol{x_k} + \boldsymbol{w_k} \qquad \text{Eq. (2)}$$

where $\boldsymbol{w_k}$ is the so-called "process noise" or state- noise vector. Because the state vector $\boldsymbol{x_k}$ (to be estimated) is a hidden "state" of the atmosphere, additional information is needed from the user's side: (a) an initial guess, $\widehat{x_0} = \left[R_{bl,0}, a_0, A_0, c_0\right]^t$, and (b) an estimate of the process-noise covariance matrix, $\boldsymbol{Q_k} = E[\boldsymbol{w_k}\,\boldsymbol{w_k^t}]$. The latter is approximated in diagonal form, $\boldsymbol{Q_k} = diag\left[\boldsymbol{\sigma_Q^2}\right]$, $\boldsymbol{\sigma_Q} = (\sigma_{Rbl}, \sigma_a, \sigma_A, \sigma_C)$, the latter built from a user-defined intensity factor, $\boldsymbol{\mu_Q}$, so that $\boldsymbol{\sigma_Q} = \mu_Q\widehat{\boldsymbol{x_0}}$. For example, σ_{Rbl} roughly models the expected standard deviation of the PBLH around its mean value. (c) The filter also requires initialization of the "a priori" error covariance matrix, $\boldsymbol{P_k^-} = E[\boldsymbol{e_k^-}\boldsymbol{e_k^-}^t]$, where $\boldsymbol{e_k^-} = \boldsymbol{x_k} - \widehat{\boldsymbol{x}}_k^-$ is the error vector and $\widehat{\boldsymbol{x}}_k^-$ is the estimated *a priori* state vector (i.e., prior to assimilating the current measurement at time $\boldsymbol{t_k}$). Such initialization, $\boldsymbol{P_0^-}$, models the expected error on the state-vector initial guess and is provided in the form, $\boldsymbol{P_0^-} = diag\left[\boldsymbol{\sigma_{e,X}^2}\right]$, with $\boldsymbol{\sigma_{e,X}} = (\sigma_{e,Rbl}, \sigma_{e,a}, \sigma_{e,A}, \sigma_{e,c})$ the user's uncertainty on the state-vector components at t_0. Likewise, the latter is passed to the filter as an *a priori* error factor, μ_P, so that $\boldsymbol{\sigma_{e,X}} = \mu_P\widehat{\boldsymbol{x_0}}$.

(ii) *The measurement model*, which relates the state-vector, x_k with the measured observables, z_k,

$$\boldsymbol{z_k} = \boldsymbol{h}(\boldsymbol{x_k}) + \boldsymbol{v_k} \qquad \text{Eq. (3)}$$

where $\boldsymbol{h}(\boldsymbol{x_k}) = \frac{A_k}{2}\left\{1 - erf\left[\frac{a_k}{\sqrt{2}}\left(R - R_{bl,k}\right)\right]\right\} + c_k$

is the erf function of Fig. 1 and $\boldsymbol{v_k}$ is the observation noise at time t_k with covariance matrix, $\boldsymbol{V_k}$ (diagonal). $\boldsymbol{V_k}$ is estimated by computing the range-dependent observation noise variance from the signal-to-noise ratio, $SNR(R)$ [8].

3. INSTRUMENTATION

The Laser Remote Sensing Unit (LRSU) of NTUA (EOLE and DIAL lidar systems) is based in Athens (37.96°N, 23.78°E, 220 m), Greece. It is equipped with an advanced 10-wavelength elastic-Raman-DIAL lidar system (Fig. 2) able to perform independent and simultaneous measurements of the vertical profiles of the aerosol backscatter coefficient (at 355, 532 and 1064 nm), of the aerosol extinction coefficient (at 355 and 532 nm) and of the water vapor and ozone mixing ratio in the troposphere (using the H_2O Raman channel at 407 nm and the Differential Absorption Lidar-DIAL technique, respectively) [11, 12].

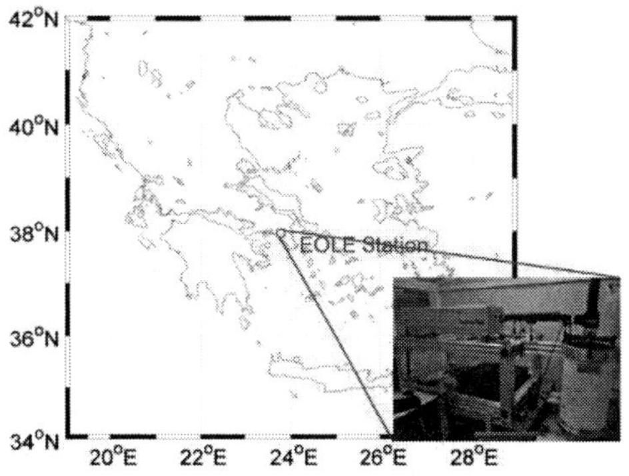

Figure 2. Map of Greece showing the location of the LRSU elastic-Raman lidar (EOLE) system, operating in Athens.

More precisely, the advanced elastic-Raman lidar system (EOLE) of LRSU (Fig. 2) is based on a pulsed Nd:YAG laser system which emits, simultaneously, high energy pulses at 355-532-1064 nm with 10 Hz repetition frequency. The laser beam is expanded by a Galilean telescope (x3), before being emitted in the atmosphere. A 300 mm diameter Cassegrainian telescope collects all elastically backscattered lidar signals, as well as those generated by the spontaneous

https://doi.org/10.1051/epjconf/201817606007

Raman effect (by atmospheric N_2 at 387-607 nm and by H_2O at 407 nm). The lidar signals are then corrected for electronic and atmospheric background noise, prior to range-corrections (RCS) pre-processing.

4. EXPERIMENTAL RESULTS

In Fig. 3 we present the monthly variability of the PBLH during daytime (12:00 UTC) and nighttime (00:00 UTC), for the entire 5 years period of measurements (2011-2016), as estimated with the EKF technique from the EOLE lidar signals. During daytime measurements the mean PBLH value is found to be ~1617±324 m, varying from 982 m (December) up to 2090 m (July). During nighttime the PBLH is found to be stable with a mean value of ~892±130 m,

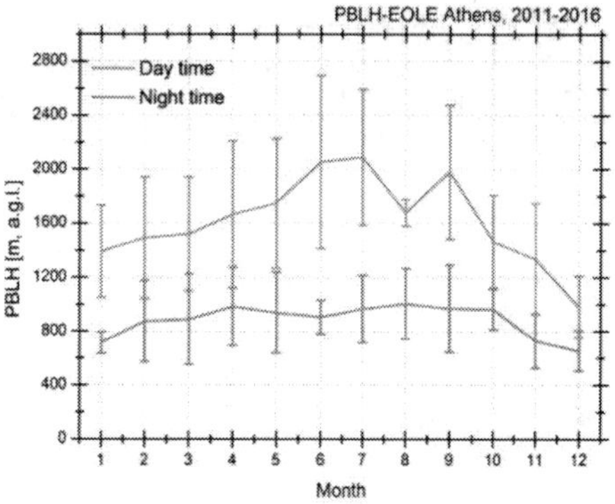

Figure 3. Monthly variability of the PBLH (during daytime and nighttime) as estimated with the EKF technique from EOLE lidar signals (2011-2016).

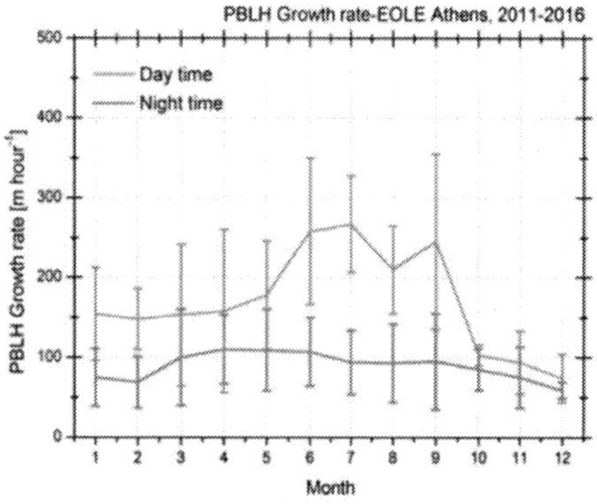

Figure 4. Monthly variability of the PBLH growth rate (during daytime and nighttime) as estimated with the EKF technique from EOLE lidar signals (2011-2016).

The daytime measurements revealed that the growth rate of the PBLH presents a maximum of ~267 m h^{-1} during summer (Fig. 4), where the highest temperature and solar radiance values (measured at 12 m above ground level, as shown in Fig. 5, orange and blue lines respectively), were recorded for the studied time period. The PBLH along with its growth rate was found to be significantly lower compared to the corresponding values revealed during dust cases. More precisely, the mean growth rate of the PBLH was found to be about 38.8 m h^{-1} lower during cases of dust particles suspended in the atmosphere over Athens, compared to the values presented without dust (Fig. 6).

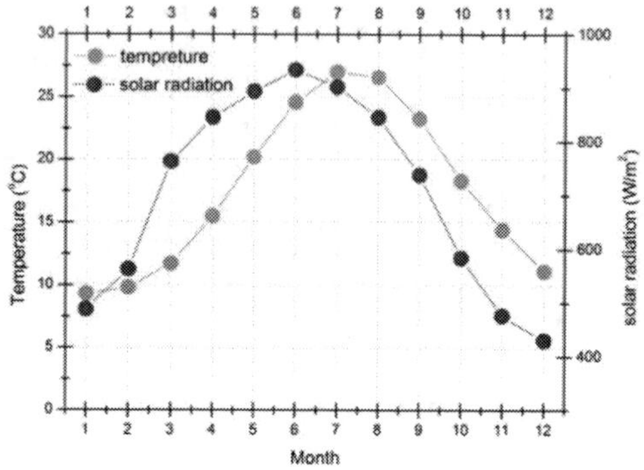

Figure 5. Monthly variability of the air temperature and solar radiation measured at 12 m above ground level in Athens (2011-2016).

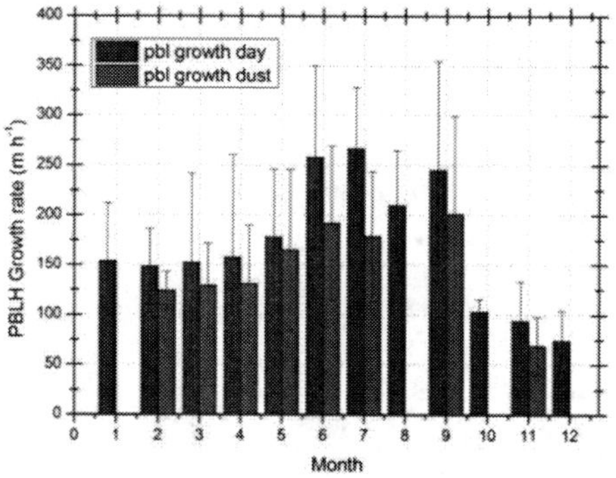

Figure 6. Monthly variability of the PBLH growth rate (during daytime) as estimated with EKF technique from EOLE lidar signals, during dust and non-dust cases over Athens (2011-2016). The error bars are computed from the standard deviation of estimated values within each month.

The values of the PBLH estimated with the EKF technique were further compared to the estimates from other methods using lidar signals (i.e. variance, gradient, inflection, threshold, wavelet covariance methods) and radiosonde data (obtained by the Hellenic National Meteorological Service, HNMS), as shown in Table 1.

This Table shows that the PBLH derived from the EKF analysis gives the best correlation with the one derived from radiosonde data (R^2=0.87 for lidar data obtained at 12:00 UTC and R^2=0.90 for lidar data averaged within a time interval of 30 min around 12:00 UTC: 12:00±0.30 UTC).

Table 1: PBLH comparison between radiosonde and lidar data based on various PBLH retrieving methods (correlation coefficient R^2 calculated at 12:00 UTC, without (middle column) and with ±30 min temporal averaging (right-hand column)).

	R^2 (12:00 UTC)	R^2 (12:00 UTC) ±30 min.
Threshold method	0.68	0.74
Gradient	0.41	0.68
Inflection point	0.26	0.62
Variance	0.43	0.74
Wavelet covariance	0.79	0.84
EKF	0.87	0.90

5. CONCLUSIONS

We studied the temporal evolution of the PBLH over Athens, Greece for the period 2011-2016. We found a mean PBLH of 1617±324 m (12:00 UTC) and 892±130 m (00:00 UTC). The PBLH growth rate was found to be ~170±64 m h^{-1} and 90±17 m h^{-1}, during daytime and nighttime, respectively. We also found that the Kalman filter follows much better than the other techniques the PBLH temporal evolution; this is corroborated when compared to the PBLH derived from radiosonde data, where it showed the best correlation (R^2=0.872 at 12:00 UTC and R^2=0.901) for 12:00±0.30 UTC). Thus, we can conclude that the EKF is the most suitable method for PBLH growth studies.

ACKNOWLEDGMENTS

The research leading to these results has received additional funding from the European Union 7th Framework Program (FP7/2011-2015) and Horizon 2020/2015-2021 Research and Innovation program (ACTRIS) under grant agreement no 262254 and 654109, respectively, as well as from Spanish National Science Foundation and FEDER funds TEC2015-63832.

REFERENCES

[1] Stull, R. B. 1988: An Introduction to Boundary Layer Meteorology, Kluwer Academic Publishers, pp. 666.

[2] IPCC: Climate Change 2013: The Physical Science Basis, Contribution of Working Group Ito the 5th Assessment Report of the Intergovernmental Panel on Climate Change. Cambridge University Press, Cambridge, United Kingdom and New York, NY, USA, pp. 1535.

[3] Menut, L., et al., 1999: Urban boundary-layer height determination from lidar measurements over the Paris area, *Appl. Opt.*, **38**, 945-954.

[4] Seibert, P., et al., 2000: Review and inter-comparison of operational methods for the determination of the mixing height, *Atmos. Environ.*, **34**, 1001-1027.

[5] Baars, H., et al., 2008: Continuous monitoring of the boundary-layer top with lidar. *Atmos. Chem. Phys.*, **8**, 7281-7296.

[6] Kalman, R.E., 1960: A new approach to linear filtering and prediction problems, Trans. American Soc. Mech. Eng. (AMSE) - *J. Basic Eng.*, **82**, Series D, pp. 35-45.

[7] Lange, D., et al., 2014: Atmospheric boundary layer height monitoring using a Kalman filter and backscatter lidar returns, IEEE Transactions on *Geos. Rem. Sensing*, **52**, 4717-4728, doi:10.1109/TGRS.2013.2284110.

[8] Lange, D., et al., 2015: Atmospheric-boundary-layer height estimation using a Kalman filter and a frequency-modulated continuous-wave radar, *IEEE Trans. Geos. Rem. Sensing*, **53**, 3338–3349.

[9] Rocadenbosch, F., et al., 1998: Adaptive filter solution for processing lidar returns: optical parameter estimation, *Appl. Opt.*, **37**, pp. 7019-7034.

[10] Rocadenbosch, F., et al., 1999: Lidar inversion of atmospheric backscatter and extinction-to backscatter ratios by use of a Kalman filter, *Appl. Opt.*, **38**, pp. 3175-3189.

[11] Kalabokas, P., et al., 2012: A study on the atmospheric concentrations of primary and secondary air pollutants in the Athens basin performed by DOAS and DIAL measuring techniques, *Science Total Environ.*, **414**, pp. 556-563.

[12] Kokkalis, P., et al., 2012: The EOLE lidar system of the National Technical University of Athens, *Proc. 26th International Laser Radar Conference*, pp. 629-632, 25-29 June 2012, Porto Heli, Greece.

IMPACT OF BIOMASS BURNING PLUME ON RADIATION BUDGET AND ATMOSPHERIC DYNAMICS OVER THE ARCTIC

Justyna Lisok[1]*, Jesper Pedersen[1], Christoph Ritter[2], Krzysztof M. Markowicz[1], Szymon Malinowski[1], Mauro Mazzola[3], Roberto Udisti[4], Iwona S. Stachlewska[1]

*[1]Institute of Geophysics, Faculty of Physics, University of Warsaw, Poland, *jlisok@igf.fuw.edu.pl*
[2]Alfred Wegener Institute for Polar and Marine Research, Germany
[3] National Research Council, Institute of Atmospheric Sciences and Climate, Italy
[4]Department of Chemistry, University of Florence, Italy

ABSTRACT

The aim of the research was to determine the impact of July 2015 biomass burning event on radiative budget, atmospheric stratification and turbulence over the Arctic using information about the vertical structure of the aerosol load from the ground–based data. MODTRAN simulations indicated very high surface radiative cooling (forcing of -150 Wm^{-2}) and a heating rate of up to 1.8 $Kday^{-1}$ at 3 km. Regarding LES results, a turbulent layer at around 3 km was clearly seen after 48 h of simulation.

1 INTRODUCTION

Lidar measurements conducted with a complex multi–wavelength Raman systems provide important information about the vertical structure of optical properties corresponding to the aerosol occurrence in the atmosphere. Lidars give high–quality wavelength–dependent profiles of particle extinction and backscattering coefficients. Such profiles may be used for an appreciable variety of further retrievals to deliver additional knowledge about the atmosphere. In this paper, we introduce one of the possibilities for reprocessing of the lidar and in–situ data to obtain information on radiation budget and atmospheric dynamics in the light of the Arctic measurements.

We focus on a strong event with a high aerosol load that was observed over Spitsbergen between 10–14 July 2015. It was characterized by a dense layer of biomass burning particles transported to the Arctic from wildfires that had spread over the Alaskan tundra at the beginning of July 2015 [1]. During that period, the observed aerosol optical properties exceeded the mean climatological summer values by a factor of at least 10. In particular, aerosol optical depth (AOD) as well as scattering and absorption coefficients at the Earth's surface increased to values of 1.2, 145 and 10 Mm^{-1} at 550 nm respectively. It was expected that such event would have a high impact on the optical properties over Spitsbergen, and thus it was chosen for further detailed study.

2 METHODOLOGY

In this section we describe the methodology applied for calculating radiation fluxes and atmospheric dynamics, using vertical profiles of single scattering albedo at ambient conditions that were obtained by the means of in–situ and lidar measurements.

2.1 Measurements

The measurements were carried out at two research stations in Ny–Ålesund on Spitsbergen. French–German (AWIPEV) station, located within the village, handled lidar measurements. Italian Gruvebadet laboratory, located approx. 1 km SW from Ny–Ålesund, maintained in–situ instruments [1].

The extinction (σ_{ext}) profiles were retrieved from signals measured by the Koldewey Aerosol Raman Lidar (KARL), equipped with a Nd:Yag laser which is emitting pulses at 355, 532, 1064 nm with 10 W each. The signals were received on a 70 cm mirror with a 1.75 mrad field of view. Geometrical compression affected the signal below 700 m a.g.l. [2]. Single wavelength M903 Nephelometer (Radiance Research) equipped with a xenon flash lamp and an opal diffuser measured scattering coefficient (σ_{scat}) at 530 nm within 10–170° integration range. Particle Soot Absorption Photometer (PSAP; Radiance Research) measured absorption coefficient (σ_{abs}) at 467, 530, 660 nm based on changes in the light attenuation indicated by the aerosol load on filter. Joint measurements of TSI Scanning Mobility Particle Sizer (SMPS 3034) and TSI Aerodynamic Particle Sizer Spectrometer (APS 3321) allowed for retrieval of aerosol size distribution in the range of 0.01–20 μm. Vertical profiles of basic thermodynamic quantities (air temperature, air pressure, relative humidity, wind speed and direction) were measured by the Vaisala radiosonde RS92.

2.2 Retrieval of aerosol single scattering properties at ambient conditions

The aerosol model for atmospheric radiative transfer calculations is built based on the following approach. Assuming constant with altitude single scattering albedo at dry conditions (ω^d), one can calculate its profile at ambient conditions (ω^a). Firstly, the value of ω^d is calculated from nephelometer and PSAP measurements at dry conditions. Secondly, by taking the lidar σ^a_{ext} profile and introducing both extinction enhancement factor (the ratio between extinction coefficient at ambient conditions to its value in dry conditions [3]) as well as Handel model (relating aerosol hygroscopicity and the relative humidity [4]), one can calculate the profile of extinction coefficient at dry conditions (σ^d_{ext}). From the σ^d_{ext} and ω^d profiles the absorption coefficient (σ^d_{abs}) profile can be retrieved. Taking into account, that σ_{abs} is weakly dependent on the relative humidity, one can assume no change of σ^d_{abs} during transition to ambient conditions (σ^a_{abs}). Finally, from σ^a_{abs} and σ^a_{ext} (lidar measurements) one can calculate the profile of ω^a at ambient conditions. Retrieval of the asymmetry parameter (g) is also possible, provided that at first, a hygroscopic growth factor related to the relative humidity is introduced [3]. Then, assuming a value of refractive index for a known particle size distribution, one can obtain g by means of Mie theory. The detailed description of the proposed approach is given in [5].

2.3 Model setups

MODTRAN v. 5.2.1 (The moderate–resolution atmospheric radiance and transmittance model) [6] was adopted to calculate profiles of the r_h. The MODTRAN solver uses discrete ordinate radiative transfer model (DISORT). In the simulations, 8–stream MODTRAN band model "M" with 17 absorption coefficients within each spectral bin and correlated–17k speed option were used. All calculations were performed assuming oceanic values of surface albedo with a Fresnel reflection included. The model calculated the clear–sky radiative fluxes and heating rates in the range of 0.3– 200 μm for two cases: the atmosphere without and with the aerosol load, referred to as "pristine" and "polluted" cases, respectively.

EULAG (The 3D nonhydrostatic anelastic Eulerian–semi–Lagrangian) model [7] was set up to perform implicit large–eddy simulation (ILES) of the two defined cases (pristine and polluted) assuming additionally clear–sky and non–condensing conditions. Initial profiles of wind speed and potential temperature θ were in both cases based on radio–soundings from Ny–Ålesund performed at 12:00 UTC on July 10 2015. The vertical velocity was initially set to 0, and a large–scale horizontal pressure gradient was applied in terms of the geostrophic wind $\mathbf{U}_g$=(10,10) ms^{-1} with the Coriolis parameter set to 1.4 x 10^{-4} s^{-1}. The r_h profiles from the MODTRAN were applied and kept constant in the simulations, which were run for 48 h.

3 RESULTS

In the following section a brief description of the obtained vertical profiles of optical and radiative properties, as well as turbulence are given. More detailed description and discussion is to be found in [5].

3.1 Impact of absorbing aerosol on radiative properties and atmospheric dynamics

Vertical profiles of aerosol and radiative properties (Fig. 1) show significant variability. Lidar extinction profile indicates a dense layer of biomass burning aerosol between 1–3.5 km, where extinction and absorption coefficients reach up to 300 and 20 Mm^{-1}, respectively. Since an appreciable variability of the relative humidity is observed (20–80 %; Fig. 1d), the particle hygroscopicity plays an important role. Thus, the noticeable rate of change of extinction coefficient at ambient conditions (black line) from 2 to 3.5 km results from the variability of the relative humidity. The corresponding extinction profile at dry conditions indicates a constant value within the layer. In comparison to the profile of potential temperature (Fig. 2b–black line), a retrieval seems to be reasonable. The vertical profile of single scattering albedo at dry conditions (Fig. 1c–blue line) is assumed constant and equal to the value obtained by in–situ measurements (0.92).

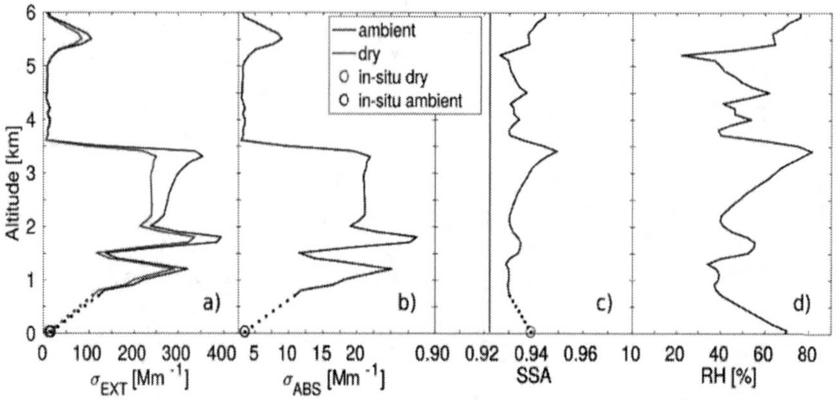

Figure 1 Vertical profiles of optical properties at 10 July 2015 12:00 UTC obtained from the measurement, and presented retrieval (sect. 2.2).

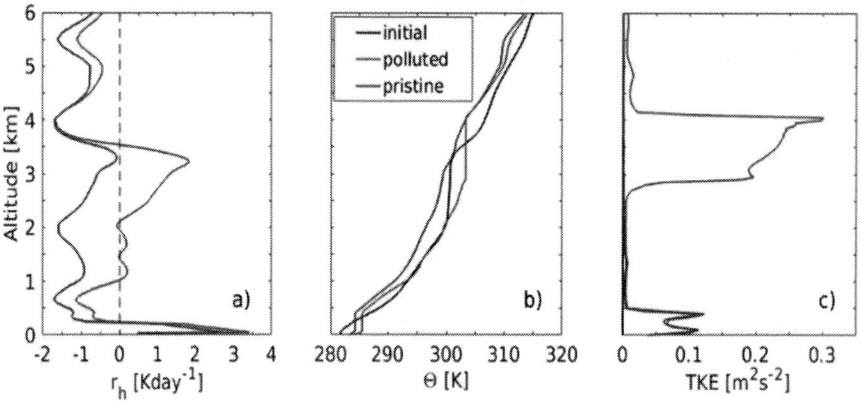

Figure 2 Vertical profiles of the heating rate (a), potential temperature (b) and turbulent kinetic energy (c). The black line represents initial conditions and the red/blue lines profiles from the polluted/pristine cases at t=48h. The heating rates are kept constant in time.

The values at ambient conditions (Fig. 1c–black line), however, range from 0.93 to 0.95. The vertical variability results from the Raoult's effect embedded in the Kohler relation used for the calculations. Thus, in comparison to dry conditions the aerosol has less absorbing properties, especially in the layers with high relative humidity (i.e. at 3.5 km). The heating rate profile calculated for the pristine case (Fig. 2a–blue line) indicates background conditions to cool the troposphere (up to –1.8 Kday^{-1}) within the first 6 km, except for the lowermost layer influenced by the surface properties. The polluted case, however shows a local heating rate (up to 1.8 Kday^{-1}) between 1 and 3.5 km caused by the biomass burning aerosol. The particular value of the heating rate is also related to the water vapor relative contamination.

As indicated by the initial profiles of potential temperature (Fig. 2b–black line), atmospheric conditions are stable at the beginning of the performed large–eddy simulations, except in the layer between 1.8 and 3.5 km where θ is nearly constant with height as a result of vertical mixing.

At time t=48 h in the polluted case, existence of a well–mixed layer is visible between 2.8 and 4 km, indicating that it is lifted with time (Fig. 2b–red line). In the pristine case, however, the initial well–mixed layer disappears over time and is not visible at t=48 h. The profile of turbulent kinetic energy from the polluted case (Fig. 2c) at t=48 h shows regions with relatively high values; up to 0.1 m^{2}s^{-2} below 0.5 km and between 0.2–0.3 m^{2}s^{-2} in the region of 2.8–4 km, with vertical velocities of as high as 1ms^{-1} (not shown for brevity). These regions correspond to layers with a constant potential temperature. On the other hand, the profile of turbulent kinetic energy calculated for the pristine case is almost entirely constant and near zero, with exception of the lowermost atmospheric layer, which is altered by the surface properties (Fig 2c–blue line). Very little vertical mixing takes place above the surface layer.

4 CONCLUSIONS

A brief presentation of the possibility of using lidar data for radiative transfer and atmospheric dynamics calculations has been given. We considered an event with high concentrations of biomass burning aerosol transported from North America to the Arctic that occurred between 10–14 July 2015 over Spitsbergen. On July 10, the radio–soundings indicated a layer with well–mixed air between 1.8–3.5 km with corresponding high values of relative humidity (60–80 %). The single scattering albedo of 0.95–0.96 and a positive heating rate of up to 3 Kday^{-1} were obtained. The EULAG simulations provided profiles of turbulent kinetic energy and potential temperature over a 48 h period, and clearly indicated the development of vertical mixing and rising of the well–mixed layer, however only in the polluted case. The layers of biomass burning aerosol seem to locally heat the surrounding air causing changes in the stability within the mid troposphere. Thus, impacting the atmospheric dynamics.

ACKNOWLEDGEMENTS

The research leading to these results has received funding from the Polish–Norwegian Research Programme operated by the National Centre for Research and Development under the Norwegian Financial Mechanism 2009–2014 in the frame of Project Contract No Pol–Nor/196911/38/2013.

EULAG simulations were performed at the Interdisciplinary Centre for Mathematical and Computational Modelling (ICM) at the University of Warsaw, under grant number G64–5.

References

[1] Markowicz K.M., Pakszys P., Ritter Ch., Zielinski T., Udisti R., Cappelletti D., Mazzola M., Shiobara M., Zawadzka O., Lisok J., Petelski T., Makuch P., Karasiński G., 2016: Impact of North American intense fires on aerosol optical properties measured over the European Arctic in July 2015, J. Geophys. Res., 121, 14,487–14512

[2] Ritter C., Neuber R., Schulz A., Markowicz K.M., Stachlewska I.S., Lisok J., Makuch P., Pakszys P., Markuszewski P., Rozwadowska A., Petelski T., Zielinski T., Becagli S., Traversi R., Udisti R., Gausa M., 2016: 2014 iAREA campaign on aerosol in Spitsbergen – Part 2: Optical properties from Raman–Lidar and in–situ observations at Ny–Ålesund, Atm. Env., 141, 1–19

[3] Zieger P., Fierz–Schmidhauser R., Gysel M., Strom J., Henne S., Yttri K.E., Baltensperger U., Weingartner E., 2010: Effects of relative humidity on aerosol light scattering in the Arctic, Atmos. Chem. Phys., 10, 3875–3890

[4] Hänel G., 1976: The properties of atmospheric aerosol particles as functions of the relative humidity at thermodynamic equilibrium with the surrounding moist air. Adv. Geophys., 19, 74–188

[5] Lisok, J., Rozwadowska, A., Pedersen, J. G., Markowicz, K. M., Ritter, C., Kaminski, J. W., Struzewska, J., Mazzola, M., Udisti, R., Becagli, S., and Górecka, I.: Impact of a Strong Biomass Burning Event on the Radiative Forcing in the Arctic, Atmos. Chem. Phys. Discuss., https://doi.org/10.5194/acp-2017-1035, in review, 2017.

[6] Berk A., Bernstein L.S., Anderson G.P., Acharya P.K., Robertson D.C., Chetwynd J.H., Adler–Golden S.M., 1998: MODTRAN cloud and multiple scattering upgrades with application to AVIRIS. Remote Sens. of Env., 65, 367–375

[7] Prusa J.M., Smolarkiewicz P.K., Wyszogrodzki A.A., 2008: EULAG, a computational model for multiscale flows. Comp. & Fluids, 37, 1193–1207

LIDAR FOR WIND AND OPTICAL TURBULENCE PROFILING

Shlomo Fastig[1]*, Omer Porat[1], Abraham Englander[1]

Detlev Sprung[2], Karin U. Stein[2], Erik Sucher[2]

*1 SOREQ NRC, Israel, *fshlomo@soreq.gov.il*

2 Fraunhofer-Institut für Optronik, Systemtechnik und Bildauswertung (Germany)

ABSTRACT

A field campaign for the comparison investigation of systems to measure wind and optical turbulence profiles was conducted in northern Germany. The experimental effort was to compare the performance of the LIDAR, SODAR-RASS and ultrasonic anemometers for the measurement of the above mentioned atmospheric parameters. Soreq's LIDAR is a fiber laser based system demonstrator for the vertical profiling of the wind and turbulence, based on the correlation of aerosol density variations. It provides measurements up to 350m with 20m resolution.

1 INTRODUCTION

Wind and turbulence height profiles data are needed for different applications: meteorology, wind energy, atmospheric model verification, aviation safety, air pollution etc.

The Atmospheric Sensing group at Soreq NRC had developed for many years techniques and systems for the remote sensing of wind turbulence and atmospheric optical refraction [1-2]. Active, laser-based systems [3] and passive technique based on video camera measurement of naturally illuminated scene [4] were developed for the measurement of the path-average crosswind and atmospheric turbulence. These techniques utilize the spatial-temporal fluctuations of light caused by the passage through the turbulent atmosphere, in order to estimate the path-average atmospheric parameters.

Wind profile measurements with LIDARs are based on transmitting short laser pulses and measuring the aerosol backscattered light. Using the time-of flight method the range resolved profiles are derived.

Doppler wind LIDARs use the frequency shift of the scattered light to derive the radial wind component. In order to measure the transverse wind components, the LIDAR beam must be scanned or transmit multiple laser beams.

Correlation wind LIDAR, are based on the measurement of the aerosol density variations. A train of laser pulses is transmitted to the atmosphere at a frequency high enough (hundreds of Hz), to match the atmospheric coherence. The measured fluctuations in the backscattered light signal from the aerosols, enables the calculation of the wind and turbulence.

This paper briefly presents the LIDAR demonstrator which was developed in Soreq and deployed for a multi-sensor comparison experiment conducted in northern Germany in collaboration with Fraunhofer IOSB [5].

2 LIDAR FOR WIND AND TURBULANCE PROFILING

Soreq's wind LIDAR is a concept demonstrator which was designed based on the following principles:

- Configuration which provides maximum working range.
- Provides measurement of the spatial-temporal variations on the imaged spot of the laser.

2.1 System Configuration

The LIDAR is constructed co-axially so that the lines of sight of the receiver and the transmitter are parallel (Figure 1). In this configuration the optical signal scattered from any distance reaches the receiver, the minimum range is determined from the forward blocking by the transmitter and the maximum range is determined from the minimum signal to noise ratio of the detector. In order to increase the working range the image plane of the receiver is fixed around the maximum range.

The receiver is composed of a $F_{\#}=5$, f=773mm Maksutov telescope. The tip of a bundle of three optical fibers is placed at the image plane of the

 https://doi.org/10.1051/epjconf/201817606009

receiver. The fibers are arranged to form a triangle and each is connected to a Si APD detector. This configuration acts as a 2D array of detectors that measure the spatial-temporal variations of the laser returns, thus enabling 2D wind vector calculation.

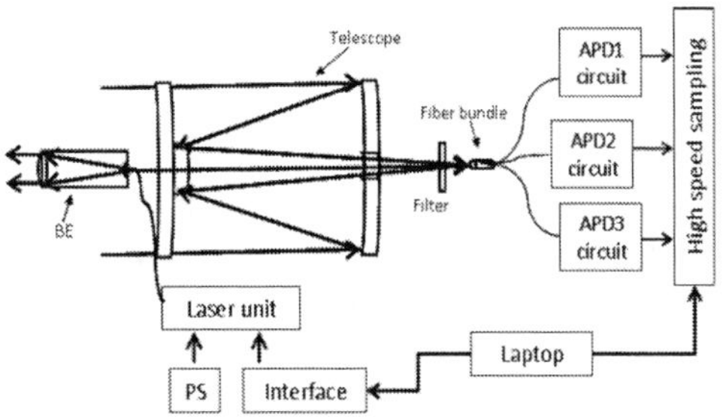

Figure 1 Block diagram of the wind LIDAR

The transmitter is a pulsed fiber laser manufactured by Raycus, transmitting 1.064μm pulses at 20 kHz repetition rate and 100nsec pulse width. The beam divergence is adjusted with a beam expander to about two-thirds of the entire field of view of the receiver (~700μRad).

The electronic signals which are created by the three detectors are sampled by a high speed digitizer – the PicoScope 5000 – at 12bit depth. The data are transferred to a laptop computer for processing. Data downloading time determines the measurement rate of the LIDAR.
The LIDAR is operated using a laptop computer which initiates the measurement procedure, synchronizes between the transmitter and the receiver and carries out data download and analysis.

Finally, a scope and a video camera are installed parallel to the laser line of sight to assist with LIDAR aiming and safe operation.

A picture of the LIDAR aimed horizontally is presented in Figure 2 and key parameters of the LIDAR demonstrator are provided in Table 1.

Table 1 Key parameters of the LIDAR

Parameter	Value	Parameter	Value
Wavelength	1.064μm	Receiver FOV	700μrad
PRF	20kHz	Operation range	50-350m
Output power	20W	Range resolution	15-20m
Pulse width	100nsec	Measurement duration	1.5sec
Beam divergence	300-500μrad	Measurement rate	1-5mins

Figure 2 The wind and turbulence profiling LIDAR

2.2 Method

The wind range profile is derived using the aerosol in the atmosphere as an extended target. The back scattered laser light measured from a range r, is used to calculate the temporal cross-correlation function between the detectors from which wind speed can be derived. The back scattered signal characteristic can either have large peaks, short duration and be sparse in time, or have small peaks with a smooth transition in time, depending on the atmospheric conditions. We believe the first case is typical for large and low density aerosols (upper graph in Figure 3), while the second one is typical for small, high density aerosols (bottom graph in Figure 3). According to the scattering type the dominant frequency of the intensity fluctuations is measured using the number of zero crossings technique [1]. This value is averaged over the three detectors, from which the wind speed is calculated.

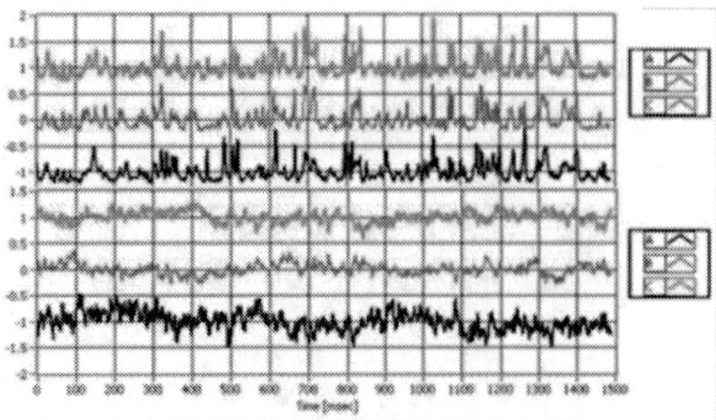

Figure 3 A series of 1500 returning pulses from one range slice as measured by the three detectors (labeled A, B and C) in the LIDAR (intensity is normalized to the series average and detector are shifted to provide clearer view). Top – data recorded in northern Germany: short large peaks, cross-correlation is clearly evident. Bottom – data recorded in central Israel: the intensity fluctuations show small peaks with smooth transition in time.

The transverse wind direction is estimated using the cross-correlation function of the intensity fluctuations measured by the detectors. For each pair of detectors the cross-correlation function is calculated (Figure 4) and the location of the peak of the correlation is determined. The ratio between the movements of the correlation function (the location of the peak) of different pairs of detector is proportional to the ratio of the wind components along the pairs. Thus, using all three pairs the transverse wind direction is given by:

$$\alpha = artan\left(\frac{3\sqrt{3}}{2}\cdot\frac{\frac{CC_{BC}}{CC_{AC}}+1}{1-\frac{CC_{BC}}{CC_{AC}}-2\frac{CC_{AB}}{CC_{AC}}}\right) \quad (1)$$

Where CC_{ij} is the location of the peak of the cross-correlation function of detectors i and j.

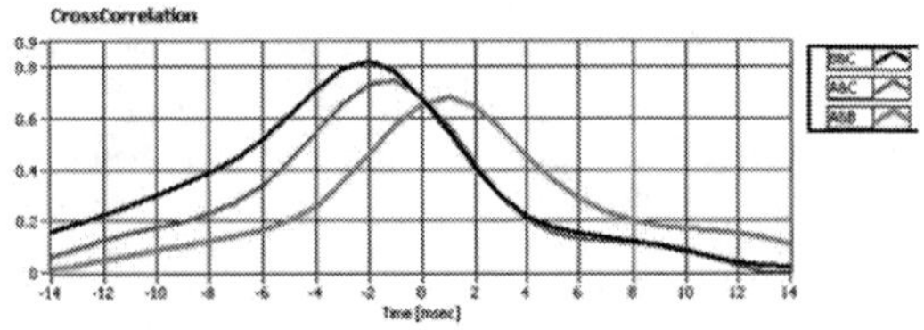

Figure 4 Temporal cross-correlation function between signals in adjacent detectors measured in a single range slice. Using the ratio between the locations of the peak of different pairs the wind direction is estimated.

The turbulence strength is calculated from the temporal variance of the intensity of the signal of each of the detectors measured over one pulse burst (1.5 secs).

The transverse wind speed and direction and the turbulence strength are calculated, as detailed above, for each available range slice to provide a range profile.

3 RESULTS

3.1 Experiment Site

The experiment took place in a area which is part of the German Federal Armed Forces Technical Center (WTD 91) near the city of Meppen (Figure 5). It is the location of the Vertical Turbulence Measurement (VerTurM) experiment – a long term study of the profiles of optical turbulence and meteorological parameters performed by Fraunhofer IOSB [7].

Figure 5 Setup of the field experiment in Germany: right – aerial photo of the site, left – Soreq's LIDAR during measurements in the site

3.2 Comparative Measurements

The measurements of vertical profiles of the wind and the optical turbulence are shown in Figures 5-7. These include the data collected on May 5th using the anemometers, scintillometer, SDAR-RASS, radiosondes and the LIDAR. The LIDAR measurements were calibrated using a reference measurement at 64m height. Generally the fit between the LIDAR and the other system for wind speed and turbulence strength is quite good at certain periods of time (around 10:00, 10:45, 12:00) within ±1.5m/s and not as good at other periods time when the LIDAR indicates stronger wind (+4-6m/s). This could be explained by the difference in the integration time of the instruments. While for the LIDAR measurement it is of the order of 1sec and thus more sensitive to wind gusts, the other instruments average over a few minutes time thus smoothing out gust effect. The wind direction fits better during the entire measurement period within $\pm 5^0$.

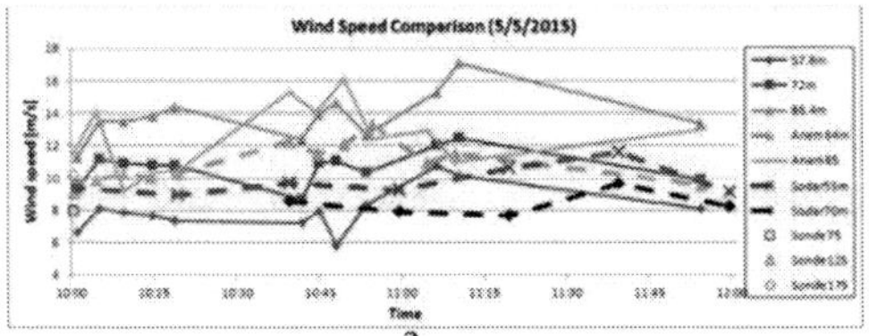

a

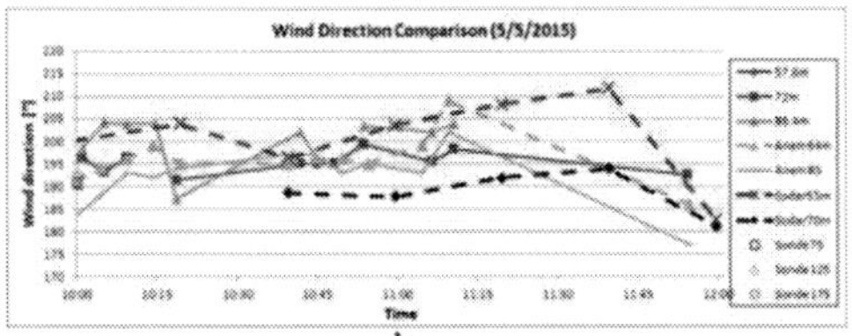

b

Figure 5 Comparison of the vertical transverse wind speed profile (a) and direction (b), as measured on May 5th by the LIDAR (solid lines) and the VerTurM instruments (annemometers, SODAR-RASS and radiosonde at 10:00, broken lines).

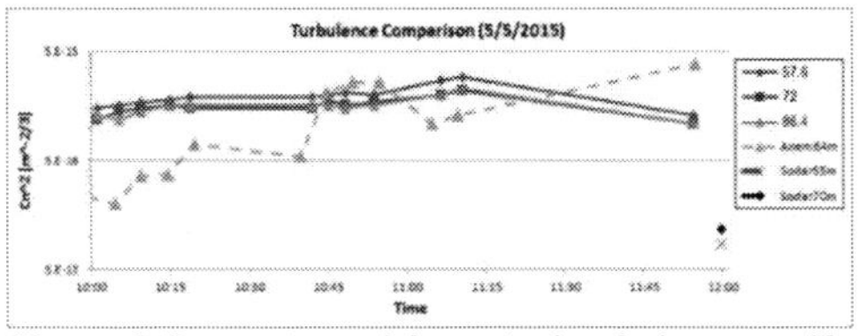

Figure 6 Comparison of the optical turbulence profile as measured on May 5th by the LIDAR (solid lines) and the VerTurM instruments (annemometers, SODAR-RASS at 12:00, broken lines).

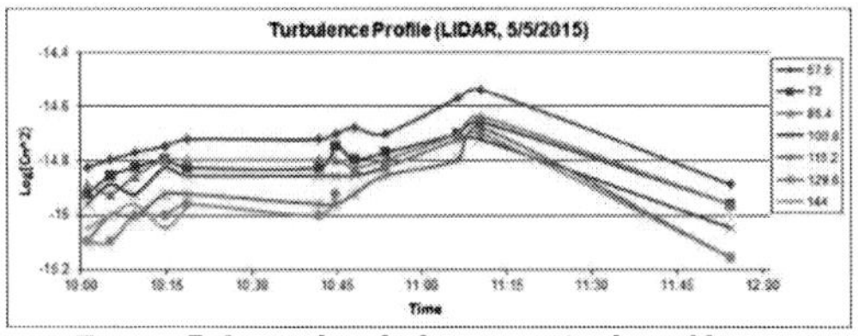

Figure 7 Optical turbulence vertical profiles as measured by the LIDAR on May 5th up to 144m.

4 CONCLUSIONS

A 3-day field experiment for comparison of the LIDAR measurements with commercially available reference systems was held in northern Germany in May 2015. The LIDAR was deployed in the site of the vertical turbulence measurement experiment (VerTurM) which comprises a SODAR-RASS, meteorological mast, scintillometer and etc. and a long-term data collection system.

Due to bad weather conditions which included fast meteorological changes from clear sun shine sky to heavy clouds, strong gusts and intense shower rains, the collected data was insufficient for a more quantitative conclusion. However, the data from two relatively calm morning measurements indicate reasonable fit, for the wind speed and the optical turbulence measurements and good fit for wind direction. Additional field experiments are required in order to fully evaluate the performance of the LIDAR at various conditions particularly at times of variable wind.

References

[1] J. Shapira, O. Porat, M. Livneh, Z. Wies, D. Heflinger, Y. Glick, A. Engel (2010): Atmospheric cross-wind and turbulence measurements using turbulence –induced scintillations, *Laser Radar Technology and Applications XV*, Proceedings of the SPIE, Volume 7684, pp. 76841L-76841L-11.

[2] A. Engel, O. Porat, J. Shapira, A. Englander (2014): Experimental evaluation of optical crosswind measurement systems, *Remote Sensing of Clouds and the Atmosphere XIX; and Optics in Atmospheric Propagation and Adaptive Systems XVII*, Proceedings of the SPIE Volume 9242, pp.92421K.

[3] G. Machavariani, J. Shapira, U. Talmud, O. Porat, Sh. Fastig (2010): Remote qualitative profiling of the cross-wind flow with lidar, *Laser Radar Technology and Applications XV*, Proceedings of the SPIE, Volume 7684, pp. 76841M-76841M-9.

[4] O. Porat, J. Shapira (2010): Crosswind sensing from optical-turbulence-induced fluctuations measured by a video camera, *Applied Optics*, Volume 49, Issue 28, pp. 5236-5244.

[5] K. Steina , E. Suchera, D. Sprunga A. Englander, S. Fastig, O. Porat (2016): Instrumental intercomparison investigating Vertical profiles of optical turbulence and wind speed in the lower atmospheric boundary layer during frontal passages in northwestern, *Optics in Atmospheric Propagation and Adaptive Systems XIX*, SPIE Proceedings Vol. 10002

EPJ Web of Conferences **176**, 06010 (2018) https://doi.org/10.1051/epjconf/201817606010
ILRC 28

STUDYING TURBULENCE BY REMOTE SENSING SYSTEMS DURING SLOPE-2016 CAMPAIGN

Gregori de A. Moreira[1,2,3,*], Juan L. Guerrero-Rascado[2,3], Jose A. Benavent-Oltra[2,3], Pablo Ortiz-Amezcua[2,3], Roberto Róman[2,3], Eduardo Landulfo[1] and Lucas Alados-Arboledas[2,3]

[1]*Institute of Energetic and Nuclear Research (IPEN), Brazil, *gregori.moreira@usp.br*
[2]*Andalusian Institute for Earth System Research (IISTA-CEAMA), Spain*
[3]*Dpt. Applied Physics, Faculty of Sciences, University of Granada, Spain*

ABSTRACT

The Planetary Boundary Layer (PBL) is the lowermost part of the troposphere. In this work, we analysed some high order moments and PBL height detected continuously by three remote sensing systems: an elastic lidar, a Doppler lidar and a passive Microwave Radiometer, during the SLOPE-2016 campaign, which was held in Granada from May to August 2016. This study confirms the feasibility of these systems for the characterization of the PBL, helping us to justify and understand its behaviour along the day.

1 INTRODUCTION

The planetary boundary layer (PBL) is defined as that *"part of the troposphere that is directly or indirectly influenced by the presence of the Earth's surface (ground or sea), and responds to surface forcings with a time scale of a few hours or less."* [1]. It is characterized by large variability in its statical stability and turbulence along the day. This behavior makes the detection of its height (PBLH) and the understanding of its turbulent characteristics very important for a wide set of studies, including pollutant dispersion and weather forecasting. However, this kind of study requires devices with high spatial and temporal resolution like as remote sensing systems (RSS) [2].

In this study, we use data from three RSS (Elastic Lidar [*EL*], Doppler Lidar [*DL*] and Microwave Radiometer [*MWR*]) acquired during SLOPE campaign, which was held in CEAMA (Andalusian Institute for Earth System Research - Granada [37.16°N, 3.61°W, 680 m asl] - Spain) from May to August/2016, in order to analyze the height and turbulence in the PBL.

2 METHODOLOGY

Below it is described the methodology applied for the three RSS used in this paper.

2.1 Microwave Radiometer (MWR)

The RPG-HATPRO Radiometer Physics GmbH is a passive microwave radiometer, which performs measurements of the sky brightness temperature with a resolution between 0.3 and 0.4 K root mean square error at 1s integration time. Beyond to providing temperature information at various heights, the MWR data were also used to identify the PBLH.

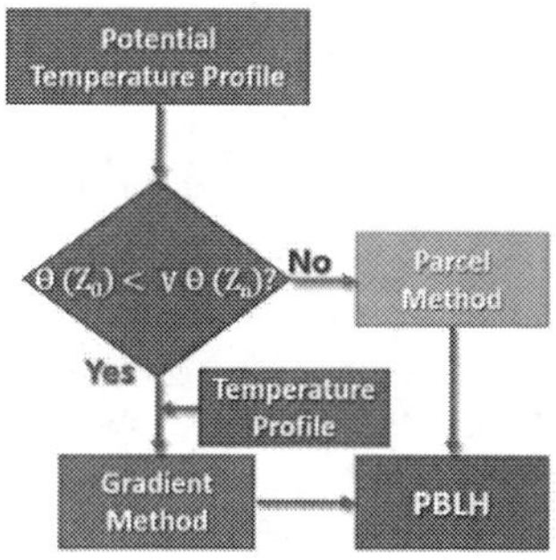

Figure 1: MWR Methodology

Figure 1 summarizes the methodology applied. It is made a comparison among surface temperature and the potential temperature profile ($\theta(Z)$) at within the firsts meters above the ground in order to identify if it has convective or stable characteristics. If it is detected as stable conditions ($(\theta(Z_0) < \forall\theta(Z_n))$ the gradient method [3] is applied on temperature profile, otherwise the Parcel Method [3] is selected.

2.2 Elastic Lidar Mulhacén (EL)

Mulhacén is an elastic-Raman lidar (Raymetrics Inc) operating at 355, 387, 408, 532 (paralell and perpendicular), 607 and 1064 nm with a temporal and spatial resolution of 2 s and 7.5 m, respectively. This system is part of EARLINET (European Aerosol Research Lidar Network) since 2005.

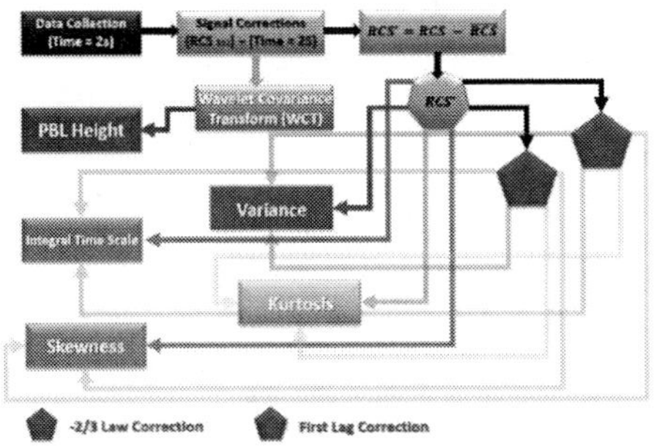

Figure 2: EL Methodology

In the Figure 2 there is a schematization of methodologies applied to study the PBL from EL data. The raw data collected are corrected for dark current, background radiation and square distance becoming the RCS (Range Corrected Signal). These data are used to identify:

- the **PBLH** by Wavelet Covariance Transform (WCT) Method [4];
- **Integral Time Scale**, which is the measure of the average size (in time) of the vortices involved in the PBL mixing processes:

$$\tau = \int_0^\infty M(t)dt \qquad (1)$$

where $M(t)$ is the auto-covariance function of RCS'

- **Variance**:

$$\sigma^2 = \frac{\sum_{i=1}^{N}(RCS'_i - \overline{RCS'})^2}{N} \qquad (2)$$

where N is the number of profiles.

- **Skewness**:

$$S = \frac{\overline{RCS'^3}}{\overline{RCS'^2}^{\frac{3}{2}}} \qquad (3)$$

- **Kurtosis**:

$$K = \frac{\overline{RCS'^4}}{\overline{RCS'^2}^{2}} \qquad (4)$$

with its respective corrections: *First Lag* and $-2/3$ *Law*, which are direved from $M(t)$ [5].

2.3 Doppler Lidar (DL)

It is a Halo Photonics model Stream Line operating at 1.5 μm with a temporal and spatial resolution of 2 s and 30 m, respectively.

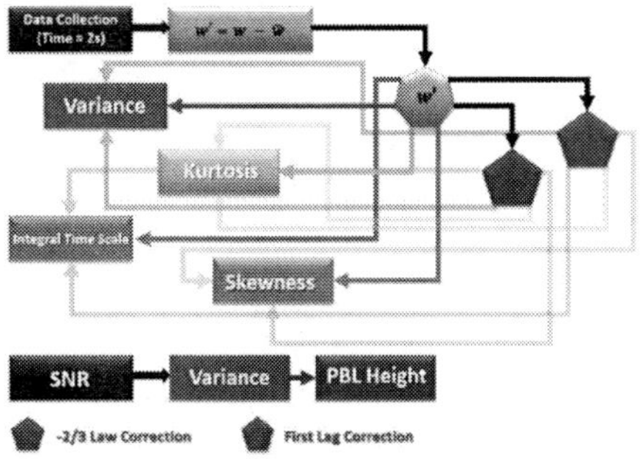

Figure 3: DL Methodology

Figure 3 shows the methodology applied to DL data, which is very similar to that presented previously for EL. The main difference is the variable used to study the high order moments, vertical wind speed (w), and the use of Signal Noise Ratio (SNR) to identify the PBLH by variance method [6], instead of using RCS in both situations.

3 RESULTS AND DISCUSSION

Figure 4 represents the integral time scale. The gray areas in the picture indicate heights where the integral time scale is lower than 2 s (DL time acquisition), therefore meaning that it is impossible to extract any information on the turbulence structures at these heights using DL. Thus, DL data provide information about turbulence throughout almost the whole PBL.

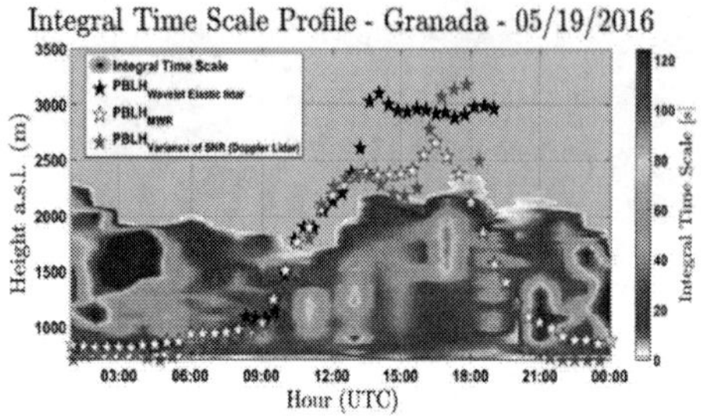

Figure 4: Integral Time Scale - DL

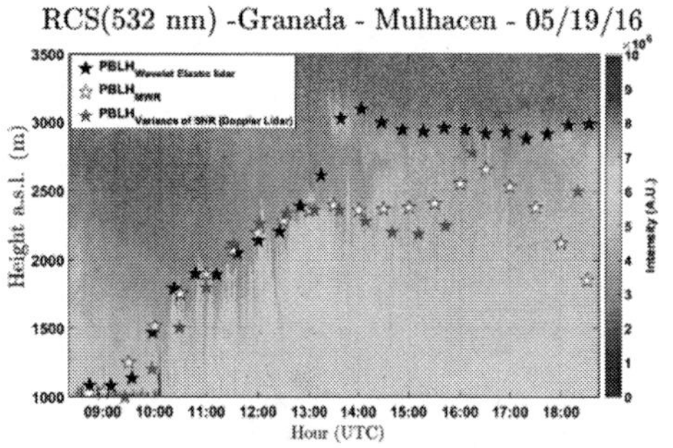

Figure 5: RCS profile - EL

In the same picture it is also possible to observe the PBLH detected by MWR, EL and DL, and we can notice that they are very similar, mainly during the PBL growing phase (09 – 13 h). In the middle of afternoon there is an ascension of an aerosol layer (Figure 5), and it causes some differences among the methods, but after some hours the PBLH obtained from EL and DL data return to coincide. During nocturnal phase PBLH from MWR and DL have a good agreement.

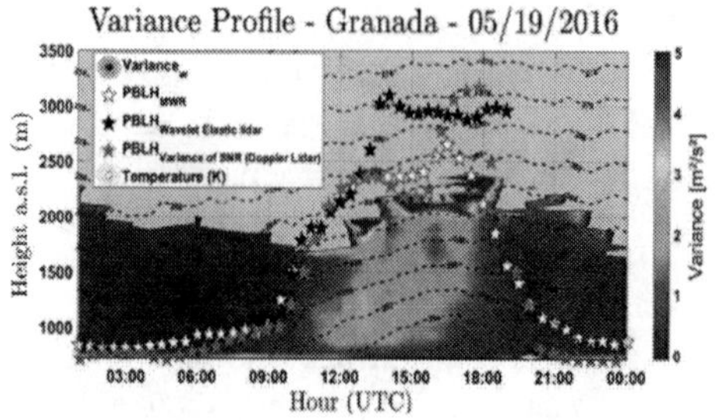

Figure 6: Variance - DL

The elevation and reduction of surface temperature directly affect the increase and decrease of the variance of *w*, as well as the PBLH (Figure 6). This occurs as expected, because the increase of temperature leads to the intensification of convective process, which culminate in the rise of PBLH and rise of vertical wind. In the same way, the reduction of temperature generates the decrease of convective process and consequently the reduction of PBLH and variance of *w*.

The maximum of RCS's variance can be used as indicator of PBLH. It represents large variation in the aerosol concentration, which is the result of the rapid mixing of air parcels located above the top of PBL (therefore with lower aerosol backscatter coefficient values) with others situated inside the convective boundary layer and consequently endowed with high aerosol concentration [1, 7]. In Figure 7 we can see that there is a good agreement among variance, WCT and MWR methods as PBLH indicator, however the variance gets confused in presence of multiple aerosol layers indicating several maxima (13 to 14 UTC).

4 CONCLUSIONS

In this work we analysed the height and turbulent structure of PBL along the time. We obtained a good agreement between the PBLH

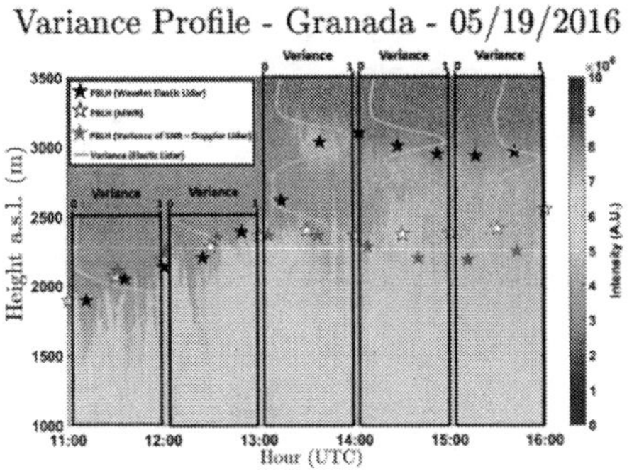

Figure 7: Variance - EL

detected by *MWR*, *EL* and *DL*. In the same way the second order moment provided by *EL* and *DL* have a high resemblance between them and they together with temperature information help us to justify and understand better the process of increasing and decreasing of PBLH. The study confirms the feasibility of RSS to studies about turbulence in the PBL.

In future studies, we intend to apply this same methodology to obtain other high order moments (Skewness and Kurtosis) and to make the combination of DL and EL data to measure particle flux.

ACKNOWLEDGEMENTS

This work was supported by the Andalusia Regional Government through project P12-RNM-2409, by the Spanish Ministry of Economy and Competitiveness through projects CGL2013-45410-R and CGL2016-81092-R and by the European Union's Horizon 2020 research and innovation programme through project ACTRIS-2 (grant agreement No 654109).The authors thankfully acknowledge the FEDER program for the instrumentation used in this work. This work was also partially funded by the University of Granada through the contract "Plan Proprio. Programa 9. Convocatoria 2013" and CNPq by scholarships from "Ciências sem Fronteiras (Convocatória 2015") and "Doutorado GD - Processo:140017/2017-8".

References

[1] To-Prof Action. Definition of terminology related to PBL, 2015.

[2] Behrendt, A., Wulfmeyer, V., Hammann, E., Muppa, S. K., and Pal, S. Profiles of second-to-fourth-order moments of turbulent temperature fluctuation in the convective boundary layer: first measurements with rotational Raman lidar, *Atmos. Chem. Phys.*, **15** (2015), 5485–5500.

[3] Coen, M. C., Praz, C., Haefele, A., Ruffieux, D., Kaufmann, P. and Calpini, B. Determination and climatology of the planetary boundary layer height above the Swiss Plateau by in situ and remote sensing measurements as well by the COSMO-2 model, *Atmos. Chem. Phys.*, **14** (2014), 13205-13221.

[4] Pal, S., Behrendt, A., and Wulfmeyer, V. Elastic-backscatter-lidar-based characterization of the convective boundary layer and investigation of related statistics, *Ann. Geophys.*, **28** (2010), 825–847.

[5] Lenschow, D. H., Wulfmeyer, V., Senff, C. Measuring Second-through Fourth-Order moments in noisy data, *Journal of Atmospheric and Oceanic Technology*, **17** (2000), 1330 - 1347.

[6] Moreira, G. de A., Marques, M. T.A., Nakaema, W., Moreira, A. C. de C. A., Landulfo, E. Planetary Boundary Layer height estimation from doppler wind lidar measurements, radiosonde and hysplit model comparison, *Óptica Pura y Aplicada*, **48** (2015), 179–183.

[7] McNicholas, C., Turner, D. D. Characterizing the convective boundary layer turbulence with a High Spectral Resolution Lidar, *Journal of Geophysical Research Atmospheres*, **119** (2014), 910 – 927.

TRACKING ATMOSPHERIC BOUNDARY LAYER IN TEHRAN USING COMBINED LIDAR REMOTE SENSING AND GROUND BASE MEASUREMENTS

Hossein Panahifar[1]* and Hamid Khalesifard[2]

[1,2]*Department of Physics and Center for Research in Climate Change and Global warming, Institute for Advanced Studies in Basic Sciences, No. 444 Prof. Sobouti Blvd., Zanjan 4513766731, Iran, *h.panahifar@iasbs.ac.ir*

ABSTRACT

The vertical structure of the atmospheric boundary layer (ABL) has been studied by use of a depolarized LiDAR over Tehran, Iran. The boundary layer height (BLH) remains under 1km, and its retrieval from LiDAR have been compared with sonding measurements and meteorological model outputs. It is also shown that the wind speed and direction as well as topography lead to the persistence of air pollution in Tehran. The situation aggravate in fall and winter due to temperature inversion.

1 INTRODUCTION

The structural parameters of atmospheric boundary layer (ABL) such as its height (BLH) is important in weather, climate and air quality models [1-2]. Despite its importance, the BLH evolution is still poorly investigated. This is as a consequence of complex processes which happen at small temporal and spatial scales.

The BLH is usually indirectly determined from Sonding. Seidel et al recommended either the parcel method or the Richardson number (RI) are the most reliable methods to retrieve BLH from sonding [3]. The LiDAR, which used for direct determination of BLH, benefits from ability to continuous sampling of the ABL. Several methods, such as the threshold method, gradient method, and variance method, have been developed to determine BLH using LiDAR measurements [4-6].

Tehran (35.5°N-35.8°N, 51.2°E-51.6°E, 1040-1700 m AMSL) is the capital of Iran with more than 12 million inhabitances. The city is the most populated city in Iran and suffering from atmospheric urban pollution. There are a few papers considering the atmospheric state of Tehran. Sabetghadam et al have shown that the changes in air quality are responsible for the long term visibility degradation [7]. Halek et al have found that the mass concentration of $PM_{2.5}$ increases almost three time in the cold seasons [8]. Khlesifard et al has shown that the city also had been subjected to dust events especially in spring and summer times [9-10].

This works aims to represents LiDAR based methods to retrieve BLH over Tehran. In Section 3 LiDAR measurements for the case study 14 Dec 2014 has been described. The result compares with the sonding observations and meteorological model output. To the best of our knowledge, this is the first attempt to determine BLH by remote sensing observations for Tehran.

2 TOPOGRAPHY, GROUND BASED MEASUREMENTS

Figure. 1 depicts the topography of Tehran, provided by the Shuttle Radar Topography Mission (SRTM) [11]. Tehran metropolitan area is extended from Alborz Mountains at the north to the Dasht-e Kavir at the south. The LiDAR facility of IASBS (35.7669°N, 51.3118°E, 1430 m AMSL) is located in northwest Tehran and has started its activity since November 2014. Data collected by this facility in combination to Radiosonding measurements (Mehrabad meteorological office) were used to investigate the boundary layer structure over the city.

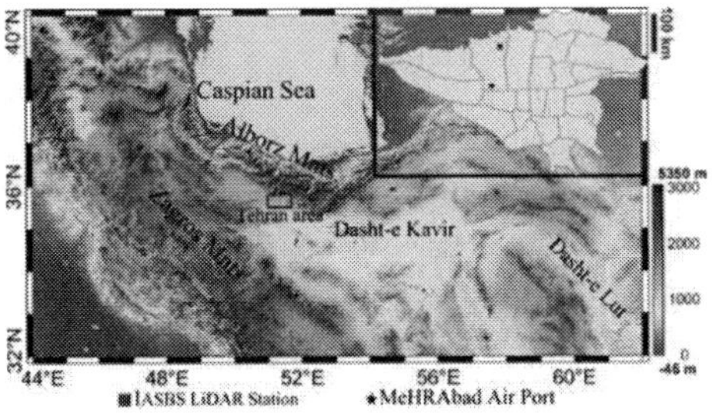

Figure 1: Topography map of Northern part of Iran provided by SRTM

The installed depolarized LiDAR benefits from a frequency doubled pulsed Nd-YAG laser with pulse duration 7 ns, energy per pulse: 40 mJ @ 532 nm as its transmitter. On the receiver side, a sensitive polarization setup after a 8" cassegrian telescope separate the received 532 nm backscattered signal to two linearly cross polarized beams. The radio-sonde (35.6900°N, 51.3112°, 1200 m AMSL, ≈8 km from IASBS Site) located in Mehrabad airport, and launched at a frequency of 2 times per day.

3 RESULTS AND DISCUSSION

Figure. 2 shows 24 hour time series for range corrected attenuated backscatter signal (RCS) on parallel channel after overlap correction. The potential temperature and relative humidity vertical profile is also shown by black and cyan line respectively. Fig. 2 very clearly depicts a nocturnal layer at early hour (00:00 till 06:00 LT) and a very stable atmosphere throughout the day. The Figure also represent development of a strong ABL due to rush-hour induced pollution (07:00 till 16:00 LT).

Figure. 3 represent the time series for depolarization signal at 532 nm on 14 Dec 2014.

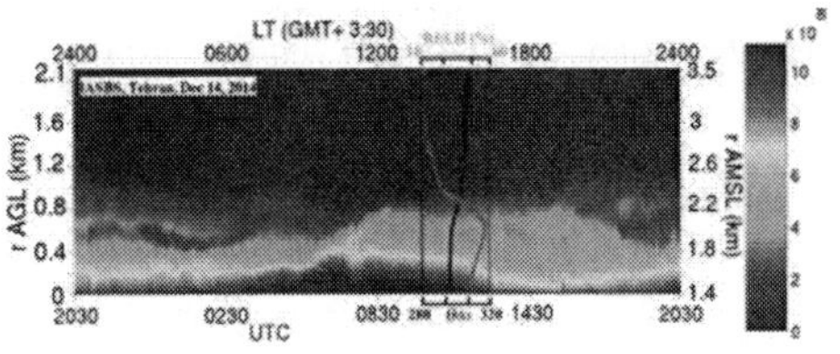

Figure 2: Time series for RCS, Potential temperature and relative humidity is shown in the box for 12:00 UTC, 14 Dec 2014

The depolarization ratio exceeds and reaches to around 25 % in second half of the day. The measured values of depolarization ratio consistent with observation of smoke ($\delta \sim 15\%$) and dusty mix ($\delta \sim 25\%$) in the atmosphere of Tehran [12].

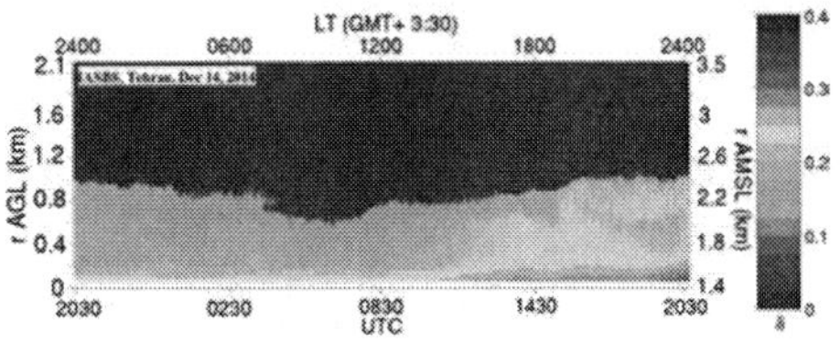

Figure 3: Time series for depolarization of LiDAR signal on 14 Dec 2014

According to HYSPLIT back trajectories (not shown) urban pollution could be the source of high $PM_{2.5}$ concentration in midday and the source of dust is Dasht-e Kavir desert.

3.1 BLH DETERMINATION WITH SONDE-LiDAR MEASUREMENTS

For radiosonde observations, the potential temperature and wind profiles are used to estimates the BLH employing Richardson (RI) number. In this method, the height at which RI is larger than the critical value ($R_{ib} > R_{ibc}$) is considered as BLH [4].

For LiDAR measurements, we took the following steps respectively to retrieve BLH from the range-corrected signal (RCS). Since high reso-

https://doi.org/10.1051/epjconf/201817606011

lution analysis of BLH requires a high signal-to-noise ratio, before any processing on RCS, an algorithm was applied to reduce the noise. The lidar profiles were also averaged over a time period 7 min. This period corresponds to a characteristic time scale of atmospheric turbulence in the boundary layer [4]. At the next step, we took first- and second-order derivative of averaged RCS with respect to the altitude to find points at which the sign of derivatives changes. For accurate detection of BLH in the case of multiple aerosol layer (Fig. 2), we considered a threshold level. The threshold value was chosen based on the analysis of several cases.

Figure. 4 illustrates the evolution of BLH for case study 14 Dec 2014. The red line identifies the BLH as determined through LiDAR approach.

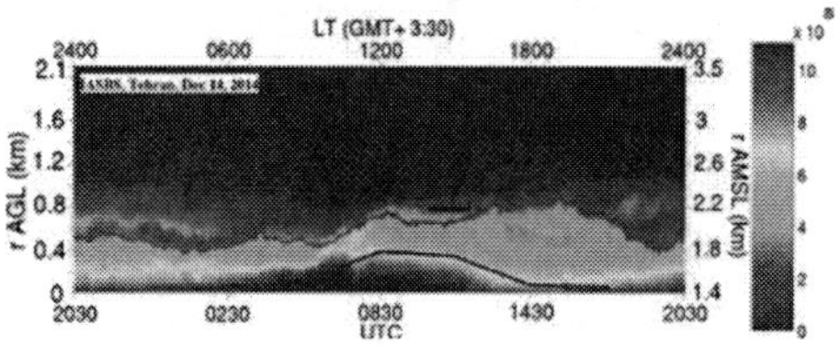

Figure 4: BLH evolution on 14 Dec 2014. The red, blue and black lines shows BLH obtained from LiDAR, NOAA model and RI.

The blue line represents the estimates obtained from NOAA HYSPLIT transport and dispersion model (GDAS 1). The black vertical line shows the BLH obtained from RI method for radiosonde launched at 12:00 UTC on Mehrabad airport. The figure covers the complete diurnal cycle of the BLH evolution. The BLH is found to grow during the day, reaching a maximum value in the early afternoon and then decaying in the late afternoon and evening (with respect to LT). Since we benefit from a high power output LiDAR with spatial resolution $\sim$ 15 m, so we also could retrieved the depth of entrainment zone precisely.

4 CONCLUSIONS

In this study, the BLH which is determined from three distinct approaches have compared for case study 14 Dec 2014 (Fig. 4). The LiDAR approach considers a method based on the identification of minima in the first- and second-order derivative of the RCS. We took a minimum which exceeds from specific threshold as the BLH. Additional estimates of the BLH are obtained from radiosonde observation and NOAA HYSPLIT meteorological model. Looking to Fig. 4, One can easily find that the differences of BLH retrieved from LiDAR and sonding observations are less than a hundred meters. While there is a considerable differences between the BLH obtained from NOAA HYSPLIT meteorological model and two other approaches.

Figure. 5 depictes average wind speed and directions for healthy (According to Tehran Air Quality Control Company) and Un-healthy days on a period of 5 years for Mehrabad airport. The majority of surface winds blows from the south when air is at unhealthy condition. On the other hand, at days with good air quality condition, the surface winds have more west-east component. So it seems that Alborz Mountain inhibit ventilation of pollutants and blocked northerly wind blowing from south. The wind speed almost remains below 3 m/se for unhealthy days.

Therefore, Topography as well as wind speed and direction, and temperature inversion could be considered as the main factors which lead to urban pollution in Tehran. The BLH also remain under 1 km especially in cold seasons.

ACKNOWLEDGEMENTS

The authors would like to express their gratitude to Ruhollah Moradhaseli for valuable discussions. The authors gratefully acknowledge the NOAA Air Resources Laboratory (ARL)

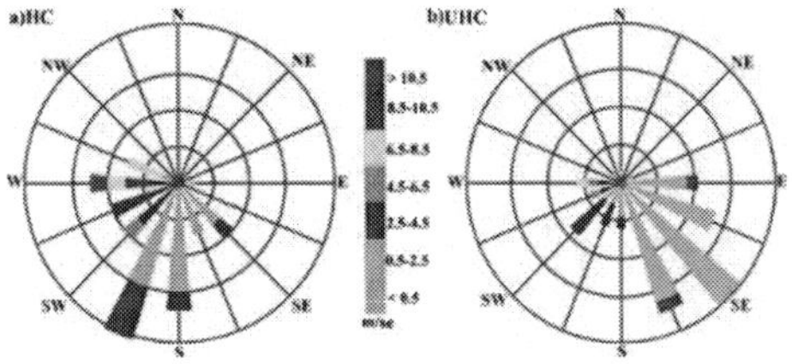

Figure 5: Windrose over Mehrabad airport over 5 year period a) Days with healthy condition (HC) b) Days with unhealthy condition (UHC)

for the provision of the HYSPLIT transport and dispersion model used in this work. Special thanks go to Larry Oolman for kindly providing some of the surface observations.

References

[1] Stull, R. B.: An Introduction to Boundary Layer Meteorology, Kluwer Academic Publishers, Norwell, MA, 1988.

[2] Zilitinkevich, S. S.: The Height of the Atmospheric Planetary Boundary layer: State of the Art and New Development. National Security and Human Health Implications of Climate Change, Springer Netherlands, 147161, 2012.

[3] Seidel, D. J., Ao, C. O., and Li, K.: Estimating Climatological Planetary Boundary Layer Heights from Radiosonde Observations: Comparison of Methods and Uncertainty Analysis, J. Geophys. Res., 115, D16113, doi:10.1029/2009JD013680, 2010.

[4] Flamant, C., Pelon, J., Flamant, P. H., and Durand, P.: Lidar Determination of the Entrainment Zone Thickness at the Top of the Unstable Marine Atmospheric Boundary Layer, Bound.-Lay. Meteorol., 83, 247284, 1997.

[5] Emeis, S., Schafer, K., and Munkel, C.: Surface-based Remote Sensing of the Mixing-layer Height-a Review, Meteorol. Z., 17,621630, 2008.

[6] Haeffelin, M., Angelini, F., Morille, Y., Martucci, G., Frey, S., Gobbi, G. P., Lolli, S., ODowd, C. D., Sauvage, L., Xueref Remy, I., Wastine, B., and Feist, D. G.: Evaluation of Mixing Height Retrievals from Automatic Profiling Lidars and Ceilometers in View of Future Integrated Networks in Europe, Bound.-Lay. Meteorol. , 143, 4975, 2012.

[7] Samaneh Sabetghadam, Farhang Ahmadi-Givi, Yahya Golestani, "Visibility trends in Tehran during 1958-2008", Atmos. Env. 62, 512-520 (2012).

[8] Halek, Farah, Mansour Kianpour-Rad, and Ali Kavousirahim. "Seasonal variation in ambient PM mass and number concentrations (case study: Tehran, Iran)." Environmental monitoring and assessment 169.1 (2010): 501-507.

[9] H. R. Khalesifard and H. Panahifar, "Monitoring Atmospheric Particulate Matters in Tehran by Synergy of LiDAR Measurement, Atmospheric Models and Satellite Data," in Light, Energy and the Environment 2015, OSA Technical Digest (online) (Optical Society of America, 2015), paper EM4A.1.

[10] Abdi Vishkaee, F., C. Flamant, J. Cuesta, L. Oolman, P. Flamant, and H. R. Khalesifard (2012), Dust transport over Iraq and northwest Iran associated with winter Shamal: A case study, J. Geophys. Res., 117, D03201, doi:10.1029/2011JD016339.

[11] USGS (2004), Shuttle Radar Topography Mission, 1 Arc Second scene SRTMu03n008e004, Unfilled Unfinished 2.0, Global Land Cover Facility, University of Maryland, College Park, Maryland, February 2000.

[12] Burton, S. P., Ferrare, R. A., Hostetler, C. A., Hair, J. W., Rogers, R. R., Obland, M. D., Butler, C. F., Cook, A. L., Harper, D. B., and Froyd, K. D.: Aerosol classification using airborne High Spectral Resolution Lidar measurements methodology and examples, Atmos. Meas. Tech., 5, 73-98, doi:10.5194/amt-5-73-2012, 2012.

https://doi.org/10.1051/epjconf/201817606012

WIND SPEED VECTOR RESTORATION ALGORITHM

Nikolay Baranov[1], Gleb Petrov[2], Ilia Shiriaev[2]

[1] *Dorodnicyn Computing Centre, FRC CSC RAS*
[2] *JSC "International Aeronavigation Systems Concern" (IANS)*

ABSTRACT

Impulse wind lidar (IWL) signal processing software developed by JSC «BANS» recovers full wind speed vector by radial projections and provides wind parameters information up to 2 km distance. Increasing accuracy and speed of wind parameters calculation signal processing technics have been studied in this research. Measurements results of IWL and continuous scanning lidar were compared. Also, IWL data processing modeling results have been analyzed.

1 INTRODUCTION

An effective means for determining the wind profile of the surface layer of the atmosphere is a Doppler wind lidar. Currently, the most widely spread lidars are operating with infrared laser beam at a wavelength of 1.54 microns. The lidar beam is scattered by aerosol particles which are present in the atmosphere.

VAD and DBS are the atmosphere scanning modes by the lidar laser beam. They provide aggregated data reception on the values of the radial projection of the wind speed in the measurement direction. This data is further converted into the value of the full vector of wind speed for all heights of scanning [1].

In a continuous circular scan the wind speed at a fixed height of its radial projection on the beam will be described by a harmonic law [1]:

$$y = -u \cdot \sin\theta \cdot \cos\varphi -$$

$$-v \cdot \cos\theta \cdot \cos\varphi - w \cdot \sin\varphi, \qquad (1)$$

where y – lidar measured value of the radial projection of the wind speed, u, v, w – full vector of wind speed components, θ azimuth of the measurement direction, φ – deviation angle from the vertical measurement direction.

The major problem with wind speed and direction remote measurement, is that the measurements are at the same height have spatiotemporal distribution. For example, when measuring the wind speed at a height of 1000 meters with the value of the deflection angle from the vertical of 20 degrees, the distance between the points of measurement can reach 730 m. In the case of spatially inhomogeneous wind field large spatial distribution of measuring points can generate additional error when dealing with the recovery of the wind velocity vector problem. Therefore, the measurement model used for the restoration of the wind velocity vector, must take into account the spatial inhomogeneity of the wind field.

2 METHODOLOGY

In this paper, the wind field model is based on the hypothesis of uncorrelated linear variable for each horizontal coordinate of the wind speed horizontal component variability. The wind speed vertical component is assumed spatially uniform throughout the volume of measurement at each height. The wind speed gradients for the directions can be set.

Let us set:

- u_0, v_0 – wind speed horizontal component spatially averaged over the measurement volume at a fixed height values of the
- u_x – gradient alteration of the u – wind speed component along the axis x,
- u_y – gradient alteration of the u – wind speed component along the axis y,
- v_x – gradient alteration of the v – wind speed component along the axis x,
- v_y – gradient alteration of the v – wind speed component along the axis y.

The usage of a gradient wind field model allows to obtain additional wind characteristics that determine horizontal wind shear at each height.

The linear model of the wind speed gradients in the directions of the horizontal plane provides for the total value of the wind speed component along each axis represents an additive mixture of the half value of the speed and gradient. For example,

 https://doi.org/10.1051/epjconf/201817606012

the value of wind speed along the x-axis components can be written as:

$$\begin{cases} u(x) = u_0 + x \cdot u_x \\ v(x) = v_0 + x \cdot v_x \end{cases}, x \in (-R..R),$$

where R – the radius of the scan circle at a fixed height.

Linear coordinates are functions of the azimuthal measurements direction angle. It provides a linear system of equations for determining the unknown parameters of the wind field, but the system is inconsistent. If you write the equations (1) for the speed radial projection, taking into account the impact of linear gradients of wind speed in the directions in the horizontal plane, you can get:

$$y = -\left(u_0 + u_x \cdot R \cdot \sin(\theta) + u_y \cdot R \cdot \cos(\theta)\right) \cdot$$
$$\cdot \sin\theta \cdot \cos\varphi -$$
$$-\left(v_0 + v_x \cdot R \cdot \sin(\theta) + v_y \cdot R \cdot \cos(\theta)\right) \cdot$$
$$\cdot \cos\theta \cdot \cos\varphi - w \cdot \sin\varphi \qquad (2)$$

It is obvious, that after the brackets removal, the coefficients of u_x and v_x are the equal, which makes it impossible for an independent analysis of wind speed gradients in the directions.

Let us analyze the possibility of the harmonic approximation of the dependence of the lined coordinates of a bell shaped function which simulates the scan beam movement in the azimuthal angle.

Consider the possibility of function bell shaped usage in the following equations:

$$\begin{cases} f_x(\theta) = R \cdot \exp(-\beta \cdot \theta^2) \\ f_y(\theta) = R \cdot \exp\left(-\beta \cdot \left(\theta - \frac{\pi}{2}\right)^2\right), \end{cases} \qquad (3)$$

where β - the normalizing index that determines the beam width, R – the radius of the scan circle at a fixed height..

The values of the azimuthal angle range from $-\pi$ to π radians. The normalization index β must provide a single value of the f_x function in 0 radians point and f_y function when the azimuth angle of $\pi/2$ radians. In addition, the values of bell shaped functions should fall to 0 within the range of the azimuth angle change. When β is close to the 1, the listed conditions are satisfied.

The wind speed components at an arbitrary point, determined by an azimuth angle θ taking into account (3) can be written as

$$\begin{cases} u = u_0 + u_x R\exp(-\theta^2) + u_y R\exp\left(-\left(\theta - \frac{\pi}{2}\right)^2\right) \\ v = v_0 + v_x R\exp\left(-\left(\theta - \frac{\pi}{2}\right)^2\right) + v_y R\exp(-\theta^2) \end{cases}$$

The equation for the wind speed radial projection in the scanning direction taking into account the uniformity of the vertical component of the wind speed is the following:

$$y = -\left(u_0 + u_x R\exp(-\theta^2) + u_y R\exp\left(-\left(\theta - \frac{\pi}{2}\right)^2\right)\right) \cdot$$
$$\cdot \sin\theta \cos\varphi -$$
$$-\left(v_0 + v_x R\exp\left(-\left(\theta - \frac{\pi}{2}\right)^2\right) + v_y R\exp(-\theta^2)\right) \cdot$$
$$\cdot \cos\theta \cos\varphi - w \cdot \sin\varphi \qquad (4)$$

For the exact restoration of the seven parameters of the numerical simulation model it is not enough to have a connection of the radial projection from the azimuthal angle. Solving the wind field parameters restoration problem by an available set of the wind speed radial projections, the solutions of the optimization problem is linearly independent on (u_0, v_0) and (u_x, u_y, v_x, v_y) parameters. The model, includes an additional optimization parameter - w, which does not have any function of the azimuth angle. Thus, the determination of the wind speed vertical component in the recovery issue, taking into account the spatial inhomogeneity is possible only with the introduction of additional modulation for the vertical component of the wind speed. This modulation is possible by changing the scan beam angle of elevation.

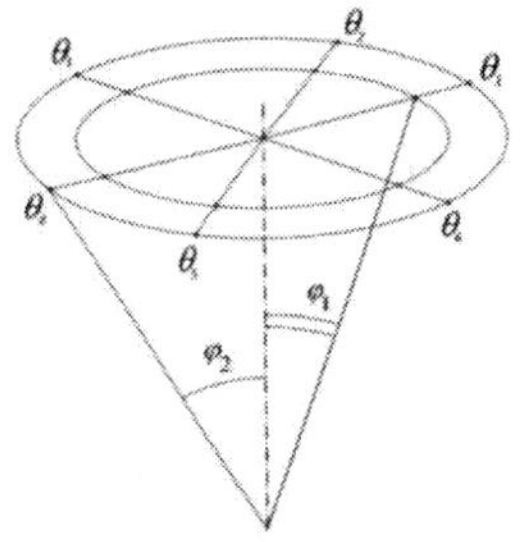

Figure 1 Lidar scan pattern

The scan algorithm should foresee the possibility of working with variable angles of elevation. The scan pattern for two fixed values of the angle of elevation and six values of azimuth are shown in the figure 1.

The equation (3) can be shown the following way:

$$y = u_0 x_{ij1} + u_x x_{ij2} + u_y x_{ij3} + v_0 x_{ij4} + v_x x_{ij5} + v_y x_{ij6} + w x_{j7}\,, \quad (5)$$

where
$$\begin{cases} x_{ij1} = -\sin\theta_i \cdot \cos\varphi_j \\ x_{ij2} = -\exp(-\theta^2)\cdot \sin\theta_i \cdot \cos\varphi_j \\ x_{ij3} = -\exp\left(-\left(\theta - \frac{\pi}{2}\right)^2\right)\cdot \sin\theta_i \cdot \cos\varphi_j \\ x_{ij4} = -\cos\theta_i \cdot \cos\varphi_j \\ x_{ij5} = -\exp\left(-\left(\theta - \frac{\pi}{2}\right)^2\right)\cdot \cos\theta_i \cdot \cos\varphi_j \\ x_{ij6} = -\exp(-\theta^2)\cdot \cos\theta_i \cdot \cos\varphi_j \\ x_{j7} = -\sin\varphi_j \end{cases}$$

With the known values of the angle of azimuth and elevation angle, the coefficients $x_{1..7}$ are permanent. The problem is reduced to finding the vector component of the wind speed values (u_0, v_0, w) and indicators of the wind field inhomogeneity (u_x, u_y, v_x, v_y), satisfying the equation (5) with the available measurement results. The result of measurements is the set of values of the radial projection of the wind speed y, obtained from the corresponding azimuth angle values and elevation, thus $x_{1..7}$. The values of the radial speed projections are attached to the same range of marks along the beam, thus the same height of scan.

For an exact solution of the equation with seven unknowns seven measurements of the radial projection of the wind speed at different values of the angle of azimuth and at least two elevation angles are necessary. This solution will be precise from a mathematical point of view, but is subject to distortion in view of the real atmospheric turbulence, as well as the presence of technical measurement system noise and interferences.

3 RESULTS

Taking into account the physical and technical aspects of the noise impact, redundant inputs are used in the experiments. In other words, the measurement is carried out over seven measurements of the wind speed radial projection. This system of equations becomes redundant and allows the plurality of solutions. To select the optimal solution the object function is introduced.

The object function is to reflect the discrepancy of the resulting function to actually measured data in the experiment. For solving this problem it is required to find such a combination of $u, u_x, u_y, v, v_x, v_y, w$ that minimizes error recovery. Since all the equations are linear, then the search for suitable solutions optimal linear regression algorithm is based on the method of least squares.

According to the obtained values of the wind speed component mean horizontal wind speed, vertical wind speed and wind direction in the horizontal plane can be calculated:

- horizontal wind speed: $H = \sqrt{u^2 + v^2}$;
- vertical wind speed: $V = w$;
- wind direction: $D = \arccos\left(\frac{v}{V}\right)$.

Furthermore, during the problem solution 4 spatial non-uniformity of the wind field are determined - the gradients of the wind speed in the horizontal plane.

From the perspective the most effective algorithm developing for space scanning, it seems the optimal to do two turns on the azimuth angle with a fixed position of the elevation. The design of an impulse wind lidar includes an xy-scanner, and such scanning algorithm can be implemented on the currently existing equipment.

Developed new model were tested on real lidar data, which were obtained during atmosphere scanning with two different elevations: 50 and 70 degrees. Wind speed vector restoration algorithm was applied to this data for three and seven parameter models with fitting function (1) and (4) respectively. Examples of results are shown in the figure 2.

Seven parameters model (red curve) improves outlier detection, prevents false filtering and better fits experimental data that decreases wind parameters restoration error, while three parameters model (gray curve) approximates data less accurately.

Statistical information for each model is shown in the table 1.

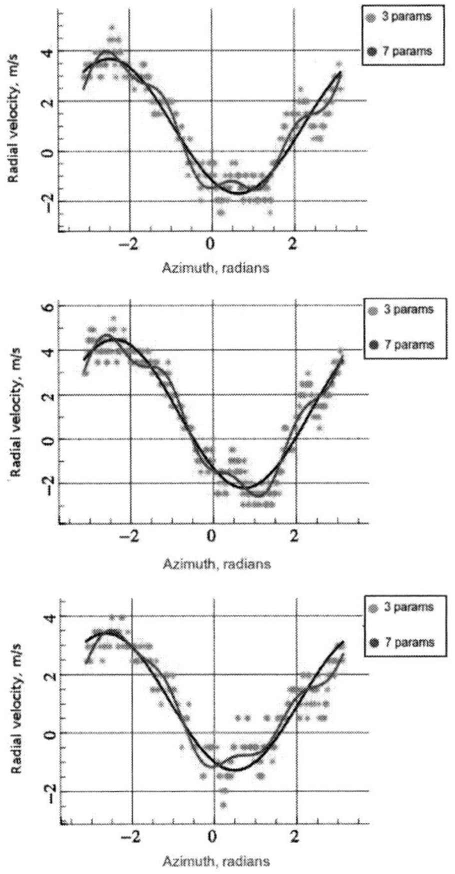

Figure 2 Comparison of three and seven parameter models

Table 1
Statistics for three and seven parameters models

Statistics	*Seven parameters model*	*Three parameters model*
R^2 adjusted	0,95	0,83
Average error	0,39	0,62
Standard error	0,48	0,74

New model lets us:

- refine description of processes occurring in the atmosphere;
- acquire additional information about wind spatial distribution.

4 CONCLUSIONS

From the mathematical modeling, and taking into account the impact of noise, it implies that the value of the elevation angle should be 20 and 40 degrees from the vertical direction, and the azimuthal direction should be evenly spaced in a circle scan. The number of fixed positions azimuth angle is 5-7. Such scanning order will ensure the scan data processing, taking into account the spatial distribution of the wind field, and in addition, will not lead to a decrease in the wind speed and direction results calculation updating frequency on the output of the algorithm.

ACKNOWLEDGEMENTS

The work was partially supported by the Russian Foundation for Basic Research, project 16-07-01072.

References

[1] Christian Werner, Dr. Claus Weitkamp, 2005: Lidar. Range-Resolved Optical Remote Sensing of the Atmosphere, Springer New York, pp 325-354.

[2] ICAO Doc 9817 - Manual on Low-Level Wind Shear.

[3] Willett H. C., Sanders F. Descriptive Meteorology, Academic Press Inc. New York. 1959.

[4] Boilley A., Mahfouf J.-F. Wind shear over the Nice C^ote d'Azur airport: case studies, Nat. Hazards Earth Syst. Sci., 13, 2223–2238, 2013.

[5] Barthelmie R.J. The effects of atmospheric stability on coastal wind climates, Meteorol. Appl. 6, 39–47. 1999.

AIRPORT LOW-LEVEL WIND SHEAR LIDAR OBSERVATION AT BEIJING CAPITAL INTERNATIONAL AIRPORT

Hongwei Zhang[1], Songhua Wu[1, 2]*, Qichao Wang[1], Bingyi Liu[1], Xiaochun Zhai[1]
[1] *Ocean Remote Sensing Institute, College of Information Science and Engineering, Ocean University of China, Qingdao 266100, China.*
[2] *Laboratory for Regional Oceanography and Numerical Modeling, Qingdao National Laboratory for Marine Science and Technology*, **Email: wush@ouc.edu.cn*

ABSTRACT

Ocean University of China lidar team operated a pulse coherent Doppler lidar (PCDL) for the low level wind shear monitoring at the Beijing Capital International Airport (BCIA) in 2015. The experiment configuration, observation modes is presented. A case study shows that the low level wind shear events at the southern end of 18R/36L runway were mainly caused by the trees and buildings along the glide path under strong northwest wind conditions.

1. INTRODUCTION

Low-level wind shear is a sharply change in wind speed or wind direction in space within 2000 feet at height. Wind shear can be broken into vertical and horizontal components, with horizontal wind shear occurring near fronts and near the coast, and vertical wind shear typically near the ground surface [[1]]. Wind shear can cause disastrous aircraft accidents by affecting aircraft airspeed during take-off and landing phases [2]. There were 33 accidents caused by turbulence encounter, covering 18 percent of the total numbers of accidents between 2014 and 2015 [3, 4]. Effective wind shear alarm systems are needed to guarantee the safety of civil aviation.

The PCDL systems have been proved to be a useful tool that can capture the small size characteristic of wind, turbulence, wake vortices, and so forth [5-8]. The PCDL is one of the most accurate optical sensing techniques by transmitting a laser beam into the atmosphere and detecting the radial backscattered signal by aerosol particles [9]. The radial velocity (Line of sight velocity, LOS) could get by analyzing the Doppler shift in the frequency of the backscattered signal [10, 11]. The LOS velocity is the component of the air motion along the laser beam, thus the wind field information could be induced [11].

Due to the variable types of low-level wind shear and its uncertainty of location and quick evolution, low-level wind shear detection requires high spatial and temporal resolution that could only be achieved by operating a PCDL. OUC's PCDL is the modified version based on a commercial WindPrint S4000 Doppler wind lidar from Seaglet Environmental Technology. The pulsed 1550 nm fiber laser source is based on the MOPA architecture with a large core fiber and excellent beam quality (M^2=1.2), whose pulsed energy is 150 μJ with a pulse repetition rate of 10 kHz and an adjustable pulse length of 100 ns to 400 ns (200 ns in this study). The software of PCDL also includes a real time display of the radial wind velocity and wind profiles production. The measurement range of the PCDL system is ± 50 m/s, while the speed measurement accuracy is 0.1 m/s. The scanner and detection range of 3000 m (4000 m at a proper aerosol concentration) enable the PCDL system to capture the wind shear occurring far away from the PCDL. The Fast Fourier Transform (FFT) spectral estimates are processed with FPGA (Field Programmable Gate Array) in real time. All these design of the PCDL system make it stable, reliable and high-integrated. The detailed specifications of the PCDL system are listed in the Table 1.

Table 1 Specifications of the pulsed coherent Doppler lidar

Qualification	Specification
Spatial Resolution (Transverse)*	< 5 m @ specified range
Radial Spatial Resolution	30 m
Measurement Range**	90 to 4000 m
LOS Velocity Update Rate	4 Hz
Wind Profile Update Rate	1 min
PPI Scan Speed	$20°s^{-1}$(max)
Scanner Positioning Accuracy	0.1°
Velocity Measurement Range	± 50 m/s
Resolution of Velocity	0.3 m/s
Wavelength	1.5 μm
Pulse Energy	150 μJ
Pulse Duration	200 ns
Pulse Repetition Frequency	10 kHz

* *Depending on the laser beam positioning accuracy.*

** *Weather condition related*

The OUC Lidar team operated the PCDL for low level wind shear study at BCIA in 2015. We have developed and carried out several low level wind shear observation campaigns to capture the wind shear events from the radial velocity measurement in PPI (Plane Position Indicator) at low elevation angles and wind profiles in 5-DBS (Doppler Beam Swing) mode.

For the study of wind shear occurring along the glide path, the PCDL was configured at the south ending of the 36L/18R runway as Figure 1 shows. Taking the 3° incidence angle of aircraft landing path into consideration, a stepwise PPI scanning pattern was performed in the field wind shear observation campaign to reveal the wind that aircrafts would encounter. The detailed designs of the glide path PPI scan are shown in Table 2.

Table 2 Scanning patterns along the glide path

No	Elevation angle	Azimuthal angle
1	3°	207° to 219°
2	2°	219° to 231°
3	1°	231° to 243°
4	0°	243° to 260°

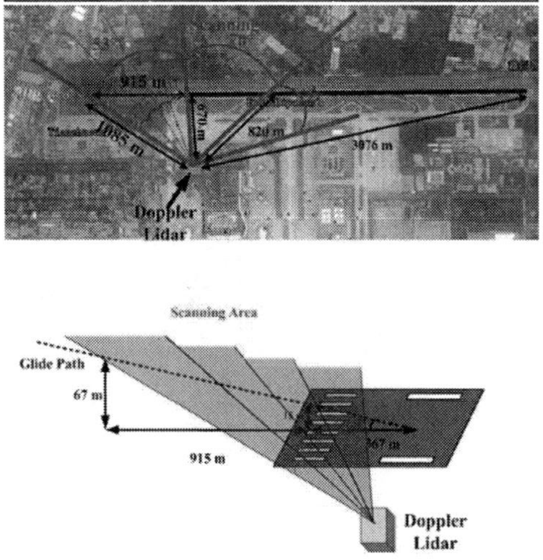

Figure 1 Location and scanning strategies of PCDL that configured at BCIA.

2. RESULTS

Wind speed profiles are effective tools to illustrate the wind speed varying with height, thus it is useful to identify the wind shear occurring at different height. We illustrated 1-min averaged wind profiles generated between 13:39 and 15:32 on Dec 15th into one figure to identify the vertical wind shear as Figure 2 shows.

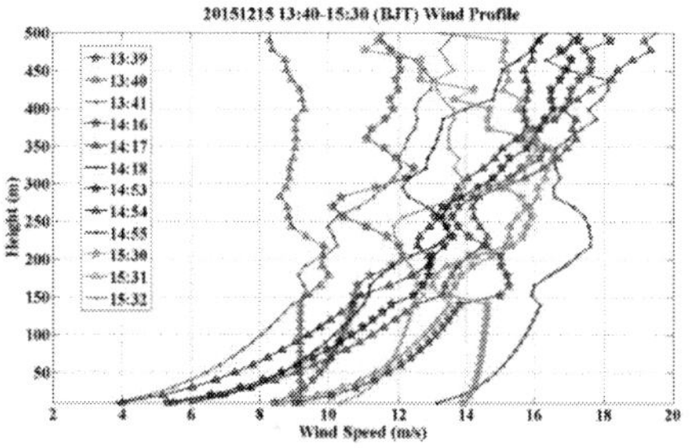

Figure 2 Wind speed profiles measured by PCDL at BCIA on Dec 15th, 2015.

Due to the influence of surface manufactures, the wind speed profiles are not perfectly match with the exponential wind profile,

$$V(h) = V_0 \left(\frac{h}{h_0}\right)^{\partial}, \qquad (1)$$

where h is the altitude of measurement volume

of PCDL, V_0 is the wind speed at the altitude of h_0 and ∂ is the Hellmann index which is related to the surface roughness and the stability of Prandtl layer. We extended the wind profiles from 10 m above ground to 500 m by taking the two lower measured points and wind profile equation (1) into consideration.

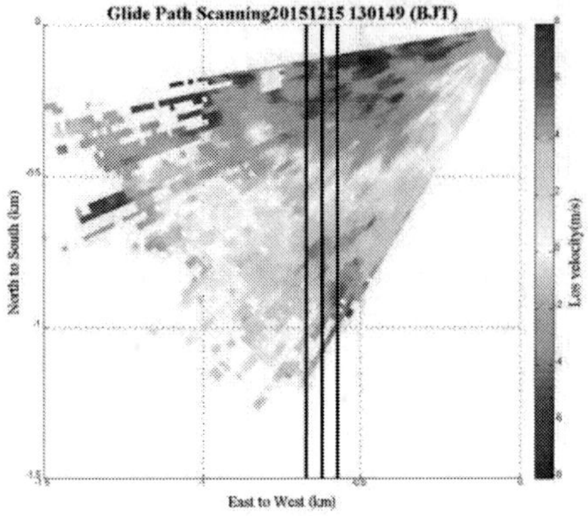

Figure 3 Glide path scanning of 36L/18R runway at BCIA on Dec 15th, 2015.

We have captured the existence of small-size airflow disturbances embedded in the prevailing wind with high spatial resolution by operating step-wise PPI scan mode as Figure 3 shows. The red and blue colors indicate positive (away from lidar) and negative (towards lidar) movement, respectively, of the atmospheric particles along the laser beam. These airflow disturbances are generally small scale with serval hundred meters at horizontal dimensions and difficulty to be captured by DBS mode. There are three black lines in Figure 3, the center one is the extended centerlines of the runway and the other two are located at about 50 m to both side of the extended centerline to take into consideration the uncertainty with the flight path of the aircraft. By checking the series of PPI scans, the airflow disturbances move along glide path with the background wind flowing over the runway corridor. They are believed to arise from disruption of the northerly airflow by the buildings above the ground surface.

For illustrating the transient and sporadic nature of the low level wind shear along the glide path, we construct the headwind profiles through the step-wise PPI scan as Figure 4 shows. The duration of a PPI scan over the runway corridor is about 30 s.

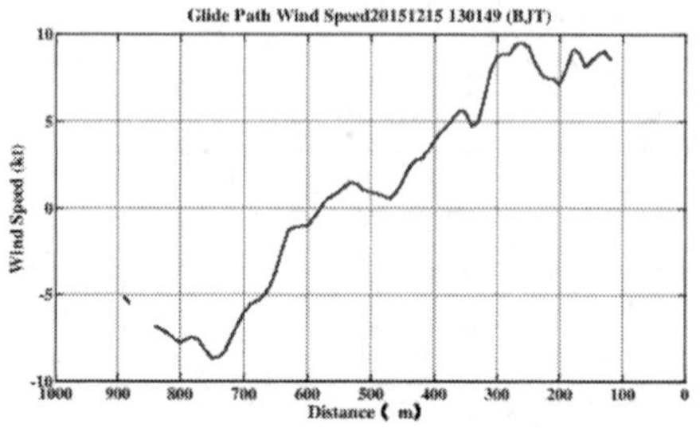

Figure 4 Wind speed profiles measured by PCDL at BCIA on Dec 15th, 2015.

Woodfield and Woods defined a severity factor I of wind shear as

$$I = \left(\frac{dV}{dt}\right)\left(\frac{\Delta V}{V_{app}}\right)^2 = \frac{1}{V_{app}}\left|\frac{\Delta V}{R^{1/3}}\right|^3, \qquad (2)$$

where $\Delta V / R^{1/3}$ is the rate of change of wind speed, ΔV is the total change of wind, V_{app} means the normal approach speed of the aircraft and R represents the ramp length.

The value of the severity factor I is 0.08 with respect to the headwind speed reduced 18 kt within 500 m. The cubic root of severity factor is used to classify the turbulence alerts into two levels: "moderate turbulence (0.3~0.5)" and "severe turbulence (>0.5)". The value in this study is 0.44, a moderate turbulence.

The reliability of algorithms have been proved by wind shear reports from the crew and the airport terminal control office. According to the pilot's report, there is wind shear event at 13:01, 15th December, 2015. The pilots have to overshoot to ensure safety.

3. DISCUSSION AND CONCLUSIONS

The PCDL is proved to be an effective remoting tool to detect the low-level wind shear under dry conditions. It is possible to capture the wind shear events with different observation modes. From the field wind shear observation campaign at BCIA, we developed algorithms to identify the wind shear and turbulence.

The method to extend the height of wind profiles with calculating the Hellmann index needs to be improved by taking surface roughness into account. Also, wind speed retrieved from the PPI scan data is related to the intercept angle between the laser beam pointing direction and the prevailing wind. When the scanning azimuthal angles are nearly correspond with or nearly opposite to the prevailing wind, the LOS wind speeds are nearly equal to the real wind speed. However, when it is nearly perpendicular to the primary wind direction, the LOS wind speed is a project component of the wind vector. If the angle of intercept is larger than 30 degrees, the radial wind data would not be utilized to construct the headwind profiles considering the increasing underestimation of the headwind component as the across angle increases.

ACKNOWLEDGEMENTS

We thank our colleagues for their kindly support during the field experiments, including Rui Wang and Xiaoqing Yu with Ocean University of China for operating the lidar at BCIA; Yilin Qi and Jiaping Yin from Seaglet Environment Technology for technical support. This work was jointly supported by the Boeing Company and Ocean University of China Research project.

References

[1] International Civil Aviation Organization, 1987: Wind Shear. CIRCULAR 186-AN/122.1987. *INTERNATIONAL CIVIL AVIATION ORGANIZATION CIRCULAR*, Montreal, Canada.

[2] James Evans and Donald Turnbull, 1989: Development of an Automated Wind shear Detection System Using Doppler Weather Radar. *Proceedings of the IEEE. Vol. 77, No. 11.*

[3] International Civil Aviation Organization (ICAO). *ICAO Safety Report 2015 Edition.*

[4] International Civil Aviation Organization (ICAO). *ICAO Safety Report 2016 Edition.*

[5] Wu, S., Yin, J., Liu, B., and Liu, J, 2015: Wind Turbine Wake Visualization by Doppler Lidar. In *Optics and Photonics for Energy and the Environment* (pp. EM3A-4). Optical Society of America.

[6] Shun, C. M., and Chan, P. W, 2008: Applications of an infrared Doppler lidar in detection of wind shear. *Journal of Atmospheric and Oceanic Technology*, 25(5), 637-655.

[7] Harris M, Young R I, Köpp F, et al, 2002: Wake vortex detection and monitoring. *Aerospace science and Technology*, 6(5): 325-331.

[8] Hon, K. K., Chan, P. W., Chiu, Y. Y., and Tang, W, 2014: Application of short-range LIDAR in early alerting for low-level wind shear and turbulence at Hong Kong international airport. *Advances in Meteorology, 2014.*

[9] Henderson, S. W., Suni, P. J., Hale, C. P., Hannon, S. M., Magee, J. R., Bruns, D. L., and Yuen, E. H. , 1993: Coherent laser radar at 2 μm using solid-state lasers. *IEEE Transactions on Geoscience and Remote Sensing*, 31(1), 4-15.

[10] Frehlich, R., and Kelley, N., 2008: Measurements of wind and turbulence profiles with scanning Doppler lidar for wind energy applications. *IEEE Journal of Selected Topics in Applied Earth Observations and Remote Sensing*, *1*(1), 42-47.

[11] Henderson, S. W., Hale, C. P., Magee, J. R., Kavaya, M. J., and Huffaker, A. V., 1991: Eye-safe coherent laser radar system at 2.1 μm using Tm, Ho: YAG lasers. *Optics letters*, *16*(10), 773-775.

[12] Wu, S., Yin, J., Liu, B., Liu, J., Li, R., Wang, X., and Zhang, K., 2014: Characterization of turbulent wake of wind turbine by coherent Doppler lidar. In *SPIE Asia Pacific Remote Sensing* (pp. 92620H-92620H). International Society for Optics and Photonics

SHIPBORNE WIND MEASUREMENT AND MOTION-INDUCED ERROR CORRECTION BY COHERENT DOPPLER LIDAR OVER YELLOW SEA IN 2014

Xiaochun Zhai[1], Songhua Wu[1,2]*, Bingyi Liu[1], Xiaoquan Song[1]

[1]*Ocean Remote Sensing Institute, Ocean University of China, China, *wush@ouc.edu.cn*
[2]*Laboratory for Regional Oceanography and Numerical Modeling, Qingdao National Laboratory for Marine Science and Technology, China*

ABSTRACT

Shipborne wind observations by the Coherent Doppler Lidar (CDL) during the 2014 Yellow Sea campaign are presented to study the structure of the Marine Atmospheric Boundary Layer (MABL). This paper gives an analysis of the correction for horizontal and vertical wind measurement, demonstrating that the combination of the CDL with the attitude correction system enables the retrieval of wind profiles in the MABL during both anchored and cruising measurement with satisfied statistical uncertainties.

1 INTRODUCTION

The vertical structure of atmospheric variables in Marine Atmospheric Boundary Layer (MABL) is significant for understanding the driving and coupling mechanisms and for parameterizing ocean-atmosphere interaction process. The Coherent Doppler Lidar (CDL) is an available tool with high spatial and temporal resolution, providing nearly continuous particle backscatter, wind profile observations in clear atmosphere [1], which is demanding for the air-sea boundary layer turbulent characteristics. Unlike the conventional wind measuring methods, CDL can only detect the Line-Of-Sight (LOS) velocity which is the projection of the horizontal and vertical velocity to the Lidar beam direction, thus it is necessary to conduct measurement at three or more different directions of the probing beam to retrieve the wind velocity vector. More complicated attitude correction should be considered when CDL is carried out at moving platform such as the ship or aircraft since the orientation of transmitting laser beam is not fixed and the speed of the ship itself and ocean wave will be stacked to the LOS velocity which has more obvious effect on vertical velocity. There have been several studies on attitude correction and error analysis of velocity measured by CDL. Wolfe et al and Pichugina et al [2] deployed the NOAA High Resolution Doppler Lidar (HSRL) along with the first use of a motion compensation system at sea in 2004. Hill et al [3] used the NOAA HSRL with a Strapped-Down System (SDS) to compensate for the orientation of the Lidar's scanning unit for the ship's motion and concluded that the attitude correction depends on the velocity of the seatainer and on the motion of the hemispheric scanner relative to the seatainer. Wulfmeyer et al corrected the vertical velocity using LOS velocity in zenith stare mode and horizontal wind derived from Velocity Azimuth Display (VAD) mode.

In order to measure the wind field and turbulence characteristics of MABL of the Yellow Sea, the experimental investigation was undertaken by Dongfanghong-2 research vessel affiliated with Ocean University of China in 2014. This paper gives a brief analysis of the correction for horizontal and vertical wind measurement using CDL with attitude correction system and robust motion correction algorithm.

2 METHODOLOGY

Ship motion turns out to be an important error source for the determination of turbulence variables using CDL. To study boundary layer dynamics, the atmospheric velocity in Earth coordinate system is required, so the correction of the Doppler frequency shift for the ship's motion relative to the Earth coordinate system is required, where the direction of the laser beam in Earth coordinate system and the LOS velocity induced by ship movement need to be determined using attitude correction system (ACS). Generally, the ACS uses Global Navigation Satellite System (GNSS) to define Earth coordinate system, and the ship speed, heading angle and earth location including the longitude and latitude in Earth coordinate system can be obtained from GNSS.

Another important part of ACS is the inertial navigation system, which records the ship motion angles including pitch, roll, laser beam azimuth and elevation in ship coordinate system.

Figure 1 shows the flowchart of data process of shipborne CDL. Specifically, the LOS velocity and Signal to Noise Ratio (SNR) can be firstly determined using Lidar data and FFT analysis. After the data pre-processing including the quality control based on SNR threshold and averaging process, the attitude transformation is then used to obtain the azimuth and elevation in each LOS vector in Earth coordinate system. The LOS velocity detected by Lidar is the atmosphere motion relative to ship coordinate system, thus the removal of the ship motion speed projection on the measured LOS is needed. Finally, the horizontal and vertical wind profiles can be retrieved using modified 4-DBS mode and zenith stare mode data, respectively.

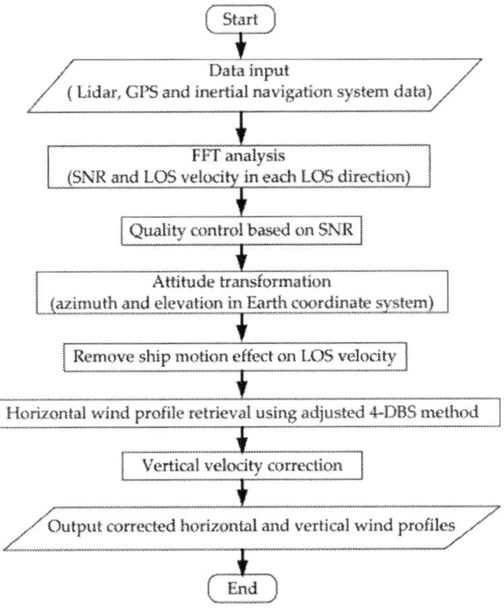

Figure 1 Flow chart of ship motion correction algorithm based on CDL

3 RESULTS

3.1 Horizontal wind evaluation

Two examples of obtained horizontal wind profiles are plotted in Figure 2 for anchored measurements and in Figure 3 for cruising observations, respectively. Meanwhile, the Lidar data (blue curves) are also compared to the radiosonde data (red curves), which were averaged over at least 10 min after the launch of the radiosonde. In Figure 2, The Root-Mean-Square-Error (*RMSE*) in speed between Lidar and radiosonde below 2 *km* are 0.49 ms^{-1} for the uncorrected measurement and 0.45 ms^{-1} for the corrected measurements, both showing consistent with the radiosonde wind speed, which is reasonable since the effect of ship speed on LOS velocity is less in anchored measurement. However, the *RMSE* in wind direction between Lidar and radiosonde are 84.43° for uncorrected measurement and 5.27° for corrected measurement. The obvious difference of the uncorrected measurement in wind direction is mainly derived from two aspects. The first one results from the definition in different coordinate systems, where the heading has an important effect on Lidar azimuth. The second aspect is that because of the experimental field limitation, the direction of GNSS master antenna is perpendicular to the ship bow, meaning that the "real" heading is the recorded heading plus 90°. Generally, the attitude correction is necessary especially for the wind direction retrieval when the ship is anchored with slight shake.

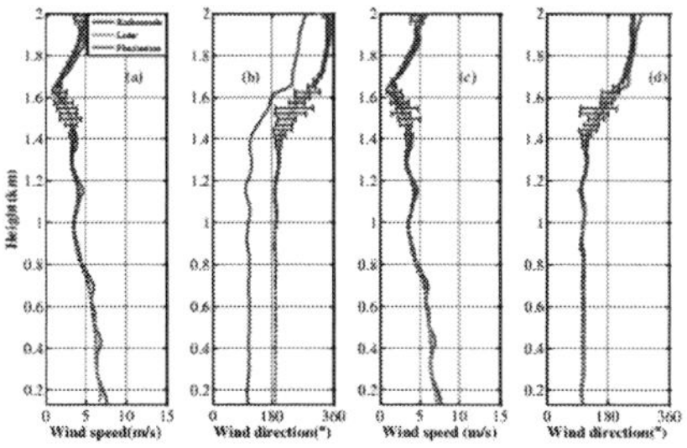

Figure 2 Anchored observation: (a) (c) wind speed and (b) (d) wind direction measured by CDL (blue line) before and after attitude correction, respectively. The simultaneous radiosonde data is shown in red line. The blue bars represent the sampling fluctuations from 15:52 to 16:02 LT (Local Time), 09 May, 2014.

Figure 3 shows the results of the cruising observation from 07:44 to 07:54 LT on 13 May 2014. In this case, the ship roll has more effect on the Lidar elevation when it points to port or starboard, on the contrary, the Lidar elevation in bow or stern direction is more sensitive to ship pitch. The *RMSE* in speed between Lidar and radiosonde data below 1.0 *km* are 4.42 ms^{-1} for uncorrected measurement and 0.88 ms^{-1} for corrected measurements, and the corresponding *RMSE* in wind direction are 48.71° and 9.52°, respectively. Therefore, the attitude correction algorithm has obviously improved the wind profile results when the ship is in cruising measurement. It is noted that the discrepancies in wind speed and direction above 1.0 *km* between radiosonde and corrected CDL may result from multipath effects at the ship's platform.

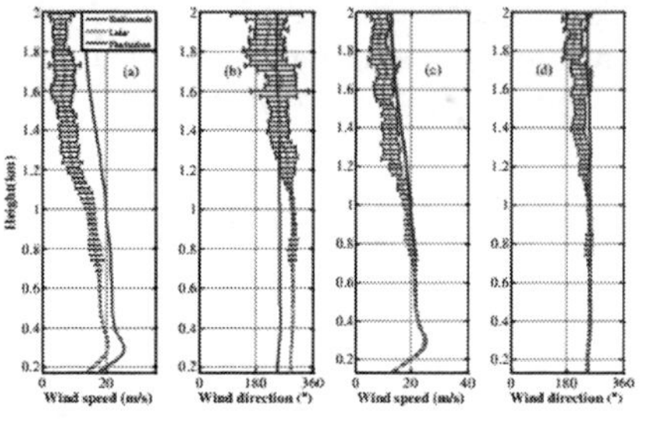

Figure 3 As figure 2, but for 07:44 to 07:54 LT 13 May, 2014 in cruising observation.

In order to assess the accuracy of the Lidar wind retrievals, results have been compared to the 11 radiosonde dataset during the experiment. Figure 4 shows scatter plots of wind speed and direction for Lidar and radiosonde using modified 4-DBS solution. Both wind speed and direction show negligible bias, demonstrating the feasibility and reliability of the modified 4-DBS method.

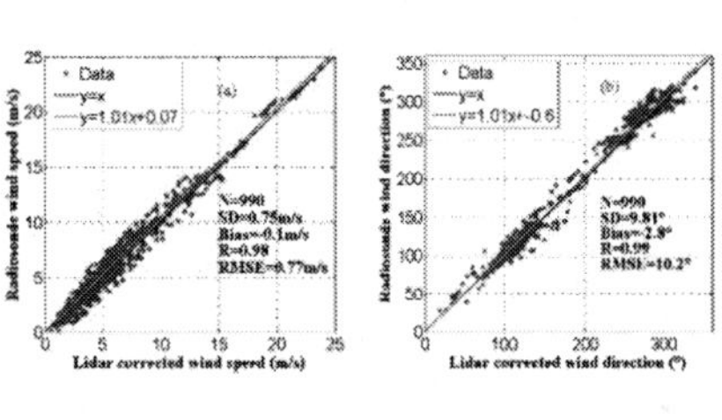

Figure 4 Comparison of (a) wind speed and (b) wind direction between CDL and radiosonde data from 09 May 2014 to 19 May 2014.

3.2 Vertical wind evaluation

The motion correction of vertical velocity is more obvious compared with the horizontal wind components. A typical measurement of 14 May 2014 is presented in Figure 5. Figure 5 (a) shows the whole series of time-height cross sections of the SNR. The MABL height is also presented in black solid circles. Figure 5 (b) shows the time series of ship heading, CDL laser beam azimuth and elevation, and horizontal wind direction at $H = 0.4$ *km*. During the zenith stare mode, the mean angle between ship heading and the laser azimuth is 66° with standard deviation of 7°, thus the projection of ship velocity on vertical velocity is always positive, the results of which are shown in Figure 6 (a). Furthermore, the estimation of the horizontal wind speed and direction (black line in Figure 5 (b)) from modified 4-DBS solution is used to substract the horizontal wind speed projection $\overrightarrow{r_g} \cdot \overrightarrow{V}$ from the relative speed measured by CDL. In this case, the $\overrightarrow{r_g} \cdot \overrightarrow{V}$ is positive and negative in downwind and headwind, respectively, causing the overestimate and underestimate of the vertical velocity, the effect of which is shown in Figure 6 (b). The resulting vertical velocity wind speed are presented in Figure 5 (c). The red and blue color indicates positive (upward) and negative (downward) movement, respectively, of the atmosphere parcels along the laser beam. It has been noted that the vertical velocity has a significant diurnal variation. Specifically, the downdrafts dominants mixing layer in the morning and amounted to about 0.5 ms^{-1}, and small-scale convective activity can be observed at the top of mixing layer. As the solar radiation strengthens, the atmospheric convection becomes more active and extends to the whole mixing layer, the strengths of updrafts and downdrafts are weakly stronger than before and the atmospheric vertical alternation becomes more frequent. The mixing layer recovers to descending motions with a continuous and long period after 13:11 LT. The difference between the corrected and uncorrected vertical velocity is shown in Figure. 6 (c), obviously showing the temporal and spatial variation of the contributions of ship motion and horizontal wind on vertical velocity.

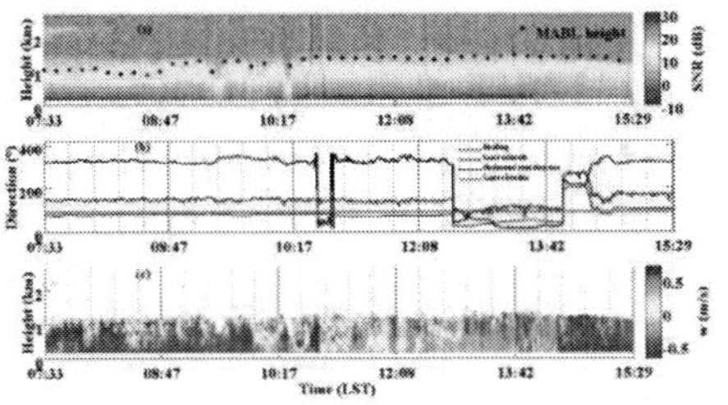

Figure 5 Example measurement from 07:33 to 15:29 14 May 2014: (a) Time-Height-Intensity of SNR and retrieved MABL height using SNR gradient method (black solid circles) (b) time series of ship heading, CDL laser beam azimuth and elevation in the Earth coordinate system, and horizontal wind direction at 0.4 km. (c) Time-Height-Intensity of vertical velocity after attitude correction.

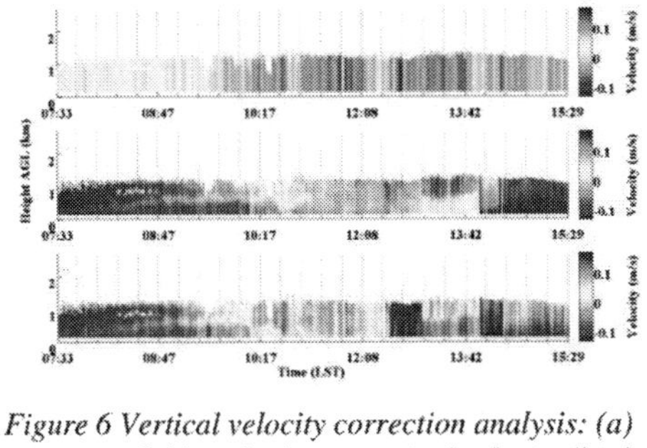

Figure 6 Vertical velocity correction analysis: (a) projection of ship velocity on vertical velocity (b) the effect of horizontal wind on vertical velocity (c) difference between vertical velocity after attitude correction and vertical velocity before attitude

4 CONCLUSIONS

Shipborne wind observation by the CDL during the 2014 Yellow Sea campaign has been presented to study the structure of the MABL. The attitude correction system of CDL consists of GNSS and inertial navigation system to directly measure the speed and the attitude of the ship. According to the transformation matrix from the product of roll, pitch and heading rotation matrix, the azimuth and elevation of the LOS velocity in the Earth coordinate system can be firstly determined. Then the removal of the ship speed projection on the measured LOS velocity is needed to obtain the "real" LOS velocity in Earth coordinate system. The horizontal wind profiles can be retrieved by a modified 4-DBS of CDL velocity data. For the case of vertical velocity, small deviations from vertical pointing due to ship motion and pointing angle knowledge errors introduces a projection of the horizontal wind on the LOS vector, thus estimations of the horizontal wind speed contribution are used to correct the vertical velocity.

The comparison of the CDL to radiosonde shows that attitude correction is essential for the wind retrieval in cruising measurement. The correlation coefficients of wind speed and direction are 0.98, 0.99, respectively, both of which shows negligible bias and demonstrates the feasibility and reliability of the modified 4-DBS method. A case study of 8-h time series observation on 14 May 2014 is presented to compare uncorrected and corrected vertical velocity, additionally showing the specific temporal and spatial variation of the contributions of ship motion and horizontal wind on vertical velocity. Generally, the combination of CDL with attitude correction system and accurate motion correction process has the potential of continuous long-term high-spatial and temporal resolution measurement for MABL thermodynamic and turbulence process.

ACKNOWLEDGEMENTS

This work was supported by the National Natural Science Foundation of China (NSFC) under grant 41471309 and 41375016 and the National High Technology Research and Development Program of China (No. 2014AA09A511).

References

[1] Wu, S. H., Liu, B. Y., Liu, J. T., Zhai, X. C., Feng, C. Z., Wang, G. N., Zhang, H. W., Yin, J. P., Wang, X. T., Li, R. Z., Gallacher, D., 2016: Wind turbine wake visualization and characteristics analysis by Doppler lidar, *Opt. Express*, **24** (10), A762-A780.

[2] Wolfe, D., Brewer, W., Tucker, S., White, A., White, D., Welsh, D., Ruffieux, D., Fairall, C., Ratterree, M., and Intrieri, J., 2007: Shipboard multisensor merged wind profiles from the New England Air Quality Study 2004, *J. Geophys. Res.: Atmos.*, 112.

[3] Hill, R. J., Brewer, W. A., and Tucker, S. C., 2008: Platform-motion correction of velocity measured by Doppler lidar, *J. Atmos. Oceanic. Technol.*, **25**, 1369-1382.

https://doi.org/10.1051/epjconf/201817606015

ESTIMATING THE PLANETARY BOUNDARY LAYER HEIGHT FROM RADIOSONDE AND DOPPLER LIDAR MEASUREMENTS IN THE CITY OF SÃO PAULO - BRAZIL

Márcia T. A. Marques[1]*, Gregori de A. Moreira[1], Maciel Pinero[2], Amauri P. Oliveira[2], Eduardo Landulfo[1]

*[1]CLA, IPEN/CNEN, Center for Lasers and Applications, Brazil, *marcia.marques@usp.br*
[2]University of São Paulo, IAG, Department of Atmospheric Sciences, Micrometeorology Group, Brazil

ABSTRACT

This study aims to compare the planetary boundary layer height (PBLH) values estimated by radiosonde data through the bulk Richardson number (BRN) method and by Doppler lidar measurements through the Carrier to Noise Ratio (CNR) method, which corresponds to the maximum of the variance of CNR profile. The measurement campaign was carried during the summer of 2015/2016 in the city of São Paulo. Despite the conceptual difference between these methods, the results show great agreement between them.

1 INTRODUCTION

The planetary boundary layer (PBL) is the region of the atmosphere directly influenced by the Earth's surface, it is also defined by the exchange of heat, moisture and momentum with the surface. The PBLH is a fundamental parameter characterizing the structure of this region and is of great interest for meteorological applications, e.g., atmospheric modelling, air quality control and dispersion of pollutants [1], [2].

PBLH estimation relies on profile data analysis or its parametrization using equations and models [2]. In this study we will work with two different methods based on profile data. The most common measurements of profile data are taken by radiosondes that are widely used, although, standard radiosonde data are only taken twice a day, which doesn't allow a more detailed analysis.

Remote sense systems (lidars and sodars) are an alternative source of profile data that can provide continuous measurement of the PBLH with high spatial and temporal resolution, ideal for studying the PBL behavior. In this study we will work with a Doppler lidar system.

The PBL can be classified in a convective boundary layer (CBL), a well-mixed layer typically as a result of surface heating during the day and a stable boundary layer (SBL), a not well mixed layer where turbulence tends to be suppressed. The PBLH has different definitions in the CBL and in the SBL due to the PBL dynamics. The PBLH detection in the SBL presents much more difficulties due to features such as intermittent and weak turbulence and radiative cooling.

2 METHODOLOGY

The measurement campaign was carried during the summer of 2015/2016, from mid-December to early March, completing 80 days.

A commercial Doppler lidar model WLS70 from Leosphere was set up to measure vertical profiles of wind velocity, wind direction and CNR from 100 to 1500 m with vertical resolution of 50 m and recording 10 minutes averaging with a temporal resolution of 3 Hz. The lidar was located at the top of the Center for Lasers and Application building at the University Campus in São Paulo (Figure 1), at 760 m above mean sea-level (23°33'38'' S, 46°44'25'' W).

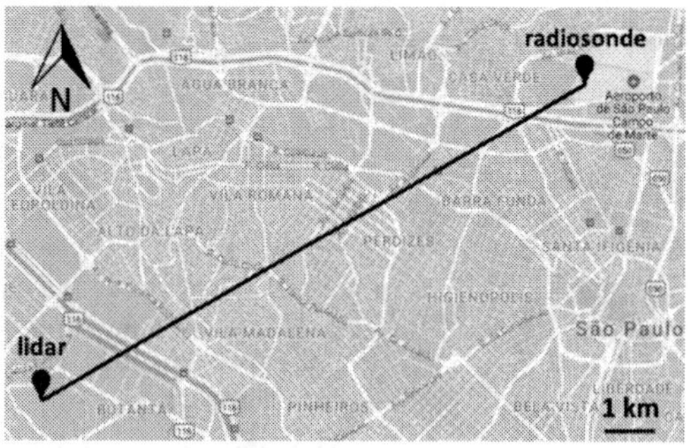

Figure 1 Study area scheme (Source: Google Maps)

 https://doi.org/10.1051/epjconf/201817606015

The radiosonde data used were launched from Campo de Marte airport (SBMT) located about 11 km northeast from the lidar measurement place (Figure 1), at 722 m above mean sea-level (23°30'46'' S, 46°38'30'' W). The radiosonde was launched twice a day at 00 and 12 UTC. In this study we only used the 12 UTC data for the CBL, which corresponds to 09 PM local time. The radiosonde has a vertical resolution of tens of meters, but the available data are only of representative heights with a vertical resolution that can vary a few hundred meters.

The atmospheric conditions affected the lidar data availability, mainly because in São Paulo city the summer is the rainy season. The rain, some power outages and unavailable radiosonde data contributed to the reduction of available days for analysis, so of an 80-day dataset, only 49 days were chosen.

2.1 CNR method

The CNR method assumes that the PBLH is demarcated by a strong gradient of aerosol content where there is a transition between the PBL with high aerosol concentration and a relatively clean free atmosphere [3].

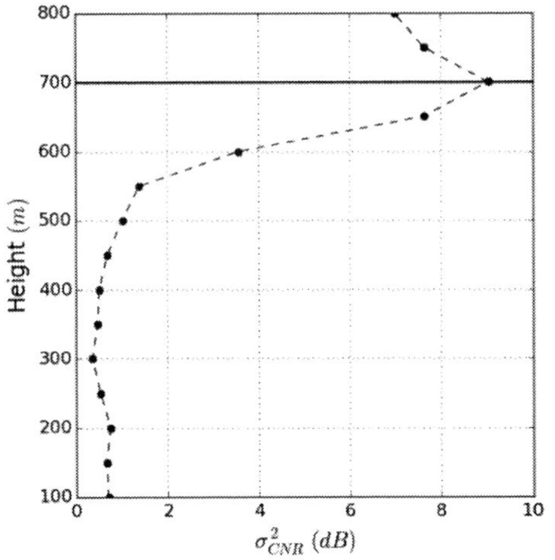

Figure 2 Example of the σ^2_{CNR} vertical profile. The horizontal solid line indicates the PBLH

Therefore, the PBLH corresponds to the height at which the variance of the CNR (σ^2_{CNR}) is maximum as exemplified in figure 2.

2.2 BRN method

The BRN is defined as the dimensionless ratio of buoyancy to shear production turbulence, expressed as:

$$BRN(z) = \frac{(g/\theta_{vs})\,(\theta_{vz}-\theta_{vs})\,z}{V_{hz}^{\,2}} \qquad (1)$$

where θ_{vz} and θ_{vs} are the virtual potential temperatures at the height z and the surface, respectively, g/θ_{vs} is the buoyancy parameter, where g is the gravitational acceleration and V_{hz} is the horizontal wind velocity at height z.

The BRN method assumes that the PBLH corresponds to the height at which the BRN reaches a critical value that represents a transition between a turbulent and a laminar flow. In the literature we can find a variety of critical values, often between 0.21 and 0.25. In this study we used a critical value of 0.17 estimated according to the methodology described by [4] based on linear fitting and statistical error minimization applied to different critical values. This methodology was applied to data from MCYTY BRAZIL program [5] that carried out 162 radiosondes launched every 3 hours collecting data with time resolution of 0.1 Hz at SMBT in 2013.

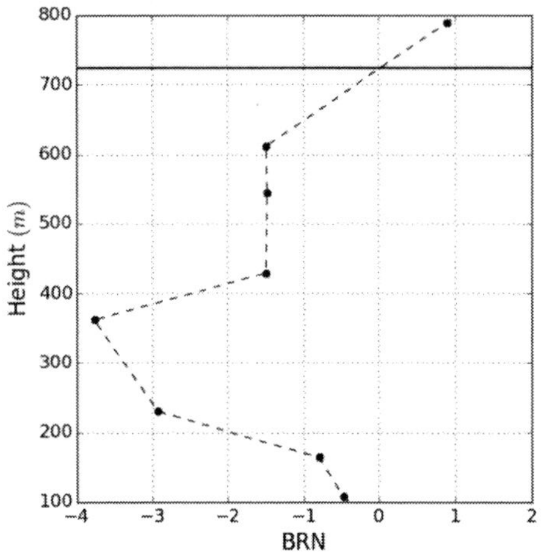

Figure 3 Example of the BRN vertical profile. The horizontal solid line indicates the PBLH

As previously mentioned, the vertical resolution of available radiosonde data is reduced, therefore, the PBLH was derived by a linear interpolation between the height where BRN < 0.17 and the height below [2].

3 RESULTS

PBLH was estimated by BRN and CNR methods for the chosen 49-day dataset at 12 UTC. Figure 4 shows the correlation between both methods with a Pearson correlation coefficient of 0.88.

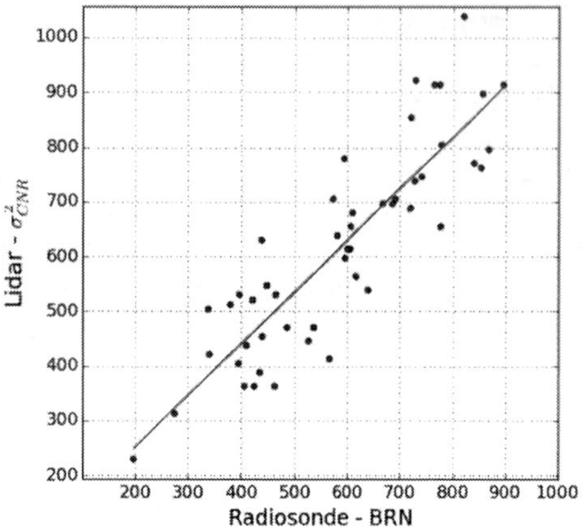

Figure 4 Correlation between PBLH estimated by both methods with a Pearson correlation coefficient of 0.88

Figure 5 shows the box plot of the estimated values for both methods. The center line of the box plot indicates the median value and the center box indicates the mean value.

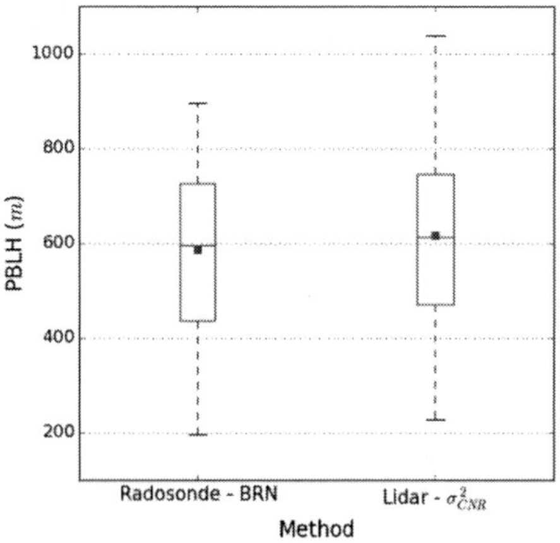

Figure 5 Box plot of PBLH estimated by both methods

The PBLH values estimated by CNR method were super estimated in comparison to the BRN method, as expected accordingly to [1]. The values were super estimated in 69% of the cases.

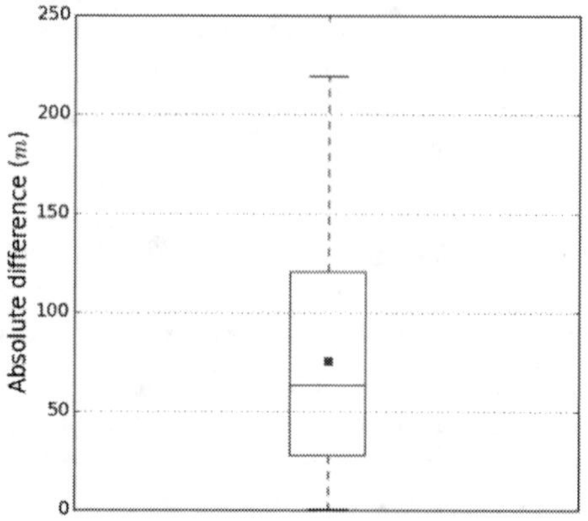

Figure 6 Box plot of the absolute difference between PBLH estimated by both methods

The average absolute difference is approximately 75 m, which represents a relative difference of 12%. Nevertheless, most absolute difference values are less than 75 m (Figure 6).

4 CONCLUSIONS

Considering these main limitations:

- Vertical resolution of 50 meters of the lidar (Figure 2) and a few hundred meters of the radiosonde (Figure 3) in which a linear interpolation between the levels of interest was carried out.

- Approximately 11 km distance between the measurements sites in a relatively complex terrain in the urban area.

- The conceptual definition of each method, which involves different definitions of the PBLH.

we were able to achieve a reasonable correlation between the methods (Figure 4). However, further investigation of PBLH behavior and the influence of other systems in the CBL can contribute to understand these differences

ACKNOWLEDGEMENTS

The author thankfully acknowledges the Micrometeorology Group from the Department of Atmospheric Science of the University of São

Paulo for its contribution, the team of the Center for Lasers and Applications for the assistance in the measurement campaign and PETROBRAS Research Center for making the lidar available.

References

[1] Seibert, P., Beyrich, F., Gryning, S., Joffre, S., Rasmussen, A., Tercier, P., 2000: Review and intercomparison of operational methods for the determination of the mixing height, *Atmos. Environ.* **34** (7), 1001-1027.

[2] Zhang, Y., Gao, Z., Li, D., Li, Y., Zhang, N., Zhao, X., Chen, J., 2014: On the computation of planetary boundary-layer height using the bulk Richardson number method, *Geosci. Model Dev.* **7** (6), 2599–2611.

[3] Lothon, M., Lenschow, D. H.; Mayor, S. D., 2009: Doppler lidar measurements of vertical velocity spectra in the convective planetary boundary layer, *Bound.-Layer Meteorol.* **132** (2), 205-226.

[4] Sorensen, J. H., Rasmussen, A., Svensmark, H., 1996: Forecast of atmospheric boundary-layer height utilized for ETEX real-time dispersion modelling, *Phys. Chem. Earth* **21** (5-6), 435-439.

[5] Oliveira, A. P., Marques Filho, E. P., Ferreira, M. J., Ribeiro, F. N. D., Soares, J., Codato, G., Mlakar, P., Boznar, M. Z., Pereira, M. M. R., Assis, E. S., Cassol, M., Escobedo, J. F., Dal Pai, A., França, J. R. A., Sánchez, M. P., Silveira, L. C., Buzati, B. C., Nofoente, J., Valdés, J., 2016: Assessing urban effects on the climate of metropolitan regions of Brazil - Implementation and preliminary results of the MCITY BRAZIL PROGRAM for the cities of São Paulo and Rio de Janeiro, (submitted to Urban Climate).

INVESTIGATION OF FEASIBILITY OF WIND TURBULENCE MEASUREMENT BY A PULSED COHERENT DOPPLER LIDAR IN THE ATMOSPHERIC BOUNDARY LAYER

Igor Smalikho[1]*, Viktor Banakh[2]

[1,2]Institute of Atmospheric Optics of Siberian Branch of the Russian Academy of Sciences, Russia,
**smalikho@iao.ru*

ABSTRACT

Feasibilities of determination of the wind turbulence parameters from data measured by the Stream Line coherent Doppler lidar under different atmospheric conditions have been studied experimentally. It has been found that the spatial structure of the turbulence is described well by the von Karman model in the layer of intensive mixing. From the lidar measurements at night under stable conditions the estimation of the outer scale of turbulence with the use of the von Karman model is not possible.

1 INTRODUCTION

Pulsed coherent Doppler lidars (PCDL) are used for the study of dynamic processes in the atmosphere, aircraft wake vortices and wind turbine wakes [1]. PCDLs can be also used for measurements of wind turbulence in the atmospheric boundary layer. A method of the azimuth structure function (ASF) for estimation of the turbulence parameters (radial velocity variance σ_r^2, turbulent energy dissipation rate ε and integral scale of the turbulence L_V) from data measured with a conically scanning PCDL has been developed in [2, 3]. In this method it is assumed that the spatial structure of the wind turbulence is described by the von Karman model. In this work we investigate of feasibility of application of the ASF method for wind turbulence measurement under different conditions of the atmospheric boundary layer.

2 MEASUREMENT STRATEGY AND MODIFIED ASF METHOD

Conical scanning is performed at a fixed elevation angle φ, using rotation of the probing beam about a vertical axis z with a constant angular speed ω_c. During one scan the azimuth angle θ varies from 0 to 2π. The components of the instantaneous wind velocity vector $\mathbf{V}=\{V_z,V_x,V_y\}$ are random functions of coordinates $\mathbf{r}=\{z,x,y\}$ and time t. Due to the turbulence anisotropy, their variances σ_z^2, σ_x^2 and σ_y^2 differ from each other, and the radial velocity variance σ_r^2 depends on the azimuth angle θ, where $\sigma_\alpha^2=<(V'_\alpha)^2>$; $V'_\alpha=V_\alpha-<V_\alpha>$; $\alpha=z,x,y,r$; $V_r=\mathbf{S}\cdot\mathbf{V}$; and $\mathbf{S}=\{\sin\varphi,\cos\varphi\cos\theta,\cos\varphi\sin\theta\}$. For the elevation angle $\varphi=\tan^{-1}(1/\sqrt{2})\approx 35.26°$ the radial velocity variance averaged over the azimuth angles $\bar{\sigma}_r^2$ is related with the kinetic energy of turbulence $E=(\sigma_z^2+\sigma_x^2+\sigma_y^2)/2$ as [4]

$$E=(3/2)\bar{\sigma}_r^2. \quad (1)$$

In this equation it is possible to take into account the averaging of the radial velocity over the sensing volume based on the theory developed in the monograph [1].

From raw lidar data we obtain an array of estimates of the radial velocity $V_L(\theta_m,R_k,n)$, where $\theta_m=m\Delta\theta$ is the azimuth angle; $m=0,1,2,...,M-1$; $\Delta\theta$ is the azimuth angle resolution; $R_k=R_0+k\Delta R$ is the distance from the lidar to the sensing volume center; $k=0,1,2,...,K$; ΔR is the range gate length; $n=1,2,3,...,N$ is the number of conical scan. Then for each height $h_k=R_k\sin\varphi$ we calculate the azimuth structure function $D_L(\psi_l)$ by the equation

$$D_L(\psi_l)=\frac{1}{N}\sum_{n=1}^{N}\frac{1}{M-l}\sum_{m=0}^{M-1-l}B^2(\psi_l,\theta_m,n), \quad (2)$$

where $\psi_l = l\Delta\theta$; $l = 1, 2, ..., L$; $B(\psi_l, \theta_m, n) = V_L'(\theta_m + \psi_l, R_k, n) - V_L'(\theta_m, R_k, n)$; $V_L'(\theta_m, R_k, n) = V_L(\theta_m, R_k, n) - \mathbf{S}(\theta_m) \cdot \bar{\mathbf{V}}(h_k)$; and $\bar{\mathbf{V}}(h_k)$ is the lidar estimate of the mean wind velocity vector obtained with the use of the sine-wave fitting [1]. According to the theory [1], the averaged lidar estimate of the structure function can be represented as $< D(\psi_l) > = D_a(\psi_l) + 2\sigma_e^2$, where $D_a(\psi_l)$ is the azimuth structure function of the radial velocity averaged over the sensing volume and σ_e is the instrumental error of the radial velocity estimate. If $R_k \cos\varphi > L_V$, the function $D_a(\psi_l)$ can be replaced by the transverse structure function $D_\perp(l\Delta y_k)$, where $\Delta y_k = \Delta\theta R_k \cos\varphi$.

Using the von Karman model for the 2-D spatial spectrum of wind velocity fluctuations and assuming that the turbulence is isotropic one can calculate the function $D_\perp(l\Delta y_k)$, which depends on two turbulent parameters: ε and L_V [3]. Taking into account that the Kolmogorov constant $C_K \approx 2$, for the von Karman model the integral scale L_V, the kinetic energy E and the dissipation rate ε have the following relation:

$$L_V = 0.38E^{3/2} / \varepsilon . \quad (3)$$

In [3] it is offered to estimate ε and L_V by the use of the least square method for the fitting of $\Delta D_\perp(l\Delta y_k) = D_\perp(l\Delta y_k) - D_\perp(\Delta y_k)$ to $\Delta D_L(\psi_l) = D_L(\psi_l) - D_\perp(\Delta\theta)$. However, if the condition $R_k \cos\varphi > L_V$ is not true, the error of estimate of the integral scale L_V can be very large. To avoid this problem, we offer the following approach for estimation of the turbulence parameters (modified ASF method). First, from measured $\Delta D_L(\psi_l)$ we estimate the turbulent energy dissipation rate ε within the inertial interval of the turbulence (when $l\Delta y_k < L_V$) [5]. Then, from the array $V_L'(\theta_m, R_k, n)$ the radial velocity variance $\bar{\sigma}_r^2$ is calculated. Taking into account that the elevation angle φ = 35.26°, we estimate E and L_V by the Eqs. (1) and (3), respectively.

3 EXPERIMENT

To investigate feasibility of determination of the wind turbulence parameters from data measured by PCDL with the use of the conical scanning at the elevation angle φ = 35.26°, we carried out an experiment on the Basic Experimental Complex of the Institute of atmospheric Optics SB RAS from 21 to 23 July 2016. The continuous measurement by the Stream Line lidar (made by HALO Photonics) was conducted at ω_c = 6°s^{-1} (duration of one scan T_{scan} = 1 min) and $\Delta\theta$ = 3° (M = 120). During the entire experiment was clear weather. At night time always the stable thermal stratification took place. Vertical profiles of the mean wind velocity and turbulence parameters were retrieved from the measured lidar data with the height resolution of 10 m. Each profile was obtained when for the averaging we used the raw lidar data measured for 30 scans (N = 30, duration of the measurement NT_{scan} = 30 min).

Figure 1 shows results of the spatiotemporal visualization of the mean wind velocity and the turbulence parameters: E, ε and L_V. In this figure it is well visible contrasts of the turbulence intensity at night and daytime. At night, due to the stability of the thermal stratification, when a low-level jet stream was formed (in a layer at altitudes of 150 - 300 m), the turbulence was very weak, and in the afternoon there was an intensive turbulent mixing. The data obtained for the kinetic energy and its dissipation rate are not contrary to the atmospheric boundary layer theory. Obtained using Eq.(3) the values of the integral scale of turbulence for the daytime (when the temperature stratification is neutral or unstable) are consistent with known experimental data [6]. However, in areas colored white in Figure 2(d) (areas where temperature stratification was stable), the integral scale of turbulence L_V takes very large values, which contradicts clearly to the theory of turbulence of the atmospheric boundary layer.

To identify the reasons of the overestimation of the integral scale L_V for the data measured by the lidar during nighttime, we compared experimental azimuth structure function $\bar{D}_L(\psi_l)$ =

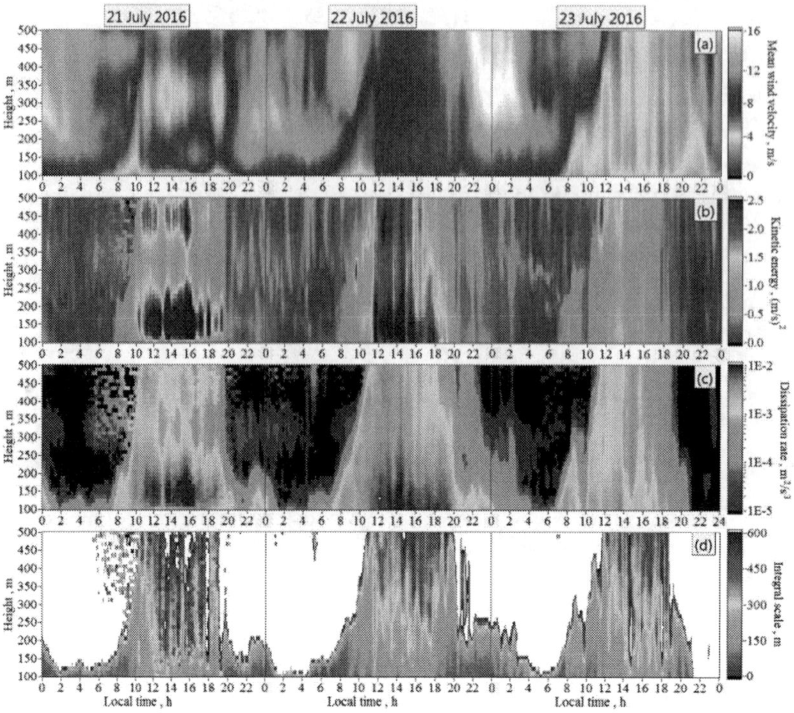

Figure 1 Spatiotemporal distribution of the mean wind velocity (a), the turbulent kinetic energy (b), the turbulent energy dissipation rate (c) and the integral scale of turbulence (d) obtained from measurements by the Stream Line lidar on the Basic Experimental Complex IAO SB RAS.

$D_L(\psi_l)-2\sigma_e^2$ with the function $D_\perp(l\Delta y_k)$ which is calculated using the von Karman model and the obtained estimates of dissipation rate ε and integral scale L_V. Figure 2 shows two examples of comparison of these functions for the cases of measurement at day and night. In Figure 2(a), there is an almost complete coincidence of the structure functions. Consequently, the turbulence structure is well described by the von Karman model and correct estimate of the integral scale of turbulence can be obtained using Eq. (3). The data in Figure 2(b) were obtained from lidar measurements at stable temperature stratification. In this case, the function $\bar{D}_L(\psi_l)$ greatly exceeds the $D_\perp(l\Delta y_k)$ at large spatial scales and we can conclude that for such conditions the turbulence structure is not described by the von Karman model. Accordingly, the correct estimate of the integral scale L_V can not be obtained by Eq. (3).

To determine the areas for which the von Karman model is applicable, we calculated parameter characterizing the degree of divergence of structure functions $\bar{D}_L(\psi_l)$ and $D_\perp(l\Delta y_k)$ for different heights and time of day by the equation

$$\gamma=\{L^{-1}\sum_{l=1}^{L}[\bar{D}_L(\psi_l)/D_\perp(l\Delta y_k)-1]^2\}^{1/2}. \quad (4)$$

Parameter γ = 0.02 for the data in Figure 2(a) and γ = 1.57 for the data in Figure 2(b).

Figure 3 shows the spatiotemporal distribution of the parameter γ for the data measured for a full day 07.22.2016. According to this figure, within a layer of 100 - 500 m and interval from 11:00 to 20:00 LT the parameter γ varies from 0.01 to 0.2 (an average of about 0.1). This means that in this period of time, wind turbulence quite well described by the von Karman model and from the data measured by the conical scanning lidar one can obtain reliable information about the integral

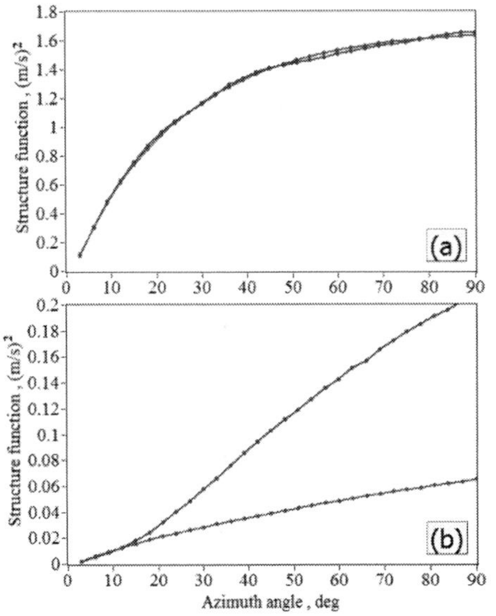

Figure 2 Structure functions $\bar{D}_L(\psi_l)$ (blue curves) and $D_{\perp}(l\Delta y_k)$ (red curves) obtained from lidar measurements at a height of 150 m on 22 July 2016 at 19:00 (a) and 03:00 (b) LT.

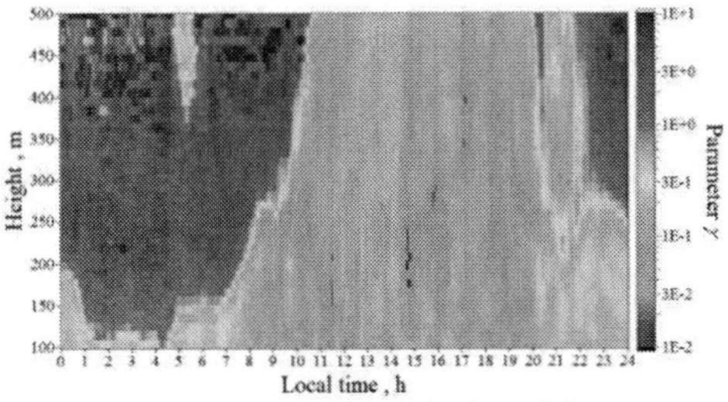

Figure 3 Spatiotemporal distribution of the parameter γ obtained from measurements by the Stream Line lidar on 22 July 2016.

scale of turbulence, in contrast to the measurements at night, in the morning and in the evening after 20:00 LT.

4 CONCLUSION

Thus, in this work the feasibility of determining the wind turbulence parameters from the data measured by a conically scanning PCDL has been investigated. From the raw data of the lidar experiment carried out for three days, we obtained a spatiotemporal distribution of the turbulence parameters with a height resolution of 10 m (in the layer 100 - 500 m) and a temporal resolution of 30 min. Analysis of the experimental results has shown that the turbulence structure is well described by the von Karman model in a layer of intensive mixing. The method described in Section 2 does not work for the data measured at night (under stable temperature stratification). Possible reason is a significant effect of horizontal inhomogeneity and nonstationarity of the mean wind at a very weak turbulence.

ACKNOWLEDGEMENTS

This study was supported by the Russian Foundation for Basic Research, Project No. 16-05-00668.

References

[1] Banakh, V.A., and Smalikho, I.N., 2013: Coherent Doppler Wind Lidars in a Turbulent Atmosphere, *Artech House*, Boston and London, 248 pp.

[2] Frehlich, R.G., Meillier, Y., Jensen, M.L., Balsley, B., and Sharman, R., 2006: Measurements of boundary layer profiles in urban environment, *J. Appl. Meteorol. Clim.* **45**(6), 821-837.

[3] Smalikho, I.N., Banakh, V.A., 2013: Accuracy of estimation of the turbulent energy dissipation rate from wind measurements with a conically scanning pulsed coherent Doppler lidar. Part I. Algorithm of data processing, *Atmos. Ocean. Opt.* **26** (5), 404–410.

[4] Eberhard, W. L., Cupp, R. E., and Healy, K. R., 1989: Doppler lidar measurement of profiles of turbulence and momentum flux, *J. Atmos. Ocean. Tech.*, **6**(10), 809–819.

[5] Smalikho, I.N., Banakh, V.A., Falits, A.V., Rudi, Yu.A., 2015: Estimation of the turbulent energy dissipation rate from the data measured by the Stream Line lidar in the surface layer of atmosphere, *Atmos. Ocean. Opt.* **28** (10), 901–905 [in Russian].

[6] Byzova, N.L., Ivanov, V.N., and Garger, E.K., 1989: Turbulence in Atmospheric Boundary Layer, *Gidrometeoizdat*, Leningrad, 265 pp. [in Russian].

 https://doi.org/10.1051/epjconf/201817606017

MIXED LAYER DEPTHS VIA DOPPLER LIDAR DURING LOW-LEVEL JET EVENTS

Brian Carroll[1]*, Belay Demoz[1], Timothy Bonin[2], and Ruben Delgado[1]

[1]*Joint Center for Earth Systems Technology/Univ. of Maryland, Baltimore County, MD, USA*
**brian.carroll@umbc.edu*
[2]*Cooperative Institute for Research in Environmental Sciences, Boulder, CO, USA*

ABSTRACT

A low-level jet (LLJ) is a prominent wind speed peak in the lower troposphere. Nocturnal LLJs have been shown to transport and mix atmospheric constituents from the residual layer down to the surface, breaching quiescent nocturnal conditions due to high wind shear. A new fuzzy logic algorithm combining turbulence and aerosol information from Doppler lidar scans can resolve the strength and depth of this mixing below the jet. Conclusions will be drawn about LLJ relations to turbulence and mixing.

1 INTRODUCTION

A low-level jet (LLJ) is a mesoscale phenomenon defined simply as a prominent wind speed peak in the lower troposphere with a significant decrease in wind speed above the peak. LLJs over land are almost exclusively nocturnal events that can persist from sunset to sunrise, reaching peak speeds anywhere from about 12 m/s to 30 m/s depending on geographic location and environmental conditions. The wind speed peak, referred to as the jet "nose", is typically below 1 km AGL due to linkages between jet formation mechanisms and the diurnal boundary layer cycle.

These jets transport atmospheric constituents such as water vapor and pollution hundreds of kilometers, and the high wind shear below the jet nose is known to generate substantial mixing during otherwise quiescent nocturnal conditions. This mixing of the residual and surface layers can lead to exceedances of air quality thresholds [1] and impact surface moisture flux. The depth of the mixed layer is also a key parameter for numerical weather prediction.

Continuous and accurate measurement of the mixed layer height, especially at night, is a challenging task. Radiosondes provide only an instantaneous snapshot of atmospheric structure. Radar wind profilers perform well during daytime convective conditions, but have trouble at night with minimum range potentially higher than a shallow boundary layer. Backscatter lidars perform well during the day, but have trouble discerning the MLH from residual layer depth outside of daytime conditions.

Scanning Doppler lidar has many strengths in approaching the problem of MLH retrieval. It can observe aerosol backscatter profiles and provide profiles of turbulence information such as vertical velocity variance. Multiple methods have been utilized in the literature. Tucker et al. [2] found vertical velocity variance to be the most accurate single measure of MLH from Doppler lidar, but all have their strengths and weaknesses (e.g. minimum range for vertical stares possibly overshooting the MLH at night). A new fuzzy logic algorithm uses all information relating to the mixed layer from a Doppler lidar scan routine to produce one best estimate for the MLH. This algorithm will be used here to study mixing and turbulence linkages to LLJs.

2 METHODOLOGY

2.1 Mixed layer height retrieval algorithm

Several independent techniques exist for estimating MLH from Doppler lidar, most of which rely on retrieving either proxies for turbulence (i.e. velocity variance) or aerosol presence. Velocity variance methods include vertical velocity variance from vertical stares, variance of velocity-azimuth-display (VAD) fit residuals, and vertical slice scan velocity variance which encompasses variations in horizontal wind. Methods that rely on aerosol presence and gradients in the aerosol profile include the Haar wavelet transform and signal-to-noise ratio variance. Each of these methods has strengths and weaknesses in terms of retrieval accuracy, and any given scan is confined to minimum and maximum

 https://doi.org/10.1051/epjconf/201817606017

measurement altitudes. For example, VAD fit residuals from low elevation conical scans can yield high-resolution information at low altitudes but the elevation angle results in rapid signal attenuation in the vertical. This can fill in the gap below a vertical stare's minimum range. In this way, combining all methods of MLH retrieval optimizes vertical range and resolution.

The new MLH algorithm presented here utilizes fuzzy logic to combine the independent methods in two phases. Gravity waves are removed before the bulk of this process by removing low-frequency variance, as they are indicative of larger-scale motions and not the mixing height. The first phase then fuzzifies each of the velocity variance method profiles, converting each variance measurement to a fuzzy membership value between zero and one via a membership function. This is demonstrated in Fig. 1. The membership value represents the variance in a format that is easily combined with the other independent methods. All membership profiles are interpolated to a 5 m vertical resolution grid and averaged together to produce an aggregate membership profile. A first guess for the MLH is obtained by finding the lowest altitude bin that has an aggregate membership value < 0.5, effectively expressing < 50% likelihood of being in the surface-connected turbulent ML.

The first guess derived from the velocity variance methods is then supplemented by the methods related to aerosol presence and gradients. The aerosol methods use peaks in profiles of Haar wavelet transforms or signal-to-noise ratio variance, that can represent the boundary between mixed layer aerosol and cleaner air above. Aerosol is used as the second method because it is less of a direct measure of mixing and hence less reliable; false boundaries are possible due to residual layers and other effects. To avoid these false peaks, the aerosol method peaks are only allowed to contribute to a final guess of the MLH if they are close to the first guess. Peaks in the wind shear profile are also included in this second phase.

A final best estimate for the MLH is obtained by applying a threshold to the updated aggregate membership profile in an identical manner to the first guess. Upper and lower confidence bounds are then determined by 0.8 and 0.2 thresholds, respectively. This is shown in Fig. 2.

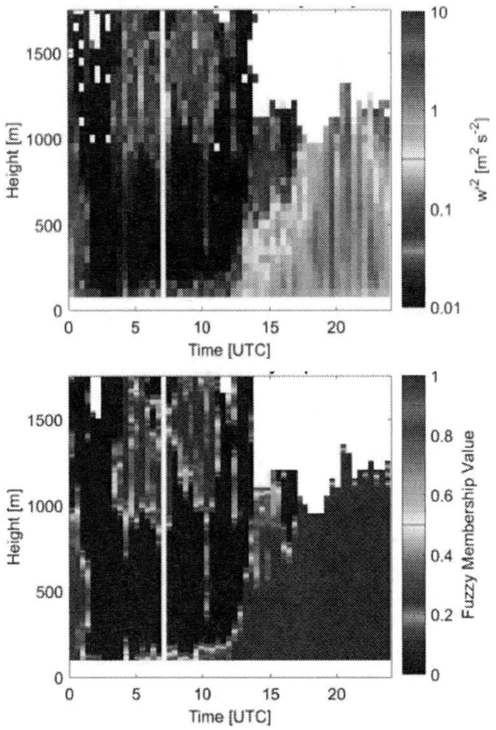

Figure 1: Variance of 45° elevation angle VAD fit residuals over a 24-hour period (top) and the corresponding fuzzy values (bottom).

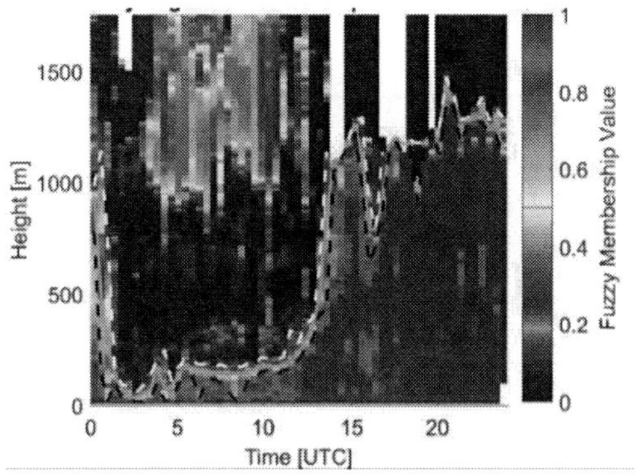

Figure 2: Final aggregate membership profiles for the same period as Fig. 1. Retrieved MLH is the cyan line, with upper and lower confidence bounds as white and black dashed lines, respectively.

2.2 Measurements

The data for this study is from the Plains Elevated Convection at Night (PECAN) campaign, which was a large multi-agency project targeting warm-season nocturnal precipitation and related phenomena over the Great Plains region of the Unites States. The Great Plains LLJ can generate or amplify mesoscale convective systems and has a large impact on the region's moisture budget [3]. It also modulates air quality via mixing as discussed previously [4]. PECAN measurements were taken from June 1st to July 15th, 2015.

The scanning Doppler lidar used for MLH retrieval during PECAN was a Leosphere WINDCUBE™ 200S owned by the University of Maryland, Baltimore County. It ran almost continuously for the six weeks of PECAN. The scan cycle period was approximately 25 minutes long and had a vertical stare, 5 conical scans between 5° and 45° elevation, and two orthogonal vertical slice scans. The lidar was co-located with several other instruments, including an atmospheric emitted radiance interferometer (AERI), radiosondes, and surface instruments. The MLH algorithm shows good agreement with the height of temperature inversions retrieved from PECAN radiosondes.

3 PRELIMINARY RESULTS

A VAD wind retrieval and the MLH algorithm have been applied to the Doppler lidar PECAN dataset to examine LLJs and their interaction with the boundary layer and turbulence. 33 LLJs were observed with a wide range of wind speed and nose height.

Current preliminary results qualitatively show some major differences in mixing below different LLJs. Two jets of very similar peak wind speed and nose height are shown in Fig. 3, plotted along with the retrieved MLH. Despite similar shear profiles below the jet nose, the June 20th case has little nocturnal mixing but the June 21st case shows substantial surface-connected mixing up to about 160 m AGL. The larger shear above the nose on June 21st may contribute to this, affecting the stability of the whole layer.

Static stability will be investigated as a key parameter to further this study. Static stability and turbulence intensity are inversely correlated, but it

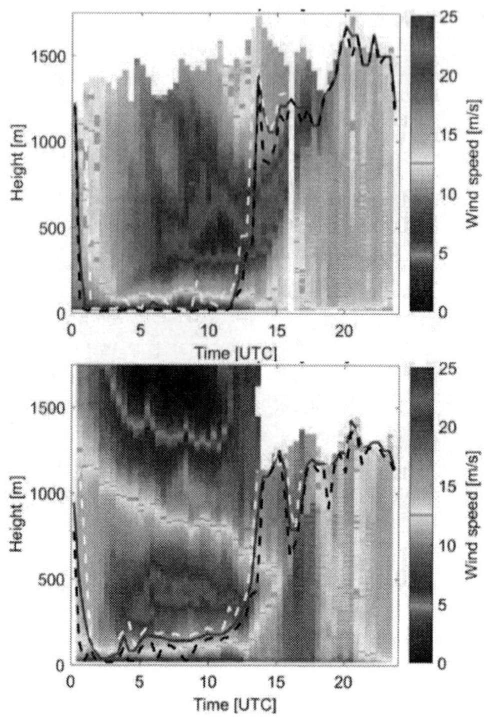

Figure 3: Wind speed from Doppler lidar for the 24 hours of June 20th (top) and 21st (bottom) 2015. Red line is retrieved MLH, with dashed line confidence bounds.

is unknown whether turbulence weakens strong stable boundary layer inversions or if preexisting weak inversions allow for turbulence production

[5]. Potential temperature profiles for stability analysis will be retrieved from radiosondes and AERI, and the bulk Richardson number profiles will be produced.

Fig. 3 also reveals very steep increases in the MLH as the jets begin to dissipate around sunrise, which is at 1130 UTC. On days without a LLJ the convective boundary layer often rises to high daytime values gradually over a few hours. This rapid jump during persistent LLJ events suggests that the sudden input of solar energy and surface buoyancy may act in conjunction with the LLJ wind shear turbulence to quickly enhance turbulence intensity throughout the lowest several hundred meters of the boundary layer. This hypothesis will be investigated further.

The MLH algorithm and its individual turbulence-related components will also be used to target the afternoon transition in relation to LLJ formation and characteristics.

It should be noted that neither of the cases in Fig. 3 is completely unique within the dataset. This study will include several other LLJ events.

4 SUMMARY

Turbulence associated with LLJs has been a topic of interest in recent years due to linkages with air quality and weather prediction. A new algorithm that detects mixed layer depth from a combination of Doppler lidar retrievals is well-suited to address topics related to LLJ turbulence, with the Doppler lidar simultaneously profiling the jet wind speed.

While high wind shear is present below all observed LLJ noses, it is not the only factor in determining whether substantial mixing will occur. Wind shear above the nose may have a controlling influence on mixing in the nocturnal boundary layer, with higher shear impacting turbulence generation throughout the layer. Static stability will also be investigated as this work continues to determine if strong inversions are suppressing mixing. Critical values for turbulence generation will be pursued for each of these factors, including a study of the Richardson number.

ACKNOWLEDGEMENTS

Financial support for this work comes from the Joint Center for Earth Systems Technology and NOAA-CREST/CCNY foundation CREST grant NA11SEC481004.3. We also acknowledge PECAN funding from the National Science Foundation award number 1503563 and from NASA.

References

[1] Delgado, R., S. D. Rabenhorst, B. B. Demoz, R. M. Hoff, 2015: Elastic lidar measurements of summer nocturnal low level jet events over Baltimore, Maryland, *J. Atmos. Chem.* **72**, 311–333.

[2] Tucker, S. C., C. J. Senff, A. M. Weickmann, W. A. Brewer, R. M. Banta, S. P. Sandberg, D. C. Law, and R. M. Hardesty, 2009: Doppler lidar estimation of mixing height using turbulence, shear, and aerosol profiles, *J. Atmos. Ocean. Tech.* **26**, 673-688.

[3] Higgins, R. W., Y. Yao, E. S. Yarosh, J. E. Janowiak, K. C. Mo, 1997: Influence of the great plains low-level jet on summertime precipitation and moisture transport over the central United States. *J. Clim.* **10**, 481–507.

[4] Klein, P. M., X.-M. Hu, M. Xue, 2014: Impacts of Mixing Processes in Nocturnal Atmospheric Boundary Layer on Urban Ozone Concentrations. *Bound.-Layer Meteor.* **150**, 107–130.

[5] Bonin, T. A., W. G. Blumberg, P. M. Klein, P. B. Chilson, 2015: Thermodynamic and Turbulence Characteristics of the Southern Great Plains Nocturnal Boundary Layer Under Differing Turbulent Regimes. *Boundary-Layer Meteorol.* **157**, 401–420.

EPJ Web of Conferences **176**, 06018 (2018) https://doi.org/10.1051/epjconf/201817606018
ILRC 28

MEASUREMENTS AT FP3 IN SUPPORT OF PECAN SCIENTIFIC Objectives Using MPL-111 Lidar

Kristen Pozsonyi[1], Natalie Midzak[2], Christina Prestine[3], Richard Clark[4]

[1]*Millersville University, now at U. Alabama in Huntsville, USA, kristenpozsonyi@yahoo.com*
[2]*Millersville University, now at U. North Dakota, USA*
[3]*Millersville University, USA*
[4]*Department of Meteorology, Millersville University, USA*

1. ABSTRACT

This paper will report on the data collected by a Sigma Space Micropulse Lidar (MPL-111), and how these measurements, when integrated with other data, helps to inform our analysis of two cases of the Great Plains nocturnal Low-Level Jet (LLJ) in the vicinity of FP3.

2. INTRODUCTION

The primary objective of the Plains Elevated Convection at Night (PECAN) campaign was to gain a better understanding of conditions and mechanisms contributing to nocturnal precipitation. Most of the summertime rainfall over the Great Plains occurs at night, often well after the maximum in daytime heating and when the nocturnal boundary exhibits static stability.

The PECAN field campaign spanned the period from 1 June 2015 to 15 July 2015. The domain for this project included Kansas, northern Oklahoma, and Southern Nebraska, but eventually spread outward to other locations across the Great Plains. PECAN was a collaborative event between eight research facilities and 14 universities. There were a total of 31 Intensive Observing Periods (IOPs) conducted during PECAN.

PECAN focused on the study of nocturnal convection initiation, the internal structure of mesoscale systems, and propagating wave-like disturbances that are spawned as a result of these convective storms. The experimental design included six fixed and four mobile ground-based facilities comprising the PECAN Integrated Sounding Arrays (PISAs), along with many more facilities. FP3 was one of six fixed PISA sites and located at Ellis, Kansas as in Fig. 1.

PECAN had five main research objectives: 1) explain the characteristics and processes that lead to pristine nocturnal convective initiation (CI) and early phases of large-scale storm systems; 2) explore the evolution of bores and their impact on maintaining MCSs; 3) investigate the initiation, structure, and evolution of the LLJ and its relation to CI; 4) study the microphysical structure of MCSs; and 5) improve the prediction of nocturnal convection, which will be useful to operational forecasting and climate models (Geerts et al. 2017).

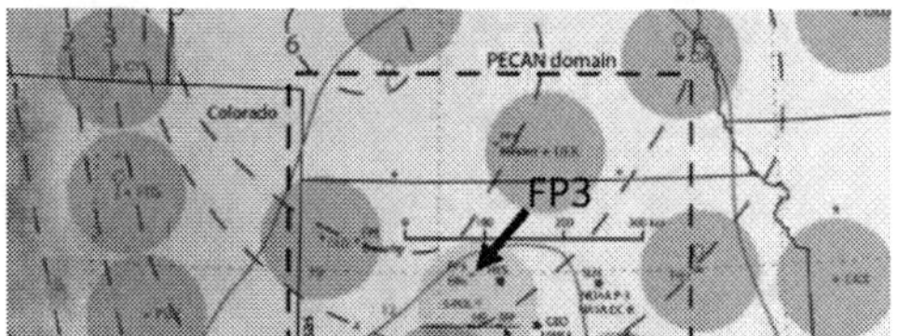

Figure 1: Map of PECAN domain and facilities

The primary instrument relevant to this paper, the Sigma Space Micropulse Lidar (MPL)-111, is part of the Millersville [University]

Atmospheric Research and Aerostat Facility (MARAF), which is a multi-capability boundary layer facility that includes tethered and free balloon atmospheric profiling systems, 10 m flux tower, Medium Frequency Acoustic Sodar (MFAS) with a radio acoustic sounding system (RASS), and a full suite of trace gas analyzers.

The measurements obtained by MARAF are used to here to report on the structure and evolution of two very different LLJs: a 15 m s^{-1} LLJ on 10 June 2015 that resided about 200-300 m above ground level (AGL), and a 30 m s^{-1} LLJ residing near 500 m AGL with a boundary layer in dynamic instability due to shear. The LLJ is known to be a recurring feature of the summertime boundary layer over the central plains, and is responsible for the influx of heat and moisture important to the initiation and maintenance of convection across the PECAN domain.

The MPL-111 was used mainly as a turnkey instrument to determine the height of the atmospheric boundary layer (ABL) and aerosol layer (AL). The MPL-111 has the ability to generate three primary products that are compiled from 30-second averages: Raw, R^2, and Normalized Relative Backscatter (NRB). The backscatter plot depicts the height of aerosols, cloud base (it serves as a ceilometer), and the height of the ABL up to five kilometers. Particular interest was given to the time-height series of the ABL and AL from the derived normalized relative backscatter. These products can be found on the PECAN field campaign website (http://catalog.eol.ucar.edu/pecan).

2.1 Site Specifications

There were multiple facilities taking part in this field campaign, including multiple mobile integrated sounding arrays and six fixed sites throughout the domain. Millersville University and NCAR brought facilities to FP3, with NCAR maintaining a 30 m instrumented tower, 915 MHz profiler, and other instruments north of I-70, and Millersville with its MARAF facility just south of I-70 (see Fig. 2.)

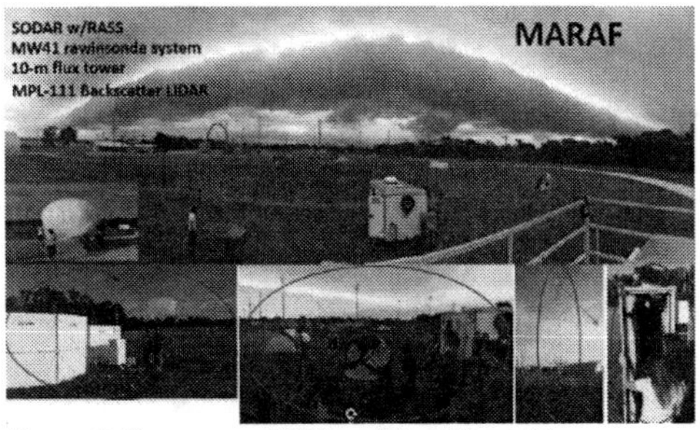

Figure 2: Panorama of FP3 and the Millersville Atmospheric Research and Aerostat Facility, which occupied the Ellis Fairground.

3. METHODOLOGY

Although FP3 accommodated an array of instrumentation, the focus for this study was on the Sigma Space MPL-111 Lidar. Measurements taken by this Lidar were gathered daily throughout the duration of this campaign. The instrument ran continuously, providing autonomous data collection using 30 meter bin resolution. The min/max range for this MPL is 150 m/25 km, however, for this campaign we only recorded data up to 5 km, focusing on the

		MPL
PERFORMANCE	range resolution	15/30/75 m (software programmable)
	minimum range	150 m
	accumulation time	1 sec - 15 min
	detection range	up to 25 km
	polarization	optional
	scanning	optional
OPTICS	laser wavelength	532 nm
	laser pulse energy	6-10 µJ @ 2500 Hz
	eye-safety	ANSI Z136.1 2000, IEC 60825
	receiver diameter	178 mm
	pump laser diode	fiber coupled, user replaceable
	detector	fiber coupled, user replaceable
DIMENSIONS	size (mm)	300x350x850
	weight (portability)	25+2 kg (2 persons)
DATA	operating system	Windows XP/Vista/7
	computer interface	USB
	data transfer	LAN Ethernet
ENVIRONMENT	temperature	Operating -10° to 40° C with NEMA 4 enclosure
	humidity	Operating 0-100% with NEMA 4 enclosure
POWER	supply	100/240V AC 50-60 Hz
	consumption	500W

Figure 3: Specifications for the MPL-111

boundary layer conditions. Other specifications for this are given in Figure 3.

https://doi.org/10.1051/epjconf/201817606018

4. RESULTS

Characteristics of the LLJ were observed with the MPL and are combined with data from the 10 m flux tower and the MFAS Sodar and upper air profiles from the RS41-SGP radiosonde to describe the vertical structure and evolution of the LLJ.

4.1 Weak Jet – 10 June 2015

One of the weaker LLJ cases that was observed during the campaign occurred on 10 June 2015 with a peak wind speed of 15 m s^{-1} around 0900 UTC. Wind profiles taken every three hours from 0000 UTC to 0900 UTC are shown in Fig. 4. Around 0300 UTC there is an abrupt acceleration of the LLJ and this is correlated with an increase in shear and a commensurate increase in the Lidar backscatter from the level of maximum winds to 1-1.2 km (see dashed line in Fig. 6). The top of the boundary layer, which is represented by the solid line and correlated with the capping inversion seen in the virtual potential temperature profiles in Fig. 5, exhibits little variability over the nighttime period. The relatively constant thickness of the two high-backscatter levels over the nighttime duration shown in Fig. 6 is consistent with a

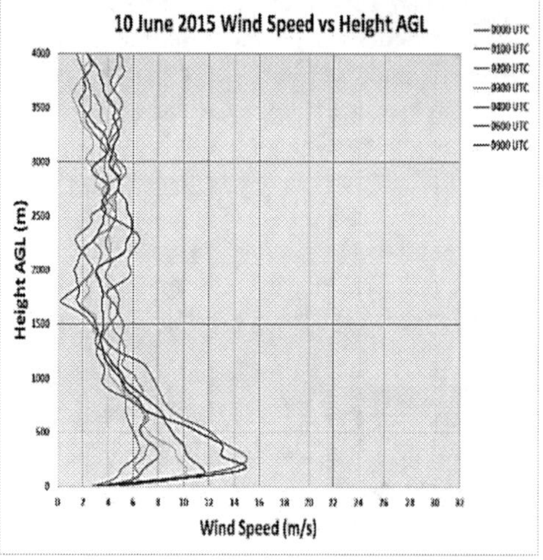

Figure 4: Wind profiles for 10 June 2015

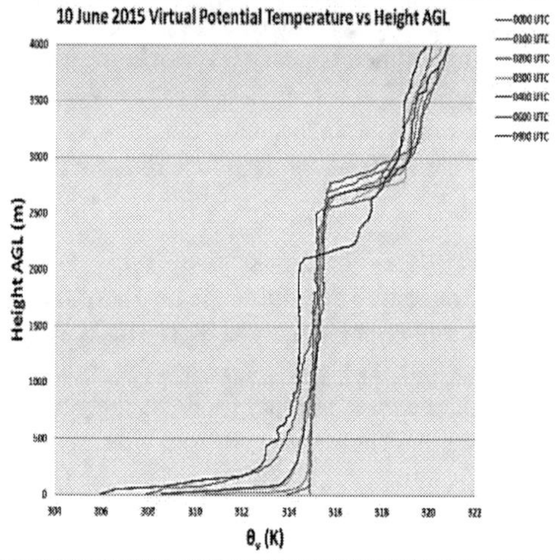

Figure 5: Virtual potential temperature profiles - 10 June 2015

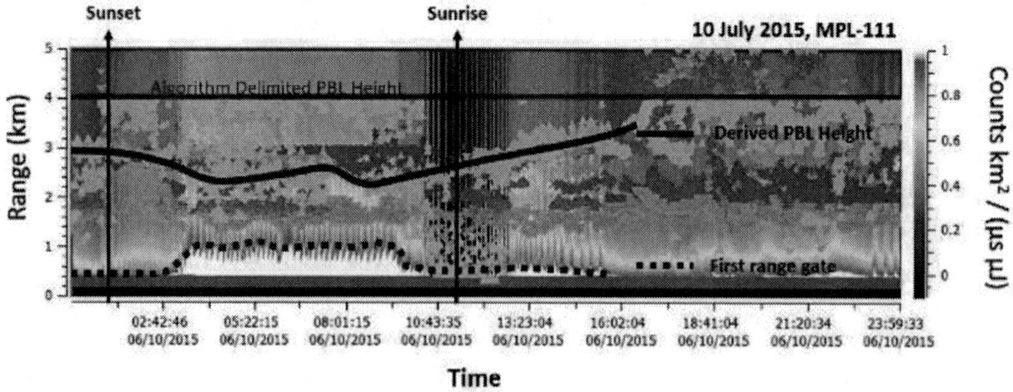

Figure 6: NRB product -10 June 2015 LLJ. The vertical lines are an artifact of the air conditioning system in the enclosure.

residual layer between 500 and 2500-3000 m that is largely unaffected by the dynamic changes and kinematic structure of this weaker LLJ.

4.2 Strong Jet – 22 June 2015

A stronger LLJ event occurred on 22 June 2015 reaching peak wind speed of 31 m s^{-1} around 1030 UTC. Wind profiles from soundings launched every 1.5 to 3 hours reveal a deep layer of maximum winds at an altitude around 500 m (Fig. 7). Strong shear both above and below the LLJ core generates dynamic instability (Richardson numbers remain less than 0.25 for the entire duration). But there is a distinct transition from high shear to lesser shear and strong static stability to weaker stability (Fig. 8) in the layer 1 – 1.2 km, giving rise to the Lidar backscatter shown in Fig. 9 (dashed line). Note that this layer remains deep and fairly constant over the event duration, with only a slight reduction in altitude with time as the LLJ weakens. The height of the top of the boundary layer, another concentrated layer of Lidar scatter and represented by the solid line, increases in altitude from 2 – 3.3 km from 0330 UTC on as the LLJ accelerates, generates turbulence above the LLJ core and produces a

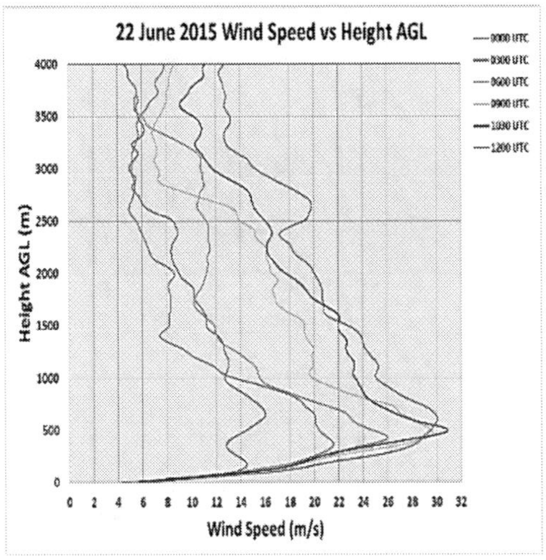

Figure 7: Wind profiles for 22 June 2015

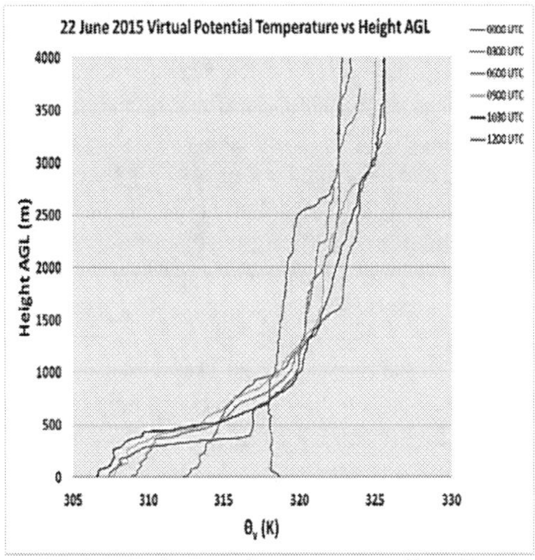

Figure 8: Virtual potential temperature profiles - 22 June 2015

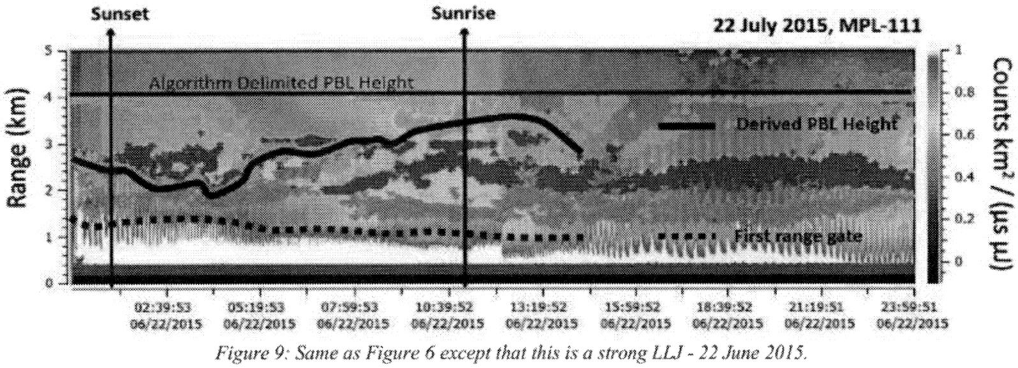

Figure 9: Same as Figure 6 except that this is a strong LLJ - 22 June 2015.

https://doi.org/10.1051/epjconf/201817606018

thickening residual layer as the LLJ continues to accelerate.

5. CONCLUSIONS

Two very different LLJ case studies were analyzed using data from upper air profiles and the MPL-111 Lidar. The MPL is a useful tool for capturing the time-height evolution of the boundary layer and filling in gaps remaining from other observations. The weaker LLJ exhibits little change once the LLJ develops, even as static stability increases. We believe that the wind is not strong enough to create the instability and turbulent mixing above the layer of maximum winds to affect the height of the boundary layer. Conversely, the strong LLJ on 22 June generated strong shear, consequent instability and turbulent eddies that moved momentum and heat throughout the residual layer owing to the weak static stability, causing the depth of the boundary layer to increase in time.

These preliminary findings can help inform future investigations on the structure and evolution of the LLJ.

6. REFERENCES

Geerts, B., D. Parsons, C.L. Ziegler, T.M. Weckwerth, M.I. Biggerstaff, R.D. Clark, M.C. Coniglio, B.B. Demoz, R.A. Ferrare, W.A. Gallus, K. Haghi, J.M. Hanesiak, P.M. Klein, K.R. Knupp, K. Kosiba, G.M. McFarquhar, J.A. Moore, A.R. Nehrir, M.D. Parker, J.O. Pinto, R.M. Rauber, R.S. Schumacher, D.D. Turner, Q. Wang, X. Wang, Z. Wang, and J. Wurman, 2017: The 2015 Plains Elevated Convection at Night Field Project. *Bull. Amer. Meteor. Soc.*, **98**, 767–786, doi: 10.1175/BAMS-D-15-00257.1.

INSECT REMOTE SENSING USING A POLARIZATION SENSITIVE CW LIDAR SYSTEM IN CHINESE RICE FIELDS

Shiming Zhu[1], Elin Malmqvist[2], Yiyun Li[1], Samuel Jansson[2], Wansha Li[1], Zheng Duan[1], Wei Fu[1], Katarina Svanberg[1,3], Joakim Bood[2], Hongqiang Feng[4], Susanne Åkesson[5], Ziwei Song[6], Baoxin Zhang[6], Guangyu Zhao[1], Dunsong Li[6], Mikkel Brydegaard[2,5,7] and Sune Svanberg[1,3,8*]

[1] *Center for Optical and Electromagnetic Research, South China Normal University, Guangzhou, China*
[2] *Combustion Physics Division, Department of Physics, Lund University, Sweden*
[3] *Lund Laser Centre, Lund University, Sweden*
[4] *Henan Academy of Agricultural Sciences, Zhengzhou, China*
[5] *Centre for Animal Movement Research, Department of Biology, Lund University, Sweden*
[6] *Guangdong Academy of Agricultural Sciences, Guangzhou, China*
[7] *Norsk Elektro Optikk A/S, Norway*
[8] *Atomic Physics Division, Department of Physics, Lund University, Sweden*
**Email sune.svanberg@fysik.lth.se*

ABSTRACT

A joint Chinese-Swedish field campaign of Scheimpflug continuous-wave lidar monitoring of rice-field flying pest insects was pursued in very hot July weather conditions close to Guangzhou, China. The occurrence of insects, birds and bats with almost 200 hours of round-the-clock polarization-sensitive recordings was studied. Wing-beat frequency recordings and depolarization properties were used for target classification. Influence of weather conditions on the flying fauna was also investigated.

1 INTRODUCTION

Optical monitoring of flying insects is an emerging field of research. There are many reasons for pursuing such endeavors. Certain insects, such as bees are important pollinators, and without their services, many crops would fail. Disease vectors, such as the deadly malaria-spreading mosquito *Anopheles*, are responsible for annual deaths in the order of one million (especially in Africa), by far outnumbering all armed conflicts, crime and terrorist actions. Finally, agricultural pest insects consume important parts of the crops intended for human consumption. The abatement through the strongly expanding use of pesticides in turn may put human health in jeopardy because of unwanted side effects, and constitutes a huge threat to the pollinators. For these, and additional reasons, effective monitoring of the whereabouts and action of flying insects is of considerable interest. Apart from sampling techniques traditionally pursued by entomologist, short-wavelength radar systems have proven effective in long-range monitoring of flying insects [1,2]. However, such systems cannot record close to the ground, and lack the spectroscopic identification ability, which is available for optical systems. Thus, optical radars, or lidar systems, could be of considerable interest. Pioneering work was pursued by Shaw *et al.* [3,4], focusing on the detection of hidden mines, found by specially trained honey bees, which aggregate in the air above the explosive, the odor of which is an attracting agent.

The Lund University lidar group started its research activity related to entomology in 2008. Using a pulsed lidar system, remotely detected fluorescence properties were first studied with mounted static insects, followed by successful field experiments [5]. Chinese agricultural insects were likewise investigated with elastic back-scattering and fluorescence techniques with the Swedish mobile pulsed lidar system brought to Hangzhou, China [6].

Passive, dark-field remote sensing of sun-illuminated flying insects observed against a black back-ground is an attractive simple way of recording flying insects. Occurrence, wing-beat frequencies (including overtones), as well as the reflectance spectra of the insects can be assessed [7]. Such work was also pursued by the Guangzhou group with collaborators [8,9]. While attractive, these experiments did not yield range resolution.

 https://doi.org/10.1051/epjconf/201817607001

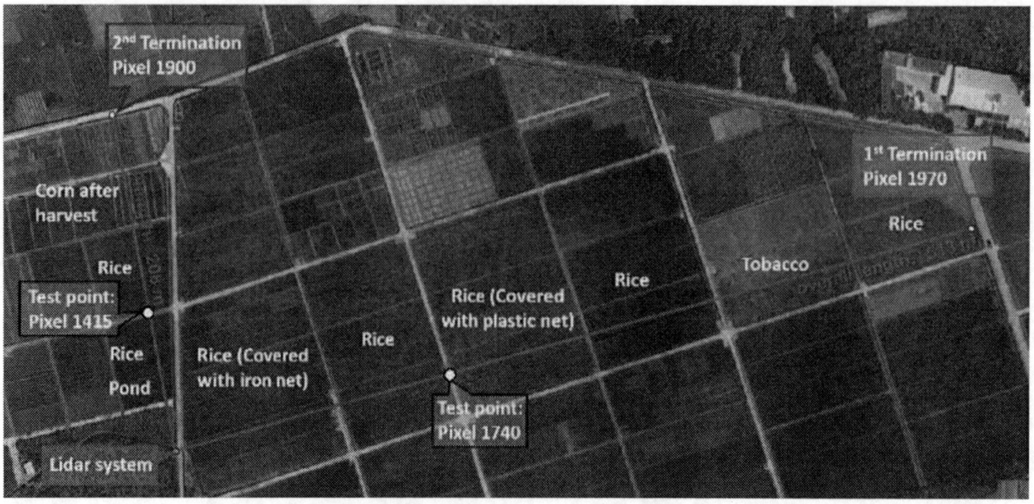

Fig. 1. Field experiment arrangement in Chinese rice fields.

In a simple CW lidar implementation using a so-called Scheimpflug arrangement, range resolution can be achieved [10,11]. Basically, a transmitted laser beam is observed by a receiving telescope placed about 82 cm separated from the transmitting telescope, and the linear array detector observing the back-scattered light is placed in a tilted position, so that close-range, as well as far-range locations can be observed in focus simultaneously. Range-resolution is high at close range while lower at far distance, where ultimately, approaching infinity, all locations fall onto the same detector pixel. On the other hand, and because of this, detected return intensities from air are basically the same for all pixels, in contrast to the customary lidar $1/R^2$ dependence. Such an arrangement has several attractive features, including reduced cost, weight and complexity, while allowing a high data sampling rate (see, e.g. [10,11]).

2 FIELD TEST AND ARRANGEMENTS

We here report on the arrangement for and results from a joint field experiment campaign pursued in rice fields at an experimental farm belonging to the Guangdong Academy of Agricultural Sciences and located in the Baiyun district north of the major Chinese city of Guangzhou. A CW Scheimpflug lidar set-up arranged for polarization sensitive recordings of flying insects was used. Polarization discrimination provides additional information on the target, e.g. distinction between specular and diffuse reflection, and assessment of, e.g. hair coverage of insects or glossy wings.

The field area with two utilized perpendicular measurement transects is shown in Fig. 1, where also the lengths of the beam paths are indicated. One path regarded crops treated with pesticides while the other, shorter one, represented non-treated conditions. The laser beams were sent at least 2.5 meters above the ground and terminated on large black beam stops. The lidar system and the beam termination are seen in Figs 2a,b. Full security was ensured by constant surveillance of the measurement area. On the order of 200 hours of around-the-clock recordings of insects and, especially in morning and evening hours, birds and bats, were taken during two very hot (24-38°C) July weeks in Guangdong. We performed extensive monitoring of diurnal variations, the influence of weather, in particular rain-fall, and also studied the insect attraction power of night-time search-light radiation crossing the lidar monitoring beam. An overview figure of the lay-out of the measurement

a  b 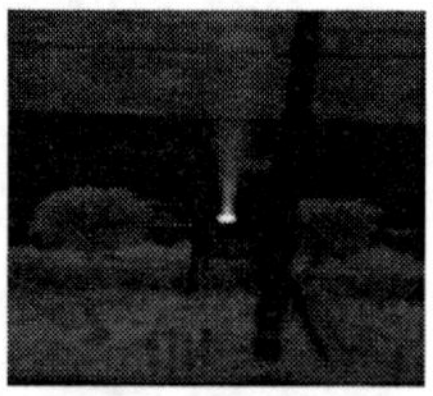

Fig. 2 a) Photograph of the CW lidar measurement system. b) CW laser beam and termination.

 https://doi.org/10.1051/epjconf/201817607001

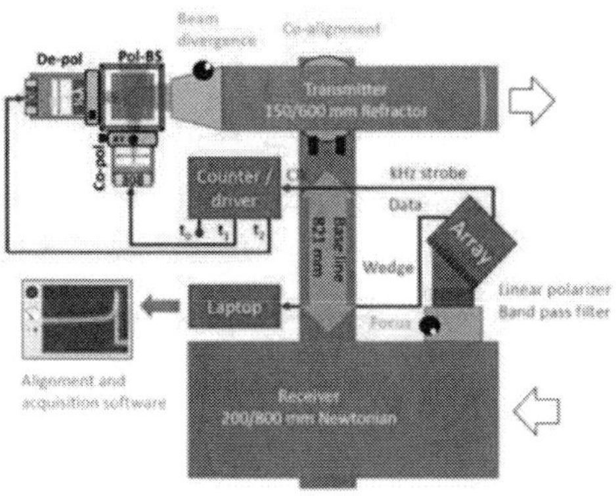

Fig. 3. Schematic diagram of the CW lidar system.

system is given in Fig. 3. The system incorporates two CW lasers operating at 808 nm, and transmitting at perpendicular linear polarizations. The beams are combined and the lasers are activated intermittently with a period of background recording added. A common detector array behind a linear polarizer is utilized. Similar systems, now with two lasers of different wavelengths, have been described – at 405 and 808 nm [12], and more recently at 980 and 1550 nm [13], and have been used, e.g. for atmospheric particle characterization.

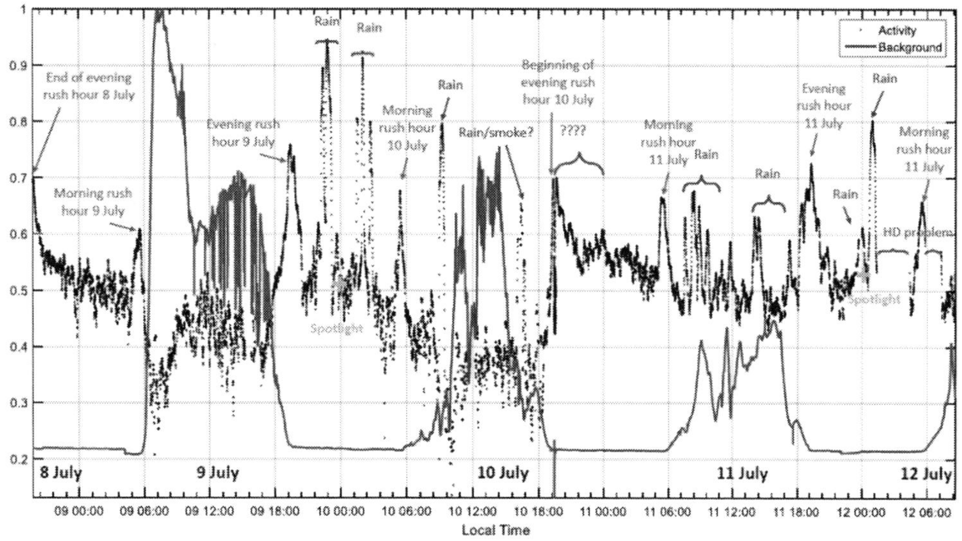

Fig. 4. Data overview for the first four field-experiment days. Atmospheric activity is on logarithmic scale.

3 MEASUREMENTS AND RESULTS

Data were recorded continuously as illustrated in Fig. 4, where the number of counts of atmospheric spikes, due to flying fauna as well as rain-drops, is displayed for the longer measuring path together with the signal background light level (the level of day-light). The plot is obtained from system read-outs, such as the one in Fig. 5, where individual spikes recorded for both co- and de-polarized light are shown. The strong echo to the far right is due to the beam termination at a range of about 510 m. The signals from high-flying migratory insects, driven down to the ground at the beginning of rain-fall, are shown in Fig. 6. Here, the short passage time of raindrops through the laser beam (less than 8 ms) is used to discriminate swiftly falling rain-drops from more slowly flying atmospheric fauna.

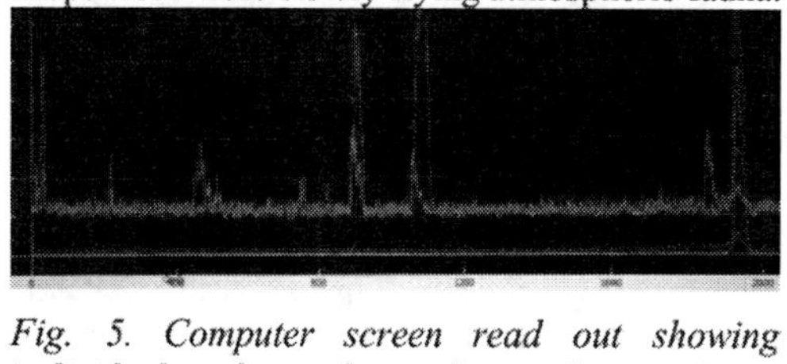

Fig. 5. Computer screen read out showing individual echoes from flying fauna. Beam termination to the right.

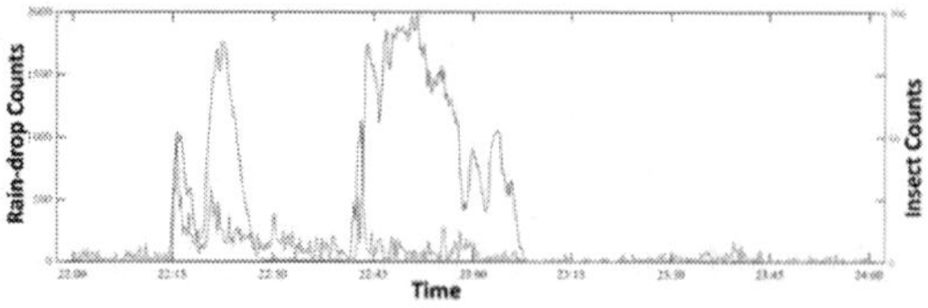

Fig. 6. Insect counts (red) and rain-drop counts (blue) during two night-time hours. High-altitude insects are forced down by the rain-fall.

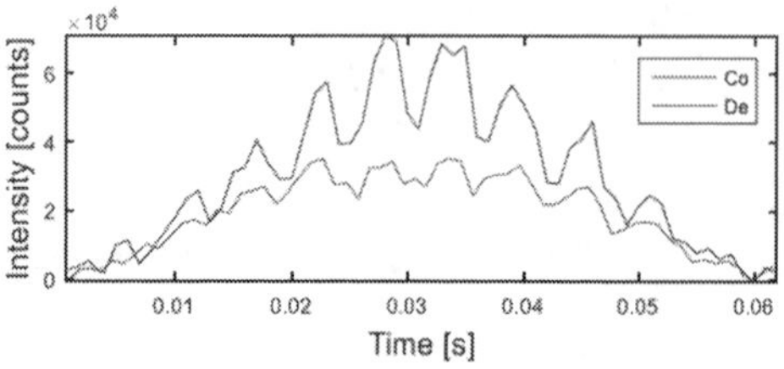

Fig. 7. Passage of an individual insect through the laser beam during 60 ms. Both co- and de-polarized signals are displayed. A considerable depolarization suggests that the insect with wingbeat frequency 160 Hz is not glossy.

For identification of insect species the wing-beat frequency and the depolarization of the back-scattered light can be utilized. Each of the hundreds of thousands of insect echo signals can be analyzed in these respects. As an example we show in Fig. 7 the detection of a rice-field insect with a given wing-beat frequency, and displaying a certain depolarization ratio. Clearly, extensive interdisciplinary collaboration between entomologists and the physicists operating the lidar system is needed for proper identification.

References

1. V.A. Drake, and D.R. Reynolds, *Radar Entomology: Observing Insect Flight and Migration* (CABI, Wallingford, Oxfordshire, Boston, MA, 2012).
2. J.W Chapman, D.R. Reynolds, A.D. Smith, Vertical-looking radar: A new tool for monitoring high-altitude insect migration. Bioscience **53**, 503 (2003).
3. J.A. Shaw, N.L. Seldomridge, D.L. Dunkle, P.W. Nugent, L.H. Spangler, Polarization lidar measurements of honey bees in flight for locating land mines. Opt. Expr. **13**, 5853 (2005).
4. E.S. Carlsten, G.R. Wicks, K.S. Repasky, J.L. Carlsten, J.J. Bromenshenk, C.B. Henderson, Field demonstration of a scanning lidar and detection algorithm for spatially mapping honeybees for biological detection of land mines. Appl. Opt. **50**, 2112 (2011).
5. Z.G. Guan, M. Brydegaard, P. Lundin, M. Wellenreuther, E. Svensson, S. Svanberg, Insect monitoring with fluorescence LIDAR techniques - Field experiments. Appl. Optics **48**, 5668 (2010).
6. L. Mei, Z.G. Guan, H.J. Zhou, J. Lv, Z.R. Zhu, J.A. Cheng, F.J. Chen, C. Löfstedt, S. Svanberg, G. Somesfalean, Agricultural pest monitoring using fluorescence lidar techniques. Applied Physics B **106**, 733 (2012).
7. A. Runemark, M. Wellenreuther, H. Jayaweera, S. Svanberg, M. Brydegaard, Rare events in remote dark field spectroscopy: An ecological case study of insects. IEEE JSTQE **18**, 1573 (2012).
8. Y.Y. Li, H. Zhang, Z. Duan, M. Lian, G.Y. Zhao, X.H. Sun, J.D. Hu, L.N. Gao, H.Q. Feng, S. Svanberg, Optical characterization of agricultural pest insects: A methodological study in the spectral and time domains, Appl. Physics B, doi 10.10007/s00340-016-6485-x (2016).
9. S.M. Zhu, Y.Y. Li, N.L. Gao, T.Q. Li, G.Y. Zhao, S. Svanberg, C.H. Lu, J.D. Hu, J.R. Huang, H.Q. Feng, Optical remote detection of flying Chinese agricultural pest insects using dark-field reflectance measurements, Acta Sinica Entomologica **59**, 1376 (2016).
10. M. Brydegaard, A. Gebru, S. Svanberg, Super resolution laser radar with blinking atmospheric particles – application to interacting flying insects, Prog. Electromagn. Res.**147**, 141 (2014).
11. M. Brydegaard, Towards quantitative optical cross sections in entomological laser radar– Potential of temporal and spherical parameterizations for identifying atmospheric fauna, PLoS One **10**, e0135231 (2015).
12. G. Zhao, E. Malmqvist, S. Török, P.-E. Bengtsson, S. Svanberg, J. Bood, *et al.*, Dual-band continuous-wave lidar system employed for particle classification, *To appear*, 2017.
13. M. Brydegaard, J. Larsson, S. Török, E. Malmqvist, S. Jansson, M. Andersson, S. Svanberg, S. Åkesson, F. Laurell, J. Bood, Short-wave Infrared Atmospheric Scheimpflug lidar, ILRC28.

EPJ Web of Conferences **176**, 07002 (2018) https://doi.org/10.1051/epjconf/201817607002
ILRC 28

MOBILE LIDAR SYSTEM FOR ENVIRONMENTAL MONITORING

Guangyu Zhao[1], Ming Lian[1], Yiyun Li[1], Zheng Duan[1], Shiming Zhu[1], and Sune Svanberg[1,2*]

[1]Center for Optical and Electromagnetic Research, South China Normal University, University City Campus, Guangzhou 510006, China
[2]Department of Physics, Lund University, P.O. Box 118, SE 221 00 Lund, Sweden
[*]Email: sune.svanberg@fysik.lth.se

ABSTRACT

A versatile mobile remote sensing system for multidisciplinary environmental monitoring tasks on the Chinese scene is described. The system includes a 20 Hz Nd:YAG laser-pumped dye laser, optical transmitting/receiving systems with a 30 cm and a 40 cm Newtonian telescope, and electronics, all integrated in a laboratory, installed on a Jiefang truck. Results from field experiments on atomic mercury DIAL mapping and remote laser-induced fluorescence and break-down spectroscopy are given.

Fig. 1. SCNU mobile remote sensing laboratory.

1 INTRODUCTION

Environmental issues attract an ever increasing interest, and measures to improve the ambient living conditions are widely launched to reduce the impact of environmental pollution, not the least on the Chinese scene, where the situation is severe (see, e.g. [1]). Laser spectroscopy has proven as a powerful methodology for environmental studies, both in the laboratory and in remote-sensing applications. We here present a new and versatile mobile laboratory, which we have constructed at South China Normal University (SCNU) in Guangzhou. While specially adapted for laser radar applications, is also allows a wide range of laboratory studies to conveniently be pursued out in the field.

2 SYSTEM DESCRIPTION

The SCNU lidar system is a mobile unit with a large-size laboratory container, 7.5m x 2.4 m x 2.4 m, integrated on a Jiefang truck. A photograph of the system is shown in Fig. 1. Fig. 2a provides a side view, Fig. 2b a top view, while Fig. 2c shows the system in its docking position at the fixed SCNU laboratory in Guangzhou. While docked its full infra-structure is available for ordinary laser spectroscopy experiments pursued also in the fixed laboratory. The mobile system, in differential absorption lidar (DIAL) configuration [2],

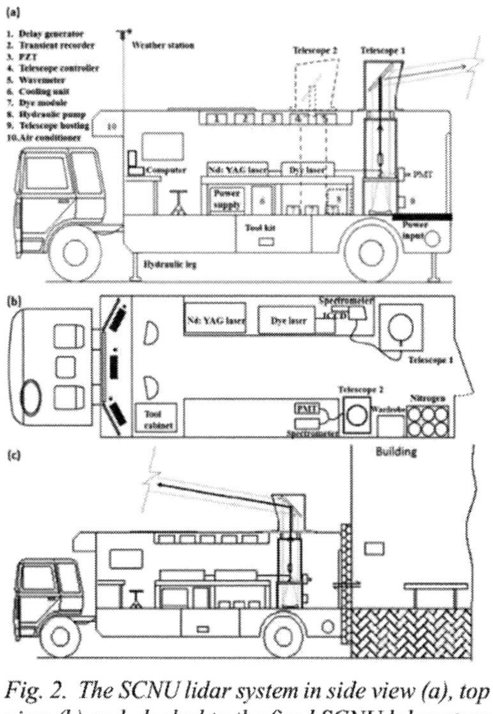

Fig. 2. The SCNU lidar system in side view (a), top view (b) and, docked to the fixed SCNU laboratory in Guangzhou (c).

operates with a 20 Hz Nd:YAG-pumped dye laser, which, e.g., for atomic mercury monitoring is operating at 508 nm, frequency doubled to the 254 nm Hg absorption line, and providing a mapping range approaching 2 km. The choice of a dye laser, instead of an all-solid-state transmitter, e.g. an optical parametric oscillator (OPO) [3], is motivated by a better adaption to realistic field operating conditions. The system incorporates two vertically looking Newtonian telescopes (diameter 40, and 30 cm, respectively), both with roof-top folding mirrors. The system is fully controlled by LabVIEW software. A full description of the system and measurement examples are given in [4].

3 ATMOSPHERIC MONITORING

Mercury is a particularly severe heavy-metal pollutant [5-8], and is also an interesting geophysical tracer gas [9,10]. China uses 500-700 tons of mercury annually and has severe mercury pollution problems in certain areas [6-8, 11]. We have used the mobile system in measurements of the vertical distribution of atmospheric mercury in the Chinese cities of Guangzhou, Zhengzhou and Xi'an. Concentrations were often found to be around 5 ng/m^3. In contrast, very high concentrations were observed in the mercury mining area of Wanshan, Guizhou province. An example of Hg DIAL measurements in this area is shown in Fig. 3, where concentrations in excess of 150 ng/m^3 were observed. Our system can readily be used for studying other important atmospheric pollutants, such as NO_2, SO_2, O_3, and NO around 440, 300, 280 and 226 nm, respectively. Examples of recordings relating to SO_2 are shown in Fig. 4.

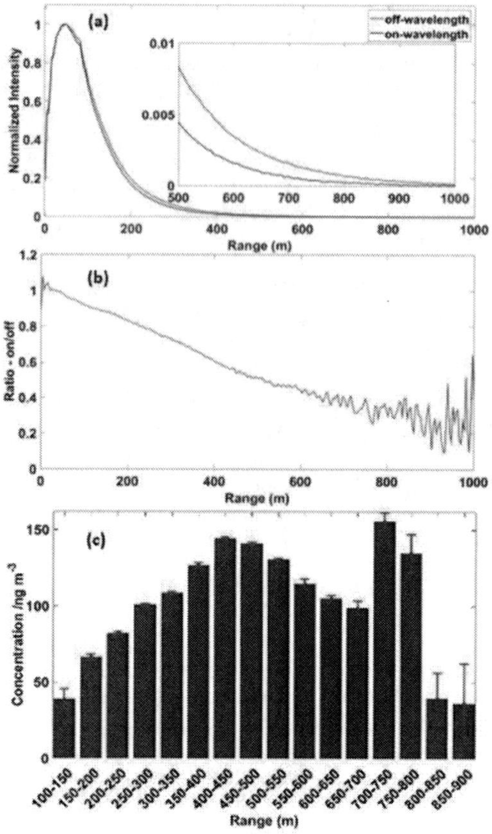

Fig. 3. Differential absorption lidar (DIAL) measurement of atomic mercury in the Wanshan abandoned mercury-mining area in China. a) on- and off-resonance lidar returns around 254 nm b) DIAL curve c) evaluated range-resolved mercury concentrations.

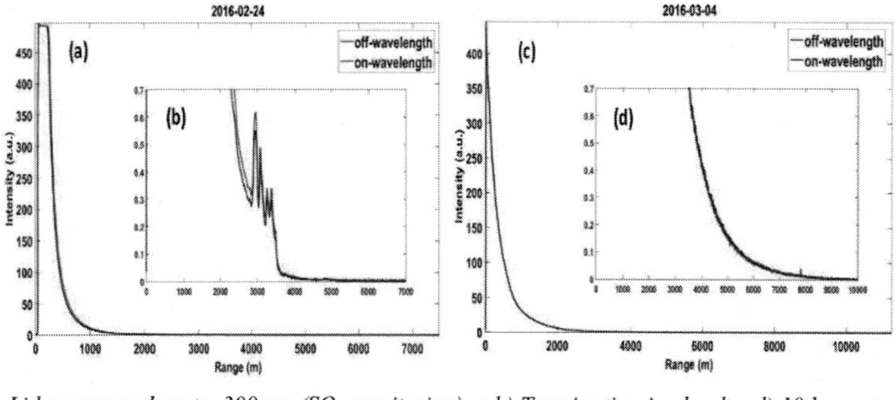

Fig. 4. Lidar curves close to 300 nm (SO_2 monitoring). a,b) Termination in cloud. c,d) 10 km return.

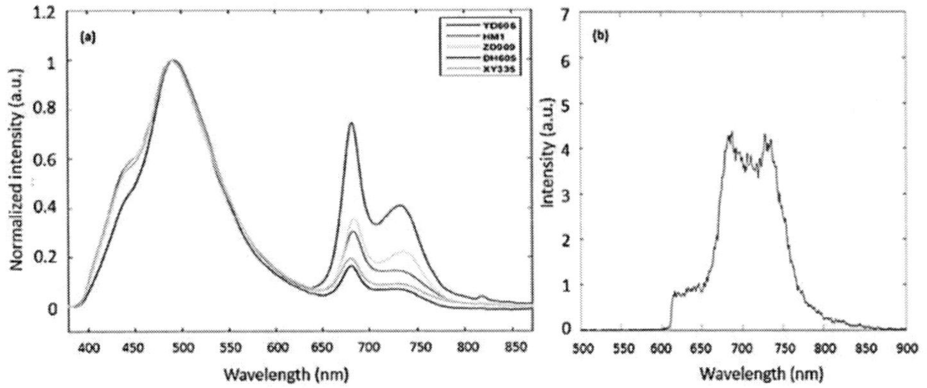

Fig. 5. Laser-induced fluorescence measurements on agricultural crops. a) shows recordings of 5 different types of maize, measured at 30 m range with 355 nm excitation. b) shows a recording of rice at 120 m distance and using 532 nm excitation.

4 REMOTE LASER-INDUCED FLUORESCENCE MEASUREMENTS

Remote laser-induced fluorescence (LIF) measurements enable the monitoring of important properties of solid and liquid targets [12,13]. We have used the new mobile laboratory in field work in Zhengzhou and Xinyang, Henan Province [14]. Examples of recordings of remote vegetation targets (maize and rice) are shown in Fig. 5. Fluorescence properties can yield information on environmental stress and the level of fertilizer available. Likewise, the status of facades of historical buildings can be assessed [15]. A spectrometer is fiber-optically coupled to the output of the lidar receiving telescope in these measurements, where harmonics of the Nd:YAG laser (at 532, 355, or 266 nm) can be utilized. For day-light recordings, influence of ambient light is suppressed by using a gated and intensified detector.

5 REMOTE LASER-INDUCED BREAK-DOWN SPECTROSCOPY MEASUREMENTS

Laser-induced break-down spectroscopy (LIBS) is a much employed technique for studying the elemental composition of samples, mostly in the laboratory environment [16]. It is also possible to perform LIBS remotely using a lidar system [17,18]. We have used our new mobile system for such measurements, and an example is shown in Fig. 6. A soil sample is measured at 15 m distance, as well as in the laboratory. A spectrometer with a gated and intensified detector was again used, in

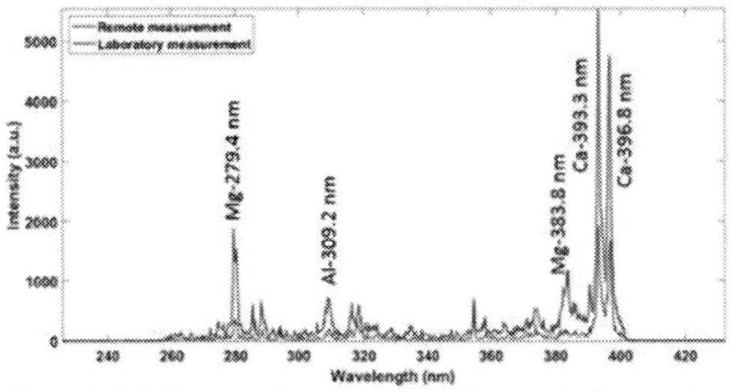

Fig. 6. LIBS recordings for a soil sample located at 15 m distance from the measurement system, as well as when kept in the laboratory are shown, with sharp spectral lines due to Mg, Al, and Ca.

order to allow delayed spectral monitoring of the recombining plasma, which then exhibits a multitude of sharp and characteristic elemental spectral lines. The availability of nutrients can be assessed, as well as the presence of, e.g. heavy-metal soil pollutants.

Further measurement examples will be discussed.

ACKNOWLEDGEMENTS

We much value the support by Prof. Sailing He, and Dr. Liang Mei, and the fruitful collaboration with colleagues at the Chinese Academy of Sciences Institute for Geochemistry, Guiyang, at the Henan Agricultural University, Zhengzhou, and the Henan and Guangdong Academies of Agricultural Sciences. This work was supported by the Guangdong Province Innovation Team Research Program (No. 201001D0104799318).

References

[1] K. He, H. Huo, and Q. Zhang, Urban air pollution in China: Current status, characteristics, and progress, *Annual Review of Energy and the Environment,* vol. 27, pp. 397-431, 2002.

[2] 2. S. Svanberg, Differential Absorption Lidar (DIAL), in *Air Monitoring by Spectroscopic Techniques*, M. Sigrist (ed.), Wiley, N.Y. 1994, p. 8

[3] P. Weibring, H. Edner, and S. Svanberg, Versatile mobile lidar system for environmental monitoring, Applied Optics **42**, 3583 (2003).

[4] G.Y. Zhao, M. Lian, Y.Y. Li, Z. Duan, S.M. Zhu, L. Mei, and S. Svanberg, Mobile lidar system for environmental monitoring, Appl. Opt., in press (2017)

[5] International Conference on Mercury as a Global Pollutant, with the latest version arranged in 2015 in Korea; http://mercury2015.com

[6] X. Fu, X. Feng, J. Sommar, and S. Wang, A review of studies on atmospheric mercury in China, Science of the Total Environment **421–422**, 73-81 (2012).

[7] D.G. Streets, J.M. Hao, Y. Wu, J.K. Jiang, M. Chan, H.Z. Tian, and X.B. Feng, Antropogenic mercury emissions in China, Atmos. Env. **39**, 7789-7806 (2005).

[8] L. Zhang and M.H. Wong, Environmental mercury contamination in China: Sources and impacts, Environment International **33**, 108-121 (2007).

[9] J. H. McCarthy, Jr., W. W. Vaughn, R. E. Learned, and J. L. Meuschke, Mercury in soil gas and air - A potential tool in mineral exploration," Geological Survey Circular 609 (US Geological Survey, Washington 1969).

[10] S. Svanberg, Geophysical gas monitoring using optical techniques: Volcanoes, geothermal fields and mines, Optics and Lasers in Engineering **37**, 245 (2002).

[11] Z.H. Dai, X.B. Feng, J. Sommar, P. Li, and X.W. Fu, Spatial distribution of mercury deposition fluxes in Wanshan Hg mining area, Guizhou province, China, Atmos. Chem. Phys. **12,** 6207 (2012).

[12] S. Svanberg, Fluorescence spectroscopy and imaging of lidar targets, Chapter 7 in T. Fujii and T. Fukuchi (Eds) *Laser Remote Sensing* (CRC Press, Boca Raton 2005) pp 433-467.

[13] H. Amann, Laser spectroscopy for monitoring and research in the ocean, Phys. Scripta **T78**, 68 (1998).

[14] G.Y. Zhao, Z. Duan, M. Lian, Y.Y. Li, R.P. Chen, J.D. Hu, S. Svanberg, and Y.L. Han, Reflectance and fluorescence characterization of maize species using laboratory measurements and lidar remote sensing, Applied Optics **55**, 5273 (2016).

[15] V. Raimondi, G. Cecchi, D. Lognoli, L. Palombi, R. Grönlund, A. Johansson, S. Svanberg, K. Barup, and J. Hällström, The fluorescence LIDAR technique for the remote sensing of photoautotrophic biodeteriogens on outdoor cultural heritage: A decade of in situ experiments, International Biodeterioration & Biodegradation (2009), doi:10.1016/j.biod.2009.03.006.

[16] D. A. Cremers, and L. J. Radziemski, *Handbook of Laser-Induced Breakdown Spectroscopy* (John Wiley & Sons Ltd., West Sussex, UK, 2006).

[17] S. Palanco, J.M. Baena, and J.J. Laserna, Open-path laser-induced plasma spectrometry for remote analytical measurements on solid surfaces, Spectrochim. Acta B **57**, 591 (2002)

[18] R. Grönlund, M. Lundqvist, and S. Svanberg, Remote imaging laser-induced break-down spectroscopy and remote cultural heritage ablative cleaning, Opt. Lett. **30**, 2882 (2005).

SYNERGY OF LIDAR AND PASSIVE REMOTE SENSOR DATA FOR RETRIEVING PROFILES OF MICROPHYSICAL PROPERTIES OF NON-SPHERICAL PARTICLES

Alexei Kolgotin[1*], Detlef Müller[2,3], Eduard Chemyakin[3], Anton Romanov[4]

[1]*Physics Instrumentation Center, Troitsk, Moscow Region, 142190, Russia, *alexeift@yahoo.com*
[2]*University of Hertfordshire, Hatfield, Hertfordshire, AL10 9AB, UK, d.mueller@herts.ac.uk*
[3]*Science Systems and Applications, Inc. (SSAI), NASA LaRC,*
1 Enterprise Parkway, Hampton, VA, 23666, USA,
[4]*The National University of Science and Technology, Leninskii av. 4, Moscow, 119049*

ABSTRACT

In this study we explore how the combination of 3 backscatter and 2 extinction lidar data with data that can be collected with ground-based and space-borne passive remote sensors, e.g. phase function coefficients which can be derived at various measurement wavelengths and scattering angles can result in improved profiles of particle microphysical properties. The algorithm is based on a light-scattering model that uses a mixture of spheres and randomly oriented spheroids.

1 INTRODUCTION

Our study addresses the question on what combination of lidar and passive remote sensing data could provide an optimum set of data for the retrieval of profiles of key particle parameters when observed from ground or space. Over the past 10 years, multiwavelength Raman/HSRL lidars have successfully demonstrated their capacity with regard to retrieving particle microphysical parameters [1, 2]. In these approaches three backscatter (β) and two extinction (α) coefficients measured with lidar at wavelengths (λ) from 355 to 1064 nm are used as input data for the retrieval of particle size distribution (PSD) as well as particle bulk parameters, i.e. effective radius, number, surface-area and volume concentrations, and the complex refractive index $m=m_R-im_I$ (CRI). The major challenge in identifying the correct solution of the underlying ill-posed inverse problem is the limited number of input data, i.e. backscatter coefficients at 355, 532, and 1064 nm, and extinction coefficients at 355 and 532 nm (3+2 data), high measurement errors of up to 10 - 20%, the wide range of particle radii r from 0.05 to 15 μm, real parts (m_R) of the CRI from 1.3 to 1.6 and imaginary parts (m_I) from $i0$ to $i0.05$. Considering all these factors, it has been shown that different solution techniques result in large retrieval uncertainties of profiles of particle properties.

One of the ways that allows us to improve the retrieval quality is to increase the number of input data. In that context passive remote sensors provide us with valuable additional information. In this study we explore how the combination of 3+2 lidar data with data that can be collected with ground-based and space-borne passive remote sensors, e.g. phase function coefficients which can be derived at various measurement wavelengths and scattering angles can result in improved profiles of particle microphysical properties. The algorithm is based on a light-scattering model that uses a mixture of spheres and randomly oriented spheroids. This light-scattering model is widely used in the analysis of passive remote sensing of particles of arbitrary shape, e.g. smoke, mineral dust, volcanic ash, and anthropogenic pollution.

2 METHODOLOGY

2.1 Mathematical model

The equation that is based on a light-scattering model that uses a mixture of spheres and randomly oriented spheroids and relates the PSD and its optical characteristics is presented in Ref. [3]. We can modify this equation as

$$\int_{r_{\min}}^{r_{\max}} k_g^{(\eta,\varepsilon)}(\lambda, m(z), r)\upsilon(r,z)dr = g(\lambda,z), \quad g=\alpha,\beta \qquad (1)$$

for the case of a height-dependent volume particle size distribution $\upsilon(r,z)$ and optical profiles $3\beta(\lambda,z)+2\alpha(\lambda,z)$, where the kernels of the integral equation

$$k_g^{(\eta,\varepsilon)}(\lambda,m(z),r) = (1-\eta)k_g(\lambda,m(z),r,\varepsilon=1) + \\ +\eta k_g(\lambda,m(z),r,\varepsilon) \quad (2)$$

depend on particle shape (ε) and the volume fraction of the spheroid particles $\eta\in[0;1]$. If the particles are spherical the respective kernel $k_g(\ldots,\varepsilon=1)$ is calculated on the basis of Mie-scattering theory [4]. The kernels for randomly oriented spheroids $k_g(\ldots,\varepsilon)$ can be computed in the way shown in Ref. [5]. In our study we assume that the aspect ratio ε (or aspect ratio distribution) in the sphere-spheroid model (η,ε) is known, but that the spheroid fraction needs to be estimated.

Our goal is to find the profile $\upsilon(r,z)$ and the profile $m(z)$, see Eq. (1), on the radius and height domains $r\in[r_{\min};r_{\max}]$ and $z\in(0;z_{\max}]$ respectively. This is an ill-posed, underdetermined problem. In order to redefine this problem we include in the solution procedure data from a potential passive sensor that provides phase function coefficients $\varphi_\theta(\lambda)$ measured at 5 scattering angles $\theta = 5°$, 30°, 60°, 90°, and 120° at 1 or 2 wavelengths, e.g. at λ=410 and/or 670 nm. Passive sensors provide column-integrated data. Therefore, we have to reformulate Eq. (1) so that it can be used for column-integrated data as well. The integration of Eq. (1) over height z allows us to find the equation that is consistent for data provided by both types of instruments

$$\int_{r_{\min}}^{r_{\max}} k_g^{(\eta,\varepsilon)}(\lambda,m,r)\upsilon(r)dr = g(\lambda)\,,\quad g=\alpha,\ \beta,\ \varphi_\theta \quad (3)$$

where

$$g(\lambda) = \int_0^{z_{\max}} g(\lambda,z)dz\,,\ \upsilon(r) = \int_0^{z_{\max}} \upsilon(r,z)dz\,, \\ g=\alpha,\ \beta \quad (4)$$

and the parameter m is equal to some intermediate value in the profile $m(z)$, so that the equation

$$\int_0^{z_{\max}}\int_{r_{\min}}^{r_{\max}} k_g^{(\eta,\varepsilon)}(\lambda,m(z),r)\upsilon(r,z)drdz = \\ \int_{r_{\min}}^{r_{\max}} k_g^{(\eta,\varepsilon)}(\lambda,m,r)\int_0^{z_{\max}} \upsilon(r,z)dzdr \quad (5)$$

is fulfilled on the basis of the average theorem.

We use Eq. (3) and (1) in our strategy in terms of two steps in order to find the particle microphysical parameters for 1) the column integrated data, and 2) the profile data, taking into account the column-integrated data as benchmark.

2.2 Retrieval approach for column-integrated data

Eq. (3) for column-integrated data is the classical integral equation of the Fredholm type of the 1st kind with unknown input parameters η and m. This equation can be solved by inversion with regularization, as shown in Ref. [3], but now data from a passive sensor (5-10φ) are added to the input lidar data (3β+2α). It means that the underlying inversion problem (3) has N_O=10-15 pieces of input information, i.e., 3β+2α + (5-10)φ. Such a relatively high number of input data allows us to modify the algorithm in Ref. [3] and take into account just a few inversion windows $[r_{\min};r_{\max}]$ located from 0.05 to 10-15 µm particle radius. The use of 5-7 inversion windows instead of 100-120 inversion windows reduces the solution space by one order of magnitude.

We find the column-integrated PSD $\upsilon(r)$ as well as effective radius r_{eff}, number n, surface-area s and volume v concentrations, and the complex refractive index m. In section 3 we mainly describe this approach and show results of numerical simulations.

2.3 Retrieval approach for profile data

Eq. (1) can be considered as a system of Fredholm integral equations of the 1st kind at a finite number of height bins z_l, l=1, 2, …, N_L, distributed along the height from 0 to $z_{\max}$. In a recent publication [6] the gradient correlation method (GCM) and proximate analysis (PA) were presented. GCM and PA allow us to find vertical profiles that contain stable solutions of the inversion problem. GCM and PA use regression equations that connect the lidar data products (optical data) and the particle microphysical parameters. It was shown in Ref. [6] that the careful estimation of the regression coefficients (RC) of the underlying regression equations permits for a comparably accurate retrieval of the PSD $\upsilon(r,z_l)$ as well as the profiles of effective radius $r_{\text{eff}}(z_l)$, number $n(z_l)$, surface-area $s(z_l)$ and volume $v(z_l)$ concentrations, and the complex refractive index $m(z_l)$. Knowledge of column-integrated data derived in step 1 of our strategy provides us with a robust estimation of the RCs. Detailed results of numerical simulation with GCM and PA can be found in Ref. [6].

3 RESULTS OF NUMERICAL SIMULATIONS

In our numerical simulations we generated synthetic optical data $3\beta+2\alpha+10\varphi$. We used two types of PSDs, i.e. one monomodal and one bimodal PSD. The monomodal PSD (MPSD) is described by the log-normal law with mean radius r_0=0.1 µm and mean width σ=1.49. The bimodal PSD (BPSD) includes the fine mode with the same parameters r_0 and σ of the MPSD and the coarse mode with r_0=1 µm and σ=1.8.

We use three values (levels) of particle light-absorption. For low light-absorbing (LA) particles the CRI is m=1.38-i0.002, for moderate light-absorbing (MA) m=1.45-i0.005, and for high light-absorbing (HA) m=1.57-i0.018. Last but not least we use three types of sphere-spheroid models (η,ε), i.e., (0,0.33), (0.7,0.33) and (0.75,033).

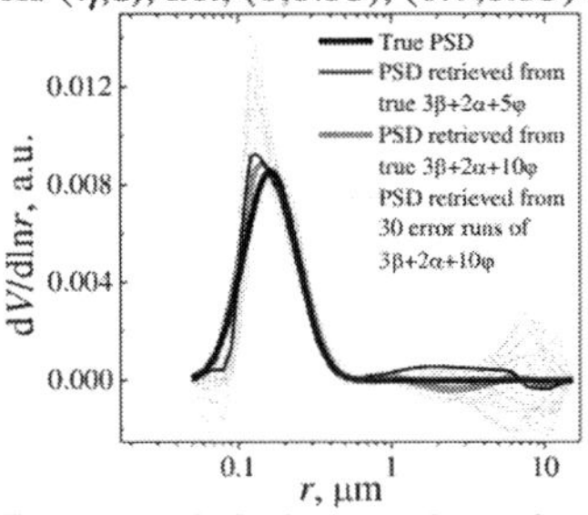

Fig. 1. Black thick curve shows the true (0.7,0.33)-HA-MPSD. The grey solid curve and the 30 grey-dotted curves represent the PSDs retrieved in scenario A and C, respectively, from the optical data set $3\beta+2\alpha+10\varphi$. The black thin curve shows the PSD retrieved in scenario A from the optical data set $3\beta+2\alpha+5\varphi$.

We generated two data sets, i.e., $3\beta+2\alpha+5\varphi$ and $3\beta+2\alpha+10\varphi$ that were used as input for the inversion problem (3). The solution space in the retrieval approach is defined on the basis of the input data that are defined by Eq. (3) as follows:

– 7 inversion windows on the radius domain r_{min}=0.05-0.10 µm and r_{max}=8-15 µm;

– 6 real parts of the CRI, equidistantly distributed between 1.35 and 1.6;

– 11 imaginary parts of the CRI, equidistantly distributed between i0 and i0.05;

– 5 spheroid fractions, equidistantly distributed between 0 and 1.

We assume the following scenarios in our simulations in order to solve the inversion problem (3):

A. The optical data $3\beta+2\alpha+5\varphi$ and $3\beta+2\alpha+10\varphi$ are error-free; the spheroid fraction η is unknown.

B. The synthetic optical data $3\beta+2\alpha+(5\text{-}10)\varphi$ have maximum errors up to 5% and average errors up to 2%; the spheroid fraction η is not known.

C. the synthetic lidar data ($3\beta+2\alpha$) have errors up to 10%, the synthetic data of the passive sensor (10φ) have errors up to 5%; the spheroid fraction η is not known.

D. We use scenario C but the spheroid fraction η is known.

We carried out inversions with noisy data (erroneous data), which implies that 30 different error runs were made in scenarios B-D.

We show the statistics of the results of our numerical simulations in Fig. 1-2 and Table 1-2. The results derived using scenario A agree well with the theoretical values. MPSD (thin black curves and BPSD (solid grey curves) in Fig. 1, 2 almost coincide, however the PSD is more stable in the case of $3\beta+2\alpha+10\varphi$ (solid grey curve).

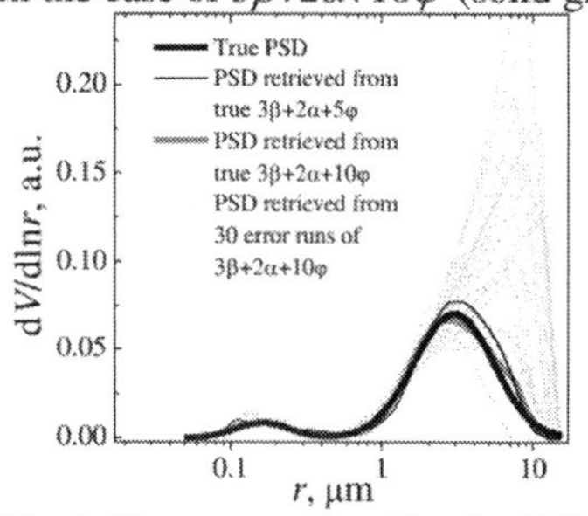

Fig. 2. The same as Fig. 1 but for (0.7,0.33)-HA-BPSD.

The uncertainties of the estimated parameters stay within a few percent. Such a quality of retrieval results cannot be obtained from a 3+2 data set even if particle depolarization coefficients are available [3]. The uncertainties of the estimated real and imaginary parts do not exceed 0.025 and i0.0025, respectively, which are half of the step sizes of our entry values, and which was never obtained in the case of using 3+2 data only [3].

The results derived from scenario B are close to the results obtained in scenario A. The uncertainties of the estimated parameters of the PSD do not exceed 50% in the worst case. The CRI is estimated with the same uncertainty in both scenarios (not shown here).

The retrieval results become worse in scenario C. For some error runs we find that the PSDs show a strong (extensive) coarse mode (see dotted lines in Fig. 2). Such outliers result in the overestimation of the volume concentration and the effective radius of up to 155% and 120%, respectively (see Table 1). The analysis of these PSDs shows that they are linked to very high values of the retrieved imaginary part, exceeding the true value by *i*0.03 and more (see for example HA-MPSD in table 2). Simultaneously, the value of the spheroid fraction cannot be retrieved and we obtain the value 0. It means that the sphere-spheroid model (η,ε) can be mimicked by a pure sphere model ($0,\varepsilon$) at the cost of overestimating the true values of the imaginary part and the volume concentration. We note that the retrieval of number concentration is more stable; the uncertainty is 80% whereas using the 3+2 data set the uncertainty is 100% and above [3]. Surface-area concentration is the most stable parameter in the retrieval (uncertainty is 53%). As we mentioned before, the uncertainty of the CRI retrieval increases as well; it is up to *i*0.04 for the imaginary part and up to 0.07 for the real part. However, our approach still allows us to distinguish the L-MA and HA particles because in the first case the uncertainty is less than *i*0.009 for the imaginary part of large particles (see LA, HA BPSD in table 2).

Table 1.Retrieval uncertainty of the PSD parameters in scenarios C and D

Scenario	r_{eff}, %	n, %	s, %	v, %
C	120	80	53	155
D	80	75	40	63

The results in scenario D shows that the uncertainties we obtained in scenario C can be reduced if we know the true sphere-spheroid model, i.e. the parameter pair (η,ε). In this case the uncertainties of r_{eff}, n, s and v decrease to 80, 75, 40 and 63%, respectively (see table 1). The real parts of the CRI can be estimated with an uncertainty of 0.08, the imaginary parts with an uncertainty of *i*0.003 for LA, MA BPSD and *i*0.009 for HA BPSD (table 2).

4 CONCLUSIONS

Preliminary simulations show that the synergy of both instruments allows us to improve the retrieval quality not only of particle size distribution parameters but particularly also allows us to improve the accuracy of the CRI retrievals. If maximum measurement errors of the $3\beta+2\alpha+10\varphi$ data set do not exceed 5% and if the average uncertainty is 2%, the uncertainties of the estimated PSD parameters do not exceed 50%, the accuracy of the CRI retrievals is 0.025 for the real part and *i*0.0025 for the imaginary part.

Table 2.Retrieval uncertainty of the CRI in scenarios C (numerator) and D (denominator) for MPSD, BPSD and different levels of particle light-absorption

Particle Absorption	m_R		m_I	
	MPSD	BPSD	MPSD	BPSD
LA	0.03 / 0.07	0.07 / 0.04	*i*0.043 / *i*0.010	*i*0.009 / *i*0.003
MA	0.07 / 0.05	0.07 / 0.08	*i*0.029 / *i*0.013	*i*0.009 / *i*0.003
HA	0.07 / 0.02	0.07 / 0.08	*i*0.030 / *i*0.019	*i*0.027 / *i*0.009

10% measurement error of $3\beta+2\alpha$ lidar data can lead to an overestimation of the volume concentration up to 155% and up to *i*0.03 of the imaginary part of the CRI. The retrieval results can be stabilized if the true sphere-spheroid model, i.e. the parameter pair (η,ε) is known.

References

[1] D. Müller, et al., 2013: Vertical profiles of pure dust (SAMUM 2006) and mixed-smoke-dust plumes (SAMUM 2008) inferred from inversion of multiwavelength Raman lidar data and comparison to AERONET retrievals, *Appl. Opt.* **52** (14), 3178-3202.

[2] D. Müller, et al., 2014: Airborne multiwavelength High Spectral Resolution Lidar (HSRL-2) observations during TCAP 2012: vertical profiles of optical and microphysical properties of a smoke/urban haze plume over the northeastern coast of the US, *Atmos. Meas. Tech.* **7**, 3487–3496.

[3] I. Veselovskii, et al., 2010: Application of randomly oriented spheroids for retrieval of dust particle parameters from multiwavelength lidar measurements", *J.Geophys.Res.* **115**, D21203.

[4] C. F. Bohren and D. R. Huffman, eds., 1983: Absorption and Scattering of Light by Small Particles, *Wiley*.

[5] O. Dubovik, et al., 2006: Application of spheroid models to account for aerosol particle nonsphericity in remote sensing of desert dust, *J.Geophys.Res.* **111**, D11208.

[6] A. Kolgotin, et al., 2016: Improved identification of the solution space of aerosol microphysical properties derived from the inversion of profiles of lidar optical data, part 2: simulations with synthetic optical data, *Appl. Opt.* in press.

RETRIEVING MICROPHYSICS OF CIRRUS CLOUDS FROM DATA MEASURED WITH RAMAN LIDAR RAMSES AND A TILTED CEILOMETER

Anatoli Borovoi[1]*, Jens Reichardt[2], Ulrich Görsdorf[2], Veronika Wolf[2,3], Alexander Konoshonkin[1], Victor Shishko[1], Natalia Kustova[1]

[1]*Institute of Atmospheric Optics, Rus. Acad. Sci., Tomsk, Russia, *borovoi@iao.ru*
[2]*Richard-Aßmann-Observatorium, DWD, Lindenberg, Germany*
[3]*Now at: Luleå University of Technology, Kiruna, Sweden*

ABSTRACT

To develop a microphysical model of cirrus clouds, data obtained by Raman lidar RAMSES and a tilted ceilometer are studied synergistically. The measurements are interpreted by use of a data archive containing the backscattering matrixes as well as the depolarization, color and lidar ratios of ice crystals of different shapes, sizes and spatial orientations calculated within the physical-optics approximation.

1 INTRODUCTION

Retrieval of microphysical properties of cirrus clouds, i.e. sizes, shapes, and orientations of ice crystals, from lidar signals is a challenging problem for the lidar community that is not satisfactorily solved yet [1]. One of the problems in the lidar studies of cirrus clouds by means of the conventional vertically-pointing lidars is that appearance of even small fraction of quasi-horizontally oriented plate-like ice crystals in the clouds leads to strong variations of the lidar signals on the background of the signals formed by the randomly oriented fraction of the crystals. This phenomenon is easily explained by the specular reflection (or specular scattering) from the horizontally oriented plates [2, 3]. In particular, the space lidar CALIPSO was specially oriented at 3° off nadir to avoid the specular scattering. To retrieve the microphysical properties of the horizontally-oriented fraction of the ice crystals of cirrus clouds, the vertically-pointing Raman lidar RAMSES [4] has been supplemented with a ceilometer tilted at 5° off zenith.

In this contribution we show that joint use of signals from both RAMSES and the ceilometer allow us to retrieve the percentage of the quasi-horizontally oriented plate-like crystals and their effective tilts near the horizon.

2 METHODOLOGY

We assume that cirrus clouds are formed by two co-existing particle fractions. The main fraction consists of randomly oriented ice crystals of arbitrary shapes, while the minor fraction is an ensemble of quasi-horizontally oriented plate-like crystals. In general, if a crystal is illuminated by linearly polarized light, the backscattering cross section for a randomly oriented crystal consists of the parallel and perpendicular components:

$$\sigma = \sigma_{\parallel} + \sigma_{\perp}. \qquad (1)$$

However a quasi-horizontally oriented plate creates, with good accuracy [5], only the parallel component which is denoted as σ_{pl}. Hence two backscattering coefficients measured by RAMSES for a given height h are equal to

$$\beta_{\perp} = c\sigma_{\perp}, \qquad (2)$$

$$\beta_{\parallel} = c\sigma_{\parallel} + C\sigma_{pl}, \qquad (3)$$

where c and C are the number densities of the main and minor fractions, respectively. Their ratio

$$k = C / c \qquad (4)$$

is an important microphysical characteristics called the plate percentage.

Also, RAMSES measures the extinction coefficient

$$a = 2cs + 2CS, \qquad (5)$$

where s and S are average areas of the shadows produced by a randomly oriented crystal and a quasi-horizontally oriented hexagonal plate wavering about the horizontal plane, respectively.

The ceilometer does not detect polarization of light, it measures the total cross section of Eq. (1) σ_{ceil} at the wavelength of 1064 nm. Being tilted at 5° off zenith, it does not detect the minor fraction. Hence its signal is equal to

$$\beta_{ceil} = c\sigma_{ceil}. \quad (6)$$

In contrast to backscatter and extinction coefficients, the depolarization ratio (δ) and the lidar ratio (L) are two optical properties that do not depend on the number density of the cloud particles and thus are characteristics of the scattering behavior of the particle ensemble and, hence, more meaningful in the context of interpreting lidar data in terms of microphysics. If all crystals in clouds are randomly oriented, these properties are equal to

$$\delta_0 = \beta_\perp^0 / \beta_\parallel^0 = \sigma_\perp / \sigma_\parallel, \quad (7)$$

$$L_0 = \alpha_0 / (\beta_\perp^0 + \beta_\parallel^0) = 2s / \sigma. \quad (8)$$

For the mixture of the randomly oriented and quasi-horizontally oriented plates we have

$$\delta = \frac{\beta_\perp}{\beta_\parallel} = \frac{c\sigma_\perp}{c\sigma_\parallel + C\sigma_{pl}} = \frac{\delta_0}{1 + k\sigma_{pl} / \sigma_\parallel}, \quad (9)$$

$$L = \frac{\alpha}{\beta_\perp + \beta_\parallel} = \frac{L_0 + 2kS / \sigma}{1 + k\sigma_{pl} / \sigma}. \quad (10)$$

As experience shows, number densities c and C [Eqs. (2)-(6)] change much more rapidly with height h in natural clouds than the microphysical proxies δ and L.

Therefore, if horizontally oriented plates were absent, $C(h) = 0$, the four experimentally measured vertical profiles of Eq. (2), (3), (5) and (6), plotted with arbitrary constant factors, should exhibit similar curve shapes, reflecting the dependence $c(h)$.

Experimental data where aligned plates were absent are presented in Figs. 1 and 2. Indeed, in Fig. 1 we see that all four profiles are similar. In addition, we see in Fig. 2 that both the depolarization ratio and lidar ratio only weakly depend on the height unlike the quantities of Fig. 1. In the following, we refer to these data as 'Case 1'.

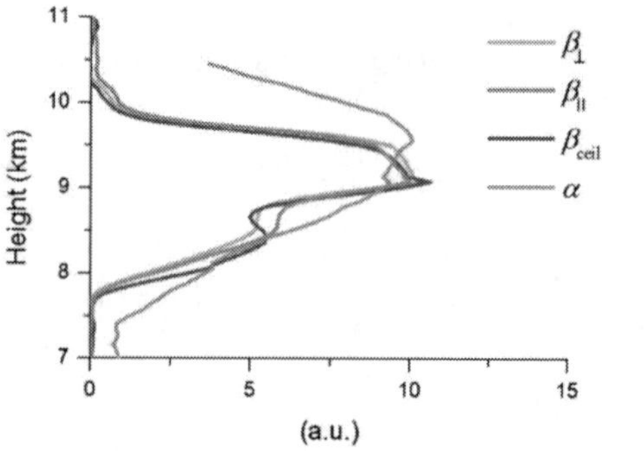

Figure 1. Case 1: Four simultaneously measured profiles of Eqs.(2), (3), (5), and (6). Profiles are scaled so that maxima have roughly the same magnitude

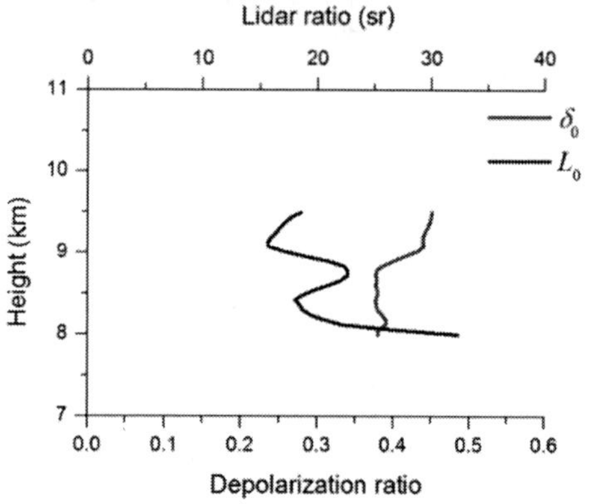

Figure 2. Case 1: Vertical profiles of the depolarization and lidar ratios

We note that the values $\delta_0 \approx 0.4$ and $L_0 \approx 20$ sr are rather typical values observed by RAMSES in cirrus clouds. According to our data archive with the backscattering matrixes of ice crystal particles [6], these values match best the results obtained for randomly oriented columns with distorted shapes.

3 RESULTS

Figures 3-5 showcase the experimental data referred to as Case 2. Here the appearance of horizontally oriented plates is revealed by the different shapes of the curves in Fig. 3. In addition, in Figs. 4 and 5, we see that the values of the depolarization and lidar ratios are close to those of Case 1 at a height of about 7 km. These observations indicate that there were only randomly oriented crystals at $h \approx 7$ km and that horizontally oriented plates appeared below.

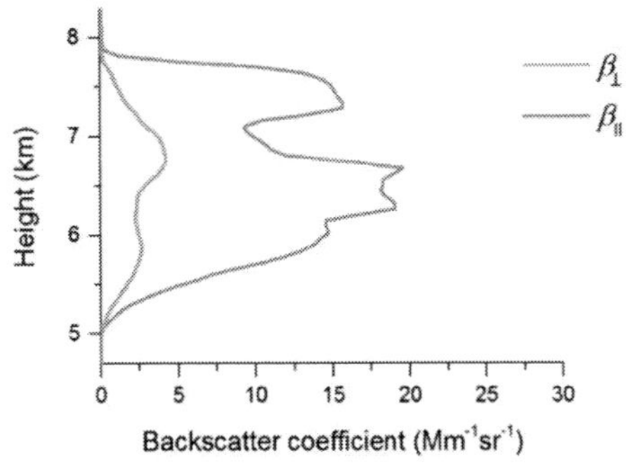

Figure 3. Case 2: Perpendicular and parallel backscattering coefficients measured simultaneously

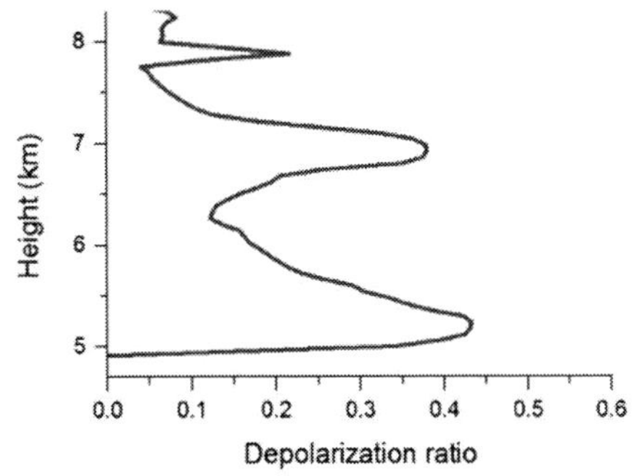

Figure 4. Case 2: Depolarization ratio

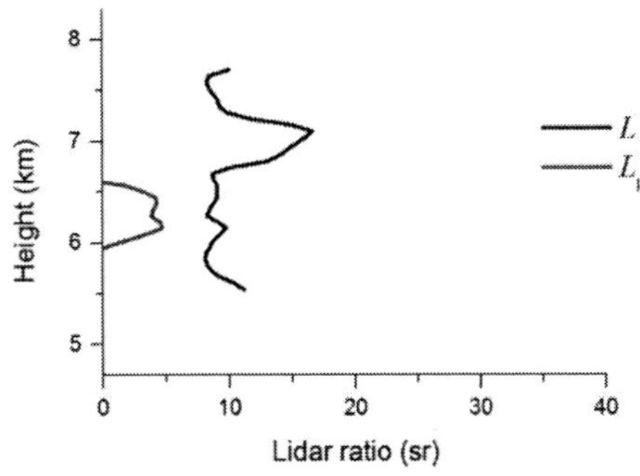

Figure 5. Case 2: Lidar ratios L and L_{pl}, the lidar ratio of the horizontally oriented plates [see defining Eq. (11)]

We can now retrieve some new microphysical characteristics of the plates in Case 2. For the first, we use the following approximation. We suppose that the microphysical characteristics of the randomly oriented fraction change only negligibly with height. So their number-density independent optical properties are considered to be equal to the values measured at 7 km, $\delta_0 \approx 0.38$ and $L_0 = 15$ sr (Figs. 4 and 5). As a result, by applying Eqs. (7)-(10), we retrieve the ratio

$$L_{pl} = 2S / \sigma_{pl}, \tag{11}$$

which is the lidar ratio of the oriented plates. The retrieved profile of L_{pl} is depicted in Fig. 5.

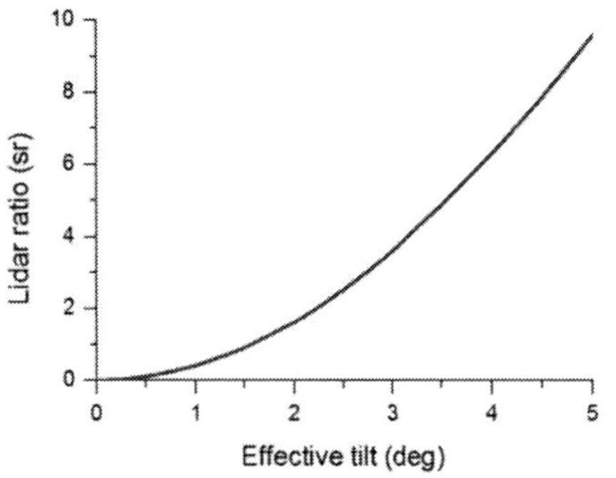

Figure 6. The lidar ratio of a quasi-horizontally oriented plate versus its effective tilt near the horizontal plane

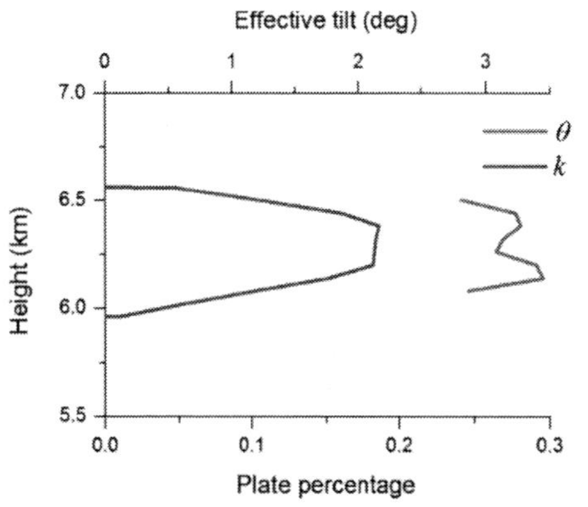

Figure 7. Effective tilt angle θ of the quasi-horizontally oriented plates and the plate percentage k retrieved from the experimental data of Figs. 3-5

The lidar ratio of Eq. (11) can be used to find the effective tilt (with respect to the horizon) of the quasi-horizontally oriented plates. For this purpose, we have calculated L_{pl} as a function of the effective tilt according to our papers [5, 7], the results are presented in Fig. 6. This relation converts the L_{pl} profile of Fig. 5 into the profile of the effective tilt θ (Fig. 7). The effective tilt angles were about 3°.

Finally, Eqs. (9) and (10) allow us to find also the quantity

$$B = 2kS / \sigma. \quad (12)$$

Here the first factor is the plate percentage k while the other factor corresponds to the ratio of plate and column sizes. This ratio should be assessed from either other experimental data or from model considerations. In our current work, we use the hypothesis that the sizes of the plates and columns are gamma-distributed [5] with their modal diameters being equal. Under this assumption, we are able to retrieve the plate percentage, which is depicted in Fig. 7 as well.

4 CONCLUSIONS

The joint use of the parallel and perpendicular backscattering coefficients together with the extinction coefficient measured by RAMSES in cirrus clouds allows us to detect and, to a certain degree, characterize the fraction of quasi-horizontally oriented plates. For such mixtures of randomly oriented crystals and quasi-horizontally oriented plates, we are able to retrieve approximately the plate percentage and the effective tilt angles for near-perfect horizontal alignment.

ACKNOWLEDGEMENTS

The work is partly supported by the grants 15-05-06100 and 16-35-60089 of the Russian Foundation for Basic Research and by the grant NS-8199.2016.5 of the President of the Russian Federation.

References

[1] Reichardt, J., Reichardt, S., Lin, R.-F., Hess, M., McGee, T. J., Starr, D. O., 2008: Optical-microphysical cirrus model, *J. Geophys. Res.* **113**, D22201.

[2] Platt, C. N. R., Abshire, N. L., McNice, G. T., 1978: Some microphysical properties of an ice cloud from lidar observation of horizontally oriented crystals, *J. Appl. Meteorol.* **17**(8), 1220-1224.

[3] Sassen, K., Benson S., A midlatitude cirrus cloud climatology from the Facility for Atmospheric Remote Sensing: II. Microphysical properties derived from lidar depolarization, 2001: *J. Atmos. Sci.* **58**(15), 2103–2112.

[4] Reichardt, J., Wandinger, U., Klein, V., Mattis, I., Hilber, B., Begbie, R., 2012: RAMSES: German Meteorological Service autonomous Raman lidar for water vapor, temperature, aerosol, and cloud measurements, *Appl. Opt.* **51**(34), 8111-8131.

[5] Borovoi, A., Konoshonkin, A., Kustova, N., Okamoto, H., 2012: Backscattering Mueller matrix for quasi-horizontally oriented ice plates of cirrus clouds: application to CALIPSO signals, *Opt. Exp.* **20**(27), 28222-28233.

[6] ftp://ftp.iao.ru/pub/GWDT/Physical_optics/Backscattering

[7] Borovoi, A., Konoshonkin, A., Kustova, N., 2014: Backscatter ratios for arbitrary oriented hexagonal ice crystals of cirrus clouds, *Opt. Lett.* **39**(19), 5788-5791.

A MOBILE SYSTEM COMBINING LIDAR AND SUNPHOTOMETER ON-ROAD MEASUREMENTS: DESCRIPTION AND FIRST RESULTS

Ioana Popovici[1,2]*, Philippe Goloub[1], Thierry Podvin[1], Luc Blarel[1], Rodrigue Loisil[1], Augustin Mortier[3], Christine Deroo[1], Fabrice Ducos[1], Stéphane Victori[2], Benjamin Torres[4]

[1]*Univ. Lille, CNRS, UMR 8518 - LOA - Laboratoire d'Optique Atmosphérique, F-59000 Lille, France*
**ie.popovici@ed.univ-lille1.fr*
[2]*Cimel Electronique, 75011 Paris, France*
[3]*Norwegian Meteorological Institute, 0313 Oslo, Norway*
[4]*GRASP-SAS, 59000 Lille, France*

ABSTRACT

The mobile system described in this paper integrates a commercial eye-safe lidar (CIMEL), a sunphotometer and in situ instruments. The system is distinguished by other transportable platforms through its capabilities to perform on-road measurements. The potential of a commercial lidar to provide reliable information on aerosol properties is investigated through comparison with a multi-wavelength Raman lidar. First results from observation campaigns in northern France are presented.

1 INTRODUCTION

Existing lidar networks consist of different systems with various configurations, from single-wavelength elastic lidars to multi-wavelength Raman lidars. Most systems are complex instruments that require regular maintenance and controlled environment for their operation, so they are predominantly installed in laboratory rooms. Therefore, their use for atmospheric profiling is limited over a fixed location. Nevertheless, aerosol distribution is highly variable in case of sudden events; e.g. pollution episodes, dust and fire outbreaks, volcano eruptions, long-range transports and dispersion of pollutants from emission sources. In these situations, field observations are important, as the spatial variability is impossible to assess from point measurements.

Mobile observations are one of the best solutions to map the extent of such events and to study the regional gradients. If lidars could be easily deployed at the time needed, the number of applications would rapidly increase, from the validation of satellite measurements and model predictions to the investigation of plumes at industrial sites.

Currently, the term *mobile* is referred to scanning lidars and transportable systems (lidar or in situ), used to perform measurements in remote places, during campaigns or simply outside laboratory rooms. Other works describe mobile laboratories performing lidar [1] and particle/gas [2] on-line measurements during vehicle's motion. Nonetheless, the capability of deploying lidars and other instruments for on-road measurements is not enough explored.

For the first time, the description of a compact mobile system to perform on-road measurements, which integrates a lidar, a sunphotometer and optical particle counters, is presented here. The uniqueness of the system lies mainly in the sunphotometer, capable to track the sun during vehicle's motion, and the synergy of joint lidar and sunphotometer measurements to derive extinction profiles along the investigated route.

In this paper, after the description of the mobile system's current configuration, we start a discussion on the overlap function used for correction and confirm the lidar quality by comparison to data from an EARLINET station. Finally, the application of the mobile system is illustrated by measurements in northern France.

2 MOBILE SYSTEM DESCRIPTION

Most mobile lidars deployed in field campaigns are bulky systems, requiring large vehicles for their installation and transportation. Compared to such systems, the mobile system described here is more compact and robust. The MAAMS (*Mobile Automated Aerosol Monitoring System*) is a light commercial vehicle (Renault Kangoo), equipped with a lidar, a sunphotometer and in situ instruments. An overview of the mobile system is given in Figure 1. The van's trunk was modified to allow the installation of instruments. Moreover,

https://doi.org/10.1051/epjconf/201817608003

the vehicle was equipped with an electrical installation and an external outlet. To continuously operate all the instruments, a 12V/300Ah AGM battery and a power inverter were installed in the van.

2.1 Instruments

The lidar included in MAAMS payload is currently a *CE370 Lidar,* an aerosol backscattering lidar emitting at 532 nm, developed and commercialized by Cimel Electronique (www.cimel.fr). The system uses a diode-pumped Nd:YAG laser with small laser pulse energy of 20 μJ and a high repetition rate of 4.7 kHz. The instrument's design consists of a shared transmitter-receiver telescope connected to the control and acquisition unit through a 10 m optical fiber. The telescope has a diameter of 200 mm and narrow field of view (55 μrad) to limit the background light. Backscattered light is detected by an avalanche photodiode detector (APD), running in photon-counting mode. The lidar sounds the atmosphere up to 15 km with a vertical resolution of 15 m. It is autonomous, lightweight and compact and operates in eye-safe mode, requiring no authorization for its operation. These features make it suitable for a mobile system.

The lidar control and acquisition unit, enclosing the optical and electronic components, is installed on vibration isolators to protect the optical alignment from shock while the vehicle is moving. Likewise, the lidar transmitter-receiver telescope is fixed to the van using a specially designed support mounted in the van's floor, in order to ensure stability during movement. Furthermore, the shared telescope design eliminates any optical alignment problems.

The sun-tracking photometer *PLASMA* [3] (*Photomètre Léger Aéroporté pour la Surveillance des Masses d'Air*), developed by LOA, is used to derive spectral aerosol optical depths (AOD) over the 0.343-1.23 μm range. PLASMA can move in elevation (0-90°) and azimuth (0-360°) and can rotate in azimuth thanks to a ring power connector, which makes it suitable for mobile applications.

The *OPC-N2* is an optical particle counter designed by Alphasense (www.alphasense.com). It measures particle size distribution from 0.38 to 17 μm within 16 channels every 6 seconds with a flow rate of 1.2 L/min.

Another particle counter, *mini-WRAS,* developed by GRIMM, has been also mounted in MAAMS. The instrument measures the particle size distribution in 40 channels, from 10 nm to 25 μm every minute with a flow rate of 1.2 L/min.

A *meteorological probe* mounted on the van is used to measure the temperature, the relative humidity and the atmospheric pressure. Finally, MAAMS is equipped with a GPS navigation system, to map the mobile measurements and a camera, used to record sky cloudiness.

Data is transferred to LOA database using a 3G communication and is processed routinely.

Figure 1. Photographs of the mobile system and the components mounted inside and outside the vehicle: (a) Lidar transmitter-receiver optical head, (b) Lidar control and acquisition unit, (c) battery, (d) power inverter, (e) optical particle counter, (f) action camera, (g) 3G antenna, (h) meteorological probe, (i) housing for PLASMA's compass, (j) protection domes for PLASMA, (k) PLASMA sunphotometers, (l) GPS

2.2 Overlap function

Figure 2 shows the overlap correction used for CE370 lidar (red curve). It is assessed using slope method from fixed vertical observations (black curve), when fair weather or homogeneous atmosphere conditions are met. A correction is possible between 150 m and 5 km, where complete overlap is reached. In order to verify the correction, we compared it with the ratio of LILAS (Multiwavelength-Raman lidar) and CE370 (mobile lidar) signals (green curve), which is a good estimate of the overlap function. A mean difference of 4% between the two overlap functions (green and black curves) is obtained for the 0.15-5 km range. LILAS, part of EARLINET, successfully went through its quality assurance chain (Rayleigh fit, telecover test, [4]) and is therefore considered as a reference in this comparison exercice. Preliminary results show a good agreement, which confirms the reliability of the function used for overlap correction and the quality of our data.

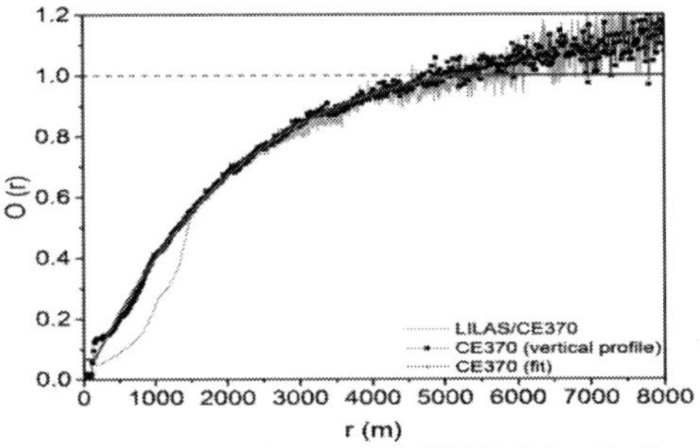

Figure 2. Overlap function of CE370 lidar (red) and ratio of Raman (LILAS) and CE370 micro-lidar signals (green).

3 RESULTS

Mobile measurements have been performed in northern France during spring and summer 2015 and 2016, as higher aerosol loadings are generally observed throughout these periods. In this section, we focus our analysis on the day of August 26th, 2016, since it is representative for strong spatial variability within the region.

The mobile system was driven from Lille (LL) to Dunkerque (DK), situated at 80 km NW from Lille. Figure 3 shows the lidar Range-Corrected Signals (RCS) recorded along the investigated route. Higher aerosol concentration is observed for Lille up to 1 km, which decreases towards the coastal region (DK). Moreover, outside Lille region, the presence of several aerosol layers up to 5 km is revealed. These layers are hardly observed in Lille due to the strong backscatter in the lower altitudes

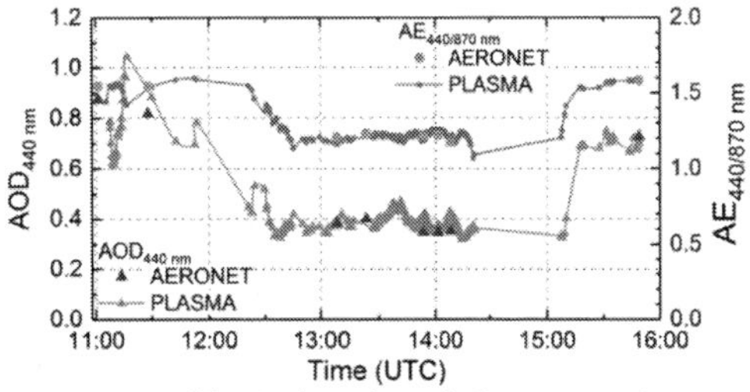

Figure 4. AOD (440 nm) and Extinction Angström Exponent (AE) values measured by PLASMA on the round-trip between Lille and Dunkerque Values from AERONET stations are also reported as reference.

The Figure 4 shows Aerosol Optical Depth (AOD) and Angstrom Exponent (AE) processed from PLASMA data using AERONET processing system. A strong variability can be seen between the two end points (LL and DK). High AOD values of 0.6-0.8 were recorded at Lille and surroundings between 11:00 and 11:30 UTC and decreased rapidly to 0.37 when arriving at Dunkerque around 13:00 UTC. Moreover, AE values decreased from 1.4-1.6 at Lille to 1.2 at Dunkerque, indicating a change in columnar microphysical properties along the road. During one hour of stationary measurements starting from 13:07 UTC, the AOD levels remained stable around 0.4. On the way back to Lille, AOD and AE values increased to 0.75 and 1.6, respectively.

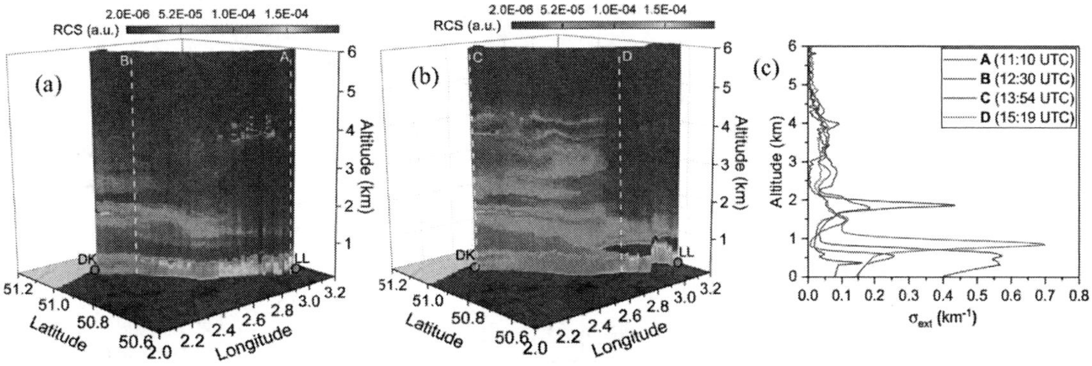

*Figure 3. 3D color-coded spatial variability of lidar range-corrected signals along: **(a)** Lille (LL)-Dunkerque (DK), **(b)** Dunkerque (DK)-Lille (LL) transects. **(c)** Extinction coefficient profiles derived from lidar-sunphotometer measurements. The white dotted lines and letters (A, B, C, D) indicate the location of the profiles chosen for analysis.*

A good consistency is observed between AOD and AE values from PLASMA and the closest AERONET stations, which demonstrate the reliability of PLASMA sunphotometer measurements. From the synergy of lidar and sunphotometer measurements, vertical extinction profiles (Figure 3 c) are derived using BASIC algorithm [5]. The profiles taken at different locations evidence the presence of several layers. The first one, up to 1 km at Lille is the boundary layer, whose height decreases to 0.6 km at DK (continental vs. coastal). A second layer is observed on A and D profiles at 1.5 km, but with a slight shift in altitude on B and C (1.8 km). A third layer is seen between 2 and 5 km. Back-trajectories and forecast dust models suggest that aerosols in this range are of Saharan origin. Several instruments along the coastal regions observed this dust intrusion. The upgrade of MAAMS with a 2λ micro-lidar (CE376) will be helpful in the analysis of such complex cases. Figure 5 presents size distributions retrieved by GRASP-AOD [6] from PLASMA AOD measurements. Column-integrated size distribution can be obtained at different points on the investigated transect. At Lille and surroundings (A, D), both fine and coarse mode are present with nonetheless a higher fine mode fraction. For Dunkerque, fine and coarse modes have equal amplitudes. Results from retrievals are in good consistency with AERONET retrievals.

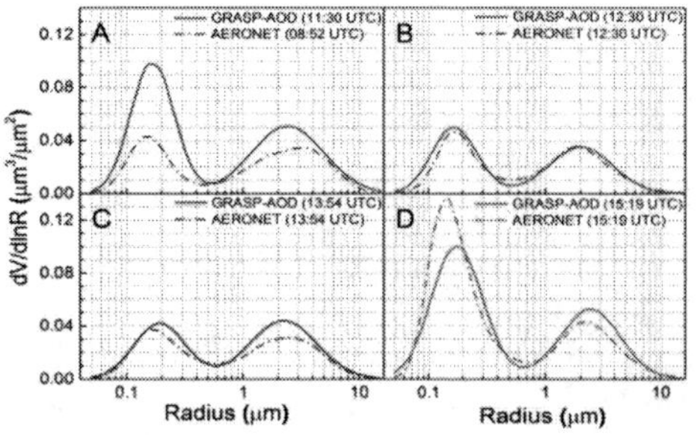

Figure 5. Size distributions retrieved by GRASP-AOD using AOD from PLASMA sun-photometer. The size distributions from the closest AERONET inversions are also represented as a reference.

4 CONCLUSIONS

The MAAMS mobile system configuration has been described, demonstrating its robustness for mobile applications. Preliminary comparisons against a Raman lidar and AERONET show good agreement between measurements and retrievals, respectively. Results for the analyzed example have shown variations of AOD betweeen 0.32 to 1.04 in 80 km range and 50 minutes time interval and highly variable aerosols vertical distribution. The presence of dust in the 2-5 km altitude range is indicated, which, added to an important event of local pollution, illustrate an interesting case of spatial and vertical variability.

ACKNOWLEDGEMENTS

The authors would like to thank to ANRT France and CIMEL Electronique for the financial support of the PhD work of Ioana Popovici. We also acknowledge the support from CaPPA project and SNO PHOTONS/AERONET from CNRS-INSU.

References

[1] Raut, J.-C., and Chazette, P., 2009: Assessment of vertically-resolved PM_{10} from mobile lidar observations, *Atmos. Chem. Phys.* 9, 8617-8638.

[2] Drewnick, F., Böttger, T., von der Weiden-Reinmüller, S.-L., Zorn, S.R., Klimach, T., Schneider, and J., Borrmann, S., 2012: Design of a mobile aerosol research laboratory and data processing tools for effective stationary and mobile field measurements, *Atmos. Meas. Tech.* 5, 1443-1457.

[3] Karol, Y., Tanré, D., Goloub, P., Vervaerde, C., Balois, J.Y., Blarel, L., Podvin, T., Mortier, A., and Chaikovsky, A., 2013: Airborne sun photometer PLASMA: concept, measurements, comparison of aerosol extinction vertical profile with lidar, *Atmos. Meas. Tech.* 6, 2383-2389.

[4] http://www.actris.net/Portals/97/Publications/quality%20standards/lidar/QA-InternalCheckups_version_121017b.pdf, EARLINET-Internal Checkups, V. Freudentalher

[5] Mortier, A., Goloub, P., Podvin, T., Deroo, C., Chaikovsky, A., Ajtai, N., Blarel, L., Tanré, D., and Derimian, Y., 2013: Detection and characterization of volcanic ash plumes over Lille during Eyjafjallajokull eruption, *Atmos. Chem. Phys.* 13, 3705-3720.

[6] Torres, B., Dubovik, O., Fuertes, D., Schuster, G., Cachorro, V.E., Lapionak, T., Goloub, P., Blarel, L., Baretto, A., Mallet, M., Toledano, C., and Tanré, D., 2016: Advanced characterization of aerosol properties from measurements of spectral optical depth using the GRASP algorithm, *Atmos. Meas. Tech. Discuss*, doi:10.5194/amt-2016-334, in review.

ACTIVE SENSOR SYNERGY FOR ARCTIC CLOUD MICROPHYSICS

Kaori Sato [1*], ***Hajime Okamoto*** [1], ***Shuichiro Katagiri*** [1] ***Masataka Shiobara*** [2], ***Masanori Yabuki*** [3], ***Toshiaki Takano*** [4],

[1]*Research Institute for Applied Mechanics, Kyushu University, Kasuga, Kasuga koen 6-1, 816-8580, JAPAN, *Email: sato@riam.kyushu-u.ac.jp*
[2] *National institute of Polar Research, 10-3, Midori-cho, Tachikawa-shi, Tokyo 190-8518, JAPAN*
[3] *Research Institute for Sustainable Humanosphere, Kyoto University. Gokasho, Uji City, Kyoto Prefecture, 611-0011, JAPAN*
[4]*Chiba University, 1-33, Yayoicho, Inage-ku, Chiba-shi, Chiba, 263-8522, JAPAN*

ABSTRACT

In this study, we focus on the retrieval of liquid and ice-phase cloud microphysics from space-borne and ground-based lidar-cloud radar synergy. As an application of the cloud retrieval algorithm developed for the EarthCARE satellite mission (JAXA-ESA) [1], the derived statistics of cloud microphysical properties in high latitudes and their relation to the Arctic climate are investigated.

1 INTRODUCTION

In recent years, space-borne lidar and cloud radar have been widely used to investigate global three-dimensional cloud properties [2][3]. Advanced research has been performed such as to investigate the relation of clouds to the observed recent-year change in sea ice concentration in the Arctic region [4][5]. Combined use of lidar and cloud radar provides reliable estimate of cloud phase and cloud microphysical properties, which are expected to further help quantify the radiative effect of each cloud phase categories in the arctic that has not been fully done. Still, ambiguity exists in the reliability of low-level cloud detection and cloud-base height estimate from space-borne lidar. In addition, multiple scattering effects on lidar backscattering coefficient and depolarization need to be considered when retrieving water cloud microphysical properties. The main objective of this paper is to use both ground-based and space-borne lidar/cloud radar to complement each other and then use the evaluated liquid and ice-phase cloud microphysics obtained by satellites for the cloud-sea ice interaction studies in the Arctic.

2 DATA AND METHODOLOGY

2.1 Observation data

Here we briefly describe the observation data. In 2013, FMCW 95GHz cloud radar FALCON-A (Chiba University) has been installed at Ny-Alesund, Svalbard and collocated measurement with Polarization Micro-Pulse Lidar (PMPL; National Institute of Polar Research) has been conducted for the first time in the arctic region near Europa. The PMPL data used in this study are obtained with vertical and temporal resolutions of 30m and 300sec, respectively. FALCON-A has 48m and 10sec, resolution. Time-height plot of the observables of PMPL and FALCON-A interpolated to 48m and 300sec resolution obtained in 2013 are shown in Fig. 1. It is seen that major parts of the cloud consist from low depolarization ratio (< 10%) regions, which indicate the presence of horizontally oriented ice particles (2D-ice) or super-cooled water.

For satellite data, Cloud-Aerosol Lidar and Infrared Pathfinder Satellite Observations (CALIPSO) and CloudSat data are used in this study. The CloudSat/CALIPSO merged data set created at Kyushu University (KU-merged data) was used, which is created by projecting the cloud-masked CALIPSO data onto the CloudSat grid with the same 240m vertical and approximately 1km horizontal resolutions [6]

2.2 Retrieval algorithm

Basically the same cloud mask and cloud particle type classification scheme developed for CloudSat/CALIPSO and EarthCARE are applied to both ground-based and satellite data [6][7]. The retrieval of ice and water cloud microphysics was performed based on the algorithm developed in

[2][8][9], but with further improvements made in the lidar forward model for liquid phase that accounts for multiple scattering in both backscattering coefficient and depolarization ratio, and also extended to take into account ice and liquid mixture as well as 2-D/3-D ice mixtures [2] within a cloud or within the same lidar and radar grid.

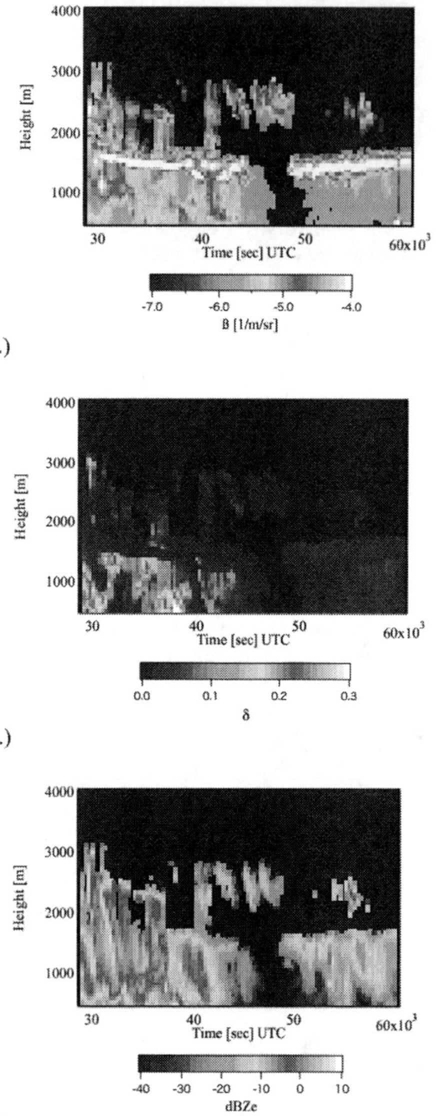

(a.)

(b.)

(c.)

Fig. 1 Time-height plot for the (a.)(b.) attenuated backscattering coefficient β and depolarization ratio δ at 532nm obtained by PMPL, and (c.) radar reflectivity factor obtained by Falcon-A.

3 RESULTS

Comparison of the vertical cloud frequency of occurrence obtained by ground-based and space-borne lidar/radar measurement was performed during satellite overpasses within the observation period 2013-2015. The cloud frequency of occurrence below (above) 2km obtained by CALIOP (PMPL) tended to underestimate the PMPL (CALIOP) value when fully attenuated pixels where not considered in the statistics. The agreement was improved after developing a new cloud mask scheme to identify fully attenuated pixels [10][11]. On the other hand, there was an overall agreement between the cloud radars when attenuation due to wet Radome is taken into account.

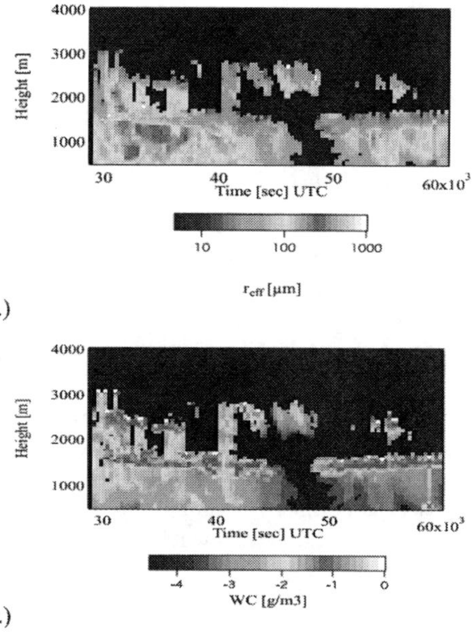

(a.)

(b.)

Fig. 2 Time-height plot for (a.) the retrieved effective radius (r_{eff}) (b.) total ice and water content (WC).

The microphysical properties retrieved for the ground-based observation are shown in Fig.2. The effective radius r_{eff} and total water content within the thin layer with high backscatter coefficient and low depolarization ratio located around 1.5km were around 20~30 μm and 10^{-2}~10^{-1} g/m^3, respectively. The particle types of the layers are identified as mixture of super cooled water and ice particles. The ice particles falling out of this region have r_{eff} around 100 μm and ice water content of 10^{-3} g/m^3. Similar features were also seen in the satellite retrieval data.

The cloud properties obtained by satellite data in the Arctic (latitude > 65 degrees) were partitioned into low-level, mid-level and high level. The water cloud coverage and its variability obtained by CALIOP was the largest at low-level throughout the year. The maximum and minimum water cloud coverage appears around May/ September and July, respectively. Relation between the seasonal variability of low-level water cloud coverage to the sea ice concentration obtained from the passive microwave radiometer Aqua/AMSER-E during the period 2006-2010 showed similar tendency as indicated in previous studies [4]. The retrieved water and ice cloud microphysical properties will be further analyzed to estimate their radiative effects in future study.

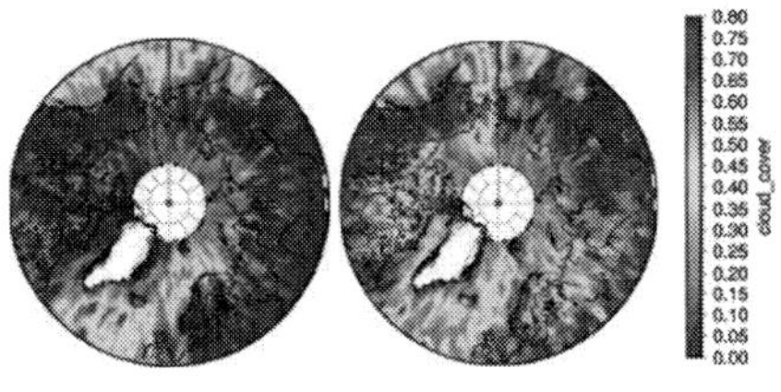

(a.) (b.)

Fig. 3 Cloud coverage of water cloud in the 0.5-1.5km altitude range obtained from CALIOP data in (a.) july and (b.) September 2008

4 CONCLUSIONS

Ground-based lidar and cloud radar data obtained at Ny-Alesund were used to evaluate vertical cloud properties retrieved from CloudSat/ CALIPSO satellite data during overpasses in the period 2013-2015. Cloud frequency of occurrence observed by both platforms showed overall agreement with each other when fully attenuated pixels were removed from the statistics. Water cloud microphysics was retrieved by taking into consideration the multiple scattering effects on lidar backscattering coefficients and depolarization ratio. Mixed phase condition was also analyzed by extending ice cloud microphysics retrieval scheme. The retrieved cloud microphysical properties will be used to investigate their radiative effects in the Arctic. The algorithms used in the studies will be adapted for the high spectral resolution lidar ATLID, and Doppler cloud profiling radar CPR onboard EarthCARE.

ACKNOWLEDGEMENTS

This study was supported by JSPS KAKENHI Grant Numbers JP17H06139 and JP15K17762, by Green Network of Excellence (GRENE) Program Arctic Climate Change Project and the Arctic Challenge for Sustainability (ArCS), by the Japan Aerospace Exploration Agency for EarthCARE Research Announcement, and by the Collaborative Research Program of Research Institute for Applied Mechanics, Kyushu University.

References

[1] Illingworth A., H. Okamoto, and others, 2015: THE EARTHCARE SATELLITE: THE NEXT STEP FORWARD IN GLOBAL MEASUREMENTS OF CLOUDS, AEROSOLS, PRECIPITATION AND RADIATION, *Bull. Amer. Meteor. Soc.*, 96, 1311-1332, doi: http://dx.doi.org/10.1175/ BAMS-D-12-00227.1

[2] Okamoto, H., K. Sato, and Y. Hagihara, 2010: Global analysis of ice microphysics from CloudSat and CALIPSO: Incorporation of specular reflection in lidar signals, *J. Geophys. Res.*, 115, D22209, doi:10.1029/ 2009JD 013383.

[3] Sato, K., and H. Okamoto, 2011: Refinement of global ice microphysics using spaceborne active sensors, *J. Geophys. Res.*, 116, D20202, doi:10.1029/2011JD015885.

[4] Kay, J. E., and A. Gettelman, 2009: Cloud influence on and response to seasonal Arctic sea ice loss, *J. Geophys. Res.*, 114, D18204, doi:10.1029/ 2009JD 011773.

[5] Palm, S. P., S. T. Strey, J. Spinhirne, and T. Markus, 2010: Influence of Arctic sea ice extent on polar cloud fraction and vertical

structure and implications for regional climate, *J. Geophys. Res.*, 115, D21209, doi:10.1029/2010JD013900

[6] Hagihara, Y., H. Okamoto, and R. Yoshida Development of combined CloudSat/CALIPSO cloud mask to show global cloud distribution, 2010, *J. Geophys. Res., 115*, D00H33, doi:10.1029/ 2009JD012344.

[7] Yoshida, R., H. Okamoto, Y. Hagihara, and H. Ishimoto, 2010: Global analysis of cloud phase and ice crystal orientation from Cloud-Aerosol Lidar and Infrared Pathfinder Satellite Observation (CALIPSO) data using attenuated backscattering and depolarization ratio, *J. Geophys. Res.,* 115, D00H32, doi:10.1029/ 2009JD012334.

[8] Sato, K., H. Okamoto and H. Ishimoto, 2016: Modeling Lidar Multiple Scattering, EPJ Web of Conferences, 119, 21005,DOI:http://dx.doi.org/10.1051/epjconf/201611921005

[9] Sato, K., H. Okamoto, M. K. Yamamoto, S. Fukao, H. Kumagai, Y. Ohno, H. Horie, and M. Abo, 2009: 95-GHz Doppler radar and lidar synergy for simultaneous ice microphysics and in-cloud vertical air motion retrieval, *J. Geophys. Res.*, 114, D03203, doi:10.1029/2008JD010222.

[10] Okamoto, H., K. Sato, T. Nishizawa, N. Sugimoto, T. Makino, Y. Jin, and A. Shimizu, T. Takano, M. Fujikawa, 2016: Development of a multiple-field-of-view multiple-scattering polarization lidar: comparison with cloud radar, *Opt. Express*, 24(26), 30053-30067, https://doi.org/10.1364/OE.24.030053

[11] Katagiri, S, H., Okamoto, K., Sato, K., Ota, 2017:Refinement of the CALIOP cloud mask algorithm, *Proc. 28th ILRC*, 2017

AEROSOL ABSORPTION PROFILING FROM THE SYNERGY OF LIDAR AND SUN-PHOTOMETRY: THE ACTRIS-2 CAMPAIGNS IN GERMANY, GREECE AND CYPRUS

Alexandra Tsekeri[1*], Vassilis Amiridis[1], Anton Lopatin[2], Eleni Marinou[1,3], Eleni Giannakaki[4,5], Michael Pikridas[6], Jean Sciare[6], Eleni Liakakou[7], Evangelos Gerasopoulos[7], Sebastian Duesing[8], Joel C. Corbin[9], Martin Gysel[9], Nicolas Bukowiecki[9], Holger Baars[8], Ronny Engelmann[8], Birgit Wehner[8], Michael Kottas[1], Dimitra Mamali[10], Panagiotis Kokkalis[1], Panagiotis I. Raptis[7], Iasonas Stavroulas[6], Christos Keleshis[6], Detlef Müller[11], Stavros Solomos[1], Ioannis Binietoglou[1,12], Nikolaos Mihalopoulos[7], Alexandros Papayannis[13], Iwona S. Stachlewska[14], Julia Igloffstein[8], Ulla Wandinger[8], Albert Ansmann[8], Oleg Dubovik[2], Philippe Goloub[2]

*[1] Institute for Astronomy Astrophysics Space Applications and Remote Sensing, National Observatory of Athens, Athens, Greece, *atsekeri@noa.gr*
[2] Laboratoire d'Optique Atmosphérique, Université de Lille, Lille, France
[3] Laboratory of Atmospheric Physics, Department of Physics, Aristotle University of Thessaloniki, Thessaloniki, Greece
[4] Finnish Meteorological Institute, Kuopio, Finland
[5] Department of Environmental Physics and Meteorology, Faculty of Physics, University of Athens, Athens, Greece
[6] The Cyprus Institute, Environment Energy and Water Research Center, Nicosia, Cyprus
[7] Institute for Environmental Research and Sustainable Development, National Observatory of Athens, Athens, Greece
[8] Leibniz Institute for Tropospheric Research, Leipzig, Germany
[9] Laboratory of Atmospheric Chemistry, Paul Scherrer Institute, Villigen, Switzerland
[10] Faculty of Civil Engineering and Geosciences, Delft University of Technology, Delft, The Netherlands
[11] School of Physics, Astronomy and Mathematics, University of Hertfordshire, Hatfield, UK
[12] National Institute of R&D for Optoelectronics, Magurele, Ilfov, Romania
[13] Physics Department, National Technical University of Athens, Athens, Greece
[14] Institute of Geophysics, Faculty of Physics, University of Warsaw, Warsaw, Poland

ABSTRACT

Aerosol absorption profiling is crucial for radiative transfer calculations and climate modelling. Here, we utilize the synergy of lidar with sun-photometer measurements to derive the absorption coefficient and single scattering albedo profiles during the ACTRIS-2 campaigns held in Germany, Greece and Cyprus. The remote sensing techniques are compared with in situ measurements in order to harmonize and validate the different methodologies and reduce the absorption profiling uncertainties.

1 INTRODUCTION

Absorption is a key climate-relevant aerosol property of particular importance for aerosol-radiation and aerosol-cloud interactions. Specifically, aerosol absorption can directly modify the global radiation budget, indirectly modify cloud properties and abundance (e.g. [1]) and thus modify the atmospheric stability in the boundary layer and free troposphere (e.g. [2]). However, the magnitude of absorption at the global scale is subject to considerable uncertainties, for aerosol particles of both anthropogenic and natural origin. While the aerosol optical depth (AOD) and aerosol particle size distribution are relatively well-constrained from measurements, uncertainties related to the SSA (e.g. [3]), and especially the vertical profile of the black carbon (BC) concentration, which is the main light-absorbing component in the visible wavelength range, contribute significantly to the overall uncertainties of the aerosol radiative effect.

The European project Aerosols, Clouds and Trace gases Research Infrastructure (ACTRIS-2 http://www.actris.eu) through its Joint Research Activity 1 (JRA1) focuses on reducing the uncertainty in the determination of the aerosol light absorption vertical profiles by combining and optimizing advanced remote sensing and in situ methods. The novel remote sensing and in situ instrumentation included in the ACTRIS research infrastructure, offers a unique opportunity for a comprehensive absorption characterization of anthropogenic and natural aerosols, within urban and remote sites of the network in Europe. The vertical distribution of the absorption coefficient and the SSA are retrieved via surface and airborne in situ measurements and remote sensing techniques, and dedicated ACTRIS JRA1 campaigns have been organized to perform closure studies between these different methods. Here we present results from the campaigns in Melpitz, Germany (May-June 2015), in Athens, Greece (December 2015 – February 2016) and in Nicosia/Agia Marina/Orounda, Cyprus (May-June 2016).

2 METHODOLOGY

The remote sensing retrieval of the absorption vertical structure is performed with the synergy of passive and active remote sensing instruments, i.e. multiwavelength lidars and sun-photometers. The sun-photometer provides the columnar properties of the particles (e.g. [4]), whereas lidars are capable of providing vertically resolved profiles of the backscatter and extinction coefficients, along with vertical profiles of the particle microphysical properties, mainly for the fine mode (e.g. [5]). The Generalized Aerosol Retrieval from Radiometer and Lidar Combined data algorithm (GARRLiC [6]), developed in the framework of ACTRIS, goes a step further using for the first time both sun-photometer and lidar measurements in the combined retrieval of the particle microphysical properties. Figure 1 shows the algorithm's input and output.

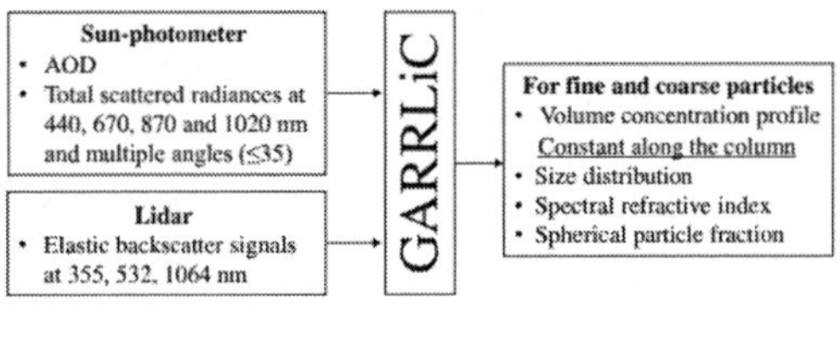

Figure 1 The GARRLiC algorithm input and output parameters

During the ACTRIS-2 JRA1 campaigns, intensive ground-based active and passive remote sensing measurements were performed with Polly^{XT} lidars and CIMEL sun-photometers. Both GARRLiC and the lidar stand-alone algorithm of Müller et al. [5] are used to produce the absorption coefficient and SSA vertical profiles. The lidar stand-alone algorithm is very useful for days with broken or high clouds, when the GARRLiC retrieval cannot be applied due to the lack of sun-photometer measurements.

For the in situ measurements of the absorption profile airborne platforms were deployed, to acquire the absorption and scattering coefficients, BC mass concentration and particle size with multi-wavelength aethalometers, nephelometers, and optical particle sizers, respectively. More specifically, in the framework of the Athens and Cyprus campaigns, a new light-weight absorption monitor, named Dual Wavelength Prototype (DWP), was used onboard unmanned aerial vehicles (UAVs), providing the absorption coefficient at 370 and 880 nm. In the Melpitz campaign, the Airborne Cloud Turbulence Observation System (ACTOS), carried by a helicopter, acquired absorption measurements with a 630 nm CAPS-PMssa monitor (Aerodyne, USA) [7]. This instrument performs calibration-free extinction coefficient measurements (with 5% accuracy). Scattering coefficients are measured from the same optical cavity via integrating nephelometry. In situ absorption (extinction minus scattering) is thus obtained, with greatest accuracy for low SSA environments. ACTOS also carried a Single channel Tri-color Absorption Photometer (STAP, Brechtel, Hayward, CA 94544), measuring the particle absorption at 450, 525 and 624 nm.

3 RESULTS

The comparison of remote sensing retrievals of the absorption coefficient with the airborne and surface in situ measurements during the ACTRIS JRA1 campaigns in Melpitz, Athens and Cyprus are shown in Fig. 2.

3.1 Melpitz campaign

During the Melpitz campaign the particle extinction coefficient was low, with AODs at 440 nm not exceeding the value of 0.2, and the SSA was generally above 0.9, with correspondingly low absorption coefficients. This means that signal-to-noise ratios for in situ and remote sensing measurements were the lowest for this campaign. Although the AOD requirements have not been quantified yet for GARRLiC, an AOD of 0.3 at 440 nm is considered sufficient for the SSA retrieval.

For the case of 24 June the AOD at 440 nm was 0.14. The GARRLiC results show large uncertainties (~7 Mm^{-1}), but they have similar slopes with the in situ absorption profiles, although they produce larger values (left plots in Fig. 2a). The lidar stand-alone retrieval has comparably large uncertainty for the case of 17 June, with an AOD at 440 nm of 0.17 (right plots in Fig. 2a). In part due to the large uncertainties, the in-situ and remote sensing results are not statistically different for this case, although the latter have larger values.

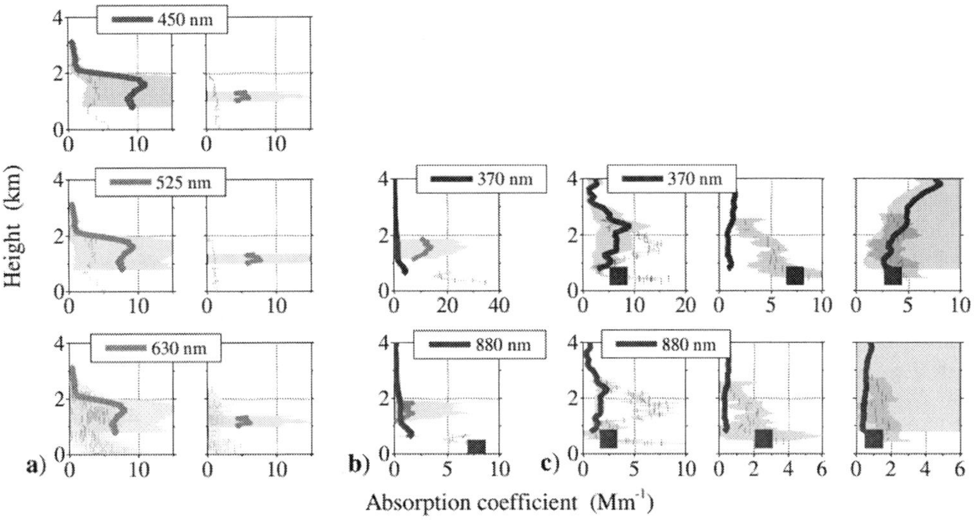

Figure 2 Absorption coefficient profiles retrieved from remote sensing measurements, from GARRLiC (colored solid lines) and lidar stand-alone algorithms (grey solid lines), and measured with airborne platforms with in situ instruments, i.e. CAPS-PMssa monitor (grey dash line) and STAP (colored dash lines) for Melpitz campaign, and the DWP (colored dashed lines) for Athens and Cyprus campaigns. The surface in situ measurements are denoted with squares. The absorption coefficients are provided at 370, 450, 525, 630 and 880 nm. The results are from the ACTRIS-2 JRA1 campaigns in: a) Melpitz, Germany, on 24 and 17 June 2015, b) Athens, Greece, on 19 January 2016 and c) Nicosia, Cyprus, on 7, 15 and 21 April 2016.

3.2 Athens campaign

During the Athens campaign there were few cloud-free days available to use for the combined lidar/sun-photometer GARRLiC retrieval and the AODs were low (<0.2 at 440 nm). Another challenge for the remote sensing retrievals is that most of the aerosol were located below ~1 km, in the lidar incomplete overlap region. In this region, there is large uncertainty in the lidar backscattered signal, which is only partially collected from the telescope [8]. We are currently working on enhancing the information in this region by utilizing the near-range measurements of the Polly^{XT} lidar, performing the signal gluing technique. Due to the low flight heights of the UAVs (up to 700 m a.g.l. for the case shown here), the retrieved profiles do not overlap. In the results of 19 January (AOD at 440 nm at 0.16) in Fig. 2b we see an agreement between the GARRLiC and the lidar stand-alone absorption coefficients at 370 nm, within their uncertainties. The agreement is better at 880 nm.

3.3 Cyprus campaign

The mandate of the Cyprus campaign was to investigate the influence of dust on absorption properties. The cases shown here (Fig. 2c) have AODs at 440 nm of 0.2-0.35. Due to higher AODs, we achieve a good agreement between GARRLiC, UAV and surface in situ measurements, especially at 370 nm, for the cases on 7 and 21 April (left and right plots, Fig. 2c). On 15 April (middle plots in Fig. 2c) we see disagreement between GARRLiC and in situ UAV measurements, for which we currently investigate the reason(s).

4 CONCLUSIONS

In this study we used remote sensing algorithms utilizing the lidar measurements (i.e., lidar stand-alone algorithm) and the combination of lidar and sun-photometer measurements (i.e., GARRLiC algorithm), along with airborne and surface in situ measurements to characterize the particle aerosol absorption vertical distribution. Data were gathered during three ACTRIS-2 JRA1 campaigns in Germany, Greece and Cyprus in 2015-2016. The lidar stand-alone and the GARRLiC retrievals agree well for a case study in Athens. The remote sensing and airborne in situ measurements were not statistically different in Melpitz, where absorption and extinction coefficients were low. The comparison in Cyprus shows very good agreement between GARRLiC and in situ measurements for most cases investigated.

The difficulties posed in retrieving the absorption profiles are associated with low AODs, the insufficient lidar information in the overlap region and possibly the uncertainties in the airborne in situ sampling. In this line specific actions need to be taken further, as for example to incorporate the lidar near-range signal in the remote sensing retrievals [e.g. 9] that will provide the missing information in the overlap region.

We aim to continue investigating the benefits and improvements brought in by the synergy of remote sensing and in situ techniques for aerosol absorption profile characterization within the framework of the ACTRIS-2 project and its experimental campaigns.

ACKNOWLEDGEMENTS

The research leading to these results has received funding from the European Union's Horizon 2020 Research and Innovation Programme ACTRIS-2 (grant agreement no. 654109). Elina Giannakaki acknowledges the support of the Academy of Finland (project no. 270108).

References

[1] Bond, T. C., et al., 2013: Bounding the role of black carbon in the climate system: A scientific assessment, J. Geophys. Res., 118, 5380–5552, doi:10.1002/jgrd.50171.

[2] Wilcox, E. M., et al., 2016: Black carbon solar absorption suppresses turbulence in the atmospheric boundary layer, *Proceedings of the National Academy of Sciences*, **113** (42), 11794-11799.

[3] Loeb, N. G., and W. Su, 2010: Direct Aerosol Radiative Forcing Uncertainty Based on a Radiative Perturbation Analysis, *J. Climate*, **23**, 5288, 2010.

[4] Dubovik, O., and M. D. King, 2000: A flexible inversion algorithm for retrieval of aerosol optical properties from sun and sky radiance measurements, *J. Geophys. Res.*, **105**, 20,673 – 20,696.

[5] Müller, D., et al., 2001: Comprehensive particle characterization from three-wavelength Raman-lidar observations, *Appl. Opt.*, **40**, 4863–4869.

[6] Lopatin, A., et al., 2013: Enhancement of aerosol characterization using synergy of lidar and sun-photometer coincident observations: the GARRLiC algorithm, *Atmos. Meas. Tech.*, **6**, 2065-2088, doi:10.5194/amt-6-2065-2013.

[7] Kebabian, et al., 2007: Optical Extinction Monitor Using CW Cavity Enhanced Detection, *Rev. Sci. Instrum.*, **78**: 063102.

[8] Wandinger U. and Ansmann A., 2002: Experimental determination of the lidar overlap profile with Raman lidar, *Appl. Opt.*, **41**, 511-514.

[9] Stachlewska, I.S., et al., 2017: Properties of arctic haze aerosol from lidar observations during iAREA 2015 campaign on Spitsbergen, *ILRC28*, this issue.

MULTI-SENSOR MEASUREMENTS OF MIXED-PHASE CLOUDS ABOVE GREENLAND

Robert A. Stillwell[1]*, Matthew D. Shupe[2,3], Jeffrey P. Thayer[1], Ryan R. Neely III[3,4], and David D. Turner[2]

*[1]Aerospace Engineering Sciences, University of Colorado, *Presenting Author, [2]NOAA/Earth System Research Laboratory, Boulder, Colorado, [3]Cooperative Institute for Research in Environmental Science, University of Colorado, [4]School of Earth and Environment, University of Leeds*

ABSTRACT

Liquid-only and mixed-phase clouds in the Arctic strongly affect the regional surface energy and ice mass budgets, yet much remains unknown about the nature of these clouds due to the lack of intensive measurements. Lidar measurements of these clouds are challenged by very large signal dynamic range, which makes even seemingly simple tasks, such as thermodynamic phase classification, difficult. This work focuses on a set of measurements made by the Clouds Aerosol Polarization and Backscatter Lidar at Summit, Greenland and its retrieval algorithms, which use both analog and photon counting as well as orthogonal and non-orthogonal polarization retrievals to extend dynamic range and improve overall measurement quality and quantity. Presented here is an algorithm for cloud parameter retrievals that leverages enhanced dynamic range retrievals to classify mixed-phase clouds. This best guess retrieval is compared to co-located instruments for validation.

1. INTRODUCTION

Understanding the occurrence of mixed-phase clouds in the Arctic is a critical building block to better specifying regional energy and ice mass budgets [1]. Approximately 7 meters of sea-level rise is locked within the Greenland Ice Sheet (GrIS) of which approximately 25 mm has been contributed since 1900 with increasing rate in recent years [2,3]. Several studies have directly linked melting to changes in surface energy budgets caused by low level liquid-only and mixed-phase (here collectively referred to as mixed-phase) clouds [4,5]. The colloidally unstable mixture of liquid and ice water within clouds is a regular occurrence in the Arctic with a complicated set of processes serving to maintain both phases [6]. In the Arctic, the liquid phase dominates the radiative contribution of clouds while the ice phase modulates the liquid via the Wegener–Bergeron–Findeisen process [7,8].

As a result of the importance of mixed-phase clouds to the surface energy and mass budgets of the GrIS, quantitative measurements of cloud properties are sought to better understand the processes driving and underlying the formation and persistence of these clouds. Historically, the unique and harsh environment of the Arctic has hampered such measurements. This work will focus on a unique set of continuous measurements made by lidar and other sensors centered at the top of the GrIS at Summit, Greenland (72°35'46" N; 38°25'1"W; 3212 m asl) since 2010 as part of the Integrated Characterization of Energy Clouds Atmospheric State and Precipitation at Summit (ICECAPS) experiment [1]. Of interest for this contribution are co-located lidars, one custom polarization lidar and one commercially available polarization sensitive micropulse lidar (MPL), a single polarization Ka-band Doppler radar, and 2 microwave radiometers with measurement bands spanning 23 to 150 GHz.

2. CLOUDS AEROSOL POLARIZATION AND BACKSCATTER LIDAR

The Clouds Aerosol Polarization and Backscatter Lidar (CAPABL) has been deployed to Summit since 2010 to measure cloud phase and horizontally oriented ice crystals (HOIC) using polarization retrievals. The retrieval of HOIC by diattenuation measurements requires a minimum of 3 polarization channels in an off zenith orientation; CAPABL makes 4 linear polarization measurements allowing for orthogonal and non-orthogonal retrievals of polarization components [9]. This set of measurements is evenly spaced in signal intensity to more judiciously account for systematic signal saturation effects. The result of orthogonal and non-orthogonal retrievals is an expanded signal dynamic range [10].

This expanded signal dynamic range is helpful in making quantitative measurements of mixed-phase clouds. Mixed-phase clouds occur predominantly low in the atmosphere (below 2-3 km above the surface) at Summit, resulting in very high dynamic range signals, often well in excess of 5 orders of magnitude in a single profile. These differences in dynamic range caused by signal saturation create tremendous variability in estimates of fractional occurrence of cloud types, resulting in errors in excess of 30-40% [10]. These errors in fractional occurrence dramatically affect the interpretation of cloud radiative effects at Summit and cause yet unquantified uncertainty in cloud radiative forcing estimates.

The focus of this work is in developing a best estimate in cloud product to identify liquid, ice, and HOIC cloud types and to verify that best estimated product (hereafter referred to as data mask) with co-located sensors.

3. CO-LOCATED SENSOR SUITE

The ICECAPS sensor suite consists of many remote sensing instruments as well as a twice-daily radiosonde program modeled on the Atmospheric Radiation Measurement (ARM) sites [11]. The following will describe the relevant instruments and data comparison in greater detail.

3.1 MICROPULSE LIDAR

Multi-sensor comparisons of data are complicated by the designed sensitivities of instrumentation. It is difficult to compare different instruments, for example radar and lidar, because they are designed to perform different tasks. It is therefore reasonable to make a first comparison between two lidar systems. CAPABL is compared to a Sigma Space V4 (MPL) [12]. Relevant specifications are given in Table 1.

Table 1: Relevant lidar specifications. Note that CAPABL uses receiver attenuation filters to better observe the troposphere and makes the power aperture product of CAPABL less than the MPL by a factor of ~50.

	CAPABL	MPL
Laser Power [W]	0.3	0.02
Receiver Attenuation [OD]	3	0
Telescope Diameter [mm]	208	178
Polarizations	4	2
Range Resolution [m]	25.98	30
Polarization Scan Res [sec]	~82	80

CAPABL and the MPL are forced to similar resolutions in post processing. Polarization retrievals and a Klett inversion are performed. MPL data is then interpolated in time and space to the CAPABL grid. Data is then classified via the scheme in [10] from 0 to 8 km.

For this comparison, pixels are either classified as clear air, cloud liquid, cloud ice, or missing because of data filtering for each system. This allows 16 possible combinations of data from CAPABL and the MPL. An example of the data overlap is given in Table 2 for July 2016 where it is noted that each column is summed and normalized. Therefore each value is the percentage of CAPABL pixels of each type.

Table 2: Pixel by pixel comparison of CAPABL (C.) and the MPL for July 2016. The green/dotted color shows agreement. Red indicates one instrument had data removed by quality control steps. Yellow indicates enhanced sensitivity by CAPABL and gray enhanced sensitivity by the MPL. Purple indicates both instruments lack data.

C. / MPL	Clear	Liquid	Ice	Filtered
Clear	69.7%	37.0%	62.2%	3.42%
Liquid	0.3%	56.3%	5.5%	0.08%
Ice	0.2%	3.7%	29.4%	0.50%
Filtered	29.9%	3.0%	3.02%	96%

The data in Table 2 indicates some striking enhancements in sensitivity, which are traced to non-orthogonal polarization retrievals. In the summer where the sun remains up 24 hours a day, the weak perpendicular polarization of the MPL is limited in range, but non-orthogonal polarization retrievals with CAPABL's parallel and 3rd channel (~45 degrees from both parallel and perpendicular) have greatly enhanced range. This enhanced ranged is due to the small difference in signal strengths between CAPABL's signals.

3.2 MILLIMETER CLOUD RADAR

A zenith pointing Ka-band (35 GHz) single polarization Doppler cloud radar (millimeter cloud radar or MMCR) is deployed in the same location as CAPABL, which is well described by [13,14]. A comparison of data products is not as simple as with an MPL because radar Doppler parameters do not directly correspond to CAPABL data masks. An example of a simple comparison is given in Fig. 1 where CAPABL cloud boundaries are plotted over MMCR reflectivity data. The cloud boundaries on the given day, Feb. 27, 2016, show good agreement in a bulk sense. CAPABL seems to have better sensitivity to high thin ice clouds than does the radar, as is expected, but attenuation is visible between approximately 21 and 24 UTC where the cloud top boundary from CAPABL does not correspond to the end of data for the MMCR.

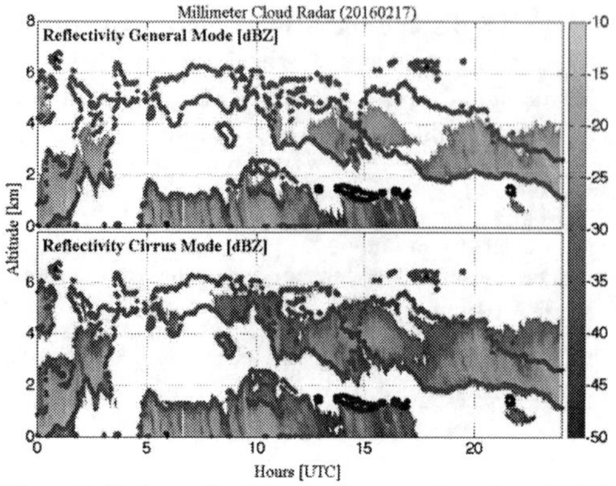

Figure 1: Radar reflectivity for its general mode and high sensitivity cirrus mode (top and bottom respectively). CAPABL derived cloud boundaries of ice and mixed-phase clouds are given by the red and black respectively.

Radar data is compared to CAPABL data in a two-step process. First, as their temporal and spatial resolutions are not the same, MMCR data is incoherently averaged in time to a similar grid as CAPABL then linearly interpolated in 2 dimensions to CAPABL's grid. Second, the radar data is grouped by the lidar data mask, then the Doppler spectral characteristics are used to create a histogram, then a cumulative distribution function of the resulting data.

A four-month comparison from July to November 2016 of data is presented in Fig. 2, which includes more than 25 million MMCR and CAPABL data pixels with temporal and spatial resolution of approximately 80 seconds and 30 meters for which both instruments had nearly 100% uptime. The MMCR data of interest are reflectivity (integral of the Doppler spectrum), Doppler velocity (first moment of the Doppler spectrum), and Doppler spectral width (second moment of the Doppler spectrum). Also plotted is liquid water path, which is described in the next section.

The cumulative distribution functions indicate what percent of the data exists at or below a given level. For example, in Fig. 2, approximately 25% of radar data identified by CAPABL as ice has a reflectivity less than -25 dBZ, whereas approximately 75% of data identified as water has a reflectivity less than -25 dBZ.

The expectations of this CAPABL/MMCR comparison should be clearly stated to put the resulting cumulative distribution functions into context. Broadly, at 35 GHz, the MMCR is more sensitive to ice than liquid water droplets via a diameter to the sixth (D^6) power scaling. By extension, the MMCR is more sensitive to liquid water droplets than clear air. One expects therefore, to see higher radar reflectivity for ice than liquid and less still for clear air. Furthermore, as ice is much larger, in diameter, at Summit than liquid water drops, one expects to see higher Doppler velocities (and given similar air motion, fall speeds) for ice. Finally, one expects Doppler spectrum width, which is dependent on a combination of microphysics and air motion, to be higher for liquid than ice or clear air due to turbulent motions that are typically associated with liquid phase clouds [15].

With these above expectations in mind, some reasonable and some striking results appear. The expected reflectivity and Doppler velocity cumulative distribution functions produce results as described above. Interestingly, Doppler velocities of ice crystals with a random orientation are higher than those with a preferential horizontal orientation. This is to be expected due to the drag forces that cause ice crystals to orient, but it is a useful indicator that polarization retrievals are reasonably separating randomly and preferentially oriented ice crystals. Additionally, the liquid spectral width curve shows interesting behavior both indicating large fractions of high spectral width particles but also a non-negligible 25% having much smaller spectral width than almost all of the ice pixels, indicating perhaps a separation between mixed phase and liquid only pixels.

3.3 MICROWAVE RADIOMETER

Comparing CAPABL data to that acquired by microwave radiometer (MWR) is also possible with the ICECAPS instrumentation. Two MWRs at Summit sample windows at 23.8, 31.4, 90, and 150 GHz. From these microwave brightness temperature observations, the liquid water path (LWP) is retrieved using optimal estimation [16].

These data are then linearly interpolated to the same time as CAPABL measurements for comparison.

CAPABL's data can be condensed into a column measurement based on the most radiatively important pixel type. A column is tagged as liquid if it contains any liquid. A column is tagged as ice if it contains ice and no liquid. A column can only be labeled as clear if no detectable cloud is present.

CAPABL column identifications can thus be used in a similar way to MMCR classification where MWR measurements are masked by CAPABL data, and then histograms and cumulative distribution functions are calculated for LWP for each mask type. This is also presented in Fig. 2 along with MMCR data previously described. Results show a very consistent agreement between CAPABL identifications of liquid clouds and MWR results.

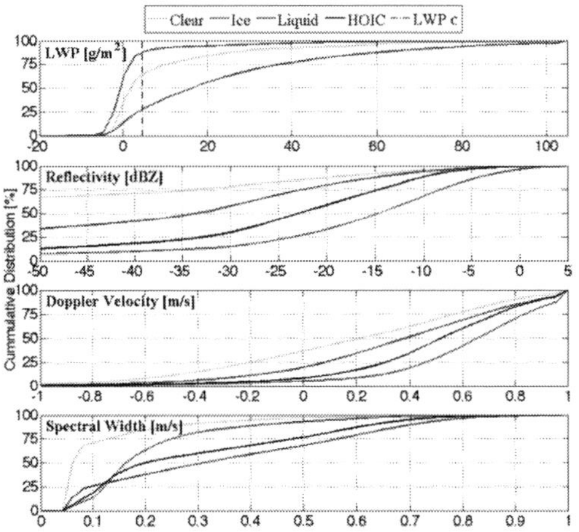

Figure 2: Cumulative distribution functions comparing MWR and MMCR data to CAPABL. The consistency of results is used in a broad sense to verify CAPABL's best estimate cloud mask.

4. DISCUSSION AND CONCLUSIONS

The data collected by CAPABL allows for an array of possible methods of calculation. Six possible methods were analyzed by [10] which showed dramatic differences in variables such as cloud fractional occurrence for different cloud phases as a function of processing type. This work has addressed that concern by combining the most critical processing types into a single cloud identification mask.

The process of creating this single mask however implicitly suggests a problem with validation. Validation of remote sensing equipment is challenging due to a lack of calibrated verification measurements. In this work, we choose to evaluate the CAPABL's phase identification algorithm by using ancillary remote sensing equipment to examine the consistency of retrievals. Here highly consistent results between CAPABL and collocated MPL, MMCR, and MWR over time periods from 1 month to 4 months are shown. One difficult is the validation of preferential orientation, because no sensors currently stationed at Summit are sensitive to such orientation except CAPABL. Differences in mean Doppler velocity measurements, related to particle fall speeds, demonstrate consistency with HOIC detection, but more work is needed to rigorously validate HOIC detections. More discussion is offered on the subject [9,10,17].

ACKNOWLEDGEMENT

This material is based upon work supported by the NSF GRFP under Grant No. DGE 1144083 and NSF Grant No. AON 1303864 and 1314156.

REFERENCES

[1] Shupe, M. D., et al.: BAMS, 94, 2013.
[2] Gregory, J. M., et al.: Nature, 428, 2004
[3] Kjeldsen, K. K., et al.: Nature, 528, 2015.
[4] Bennartz, R., et al.: Nature, 49, 2013.
[5] Tan, I., et al.: Science, 352, 2016.
[6] Morrison, H., et al.: Nature, 5, 2012.
[7] Curry, J. A., et al.: J. of Climate, 9, 1996.
[8] Shupe, M.D., and J. M. Intrieri, J. Climate, 17, 2004
[9] Neely, R.R., et al.: JTECH, 2013.
[10] Stillwell, R.A., et al.: Atmos. Meas. Tech. Discuss., in review, 2016.
[11] Turner, D.D., and R.G. Ellingson, Eds.: Meteor. Monograph, No 57, AMS, 2016
[12] Campbell, J.R., et al.: JTECH, 19, 2002
[13] Moran, K.P., et al.: BAMS, 79, 1998.
[14] Clothiaux, E.E., et al.: JTECH, 16, 1999.
[15] Verlinde, J., et al., J. Geophys. Res. Atmos., 2013
[16] Cadeddu, M.P. et al., Atmos. Meas. Tech., 6, 2013
[17] Cole, S. et al.: Submitted to ACP, in review, 2016.

LIDAR&RADIOMETER INVERSION CODE (LIRIC) FOR SYNERGETIC PROCESSING OF EARLINET, AERONET AND CALIPSO LIDAR DATA

Anatoli Chaikovsky*, Ludmila Chaikovskaya, Natalia Denishchik-Nelubina, Anton Fedarenka, Sergey Oshchepkov

[1]*Institute of Physics, National Academy of Sciences, Belarus, *chaikovl@dragon.bas-net.by*

ABSTRACT

Lidar&Radiometer Sounding (LRS) technique is used to obtain the "snapshot" of aerosol concentration profiles over the East European region during dust event, 07-08.04.2016 from EARLINET, AERONET and CALIPSO data. Aerosol concentration profiles retrieved from space and ground-based lidars are compared.

1 INTRODUCTION

In the recent years, combined lidar and radiometer sounding (LRS) technique becomes one of the basic tools in comprehensive investigations of the atmosphere aimed at retrieving altitude distributions of aerosol optical parameters and aerosol mode concentrations of an inhomogeneous aerosol layer [1,2].

However, the insufficient number of coincident lidar and sun-radiometer stations restrict employment of LRS technique for monitoring atmospheric aerosol. Nowadays, more than 400 active AERONET sun-radiometer stations provide regular observations in the different geophysical regions. At the same time, the number of lidar systems with required technical characteristics for providing LRS measurements and collocated with AERONET sun-radiometer systems is an order of magnitude smaller. Furthermore, the number of combined lidar and radiometer stations is expected not to be increased significantly in the near future.

Using data, measured by satellite multiwavelength lidars, like the CALIOP, working in the vicinity of AERONET station, for retrieving aerosol parameters becomes a real possibility to widen the LRS-area of applications. The most promise variant of complex LRS-experiment is carrying out coincident ground-based and satellite multiwavelength lidar measurements and AERONET observations.

This paper presents the results of development of the algorithm and software package for retrieval of aerosol mode concentration profiles from CALIPSO and AERONET data.

2 SET UP FOR THE PROBLEM

Initially algorithm and software LIRIC [1] were developed for the processing data of combined ground-based lidar and radiometer measurements. We assume that lidar provides measurements of backscatter signal at three wavelengths (355, 532 and 1064 nm) or at two wavelengths with both cross- and parallel polarization components of the lidar signal for a third wavelength.

The aerosol model is formed as a mixture of $k = 1,..,K$ aerosol modes which consists of spherical and spheroid particles with volume size distributions, $dV_k(r)/d\ln r$, and refractive index n that are the same for all particles. The part of spherical particles in the aerosol mixture ("sphericity") doesn't depend on particle sizes.

The aerosol layer is defined with two sets of parameters:

a) integral parameters characterizing the aerosol layer as a whole: size distributions of aerosol modes, their total content (column concentrations), refractive index and "sphericity";
b) concentrations, $c_k(h)$, of aerosol modes changing over altitude.

Aerosol particle size distribution are characterized by two fractions (fine and coarse) with bounding radius of ~0.5 micrometers, as it is the convention in AERONET retrievals. Particle size distribution of each fraction and refractive index is considered to be independent of the altitude.

Spherical and non-spherical particles of the coarse mode are considered as two different fractions if measurements of couple of cross/parallel polarization components are available.

CALIOP provides measurements of lidar signal at two wavelengths (532 and 1064 nm) and intensity of the cross polarization signal component at wavelength of 532 nm. We have shown in [1] that retrieved concentration profiles, $c_k(h)$, for two-, and three aerosol models, after elimination of the 355–lidar signal from the input data set undergo insignificant changes. Thus, the CALIOP data are expected to be sufficient to retrieve aerosol parameters by means of LRS algorithm and LIRIC program package.

Geometry of the combined ground-based radiometer and space lidar experiment are different from that for two ground-based systems. These differences are taken into account in the algorithm.

3 ALGORITHM

For the data processing of combined AERONET/CALIPSO experiments we transform lidar signals measured by CALIOP at the level h_∞ into signals of virtual system "Lidar-V" at the ground level h_0 (Fig. 1). ,

Range-corrected CALIOP lidar signals, $S^*_{C,j}(R_i) = P^*_{C,j}(\lambda_j, R_i)R_i^2$, can be written by the following equation [3]:

$$S^*_{C,j}(R_i) = E_{C,j}\xi_{C,j}\beta_j(\lambda_j, R_i)T_j^2(\lambda_j, R_i) = E_{C,j}\xi_{C,j}\beta_j(\lambda_j, R_i)\frac{T_j^2(\lambda_j, R_{h_0})}{T_j^2(\lambda_j, h_i)} \quad (1)$$

where j is the number of receiving channel; R_i is the range from the satellite to the sampled volume; $P^*_{C,j}(R_i)$ is the CALIOP measured signal at the wavelength λ_j after background subtraction and artifact removal; $E_{C,j}$ is the average CALIOP laser energy; $\xi_{C,j}$ is the CALIOP system parameter; $\beta_j(\lambda_j, R_i)$ is the backscatter coefficient at the range R_i; $T_j(\lambda_j, R_i)$ is transmittance from CALIOP, level, h_∞, to the scattering volume at the range R_i.

A range corrected signal of the virtual lidar,

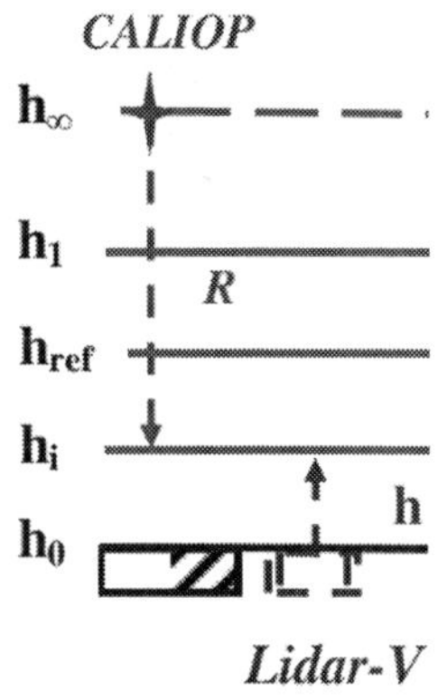

Figure 1 The scheme of lidar measurements

$S^*_{V,j}(h_i) = P^*_{V,j}(\lambda_j, h_i)h_i^2$ can be written in the same manner:

$$S^*_{V,j}(h_i) = E_{V,j}\xi_{V,j}\beta_j(\lambda_j, h_i)T_j^2(\lambda_j, h_i). \quad (2)$$

Using Eq. (1) and Eq. (2) the relation between $S^*_{V,j}(h_i)$ and $S^*_{C,j}(R_i)$ can be defined as follows:

$$S^*_{V,j}(h_i) = \frac{E_{V,j}\xi_{V,j}S^*_{C,j}(h_i)T_j^4(\lambda_j, h_i)}{E_{C,j}\xi_{C,j}T_j^2(\lambda_j, R_{h_0})}. \quad (3)$$

Given Eq. (3) the relation between the measured $\tilde{\mathbf{L}}^*$ and aerosol mode concentration, $c_k(h)$, can be expressed by the vector equation:

$$\tilde{\mathbf{L}}^* = \tilde{\mathbf{L}}\left(\lambda, c_k(h), \mathbf{a}_k, \mathbf{b}_k\right) + \mathbf{\Delta}_L, \quad (4)$$

In this equation the components of $\tilde{\mathbf{L}}^*$ are defined as follows:

$$\tilde{L}^*_{V,j}(h_i) = \frac{S^*_{C,j}(h_i)}{S^*_{C,j}(h_{ref})}\exp\left(2\delta_{h,h_{ref}}\tau_{r,j}(h_i, h_{ref})\right) \quad .(5)$$

$\delta(h, h_{ref})$ is the utility function:

$$\delta_{h_i,h_{ref}} = \begin{cases} 1 & if \quad h_i \le h_{ref} \\ -1 & if \quad h_i > h_{ref} \end{cases}. \quad (6)$$

Modeled lidar function $L_{V,j}(\ldots)$ can be written in the following:

$$\hat{L}_{V,j}(h_i) = \frac{\beta_j(h_i)\exp\left(-2\delta_{h,h_{ref}}\tau_{a,j}(h_i,h_{ref})\right)}{\beta_{r,j}(h_{ref})\hat{R}_j(\lambda_j,h_{ref})} \quad ;(7)$$

$\tau_{r,j}(\lambda_j,h_i,h_{ref})$ is the molecular optical thickness related to the range of (h,h_{ref}); $\tau_{a,j}(\lambda_j,h_i,h_{ref})$ is the aerosol optical thickness; $\mathbf{\Delta}_L$ is the vector of measurement uncertainties.

Equations (4)-(7) determine procedure of the lidar data processing for retrieving aerosol parameters from combined AERONET/CALIOP data. The basic difference between Eqs. (4) – (7) and appropriate equations for processing data of two ground-based systems is opposite sign in the exponential rate of eq. (5) and (7). This feature provides easy correction of the software LIRIC for the data processing combined AERONET/CALIOP.

4 CASE STUDY OF COINCEDENT EARLINET, AERONET AND CALIPSO DATA PROCESSING

Data of coincident EARLINET, AERONET and CALIPSO measurements in the East European region during Saharan dust events, 07-08.04.2016 were used to test LRS technique and software for retrieving aerosol parameters from ground-based radiometer and satellite lidar systems.

The map of dust optical thickness, AERONET stations and CALIPSO trajectories are shown in figure 2. The AERONET active systems in the East European region which data can be combined with CALIOP data during 07-08.04.2016 were stations in Kyiv, Minsk, Toravere, Ersa (07.04.2016) and Athens-NOA (08.04.2016). Simultaneously, the EARLINET station in Minsk provided ground-based lidar observations.

The results of the EARLINET, AERONET and CALIPSO data processing are presented in Fig. 4 and Fig. 5. Retrieved profiles of the aerosol mode concentration correspond to the forecast of the dust spatial distribution.

Figure 2 The map of dust optical thickness, (07-04-2016, NAAPS), ground-based stations and CALIPSO trajectories

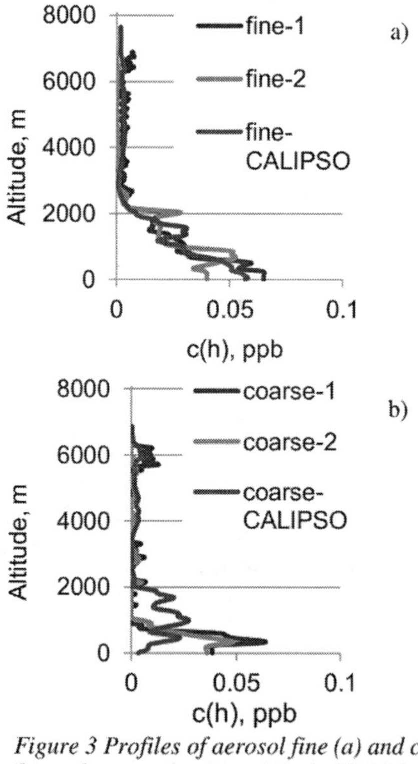

Figure 3 Profiles of aerosol fine (a) and coarse (b) mode concentrations, Minsk, 07-04-2-16; 1 - ground-based lidar, 09:17:30; 2 -ground-based lidar, 10:11:46; CALIPSO,11:00:47

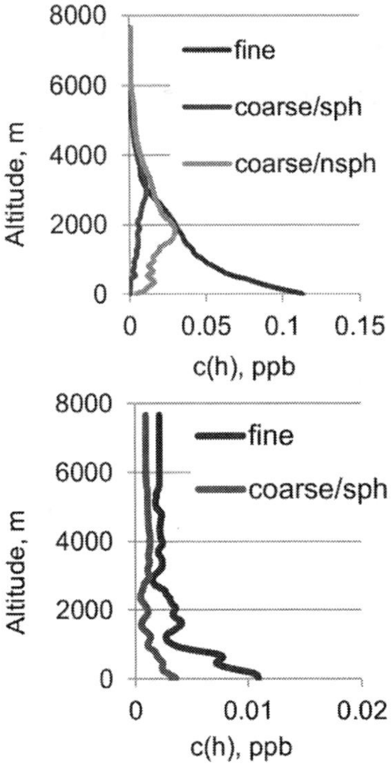

Figure 4 Profiles of aerosol mode concentrations; a) Kyiv (CALIPSO), 07-04-2016; 10:59:37; b) – Toravere (CALIPSO), 07-04-2016; 11:02:27

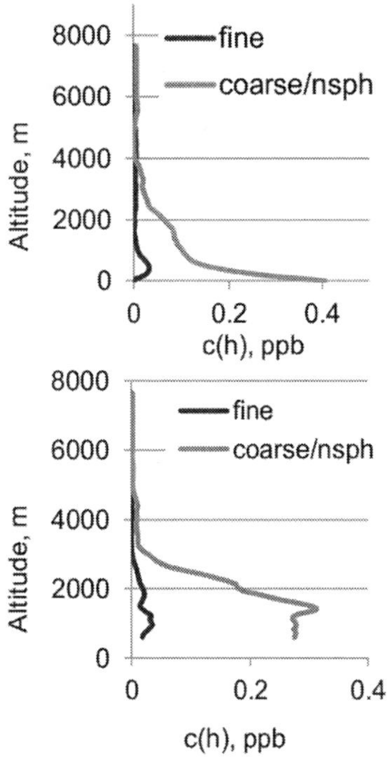

Figure 4 Profiles of aerosol mode concentrations; a) Ersa (CALIPSO), 07-04-2016; 12:13:16; b) – Athens-NOA (CALIPSO), 08-04-2016; 11:39:19

Selected CALIPSO trajectory segment for Minsk site was moved to the South because of cloudy atmosphere in the Minsk region. Nevertheless, profiles of fine aerosol particles retrieved from ground-based lidar are very close to those from CALIOP. Coarse mode of aerosol particles was more variable.

Results of the case study demonstrate realizability of the LRS-technique for retrieving aerosol parameters from combined AERONET/CALIPSO data.

ACKNOWLEDGEMENTS

The financial support by the European Union's Horizon 2020 research and innovation programme (ACTRIS-2, grant agreement no. 654109) is gratefully acknowledged. The investigation was supported by the Belarusian Republican Foundation for Fundamental, Research Agreement No. F15SB-023. We acknowledge the AERONET community, Atmospheric Science Data Center and the Naval Research Laboratory for the provision of the radiometer and CALIPSO data and NAAPS information.

References

[1] Chaikovsky, A., et al., 2016: Lidar-Radiometer Inversion Code (LIRIC) for the retrieval of vertical aerosol properties from combined lidar/radiometer data: development and distribution in EARLINET, *Atmos. Meas. Tech.*, **9**(3), 1181–1205.

[2] Lopatin, A., et al., 2013: Enhancement of aerosol characterization using synergy of lidar and sun – photometer coincident observations: the GARRLiC algorithm, *Atmos. Meas. Tech.* **6**, 2065-2088.

[3] CALIOP Algorithm Theoretical Basis Document; Calibration and Level 1 Data Products. PC-SCI-201, Release 1.0, 7 April 2006, 66 ph.

https://doi.org/10.1051/epjconf/201817608008

OPTICAL INSTRUMENTS SYNERGY IN DETERMINATION OF OPTICAL DEPTH OF THIN CLOUDS

Daniela Viviana Vlăduțescu[1]*, Stephen E. Schwartz[2], Dong Huang[3]

[1]*New York City College of Technology of the City University of New York, USA*
vivianavladutescu@yahoo.com,
[2]*Environmental and Climate Sciences, Brookhaven National Laboratory, NY, USA*
[3]*Science Systems and Applications Inc., Greenbelt, MD, USA*

ABSTRACT

Optically thin clouds have a strong radiative effect and need to be represented accurately in climate models. Cloud optical depth of thin clouds was retrieved using high resolution digital photography, lidar, and a radiative transfer model. The Doppler Lidar was operated at 1.5 μm, minimizing return from Rayleigh scattering, emphasizing return from aerosols and clouds. This approach examined cloud structure on scales 3 to 5 orders of magnitude finer than satellite products, opening new avenues for examination of cloud structure and evolution.

1 INTRODUCTION

Considering the substantial influence of clouds on short- and longwave irradiance at the top of the atmosphere and the necessity of accurately representing radiative and other effects of clouds in models, we initiated a project to characterize clouds by high-resolution digital photography looking vertically upward from the surface; the present analysis is limited to thin, boundary-layer cumulus clouds as determined by lidar vertical profiles. Cloud properties are characterized by cloud contribution to zenith radiance. The approach presented here, has the strength of yielding a quantitative measure of the effect of clouds on radiance at the surface and affords the prospect of inverting the measured radiance to infer cloud properties. High spatial resolution is achieved with a commercially available digital camera (Fujifilm FinePix, model S1) having a nominal resolution of 6 μrad, which, for a cloud 1 km above the surface, corresponds to a spatial resolution of 6 mm. In addition to resolution, photography from the surface looking upward affords the advantage over satellite imagery in quantitative interpretation of measured signal in that the background radiance is the black of outer space, with contributions to measured radiance only from light scattering by air (Rayleigh scattering), aerosol particles, and cloud drops, with minimal confounding contribution to the signal from surface-leaving radiance.

A Doppler Lidar was used to determined valuable context of the images. The lidar (wavelength 1.5 μm; range gate, 30 m; integrating time, 1 s; full angle field of view, 66 μrad; Pearson et al., 2009) provided time-height profiles of back-scattered intensity and vertical velocity.

A radiation transfer model was used to invert normalized zenith radiance (NZR, zenith radiance per downwelling irradiance) obtained in each of two color channels, red and blue, of the digital images to yield shortwave cloud optical depth (COD, the vertical integral of the scattering coefficient) for clouds of COD less than about 2 and report initial findings to show the capabilities of this approach for characterization of cloud radiance and its variation on fine spatial scales.

Other key measurements were with a multi-filter rotating shadow band radiometer (MFRSR) [Harrison et al., 1994] for cloud detection and aerosol optical depth; and Cimel sun photometer [Holben et al., 1998] for aerosol optical depth.

2 METHODOLOGY

The short-range spatial and temporal variation of cloud contribution to zenith radiance has been extensively studied at Brookhaven National Laboratory (Upton, Long Island NY, June-August, 2014), City College of the City University of New York, CCNY/CUNY (New York City, May, 2015; Figure 1 *a*), and at the Atmospheric Radiation Measurement (ARM) Southern Great Plains (SGP) site in north central

Oklahoma (36.6°N, 97.5°W; altitude 317 m, July 2015) (Figure 1 *b-e*) (Stokes and Schwartz, 1994).

All of these studies revealed frequent occurrence of clouds exhibiting similar short-range structure and frequently consisting of an intimate mixture of blue sky with cloud and/or transmittance of blue through thin clouds. Based on these observations we have reason to believe that the high temporal and spatial variation reported here may be quite general. In this regard we present here one case, selected from the measurements performed at the ARM SGP site on July 31st, 2015 (Fig.1, Schwartz et al., 2017).

Our research focus was on optically thin clouds, which are radiatively quite important, (typical local and instantaneous TOA shortwave radiative effect for COD = 1, about 80 W m^{-2}) and are underrepresented in models and analysis. Our quantitative analysis of the images has been restricted thus far to optically thin clouds, optical depth $\lesssim$ 2. This low COD value has also been selected to avoid the ambiguity problem noted later in this section (Figure 2).

The cloud layers selected for analysis are indicated by the lidar (Pearson, 2009) in regions of the images (see Figure 1*e*) of intense return (attenuated backscatter coefficient ~$10^{-3.5}$ sr^{-1} m^{-1}), for which the vertical profiles of returned intensity showed a strong peak at cloud altitude 2 km above the surface. A quantity determined for these clouds was the effective COD, ECOD, which was determined for each image collected with the digital camera, on a pixel-by-pixel basis by interpolation in the relation between COD and NZR shown in Figure 2. The dependence of COD on NZR given in that figure rests on plane-parallel calculations, and thus because of 3D effects the ECOD so determined for a given pixel may depart from the actual COD, the local vertical integral of the scattering coefficient.

Additionally, the optical depth was also measured by a multifilter rotating shadowband radiometer (MFRSR) and by a Cimel sun photometer. The MFRSR returned signal during much of the period 16:00 - 17:00, with optical depth at 500 nm 0.4 to 0.5, but with excursions up to 1.5 earlier in the hour and 1.0 at 16:34:00; the high excursions were characterized by low Ångström exponent (0 - 0.2) indicative of optically thin clouds.

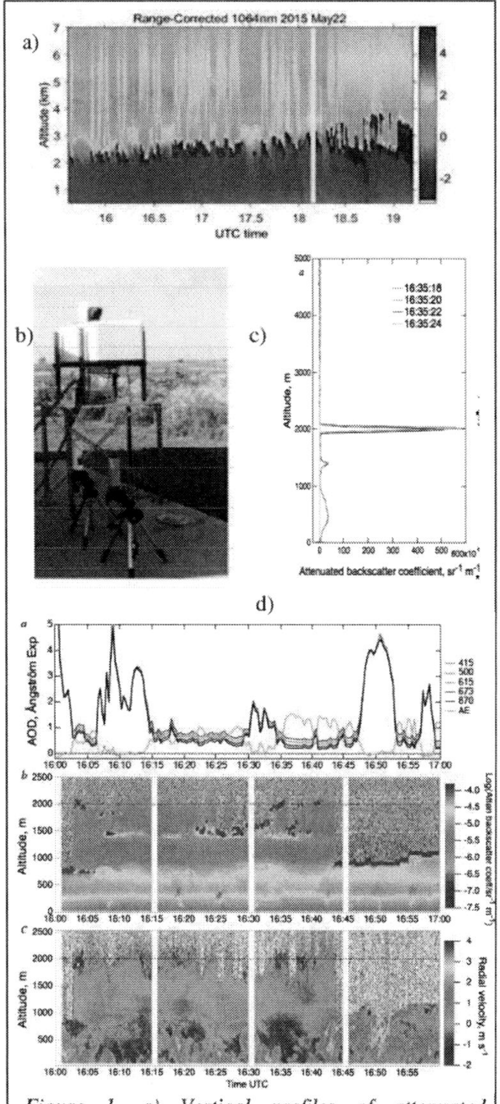

Figure 1. a) Vertical profiles of attenuated backscatter signal at 1064 nm from the CCNY/CUNY Lidar; b) Doppler Lidar and cameras at ARM SGP Site in Oklahoma c) Vertical profiles of range-corrected attenuated backscatter collected by the Doppler Lidar at times indicated on the figure, corresponding to cloud at 2km. d) MFRSR retrieved AOD and Angstrom exponent and Lidar vertical profiles between 16:00 - 17:00 UTC (09:30 - 1030 local sun time) on 07/31/15; (b-d modified from Schwartz et al., 2017.

Additional perspective to the atmospheric state was provided by radiosonde (not shown here), which showed a region of high relative humidity between 1000 and 2000 m altitude above the surface, consistent with cloud height determined by lidar. Based on the sounding, the atmosphere is neutral to slightly stable at the altitude of the aerosol and cloud layers, supporting interpretation of the clouds as weakly dissipative, an interpretation supported also by the slightly negative vertical velocities (~ 1 m s^{-1}) indicated by the Doppler Lidar (Schwartz et al., 2017).

The dependence of zenith radiance on COD, shown in Figure 2, was inverted at a given solar zenith angle to yield COD as a function of NZR for 640 and 460 nm, and cos(SZA) = 0.85. At low NZR the relation is approximately linear suggesting the possibility of sensitive retrieval of COD from NZR.

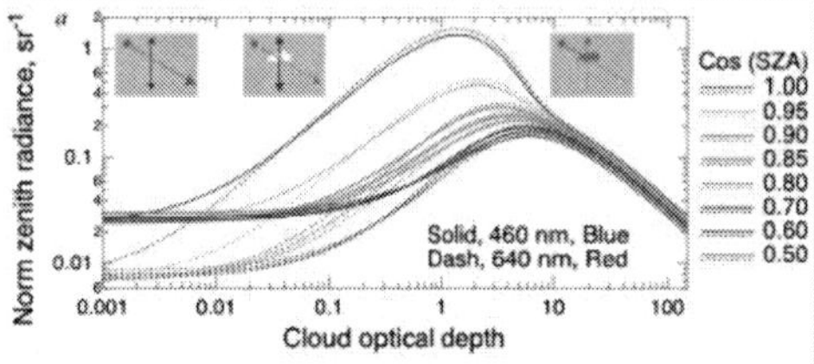

Figure 2. Double-logarithmic plot of dependence of normalized zenith radiance NZR on cloud optical depth COD at 460 nm, (solid curves) and red 640 nm, (dashed curves) wavelengths for liquid water clouds for indicated values of the cosine of the solar zenith angle (SZA). Radiance is normalized to incident hemispheric solar irradiance at each wavelength. Inset sketches show scattering processes contributing to zenith radiance. From Schwartz et al., 2017.

The normalized zenith radiance NZR, denoted R_λ^{nz} (unit: sr^{-1}), at a given wavelength *l* as the ratio of downwelling zenith radiance $I_\lambda^z(\mu_0)$ (W m^{-2} nm^{-1} sr^{-1}) to incident downwelling irradiance at the top of the atmosphere (W m^{-2} nm^{-1}), is defined as

$$R_\lambda^{nz}(\mu_0)=\frac{I_\lambda^z(\mu_0)}{\mu_0 F_\lambda}, \quad (1)$$

where μ_0 is the cosine of the solar zenith angle SZA and F_λ is the direct normal solar irradiance at the top of the atmosphere; this definition differs by a factor of π from that of Marshak et al. (2009). Here we specialize to the red and blue channels, having peak sensitivities at 640 and 460 nm, respectively, in the camera employed.

As the inference of COD from NZR is based on a radiative transfer (RT) model that assumes horizontal homogeneity, the retrieval is restricted to clouds that meet this requirement. In the present study use of this model for retrieval of COD gains further justification by the restriction of the retrieval to situations of single, optically thin layer clouds, in which the downwelling radiance is dominated by single scattering, as manifested by the near linearity of the relation between COD and NZR shown in Figure 2 for COD $\lesssim$ 2. This restriction also avoids ambiguity arising from the double valued dependence of COD in the retrieval of COD from NZR (so-called ambiguity problem). The lidar profiles confirm this situation in the cases selected.

The normalized zenith radiance R_s is obtained for both the red and blue channels on a pixel-by-pixel basis from the count rate in each pixel as

$$R_s = R_{min} + \frac{C - C_{min}}{C_{max} - C_{min}}(R_{max} - R_{min}) \quad (2)$$

Here C denotes red or blue counts in a given pixel. C_{min} and C_{max} denote the minimum and maximum counts, corresponding to cloud-free and maximum-brightness cloud pixels, respectively; determination of C_{min} and C_{max} thus requires identification of a nominally cloud-free region and of a region of maximum cloud brightness, respectively, in a single image or in temporally proximate images. R_{min} and R_{max} similarly denote the NZR determined from the RT calculations in the absence of cloud and at the maximum brightness as COD is increased, respectively, at the pertinent solar zenith angle. The calibration obtained in this way is expected to be transferrable to temporally proximate images with approximately constant solar geometry. Using a cloud-free region in the image itself to determine C_{min} implicitly accounts for contribution to zenith radiance from aerosols, the contributions to radiance from clouds and aerosols being nearly additive at the low optical depths for which the method is applicable. The ability to determine NZR from the camera counts for each pixel (and separately for the red and blue channels) in turn permits determination of COD from the data for each channel on a pixel-by-pixel basis.

3. RESULTS

The lidar measurements for the day of analysis indicate that the Planetary Boundary Layer signal was temporally much more uniform than that from the clouds, consistent with an aerosol layer (or nascent cloud layer) at about 1.5 km that briefly developed into a thin cloud; the slight variation in aerosol signal, which dominated the return over the 16:35 - 16:40 period (Figure 1 *c & d*) is attributed more to variation in relative humidity than to variation in aerosol amount. We analyzed situations where the transmittance through the cloud layers was sufficiently high that it can be inferred that the return from above 2 km is indicative of the absence of any significant aerosol or clouds above 2 km; this conclusion is also supported by the sounding. An image that satisfied all these requirements, taken at 16:35:20 and examined in Figure 3, exhibits considerable variation in ECOD on a variety of spatial scales. We find that such variation in radiance at all scales examined, with coherent spatial structures at scales as low as 10 cm. We attribute this variation to variation in cloud optical depth (Figure 3 *f*) on such scales, that we hypothesize to be due to local turbulence. Should this turn out to be the case, then it suggests that high resolution imaging of clouds from the surface may be a means of remotely examining the variation of such turbulence at scales approaching the Kolmogorov scale. Such imaging may thus provide an approach to relating these variations to atmospheric turbulence, which we hypothesize is responsible for the fluctuations in radiance observed on the centimeter scale.

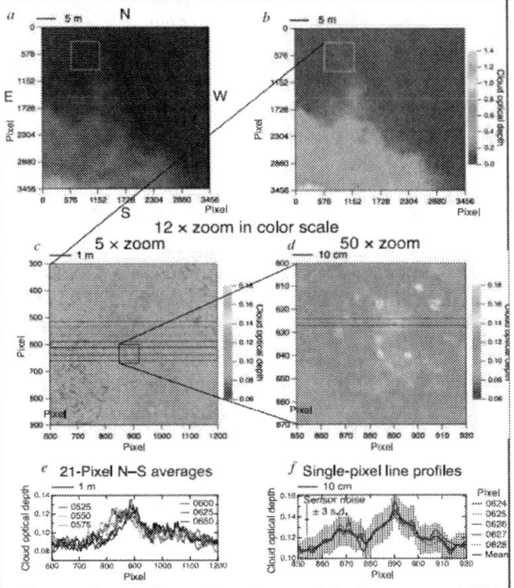

Figure 3. a).Natural color zenith image at ARM SGP at 16:35:20 UTC; axes denote pixel number; N, E, S, W denote cardinal directions; scale bar is evaluated for cloud height 1.5 km.. b) False color image (FCI) of ECOD obtained from red-channel image. c) FCI of ECOD for sub-region of image denoted by yellow box in a and b; note 12-fold zoom in false color scale. d) FCI of ECOD for black square in c. (e) Mean ECOD calculated over 21 line profiles centered at indicated N-S pixel number bounded by pairs of horizontal lines in corresponding colors in c. f) Single-pixel line profiles of COD at N-S pixel number indicated by colors of lines in e; uncertainty bar denotes ± 3 standard deviations of noise in retrieval of ECOD. Modified from Schwartz et al., 2017.

3 CONCLUSIONS AND FUTURE WORK

A method is presented here to infer cloud optical depth COD on a pixel-by-pixel from high-resolution digital photographic images by scaling the counts in individual pixels to COD via RT calculations. We find that there is commonly considerable variation in radiance at all scales examined, with coherent spatial structures at scales as low as 10 cm. Future work will include further validation and comparison based on lidar retrieved COD.

References

[1] Pearson, G., F. Davies, and C. Collier, 2009: An Analysis of the Performance of the UFAM Pulsed Doppler Lidar for Observing the Boundary Layer, *J. Atmos. Ocean. Tech.*, **26**(2), 240-250

[2] Stokes, G. M., and S. E. Schwartz, 1994, The Atmospheric Radiation Measurement (ARM) Program: Programmatic Background and Design of the Cloud and Radiation Test Bed, *Bull. Amer. Meteorol. Soc.*, **75**(7), 1201-1221

[3] Schwartz S. E., Huang D. and Vladutescu D. V. 2017: High-Resolution Photography of Clouds from the Surface: Retrieval of Optical Depth of Thin Clouds down to Centimeter Scales. *J. Geophys. Res. - Atmos.*, Paper #2016JD025384, in press

[4] Marshak, A., Y. Knyazikhin, J. C. Chiu, and W. J. Wiscombe (2009), Spectral invariant behavior of zenith radiance around cloud edges observed by ARM SWS, Geophys. Res. Lett., 36, L16802

GROUND BASED AND AIRBORNE ATMOSPHERIC MEASUREMENTS NEAR BUCHAREST

Anca Nemuc[1]*, Andreea Boscornea[2], Livio Belegante[1], Jeni Vasilescu[1], Sorin Vajaiac[2], Dragos Ene[1], Luminita Marmureanu[1], Simona Andrei[1]

[1]*National Institute of R&D for Optoelectronics, Romania, *Email: anca@inoe.ro*
[2]*National Institute for Aerospace Research "Elie Carafoli"*

ABSTRACT

This paper presents the results from a coordinated approach for atmospheric investigation, exploring synergies between different techniques. A wide range of instruments have been used during an intensive measurement period both from ground (lidar, sunphotometer, aethalometer and Aerosol Chemical Speciation Monitor) and airborne (aerodynamic particle sizer, the Picarro gas analyzer and the NO_2 CAPS analyzer) in 2016 over Magurele, 6 km South West of Bucharest.

1 INTRODUCTION

Despite all the efforts made in recent years for improving the understanding of aerosol characteristics more detailed airborne and ground based observations of aerosol particles are needed, due to their high spatial and temporal variability[1]. A lot of experimental techniques are available today to perform measurements from ground, *in situ,* airborne or from satellites. Unfortunately, there are some problems in correlating these data and systematic validation of the measured data is necessary. Data collection and handling procedures are dramatically different in terms of measurement and response time, resolution and accuracy, derived parameters are inter-dependent. Airborne platforms enable new measurement geometries, leading for instance to a high sensitivity in the free troposphere. The fundamental challenge for satellites measuring air quality is to distinguish between pollution near the surface and pollution in the higher atmosphere. Scientific questions are still under debate about the vertical distribution of pollutants. Measurements from aircraft, in combination with ground-based measurements, offer a key perspective that makes such distinctions easier to make. In an ideal campaign the ground-based instruments offer a critical view of the same patches of air that an aircraft are monitoring from above (e.g.[2]). The Airborne ROmanian Measurements of Aerosols and Trace gases (AROMAT) campaigns were held in Romania in September 2014, August 2015 and 2016. They were supported by ESA in the framework of its Copernicus Earth observation programme, to prepare the validation programme of the future Atmospheric Sentinels.

A diversity of resources have been used during the campaign to measure air pollution in a pre-urban environment focusing both on the ground level concentrations and on the vertical homogeneity of the pollution plumes.

This paper presents measurements from the three days intensive period measurements 30.05-01.06 2016 over Magurele, 6 km South West of Bucharest.

2 METHODOLOGY

2.1 Instruments

During the AROMAT 2016 the following instruments were used:

- Aerosol lidar for dynamics of the atmospheric structure (including the Planetary Boundary Layer-PBL), backscatter coefficient and extinction coefficient profiles at several wavelengths[3], [4]
- Aethalometer AE33 measurements for mass concentration of Black Carbon particles at ground level [5]
- Sun photometer for Atmospheric Optical Depth(AOD) at 7 wavelengths[6]

- Gas analyzers for NOx, NO,NO_2, SO_2, O_3, THC, CH_4 and CO concentrations
- Aerosol Chemical Speciation Monitor for quantitative particle mass loading and chemical composition in real-time for non-refractory sub-micron aerosol particles measurements [7]
- Airborne Aerodynamic Particle Sizer
- Airborne Picarro gas analyzer- The Picarro gas analyzer, model G2401-mc, for simultaneous and continuous measurements of four most common greenhouse gases: water vapour, carbon monoxide, carbon dioxide and methane, based on near–infrared cavity ring down spectroscopy technology
- and the NO_2 CAPS analyzer- The Cavity Attenuated Phase Shift (CAPS) nitrogen dioxide (NO_2) AS32M monitor from Environment S.A. for the NO_2 volume mixing ratio using its broadband absorption in the blue and the CAPS principle [8]

2.2 Meteorological context

During 30.05 and 01.06.2016 south-eastern Europe was influenced by a cyclonic regime within the lowest part of the atmosphere, while in the upper troposphere two cyclonic areas were active both sides of Romania (west and east). The western cyclonic nucleus with its south-western circulation, has generated the advection of warm and humid air from Mediterranean Sea, while the diurnal heating added its energy to generate the afternoon instability.

3 RESULTS

3.1 Ground based measurements

The ground based measurements indicate the influence of local pollution source. The wind direction data at ground level shows SW influence, therefore the rural and local predominance.

During the first day of 2016 campaign a decrease of the Ängstrom exponent have been observed from the sun photometer measurements, highlighting an increase of the size particles in the upper layers above Magurele, most probably dust[4], but with no significant variation of the AOD. AOD daily average at 500nm is 0.204, while the Ängstrom exponent has a daily average of 1.096.

During 30.05-02.06 the aerosol load above Magurele decreased as it is shown by the lidar measurements; the AOD maximum reached in the first day was about 0.38 at 532nm while the minimum was 0.1 on the 1st of June, in good agreement with the ones derived from sunphotometer measurements (Figure 1).

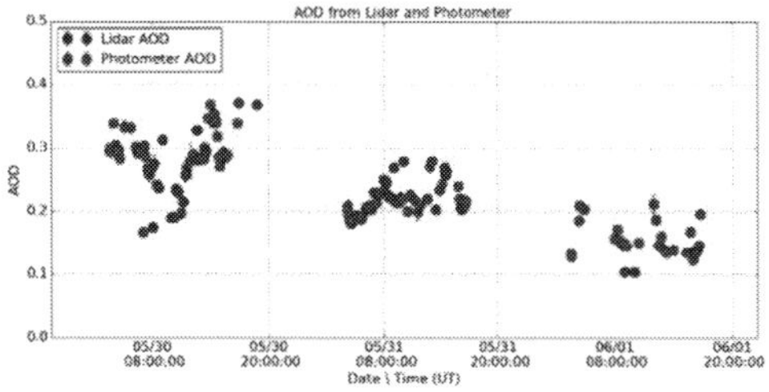

Figure 1 AOD from lidar and sun photometer during the campaign

The PBL height from lidar varied between 600m to about 2000m. A second layer is also visible in all days. The ECMWF prognosis is completely missing the PBL height during the night (depicted as very close to zero-red in Figure 2) while during the day is between the values of the first layer detected in the lidar data and PBL height (Figure 2).

The forecast dust model (DREAM, available online at http://www.bsc.es/projects/earthscience/visor/sub_fc8.php?dom=med8&type=dld&vyear=2016&vmonth=06&vday=01&vhour=12) shows both the altitude and the amount of dust particle reaching Bucharest-Magurele area. Hysplit back trajectories show the air masses arriving at 4 km above the measurement site coming from North Sahara after a 48 hours travel across Mediterranean Sea and South East Europe (not shown) [9].

We can conclude that the long range transported mineral dust particles were present at 3-4km altitude above Bucharest-Magurele on May 30, 2016.

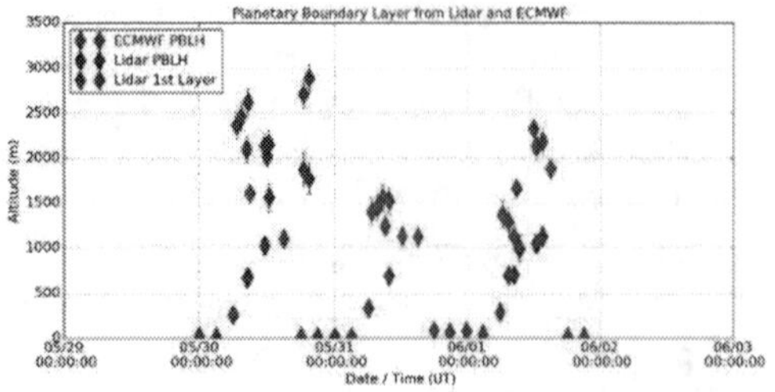

Figure 2 Planetary Boundary layer height and the first layer altitude derived from lidar measurements along with the ECMWF forecast

Similar behaviour of the concentrations for NO and NO_2 with peak values at about 20ppb at the same time of the day are highlighting traffic influence.

The O_3 and SO_2 concentrations patterns show a diurnal cycle and maximum values during the day reaching about 90 ppb and 3 ppb respectively.

CH_4 and THC concentrations at ground level were almost constant at about 2.5 ppm with some peaks close to mid-night of about 4ppm.

High concentrations for entire period of measurements are evidenced for fossil fuel black carbon (representing approx. 83% from total BC). Most probably source of fossil fuel BC is traffic. The average concentrations for fossil fuel are 880ng/m^3 since for biomass burning the average concentration is 176ng/m^3, normal values for this time of the year.

Low values of BC biomass burning is characteristic for the 30 May 2016, there are no sources of wood burning. Highest concentrations of BC fossil fuel characterize the first part of the day, almost double than the values for the evening period. The aerosol chemical composition on 30 May has been influenced only by traffic sources.

Aerosol chemical composition of particle less than 1µm depicted 5 main species: Organics, Sulphate, Nitrate, Ammonium and Chloride. Highest proportions are represented by Organics (~48%) and Sulphate (~30%). Higher concentrations at the beginning of the campaign are pin pointed, decreasing during the next days. The average concentrations are 8.11±3.48 µg/m^3 on 30 May, 6.09±2.00 µg/m^3 on 31st of May and 6.11±1.65 µg/m^3 on 1st of June. Much lower concentration can be observed between the night of 30 May and morning of 31 May, characterized by hydrocarbon organic aerosols, coming from fresh sources – local and traffic sources.

4 Airborne *in situ* measurements

A detailed flight strategy has been done before the campaign. After take-off from Strejnicu aerodrome (Prahova county), the BN2 aircraft had a North-West-South course around Bucharest Metropolitan Area, directing to the South-Western Bucharest, were the observation area was: Bragadiru and Magurele. After performing the sounding (at three different altitude levels: 300, 500 and 1000m), the aircraft returned to Strejnicu aerodrome .The next plot (Figure 3) show the aerodynamic Particle Sizer data for the altitude levels of 341m. Because of the low number of data points (every minute one data point) no interpolation method could be used. Therefore, the plot contains colored points corresponding to every measured point of total aerosol concentration for the flight trajectory. The position of orange and red dots for all altitude levels correspond to identified source locations (traffic or industrial sources). It is interesting to observe that in all three altitude levels in the upper left side and in the right side of the scanned area the presence of a strong source is identified.

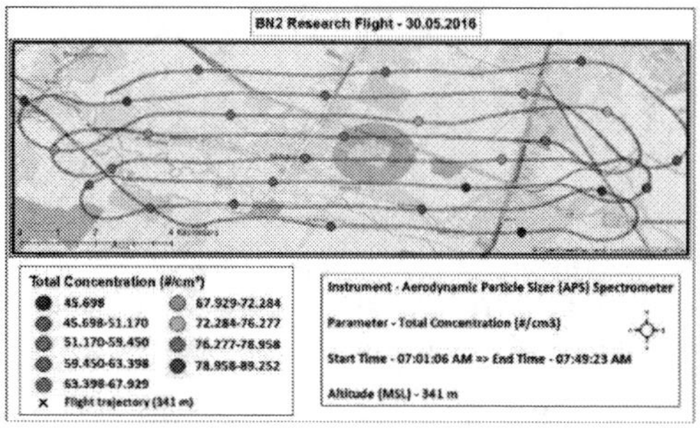

Figure 3 Aerosol total concentration recorded during flight perfomed on 30.06.2016 at an altitude level of 341 m

5 CONCLUSIONS AND REMARKS

This paper presents the measurements results from AROMAT 2016 campaign held in Romania during 30 May - 2 June. We explored the suitability of using the existing Romanian ground based and airborne instruments to assess the pollution sources in a region close to south west Bucharest.

Synergies between different techniques gave us useful perspective. We discovered also drawbacks to be corrected for future campaigns.

This measurement campaign carried out in early summer, has delivered these first results, showing in particular that a proportion of the pollution came from long range transported aerosols of mineral dust. Highly localized sources could be found and chemically characterized.

Supplemented with satellite observations, these campaigns could provide the missing data in an explanatory mechanism for the interaction between atmospheric chemistry, aerosols, formation and dispersal of low clouds, radiation budget and precipitation.

The goal of future campaigns could be to use these types of data to improve regional air quality models, which will support environmental policies for sustainable development of the region.

ACKNOWLEDGEMENTS

This work has been supported by the European Union's Horizon 2020 Research and Innovation Programme, under the Grant Agreement no 692014 – ECARS, by the National Core Program-PN16.40.01.01, by ESA contract 4000113511/NL/FF/gp- AROMAT. Also this project has received funding from the European Union's Horizon 2020 research and innovation programme under grant agreement No. 654109, ACTRIS 2.

The authors gratefully acknowledge the NOAA Air Resources Laboratory (ARL) for the provision of the HYSPLIT transport model website (http://www.ready.noaa.gov) used in this publication.

References

[1] Groß, S., Esselborn, M., Abicht, F., Wirth, M., Fix, A., and Minikin, A.: Airborne high spectral resolution lidar observation of pollution aerosol during EUCAARI-LONGREX, Atmos. Chem. Phys., 13, 2435–2444, doi:10.5194/acp-13-2435-2013, 2013

[2] Whiteman, D. N and Coauthors: Airborne and ground-based measurements using a high-performance Raman lidar. J. Atmos. Oceanic Technol., 27, 1781–1801, doi:10.1175/2010JTECHA139, 2010

[3] Nemuc, A., Vasilescu, J., Talianu, C., Belegante, L., and Nicolae, D.: Assessment of aerosol's mass concentrations from measured linear particle depolarization ratio (vertically resolved) and simulations, Atmos. Meas. Tech., 6, 3243-3255, doi:10.5194/amt-6-3243-2013, 2013

[4] Nicolae, D., Nemuc, A., Müller, D., Talianu, C., Vasilescu, J., Belegante, L., & Kolgotin, A. : Characterization of fresh and aged biomass burning events using multiwavelength Raman lidar and mass spectrometry. Journal of Geophysical Research: Atmospheres, 118(7), 2956–2965. http://doi.org/10.1002/jgrd.50324, 2013

[5] Drinovec, L., Mocnik, G., Zotter, P., Prévôt, A. S. H., Ruckstuhl, C., Coz, E., Rupakheti, M., Sciare, J., Müller, T., Wiedensohler, A., and Hansen, A. D. A.: The "dual-spot" Aethalometer: an improved measurement of aerosol black carbon with realtime loading compensation, Atmos. Meas. Tech., 8, 1965–1979, doi:10.5194/amt-8-1965-2015, 2015.

[6] Holben, B. N., Eck, T. F., Slutsker, I., Tanré, D., Buis, J. P., Setzer, A., Smirnov, A. (). AERONET—A Federated Instrument Network and Data Archive for Aerosol Characterization. *Remote Sensing of Environment*, *66*(1), 1–16. http://doi.org/10.1016/S0034-4257(98)00031-5, 1998

[7] Jayne, J. T., Leard, D. C., Zhang, X., Davidovits, P., Smith, K. A., Kolb, C. E., & Worsnop, D. R. (). Development of an Aerosol Mass Spectrometer for Size and Composition Analysis of Submicron Particles. *Aerosol Science and Technology*, *33*(1–2), 49–70, 2000

[8] Kebabian, P. L., Herndon, S. C., & Freedman, A.: Detection of nitrogen dioxide by cavity attenuated phase shift spectroscopy, *Analytical Chemistry*, *77*(2), 724–728, 2005

[9] Stein, A.F., Draxler, R.R, Rolph, G.D., Stunder, B.J.B., Cohen, M.D., and Ngan, F., . NOAA's HYSPLIT atmospheric transport and dispersion modeling system, Bull. Amer. Meteor. Soc., **96**, 2059-2077, 2015

TEMPERATURE INTER-COMPARISON EFFORT IN THE FRAMEWORK OF HYDROLOGICAL CYCLE IN THE MEDITERRANEAN EXPERIMENT – SPECIAL OBSERVATION PERIOD (HYMEX-SOP1)

Benedetto De Rosa[1], Paolo Di Girolamo[1], Donato Summa[1], Cyrille Flamant[2], Olivier Bousquet[3], , Marco Cacciani[4], Dario Stelitano[1]

[1] Scuola di Ingegneria, Università degli Studi della Basilicata, 85100 Potenza – Italy, emails: benedetto.derosa@unibas.it; digirolamo@unibas.it;donato.summa@unibas.it
[2] LATMOS/IPSL, UPMC Univ. Paris 06 Sorbonne Universités, UVSQ, CNRS, Paris, France, email: cyrille.flamant@latmos.ipsl.fr
[3] Météo-France & Laboratoire de l'Atmosphère et des Cyclones, DIROI / CRC, 97490 Sainte Clotilde Cedex, France, email: olivier.bousquet@meteo.fr
[4] Dipartimento di Fisica, Università di Roma "La Sapienza", Piazzale Aldo Moro, n. 2, 00100 Roma – Italy, email: marco.cacciani@uniroma1.it

ABSTRACT

Accurate measurements of the vertical profiles of atmospheric temperature are necessary to advance the knowledge of dynamics-thermodynamics-radiative interaction mechanisms triggering convection, and ultimately improve weather forecasting capabilities. Comprehensive inter-comparisons between different remote sensing and in-situ sensors have to be carried for the purpose of obtaining accurate error estimates for these sensors. This paper reports results obtained in the frame of the Hydrological Cycle in the Mediterranean Experiment – Special Observation Period (HyMeX-SOP1).

1 INTRODUCTION

The Hydrological Cycle in the Mediterranean Experiment aims at a better understanding, of the hydrological cycle in the Mediterranean, with specific emphasis on the observation and modeling of extreme weather events.

Heavy precipitation and flash-flooding events affect the Mediterranean coastline as a consequence of specific forcing mechanisms (orography) and abundant evaporation from the sea surface. These events, which may cause hundreds of millions of euros in damages, are usually associated with increasing levels of humidity within the boundary layer and the lower part of the free troposphere The role that played by water vapour on meteorological process is only partially understood due to its highly variable distribution in space and time. This characteristic is partially determined by the fact that water vapour pressure varies significantly with temperature. Therefore accurate and high space-temporal resolution measurements of the vertical profiles temperature are even more necessary in these areas. A proper understanding addressing of these

The improvement of weather forecasting skills, especially in the presence of extreme events, requires an appropriate comprehension of the key processes influenced by the three-dimensional distribution of atmospheric water vapor and temperature. As a consequence, accurate and high-resolution global-scale measurements of the vertical profiles of atmospheric temperature, besides those of water vapour mixing ratio, are highly necessary. Unfortunately, none of the presently available measurement techniques for tropospheric temperature profiling is able to achieve the highly demanding observational requirements, in terms of measurement accuracy and precision, capable to produce a significant impact on mesoscale weather prediction. Some observational networks, such as the network of upper air radiosounding stations, are characterized by a sufficient density of sensors, high accuracy and vertical resolution, but lack in temporal resolution. Other networks, on the other hand, such as the global network of ground-based microwave radiometers, are characterized by a sufficient density of sensors, high accuracy and temporal resolution, but lack in vertical resolution.

These limitations can be in principle overcome based on the implementation of a network of lidar systems exploiting the rotational Raman technique.

In this paper, we report results from an inter-comparison study involving a temperature Raman lidar and other in-situ and remote sensing sensors with the purpose of obtaining accurate error estimates for all the involved instruments. More specifically, Raman lidar temperature profiles are compared to those collected by an in-situ temperature sensor deployed on-board the French research aircraft ATR42, operated by the Service des Avions Instrumentés pour la Recherche en Environnement (SAFIRE), as well as to those collected by a microwave radiometer and radiosondes.

Measurements reported in this paper were carried out in the frame of HyMeX-SOP1. Within this experiment a major field campaign, SOP1, took place in the period September–November 2012 over the northwestern Mediterranean Sea and its surrounding coastal regions in France, Italy and Spain. In the frame of HyMeX-SOP1, the University of BASILicata ground-based Raman Lidar system (BASIL) was deployed in the Cévennes-Vivarais site (Candillargues, Southern France, Lat: 43°37' N, Long: 4° 4' E, Elev: 1 m) and operated from 5 September to 5 November 2012.The inter-comparison effort described in this work allows obtaining accurate error estimates for the involved sensors. For this purpose, simultaneous and co-located profiles from different sensors' pairs are considered. These are used to compute the mutual BIAS and root-mean-square (RMS) deviation between the different sensors' pairs as a function of height. Similar sensors' intercomparison studies involving lidar systems had already been reported in the past [1, 2, 3], but only for water vapour measurements and not for temperature measurements. So this study is a first attempt in trying to fill this intercomparison gap.

2 METHODOLOGY

The Raman lidar system BASIL has been described in detail in a variety of previous papers (among others, [4, 5,6]). During HyMeX-SOP 1, BASIL was calibrated based on the comparison with radiosondes launched directly from the lidar site in Candillargues (approx. 100-150 m away).

In order to carry out an appropriate error analysis for the different sensors involved in the intercomparison effort, the mutual BIAS and root-mean-square (RMS) deviation of simultaneous and co-located data from the different sensors' pairs have been calculated. Expressions to compute these quantities have been reported by several authors [1, 2, 3]. Specifically:

$$BIAS = \frac{1}{N}\sum_{i=1}^{N} BIAS_i = \frac{2}{N}\sum_{i=1}^{N}\left\{\frac{\sum_{z=z_1}^{z_2}[q_1(z)-q_2(z)]}{\sum_{z=z_1}^{z_2}[q_1(z)+q_2(z)]}\right\} \quad (1)$$

$$RMS = \frac{1}{N}\sum_{i=1}^{N} RMS_i = \frac{2}{N}\sum_{i=1}^{N}\left\{\frac{\sqrt{N_z\sum_{z=z_1}^{z_2}[q_1(z)-q_2(z)]}}{\sum_{z=z_1}^{z_2}[q_1(z)+q_2(z)]}\right\} \quad (2)$$

where $q_1(z)$ and $q_2(z)$ represent the water vapor mixing ratio values at altitude z for sensor 1 and sensor 2, respectively, z_1 and z_2 are the lower and upper altitude of the considered altitude interval, respectively, and Nz is the number of data points for each sensor in this interval. In the present work, our attention was focused on altitude region 0.5 – 5.5 km, with an interval width of 0.5 km. Index i, with values from 1 to N, denotes the intercomparison sample, with N being the total number of possible comparisons for each pair of sensors. To apply expressions (1) and (2) a common altitude array has to be considered for each pair of sensors. Expressions (1) allow to compute relative BIAS and RMS deviation. Absolute BIAS and RMS deviation are obtained from expressions (1) and (2) by multiplying these expressions by the mean of the two sensors' data:

$$2\sum_{z=z_1}^{z_2}\{q_1(z)+q_2(z)\}\Big/N_z \quad (3)$$

3 RESULTS

Comparisons reported in this paper were carried out in the time period from 11 Sept. to 5 Nov. 2012. As an example of BASIL measurement capabilities, figure 1 illustrates the time evolution of temperature over a 3,3 h time interval from

00:00 to 3:30 UTC on 30 October 2012 . The figure is generated using a succession of 44 consecutive profiles (integration time Δt=5 min, vertical resolution Δz=120 m).

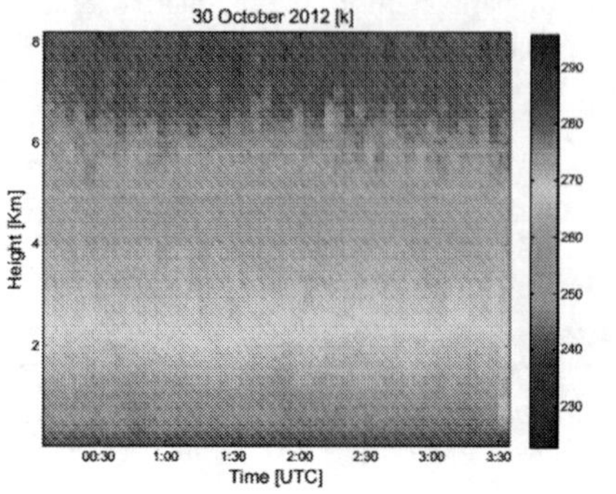

Figure 1 Time evolution of temperature from 00:00 UTC to 3:30 UTC on 30 October 2012.

A specific intercomparison effort was carried between BASIL and the airborne in-situ temperature sensor (a platinum resistance wire sensor). In-situ temperature measurements from measurements are usually collected during the aircraft take-off and landing phases.

On 12 September and 2 October 2012 the aircraft preformed specific ascending and descending loops in the proximity of the lidar site. Figure 2 illustrate the intercomparison between BASIL and the temperature in-situ sensor (hereafter represented with the short name IS). An integration time of 5 min has been considered for BASIL data.

The figure reveals a quite good agreement between these two sensors BASIL and the aircraft in situ sensors, obtained with a, is, with few specific exceptions. Specifically, absolute BIAS does not exceed 2 K in the free troposphere.

The higher BIAS values in the lowest 2 km are most probably associated with the temperature variability in the planetary boundary layer (it is to be kept in mind that the typical distance between the profiles' pairs is 20-30 km) and with the systematic error affecting Raman lidar temperature measurements in the overlap region (in the lowest 1 km).

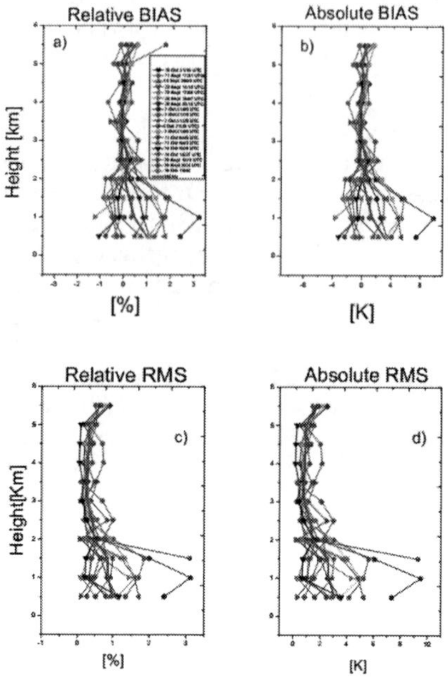

Figure 2 Relative (a) and absolute (b) BIAS and relative(c) and absolute(d) RMS deviation of BASIL vs. IS. Considered case studies are from period 11 September - 18 October 2012.

Figure 3 illustrates the vertical profiles of mean BIAS and RMS deviation for the different sensors' pairs. The figure reveals that the BIAS between BASIL and the radiosonde does not exceed 1.2 K or 0.75 % above the planetary boundary layer, while the BIAS between BASIL and IS does not exceed 0.5 K or 0.25% above the planetary boundary layer. Again, deviations observed in the lower portion of the boundary layer are primarily due to the differences between the air masses sounded by the two sensors. This is also testified by the large values of RMS deviations in this same altitude range. Comparisons between BASIL and the microwave radiometer (MWR) were possible in the period 11 October-05 November 2012 (the MWR was deployed at a distance of 150-200 m from BASIL). During this period 348 night-time clear sky occurrences were identified and considered. The mean absolute BIAS and RMS deviation within the planetary boundary layer are found to have values not exceeding 1K and 2 K, respectively.

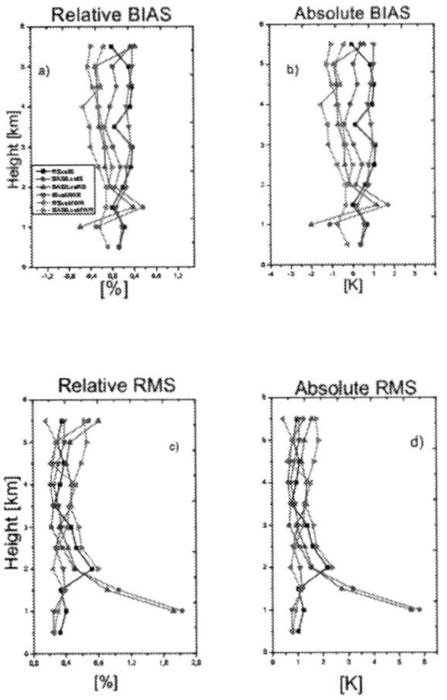

Figure 3 Average values of Relative BIAS, absolute Bias, relative RMS and absolute RMS deviation for all sensors

Vertically averaged values of BIAS and RMS deviation for all sensors' pairs are listed in Table1. The mean absolute BIAS of BASIL vs. RS –is 0.25 K, while the mean absolute BIAS of BASIL vs. MWR and IS are 0.74 K and 0.28 K, respectively. The mean absolute BIAS of RS vs. MWR and IS are -0.74 K and 0.610 K, respectively. Finally, the mean absolute BIAS of MWR vs. IS is 0.74 K. More results from this study will be illustrated at the Conference.

Table 1 Intercomparison between all sensor's pairs

	BIAS [%]	BIAS [K]	RMS [%]	RMS [K]
RS vs. IS	0,20	0,61	0,40	1,19
BASIL vs. IS	0,12	0,28	0,43	1,17
BASIL vs. RS	-0,06	-0,25	0,53	1,48
IS vs. MWR	0,26	0,74	0,32	0,94
RS vs. MWR	-0,25	-0,74	0,32	0,94
BASIL vs.MWR	-0,23	-0,62	0,60	1,71

References

[1] Whiteman, D.N., Demoz, B.B., Di Girolamo, P., Come,r J., Veselovskii, I., Evans, K., Wang, Z., Sabatino, D., Schwemmer, G., Gentry, B., Lin, R.-F., Behrendt, A., Wulfmeyer, V., Browell, E., Ferrare, R., Ismail, S.,Wang,J., 2006. Raman lidar measurements during the International H2O Project. Part II: Case studies. *J. Atmos. Oceanic Technol.*, **23**, 170–183.

[2] Behrendt, A., Wulfmeye,r V., Di Girolamo, P., Kiemle, C., Bauer, H.-S., Schaberl, T., Summa, D., Whiteman, D.N., Demoz, B.B., Browell, E.V., Ismail, S., Ferrare, R., Kooi, S., Ehret, G,, Wang, J., 2007a. Intercomparison of water vapor data measured with lidar during IHOP 2002. Part I: Airborne to ground-based lidar systems and comparisons with chilled-mirror hygrometer radiosondes. *J. Atmos. Oceanic Technol.*,**24**, 3–21.

[3] Bhawar, R.,Di Girolamo,P., Summa,D., Flamant, C., Althausen, D., Behrendt, A., Kiemle, C., Bosser, P., Cacciani, M., Champollion, C., Di Iorio, T., Engelmann, R., Herold, C., Müller, D., Pal,S., Wirth, M., Wulfmeyer, V., 2011: The Water Vapour Intercomparison Effort in the Framework of the Convective and Orographically-Induced Precipitation Study: Airborne-to-Ground-based and airborne-to-airborne Lidar Systems, *Quarterly Journal of the Royal Meterological Society*, **137**, 325-348.

[4] Di Girolamo, P., Summa, D., Ferretti, R.: Multiparameter Raman Lidar Measurements for the Characterization of a Dry Stratospheric Intrusion Event. *J. Atm. Ocean. Tech..*, **26**, 1742-1762, doi:10.1175/2009JTECHA1253.1, 2009a.

[5] Di Girolamo, P., Flamant, C., Cacciani, M., Richard, E., Ducrocq, V., Summa, D., Stelitano, D., Fourrié, N., and Saïd, F., (2016), Observation of low-level wind reversals in the Gulf of Lion area and their impact on the water vapour variability, *Quarterly Journal of the Royal Meterological Society.*, **142** (Suppl 1), 153–172, doi:10.1002/qj.2767.

[6] Di Girolamo, P., Cacciani, M., Summa,D., Scoccione, A., De Rosa, B., Beherendt A., and Wulfmeyer, V., (2017), Characterisation of boundary layer turbulent processes by the Raman lidar BASIL in the frame of HD(CP) 2 Observational Prototype Experiment, *Atmos. Chem.Phys.*,**17**, 745-767, doi:10.5194/acp-17-745-2017.

Rehearsal for Assessment of atmospheric optical Properties during biomass burning Events and Long-range transportation episodes at Metropolitan Area of São Paulo-Brazil (RAPEL)

Fábio J. S. Lopes[1,*], Juan Luis Guerrero-Rascado[2,3], Jose A. Benavent-Oltra[2,3], Roberto Román[2,3], Gregori A. Moreira[1,2,3], Marcia. T. A. Marques[1], Jonatan J. da Silva[1], Lucas Alados-Arboledas[2,3], Paulo Artaxo[4] and Eduardo Landulfo[1]

[1]*Center for Lasers and Applications (CLA), Nuclear and Energy Research Institute (IPEN), Brazil, *fabiolopes@usp.br*
[2]*Institute for Earth System Research (IISTA-CEAMA), Spain*
[3]*Dpt. Applied Physics, Faculty of Sciences, University of Granada, Spain*
[4]*Institute of Physics, University of São Paulo, Brazil*

ABSTRACT

During the period of August-September 2016 an intensive campaign was carried out to assess aerosol properties in São Paulo-Brazil aiming to detect long-range aerosol transport events and to characterize the instrument regarding data quality. Aerosol optical properties retrieved by the GALION - LALINET SPU lidar station and collocated AERONET sunphotometer system are presented as extinction/backscatter vertical profiles with microphysical products retrieved with GRASP inversion algorithm.

1 INTRODUCTION

According to the Intergovernmental Panel on Climate Change (IPCC) [1], anthropogenic aerosols are responsible for a negative radiative forcing for climate, widely contributing to cooling or warming the atmosphere through different processes such as aerosol-radiation and aerosol-cloud interactions. However, due to lack of observations and the difficult in measurements of important aerosol parameters, considerable uncertainties remain in assessments of long-term trends of global properties of aerosols. At this point, an important issue must be addressed: it is necessary to increase the number of observational stations and the experimental field campaigns to increase the global observation of aerosol optical and physical properties, and consequently, improve the climate numeric model validation. The development of aerosol measurement networks, with rigorous and systematic protocols of measurements is an important step to increase the reliability of aerosol measurements. In this context, the Latin America Lidar network (LALINET) [2], has been monitoring the vertical distribution of particle optical properties since 2001, and started to be operative with systematic measurements since 2013. LALINET has the purpose of developing an accurate database of vertical distribution of aerosol over South America providing data for air quality, information of aerosol microphysical properties and other atmospheric components such as water vapour and clouds, also, to provide quality assurance data for atmospheric satellite missions such as CALIPSO [4], and the future ESA's mission ADM-Aeolus and EarthCARE [3]. Focusing on the validation processes of ESA's mission, LALINET is working to implement quality assurance standards based on EARLINET procedures to increase the capability to provide a reliable dataset. During August and September of 2016 the LALINET São Paulo station (SPU) started a rehearsal performing a pre-project field experiment to gather a high quality dataset to assess the atmospheric optical properties during biomass burning events and long-range transportation episodes at Metropolitan Area of São Paulo (MASP)-Brazil (RAPEL campaign). During RAPEL were detected at least 5 episodes of biomass burning (BB) transportation. In this work the aerosol properties and microphysical properties using the GRASP (Generalized Retrieval of Atmosphere and Surface Properties) inversion algorithm [5] were retrieved.

2 INSTRUMENTATION AND METHODOLOGY

An elastic and Raman lidar system (MSP-Lidar I) installed at the Nuclear and Energy Research Institute (IPEN) was employed to measure the particle extinction and backscatter coefficients and, thus, the respective particle (extinction-to-backscatter) lidar ratio at 532 and 355 nm. The configuration of SPU lidar station for RAPEL campaign were composed by four detection channels, 355, 387, 532 and 607 nm, all in both AN and PC detection modes. The scheduled measurements in this particular campaign covered the period from 11th August to 30th September. Measurements were performed from Monday to Friday during daytime, and Monday and Thursday at nighttime. In total, the number of scheduled days was 37, split into 32 days with performed measurements (86.5%). The RAPEL campaign was coordinated during the Southern hemisphere winter season, considered the dry season at the Southeastern part of Brazil, where SPU lidar station is located. During the winter/dry season São Paulo city can experience high pollution episodes due poor dispersion conditions and aerosol biomass burning transportation from different parts of Brazil and South America continent. During the campaign there were at least 5 possible days with biomass burning transportation occurrence. The selected days with lidar measurements to be studied were analyzed to derive optical properties at the available wavelengths and particle backscatter coefficient profiles. In addition, ancillary information was used: optical and microphysical column integrated aerosol properties derived from a co-located AERONET sunphotometer, radiosoundings launched at Campo de Marte station, backtrajectories at 6 different heights to investigate the air masses origin from HYSPLIT model.

3 RESULTS

As a first step, AERONET aerosol optical depth (AOD) retrievals for cloud-free days during the campaign period are derived. The mean values and the standard deviation of AOD at 532 nm retrieved from AERONET were 0.22 ± 0.15 during August to September of 2016. For the RAPEL period, the number of focus fires in the brazillian territory increased around 65% compared to the wet season, according to brazillian National Institute For Space Research - INPE (http://www.inpe.br/queimadas/). Figure 1 shows the range corrected signal (RCS) at 532 nm measured by SPU station during a BB case on August 16th 2016. It can be seen that most part of aerosol are confined into the atmospheric boundary layer (ABL) region,which should be from local sources. However, there were some decoupled aerosol plumes above the PBL, between the altitude of 2 to 3.5 km (a.g.l). Applying the Klett-Fernald-Sasano inversion method by tuning the initial lidar ratio assumption with the AOD values retrieved from AERONET, particle backscatter profiles at 532 nm were retrieved, as can be seen in figure 1 (in red). For the mean profiles between 19:00 to 19:30 UTC and 19:30 to 20:00 UTC, a lidar ratio of $70 \pm 14 sr$ was obtained for both profiles. In order to investigate the aerosol plumes between 2 to 3.5 km detected by the SPU station, HYSPLIT trajectory model was used to calculate backward trajectories and derive information about from where, when and which altitude aerosols layers were transported to São Paulo. Five-day back-trajectories of air-masses starting at the SPU station were calculated using the GDAS database from Global Data Assimilation

System, for six different altitudes ranging from 1000 to 3800 m above ground level (a.g.l.). The back-trajectories starting at 19 UTC and with altitude level from 2000 to 2500 m a.g.l. came originally from North-Western region of Brazil, and the back-trajectories with altitude between 3000 to 3500 m a.g.l. came from Bolivia region. Both locations have intense burning events during this period of the year, thus it can be expected that BB aerosols were advected from this source region to São Paulo, as depicted in figure 2.

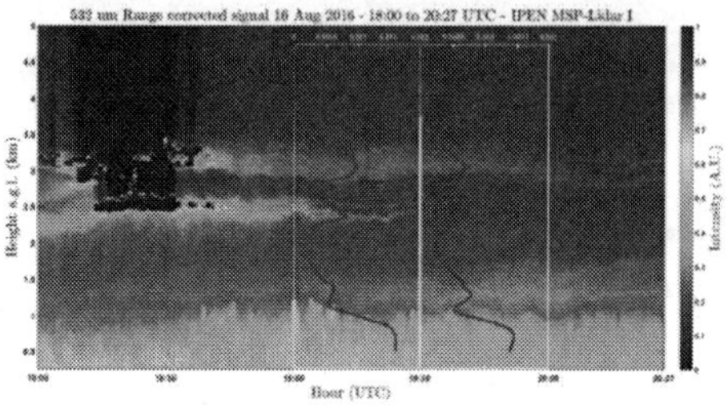

Figure 1: 532 nm range corrected signal on August 16th 2016 measured by SPU Station and the particulate backscatter profiles (red) retrieved during RAPEL.

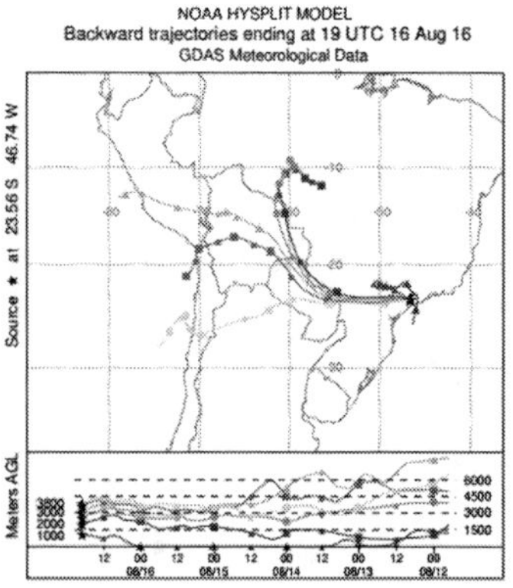

Figure 2: Five-day backward trajectories ending at SPU station at 19 UTC, at different levels 1000, 2000, 2500, 3000, 3500 and 3800 m a.g.l.) on August 16th 2016.

In order to derive complementary information of the aerosol load at São Paulo atmosphere combining the SPU lidar and AERONET retrieved data, GRASP inversion algorithm were used to retrieve optical and microphysical parameters on the vertical distribution and compared with lidar measurements, using as input the AOD and sky radiances at 440, 675, 870 an 1020 nm, and also the RCS at 355 and 532 nm. Figure 3 shows that the aerosol size distribution values obtained from GRASP and AERONET are in good agreement. It also shows that the size of the most part of aerosol presented in the atmosphere for this measurement days was fine mode aerosol, which is a characteristic of BB aerosol type.

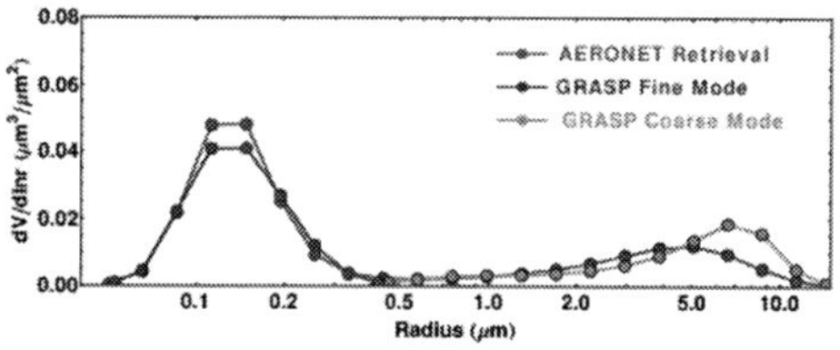

Figure 3: GRASP (blue and cyan) and AERONET (red) size distribution for August 16th 2016 (19:15 UTC) over SPU station.

Particle backscatter coefficient at 532 nm derived from GRASP shows that aerosol fine mode is predominant in the profile for this measurements, as can be seen in figure 4 (left panel). It also can be seen a good agreement between 532 nm total backscatter coefficient derived from GRASP and that on retrieved by the SPU station using the Klett-Fernald method. The 532 nm lidar ratio profile retrieved by GRASP inversion, right panel of figure 4, shows a mean lidar ratio of 75 sr for the aerosol plumes between 2 to 3.5 km of altitude, which is the typical value for aerosol from BB type. The vertical distribution of single scattering albedo shows values between 0.70 and 0.85, which can be considered low values for Brazilian BB aerosol type. However, considering that the aerosol plume were transported from the North-Western region of Brazil to São Paulo, their properties may have been modified due the increase of humidity at the São Paulo atmosphere and this hypothesis will be investigated.

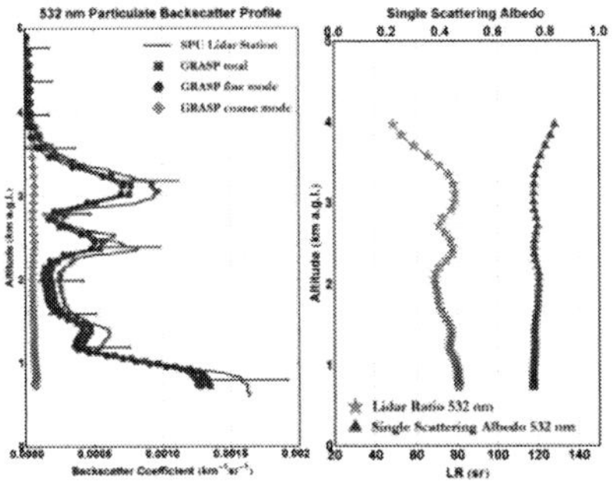

Figure 4: 532 nm particle backscatter profile and retrieved from GRASP (left panel) and 532 nm lidar ration and single scattering albedo profiles for August 16th *2016 (19:30 UTC) over SPU station.*

4 CONCLUSIONS

The work presented a case study of of biomass burning transportation detected during the RAPEL campaign in the framework of a pre-project field experiment in order to obtain a high quality data set to assessment of particle properties. Particle backscatter coefficient at 532 nm and HYSPLIT back-trajectories indicated the presence of aerosol plumes transported from Central-Western region of Brazil to São Paulo. Combination of sun/sky photometer and lidar measurements applied to GRASP algorithm show to be a power tool to derive microphysical parameters of aerosol on vertical scale. Also, GRASP results were in agreement with AERONET and lidar retrievals. GRASP 532 nm particle backscatter profile shows the predominance of fine mode aerosol type, and the lidar ratio profile retrieved from GRASP confirmed the presence of BB aerosol type.

ACKNOWLEDGEMENTS

The authors would like to thank for the support given by the following funding agencies: FAPESP (Projects 2015/12793-0 and 2016/09151-0), CNPq (Universal project 471798/2013-4). This work was supported by the University of Granada through "Plan Propio. Programa 9 Convocatoria 2013, by the Andalusia Regional Government through project P12-RNM-2409, by the Spanish Ministry of Economy and Competitiveness through projects CGL2013-45410-R and CGL2016-81092-R and by the European Union's Horizon 2020 research and innovation programme through project ACTRIS-2 (grant agreement No 654109).

References

[1] T. F. Stocker et al., 2013: Climate Change 2013: The Physical Science Basis. Contribution of Working Group I to the Fifth Assessment Report of the Intergovernmental Panel on Climate Change, *Cambridge University Press*, Cambridge, United Kingdom and New York, NY, USA, 1535 pp. doi:10.1017/CBO9781107415324.

[2] J. L. Guerrero-Rascado et al., 2016: Latin American Lidar Network (LALINET) for aerosol research: Diagnosis on network instrumentation, *J Atmos Sol-Terr Phy*, **138-139**, 112-120. Doi:10.1016/j.jastp.2016.01.001.

[3] A. Hélière et al., 2016: Development status of the EarthCARE Mission and its atmospheric Lidar, *Proc. SPIE, Earth Observing Systems XXI*, **9972**, 997215. Doi:10.1117/12.2235412.

[4] F. J. S. Lopes et al., 2013: Evaluating CALIPSO's 532 nm lidar ratio selection algorithm using AERONET sun photometers in Brazil, *Atmos. Meas. Tech.*, **6**, 3281-3299. Doi:10.5194/amt-6-3281-2013.

[5] O. Dubovik et al., 2011: Statistically optimized inversion algorithm for enhanced retrieval of aerosol properties from spectral multi-angle polarimetric satellite observations, *Atmos. Meas. Tech.*, **4**, 975-1018. Doi:10.5194/amt-4-975-2011.

REMOTE SENSING OF AEROSOLS BY SYNERGY OF CALIOP AND MODIS

Rei Kudo[1]*, Tomoaki Nishizawa[2], Akiko Higurashi[2], and Eiji Oikawa[3]

[1]*Meteorological Research Institute, JMA, Japan, *reikudo@mri-jma.go.jp*
[2]*National Institute for Environmental Studies, Japan*
[3]*National Institute of Information and Communication Technology*

ABSTRACT

For the monitoring of the global 3-D distribution of aerosol components, we developed the method to retrieve the vertical profiles of water-soluble, light absorbing carbonaceous, dust, and sea salt particles by the synergy of CALIOP and MODIS data. The aerosol product from the synergistic method is expected to be better than the individual products of CALIOP and MODIS. We applied the method to the biomass-burning event in Africa and the dust event in West Asia. The reasonable results were obtained; the much amount of the water-soluble and light absorbing carbonaceous particles were estimated in the biomass-burning event, and the dust particles were estimated in the dust event.

1 INTRODUCTION

Aerosols have significant impacts on the Earth's radiation budget. It is essential to investigate the spatial and temporal variability of the microphysical and optical properties of aerosols.

The remote sensing from the space has a role to monitor the aerosols in the global scale. The satellite-borne passive sensor, such as MODIS, provides the global distribution of the columnar properties of the aerosols [1]. The active sensor onboard the satellite, such as CALIOP, can capture the vertical profile of aerosols [2]. A synergistic remote sensing using both the passive and active sensors makes up for each other's weak points and have an ability to provide the better result of aerosols. For example, the aerosol vertical profile is assumed in the retrieval using the passive sensor although the radiance observed by the passive sensor depends on the vertical profile. The active sensor can provide the information of the vertical profile.

We are developing a new method to retrieve the vertical profiles of aerosol components by the synergy of CALIOP and MODIS. The data and the retrieval algorithm are described in Sect. 2. Two case studies are shown in Sect. 3. Finally, the results are summarized in Sect. 4.

2 METHODOLOGY

2.1 Data

The CALIOP data used in this study is the attenuated backscatter at 532 and 1064 nm, the depolarization ratio at 532nm, and the vertical feature mask of the CALIPSO Lidar Level 1B and 2 Products Version 3. The vertical resolution is 120 m. The horizontal resolution is 1 km. The MODIS data is the radiances in band 1 (620-670 nm) and 2 (841-876 nm) of Level 1B Calibrated Radiances (MYD021KM). The horizontal resolution is 1 km. We selected the radiance scattered from a point near the CALIOP observation and used it for the retrieval.

We used the auxiliary data in the forward modeling of the aerosol physical and optical properties, the CALIOP and MODIS data. The data is the pressure, the temperature, the relative humidity, and the surface wind speed on the ocean derived from the GEOS-5 meteorological data product. In calculating the radiance on the land, the surface albedo of the MODIS global albedo product (MCD43C3) was used.

2.2 Retrieval algorithm

We assumed four aerosol components, water-soluble (WS), light absorbing carbonaceous (LAC), dust (DS), and sea salt (SS). These are the major components in the atmosphere and have different particle sizes and light absorption characteristics. The vertical profiles of the extinction coefficients of four components are retrieved by simultaneously optimizing their vertical profiles of the dry volume concentrations, and their vertically mean dry mode-radii to the CALIOP and MODIS data. The optimization is conducted by the method developed by Kudo et al. [3]. This method optimizes the parameters by minimizing the objective function defined as

$$f(\mathbf{x}) = \frac{1}{2}\left(\mathbf{y}^{obs} - \mathbf{y}(\mathbf{x})\right)^T \left(\mathbf{W}^2\right)^{-1}\left(\mathbf{y}^{obs} - \mathbf{y}(\mathbf{x})\right) + \frac{1}{2}\mathbf{y}_a(\mathbf{x})^T\left(\mathbf{W}_a^2\right)^{-1}\mathbf{y}_a(\mathbf{x}) \quad , (1)$$

where vector $\mathbf{y}_{obs}$ is the CALIOP and MODIS observation data, vector $\mathbf{x}$ comprises the dry volume concentrations and the vertically mean dry mode-radii of WS, LAC, DS, and SS, vector $\mathbf{y(x)}$ is the calculated values by the forward model, matrix $\mathbf{W}$ is the standard errors in the observations, vector $\mathbf{y}_a$ is the constraints to $\mathbf{x}$, and matrix $\mathbf{W}_a$ is parameters to determine the strength of the constraints.

Our developed forward model consists of three parts. In the first part, the aerosol optical properties are calculated from the dry volume concentrations and the vertically mean dry mode-radii. Other two parts are the calculations of the CALIOP and MODIS data.

The particle shape of WS, LAC, and SS were assumed to be spherical. The DS was assumed to be the randomly oriented spheroids [4]. The LAC was assumed to be a core-gray shell model, which is an internal mixture model, and its optical properties are better than those of the core-shell model and the homogeneous internal mixture model [5]. We defined the WS as the shell around a core of LAC, and the shell-core ratio was fixed by 1.5. This lies among the values observed in the A-FORCE aircraft campaign [6].

The size distribution of each component was assumed to be a lognormal function, and the values of the standard deviations were given from the database of OPAC [7]. The vertically mean dry mode-radii of WS, LAC, and DS are parameters to be optimized. We assumed that the dry mode-radii of WS and LAC are same. The dry mode-radius of SS was parameterized by the surface wind speed on the ocean [8].

The refractive index of WS and SS were given form the OPAC [8]. The refractive index of LAC was given form the measurements reported in [9]. The refractive index of DS is given from the result of the ADEC project [10].

Before calculating the optical properties from the size distribution and the refractive index, the volumes and mode radii of WS, SS, and the shell of LAC are increased by the hygroscopic growth depending on the relative humidity of the ambient condition. Their growth factors and the refractive index for the particle mixed with the water were given from the database of OPAC [8].

The optical properties of WS and SS were calculated using Mie theory. The optical properties of LAC were calculated using the stratified sphere Mie code [11]. The optical properties of DS are calculated by using the data table developed by [4].

The attenuated backscatter coefficients for co- and cross polarization are calculated by the lidar equation. The radiances are calculated by PSTAR, a vector radiative transfer model developed by [12].

We constrained the range that the values of $\mathbf{x}$ can be by some methods. The smoothness of the vertical profiles of WS, DS, and SS are given by

$$y_a(\mathbf{x}) = \ln\left(V_j(z_{i-1})\right) - 2\ln\left(V_j(z_i)\right) + \ln\left(V_j(z_{i+1})\right), \quad (2)$$

where V is the dry volume concentration at the altitude z for j = WS, LAC, DS, and SS.

We assumed that the vertical profile of LAC has a similar shape to that of WS and introduced the shape constraint for LAC by,

$$y_a(\mathbf{x}) = \ln\left(V_{LAC}(z_i)/V_{LAC}(z_{i+1})\right) - \ln\left(V_{WS}(z_i)/V_{WS}(z_{i+1})\right). \quad (3)$$

In general, it is considered that the aerosol optical depth (AOD) of LAC is smaller than that of WS. This inequality was introduced by using the penalty function,

$$y_a(\mathbf{x}) = \begin{cases} -\ln\left(1 - \tau_{LAC}/\tau_{WS}\right) & for \quad \tau_{LAC} < \tau_{WS} \\ +\infty & otherwise \end{cases}, \quad (4)$$

where τ is the aerosol optical depth.

When the AOD of LAC approaches to the AOD of WS, the penalty function rapidly increases.

The retrieval of the dry mode radii of WS, LAC, and DS is very challenging. In the numerous tests of our algorithm using the simulated and actual data of CALIOP and MODIS, the mode-radii can be retrieved from the noiseless data of CALIOP and MODIS. The retrieval of the mode-radii was influenced by the noise of CALIOP data. For the stable retrieval, the a priori values of the dry mode-radii for WS, LAC, and DS are given by,

$$y_a(\mathbf{x}) = \ln(r_j) - \ln(r_{a,j}), \quad (5)$$

where r is the dry mode-radii, and r_a is a priori values for j = WS, LAC, and DS.

Based on the above mentioned forward model and constraints, the dry volume concentrations and the vertically mean dry mode-radii of WS, LAC, DS, and SS are optimized to the CALIOP and MODIS data. Finally, the extinction coefficients of each component are output.

3 RESULTS

Two case studies of the biomass-burning and dust events are presented. Figures 1 and 2 show the case of the biomass-burning event in Africa. The CALIPSO satellite passed by the region where the fires are active (Figure 1). The retrieved extinction coefficients are characterized by the large values of WS and LAC (Figure 2). In general, the large amount of the fine particles and the strong light absorbing property are observed in the biomass-burning event.

Figures 3 and 4 show the case of the dust event in West Asia. The CALIPSO passed by the desert region (Figure 3). The retrieved extinction coefficient was characterized by the large value of WS and DS.

4 CONCLUSIONS

We developed the synergistic method to retrieve the vertical profiles of aerosol components from the CALIOP and MODIS data. The products are the extinction coefficients of WS, LAC, DS, and SS. Two case studies for the biomass-burning and dust events showed the consistent results with each scene. We will apply the method to the long-term data of CALIOP and MODIS to investigate the global 3-D distributions of aerosol components.

We plan to apply our method to the high spectral resolution lidar and the imaging spectral radiometer onboard the EarthCARE satellite (Earth Clouds, Aerosols and Radiation Explorer) [13]. The lidar ratio derived from the high spectral resolution lidar is expected to improve the retrieval of LAC because the lidar ratio significantly depends on the light absorption particles.

ACKNOWLEDGEMENTS

This work was supported by the Japan Society for the Promotion of Science KAKENHI grant no. 15H02808.

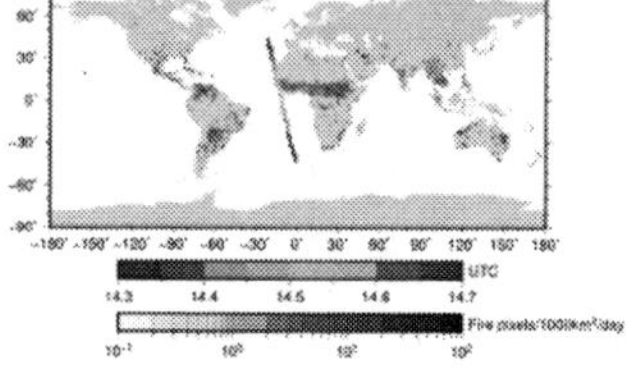

Figure 1 The orbit of CALPSO in 29 Jan, 2010, and the fire activity from 25 Jan to 2 Feb, 2010. The color of the orbit is time (UTC). The fire activity is based on data of MODIS active-fire product.

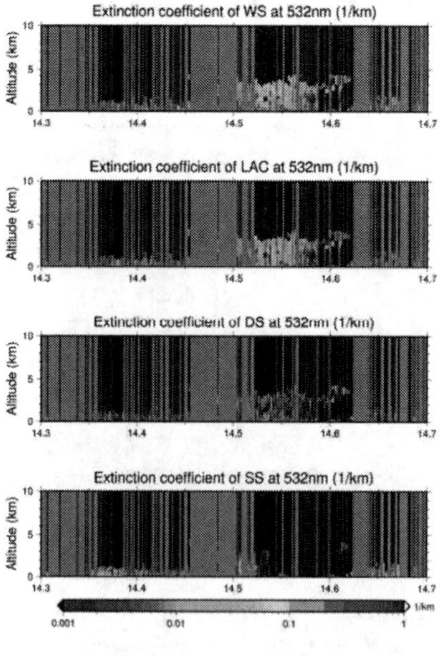

Figure 2 Retrieved extinction coefficients of WS, LAC, DS, and SS in the biomass-burning event. The gray

region indicates the cases that include clouds. Our method cannot be applied to such cases.

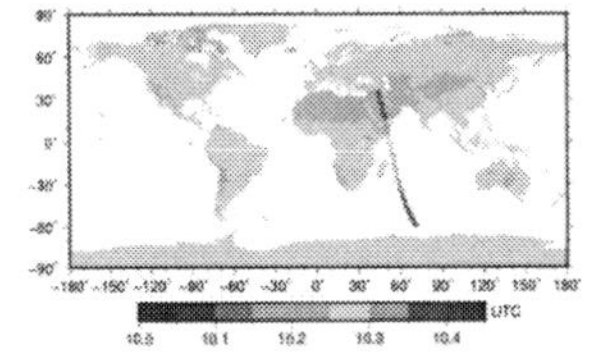

Figure 3 The orbit of CALPSO in 30 Jan, 2010. The color of the orbit is time (UTC). The ochre region indicates barren or sparsely vegetated region.

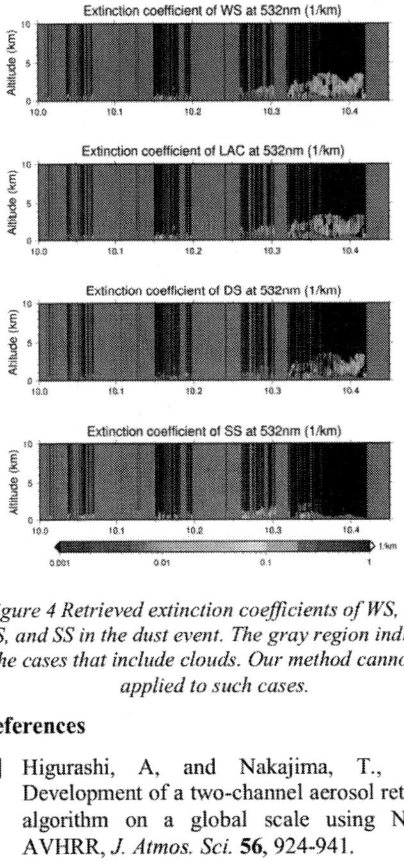

Figure 4 Retrieved extinction coefficients of WS, LAC, DS, and SS in the dust event. The gray region indicates the cases that include clouds. Our method cannot be applied to such cases.

References

[1] Higurashi, A, and Nakajima, T., 1999: Development of a two-channel aerosol retrieval algorithm on a global scale using NOAA AVHRR, *J. Atmos. Sci.* **56**, 924-941.

[2] Winker, D. M., et al., 2009: Overview of the CALIPSO mission and CALIOP data processing algorithms, *J. Atmos. Oceanic Technol.* **26**, 2310-2323.

[3] Kudo, R., et al., 2016: Vertical profiles of aerosol optical properties and the solar heating rate estimated by combining sky radiometer and lidar measurements, *Atmos. Meas. Tech.* **9**, 3223-3243.

[4] Dubovik, O., et al., 2006: Application of spheroid models to account for aerosol particle nonsphericity in remote sensing of desert dust, *J. Geophys. Res.* **111**, D11208, doi:10.1029/2005JD006619.

[5] Kahnert, M., et al., 2013: Models for integrated and differential scattering optical properties of encapsulated light absorbing carbon aggregates, *Opt. Express* **21**, 7, 7974-7993.

[6] Matsui, H., et al., 2013: Development and validation of a black carbon missing state resolved three-dimensional model: Aging process and radiative impact, *J. Geophys. Res.* **118**, 2304-2326.

[7] Hess, M., et al., 1999: Optical properties of aerosols and clouds: The software package OPAC, *Bull. Amer. Metor. Soc.* **79**, 831-844.

[8] Erickson, D. J., and Duce, R. A., 1988: On the global flux of atmosphere sea salt, *J. Geophys. Res.* **93**, 14079-14088.

[9] Chang, H., and Charalampopoulos, T. T., 1990: Determination of the wavelength dependence of refractive indices of flame soot, *Proc. R. Soc. Lond.* A **430**, 577-591.

[10] Aoki, T., et al., 2005: Sensitivity experiments of direct radiative forcing caused by mineral dust simulated with a chemical transport model, *J. Meteor. Soc. Jpn.* **83A**, 315-331.

[11] Toon, O. B., and Ackermann, T. P., 1981: Algorithms for the calculation of scattering by stratified spheres, *Appl. Opt.* **20**, 3657-3660.

[12] Ota, Y., et al., 2010: Matrix formulations of radiative transfer including the polarization effect in a coupled atmosphere-ocean system, *J. Quant. Spect. Rad. Trans.* **111**, 6, 878-894.

[13] Illingworth, A. J., et al., 2015: The EARTHCARE satellite: the next step forward in global measurements of clouds, aerosols, precipitation, and radiation, Bull. Am. Meteor. Soc., 96, 1311-1332.

https://doi.org/10.1051/epjconf/201817608013

BIOMASS BURNING AEROSOLS CHARACTERIZATION FROM GROUND BASED AND PROFILING MEASUREMENTS

Cristina Marin[1,2,*], Jeni Vasilescu[1], Luminita Marmureanu[1], Dragos Ene[1], Liliana Preda[2], Mona Mihailescu[2]

*[1]National Institute of Research and Development for Optoelectronics, *cristina.marin@inoe.ro*
[2] University Politehnica of Bucharest, Faculty of Applied Sciences, Bucharest, Romania

ABSTRACT

The study goal is to assess the chemical and optical properties of aerosols present in the lofted layers and at the ground. The biomass burning aerosols were evaluated in low level layers from multi-wavelength lidar measurements, while chemical composition at ground was assessed using an Aerosol Chemical Speciation Monitor (ACSM) and an Aethalometer. Classification of aerosol type and specific organic markers were used to explore the potential to sense the particles from the same origin at ground base and on profiles.

1 INTRODUCTION

Biomass burning aerosol (BBA) represents animportant part of the total aerosol emission, having both natural and anthropogenic sources. Zhang et al. [1] estimated that BBA represents 13% of the $PM_{2.5}$, having seasonal variability, with greater fraction in the winter. Biomass burning emissions include BBA, which is primary composed of black carbon (BC) and organic carbon (OC); and important trace gases such as CO_2, CO, SO_2[2]. The BBA effects include changes in the Earth radiative budgets [2] and human health [3].

Giving the spatial and temporal variability of aerosol distribution and composition, it is necessary to use assemble of different types of observation from passive and active techniques measurements, in order to achieve a good characterization of the aerosol.A major step forward and a challenge are represented nowadays by automatic aerosol classification using optical parameters. Lidar (Light detection and ranging) measurements provide microphysical and optical properties which permit aerosol classification or clustering [4].

Good options for automatic pattern recognition are neural networks, due to their unique mathematical approach. These have been already tested and extensively used in many domains related to classification such as protein structure, organic disorders, medicine and pharmacy [5]. The aerosol classification using only aerosol intensive optical parameters derived from multiwavelength lidar data as input, permit a fast identification of layers and their pattern[6]. New algorithms combine lidar profiles with column integrated measurements in order to retrieve aerosol concentration profiles such as: Liric[7].

For ground based measurements, spectral analysis [8]and mass spectrometry [9]are important tools to investigate aerosol composition, special ion markers being used for BBOA (biomass burning organic aerosol) assessment.

In this paper we evaluate the possibility of sensing the particles from the same origin in lofted layers measured by the lidar, and at ground base, by analyzing the chemical composition.

2 METHODOLOGY

Remote sensing and in-situ instruments present at the Bucharest, Romania site (26.029°E, 44.048°N, 93m above sea level) and corresponding measurements are used in order to characterize from both optical and chemical point of view the aerosol, from May to September 2016. The multiwavelenght lidar system, RALI, part of EARLINET is based on a 3 wavelength Nd:YAG laser and detection on 12 channels using both analogue and photoncounting [6].

In this paper, an algorithm for automatic aerosol classification from multiwavelength lidar data, based on Artificial Neural Networks, have been applied. The lidar data are used to calculate the intensive properties such as Ångstrom exponent (AE), color ratios, color indexes and lidar ratios, which are further presented to highly trained neural networks which pick up the most probable type of aerosols from several already defined

https://doi.org/10.1051/epjconf/201817608013

classes[6]. A good classification process is strongly dependent on the physical content of the optical inputs and on their uncertainty.

The processing algorithm based on remote sensing measurements, used in this paper is LIRIC (Lidar/Radiometer Inversion Code)[7]. It combines pre-processed lidar, and column AERONET sunphotometer inversion data[7] in order to retrieve aerosols volume concentration profile for fine and coarse mode. The algorithm uses elastic lidar channels at 1064 nm, 532nm, 355 nm and when available the depolarization measurements (channel 532 cross). In this case, the aerosol in the coarse mode is divided into spherical coarse and spheroid coarse, besides the fine mode. The separation between the fine and coarse mode is made by finding the minimum radii from the AERONET size distribution in the 0.194-0.576 μm range. Also the separation for the coarse mode is done using the sphericity parameter.

At ground base, chemical characterization for aerosol is achieved with an ACSM[10]. Using mass spectrometry principles, aerosol species are divided into organics (Org), ammonium(NH_4), nitrate(NO_3), chloride (Chl)and sulfate(SO_4).The presences of BBA from mass spectrometry is evaluated using the f60 specific marker(ratio between m/z60 concentration and organics concentration), due to signal at m/z60 is attributed to levoglucosan and other "levoglucosan-like" species resulted from the pyrolysis of cellulose[9]. For a qualitatively evaluation of the aerosol degree of oxidation, two other markers are used: f43 and f44 (computed similarly as f60). An increase in f44 and a decrease in f43 indicates changes in aerosol volatility, thereby indicates the presence of oxidized aerosol [9].

Light absorption at seven wavelengths by aerosol particles is the principle of the Aethalometer[8]. Source apportionment of BC can be retrieved based on wavelength dependency of the absorption coefficient for different types of aerosols [8]. For the current paper, source apportionment for biomass burning and fossil fuel has been evaluated.

To assess the possible origin of upper layers we used forward dispersion model FLEXPART[11], based on fire emission rate estimated from Modis orbital satellites. FLEXPART uses meteorological data as input for the Lagrangian particle dispersion model and it can provide backward and forward simulations of particle trajectories.

3 RESULTS

Five months, from May to September 2016, of lidar, mass spectrometry and absorption measurements have been analyzed. At ground the measurements are characterized by higher average concentration for organics (5.4 $\mu g/m^3$) than the other chemical species (1.7 $\mu g/m^3$ for NH_4, 0.77 $\mu g/m^3$for NO_3 and 2.12 $\mu g/m^3$ for SO_4), while spectral analysis retrieves average concentrations of 0.08 $\mu g/m^3$ for BBA and 1.15 $\mu g/m^3$for fossil fuel BC. Two major peaks for BBA concentration are noticeable on 12th of July (2.52 $\mu g/m^3$) and 30th of September (2.01 $\mu g/m^3$).

The lidar retrievals and automatic classification algorithm highlight that fine and low depolarizing particles are predominant components of the lower layers during the measurements period, representing approximately 70% of the analyzed cases. During the five months of measurements, only few cases of proper meteorological conditions which permit aerosol mixing and lofted layers descending to the ground have been encountered. One of these study cases is identified on 29th September 2016, and is further analyzed.

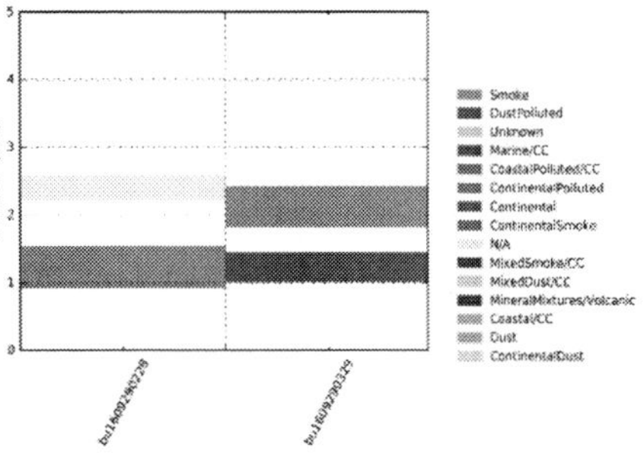

Figure 1.Types of aerosols retrieved using automated classification algorithm on 29th September 2016

During 28th to 30th September 2016 the BBA aerosols have been present in lofted layers and at the ground. The automatic classification algorithm pin points the presence of smoke, continental and continental polluted as predominant components characteristic for the lower layers (*Figure 1*).

For the LIRIC algorithm computed for 29th September 2016 only elastic channels without depolarization measurements have been used. The vertical distribution of coarse and fine aerosol and vertical profiles of backscatter coefficient are illustrated in *Figure 2*. Fine particles are dominant on the entire vertical profile, but a layer of fine aerosol is emphasized from 1 to 2 km. This result is in accordance with the layer of BBA retrieved by the automatic classification algorithm.

At ground level, the average mass concentrations measured for organics, sulfate, nitrate and ammonium were 7.54 μg/m^3, 1.82 μg/m^3, 1.6 μg/m^3, and 2.6 μg/m^3 respectively.Around 70% of the total aerosol sampled exceeded the background level of f60 marker for BBA, while theaging markers f44 and f43 indicates mainly the presence of particles with low volatility, and also semi-volatility attributed to oxidized aerosol. A special interest is given to last aerosol sampled on 29th September, 22.10 UTC, characterized by high organics concentration, and semi-volatile BBA aerosol. The sampled aerosols are classified as oxygenated BBA, which are characteristic for BBA or smoke aerosols emitted in the atmosphere by vegetation fires and could by associated with long-range transport.

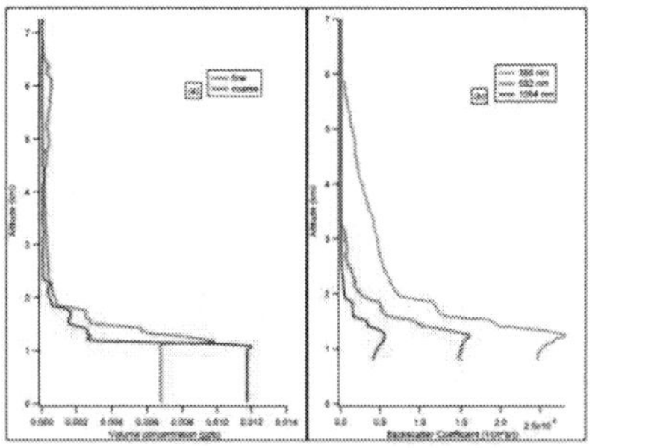

Figure 2. a) LIRIC retrieval for 29th September. Altitude vs. volume concentration for fine (blue) and coarse (red) aerosol; b) Vertical profiles of backscatter coefficient at 355 nm (purple), 532 nm (green) and 1064nm (red) derived from lidar measurements for 29th September 2016

The organic fraction at m/z60 and black carbon (BC) emitted from biomass burning retrieved from Aethalometer (based on differential abortion at 370 and 880 nm) have the same appearances highlighting the presence of BBA (*Figure 3*). The time series, presented in UTC time, illustrates a diurnal trend, with higher values in the afternoon (17-18 UTC), and minimum values at noon.

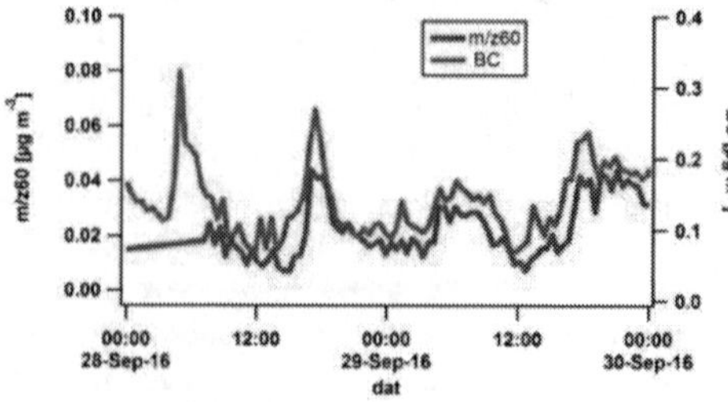

Figure 3. Time series for m/z60 (red) and BC (green) mass concentration

The atmospheric layers from long range transport in the sample site were evaluated using the FLEXPART model. Specific sets of particles for BBA, based on a mixture between organics and BC was used to compute the dispersion from sources. The model was set-up assimilating Aqua and Terra fire spot satellite data, and using fire radiative power (FRP) retrieval as input in the model that was transformed in emission rate, using emission coefficient established by Ichoku et al., 2014[12]. The period assimilated was between 27th and 29th September when hot spots with high degree of confidence and significant value of FRP was taking into account.

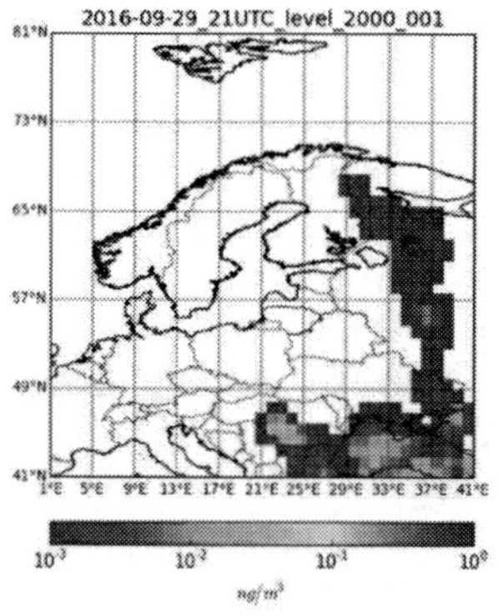

Figure 4. BBA dispersion retrieved from FLEXPART, on 29th September 2016, at 2000m altitude

Important satellite thermal anomalies have been detected in the south part of Europe that influence the upper layers at 2000 m as shown in *Figure 4.* The output model also underlines influences from long range transport at ground base100m (not shown).

4 CONCLUSIONS

Independent assessments of biomass burning aerosols presence at the Bucharest site during May-September 2016 have been assessed using in situ and remote sensing measurements. Chemical composition, specific markers and differentiated spectral absorption pin pointed the BBA characterizes, including oxidation state and potential source highlighting the possible long-range transport. In the same time the low level layers of aerosols have been evaluated from multi-wavelength lidar measurements. The LIRIC algorithm identified the fine particle volume concentrations, while automatic classification algorithm pin pointed the predominant component.The 29th September 2016 case is intensely analyzed proving the potential to sense the same BBA particles at ground base and on profiles.

ACKNOWLEDGEMENTS

The research leading to these results has received funding from the European Union's Horizon 2020 Research and Innovation Programme, under Grant Agreement no 692014-ECARS and from the National Core Program- PN 16 40 01 01.

References

[1] Zhang, X., Hecobian,A., Zheng, M., Frank, N. H., Weber, R. J., 2010: Biomassburning impact on PM2.5 over thesoutheastern US during 2007: integrating chemically speciated FRM filter measurements, MODIS fire counts and PMF analysis, *Atmos. Chem. Phys*. **10**, 6839–6853.

[2] Jacobson, M. Z., 2014: Effects of biomassburning on climate, accounting for heat and moisture fluxes, black and brown carbon, and cloud absorption effects, *J. Geophys. Res. Atmos.*, **119**, 8980–9002.

[3] Sigsgaard, T., et al., 2015: Health impacts of anthropogenic biomass burning in the developed world, *EurRespir J.*,**46**(6):1577-1588.doi: 10.1183/13993003.01865-2014.

[4] Groß, S., Esselborn, M., Weinzierl, B., Wirth, M., Fix, A., Petzold, A., 2013: Aerosol classification by air borne high spectral resolution lidar observations, *Atmos. Chem. Phys.*, **13**, 2487–2505.

[5] Li, H., Liu, C., Burge, L., Southerland, W., 2012: Classification of Protein 3D StructuresUsing Artificial Neural Network, *IJMLC*,**2**(6).

[6] Nicolae, D., Vasilescu, J., Talianu, C., Dandocsi, A., 2016: Independent retrieval of aerosol type from lidar,*EPJ Web of Conferences*, http://dx.doi.org/10.1051/epjconf/201611918002.

[7] Chaikovsky, A.,et al., 2016: Lidar-Radiometer Inversion Code (LIRIC) for theretrieval of vertical aerosol properties from combined lidar/radiometer data: development and distribution in EARLINET, *Atmos. Meas. Tech.*, **9**, 1181–1205.

[8] Lewis,K., Arnott, W.P., Moosmüller, H., Wold, C.E., 2008: Strong spectral variation of biomass smoke light absorption and single scattering albedo observed with a novel dual-wavelength photoacoustic instrument, *J. Geophys. Res*, **113**, D16.

[9] Adler, G.,Flores, J.M., AboRiziq, A., Borrmann, S., Rudich, Y., 2011: Chemical, physical, andopticalevolution of biomassburningaerosols: a case study, *Atmos. Chem. Phys.*, **11**, 1491–1503.

[10]Ng, N.L., Herndon, S. C., Trimborn, A., Canagaratna, M.R., Croteau, P.L., Onasch, T.B., Sueper, D, Worsnop, D.R., Zhang, Q., Sun, Y.L., Jayne, J.T., 2011: An Aerosol Chemical Speciation Monitor (ACSM) for Routine Monitoring of the Composition and Mass Concentrations of Ambient Aerosol, *Aerosol Sci. Technol.*, **45**, 780–794.

[11]Stohl, A., Forster, C., Frank, A.,Seibert, P.,Wotawa, G., 2005:Technical note: The Lagrangian particle dispersion model FLEXPART version 6.2, *Atmos. Chem. Phys.*, **5**, 2461-2474.

[12]Ichoku, C., and Ellison, L., 2014: Global top-downsmoke-aerosol emissions estimation using satellite fire radiative power measurements, *Atmos. Chem. Phys.*, **14**(13), 6643–6667.

Aerosol Characterization from Active and Passive Ground Measurements and Satellite Data

Dandocsi Alexandru[1,2], Sapartoc Georgiana[2], Preda Liliana[2], Stan Cristina[2], Radu Cristian[1]

*[1]National Institute of R&D for Optoelectronics, Romania, *alexandru.dandocsi@inoe.ro*
[2]University Politehnica of Bucharest, Romania

ABSTRACT

One year records of AErosolROboticNEtwork (AERONET) sun photometer measurements were analyzed to investigate the seasonal and daily variations of columnar aerosol optical depth. Some irregularities of this time series are associated with aerosol intrusions. The aerosol layers indicated by these irregularities are identified and characterized using the extensive optical data from coincident CALIPSO satellite observations and ground based LIDAR.

INTRODUCTION

Long term Aerosol Optical Depth studies show interesting influences of natural and anthropogenic aerosol sources. Studies shows daily frequencies as well as weekly or seasonal [1] as well as natural influences like diurnal cycle of the planetary boundary layer (PBL).

Location of the site is in Magurele, a suburban area of Bucharest (26.029E; 44.048N, 93m above sea level). In such location, one can expect to find the urban influence as well as the natural influences of continental climate.

In this study, we use time series analysis to determine the time periods specific to our location and identify some irregularities of the measured AOD values. These irregularities can be explained with the presence of clouds and/or long range transported aerosol layers.

One year record of level 1.5 AOD data from AERONET were used in this analysis and validate the results with level 2 lidar data from Calipso and measurements from a ground based Lidar. Both sun photometer and the ground based Lidar are collocated while the satellite measurements were selected around a 50 km region around the site.

1 METHODOLOGY

One year measurements of the AOD from CIMEL's sun photometer were used in this study, during 2014. The AOD retrieval is computed based on the Beer-Lambert-Bouger equation [2].

The data is measured at various time stamps because of the weather conditions and has missing values. To apply the decomposition algorithm, which gives the frequencies in the AOD data, the time series has to be continuous. In order to identify the small periods and to reduce the error in the same time, we used a sampling rate of 1 for each 12 hours, by averaging the measurements done in first and last 12 hours for each day. Summer time is the most suitable period in terms of AOD because only a specific range of isolated data are missing and it can be easily interpolated using a shape-preserving piecewise cubic interpolation without changing the time evolution. The data shows an almost linear trend probably due to shorter frequencies which can be associated with seasonal variations. Applying the detrend, we estimated a power spectrum density which gives the frequencies of the series.

Removing the frequencies caused by the natural and the anthropogenic influences of the atmosphere, we obtained the irregularities in the AOD measurements. These irregularities can be explained with the presence of long range transported aerosols and/or clouds. To validate this statement we used satellite lidar data and ground based lidar measurement for the given location.

Backscatter and extinction profiles from a multiwavelength Lidar measurements, described by [3], were used to highlight the presence of aerosol layers or clouds and to compare with the measurements from Calipso. Both Lidar and Calipso measurements were used to validate the presence of the irregularities in the AOD data. Calculation of the profiles was made with a good

 https://doi.org/10.1051/epjconf/201817608014

accuracy using an assumed lidar ratio and a calibration in the far range **[4]**,**[5]**.

Range corrected signals were used to show an overview of the atmosphere while backscattered profiles were used to indicate the presence of aerosols, using the gradient method described by **[6]**.

From the satellite data point of view, profiles from Lidar data from Level 2, were used. Selection of the profiles was made with a location constraint, only data within 50 km range from the fixed location were considered.

2 RESULTS

One year measurements of the AOD, 500 nm filter, aerosol related from the sun photometer is shown in Figure 1.Sun photometer measurements for the AOD have a maximum 5% error [7].

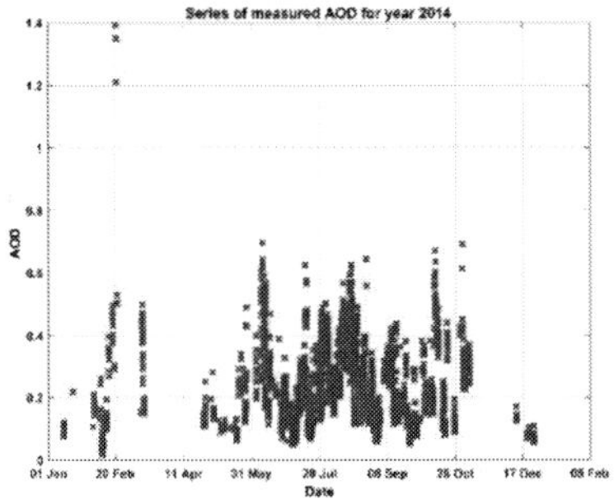

Figure 1 AOD measurements from sun photometer for 2014

From this time series, the most continuous measurements are during summer time for which we made interpolated the missing values using a cubic function. The power spectrum of this series is presented in Figure 2 and shows the following characteristic periods: 20 days, 10 days, 5 days and some smaller components to 6.7 days, 2.9 and 2.4 days. These latter frequencies are most probably linked to urban aerosol due to weekly traffic variations [8]

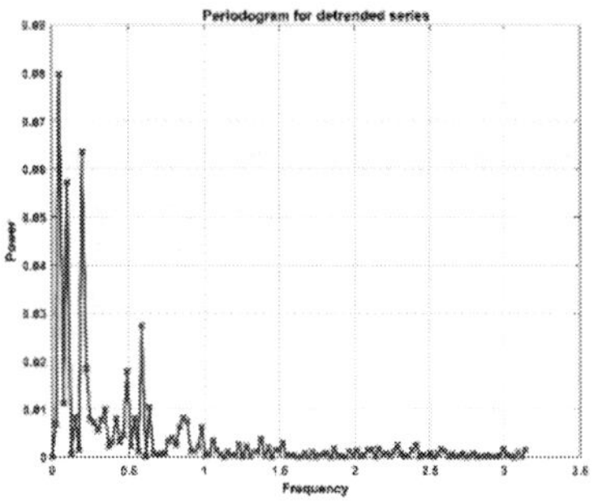

Figure 2 Periodogram for the AOD data

Eliminating the frequencies from the natural and anthropogenic influences we focused on one case (Figure 3)

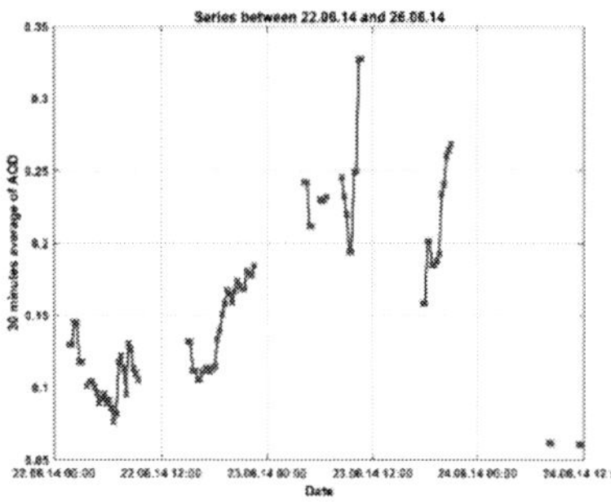

Figure 3 Case of 23.06 - 25.06 a half an hour mean data

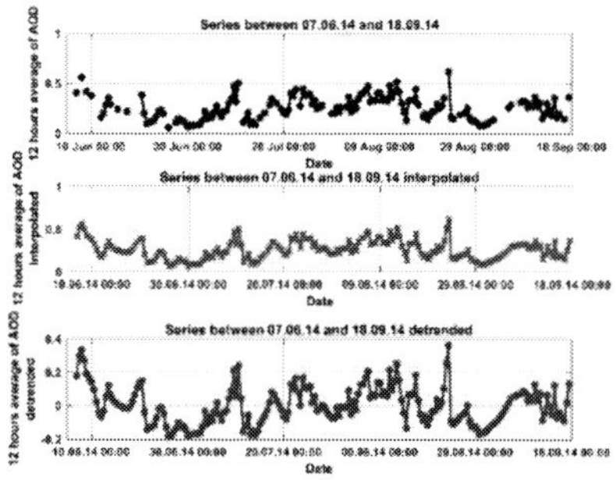

Figure 4 Series of AOD averaged for 12 hours (I), interpolated (II) and detrended (III)

In this case, range corrected signal quicklook (Figure 4) shows the planetary boundary layer in the lower region (between 0.5 and 1 km) while in the free one can expect relatively higher values in the backscatter profile due to the presence of a diffuse aerosol layer..

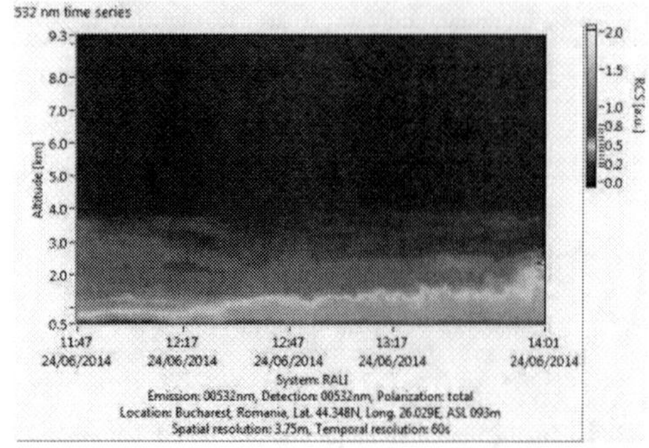

Figure 4 Range Corrected signal for 24th of June 2014

Visualization of the backscatter coefficient profiles at 532 nm channel from the ground based lidar system and Calipso can be seen in Figure 5. Both systems are in a good agreement showing higher values of backscatter coefficient in the 3 km – 4 km region.

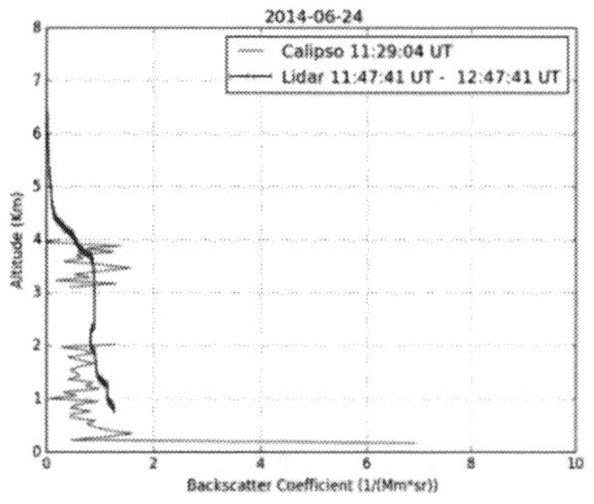

Figure 5 Backscatter coefficient from both Calipso and Lidar measurements from 24th of June 2014

CAD score for the Calipso profiles show values between -97 and -100 which indicates the presence of aerosols with high confidence.

This region is due to long range transport aerosol layer and can be validated using the backwards trajectories from the HYSPLIT model which shows the source (Figure 6) for 2, 3 and 4 km altitudes sources.

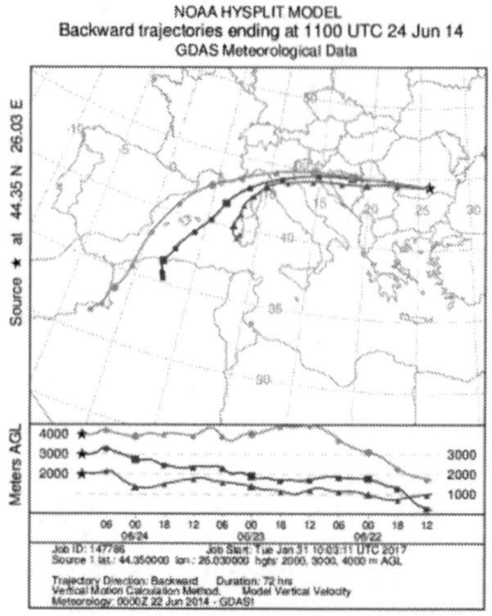

Figure 6 HYSPLIT backwards trajectories for 24.06.2014

AOD values obtained from the Lidar profile have a value of 0.1908±0.0003 while from the Calipso extinction coefficient profile the values has a value of 0.1568±0.0629. The AOD values obtained from the sun photometer in same time period shows a mean value of 0.2028±0.0026 and a variability between 0.1848 and 0.2228. As expected, the values are lower than both lidar systems, ground based and from Calipso because neither lidar systems have the information for all the atmospheric column [9].

Earlier that case, another range corrected signal quicklook (Figure 7) show similar atmospheric condition with a long range transporter aerosol layer which can be validated by running the backwards trajectories provided by HYSPLIT model (not shown).

Figure 7 Range Corrected Signal from ground based lidar

The aerosols shown in the 3 km region acted as cloud nuclei in the last time period of the measurement, the radiation from the lidar system being highly backscattered.

3 CONCLUSIONS

Using a high rate of sampling for AOD data measured in the summer of 2014, we identify long periods variations, 20 and 10 days driven by local climatic conditions and shorts periods of AOD variations (5, 2.9 and 2.4days) due to induced air pollution and due to intense weekly traffic of Bucharest.

The irregularities shown in the AOD time series from sun photometer data can indicate extreme cases of long range transport of aerosol layer and clouds formation.

In the case presented in this paper, satellite data from Calipso and ground based Lidar measurements were used to validate not only the Planetary Boundary Layer height, but also the presence of long range transported aerosols.

Also, the values obtained for AOD from both Calipso and ground based lidar systems and sun photometer data are in a good agreement showing that the validation made with both lidar systems was a suitable one.

Earlier ground based lidar measurement of that case showed that the aerosol layer found acted as nuclei for cloud formation.

ACKNOWLEDGEMENTS

Satellite data were obtained from the NASA Langley Research Center Atmospheric Science Data Center.

This work has been supported by the European Union's Horizon 2020 Research and Innovation Programme, under the Grant Agreement no 692014 – ECARS, by the National Core Program-PN16.40.01.01/2016, by ESA contract 4000113511/NL/FF/gp- AROMAT. Also this project has received funding from the European Union's Horizon 2020 research and innovation programme under grant agreement No 654109, ACTRIS 2.

References

[1] Xia, X., T. F. Eck, B. N. Holben, G. Phillippe, and H. Chen 2008, *Analysis of the weekly cycle of aerosol optical depth using AERONET and MODIS data*, J. Geophys. Res., 113, D14217, doi:10.1029/2007JD009604

[2] Dubovik, O. and M. D. King, 2000: A flexible inversion algorithm for retrieval of aerosol optical properties from Sun and sky radiance measurements," *J. Geophys. Res.*, **105**, 20 673-20 696

[3] Nemuc, A., Nicolae, D., Talianu, C., Carstea, E., Radu, C. (2009). Dynamic of the lower troposphere from multiwavelengthlidar measurements. *Rom. Rep. Phys.*, 61, 313-323;

[4] Fernald, F.G., Herman, B. M., and Reagan, J. A. (1972). *Determination of aerosol height distribution of Lidar* J. Appl. Meteorol, 11, 482 – 489;

[5] Klett, J.D. 1981, *Stable Analytical Inversion Solution for Processing Lidar Returns,* Appl. Opt, 20, 211 – 220;

[6] Belegante, L., et al., 2004, *Retrieval of the boundary layer height from active and passive remote sensors. Comparison with a NWP model,*.ActaGeographica, 62, 2, 276 – 289;

[7] Dubovik, O., B. N. Holben, T. F. Eck, A. Smirnov, Y. J. Kaufman, M. D. King, D. Tanré, and I. Slutsker, "*Variability of absorption and optical properties of key aerosol types observed in worldwide locations*", J. Atmos. Sci.,59, 590-608, 2002b.

[8] Dominique Ba¨umer and Bernhard Vogel, 2007, *An unexpected pattern of distinct weekly periodicities in climatologica lvariables in Germany,* Geophysical Research Letters, vol. 34, L03819, doi:10.1029/2006GL028559

[9] A. Nemuc, I. Binietoglou, S. Andrei, A. Dandocsi, H. Stefanie, 2016, *Multiyear aerosol study based on lidar & sun photometer measurements in Romania,* The European Physical Journal Conferences 119:24001, DOI: 10.1051/epjconf/201611924001

A Calibration of the MeteoSwiss RAman Lidar for Meteorological Observations (RALMO) Water Vapour Mixing Ratio Measurements using a Radiosonde Trajectory Method

Shannon Hicks-Jalali[1*], R.J. Sica[1,2], Alexander Haefele[2,1] and Giovanni Martucci[2]

[1]*Department of Physics and Astronomy, The University of Western Ontario, London, Canada,*
**shicks26@uwo.ca*
[2]*Federal Office of Meteorology and Climatology, MeteoSwiss, Switzerland*

ABSTRACT

With only 50% downtime from 2007-2016, the RALMO lidar in Payerne, Switzerland, has one of the largest continuous lidar data sets available. These measurements will be used to produce an extensive lidar water vapour climatology using the Optimal Estimation Method introduced by Sica and Haefele (2016). We will compare our improved technique for external calibration using radiosonde trajectories with the standard external methods, and present the evolution of the lidar constant from 2007 to 2016.

1 INTRODUCTION

Water vapour is the dominant greenhouse gas and it plays a prominent role in atmospheric chemistry and dynamics. Water vapour distribution is controlled by and controls upper troposphere and lower stratosphere (UTLS) dynamics. Water vapour density is extremely variable both spatially and temporally and can exceed variations over 4 orders of magnitude between the ground and the tropical tropopause [1]. There are few long-term water vapour data sets with high enough accuracy and precision to provide the information necessary to discern cyclical and persistent behaviour. The RALMO has 9 years of measurements with an average of 50% uptime. This makes it uniquely capable of detecting long-term variations as well as diurnal, seasonal, and annual cycles. The RALMO is also supported with twice-daily radiosonde and sporadic lyman-α hygrometer soundings for measurement validation and redundancy. The RALMO water vapour mixing ratio measurements will be processed using the OEM introduced in [2].

In preparation for a multi-year water vapour mixing ratio climatology, it is first necessary to determine the evolution of the RALMO water vapour calibration factor. One cause of the change in the calibration factor is the accelerated deterioration of the water vapour photomultiplier tube (PMT) with respect to the nitrogen PMT [3].

For our purposes the calibration factor is defined as:

$$C_w = 0.781 \frac{\eta_{N_2}}{\eta_{H_2O}} \frac{O_{N_2}(z)}{O_{H_2O}(z)} \frac{\sigma_{N_2}(z)}{\sigma_{H_2O}(z)} \quad (1)$$

where 0.781 is the conversion value between the fraction of nitrogen molecules in air, η_{N_2,H_2O} is the system efficiency for the nitrogen and water vapour channels, $O_{N_2,H_2O}(z)$ is the overlap function for both channels,and $\sigma_{N_2,H_2O}(z)$ is the Raman cross-section for each molecular species.

2 METHODOLOGY

We have developed an improved lidar calibration method using Vaisala RS92 radiosondes based on the radiosonde tracking methods presented in [4] and [5]. However, unlike Whiteman et al. and Adam et al., we consider the wind speed and direction for each altitude bin. This calibration technique systematically accounts for the displacement of the radiosonde from the lidar, thereby increasing the validity of the calibration when other instruments such as a microwave radiometer are not available. This paper presents a new calibration procedure that is tested on a single night's measurements. We are in the process of automating the procedure and applying it to 120 nighttime radiosonde flights.

All RS92 data is processed using the GCOS Reference Upper-Air Network (GRUAN) "gt92" processing tools [6] and only "FLEDT" files are used. One-minute lidar scans from the water vapour and nitrogen channels corresponding to $\pm$ 2 hours around the radiosonde launch are selected. Water vapour and nitrogen scans are selected based on nitrogen signal to noise ratios (SNRs) at 12 km. If the SNR of the nitrogen scan is below 1, the nitrogen scan and the corresponding water vapour scan are not used. Additionally, if the background in the nitrogen scan is higher than 0.01 photons, where the background is defined by the photocounts above 40 km, then both the nitrogen scan and water vapour scan are unusable. The radiosonde is interpolated onto the lidar altitude grid with a bin size of 3.75 meters. Molecular transmission is calculated using Nicolet [7] cross-sections for 407 nm and 387 nm. Number density is determined using radiosonde pressure and temperature.

The 2D air parcel trajectories are initialized from the radiosonde position at each altitude (Figure 1). The trajectories extend $\pm$ 2 hours in the direction of the wind, where their length is determined by the wind speed measured by GPS tracking. In most cases the air parcel trajectory does not pass directly over the field of view of the lidar; therefore, we define a region around the lidar where we assume that the mixing ratio is horizontally homogeneous. In order to maintain significant SNR values, we have chosen a 2 km radius around the lidar as the homogeneous region. The next step is to determine how long the air parcels remain inside the homogeneous region. The lidar measurements are integrated only for the period of time in

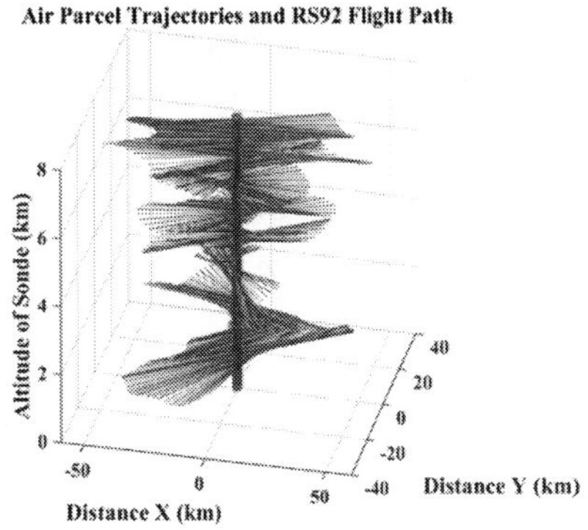

Figure 1: 4-hour 2D Air Mass Trajectories initialized from the radiosonde position (red line). The length of the trajectories is determined by the wind speed measured by the radiosonde GPS tracking. The blue cylinder represents the assumed horizontally homogeneous region around the lidar. We have chosen the radius to be 2 km.

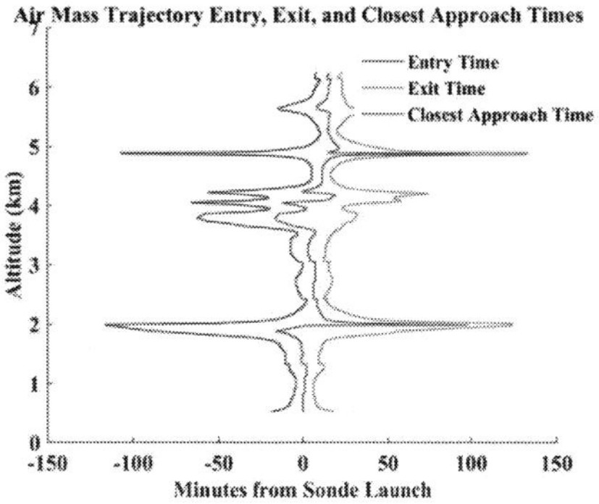

Figure 2: Air Mass trajectory entry (blue), exit (green), and closest approach (red) times. Entry and exit times correspond to the time when the air mass enters or exits the homogeneous region. The time of closest approach is the time at which the air mass is closest to the lidar. Integration times are determined by the exit and entry if the time spent inside the homogeneous region is less than 30 minutes. Otherwise, the integration time is 30 minutes around the time of closest approach.

which the air parcel remains inside the homogeneous cylinder (Figure 2), with a maximum of 30 minute integration in order to assume temporal homogeneity. Occasionally, due to wind direction, an air parcel does not pass through this region. In this case, there are no measurements for that particular altitude. Naturally, this process results in uneven integration times as a function of altitude. All pre-processing for saturation and background removal is done on a per scan basis before integration. After integration the mixing ratio is calculated using Eq. (2) where N_{H_2O} and N_{N_2} are the corrected water vapour and nitrogen photon counts, respectively, and T_{H_2O} and T_{N_2} are the molecular transmissions.

$$w = \frac{N_{H_2O}}{N_{N_2}} \cdot \frac{T_{H_2O}}{T_{N_2}} \tag{2}$$

We define the calibration region to be above 3 km to avoid signal saturation and using analog data. Additionally, we remove low SNR data by limiting the profiles to a region where the water vapour SNR is above 5. Finally, a standard least-squares fitting is done over the entire calibration region.

We then compare this trajectory method with the standard calibration method. The standard method integrates over a time interval that is constant with altitude after the radiosonde is launched. For our purposes, we chose to integrate for 30 minutes after the radiosonde is launched. All pre-processing of the data and data corrections are the same as the Trajectory Method.

3 RESULTS

A preliminary comparison of the standard calibration method with the trajectory method shows that the two produce very similar results (Figure 3). The standard method produces a calibration factor of 31.84 ± 0.09 and a regression coefficient of 0.99. The trajectory method produces a smaller calibration factor of 31.06 ± 0.10 and a regression coefficient of 0.99. The difference between the two calibration coefficients is statistically significant by a factor of 7 sigma. The larger fitting error of the trajectory method is likely due to the data gap at 3.8 km. This gap is caused by time of closest approach of the air parcels arriving before the launch when all of the data was masked by the vetting procedure prior to calibration. Differences between the two results are likely due to the trajectory method using data beyond the 30-minute window of the standard method. Figure 4 shows a 133% change in the water vapour content between 4 and 6 km over the 120 minutes of measurements.

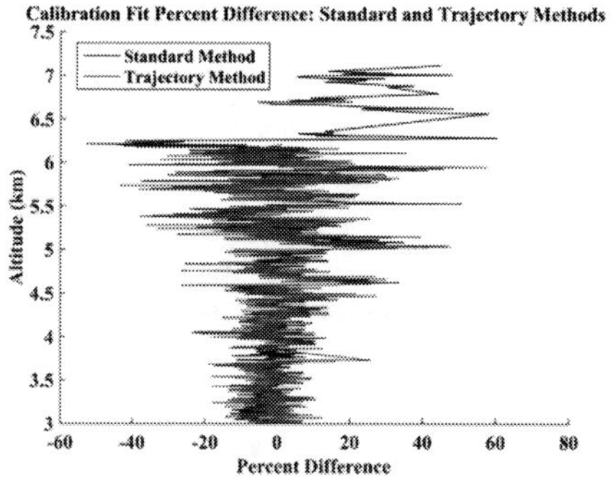

Figure 3: The percent difference between the RS92 mixing ratio profile and lidar calibrated profile using both the standard (black) and trajectory (red) methods.

4 CONCLUSIONS

Further comparisons are needed using all calibration nights to accurately determine the success of the trajectory method. These preliminary results suggest that the traditional method produces a larger fitting region due to larger SNR values at higher altitudes; however, the trajectory method produces a profile closer to the radiosonde (Figure 3).

Both the traditional and trajectory methods

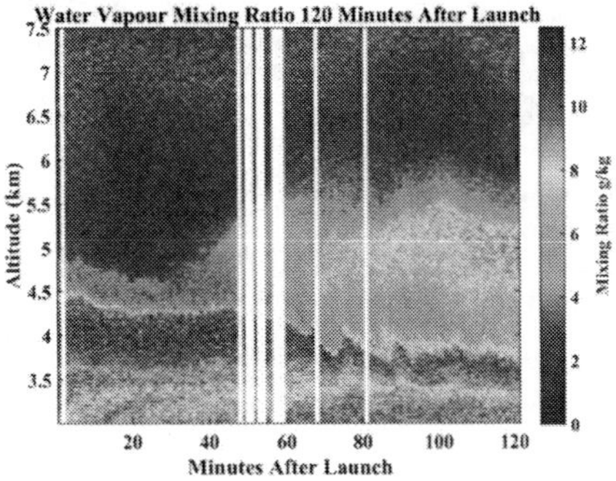

Figure 4: Lidar water vapour mixing ratio after RS92 launch. The white lines are scans that were filtered by the vetting procedure prior to calibration.

have limitations. The traditional method's results may be biased depending on the wind direction and speed. Low windspeeds allow the radiosonde to stay close to the lidar for the entire half-hour integration, in which case the traditional method would be valid. However, high wind speeds could carry the radiosonde farther from the lidar causing it to measure different air parcels thereby biasing the result. Without measuring the trajectory of the radiosonde, it is impossible to know whether or not the traditional method is acceptable. The trajectory method is primarily limited by short integration periods at higher altitudes due to high wind speeds. This could be compensated by allowing the radius of the homogeneous region to increase with altitude (i.e. using a cone instead of a cylinder); however, this remains a topic of further exploration.

We will expand upon this single-case result by testing the method over 120 calibration nights and will then determine a function representative of the calibration factor's evolution to use in the forthcoming water vapour mixing ratio OEM climatology.

ACKNOWLEDGEMENTS

This project has been supported by the Natural Sciences and Engineering Research Council (NSERC), the NSERC CREATE Training Program in Arctic Atmospheric Science and the Canadian Network for the Detection of Atmospheric Change, Canadian Space Agency, and MeteoSwiss.

References

[1] Andrew Gettelman and Michaela I Hegglin. Chapter 7: Upper Troposphere and Lower Stratosphere. *SPARC Report No.5*, volume 5, chapter 7. World Climate Research Program, 2010.

[2] R.J. Sica and Alexander Haefele. Retrievalof water vapor mixing ratio from a multiple channel Raman-scatter lidar using an optimal estimation method.*Applied Optics*, **55** (4), 2016.

[3] T. Dinoev et al. Raman Lidar for Meteorological Observations , RALMO Part 1: Instrument Description. *Atmos. Meas. Tech.*, 6:1329–1346, 2013.

[4] D.N. Whiteman et al. Analysis of Raman lidar and radiosonde measurements from the AWEX-Gfield campaign and its relation to Aqua validation.*JGR*,111 (November 2003):1–15, 2006

[5] M. Adam et al. Water vapor measurements by Howard university Raman lidar during the WAVES 2006 campaign. *Journal of Atmospheric and Oceanic Technology*, 27 (1):42-60, 2010

[6] R.J. Dirksen et al. Reference quality upper-air measurements: GRUAN data processing for the Vaisala RS92 radiosonde. *Atmos. Meas. Tech.*, 7(12):4463–4490, 2014

[7] Marcel Nicolet. On the molecular scattering in the terrestrial atmosphere : An empirical formula for its calculation in the homosphere.*Planetary and Space Science*, 32(11):1467–1468, 1984

WATER VAPOUR INTER-COMPARISON EFFORT IN THE FRAMEWORK OF THE HYDROLOGICAL CYCLE IN THE MEDITERRANEAN EXPERIMENT – SPECIAL OBSERVATION PERIOD (HYMEX-SOP1)

Donato Summa[1], Paolo Di Girolamo[1], Cyrille Flamant[2], Benedetto De Rosa[1], Marco Cacciani[3], Dario Stelitano[1]

[1] *Scuola di Ingegneria, Università degli Studi della Basilicata, Italy , *Email: donato.summa@unibas.it*
[2] *LATMOS/IPSL, UPMC Univ. Paris 06 Sorbonne Universités, UVSQ, CNRS, France*
[3] *Dipartimento di Fisica, Università di Roma "La Sapienza", Italy*

ABSTRACT

Accurate measurements of the vertical profiles of water vapour are of paramount importance for most key areas of atmospheric sciences. A comprehensive inter-comparison between different remote sensing and in-situ sensors has been carried out in the frame work of the first Special Observing Period of the Hydrological cycle in the Mediterranean Experiment for the purpose of obtaining accurate error estimates for these sensors. The inter-comparison involves a ground-based Raman lidar (*BASIL*), an airborne DIAL (*LEANDRE2*), a microwave radiometer, radiosondes and aircraft in-situ sensors.

1. INTRODUCTION

The ability to accurately forecast high-impact weather events, especially in the Mediterranean region, is limited by the present poor capability to properly simulate the contribution of very fine-scale processes and their non-linear interactions with the larger scale processes [1]. The improvement of weather forecasting skills requires an appropriate comprehension of the key processes influenced by the three-dimensional distribution of atmospheric water vapor and temperature. As a consequence, accurate and high-resolution global-scale measurements of the vertical profiles of water vapour mixing ratio and temperature are highly demanded [2]. Space and time resolution must be sufficiently high to allow resolving the fine structure of the atmospheric boundary layer.

Unfortunately, none of the currently available operational observational techniques for measuring water vapour and temperature tropospheric profiles (radiosounding, microwave radiometry, Global Navigation Satellite System) is characterized by the necessary measurement capability and performances (Wulfmeyer et al., 2015 [3]). Only lidar systems exploiting the differential absorption (DIAL) and Raman techniques have the potential to overcome these limitations, thus providing high-precision low-bias water vapour and temperature profile measurements with sufficiently high vertical and temporal resolution.

For the purpose of characterizing the performances of lidar systems and demonstrating their capabilities in terms of measurement precision and accuracy, comprehensive inter-comparison efforts involving lidars and other sensors measuring the same atmospheric parameters are required. This paper illustrates results from a specific inter-comparison effort dedicated to water vapour sensors carried out in the frame work of the first Special Observing Period of the Hydrological cycle in the Mediterranean Experiment (HyMeX-SOP1) [4].

2. METHODOLOGY

Different water vapour sensors were operated in the frame of HyMeX-SOP1. The University of *BASIL*icata ground-based Raman Lidar system (*BASIL*) was deployed in the Cévennes-Vivarais site (Candillargues, Southern France, Lat: 43°37' N, Long: 4° 4' E, Elev: 1 m) and operated between 5 September and 5 November 2012, collecting more than 600 hours of measurements, distributed over 51 measurement days and 19 intensive observation periods (IOPs). *BASIL* is capable to perform high-resolution and accurate measurements of atmospheric temperature and water vapour, both in daytime and night-time, based on the application of the rotational and vibrational Raman lidar techniques in the UV [5, 6, 7, 8, 9].

This measurement capability makes *BASIL* a key instrument for the characterization of the water vapour inflow in Southern France, which is an important ingredient of heavy precipitation events in the North-western Mediterranean basin. *BASIL* makes use of a Nd:YAG laser source capable of emitting pulses at 355, 532 and 1064 nm, with single pulse energies at 355nm of 500 mJ.

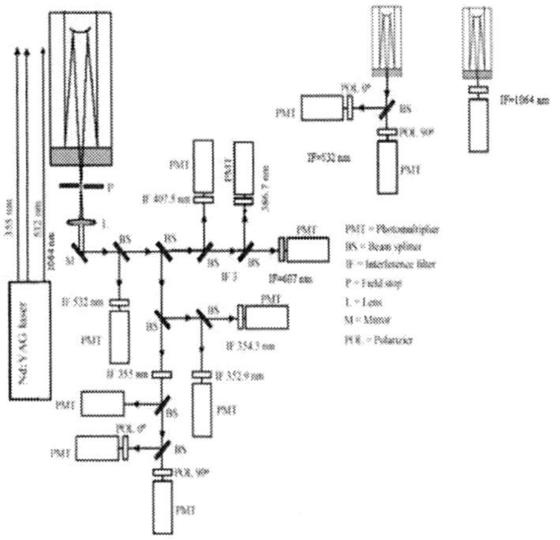

Figure1: *Block diagram of the system setup.*

The receiver includes a Newtonian telescope in (50-cm diameter primary mirror). Data are acquired with a vertical and temporal resolution of 7.5-30m and 10 sec, respectively. A block diagram of the system is illustrated in figure 1.

The airborne water vapour DIAL system LEANDRE 2, embarked on-board the French research ATR42 operated by the Service des Avions Français Instrumentés pour la Recherche en Environnement, is capable of profiling water vapour mixing ratio above or beneath the aircraft.

Details concerning the design and the signal processing procedure of LEANDRE2 are given in Bruneau *et al.* (2001a [10], 2001b [11]). Nadir profiles from *LEANDRE2* were considered in this study in the comparison with other sensors.

The ATR42 is also equipped with a variety of *in situ* thermodynamic sensors (temperature/pressure/humidity) as well as with sensors for turbulence measurements and aerosol/cloud microphysics probes. The *in situ* thermodynamic sensors present on-board were considered in this study. More specifically, the humidity sensor is an hygrometer for water vapour mixing ratio and relative humidity measurements, developed by Aerodata (Braunschweig, Germany), whose sensing element is a capacitive relative humidity sensor used in combination with a temperature sensor. The temperature sensor is platinum resistance wire sensor, developed by Rosemount.

A radiosonde launching facility was also set-up in the Cévennes-Vivarais site in Candillargues. The launched type of radiosonde was the RS92, manufactured by the Finnish company Vaisala. Radisondes were launched primarily during the intensive observation periods.

These sondes provide vertical profiles of atmospheric pressure, temperature, humidity and wind (direction and speed), both during the ascent and the descent phase.

Additionally, a micro-wave radiometer (MWR) was deployed in Candillargues, approx. 50 m away from the radiosonde launching facility and 150-200 m away from *BASIL*. The MWR, operated continuously from 11 October to 05 November 2012.

3. RESULTS

The inter-comparison effort described in this work allows obtaining accurate error estimates for the involved sensors. For this purpose, simultaneous and co-located profiles from different sensors' pairs are considered.

These are used to compute the mutual bias and root-mean-square (RMS) deviation between the different sensors' pairs as a function of height. Expressions to compute these quantities are:

$$BIAS_{i,\mathrm{relative}}(z_1,z_2) = \frac{2\sum_{z=z_1}^{z_2}\{q_1(z)-q_2(z)\}}{\sum_{z=z_1}^{z_2}\{q_1(z)+q_2(z)\}} \qquad (1)$$

$$RMS_{i,\mathrm{relative}}(z_1,z_2) = \frac{2\sqrt{N_z\sum_{z=z_1}^{z_2}\{q_1(z)-q_2(z)\}^2}}{\sum_{z=z_1}^{z_2}\{q_1(z)+q_2(z)\}} \qquad (2)$$

where i is an index denoting the intercomparison sample, $q_1(z)$ and $q_2(z)$ are the water vapour mixing ratio values from the two sensors at height z, z_1 and z_2 are the lower and upper boundary of the considered height interval, respectively, and N_z is the total number of data points for each sensor in the interval $[z_1, z_2]$. Profiles of mean bias and RMS deviations are computed considering the total number of possible intercomparisons for each sensors' pair.

Table 1: Intercomparison BASIL and LEANDRE2: the first and second column identify date and time of the intercomparison, respectively; the third column identifies the flight number and IOP; the fourth column identifies the minimum distance between BASIL and the ATR42 footprint (expressed in km), the fifth and sixth column identify the relative bias and RMS deviation

Date (2012)	Time (UTC)	Flight / IOP	Min. Dis. (km)	Mean bias (%)	Mean RMS (%)
11 Sept.	11:31	34 / IOP 1	0.13	-20	32
23 Sept.	14:19	39 / IOP 5-6	28.02	-21,80	45,20
23 Sept.	17:06	39 / IOP 5-6	61.05	-22,90	33,60
26 Sept.	06:16	40 / IOP 7a	26.21	-6,4	21,60
26 Sept.	09:24	40 / IOP 7a	34.27	6,45	20,10
28 Sept.	15:07	41 / IOP 8	28.29	-6,67	30,20
28 Sept.	20:15	41 / IOP 8	63.42	-11,60	29,70
02 Oct.	21:00	42 / IOP 9	6.17	-1,06	11,30
02 Oct.	21:10	42 / IOP 9	7.13	-2,54	13,20
02 Oct.	21:20	42 / IOP 9	7.92	2	11,40
02 Oct.	21:25	42 / IOP 9	8.38	-3,82	12,60
02 Oct.	21:30	42 / IOP 9	7.83	0,15	9,20
02 Oct.	21:35	42 / IOP 9	7.66	-2,19	9,4
11 Oct.	06:45	43 / IOP 12a	145.1	-4,28	37,10
11 Oct.	09:43	43 / IOP 12a	38.78	-1,66	21,20
12 Oct.	01:18	44 / IOP 12b	28.45	-6,46	11,90
12 Oct.	06:57	45 / IOP 12b	38.17	-1,78	8,10
14 Oct.	08:28	46 / IOP 13	28.00	4,30	24,60
14 Oct.	15:27	47 / IOP 13	22.25	16,10	30,40
15 Oct.	05:24	48 / IOP 13	27.59	14,50	11,60
18 Oct.	15:43	51 / IOP 14	26.39	16.1	26,10
18 Oct.	19:02	51 / IOP 14	38.29	7,41	19,1
20 Oct.	10:03	52 / IOP 15a	37.58	3,93	36

(expressed in %) between the two sensors, respectively.

To apply equation (1) and (2) a common height array has to be considered for each sensors' pair. Equations (1) and (2) allow computing the relative, or percentage, bias and RMS deviation. These quantities allow quantifying the mutual performances of the two compared sensors, i.e. the performance of one sensor with respect to the other. This approach attributes equal weight to the data reliability of each sensor and it is particularly appropriate when none of the two compared sensors can *a priori* be assumed to be more accurate than the other one [12].

Comparisons between *BASIL* and *LEANDRE2* were possible on a variety of case studies. This effort benefitted from the dedicated ATR42 flights in the frame of the EUFAR Project "WaLiTemp". In addition to these dedicated flights, LEANDRE 2 data from additional HyMeX flights were used for intercomparison purposes (see complete list in Table 1). Due to eye-safety issues, LEANDRE 2 has been acquiring nadir profiles only once the ATR was flying higher than 2 km above surface.

Specifically, Table 1 lists the date and time of the intercomparison, the number of the flight and corresponding IOP number, the minimum distance between BASIL and the ATR42 footprint, as well as the relative bias and RMS deviation (expressed in %) between the two sensors, respectively. The table reveals that single bias values do not exceeding 20 %, with an average value of -2.65%, while single RMS values are not exceed 45% with an average value of 21%.

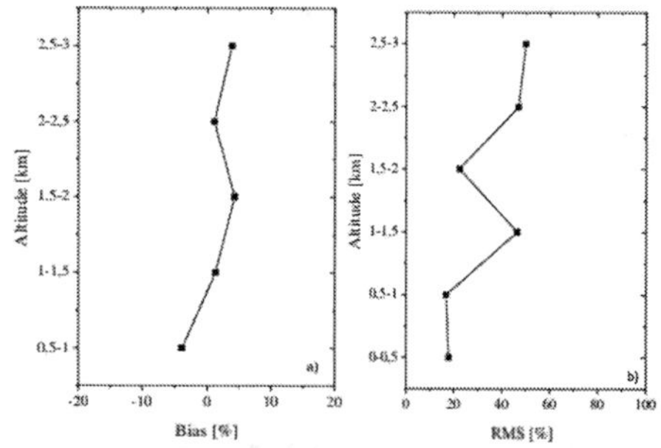

Figure 2: Relative bias (panel a) and RMS deviation (panel b) of BASIL vs. the water vapour in-situ sensor (expressed in %)

Intercomparison between BASIL and other sensors were also carried out. Figure 2 illustrates the mean bias (left panel) and RMS deviation (right panel) of BASIL vs. the water vapour in-situ sensor on-board the ATR42 aircraft. Mean bias and RMS deviation between these two sensors are -1.17% and 33.41 %, respectively, while mean bias and RMS deviation of BASIL vs. radiosondes are -1.8% and 21%, respectively. Additional results considering with all possible sensors' pairs will be illustrated and presented.

ACKNOWLEDGEMENTS

This work is a contribution to the HyMeX programme supported by MISTRALS and ANR IODA-MED Grant ANR-11-BS56-0005. This research effort was supported by the European Commission under the European Facility for Airborne Research programme of the seventh Framework Programme (Project WaLiTemp).

References

[1] Kalthoff, N., M. Kohler, C. Barthlott, B. Adler, S.D. Mobbs, U. Corsmeier, K. Träumner, T. Foken, R. Eigenmann, L. Krauss, S. Khodayar and P. Di Girolamo, 2011. The dependence of convection-related parameters on surface and boundary-layer conditions over complex terrain, *Q. J. Roy. Meteor. Soc.* **137**, 70-80, doi: 10.1002/qj.686.

[2] Wulfmeyer V., H. Bauer, P. Di Girolamo, C. Serio, 2005: Comparison of active and passive water vapour remote sensing from space: An analysis based on the simulated performance of IASI and space borne differential absorption Lidar, *Remote Sens. Environ.* **95**, 211-230, doi: 10.1016/j.rse.2004.12.019.

[3] Wulfmeyer, V., R. M. Hardesty, D. D. Turner, A. Behrendt, M. P. Cadeddu, P. Di Girolamo, P. Schlüssel, J. Van Baelen, and F. Zus, 2015: A review of the remote sensing of lower tropospheric thermodynamic profiles and its indispensable role for the understanding and the simulation of water and energy cycles, *Rev. Geophys.* **53**, 819–895, doi:10.1002/2014RG000476.

[4] Di Girolamo, P., C. Flamant, M. Cacciani, E. Richard, V. Ducrocq, D. Summa, D. Stelitano, N. Fourrié and F. Saïd, 2016: Observation of low-level wind reversals in the Gulf of Lion area and their impact on the water vapour variability, *Q. J. Roy. Meteor. Soc.* **142 (Suppl 1)**, 153–172, doi: 10.1002/qj.2767.

[5] Di Girolamo, P., D. Summa, R. Ferretti, 2009: Multiparameter Raman Lidar Measurements for the Characterization of a Dry Stratospheric Intrusion Event, *J. Atmos. Ocean. Tech.* **26**, 1742-1762, doi: 10.1175/2009JTECHA1253.1.

[6] Summa D., Di Girolamo P., Stelitano D. and Cacciani M., 2013: Characterization of the Planetary boundary layer height and structure by Raman lidar: Comparison of different approach, *Atmos. Meas. Tech.* **6**, 3515-3525, doi: 10.5194/amt-6-3515-2013.

[7] P. Di Girolamo, D. Summa, B. De Rosa, M. Cacciani, A. Scoccione, A. Behrendt, V. Wulfmeyer, 2017: Characterization of Boundary Layer Turbulent Processes by Raman Lidar: Demonstration of the Measurement Capabilities of the Raman Lidar System BASIL, *Atmos. Chem. Phys.* **17**, 745-767, doi:10.5194/acp-17-745-2017.

[8] P. Di Girolamo, P., A. Behrandt, and V. Wulfmeyer (2006). Spaceborne profiling of atmospheric temperature and particle extinction with pure rotational Raman Lidar and of relative humidity in combination with differential absorption Lidar: performance simulations, *Appl. Opt.* **45**, 2474-2494, ISSN: 0003-6935, doi: 10.1364/AO.45.002474.

[9] Di Girolamo P., D. Summa, R.-F Lin, T. Maestri, R. Rizzi, G. Masiello, 2009: UV Raman Lidar measurements of relative humidity for the characterization of cirrus cloud microphysical properties, *Atmos. Chem. Phys.* **9**, 8799-8811, doi: 10.5194/acp-9-8799-2009.

[10] Bruneau D, Quaglia P, Flamant C, Meissonnier M, Pelon J., 2001: Airborne lidar LEANDRE II for water-vapor profiling in the troposphere. I: System description, *Appl. Opt.* **40**, 3450–3461.

[11] Bruneau D, Quaglia P, Flamant C, Pelon J., 2001b: Airborne lidar LEANDRE II for water-vapor profiling in the troposphere. II: First results. *Appl. Opt.* **40**, 3462–3475.

[12] Bhawar, R., P. Di Girolamo, D. Summa, C. Flamant, D. Althausen, A. Behrendt, C. Kiemle, P. Bosser, M. Cacciani, C. Champollion, T. Di Iorio, R. Engelmann, C. Herold, Müller, D., S. Pal, M. Wirth, V. Wulfmeyer, 2011: The Water Vapour Intercomparison Effort in the Framework of the Convective and Orographically-Induced Precipitation Study: Airborne-to-Ground-based and airborne-to-airborne Lidar Systems, *Q. J. Roy. Meteor. Soc.* **137,** 325–348, doi:10.1002/qj.697.

THE ANALYSIS OF A COMPLEX FIRE EVENT USING MULTI-SPACEBORNE OBSERVATIONS

Simona Andrei[1]*, Emil Carstea[1], Luminita Marmureanu[1], Dragos Ene[1], Ioannis Binietoglou[1], Doina Nicolae[1], Dimitra Konsta[2], Vassilis Amiridis[2], Emmanouil Proestakis[2]

[1]*National Institute of R&D for Optoelectronics, Bucharest-Magurele, 077125, Romania,*
**Email:simona.andrei@inoe.ro*
[2]*National Observatory of Athens, Athens, Greece*

ABSTRACT

This study documents a complex fire event that occurred on October 2016, in Middle East belligerent area. Two fire outbreaks were detected by different spacecraft monitoring instruments on board of TERRA, CALIPSO and AURA Earth Observation missions. Link with local weather conditions was examined using ERA Interim Reanalysis and CAMS datasets. The detection of the event by multiple sensors enabled a detailed characterization of fires and the comparison with different observational data.

1 INTRODUCTION

Oil and sulfur fires have a major deleterious impact on the environment and human health. Crude oil burns produce smoke containing carbon dioxide, carbon monoxide, sulfur dioxide, oxides of nitrogen, volatile organic hydrocarbons, ozone, acidic aerosols, soot, droplets of unburned oil and PM_{10}. Sulfur fires generate fumes of sulfur dioxide and sulfur trioxide, and when these react with moisture they produce sulfurous and sulfuric acids [1]. Sulfur dioxide (SO_2), in particular, is an important pollutant gas with significant impact on the environment. It is a precursor of sulfate aerosols that can affect air quality and climate. Although SO_2 has a relatively short atmospheric lifetime, sizable transient SO_2 plumes can travel into remote oceanic areas [2].

Sulfur fires were documented as some of the largest sources of SO_2 fumes released into the atmosphere. In June 2003, a fire at a sulfur mine and processing facility near Mosul, Iraq, released about 600 kilotons of SO_2, making it the largest non-volcanic release of SO_2 that has ever been observed with satellites [3].

Major oil fires were reported in the 1990s, in the Persian Gulf War, when Iraqi troops ignited over 700 oils well in Kuwait and consumed roughly 4.6 million barrels of oil a day. Similar events occurred over several months in 2016, periodic oil fires being recorded by satellite imagery. These events produced a thick, dark smoke [4].

On October 2016, a complex fire event was registered, combining fires at the Al-Mishraq sulfur plant (the same that produced the event in 2003) and at the nearby Qayyarah oil field, when ISIS forces that retreated from Mosul torched the plants. The complex event lasted for 6 days and spread a sulfur plume across northern Iraq, Syria and Turkey [5], [6] exposing the local population to substantial risks [7].

Taking into consideration the negative effects of oil and sulfur fires it is highly important to study the events using state of the art satellite data. The present study is focused on the analysis of a prolonged fire event occurred during 20th and 25th of October, 2016 over the belligerent area of Mosul-Iraq (35.99°N, 43.34°E). Using different spacecraft monitoring instruments onboard of TERRA satellite, two distinct fire outbreaks have been spotted: one with dark plume due to oil wells burning and one with white plume coming from a burning sulfur plant (Fig.1).

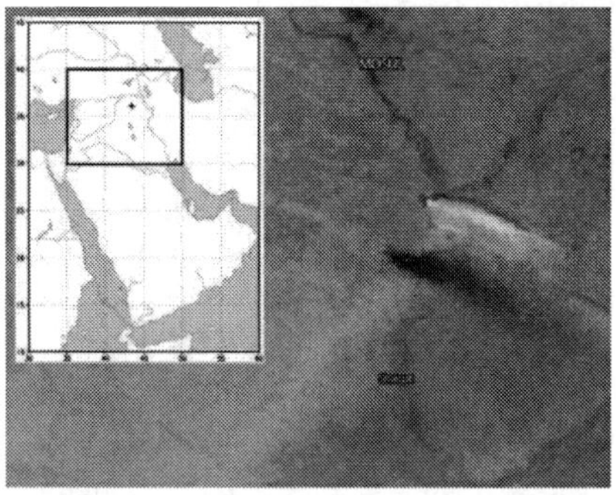

Figure 1 Satellite image of fire event on 22nd October 2016, (https://worldview.earthdata.nasa.gov/)

2 METHODOLOGY

The analysis has implied the use of multiple data from the following instruments: Multi-angle Imaging SpectroRadiometer (MISR), Cloud-Aerosol Lidar with Orthogonal Polarization (CALIOP), Ozone Monitoring Instrument (OMI). The link with the local weather conditions has been examined by the use of Copernicus Atmosphere Monitoring Service (CAMS) datasets.

Onboard of TERRA, the Multi-angle Imaging SpectroRadiometer instrument uses 9 push-broom cameras, which allow the system to image every scene at 9 viewing zenith angles in line with the ground track of the satellite in 7 minutes. [8]. The instrument has been used to access the magnitude and variability of smoke plume heights.

The retrievals from MISR have been validated by/compared with the spaceborne lidar retrievals coming from CALIPSO (Cloud-Aerosol Lidar and Infrared Pathfinder Satellite Observations).

CALIPSO carries the Cloud-Aerosol Lidar with Orthogonal Polarization (CALIOP), which provides vertical resolved aerosol and cloud observations. CALIOP measures the backscatter signal at 532 nm and 1064 nm, and polarization at 532 nm. Cloud and aerosol layers are identified using the level 2 retrievals and through the cloud aerosol discrimination algorithm, based on the probability distribution functions of altitude–and-latitude dependent parameters (integrated colour ratio, layer-integrated volume depolarization ratio, mean attenuated backscatter coefficient) [9].

The Ozone Monitoring Instrument (OMI) has been used to detect the concentration of SO_2 over the plume. OMI, onboard NASA Earth Observing System Aura satellite, observes aerosols and trace gases, including SO_2. Data from the first and last 10 cross-track positions were excluded from the analysis to limit the across-track pixel width to about 40 km [10], [11].

Weather conditions during the event were examined using ERA Interim Reanalysis and Copernicus Atmosphere Monitoring Service (CAMS) datasets [12], [13]. Boundary layer height, obtained from the European Centre for Medium-Range Weather Forecast (ECMWF), was compared with the injection height measured & computed by MISR [8]. Concentration of SO_2 from OMI was compared/validated using SO_2 from CAMS.

3 RESULTS

The cameras of Multi-angle Imaging SpectroRadiometer have detected the fire event on October 21st, 2016. Digitized plumes (Figure 2 a, b) indicate the dispersion along W-NW air circulation within the troposphere.

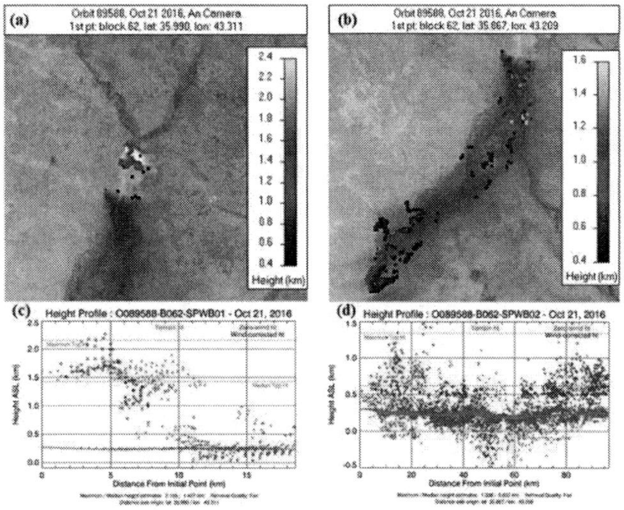

Figure 2 Digitized plumes of SO_2 (a), and oil well (b) in nadir view, height profile of SO2 plume (c), and height profile of oil plume (d) detected with MISR, on October 21st, 2016

The corresponding plumes height generated by MINX are shown in Figure 2 c, d. Red points represent zero-wind heights, blue points wind-corrected heights, and the green line represents the terrain heights with sea level as references.

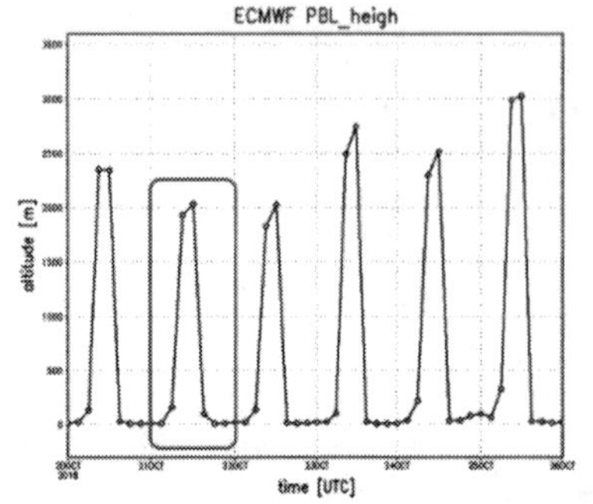

Figure 3 Time series of PBL height from ECMWF during the fire event

Time series of planetary boundary layer (PBL) height from ECMWF (Figure 3) shows that the maximum value of PBL on October 21st has reached 2 km at the afternoon hours. Comparing the altitude where the smoke particles were detected by MISR (Figure 2 c, d) with the PBL height values corresponding to the morning hours (06UTC) when TERRA overpassed Mosul area, it is obvious that the smoke injection dispersed within the free atmosphere and affected a large area, as it was mentioned by WHO reports.

CALIOP is rarely able to detect specific targets such as aerosol source due to its narrow field-of-view [14]. However, by the time of the event CALIPSO overpass the area of Mosul and an aerosol layer can be seen in the PBL, using the CALIOP Vertical Feature Mask on October 21st 2016, 23:15 UTC (Fig.4). The layer is localized between near surface and 2.5 km altitude and is classified as a mix of dust and pollution, (2= dust, 5=polluted dust) with traces of smoke (6 = smoke) [15].

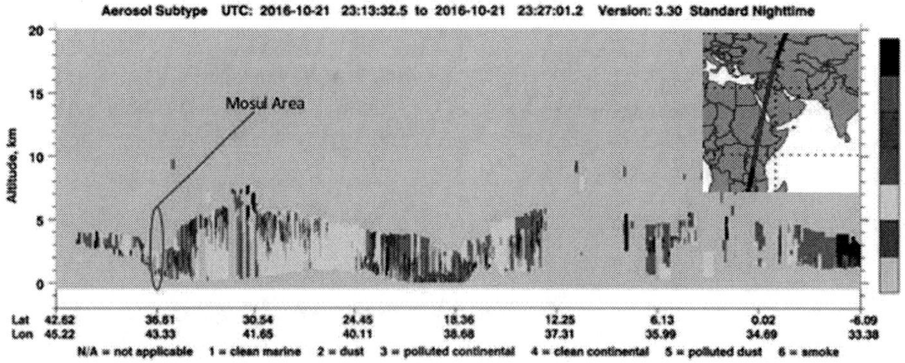

Figure 4 Vertical Feature Mask of aerosol layers detected with CALIOP over Mosul, Iraq

OMI was used to detect the column integrated SO_2 over the dispersion area. The data measured by OMI from 20 to 25 Oct 2016 reveals an average maximum value of 7 DU (Figure 5 a). CAMS analysis (Figure 5 b) emphasizes the initiation of the fire event, showing an injection of SO_2 throughout the atmospheric column.

4 CONCLUSIONS

The detection of smoke by multiple instruments has enabled the detailed characterization of the fire event and the comparison between different observational datasets. CALIOP was able to detect the vertical details of the aerosol layers highlighting both the dust and smoke pollution. Complementary data from the OMI quantitatively estimated the SO_2 released in the atmospheric column.

ACKNOWLEDGEMENTS

The research leading to these results has received funding from ANCS Core program, project no. 16.40.01.01 and the European Union's Horizon 2020 Research and Innovation Programme, under Grant Agreement no 692014 – ECARS.

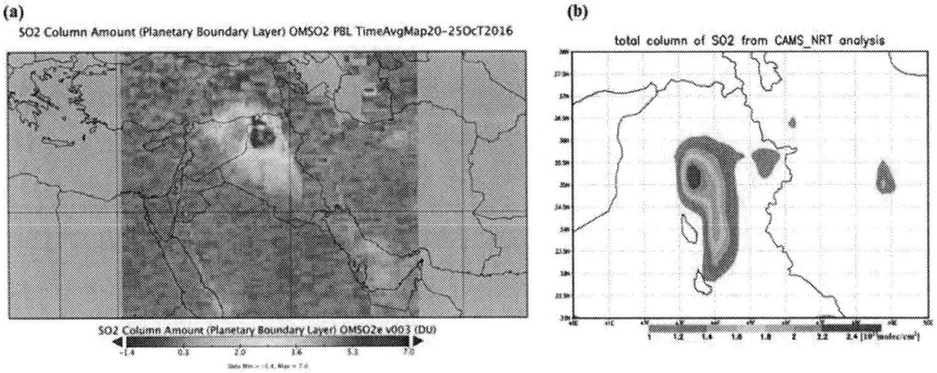

Figure 5 Time averaged map of SO_2 detected with OMI (a), SO_2 total column derived from CAMS (b)

References

[1] http://www.emro.who.int/images/stories/eha/WHO_health_advice_oil_and_sulphur_fires.pdf?ua=1.

[2] Li, C., Joiner, J., Krotkov, N. A., Bhartia, P. K., 2013: A fast and sensitive new satellite SO2 retrieval algorithm based on principal component analysis: Application to the ozone monitoring instrument, *Geophys. Res. Lett.*, **40**, 6314–6318.

[3] Carn, S.A, Krueger, A.J., Krotkov, N.A., Gray, M. A., 2004: Fire at Iraqi sulfur plant emits SO2 clouds detected by Earth Probe TOMS, *Geophys. Res. Lett.*, **31**, L19105.

[4] http://earthobservatory.nasa.gov/IOTD/view.php?id=88666.

[5] http://earthobservatory.nasa.gov/NaturalHazards/view.php?id=88994.

[6] http://reliefweb.int/sites/reliefweb.int/files/resources/mosul_env_health_hazards_report_final_8nov.pdf.

[7] http://www.emro.who.int/irq/iraq-news/who-and-ministry-of-health-manage-over-1000-cases-of-suffocation.html.

[8] Nelson, D. L., Chen, Y., Kahn, R. A, Diner D. J., Mazzoni, D., 2008: Example applications of the MISR INteractive eXplorer (MINX) software tool to wildfire smoke plume analyses, *Proc. SPIE*, **7089** 708909-2.

[9] Gkikas, A., Basart, S., Hatzianastassiou, N., Marinou, E., Amiridis V., Kazadzis, S., Pey, J., Querol, X., Jorba, O., Gassó, S., Baldasano, J.M., 2016: Mediterranean intense desert dust outbreaks and their vertical structure based on remote sensing data, *Atmos. Chem. Phys.*, **16**, 8609–8642.

[10] Fioletov, V. E., McLinden, C. A., Krotkov, N., Yang, K., Loyola, D. G., Valks, P., Theys, N. Van Roozendael, M., Nowlan, C. R., Chance, K., Liu, X., Lee, C., Martin, R. V., 2013: Application of OMI, SCIAMACHY, and GOME-2 satellite SO_2 retrievals for detection of large emission sources, *J. Geophys. Res. Atmos.*, **118**, 11399-11418.

[11] Fioletov, V. E., McLinden, C. A., Krotkov, N, Li, C., 2015: Lifetimes and emissions of SO_2 from point sources estimated from OMI. *Geophys. Res. Lett.* **42**, 1969–1976.

[12] Dee DP, Uppala SM, Simmons AJ, Berrisford P, Poli P, Kobayashi S, Andrae U, Balmaseda MA, Balsamo G, Bauer P, Bechtold P, Beljaars ACM, van de Berg L, Bidlot J, Bormann N, Delsol C, Dragani R, Fuentes M, Geer AJ, Haimberger L, Healy SB, Hersbach H, H´olm EV, Isaksen L, K°allberg P, K¨ohler M, Matricardi M, McNally AP, Monge-Sanz BM, Morcrette J-J, Park B-K, Peubey C, de Rosnay P, Tavolato C, Th´epaut J-N, Vitart F. 2011. The ERA-Interim reanalysis: configuration and performance of the data assimilation system. *Q. J. R. Meteorol. Soc.* **137**: 553–597.

[13] https://atmosphere.copernicus.eu/sites/default/files/repository/SPP%20November%202016_2.pdf.

[14] Winker, D. M., W. H. Hunt, and C. A. Hostetler, 2004: Status and Performance of the CALIOP lidar, *Proc. SPIE*, **5575**, 8-15.

[15] Omar, A., D. Winker, C. Kittaka, M. Vaughan, Z. Liu, Y. Hu, C. Trepte, R. Rogers, R. Ferrare, R. Kuehn, and C. Hostetler, 2009: The CALIPSO Automated Aerosol Classification and Lidar Ratio Selection Algorithm, *J. Atmos. Oceanic Technol.* **26**, 1994-2102.

https://doi.org/10.1051/epjconf/201817608018

CONSTRAINING LIDAR STAND-ALONE RETRIEVALS WITH LUNAR PHOTOMETRY MEASUREMENTS

Pablo Ortiz-Amezcua[1,2]*, Juan Luis Guerrero-Rascado[1,2], Jose Antonio Benavent-Oltra[1,2], Roberto Román[1,2], Christine Böckmann[3], Lucas Alados-Arboledas[1,2]

[1]*Andalusian Institute for Earth System Research (IISTA-CEAMA), Granada, Spain, *portizamezcua@ugr.es*
[2]*Department of Applied Physics, University of Granada, Spain*
[3]*Institute for Mathematics, University of Potsdam, Germany*

ABSTRACT

This study combines atmospheric optical information measured with lidar and nocturnal photometers in order to find configurations that allow for the retrieval of particle microphysical properties without "3+2" lidar setups. It has been carried out using data measured at the EARLINET Granada station during the experimental campaign SLOPE in the framework of ACTRIS-2 project.

1 INTRODUCTION

In the last decades, lidar technique has become a key methodology to obtain vertical profiles of atmospheric aerosol properties at several wavelengths with high spatial and temporal resolution. Multiwavelength Raman lidar measurements allow for stand-alone microphysical retrievals, i. e., the calculation of particle microphysical properties using only Raman lidar data, but the minimum requirement is having a setup with 3 profiles of backscatter coefficient (355, 532, 1064 nm) and 2 of extinction coefficient (355, 532 nm), the so-called "3+2" configuration [1, 2]. However, there are a lot of lidar systems, in particular inside EARLINET [3] and LALINET [4] networks, that do not fulfill this requirement.

The aim of this work is to find alternative setups that, in combination with lunar photometry measurements, can provide reasonably good inputs for microphysical retrievals

2 LOCATION, INSTRUMENTATION AND METHODOLOGY

The measurements have been performed at Granada (37.16° N, 3.61°W, 680 m asl), a medium-size city in Southeastern Spain affected by air masses mainly coming from the Atlantic Ocean, Europe, North Africa and the Mediterranean Basin. The data for this study were obtained in the context of SLOPE, a field campaign carried out in summer 2016 to gather data useful for testing the retrieval schemes through inversion of remote sensing observations.

2.1 Multiwavelength Raman lidar and microphysical algorithm

The vertical profiles of particle optical properties were obtained using signals from the multiwavelength Raman lidar system MULHACEN (LR331D400, Raymetrics Inc.), configured in a monostatic biaxial alignment pointing vertically to the zenith. A pulsed Nd:YAG laser with emission at wavelengths of 355, 532 and 1064 nm is used as a radiation source, and the backscattered signal is collected by a Cassegranian telescope and splitted to detect elastic signals at 355, 532 (in parallel and perpendicular polarizations) and 1064 nm and Raman shifted signals at 387, 607 and 408 nm.

Night-time lidar signals are used to obtain particle backscatter (β) and extinction (α) coefficients using the Raman algorithm [5]. The resulting set of 3 β (at 355, 532 and 1064 nm) and 2 α (at 355 and 532 nm) is employed as input to obtain particle microphysical properties with a regularization method. The used algorithm (UP) was developed at the University of Potsdam [2] in the framework of EARLINET [6], and is based on explicitly solving the equations that relate the particle microphysical and optical properties by means of Truncated Singular Value Decomposition (TSVD) regularization. The resulting particle properties, namely volume concentration (v_c), effective radius (r_{eff}), single scattering albedo (*SSA*) and real and imaginary parts of refractive index (RRI, IRI), constitute the reference values in the present work.

2.2 Sun/sky/lunar photometer

The nocturnal aerosol optical depth (AOD) is measured with the sun-sky-lunar CE318-T photometer (Cimel Electronique) [7], developed to extend the performance of the common sun-photometers in AERONET (Aerosol Robotic Network) [8] to both daytime and night-time measurements.

This instrument performs the daytime AERONET routine measurements, but also night-time direct measurements of the lunar irradiance at nominal wavelengths of 1640, 1020, 870, 670, 500 and 440 nm. A sequence of three measurements of the radiation reflected by the Moon is taken every 30 s at each wavelength. AOD is obtained from these irradiances following the method of [7], which uses the ROLO model [9] to calculate the lunar extraterrestrial irradiance. The nocturnal measurements can be taken during 50% of the Moon cycle, extending the diurnal measurements during this period.

2.3 Simulation of missing lidar channels with spectral AOD

In order to test the accuracy of the microphysical retrievals for the combination of "incomplete" lidar systems with lunar photometer, several possible configurations were simulated and compared with the reference 3+2 results. To this end, the missing lidar information was reconstructed using columnar Angström Exponent (*AE*) of spectral AOD's from photometer measurements, and following Angström equation for wavelengths λ_1 and λ_2:

$$X(\lambda_2) = X(\lambda_1)\left[\frac{\lambda_2}{\lambda_1}\right]^{-AE_{\lambda_1,\lambda_2}} \quad (1)$$

where X stands for α or β.

With this in mind, the following lidar configurations were chosen and completed with lunar photometer measurements:

- ***3+1 (355):*** three β at 355, 532, 1064 nm and one α at 355 nm are used. Missing α_{532} profile is generated from α_{355} profile and $AE_{440,670}$ from photometer using Eq. (1).
- ***3+1 (532):*** three β at 355, 532, 1064 nm and one α at 532 nm are used. Missing α_{355} profile is generated from α_{532} profile and $AE_{440,670}$ from photometer.
- ***2+2:*** two β and two α at 355 and 532 nm are used. Missing β_{1064} profile is generated combining β_{532} profile and $AE_{670,1640}$ in Eq. (1).

3 RESULTS

Two night-time cases measured during SLOPE field campaign at Granada were selected to test the proposed methodology.

3.1 Case A

It corresponds to a measurement on 18th May 2016 at 22:00-22:30 UTC. According to satellite information and transport models (not shown here), the event is classified as biomass burning particles from North American forest fires.

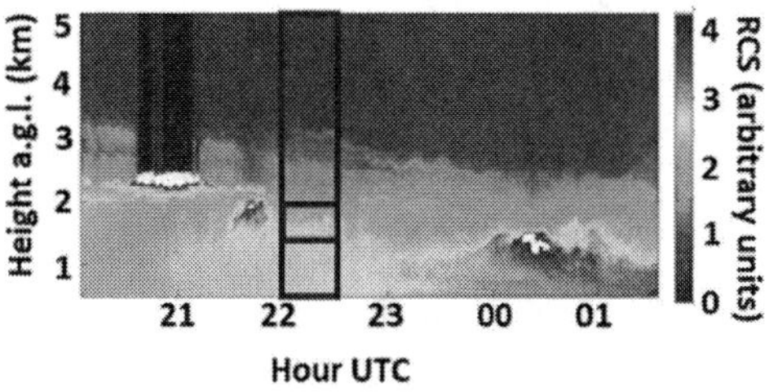

Figure 1: Time evolution of the lidar range corrected signal (RCS) at 532 nm for case A. Brown boxes indicate the period and region selected for microphysical analysis

The time evolution of the lidar range corrected signal (RCS) at 532 nm (Fig. 1) shows a multi-layered aerosol structure up to 3.5 km above ground level (agl). A region from 1.4 to 2 km agl (brown box) was selected to perform the microphysical analysis.

Simultaneous lunar photometer measurements indicated AOD_{440} of 0.28. The spectral AOD was used to simulate the different configurations for microphysical retrievals, and the results (Fig. 2) showed similar values to the reference set (3+2) retrievals, since most of the differences among configurations were inside the uncertainty bars (which indicate a statistical standard deviation of the retrieval).

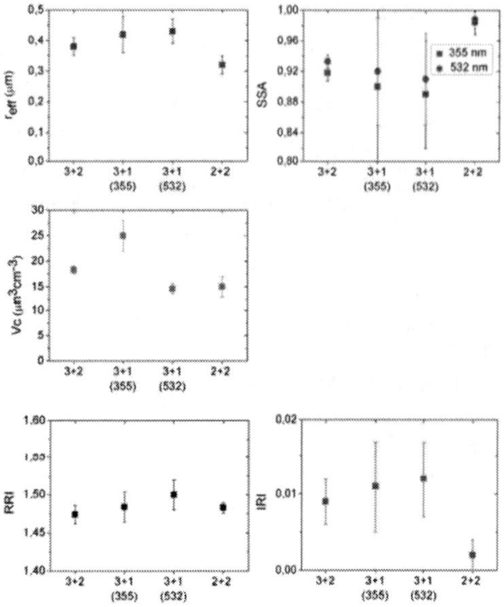

Figure 2: Effective radius (r_{eff}), single scattering albedo (SSA), volume concentration (V_c), real and imaginary parts of refractive index (RRI, IRI) retrieved for Case A using the complete lidar setup (3+2) and different combinations with lunar photometer

3.2 Case B

The second case from SLOPE, with different aerosol type, corresponds to an event of local pollution at Granada on 17th June 2016 at 20:30-21:00 UTC.

Figure 3: Time evolution of the lidar RCS at 532 nm for case B. Brown boxes indicate the period and region selected for microphysical analysis

The particle microphysical properties were calculated for an aerosol layer between 1.5 and 2 km agl (brown box in Fig. 3, corresponding to time evolution of RCS at 532 nm).

For this case, the microphysical retrievals resulting from the different combinations made with lunar photometer spectral AOD (Fig. 4) also agreed well with the 3+2 setup. However, the deviations are slightly higher than those for case A, what can be attributed to a lower AOD, taking the value of 0.10 at 440 nm.

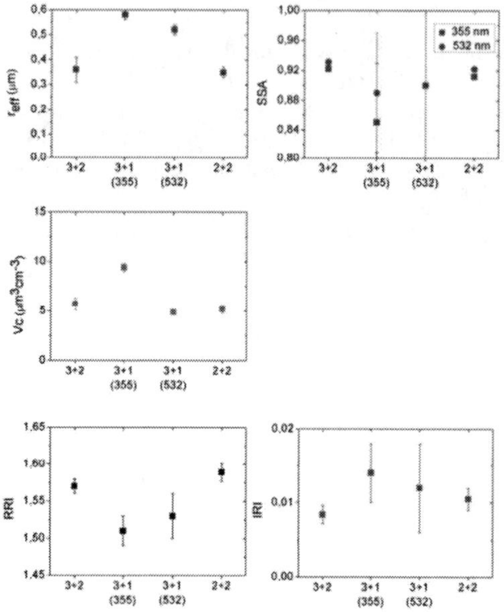

Figure 4: Microphysical properties (same as Fig. 2) for Case B with different lidar + lunar photometer setups.

The deviation to the 3+2 retrieval of all simulated configurations was averaged for each microphysical property and for cases A and B (Table 1). It can be observed that the parameter with most sensitivity to the changes for the studied cases is IRI, and that SSA and RRI are almost unaffected by the variations.

Table 1: Deviations from the 3+2 configuration retrieval, averaged for all photometer configurations in cases A and B

Microphysical Properties	Mean Deviation (%)	
	Case A	Case B
r_{eff}	13	36
SSA	4	4
Vc	25	30
RRI	1	3
IRI	44	45

4 CONCLUSIONS AND OUTLOOK

This work explores the combination of an "incomplete" lidar system (without 3+2 setup) with the night-time AOD from lunar photometer, finding that it can be used to derive alternative input setups for a lidar stand-alone regularization algorithm to obtain particle microphysical properties. Optimistic results have been found for two test cases: there are small differences among different datasets (sometimes they are inside the standard deviation bars). Comparing the two cases, slightly better results are obtained for a biomass burning case, which presented higher AOD.

In future work, more cases and more configurations will be analyzed in order to extract a general behavior about the best configurations and the relationship with the AOD.

ACKNOWLEDGEMENTS

This work was supported by the Andalusia Regional Government through project P12-RNM-2409, by the Spanish Ministry of Economy and Competitiveness through projects CGL2013-45410-R and CGL2016-81092-R and by the European Union's Horizon 2020 research and innovation programme through project ACTRIS-2 (grant agreement No 654109). The authors thankfully acknowledge the FEDER program for the instrumentation used in this work.. This work was also partially funded by the University of Granada through "Plan Propio. Programa 9. Convocatoria 2013".

References

[1] Müller, D., Wandinger, U., Ansmann, A., 1999: Microphysical particle parameters from extinction and backscatter lidar data by inversion with regularization: Theory, *Appl. Optics* **38**, 2346-2357.

[2] Böckmann, C., 2001: Hybrid regularization method for the ill-posed inversion of multiwavelenght lidar data to determine aerosol size distributions, *Appl. Optics* **40**, 1329-1342.

[3] Pappalardo, G., Amodeo, A., Apituley, A., Comeron, A., Freudenthaler, V., Linné, H., Ansmann, A., Bösenberg, J., D'Amico, G., Mattis, I., Mona, L., Wandinger, U., Amiridis, V., AladosArboledas, L., Nicolae, D., Wiegner, M., 2014: EARLINET: towards an advanced sustainable European aerosol lidar network, *Atmos. Meas. Tech.* **7**, 2389-2409.

[4] Guerrero-Rascado, J. L., Landulfo, E., Antuña, J. C., Barbosa, H. M. J., Barja, B., Bastidas, A. E., Bedoya, A. E., da Costa, R., Estevan, R., Forno, R. N., Gouveia, D. A., Jimenez, C., Larroza, E. G., Lopes, F. J. S., Montilla-Rosero, E., Moreira, G. A., Nakaema, W. M., Nisperuza, D., Alegria, D., Manera, M., Otero, L., Papandera, S., Pallota, J. V., Pawelko, E., Quel, E. J., Ristori, P., Rodrigues, P. F., Salvador, J., Sanchez, M., Silva, A., 2016: Latin American Lidar Network (LALINET): diagnosis on network instrumentation, *J. Atmos. Sol. Terr. Phys.* **138-139**, 112-120.

[5] Ansmann, A., Wandinger, U., Riebesell, M., Weitkamp, C., Michaelis, W., 1992: Independent measurement of extinction and backscatter profiles in cirrus clouds by using a combined Raman elastic-backscatter lidar, *Appl. Opt.* **31**, 7113-7131.

[6] Müller, D., Böckmann, C., Kolgotin, A., Schneidenbach, L., Chemyakin, E. V., Rosemann, J., Znak, P., Romanov, A., 2016: Microphysical particle properties derived from inversion algorithms developed in the framework of EARLINET, *Atmos. Meas. Tech.* **9**, 5007-5035.

[7] Barreto, A., Cuevas, E., Granados-Muñoz, M. J., Alados-Arboledas, L., Romero, P. M., Gröbner, J., Kouremeti, N., Almansa, A. F., Stone, T., Toledano, C., Román, R., Sorokin, M., Holben, B., Canini, M., Yela, M., 2016: The new sun-sky-lunar Cimel CE318-T multiband photometer – a comprehensive performance evaluation, *Atmos. Meas. Tech.* **9**, 631-654.

[8] Holben, B., N., Eck, T. F., Slutsker, I., Tanré, D., Buis, J. P., Setzer, A., Vermote, E., Reagan, J. A., Kaufman, Y. J., Nakajima, T., Lavenu, F., Jakowiak, I., Smirnov, A., 1998: AERONET – A Federated Instrument Network and Data Archive for Aerosol Characterization, *Remote Sens. Environ.* **66**(1), 1-16.

[9] Kieffer, H. H., Stone, T. C., 2005: The spectral irradiance of the moon, *Astron. J.* **129**, 2887-2901.

LIDAR TRANSMITTER OFFERS "NON-DIFFRACTING" PROPERTY THROUGH SHORT DISTANCE IN HIGHLY-DENSE RANDOM MEDIA

Xiafukaiti Alifu*, Peng Ziqi and Tatsuo Shiina

Chiba University 1-33 Yayoi-cho, Inage-ku, Chiba-shi, Chiba 263-8522 JAPAN
**xapkat_alip@chiba-u.jp*

ABSTRACT

Non-diffracting beam (NDB) is useful in lidar transmitter because of its high propagation efficiency and high resolution. We aimed to generate NDB in random media such as haze and cloud. The laboratory experiment was conducted with diluted processed milk (fat: 1.8%, 1.1μmφ). Narrow view angle detector of 5.5mrad was used to detect the forward scattering waveform. We obtained the central peak of NDB at the propagation distance of 5cm ~ 30cm in random media by adjusting the concentration of <10%.

1. INTRODUCTION

Aerosols and dusts have interactions between the ground and the cloud. They actively float in the low-altitude atmosphere. To observe them, we have developed an inline type micro-pulse lidar and a high-resolution polarization lidar, which have conducted polarization measurement in low-altitude atmosphere [1-3]. The lidar transmitted beam catches strong scattering, hence it cannot penetrate at long distance in random media such as cloud and haze. Having a long distance penetration in random media is important for studying the remote sensing techniques such as deeper sensing in cloud structure and target recognition through haze [4-5].

In general, beams spread by its diffraction effect. However, in the case of a non-diffracting beam (NDB), such as Bessel beam, it can propagate over a long distance and keep its narrow beam width, because it does not diffract (self-formation) [6-8]. In previous study, it has been confirmed that an annular beam could change its beam shape into quasi-NDB after being scattered several times in random media (10-30cm) [9]. But, variation of NDB in different receiving distance was not clear. The objective of this research is to obtain variation of NDB in different receiving distances to analyze the components of the scattering waveform, and generate NDB at short distances (5cm) in random media with higher concentration.

2. ANNULAR BEAM

In this research, the transmitted light of lidar is an annular beam, which is converted from a Gaussian beam by a pair of axicon prism. Figure 1 shows the transformation from the Gaussian beam to the annular beam using the axicon pair.

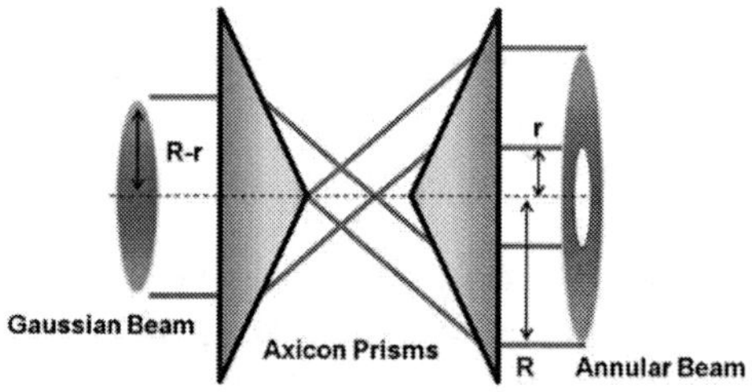

Figure 1 Transformation from the Gaussian beam to the annular beam using the axicon pair.

The Gaussian beam distribution g(r) is expressed in Eq.1.

$$g(r) = \frac{1}{\pi h^2} \exp\left[-\left(\frac{r}{h}\right)^2\right] \qquad (1)$$

Where h is a half width at the point where the intensity equals $1/e^2$ of the central peak value, and r represents radial point of the plane perpendicular to the propagation direction of light. The converted annular beam is expressed in Eq.2. R is the external radius of the annular beam.

$$a(r) = \sqrt{\frac{R-r}{r}}\, g(R-r) \qquad (2)$$

In our laboratory experiment, we applied the annular beam as the lidar transmitted light, which has an external diameter of 40mm with a ring width of 3mm. When the annular beam propagates in the atmosphere, as the propagation range increased, the energy is focused in the center due to a non-diffraction effect. Figure 2 shows the intensity distributions of the annular beam and NDB. This NDB was generated by propagating the annular beam at a long distance of 210m in air, which is calculated by Fresnel equation. The center peak of NDB is narrow, and it has an advantage on optical sensing with high resolution.

3. SYSTEM SETUP

Figure 3 shows the experiment setup for the annular beam propagation through a random media. The experiment parameters are summarized in Table 1. A DPSS pulse laser with peak power of 4.6kW, 532nm wavelength, 2ns pulse width and 10 kHz repetition frequency was used. The ND filter was applied to adjust the intensity of the transmitted light. A diluted processed milk (milk fat: 1.8%, 1.1μmφ) with 0% ~ 6% concentration was used as a random media in this experiment. The media tanks of 20cm(W)x20cm(H)x5cm ~ 30cm(L) were used for the evaluating the propagation property of light in a short distance. The intensity distributions of the propagated light in random media were recorded by a receiving unit, which includes a collimating lens, a multimode optical fiber and a photomultiplier tube . This receiving unit had a narrow viewing angle of 5.5mrad. The forward-scattered light in a random media is received selectively. The photomultiplier tube and a sampling oscilloscope were used to detect the weak signals from the media of definite concentration.

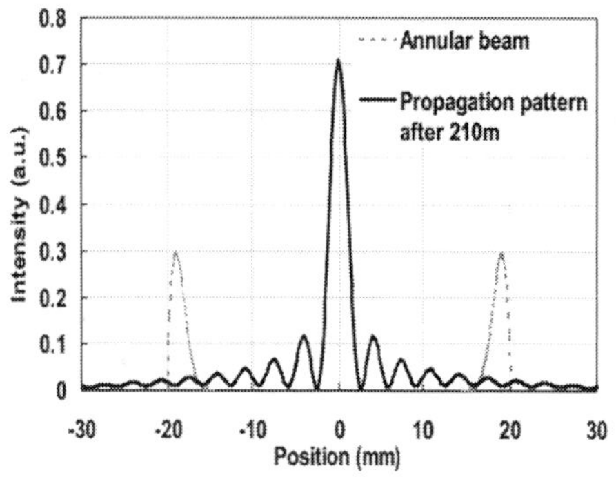

Figure 2 Intensity distributions of annular beam and its propagation after 210m.

Table 1 Parameters for the experimental setup

Light source	DPSS laser CryLas, 1Q532-1: Wavelength: 532nm; Peak power: 4.6kW; Pulse Width: 2ns; Repetition: 10kHz.
Annular beam converter	Axicon prims: Zenith angle: 150° (±10'); Core diameter: 50.8mm; Annular beam: External diameter : 42mm
Random media	Media tank: Material: tempax glass; Size: W×H×L 20cm×20cm× (1~30)cm; Media: Diluted processed milk: Milk fat: 1.8%; Fat size: 1.1μm;
Receiver	Optical elements: Optical Fiber: Multiple mode Core Diameter: 50μm; Collimating lens: Lens aperture:6mm View angle: 5.5mrad;

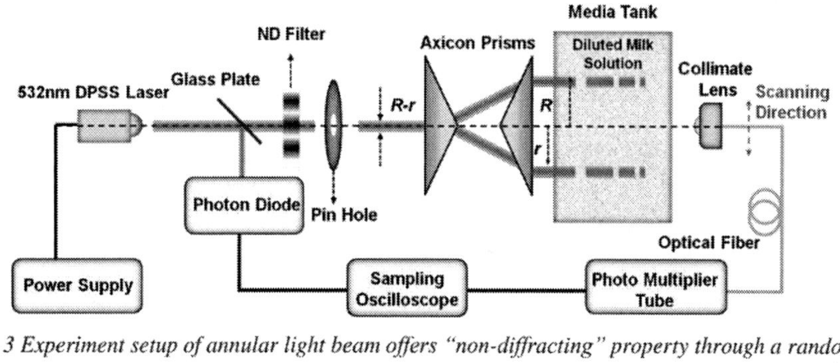

Figure 3 Experiment setup of annular light beam offers "non-diffracting" property through a random media.

4. RESULTS AND DISCUSSION

When the annular beam propagated at 20cm distance in milk diluted solution of 0.6% concentration, the forward scattering waveforms were measured at 5cm and 10cm away from the back of the media tank. Figure 4 shows the measured receiving waveforms. The center of the waveform has a peak (6mm in diameter) which we call NDB. The surrounding curve of the waveform is constructed by the multiple scattering lights (the intensity was normalized at position of 20mm). The central intensity ratio can be calculated by the ratio of the maximum intensities between NDB and multiple scattering lights. As a result, the central intensity ratio was 1.2% in both waveforms. Curvature radius of surrounding part of scattering waveforms was 9.6cm when the receiving distance was 5cm. It changed to 10.5cm and its intensity was declined overall when the receiving distance was 10cm. We confirmed that surrounding part of scattering waveforms spread out while center peak of NDB did not change its shape as the generated waveform propagated in air. Furthermore, according to the calculation, the generated waveform keeps its peak of NDB up to the propagation of 100cm in the air. This result proves that the waveforms are composed of NDB in center part, which was constructed from forward scattering and the surrounding part, which was constructed from the multiple scattering, respectively.

To generate and control NDB in short-distance random media, we set the media tank of 5cm and measured the scattering waveforms at the concentration range of 2% ~ 6%. Figure 5 shows the measured waveforms at media concentration range of 3.5% ~ 5%. At the beginning, a small peak appeared at the center of 3.5% concentration, the central intensity ratio became high as the concentration increases. The highest peak of NDB appeared at 5% concentration, and the central intensity ratio was 1.8%. In the previous study, the highest peak of NDB appeared at concentrations of 0.4%, 0.6% and 1% in the media tank of 30cm, 20cm and 10cm respectively. But the highest peak of NDB appeared at the propagation distance of 5cm. In this case, the center intensity ratio was 50% higher than that at 20cm propagation distance. Figure 6 shows the variation of the central intensity ratio in different

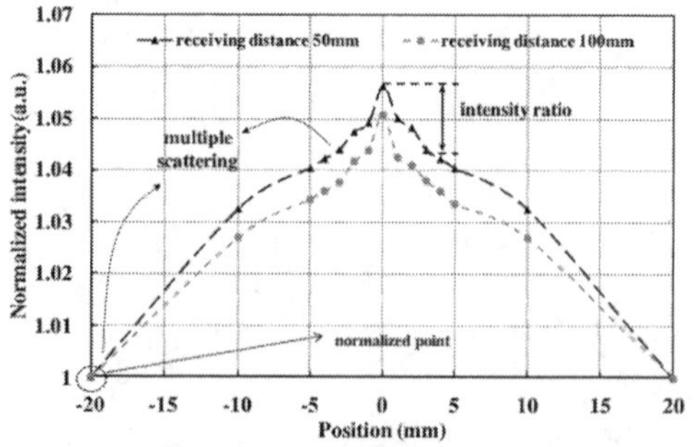

Figure 4 Scattering waveforms of annular beam through 20cm long-distance random media by different receiving distance.

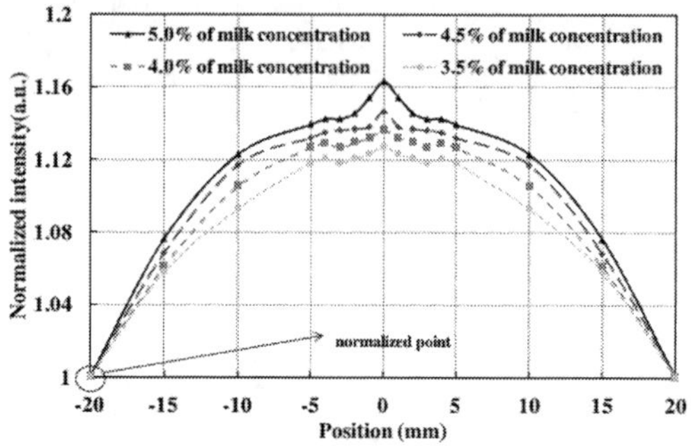

Figure 5 Scattering waveforms of annular beam at concentration range of 3.5% ~ 5% in random media by 5cm propagation.

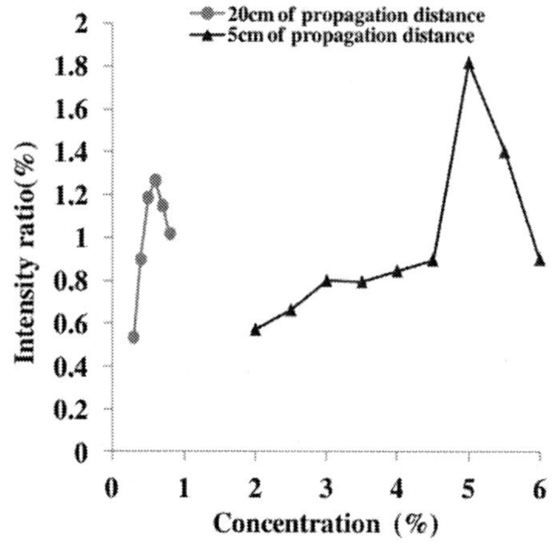

Figure 6 Analysis of intensity ratios at concentration increases by different propagation distance.

concentration when the propagation distance is 5cm and 20cm. For the case of 20cm propagation, the NDB was generated at concentrations range of 0.3% ~ 0.8%. The highest central intensity ratio of NDB was obtained at 0.6% concentration. The non-diffracting intensity start to decline when the concentration is higher than 0.6%. In the case of 5cm propagation, the NDB can be obtained at the concentration range of 2% ~ 6%. The highest central intensity ratio of NDB was obtained at 5% concentration. Same to the case of 20cm propagation, the non-diffracting intensity started to decline after this concentration in the 5cm propagation. The concentration range of NDB appeared at propagation distance of 5cm was 8times wider than that at propagation distance of 20cm. Since, the center part was subjected to stronger forward scattering at higher concentration in a short-distance random media. As a result, the non-diffracting field is offered the high center intensity in short distance [9].

5. CONCLUSIONS

This paper presents the propagation characteristics of NDB in different receiving distance to analyze the components of the scattering waveform and generation of NDB at short distances (5cm) in random media with higher concentration. The experiment of different receiving distance demonstrated that NDB in center part is composed of forward scattering and surrounding part is composed of multiple scattering. According to the calculation, we have confirmed that the generated waveform keeps its peak of NDB up to the propagation distance of 100cm in air. Moreover, we have successfully generated the NDB in the following condition; concentration of 2% ~ 6% and short-distance of 5cm in random media. The highest center intensity ratio of 1.8% was obtained by high concentration media (5%). By comparing the results from two different propagations (5cm and 20cm), the concentration range of existing NDB is wider at shorter propagation distance. The maximum intensity ratio of NDB is also higher in shorter distance, because the center part is subjected to stronger forward scattering in the higher concentration range media. This finding could be useful reference for more shorter-distance propagation (<5cm) in future studies. In the future, the method which generates the NDB in random media by using annular beam can be applied for getting the deeper information of high concentration cloud and penetrating the heavy cloud as optical sensing field.

References

[1] Tatsuo Shiina, Eiji Minami, Masafumi Ito, and Yasuyuki Okamura, 2002: Optical circulator for an in-line-type compact lidar, *Appl. Opt.*, Vol. 41, pp. 3900-3905

[2] Tatsuo Shiina, Kei Yoshida, Masafumi Ito, and Yasuyuki Okamura, 2005: In-line type micropulse lidar with an annular beam - experiment, *Appl. Opt.*, Vol. 44, pp. 7407-7413

[3] Tatsuo Shiina, Kazuo Noguchi, and Tetsuo Fukuchi, 2012: Polarization-independent optical circulator for high accuracy Faraday depolarization lidar, *Appl. Opt.*, Vol. 51, pp. 898-904

[4] Michael I. Mishchenko, Janna M. Dlugach, Maxim A. Yurkin, Lei Bi, Brian Cairns, Li Liu, R. Lee Panetta, Larry D. Travis, Ping Yang, Nadezhda T. Zakharova 2016: First-principles modeling of electromagnetic scattering by discrete and discretely heterogeneous random media , *Physics Reports*, Vol.632, pp.1-75

[5] Yanhui Li and Zhensen Wu, 2010: Targets recognition using subnanosecond pulse laser range profiles, *Opt. Express*, Vol. 18, pp. 16788-16796

[6] J. Durnin and J. J. Miceli JR., 1987: Nondiffracting beams, *Optics News*, Vol. 13(12), pp. 27-27

[7] J. Durnin, J. H. Eberly, and J. J. Miceli, 1988: Comparison of Bessel and Gaussian beams, *Opt. Lett.*, Vol. 13, pp. 79-80

[8] Tatsuo Shiina, Kei Yoshida, Masafumi Ito, Yasuyuki Okamura, 2007: Long-range propagation of annular beam for lidar application, *Optics Communications*, Vol. 279, pp. 159–167

[9] Ziqi Peng and Tastuo Shiina, 2017: Generation control of a non-diffractive beam in random media by adjusting concentration, *Optics Communications*, Vol. 391, pp. 94–99

SIBERIAN LIDAR STATION: INSTRUMENTS AND RESULTS

Gennadii G. Matvienko[1,2], Yurii S. Balin[1], Sergey M. Bobrovnikov[1], Oleg A. Romanovskii[1,2], Grigirii P. Kokhanenko[1], Svetlana V. Samoilova[1], Ioganes E. Penner[1], Evgenii V. Gorlov[1], Victir I. Zharkov[1], Sergey A. Sadovnikov[1], Semen V. Yakovlev[1,2], Oleg E. Bazhenov[1], Sergey I. Dolgii[1], Andrey P. Makeev[1], Alexey A. Nevzorov[1], Alexey V. Nevzorov[1], Olga V. Kharchenko[1]

[1]*V.E. Zuev Institute of Atmospheric Optics, Tomsk, Russia, *roa@iao.ru*
[2] *National Research Tomsk State University, Tomsk, Russia*

ABSTRACT

The Siberian Lidar Station created at V.E. Zuev Institute of Atmospheric Optics and operating in Tomsk (56.5° N, 85.0° E) is a unique atmospheric observatory. It combines up-to-date instruments for remote laser and passive sounding for the study of aerosol and cloud fields, air temperature and humidity, and ozone and gaseous components of the ozone cycles. In addition to controlling a wide range of atmospheric parameters, the observatory allows simultaneous monitoring of the atmosphere throughout the valuable altitude range 0–75 km. In this paper, the instruments and results received at the Station are described.

1 INTRODUCTION

Long-term routine measurements by lidar facilities are very effective. The analysis of the measurement data allows the development of empirical models of atmospheric parameters; determination of their seasonal, quasi-two-year, and other periodic cycles of variations; the study of causes that determine the dynamics and correlations between the parameters measured; detection of short-term atmospheric disturbances of slow variations under the action of natural and anthropogenic factors, and, finally, detection and forecast of trends in atmospheric variations with estimation of their effects on possible climate changes. The Siberian Lidar Station (SLS) is a unique atmospheric observatory created at V.E. Zuev Institute of Atmospheric Optics. It operates in Tomsk (56.5° N, 85.0° E) and combines up-to-date instruments for remote laser and passive sounding of aerosol and cloud fields, air temperature and humidity, and ozone and gaseous components of the ozone cycles. In addition to controlling a wide range of atmospheric parameters, the observatory allows simultaneous monitoring of the atmosphere throughout the valuable altitude range 0–75 km. The possibility of simultaneous complex measurements of key climate forcing and ecological parameters of the atmosphere is a feature of measurements at SLS. It is provided by use of different laser sounding techniques in combination with spectrophotometric measurements This paper describes the monitoring of such atmospheric parameters as aerosols, temperature, clouds, ozone, gaseous components of ozone cycles at SLS.

2 GENERAL SCEMATIC OF THE SLS MEASURING COMPLEX

Experiments on the sounding of different atmospheric parameters (aerosols, temperature, ozone, humidity) have been carried out at the High-Altitude Sensing Station since 1991 using lidar channels with receiving mirrors of 2.2, 0.5, and 0.3 m in diameters. The name "Siberian Lidar Station" was introduced in 1995, during preparation of documents for inclusion of the Station into the List of Unique Research and Experimental Systems of National Significance of Russian Federation (RF). For a long time (from 1986), SLS was the only station of routine lidar monitoring of the stratosphere not only on a great territory of Siberia, but in the whole Russia. Only in 2005, measuring lidar facilities were elaborated and then started operating in the Asian Russia, in Vladivostok and Surgut, within the CIS-LiNet network [1].

The main SLS components are mounted in a specially equipped four-storey building with shaft and hatch (Figure 1). Laser sources are located at vibration-resistant benches around the shaft at the ground level. The key component of SLS is a receiving mirror with a diameter of 2.2 m and a focal length of 10 m; it is located on the same level inside the shaft on a vibro-isolated foundation. The focal plane of this mirror is formed at the level of the 3rd floor. The multichannel operation mode of the lidar with the receiving mirror of 2.2 m in diameter is possible due to the slight inclination of some sounding laser beams relative to the vertical. Thus, there is the possibility of simultaneous reception of up to nine optical signals. Additional receiving mirrors with smaller apertures and diameters of 0.3 and 0.5 m are mounted on the 3rd floor. They provide for the optimal lidar signal detection mode in different sounding altitude ranges due to different base distances between the axes of the receiving and transmitting antennas of the lidar channels. Lidar signals are detected in the photocurrent pulse counting mode. The sounding is carried out in the darkness hours.

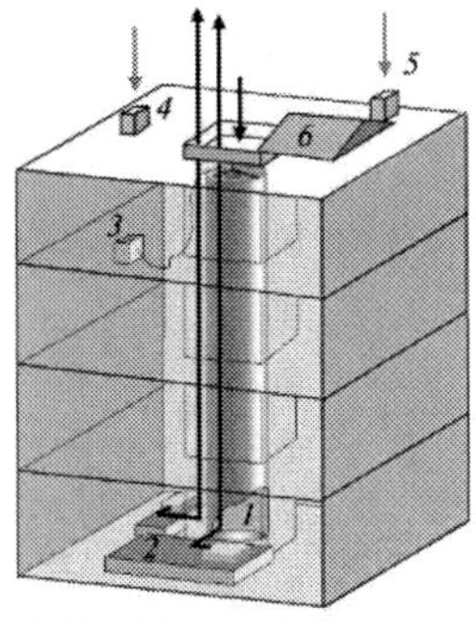

Figure 1 SLS building with the master receiving telescope in the shaft: 2.2 m mirror (1); laser benches (2); receiving and recording equipment (3); M-124 ozonometer (4); twilight spectrophotometer (5); hatch cover (6)

3 SLS STATIONARY MULTICHANNEL LIDAR COMPLEX

The stationary multichannel lidar complex of SLS has been developed with the use of laser sources of different types generating in the 266–1064 nm spectral region and receiving telescopes with 2.2, 0.5, and 0.3 m diameter mirrors. Different sounding techniques allow the complex to receive information about the vertical distribution of some radiation significant atmospheric parameters, i.e., stratospheric ozone and aerosols, air temperature in the region from the tropopause to ~ 80 km, probably optical and geometrical parameters of high-level clouds, and humidity. The use of passive spectrophotometric sounding techniques allows measurements of the total ozone and total NO_2 and their vertical distribution. Table 1 includes the main specifications of the laser sources and receiving optics of the complex.

Table 1 – Specifications of the SLS lidar complex

Lidar parameter	Measured parameter								
	Aerosols					Temperature		Ozone	
Transmitter									
Sounding wavelength λ, nm	511	532	355	683	1064	532		308 353	299 341
Pulse energy, mJ (corr. λ)	1	100	40	60	200	150		100 50	20
Pulse frequency, Hz	2500	15	15	15	15	20		50-200	15
Divergence, mrad	~01					0.1		0.1–0.3	0.1–0.3
Receiver									
Sounding technique	Elastic backscattering						RS by N_2	DIAL	
Wavelength, nm	511	532	355	683	1064	532	607	308 353	299 341
Mirror diameter, m	2.2	0.3	0.3	0.3	2.2	2.2	2.2 0.3	0.5	0.5
Focal length, m	10	1	1	1	10	10	10 1	1.5	1.5
Interference filters									
Maximal transmittance, %	65	62	60	68	60	62	70	56 60	42 50
Bandwidth at the level 0.5T, nm	1.8	2	5	2	3.9	2	1.8	4.5 6	3.6 4

Today, the SLS lidar measurement complex includes two independent lidar systems (Figure 2): a system for measuring parameters of the stratospheric aerosol layer and the middle atmosphere temperature (a) and a lidar for sounding the vertical distribution of the ozone concentration over the stratosphere (b). Figure 3a shows that the system developed provides for lidar signal recording from up to the upper boundary of the stratospheric aerosol layer, ~30 km. Figure 3b shows the profiles of the scattering ratio R(H), which exemplify the three-frequency stratospheric aerosol sounding results under the presence of traces of volcanogenic aerosols emitted during the explosions of Okmok and Kasatochi volcanoes.

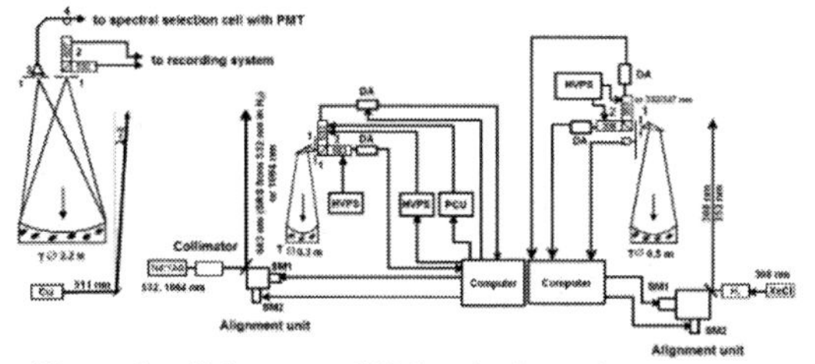

Figure 2 – Schematic of SLS multichannel measuring complex: field stops (1); spectral selection cell with PEM (2); focon (3); light guide (4); mechanical obturator (5); solid laser (Nd:YAG); eximer laser (XeCl); Cu vapor laser (Cu); SRS conversion cell filled with hydrogen (H_2); collimator (C); discriminator amplifier (DA); high-voltage power supplies (HVPS); step motors (SM)

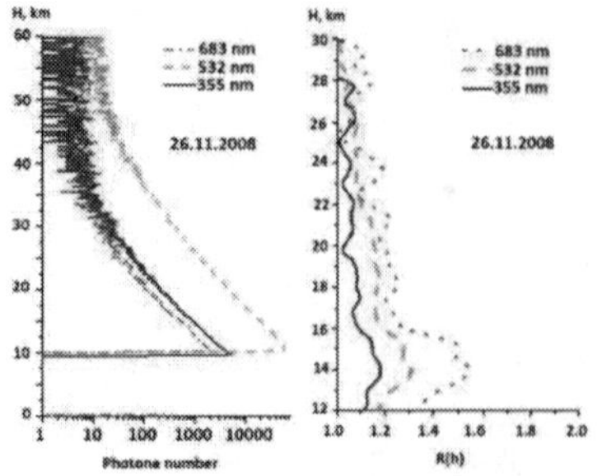

Figure 3 – Lidar signals at three sounding wavelengths (a) and altitude profiles of the scattering ratio retrieved for November 11, 2008 (b)

4 MULTIWAVELENGTH POLARIZATION LIDAR LOZA-S

The stationary LOZA-S lidar has been designed and created at the Institute of Atmospheric Optics. It ensures simultaneous recording of lidar signals of elastic and Raman scattering (RS) when irradiating a medium at wavelengths of 1064, 532, and 355 nm. The lidar (Figure 4) is based on a LOTIS-2135 Nd:YAG laser and a receiving Cassegrain mirror telescope with a diameter of 300 mm. Lidar specifications are given in Table 2.

Figure 4 – LOZA-S lidar

Table 2 – Stationary lidar LOSA-S

Coordinates:	56.48 N, 85.05 E;
Altitude range, approx.	0.1–20 km
Spatial resolution	1.5–12 m
Radiation wavelength:	355, 532, 1064 nm
Pulse energy (typ.):	160, 400, 700 mJ
Pulse frequency (typ.):	10 Hz
Beam divergence:	0.2 mrad
Telescope:	Cassegrain
Objective diameter:	300 mm
Focus length:	1500 mm

The LOZA-S lidar was used for routine measurements of the vertical aerosol distribution in Tomsk. On the basis of these measurements, the altitude distribution of tropospheric aerosol layers was analyzed, including the boundary layer, which consists of internal mixing layer, mixing layer, and entrainment layer, and the middle tropospheric layer. Figure 4 exemplifies lidar signals detected in day- and nighttime; RS signals were recorded in the photon counting mode in darkness hours.

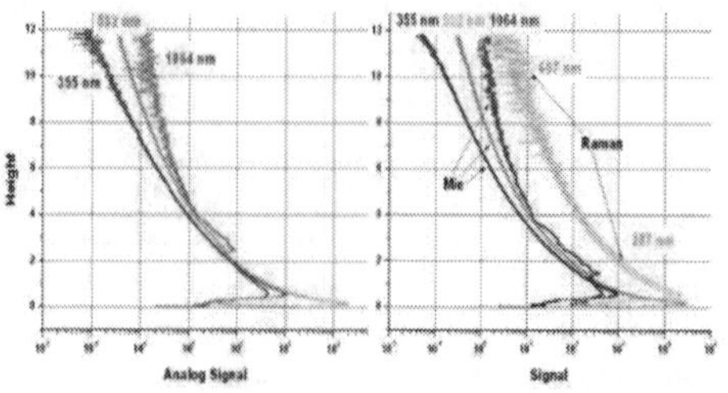

Figure 4 – LOZA-S lidar signals when sounding in day (left) and nighttime (right)

5 REMOTE MEASUREMENTS OF THE TEMPERATURE IN THE STRATOSPHERE AND MESOSPHERE

Figure 5 shows the optical schematic of the temperature channel of the complex mean pure rotational Raman scattering (PRRS) lidar of SLS. It was used for testing the technique for discrimination of the regions of pure rotational SR spectrum with the help of a double monochromator. A Powerlight 9030B Nd:YAG laser is used as a source of exciting radiation.

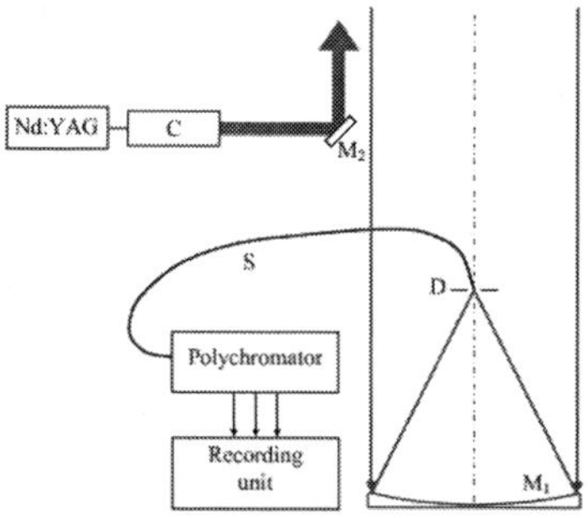

Figure 5 Schematic structure of the temperature channel of the PRRS lidar: main mirror of the receiving telescope of 2.2 m in diameter (M_1), receiver's deflecting mirror (M_2), field stop (D), light guide (S), collimator (C)

Figure 6 shows the lidar response of the pure rotational RS spectrum in the presence of a deep ice crystal cloud. A high level of suppression of the unbiased scattering provides for a high quality of the Raman signal at the lower boundary and inside the cloud.

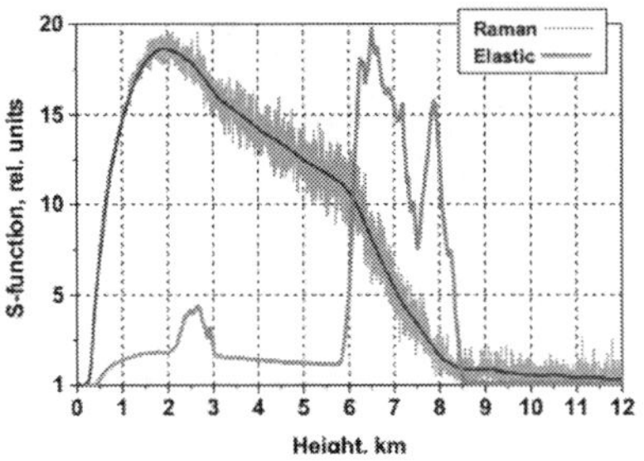

Figure 6 Lidar response of the pure rotational RS spectrum in the presence of a deep ice crystal cloud

6 LIDAR FOR OZONE SENSING

The sounding wavelengths selected 299/341 nm were used in the ozone lidar designed. The 4th harmonic 266 nm of the fundamental frequency of a Nd:YAG laser (LS-2134UT laser, LOTIS TII company, Minsk) is used as a laser radiation source, which is then converted in hydrogen in the 1st (299 nm) and 2nd (341 nm) Stokes components. Ozone profiles retrieved from lidar sounding data are compared with the IASI/MetOp satellite measured profiles. The range of lidar signal detection was from 5 to 18 km. The lidar measurements show the ozone behavior in more detail. Figure 7 shows all the considered cases of comparison of ozone profiles measured at the SLS and retrieved from IASI data. The measurements are reduced to the total altitude range 6–15 km for convenience. Increased ozone concentrations measured at the SLS as compared to IASI are seen in the range from 8.5 to 12.5 km throughout the whole observation period.

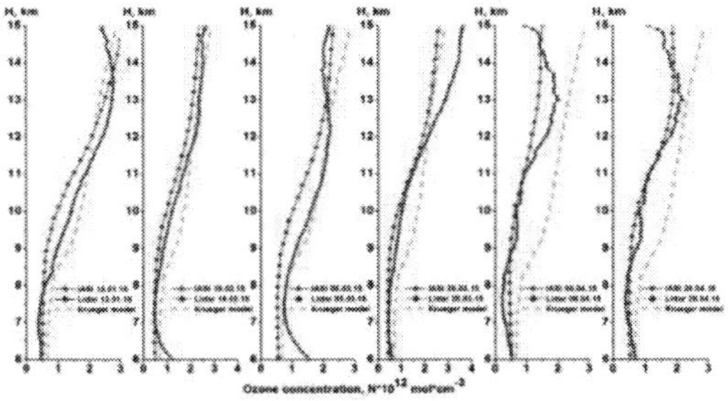

Figure 7 Comparison of vertical profiles of ozone

7 OPO LIDAR SYSTEM

SLS includes an OPO lidar for remote gas analysis of the atmosphere in the 3–4 μm spectral range. A pumping wavelength of 1064 nm is transformed by means of OPO based on nonlinear KTP and KTA crystals. Crystal turn allows turning the radiation wavelength in the 3–4 μm spectral range. The main OPO lidar specifications are given in Table 3.

Table 3 Specification of the experimental setup

Laser maximal pulse energy, mJ	9
Laser pulse length, ns	10–13
Laser pulse frequency, Hz	10
Laser beam divergence, mrad	2
Main mirror diameter, mm	300
Focal length, mm	120
Photodetector spectral range, μm	1–4
Diameter of photodiode detection area, μm	300
Input rise time, ns	50

Energy parameters of a laser system have been measured in experiments. The sounding path is 20 m long. Backscattered radiation is gathered by a receiving telescope and focused to the PD38-03-PR (IBSG Co., St-Peterburg, Russia) photodiode detection area. The level of signals with the use of the above optical elements allows a suggestion about possibilities of sounding the atmosphere along a surface tropospheric path up to 1 km long. Figure 8 shows lidar signals: minimum signals (black curve), maximum signals (blue curve), result of averaging of lidar signals (red curve) in the 3 – 4 μm spectral region.

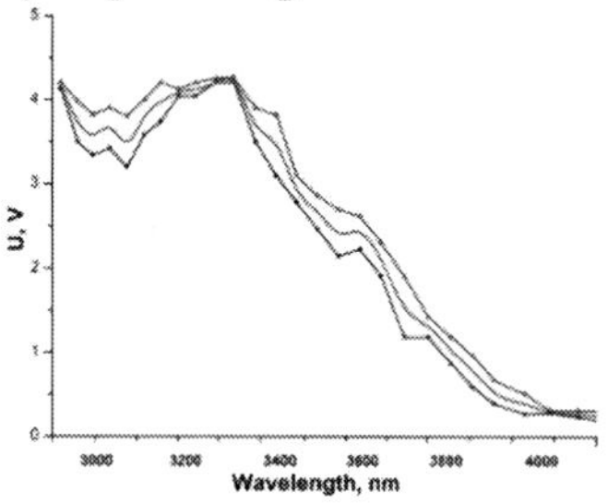

Figure 8 OPO lidar signals

ACKNOWLEDGEMENTS

The work was supported by the Russian Science Foundation (Agreement 15-17-10001 in part of the ozone lidar measurements) and President of the Russian Federation (grant MK-1367.2017.5 in the part of the experiments on recording lidar signals of the broadband mid-IR lidar).

References

[1] Chaykovskii A.P., Ivanov A.P., Balin Yu.S., El'nikov A.V., Tulinov G.F., Plusnin I.I., Bukin O.A., Chen B.B., 2005: CIS-LiNet lidar network for monitoring aerosol and ozone: methodology and instrumentation, *Atmospheric and oceanic optics*, 18 (12), 958-964.

RECENT DEVELOPMENTS WITH THE ASIAN DUST AND AEROSOL LIDAR OBSERVATION NETWORK (AD-NET)

Nobuo Sugimoto[1]*, Atsushi Shimizu[1], Tomoaki Nishizawa[1], Yoshitaka Jin[1]

[1]*National Institute for Environmental Studies, Japan, *nsugimot@nies.go.jp*

ABSTRACT

Recent developments of lidars and data analysis methods for AD-Net, and the studies using AD-Net are presented. Continuous observation was started in 2001 at three stations using polarization-sensitive Mie-scattering lidars. Currently, lidars, including three multi-wavelength Raman lidars and one high-spectral-resolution lidar, are operated at 20 stations. Recent studies on validation/assimilation of chemical transport models, climatology, and epidemiology of Asian dust are also described.

1 INTRODUCTION

Continuous operation of AD-Net was started in 2001 with three stations (Tsukuba, Nagasaki, and Beijing). The number of stations increased in various research and international cooperation programs. Currently 21 lidars are operated in East Asia (Japan, (China), Korea, Mongolia, Thailand). The lidars at 12 stations are two-wavelength (1064 nm and 532 nm) polarization (532 nm) backscattering lidars. The lidars at 6 stations have additional nitrogen vibrational Raman receivers (607 nm). The lidars at 3 stations are multi-wavelength Raman lidars (backscattering at 1064 nm, 532 nm, and 355 nm, extinction at 532 nm and 355 nm, depolarization at 532 nm and 355 nm). Additionally, continuous observation was recently started at NIES in Tsukuba with a high-spectral-resolution and Raman lidar (HSR at 532 nm, Raman at 355 nm, depolarization at 532 nm and 355 nm) [1].

The lidar data from AD-Net are transferred to NIES. The backscattering lidar part of the data are automatically processed in realtime.

2 AD-NET DATA PRODUCTS

Basic real-time products of AD-Net are the attenuated backscattering coefficient at 532 nm and 1064 nm and the volume depolarization ratio. The extinction coefficient estimates for non-spherical and spherical aerosols are also provided in realtime [2, 3]. Table 1 lists the realtime standard AD-Net products, and Figure 1 presents an example of the data.

Table 1 AD-Net standard data product

Level	Product	Used lidar channels	Other data used
1	Attenuated backscattering coefficient at 532 nm	532 nm co- & cross-pol.	No
1	Attenuated backscattering coefficient at 1064 nm	1064 nm	No
1	Volume depolarization ratio at 532 nm	532 nm co- & cross-pol.	No
2	Dust extinction coefficient estimate at 532 nm + cloud mask	532 nm co- & cross-pol + 1064 nm	S_1=50 sr assumption, molecular profile
2	Spherical aerosol extinction coefficient estimate at 532 nm + cloud mask	532 nm co- & cross-pol + 1064 nm	S_1=50 sr assumption, molecular profile

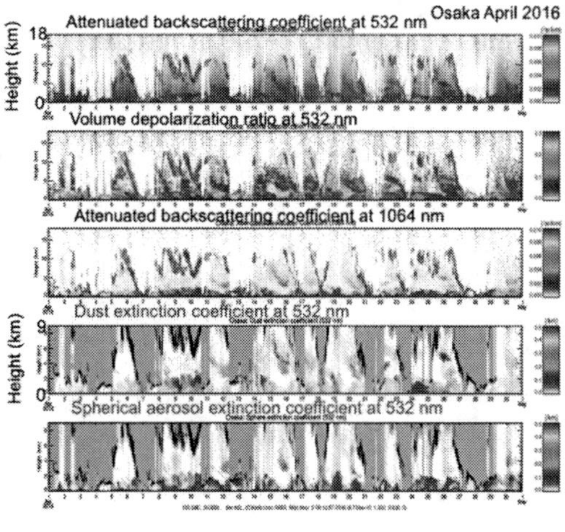

Figure 1 AD-Net standard data products (example for Osaka, April 2016).

The data are updated every hour and provided also in the NetCDF format through the AD-Net www page (http://www-lidar.nies.go.jp/AD-Net/).

Data from Raman lidars and a HSRL are also transferred to NIES and processed. Figure 2 shows an example of the data from the HSRL in Tsukuba. A near real-time data analysis system for multi-parameter lidars was developed to derive the backscattering coefficient at 1064 nm, 532 nm, and 355 nm, the extinction coefficient at 532 nm and 355 nm, and the depolarization ratio at 532 nm and 355 nm. Also, data analysis method for deriving concentration of aerosol components was developed [4, 5].

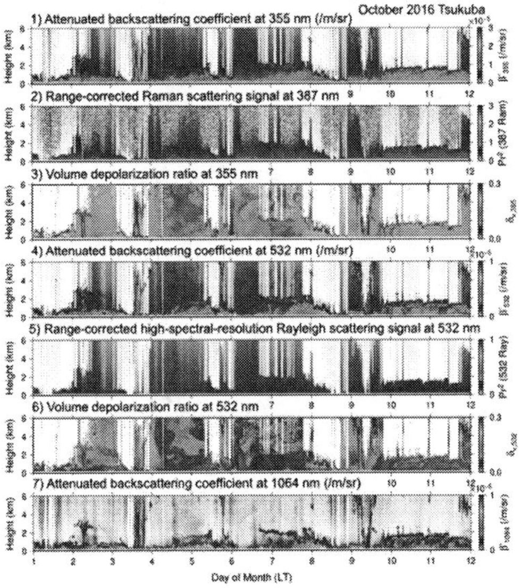

Figure 2 Continuous 532-nm HSR and 355-nmRaman lidar observation in Tsukuba.

3 APPLICATIONS OF AD-NET DATA

AD-Net data are used in various applications. The real-time data (extinction coefficient estimates for non-spherical dust) are used for nowcast of Asian dust on the www page of Japanese Ministry of the Environment for the public. The data are also used for real-time validation of dust transport models.

Data assimilation of chemical transport models (CTMs) is one of the primary purposes of AD-Net. A 4D-Var data assimilation system based on regional chemical transport model (CFORS) was developed at Kyusu University and applied to analysis of several dust events successfully [6, 7]. The dust extinction coefficient was used in the data assimilation, but the assimilated model reproduced both optical and surface observations not used in assimilation (e.g. AOD and PM10) very well.

AD-Net data are not regularly used in data assimilation of aerosol climate models, so far. One of the reason is the lidar ratio problem with backscattering lidar data analysis. To improve the situation, we considered methods for using backscattering lidar data validation/assimilation of chemical transport models without using assumptions in lidar data analysis.

3.1 Quantitative use of backscattering lidar data in validation/assimilation of chemical transport models

A method is to use the attenuated backscattering coefficient itself to compare with models. However, the method may have a problem for ground-based lidars. It is usually difficult to reproduce near-surface aerosol distribution accurately with models, and the modeled attenuated backscattering coefficient profile may contain large error.

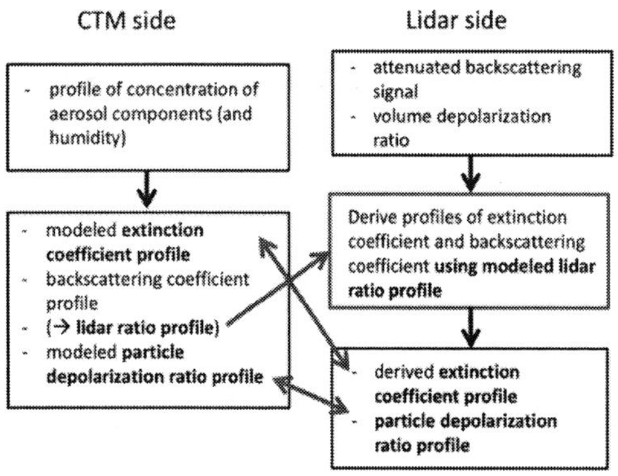

Figure 3 Validation of chemical transport model using the extinction coefficient profile derived from backscattering lidar data using the modeled lidar ratio.

An alternative method is using the extinction coefficient profile derived using modeled lidar ratio profile. Figure 3 shows the concept. If the chemical transport model does not have optical models for aerosol components, we may calculate from the modeled mass concentrations using external module.

We tested the method using modeled lidar ratio with the AD-Net lidar in Fukuoka and the MASINGAR mk-2 model of Meteorological Research Institute of Japan. Figure 4 shows the results of the model calculation.

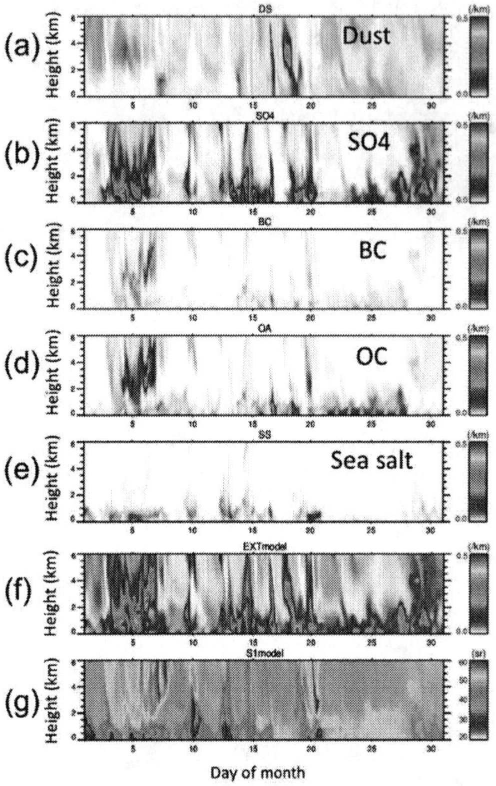

Figure 4 (a)-(f): Extinction coefficient time-height indications for (a) dust, (b) sulfate, (c) black carbon, (d) organic carbon, (e) sea salt, and (f) total, calculated by MASINGER mk-2 for Fukuoka, April 2015. (g): lidar ratio calculated by the model.

The lidar ratio of the aerosol components in the optical models used in this calculation is as follows. Dust: 48 sr (spheroid model, mode radius 2 μm), SO_4: 50 sr (GEOS-Chem, Rh=60% (dry radius 0.08 μm)), BC: 101 sr (OPAC), OA: 52 sr (OPAC water-soluble model, Rh=60%), sea salt: 20 sr (OPAC, Rm=3μm, Rh=60%). The particle depolarization ratio of dust is 0.35 and others is 0.00.

Figure 5 shows the extinction coefficient profiles derived using the modeled lidar ratio and the constant lidar ratio. The difference between them are also shown. Figure 6 compares the modeled and derived dust extinction coefficients.

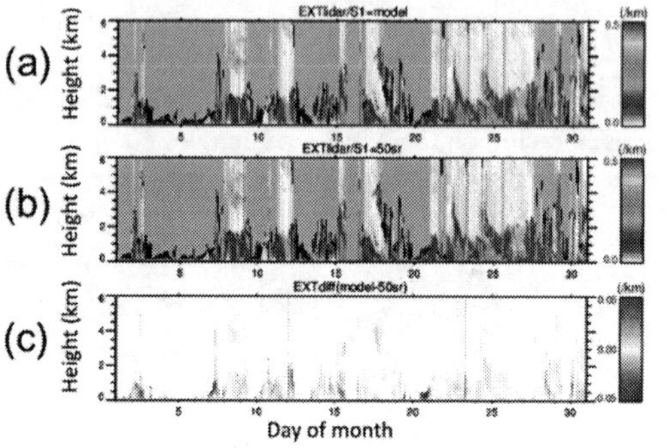

Figure 5 Extinction coefficient derived from the backscattering lidar data using (a) modeled lidar ratio, (b) constant lidar ratio (S_1=50 sr), and (c) the difference (a-b).

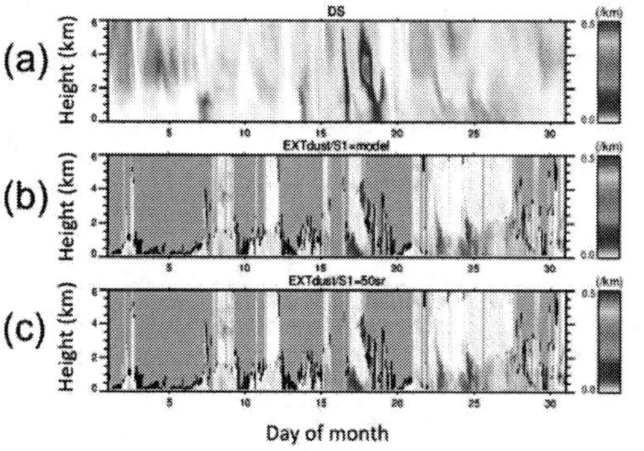

Figure 6 (a): Modeled dust extinction coefficient, (b)(c): lidar derived dust extinction coefficient using (a) modeled lidar ratio, (b) constant lidar ratio.

We showed that backscattering lidar data can be used quantitatively in validation/assimilation of CTM. The result also showed that the difference between the extinction coefficient derived with the modeled lidar ratio and the constant lidar ratio was not large except for the sea-salt dominant case. It was much smaller than the difference between the modeled extinction coefficient. We think we may use backscattering lidar extinction coefficients derived with a constant lidar ratio in applications such as dust event analysis.

3.2 Epidemiology using AD-Net data

The dust and spherical aerosol extinction coefficients provided as standard AD-Net data are useful for epidemiology of Asian dust. We calculate daily near-surface dust and spherical aerosol extinction coefficients and provide to the researchers in epidemiology of aerosols. Statistically significant effects of Asian dust were found in asthma and other respiratory symptoms. So far, about 10 papers are published using AD-Net data [8-10]. A study for providing a qualitative index of internal mixing of dust and air pollution particles is being conducted [11].

4 CONCLUDING REMARKS

AD-Net celebrated 15th anniversary in 2016. Analysis of major aerosol events and climatology of Asian dust and anthropogenic aerosols are under way using long-term AD-Net data.

Development of multi-parameter lidars and data analysis methods are also conducted for better aerosol characterization. One of our major targets is assimilation of multi-parameter lidars with a CTM that resolves aerosol components.

ACKNOWLEDGEMENTS

We thank Ichiro Matsui and all persons involved in AD-Net observations and data analysis. We thank Takashi Maki and Taichu Y. Tanaka of MRI for the MASINGAR mk-2 simulation results.

References

[1] Shimizu A., Nishizawa, T., Jin, Y., Sugimoto, N., 2016: Evolution of a lidar network for tropospheric aerosol detection in East Asia, *Opt. Eng.* **56** (3), 031219.

[2] Sugimoto, N., Uno, I., Nishikawa, M., Shimizu, A., Matsui, I., Dong, X., Chen, Y., Quan, H., 2003: Record heavy Asian dust in Beijing in 2002: Observations and model analysis of recent events, *Geophys. Res. Lett.* **30** (12), 1640.

[3] Shimizu, A., Sugimoto, N., Matsui, I., Arao, K., Uno, I., Murayama, T., Kagawa, N., Aoki, K., Uchiyama, A., Yamazaki, A., 2004: Continuous observations of Asian dust and other aerosols by polarization lidar in China and Japan during ACE-Asia, *J. Geophys. Res.*, **109**, D19S17.

[4] Nishizawa, T., Sugimoto, N., Matsui, I., Shimizu, A., Hara, Y., Uno, I., Yasunaga, K., Kudo, R., Kim, S.-W., 2017: Ground-based network observation using Mie-Raman lidars and multi-wavelength Raman lidars and algorithm to retrieve distributions of aerosol components, *J. Quantitative Spectroscopy & Radiative Transfer* **188**, 79-93.

[5] Hara Y., Nishizawa, T., Sugimoto, N., Matsui, I., Pan, X., Kobayashi, H., Osada, K., Uno, I., 2017: Optical properties of mixed aerosol layers over Japan derived with multi-wavelength Mie-Raman lidar system, *J. Quantitative Spectroscopy & Radiative Transfer* **188**, 20-27.

[6] Yumimoto, K. et al., 2008: Adjoint inversion modeling of Asian dust emission using lidar observations, *Atmos. Chem. Phys.* 8, 2869–2884.

[7] Hara, Y. et al., 2009: Asian dust outflow in the PBL and free atmosphere retrieved by NASA CALIPSO and an assimilated dust transport model, *Atmos. Chem. Phys.* **9**, 1227-1239.

[8] Kanatani, K. et al., 2016: Effect of desert dust exposure on allergic symptoms: a natural experiment in Japan, *Ann. Allergy, Asthma Immunol.* **116** (5), 425–430.

[9] Higashi, T. et al., 2014: Effects of Asian dust on daily cough occurrence in patients with chronic cough: a panel study, *Atmos. Environ.* **92**, 506–513.

[10] Watanabe, M., et al., 2015: Association of sand dust particles with pulmonary function and respiratory symptoms in adult patients with asthma in western Japan using light detection and ranging: a panel study, *Int. J. Environ. Res. Public Health* **12** (10), 13038–13052.

[11] Sugimoto, N. et al. 2015: Detection of internally mixed Asian dust with air pollution aerosols using a polarization optical particle counter and a polarization-sensitive two-wavelength lidar, *J. Quant. Spectrosc. Radiat. Transfer* **150**, 107–113.

LALINET STATUS - STATION EXPANSION AND LIDAR RATIO SYSTEMATIC MEASUREMENTS

Eduardo Landulfo[1,*], Fabio Lopes[1], Gregori Arruda Moreira[1], Jonatan da Silva[1], Pablo Ristori[2], Eduardo Quel[2], Lidia Otero[2], Juan Vicente Pallota[2], Milagros Herrera[2], Jacobo Salvador[2], Juan Lucas Bali[2], Eliam Wolfram[2], Paula Etala[3], Albane Barbero[3], Ricardo Forno[4], Maria Fernanda Sanchez[4], Henrique Barbosa[5], Diego Gouveia[5], Amanda Vieira Santos[5], Judith Hoelzemann[6], Jose Henrique Fernandez[6], Anderson Guedes[6], Antonieta Silva[7,8], Boris Barja[9], Felix Zamorano[9], Raul Perez Legue[9], Alvaro Bastidas[10], Maribel Vellejo Zabala[10], Juan Velez[10], Daniel Nisperuza[10], Elena Montilla[11], Rene Estevam Arredondo[12], Juan Carlos Antuña Marrero[12], Alberth Rodriguez Vega[12], Lucas Alados-Arboledas[13,14], Juan Luis Guerrero-Rascado[13,14], Nobuo Sugimoto[15] and Jin Yoshitaka[15]

[1]*Instituto de Pesquisas Energéticas e Nucleares, Centro de Lasers e Aplicações, Av. Prof. Lineu Prestes, 2242, São Paulo, 05508000, Brazil, *elandulf@ipen.br*
[2]*CITEDEF, CEILAP, San Juan Bautista de la Salle 4397, Buenos Aires, B1603ALO, Argentina*
[3]*Servicio Meteorologico Nacional, 25 de Mayo 658, C1002ABN CABA, Argentina*
[4]*Laboratorio de Física de la Atmosfera, Universidad Mayor de San Andrés, La Paz, Bolivia*
[5]*Dep. de Fisica Aplicada, U. de São Paulo, Rua do Matão, 1371, 05508-090, São Paulo, Brazil*
[6]*CCET, U. Federal do Rio Grande do Norte, Av Sen Salgado Filho, 59078900, RN, Brazil*
[7]*Center for Optics and Photonics (CEFOP), Universidad de Concepción - Chile*
[8]*Departamento de Ciencias Físicas, Universidad de La Frontera, Temuco, Chile*
[9]*Atmospheric Research Laboratory (LIA), UMAG, Punta Arenas, Chile*
[10]*Grupo de Investigación Láseres y Espectroscopia, Universidad Nacional de Colombia, Medellín, Colombia*
[11]*Escuela de Ciencias, Universidad EAFIT, Medellín, Colombia*
[12]*Grupo de Óptica Atmosférica de Camagüey, CMC, INSMET, Cuba*
[13]*Andalusian Institute for Earth System Research (IISTA-CEAMA), Spain*
[14]*Dpt. Applied Physics, Faculty of Sciences, University of Granada, Spain*
[15]*National Institute for Environmental Studies, Tsukuba, Japan*

ABSTRACT

LALINET is expanding regionally to guarantee spatial coverage over South and Central Americas. One of the network goals is to obtain a set of regional representative aerosol optical properties such as particle backscatter, extinction and lidar ratio. Given the North-South extension and influence of distinct airmass circulation patterns it is paramount to distinguish these optical parameters in order to gain better perfomance in radiation transfer models. A set of lidar ratio data is presented.

1 INTRODUCTION

Latin America Lidar Network is a federative fully operative lidar for aerosol research in South and Central America. Its operational cycle concerns probing the atmosphere on regular basis to retrieve vertically and temporally resolved aerosol optical and microphysical properties(Luis Guerrero-Rascado et al., 2016). Most of the instruments based in the network were customized and produced configurations suitable for the aerosol observation task. The network was officially established in 2013 by recognition as being part of **GAW-GALION** programs under the supervision of World Meteorological Organization(Antuña-Marrero et al., 2016). The present goals of LALINET are:

1. Consolidate the measurement and data acquisition protocols

2. Establish a QA/QC routine among all stations

3. Improve and establish an unified data analysis routine common to all stations, e.g., Single Calculus Chain(D'Amico et al., 2015)

4. Create a scientifically signficant distributed database,e.g., lidar ratio, particle extinction, backscatter, angstrom exponents and particle depol. regional values

5. Guarantee the continuity of LALINET's biannual workshops by the intensification its role as a mechanism of transfer of knowledge, evaluation of joint actions conducted and agreemnts on how to continue under the new evolving situations. In 2018 second semester, the host site should be in Medellin, Colombia

Yet regarding the vast majority of stations share a collocated AERONET station the use of the GRASP approach is also envisaged(Torres et al., 2016) in order to obtain particle microphysical properties. Currently LALINET is seeking its expansion to better cover up a region roughly twice the size of EARLINET(Mona, L. et al., 2016) and therefore a region where more circulation patterns take place. A careful establishment of a regional distributed aerosol optical parameter database pose a challenge wich could be tackled with the network expansion. In this matter 2 stations were deployed as seen in table 1. One in Natal, Brazil, which should be focused on long-range aerosol transport from Saharan dust. The other in Punta Arenas in Chile, which is also part of SAVER.NET(Quel, Eduardo et al., 2016) which merges the efforts of LALINET and AD-NET(Nishizawa, Tomoaki et al., 2016).

1.1 LALINET Lidar Stations

Up to date there are 14 operative stations, see Table 1, distributed along the South american continent. The vast majority has collocated AERONET sunphotometer stations which helps in the lidar data retrieval of the lidar ratio by means of a closure between the two instruments and further on the implementation of data integration algorithms such as GRASP.

2 METHODOLOGY

LALINET measurement protocol follows an twice a week measurements on Mondays and Thursdays, even though pne station operates continuously (MAO) and with other schedules for other proposes. The products obtained are aerosol backscattering and extinction profiles, Lidar ratios (average and profiles when raman channels are operative). Since the vast majority of stations have collocated AERONET sunphotometers it is possible to obtain microphysical products. Some of the station also have polarization-sensitive channels which allows information on shape and identification of aerosols.

3 RESULTS

3.1 Calbuco Eruption

On April 22, 2015, the Calbuco volcano in Chile (41.33^{o}S, 72.62^{o}W) erupted after 43 years of inactivity, followed by an injection of pyroclastic material dispersed into the atmosphere on weeks after the event. At first a threat-alarm to aviation traffic and air quality over a large area was released, and the stations on the southermost part of SA were triggered on daily monitoring basis in Argentina, Brazil, and Chile. Below we find a temporal serie of range corrected signal at 532 nm the quick look of the

*Table 1: LALINET Lidar Stations - The boxes in green are newly added stations in 2016. *Temuco station is the previous Concepcion Station.*

ST.	ID	LAT(S) LON(W)	Channels (nm)
Buenos Aires	*AEP*	34.56°*S* 58.42°*W*	*1064, 532^p & 355^p*
Buenos Aires	*VMA*	34.56°*S* 58.51°*W*	*1064, 607, 532^p, 387 & 355^p*
Neuquen	*NQN*	38.95°*S* 68.13°*W*	*1064, 532^p & 355^p*
Barilo-che	*BRC*	41.15°*S* 71.16°*W*	*1064, 607, 532, 387 & 355*
Commo doro	*CDR*	45.79°*S* 67.46°*W*	*1064, 532 & 355*
Gal-legos	*RGL*	51.61°*S* 69.31°*W*	*1064, 532^p & 355^p*
Punta Arenas	*PAR**	53.13°*S* 70.88°*W*	*1064, 607, 532^p, 408, 387 & 355^p*
S. Paulo	*SPU*	23°13′ 46°28′	*1064, 607, 532, 408, 387 & 355*
S. Paulo	*SPT*	*VAR*	*607, 532*
Man-aus	*MAO*	02.60°*S* 60.21°*W*	*408, 387, 355*
Natal	*NAT*	05.82°*S* 35.20°*W*	*1064, 532^p & 355^p*
*Temuco**	*TMU**	38.74°*S* 72.62°*W*	*1064, 532^p & 355^p*
Medellin	*MED*	06.26°*N* 75.58°*W*	*532 & 355*
La Paz	*LPZ*	16.54°*S* 68.07°*W*	*1064, 532^p & 355^p*

lidar measurements conducted at AEP station, in Buenos Aires, Argentina. The volcanic layer detected on April 24 shows a persistency along the entire day ranging between 5 and 7 km Later on there is a subsidency of this layer not mixing however with the urban mixing layer below the PBL around 1.5 - 2.0 km. This station, to the east of the Calbuco, measured a tropospheric aerosol layer at about 6 km a.g.l. which subsided along the day of measurement.

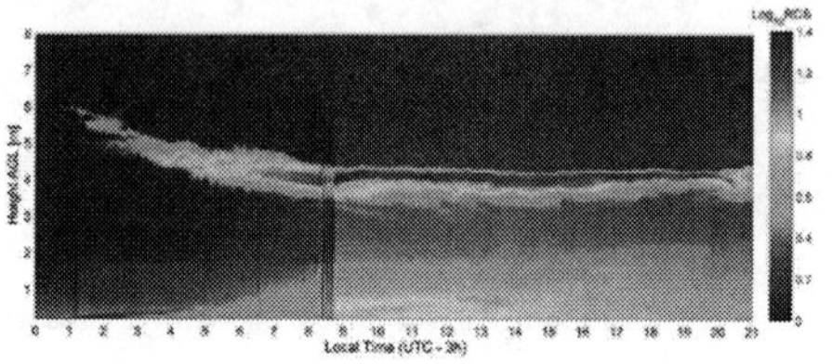

Figure 1: RCS quicklook obtained by AEP station, in Buenos Aires, Argentina, 2 days after the Calbuco Eruption allowing a consistent tracking of the pyroclastic material ejected by the volcano spreading over south of South America.

3.2 Lidar Ratio Evaluation

Aiming to create a consistent lidar ratio climatology regionally distributed we have evaluated the yearly mean values of some LALINET stations which included the results obtained from procotol measurements taken routineously. Some of the values are from intensive campaigns taken previously (BUE and PAR) using lidar/sunphotometer sinergy in clear sky conditions. All LR"s are at 532 nm except MAO (355 nm). The standard deviations at each value are about $10 - 15\%$.

4 CONCLUSIONS

LALINET is in expansion and working to reach good standards and practices for regional monitoring of aerosols and trace gases in South and Central America. The network now comprehends 14 operative stations. The network has been capable of monitoring events under

Figure 2: Lidar Ratios at some of LALINET's stations. These values should be updated and extended to other stations in order to have a regional representative values.

a routine and obtaining aerosol optical properties with regional representation and sparse events such as volcanic eruptions. Network capabilities are also under scrunity to comply with QA/QC protocols.

ACKNOWLEDGEMENTS

This work was also partially funded by the University of Granada through the contract *Plan Propio. Programa 9. Convocatoria 2013*, additional support was given by Universidad Nacional Sede Medellín, NASA-AERONET. FAPESP and CNPq support the brazilian stations.

References

Antuña-Marrero, J. C., Landulfo, E., Estevan, R., Barja, B., Robock, A., Wolfram, E., … Guerrero-Rascado, J. L. (2016). LALINET: The first Latin American-born regional atmospheric observational network. *Bulletin of the American Meteorological Society.*

D'Amico, G., Amodeo, A., Baars, H., Binietoglou, I., Freudenthaler, V., Mattis, I., … Pappalardo, G. (2015). Earlinet single calculus chain – overview on methodology and strategy. *Atmospheric Measurement Techniques*, *8*(11), 4891–4916.

Luis Guerrero-Rascado, J., Landulfo, E., Carlos Antuna, J., Jorge Barbosa, H. d. M., Barja, B., Efrain Bastidas, A., … Silva, A. (2016). Latin American Lidar Network (LALINET) for aerosol research: Diagnosis on network instrumentation. *JOURNAL OF ATMOSPHERIC AND SOLAR-TERRESTRIAL PHYSICS*, *138*, 112-120.

Mona, L., Arboledas, L. Alados, Amiridis, V., Amodeo, A., Apituley, A., Balis, D., … Pappalardo, G. (2016). Earlinet: 12-year of aerosol profiling over europe. *EPJ Web of Conferences*, *119*, 19002.

Nishizawa, Tomoaki, Sugimoto, Nobuo, Matsui, Ichiro, Shimizu, Atsushi, Higurashi, Akiko, & Jin, Yoshitaka. (2016). The asian dust and aerosol lidar observation network (ad-net): Strategy and progress. *EPJ Web of Conferences*, *119*, 19001.

Quel, Eduardo, Sugimoto, Nobuo, Otero, Lidia, Jin, Yoshitaka, Ristori, Pablo, Nishizawa, Tomoaki, … Mizuno, Akira (2016). Aerosols monitoring network to create a volcanic ash risk management system in argentina and chile. *EPJ Web of Conferences*, *119*, 19006.

Torres, B., Dubovik, O., Fuertes, D., Schuster, G., Cachorro, V. E., Lapionak, T., … Tanré, D. (2016). Advanced characterization of aerosol properties from measurements of spectral optical depth using the grasp algorithm. *Atmospheric Measurement Techniques Discussions*, *2016*, 1–47.

STATUS OF THE NASA MICRO PULSE LIDAR NETWORK (MPLNET): OVERVIEW OF THE NETWORK AND FUTURE PLANS, NEW VERSION 3 DATA PRODUCTS, AND THE POLARIZED MPL

Ellsworth J. Welton[1]*, Sebastian A. Stewart[2], Jasper R. Lewis[3], Larry R. Belcher[2], James R. Campbell[4], and Simone Lolli[3]

*[1]NASA Goddard Space Flight Center, USA. *Ellsworth.J.Welton@nasa.gov*
[2]Science Systems and Applications Inc., Goddard Space Flight Center, USA
[3]University of Maryland Baltimore County, Goddard Space Flight Center, USA
[4]Naval Research Laboratory, Marine Meteorology Division, USA

ABSTRACT

The NASA Micro Pulse Lidar Network (MPLNET) is a global federated network of Micro-Pulse Lidars (MPL) co-located with the NASA Aerosol Robotic Network (AERONET). MPLNET began in 2000, and there are currently 17 long-term sites, numerous field campaigns, and more planned sites on the way. We have developed a new Version 3 processing system including the deployment of polarized MPLs across the network. Here we provide an overview of Version 3, the polarized MPL, and current and future plans.

1 INTRODUCTION

The NASA Micro-Pulse Lidar Network (MPLNET) [1] is a global federated network of polarized Micro-Pulse Lidar (MPL) systems designed to measure aerosol and cloud vertical structure continuously, over long time periods required to contribute to climate change studies and provide ground validation for satellite sensors and related aerosol modeling and forecasting efforts. MPLNET began in 2000, and there are currently 17 long-term and numerous short-term field campaign sites. Seven of the long-term sites have 10+ years of data, and many more have 5+ years. There are 6 more sites in planning stages towards operational status by end of 2017 and several more at proposal stage. Most MPLNET sites are co-located with the NASA Aerosol Robotic Network (AERONET) [2], producing both column and vertically resolved data on aerosol and cloud properties and the evolution and structure of the planetary boundary layer (PBL). MPLNET is also a member of the WMO Global Atmospheric Watch (GAW) Aerosol Lidar Observation Network (GALION) [3].

MPLNET has been running the Version 2 processing system [4] since 2006. We have developed an enhanced and more mature Version 3 (V3) processing system, set for release in spring 2017. The new website is already active at http://mplnet.gsfc.nasa.gov. V3 introduces a new polarized MPL, already deployed throughout the network. In addition to the new polarized data products, vast improvements to all products will be provided including new cloud optical data and a more robust mixed layer depth product to examine the PBL. The entire V2 data archive is available from the new V3 website, and new V2 data will continue to be produced through 2017. This presentation will provide an overview of MPLNET Version 3, the polarized MPL, and current and future plans.

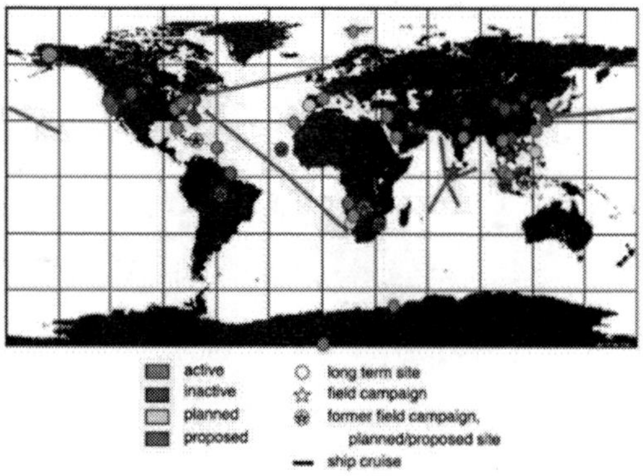

Figure 1. MPLNET Sites and Deployments

2 THE POLARIZED MPL

The original MPL design from the 1990s was depolarized. A solution to polarize the MPL was presented by Flynn et al [5] in the early 2000s, and many instruments were sold by the manufacturer (Sigma Space Corporation). This version of the polarized MPL was not used within MPLNET because a full performance and sensitivity study was not completed. The design showed promise but was not ready to integrate into MPLNET. In addition, the design utilized a liquid crystal retarder (LCR) to switch polarization states, but the nematic LCR is limited to millisecond switching speeds. In practice the MPL was switched much slower which could affect cloud observations.

Several years ago, the MPLNET team began working with Sigma Space to redesign the polarized MPL to utilize a ferroelectric liquid crystal (FLC) rotator, which is capable of switching the emitted polarization state of the transmitted MPL pulses at microsecond data rates. The new design produces similar data, and data processing techniques described by Flynn et al [5] remain applicable. However, the operation of the MPL using a rotator versus a retarder requires a different assembly methodology and calibration process. Stewart et al [6] (this issue) present an overview of the new polarized MPL design and present a methodology for calibrating the FLC component. A more thorough presentation of this polarized MPL, its performance characteristics, system bias, and calibration procedures will be provided in a forthcoming publication.

In summary, the new calibration protocols account for the optical properties of the MPL components, the most critical being the polarizing beam-splitter (PBS) and the FLC. The window used for data collection may also be important if stressed due to mechanical/temperature fluctuations. The diattenuation, retardance, and alignment/rotation angles of the PBS and FLC are critical parameters. Once obtained (see [6]), the MPL can be calibrated to reduce systematic bias in the measured signals and resulting volume depolarization ratio to within fractions of a percent. As a result, the new polarized MPL is capable of measuring the molecular volume depolarization background with sufficient time averaging to reduce signal noise (~1 hour at night for less turbid conditions). Without careful application of such calibrations, large systematic bias (up to ~30%) may be present in the signals, especially for returns from weakly depolarizing layers (e.g. molecular background). The affects are highly non-linear, with low bias in signals from strongly depolarizing layers. Thus the new MPL is easily capable of detecting the presence of non-spherical particles (dust, ice clouds), but quantitative analysis requires the polarization calibrations. Figure 2 presents examples of the new MPL signals and depolarization ratios measured at NASA Goddard Space Flight Center from May 20 – 24, 2015.

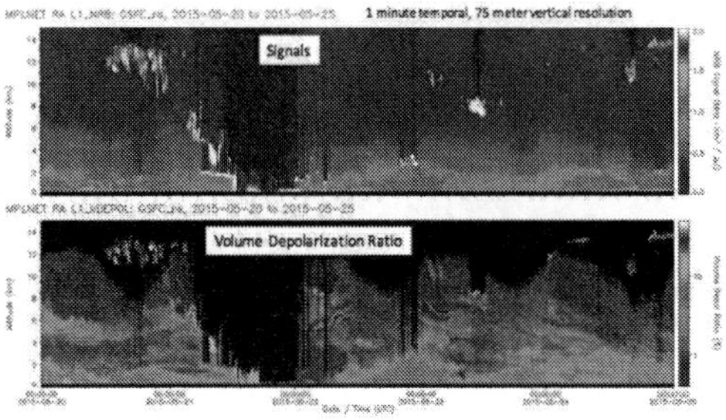

Figure 2. MPLNET V3 Level 1 Products

In addition to calibrating the optical components, ferroelectric materials are temperature dependent by nature. The FLC is subject to both solar background and internal heat within the MPL, which can change the location of the fast axes, and the resulting rotation angle (studies indicate approximately 0.3° change per 1° C). A temperature correction has been developed as part of the new calibration procedures, but is a second order affect unless the FLC temperature changes by several degrees (this is uncommon). For typical operating temperatures, the expected bias from temperature dependent FLC effects is ~0.5% at most. The temperature calibrations are more time consuming and difficult to determine and apply per MPL. If the basic polarization component calibrations are applied to the MPL, but the temperature dependent effects are ignored, then for normal operating temperatures it is possible to accurately measure volume depolarization ratios ≥ 1.5%. For more weakly depolarizing aerosols (such as sulfate), the temperature corrections must be applied to produce quality data with minimum of systematic bias.

https://doi.org/10.1051/epjconf/201817609003

3 VERSION 3 PROCESSING SYSTEM

The new V3 processing system is a completely new version of MPLNET. In addition to the new polarized MPL, all data communications, archiving, processing code, and the website are new. Some V3 algorithms are based on mature V2 processing, but the code and processing structure are new. Some of the important new changes in V3 include:

- CF compliant NETCDF4 data formats
- QA flags at all data levels
- New product structure and definitions
- NASA GEOS-5 Meteorology
- Online data ordering and delivery
- Near-real-time (NRT) Level 1.5 products are now quality assured
- Custom data products and delivery to aerosol forecasting centers

Both V2 and V3 data will be provided from the new website. Older V2 data will be provided in the new V3 data file formats. Table 1 provides definitions and descriptions of the new V3 products levels.

Signal data (NRB) are processed using mature V1 and V2 algorithms [7,8]. The new polarized MPL and calibration requirements will be presented in a future paper. The V3 cloud products (CLD) are new, and generated from algorithms presented by Lewis et al [9], including improved cloud detection versus V2 and new thin cloud optical depth and phase information. The new V3 planetary boundary layer products (PBL) include mixed layer depths presented by Lewis et al [10]. Finally, the aerosol algorithms remain unchanged from V2 [11, 12], the only differences are in the operational implementation (online algorithm theoretical basis documents, ATBD, will be available on our website soon).

L1 products are automated and generated in NRT (latency for most all of network is ~1.5 hours). L1 processing will utilize forecast meteorology from the NASA GEOS-5 model [15]. Quality assurance screens will be applied to all L15 data. In V2, this was only done for L2 products. This change follows the same protocol developed for AERONET V3 data. L15 processing will also occur in NRT and browse imagery will be available on our website, but data files will not be publicly downloadable until the final assimilated GEOS-5 data are available for reprocessing (typically ~ 24 hour delay). The NRT L15 data files will be made available to operational aerosol forecasting centers under agreement.

Table 1. V3 Product Level Definitions

V3 Product	Definitions
NRB	Lidar signals, volume depolarization ratios, and diagnostics
CLD	Cloud heights and retrievals.
PBL	PBL height and estimated AOD
AER	Aerosol heights and retrievals for coincident, co-located sunphotometer observations.
CAER	Aerosol heights and retrievals (continuous). Less accurate than AER product.

V3 Product Levels	Descriptions
L1_NRB	NRT, not screened, initial calibration, automated
L1_CLD	
L1_PBL	GEOS5 Forecast NRT, reprocessed next day with GEOS5 assimilated, AERONET L15 AOD
L1_AER	
L1_CAER	
L15_NRB	NRT, screened, initial calibration, automated
L15_CLD	
L15_PBL	GEOS5 Forecast NRT, reprocessed next day with GEOS5 assimilated, AERONET L15 AOD
L15_AER	
L15_CAER	
L2_NRB	Not NRT, screened, post calibration
L2_CLD	
L2_PBL	GEOS5 assimilated, AERONET L2 AOD
L2_AER	
L2_CAER	

Figure 3 presents examples of V3 data products retrieved from data collected at NASA Goddard Space Flight Center for the same time span as shown in Figure 2 (May 20 – 24, 2015). The top pane displays V3 cloud heights (gray) and identification of ice clouds (purple). The next pane presents estimates of thin cloud optical depth from [9], where the results are bounded using lidar ratios of 20 and 30 sr (dark and light blue respectively). The following pane shows retrievals of aerosol extinction [11,12] and an overlay of the new V3 mixed layer depth (brown). The bottom pane displays the column aerosol optical depth (AOD), from AERONET (green), the continuous V3 aerosol product (CAER, red), and the estimate of the PBL AOD (brown). All MPLNET aerosol retrievals are constrained by the co-located AERONET data [11].

4 CONCLUSIONS

An overview of the new V3 MPLNET processing system has been presented. V3 data products will be released in 2017, and joint processing of V2 data will continue through at least 2017. All V2 products will be available from the new website, including after joint processing of V2 ends. The size and distribution of the network and the planned addition of new sites in the near term were shown. Additional growth of the permanent, long-term sites in the network requires a funding increase to accommodate new personnel. However, MPLNET will continue to participate in short-term field campaigns worldwide as well as the WMO GALION project.

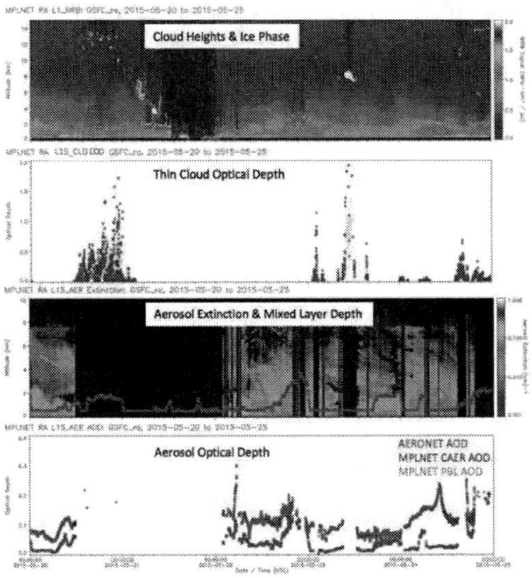

Figure 3. MPLNET V3 Cloud, Aerosol and PBL Products

ACKNOWLEDGEMENTS

The MPLNET project is funded by the NASA Radiation Sciences Program and Earth Observing System. We also thank Justin Fisher, Adam Slagel, and Phillip DeCola at Sigma Space Corporation for helpful collaborations developing and testing the new polarized MPL.

References

[1] Welton, E. J., et al, 2001: Global monitoring of clouds and aerosols using a network of micro-pulse lidar systems, in Lidar Remote Sensing for Industry and Environmental Monitoring, U. N. Singh, T. Itabe, N. Sugimoto, (eds.), Proc. SPIE, 4153, 151-158.

[2] Holben B.N., et al, 1998: AERONET - A federated instrument network and data archive for aerosol characterization. Rem. Sens. Environ., 66, 1-16.

[3] Bosenberg, J., et al, 2007: Plan for the implementation of the GAW Aerosol Lidar Observation Network GALION. WMO GAW Report No. 178, WMO/TD-No. 1443.

[4] Welton E.J., et al., 2010: The NASA Micro Pulse Lidar Network (MPLNET): summary of the last 10 years, current status, and future plans, Proceedings of the 25th Inter. Laser Radar Conf., 1, 875-878.

[5] Flynn, C., et al, 2007: Novel polarization-sensitive micropulse lidar measurement technique. Optics Express, 15, 2785-2790.

[6] Stewart, S.A., et al, 2017: Experimental Characterization of the Ferroelectric Liquid Crystal (FLC) Modulator in the Polarized Micro Pulse Lidar. Proceedings of the 28th Inter. Laser Radar Conf., this issue.

[7] Campbell, J.R., et al, 2002: Full-time, Eye-Safe Cloud and Aerosol Lidar Observation at Atmospheric Radiation Measurement Program Sites: Instrument and Data Processing. J. Atmos. Oceanic Technol., 19, 431-442.

[8] Welton, E.J., and J.R. Campbell, 2002: Micro-pulse Lidar Signals: Uncertainty Analysis. J. Atmos. Oceanic Technol., 19, 2089-2094.

[9] Lewis, J.R., et al, 2016: Overview of MPLNET Version 3 Cloud Detection. J. Atmos. Oceanic Tech., 33, 2113 – 2134, doi: 10.1175/JTECH-D-15-0190.1.

[10] Lewis, J.R., et al, 2013: Improved boundary layer depth retrievals from MPLNET, J. Geophys. Res., 118, 9870-9879, doi:10.1002/jgrd.50570.

[11] Welton, E.J., et al, 2000: Ground-based Lidar Measurements of Aerosols During ACE-2: Instrument Description, Results, and Comparisons with other Ground-based and Airborne Measurements. Tellus B, 52, 635-650.

[12] Welton, E.J., et al, 2002: Measurements of aerosol vertical profiles and optical properties during INDOEX 1999 using micro-pulse lidars. J. Geophys. Res., 107, 8019, doi:10.1029/2000JD000038.

ACTRIS

Aerosol, Clouds and Trace gases Research Infrastructure

Gelsomina Pappalardo

*Consiglio Nazionale delle Ricerche - Istituto di Metodologie per l'Analisi Ambientale CNR-IMAA, C.da S. Loja, I-85050 Tito Scalo, Potenza, Italy, *Email: gelsomina.pappalardo @imaa.cnr.it*

ABSTRACT

The Aerosols, Clouds and Trace gases Research Infrastructure (ACTRIS) is a distributed infrastructure dedicated to high-quality observation of aerosols, clouds, trace gases and exploration of their interactions. It will deliver precision data, services and procedures regarding the 4D variability of clouds, short-lived atmospheric species and the physical, optical and chemical properties of aerosols to improve the current capacity to analyse, understand and predict past, current and future evolution of the atmospheric environment.

1 INTRODUCTION

Short-lived atmospheric components – aerosols, clouds, trace gases – have a residence time in the atmosphere from hours to few weeks, which differentiates them from long-lived greenhouse gases. The short lifetimes make their concentrations highly variable in time and space and involve fast processes. They are recognised to be among the most significant anthropogenic pollutants affecting Earth's radiation balance and the largest source of uncertainty in terms of radiative forcing impact [1]. In parallel, short-lived atmospheric compounds have recognized adverse health effects at concentrations typically found across Europe and potentially lead to more than 400.000 premature deaths annually in the EU28 [2]. Information on concentrations and distributions of aerosols and trace gases is therefore required to reduce air pollution and related adverse effects on health and ecosystems [3]. ACTRIS (www.actris.eu) addresses these challenges by operating at National Facilities a combination of near-surface and remote-sensing systems including: near surface measurements of aerosols and short-lived trace gases, vertically resolved measurements of aerosols, vertically resolved measurements of clouds and precipitation, profile and column observations of short-lived trace gases and ancillary measurements of meteorological and radiation quantities. ACTRIS also includes exploratory platforms at the national level. The observation platforms are often components of European or international networks [4]. ACTRIS relies on appropriate Central Facilities – Calibration Centres, Data Centre, Head Office – that ensure compliance with standard operating procedures and/or quality protocols to provide harmonized, reliable, and documented observational data. The data curation and storage services are handled by a dedicated Central Facility, the ACTRIS Data Centre.

ACTRIS serves a vast community of users working on observations, experiments, models, satellite data, analysis and predicting systems and offers access to advanced technological platforms for exploration of the relevant atmospheric processes in the fields of climate change and air quality.

2 ACTRIS Research Infrastructure

The Research Infrastructure ACTRIS is the pan-European initiative that consolidates strategies amongst European partners for observations of aerosols, clouds, and trace gases. The infrastructure is unique in providing the 4D-variability of clouds and of the physical, optical and chemical properties of short-lived atmospheric species, in particular those involved in climate forcing.

The primary objective of ACTRIS is to provide the 4D-variability of clouds and of the physical, optical and chemical properties of short-lived atmospheric species, from the surface throughout

https://doi.org/10.1051/epjconf/201817609004

the troposphere to the stratosphere, with the required level of precision, coherence and integration. The second objective is to provide effective access to this information and the means to more efficiently use the complex and multi-scale ACTRIS parameters serving a vast community of users working on models, satellite retrievals, and analysis and forecast systems. The third objective is to raise the level of technology used in the RI and the quality of services offered to the community of users, involving partners from the private sector. Finally, the fourth objective of ACTRIS is to promote training of operators and users and enhance the linkage between research, education and innovation in the field of atmospheric science.

ACTRIS is composed of several connected elements:

- National Facilities (NFs) (observation platforms, stationary and mobile exploratory platforms hosted and operated at the national level) located both within Europe and globally, documenting with high precision and reliability the 4D-variability of short-lived atmospheric species, in particular those involved in climate forcing and the processes driving their life-cycle;
- Central Facilities (CFs): Calibration Centres, Data Centre, Head Office hosted and operated at the European level.

ACTRIS engages the National Facilities with highly standardized and coordinated procedures. The Central Facilities are required to ensure that all data are produced, treated and quality-controlled with specified procedures, properly archived for long-term usage, and accessible in a timely manner to all users. Central Facilities are responsible for the definition of ACTRIS data products, including definition of quality requirements, format and nature of delivery (near real time or consolidated), flagging and development of value-added products. Central Facilities are fundamental to provide the access to the ACTRIS services, organising the right level of training and education, both within and outside the RI, and delivering tailored services for various users, scientific community, space agencies, COPERNICUS [5] and the private sector.

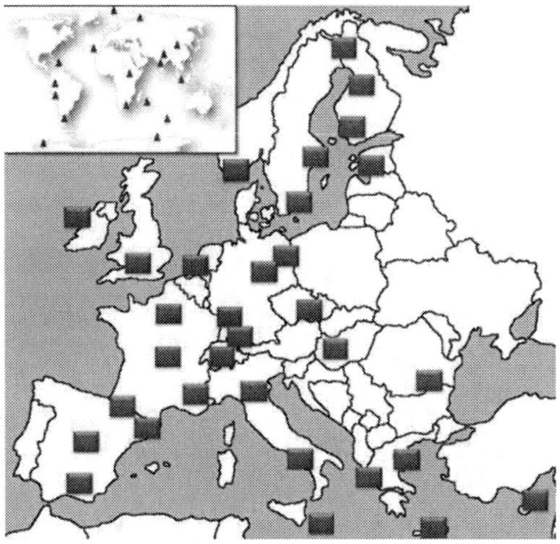

Figure 1 Distribution of ACTRIS observation stations

More specifically, Calibration Centres provide the infrastructure with: a) Standard operation procedures and best practices; b) Quality-assurance criteria; c) Intercomparison of instruments and procedures; d) Development and testing of new instrumentation for variables that are not yet accessible at the required sensitivity/accuracy; e) Training of RI managers, staff, data providers, and users.

The data curation and repository services, including data management and necessary development work on, workflow documentation, metadata issues, etc., are handled by the ACTRIS Data Centre (DC). All primary data and added-value data products are made available to the users via the ACTRIS data portal. ACTRIS follows the open research data initiative of the European Union, making data discoverable (DOI identification), accessible (via open licence) and assessable (via necessary documentation and description). These data products are made interoperable, adhere to specific quality standards and are free of charge.

Besides data services and access to data products, ACTRIS will also offer physical and remote access to Calibration Centres and physical access to selected National Facilities. Access will be handled by the ACTRIS Service and Access Management Unit (SAMU) which is the entry point to coordinate access to the users to the different services granted through a per review process. SAMU is integrated within the ACTRIS Head Office (HO) which coordinates the research infrastructure operations, facilitates the strategic decision-making, and represents ACTRIS.

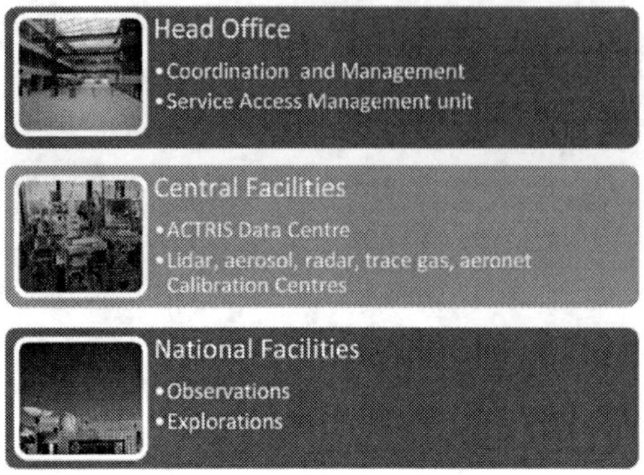

Figure 2 ACTRIS Structure

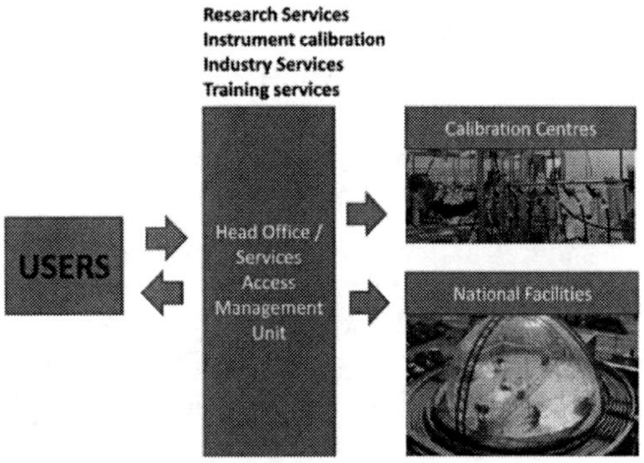

Figure 3 Physical access related ACTRIS services

3 Towards the Implementation

ACTRIS, as research infrastructure, was listed in 2016 among the ESFRI projects (www.esfri.eu) [6], and has just started the preparation phase but it results from long-term collaborative work of the atmospheric science community through a series of European projects (EARLINET, EUSAAR, EARLINET-ASOS, CloudNet, EUROCHAMP 2020, ACTRIS 1 and ACTRIS 2) that started in 2000 [7, 8, 9]. The gradual shift towards the construction phase of the research infrastructure is foreseen in 2019- 2021 planning the commissioning phase in 2021-2022. ACTRIS will be fully operational in 2025.

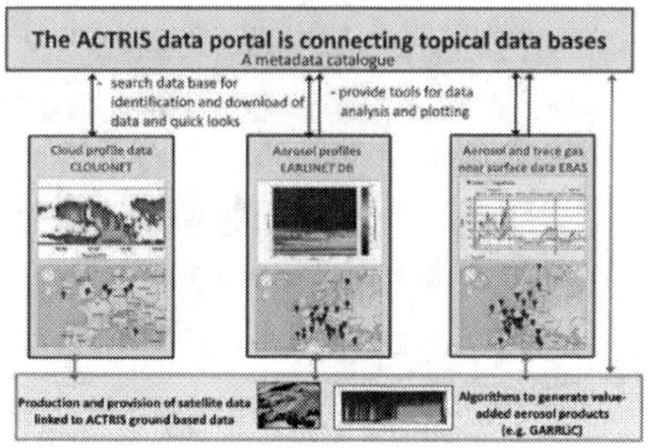

Figure 4 ACTRIS data portal

The implementation of ACTRIS as a research infrastructure is funded through the ACTRIS Preparatory Phase project devoted to develop the organizational, operational and strategic frameworks of the RI. The work includes legal, governance, financial, technical, strategic, and administrative aspects. The main outcomes of PPP are signature-ready documents for establishment of a legal entity with well-defined operations and a sound business plan.

More technical and scientific aspects are supported through the ACTRIS 2 project which addresses the scope of integrating state-of-the-art European ground-based stations for long-term observations of aerosols, clouds and short lived gases. It consolidates and improves services offered within FP7 funded Integrated Infrastructures Initiative ACTRIS (2011-2015). ACTRIS-2 takes up the overarching objectives of ACTRIS to further integrate the European ground-based stations and to construct a user-oriented RI, unique in the EU-RI landscape, for aerosols, clouds, and short-lived gas-phase species. One of the key activities within the project is the provision of access to a large number of ACTRIS-2 services, implemented through its virtual and trans-national access programme (TNA), comprising wide and open access to high-quality information, tools and services through the ACTRIS Data Centre, hands-on access to a large number of ground-based observing platforms and advanced instrumentation, and physical and remote access to thematic calibration centres for instrument testing and development, quality assurance, training and best practices, and international collaboration.

Eighteen world-class observational facilities, unique in Europe and representing a variety of environments and air mass types, offer access to comprehensive measurement programmes and cutting-edge research, scientific campaigns, intercomparisons, and training capabilities for exploitation of state-of-the art resources, knowledge, and expertise. The observational sites comprise historical sites, operational since decades with long records and experience of access provision, and new sites for novel research opportunities.

The calibration facilities constitute multi-national installations and are globally unparalleled, being beneficial not only to the entire scientific communities, but also to users from the industrial sector and operational networks, and offer access and services to:

- the Lidar Calibration Centre (LiCal) for testing and calibration of lidars and ceilometers, including characterization and optimization of single components and assessment of whole system performances;
- the AERONET-Europe Calibration facility for calibration and standards maintenance service for sun/sky/polar/moon photometers, complementing the NASA AERONET programme for routine, real-time, and high-quality aerosol column integrated properties;
- the European Centre for Aerosol Calibration (ECAC) for calibration and quality assurance of high quality physical, optical, and chemical in-situ aerosol measurements, instrument audits and capacity building.

The Transnational Access program for ACTRIS and the various opportunities are described in more detail at www.actris.eu.

ACKNOWLEDGEMENTS

21 EU or associated Countries are supporting ACTRIS with a total of more than 120 research performing organizations involved.

The financial support for ACTRIS by the European Union's Horizon 2020 research and innovation programme under grant agreement no. 654169 and no. 739530 is gratefully acknowledged.

References

[1] IPCC 2013 WGI AR5 Chapter.

[2] European Environmental Agency, "Air quality in Europe — 2014 report", EEA Report No 5/2014.

[3] WHO, Review of Evidence on Health Aspects of Air Pollution – REVIHAAP Project, Technical Report, World Health Organization Geneva, 2013, http://www.euro.who.int/ data/assets/pdf file/0004/193108/REVIHAAP-Final-technical-report-final-version.pdf.

[4] Lund Myhre, C, and Baltensperger, U, et al., 2012: Recommendations for a Composite Surface-Based Aerosol Network, GAW Report n. 207, www.wmo-gaw-wcc-aerosol-physics.org/files/gaw-207.pdf.

[5] COPERNICUS www.copernicus.eu.

[6] ESFRI roadmap 2016 - Strategy Report on Research Infrastructures, www.esfri.eu.

[7] Pappalardo, G., et al., 2014: EARLINET: towards an advanced sustainable European aerosol lidar network, Atmos. Meas. Tech., 7, 2389-2409, doi:10.5194/amt-7-2389-2014, 2014.

[8] Laj P., et al., 2009: Measuring Atmospheric Composition Changes, *Atmospheric Environment*, Vol. 43, No. 33, pp. 5351-5414, doi:10.1016/j.atmosenv.2009.08.020.

[9] Illingworth, A. J. et al., 2007: Cloudnet Continuous Evaluation of Cloud Profiles in Seven Operational Models Using Ground-Based Observations, *BAMS*, doi: http://dx.doi.org/10.1175/BAMS-88-6-883.

 https://doi.org/10.1051/epjconf/201817609005

AEROSOL OBSERVATION USING MULTI-WAVELENGTH MIE-RAMAN LIDARS OF THE AD-NET AND AEROSOL COMPONENT ANALYSIS

Tomoaki Nishizawa[1*], Nobuo Sugimoto[1], Atsushi Shimizu[1], Itsushi Uno[2], Yukari Hara[2], and Rei Kudo[3]

*[1]National Institute for Environmental Studies, 16-2 Onogawa, Tsukuba, 305-8506, Japan, *nisizawa@nies.go.jp*
[2]Research Institute for Applied Mechanics, Kyushu University, 6-1 Kasugakoen, Kasuga 816-8580, Japan
[3]Meteorological Research Institute, Japan Meteorological Agency, 1-1 Nagamine, Tsukuba 305-0052, Japan

ABSTRACT

We deployed multi-wavelength Mie-Raman lidars (MMRL) at three sites of the AD-Net and have conducted continuous measurements using them since 2013. To analyze the MMRL data and better understand the externally mixing state of main aerosol components (e.g., dust, sea-salt, and black carbon) in the atmosphere, we developed an integrated package of aerosol component retrieval algorithms, which have already been developed or are being developed, to estimate vertical profiles of the aerosol components. This package applies to the other ground-based lidar network data (e.g., EARLINET) and satellite-borne lidar data (e.g., CALIOP/CALIPSO and ATLID/EarthCARE) as well as the other lidar data of the AD-Net.

1. INTRODUCTION

To assess the effects of aerosols on the atmospheric environment and climate, it is essential to observe the optical and microphysical properties and chemical composition of aerosols and to understand their spatial distribution and temporal variation. We deployed multi-wavelength Mie-Raman lidars (MMRL) at the Hedo (26.87°N, 128.25°E), Fukuoka (35.52°N, 130.47°E), and Toyama (36.70N, 140.12E) **(Fig. 1)**, which are the AD-Net **[1]** observation sites in Japan and have conducted continuous measurements in collaboration with Kyushu University (Japan) and the University of Toyama (Japan) since 2013. The main objectives of these observations are to evaluate the extinction coefficients of major aerosol components (dust (DS), sea-salt (SS), black carbon (BC), and air pollution particles without BC (AP)) in the atmosphere from the MMRL measurements, to provide the aerosol component data for data assimilation analyses as the input data, and to contribute to evaluating the emission of the aerosol components and building aerosol-assimilated data for each region of East Asia.

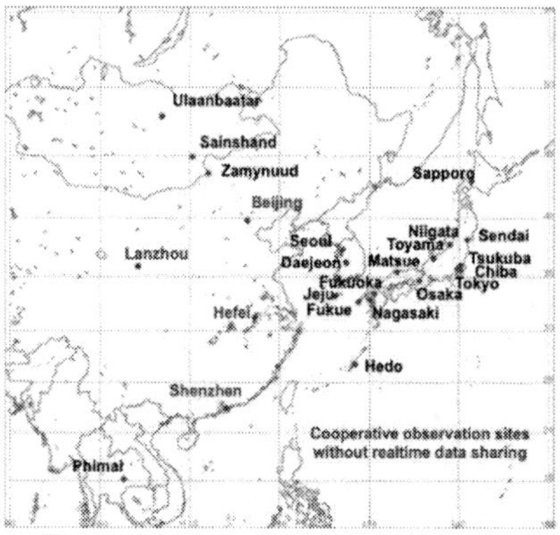

Fig. 1 Observation sites in AD-Net

In this paper, we report the MMRL system and analysis results of aerosol optical properties derived from the MMRL measurements (Section 2). Furthermore, we developed an algorithm to retrieve extinction coefficients of the aerosol components using the MMRL data. We report the developed algorithm and the primary results (Section 3).

2. AEROSOL OBSERVATION USING MULTI-WAVELENGTH MIE-RAMAN LIDARS (MMRL)

MMRL provides particle extinction (α) and backscatter (β) coefficients at 355 and 532nm, particle depolarization ratio (δ) at 355 and 532nm, and the attenuated backscatter coefficient at 1064nm (i.e., 2α(355,532) + 3β (355,532,1064) + 2δ(355,532) data). The details of the system, method of retrieving aerosol optical properties, and the measurement and retrieval accuracies are given in Nishizawa et al. **[2].** Here, we present the outlines.

The MMRL system employs a commercial compact Nd:YAG laser with second-harmonic and third-harmonic generators and a rigid biaxial configuration. Backscattered light is collected with a vertically mounted telescope. The collected light is split among the receivers at 355, 387, 532, 607, and 1064nm using dichroic beam splitters. At 532 and 355nm, the co-polarized and cross-polarized components of the total backscatter light split with a cube polarizer are detected using two PMTs. At 607 and 387nm, nitrogen Raman backscatter signals are detected using a PMT. At 1064nm, the total backscatter light is detected using an APD. Band-path filters are set for all the detectors in order to reduce contamination from background light. We conducted photon-counting measurements for Raman backscatter and analog measurements for elastic backscatter. No Raman channel data are available in the daytime due to strong sunlight. The lidar continuously follows a work cycle of 5min on and 10min off in order to maintain laser life. The signals are accumulated for 3000 laser shots, corresponding to 5min, in the 7.5m (6m) range bin up to 30km (24km) for Raman (elastic) backscatter signals, and recorded every 15min for both elastic and Raman signals.

The aerosol optical properties (i.e., α_a, β_a, and δ_a) are derived every 30m and 15min using a conventional retrieval method **[e.g., 3]**. We first average the measured signals over every 30m range to match the vertical resolution between the Raman signal data and the elastic signal data. Next, we identify an aerosol-rich layer as well as a molecule-rich layer and a cloud-rich layer using the averaged β_{1064} **[e.g., 4]**. To remove data contaminated by clouds or rain and to improve the signal-to-noise ratio (SNR) of the measured signals, we apply a moving average in window ranges (half range) of 30m vertical and 45min temporal to the signal data after we remove the data at and above any slab-layer identified as a cloud or rain layer **(Fig. 2)**. α_a, β_a, and δ_a are then derived using the moving-averaged data. Finally, we also apply a moving average in window ranges (half width) of 60m vertical resolution and 30min temporal resolution to the derived α_a, β_a, and δ_a to reduce the impact of signal noise **(Fig. 3)**. The lidar ratio of aerosols (S_a), defined as the extinction-to-backscatter ratio ($S_a = \alpha_a / \beta_a$), is derived using the moving-averaged α_a and β_a.

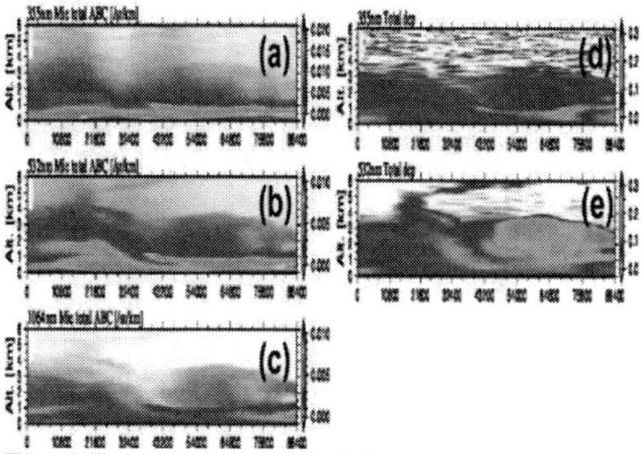

Fig. 2 Example of time-height cross sections of moving-averaged attenuated backscatter coefficients at 355nm (a), 532nm (b), and 1064 nm (c), and total depolarization ratio at 355nm (d) and 532nm (e) after removing the data contaminated by clouds or rain measured at Fukuoka on 21 Mar. 2015.

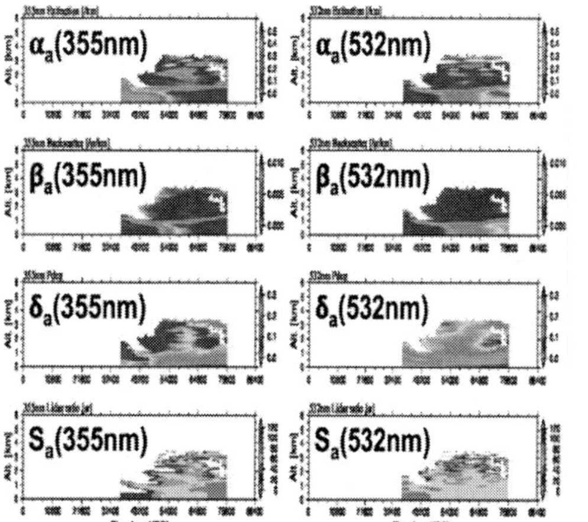

Fig. 3 Example of time-height cross sections of derived α_a, β_a, δ_a, and S_a at 355 and 532nm measured at Fukuoka on 21 Mar. 2015.

Using the α_a, β_a, δ_a, and S_a data at Fukuoka from January 2015 to May 2015, we characterized aerosol optical properties for several remarkable aerosol events (e.g., air pollution rich or mineral dust rich events) **[5]**. The aerosol events were basically extracted from ground-based sampling and chemical measurements. The mean S_a for air pollution cases (5-day data) was 57±4sr at 355nm and 53±8sr at 532nm. The mean δ_a was 4±1% at 355nm and 6±2% at 532nm. These values are consistent with previous observations using lidar for air pollution events **[e.g., 6, 7]**. For mineral dust cases (6-day data), the mean S_a was 50±7sr at 355nm and 54±9sr at 532nm; The mean δ_a was 13±8% at 355nm and 16±6% at 532nm. These values are close to those for the "Dusty-mixture" categorized in Illingworth et al. **[8] and** imply mixture of mineral dust with air pollution particles. We will improve this aerosol characterization analysis by increasing data samples for noteworthy aerosol events.

3. AEROSOL COMPONET RETRIEVAL USING AD-NET LIDARS

We developed some algorithms to identify the aerosol components and retrieved their vertical distributions in an extinction coefficient assuming an external mixture of some aerosol components and prescribing their microphysical and optical properties from AD-Net lidar measurements. Sugimoto et al. **[9]** used depolarization ratios for classifying non-spherical DS and spherical aerosols (i.e., AP+BC+SS) and retrieved their extinction coefficients using 1β(532) + 1δ(532) lidar data (1β+1δ method). Nishizawa et al. **[10]** used spectral (color) ratios of backscatter coefficients for classifying large aerosols (e.g., SS and DS) and small aerosols (e.g., AP and BC) as well as depolarization ratio and retrieved DS, SS, and small aerosols (i.e., AP+BC) using 2β(532, 1064) + 1δ(532) data (2β+1δ method). Nishizawa et al. **[11]** used lidar ratios for classifying BC with strong light absorption and weak light absorption aerosols (e.g., SS, AP) as well as the depolarization ratio, and retrieved DS, BC, and weak light absorption aerosols (i.e., AP+SS) using 1α+1β+1δ data at 532 or 355nm (1α+1β+1δ method). Nishizawa et al. **[2]** used all the depolarization ratio, color ratio, and lidar ratio and retrieved DS, SS, BC, and AP using 1α(532) + 2β(532, 1064) + 1δ(532) data (1α+2β+1δ method).

3.1 Algorithm using all the seven MMRL channels (2α+3β+2δ method)

The MMRL has seven channels (i.e., 2α + 3β + 2δ data). For developing the 1α+2β+1δ method, we are developing an algorithm to retrieve mode-radii (R_m) of DS, AP, and SS using color ratios of extinction, backscatter and depolarization ratio as well as extinction coefficients of the four aerosol components (σ) (DS, SS, BC, and AP). In this algorithm, we assume that (1) the size distribution for each aerosol component is log-normal; (2) DS is spheroidal and the others are spherical; and (3) the refractive index and standard deviation of DS, SS, AP, and BC are fixed and the mode radius of BC are fixed (see **Table 1**). This estimation of the seven parameters (i.e., σ for DS, SS, BC, and AP and R_m for DS, SS, and AP) using seven channel lidar data amounts to solving seven-order non-linear simultaneous equations, and thus we use the Gauss-Newton method with the line search based on the least-square method **[2]**. We will report the primary results in the conference.

3.2 Integrated package of aerosol component retrieval algorithms

We developed an integrated package of aerosol component retrieval algorithms, which includes developed algorithms (i.e., 2β+1δ, 1α+1β+1δ, and 1α+2β+1δ methods) and a new developed one (i.e., 2α+3β+2δ method), to estimate vertical profiles of the aerosol components. In this package, the same analytical method (i.e., Gauss-Newton method with the line search) is used for all retrievals and the optical model of each aerosol component consistent for all the retrieval methods are used. The primary results are depicted in **Fig. 4.** The 2β+1δ method was applied to 2β(532,1064) + 1δ(532), the 1α+1β+1δ method was applied to 1α(532) + 1β(532) + 1δ(532), and the 1α+2β+1δ method was applied to, 1α(532) + 2β(532,1064) + 1δ(532) of the MMRL data at Fukuoka on 21 Mar. 2015. Note that the 2β+1δ method can apply to daytime data because it does not use Raman channel data. The optical model for each aerosol component used in this estimation is given in Table 1. When the 2α+3β+2δ method is applied, the mode-radii for

DS, SS, and BC are estimated (not fixed).

This package applies to the other ground-based lidar network data (e.g., EARLINET) and satellite-borne lidar data (e.g., CALIOP/CALIPSO and ATLID/EarthCARE) as well as the other lidar data of the AD-Net.

Table 1 Optical model for each aerosol component used in the component retrieval.

	AP	**DS**	**BC**	**SS**
Effective radius [μm]	0.11	1.44	0.04	2.1
Standard deviation	1.6	2.2	2.0	2.1
Refractive index (532nm)	1.43 $3\cdot10^{-3}$	1.51 $3\cdot10^{-3}$	1.75 0.44	1.37 $4\cdot10^{-9}$
S_a [sr] (532, 355nm)	52, 70	47, 79	98, 97	19, 16
δ_a (532,355nm)	0.02, 0.02	0.30, 0.28	0.02, 0.02	0.02, 0.02
Backscatter color ratio (1064/532, 355/532)	0.35, 1.68	0.98, 0.57	0.24, 1.74	0.66, 1.13

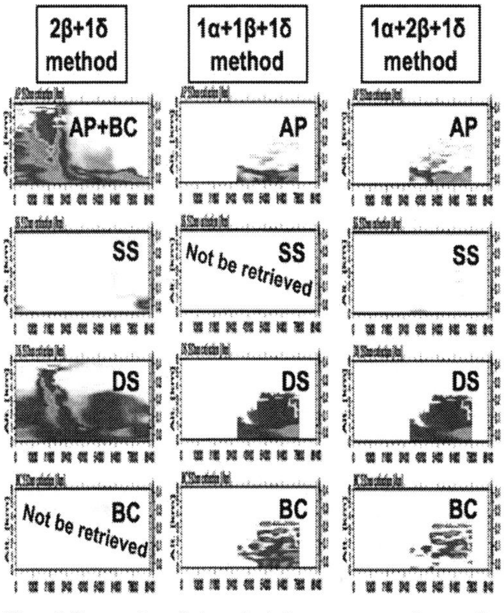

Fig. 4 Example of time-height cross sections of σ at 532nm retrieved by the 2β+1δ method (left column), 1α+1β+1δ method (middle column), and 1α+2β+1δ method using data measured at Fukuoka on 21 Mar., 2015.

EarthCARE is an ongoing aerosol, cloud, and radiation observation project conducted by ESA, JAXA, and NICT. A 355nm high-spectral resolution lidar (HSRL) ATLID providing 1α+1β+1δ at 355nm and multi-spectral imager MSI are aboard EarthCARE sattelite **[8]**. We are developing an algorithm to estimate extinction coefficients of BC, SS, DS, and AP from the ATLID and MSI measurements in collaboration with JAXA. We are also improving AD-Net by introducing multi-wavelength Raman lidars and HSRLs with measurement capability at 355nm to assist with the EarthCARE validation observation. The data analysis of a multi-channel lidar with more channels than the ATLID is useful for validating the ATLID-derived extinction coefficients of aerosol components because the algorithm using the multi-channel lidar data makes fewer assumptions and is more accurate. This developed package will be used for this validation study.

ACKNOWLEDGEMENTS

This research was partly supported by a Grant-in-Aid for Scientific Research (25220101) from the Japan Society for the Promotion of Science (JSPS) and the first EarthCARE algorithm and validation Research Announcement from JAXA.

References

[1] Shimizu, A., et al., 2016: *Opt. Eng.*, **56**, 031219, doi:10.1117/1.OE.56.3.031219.
[2] Nishizawa, T., et al., 2017: *J. Quant. Spec. Rad. Trans.*, 188, 79-93.
[3] Xie, C., et al., 2008: *Appl. Opt.*, **47**, 4945-4951.
[4] Shimizu, A., et al., 2004: *J. Geophys. Res.*, **109**, D19S17, doi:10.1029/2002JD003253.
[5] Hara, Y., et al., 2017: *J. Quant. Spec. Rad. Trans.*, 188, 20-27.
[6] Noh Y. M., et al., 2018: *Atmos. Env.*, 42, 2208-2224
[7] Muller D., et al., 2007: *J. Geophys. Res.*, 112, D16202, doi:10.1029/2006JD008292.
[8] Illingworth, A. J., et al., 2016: *BAMS*, doi:10.1175/BAMS-D-12-00227.1
[9] Sugimoto, N., et al., 2003: *Geophys. Res. Let.*, **30**, 1640, doi:10.1029/2002GL016349.
[10] Nishizawa, T., et al., 2011: *J. Quant. Spec. Rad. Trans.*, 112, 254-267.
[11] Nishizawa, T., et al., 2008: *Trans. Geosci. Rem. Sens.*, **41**, 1743-54.
[12] Nishizawa, T., 2013: J. Rem. Sens. Soc. Jap., 33, 367-376 (in Japanese)

EPJ Web of Conferences **176**, 09006 (2018) https://doi.org/10.1051/epjconf/201817609006
ILRC 28

The UK Lidar-sunphotometer operational volcanic ash monitoring network

Mariana Adam[1]*, Joelle Buxmann[1], Nigel Freeman[1], Andrew Horseman[1], Christopher Salmon[1], Jacqueline Sugier[1], Richard Bennett[1]

*[1]The Met Office, United Kingdom, *Email: mariana.adam@metoffice.gov.uk*

ABSTRACT

The Met Office completed the deployment of ten lidars (UV Raman and depolarization), each accompanied by a sunphotometer (polarized model), to provide quantitative monitoring of volcanic ash over UK for VAAC London. The lidars provide range corrected signal and volume depolarization ratio in near-real time. The sunphotometers deliver aerosol optical depth, Ångstrom exponent and degree of linear polarization. Case study analyses of Saharan dust events (as a proxy for volcanic ash) are presented.

1 INTRODUCTION

The Met Office hosts and runs the London Volcanic Ash Advisory Centre (VAAC) as part of nine VAACs worldwide [1]. The VAAC London advisories are based on input from ground-based, satellite-based and aircraft-based observations as well as dispersion models forecast [2]. The ceilometer network is used operationally since 2012 and it produces range corrected signal (*RCS*) in near-real time (NRT) [1]. The procurement of lidars and sunphotometers started in 2013 and the lidar-sunphotometer network became operational in 2016 (for more details, including the data flow to provide NRT visualization, see [1]). The ultimate goal for the volcanic ash (VA) network is to provide for VAAC the location (altitude of the boundaries) of the VA plumes and estimates of the VA concentrations within the plumes.

2 LIDAR-SUNPHOTOMETER OPERATIONAL NETWORK

The Met Office VA Lidar-sunphotometer operational network contains nine fixed locations and a mobile unit (Fig. 1). The lidars (Raymetrics LR111-300) operate in UV and have N_2 Raman and depolarization (polar and co-polar) channels. The method used for depolarization calibration for our lidar systems is described elsewhere [3]. The lidars provide *RCS* and *VDR* (volume depolarization ratio) in NRT. The sunphotometers (Cimel CE318-NE DPS9) provide the aerosol optical depth (*AOD*) and Ångstrom exponent (*AE*) in NRT. The data is also submitted to AERONET [4] with one day delay. In the near future, the degree of linear polarization will be added to NRT products. To assure the eye-safety, the lidars operate at ~50mJ/pulse and 20 Hz.

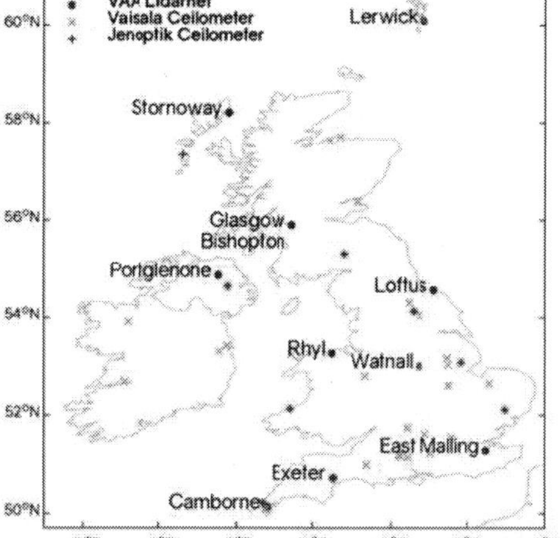

Figure 1.The Met Office Lidar-sunphotometer operational network. Also shown the location of the operational ceilometers.

As a consequence of its geographical location, clouds (i.e. low level clouds or cloud-topped boundary layer) are typically present over the UK. Within these conditions, clear skies or skies with cirrus (Ci) clouds are less frequent.

On the other hand, UK atmosphere is clearer than continental Europe (less pollution). Thus, the network is largely utilized to monitor long-range transport of pollution, such as volcanic ash or Saharan dust clouds.

In the following sections, we describe two events observed by the UK VA network: biomass burning event at the end of May 2016; and a volcanic ash event early January 2017 interestingly also associated with buoyancy internal gravity waves.

2.1 Biomass burning pollution event

Smoke layers of Canadian forest fires were observed by Lidar systems over centre-Southern UK between 25th and 31st of May 2016 at Exeter, East Malling and Watnall and between 23rd and 31st of May at Camborne. Figure 2 shows *RCS* and *VDR* collected at Camborne on May 24th, 2016. Note that *VDR* is shown in log10 scale to emphasize the depolarization layers.

The pollution event was observed in all four locations at high altitude (between ~5 km and ~10 km). On several occasions, the pollution layers coexisted along the Ci clouds. In the UV region the molecular component of the backscatter signal is important and thus, weak aerosol pollution layers are not always clearly observed in *RCS* (Fig. 2, upper plot). However, the signature of the non-spherical particles can be seen in *VDR* (Fig. 2, middle plot). High AOD were also observed (up to 0.45), especially between 15:00 and 18:00. In this period *AE* increases (from 0.5 to 0.9), which indicates the relevance of small particles.

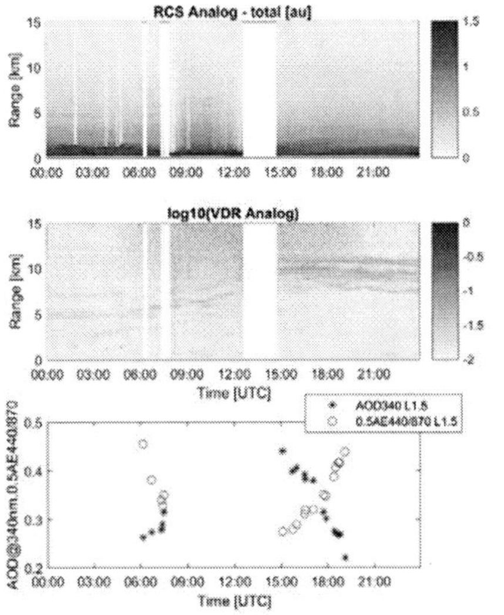

Figure 2. RCS (upper), VDR (middle) and AOD and AE (lower) at Camborne on 24/05/2016 during a biomass burning event.

Consequently, it looks like the contribution of the thin pollution layers is insignificant. On May 26, the pollution layers can still be seen at 8-10km at all four locations (Figs. 3-6). No sunphotometer data was available at East Malling and Watnall.

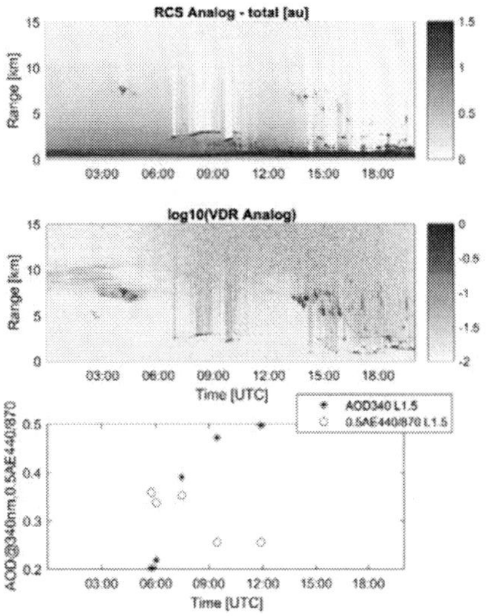

Figure 3. RCS (upper), VDR (middle) and AOD and AE (lower) at Camborne on 26/05/2016 during a biomass burning event.

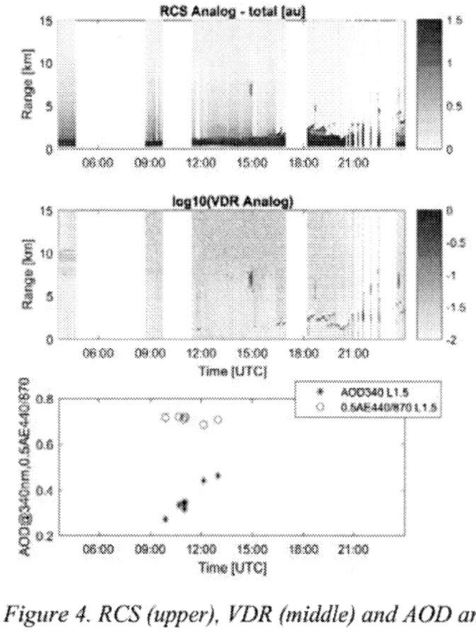

Figure 4. RCS (upper), VDR (middle) and AOD and AE (lower) at Exeter (van) on 26/05/2016 during a biomass burning event

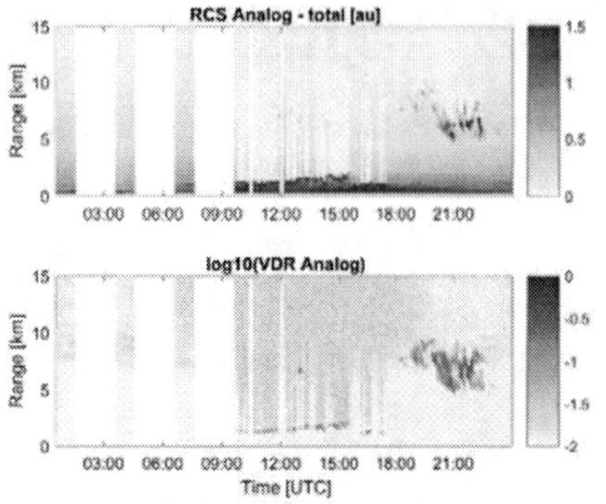

Figure 5. RCS (upper) and VDR (middle) at East Malling on 26/05/2016 during a biomass burning event

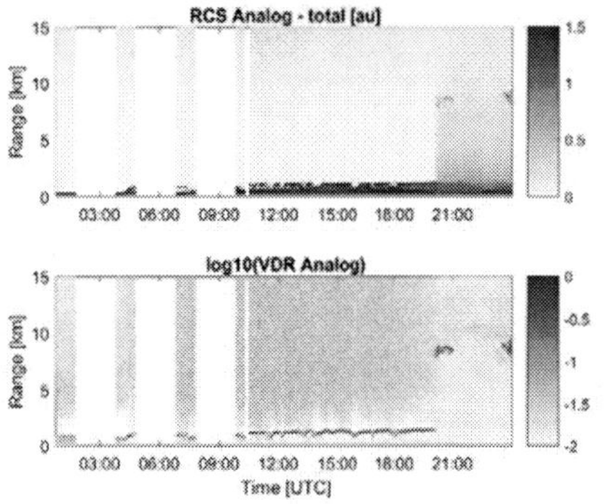

Figure 6. RCS (upper) and VDR (middle) at Watnall on 26/05/2016 during a biomass burning event

For Camborne and Exeter, high *AOD* values (up to ~0.5) can still be seen but most probably they are due to PBL particles contribution. The full analysis of the particles optical properties will be shown during conference.

2.2 Buoyancy internal gravity waves

A weak pollution event took place over central UK towards the end of December 2016 and early January 2017. Preliminary examination suggests that the layer originates from western Alaska and it could have as source the eruption from Bogoslov volcano during December [6]. Continuous measurements were taken at three locations including Glasgow Bishopton in Scotland. On January 3rd, a pollution layer is seen descending towards the planetary boundary layer (PBL). On January 4th and 5th, a very thin depolarizing layer is noticeable at the top of PBL. The assumption is that this depolarizing layer originates from the descendent pollution layer. Figures 7 and 8 show *RCS* and *VDR* zoomed over first two km above the ground for January 4 and January 5. The presence of LLC (see Figs 7-8) as

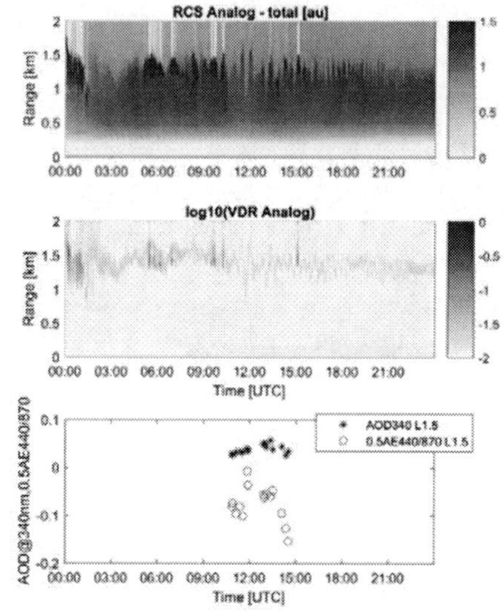

Figure 7. RCS (upper), VDR (middle) and AOD and AE (lower) at Glasgow Bishopton on 04/01/2017after a pollution event. Zoom over [0, 2] km.

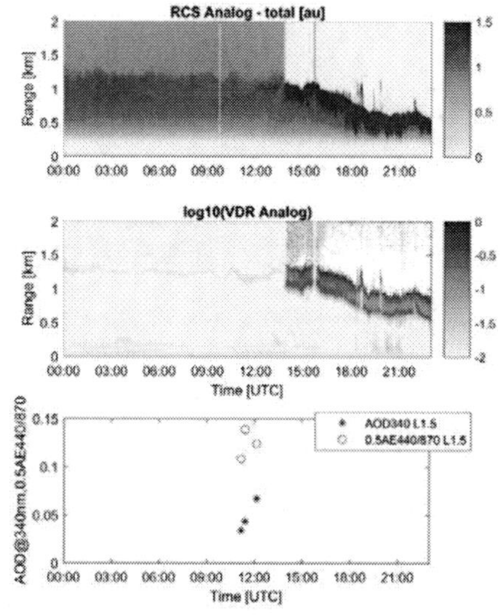

Figure 8. RCS (upper), VDR (middle) and AOD and AE (lower) at Glasgow Bishopton on 05/01/2017after a pollution event. Zoom over [0, 2] km.

well as of Ci clouds (from ~15:00UTC on 04/01 until ~07:00UTC on 05/01) and middle clouds (after ~12:00UTC on 05/01) limited the retrieval of *AOD*. Sporadic *AOD* values show small values (below 0.1 for 340nm). The Met Office NRT for *AOD* and *AE* for these days are not in agreement with AERONET [4], for reasons currently unknown. This is under investigation and an update will be given during IRLC.

Figure 9 shows the MSG satellite imagery [5] over Scotland, which confirms the presence of the gravity waves.

Figure 9. Satellite imagery (MSG) at14:15 on 04/01/2017.

3 CONCLUSIONS

An operational network of ten lidars and sunphotometers was successfully set up over the UK by the Met Office in 2016 to monitor volcanic ash. NRT plots with *RCS* and *VDR* are used by the VAAC London, along with *AOD* and *AE* from sunphotometers. The algorithms to retrieve the particle optical properties are under development

Detailed analysis of the events described here, including the aerosol particles optical properties and optical and microphysical properties as retrieved by AERONET, will be presented at ILRC.

ACKNOWLEDGEMENTS

We would like to thank all the Met Office teams involved in the VA Lidar-sunphotometer project over the last two years, especially to our engineers' teams. We are grateful to Civil Aviation Authority and Department for Transport for funding this project.

References

[1] Adam, M., M. Turp, A. Horseman, C. Ordóñez, J. Buxmann, J. Sugier, 2015: From operational ceilometer network to operational Lidar network, proceedings *27 ILRC*, S14.06-163.

[2] Marenco, F., J. Kent, M. Adam, J. Buxmann, P. Francis, J. Haywood, 2015: Remote Sensing of volcanic ash at the Met Office, proceedings *27 ILRC*, PS A4.03-015,

[3] Buxmann, J., Adam, M., Georgoussis, G., Louridas, A., Horseman, A., Sugier, J. 2017: Calibration of the depolarization channel of UV Raman Lidars, proceedings *28 ILRC*.

[4] http://aeronet.gsfc.nasa.gov/new_web/index.html

[5] http://research.metoffice.gov.uk/research/nwp/satellite/imagery/msg.html

[6] https://watchers.news/2016/12/27/bogoslov-volcano-alaska-december-2016-eruptions/

EPJ Web of Conferences **176**, 09007 (2018) https://doi.org/10.1051/epjconf/201817609007
ILRC 28

CONSISTENCY OF THE SINGLE CALCULUS CHAIN FOR CLIMATOLOGICAL STUDIES USING LONG-TERM MEASUREMENTS FROM THESSALONIKI LIDAR STATION

Nikolaos Siomos[1,*], Kalliopi A. Voudouri[1], Maria Filioglou[1,2], Eleni Giannakaki[2,5], Vasilis Amiridis[3], Giuseppe D'Amico[4] and Dimitris S. Balis[1]

[1]*Laboratory of atmospheric physics, Physics Department, Aristotle University of Thessaloniki, Greece, *Email: nsiomos@physics.auth.gr*
[2]*Finnish Meteorological Institute, Atmospheric Research Centre of Eastern Finland, Kuopio, Finland*
[3]*Institute for Astronomy, Astrophysics, Space Applications and Remote Sensing, National Observatory of Athens, Athens, Greece*
[4]*Consiglio Nationale delle Ricerche, Instituto di Metodologie per l' Analisi Ambientale (CNR - IMAA), Tito Scalo, Potenza, Italy*
[5]*Department of Environmental Physics and Meteorology, Faculty of Physics, University of Athens, Greece*

ABSTRACT

The long term analysis of 15 years of lidar data derived from a Raman lidar at Thessaloniki is presented here. All measurements have been processed with the latest version 4 of the EARLINET Single Calculus Chain algorithm and are compared with the results from the current operational retrieval algorithm. In this paper we investigate the consistency between the EARLINET database and SCC for the case of Thessaloniki and we identify the issues to be considered when switching from current operations to SCC.

1 INTRODUCTION

Thessaloniki lidar system (THELISYS), is operational for the detection of aerosol particles since 2000 in the framework of the European Aerosol Network EARLINET [1]. The primary setup of the lidar system included two elastic backscatter channels at 355nm and 532nm and a nitrogen Raman channel at 387nm. A second Raman channel at 607nm was added in 2008. In 2011, two channels for the measurement of the cross and parallel polarized signal at 532nm were added. Finally, a third elastic backscatter channel at 1064nm was added in 2012. All the measurements processed within the study period 2001-2015 have been uploaded in the EARLINET database (https://data.earlinet.org). The total number of uploaded files for the whole period is 1831. A number of 527 files correspond to the elastic backscatter profiles at 355nm, 723 files to the elastic backscatter profiles at 532nm, 182 files to the elastic backscatter profiles at 1064nm, 290 files to Raman extinction and backscatter profiles at 355nm and 109 files to Raman extinction and backscatter profiles at 532nm. All the raw data have been converted to the network's standard format in order to be reprocessed with the SCC v4.0 algorithm. The SCC structure is briefly described in section 2.

2 The Single Calculus Chain

An automated tool for the processing of lidar data from raw signals up to the final products was developed within EARLINET. The SCC algorithm [2] provides high quality standardized aerosol optical products in near real time along within the entire EARLINET network (Pappalardo et al. 2014 [1]). Briefly, the SCC consists of different modules and uses raw lidar data as input in a standard format (NetCDF). The raw lidar signals are initially processed by the Preprocessor module (ELPP: EARLINET Lidar PreProcessor). After the applied system-dependent corrections to the raw signals, the aerosol backscatter and extinction coefficient profiles are derived by the optical processing module (ELDA: EARLINET Lidar Data Analysis). Also, with the latest version of SCC the calculation of the particle linear depolarization ratio is possible, by the new module ELDEC (EARLINET Lidar Depolarization ratio). All these processes are monitored by the SCC daemon which runs continuously in the background. The software is installed on a centralized server and a web interface is available giving the possibility of uploading the data and selecting and modifying any of the input SCC parameters. The output data are also in the NetCDF format accepted by the EARLINET database.

3 METHODOLOGY

In the first part of the study, the SCCv.4 Raman extinction product at 355nm is evaluated for Thessaloniki. The evaluation procedure is based on a comparison of the aerosol optical depth (AOD) values derived by the SCCv.4 and the operational algorithm. AOD data from the Aerosol Robotic Network (AERONET) for Thessaloniki are also compared against the AOD of the two algorithms. The two lidar datasets include data from the period 2001-2015, processed by both the operational algorithm and the SCC. A common overlap correction was applied for the SCC Raman analysis. In the archived data Raman inversion, however, the applied overlap function is derived individually for each measurement [3]. Since the two algorithms don't always produce profiles with the same height range we use only the common part of the two profiles for the AOD integration. The values in the first point of the common height range are assumed to be constant down to the ground. The sunphotometer dataset includes the level 1.5 AOD at 340nm for Thessaloniki. This product is available for most of the period 2005-2015. Since the raman AOD is a nighttime product while the sunphotometer AOD is a day time product we use the closest sunphotometer measurement. In the second step, we examine the ability of the SCC algorithm to reproduce a seasonal climatology similar to the ones generated by the operational algorithm and the sunphotometer retrievals for the AOD at 355nm. Since a longer common period of measurements is important for a climatologic study, instead of using the sunphotometer AOD at 340nm, that is directly comparable with the lidar AOD at 355nm, we prefer to convert the sunphotometer's 440nm AOD using the angstrom 440-675nm that are available for the whole period 2003-2015. For the climatological comparison we preferred to use the daily mean AOD from the sunphotometer. We calculate and compare the seasonal averages, the annual cycle and the AOD trends for the three datasets.

4 RESULTS

The evaluation of the SCCv.4 Raman AOD at 355nm against the operational algorithm is presented in figure 1. A high correlation of 0.89 is derived with a least square fit slope at 0.78. The AOD seems to be underestimated by the SCC algorithm for the majority of the data points. In figure 2 and 3 the SCC AOD and the operational algorithm AOD are compared with the sunphotometer AOD. The results are similar for both lidar datasets. The correlation is close to 0.6 and the least square fit slope is the same for both comparisons. We have to mention here that figure 1 is not directly comparable with figures 2 and 3 since the sunphotometer data weren't available in the period 2001-2005. The annual cycle of the period is pre-

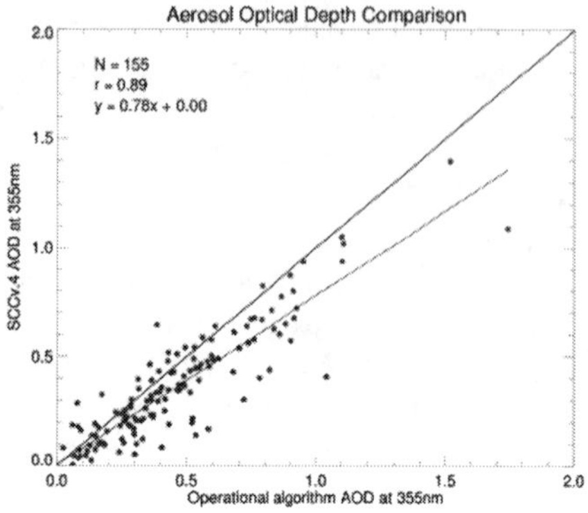

Figure 1: Comparison of the SCC v.4 AOD at 355nm with the operational algorithm AOD at 355nm.

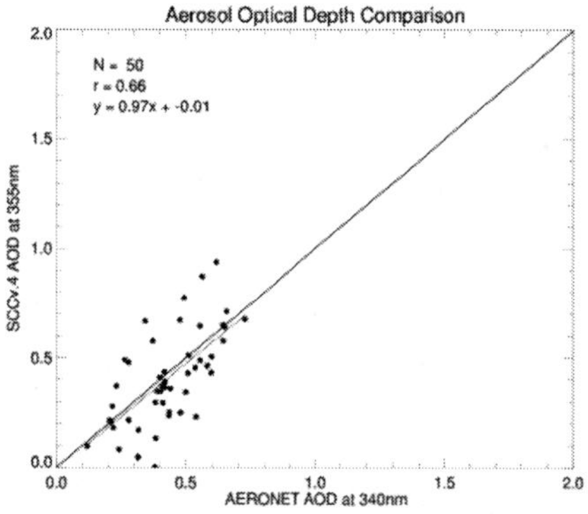

Figure 2: Comparison of the SCC v.4 AOD at 355nm with the sunphotometer AOD at 340nm.

sented in figure 4. The maximum values appear in summer and the minimum ones in winter in all datasets. The lidar AODs are usually underestimated compared to the sunphotometer AOD with the exception of the summer AOD of the operational algorithm. There is also an offset between the two lidar products

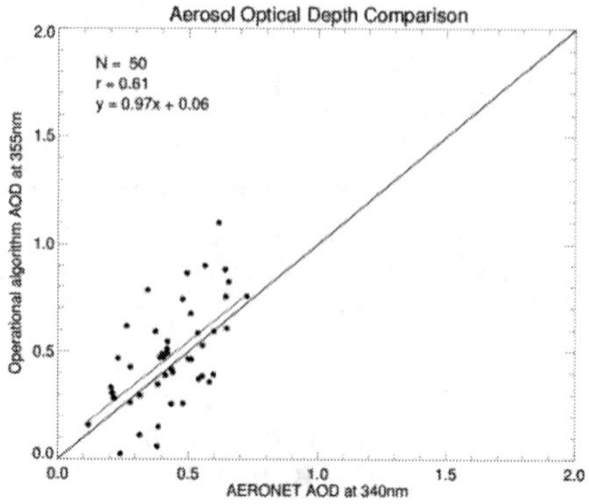

Figure 3: Comparison of the operational algorithm AOD at 355nm with the sunphotometer AOD at 340nm.

for all the seasons. Figure 5 contains the comparison of the AOD seasonal averages between the sunphotometer, the SCCv.4 and the operational algorithm datasets for the period 2003-2015. The least square fit line has been calculated for each timeseries. The sunphotometer values are considered as reference data. Both the operational algorithm and the SCCv.4 seasonal trends are similar to the sunphotometer one. There is a negative offset for both lidar

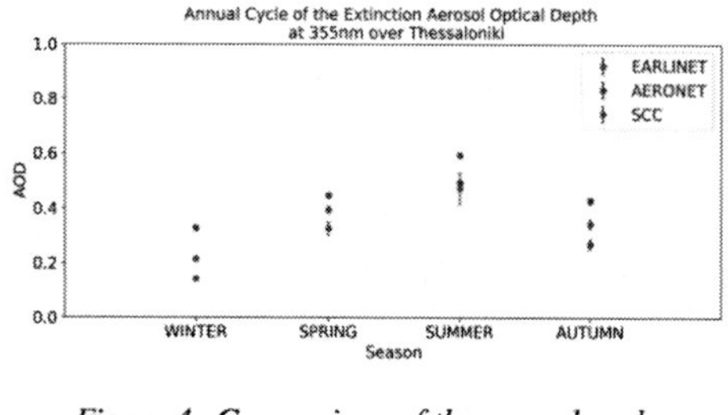

Figure 4: Comparison of the annual cycle calculated with the three datasets.

products which is larger for the SCC. Overall the SCC lidar AODs tend to be underestimated compared to the sunphotometer and the operational algorithm. This is compatible with figure 1 where the SCC AOD seems underestimated for most data points. It is important to mention

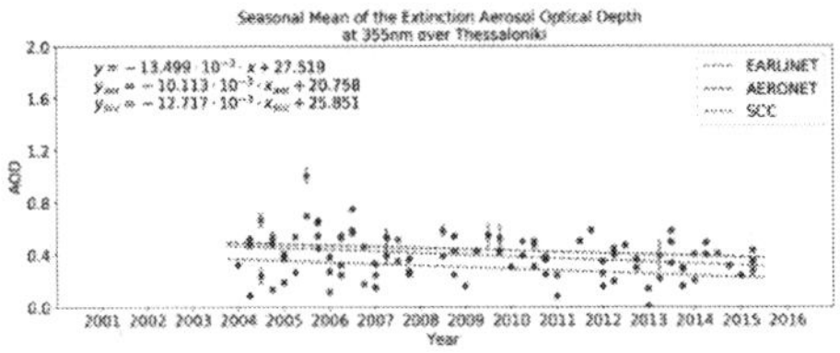

Figure 5: Comparison of the seasonal averages calculated with the three datasets.

though that the AOD is calculated in different ways between figures 1,2,3 and figure 4,5 (see section 3). This behavior could be attributed to the fact that each algorithm applies the overlap function in a different way and this affects the AOD calculation [4] (section 3).

5 CONCLUSIONS

The SCCv.4 algorithm has been evaluated for the AOD at 355nm for Thessaloniki. The agreement with the standard algorithm seems promising with a correlation of 0.89. The comparison with the sunphotometer gives similar results for the two algorithms. In both cases the correlation is close to 0.6 and the least square fit slopes are the same. The seasonal variability of the AOD at 355nm in the period 2003 to 2015 is reproduced adequately by the lidar products despite the fact that the data availability for the lidar is quite lower than the sunphotometer. The trends are similar and the minimum and maximum values in the annual cycle occur in the same seasons. The overall underestimation of the SCCv.4 results is mostly connected to differences in the overlap function that is applied in each algorithm. In the future additional optimizations in the SCC products, such as the addition of radiosonde data for each measurement are going to be implemented and the effect of the overlap function in the climatological results will be further examined.

ACKNOWLEDGEMENTS

This work has been conducted in the framework of EARLINET (EVR1 CT1999-40003), EARLINET-ASOS (RICA-025991) and ACTRIS-2 funded by the European Commission. This project has received funding from the European Union's Horizon 2020 research and innovation programme under grant agreement No 654109.

References

[1] G. Pappalardo, A. Amodeo, A. Apituley, A. Comeron, V. Freudenthaler, H. Linné, A. Ansmann, J. Bösenberg, G. D'Amico, I. Mattis, L. Mona, U. Wandinger, V. Amiridis, L. Alados-Arboledas, D. Nicolae, and M. Wiegner. Earlinet: towards an advanced sustainable european aerosol lidar network. *Atmospheric Measurement Techniques*, 7(8):2389–2409, 2014.

[2] G. D'Amico, A. Amodeo, H. Baars, I. Binietoglou, V. Freudenthaler, I. Mattis, U. Wandinger, and G. Pappalardo. Earlinet single calculus chain – overview on methodology and strategy. *Atmospheric Measurement Techniques*, 8(11):4891–4916, 2015.

[3] Ulla Wandinger and Albert Ansmann. Experimental determination of the lidar overlap profile with raman lidar. *Appl. Opt.*, 41(3):511–514, Jan 2002.

[4] Voudouri, K., Siomos, N., Giannakaki, E., Amiridis, V., d'Amico, G., and Balis, D. S. Comparison of aerosol backscatter and extinction profiles based on the earlinet database and the single calculus chain for thessaloniki greece (2001–2014). *EPJ Web of Conferences*, 119:23024, 2016.

ATHLI16: THE ATHENS LIDAR INTERCOMPARISON CAMPAIGN

Aldo Amodeo[1*], Giuseppe D'Amico[1], Aldo Giunta[1], Nikolaos Papagiannopoulos[1], Alex Papayannis[2], Athina Argyrouli[2], Maria Mylonaki[2], Georgios Tsaknakis[2], Panos Kokkalis[2], Ourania Soupiona[2], Chris Tzanis[3]

[1]*National Research Council - Institute of Methodologies for Environmental Analysis (CNR-IMAA), Italy,*
**aldo.amodeo@imaa.cnr.it*
[2] *National Technical University of Athens, Laser Remote Sensing Unit, Greece*
[3] *National and Kapodistrian University of Athens, Department of Physics, Greece*

ABSTRACT

The results of the ATHLI16 (ATHens Lidar Intercomparison) campaign, held in Athens from 26/09 to 07/10 2016 are presented. The campaign was performed within the Lidar Calibration Centre activities (EU H2020 ACTRIS-2 project) to assess the performance of the EOLE lidar system (NTUA, Athens, Greece), operating within EARLINET, by comparing against the EARLINET reference lidar system MUSA (CNR-IMAA, Potenza, Italy). For both lidars only products retrieved by the EARLINET Single Calculus Chain have been compared.

1 INTRODUCTION

EARLINET, the European Aerosol Lidar Network [1], was established in 2000 with the aim to provide a statistical significant and high quality database of aerosol profiles at European scale. After the first phase, from 2000 to 2003, in which the network was established as a research project funded by EU, EARLINET was funded again by EU under the projects EARLINET-ASOS (Advanced Sustainable Observation System) and ACTRIS (Aerosol, Clouds, and Trace gases Research Infrastructure), and presently it is funded in the frame of the EU project ACTRIS-2 (EU H2020 project) [www.actris.eu]. The network has been growing continuously over the years and several new stations have joined. At present, it consists of 28 advanced lidar stations distributed over 17 European countries. Most of the network instruments have been developed individually and therefore they are not standard systems. In order to guarantee the quality of the data from all the lidar systems, a quality assurance (QA) strategy has been developed both at algorithm and instrument levels. In particular, at instrument level, a direct intercomparison has been planned for all the network lidars against reference systems. Such intercomparison is made every time a new station joins the network or when a station performs a substantial upgrade on an already intercompared system. The first intercomparison campaign was carried out in 2009 (EARLI09 - EARlinet Lidar Intercomparison campaign 2009), involving 11 co-located lidar systems from nine countries [2]. This campaign was particularly important because it allowed to fix the quality standards for the EARLINET systems and to define five reference lidar systems to be used in the future intercomparison campaigns. After that, further six campaigns have been carried out, involving further eleven lidar systems and more are planned in the near future.

In this paper, the ATHLI16 (ATHens Lidar Intercomparison 2016 measurement campaign) is presented. The campaign was held in Athens (Greece) from 26 September to 7 October 2016, within the services provided by the Lidar Calibration Centre (LiCal) [3]. LiCal is a multi-installation facility located in Romania (INOE), Germany (LMU), and Italy (CNR-IMAA), established in the frame of the ACTRIS-2 project and offering a wide range of services to test and calibrate lidars and ceilometers, starting from the characterization and optimization of single components, to the assessment of the whole system's performance, and training of instrument operators. The ATHLI16 campaign allowed to check the performance of EOLE, the lidar system running in Athens EARLINET station, through the comparison with the EARLINET reference lidar system MUSA (MUltiwavelength lidar System for Aerosol measurements), operated by CNR-IMAA in Potenza, Italy.

https://doi.org/10.1051/epjconf/201817609008

2 METHODOLOGY

The used methodology has been described in detail by Wandinger et al. [2]. The goals of these campaigns are the verification of the performance of the systems in different atmospheric conditions, the identification of the main system deficiencies (e.g. optical misalignment, detector saturation, experimental setup, electronic problems) leading to systematic errors in the aerosol products, and to figure out technical solutions to fix instrumental problems. As a first step, a set of tests are performed for each optical channel to check the optical alignment (Rayleigh-fit and telecover tests), the trigger delay, the dark current (to evidence electronic distortions), the detector saturation and several instrumental parameters [4]. As a second step, the range corrected signals (RCS) are compared in terms of relative differences in the PBL, in the layers and on the whole measurement range. The agreement of the signals is evaluated according to the EARLINET standards [2, 5].

The EOLE Raman lidar system is located at the National Technical University of Athens (NTUA) (37.97° N, 23.79° E, 212 m a.s.l.) [6]. The MUSA Raman lidar system [7] was located at an altitude of 200 m a.s.l. The difference in altitude between the locations of the two systems (12 m) was taken into account in the data analysis and comparison. Both EOLE and MUSA are able to measure aerosol backscatter coefficient at 1064, 532 and 355 nm, and aerosol extinction at 532 and 355 nm, using 607 and 387 nm Raman signals. For both systems, the infrared radiation is acquired in analog mode using an APD detector, the 355 and 532 nm backscattered radiation is acquired both in analog and photon-counting modes, while the Raman backscattered radiation at 387 and 607 nm is detected in photon-counting mode only for EOLE and both in analog and photon-counting modes for MUSA. MUSA and EOLE laser repetition rates are 20 and 10 Hz respectively, and the range resolutions were 3.75 and 7.5 m, therefore the comparison was carried out reducing all the lidar signals at the lowest resolution (7.5 m).

In the comparison, the atmospheric molecular backscatter signals were also used. They were retrieved from the pressure and temperature vertical profiles provided each day by a radio-sounding launched by the Hellenic National Meteorological Service (HNMS) at about 12:00 UTC from a site located 12 km far away from measurement site and at an altitude of 10 m a.s.l.

A total of 6 daytime and 7 nighttime measurement sessions were performed during the campaign. Each session had a duration of about 2-3 hours allowing to select a sufficient long time window for the comparison (at least 30 minutes) in which the atmospheric conditions and layering were quite stable.

The range corrected signals were processed using the Single Calculus Chain (SCC) [8, 9], the centralized analysis tool developed within EARLINET in order to process in an automatic way the data measured by the network stations, and a set of software tools, specially developed for the intercomparison campaigns [2].

Whenever a problem was detected, a possible reason was identified, and the corresponding correction was applied in the next measurement session or in the signal processing.

3 RESULTS

We present an example of comparison of measurements performed in nighttime conditions on 26 September 2016 from 18:30 to 20:10 UTC. Fig. 1 shows the time evolution of the RCS at 1064 nm measured by MUSA for the whole measurement session. The red square indicates the time window (18:44 - 19:44 UTC) where the RCS of both systems were averaged for the comparison. The selected time window was characterized by quite stable atmospheric conditions.

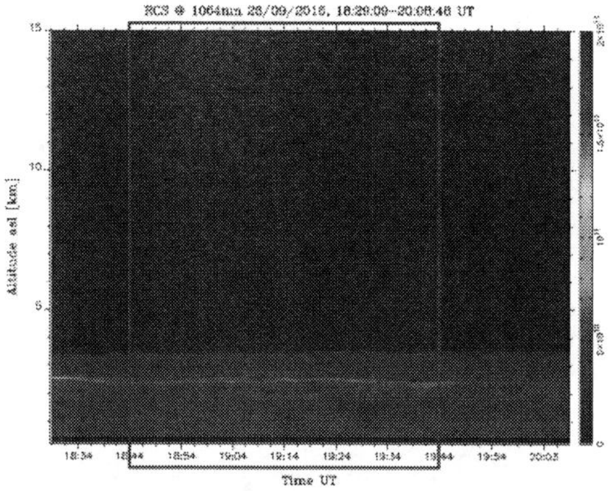

Figure 1 Time evolution of the RCS at 1064 nm, measured on 26 September 2016 by MUSA (18:30 - 20:10 UTC) in Athens during ATHLI16.

The comparison allowed us to identify two technical problems on the EOLE infrared channel.

The first one was an incorrect setting of the trigger delay (i.e. the delay between the electronic start of the measurement and the effective time of the laser light emission) which was corrected by minimizing the vertical scale shift between EOLE and MUSA RCS. The second problem on the EOLE infrared channel was the presence of electronic distortions which was fixed by subtracting the dark current signal from the measured lidar signals. We concluded that this correction was particularly significant at 1064 nm.

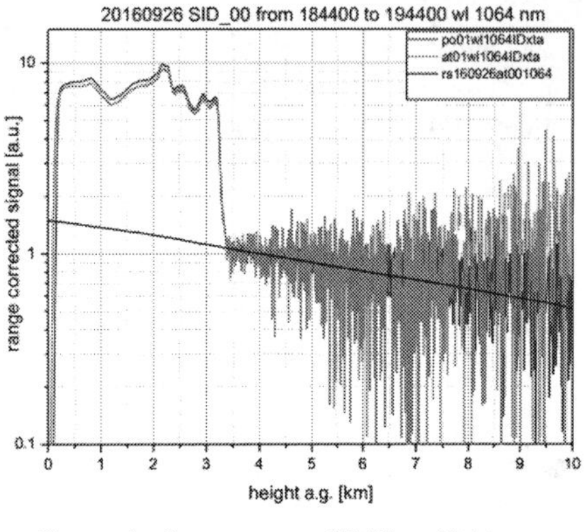

Figure 2. Comparison of RCS at 1064 nm.

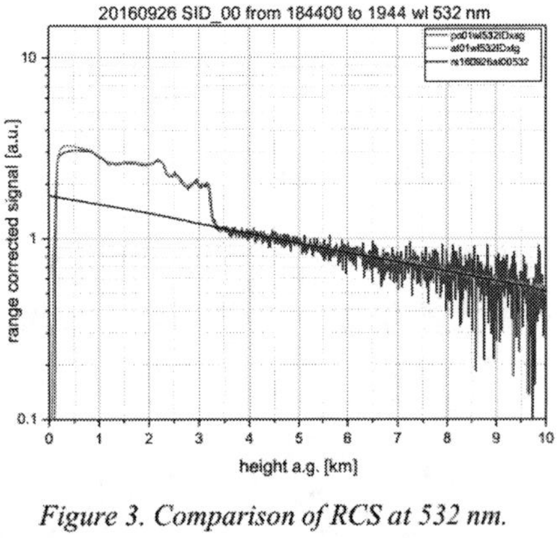

Figure 3. Comparison of RCS at 532 nm.

Figs. 2-6 show the comparison of the RCS at 1064, 532, 355, 607 and 387 nm measured by MUSA (red line) and EOLE (green line) and averaged in the time window shown in Fig. 1. In all the figures it is also reported the molecular backscattering (black line) at the corresponding wavelength. The signals and the molecular backscattering have been normalized in the aerosol free region. A large urban aerosol layer up to about 3.5 km is present.

From the comparison, it is evident the general good agreement between the elastic signals of EOLE and MUSA systems.

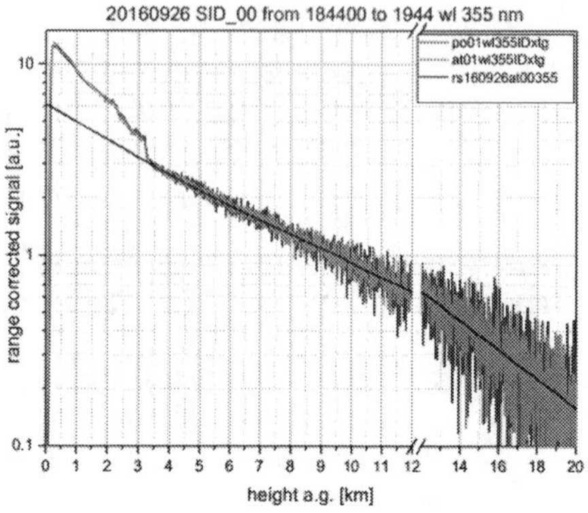

Figure 4. Comparison of RCS at 355 nm.

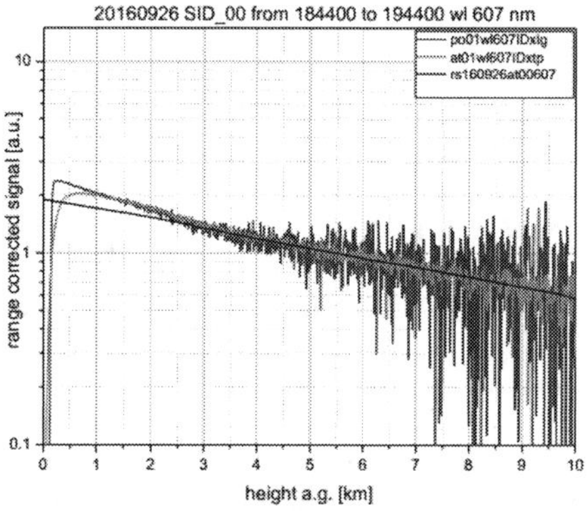

Figure 5. Comparison of RCS at 607 nm.

The relative differences between EOLE and MUSA RCS at 1064 nm are within 6% in the range where most of the aerosol is resided (up to about 3 km). Above this range the relative differences are larger (20% below 5 km) because the aerosol load is quite low and as a consequence the signals of both lidars are quite noisy. The signals at 532 (355) nm agree within 2% (3%) up to 3.5 km and within 5% (10%) in the range 3.5-6 km.

The Raman signals show good agreement above 1.2 km and 1.5 km at 607 and 387 nm respectively. The large discrepancy at lower ranges is related to the different acquisition mode used by the two lidars: EOLE Raman signals are acquired using photon-counting mode while MUSA ones are detected in both analog and photon-counting mode and the corresponding glued signal is provided. As a consequence the EOLE Raman signals are saturated at low ranges whereas for MUSA system the analog signals start to play an important role.

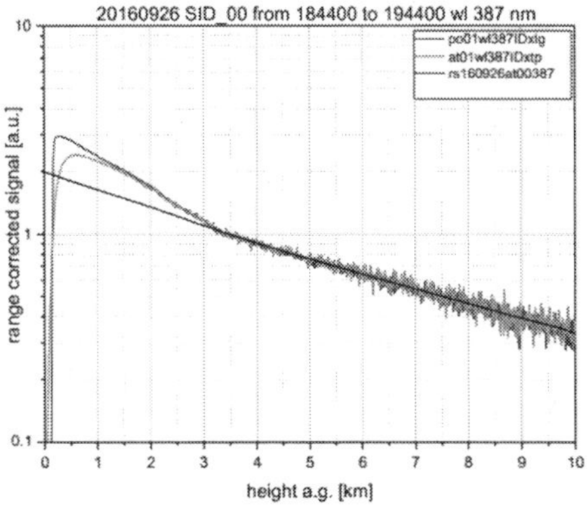

Figure 6. Comparison of RCS at 355 nm.

We also calculated the mean relative deviations, averaged within the large aerosol layer present between 200 and 3500 m (see Tab 1). The results show a good agreement, but with large standard deviations due to the signal noise.

Table 1. Mean relative difference between EOLE and MUSA RCS, calculated within the aerosol layer between 0.2 and 3.5 km.

λ (nm)	**0.2 - 3.5 km**
1064	-0.05 (0.02)
532	0.01 (0.03)
355	-0.0003 (0.0276)
607	-0.02 (0.09)
387	-0.05 (0.08)

4 CONCLUSIONS

The lidar intercomparison campaign ATHLI16 was addressed to check the quality of the EOLE system through the comparison with the EARLINET reference lidar system MUSA. The results show a good performance of the EOLE system in terms of RCS comparison according to the EARLINET standards [2, 5]. The discrepancy at low ranges of Raman signals will be solved by upgrading the Raman optical channels also with analog acquisition. The intercomparison evidenced the importance of subtracting the dark to the 1064 nm measurement regularly in order to minimize electronic distortions. For the same optical channel, the correct trigger delay value was determined as well. We also tried, as a preliminary stage, to compare the optical products (aerosol backscatter and extinction), retrieved using the SCC, obtaining promising results (examples will be presented at the conference).

ACKNOWLEDGEMENTS

This project has received funding from the European Union's Horizon 2020 research and innovation programme under grant agreement No 654109.

References

[1] Pappalardo, G., et al., 2014: EARLINET: towards an advanced sustainable European aerosol lidar network, *Atmos. Meas. Tech.*, **7**, 2389-2409, doi:10.5194/amt-7-2389-2014.

[2] Wandinger, U., et al., 2016: EARLINET instrument intercomparison campaigns: overview on strategy and results. *Atmos. Meas. Tech.*, **9**, 1001–1023, doi:10.5194/amt-9-1001-2016.

[3] Belegante, L., et al., 2017: The lidar calibration centre: a multi installation facility in the framework of ACTRIS. In: *Proc. 28th International Lidar Aerosol Conference (28th ILRC)*. This volume.

[4] Freudenthaler, V., et al.: Internal quality assurance tools, *Atmos. Meas. Tech. Discuss.*, in prep.

[5] Matthias, V., et al., 2004: Aerosol lidar intercomparison in the framework of the EARLINET project. 1. Instruments, *Appl. Opt.*, **43**, 961–976, doi:10.1364/AO.43.000961.

[6] Kokkalis, P., et al., 2012: The EOLE lidar system of the National Technical University of Athens. In: *Proc. 26th International Laser Radar Conference (26th ILRC)*, pp. 629–632.

[7] Madonna, F., et al., 2011: CIAO: the CNR-IMAA advanced observatory for atmospheric research, *Atmos. Meas. Tech.*, **4**, 1191-1208, doi:10.5194/amt-4-1191-2011.

[8] D'Amico, G., et al., 2016: EARLINET Single Calculus Chain – technical – Part 1: Pre-processing of raw lidar data, *Atmos. Meas. Tech.*, **9**, 491-507, doi:10.5194/amt-9-491-2016.

[9] Mattis, I., et al., 2016: EARLINET Single Calculus Chain – technical Part 2: Calculation of optical products, *Atmos. Meas. Tech.*, **9**, 3009-3029, doi:10.5194/amt-9-3009-2016.

South American Aerosol Tracking - LALINET

Eduardo Landulfo[1,*], Fabio Lopes[1], Pablo Ristori[2], Eduardo Quel[2], Lidia Otero[2], Ricardo Forno[3], Maria Fernanda Sanchez[3], Henrique Barbosa[4], Diego Gouveia[4], Amanda Vieira Santos[4], Alvaro Bastidas[5] and Daniel Nisperuza[5]

[1]*Instituto de Pesquisas Energéticas e Nucleares, Centro de Lasers e Aplicações, Av. Prof. Lineu Prestes, 2242, São Paulo, 05508000, Brazil, *elandulf@ipen.br*
[2]*CITEDEF, CEILAP, San Juan Bautista de la Salle 4397, Buenos Aires, B1603ALO, Argentina*
[3]*Laboratorio de Física de la Atmosfera, Universidad Mayor de San Andrés, La Paz, Bolivia*
[4]*Dep. de Fisica Aplicada, U. de São Paulo, Rua do Matão, 1371, 05508-090, São Paulo, Brazil*
[5]*Grupo de Investigación Láseres y Espectroscopia, Universidad Nacional de Colombia, Medellín, Colombia*

ABSTRACT

LALINET lidar stations were used to track down aerosols generated over Amazon region and transported over the continent. These data were merged with collocated Aeronet stations in order to help in their identification together with HYSPLIT simulations. The results show potential indication of how aerosol can age in their long transport over regions South and Westward from the source areas by change of their optical properties.

1 INTRODUCTION

The transport of aerosols in South America has as one of its main sources those generated in the Amazon and adjacent regions[3]. The circulation pattern is well known and can make this particles reach most countries in the continent.During the transport which may last several days, 4-8 days, the aerosols which do not mix or are not removed may change their optical properties and affect the air quality and mix with those aerosols present in the areas they are detected. A consistent tracking of these changes might be performed by sattelites such as CALIOP and/or MODIS or by an array of instruments[4, 6, 1] in order to increase modelling reliability .Aerosols from Amazon may change locally by water uptake or condensation of organics as suggested by Johnson *e*t al.[2] or during transport into remote areas when more complex processes take place such as coagulation and gas-to-particle conversion and aerosol-cloud-interaction[5] which should be taken into account as a valued source for potential aerosol aging studies.

2 METHODOLOGY

For this study 5 LALINET stations were used namely: Manaus(BR), Medellin(CO), La Paz(BO), Buenos Aires(AR) and São Paulo(BR), the station information are given in table below. Their location span over the entire continent and could be affected by transport from biomass burning activities. Collocated with these stations are AERONET sunphotmeter stations which synergizes with the local lidar instruments. Besides HYSPLIT simulations were made on specific days for arriving air masses at 3000 km, a.g.l., to help in identifying potential aerosols sources present on the lidar routine measurements carried on by the network. This first approach is to comprehend the overall pattern circulation present on the station sites and also to better undestand possible *a*ging effects, at first quantitatively.

 https://doi.org/10.1051/epjconf/201817609009

Table 1: LALINET Lidar Stations

ST.	ID	LAT(S) LON(W)	Channels (nm)
Buenos Aires	*AEP*	$34^{o}33'$ $58^{o}25'$	*1064, 532^{p} & 355^{p}*
S. Paulo	*SPU*	$23^{o}13'$ $46^{o}88'$	*1064, 607, 532, 408, 387 & 355*
Man-aus	*MAO*	$02^{o}60'$ $60^{o}21'$	*408, 387, 355*
Medellin	*MED*	$-06^{o}16'$ $75^{o}35'$	*532 & 355*
La Paz	*LPZ*	$16^{o}32'$ $72^{o}04'$	*1064, 532^{p} & 355^{p}*

3 RESULTS

3.1 Sunphotometric Results

The stations have collocated AERONET stations which provide informations on the aerosol optical properties such as **A**erosol **O**ptical **D**epth (**AOD**) and **A**ngström **E**xponent (**AE**). In figures 1 and 2 one can see the AOD statistics in 2016 and AOT $\times$ AE for the same period in the 5 LALINET stations. The MED station is the one presenting the largest variability and certainly influence by aerosos from local and remote sources. The plots show also a large variability in MED station, while MAO,LPZ, SPU and BUE are those which present less change with a mean value of 0.19 ± 0.11 , 0.10 ± 0.09 , 0.19 ± 0.12 and 0.11 ± 0.12, respectively. On other hand as forementioned, MED has an AOD mean value of 0.43 ± 0.42. When looking at size range one checks the Angström Exponent values and variability: $AE_{SP} = 1.50 \pm 0.21$, $AE_{MAO} = 1.26 \pm 0.34$, $AE_{BUE} = 1.25 \pm 0.47$, $AE_{LPZ} = 0.82 \pm 0.37$, $AE_{MED} = 1.19 \pm 0.47$.

Among many cases when high values of AOD

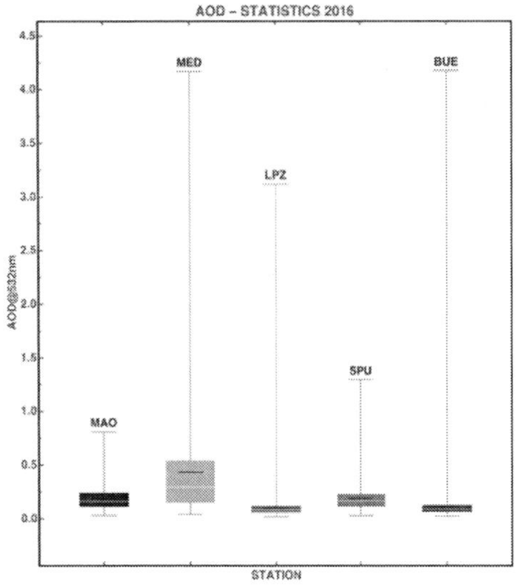

Figure 1: AOD Statistics for 2016among LALINET stations.

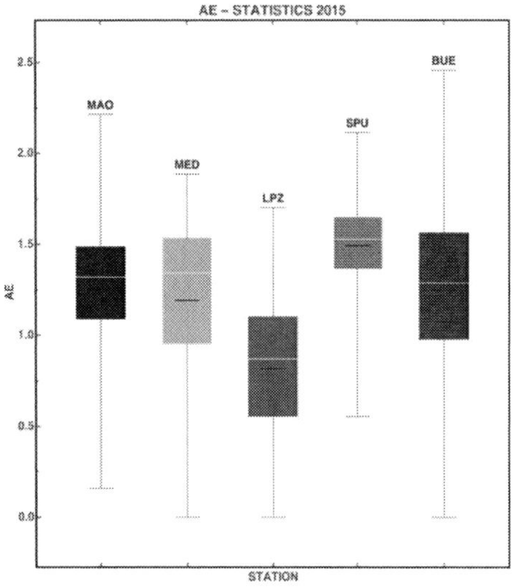

Figure 2: AE Statistics for 2016 among LALINET stations

were present in all the stations we have selected some specific days, namely August 25, September 10, September 19, November 28, in 2016, when air mass transport into the stations are shown by HYSPLIT simulations. The chosen arrival altitude was 3000 m to minimize is this first study Planetary Boundary Layer (PBL) and uncomplete overlap effects. It is important to note that the different circulation patterns playing a role in the continent show a potential variety of aerosols from Atlantic/African origin to in-continental sources, dust, marine/clear and biomass burning. Also it is noteworthy that some pair (threewise as well) of station could be used to track the aerosol properties and possible aging and changes in their optical properties. For example : MAO-LPZ-MED or BUE-SPU groups. Of course this approach could be extended to other LALINET stations thus enhancing the regional coverage. Also the aerosol typing according to the source region allows to identify saharan dust and biomass burning events in the case o MAO, LPZ and MED. When one looks at the Range Corrected Sig-

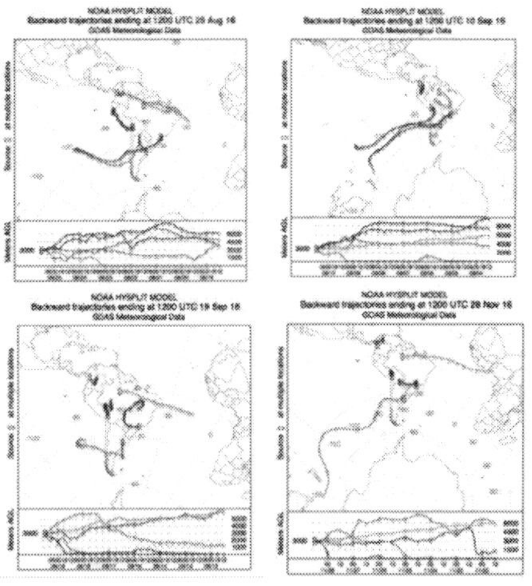

Figure 3: HYSPLIT backtrajectory simulation over LALINET stations arriving on September 19, 2016.

nal data from the stations it is possible to verify the layering structure over the lidar sites being supported and/or validating the HYSPLIT simulations in the instances shown. The RCS plots corrspond to the Monday/Thursday routine measurements taken by the network. Begining by the AER station in Buenos Aires on November, 28, a mixing of Pacific marine and continental (dust) is present in the Argentinian city up to 6 km with a clear enhancement of signal above 2 km. The bolivian site, the cleanest site, show the presence of aerosols up to 3 - 4 km a.g.l. but this station is in the Andean mountain range and close to La Paz. Both Manaus and São Paulo stations are represented on August, 25, exemplifying the two different circulation patterns present in the continent, while Manaus is influenced by a wersterly air mass bringing diffent types of aerosols, São Paulo, the largest urban site, shows a heavy load of aerosols with a well defined PBL and a transported air mass originating in the Pampas region in the southern part of the continent. Finally the Medellin station on September, 19, shows that even though air masses show advection from local areas the presence of aerosols aloft the PBL plus the presence of low-middle clouds poses a challenge in studiyng aerosols in the areas with largest values of AOD's.

4 CONCLUSIONS

This first study on identifying the local and transported aerosol layers in 5 different LALINET stations presented here show an important step towards the aerosol optical properties based on their potential sources, while this procedure is not new, it is for South America. By the extension of the timeseries from only one year in this study, 2016, to a longer period will show reveal important patterns in the aerosol study. Also the inclusion of other netwrok station not presented here for the sake of brevity will improve the results.

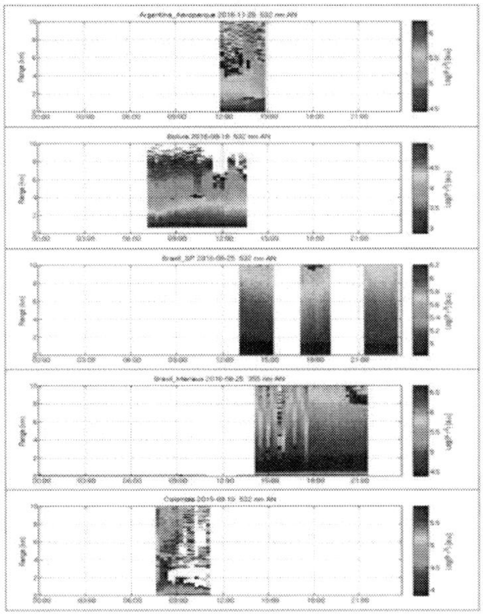

Figure 4: Range Corrected Signal for the five LALINET stations. The dates correspond to the Monday/Thursday measurement protocol.

ACKNOWLEDGEMENTS

This work was also partially funded by the University of Granada through the contract *Plan Propio. Programa 9. Convocatoria 2013*, additional support was given by Universidad Nacional Sede Medellín, NASA-AERONET. We are also indebted with NIES science partner, namely Dr. Yoshitaka Jin and Nobuo Sugimoto on the data transfer and storage in their facilties.

References

[1] Y. Hara, K. Yumimoto, I. Uno, A. Shimizu, N. Sugimoto, Z. Liu, and D. M. Winker. Asian dust outflow in the pbl and free atmosphere retrieved by NASA CALIPSO and an assimilated dust transport model. *Atmospheric Chemistry and Physics*, 9(4):1227–1239, 2009.

[2] B. T. Johnson, J. M. Haywood, J. M. Langridge, E. Darbyshire, W. T. Morgan, K. Szpek, J. K. Brooke, F. Marenco, H. Coe, P. Artaxo, K. M. Longo, J. P. Mulcahy, G. W. Mann, M. Dalvi, and N. Bellouin. Evaluation of biomass burning aerosols in the HadGEM3 climate model with observations from the SAMBBA field campaign. *Atmospheric Chemistry and Physics*, 16(22):14657–14685, 2016.

[3] E. Landulfo, S. R. Freitas, K. M. Longo, S. T. Uehara, and P. Sawamura. A comparison study of regional atmospheric simulations with an elastic backscattering lidar and sunphotometry in an urban area. *Atmospheric Chemistry and Physics*, 9(18):6767–6774, 2009.

[4] Jianfei Peng, Min Hu, Zhaoheng Gong, Xudong Tian, Ming Wang, Jing Zheng, Qingfeng Guo, Wei Cao, Wei Lv, Weiwei Hu, Zhijun Wu, and Song Guo. Evolution of secondary inorganic and organic aerosols during transport: A case study at a regional receptor site. *Environmental Pollution*, 218:794 – 803, 2016.

[5] Jeffrey S. Reid, Thomas F. Eck, Sundar A. Christopher, Peter V. Hobbs, and Brent Holben. Use of the ångstrom exponent to estimate the variability of optical and physical properties of aging smoke particles in Brazil. *Journal of Geophysical Research: Atmospheres*, 104(D22):27473–27489, 1999.

[6] F. Waquet, F. Peers, P. Goloub, F. Ducos, F. Thieuleux, Y. Derimian, J. Riedi, M. Chami, and D. Tanré. Retrieval of the Eyjafjallajökull volcanic aerosol optical and microphysical properties from POLDER/PARASOL measurements. *Atmospheric Chemistry and Physics*, 14(4):1755–1768, 2014.

https://doi.org/10.1051/epjconf/201817609010

VALIDATION AND DEPLOYMENT OF THE FIRST LIDAR BASED WEATHER OBSERVATION NETWORK IN NEW YORK STATE: THE NYS MESONET PROJECT

L. Thobois[1]*, J. Freedman[2], P. Royer[1], J. Brotzge[2], E. Joseph[2]

[1]*Leosphere: Lidar Environmental Observations, Orsay, France, *lthobois@leosphere.com*
[2] *University of Albany, Albany, USA*

ABSTRACT

The number and quality of atmospheric observations used by meteorologists and operational forecasters are increasing year after year, and yet, consistent improvements in forecast skill remains a challenge. While contributing factors involving these challenges have been identified, including the difficulty in accurately establishing initial conditions, improving the observations at regional and local scales is necessary for accurate depiction of the atmospheric boundary layer (below 2km), particularly the wind profile, in high resolution numerical models. Above the uncertainty of weather forecasts, the goal is also to improve the detection of severe and extreme weather events (severe thunderstorms, tornadoes and other mesoscale phenomena) that can adversely affect life, property and commerce, primarily in densely populated urban centers.

This paper will describe the New York State Mesonet that is being deployed in the state of New York, USA. It is composed of 126 stations including 17 profiler sites. These sites will acquire continuous upper air observations through the combination of WINDCUBE Lidars and microwave radiometers. These stations will provide temperature, relative humidity & "3D" wind profile measurements through and above the planetary boundary layer (PBL) and will retrieve derived atmospheric quantities such as the PBL height, cloud base, momentum fluxes, and aerosol & cloud optical properties. The different modes and configurations that will be used for the Lidars are discussed. The performances in terms of data availability and wind accuracy and precision are evaluated. Several profiles with specific wind and aerosol features are presented to illustrate the benefits of the use of Coherent Doppler Lidars to monitor accurately the PBL.

1 INTRODUCTION

Despite the constant effort to improve weather forecasts over the last few decades, depiction of the evolution of the planetary boundary layer (PBL) remains a challenge. Today, there is a growing interest focusing on local and regional scales. Indeed, for severe weather, weather forecasts are crucial to drastically reduce the impact on human lives and economical activities where weather risk exposure is the highest (urban, business, industrial area). The forecast goal is to provide information regarding weather risk to support local authorities and decision makers. To reach this objective, numerical weather prediction (NWP) models must be coupling increased spatial resolution with improved data assimilation methods. These models, however, require more comprehensive local observations (e.g. the 3D wind field) as initial conditions. But existing weather observation networks were designed for national coverage and not for providing dense observations to high resolution models. New sensors are then required to provide accurate 3D observations with high spatial resolution from the surface through the top of the PBL. Wind Lidar technologies developed since the 1990s have expanded vastly in the 2000s with the introduction of commercial off-the-shelf fiber optics components developed for the telecommunication industry. This Lidar technology now incorporates cost effective sensors thanks to industrial manufacturing. Today, Coherent Pulsed Doppler Lidar (CDL) is being evaluated globally in a wide variety of projects to determine its benefits for weather observing networks [1][2].

In this paper, the incorporation of CDL into the operations of the New York State Mesonet (NYSM) is presented. The capabilities of CDLs used in this network are described in terms of wind and aerosol/cloud measurements. Their potential benefits for future observing networks are demonstrated with a few examples of wind and aerosol cases.

2 THE NEW YORK STATE MESONET

In 2014 Governor Cuomo, with funding from the Federal Emergency Management Agency, initiated the New York State Early Warning Weather Detection System. The centerpiece of this system is the NYSM, an advanced, statewide weather station network explicitly designed to enhance local data collection for improved local weather monitoring and prediction. It consists of a network of 126 weather stations that provide surface temperature, relative humidity, wind speed and direction, precipitation, solar radiation, atmospheric pressure, photographic images and soil moisture and temperature at three depths (5, 25, and 50 cm). In addition, seventeen of the sites are being outfitted with WINDCUBE coherent Doppler lidars (CDL), Microwave radiometers and sun photometers providing 3D wind, turbulence, temperature, and moisture profiles in the vertical within and above the PBL, and cloud and aerosol information, as well as mixing layer and residual layer height. Upon completion in 2017, real time data along with graphical products will be available to the public via a website (http://nysmesonet.org).

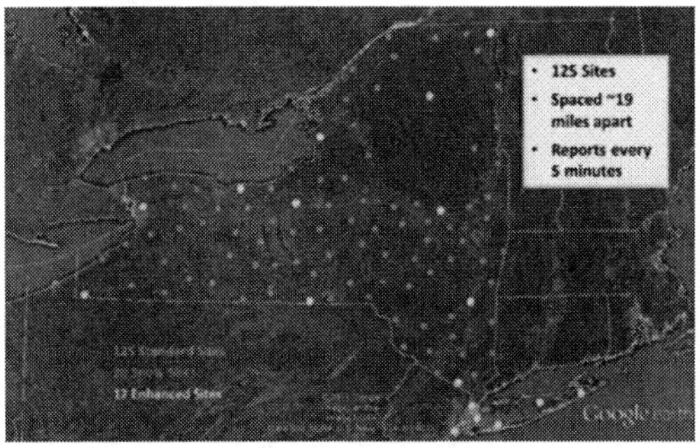

Figure 1 NYS MESONET deployment

3 PROFILER STATIONS WITH CDLs

Thanks to the rapid development of fiber based Lidar technology, more than a thousand of CDLs are now in use worldwide as operational instruments for wind energy, meteorology and air traffic applications. CDLs have been chosen for the NYSM to provide one of the most important parameters to be assimilated into NWP model wind profiles. The CDLs of the NYSM are LEOSPHERE WINDCUBE100S scanning CDLs that are configured in Doppler Beam Swinging (DBS) scenario to provide vertical profiles of horizontal wind speeds, vertical wind speed and wind direction from the ground up to the top of the PBL (1-2km). The physical resolution chosen is 25m corresponding to pulse duration of 100 ns and Doppler spectra discretized by 32 points. The accumulation time is 1s, and the update rate of final products is 5 min. In addition, the signal reflectivity of CDLs will be used to provide aerosol and cloud functionalities: detection of aerosol/cloud layers, retrieval of PBL heights, and backscatter coefficient [3][4].

Figure 2 – Installations of WINDCUBE100S lidar and MP3000a radiometer at University of Albany

4 PRELIMINARY RESULTS

Monitoring the PBL with CDLs

As an illustration, the measurements performed by one WINDCUBE100S CDL during two summer days with moderate convection and few clouds and aerosols layers above the boundary layer are presented on Figure 3. The observations show rapid changes in vertical velocity during the late morning and afternoon (local time ~ 1400 – 1800 UTC)—evidence of convective plumes and corresponding subsidence) and significant variations of wind speed on these two days. The vertical wind speed allows monitoring the convection from the surface and up to the top of the mixing layer. The backscatter coefficient allows the determination of different vertical layers including the residual layer of the boundary layer, clouds and other aerosol layers. In the case of relatively thin cloud layers, the base and top of the multiple cloud layers can be determined.

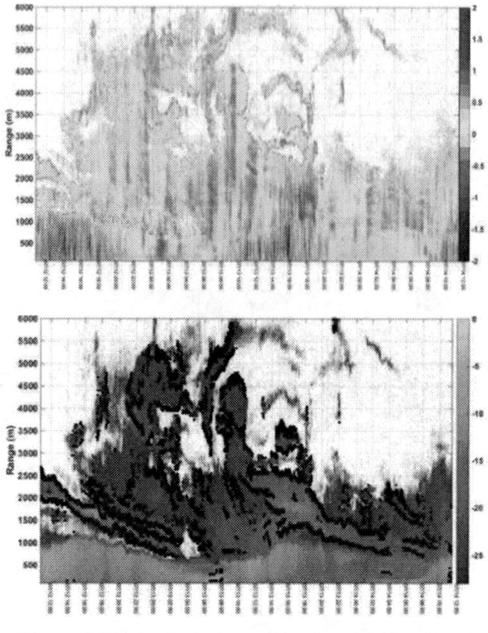

Figure 3 Vertical wind speed (top) and Lidar signal with detected layers (bottom) measured by a CDL

Comparison of CDLs with Radiosondes

The CDLs that will compose the NYS Mesonet are evaluated in terms of wind accuracy against radiosondes launched twice daily (0000 and 1200 UTC) at the National Weather Service Forecast Office in Albany, NY. Figure 4 shows one example showing excellent agreement between the CDL and radiosonde-derived horizontal wind speed profiles for the 26 August 2015. When evaluating the statistical accuracy and precision of CDL with radiosonde, excellent correlation is obtained in the higher part of the boundary layer (above 500m; Figure 5 (bottom)) with less agreement at lower levels due to turbulence induced by the presence of nearby buildings.

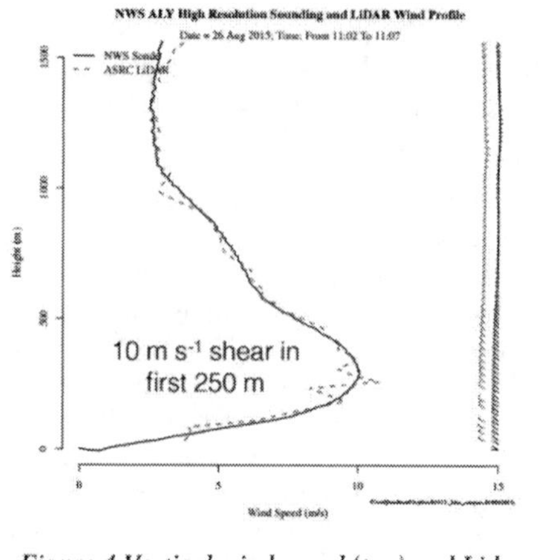

Figure 4 Vertical wind speed (top) and Lidar signal with detected layers (bottom) measured by a CDL

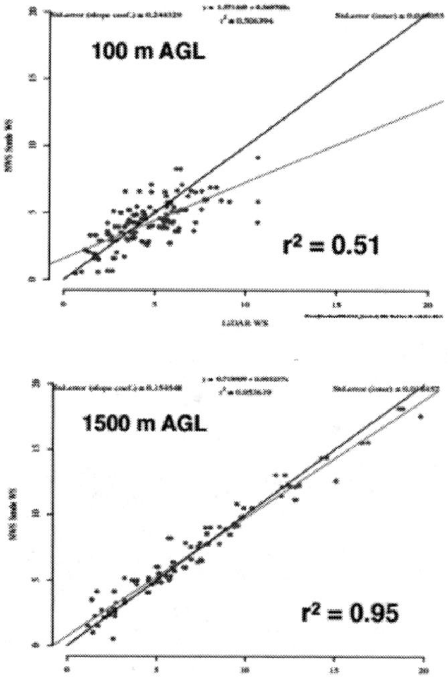

Figure 5 Comparison of horizontal wind speeds between CDL and radiosondes at 100m (top) and 1500m (bottom)

The CDL can also be used to observe rapid changes in the wind speed profile (Figure 6), a characteristic of special interest to the PBL community (e.g., the evolution of low level jets) and the aviation industry (timely detection of changes in low level wind shear).

https://doi.org/10.1051/epjconf/201817609010

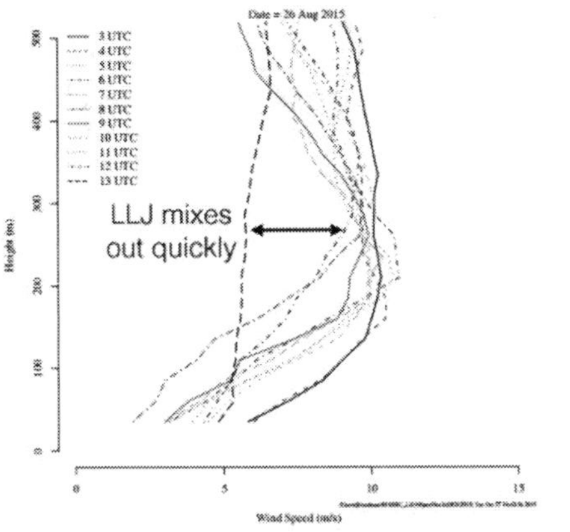

Figure 6 Evolution of wind profiles during one day with a low level jet

Comparison between CDLs

The CDLs of the NYS Mesonet are also compared with each other to validate the consistency of their data and ensure unbiased measurements at the different sites of the network. Wind speed, Lidar signal strength and data availability are the main parameters that are studied. As an example, wind speed profiles of three co-located CDLs show very good agreement (Figure 7). Advanced inter-comparison studies are currently on-going.

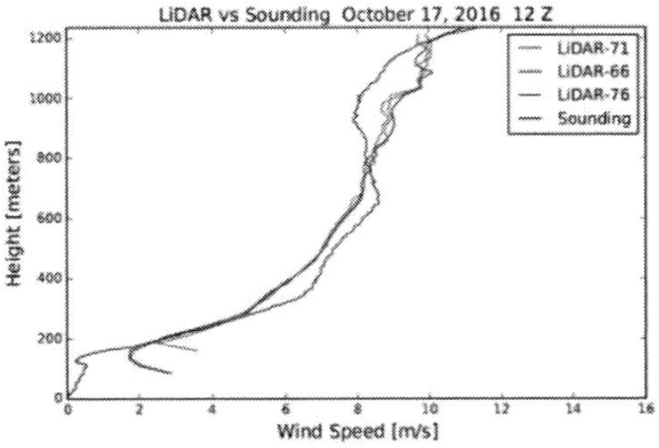

Figure 7 Comparison of wind profiles during one day for three CDLs with the same configuration

5 CONCLUSIONS

For the first time, a regional network equipped with CDLs is currently under deployment in the state of New York. The CDLs will be used as a multi-function sensor to provide wind and turbulence measurements but also aerosol and cloud observations usually performed by ceilometers or aerosol Lidars. The measurements of each CDL will be validated in terms of accuracy and precision and compared with a reference CDL. Advanced studies are currently on-going to evaluate their performance and data reliability. Data and derived products from CDLs will be used to improve weather analysis and prediction in combination with MW radiometers that allows the continuous tropospheric profiling of wind, temperature and humidity. A focused effort will be made to assimilate conventional data and derived products like PBL and turbulence into high resolution weather forecast models (NWP) to improve local forecasts.

References

[1] Illingworth, A. et al., Exploiting Existing Ground-Based Remote Sensing Networks to Improve High-Resolution Weather Forecasts, Bulletin of the American Meteorological Society 2015 96:12, 2107-2125

[2] Wulfmeyer, V. et al., A review of the remote sensing of lower tropospheric thermodynamic profiles and its indispensable role for the understanding and the simulation of water and energy cycles, Rev. Geophys., 53, 819–895, doi:10.1002/2014RG000476.

[3] O'Connor, E. et al.: Observing wind, aerosol particles, cloud and precipitation: Finland's new ground-based remote-sensing network, Atmos. Meas. Tech., 7, 1351-1375, doi:10.5194/amt-7-1351-2014, 2014.

[4] Hardesty, R. et al., One Year of Doppler Lidar Observations Characterizing Boundary Layer Wind, Turbulence, and Aerosol Structure During the Indianapolis Flux Experiment. In AGU Fall Meeting Abstracts (Vol. 1, p. 03).

SAVER.NET LIDAR NETWORK IN SOUTHERN SOUTH AMERICA

Pablo Ristori[1]*, Lidia Otero[1-5], Yoshitaka Jin[2], Boris Barja[3], Atsushi Shimizu[2], Albane Barbero[4], Jacobo Salvador[1-5], Juan Lucas Bali[5-6], Milagros Herrera[5], Paula Etala[4], Alejandro Acquesta[6], Eduardo Quel[1], Nobuo Sugimoto[2], Akira Mizuno[7]

[1] *Lidar Division, CITEDEF and UNIDEF (MINDEF - CONICET)*
**Email: pablo.ristori@gmail.com*
[2] *National Institute for Environmental Studies (NIES), Japan*
[3]*Universidad de Magallanes (UMAG), Punta Arenas, Chile*
[4]*National Weather Service, Argentina*
[5]*National Scientific and Technical Research Council (CONICET), Argentina*
[6] *Crisis Modelling and Management Department, CITEDEF, Argentina*
[7]*Atmospheric Environment Division, Solar-Terrestrial Environment Laboratory, Nagoya University, Japan*

ABSTRACT

The South American Environmental Risk Management Network (SAVER-Net) is an instrumentation network, mainly composed by lidars, to provide real-time information for atmospheric hazards and risk management purposes in South America. This lidar network have been developed since 2012 and all its sampling points are expected to be fully implemented by 2017. This paper describes the network's status and configuration, the data acquisition and processing scheme (protocols and data levels), as well as some aspects of the scientific networking in Latin American Lidar Network (LALINET). Similarly, the paper lays out future plans on the operation and integration to major international collaborative efforts.

1 INTRODUCTION

In recent years, volcanic activity in the Andes Mountain Range have impacted many countries in the world, primarily Argentina and Chile. These hazards promoted mainly two atmospheric risk management projects (Ministry of Defense – Special Project N°31554/11 and the SAVER.Net Project.). The first project focused on the development of five lidar deployable stations. The stations were supplemented with ground base instrumentation to monitor atmospheric aerosols and gas constituents (SO_2, BrO, N_2O, O_3, HCHO, H_2O, by means of MAX-DOAS), solar radiation and meteorological variables. The main goal is the detection of suspended volcanic ash. For this purpose, stations have been strategically distributed across latitudes covering from Patagonia 51°37'S in the South to the central Argentinean region in Buenos Aires from 31°33'S as the northern boundary.

The second project is funded by the government of Japan and provides support for development and assistance to third countries based on a joint funding provided by: Japan International Collaboration Agency (JICA) and the Japan Science and Technology Agency (JST). This support created a Science and Technology Research Partnership for Sustainable Development (SATREPS) to address global issues with practical benefits on both local and global society. The overarching objective of SAVER.Net is to develop an atmospheric hazard and risk management system in Southern South America; focused on aerosol detection and identification as well as monitoring of stratospheric ozone ultraviolet (UV) radiation levels.

This multinational support for science and technological development geared joint efforts between participating institutions: CITEDEF, NIES, UMAG, SMN and the Chilean Meteorological Service (DMC). A common goal across organizations is to use lidar data for providing risk management information.

2 Hardware description

The Lidar Division at CITEDEF installed the instrumentation on the stations, developing and constructing the lidars on standard 20 ft. containers. These containers are equipped with open hatches at the rooftop, two air conditioning systems, a local area network and electricity connectivity. The roof on each container offers future expansion possibilities for installing additional instrumentation. The container includes office space since the lidar fills out one half of the container divided from the other half by a door.

The lidar network has nine instruments in total from which eight of them can measure depolarization at 532 nm and 355 nm; and one of them contains a telescope fiber-coupled to the spectroscopic box. Four of these instruments are prepared to measure inelastic Raman backscattering returns. And, two of them have high spectral resolution capabilities (HSRL) following the emission and detection scheme described by [1]. Table 1 shows the measurement capabilities.

Table 1 SAVER.Net and PE31554/11 Lidar Receiver and detection channels: p and s are parallel and perpendicular polarization; h is for High Spectral Resolution

Site / λ [nm]	355	387	408	532	607	1064
Bariloche	yes	yes	yes	yes	yes	yes
Aeroparque	p, s	no	no	p, s	no	yes
Cdo. Rivadavia	p, s	no	no	p, s	no	yes
Neuquén	p, s	no	no	p, s	no	yes
Río Gallegos	p, s	no	no	p, s	no	yes
Tucumán	p, s	no	no	p, s	no	yes
Pta. Arenas	p, s	yes	yes	p, s, h	yes	yes
Pilar-Córdoba	p, s	yes	yes	p, s, h	yes	yes
Villa Martelli	p, s	yes	yes	p, s, h	yes	yes

The Lidar emission sub-system in all cases is a Nd:YAG laser (at 1064 nm) doubled (at 532 nm) and tripled (at 355 nm). HSRL Lidars and the one installed in Punta Arenas are based on Continuum Surelite II – 10 Hz lasers. However, the lidar in the Bariloche site is based on a Quantel Brilliant B 20 Hz laser and the remaining of them are all based on Continuum Surelite I – 30 Hz lasers. See Table 2 for details on laser model, repetition rate and energy per pulse at 1064 nm.

Table 2 SAVER.Net and PE31554/11 Lasers: Quant. Stands for "Quantel", Cont., for "Continuum", and the Energy rating is for the fundamental wavelength

	Laser	Rep.rate	Energy
Bariloche	Quant. Brilliant B	20 Hz	360 mJ
Aeroparque	Cont. Surelite I	30 Hz	380 mJ
Comodoro. Rivadavia	Cont. Surelite I	30 Hz	380 mJ
Neuquén	Cont. Surelite I	30 Hz	380 mJ
Río Gallegos	Cont. Surelite I	30 Hz	380 mJ
Tucumán	Cont. Surelite I	30 Hz	380 mJ
Punta Arenas	Cont. Surelite II	10 Hz	675 mJ
Pilar-Córdoba	Cont. Surelite II	10 Hz	675 mJ
Villa Martelli	Cont. Surelite II	10 Hz	675 mJ

Figure 1 shows the lidar system installed in Punta Arenas, Chile (see Table 1).

Figure 1 Lidar system. Top shelf is the laser emitter, left side is the telescope receiver and first shelf is the detection and acquisition system.

The lidar receiver unit has a 200mm *f/5* Newtonian telescope connected to spectroscopic box except at Bariloche station where the telescope is 200mm *f/10* Cassegrain.

The detection units have Hamamatsu H10721 photomultiplier modules optimized for UV (-110) and for visible detection (-20). Infrared detection is done using Licel detector modules based on a 3mm effective area APD Si Detector thermoelectrically cooled. Transient recorders with split analog and photocurrent detection acquires backscatter signals. The analog inputs record the signals for all the elastic channels, except the depolarization channel at 355 nm. In this case, the signal is acquired in photon counting mode.

Lidars that are only elastic can be easily upgraded to include nighttime Raman detection in photon counting mode.

3 LIDAR OPERATION DESCRIPTION

The lidars were prepared to operate continuously under all weather conditions in the span of climate conditions from northern Argentina to sub Antarctic regions. Except the HRSLs that measure continuously, the other systems measure every 15 minutes during at least one-minute integrating profiles every 10 seconds (300 shots). This operation mode extends the flash lamp lifetime to one year for the lidar emission sub-system, with the exception of HSRL which requires six flash lamps per year due to their continuous operation.

The Lidar instruments are intended for continuous mode operation but also for intensive observing mode when specific hazardous events occur and then continuous profiling is required. This versatility allows for complementary operations of atmospheric modeling since datasets and level process data can be accessed remotely. Each lidar upload the data to a remote SFTP server every 15 minutes to assure almost real time transfer.

A web server manages and centralize the data through an interactive portal where quick-looks are refreshed illustrating backscattering profiles and lidar retrieved aerosol parameters. The actual quick looks and NetCDF profiles are processed at NIES and have the same format as AD-Net profiles. The online data processing algorithms are described in [2].

4 LIDAR OPERATION SITES

The lidar site location followed several criteria. Some of them are:

- The operation of the lidar should be performed by members of the SAVER.Net Project (in this case CITEDEF, SMN, UMAG)
- The site location should be as close as possible to main airports.
- The chosen region for installation should contribute improved knowledge on aerosol's optical and microphysical properties as well as their spatial distribution.
- The lidar network contours the Andean volcanic belt of Argentina and Chile in Patagonia were several active volcanoes can be found.

Detailed location of the lidar sites can be seen on Table 3 and figure 2

Table 3 SAVER.Net and SP 31554/11: Operation sites

Lidar Station	Latitude	Longitude
Bariloche	41° 08' 50"S	71° 9' 51"W
Aeroparque	34° 33' 51"S	58°25' 02"W
Cdo. Rivadavia	45° 47' 32"S	67°27' 46"W
Neuquén	38° 57' 08"S	68° 08' 13"W
Río Gallegos	51° 36' 42"S	69° 18' 26"W
Tucumán	26° 47' 14"S	65° 12' 25"W
Pta. Arenas	53° 08' 04"S	70° 52' 49"W
Pilar-Córdoba	31° 40' 03"S	63° 52' 58"W
Villa Martelli	34° 33' 21"S	58° 30' 23"W

With the exception of CEILAP at CITEDEF Villa Martelli (Argentina) and Punta Arenas (Chile), the sites are collocated to radiosondes launches on SMN sites.

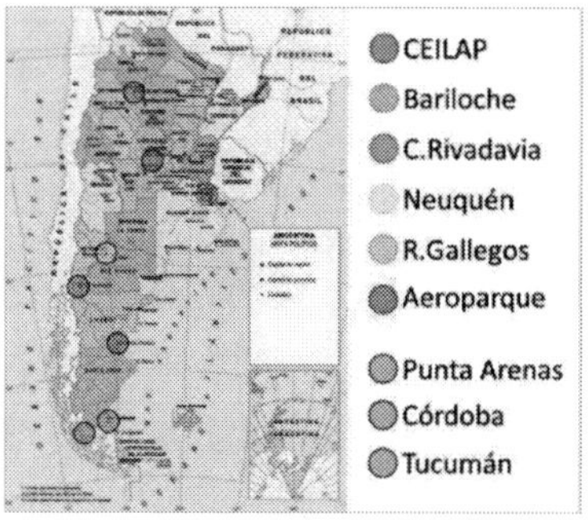

Figure 2 Lidar operation sites. To complete the network, Córdoba and Tucumán will be installed on the first quarter of 2017

5 DATA ANALYSIS

One of the main goals of SAVER.Net is to provide operational aerosol information supporting governmental agencies operations for hazards and risk analysis and prevention. For research purposes, lidar datasets are combined to other instrumental datasets (e.g. co-located sun-photometers, radiometers and meteorological stations).

The algorithms to provide lidar information were originally written in MatLab and IDL. Recently, the entire processing platform was upgraded to Python as single open source language to ensure wide distribution and sharing process. The algorithms will deliver in pseudo-real time dynamic structures like atmospheric boundary layer and cloud height and aerosol layers. They will also differentiate spherical from non-spherical backscatters.

In terms of retrieval of aerosol concentration, the network will implement those known and current available inversion methodologies [3-8] that can provide an operative solution in pseudo-real time.

By project ends in March 2018 the groups involved in the network will develop a web based platform to distribute datasets and products tailored to risk managers, governmental agencies involved in hazard mitigation and risk assessment [9].

6 LIDAR OPERATION, TRAINING, AND MAINTENANCE

The instruments are designed to measure continuously but the operator should be ready to deal with shutting down (e.g., to prevent damage due to hail or strong storm) and similarly to restarting the system (e.g., after a blackout lasting longer than the UPS capacity) when necessary. Instruments may also require optical realignment and signal optimization as well as basic maintenance. For this reason, researchers involved leading the lidar development have started training local lidar operators so that after installing and repairing; realigning and other simple routinely measures can be developed by on-site operators.. Furthermore, short courses on lidar signal processing are being prepared to train operators, forecasters and managers on demand of high level data product. Training courses are continuously evolving and improving to cover the majority of data-network users. Recent efforts between the lidar division at CEILAP-CITEDEF and the SMN have focused on preparing distance learning courses in specific topics of SMN interest. Finally, an operation handbook is being prepared for basic on-site operation and maintenance rules and procedures.

7 CONCLUSIONS AND PERSPECTIVES

Currently seven of nine lidars of the SAVER.Net are operating across different regions of Argentina and Chile. By March 2017 it is expected to have the last two lidars installed on their final locations. This final milestone will fulfill an unprecedented effort to provide a fully operational monitoring network from which several governmental agencies will benefit. After network completion analysis of lidar datasets will be implemented for aerosols transport from local and regional transport events in Argentina and Chile. This way SAVER.Net will contribute and integrate LALINET and other international efforts.

ACKNOWLEDGEMENTS

The researchers would like to acknowledge the support of the National Weather Service Lidar Operators, the Japan Science and Technology Agency (JST) / Japan International Cooperation Agency (JICA), the Science and Technology Research Partnership for Sustainable Development (SATREPS), CITEDEF and the Ministry of Defense of Argentina. The authors would like to thank the collaboration of Yuji Misu, Francisco González, Juan Carlos Dwornickzak, Osvaldo Vilar, Marcelo Ferrari, Rodrigo Videla, Mario Proyetti, Sebastián Papandrea, Guillermo Acosta, Fernando Chouza, Andrea Pereyra, Evangelina Martorella, Raúl D'Elía, Jorge Codnia, Laura Azcárate, Francisco Manzano, Claudio Libertelli, Hernán Ciminari, Carlos Pacheco, Fabián Jeric, Joaquín Miranda, Juan Carlos Condori, Marcelo Bustos, Martín Laterra, Raúl Rodríguez, and the CITEDEF machine shop for their collaboration on the construction and update of the lidar systems and related subsystems.

REFERENCES

[1] Z. Liu, I. Matsui, and N. Sugimoto: High-spectral-resolution lidar using an iodine absorption filter for atmospheric measurements, *Opt. Eng.*, **38**, 1999.

[2] Shimizu A., Nishizawa T., Jin Y., Kim S-W., Batdorj D. and Sugimoto N.: Evolution of a lidar network for tropospheric aerosol detection in East Asia. *Opt. Eng.*, **56(3)**, 31219-1-12, 2017

[3] Kudo R., Nishizawa T. and Aoyagi T: Vertical profiles of aerosol optical properties and the solar heating rate estimated by combining sky radiometer and lidar measurements, AMT, **9(7)**, 3223-3243, 2016.

[4] Lopatin A: Enhanced Remote Sensing of atmospheric aerosol by joint inversion of active and passive remote sensing observations, PhD thesis.

[5] J. Wagner, A. Ansmann, U. Wandinger , P. Seifert , A. Schwarz , M. Tesche ,A. Chaikovsky , and O. Dubovik: Evaluation of the Lidar/Radiometer Inversion Code (LIRIC) to determine microphysical properties of volcanic and desert dust, *AMT*, **6(7)**, 1707-1724, 2013.

[6] Ansmann A., Seifert P., Tesche M. and Wandinger U.: Profiling of fine and coarse particle mass: case studies of Saharan dust and Eyjafjallajökull/Grimsvötn volcanic plumes, *Atm. Chem. Phys.* , **12(20)**, 9399-9415, 2012

[7] Ristori, P., L. Otero, J. Fochesatto, P.H. Flamant, E. Wolfram, E. Quel, R. Piacentini, and B. Holben. 2003: "Wavelength Dependence and Variability of Atmospheric Aerosols in the Buenos Aires area using Sun Photometer Measurements", *Optics and Laser in Engineering, 40*, 91-104.

[8] Otero, L., P. Ristori, G. J. Fochesatto, E. Wolfram, J. Porteneuve, P.H. Flamant, and E. Quel. "First Aerosol Measurements with a Multiwavelength Lidar System at Buenos Aires, Argentina". 2004: Peer reviewed articles presented at the *22nd International Laser Radar Conference,* Matera, Italia, ESA SP561, Vol. II, ISBN 92-9092, ISSN 0379-6566, 769-772.

[9] Webley P., D. Atkinson, R. Collins, K. Dean, G. J. Fochesatto, K. Sassen, C. Cahill, C. Flynn and K. Mazutani. 2008: "Predicting and Validating the tracking of a volcanic ash cloud during the 2006 eruption of Mt. Augustine Volcano". *Bull. Amer. Meteor. Soc.* 1647-1658.

AN AUTOMATIC AEROSOL CLASSIFICATION FOR EARLINET: APPLICATION AND RESULTS

Nikolaos Papagiannopoulos[1]*, Lucia Mona[1], Vassilis Amiridis[2], Ioannis Binietoglou[2,3], Giuseppe D'Amico[1], P. Guma-Claramunt[1], Anja Schwarz[4], Lucas Alados-Arboledas[5,6], Aldo Amodeo[1], Arnoud Apituley[7], Holger Baars[4], Daniele Bortoli[8], Adolfo Comeron[9], Juan Luis Guerrero-Rascado[5,6], Panos Kokkalis[2,10], Doina Nicolae[3], Alex Papayannis[10], Gelsomina Pappalardo[1], Ulla Wandinger[4], and Matthias Wiegner[11]

*[1]Consiglio Nazionale delle Ricerche, Istituto di Metodologie per l'Analisi Ambientale (CNR-IMAA), Tito Scalo (PZ), Italy, *nikolaos.papagiannopoulos@imaa.cnr.it*

[2] IAASARS, National Observatory of Athens, Athens, Greece

[3]National Institute of R&D for Optoelectronics (INOE), Magurele, Romania

[4]Leibniz Institute for Tropospheric Research (TROPOS), Leipzig, Germany

[5]Instituto Interuniversitario de Investigación del Sistema Tierra en Andalucía (IISTA-CEAMA), Granada, España

[6]Dpto. Física Aplicada, Universidad de Granada, Granada, España

[7]Royal Netherlands Meteorological Institute (KNMI), De Bilt, the Netherlands

[8]Earth Science Institute – ICT, Évora, Portugal

[9]Dept. of Signal Theory and Communications, Remote Sensing Lab. (RSLab), Universitat Politècnica de Catalunya, Barcelona, Spain

[10]Laser Remote Sensing Unit, Physics Dept., National Technical University of Athens, Greece

[11]Ludwig-Maximilians-Universität (LMU), Meteorologisches Institut, Munich, Germany

ABSTRACT

Aerosol typing is essential for understanding the impact of the different aerosol sources on climate, weather system and air quality. An aerosol classification method for EARLINET (European Aerosol Research Lidar Network) measurements is introduced which makes use the Mahalanobis distance classifier. The performance of the automatic classification is tested against manually classified EARLINET data. Results of the application of the method to an extensive aerosol dataset will be presented.

1 INTRODUCTION

A thorough characterization of atmospheric aerosol particles is a key factor for accurate climate modeling [1]. The vertical profile of atmospheric particles is particularly important for the study of aerosol transport, and of the radiative forcing. Lidar measurements provide vertically resolved information on the aerosol distribution as well as on their optical properties. These parameters obtained can be used to classify into several aerosol types the various layers in the vertical.

Automatic procedures for aerosol typing combine information and based on a set of pre-specified clusters, to determine the aerosol type. Different methods for retrieving aerosol types are applied depending on the lidar abilities. The spaceborne CALIOP (Cloud-Aerosol Lidar with Orthogonal Polarization) lidar utilizes a decision-tree that takes into account lidar-derived information but also external information on geographical location, surface type, and season [2]. In contrast [3] performed an objective, multi-dimensional analysis that exclusively makes use of airborne lidar measurements. Recently, a novel method was implemented using artificial neural networks based on optical data provided by multi-wavelength Raman lidars measurements and advanced aerosol models in order to categorize aerosols [4].

EARLINET provides profiling data on a continental scale. The majority of the measuring stations operate multi-wavelength Raman lidars with the ability of linear particle depolarization ratio measurements. Based on these characteristics, EARLINET constitutes an excellent basis to distinguish different aerosol types.

Here, we apply the Mahalanobis distance

classifier to EARLINET data. Previous to classifying a sensitivity analysis is performed to identify the intensive parameters most adequate for the classification. Examples of aerosol classification are presented, followed by a short discussion of these results.

2 AUTOMATIC AEROSOL TYPE CLASSIFICATION

The Mahalanobis-based classification found great applicability in aerosol studies. For instance, the algorithm developed by [3] is lidar stand-alone classification using four lidar intensive properties with the aim of classifying aerosols into eight types. A slightly different algorithm was introduced by [5] that incorporates uncertainties in the input properties. Recently, [6] used the same classifier to produce an aerosol classification scheme based on long term AERONET data.

2.1 Methodology

The aerosol classification is performed in two parts. First, specific samples of known aerosol types are used to make the model distributions, this is called the training phase. Second, a test dataset is classified by comparison with these models and it is called the testing phase.

The calculated classifier assigns any given N-dimensional point (x_1, x_2, ..., x_N) to the cluster that has minimum Mahalanobis distance (D_M) from that point. A cluster is defined by its mean (μ_1, μ_2, ..., μ_N) and its covariance matrix, S. In N dimensions,

$$D_M(x) = \sqrt{(x-\mu)^T S^{-1}(x-\mu)} \tag{1}$$

2.2 Training phase

For the training phase we used EARLINET network-wide typing results already published in literature [7, 8] for a total of 64 samples. The clusters were chosen for the aerosol types to be physically meaningful and mark the sources. The 7 selected aerosol types are: pure dust (D), polluted dust (PD = Dust + Smoke), mixed dust (MD = Dust + Marine), clean continental (CC), polluted continental (PC), smoke (S), and mixed marine (MM). However, aerosol types with similar characteristics were merged to identify whether this merging enhances the correct prediction. The S and PC were merged into a single category of small, absorbing particles. Further, the pure dust and the dust mixtures (PD and MD) were integrated into one.

Next, we performed a sensitivity analysis to identify which wavelengths and intensive parameters provide the adequate information. In contrast to the work of [3] we used three aerosol intensive properties due to lack of depolarization ratio profiles: the lidar ratio (S_{aer}), the backscatter/extinction related Ångström exponent (B-AE/E-AE), and the ratio of the lidar ratios (RS_{aer}). Two statistical parameters that highlight the strength of the selected classifiers to discriminate between clusters are used: the total and the partial Wilk's lambda (for more info check [7]). The lowest Wilk's total lambda was found to be 0.03 for the set B-AE (IR, UV), S_{aer} (VIS), and RS_{aer}; the B-AE has the most weight in the classification.

Figure 1 shows the characteristics of the training dataset in terms of the above mentioned intensive properties. The crosshairs indicate the standard deviation of the mean cluster properties. The 2-σ ellipses are calculated using the eigenvalues and eigenvectors of the covariance matrix. For the pair S_{aer} and B-AE, the various aerosol clusters tend to populate specific areas of the graph whereas for the pairs S_{aer}, RS_{aer} and B-AE, RS_{aer} no evident grouping is observed. It might be noted that the parameters do not give a good separation of the clusters, however, this is, of course, the benefit of using multivariate analysis.

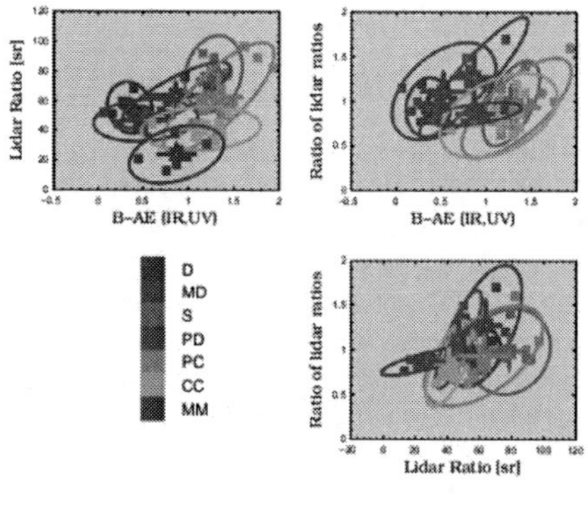

Figure 1 Pre-specified clusters delineated by two-sigma ellipses.

3 RESULTS

EARLINET data collected during the ACTRIS Summer 2012 campaign were chosen to test the automatic typing algorithm. The description of the aerosol type distribution over Europe during the campaign was obtained through a combined use of advanced lidar measurements, backward trajectory analyses and model outputs [9]. The test dataset comprises of 47 samples, 21 of which yield depolarization values.

In Tab. 1, the agreement between the manual and automatic procedures is 64 %, the samples wrongly flagged are 34 %, and the remaining 2 % is not assigned due to the low confidence (D_M < 4.0 for 3 d.f.). The aerosol types that performed worse are the smoke and polluted continental categories owing to the similarities in the intensive properties. Given the noticeable signature of dust particles the agreement is limited to 54 % which can be assigned to the lack of depolarization measurements.

In the light of these discrepancies, first, we introduce a combined smoke and polluted continental category, as already discussed in Sect. 2.1, and the results are shown in the same table. The agreement increased by 13 %. Second, when dust is represented by a single cluster (the pre-specified clusters are limited to four), the agreement reached 87 %.

Table 1 Agreement in percentage of the automatic classification results with manually classified data

# of clusters	Correct Pred.	Wrong Pred.
7	64 %	36 %
6	77 %	23 %
4	87 %	13 %

The current status of EARLINET does not allow incorporating the particle linear depolarization ratio (δ_{aer}) into the automatic typing scheme. Whereas, for [3] the depolarization showed to be a robust means to discriminate the various aerosol types. Therefore, we used literature values for particle linear depolarization ratio (Tab. 2) in order to implement this parameter into the automatic algorithm. Values within the aerosol type range were randomly assigned to each sample and the Wilk's lambda distribution was calculated. Wilk's total lambda is 0.01. For B-AE is 0.41, for S_{aer} is 0.36, for RS_{aer} is 0.48 and for δ_{aer} is 0.17. δaer for this dataset indicates that this variable has the most weight in the classification.

Table 2 The range of the linear particle depolarization (δ_{aer}) ratio for each of the aerosol types along with the bibliographic reference.

Type	δ_{aer}	References
Dust	0.28-0.33	[10]
Poll. Dust	0.12-0.23	[10]
Mixed Dust	0.13-0.17	[10]
Poll. Continent.	0.03-0.07	[11]
Clean Continent	0.01-0.07	[2]
Smoke	0.04-0.09	[11]
Mixed Marine	0.01-0.05	[10]

The Mahalanobis distance threshold, in this case, for the type inference is 4.3 (4 d.f.). Table 3 shows the agreement between the manual and the automatic procedures is 67 % (64 % without δ_{aer}). The agreement for the combined smoke and polluted continental types, now, becomes 86% (77 % without δ_{aer}); whereas for the combined dust type is 81 % (87 % without δ_{aer}). It becomes evident that the depolarization ratio increases the ability for predicting correctly the aerosol type when considering 7 clusters and 6 clusters, while it performs worse in case of 4 clusters, however the prediction rate is rather high.

Table 3 Agreement in percentage of the automatic classification results with manually classified data.

# of clusters	Correct Pred.	Wrong Pred.
7	67 %	33 %
6	86 %	14 %
4	81 %	19 %

4 CONCLUSIONS

An automatic classification for EARLINET data method based on Mahalanobis distance was presented. A Wilk's lambda analysis was performed and the three best performing classifying parameters were the S_{aer} at 532 nm; the RS_{aer} and the B-AE (IR, UV). The prediction of the automatic classification showed positive results when compared against manually classified EARLINET data. The performance was further enhanced when the aerosol clusters were merged according to aerosol types' similar characteristics. The fewer aerosol clusters make a coarse aerosol identification and do not represent the wide aerosol spectrum. However, the positive learning success indicates that the user can select the number of clusters depending on the application. Here, we think that six aerosol clusters cover the aerosol situation with a configuration of 3 backscatter and 2 extinction coefficient profiles. Accordingly, the algorithm was trained with depolarization data that were randomly generated based on literature values, and showed the strengthening of correct prediction.

Specifically, for the dust subtype the prediction rate was higher than 50 %, even though the spectral signature is easily identifiable, the paucity of depolarization measurements makes the classification trivial as polluted dust and mixed dust reflect similar behavior with the dust type. For Polluted Continental and Smoke categories the algorithm performed the worse, this was expected as the two subtypes are overlapping and is highly difficult to separate them.

The manageability of the algorithm regarding the training dataset, the number of the clusters (i.e., aerosol types) and the classifying parameters (i.e., available intensive parameters) makes the method easily adaptable and handled by individual users. For instance, the training dataset can be easily enlarged with high quality data coming from a multitude of EARLINET stations and a longer time record. Moreover, new classifying parameters, as particle linear depolarization ratio at more wavelengths and aerosol extinction coefficient in the infrared, can be easily added as the observing capacity increases. Future work includes an extended EARLINET testing dataset for this study.

ACKNOWLEDGEMENTS

This project has received funding from the European Union's Horizon 2020 research and innovation programme under grant agreement No 654109.

References

[1] Laj, P. et al.: Measuring atmospheric composition change, Atmos. Environ., 43, 5351-5414, 10.1016/j.atmosenv.2009.08.020, 2009.

[2] Omar, A. et al.: The CALIPSO Automated Aerosol Classification and Lidar Ratio Selection Algorithm, J. Atmos. Ocean. Tech., 26, 1994-2014, 10.1175/2009JTECHA1231.1, 2009.

[3] Burton, S.P. et al.: Aerosol classification using airborne High Spectral Resolution Lidar measurements – methodology and examples, Atmos. Meas. Tech., 5, 73–98, 10.5194/amt-5-73-2012, 2012.

[4] Nicolae, D. et al.: Using artificial neural networks to retrieve the aerosol type from multi-spectral lidar data, European Geosciences Union, General Assembly, Vol. 17, EGU2015-9793, 2015.

[5] Russell, P.B. et al.: A multiparameter aerosol classification method and its application to retrievals from spaceborne polarimetry, J. Geophys. Res. Atmos., 119, 9838–9863, 10.1002/2013JD021411, 2014.

[6] Hamill, P. et al.: An AERONET-based aerosol classification using the Mahalanobis distance, Atmos. Env., 140, 213-233, 10.1016/j.atmosenv.2016.06.002, 2016.

[7] Pappalardo, G. et al.: EARLINET correlative measurements for CALIPSO: First intercomparison results, J. Geophys. Res., 115, D00H19, 10.1029/2009JD012147, 2010.

[8] Papagiannopoulos, N. et al.: CALIPSO climatological products: evaluation and suggestions from EARLINET, Atmos. Chem. Phys., 16, 2341-2357, 10.5194/acp-16-2341-2016, 2016a.

[9] Papagiannopoulos, N. et al.: Aerosol classification using EARLINET measurements for an intensive observational period, European Geosciences Union, General Assembly, Vol. 18, EGU2016-16026, 2016b.

[10] Groß, S. et al.: Characterization of Saharan dust, marine aerosols and mixtures of biomass-burning aerosols and dust by means of multi-wavelength depolarization and Raman lidar measurements during SAMUM 2. Tellus B, 63, 706-724, 10.3402/tellusb.v63i4.16369, 2011.

[11] Burton, S.P. et al.: Separating mixtures of aerosol types in airborne High Spectral Resolution Lidar data, Atmos. Meas. Tech., 7, 419-436, 10.5194/amt-7-419-2014, 2014.

https://doi.org/10.1051/epjconf/201817609013

POLLY^{NET} - AN EMERGING NETWORK OF AUTOMATED RAMAN-POLARIZARION LIDARS FOR CONTINUOUS AEROSOLPROFILING

Holger Baars[1,*], Dietrich Althausen[1], Ronny Engelmann[1], Birgit Heese[1], Albert Ansmann[1], Ulla Wandinger[1], Julian Hofer[1], Annett Skupin[1], Mika Komppula[2], Eleni Giannakaki[2], Maria Filioglou[2], Daniele Bortoli[3], Ana Maria Silva[3], Sergio Pereira[3], Iwona S. Stachlewska[4], Wojciech Kumala[4], Dominika Szczepanik[4], Vassilis Amiridis[5], Eleni Marinou[5], Michail Kottas[5], Ina Mattis[6] and Gerhard Müller[6]

[1]*Leibniz Institute for Tropospheric Research (TROPOS), *baars@tropos.de*
[2]*Finnish Meteorological Institute, Kuopio, Finland*

[3]*Évora University, Institute for Earth Sciences, Évora, Portugal*
[4]*University of Warsaw, Faculty of Physics, Institute of Geophysics, Poland*
[5]*National Observatory of Athens, Greece*
[6]*Deutscher Wetterdienst, Observatorium Hohenpeißenberg, Germany*

ABSTRACT

Polly^{NET} is a network of portable, automated, and continuously measuring Raman-polarization lidars of type Polly operated by several institutes worldwide. The data from permanent and temporary measurement sites are automatically processed in terms of optical aerosol profiles and displayed in near-real time at *polly.tropos.de*. According to current schedules, the network will grow by 3-4 systems during the upcoming 2-3 years and will then comprise 11 permanent stations and 2 mobile platforms.

1 INTRODUCTION

Motivated by the urgent need for robust Raman-polarization lidars that are easy to operate and allow aerosol typing, an automated portable lidar system, called Polly^{XT}, has been developed at the Leibniz Institute for Tropospheric Research (TROPOS) with international partners during the last decade [1, 2]. The aim was to develop a sophisticated automated Raman-polarization lidar for scientific purpose, but with the advantages of an easy-to-use and well-characterized instrument with same design, same automated operation, and same centralized data processing delivering near-real-time data products. By the end of the year 2016, ten Polly systems have been constructed and are employed worldwide. As the number of Polly lidars and measurement sites has increased, an independent, voluntary, international network of cooperating institutes, the so-called Polly^{NET} [3], has evolved as an additional contribution to the worldwide aerosol observational efforts like for example the European Aerosol Research Lidar Network (EARLINET) and the Asian Dust Network (AD-NET). In this contribution, we present the basic idea behind Polly^{NET}, its stations, the automated data analysis efforts and future plans.

2 POLLY LIDAR SYSTEMS

All Polly lidar systems are designed for automatic and unattended operation in 24/7 mode (continuously 24 h a day, 7 days per week). As the capabilities have continuously expanded, all lidars differ slightly from each other. Nevertheless, they feature a similar design, the same data format, and benefit from unified calibration and quality assurance routines. An overview of the different systems including their capabilities and different characteristics can be found in [2]. The latest developed Polly^{XT} (with extended capabilities) is a so-called $3+2+2+1+2+2$ multiwavelengths lidar with near-range capabilities (3 elastic, 2 Raman, 2 depolarization, 1 water-vapor, 2 near-range elastic, and 2 near-range Raman channels). The lidar emits light at 1064, 532, and 355 nm at an energy of 180, 110, 60 mJ per pulse. The primary mirror of the receiver unit has a diameter of 30 cm. The data of all channels is acquired with a vertical resolution of 7.5 m in temporal steps of 30 s (laser repetition rate 20 Hz). Below about 600–800 m above the lidar, the overlap of the laser beam and the receiver field of view is incomplete and needs to be corrected. As a consequence, a second detection unit together with a near-range telescope was added to the system to detect the elastic and inelastic backscatter in the UV and VIS in the lowermost height range above the system. As a result, the determined overlap function of the far-range channels can be verified and backscatter and extinction profiles can be obtained down to about 100 m a.g.l. Further system details are given by [1] and [2].

3 POLLY^{NET}

Until now, the Leibniz Institute for Tropospheric Research (TROPOS), the Finnish Meteorological Institute (FMI), the National Institute of Environmental Research in Korea (NIER), the Évora University in Portugal (CGE), the University of Warsaw in Poland (UW), the German Meteorological Service (DWD) and the National Observatory of Athens in Greece (NOA) contribute actively to Polly^{NET} by operating Polly systems. Each group contributes with its expertise and knowledge to the network and to joint scientific projects.
Within Polly^{NET}, Polly lidar measurements have been performed at 30 locations in Europe, the Amazon rain forest, Southern Chile, South Africa, India, China, Korea, Tajikistan, Israel and over the Atlantic Ocean. Very different aerosol types and aerosol mixtures have been observed. An overview of findings from these observations is given in [3].
Figure 1 shows the current and planned measurement sites within Polly^{NET}. Currently, measurements are performed at Kuopio, Finland (FMI); Leipzig, Germany; Limassol, Cyprus; Haifa, Israel (TROPOS); Warsaw, Poland (UW), Hohenpeißenberg, Germany (DWD); Evora, Portugal (CGE); and Banegnyeong Island, Korea (NIER). Most of these stations operate the lidar continuously. Additionally, regular shipborne observations are performed twice a year by TROPOS onboard the German research vessel Polarstern crossing the Atlantic Ocean latitudinal.
Data from the permanent locations as well as from measurement campaigns are centrally collected via internet, processed, and displayed in near-real time at *polly.tropos.de*.
In future, new permanent stations are scheduled for Cape Verde (TROPOS), Finokalia, Crete, Greece (NOA), Cyprus (TROPOS), Tel Aviv, Israel (TAU) and Dushanbe, Tajikistan (TROPOS).

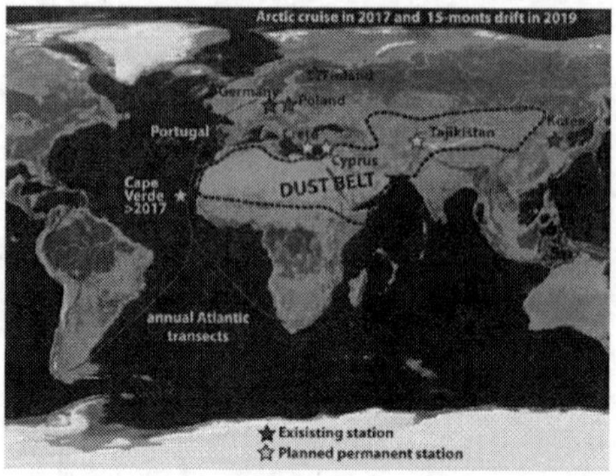

Figure 1: Map of current and planned Polly^{NET} stations.

4 AUTOMATED DATA ANALYSIS

Polly systems are designed to operate continuously, i.e., accumulate up to 2880 raw files per day. Naturally, a robust automatic data analysis algorithm is necessary to make use of such an amount of data. Thanks to equal system

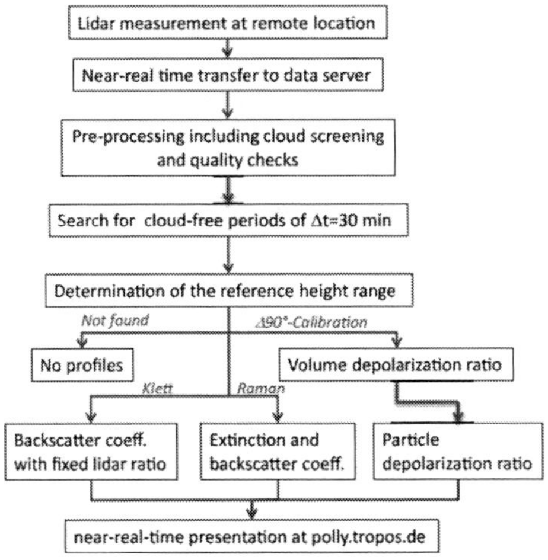

Figure 2: Schematic description of the steps for the automatic data analysis chain.

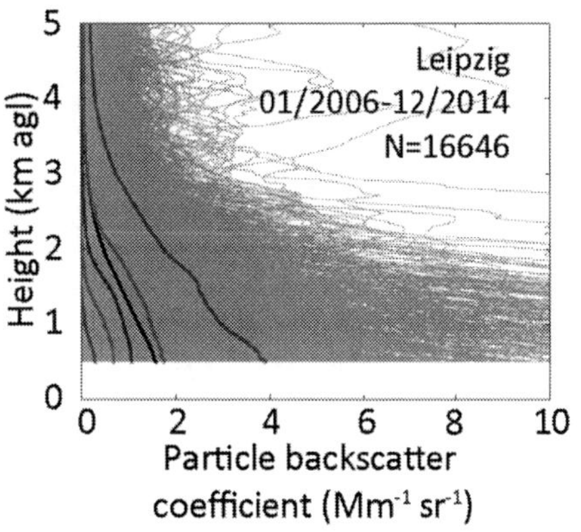

Figure 3: Particle backscatter coefficients profiles at 532 nm at Leipzig from 2006 to 2014. 16646 individual 30-min profiles were used to compute the median (blue), mean (black), 25 and 75 % percentile (purple), and 5 and 95 % percentile (red) profiles.

setup and data format this could be achieved within Polly^{NET}. The complete procedure for the automatic determination of aerosol optical profiles is illustrated in Fig. 2 and in detail described in [3].

After pre-processing including cloud-screening, optical profiles at 355, 532, and 1064 nm are calculated using the well-known retrieval algorithms with the help of an automated Rayleigh fit procedure for reference height finding. The choice between the preferred Raman method and the Klett method is based on the signal-to-noise ratio (SNR) in the channels detecting the Raman scattered light (387 and 607 nm). Then, either profiles of the backscatter and extinction coefficient (Raman method) or the backscatter coefficient only (Klett method) are determined. From the profile of the volume linear depolarization ratio and the backscatter coefficient profile, the particle linear depolarization ratio is calculated. For high quality in the depolarization measurements, the Δ90° calibration [4] is automatically performed three times a day since 2012 [2].

Finally, the vertical aerosol profiles (backscatter, extinction, and depolarization) are stored centrally and displayed at *polly.tropos.de* in near-real time without any manual intervention. For research on aerosols only, a post-processing is applied, e.g., screening for cirrus clouds and discarding contaminated measurements. From those screened profiles, further studies like, e.g., a statistical analysis of the vertical aerosol distribution at the Polly^{NET} stations can be performed. Such an analysis based on the backscatter coefficient profiles at 532 nm is shown for a nine-year data set of Leipzig in Fig. 3. In addition to the 16646 individual 30-min profiles, the mean (black), the median (blue), the 25- and 75 %-percentile (purple), and the 5- and 95 %-percentile (red) profiles are plotted. The median and extreme profiles may be of particular interest, if representative aerosol profiles at specific locations are needed as a priori input for models or data retrieval algorithms.

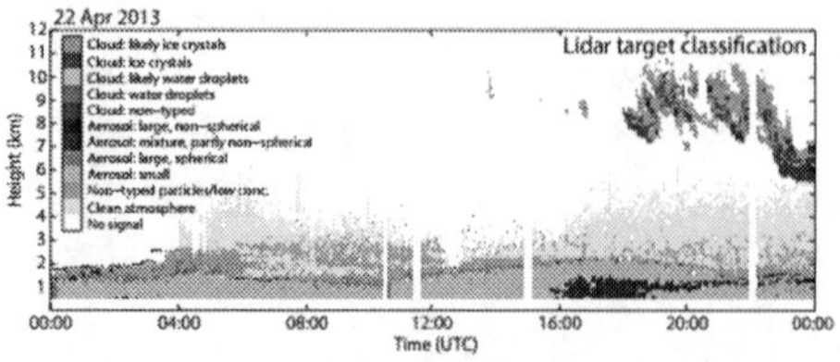

Figure 4: Example of target classification from Polly lidar observations near Jülich, Germany on 22 April 2013.

5 CONCLUSIONS AND OUTLOOK

A powerful global network of automated and continuous measuring Raman-polarization lidar systems has been established with Polly^{NET}. Permanent as well as temporary sites form a network of Polly instruments.
The lidar data collected within Polly^{NET} since 2006 is analyzed with an automated data processing procedure. All analyzed data are shown online at *polly.tropos.de* and are available on demand for further use in agreement with the respective station representatives.
In future, Polly^{NET} will expand. New stations with new systems in Cape Verde, Cyprus, and Tajikistan are planned by TROPOS. NOA plans to install a permanent site in Crete. Also the data processing chain will be further developed and merged with ACTRIS (The European Research Infrastructure for the observation of Aerosol, Clouds, and Trace gases, *www.actris.eu*). E.g., the combination with Cloudnet (continuous measurements with cloud radar, microwave radiometer, and ceilometer, *www.cloud-net.org*) might be a powerful tool to investigate aerosol-cloud-interaction. A high-resolution target categorization based on absolute calibrated lidar signals as exemplary shown in Fig. 4 have been developed for such purposes [5].

ACKNOWLEDGEMENTS

The authors acknowledge support through ACTRIS under grant agreement no. 262254 and BACCHUS under grant agreement no. 603445 of the European Union Seventh Framework Programme (FP7/2007-2013), ACTRIS-2 under grant agreement no. 654109 from the European Union's Horizon 2020 research and innovation programme, and "High Definition Clouds and Precipitation for Climate Prediction – HD(CP)2" (FKZ: 01LK1209C and 01LK1212C) funded by Federal Ministry of Education and Research in Germany (BMBF). Many improvements were triggered by the fruitful discussions and network activities within EARLINET. We appreciate the contributions of the individuals that have been involved in supporting, enabling, and maintaining Polly measurements.

References

[1] Althausen, D., et al.: Portable Raman Lidar PollyXT for automated profiling of aerosol backscatter, extinction, and depolarization, J. Atmos. Ocean. Tech., 26, 2366–2378, 2009.

[2] Engelmann, R., et al.: The automated multiwavelength Raman polarization and water-vapor lidar Polly^{XT}: the neXT generation, Atmospheric Measurement Techniques, 9, 1767–1784, 2016.

[3] Baars, H., et al.: An overview of the first decade of Polly^{NET}: an emerging network of automated Raman-polarization lidars for continuous aerosol profiling,Atmospheric Chemistry and Physics, 16, 5111–5137, 2016.

[4] Freudenthaler, V.: About the effects of polarising optics on lidar signals and the Δ90 calibration, Atmos. Meas. Tech., 9, 4181-4255, doi:10.5194/amt-9-4181-2016, 2016.

[5] Baars, H., et al.: Target categorization of aerosol and clouds by continuous multiwavelength-polarization lidar measurements, Atmos. Meas. Tech., 10, 3175-3201, https://doi.org/10.5194/amt-10-3175-2017, 2017.

EPJ Web of Conferences **176**, 09014 (2018) https://doi.org/10.1051/epjconf/201817609014
ILRC 28

EARLINET SINGLE CALCULUS CHAIN: NEW PRODUCTS OVERVIEW

Giuseppe D'Amico[1,*], Ina Mattis[2], Ioannis Binietoglou[3,5], Holger Baars[4], Lucia Mona[5], Francesco Amato[1], Panos Kokkalis[5], Alejandro Rodríguez-Gómez[6], Ourania Soupiona[7] and Voudouri Kalliopi-Artemis[8]

[1]*Consiglio Nazionale delle Ricerche, Istituto di Metodologie per l'Analisi Ambientale (CNR-IMAA), Potenza Italy, *Email:giuseppe.damico@imaa.cnr.it*
[2]*Deutscher Wetterdienst, Meteorologisches Observatorium Hohenpeißenberg, Germany*
[3]*National Institute of R&D for Optoelectronics INOE, Bucharest, Romania*
[4]*Leibniz Institute for Tropospheric Research (TROPOS), Leipzig, Germany*
[5]*National Observatory of Athens, Athens, Greece*
[6]*BarcelonaTech University (UPC), Barcelona, Spain*
[7]*Laser Remote Sensing Unit, National Technical University of Athens, Physics Dept., Greece*
[8]*Laboratory of Atmospheric Physics, Aristotle University of Thessaloniki, Greece*

ABSTRACT

The Single Calculus Chain (SCC) is an automatic and flexible tool to analyze raw lidar data using EARLINET quality assured retrieval algorithms. It has been already demonstrated the SCC can retrieve reliable aerosol backscatter and extinction coefficient profiles for different lidar systems. In this paper we provide an overview of new SCC products like particle linear depolarization ratio, cloud masking, aerosol layering allowing relevant improvements in the atmospheric aerosol characterization.

1 INTRODUCTION

The SCC is the standard EARLINET (European Aerosol Research Lidar Network) tool for the analysis of lidar data providing high-quality aerosol properties at network level. The SCC has been installed on a centralized server where raw data measured by all EARLINET systems can be submitted and automatically analyzed. Moreover, the SCC is highly configurable and can be easily adapted to analyze data coming from new lidar systems. In particular, using a user-friendly web interface it is possible to change all instrumental and configuration parameters to be used in the analysis. The products of the SCC are all quality certified according to the EARLINET quality assurance program. The SCC can provide different levels of output: pre-processed signals (range-corrected lidar signals corrected for all instrumental effects) and aerosol optical products (aerosol backscatter and extinction coefficient profiles). The SCC comprises two main calculus modules: the ELPP (EARLINET Lidar Pre-Processor) module which accepts as input the raw lidar data and produces the preprocessed signals [1] and the ELDA (EARLINET Lidar Data Analyzer) module which takes as inputs the outputs of the ELPP module and generates optical products like particle extinction and backscatter coefficient [2]. The actions of the two modules are automatically synchronized and coordinated by another module called SCC daemon. All parameters required by the ELPP and ELDA modules are stored in an efficient way in an MySQL database. The SCC has been validated by [2] using synthetic lidar signals used during the EARLINET algorithm inter-comparison exercise and by [3] using real lidar data. Such validations show good SCC performance in analyzing specific measurements of several co-located lidar systems as well as in case of statistical analysis made by using long-term data set.

The development of the SCC is continuing and new important features are going to be included in the framework of the ACTRIS-2 project. In this paper we provide an overview on these new SCC products.

2 NEW SCC PRODUCTS

The SCC can deliver quality assured particle extinction and backscatter coefficient profiles.

It is well known that this set of optical parameters can already provide a quite complete characterization of atmospheric aerosols especially in case of multi-wavelength Raman lidar measurements. Anyway, to further improve the SCC capabilities in characterizing atmospheric particles, the following new products have been included: 1) particle linear depolarization ratio, 2) cloud masking, and 3) aerosol layering. The calculation of particle linear depolarization ratio allows a better differentiation between spherical and not spherical particles making easier to discriminate for example dust or ash from marine or sulfate aerosol types. The cloud masking is another important improvement because the quality of the SCC optical products cannot be assured, if there are signatures of low-level clouds in the raw lidar time series. Without a cloud masking module it is an user responsibility to not submit raw datafile contaminated by low-level clouds and this makes the process of submission to the SCC not fully automatable. Finally, the automatic characterization of the atmospheric aerosol layers in terms of their optical and geometrical properties, makes feasible the provision of the SCC products to external algorithms related for example to micro-physical retrievals or aerosol typing.

2.1 Particle Linear Depolarization Ratio

The Volume Linear Depolarization Ratio (VLDR) and Particle Linear Depolarization Ratio (PLDR) are of great importance for the characterization of atmospheric aerosols especially in terms of spherical and not-spherical scatterers. Before the implementation of the calculation of the PLDR and VLDR products in the SCC it is required to define quality assured procedures for the retrieval of these important parameters. Despite the apparent simplicity, there are several instrumental effects that need to be taken into account to deliver reliable PLDR values. Such effects are highly system dependent and require dedicated measurements to fully characterize all the optics of lidar from polarization point of view. In what it follows, we briefly summarize the basic aspects regarding the implementation of the PLDR and VLDR in the SCC. The full theoretical background behind what has been implemented is described in details by [4]. The first critical point is the calibration of the polarization sensitive channels involved in the calculation of the VLDR (and consequently of the PLDR). As, in general, the total efficiency of the polarization channels perpendicular and parallel to the direction of the polarization of the emitted light (supposed to be linear polarized) are different, accurate values of the gain ratio η^* of these two channels are required. The calibration procedures implemented in the SCC are based on $+45$ or $\Delta 90$ calibration methods [4] and are made by submitting to the SCC a raw calibration dataset. Calibration based on Rayleigh method is not implemented because it is considered not enough accurate (small strongly depolarizing aerosol layer within the assumed aerosol-free atmospheric region could lead to major deviations from the assumed molecular depolarization value). Once the value of η^* is determined it is possible to calculate the "apparent" VLDR:

$$\delta^* = \frac{K}{\eta^*}\frac{I_R}{I_T} \tag{1}$$

where the factor K is a calibration correction factor defined in [4] and I_R and I_T are the intensity or the radiation reflected and transmitted by the polarization sub-system. For ideal systems for which the cross polarized channel is reflected and the parallel one is transmitted, δ^* is equal to the VLDR δ. For real systems the relationship between δ^* and δ is more complex as it needs to include cross-talk or di-attenuation effects due to all lidar optics and it is given by:

$$\delta = \frac{\delta^*(G_T + H_T) - (G_R + H_R)}{(G_R - H_R) - \delta^*(G_T - H_T)} \tag{2}$$

where G_T, H_T, G_R, H_R are four polarization cross-talk correction parameters characterizing both transmitted and reflected polarization channels. Such parameters can be calculated (for details see [4]) and stored in the SCC database as channel options. Finally, when the

https://doi.org/10.1051/epjconf/201817609014

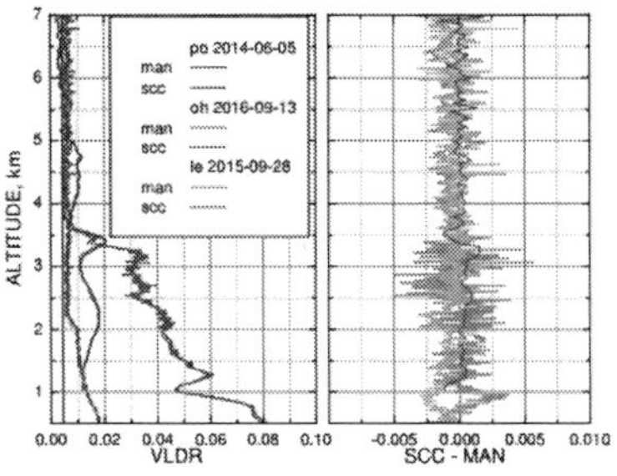

Figure 1: Comparison of VLDR profiles which were derived manually (bold lines) and by SCC (thin lines) for several EARLINET systems.

VLDR is known, it is possible to calculate the PLDR using the following formula:

$$\delta_a = \frac{(1+\delta_m)\delta R - (1+\delta)\delta_m}{(1+\delta_m)R - (1+\delta)} \qquad (3)$$

where δ_m is the theoretical value for the molecular linear depolarization ratio and R is the backscatter ratio. In Fig. 1 is shown a comparison between the manually retrieved VLDR profiles and the corresponding ones calculated by the SCC for several lidar systems. The red (blue) profiles were measured in Potenza (Leipzig) by the MUSA (PollyXT) system. The green profiles were measured in Hohenpeißenberg by the EARLINET reference lidar POLIS (light green, bold - manually retrieved) and by RALPH system (dark green, thin - SCC retrieved). Differences between the manually and corresponding SCC retrieved profiles are shown in the right panel.

2.2 Cloud masking

The aim of the cloud masking module, still in a testing phase, is to separate cloudy and cloud-free regions from raw, uncalibrated lidar signals. The module treats each submitted measurement file independently, not considering previous analysis, as we cannot be sure that the instrument calibration constant did not change even in a small time period. This approach does not allow us to apply quantitative,

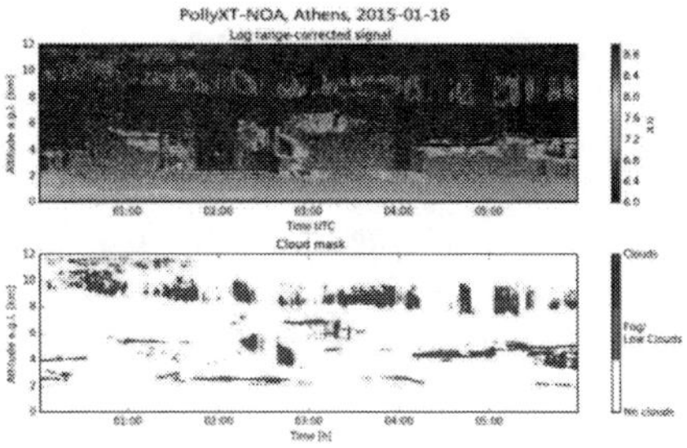

Figure 2: Example output of the cloud masking algorithm on PollyXT lidar measurements.

physically meaningful thresholds, for cloud detection. Therefore the aim of this cloud masking module is to provide an initial estimate of cloud-free regions, and it is not intended for the analysis of cloud properties. Before applying the cloud masking procedure, all signals are pre-processed and normalized using the median value of the logarithm of the range corrected signal in the first few kilometers of the atmosphere. In this way signals from different systems have comparable magnitude, helping the stability of the cloud detection. To separate cloudy and non-cloudy regions we examine two features: a) the presence of sharp edges in the signal, quantified by the Sobel operator, and b) the large time variability of the signal, quantified by the standard deviation of the signal in a moving window. Presently, we assume that we can adequately separate cloudy and cloud-free regions in this feature space using a linear decision boundary. To find the optimal boundary, we constructed a "training" set of measurements where we manually apply a cloud mask, based on visual inspection. Then we use a stochastic linear descent algorithm with logistic loss function to find the optimal linear relationship between the two features that is able to separate the two classes of data.

2.3 Aerosol Layering

Aerosol layer detection can be done on high-resolution pre-processed data or on the vertical profiles of the aerosol optical properties. Both approaches will be implemented in the SCC as separate modules. As first step, the assessment of the performances of the aerosol layer detection based on optical properties is planned. In particular, for a pilot study, a prototype of the detection algorithm has been developed. To allow the application of automatic procedures during both day and night time conditions, only the backscatter coefficient profiles and the related Ångström exponents are considered. The current algorithm first checks at which altitudes a significant aerosol load is present (by defining proper threshold on the value of backscatter coefficients); then, within these ranges, significant gradients of the backscatter coefficient profile are identified and from that layer boundaries are defined. Finally, it is checked if the intensive aerosol properties (Ångström exponent) within the defined layers are nearly constant. If not, the layer is divided into sub layers of nearly constant Ångström exponents. Fig. 3 shows an example of the detection algorithm on a Leipzig measurement case where two distinct aerosol layers were automatically detected. In the near future it is planned to extend the procedure to other intensive properties like lidar ratio and particle depolarization ratio.

3 CONCLUSIONS

We have provided an overview of the new SCC products like PLDR, VLDR, cloud masking, and aerosol layering. These new products together with the ones already available (aerosol extinction and backscatter coefficients) allows for a better characterization of atmospheric aerosols. In particular, the quality assured procedures used to calculate the PLDR allows a reliable classification of scatterers in terms of their sphericity. The cloud masking module provides a big improvement in terms of SCC near-real time products availability. The aerosol layering products are useful for a large number of application like micro-physical retrievals or aerosol typing schemes. Finally, the flexibility characterizing the implemented products allows the SCC to be easily extended

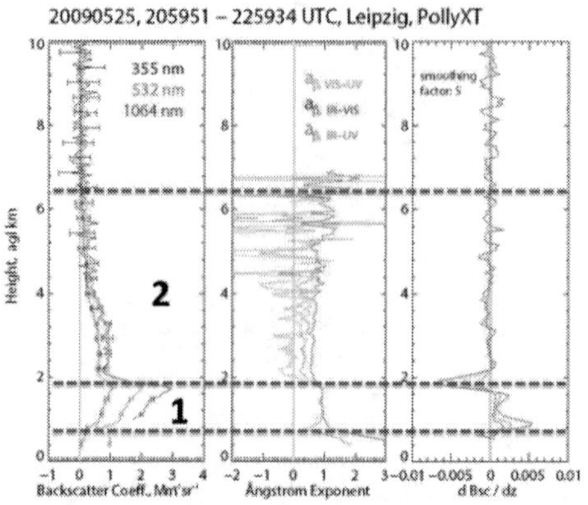

Figure 3: Automatic layer detection algorithm applied on a Leipzig measurement case.

to GALION (GAW Aerosol LIdar Observation Network) to evaluate lidar data of networks different from EARLINET.

ACKNOWLEDGMENTS

The financial support for EARLINET in the ACTRIS Research Infrastructure Project by the European Union's Horizon 2020 research and innovation programme under grant agreement no 654169 is gratefully acknowledged.

References

[1] D'Amico, G. et al., 2016: EARLINET Single Calculus Chain – technical – Part 1: Preprocessing of raw lidar data, *Atmos. Meas. Tech.*, **9**, (2), 491-507.

[2] Mattis, I. et al., 2016: EARLINET Single Calculus Chain – technical – Part 2: Calculation of optical products, *Atmos. Meas. Tech.*, **9**, (7), 3009-3029.

[3] D'Amico, G. et al., 2015: EARLINET Single Calculus Chain – overview on methodology and strategy, *Atmos. Meas. Tech.*, **8**, (11), 4891-4916.

[4] Freudenthaler, V. 2016: About the effects of polarising optics on lidar signals and the Δ90 calibration, *Atmos. Meas. Tech.*, **9**, 4181-4255.

https://doi.org/10.1051/epjconf/201817609015

OPTICAL MEASUREMENT OF ASIAN DUST OVER DAEJEON CITY IN 2016 BY DEPOLARIZATION LIDAR IN AD-NETWORK

Park Chan Bong[1*], Atsushi Shimizu[2], Nobuo Sugimoto[2]

[1]*Dept. of Electronic Engineering, Mokwon University, 88 Doan-bukro, Seo-Gu, Dae-jeon, Korea, cbpark@mokwon.ac.kr*

[2]*National Institute for Environmental Studies, 16-2, Onogawa, Tsukuba, Ibaraki, 305-8506, Japan, nsugimot@nies.go.kr*

ABSTRACT

Long-term sustained heavy Asian Dust below 3 km was measured in the period of April 22-26 by 2-wavelengths depolarization lidar system. As the comparison results of Daejeon station with other stations in the AD-Net, similarly formed Asian dust had been measured at Nigata, Toyama, Matsue, and Sendai. The route of the dust was examined by HYSPLIT. More than 80 μg / m^3 mass concentrations derived from the lidar measurements were compared with Air-Korea PM10 data.

1 INTRODUCTION

In every spring season, a huge amount of Asian dust generated in inner China, Mongolia and Taklimakan areas are transported and widely dispersed to East Asia region including Korean peninsula and Japan. As one of the major continental aerosols, the Asian dust caused great effects on radiation budget and climate changes, depending on their concentration, composition, and height distribution in the atmosphere. Optical properties, spatial distribution, and long-range transportation mechanism of the dust have been studied using ground based measurement profiles, satellite images and numerical modeling. Among them, lidar measurements method provides a useful and simple tool that can directly measure the vertical profiles of the aerosols and dust up to the upper troposphere with higher spatial and temporal resolutions [1][2]. Many studies about the origins and transportation route of the dust in free troposphere have considered several dominant sources of the Asian dust such as Mongolia, Taklimakan, and Loess area in east China. Recently, to know the contributions from unknown desert sources and understanding of transportation route, many lidar stations have been constructed in the East Asia and collaboration measurements have been performed. Since 2001, the Asian dust Network (AD-Net) has been expanding the stations. At present, more than 20 lidar stations in Japan, Korea, and China are located in the Network [3]. For collaboration measurements, Daejeon station in Korea has been involved in the Network since 2010. Because of the dust frequently mixed with pollutant gases or water vapor during their transportation, the optical characteristics of the dust are also changed. A polarization lidar technique is very useful tool to distinguish the dust aerosol, spherical aerosols, and clouds according to their shapes [3]. However, uncertainties of the dust aerosols (distribution, amount, characteristics) are still remained. To analyze the contribution of the dust and the spherical aerosol on the calculation of the backscattering coefficient, method using depolarization ratio was developed and published.

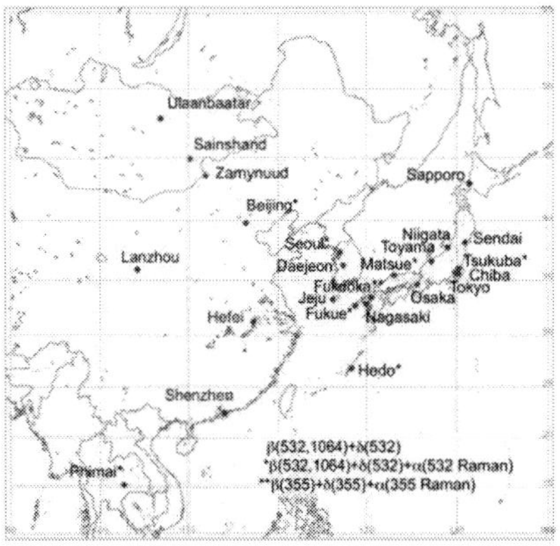

Figure 1 Stations in the AD-Net

2 LIDAR SYSTEM

The Lidar system in Daejeon station as a one compact package unit consisted of transmitter and receiver modules. The lidar system use the

https://doi.org/10.1051/epjconf/201817609015

second-harmonic Nd:YAG laser and 300 mm diameter telescope. The transmitting output energy at each wavelength of Nd:YAG laser is 65 mJ at 532 nm, 190 mJ at 1064 nm. The receiving system has Newtonian telescope with 30 cm diameter and four polarization channels that are 532(Vertical) /532 (Parallel) and 1064(Vertical) /1064(Parallel). PMT and APD detectors are allocated to the 532 nm and 1064 nm channels respectively. For the depolarization alignments, the polarizing beam splitter (PBS) is used only for the separate of the polarization (Vertical and Parallel of the receiving beam). Using the return signal at 532 nm and 1064 nm we calculate the aerosol backscattering coefficient using the Fernald method [4].

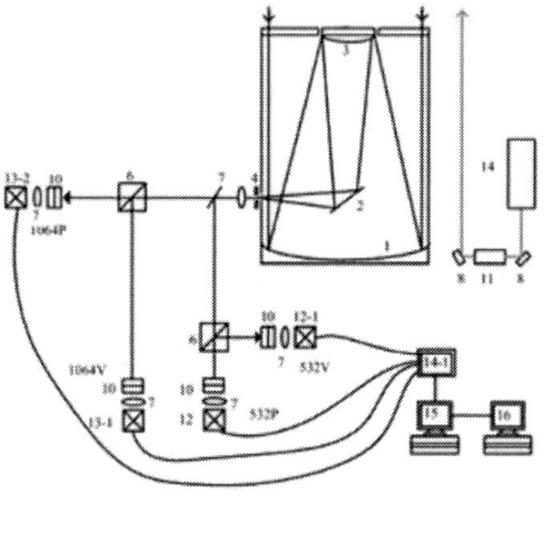

Figure 2 Block diagram of ACA Lidar system

(1. main mirror, 2. telescope reflection mirror, 3. secondary mirror, 4. Iris, 6. polarization 7. focusing lens, 8. transmitter reflection mirror, 10. interference filter, 11. beam expander, 12. PMT 13. APD, 14.ND:YAG laser, 14.1 Digital OSC, 15. PC1, 16. PC2)

At the first step in the calculation of, the value of $\beta_a(z_b)$ is given as the boundary value at the boundary altitude z_b. Usually, the a priori value of 0%–3% of the Rayleigh backscattering coefficient at the altitude of 6–10 km is used as the boundary $\beta_a(z_b)$ value, and the boundary altitude z_b and the extinction/backscatter ratio of aerosols S_a are assumed to be constant with time and altitude. Figure 2 show a block diagram of the two wavelengths dual polarization lidar system and Table I is a specification of the lidar system.

Table 1 Specification of Lidar System

Transmitter
Laser : Surelite I-10
Max. output : 470 mJ @1064 nm
220 mJ @532 nm
Beam div : 0.6 mrad
Repetition : 10 Hz
SHG : type I for 1064 nm and 532 nm
Receiver
Telescope : Newtonian type
1st mirror, 300 mm
FOV, 0.3-6.4 mrad (1 mrad)
Receiving channel : 1064 nm V and 1064 nm P
532 nm V and 532 nm P
Detector : PMT , Hamamatsu-H6780
APD , Licel-APD-1.5
Data Acquisition : LeCroy-LT264, 4ch,
350 MHz, 1 GS/sec, 8 bits
Operation mode : 5 min-on, 10 min-off
Range resolution : 6 m

3 MEASUREMENTS

In a spring season in 2016, we measured very heavy dust layers in the lower troposphere by the lidar during a time of April 22th (06:00, UTC) ~ April 26th (22:00, UTC) at Daejeon station in Korea. Figure 3 show Time-Height Indication (THI) plots of 532 nm backscatter coefficient (a), 1064 nm backscatter coefficient (b), depolarization ratio of 532 nm (c), and ratio of backscatter coefficient between 1064 nm and 532 nm (d). Higher backscatter coefficients and depolarization ratios measured at each wavelength of 532 nm and 1064 nm below 2 km were observed during that period. The maximum backscatter coefficient of 532 nm and 1064 nm showed over $1x10^{-5}$(/m/sr). The value of ratio of backscatter coefficient between

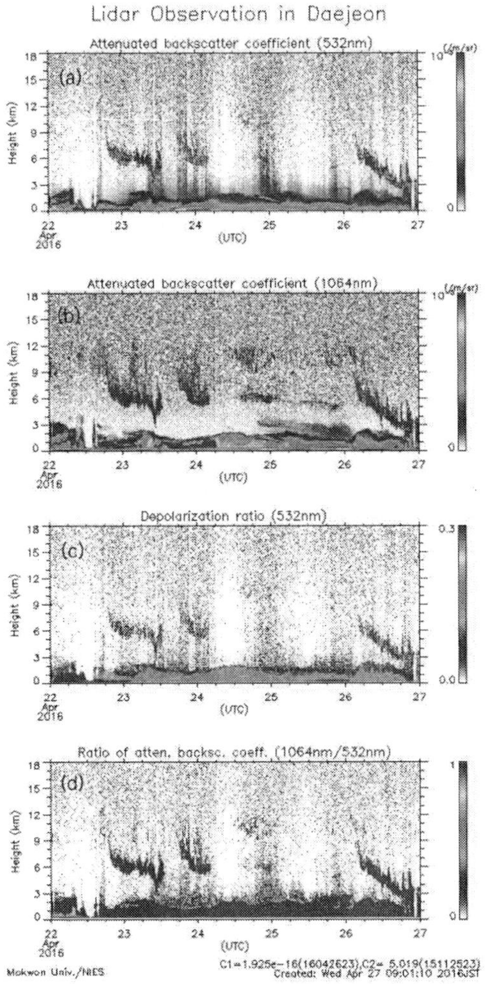

Figure 3 THI plots of 532 nm backscatter coefficient (a), 1064 nm(b), 532 nm depolarization ratio(c), and ratio of backscatter coefficient between 1064 nm and 532 nm(d).

1064 nm and 532 nm (d) dependent on particle size showed near 1 and indicate that larger particles were located in center of the dust layer. The maximum depolarization ration showed about 20~25 % at 532 nm.

4 COMPARISON

As the comparison results of Daejeon station with other stations in the AD-Net, similarly formed

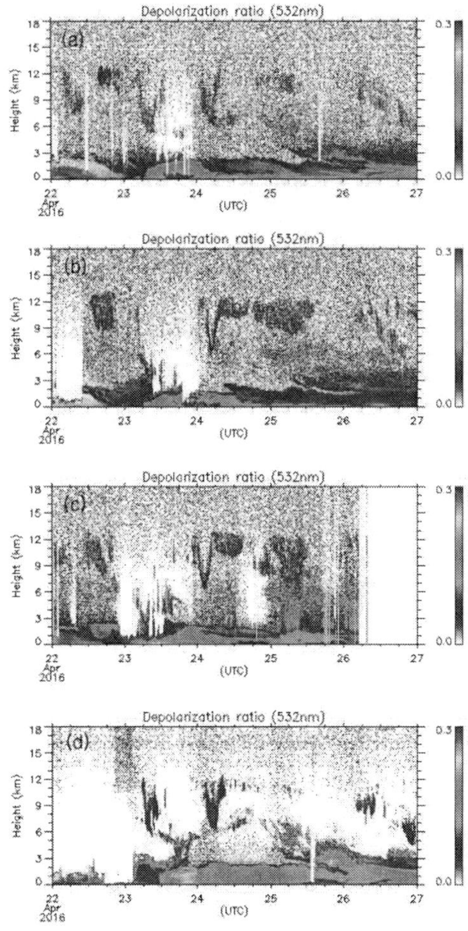

Figure 4 THI plots of 532 nm depolarization ratio measures at Nigata(a), Toyama(b), Matsue(c), and Sendai(d).

Asian dust with 10~25% depolarization ratio had been measured at Nigata, Toyama, Matsue, and Sendai in Japan during that period of April 23th ~ April 26th. However, this kind of long-term continuously measured dust was not observed at other stations in Japan such as Fukue, Nagasaki, Osaka, Tsukuba, Tokyo, and Chiba. Figure 5 show the Time-Height Indication(THI) plots of depolarization ratio of 532 nm at Nigata (138.94E, 37.84N), Toyama(137.10E, 36.70N), Matsue (133.01E, 35.48N), and Sendai (140.83E, 38.26N). To know the transported route of the dust layer, backward and forward trajectory analysis from Daejeon station was performed. Figure 5 show the results of the backward and forward trajectory analysis. NOAA HYSPLIT MODEL using GDAS meteorological data was used for this Analysis [5]. The forward trajectory below 2 km over Daejeon station started on April 22, 06:00 with running time of 72 hrs. The result of this analysis clearly showed that the dust transported route passed Nigata, Toyama, and Matsue. The backward trajectory started from Sendai on April 22, 18:00 also showed the dust layer over passed Daejeon area.

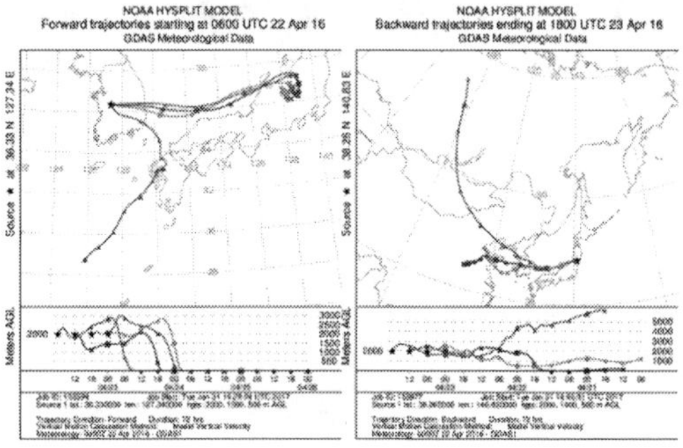

Figure 5 Forward Trajectory started from Daejeon station on April 22, 06:00 and Backward Trajectory started from Sendai station on April 22, 18:00.

The mass concentration of aerosol was calculated from averaged AOD values in PBL altitudes with mass conversion factor of $3.6 \pm 1.2\ m^2\ g^{-1}$. Figure 6 show the mass concentration variation in April, 2016. More than 80 $\mu g / m^3$ mass concentrations were observed during that period of Asian dust measurements. Its results was also coincides well with the report of PM10 by KMA in Korea.

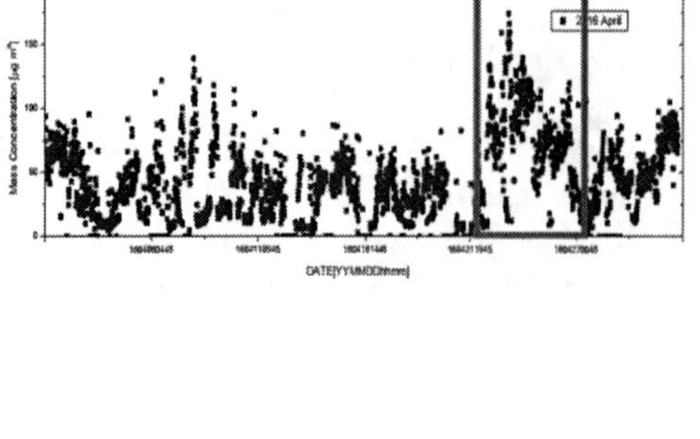

Figure 6 Mass concentration variations in April, 2016 derived from averaged AOD value in PBL (red box shows the period of the Asian dust measurements).

5 CONCLUSIONS

The monitoring of Asian Dust has been performed with 2-wavelengths depolarization lidar in Daejeon station. Strong and long term sustained Asian dust had been measured during a period from April 22th to April 26th. The maximum backscatter coefficient over 10^{-5}(/m/sr) with depolarization ratio over 25 % were observed. At the time period within a few days, similar dust events were measured consequently at Nigata, Toyama, Matsue, and Sendai in Japan. The backward and forward trajectory analysis by NOAA HYSPLIT Model showed coincidence with the results of lidar measurement in the station of AD-Net.

ACKNOWLEDGEMENTS

The research was supported by 'Software Convergence Technology Development Program', through the Ministry of Science, ICT and Future Planning (S0177-16-1059).

References

[1] Sugimoto N, A. Shimizu, I. Matsui, I. Uno, K Arao, X. Dong, S. Zhao, J. Zhou, and C. H. LEE, "Study of Dust Transport Using a Network of Continuously Operated Polarization Lidars", Water, Air, and Soil Pollution: Focus 5, (2005), pp. 145-157.

[2] Liu X, Y. Cheng, Y. Zhang, J. Jung, N. Sugimoto, S.Y. Chang, Y. J. Kim, S. Fan, L. Zeng, "Influences of relative humidity and particle chemical composition on aerosol scattering properties during the 2006 PRD campaign", Atmospheric Environment, vol. 42, No. 7, (2008), pp. 1525-1536.

[3] Shimizu A, K Arao, T. Murayama, N. Kagawa, K. Aoki, A. Uchiyama, and A. Yamazaki, "Continuous observations of Asian dust and other aerosols by polarization lidars in China and Japan during ACE-Asia", Journal of Geophysical Research, vol. 109, (2004), pp. D19S17.

[4] Fernald F.G, "Analysis of atmospheric LIDAR observations: Some comments", Applied Optics, vol. 23, No. 5, (1984), pp. 652–653.

[5] NOAA Air Resources Laboratory, http://ready.arl.noaa.gov/ HYSPLIT .php

EARLINET DATABASE:

NEW DESIGN AND NEW PRODUCTS FOR A WIDER USE OF AEROSOL LIDAR DATA

Lucia Mona[1]*, Giuseppe D'Amico[1], Francesco Amato[1], Holger Linné[2], Holger Baars[3], Ulla Wandinger[3], Gelsomina Pappalardo[1]

[1]*Istituto di Metodologie per l'Analisi Ambientale CNR-IMAA, C.da S. Loja, I-85050 Tito Scalo, Potenza, Italy, *Email: lucia.mona@imaa.cnr.it*

[2]*Max-Planck-InstitutfürMeteorologie, Hamburg, Germany*

[3]*3Leibniz Institute for Tropospheric Research, Leipzig, Germany*

ABSTRACT

The EARLINET database is facing a complete reshaping to meet the wide request for more intuitive products and to face the even wider request related to the new initiatives such as Copernicus, the European Earth observation programme. The new design has been carried out in continuity with the past, to take advantage from long-term database. In particular, the new structure will provide information suitable for synergy with other instruments, near real time (NRT) applications, validation and process studies and climate applications.

1 INTRODUCTION

EARLINET (European Aerosol Research Lidar Network) is the aerosol vertical profile component of ACTRIS (Aerosol, Cloud and Trace gases Research infraStructure). EARLINET aerosol profile data are collected at each station applying measurement procedures compliant with the recommendations provided by the ACTRIS aerosol profile community. Data in the EARLINET DB are reported in netCDF format in accordance with the reporting procedures (available at http://www.earlinet.org/).

The EARLINET database contains at the moment quality assured aerosol extinction and backscatter profiles, reported in separated files (called e- and b-files). The main concept behind the current EARLINET database is to provide basic aerosol lidar products at the best available resolution.

A strong link with EARLINET data users has been established since the beginning of EARLINET, in which network members acted as mentors and guides for users to assure a correct use and interpretation of the lidar data. This big effort for the EARLINET community has been an excellent opportunity to improve the knowledge about the user needs.

The main result of this learning process was to realize that the EARLINET database is under-exploited, because although it contains a lot of information needed for several applications, the way these data are provided are not easily understandable to many external (non-lidar) users.

This awareness was the seed for the development of new products for making EARLINET data more visible, effective and easier to handle to potential users of different communities.

2 NEW EARLINET DATABASE

The new EARLINET database follows a general concept used for most databases in the atmospheric field: a level structure going from low level and fast delivered data towards more advanced and correspondingly later released data.

The new structure of the EARLINET database is schematically reported in Figure 1. Level 1 contains pre-processed lidar data, i.e. a step in between the raw signal (Level 0 data stored at each station) and the optical properties, where all instrumental corrections have been already implemented. The Level 1 data are the base for the retrieval of the optical properties contained in the Level 1.5 products. The Level 1.5 datasets are not quality checked, except for format aspects, and therefore released as soon as data originators submit them to the database. Afterwards, all the Level 1.5 data pass through the quality check procedures described in section 2.1. The optical data that passed such quality checks are promoted to Level 2 product. Finally, the Level 3 data contain climatological datasets retrieved from the Level2optical products.

The increase of quality checks (QC) from the Level 1 to Level 3 implies a corresponding increase in the delivery time of the data which ranges from potential NRT for Level 1 and 1.5 to annual and bi-annual release of Level2 and Level 3 data.

The new structure of the EARLINET database will take advantage from the centralized data processing through the common standardized automatic analysis software developed within EARLINET, the Single Calculus Chain (SCC) [1] , even if it will be still allowed the submission to the database of data analyzed by using specific QA and documented software. The usage of the SCC assures the usage of QA retrieval algorithms and the full traceability of all EARLINET products.

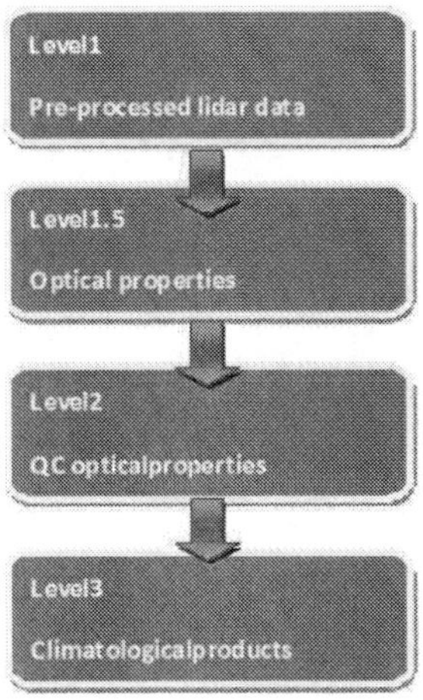

Figure 1Layout of the new structure of the EARLINET database.

2.1 Level 1 products

The Level 1 data contain the pre-processed lidar data, i.e. range-corrected signals derived from the raw lidar data measured by each station for each detection channel of the lidar system corrected for several instrumental effects like dead-time, partial overlap between laser beam and receiving system etc. The input parameters needed to correct for these effects are measured applying the QA procedures through the Lidar Calibration centre (http://lical.inoe.ro). Level 1 products are used internally to deliver aerosol optical products (Level 1.5). At same time, they could be used by external processing modules to calculate combined/advanced products, for example in combination with Cloudnet data. In addition, the ICARE thematic center (http://www.icare.univ-lille1.fr/) will automatically process the L1 products combining them with coincident AERONET data through the GARRLiC (Generalized Aerosol Retrieval from Radiometer and Lidar Combined data) algorithm.

The standardized and quality-assured pre-processed lidar data have been also used to develop a homogeneous network-wide, open and freely accessible quick-look database (high-resolution images of time-height cross sections). The quick-look database is now in the testing phase and will be externally accessible starting from June 2017.

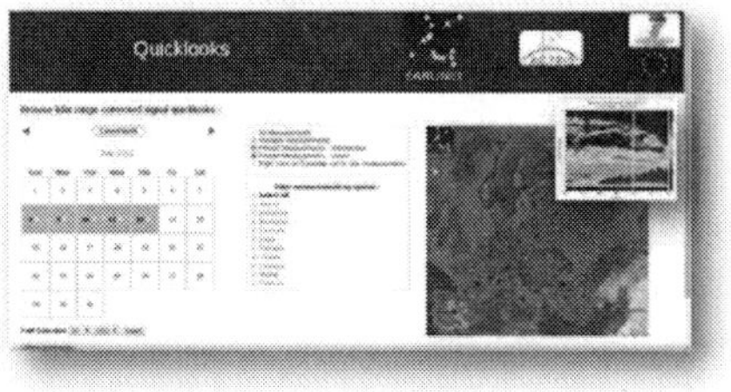

Figure 2 Web interface of the new EARLINET quicklook database.

Furthermore, the results of cloud/aerosol masking will be graphically reported in these quick-looks.

2.2 Level 1.5 products

In order to foster the fast use of the EARLINET data which is essential for assimilation purposes and hazards, and to clarify the difference between quality-checked and not quality-checked, an intermediate step between the pre-processed and quality-checked data has been added: Level 1.5. The data products contained in there are the same aerosol optical products planned for the Level 2 (see next section) with the difference that for Level 2. The only difference is no quality check procedures were applied except of some basic format checks. These technical checks are applied to the data to assure the consistency between the file contents and the file structure as defined in the EARLINET database.

The Level 1.5 products allow for a NRT use of lidar data. EARLINET already proved its capability of providing data in NRT using the SCC during a 72-h experiment carried out in the summer 2012 ACTRIS campaign [2].

2.3 Level 2 products

Level 1.5 data submitted to the EARLINET database are subject to centralized quality check procedures. In addition to some technical quality checks, scientific QCs are applied to assure the validity of the measured parameters from physical point of view. The data submitter receives feedback about the outcome of the QC. Data compliant to all the QC requirements are made public available as Level2 data. Level 2 data will be regularly published with primary Digital Object Identifiers (DOIs) in continuity with previous volume publications on the CERA database [3].

Three different data products are present in Level 2 (and Level 1.5): single-wavelength profiles, multi-wavelength profiles and layer products.

Single-wavelength profiles

In continuity with the past, the new EARLINET database will report the extinction and backscatter profiles at their best vertical resolution in separated products. This allows for a consistent extension of the previous database version already published with DOI on the CERA database [3].

Multi-wavelength profiles

In addition to the standard (single-wavelength) optical property profiles, a new product is envisaged in order to take the most from the multi-wavelength capabilities widely available within the network. The multi-wavelength profile product contains the vertical profiles of all the optical properties measured in the same temporal window and with the same vertical resolution. This allows investigation of the aerosol type and, if applicable, mixing [e.g., 4] as well as the determination of aerosol microphysical properties [e.g., 5].

Layer products

Building on the previous experiences in providing tailored datasets to different communities [6, 7, 8], specific products focused on aerosol layer properties have been introduced. For each Level 2 product, a corresponding Level 2 Layer Product is generated reporting the relevant layer information such as the base height, top height and thickness of each identified layer as well as mean, median, standard deviation and mean statistical error for each measured optical property. Integrated quantities within each layer and the corresponding columnar ones are reported as well for extensive optical properties. Finally, preliminary information about the aerosol typing is provided whenever available [9, 10].

2.4 Level 3 products

The Level 3 standard product contains climatological datasets obtained as aggregated products from the Level 2 aerosol optical products. Data will be aggregated into monthly, seasonal and annual datasets for both profiles and vertically integrated quantities. Information about the number of collected samples, mean, median and standard deviation of the properties, as well as mean statistical error for each property are reported. Metrics of the comparison against reference datasets (e.g. AERONET for AOD) are included whenever available, in order to provide information about data representativeness. At the moment, a bi-annual release of the Level 3 standard dataset is envisaged. The Level 3 standard product is the result of the cooperation between EARLINET and AEROCOM (Aerosol Comparisons between Observations and Models project) carried out during the previous ACTRIS project, responding to need of the global aerosol modeling community for aggregated data together with information about their representativeness.

2.5 Advanced secondary data sets

Additional advanced products will be designed step by step following the specific needs of specific studies.

At the moment, two secondary datasets are available at www.earlinet.org and linked to the ACTRIS data portal: the Eyjafjallajökull 2010 – EARLINET database, and the EARLINET 72h operational exercise dataset. The first reports the four-dimensional (4-D) distribution of the Eyjafjallajökull volcanic cloud in the troposphere over Europe as observed by EARLINET [8]. The second instead collects the SCC products obtained during the 72h EARLINET exercise for near-real time data provision [2].

A new dataset will be provided as results of the collaboration with ICARE for the lidar+photometer data processing through GARRLiC. This dataset will provide vertical profiles of aerosol microphysical properties.

Further advanced products can be the results of the specific case studies and measurement campaigns, and will be provided as additional datasets where aerosol profile data are combined to different platform data and analysis.

3 CONCLUSIONS

The EARLINET database is facing a complete redesign with the implementation of a new workflow and delivering new products as response to the growing demand of aerosol vertical profiling information. The complete switch to the new version is foreseen by the end of ACTRIS-2 project in May 2018, but many new products and procedures are already in place. Advances will be presented at the conference. In particular, the climatological products and their potential applications will be presented as result of the implemented QC and averaging schemes, underlying their scientific content in terms of aerosol climatological analysis over Europe in the last 15 years.

ACKNOWLEDGEMENTS

The financial support for this activity in the ACTRIS Research Infrastructure Project by the European Union's Horizon 2020 research and innovation programme under grant agreement no. 654169 is gratefully acknowledged.

References

[1] D'Amico G., Amodeo, A., Baars, H., Binietoglou, I., Freudenthaler, V., Mattis, I., Wandinger, U., and Pappalardo, G., 2015: EARLINET Single Calculus Chain – overview on methodology and strategy, Atmos. Meas. Tech., 8, 4891-4916, doi:10.5194/amt-8-4891-2015.

[2] Sicard, M., et al., 2015: EARLINET: potential operationality of a research network,EARLINET: potential operationality of a research network, *Atmos. Meas. Tech.*, **8**,4587-4613, doi:10.5194/amt-8-4587-2015.

[3] The EARLINET publishing group 2000–2010, 2014: EARLINET all observations (2000–2010), World Data Center for Climate(WDCC), doi:10.1594/WDCC/EN_all_measurements_2000-2010.

[4] Mona, L., Amodeo, A., D'Amico, G., Giunta, A., Madonna, F., and Pappalardo, G., 2012: Multiwavelength Raman lidar observations ofthe Eyjafjallajökull volcanic cloud over Potenza, southern Italy,*Atmos. Chem. Phys.*,**12**, 2229–2244, doi:10.5194/acp-12-2229-2012.

[5] Veselovskii, I., A. Kolgotin, D. Müller, and D. N. Whiteman, 2005: Information content of multiwavelengthlidar data with respect to microphysical particle properties derived from eigenvalue analysis, *Appl. Opt.*,**44** (25), 5292–5303.

[6] Wandinger, U., Hiebsch, A., Mattis, I., Pappalardo, G., Mona, L., and Madonna, F., 2011: Aerosols and Clouds: long-term Database from Spaceborne Lidar Measurements, Tech. rep., ESA Publications Division, final report, ESTEC Contract 21487/08/NL/HE, Noordwijk, the Netherlands.

[7] Amiridis, V., et al., 2015: LIVAS: a 3-D multi-wavelength aerosol/cloud database based on CALIPSO and EARLINET, *Atmos. Chem. Phys.*,**15**, 7127-7153, doi:10.5194/acp-15-7127-2015.

[8] Pappalardo, G., et al., 2013: Four-dimensional distribution of the 2010 Eyjafjallajökull volcanic cloud over Europe observed by EARLINET, *Atmos. Chem. Phys.*,**13**, 4429-4450, doi:10.5194/acp-13-4429-2013.

[9] Papagiannopoulus et al., 2017: An automatic aerosol classification for EARLINET: application and results, ILRC 2017.

[10] Nicolae et al.,2016: Independent Retrieval of Aerosol Type From Lidar, EPJ Web of Conferences 119, 18002 (2016), https://doi.org/10.1051/epjconf/201611918002.

STUDIES ON MINERAL DUST USING AIRBORNE LIDAR, GROUND-BASED REMOTE SENSING, AND IN SITU INSTRUMENTATION

Franco Marenco[1*], Claire Ryder[2], Victor Estellés[3], Sara Segura[3], Vassilis Amiridis[4], Emmanouil Proestakis[4], Eleni Marinou[4], Alexandra Tsekeri[4], Helen Smith[5], Zbigniew Ulanowski[5], Debbie O'Sullivan[1], Jennifer Brooke[1], Yaswant Pradhan[1], and Joelle Buxmann[1]

[1] *Met Office, United Kingdom, *Email: franco.marenco@metoffice.gov.uk*
[2] *University of Reading, United Kingdom*
[3] *University of Valencia, Spain*
[4] *National Observatory of Athens, Greece*
[5] *University of Hertfordshire, United Kingdom*

ABSTRACT

In August 2015, the AER-D campaign made use of the FAAM research aircraft based in Cape Verde, and targeted mineral dust. First results will be shown here. The campaign had multiple objectives: (1) lidar dust mapping for the validation of satellite and model products; (2) validation of sunphotometer remote sensing with airborne measurements; (3) coordinated measurements with the CATS lidar on the ISS; (4) radiative closure studies; and (5) the validation of a new model of dustsonde.

1 INTRODUCTION

From 6 to 25 August 2015, the FAAM BAe-146 research aircraft was based in Praia on the Santiago island (Cape Verde). A ground-based site with several types of instruments was established at the base airport, including two sun-photometers (PREDE-POM01 and CIMEL CE318-NEDPS9 dual-polarization). Additional sun-photometer measurements were available through the permanent AERONET station Cabo Verde in Sal Island. A number of dustsondes of a new type [1] were also launched from Sal.

In this season, mineral dust transport from the Sahara is frequent (see Figure 1). Sixteen research flights were carried out, 10 of which targeted the role of dust as primary heterogeneous ice nuclei (ICE-D campaign). Six more flights, described in this paper and listed in Table 1,

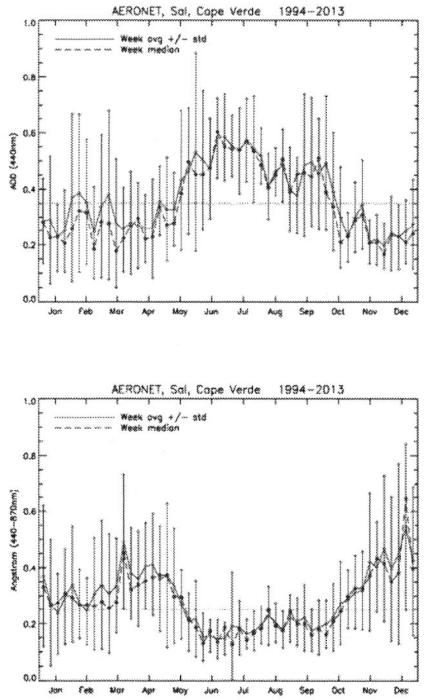

Figure 1. Yearly cycle of aerosol optical depth and Ångstrom exponent, as derived from 20 years AERONET observations (1994-2013) at Sal, Cape Verde.

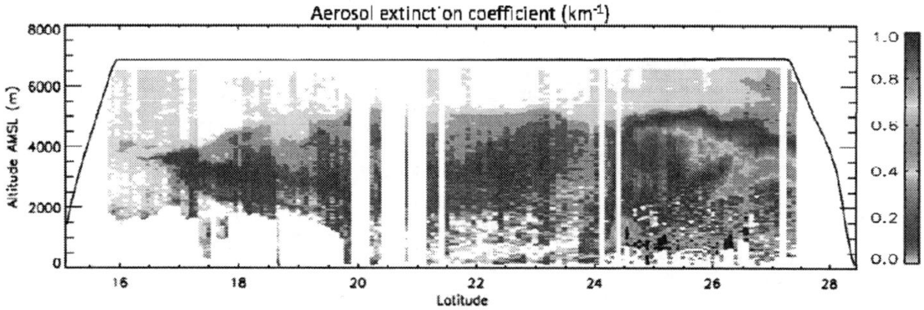

Figure 2. Aerosol extinction coefficient measured by airborne lidar in flight B923(flight from Cape Verde to the Canary Islands).

aimed at a series of objectives in the field of aerosol science and remote sensing (AER-D campaign). The aims of AER-D were multiple: (1) lidar dust mapping for the validation of satellite and model products; (2) validation of sunphotometer remote sensing with airborne measurements; (3) coordinated measurements with the CATS lidar on the ISS; (4) radiative closure studies; and (5) the validation of a new model of dustsonde.

Table 1. Summary of the AER-D research flights

B920	7 Aug.	CATS underflight
B923	12 Aug.	Dust mapping
B924	12 Aug.	Dust mapping
B928	16 Aug.	SAVEX-D (Santiago)
B932	20 Aug.	Dust mapping
B934	25 Aug.	SAVEX-D (Sal)

2 LIDAR SYSTEM

The aircraft carried a nadir-pointing ALS450 elastic backscatter lidar, operating at 355 nm, manufactured by Leosphere. Signal inversion to aerosol extinction coefficient is done following reference [2].

3 DUST MAPPING

The dust mapping flights were designed to map the spatial distribution and vertical structure of the dust layer with the on-board lidar, by flying at high altitude over a long distance. Our main goal is the development of a dust product based on SEVIRI, based on infrared channels and a 1D-VAR retrieval scheme [3], and these flights will yield a good opportunity for its validation. We anticipate also other comparisons, such as e.g. with the Met Office Unified Model dust prediction.

On 12 August the advection of a a large dust load was predicted over the Canary islands, and a double flight was carried out with a refuel at Fuerteventura, and on 20 August the experiment was repeated but with a shorter, single flight towards the Northern border of Mauritania on the coast. Figure 2 shows the cross-section of the dust extinction coefficient determined with the airborne lidar during the first of these flights. Whereas the flight starts with little or no dust near Cape Verde, the dust layer is seen to increase in depth initially, and then in optical depth. North of 24N, a dust "front" is encountered and its structure is mapped. This "front" is clearly identifiable in the visible and infrared satellite imagery. A similar pattern is observed in flights B924 and B932. Small sections of the mapping flights were also devoted to in situ measurements, by bringing the aircraft down into the dust layer, and they revealed very large particles (up to ~ 100 μm in diameter). The AOD in the high dust region reached ~2, approaching the limits of the lidar system itself due to the large signal extinction.

4 CATS UNDERFLIGHT and SAVEX-D

The first of our flights was aimed at a comparison of aircraft measurements to observations from the CATS lidar on board the International Space Station. This is a backscatter lidar operating at 1064 nm and 532 nm, launched in 2015, and with a processing chain similar to CALIOP's [4]. The aim of the SAVEX-D flights, instead, was the

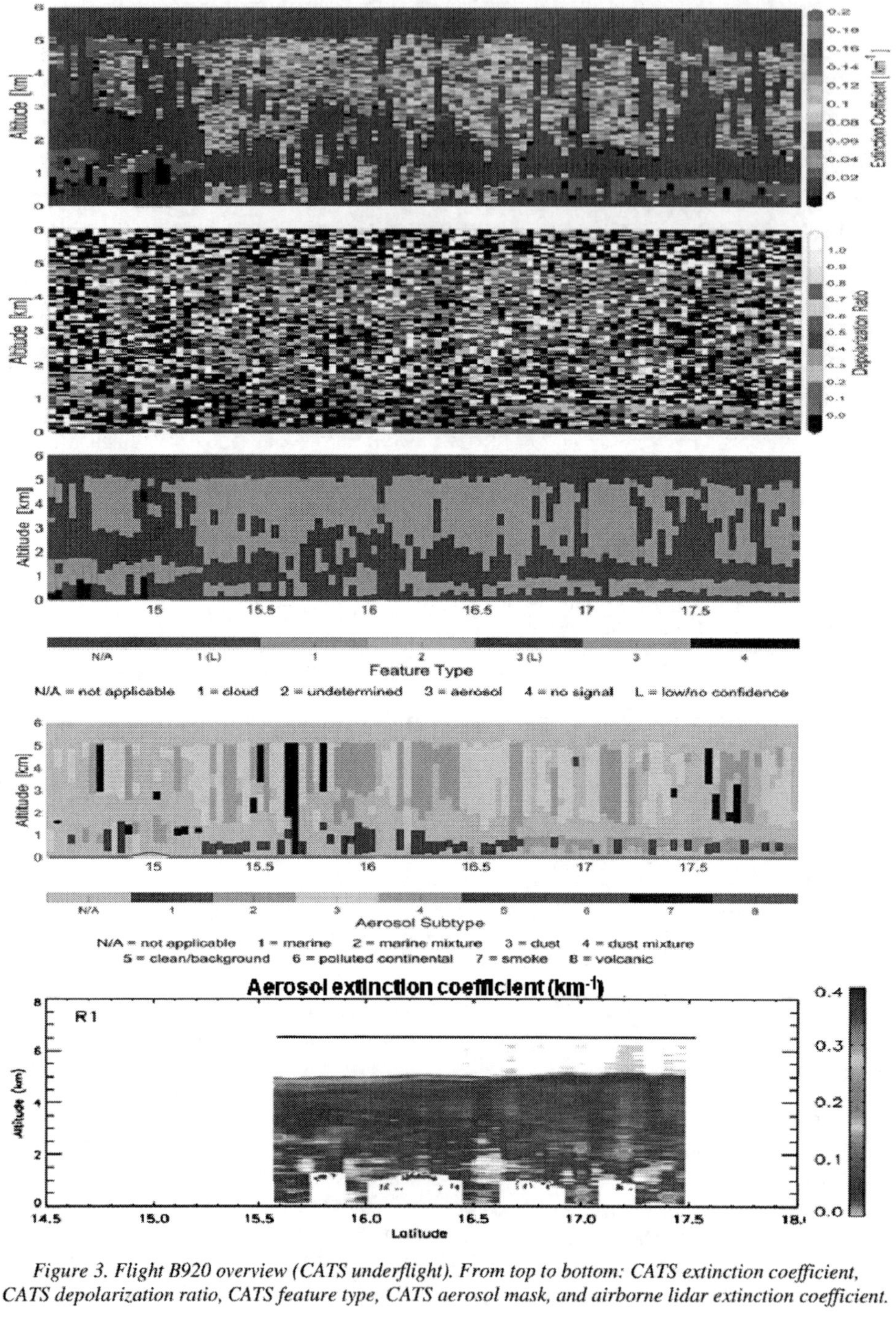

Figure 3. Flight B920 overview (CATS underflight). From top to bottom: CATS extinction coefficient, CATS depolarization ratio, CATS feature type, CATS aerosol mask, and airborne lidar extinction coefficient.

validation of SKYNET and AERONET ground-based sunphotometer retrievals of columnar aerosol microphysical properties, and in particular the size-distribution [5]. The track for these flights was confined within a small area, and the airplane sampled the atmosphere at a series of levels within the dust layer (to measure in situ properties) as well as above (to provide a lidar overview of the scene). As an example, Figure 3 compares the CATS L2 products with the scene depicted by the airborne lidar, showing great consistency.

5 RADIATIVE CLOSURE, DUST PROPERTIES, AND NEW MODEL DUSTSONDES

In this presentation, we will also show first results concerning our radiative closure studies, the detailed characterization of dust properties, and the comparison of the airborne data with the new dustsondes.

6 CONCLUSIONS

AER-D has been a very successful campaign, despite relying on a small number of flights. The measurements provide insight on the vertical structure of the Saharan Air Layer and on its microphysical properties; they provide a small but detailed dataset on the dust spatial distribution, designed for satellite and modeling cal/val exercises; and they provide in situ and remote sensing information for the validation of CATS retrievals, sunphotometer retrievals, and the measurements from the new model of dust sondes.

The data analysis is still preliminary at this stage but progress is being made quickly and we anticipate a number of publications in the years to come.

ACKNOWLEDGEMENTS

Airborne data was obtained using the BAe-146 Atmospheric Research Aircraft flown by Directflight Ltd. and managed by the Facility for Airborne Atmospheric Measurements (FAAM), which a joint entity of the Natural Environment Research Council (NERC) and the Met Office. R. Cotton, T. Choularton, M. Gallagher, K. Bower, J. Taylor, J. Crosier, G. Lloyd, D. Lui, P. Rosenberg, J. McQuaid, B. Murray, H. Price, FAAM, and the full team taking part in the ICE-D campaign are gratefully acknowledged for their assistance and the provision of datasets. The SAVEX-D project and the participation of Sara Segura and Victor Estellés were possible thanks to EUFAR TNA (European Union Seventh Framework Programme grant agreement 312609) and projects PROMETEUII/2014/058 and GV/2014/046 from the Valencia Autonomous Government. Copernicus Atmosphere Monitoring Services and ECMWF are gratefully acknowledged for the provision of forecast products used for flight planning. Axel Wellpott is thanked for his help in predicting the ISS orbit and other satellites for flight planning. We thank Didier Tanré for the effort in establishing and maintaining the Sal AERONET station. John Yorks, Ed Nowottnick, and NASA are kindly thanked for their assistance during the campaign and for the provision of CATS data.

References

[1] Smith, H.R., J. Ulanowski, P. Kaye, E. Hirst, W. Stanley, and A. Weiser, 2016: A Universal Cloud and Aerosol Sounding System (UCASS) based on wide angle light scattering, *presented at the Focus Meeting on Recent Advances in Aerosol Measurement Technology*, Aerosol Society, 5 April 2016.

[2] Marenco, F., Nadir airborne lidar observations of deep aerosol layers, *Atmos. Meas. Tech.* **6**, 2055-2064, 2013

[3] Francis, P.N., M.C. Cooke, and R.W. Saunders, 2012: Retrieval of physical properties of volcanic ash using Meteosat: A case study from the 2010 Eyjafjallajökull eruption, *J. Geophys. Res.* **117**, doi:10.1029/2011JD016788.

[4] Yorks, J.E., M.J. McGill, S.P. Palm, D.L. Hlavka, P.A. Selmer, E P. Nowottnick, M.A. Vaughan, S.D. Rodier, and W.D. Hart, 2016: An overview of the CATS level 1 processing algorithms and data products, *Geophys. Res. Lett.* **43**, 4632–4639.

[5] Estellés, V., Campanelli, M., Utrillas, M.P., Expósito, F., and Martínez-Lozano, J.A., 2012: Comparison of AERONET and SKYRAD4.2 inversion products retrieved from a Cimel CE318 sunphotometer, *Atmos. Meas. Tech.* **5**, 569-579.

ATMOSPHERIC NO_2 PROFILES MEASURED WITH LIDAR DURING THE CINDI-2 CAMPAIGN

A.J.C. Berkhout[1]*, L.F.L. Gast[1], G.R. van der Hoff[1], D.P.J. Swart[1], M. Hoed[2], M. Allaart[2]

[1]National Institute for Public Health and the Environment (RIVM), the Netherlands,
**stijn.berkhout@rivm.nl*
[2]Royal Netherlands Meteorological Institute (KNMI), the Netherlands

ABSTRACT

From 12 to 28 September 2016, the CINDI-2 NO_2 intercomparison campaign took place at the Cabauw research site in the Netherlands. Aimed principally at MAX-DOAS instruments, other techniques participated as well. The RIVM mobile lidar measured atmospheric NO_2 profiles on six campaign days. Results show the development of NO_2 concentrations during the day, as well as layered structures in the profiles. These were compared with profiles from other instruments.

1 INTRODUCTION

Measurements of atmospheric trace gases from satellites offer global coverage but need validation from ground-based data. The CINDI-2 campaign was set up to intercompare a suite of MAX-DOAS instruments. These instruments are to provide data for the validation of the upcoming Sentinel 5 and TROPOMI satellite missions, as profile information is highly relevant in satellite retrieval algorithms. In addition to slant column density measurements of NO_2, HCHO, O_3, and O_4, profile measurements and intercomparisons were included in the campaign objectives as well. CINDI-2, short for Cabauw Intercomparison of Nitrogen Dioxide Measuring Instruments 2, was the successor to the successful CINDI campaign held in 2009 [1]. An overview paper of this campaign is presented at this Conference by Arnoud Apituley.

2 METHODOLOGY

2.1 The CINDI-2 campaign

The campaign took place at the Cabauw Experimental Site for Atmospheric Research (CESAR), 51.971°N, 4.927°E, 0.7 m below mean sea level (Figure 1). A 213 m high meteo mast and a suite of meteorological, atmospheric and air quality instruments are always operational on the site. There are no sources of air pollution in the immediate vicinity. Depending on the wind direction, air at the site comes from the sea, or from densely populated and heavily industrialised areas, leading to a variety of conditions.

Figure 1. The Lidar vehicle at the site in Cabauw, next to the meteo mast.

Participants included 35 MAX-DOAS instruments from institutes from all over the world. Other measurement techniques, such as in-situ instruments and Lidars, participated as well. A profiling task team was set up, which aims to retrieve profile information from MAX-DOAS measurements and perform intercomparisons between these and other techniques.

2.2 The NO_2 Lidar

For this measurement campaign, the RIVM mobile Lidar system was used. This system is extensively described in [2], [3] and [4]. It measures either NO_2, SO_2 [3], NH_3 [4] and benzene [5]. It uses the DIAL technique to measure concentration profiles of NO_2 in the atmosphere.

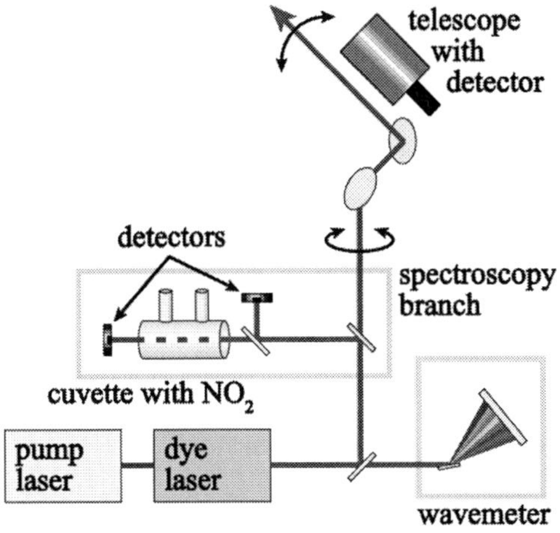

Figure 2. Schematic representation of the mobile Lidar DIAL system.

The lidar instrument (Figure 2) uses a Spectra-Physics Quanta-Ray Pro 250 pulsed Nd:YAG laser pumping a Sirah PrecisionScan PRSC-D-30 dye laser. The Exalite 411 dye used in this laser produces alternating wavelength pairs of 413.5382 nm (on) and 413.3775 nm (off), both wavelengths in vacuum. A wavemeter checks the wavelength of each pulse, if it has drifted too much, the dye laser is retuned. Output power of the dye laser is 30 mJ pulse[1] at 30 Hz. The spectroscopy branch shown in Figure 2 is used to determine the effective molecular cross section of the gas being measured. The instrument is housed in a fully self-supporting measurement vehicle.

2.3 Lidar measurements of NO_2 profiles

When measuring NO_2, the measurement range of the lidar runs from 300 m to about 2.5 km. At ambient concentration levels, typically only one or two concentration values are derived over this interval. To arrive at a higher vertical resolution, measurements at eight different elevation angles are combined (see Figure 3).

Profiles were obtained consisting of about six or seven values inside the planetary boundary layer, and, depending on atmospheric conditions, one above this layer. The vertical resolution of a profile varies (see the vertical bars illustrating this in Figure 3). It ranges from 7.5 m in the lowest part of the profile to 800 m above the boundary layer.

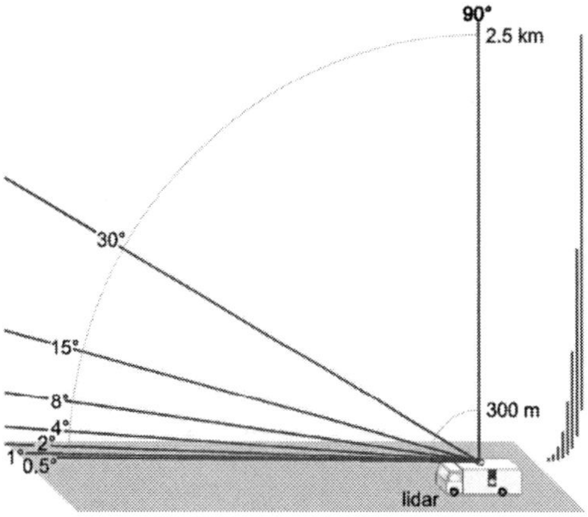

Figure 3. Construction of a vertical NO2 concentration profile from a set of measurements at different elevations. The vertical bars on the right illustrate the altitude intervals to which the measured NO2 concentrations are assigned.

Measuring a single sequence of elevation angles took about 84 s. Usually, 20 sequences were averaged to arrive at a single profile in about 28 min.

Lidar measurements were planned to yield as much information as possible. This meant that measurements were taken:

- on clear days, to facilitate intercomparison with satellite data;
- starting early in the morning (preferably around 6:30 local time), to capture the build-up of the atmospheric boundary layer;
- at times that coincided with balloon sonde launches (see below).

2.4 Balloon sondes

On some of the campaign days, one or more balloon sondes with fast NO_2 sensors [6] were launched. These reached altitudes of typically 10 km.

3 RESULTS

3.1 Example profile.

An example profile, measured with the Lidar on 15 September 2016, is shown in Figure 4. The horizontal error bars indicate the uncertainty in the measurement. The vertical bars indicate the altitude intervals over which the data points were collected. Note that the altitude axis is logarithmic, to better resolve the higher vertical resolution at lower altitudes.

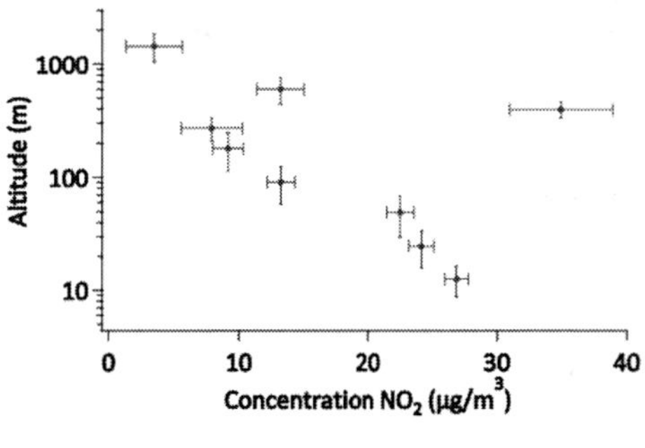

Figure 4. Example profile, measured on 15 September 2016 between 5:34 and 6:01 UTC. Local time is UTC + 2 hours.

Concentrations of NO_2 are highest at low altitudes and decrease with increasing altitude. Above the planetary boundary layer (~1000 m) concentrations approach zero. This is as expected, as most sources are at or near the ground. The high concentration layer at 500 m altitude is therefore unexpected. It is, however, consistently found in other time intervals as well during several hours. An independent alternative instrument, a balloon sonde launched that day, also measures this layer.

3.2 All profiles on one day

The development of the concentration during 15 September 2016 is shown in Figure 5. The development of the layer around 500 m is clearly visible.

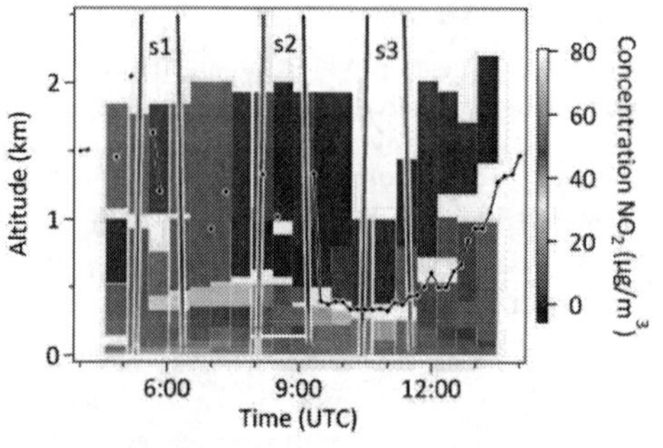

Figure 5. Example day: Thursday 15 September 2016. The traces marked s1 to s3 are balloon sonde launches on that day (see text). The black points indicate the height of the planetary boundary layer, retrieved from a ceilometer on site. Local time is UTC + 2 hours.

Figure 6 shows the same day, but zoomed in on the lower altitudes during the morning. The development of the boundary layer is clearly visible, as is the influence of the weekday morning rush hour peak on the NO_2 concentrations. Note that sunrise on this day was at 5:17 UTC.

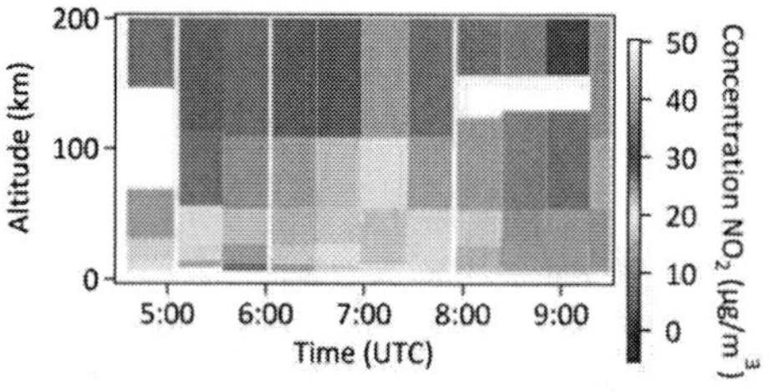

Figure 6. Example day: Thursday 15 September 2016, zoomed in.

3.3 Comparison with balloon sonde

In Figure 5, the traces marked s1, s2 and s3, show the lower part of the data of the three balloon sonde launches of this day. The colours of the traces indicate the concentrations measured by the sondes at the respective times and altitudes. Note that the sondes covered a considerable distance. When they measured the second part of the traces, they were located about 30 km from Cabauw.

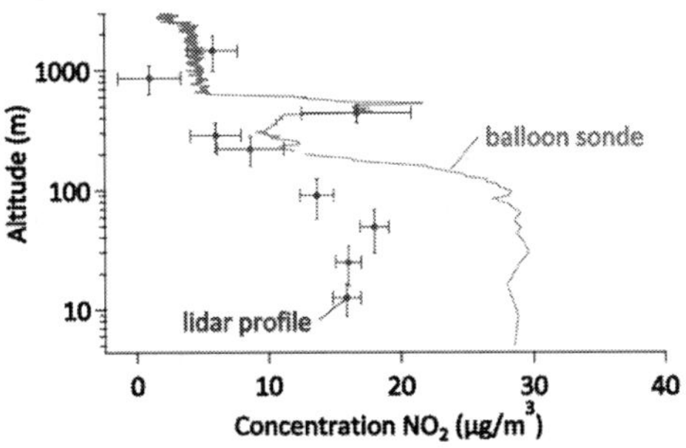

Figure 7. Lidar profile, measured on 15 September 2016 between 7:56 and 8:24 UTC, and balloon sonde, measured between 8:04 and 8:11 UTC.

Figure 7 shows the Lidar profile measured between 7:56 and 8:24 UTC, and the profile from the ascending balloon sonde, measured between 8:04 and 8:11 UTC. Lidar and sonde measure different concentrations, but the shapes of the profiles are similar. Especially striking is how well both instruments agree on the presence of the layer of elevated concentrations around 500 m altitude.

3.4 All profiles from the campaign.

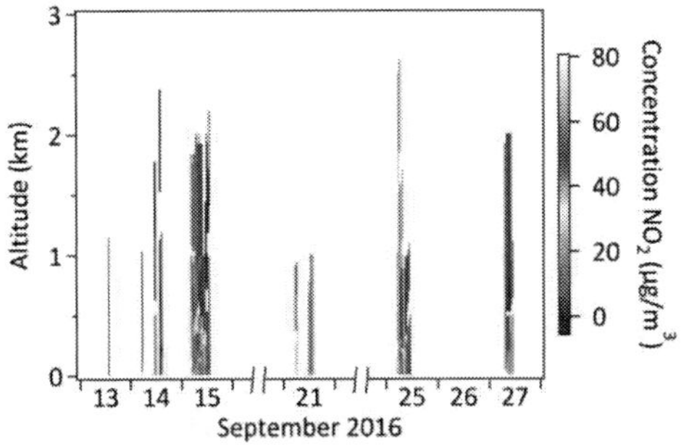

Figure 8. All profiles of the campaign.

All profiles of the campaign are shown in Figure 8. The profile of 21 September 2016 shows a similar patterns as 15 September, but with higher concentrations in the morning. On 27 September the pattern is similar as well, but the development of the rush hour peak is delayed somewhat. On 25 September, concentrations remain much lower all day long. The gaps in the day profiles on 14 and 21 September are due to instrumental problems, the days without measurements are due to unfavourable weather conditions.

4 CONCLUSIONS

The lidar system is well suited to mapping the atmospheric profile of NO_2. The development of the atmospheric boundary layer and the dynamics of the NO_2 concentration are clearly shown. At the time of writing, comparisons with only one other instrument were made, more intercomparisons will follow.

ACKNOWLEDGEMENTS

The authors wish to thank CINDI-2 organising committee for their work on this campaign. They also wish to thank Henk Klein Baltink (KNMI) for retrieving the planetary boundary layer height.

The CINDI-2 campaign was funded by the Netherlands Space (NSO) and the European Space Agency (ESA).

References

[1] Piters, A.J.M., K.F. Boersma, *et al.*, *The Cabauw Intercomparison campaign for Nitrogen Dioxide measuring Instruments (CINDI): Design, execution, and early results.* Atmospheric Measurement Techniques, 2012. **5**(2): p. 457-485.

[2] Volten, H., E.J. Brinksma, *et al.*, *NO2 lidar profile measurements for satellite interpretation and validation.* Journal of Geophysical Research D: Atmospheres, 2009. **114**(24).

[3] Berkhout, A.J.C., D.P.J. Swart, G.R. van der Hoff, and J.B. Bergwerff, *Sulphur dioxide emissions of oceangoing vessels measured remotely with Lidar.* RIVM, Bilthoven, Research report 609021119, 61 pp., 2012.

[4] Berkhout, A.J.C., G.R. van der Hoff, *et al.*, *Measuring ammonia emissions from manured fields.* RIVM, Bilthoven, Research report 680150003, 80 pp., 2008.

[5] Berkhout, A.J.C., G.R. Van Der Hoff, and L.F.L. Gast. *Lidar Measurements of Industrial Benzene Emissions.* in *27th International Laser Radar Conference, ILRC 2015.* 2016. EDP Sciences.

[6] Sluis, W.W., M.A.F. Allaart, A.J.M. Piters, and L.F.L. Gast, *The development of a nitrogen dioxide sonde.* Atmospheric Measurement Techniques, 2010. **3**(6): p. 1753-1762.

AEROSOL ABSORPTION MEASUREMENTS AND RETRIEVALS IN SHADOW2 CAMPAIGN

Qiaoyun HU[1*], *Philippe GOLOUB*[1], *Thierry PODVIN*[1], *Igor VESELOVSKIY*[2,3], *Anton LOPATIN*[1], *Oleg DUBOVIK*[1], *Benjamin TORRES*[4], *Laura REVILINI*[1], *Suzanne CRUMEYROLLE*[1], *Tatsiana LAPIONAK*[1], *Christine DEROO*[1]

[1] Laboratoire d'Optique Atmosphérique, Université de Lille 1, Villeneuve d'Ascq, France
[2]Physics Instrumentation Center of GPI, Troitsk, Moscow, Russia
[3]Joint Center for Earth Systems Technology, UMBC, Baltimore, MD, USA
[4]GRASP-SAS-SME, Lille, France
**Email:* qiaoyun.hu@ed.univ-lille1.fr

ABSTRACT

Dust, maritime and dust-smoke mixture events observed during SHADOW2 (**SaHAran Dust Over West Africa**) field campaign are selected and analyzed by using Raman and GARRLiC retrievals.

The derived aerosol optical and microphysical properties will be shown. Dust absorption profile and on ground level are derived from GARRLiC retrievals and Aethalometer measurements, respectively. Our results provide a closer insight about dust absorbing properties.

1 INTRODUCTION

Mineral dust is the most abundant constitute in the atmosphere. Originated mostly from the arid and half-arid areas, every year large amounts of mineral dust are emitted into the atmosphere. Mineral dust plays a crucial role in the radiative forcing because it absorbs and scatters the solar radiation thus directly influencing the radiation budget. Additionally, mineral dust may change the cloud condensation nuclei process, thus affecting indirectly the radiation arriving to the ground [1].

Plenty of researches have focused on the quantification of the absorption of mineral dust, which is a key factor for estimating its radiative forcing effect. Generally, aerosol absorption is obtained from laboratory measurements or retrievals of remote sensing measurements in ambient air. Both techniques have their limitations. The widely used Aethalometer for measuring absorption coefficient has drying effects on the particles that are taken into the chamber. And the high irregularity of the shape of mineral dust poses challenges to the modeling [2], thus resulting in uncertainties for the retrieval.

To investigate the mineral dust properties near the main source, the SHADOW2 campaign was conducted at M'bour, Senegal, in March, April, December 2015 and January of 2016. Both in-situ and remote sensing instruments are deployed in the campaign; airborne measurements were performed as well, so that intensive parameter sets were achieved.

In this paper, we select two typical and representative cases; analyze the measurements from Multi-wavelength Raman Lidar (LILAS) and Aethalometer. Optical properties from Raman retrievals, absorption coefficients (ABS) and SSA (**S**ingle **S**cattering **A**lbedo) derived from Aethalometer (ABS) and GRASP/GARRLiC (ABS and SSA) retrieval are shown and analyzed. The results show the potentials and our first try to quantify the absorption of mineral dusts.

2 INSTRUMENT&METHODOLOGY

M'bour site is affiliated to the AERONET network so the standard AERONET protocol is performed regularly. Apart from the sun photometer, multiple instruments were deployed in the campaign. Nephelometer and Aethalometer measured the scattering coefficient and absorption coefficient, respectively. It should be noted that, in the campaign, the Aethalometer was equipped with a sampling head that took only particles with diameter of 1 μm or smaller. The OPC (**O**ptical **P**article **C**ounter) and GRIMM measured the particle size distribution on the ground level and also aloft when airborne measurements were conducted. Two Lidars: LILAS and a single wavelength CIMEL Lidar were in operation to monitor the aerosol temporal and vertical variations. Other instruments, such as wind Lidar, Fluxmeter, Radiometer, etc. were also operating

 https://doi.org/10.1051/epjconf/201817610003

simultaneously. And here we emphasize the measurements from Aethalometer and LILAS.

The Aethalometer measures the absorption at 7 wavelengths: 370, 470, 520, 590, 660, 880 and 950 nm. LILAS has three emitting wavelengths: 355, 532 and 1064 nm. The receiving wavelengths consist of three elastic channels that are the same as the emission and three Raman channels: 387, 530 and 408 nm. In addition, 355 nm and 532 nm channels are operated with both parallel and perpendicular channels, which enable to measure the linear depolarization ratio of aerosols at both wavelengths.

GRASP/GARRLiC retrieval uses the AOD/TOD and Almucantar measurements from sun photometer (at 440, 670, 870 and 1020 nm) and 3 Lidar profiles (elastic channels) from LILAS. The retrieved parameter sets are aerosol size distribution, aerosol vertical concentration, and complex refractive indices (called CRI hereafter) of both fine and coarse mode aerosols. The vertical absorption profile can be calculated from the aforementioned parameters. To show the comparison between the Aethalometer measurements, the absorption of fine mode particles is specifically calculated in the case studies.

3 CASE STUDY

In March and April 2015, dust was the most frequently observed aerosol type, and maritime aerosols and smoke occurred from time to time. Whereas, mixture of dust and smoke occurred relatively more frequent in December 2015 and January 2016.

3.1 11 April, 2015

Figure 1 shows the extinction, backscattering (Figure 1(a)) and particle linear depolarization ratio (called PDR hereafter) profiles retrieved from Lidar measurements for the time interval of 00:00-01:30 UTC, 11 April 2016. The AOD during this period is around 1.0 at 532 nm. The extinction profile does not show obvious spectral dependence. Whilst, the backscatter coefficient at 532 nm is obviously stronger in amplitude compared with 355 nm. The PDR is of values between 30% and 35%, which indicates the presence of irregular particles. Additionally, the Ångström Exponent is -0.2 ~ 0 (not shown in the figure), from which we can tell there are bigger particles in the atmosphere. Relative humidity (RH) profile increases slightly with height. An insignificant peak is observed at 2.0 km. It needs to be mentioned that, the RH is calibrated by radio sounding measurements in Dakar, which is not collocated with Lidar, so that the absolute values of RH may not be accurate, but the variation of the its amplitude represent qualitatively the vertical and temporal variations of RH. The back trajectory (Figure 2(a)) shows the air mass at 2 km and 3 km comes from Mauritania and Mali. Around 0.8 km, the air mass originated from the North Atlantic Ocean, and the air mass crossed Western Sahara and Mauritania before reaching the site.

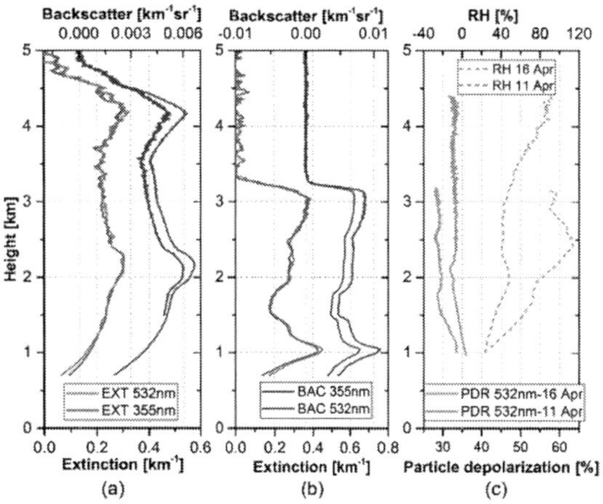

Figure 1. (a) Extinction and backscatter profiles retrieved from LILAS. Profiles are averaged between 00:00-01:30 UTC, 11 April 2015. (b) Similar plots to (a), but profiles are averaged between 05:00-06:00 UTC, 16 April 2015. (c) PDR and RH profiles corresponding to the time intervals in (a) (orange) and (b) (cerulean).

3.2 16 April, 2015

Maritime aerosols mixed with dust were observed on 16 April 2016. For the time interval of 05:00-06:00, the AOD at 532nm is around 0.9. The extinction profiles show very weak spectral dependence, like in the previous case. However, the backscatter coefficient at 355 nm is augmented and stronger than that at 532 nm. The PDR at 532 nm is ~ 27% and Ångström Exponent varies from 0 to 0.1, between 1.2 km and 3.2 km.

Below 1.2 km, the PDR is between 30% and 35% and Ångström Exponent between -0.1 and 0. This difference suggests different aerosol types in the two layers. The RH of 16 April is much higher compared with 11 April and increases with height.

Referring to the back trajectories, in the upper layer air mass came originally from the surface of North Atlantic Ocean and then landed from Western Sahara. In the lower layer, air mass came from the continent, which is usually transported dust.

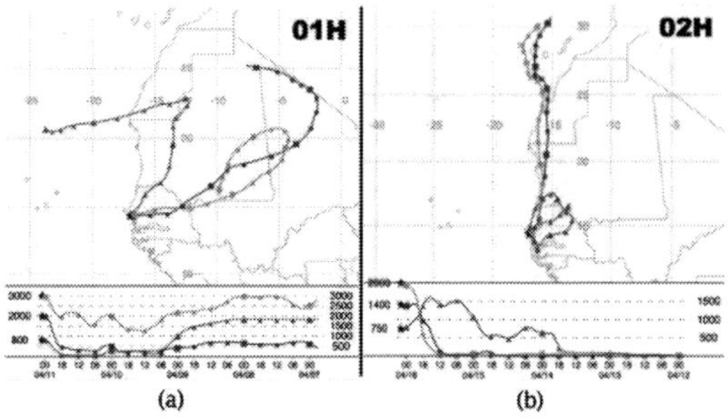

Figure 2. 100-hour back trajectories at (a) 01:00 UTC, 11 April 2015 and (b) 02:00 UTC, 16 April 2015.

4 GRASP/GARRLiC RETRIEVAL

GRASP/GARRLiC retrievals are performed around 09:30, 11 April 2015 and 08:02, 16 April 2015, and about half an hour Lidar profiles are averaged as the input of GRASP/GARRLiC.

4.1 Size distribution and concentration

Figure 3(a) shows a bimodal size distribution with a pronounced coarse mode in both cases. For 16 April, the median sizes of coarse and fine mode are 1.93 μm and 0.11 μm. The ratio of concentration is: $C_{vc}/C_{vf} \approx 8.74$. For 11 April, the median sizes of coarse and fine mode are 2.44 μm and 0.12 μm, respectively. The ratio of concentration: $C_{vc}/C_{vf} \approx 12.59$. It is noteworthy that, from 04:00 UTC, 16 April, the wind direction changed from northerly to northeasterly, thus bringing more dust aerosols, which may, to a certain extent, dilute the marine aerosols. The insert of Figure 3 shows the vertical concentration of fine and coarse mode aerosols. On 11 April, the aerosol layer is up to 4.5 km and most aerosols concentrated below 3 km on 16 April. As we have mentioned that, on 16 April, the aerosol type below 1.2 km differs from the upper layer, since continental wind injected to the lower layer may bring more dust particles. The retrieval shows a sharp peak of the concentration of coarse mode at ~0.9 km, this could qualitatively support our inference.

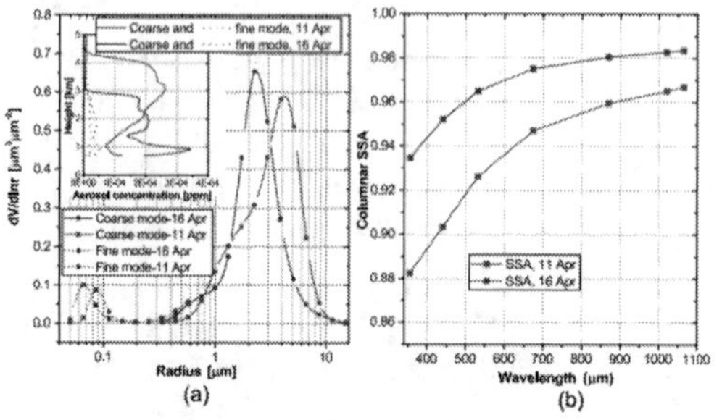

Figure 3. (a) Size distribution and aerosol vertical concentration (insert) retrieved from GRASP/GARRLiC and (b) Columnar SSA from GRASP/GARRLiC for 16 April (blue) and 11 April retrieved (red).

4.2 Single Scattering Albedo and absorption

In Figure 3(b), the columnar SSA increases with wavelength in both cases, and on 16 April, aerosols are apparently less absorbing than on 11 April. Studies have revealed that clean maritime aerosols have much lower absorption than mineral dust [3]. When mixed with each other, the optical properties of the mixture depend on the fraction of each component and the mixing mode as well. Previous researchers also obtained a lifted SSA of mineral dust when mixed with maritime aerosols [4]. The profile of SSA (Figure 4(a)) has no obvious vertical variation for both cases. A minor increase of SSA at 355 nm and 532 nm on 11 April is due to the change of the proportion of fine and coarse mode.

The maximum absorption appears at ~0.8 km and ~3.0 km for 11 April and 16 April, respectively. Above 1 km, the absorption on 11 April is 1~2 times stronger than 16 April. Around 2 km, the particle concentration in the two cases is close, however, the absorption on 11 April is ~30% higher than that on 16 April. Below 1 km, the particle concentration of 16 April exceeds that on 11 April, and the absorption is superior.

The fine mode absorption profiles are plotted to compare with the in-situ absorption, since the

Aethalometer samples particles with diameter of 1 μm or smaller. On 11 April, the absorption of fine mode particles is very low and shows almost no spectral dependence, while the fine mode absorption on 16 April is 2-3 times stronger and has a more pronounced spectral dependence. Similar tendencies are also observed from the in-situ measurements.

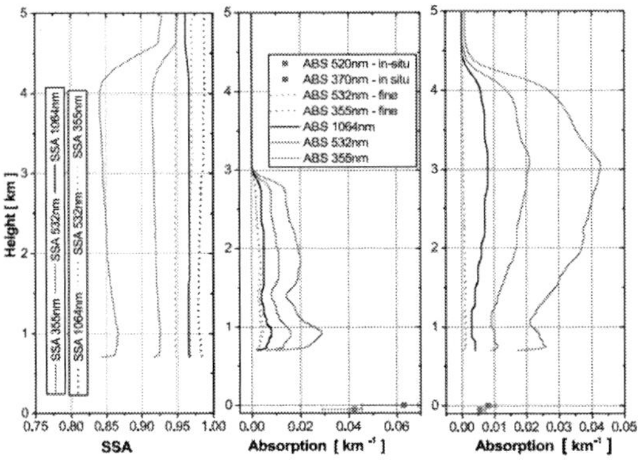

Figure 4. (a) SSA profile for 11 April (solid line) and 16 April (short dashed line). (b) Absorption coefficient profile for 16 April. (c) The same plot as (b), but for 11 April. In (b) and (c), the points on the ground level represent the in-situ Aethalometer measurements.

4.3 Complex Refractive Indices

The real parts of CRI on 11 April are higher than 16 April (Figure 5). The reported values of real part of CRI vary with the origins of mineral dust. For instance, 1.55 ± 0.03 is reported for Bahrain-Persian Gulf, 1.48 ± 0.05 for Cape Verde. Here, we retrieved:

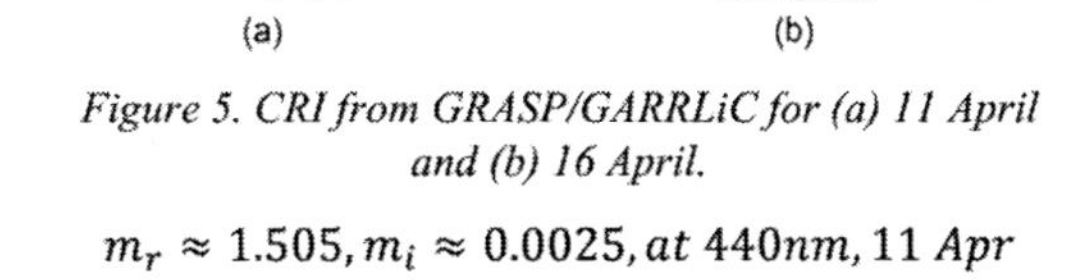

Figure 5. CRI from GRASP/GARRLiC for (a) 11 April and (b) 16 April.

$$m_r \approx 1.505, m_i \approx 0.0025, at\ 440nm, 11\ Apr$$

$$m_r \approx 1.417, m_i \approx 0.0008, at\ 440nm, 16\ Apr$$

On 16 April, aerosols are 'softer' and less absorbing, which is in agreement with our inference that dust and maritime aerosols are observed. Due to the low concentration of fine mode particles, the retrieved fine mode properties may have high uncertainties.

5 Conclusions

This study shows the absorption measurements and retrievals for selected two cases, derived columnar and vertical resolved SSA, absorption coefficients and ground-level absorption are shown and compared. The results are qualitatively in agreement with the expecting aerosol properties and suggest the potential of quantifying the vertical absorption. Uncertainties and validation still need to be studied carefully in future work.

The authors acknowledge the use of GRASP inversion algorithm software and the support from CaPPA project, ESA and H2020/ACTRIS2.

Reference

[1] Haywood, J., & Boucher, O. (2000). Estimates of the direct and indirect radiative forcing due to tropospheric aerosols: A review. Reviews of geophysics, 38(4), 513-543.

[2] Veselovskii, I., Dubovik, O., Kolgotin, A., Lapyonok, T., Di Girolamo, P., Summa, D., ... & Tanré, D. (2010). Application of randomly oriented spheroids for retrieval of dust particle parameters from multiwavelength lidar measurements. Journal of Geophysical Research: Atmospheres, 115(D21).

[3] Dubovik, O., Holben, B., Eck, T. F., Smirnov, A., Kaufman, Y. J., King, M. D., ... & Slutsker, I. (2002). Variability of absorption and optical properties of key aerosol types observed in worldwide locations. Journal of the atmospheric sciences, 59(3), 590-608.

[4] Cattrall, C., Carder, K. L., & Gordon, H. R. (2003). Columnar aerosol single - scattering albedo and phase function retrieved from sky radiance over the ocean: Measurements of Saharan dust. Journal of Geophysical Research: Atmospheres, 108(D9).

LIDAR PROFILING IN THE LOWER TROPOSPHERE: EXPERIENCE FROM PECAN

BELAY B. DEMOZ, RUBEN DELGADO, BRIAN CAROLL, KEVIN VERMEESCH
Joint Center for Earth Systems Technology, University of Maryland Baltimore County, Baltimore, MD 21250 USA

DAVID N. WHITEMAN
NASA/Goddard Space Flight Center, Greenbelt, MD 20771 USA

RICARDO SAKAI, SIUM TESFAY, LORENZA COOPER
Howard University, Washington, DC USA

ABSTRACT

Results from the PECAN (Plains Elevated Convection at Night) campaign are discussed. In particular, the utility of simple backscatter lidars/ceilometers in quantifying atmospheric dynamics parameters and variables as well as evolution of the lower tropospheric dynamics are made. Cases of bore wave dynamics and the potential of these events in lofting of low level, moist, airmass and its consequence in thunderstorm initiation are made. A suite of thermodynamic profiling instruments are combined and compared to describe and visualize lower tropospheric dynamic evolution.

1. Introduction

PECAN (Plains Elevated Convection at Night) campaign, designed to advance the understanding of "continental, nocturnal, warm-season precipitation" in the Southern Great Plains, brought together a suite of remote sensing and in-situ instruments. Although PECAN's focus was on nocturnal convection during stable boundary layer (SBL) conditions with large CAPE (Convectively Available Potential Energy) located above the SBL and when nocturnal low-level jets (NLLJ) are present, it offered a good opportunity for inter-comparison of lidar-radar-microwave and other profilers and assessment of their utility in atmospheric dynamics studies. PECAN's focus in these conditions stems from the fact that in this region thunderstorms are most common after sunset, and much of the heavier precipitation resulting from a variety of mesoscale convective systems (MCSs) tends to be correlated with the intensity of the NLLJ (e.g., Arritt et al. 1997; Tuttle and Davis 2006). An accurate prediction and an in-depth understanding of nocturnal convection in this region are lacking. In particular, the SBL structure leading to the initiation of nocturnal elevated convection (~900 – 600 hPa) is poorly understood and not well observed (e.g., Wilson and Roberts 2006). Operational observation of the thermodynamics of this atmospheric region is then better done with profilers – lidar being among the primary instruments.

Several Fixed PECAN (FP) Sites were deployed during the experiment. We report on data from a single site in Greensburg, Kansas, referred to as Fixed PECAN site #2 (FP2). FP2 served as a hub for Profiling Microwave Radiometers, wind and water vapor lidars, radio sounding unit, a surface station, and a network of ceilometers. FP2 anchored the southwest part of the instrumentation for the PECAN domain. The instrumentation was organized by teams from Howard University, NASA/GSFC and UMBC as well as other collaborators. In addition, the Naval Postgraduate School (NPS), located in Monterey, CA, also deployed a 16 m main tower, a 6 m and 3 m supporting towers, a SODAR, a tethered balloon, and ceilometers. Moreover, NOAA/NSSL deployed an AERI system for part of the experiment.

We will discuss lidar and collaborating data from selected cases and demonstrate the utility of even the less sophisticated ceilometer-lidar in investigating storm and atmospheric dynamics.

2. Undular Bore: 3 June 2015

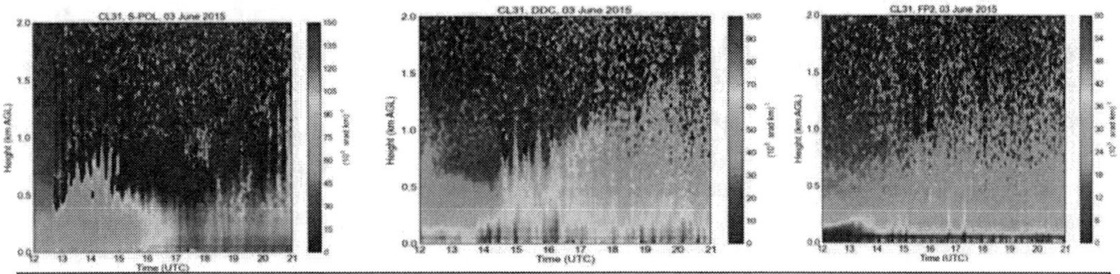

Figure 1: A plot of the aerosol backscatter profile obtained over three different sites by Vaisala CL31 ceilometer during the June 3, 2015 PECAN case over southwest Kansas, KS.

A wave train of roll clouds was visible early morning covering the entire southwest Kansas using satellite and visual (photos) observations (not shown). These waves resemble the classic bore-soliton conditions described by Christie (1989). The water vapor density, temperature and relative humidity at FP2 as well as aerosol backscatter signatures from three of the PECAN network sites (located at the central radar location SPOL, Dodge city, Kansas, and FP2) for this case are shown in fig.1 above. The figures show the following wave characteristics (i) a progressively decaying wave pattern, undulations barely visible over the FP2 site but a strong cloud signature; (ii) about 100 minutes difference between the passage of the wave at the SPOL site and Dodge City, Kansas with additional 40min delay before reaching FP2; and (iii) the lift associated with the first cloud was higher at FP2 than the other sites despite the fact that the wave undulations were decreasing and dying; an indication of the role played by the evolving early morning boundary layer. In these plots, note that the backscatter false color depictions shown are relative and that scaling of individual profiles should not be compared. Although the manufacturer of these ceilometers has a calibration technique that attempts to correct and quantify the profiles and provide outputs in Str-1 per km^{-1}, we have not found an independent confirmation of the validity of this calibration. There are however independent comparison experiments of one or more manufacturers of such lidars and we hope results will be published soon discussing the quantitative value of the profilers. This doesn't diminish the value of the relative measurements in the study of atmospheric dynamics, as being attempted in this work.

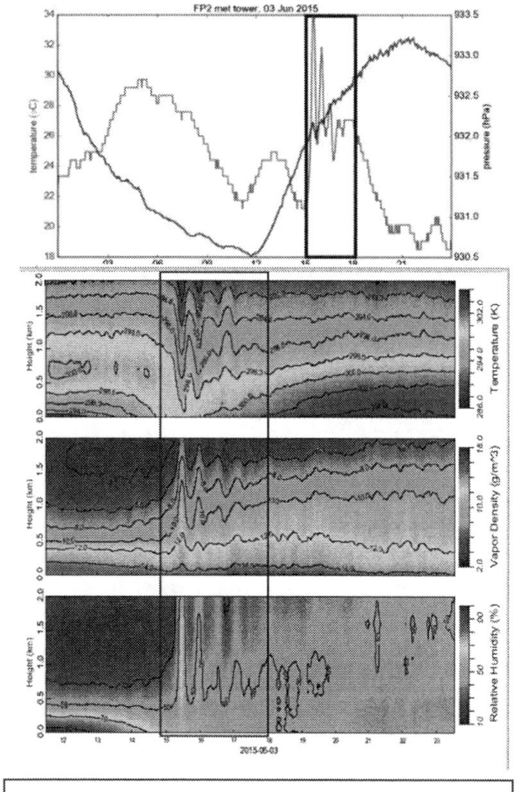

Figure-2: (A) Surface temperature and pressure at FP2 on 3 June 2015. Corresponding profiles of microwave measured Temperature, vapor density, and relative humidity are also shown. Passage time of the bore wave is indicated by black-box on the plots.

In figure-2, we present a detailed look at the undulations captured by the CL31 lidar at FP2 (the weakest aerosol signature of the three sites in fig.1) and the resulting thermodynamic signatures as captured by surface meteorological stations and captured by the microwave radiometric profilers. Shown in fig.2 are plots for the water vapor density (g/m^3), Temperature (K) and relative Humidity (%). Note that the seemingly weak perturbations captured by a simple ceilometer are an indication of intense water vapor and temperature modulations caused by the strong atmospheric dynamics of the bore/soliton on this day. This moisture-temperature distribution is key to understanding storm initiation and moistening of the upper tropospheric region. The pressure perturbations and subsequent atmospheric oscillations in temperature, moisture and aerosol backscatter profiles in this case reveal the entire sub-cloud layer. Surface signatures also are clearly revealed in the pressure-temperature perturbations and thus the dynamics has a clear surface signal and are hence not of an "elevated" nature. This is not always true – there are cases where the strong dynamic oscillations in the thermodynamic profiles are less evident or "weak at best" in the surface signal.

Such an example is a case of an undular bore that occurred on 14 July 2015. On this date, shown in Fig. 3, the wave oscillation occurred at night where there was no developing boundary layer. This case also had two periods of pressure jumps, at 5:30UTC and 7:00 UTC. The magnitudes of the pressure jumps recorded were similar but the resulting atmospheric water vapor perturbation was very different. As can be seen from figure 3, the low level (below 1km) water vapor mixing ratio and temperature did not show that strong perturbations, but the upper level 2-3km water vapor wave reveal a strong dynamics. This case revealed a very strong and persistent lofting of the warm-moist near-surface airmass up to almost 3km in elevation. This lifting of the moist layer was much larger than the previous case discussed (~0.5km).

These two cases did not initiate a lasting convection, a subject of PECAN, but fall in the continuum of bore-wave dynamics that were observed. During PECAN a case was observed (not shown here) that was likely part of a bore-wave system that initiated severe convection. A detailed presentation of these two cases and the additional case that may have triggered an intense convection during PECAN will be discussed.

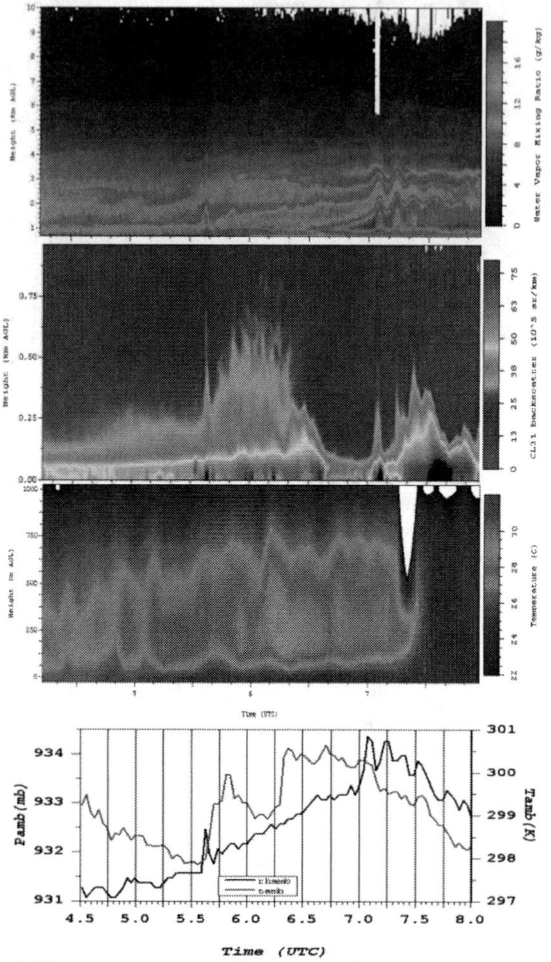

Figure-3: Profiles of (A) water vapor mixing ratio measured by the NASA/GSFC ALVICE Raman lidar (B) Vaisala CL31ceilometer measured backscatter (C) Temperature profile measured by the MTP and (D) Surface temperature and Pressure made at FP2 on 14 July 2015.

In addition to the discussion above and a description of the PECAN ceilometer network design, a progress report on the NOAA-NWS effort to create a national network of ceilometers from the Automated Surface

Observing System (ASOS) network (which consists of close to 1000 lidar-ceilometers nationwide) will be presented and reported.

3. LLJ

An additional objective of the PECAN experiment was investigation of the character and evolution of the plains low-level jet (LLJ). The LLJ is defined simply as lower-tropospheric maximum in the vertical profile of the horizontal wind, and occurring in the boundary layer at night. Of practical interest is their impact on the transport of moisture although it should be noted that a theoretical framework explaining them is still not quite settled yet (Du and Rottuno, 2014; Shapiro et al. 2016). Notable is the large amount of vertical wind shear associated with LLJ and the observation that they are typically super-geostrophic by a large (>50%) amount with no appreciable vertical wind. There are a number of specific geographic locations that have been identified as especially favourable for LLJ development and their impact. The Southern Great Plains (SGP) region of the United States is one such region where the LLJ events have been studied extensively because of their impact on the development of nocturnal severe storms and their role in precipitation (Tollerud et al., 2004, Geerts et al. 2016).

During PECAN, several cases of LLJ were observed at FP2 and the distributed network stations. A summary of findings and progress in analysis to date of lidar observations made will be made.

4. Conclusions

A summary of FP2 measurements and lidar instrumentation during the PECAN experiment is discussed. Utility of simple backscatter lidars, CL31 Vaisala ceilometer, were used to diagnose lower atmosphere dynamics. In particular, the pressure perturbations that would result in water vapor lofting and thus may lead to convection initiation and often referred to as bore-solitary waves are discussed. Detailed data from two case studies, including surface meteorology, co-located microwave radiometers, Raman and Doppler wind lidars are presented. Preliminary findings indicate that the ceilometer profile data is an excellent indicator of the tropospheric dynamics. When combined with surface meteorology data, these ceilometers could lead to a useful documentation and statistics of the occurrence of bore events. These PECAN data sets are being used as a demonstration project for the proposed NOAA/NWS – ASOS ceilometer network. A detailed discussion and report of these and related activity will be presented at the meeting.

5. Acknowledgement

This research was supported in part by NSF (Award Number 1503563) and NASA Earth Science Directorate's weather research Program under the direction of Dr. Jack Kaye.

6. References

Geerts, B., et al., 2016: The 2015 Plains Elevated Convection At Night (PECAN) field project. *Bull. Amer. Meteor. Soc.* doi:10.1175/BAMS-D-15-00257

Tollerud et al., 2004: Multiscale analyses of moisture transport by the central plains low-level jet during IHOP. *20th Conf. On Weather Analysis and Forecasting, Seattle, WA, Amer. Meteor. Soc.*, CD-ROM, P4.21.

Shapiro, A., E. Fedorovich, and S. Rahimi, 2016: A unified theory for the Great Plains nocturnal low-level jet. *J. Atmos. Sci.*, 73, 3037-3057.

Du Y., and R. Rotunno (2014): A simple analytical model of the nocturnal low-level jet over the Great Plains of the United States. *J. Atmos. Sci.*, 71, 3674-3683.

Arritt, R. W., T. D. Rink, M. Segal, D. P. Todey, C. A. Clark, M. J. Mitchell, and K. M. Labas, 1997: The Great Plains Low-Level Jet during the Warm Season of 1993. *Mon. Wea. Rev.*, 125, 2176–2192.

Tuttle, J. D., and C. A. Davis, 2006: Corridors of warm season precipitation in the central United States. *Mon. Wea. Rev.*, 134, 2297–2317.

Wilson, J. W. and R. D. Roberts, 2006: Summary of convective storm initiation and evolution during IHOP: observational and modeling perspective. *Mon. Wea. Rev.*,134, 23-47.

Christie, D. R., 1989: Long nonlinear waves in the lower atmosphere. *J. Atmos. Sci.*, **46,** 1462–1491.

EPJ Web of Conferences **176**, 10005 (2018) https://doi.org/10.1051/epjconf/201817610005
ILRC 28

AEROSOL PROFILING DURING THE LARGE SCALE FIELD CAMPAIGN CINDI-2

Arnoud Apituley[1,*], Michel Van Roozendael[2], Andreas Richter[3], Thomas Wagner[4], Udo Friess[5], Francois Hendrick[2], Karin Kreher[6], Jan-Lukas Tirpitz[5] and CINDI-2 Participants[7]

[1]*Royal Netherlands Meteorological Institute (KNMI), The Netherlands, *apituley@knmi.nl*
[2]*BIRA-IASB, Brussels, Belgium*
[3]*IUP-Bremen, Germany*
[4]*MPIC, Mainz, Germany*
[5]*IUP-Heidelberg, Germany*
[6]*Bodeker Scientific, Alexandra, New Zealand*
[7]*Various countries world-wide*

ABSTRACT

For the validation of space borne observations of NO_2 and other trace gases from hyperspectral imagers, ground based instruments based on the MAXDOAS technique are an excellent choice, since they rely on similar retrieval techniques as the observations from orbit. To ensure proper traceability of the MAXDOAS observations, a thorough validation and intercomparison is mandatory. Advanced MAXDOAS observation and retrieval techniques enable inferring vertical structure of trace gases and aerosols. These techniques and their results need validation by e.g. lidar techniques.

For the proper understanding of the results from passive remote sensing techniques, independent observations are needed that include parameters needed to understand the light paths, i.e. in-situ aerosol observations of optical and microphysical properties, and essential are in particular the vertical profiles of aerosol optical properties by (Raman) lidar.

The approach used in the CINDI-2 campaign held in Cabauw in 2016 is presented in this paper and the results will be discussed in the presentation at the conference.

1 INTRODUCTION

Ground based and spaceborne hyperspectral instruments for observations of NO_2 and other trace gases rely on similar retrieval techniques. In both cases, retrievals take into account the light path of scattered sunlight though the entire atmosphere (see Fig.1). In case of high aerosol load close to the surface, this will dominate the light path and therefore bias the sensitivity of the measurement to this part of the atmosphere. Since MAXDOAS instruments are relatively low cost and can be operated autonomously almost anywhere, they are credible candidates to form a world-wide ground based reference network for satellite observations. To ensure proper traceability of the MAXDOAS observations, a thorough intercomparison is mandatory.

With the imminent launch of Sentinel-5 Precursor/TROPOMI (6), with a nadir pixelsize of 3.5 × 3.5 km^2, and recent developments in MAXDOAS instruments there was a need for

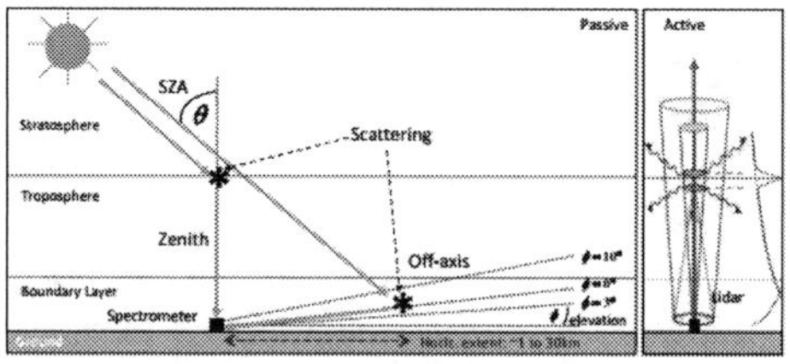

Figure 1: Schematic of light paths for MAXDOAS (left) and for lidar (right). Note that the effective footprint of the passive MAXDOAS observation extends multiple kilometers (similar to satellite observations), while the lidar footprint is much smaller.

https://doi.org/10.1051/epjconf/201817610005

Figure 2: Aerial view of the CESAR site near Cabauw, showing the location of the main instrumentation of the campaign.

CINDI-2. This campaign was completed in September 2016 and had the goals to: 1) Asess the consistency of slant column measurements of key target species (NO_2, O_3, O_4 and HCHO) relevant for the validation of S5P and the future ESA atmospheric Sentinels, from a large number of DOAS and MAXDOAS instruments from all over the world, and 2) study the relationship between remote-sensing column and profile measurements of those species and reference measurements of the same species.

Figure 3: Picture of the CINDI-2 observation site. The MAXDOAS instruments were placed on the roof of temporary containers situated at the Remote Sensing Site of the CESAR Observatory where the Raman lidar and ceilometer are permanently located.

2 CAMPAIGN DESIGN

A feature of recent MAXDOAS developments is the use azimuthal scanning, in addition to elevation scanning such as in e.g. the PANDORA type of instruments (3). This, and the number of participating instruments, that expanded to 42, posed a challenge to the design of the CINDI-2 campaign.

The Cabauw Experimental Site for Atmospheric Research (CESAR) site in centre of The Netherlands (1) was the stage of the Cabauw Intercomparison of Nitrogen Dioxide Measuring Instruments (CINDI) in June-July 2009 (4) and again for the second campaign, CINDI-2, in 2016. Cabauw was chosen because the flat terrain offers a free view of large parts of the horizon, needed to accommodate the viewing geometry of the MAXDOAS observations. The location is under influence of both clean as well as polluted airmasses. This gives a wide range of possible trace gas concentrations and mixtures.

Furthermore, at CESAR a wide range of observations are routinely carried out that fulfil the requirement to provide the background necessary for unraveling the differences between the observations from different MAXDOAS instruments that can be quite diverse in design and data treatment. These observations include parameters needed to understand the light paths, i.e. in-situ aerosol observations of optical and microphysical properties, as well as vertical profiles of aerosol optical properties by (Raman) lidar (2). In addition, vertical profiles of NO_2 were provided by the unique NO_2 sonde (5), and a NO_2 lidar system (7). In situ observations of aerosol scattering and absorbing properties, as well as microphysical properties were made in collaboration with the ACTRIS-2 project. The placement of the instruments on the Cabauw site is schematically shown in Fig.2 and a picture of the remote sensing site is shown

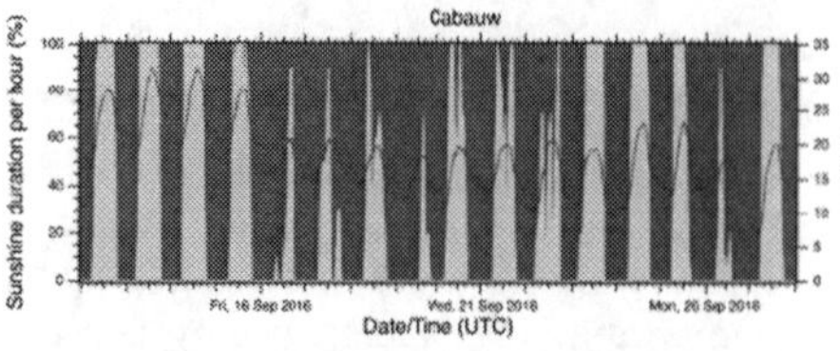

Figure 4: General meteorological conditions during the campaign main intercomparison phase between 12 and 28 Sept. 2016.

in Fig.3. Here, we focus on the aerosol profile. The NO_2 profiles are discussed in a paper by S. Berkhout at ILRC28.

3 FIRST RESULTS

Due to favourable weather conditions for making MAXDOAS and lidar observations, i.e. absence of clouds as shown in Fig.4, a very interesting dataset has been collected, where most of the instruments were able to provide data.

An example of an aerosol profiling case is shown in Fig.5 and 6. Fig.5 shows the (preliminary) MAXDOAS aerosol extinction profiles retrieved for 477 nm. The extinction peaks between 0.5 and 1.5 km above ground while some structure in the aerosol profile is present for the results from all groups below 1 km. The peak value of the extinction is about 0.3 km^{-1}. Lidar data for the same day are shown in Fig.6, where the top panel shows the 24-h aerosol profile from the Lufft CHM15k ceilometer and the bottom left in the figure shows the Raman lidar range corrected signal. The lower right shows the Raman lidar backscatter and extinction profiles, as well as the lidar-ratio and Ångström coefficient. From the lidar extinction profile at 355 nm it can be seen that extinction profile peaks at $\approx$ 2.5 km at a value of $\approx$ 0.3 km^{-1}. So, while the peak extinction values are very similar, the shape of the profile shows marked differences. The analysis of the CINDI-2 data is ongoing and the latest stage of development will be shown at the conference.

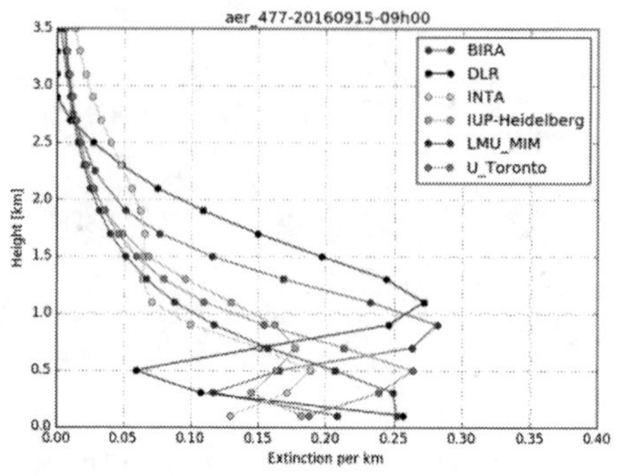

Figure 5: Example of MAXDOAS aerosol profile retrieval at 477 nm for 15 Sept. 2016 around 9 UTC from various participating groups. Note that these are preliminary results.

4 CONCLUSIONS

The CINDI-2 campaign conducted in Sept. 2016 in Cabauw was successful due to well functioning instruments and thanks to favourable weather conditions. A valuable dataset was collected. Preliminary data analysis shows that the observations are suitable for study of the (aerosol) profile retrieval from MAXDOAS observations. The lidar measurements carried out the the facility are key data for the evaluation of the results.

ACKNOWLEDGEMENTS

The CINDI-2 campaign was supported by ESA, the Netherands Space Office (NSO) and ACTRIS-2 (H2020-INFRAIA 654109).

References

[1] A. Apituley, et al. Overview of research and networking with ground based remote sensing for atmospheric profiling at the cabauw experimental site for atmospheric research (cesar) – the netherlands. In *Proceedings IGARSS 2008,*

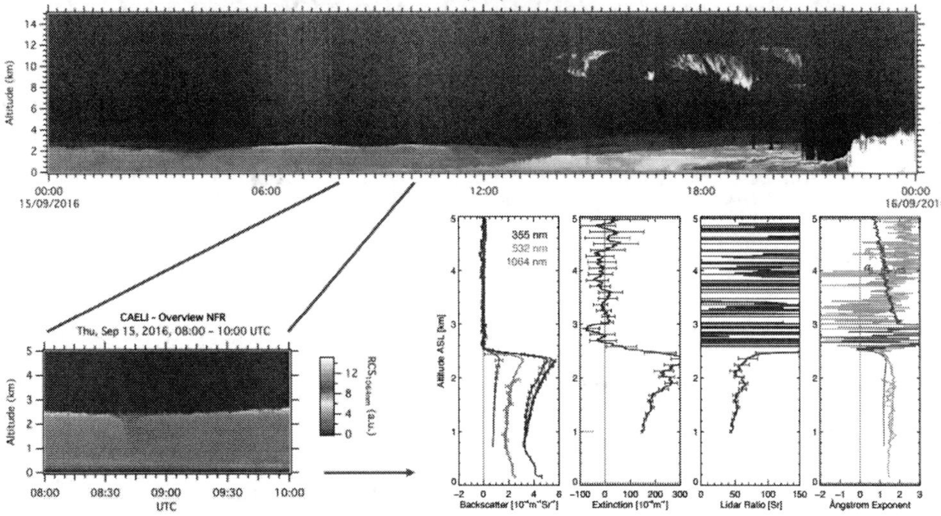

Figure 6: Example of a diurnal CHM15k ceilometer observation for 15 Sept. 2016 (top) and a daytime multi-wavelength Raman lidar observation between 8-10 UTC (bottom). In the lower-right panel, the retrieved backscatter and extinction profiles are shown. Note that these are daytime measurements.Therefore, only the 355 nm Raman extinction profile is available. The lidar ratio above the aersol layer that extends to about 2.5 km appears as noise, since the extinction fluctuates around zero.

Boston, Massachusetts, volume III, pages 903–906, 2008.

[2] A. Apituley, et al. Performance Assessment and Application of Caeli – A high-performance Raman lidar for diurnal profiling of Water Vapour, Aerosols and Clouds. In A. Apituley, H. Russchenberg, and W. Monna, editors, *Proceedings of the 8th ISTP*, pages S06–O10, 2009.

[3] J. Herman, et al. NO2 column amounts from ground-based Pandora and MFDOAS spectrometers using the direct-sun DOAS technique: Intercomparisons and application to OMI validation. *Journal of Geophysical Research: Atmospheres*, 114(D13):n/a–n/a, 2009. D13307.

[4] A. J. M. Piters, et al. The Cabauw Intercomparison campaign for Nitrogen Dioxide measuring Instruments (CINDI): design, execution, and early results. *Atm. Meas. Tech.*, 5(2):457–485, 2012.

[5] W. W. Sluis, et al. The development of a nitrogen dioxide sonde. *Atmospheric Measurement Techniques*, 3(6):1753–1762, 2010.

[6] J. Veefkind, et al. TROPOMI on the ESA Sentinel-5 Precursor: A GMES mission for global observations of the atmospheric composition for climate, air quality and ozone layer applications. *Remote Sensing of Environment*, 120(0):70 – 83, 2012.

[7] H. Volten, et al. NO2 lidar profile measurements for satellite interpretation and validation. *J. Geophys. Res.*, 114, 12 2009.

GROUND-BASED AND SATELLITE OPTICAL INVESTIGATION OF THE ATMOSPHERE AND SURFACE OF ANTARCTICA

Aleksey Malinka[1]*, Luc Blarel[2], Ludmila Chaikovskaya[1], Anatoli Chaikovsky[1], Natalia Denishchik-Nelubina[1], Sergei Denisov[1], Vladimir Dick[1], Anton Fedaranka[1], Philippe Goloub[2], Iosif Katsev[1], Michail Korol[1], Aleksandr Lapyonok[2], Thierr Podvin[2], Alexander Prikhach[1], Vadim Svidinsky[1], Eleonora Zege[1]

[1]*Institute of Physics, National Academy of Sciences, Belarus, *mal@light.basnet.by*
[2] *Laboratoire d'Optique Atmosphérique, Universite de Lille, France*

ABSTRACT

This presentation contains the results of the 10-year research of Belarusian Antarctic expeditions. The set of instruments consists of a lidar, an albedometer, and a scanning sky radiometer CIMEL. Besides, the data from satellite radiometer MODIS were used to characterize the snow cover. The works focus on the study of aerosol, cloud and snow characteristics in the Antarctic, and their links with the long range transport of atmospheric pollutants and climate changes.

1 INTRODUCTION

Recent knowledge on the Earth climate formation undoubtedly shows that Antarctica is one of the key points in the global climate. The polar caps, especially the ice cap in Antarctica that is covered with a snow blanket all-the-year-round, play a key role in the Earth radiation budget. The pollution of snow, e.g., from industrial sources, drastically affects the snow radiation regime and glaciers melting. So, environmental monitoring in Antarctica becomes a global challenge. It's being carried out by the world community, in compliance with the provisions of the international conventions.

Since 2006, in the frame of the Belarusian State Program, the scientific team of the National Academy of Sciences of Belarus has been developing the equipment for monitoring the Antarctic environment, the methods and algorithms for processing satellite and ground-based measurement data, has been performing regular ground-based observations at the Belarus station *Mount Vechernyaya* (67.66S, 46.16E) and at the Russian stations *Molodezhnaya* (67.67S, 45.85E) and *Progress* (69.38S, 76.38E).

The present paper summarizes results of the 10-year research of the Belarusian Antarctic Expedition [1].

2 EQUIPMENT

Nowadays, three types of the ground-based measurements have been implemented at *Mount Vechernyaya*: spectral radiometric measurements of the snow surface parameters, multi-wavelength lidar observations of the vertical structure of aerosols and clouds, and sun-radiometric measurements of the column aerosol characteristics.

2.1 Spectral albedometer *AS-A*

The spectral albedometer *AS-A* provides measurements of the spectral reflection of different types of the Earth surface. The albedometer measures both incident and reflected radiation fluxes. *AS-A* has an operating range of 350 – 1000 nm.

2.2 Multi-wavelength radiometer *CIMEL*

The multi-wavelength scanning sun/sky-radiometer *CE-318N (CIMEL)* is the basic instrument in the global radiometric network AERONET [2]. It registers two types of radiation: the direct solar (at 340, 380, 440, 500, 670, 940, and 1020 nm) and sky-scattered (at 440, 670, 870, and 1020 nm) radiation in the principle solar plane and in the almucantar.

Data processing results in retrieval of the aerosol parameters averaged over an atmospheric layer: the aerosol optical thickness (AOT), the Angström exponent, the water vapor content, the aerosol particles size distribution, the refractive index (its real and imaginary parts), and the scattering phase function [3]. The obtained observation data are accessible online (http://aeronet.gsfc.nasa.gov/).

2.3 Lidars

The multi-wavelength lidars provide measurements of the altitude distributions of the aerosol and cloud optical parameters. Up to now, two lidars have been developed for the Antarctic research: the mobile two-wavelength (532 and 1064 nm) lidar and the stationary multi-wavelength polarization Raman lidar. The mobile lightweight lidar was designed to operate on open ground. The multi-wavelength polarization Raman lidar provides measurements of the 355 and 1064 nm backscatter signals, the cross and parallel polarized components at 532 nm, and the Raman signals from atmospheric nitrogen and water vapor in the UV and visible spectral range.

3 TECHNIQUES

Parameters of the atmospheric aerosol and Earth surface are retrieved from the data of combined multisensory experiments, including the ground-based and satellite, the radiometric and lidar observations.

3.1 LIRIC

Algorithm LIRIC [4] was developed for processing the data of co-located and simultaneous lidar and radiometer soundings (LRS) at the EARLINET and AERONET network stations. It provides sequential inversion of the combined lidar and radiometric data.

The LRS-technique was evolved to retrieve the aerosol mode concentration profiles from the collocated space (lidar CALIOP http://www-calipso.larc.nasa.gov/) and ground-based (AERONET) data. The employment of the lidar satellite information has appreciably extended both the area of investigation and the number of potential users.

3.2 SGSR

Satellite remote sensing is an important tool for snow cover monitoring, especially over difficult-to-access polar regions. However, as a snow pack is a medium with strong multiple scattering by irregularly shaped grains, the classical approaches cannot be applied in algorithms to process the satellite data. An algorithm that takes these features of snow into account has been developed by the team of the Institute of Physics [5]. The Snow Grain Size and Rocks (SGSR) algorithm takes the signals in the channels of the satellite radiometer MODIS (http://modis.gsfc.nasa.gov/) as an input and gives the maps of the effective snow grain size and the fraction of rocks as an output.

4 RESULTS

4.1 Column aerosol parameters

Since 2008, the seasonal radiometric observations have been carried out at the AERONET station *Vechernyaya* and *Progress* in a coastal region of Antarctica. A specific feature of the Antarctic aerosol is the small concentration of suspended particles and, consequently, the small values of the aerosol optical thickness (AOT).

Figure 1 shows the seasonally averaged spectra of AOT retrieved from the direct solar light measurements. Values of AOT are quite small: 0.02-0.03 for $\lambda = 500$ nm. In the IR range, the AOT values fall down to 0.01 and less.

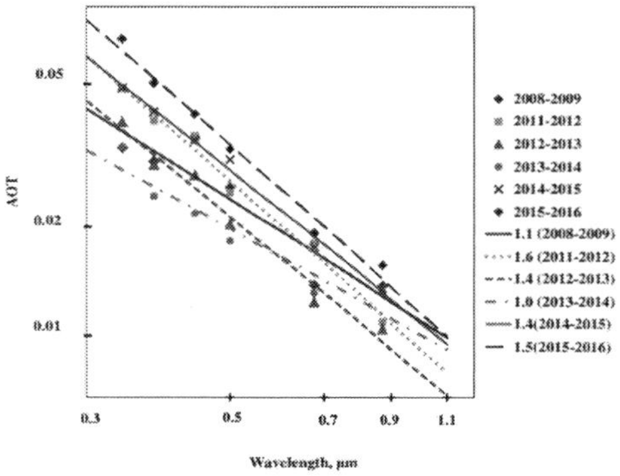

Figure 1 Spectra of AOT in Antarctic coastal zone

Average parameters of aerosol size distributions for the Antarctic coastal zone and the continental region in the Northern Hemisphere (Minsk, Belarus) are compared in Table 1. The column concentration of aerosol particles is considerably less and the content of the coarse particle mode is relatively small in the Antarctic costal region.

Table 1 Columnar concentrations of the aerosol modes

	Antarctic coast	Minsk, Belarus
Total content of aerosol, μm	0.0045	0.059
Fine/course mode, %	82/18	54/46

4.2 Aerosol vertical distribution

Information about the aerosol vertical structure is obtained from the lidar measurements. The measurements have been carried out at the *Mount Vechernyaya* station. As the aerosol scattering is very small, as compared to the Rayleigh one, the wavelength of 1064 nm was used to characterize the aerosol layer stratification (Figure 2).

The LRS-experiment that used both the ground-based and satellite lidar measurements, in addition to the CIMEL data, was made for sounding the vertical structure of aerosol at the coastal zone of Antarctica. Some results are presented in Figure 3. The ground-based lidar measurements were made at 15:00 UTC, December 23, 2012 (67.66S, 46.16E). The aerosol layer with the dominant fine particle mode is detected at the height 3-5.5 km.

Because of clouds in the coastal part of satellite trajectory the CALIOP data were taken from the sector 70.165S, 59.182E - 66.275S, 54.502E, 11:21 UTC, December 23, 2012 which passes over the continental region of Antarctica with the height above sea level 1 – 3.5 km.

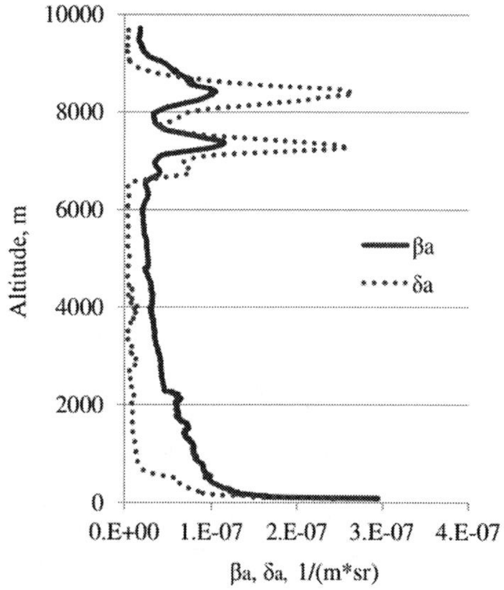

Figure 2 Vertical profile of the backscattering coefficient β_a and its standard deviation δ_a at 1064 nm (Mount Vechernyaya, 2014 – 2015)

The retrieval of aerosol characteristics from CALIOP data was carried out under the assumption that the characteristics of the fine and coarse fractions of the aerosol model were identical to parameters determined from radiometric measurements at the *Mount Vechernyaya* station. Moreover, 30% deviations of the total content of aerosol fractions in the atmospheric layer from their values at the *Mount Vechernyaya* station were allowed. The results of the retrieval are shown in Fig. 3. The horizontal lines represent the uncertainty of the fraction concentrations calculated under the assumptions made.

The additional information about the geophysical processes was taken from the results of modeling of the air-mass back trajectories HYSPLIT [6] and from the fire map of the South African region provided by FIRMS, NASA [7].

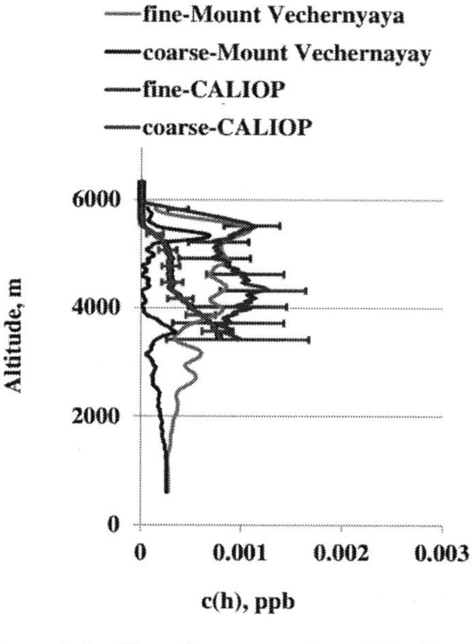

Figure 3 Profiles of concentration of the fine and coarse aerosol particles retrieved from the ground-based lidar and CALIOP data

4.3 Snow microphysical characteristics from satellite data

Figure 3 presents the example of the snow grain size and the rock fraction distributions in the region of *Vechernyaya* retrieved from the MODIS data. The seasonal series of such maps were

https://doi.org/10.1051/epjconf/201817610006

calculated to study the changes of snow characteristics in the *Vechernyaya* region. In particular, the phenomenon of a sudden decrease of the snow grain size (down to 50-100 μm) after a snowfall event was clearly observed. Between snowfalls the grain size slowly increases to 200-300 μm, due to the grain growth induced by the snow metamorphism.

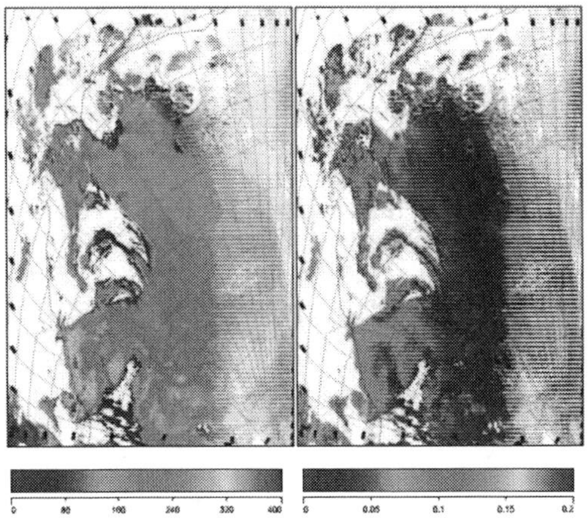

Figure 3 Map of the snow grain size (left, the scale is in micrometers) and rock fraction (right) retrieved from the MODIS-Terra data (January 01, 2015, MOD021KM.A2015001.0755.006.2015002145822.hdf)

4.4 Field measurements of the surface albedo spectra

Field measurements of the surface reflection spectra in the region of the *Vechernyaya* station were aimed at gathering the database of the surface reflectance spectra.

5 CONCLUSIONS

The technique of combined ground-based and satellite remote sensing creates the methodological basis for our investigations of the atmospheric aerosol and Earth surface in the Antarctic. The employment of the satellite information allows one to extend the region of investigation. The joint measurements with the different experimental techniques, as well as the special inversion algorithms for processing combined data, aim at obtaining supplementary information about the atmosphere and surface properties, as against the results of independent measurements (the so called synergetic effect).

ACKNOWLEDGEMENTS

The investigation was supported by: the State Program of Polar Research of Belarus, the EU research and innovation Programme *Horizon-2020* (ACTRIS-2, grant agreement no. 654109), and the Belarusian Republican Foundation for Fundamental, Research Agreement No. F15SB-023. We acknowledge the NASA/GSFC/Earth Science Data and Information System (ESDIS), the NOAA Air Resources Laboratory (ARL), and all the scientists, engineers, and technicians from Belarus, Russia and France, thanks to whom this study became possible.

References

[1] Chaikovsky, A.P., et al., 2016: Optical studies of the atmosphere and surface in Antarctic, *Belarus in Antarctic: on the 10th anniversary of the beginning of scientific and expeditional research, acad. V.F. Loginov ed., Belaruskaya navuka,* Minsk, 50-101.

[2] Holben, B.N., et al., 1998: AERONET-A federated instrument network and data archive for aerosol characterization, *Rem. Sens. of Env.* **66**, 1-16.

[3] Dubovik, O., and King, M., 2000, :A Flexible Inversion Algorithm for Retrieval of Aerosol Optical Properties from Sun and Sky Radiance Measurements, *J. Geophys. Res.* **105** (D16), 20673-20696.

[4] Chaikovsky, A., et al., 2016: Lidar-Radiometer Inversion Code (LIRIC) for the retrieval of vertical aerosol properties from combined lidar/radiometer data: development and distribution in EARLINET, *Atmos. Meas. Tech.*, **9** (3), 1181–1205.

[5] Zege, E.P., et al., 2011: Algorithm for retrieval of the effective snow grain size and pollution amount from satellite measurements, *Rem. Sens. of Env.* **115** (10), 2674-2685.

[6] Stein, A.F., Draxler, R.R, Rolph, G.D., Stunder, B.J.B., Cohen, M.D., and Ngan, F., 2015: NOAA's HYSPLIT atmospheric transport and dispersion modeling system, *Bull. Amer. Meteor. Soc.* **96**, 2059-2077.

[7] Fire Information for Resource Management System (FIRMS), https://earthdata.nasa.gov/earth-observation-data/near-real-time/firms (September 2016).

TOLNET OZONE LIDAR INTERCOMPARISON DURING THE DISCOVER-AQ AND FRAPPÉ CAMPAIGNS

Michael J. Newchurch[1]*, Raul J. Alvarez II[2], Timothy A. Berkoff[3], William Carrion[4], Russell J. DeYoung[3], Rene Ganoe[5], Guillaume Gronoff[3], Guillaume Kirgis[2,6], Shi Kuang[1], Andy O. Langford[2], Thierry Leblanc[7], Thomas J. McGee[8], Denis Pliutau[5], Christoph Senff[26], John T. Sullivan[8], Grant Sumnicht[5], Laurence W. Twigg[5], Lihua Wang[1]

*[1]University of Alabama in Huntsville, USA, *mike@nsstc.uah.edu*
[2]NOAA Earth System Research Laboratory, USA
[3]NASA Langley Research Center, USA
[4]Coherent Applications Inc., USA
[5]Science Systems and Applications, Inc, USA
[6]University of Colorado, USA
[7]California Institute of Technology, USA
[8]NASA Goddard Space Flight Center, USA

ABSTRACT

The Tropospheric Ozone Lidar Network (TOLNet) is a unique network of lidar systems that measure atmospheric profiles of ozone and aerosols, to contribute to air-quality studies, atmospheric modeling, and satellite validation efforts. The accurate characterization of these lidars is of critical interest, and is necessary to determine cross-instrument calibration uniformity. From July to August 2014, three lidars, the TROPospheric OZone (TROPOZ) lidar, the Tunable Optical Profiler for Aerosol and oZone (TOPAZ) lidar, and the Langley Mobile Ozone Lidar (LMOL), of TOLNet participated in the "Deriving Information on Surface conditions from Column and Vertically Resolved Observations Relevant to Air Quality" (DISCOVER-AQ) mission and the "Front Range Air Pollution and Photochemistry Éxperiment" (FRAPPÉ) to measure sub-hourly ozone variations from near the surface to the top of the troposphere. Although large differences occur at few individual altitudes in the near field and far field range, the TOLNet lidars agree with each other within ±4%. These results indicate excellent measurement accuracy for the TOLNet lidars that is suitable for use in air-quality and ozone modeling efforts.

1 INTRODUCTION

The Tropospheric Ozone Lidar Network (TOLNet) is designed primarily to provide time and height measurements of ozone from near the surface to the top of the troposphere at multiple locations for satellite validation, model evaluation, and scientific research (http://www-air.larc.nasa.gov/missions/TOLNet/). Particularly, these high-fidelity ozone measurements can serve to validate NASA's first Earth Venture Instrument mission, Tropospheric Emissions: Monitoring Pollution (TEMPO), planned to launch in 2020 or 2021. A second objective of TOLNet is to identify a brassboard ozone lidar instrument that would be suitable to populate a network to address increasing desire for ozone profiles by air-quality scientists and managers within the modeling and satellite communities.

TOLNet currently consists of five ozone lidars across the United States and one lidar in Canada: the Jet Propulsion Laboratory Table Mountain tropospheric Ozone (JPL-TMO) differential absorption lidar (DIAL), the TOPAZ lidar at NOAA's Earth System Research Laboratory (ESRL), the Rocket-city Ozone Quality Evaluation in the Troposphere (RO_3QET) lidar at University of Alabama in Huntsville, TROPOZ DIAL of NASA's Goddard Space Flight Space Center (GSFC), and LMOL at NASA's Langley Research Center (LaRC), and AMOLITE lidar at Environment Canada. TOLNet also includes a modeling component at NASA Ames Research Center (ARC) and a data center at LaRC.

2 LIDARS

2.1 TROPOZ at NASA's GSFC

The transmitter for TROPOZ consists of two 50-Hz Nd:YAG- lasers used to pump two Raman cells filled with Deuterium and Hydrogen gases, respectively, to generate two outgoing wavelengths at 289 and 299 nm. The typical pulse energies are 12 mJ at 299 nm (off-line) and 16 mJ at 289 nm (on-line) [1]. The receiving system for the TROPOZ consists of a 45-cm-diameter Newtonian telescope for measuring far field and four smaller 2.5-cm refracting telescopes to measure near field. The 45-cm telescope has a 1.0-mrad field of view (FOV), and the 2.5-cm telescopes have a much wider FOV at 10.0 mrad. In each channel, narrowband interference filters with 1-nm bandwidth are implemented to decrease the amount of ambient solar light, which improves the signal-to-noise ratio. The fundamental range resolution for data acquisition system is 15 m (100 ns). The relatively powerful lasers and large telescope result in ozone retrievals up to 16 km during daytime hours.

2.2 TOPAZ at NOAA's ESRL

The TOPAZ lidar is an up-looking, scanning instrument in a truck modified from an airborne DIAL system [2,3]. The lidar transmitter is based on a Ce:LiCAF laser that is pumped by a quadrupled Nd:YLF laser to produce three UV wavelengths at 333 Hz repetition rate tunable from 283 nm to 310 nm. The actual wavelengths during the DISCOVER-AQ 2014 were 287, 291, and 294 nm. Compared to the conventional two-wavelength DIAL, the three-wavelength configuration can potentially minimize the aerosol interference by using the dual-DIAL retrieval technique [4] without assuming lidar ratio and Angström exponent.

The backscattered light is collected with a 50-cm diameter Newtonian telescope and then split at a 1:9 ratio into near- and far-field detection channels. The FOVs of the near- and far-field channels are controlled by different-size apertures resulting in full overlap at distances of ~300 m and ~800 m, respectively. Both channels use gated photomultipliers operated in analog mode with solar interference filters during daytime. Compared to a photoncounting signal, an analog signal is able to keep high linearity for strong signals and, therefore, is particularly suitable for near-range measurement. The two-axis scanner on the truck permits pointing the laser beam at several shallow elevation angles at a fixed, but changeable azimuth angle, typically at 2^{O}, 10°, and 90° elevation angles that are repeated approximately every 5 minutes. The ozone profiles at these three angles are spliced together to create composite vertical profiles extending from 10 m to about 2.5 km AGL. The range resolution of the signal recording system is 6 m. The TOPAZ lidar has a unique capability to measure the near-surface ozone above 10 m due to its scanning configuration.

2.3 LMOL at NASA's LaRC

The transmitter of LMOL consists of a diode-pumped Nd:YLF laser pumping a Ce:LiCAF tunable UV laser to obtain two lasers typically at 287.1 and 292.7 nm with a pulse energy of 0.2 mJ at 500 Hz for each wavelength. The lidar receiver system consists of a 40-cm telescope with a 1.4-mrad FOV to measure far field and another 30-cm telescope with same FOV to measure near field [5]. The raw lidar signals are recorded with a 7.5-m range resolution. The LMOL data acquisition system operates in both analog and photoncounting modes. In this study, LMOL measures ozone between 0.7 and 4.5 km. Ozone measurements for DISCOVER-AQ result from LMOL's first ever remote deployment.

2.4 Measurement Uncertainty Budget

Table 1 presents the estimation of measurement uncertainty budget for 5-min integration time for above three lidars. Statistical errors [6] arising from signal and background noise fluctuations are random errors and may be improved by additional average or smoothing. The uncertainty arising from aerosol interference cloud be the largest systematic error source and can be minimized by appropriate correction algorithm [1].

3 RESULTS

Prior to the campaign, the three TOLNet lidars were deployed next to the Boulder Atmospheric Observatory (BAO) tower (40.050^{O}N, 105.003^{O}W) to take simultaneous measurements. They were within 600 meters of each other and were very near the same elevation.

Unlike stratospheric ozone lidars that focus on integrating hours of observations, tropospheric ozone lidars need to detect ozone variations with timescales on the order of minutes, considering ozone's shorter lifetime, smaller-scale transport, mixing processes within the PBL and free troposphere. Therefore, we processed all lidar data on a 5-min temporal scale (signal integration time). Rayleigh correction was performed with the same atmospheric profile from ozonesonde. Because the three lidars have different fundamental range resolutions, retrieved ozone number density values are internally interpolated on the same altitude grids with a 15-m interval for comparison.

Table 1. Estimated uncertainties for TROPOZ, TOPAZ and LMOL measurements below 2 km AGL for 5-min integration time.

Sources	Uncertainty
Statistical error	<8%
Aerosol interference	<12%
Interference by SO_2, NO_2, O_2 dimmer	<1.5%
Differential Rayleigh scattering	<1%
Total	<15%

TOPAZ and TROPOZ lidars were operated simultaneously at BAO during 1300 to 2135 UTC (6 hours ahead of local time, Mountain Daylight Time) on July 11 under a partly cloudy sky condition. Some bad data points due to cloud interference were filtered out. Ozone curtains from both lidars (Figures 1a and b) show a significant (~30%) ozone increase in the early afternoon. A total of 7655 TOPAZ and TROPOZ coincident pairs are constructed between 0.6 and 2 km AGL over this time period. The measurement differences between the two lidars are mostly within ±5% at individual grids (Figure 1, c). The ozone averages measured by TROPOZ and TOPAZ were 126.8 and 128.2x10^{16} molec·m^{-3}, respectively, with the same standard deviations, 18.8x10^{16} molec·m^{-3}. The mean relative difference (or normalized bias) is calculated by averaging the relative difference, (TROPOZ-TOPAZ)/TOPAZ (the denominator was arbitrarily chosen), for all coincident pairs. The -0.9% mean relative difference suggests an excellent agreement of the averaged ozone column within 6.5 hours between TOPAZ and TROPOZ retrievals.

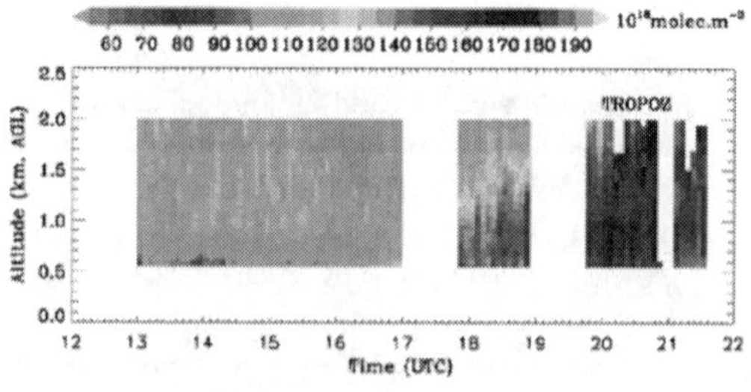

(a)

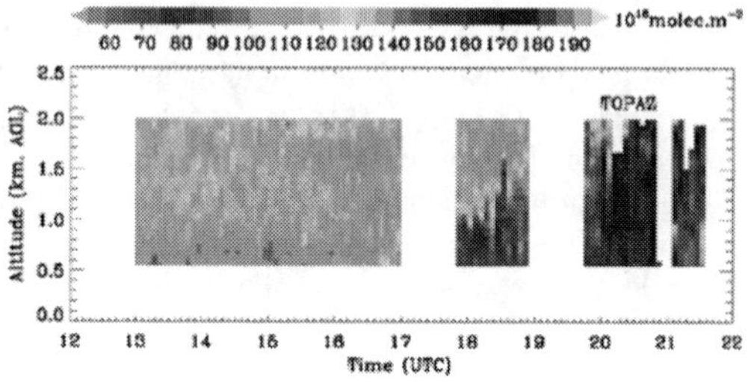

(b)

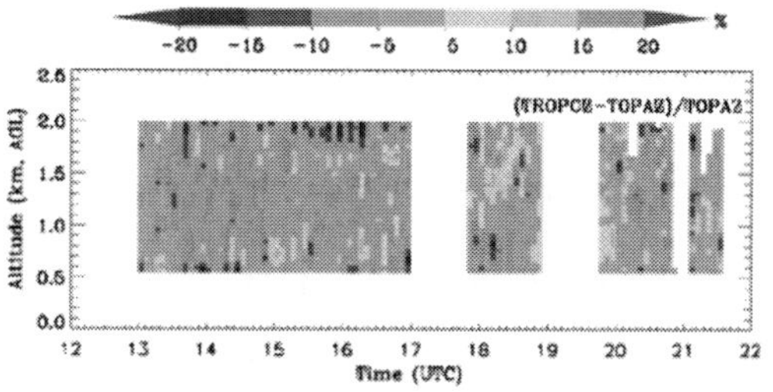

(c)

Figure 1. Comparisons of ozone measured by TROPOZ and TOPAZ. (a) Ozone number densities measured by TROPOZ. (b) Ozone number densities measured by TOPAZ. (c) Their relative percent differences, (TROPOZ-TOPAZ)/TOPAZ.

Figure 2 shows TOPAZ/LMOL intercomparison for the data taken on July 16, 2014 with 1902 coincident pairs from 0.9 to 2 km and between 1340 to 1730 UTC on this day. The mean ozone values were 97.6 and 101.4 x10^{16} molec·m^{-3} from LMOL and TOPAZ, respectively. Both retrievals coincidentally have the same standard deviation, 15.3 x10^{16} molec·m^{-3}. LMOL measured ozone 3.6% lower than TOPAZ on average with significantly fewer pairs than those from the TROPOZ/TOPAZ comparison. In spite of these measurement differences, the generally random distribution of the relative differences, in Figure 1(c) and 2(c), during the large ozone variation days, suggest overall consistent measurements from all three lidars.

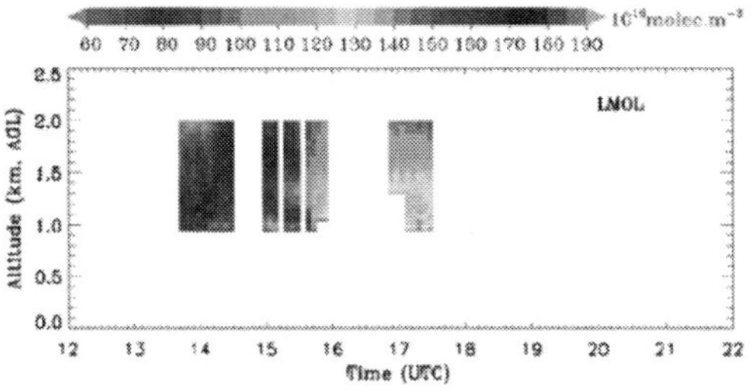

(a)

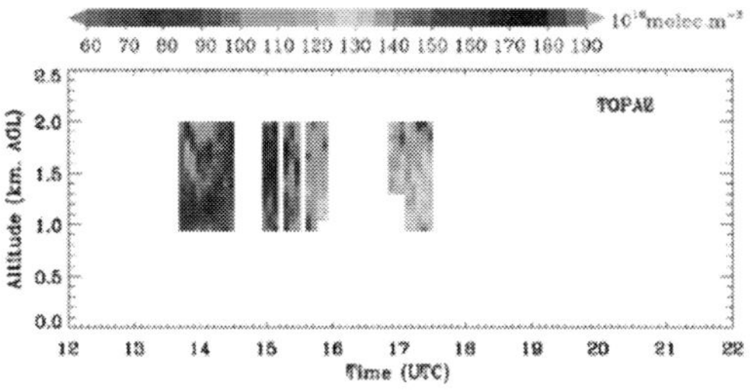

(b)

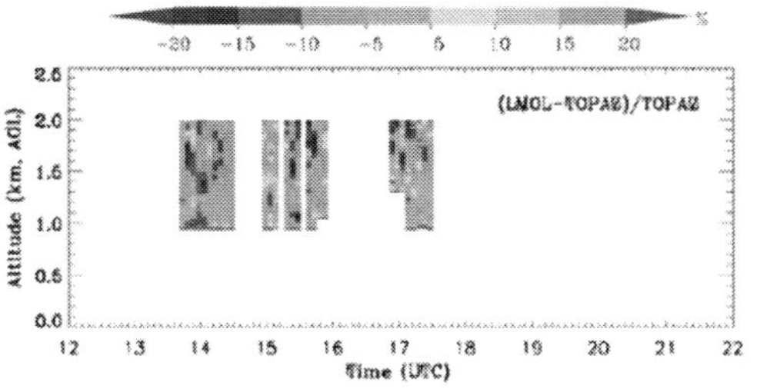

(c)

Figure 2. Comparisons of ozone number densities measured by LMOL and TOPAZ. (a) LMOL-measured ozone. (b) TOPAZ-measured ozone. (c) Their relative percent differences, (LMOL-TOPAZ)/TOPAZ.

4 CONCLUSIONS

In this study, intercomparisons between three TOLNet ozone lidars (TROPOZ at NASA's GSFC, TOPAZ at NOAA' ESRL, and LMOL at NASA's LaRC) by using coincident data during the 2014 DISCOVER-AQ and FRAPPÉ campaigns. On average, TROPOZ and TOPAZ reported very similar ozone (within 0.9%) while LMOL reported ozone slightly lower (3.6%) than TOPAZ. The three lidars measured consistent ozone variations evidenced by the lidar curtains and the distribution of their relative difference.

ACKNOWLEDGEMENTS

This work is supported by the TOLNet program developed by NASA's Science Mission Directorate. The views, opinions, and findings contained in this report are those of the authors and should not be construed as an official NOAA, NASA, or the U.S. Government position, policy, or decision.

References

[1] Sullivan, J. T., McGee, T. J., DeYoung, R., Twigg, L. W., Sumnicht, G. K., Pliutau, D., Knepp, T., Carrion, W., 2015: Results from the NASA GSFC and LaRC Ozone Lidar Intercomparison: New Mobile Tools for Atmospheric Research, *J. Atmos. Oceanic Technol.*, **32**(10), 1779–1795.

[2] Alvarez, R.J., Senff, C.J., Langford, A.O., Weickmann, A.M., Law, D.C., Machol, J.L., Merritt, D.A., Marchbanks, R.D., Sandberg, S.P., Brewer, W.A. and Hardesty, R.M., 2011: Development and application of a compact, tunable, solid-state airborne ozone lidar system for boundary layer profiling, *J. Atmos. Oceanic Technol.*, **28**(10), 1258-1272.

[3] Senff, C.J., Alvarez, R.J., Hardesty, R.M., Banta, R.M. and Langford, A.O., 2010: Airborne lidar measurements of ozone flux downwind of Houston and Dallas, *J. Geophys. Res.*, **115**, D20307.

[4] Kovalev, V.A. and Bristow, M.P., 1996: Compensational three-wavelength differential-absorption lidar technique for reducing the influence of differential scattering on ozone-concentration measurements, *Appl. Opt.*, **35**(24), 4790-4797.

[5] De Young, R., Carrion, W., Ganoe, R., Pliutau, D., Gronoff, G., Berkoff, T., Kuang, S., 2017: Langley mobile ozone lidar: ozone and aerosol atmospheric profiling for air quality research, *Appl. Opt.* **56**(3), 721-730.

[6] Papayannis, A., Ancellet, G., Pelon, J., Mégie, G., 1990: Multiwavelength lidar for ozone measurements in the troposphere and the lower stratosphere, *Appl. Opt.*, **29**(4), 467–476.

EPJ Web of Conferences **176**, 11001 (2018) https://doi.org/10.1051/epjconf/201817611001
ILRC 28

REMOTE SENSING OF LOW VISIBILITY OVER OTOPENI AIRPORT

Livius Buzdugan[1,2], Denisa Urlea[1,2], Paul Bugeac[1,2], Sabina Stefan[1]

[1] *University of Bucharest, Faculty of Physics, P.O.BOX: MG-11, Magurele, Bucharest, Romania*

(buzdugan.livius@gmail.com)

[2] *Romanian Air Traffic Services- ROMATSA, 10 Ion Ionescu De la Brad Str., Bucharest, Romania*

ABSTRACT

The paper is focused on the study of atmospheric conditions determining low vertical visibility over Henri Coanda airport. A network of ceilometers and a Sodar were used to detect fog and low level cloud layers. In our study, vertical visibility from ceilometers and acoustic reflectivity from Sodar for November 2016 were used to estimate fog depth and top of fog layers, respectively. The correlation between fog and low cloud occurrence and the wind direction and speed is also investigated.

1. INTRODUCTION

The prediction of low visibility conditions is of paramount importance for operational weather forecast services. Adverse visibility conditions can strongly reduce the efficiency of a terminal area traffic flow. The application of Low Visibility Procedures (LVP) reduces airport efficiency for takeoffs and landings by a factor of two. Costly delays and flight cancellations ensue [1].

At Bucharest Henri Coanda (OTP) international airport, LVP are applied when visibility is less than 600 m (2,000 feet) or the ceiling is below 60 m (200 feet).

The goal of this work is to test how the use of Sodar and CL31 VÄÏSÄLA Ceilometer measurements can help the detection of the top of fog layers which could in turn be used for the initialisation of a high resolution numerical model dedicated to fog forecasting.

The Sodar and Ceilometers are operated at Otopeni Airport close to Bucharest, capital of Romania.

The site, meteorological conditions and the instruments are presented in Section 2.

The capabilities of instruments to detect fog and the link between fog and low cloud occurrence the wind direction and speed are investigated and results are presented in Section3. Section 4 is reserved to conclusions.

2. METHODOLOGY

2.1 Instruments and Measuring Sites

The geographical and climatological characteristics of the site arise from Henri Coanda airport being situated nearby the city of Bucharest, the largest in Romania with strong urban and plain climatology.

The boundary layer conditions accompanying fog and low cloud are studied using 4 ceilometers and a Sodar installed at Bucharest Henri Coanda airport. The ceilometers are used operationally only to report cloud base height and vertical visibility [3] and to detect and evaluate the fractional extent of cloud layers [3, 5]. They are integrated in the local automated weather observing system (AWOS).

The data used are gathered from the ceilometer and Sodar which are co-located at approximatively 1 km east of the eastern threshold of one of the runways of Bucharest Henri Coanda airport (44.58N, 26.13E, 92m AMSL). The Ceilometer is installed on the roof of a building formerly hosting a radio-navigation facility, at 4m height and owned and operated by ROMATSA (the Romanian Air Traffic Services Administration). The Sodar is also operated by ROMATSA, but owned by INOE (National Institute of Opto-Electronics).

The ceilometer is VÄÏSÄLA CL31 on single-lens technology, 910nm (near-infrared) wavelength, operated with a 10m vertical resolution and 15s temporal resolution. Its laser diodes are pulsed with a repetition rate of 10 kHz [3]. The ceilometer has been set up for a maximum range of 7700m AMSL. It is able to detect three cloud

layers simultaneously. If the cloud base is obscured due to precipitation or ground-based fog, CL31 reports vertical visibility (VV) [3].
By assuming a linear relationship between backscatter and extinction coefficient and by assuming that the Lidar ratio (k) is constant over the observation range, it is possible to invert the backscatter profile, basically obtaining an extinction coefficient profile which would produce the measured backscatter profile [3].

It has been discovered that in many cases, k can be assumed to equal 0.03, tending to be lower (down to to 0.02) in high humidity, and higher (up to 0.05) in low humidity conditions. However, in precipitation conditions, k can have a wider range of values [3].

An estimate of vertical visibility can easily be calculated from the extinction coefficient profile because of the straightforward extinction coefficient-to-visibility relationship, assuming a constant contrast threshold. Visibility is simply the height where the integral of the extinction coefficient profile, starting from the ground, equals the natural logarithm of the contrast threshold, sign disregarded [3].

Tests and research have, however, shown that the 5% contrast threshold widely used in meteorological visibility measurements is unsuitable for vertical visibility measurement [3].

ceilometer CL31 uses a contrast threshold value which has been found to give vertical visibility values closest to those reported by ground-based human observers. A safety margin is obtained with regard to pilots looking down in the same conditions since the contrast between objects, especially runway lights, is much more distinct on the ground [3].

For the purpose of this study, the vertical visibility reported by the ceilometers has been used only to identify the cases with dense, ground-based fog.

The Sodar is a Doppler monostatic METEK PCS 2000-24.
For a Sodar the source of the received signal is the scattering eddy – essentially a density fluctuation - moving along with the air. It can be shown that in monostatic mode the backscattering is due to random temperature fluctuations only [2].

The change in frequency produced by a scatterer (Doppler shift) is proportional to the rate of change of the distance between receiver and scatterer and the initial frequency, so for a known transmitted frequency, a radial velocity - corresponding to motions along the beam - can be calculated by measuring the received backscattered wave frequency. Thus the determination of the three-dimensional wind vector requires at least 3 beams with different orientations [1,4].
The METEK PCS 2000-24 Sodar uses a phased-array antenna consisting of an array of 24 loudspeakers for sequential monostatic soundings of the three-dimensional wind field. The system software controls the operation of each beam. Electronic phase shifters apply a different phase shift to each antenna element to steer the beam. The shifted phase of each element causes the waves to interfere constructively, giving the maximum gain in the desired direction. Five beams are achieved, allowing cross checking among the wind components. [4].
The Sodar was parameterized to a vertical resolution of 20 m and a first range gate (bottom altitude) of 40 m. The temporal resolution was set to 10 min (integration time).

2.2. Data and methods

In order to find the days with fog and low cloud, the ceilometers measurements for November 2016 have been investigated simultaneously with METAR reports from Bucharest Henri Coanda airport, stored by the local AWOS system. The month of November 2016 was investigated, as several fog situations occurred on the 4th, 5th, 12th, 19th and 26th.

The cloud base heights and vertical visibility from ceilometers measurements were stored and retrieved using the software running on the AWOS operated by ROMATSA. The stored data consist of a text file for each day and parameter (cloud base-heights and vertical visibility) with temporal resolution of 15s. These data were processed with a software to determine time series of measurements with cloud base within specific intervals and with vertical visibility reported.
It was noticed that dense fog or precipitation is required for the ceilometer to output vertical visibility values instead of cloud base heights.

https://doi.org/10.1051/epjconf/201817611001

Dense fog conditions were declared on site at a time frame when vertical visibility was reported by the ceilometer, only if confirmed by METAR (SPECI) reports or by values of MOR below 1000 meters measured by a nearby transmissometer at 2,5m above the ground.
The Sodar reflectivity and wind (direction and speed) profiles were retrieved using the METEK Graphics software, as text files for the whole month, with a temporal resolution of 10 minutes.
An enhanced turbulent activity with temperature fluctuations is expected at the top of the fog layers due to radiative cooling and wind shear and therefore, Sodar reflectivity at the top of a fog layer is expected to exhibit a local maximum [1].
To check this hypothesis, a monthly vertical reflectivity profile was computed by using all the available reflectivity data and compared with a vertical profile of average reflectivity values from time frames with dense fog conditions, as previously defined – covering approximately 12 hours.
Another data set, of low cloud conditions, was defined by measured cloud base heights not higher than 300 meters. The average Sodar reflectivity profile for these conditions – covering approximately 135 hours - was computed.
To investigate if an enhanced turbulence due to increased wind speed, wind shear or an increased vertical thermal gradient was the underlying cause of any Sodar reflectivity maxima, the monthly mean Sodar wind speed and direction profiles and the characteristic wind speed profile for low cloud and fog were also computed.

3. RESULTS

In this study, data gathered during November 2016 from a ceilometer co-located with a Sodar at Bucharest Henri Coanda (OTP) airport, were evaluated to test how the use of Sodar measurements can help the detection of the top of fog layers which could in turn be used for the initialisation of a high resolution numerical model dedicated to fog forecasting. The link between fog and low cloud occurrence and the wind direction and speed was also investigated.
In Figure 1, the mean vertical profile of Sodar reflectivity in dense fog situations displays a maximum at heights of 40-100 meters and below and values significantly lower than the general monthly means at heights above 140 meters. This can be explained by the fact that dense fog situations were observed mostly during nighttime hours, with surface temperature inversions generating a significant amount of temperature fluctuations in the inversion layer even at low wind speeds. In Figure 2, the relatively low wind speeds characteristic to the fog events emphasize the decisive role of the temperature inhomogeneities in the increase of reflectivity.
A maximum of the Sodar reflectivity clearly positioning the height of the fog layer top is not evident, possibly due to the relatively high value of the lowest range gate (40 m). Nevertheless, the mean height of the fog top can be traced between 40 and 120 meters, the low values of Sodar reflectivity above 120 meters being explained by the nearly isothermal conditions above the fog top and also by the attenuation of sound in the highly reflective layer below.

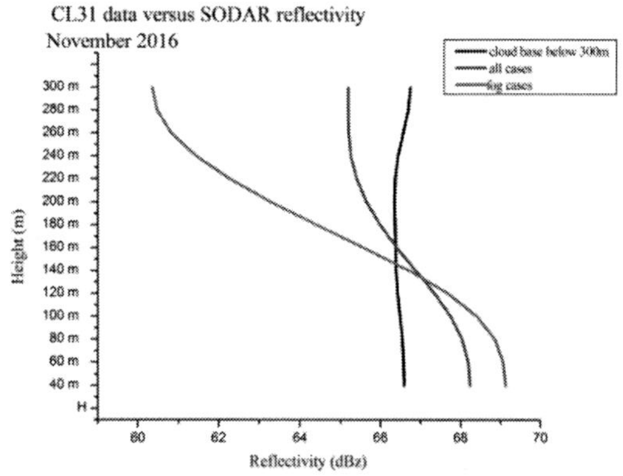

Figure 1. Vertical profiles of mean Sodar reflectivity in low cloud situations as compared to fog situations and monthly average, OTP, November 2016

When compared to the fog situations, the low cloud (base heights not higher than 300 meters, Stratus type cloud) situations exhibit a nearly-constant vertical profile of Sodar reflectivity up to the maximum cloud base height considered (300 meters). This is consistent with a well-mixed layer and a nearly constant vertical gradient of temperature within the analysed layer.
In Figure 3, the vertical profile of wind speed in the low cloud situations is similar to the monthly mean profile of the wind speed. This leads to the conclusion that the change in the reflectivity

profile associated with the low clouds exhibited in Figure 1 cannot be accounted for by changes in the wind speed profile – but rather by changes in the temperature profile associated with these clouds, during the analysed period.

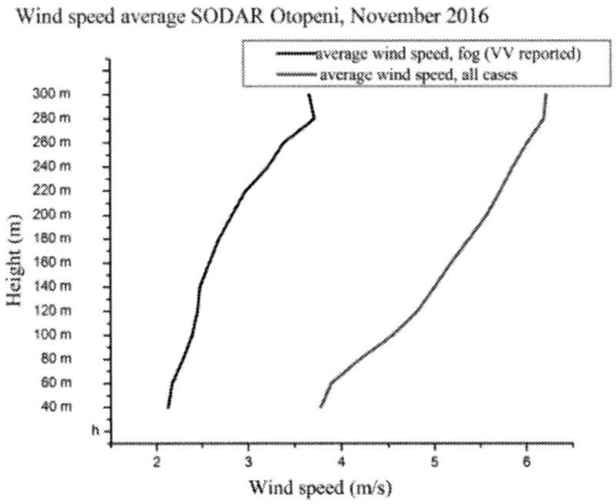

Figure 2. Vertical profiles of wind speed in fog situations as compared to monthly average, OTP, November 2016

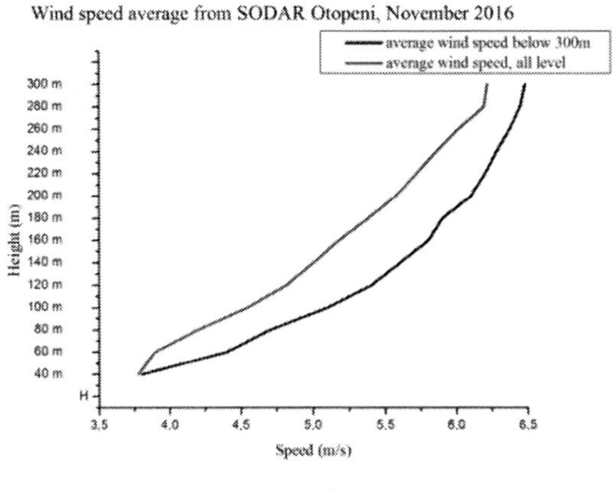

Figure 3. Vertical profiles of wind speed in low cloud situations as compared to monthly average, OTP, November 2016

4. CONCLUSIONS

The paper proves that the CL31 ceilometer and a Sodar could be used to provide information on the boundary layer conditions accompanying fog and low cloud and locating the top of the fog layers.

A higher resolution of the Sodar (10 meters) and a lower first range gate could improve the accuracy of the latter determination.

ACKNOWLEDGEMENTS

The authors gratefully acknowledge ROMATSA for the provision of the data used in this scientific paper.

References

[1] Dabas, A., Remy, S., Bergot, T (2012) Use of a Sodar to Improve the Forecast of Fogs and Low Clouds on Airports, *Pure Appl. Geophys.* vol 169, pg. 769–781, doi: 10.1007/s00024-011-0334-y

[2] LITTLE, C.G. (1969), Acoustic methods for the remote probing of the lower atmosphere, *Proc. IEEE.* vol 57, 571–578

[3] VÄÏSÄLA Ceilometer CL31user's guide

[4] PCS.2000-24/64/MF/LP User Manual, METEK Gmbh, 2012

[5] Ungureanu, I. S. Stefan, and D. Nicolae (2010) Investigation of the cloud cover and Planetary Boundary Layer (PBL) characteristics using Ceilometer CL 31. *Romanian Reports in Physics*, vol 62, nr. 2, pg. 396-404

EPJ Web of Conferences **176**, 11002 (2018) https://doi.org/10.1051/epjconf/201817611002
ILRC 28

THE LESSON LEARNT DURING INTERACT - I AND INTERACT - II ACTRIS MEASUREMENT CAMPAIGNS

Marco Rosoldi [1]*, Fabio Madonna[1], Gelsomina Pappalardo[1], Joshua Vande Hey[2], Yunhui Zheng[3], VAISALA Team[4]

[1]*Consiglio Nazionale delle Ricerche, Istituto di Metodologie per l'Analisi Ambientale (CNR-IMAA), Italy,*
**marco.rosoldi@imaa.cnr.it*
[2] *University of Leicester, Department of Physics and Astronomy, Earth Observation Science Group, UK*
[3]*Sigma Space Corporation, USA*
[4]*VAISALA, Finland*

ABSTRACT

The INTERACT-II (INTERcomparison of Aerosol and Cloud Tracking) campaign, performed at the CNR-IMAA Atmospheric Observatory (760 m a.s.l., 40.60° N, 15.72° E), aims to evaluate the performances of commercial automatic lidars and ceilometers for atmospheric aerosol profiling, through the comparison with Potenza EARLINET (European Aerosol Research Lidar NETwork) lidars. The results of the campaign and the overall lesson learnt within INTERACT-I and INTERACT-II ACTRIS campaigns will be presented.

1 INTRODUCTION

Knowledge of aerosol spatio-temporal distribution in troposphere is essential to improve our understanding of climate and air quality. For this purpose, global scale high resolution continuous measurements of tropospheric aerosols are needed. Global coverage high resolution networks of ground-based low-cost and low-maintenance remote sensing instruments, such as commercial automatic lidars and ceilometers, can strongly contribute to this scientific mission. Therefore, it is very interesting for scientific community to understand to which extent these instruments are able to provide reliable aerosol measurements and fill in the geographical gaps of existing networks of advanced lidars, like EARLINET (European Aerosol Research LIdar NETwork). The INTERACT-II (INTERcomparison of Aerosol and Cloud Tracking) campaign, carried out at CIAO (CNR-IMAA Atmospheric Observatory) in Tito Scalo, Potenza, Italy (760 m a.s.l., 40.60°N, 15.72°E), aims to evaluate the performances of commercial automatic lidars and ceilometers for tropospheric aerosol profiling. The campaign has been performed in the period from July 2016 to January 2017 in the framework of ACTRIS-2 (Aerosol Clouds Trace gases Research InfraStructure) H2020 research infrastructure project. Besides the commercial ceilometers operational at CIAO, a Vaisala CT25K and a Lufft CHM15k (not operated from July to September due to the misalignment between laser and detector), the performance of a CL51 Vaisala ceilometer, a Campbell CS135 ceilometer and a Sigma Space mini-Micro Pulse Lidar (MPL) are assessed using the EARLINET multi-wavelength Raman lidars operative at CIAO as reference. In the following sections, the methodology adopted for the intercomparison is described along with the first results.

2 METHODOLOGY

The signals of the multi-wavelength Raman lidar operative at CIAO, MUSA, are processed using the EARLINET Single Calculus Chain (SCC) also with the aim to improve the data consistency and comparability [1, 2]. The SCC provides the pre-processed range corrected signals (RCS) and the profiles of aerosol extinction coefficient at 355 and 532 nm and backscattering coefficient at 355, 532 and 1064 nm, using both Raman and elastic signals. MUSA is equipped with two channels detecting the co- and cross polarized components of the elastically backscattered radiation at 532 nm, in order to measure the particle depolarization at that wavelength. The automatic mini-MPL detects the co- and cross polarized components of the elastically backscattered radiation at 532 nm and provides continuous measurements of particle backscattering coefficient and depolarization ratio profiles. In a first stage, the comparison between MUSA and mini-MPL is focused on the signals collected by both the instruments. Because of the different polarization setup of MUSA [3] and mini-MPL [4], total signals calculated from the combination of the respective co- and cross-polarized channels are compared. In particular, RCS are compared, which are proportional to attenuated backscatter coefficient profiles used in the first INTERACT campaign for the comparison of ceilometers and MUSA [5]. Mini-MPL RCS is normalized on the MUSA RCS in a vertical range of about 1.5 km between 8 and 10 km above sea level (a.s.l.), identified as an aerosol free region from the quicklooks of the lidar time series.

The performances of CL51 and CS135 are assessed by comparing their attenuated backscatter coefficient profiles with those of MUSA, following a similar approach used in the first INTERACT campaign. To compare attenuated backscatter coefficient profiles at 905 and 1064 nm, obtained from the ceilometers and MUSA respectively, the spectral dependency of the attenuated backscatter coefficient is considered, using the backscatter related Angstrom exponent at 1064–532 nm retrieved from the processing of MUSA measurements. This is assumed to be the best approximation of the 1064–905 nm backscatter-related Angstrom exponent. Since ceilometers are characterized by a low signal to noise ratio (SNR) in the upper troposphere, their attenuated backscatter coefficient profiles are normalized on the corresponding MUSA profiles in the lower troposphere, in order to obtain a stable normalization. Typically, CL51 profiles are normalized in a vertical region between 4 and 5 km a.s.l.

All the time series considered in this comparison refer to night time measurements. The profiles from all the instruments are compared over a vertical resolution of 60 meters and a temporal integration ranging between 1 and 2 hours, depending on the observed atmospheric scenario. No vertical smoothing is applied to the data processing and all the profiles are cut in lower part of the atmosphere, below 1300 m a.s.l., to remove overlap effects.

A typical comparison between RCS provided by MUSA and mini-MPL is shown in the top panel of Figure 1. It is related to the observations collected on 29 August 2016 from 19:16 to 20:47 UT. The quicklooks of RCS time series (not reported) show a thick aerosol layer between 2 and 3 km a.s.l. The comparison shows a good agreement in the detection of the vertical structure of the atmospheric aerosol, though above 1.7 km a.s.l. a sort of bias affects most of the mini-MPL profile. This bias could be due to wrong corrections applied in the mini-MPL signal pre-processing, such as pile-up or trigger delay. Further investigation is required.

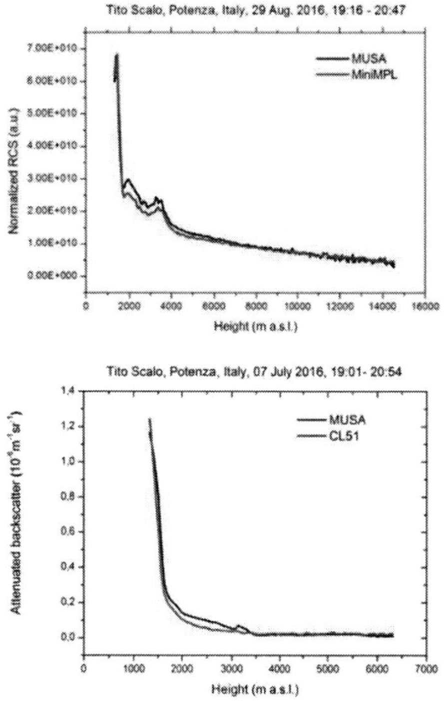

Figure 1. Comparison between range corrected signals (RCS) provided by MUSA and mini-MPL on 29 August 2016 in the time interval from 19:16 to 20:47UT (top panel) and attenuated backscatter coefficient profiles provided by MUSA and CL51 on 07 July 2016 in the time interval from 19:01 to 20:54 UT (bottom panel).

In the bottom panel of Figure 1, a typical comparison between the attenuated backscatter coefficient profiles of MUSA and CL51 is also reported. This refers to the observations of 7 July 2016 from 19:01 to 20:54 UT. The quicklooks of RCS time series (not reported) show the nocturnal boundary layer structure and above the presence of a residual layer up to an altitude higher than 3 km a.s.l. MUSA and CL51 profiles show that CL51 underestimates the attenuated backscatter coefficient value through the troposphere up to an altitude of about 3.4 km a.s.l. The differences are lower than 50% and they are probably due to a poor SNR of the ceilometer. The compared profiles also show different shapes in the investigated atmospheric region.

3 RESULTS

A first statistical analysis of simultaneous observations of all the instruments during the first half of the inter-comparison period, from July to September, is performed to investigate the differences of the mini-MPL and ceilometers with respect to MUSA. The top panel of Figure 2 shows the comparison of the frequency count distributions (fcds) of RCS values obtained from MUSA and mini-MPL. These distributions refer to 8 simultaneous measurement time series. The bottom panel of Figure 2 shows the fcds of attenuated backscatter coefficient values obtained from MUSA and CL51 and related to 13 simultaneous measurement time series. MUSA fcds are considered as the reference. The comparison of fcds confirms the overall good agreement between MUSA and mini-MPL, with some tendency of mini-MPL to overestimate MUSA for RCS values lower than 4.0 x 10^{10}, while for higher values mini-MPL underestimates MUSA. On the other hand, CL51 overestimates MUSA for the lowest values of attenuated backscatter coefficient, up to 2 x 10^{-7} m^{-1} sr^{-1}, underestimates MUSA for values higher than 2 x 10^{-7} m^{-1} sr^{-1} up to 9 x 10^{-7} m^{-1} sr^{-1} and is in good agreement with MUSA for values higher than 9 x 10^{-7} m^{-1} sr^{-1}. Moreover, fcds calculated in the vertical region above 4000 m a.s.l. (not reported) show that CL51 is not able to detect values of attenuated backscatter coefficient higher than 2 x 10^{-7} m^{-1} sr^{-1} at these altitude levels. The discrepancies between MUSA and CL51 are probably due to the ceilometer electronic distortions, reduced dymanic range, or low SNR, and not to calibration errors. Indeed, calibration errors should cause overestimations or underestimations of CL51 with respect to MUSA for all the values of fcds, though the effects of these errors might compensate over the whole dataset, so decreasing the discrepancies between fcds. In a different way, the ceilometer electronic distortions and low SNR can differently affect the different regions of fcds. These effects only partially depend on the water vapor absorption at the ceilometer wavelength, whose influence has been quantified by co-located Raman lidar and radiosounding water vapor measurements as input to radiative transfer models.

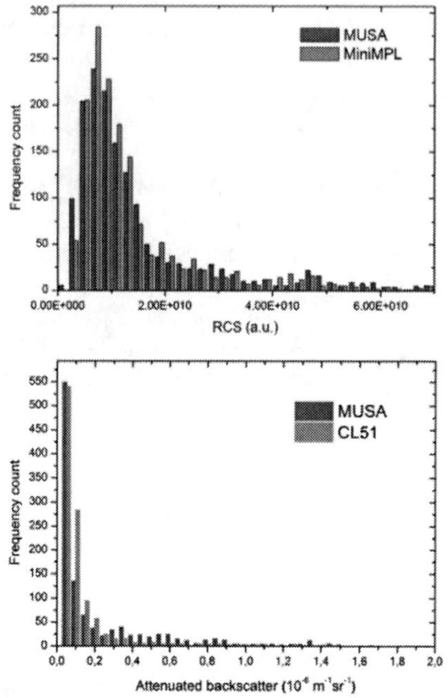

Figure 2. Frequency count distributions of RCS values retrieved from simultaneous observations of MUSA and mini-MPL (top panel) and attenuated backscatter coefficient values retrieved from simultaneous observations of MUSA and CL51 (bottom panel).

Figure 3 shows the profiles of the average percentage difference between MUSA and mini-MPL values of RCS (top panel) and between MUSA and CL51 values of attenuated backscatter coefficient (bottom panel). The vertical bars are the standard deviations of average percentage differences. The profiles show that mini-MPL underestimates MUSA below the normalization region, with differences limited to 10 %. The variability of these differences ranges from ±10-15% in the lower troposphere to about ±5% in the free troposphere (up to 8000 m a.s.l.). This also confirms the good stability of mini-MPL in its operation in the considered time period. With the exception of a small vertical region, up to about 1500 m a.s.l., where CL51 overestimates MUSA with differences lower than of 40%, and some uncertain spikes, CL51 underestimates MUSA up to an altitude of about 3700 m a.s.l., with maximum percentage differences around 50%, probably due to the low SNR of CL51. Up to an altitude of about 2500 m a.s.l., differences are limited to 30 %. Above 3700 m a.s.l., CL51 overestimates MUSA, with differences oscillating around 50% up to an altitude of about 5000 m a.s.l. and drastically increasing at higher altitudes. These overestimations are likely due to the low SNR of CL51 and the large variability of these differences reveals also some instability of the ceilometer in the considered time period.

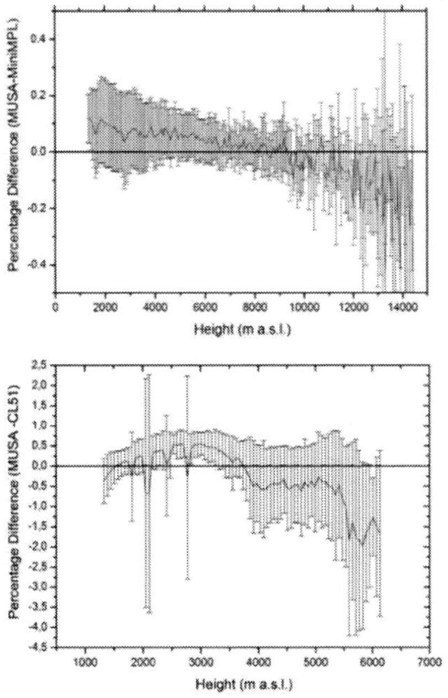

Figure 3. Profiles of the average percentage difference between MUSA and mini-MPL values of RCS (top panel) and MUSA and CL51 values of attenuated backscatter coefficient (bottom panel). The vertical bars are the standard deviations of average percentage differences.

4 CONCLUSIONS

A first statistical analysis of simultaneous observations of all the instruments involved in INTERACT – II during the first half of the inter-comparison period is performed. This analysis reveals that, among the considered devices, the mini-MPL shows the best performances with discrepancies limited to 10 % throughout the troposphere. In agreement with the outcome of INTERACT–I, ceilometers have fairly good performances in aerosol profiling in the lower troposphere, up to an altitude of about 2000 m above the ground, but they are limited at higher altitudes. The above preliminary analysis will be consolidated analyzing the cases collected in the period from October to December, including also the Lufft ceilometers data, and comparing mini-MPL and MUSA profiles of aerosol optical properties. Further analysis is ongoing also to assess the stability of the considered lidar technologies with respect to variation of working and environment temperature, aerosol loading and laser operations. Finally, the results of INTERACT-II will be also compared to those reported in literature from other international experiments carried out to assess the ceilometer performances for aerosol profiling.

ACKNOWLEDGEMENTS

The financial support by the European Commission for ACTRIS-2 in HORIZON 2020 research and innovation programme under grant agreement no. 654109 is gratefully acknowledged. The authors would also like to thank Sigma Space Corporation, Vaisala and Campbell Scientific for contributing the mini-MPL, CL51 and CS135, respectively, for INTERACT - II campaign.

References

[1] D'Amico, G., et al., 2016: EARLINET Single Calculus Chain – technical Part 1: Pre-processing of raw lidar data, *Atmos. Meas. Tech., 7, 1979–1997.*

[2] Mattis, I., et al., 2016: EARLINET Single Calculus Chain – technical Part 2: Calculation of optical products, *Atmos. Meas. Tech., 9, 3009–3029.*

[3] Freudenthaler, V., et al., 2009: Depolarization ratio profiling at several wavelengths in pure Saharan dust during SAMUM 2006, *Tellus B, 61, 165–179.*

[4] Flynn, C.J., et al., 2007: Novel polarization-sensitive micropulse lidar measurement technique, *Optics Express 15, 6,2785-2790.*

[5] Madonna, F., et al., 2015: Ceilometer aerosol profiling versus Raman lidar in the frame of the INTERACT campaign of ACTRIS, *Atmos. Meas. Tech., 8, 2207–2223.*

NOVEL LIDAR ALGORITHMS FOR ATMOSPHERIC SLANT-RANGE VISIBILITY, PLANETARY BOUNDARY LAYER HEIGHT, METEOROGICAL PHENOMENA AND ATMOSPHERIC LAYERING MEASUREMENTS

Alexandros Pantazis[1*], Alexandros Papayannis[1], and Georgios Georgoussis[2]

[1]*National Technical University of Athens, Laser Remote Sensing Laboratory, Physics Department, 15780 Zografou, Greece, *Email: alexpant@mail.ntua.gr*
[2]*Raymetrics S.A., Spartis 32, 14452 Metamorfosi, Athens, Greece*

ABSTRACT

In this paper we present a development of novel algorithms and techniques implemented within the Laser Remote Sensing Laboratory (LRSL) of the National Technical University of Athens (NTUA), in collaboration with Raymetrics S.A., in order to incorporate them into a 3-Dimensional (3D) lidar. The lidar is transmitting at 355 nm in the eye safe region and the measurements then are transposed to the visual range at 550 nm, according to the World Meteorological Organization (WMO) and the International Civil Aviation Organization (ICAO) rules of daytime visibility. These algorithms are able to provide horizontal, slant and vertical visibility for tower aircraft controllers, meteorologists, but also from pilot's point of view. Other algorithms are also provided for detection of atmospheric layering in any given direction and vertical angle, along with the detection of the Planetary Boundary Layer Height (PBLH).

1 THE LADDER TECHNIQUE

There has been a lot of questioning on whether backscattering $\beta_{aer}(\lambda,R)$ and extinction $\alpha_{aer}(\lambda,R)$ coefficients can be derived from lidar slant range measurements. The well-known lidar equation for the power received at wavelength λ from a distance R, is:

$$P(\lambda,R)=P_{OL}\cdot\frac{A_0}{R^2}\cdot\beta(\lambda,R)\cdot\eta(\lambda)\cdot\xi(R)\cdot\Delta R\cdot\exp^{-2\int_0^R\{\alpha_{Mie}(\lambda,r)+\alpha_{Ray}(\lambda,r)\}dr} \quad (1)$$

where, P_{OL} is the power of the transmitted laser pulse beam, A_0 is the diameter of the receiver's telescope, ΔR is the space resolution capability of the lidar, ξ(R) is the geometrical form factor, η(R) is the lidar opto-electronic efficiency coefficient for Mie and Rayleigh scattering, accordingly. According to Klett [1, 2] solved this equation for $\alpha_{aer}(\lambda,R)$ and $\beta_{aer}(\lambda,R)$ in a way where:

$$a_{aer}(\lambda,R)=\frac{\exp[S(R)-S(R_F)]}{\frac{1}{a(\lambda,R_F)}+2\int_R^{R_F}\exp[S(R')-S(R_F)]dR'}-a_{Ray}(\lambda,R) \quad (2)$$

and

$$\beta_{aer}(\lambda,R)=\frac{\exp[S'(R)-S'(R_F)]}{\frac{1}{\beta(\lambda,R_F)}+2\int_R^{R_F}\frac{1}{C(R')}\exp[S'(R')-S'(R_F)]dR'}-\beta_{Ray}(\lambda,R) \quad (3)$$

where,

$$S'(R)-S'(R_F)=S(R)-S(R_F)-\frac{16\pi}{3}\int_R^{R_F}\beta_{Ray}(1-\frac{3}{8\pi\cdot C(R')})dR' \quad (4)$$

$$S(R)\equiv\ln[P'(\lambda,R)\cdot R^2]=RSLS \quad (5)$$

where, R_F is a reference height where the values of $\alpha(\lambda,R)$ and/or $\beta(\lambda,R)$ are known (e.g. for a purely molecular atmosphere), RSLS is the Range-Squared Lidar Signal, with atmospheric and electronic noise background (BG) has been removed with:

$$P'(\lambda,R)=P(\lambda,R)-BG \quad (6)$$

Also, $C(\lambda,R)$ is the lidar ratio:

$$C(\lambda,R) = \alpha_{aer}(\lambda,R) / \beta_{aer}(\lambda,R) \quad (7)$$

It is obvious that in vertical pointing lidars there is a reference height where we can set the values of $\alpha(\lambda,R)$ and/or $\beta(\lambda,R)$ equal to the known molecular ones, following the Boundary Point Solution techniques as shown by Klett [10]. Thus, any received lidar signal can be calibrated according to this reference value. Therefore, the reference height R_F in Eqs. (2), (3) and (4) can be set as the vertical height at which the atmosphere is purely molecular and the signal to noise ratio is still quite high (e.g. >2-3).

By keeping the values of $\alpha(\lambda,R)$ and/or $\beta(\lambda,R)$ at the R_F, with the novel LADDER technique, we retrieve $\alpha_{aer}(\lambda,R)$ and/or $\beta_{aer}(\lambda,R)$ with the 3D-lidar by making measurements like "stair steps" from vertical to horizontal, every 1°, like "walking down a ladder". In this path, we take as R_F at Eq. (2) and (3) the distance, at which we abstract one or two bins (optional) for every new measurement and we keep the last value of $\alpha_{aer}(\lambda,R)$ and/or $\beta_{aer}(\lambda,R)$, at which R_F was previously made, as the new calibrating values for $\alpha_{aer}(\lambda,R)$ and/or $\beta_{aer}(\lambda,R)$. So, if R_F, $\alpha_{aer}(\lambda,R)$, $\beta_{aer}(\lambda,R)$ are the last known values and $R_{F\text{-}new}$, $\alpha_{aer\text{-}new}(\lambda,R)$, $\beta_{aer\text{-}new}(\lambda,R)$, the new ones, then we have:

$R_{F\ new} = R_F$ - 1bin(=7.5 m) (2 or more bins, optional) (8)

at which, according to the old values $\alpha_{aer}(\lambda,R)$, $\beta_{aer}(\lambda,R)$ the new values $\alpha_{aer\text{-}new}(\lambda,R)$, $\beta_{aer\text{-}new}(\lambda,R)$ are calibrating the next received signal. In this way, the measurement of every received signal and $\alpha(\lambda,R)$ and/or $\beta(\lambda,R)$ is made, taking into account the old (last) known values of these coefficients/received signals and by reducing R_F at the same time, there is being a continually "calibration" of the values of $\alpha(\lambda,R)$ and $\beta(\lambda,R)$, from the old to the new position:

$$R_{F\text{-}new} < R_F \qquad (9)$$

So, during measurements, we reset the values of the signal received for every "drop" we make towards to the horizontal position and in this way, we are able to keep the values of the signal and $\alpha(\lambda,R)$ and/or $\beta(\lambda,R)$ to the "original" ones and we avoid the ambiguities by presuming a homogeneous atmosphere in every measurement, especially within long distances.

Known methods like the slope technique and/or the ratio method, are mainly using this kind of assumption (homogeneous atmospheres) to provide slant range atmospheric parameters, but in this way, the error of the measurement increases according to the unsustainability of the atmospheric conditions, by means of distance R. Here, by using the LADDER technique (Fig. 1), we saw that by resetting the values of $\alpha(\lambda,R)$ and $\beta(\lambda,R)$, we are able to use the main lidar Eq. (1), like if lidar was in vertical position and molecular measurements were made to determine $\alpha_{aer}(\lambda,R)$ and $\beta_{aer}(\lambda,R)$ at lower heights.

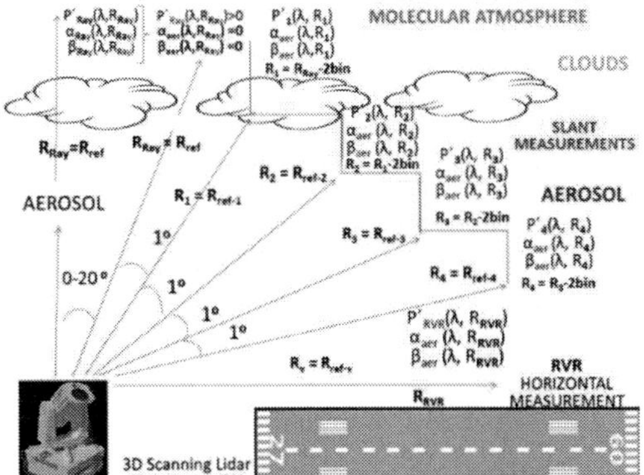

Figure 1: *The LADDER technique working principle (RVR-Runway Visual Range).*

2 NAVIS-T/P ALGORITHM

After the use of denoising filters and the choice of the slant range measurement technique, the NAVIS (Novel Algorithm for VISibility measurement) – T (Tower of the Airport)/P (Pilot) is applied for further signal processing. This algorithm is able to determine the visibility of tower towards to the pilot and vice versa, according to WMO and ICAO rules of daytime visibility, at the airports. According to [4,5,7] and as modified by Wright et al. [6], the values of $a_{aer}(\lambda, R)$ and/or $\beta_{aer}(\lambda,R)$ are shown as a function of wavelength, under different atmospheric conditions (cumulus clouds, haze, etc.). A maritime model has been used in these calculations with MAPP (Modular Atmospheric Propagation Program) developed by SRI (Stanford Research Institute), because of the global nature of its application [6]. Additionally, Measures [7] also provides another figure where a number of different atmospheric conditions are taken into account (fog, haze, etc.) with extinction coefficient versus visibility at 550 nm at daylight.

So, a combination of these two figures is been incorporated into the NAVIS Algorithm. Then, the AOD (Aerosol Optical Depth) (τ) (Eq. 13) is been calculated from the lidar - Tower point of view and from the maximum range of the lidar signal, towards the position of lidar - Tower (Pilot's point of view) using the trapezoid method:

$$\tau(0,R) = \int_0^R a(\lambda,R')dR' \qquad (13)$$

where, R′ is the distance where $\alpha(\lambda,R)$ has been measured through a distance resolution of 1 bin (i.e. 7.5 m). By using the values of $a_{aer}(\lambda, R)$ and/or $\beta_{aer}(\lambda,R)$ provided in [4,5,6,7], under different atmospheric conditions (cumulus clouds, haze, etc.), the visibility (Tower's and Pilot's) is been calculated and a categorization of atmospheric conditions–layering is been made at slant and/or vertical position of the lidar. Using a 3D scanning lidar system, a profile of fully atmospheric layering at different angles, azimuth and vertical, can be provided in near-real time, plus the daytime visibility, according to WMO and ICAO rules, using the empirical and well qualified, Koschmieder law (at 550 nm):

$$Vis=3/\alpha \text{ (for } \varepsilon=0.05 \text{ at 550 nm)} \qquad (14)$$

$$Vis=3.912/\alpha \text{ (for } \varepsilon=0.02 \text{ at 550 nm)} \qquad (15)$$

where, Vis is the visibility in m or km and ε is a pure number, showing a contrast threshold, as a difference of the self-luminance of any object and the general luminance of the area viewed from a standing position.

https://doi.org/10.1051/epjconf/201817611003

It is well known that the human visual range is ~400-700 nm, with "peaks" of visibility at 555 nm for daytime and 515 nm for night conditions, according to CIMO (Commission for Instruments and Methods of Observation) [8] for relative luminous efficiency of monochromatic radiation. By using the visibility versus the aerosol extinction coefficient at 550 nm [5], we are able to have a good approximation of the above required values at daytime, at 355 nm.

Another way to do that is by using Eq. (16) for available measurements that have been made at any location with lidar at any wavelength, categorized by season and chemical atmospheric composition (AAE – Absorption Ångström Exponent, which determines the wavelength dependence of absorption and EAE – Extinction Ångström Exponent which is an indicator of the particle size of the atmosphere [9]), together with certain meteorological conditions, like the relative humidity (RH), temperature, rain, etc. and geographical location. The idea is that through measurements of different wavelengths of an area at different locations of interest, anyone would be able to acquire the visibility and meteorological conditions with good approximation, at a single wavelength (i.e. 355 nm) (see Eqs. 17, 18), deployable at any airport:

$$\alpha(\lambda_1)\,/\,\alpha(\lambda_2) = \beta(\lambda_1)\,/\,\beta(\lambda_2) \tag{16}$$

in order to be able to acquire:

$$\beta(\lambda_1) = C * \beta(\lambda_2) \tag{17}$$

$$\alpha(\lambda_1) = C * \alpha(\lambda_2) \tag{18}$$

where, λ_1, λ_2 are different wavelengths, for example λ_1=355 nm, where measurements have been taken and λ_2=550 nm, at which the visibility and categorization of the atmospheric conditions have been described by Measures [7]; C is a constant that is defined by different values, according to the calculations made at different seasons, chemical compositions of the atmosphere, at any geographical location using Eqs. 16, 17 and 18.

3 METCON ALGORITHM

In this algorithm METCON (METeorological CONditions), we make use of the NAVIS algorithm, to provide meteorological conditions like clouds, haze, fog, etc., versus the distance R, for any vertical or slant range lidar measurements. The same 3D scanning lidar and the above-mentioned techniques and data, can be used to characterize the atmospheric conditions (fog, haze, etc.), plus the visibility. Fig. 2 shows a case study of the NAVIS and METCON algorithms, where the tower visibility from lidar's vertical pointing direction is found to be 940 m and the pilot's visibility from a 3000 m height (distance) is of 1340 m, in full agreement with Measures [7].

a)

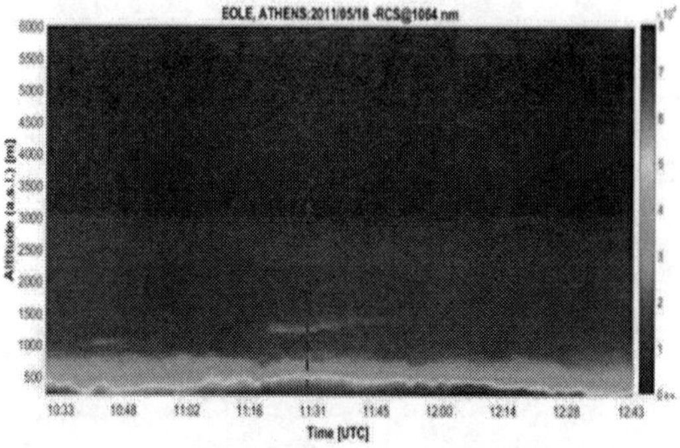

b)

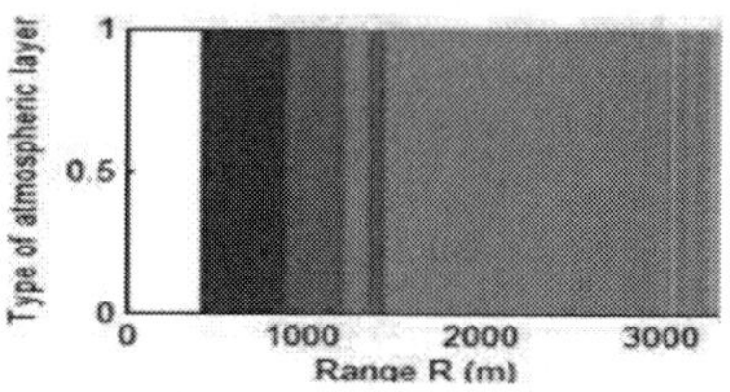

***Figure 2**: a) RCS signal acquired at 1064 nm by the LRSU-NTUA (16-05-2011), b) NAVIS and METCON algorithms applied in 3D lidar data pointing to the vertical (355 nm) at 11:30:10 UTC, starting at 417.5 m above ground level up to 3000 m. Where Blue is "Light Fog", Purple is "Thin Fog", Light Blue is "Light Haze" and Green is "Haze" [7]. Tower visibility is 940 m and pilot's visibility is 1340 m from the range of 3000 m.*

4 VASPAT-L/D ALGORITHM

The VASPAT-VARiable SPAce Time (VASPAT) algorithm is used to determine aerosol layering (L) and distribution (D), based on the existence of the $\beta_{aer}(\lambda,R)$ and $\alpha_{aer}(\lambda,R)$ coefficients, in 3D of the current atmosphere. VASPAT-L/D is pre-required for the next presented algorithm VASPAT-PBLH.

In VASPAT-L/D we take the values of $\alpha_{aer}(\lambda,R)$ or $\beta_{aer}(\lambda,R)$ versus height or slant range distance (R); we then create "space filters" of different number of bins, in order to create different "tools", to further process the $\alpha_{aer}(\lambda,R)$ and/or $\beta_{aer}(\lambda,R)$ profiles. In this work we used 12 different space filters (G2 to G269) from 15 m (2 bins) to 2017.5 m (269 bins), a good space filter for the RVR (Runway Visual Range).

Afterwards, we check if the measured value of $\alpha_{aer}(\lambda,R)$ in any of these filters is greater than $4*10^{-5}$ m^{-1} [7].

Then, we check if the ratio of two consecutive average values of $\alpha_{aer}(\lambda,R)$ in any of these "space filters" from a package of bins to the next package of equal number of bins are greater than 10%. As next step, the algorithm calculates the derivatives of all space filters (dG3/dR), at every distance (R), for vertical or slant measurements and we do the same to produce the second order derivatives (d^2G3/dR^2). A series of calculations is thus been made to provide the atmosphere aerosol layering (VASPAT-L) so that we will able to have a clear vision of the 3D spatial distribution of $\alpha_{aer}(\lambda,R)$ and/or $\beta_{aer}(\lambda,R)$ around airport sites (VASPAT-D).

5 VASPAT-PBLH ALGORITHM

At first, we place the 3D scanning lidar pointing to the vertical or up to $\pm 20^o$ around the vertical. Then, we start measuring for at least 15-20 min in order to acquire at least 15 lidar data files. Then, we proceed to time filters. Keeping the results provided by VASPAT-L/D, we can make use of time filters after the use of the above mentioned space filters. In this case our goal is to find the PBLH with a time slot corresponding to 15-20 lidar data files (e.g. each data file is acquired within 1.5 min.) Then, we keep track of the layering (L) and distribution (D) calculated values, and then, we make use of the layer "endings" ($\alpha_{aer}(\lambda,R)$ negative 1st derivative and positive 2nd derivative). Then, we check which is the most common time depending "ending" layer in all times series within for e.g. 15-20 min period of measurements, which finally, determines the most probable PBLH.

6 INFRASTRUCTURE

The lidar system used for this study is the EOLE 10-wavelength elastic-Raman-DIAL lidar system which is located in Athens (37.96°N, 23.78°E, 220 m) at LRSL-NTUA. It can perform independent measurements of $\alpha_{aer}(\lambda,R)$ and $\beta_{aer}(\lambda,R)$ (at 355, 532 and 1064 nm), as well as the water vapor and ozone mixing ratio in the troposphere [11]. The algorithms presented in this paper were qualified and tested (at 355 nm and verified by the data obtained at 1064 nm) based on EOLE data [11]. These algorithms will be then implemented in the software of the 3-dimensional scanning lidar system constructed by Raymetrics S. A., based on 355 nm with co-polar and cross-polar detection using a 387 nm N_2 Raman detection channel, designed for aviation, meteorology and environmental applications for vertical and slant range measurements.

7 CONCLUSIONS

In this paper we presented several novel algorithms and techniques applied in vertical and slant range lidar measurements to calculate horizontal, slant and vertical visibility for tower aircraft controllers, meteorologists, and for the pilot's point of view, as well as for the detection of atmospheric layering and distribution of $\alpha_{aer}(\lambda,R)$ and/or $\beta_{aer}(\lambda,R)$ in 3dimensions, along with the detection of the Planetary Boundary Layer Height. We also provided a typical example based on real lidar data obtained over Athens by the EOLE system.

References

[1] Klett, J., 1981: Stable analytical inversion for processing lidar returns, *Appl. Opt.*, **20**, 211-220.

[2] Klett, J., 1985: Lidar inversion with variable backscatter extinction ratios, *Appl. Opt.*, **24**, 1638-1643.

[3] ICAO, 2005: Manual of runway visual range observing and reporting practices, Doc 9328 AN/908.

[4] Deirmendjian, D., 1964: Scattering and polarization properties of water clouds and hazes in the visible and infrared, *Appl. Opt.*, **3** (2), 187-196.

[5] Deirmendjian, D., 1969: Electromagnetic scattering on spherical polydispersions, *American Elsevier Publishing Co.*, Inc., New York.

[6] Wright, W. L., et al., 1966: A preliminary study of air pollution measurement by active remote sensing techniques, NASA, CR-132724 C89.

[7] Measures, R. M., 1992: Laser remote sensing. Fundamentals and applications, Krieger, sys No 9247, MEA 621.3678.

[8] CIMO (Commission for Instruments and Methods of Observation (WMO-No. 807).

[9] Russell, P. B., et al., 2010: Absorption Ångström exponent in AERONET and related data as an indicator of aerosol composition, *Atmos. Chem. Phys.*, **10** (3), 1155-1169.

[10] Kovalev, V.A., and Eichinger, W. E.,, 2004: Elastic Lidar, Wiley Interscience.

[11] Kokkalis, P., et al., 2012: The EOLE lidar system of the National Technical University of Athens, *Proc. 26th International Laser Radar Conference*, pp. 629-632, 25-29 June 2012, Porto Heli, Greece.

STUDIES OF SAHARAN DUST INTRUSIONS OVER BUCHAREST USING CEILOMETER'S MEASUREMENTS AND SATELLITE DATA

Denisa Urlea[1,2*], Andreea Boscornea[3], Sorin Nicolae Vâjâiac[3], Florica Ţoancă[4], Nicu Barbu[2], Sabina Ştefan[1], Ionuţ Bunescu[3,5]

[1] *University of Bucharest, Faculty of Physics, Măgurele - Ilfov, Romania*
[*]denisaurlea@gmail.com
[2] *Romanian Air Traffic Services, Bucharest, Romania*
[3] *National Institute for Aerospace Research "ElieCarafoli", Bucharest, Romania*
[4] *National Institute of Research & Development for Optoelectronics, Măgurele - Ilfov, Romania*
[5] *POLITEHNICA University of Bucharest, Faculty of Aerospace Engineering, Bucharest, Romania*

ABSTRACT

Three case studies of Saharan dust intrusions over southern Romania were performed. For these studies the database from the ceilometers located at Magurele and Strejnic was used. In addition, the meteorological conditions were analyzed using the WLK Catalogue based on the Objektive Wetterlagenklassifikation classification of the weather types [1]. This catalogue uses information from three basic tropospheric levels: 925, 700 and 500 hPa, and information on the precipitable water content over the entire atmosphere column. Geopotential fields at 925hPa and 500hPa are used for establishing the cyclonicity or anticyclonicity, while the U and V components of wind at 700hPa for establishing the dominant direction of the wind flow. For better understanding of the atmospheric parameters we performed HYSPLIT dispersion and trajectories analysis in conjunction with DREAM model output data.

1. INTRODUCTION

The desert dust aerosol is an important factor of impact for the European regional climate. The most intense and persistent activity occurs in the southern part of the Tell Atlas Mountain, Tunisia and North-East of Algeria having a maximum on April-May often extending to August-September [2] and follows a Mediterranean path to Central and Eastern Europe [3]. Romania can be considered as being relatively close to the source region taking into account that over Belgrade were found 3.5 up to 9.8 μm diameter Saharan dust particles over two years of studies [4] and a major event took place over Germany in April, 2014 [5].
This study is exploring the capabilities of CL31 VÄÏSÄLA ceilometer to monitor, evaluate and describe the aerosols, focusing on Saharan dust intrusions over southern Romania, particularly at Magurele and Strejnic, two sites nearby Bucharest. The geographical and meteorological conditions are presented in Section 2.1. Three months were investigated (July, August and September 2016) of coincident data at the two locations and the matching dates with possible dust intrusions forecasted by the DREAM model. All data and images were derived from the BSC-DREAM8b and the NMMB/BSC-Dust models, operated by the Barcelona Supercomputing Center (http://www.bsc.es/ESS/). Three study cases were selected for the analysis. Data and methods used are presented in Section 2.2. In Section 3 we discuss the vertical profiles from simultaneous ceilometers aerosol backscatter coefficients, satellite imagery and synoptic meteorological data. The concluding remarks end our paper.

2. METHODOLOGY

2.1 Instruments and Measuring Sites

In this paper we studied three Saharan dust intrusion cases over Romania using two ceilometers. These instruments are traditionally been used only to report cloud base height and vertical visibility [6]. Ceilometers are also useful in studying the atmospheric phenomena by using backscatter profiles and time series images [7], to evaluate the cloud fraction [8], the aerosol layers height in the atmosphere as Saharan dust intrusions or volcanic ash [9, 10] and the PBL (Planetary Boundary Layer) dynamics studies [9].
We used data collected by two ceilometers located in southern part of Romania, one situated at Bucharest-Magurele (26.029E, 44.38N, 93m amsl) and the other at Strejnic (44.55N and 25.57E, 252m asl). The ceilometer located at Magurele is installed

on the roof of the Physics Faculty (20m above ground level) and managed by the University of Bucharest. The ceilometer installed at Strejnic is managed by National Institute of Aero-Spatial Research. Both instruments are VÄÏSÄLA CL31 type on single-lens technology, 910nm wavelength, with 20m vertical resolution and 2s temporal resolution. Its laser diodes are pulsed with a repetition rate of 10 kHz at high energy [11, 12]. Apart from ceilometer located at Strejnic which has the vertical range of determination up to only 4500m, the Magurele ceilometer is calibrated up to 7000m.

The geographical and climatological characteristics are different for the two locations. Magurele is situated nearby Bucharest, the largest town in Romania with strong urban and plain climatology. Strejnic is located at a higher altitude, nearly 200m upper and has a climatology with mountain influences. Therefore, it is important to have both locations experiencing the same synoptic pattern at the same time. In order to increase the ceilometers' data accuracy we investigated the dates when the columnar air characteristics over the two locations were similar.

2.2. Data and methods

In order to find the days with Saharan dust intrusions, the ceilometers' measurements for months July, August and September 2016 were investigated simultaneous with daily dust intrusions forecasted by Dust Regional Atmospheric Model (DREAM).

Three days with coincident data were selected as study cases for analysis: July 17th (case I), September 10th (case II) and September 17th (case III).

The ceilometers' measurements were processed using the Cloud View software. The output data consist of a text file for each day, every files includes cloud base-heights and aerosol layer heights with temporal resolution of 16s. These data were processed with LabView software to determine time series of the backscatter signals and the backscatter coefficient vertical profiles.

In order to have a good evaluation of the aerosols presence and characteristics, and because it is required another data source as satellite or lidar instrument to have a back-up for the investigation [11] we used in addition coincidental synoptic data, aviation meteorological and satellite data, confirming the originating dust storm.

To evaluate the atmospheric parameters and meteorological conditions we used multiple satellite products. The EUMETSAT analysis was extracted from the cloud product diagnosis Cloud Mask (CMa) from NWCSAF (Nowcasting and very Short Range Forecasting), which allows to identify the cloud types and cloud top temperature and height. In order to localize the source, we used the Single Scattering Albedo (SSA) at 550nm (it was found to be higher than 0.95 for dust [13]) provided by MODIS (Moderate Resolution Imaging Spectroradiometer) facilities, spectrometers abroad Terra and Aqua satellites. In order to find the dust source and monitor the plume dispersing path we used Absorbing Aerosol Index (AAI) from the Global Ozone Monitoring Experiment–2 (GOME-2). The Absorbing Aerosol Index (AAI) indicates the presence of elevated absorbing aerosols in the Earth's atmosphere. The aerosol types that are mostly seen in the AAI are desert dust and biomass burning aerosols.

The two ceilometers dataset were organized into an inter-comparison study in order to increase the data accuracy.

The air mass characteristics were evaluated performing vertical profiles from recomputed soundings with HYSPLIT computing facilities (GDAS, 0.5 deg) and Baneasa atmospheric sounding provided by the Romanian National Meteorological Administration.

3. RESULTS

Data from two ceilometers, were evaluated for three dust intrusion cases in order to determine ceilometer's capabilities to detect aerosol layers. The study was thought as an intercomparison between the data gathered from the two ceilometers with coincidental data and this is why the outcome consists in only three study cases. The analysis for the three cases started with the prognosis of DREAM and continues using all the announced data in Section 2.

For the first case the DREAM model is displaying a vertical profile of dust concentration around 45μg/m^3 value which is not exceeding the threshold value of 50μg/m^3. For the other two cases we found 48μg/m^3 to 50μg/m^3 at heights ranging from 1500m to 3500m. The cross-section data from

DREAM output is displaying values of 20-40μg/m^3 around 3000m height (not shown) for the all three cases.

In order to locate the Saharan dust plume source we performed HYSPLIT trajectory model. The trajectories obtained are passing nearby the regions where dust/sand storms were certified as being active by the MODIS output data with Deep Blue Aerosol retrieval in two of the three cases with AOT values ranging from 1.17 to 1.52.

Other sources, as synoptic and aeronautical data were used i.e.: confirmation on METAR telegrams for 15 September 2016 (case III) as blowing dust event for more than six hours at Biskra Airport (34.51N 05.44E).

EUMETSAT Dust Channel product with superimposed 500hPa contour field (ECMWF) (not shown) and WLK map product (not shown) are consistent with the dust trajectory displayed by the HYSPLIT trajectories model presented for the all dust events. The dust layers over the two stations were thin and sometimes dispersed. Therefore, because it was difficult to discern between aerosol layers and clouds, we payed attention especially to color and in the intensity of backscattering signals in units of $m^{-1} \cdot sr^{-1}$.

From the two time series obtained from the two ceilometers is clearly shown the presence of the low level clouds, which are confirmed by the analysis provided by the cloud product diagnosis Cloud Mask (CMa) from NWCSAF (Nowcasting and very Short Range Forecasting), EUMETSAT (not shown). This particular day (17.07.2016) was characterized by stratus clouds in the morning and broken clouds during daytime.

Study case I data are shown in Figure 1 where the red line is representing the backscatter vertical profile from Strejnic ceilometer. The profile is selected in a 30 minutes interval from 05:00 UTC to 05:30 UTC.

The vertical profile (Figure 1) gives us a more precise value for the maximum of the backscatter coefficient: $\beta = 1 \cdot 10^{-4} m^{-1} sr^{-1}$ at 2700m height. Grater values of $4 \cdot 10^{-4} m^{-1} sr^{-1}$ or the backscatter coefficient are displayed in lower levels, as 500-1100m, consistent with the EUMETSAT CMa product (not shown) which displayed Stratus/Fog presence at 06:00 UTC in the Strejnic region.

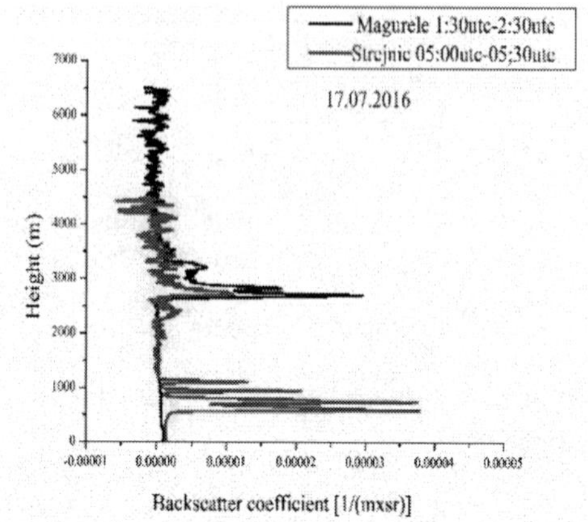

Figure 1. Superimposed backscatter vertical profiles from Magurele and Strejnic ceilometers

Magurele ceilometer backscatter coefficient vertical profile is represented in black line, showing consistency with Strejnic vertical profile i.e values of $1 \cdot 10^{-4} m^{-1} sr^{-1}$ at 3000m as the plume was dispersed. On the backscatter coefficient time series one can see a dispersed signal of the suspected dust intrusion between 2700m and 3200m (Figure not shown).

For the second and third study case there were obtained dispersed signals on time series backscatter coefficient image and values of beta below $0.5 x 10^{-4} m^{-1} sr^{-1}$ on backscatter coefficient vertical profiles situated at heights ranging 2000 to 3500m (Figures not shown). Performing HYSPLIT model soundings facilities we have obtained similar columnar air characteristics for the two locations.

For the all three study cases we found the same similarities between levels of clouds or level of temperature inversion obtained from soundings and the correspondent levels obtained from CL31 measurements.

These similarities allow us to consider the measurements of the two ceilometers being consistent with the hypotheses that the Saharan dust intrusion was manifested on the two locations and ceilometers' measurements are correct and significant.

4. CONCLUSIONS

The CL31 ceilometer was found to be a useful instrument for determining the atmosphere proprieties. Although the Saharan dust plume discussed in this paper was a faint one, we have good reason to say that the ceilometer is capable to sense the aerosol even in little quantities.

Usage of two ceilometers with coincidental data would allow us to discern between the clouds and aerosol's layers within the same synoptic conditions but keeping in mind the existing geographical difference between the ceilometer's locations.

This study has shown the importance of having some other tools for backup when performing aerosol detection.

ACKNOWLEDGEMENTS

The authors gratefully acknowledge the NOAA Air Resources Laboratory (ARL) for the provision of the HYSPLIT transport and dispersion model (http://www.ready.noaa.gov) used in this publication. The authors gratefully acknowledge the permission to use the EUMETSAT satellite products with ECMWF data overlaid delivered by EUMeTrain e-PORT PRO, access permitted by courtesy of Mr. Andreas Wirth.

References

[1] Dittmann E, Barth S, Lang J, Muller-Westermeier G., 1995: Objektive Wetterlagenklassifikation (Objective weather type classification). *Ber. Dt. Wettrd.* **197**, Offenbach a. M., Germany.

[2] Prospero, J. M., P. Ginoux, O. Torres, S. E. Nicholson, and T. E. Gill, 2002: Environmental characterization of global sources of atmospheric soil dust identified with the nimbus 7 Total Ozone Mapping Spectrometer (TOMS) absorbing aerosol product, *Rev. Geophys.*, **40**(1), 1002, doi: 10.1029/2000RG000095.

[3] Israelevich, P., Ganor, E., Alpert, P., Kishcha, P., and Stupp, A., 2012: Predominant transport paths of Saharan dust over the Mediterranean Sea to Europe, *J. Geophys. Res.: Atmospheres*, **117**(D2).

[4] Vukmirovic, Z., Unkaševic, M., Lazic, L., Tošic, I., Rajšic, S., Tasic, M., 2004: Analysis of the Saharan dust regional transport, *J. Meteorol. and Atmos. Phys.*, **85**(4), 265-273.

[5] Forstner, J., B. Ritter, B. Vogel, D. Rigier, D. Majewski, H. Vogel, 2016: Assessment of the mineral dust related reduction of photovoltaic power generation in Central Europe with COSMO-ART and ICON-ART Conf paper, Atmospheric Dust-DUST2016.

[6] Illingworth, A. J., RJ Hogan, E.J. O'Connor, D. Bouniol, J. Delanoë, J. Pelon, A. Protat, M.E. Brooks, N. Gaussiat, D.R .Wilson, D.P .Donovan, H.K .Baltink, G.J. van Zadelhoff, J.D. Eastment, JWF Goddard, CL Wrench, M Haeffelin, OA Krasnov, HWJ Russchenberg, J. M. Piriou, F. Vinit, A. Seifert, A.M. Tompkins, U. Willén, 2007: Cloudnet-continuous evaluation of cloud profiles in seven operational models using ground-based observations, *Bull. Amer. Meteorol. Soc.,* 88, 883–898, doi: 10.1175 BAMS-88-6-883.

[7] Ungureanu, I., S. Stefan, and D. Nicolae, 2010: Investigation of the cloud cover and Planetary Boundary Layer (PBL) characteristics using ceilometer CL 31. *Rom. Rep. Phys.*, **62**(2), 396-404.

[8] Stefan, S., I. Ungureanu, C. Grigoras, 2014: A survey of cloud cover over Magurele, Romania, using ceilometer and satellite data. *Rom. Rep. Phys.*, **66**(3), 812-822.

[9] Münkel, C., Eresmaa, N., Räsänen, J., Karppinen, A., 2007: Retrieval of mixing height and dust concentration with lidar ceilometer. *Boundary-Layer Meteorol.*, **124**(1), 117-128.

[10] Nemuc, A, I. S. Stachlewska, J. Vasilescu, A. Gorska, D. Nicolae, C. Talianu, 2014: Optical Properties of Long-Range Transported Volcanic Ash over Romania and Poland during Eyjafjallajokull Eruption in 2010, *Acta Geophys.*, 61, 350-266. doi:102478/s11600-013-0180-7.

[11] Kotthaus S., E.O'Connor, C. Münkel, C. Charlton-Perez, M. Gabey, C.S.B. Grimmond, 2016: Recommendation for processing atmospheric attenuated backscatter profiles from VÄÏSÄLA CL31 Ceilometers. *Atmos. Meas. Tech. Discuss*, doi: 10.5194/amt-2016-87.

[12] VÄÏSÄLA Ceilometer CL31User's Guide.

[13]. Levy, R. C., Remer, L. A., Kleidman, R. G., Mattoo, S., Ichoku, C., Kahn, R., Eck, T. F., 2010: Global evaluation of the Collection 5 MODIS dark-target aerosol products over land. *Atmos. Chem. Phys.*, **10**(21), 10399.

RESEARCH AND DEVELOPMENT OF COMMERCIAL LIDAR SYSTEMS IN ROMANIA: CRITICAL REVIEW OF THE ESYRO LIDAR SYSTEMS DEVELOPED BY SC ENVIROSCOPY SRL (ESYRO)

Marius Mihai Cazacu[1, 2, 3], Ovidiu Tudose[4], Dragos Balanici[1] and Ioan Balin[1]

[1] *SC EnviroScopY SRL, Iasi - Romania, *ioan.balin@enviroscopy.com*
[2] *Gheorghe Asachi Technical University of Iasi, Department of Physics, 700050 Iasi, Romania*
[3] *Alexandru Ioan Cuza University of Iasi, Faculty of Physics, Atmosphere Optics, Spectroscopy and Lasers Laboratory, 700506 Iasi, Romania*
[4] *SC Inoesy SRL, Iasi - Romania*

ABSTRACT

This paper is shortly presenting the two basic lidar system configurations respectively a micro-lidar and a multi-wavelength lidar systems developed by SC EnviroScopY SRL (**ESYRO**) from Iasi – Romania in the last decade. Furthermore in addition to the comparative analysis of the two technical configurations the examples of various tests and the capability of the two systems to perform are here presented. Measurements samples of aerosols, clouds, PBL, depolarization and Saharan dust are also illustrated.

1 INTRODUCTION

SC EnviroScopY SRL starts from its creation as start-up in 2006 to develop commercial aerosols lidar systems in Romania. According to size, the atmospheric aerosols have a large range, from nanometric particles (a couple of molecules) to particles larger than 10 μm. Aerosols influence the air quality and visibility, the net heat balance received by Earth's crust directly by reflecting the solar radiation back into space and indirectly by modifying the solar radiation absorption and reflection coefficients of the various cloud formations [5, 7]. Therefore, vertically measurements by lidars of physical and optical parameters of the aerosols are still of great interest. In addition the vertically monitoring of regional air pollution in order to complement the ground-based stations is nowadays clearly confirmed both by the key information concerning the atmospheric dynamics (as planetary boundary layer height and its variability) and the regional or long-range transport aerosols load estimation. Furthermore, the interaction between the aerosols (regional/global) as a trigger of regional pollution and meteorology (i.e. extreme events as hail and strong thunderstorms) is still not well known and difficult to assess without high-resolution fast atmospheric information [1].

In this context, from 2010, a new lidar network at Romanian national level was initiated under the development in the framework of Romanian LIdar NETwork (ROLINET) research project. One year later, the Romanian Atmospheric 3D research Observatory – RADO was founded. RADO [13] is a distributed atmospheric research infrastructure based on the collaboration between: National Institute of R&D for Optoelectronics (INOE), "Babes-Bolyai" University of Cluj-Napoca (UBB), "Alexandru Ioan Cuza" University of Iasi (UAIC), "Politehnica" University of Timisoara (UPT), University of Bucharest (UB) and National Administration for Meteorology (ANM) (RADO - 2015). Numerous coordinated experimental campaigns within Observation Network that is based on 5 existing lidar stations, which operate as the ROLINET, occurred [3, 4, 8]. Four of them are equipped with elastic backscatter lidars, having a dynamic range from 500 m to 15 km and a spatial resolution of 3.75 m and a multi-wavelength Raman lidar (3 elastic + 2 Nitrogen Raman + 1 water vapor channels) that is used at the INOE coordinator site, along with a tropospheric ozone lidar. All stations operate AERONET sunphotometers and ground-level in situ instruments, such as particle counters, gas analysers and weather stations in order to complement lidar observations. ESYRO LIDAR observations combined with RADAR data may help into the improvement of efficiency of anti-hail active intervention with AgI pyrotechnic charges i.e LIRA national project (2014-2015) coordinate by ESYRO in Romania, *PNCDI II – Contract No. 144 /2014* [14].

 https://doi.org/10.1051/epjconf/201817611005

2 ESYRO LIDAR systems

2.1 The ESYRO microLIDAR system

The ESYRO microlidar system (Fig.1) transmitter is based on micropulse class 3B laser emitting at 532 nm ±0.1 nm. The pulse rate frequency is at 7.2 kHz and the pulse duration is 0.5 - 1ns with energy of 3μJ /pulse. The average power is about 30mW but the pulse peak power is around 6kW. The laser beam of 0.2-0.5 mm diameter is 20x expanded in order to reduce the beam divergence to 2-4 mrad. The microlidar receiver is based on a Cassegrain 20 cm telescope of 2 m focal distance and 3D scan available on which the transmitter module (laser + beam expander) is mounted. The lidar signal acquisition is made by an opto-mechanical detection module which is composed by a variable diaphragm, collimation lens and an interferential filter. This module is coupled with a photomultiplier operating in photon counting mode. The data acquisition is due to a high-speed electronic time of fly module triggered by a photodiode. This detection setup is assuring 3.75 m spatial resolution. The data treatment is computer assisted by labVIEW-dedicated routines for data acquisition alignment, dead time correction, SNR analysis, averaging, PBL (Planetary Boundary Layer) height determination and more.

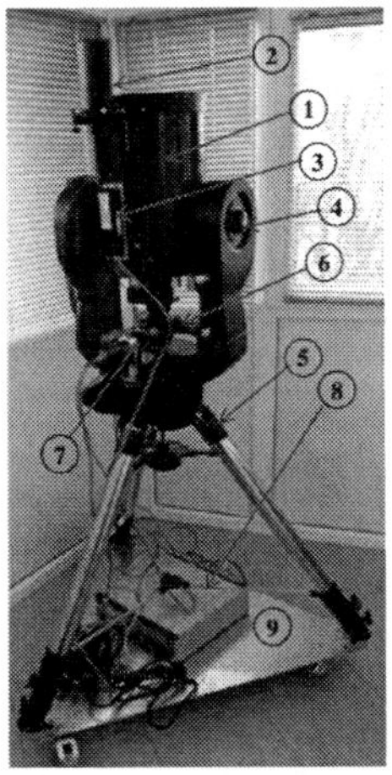

(1) Astronomic Telescope
(2) Beam expander
(3) MicroPuls Laser
(4) 3D Mount
(5) Telescope tripod
(6) Opto – mechanical module
(7) Photodetector
(8) Acquisition module
(9) Transport platform

Figure 1: The microLIDAR system setup

The system is installed on a mobile platform based on a tripod. These system performances are comparable with the microlidars systems presented in [6, 12].

2.2 The ESYRO ESYLIDAR system

The new ESYLIDAR system designed as being a modular system, mobile, easy technically upgradable (multi-angle, multi-channels) for various applications, is used at the Iasi, Cluj-Napoca and Timisoara monitoring sites. The first version was based on a coaxial UV (355 nm), VIS (532 nm) and NIR (1064 nm) emission of a Nd:YAG laser with a variable repetition rate up to 30 Hz. The initial divergence of the 6 mm laser beam of 0.75 mrad was 5 times improved, by using a 3λ UV-VIS-NIR beam expander resulting in a beam of 30 mm diameter and a final divergence of 0.15 mrad [2, 10]. The ESYLIDAR receiver is based on a Newtonian configuration telescope being equipped with a 406 mm diameter primary mirror and a combined focal length of 1829 mm. ESYLIDAR system is presented in [9].

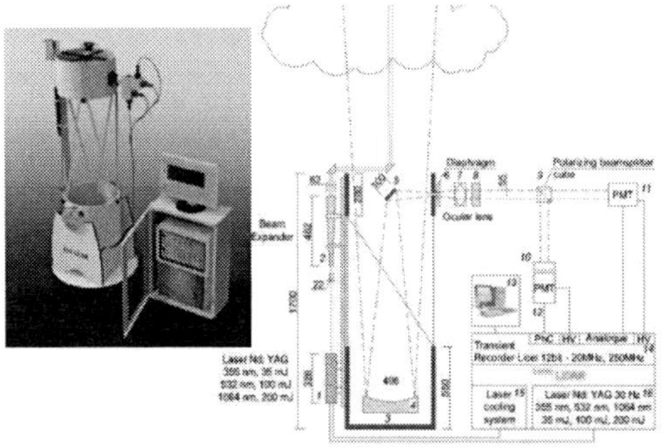

Figure 2: Schematic overview of the on axis ESYLidar system configuration: 1. Laser source Nd:YAG, 2. Beam expander, 3. Newtonian telescope, 4. Main telescope mirror, 5. Secondary telescope mirror, 6. Adjustable circular diaphragm, 7. Eye-piece, 8. Interference filters, 9. Polarizing beam splitter cube, 10. Optical filter, 11. Photomultiplier (in analog regime), 12. Photomultiplier (in photon-counting mode), 13. Computer for data analysis, 14. Acquisition board, and analog to digital convertor 15. Laser cooling unit, 16. Laser power

The detection module is further equipped with a lens assembly, filters and diaphragms which limit the acceptable spectrum, focus and select the reflected spectrum reaching the photomultipliers to make the most out of the useful signal produces by the laser-matter interaction [2, 9]. All resulting optical parameters were used in the final technical configuration of the on-axis lidar system that is represented in Fig.2.

The upgraded version was initiated to improve the fixation/alignment system by decreasing the distance between the main optical axis of the emission module and the main axis of the telescope (from the initial 360 mm to 320 mm). This reduced distance changes the overlap factor of the lidar system, factor describing the overlapping radiation ratio at emission and reception. The circular diaphragm located in the reception module can be reduced in size during sunny and bright days (necessary to maintain a good signal to noise ratio at the) and to maintain the photomultipliers in the linear response region for the wavelength of interest. By modifying the circular diaphragm, the overlapping factor is changing. To preserve the correct lidar signal the declination angle has to be adjusted. Accordingly, for a variation of the diameter of the diaphragm from 12 mm to 3 mm, the declination angle has to be changed from -0.5 mrad to 0.35 mrad (the minus sign denotes the declination orientation), thus the altitude where the overlapping factor becomes 1 can vary between 700 and 950 m. This system is comparable with one presented in [1].

4 MEASUREMENT SAMPLES

A RCS (range corrected signal) obtained with the ^{micro}LIDAR system for more than 2h of continuous measurements is shown in Fig.3.

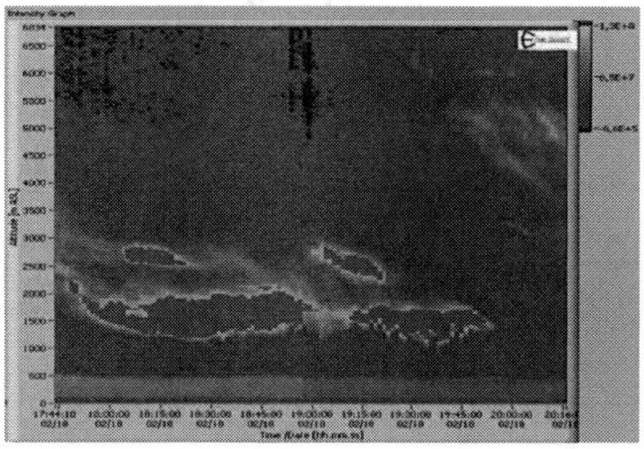

Figure 3: Sample of RCS from ^{micro}LIDAR data

In addition on the alignment stability we noticed the system is able to detect simultaneously a very low level of PBL winter time inversion at 500 m altitude, a complex group of tropospheric clouds from 1000 to 3000 m and some cirrus clouds at 6500 -7000 km with 7.5m spatial resolution. More results analysis are presented in [2].

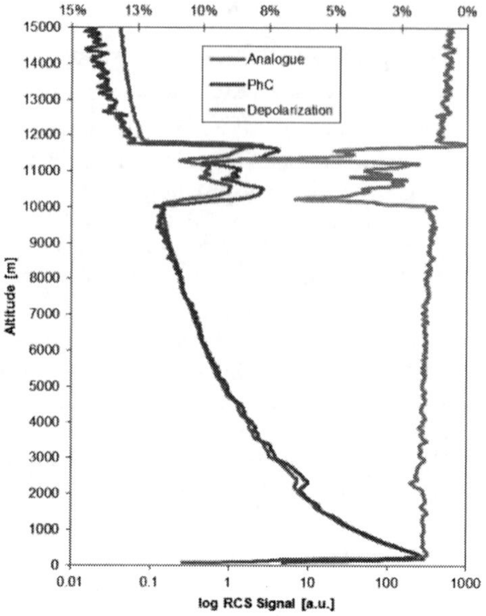

Figure 4: Sample of RCS from ^{ESY}LIDAR data [nighttime profile, 7.5 m spatial resolution, 1 min integration time]

A sample measurement obtained with ^{ESY}LIDAR system is shown in Fig.4. We can notice the 532 nm two channels (A for elastic direct signal and PhC for cross depolarization signal) as well as the depolarization ratio profile. ^{ESY}LIDAR may be considered as a powerful lidar system able to profile the atmosphere up to lower stratosphere i.e.15 km (day) - 20 km (night) and a high spatial (7.5 m) and temporal (1 min) resolution. In Fig. 4 from 10 to 12km tropospheric cirrus clouds are detected with ice crystal contents up to 8%. Thus one may also use this system for Saharan dust and o volcano ash plumes monitoring. Typically, the lidar data are correlated with complementary techniques such as the AERONET sunphotometers, the Calipso lidar system, and with theoretical models (DREAM, HYSPLIT). AErosol RObotic NETwork (AERONET) is a NASA network for monitoring and characterizing atmospheric aerosols by ground-based sun photometer. Beginning with the 7th of May 2012, the monitoring station LOA-SL in Iasi, Romania (Lat: 47.19° N, Long: 27.55° E) is active in this network providing quantitative values for various types of aerosols. Based on this synergy many measurements campaigns and case studies were performed as shown in the [2–4, 8–11].

5 CONCLUSIONS

Two operational lidar systems, respectively a ^{micro}LIDAR and the ^{ESY}LIDAR, developed by SC EnviroScopY SRL (ESYRO) based in Iasi, RO were presented together with sample relevant measurement tests. A ^{micro}LIDAR system is still in use by National Meterological Service while ^{ESY}LIDAR systems are operating at UBB, UAIC and UPT universities. A depolarization version of ^{ESY}LIDAR system was operated during LIRA project 2014-2016 by ESYRO [14]. The performances of these two systems are comparable with the up to date academic systems but they are not enough designed yet to be commercially rentable and more efforts are ongoing to improve them.

ACKNOWLEDGEMENTS

Romanian Research Ministry (A.N.C.S.I.) for co-founding the research and development and to academic research partners as INOE, UAIC, UPT, UB and UBB.

References

[1] Balin, I., 2004: Measurement and analysis of aerosols, cirrus-contrails, water vapor and temperature in the upper troposphere with the Jungfraujoch LIDAR system. Thèse École polytechnique fédérale de Lausanne, no 2975.

[2] Cazacu, M.M., 2010: Contributions to the implementation of the first national LIDAR network for atmospheric aerosols optical characterization. PhD Thesis, Alexandru Ioan Cuza University of Iasi, Romania.

[3] Cazacu, M.M., Timofte, A., Talianu, C., Nicolae, D., Danila, M.N., Unga, F., Dimitriu, D.G., Gurlui, S., 2012: Grimsvotn Volcano: atmospheric volcanic ash cloud investigations, modelling-forecast and experimental environmental approach upon the Romanian area. *J. Optoelectron. Adv. Mater.* **14** (5–6), S. 517–522.

[4] Papayannis, A., Nicolae, D., Kokkalis, P., Binietoglou, I., Talianu, C., Belegante, L., Tsaknakis, G., Cazacu, M.M., Vetres, I., Ilic, L., 2014: Optical, size and mass properties of mixed type aerosols in Greece and Romania as observed by synergy of lidar and sunphotometers in combination with model simulations: A case study. *Sci. Total Environ.* **500–501** S. 277–294.

[5] Seinfeld, J.H., Pandis, S.N., 2006: Atmospheric Chemistry and Physics: From Air Pollution to Climate Change, 2nd Edition - John Wiley & Sons, Inc. New Jersey.

[6] Spinhirne, J.D., 1993: Micro Pulse Lidar. *IEEE Trans. Geosci. Remote Sens.* **31** (1), S. 48–55.

[7] Stefan, S., Nicolae, D., Caian, M., 2008: Secretele aerosolului atmosferic în lumina laserilor. Ed. Ars Docendi.

[8] Timofte, A., Cazacu, M.M., Radulescu, R., Belegante, L., Dimitriu, D.G., Gurlui, S., 2011: Romanian lidar investigation of the Eyjafjallajokull volcanic ash. *Environ. Eng. Manag. J.* **10** (1), S. 91–97.

[9] Tudose, O.-G., 2013: Contributions to the study of atmospheric aerosols optical properties using remote sensing techniques. PhD Thesis, Alexandru Ioan Cuza University of Iasi, Romania.

[10] Tudose, O.-G., Cazacu, M.-M., Timofte, A., Balin, I., 2011: ESYROLIDAR system developments for troposphere monitoring of aerosols and clouds properties. *Proc. SPIE.* **8177** S. 817716-817716–10.

[11] Unga, F., Cazacu, M.M., Timofte, A., Bostan, D., Mortier, A., Dimitriu, D.G., Gurlui, S., Goloub, P., 2013: Study of tropospheric aerosol over Iasi , Romania , during summer of 2012. *Environ. Eng. Manag. J.* **12** (2), S. 297–303.

[12] Cheng, Y.S.A., Guo, Y., Zhu, J., 2004: Micropulse lidar system. *United States Pat. no. 6.717.655.*

[13] 2015: RADO - Romanian Atmospheric 3D research Observatory http://environment.inoe.ro/category/66/rado.

[14] LIRA: http://lira.inoe.ro/?lang=en

AUTHOR INDEX

AUTHOR INDEX

AUTHOR INDEX

AUTHOR INDEX

AUTHOR INDEX

AUTHOR INDEX

AUTHOR INDEX

AUTHOR INDEX